Biosystems - Investigated by
Scanning Probe Microscopy

Harald Fuchs · Bharat Bhushan
Editors

Biosystems Investigated by Scanning Probe Microscopy

Springer

Editors
Prof. Dr. Harald Fuchs
Center for Nanotechnology
& Universität Münster
Physikalisches Institut
Wilhelm-Klemm-Str. 10
48149 Münster
Germany
fuchsh@uni-muenster.de

Prof. Bharat Bhushan
Ohio State University
Nanotribology Lab. Information
Storage & MEMS/NEMS
201 West 19th Ave.
Columbus OH 43210-1142
W390 Scott Laboratory
USA
bhushan.2@osu.edu

Partly published in *Applied Scanning Probe Methods*, Bhushan, B., Fuchs, H. (Eds.)

Vol. II, ISBN 978-3-540-26242-8
Vol. III, ISBN 978-3-540-26909-0
Vol. IV, ISBN 978-3-540-26912-0
Vol. V, ISBN 978-3-540-37315-5
Vol. VI, ISBN 978-3-540-37318-6
Vol. VII, ISBN 978-3-540-37320-9
Vol. VIII, ISBN 978-3-540-74079-7
Vol. IX, ISBN 978-3-540-74082-7
Vol. X, ISBN 978-3-540-74084-1
Vol. XII, ISBN 978-3-540-85038-0
Vol. XIII, ISBN 978-3-540-85048-9

ISBN 978-3-642-02404-7 e-ISBN 978-3-642-02405-4
DOI 10.1007/978-3-642-02405-4
Springer Heidelberg Dordrecht London New York

Library of Congress Control Number: 2009940804

Cover design: WMXDesign GmbH, Heidelberg

Printed on acid-free paper

Springer is part of Springer Science+Business Media (www.springer.com)

Preface

Scanning probe microscopy techniques have significantly contributed to the rapid development of nanotechnology. Over several years representative developments were collected in the Springer Series 'Applied Scanning Probe Methods I–XIII' giving an overview on recent research and development activities in scanning probe methods as well as their application in various technology fields. Researchers from all over the world displayed their recent development in different fields including instrumentation and applications in material science, electronics/electro-optics, biology and medicine. The atomic force microscopy techniques such as static and dynamic AFM form a broad basis of these contributions due to their easy applicability in non-vacuum environments, e.g. air and liquids. This establishes an excellent basis for the study of bio-materials including biopolymers such as DNA, proteins, membranes and even complete cells in vitro. Static and Dynamic Force Microscopy helped to better understand the variety of molecular and physiological phenomena at the nanometer scale and opened a new window into a nano world for biological specimen. As a natural step representative contributions in this field previously published in the Springer series 'Applied Scanning Probe Methods I–XIII' were compiled in this book. Readers working in the field of nanobioscience will have a comprehensive overview of recent results obtained with scanning probes on biosystems and will also get many indications of their potential for future developments. The volume 'Biosystems – Investigated by Scanning Probe Microscopy' contains in total 27 selected articles organized in four topical parts, starting with part I on instrumentation and methods, followed by part II on applications of AFM on biomolecules and part III on AFM of biological membranes, cells and tissue, and completed by part IV on functional bio(-inspired) surfaces. The table-of-contents thus displays a hierarchical structure of the book ranging from instrumental and theoretical development to the application of AFM on biological systems with increasing complexity.

We wish to thank Springer, in particular Mrs. Hertel and Mrs. Siek, for their encouragement and their initiative to set up this volume.

Münster, Germany — Harald Fuchs
Columbus, OH, USA — Bharat Bhushan

Contents

Part III
AFM of Biological Membranes, Cells and Tissue

List of Contributors

Rehana Afrin Department of Life Science, Tokyo Institute of Technology, Nagatsuta, Midori-ku, Yokohama 226-8501, Japan; e-mail: rafrin@bio.titech.ac.jp

David Alsteens Unité de Chimie des Interfaces, Université catholique de Louvain, Croix du Sud 2/18, 1348 Louvain-la-Neuve, Belgium;
e-mail: david.alsteens@uclouvain.be

Matthias Amrein Department of Cell Biology and Anatomy, Faculty of Medicine, University of Calgary, Canada; e-mail: mamrein@ucalgary.ca

Boris Anczykowski nanoAnalytics GmbH, Gievenbecker Weg 11, 48149 Münster, Germany; e-mail: anczykowski@nanoanalytics.de

Guillaume André Unité de Chimie des Interfaces, Université catholique de Louvain, Croix du Sud 2/18, 1348 Louvain-la-Neuve, Belgium;
e-mail: guillaume.andre@uclouvain.be

Hideo Arakawa Bio Nanomaterials Group, National Institute for Materials Science, 1-1, Namiki, Tsukuba 305-0044, Japan;
e-mail: ARAKAWA.Hideo@nims.go.jp

Friedrich Aumayr Institut für Allgemeine Physik, Technische Universität Wien, Wiedner Hauptstraße 8-10/134, 1040 Wien, Austria;
e-mail: aumayr@iap.tuwien.ac.at

François Bessueille LSA, Université Lyon I, 43 Boulevard du 11 Novembre 1918, 69622 Villeurbanne Cedex, France; e-mail: francois.bessueille@univ-lyon1.fr

Bharat Bhushan Nanotribology Laboratory for Information Storage and MEMS/NEMS (NLIM), W 390 Scott Laboratory, 201 W. 19th Avenue, Ohio State University, Columbus, Ohio 43210-1142, USA;
e-mail: bhushan.2@osu.edu

Lilia A. Chtcheglova Institute for Biophysics, Johannes Kepler University of Linz, Altenbergerstr. 69, 4040 Linz, Austria; e-mail: lilia.chtcheglova@jku.at

David T. Cramb Department of Chemistry, University of Calgary, Canada;
e-mail: dcramb@ucalgary.ca

Etienne Dague Unité de Chimie des Interfaces, Université catholique de Louvain, Croix du Sud 2/18, 1348 Louvain-la-Neuve, Belgium;
e-mail: dague@cifa.ucl.ac.be

Manfred Drack GrAT Center for Appropriate Technology, Technische Universität Wien, Wiedner Hauptstraße 8-10/965, 1040 Wien, Austria; e-mail: md@grat.at

Yves F. Dufrêne Unité de Chimie des Interfaces, Université catholique de Louvain, Croix du Sud 2/18, 1348 Louvain-la-Neuve, Belgium; e-mail: dufrene@cifa.ucl.ac.be

Vincent Dupres Unité de Chimie des Interfaces, Université catholique de Louvain, Croix du Sud 2/18, 1348 Louvain-la-Neuve, Belgium; e-mail: vincent.dupres@uclouvain.be

Andreas Ebner Institute for Biophysics, Johannes Kepler University of Linz, Altenbergerstr. 69, 4040 Linz, Austria; e-mail: andreas.ebner@jku.at

Robert H. Eibl Plainburgstr. 8, 83457 Bayerisch Gmain, Germany; e-mail: robert eibl@yahoo.com

Abdelhamid Errachid Laboratory of NanoBioEngineering, Barcelona Science Park, Edifici Modular, C/Josep Samitier 1–5, 08028 Barcelona, Spain; e-mail: aerrachid@pcb.ub.es

Horacio D. Espinosa Department of Mechanical Engineering, Northwestern University, 2145 Sheridan Rd., Evanston, IL 60208-3111, USA; e-mail: espinosa@northwestern.edu

Eric Finot Institut Carnot de Bourgogne, UMR 5209 CNRS, Université de Bourgogne, France; e-mail: efinot@u-bourgogne.fr

Grégory Francius Unité de Chimie des Interfaces, Université catholique de Louvain, Croix du Sud 2/18, 1348 Louvain-la-Neuve, Belgium; e-mail: gregory.francius@uclouvain.be

Friedrich Franek Austrian Center of Competence for Tribology, Viktor Kaplan-Straße 2, 2700 Wiener Neustadt, Austria and Institut für Sensor- und Aktuatorsysteme, Technische Universität Wien, Floragasse 7/2, 1040 Wien, Austria; e-mail: franek@ac2t.at

Harald Fuchs Center for Nanotechnology (CeNTech) and Institute of Physics, University of Münster, Gievenbecker Weg 11, 48149 Münster, Germany; e-mail: fuchsh@uni-muenster.de

Ille C. Gebeshuber Austrian Center of Competence for Tribology, Viktor Kaplan-Straße 2, 2700 Wiener Neustadt, Austria and Institut für Allgemeine Physik, Technische Universität Wien, Wiedner Hauptstraße 8-10/134, 1040 Wien, Austria; e-mail: ille@iap.tuwien.ac.at

Christoph Gerber National Competence Center of Research in Nanoscale Science, Basel, Klingelbergstrasse 82, 4056 Basel, Switzerland; e-mail: Christoph.Gerber@unibas.ch

Yann Gilbert Unité de Chimie des Interfaces, Université catholique de Louvain, Croix du Sud 2/18, 1348 Louvain-la-Neuve, Belgium;
e-mail: gilbert@cifa.ucl.ac.be

Hermann J. Gruber Institute for Biophysics, Johannes Kepler University of Linz, Altenbergerstr. 69, 4040 Linz, Austria; e-mail: hermann.gruber@jku.at

Pietro Giuseppe Gucciardi CNR-Istituto per i Processi Chimico-Fisici, Salita Sperone c.da Papardo, 98158 Messina, Italy; e-mail: gucciardi@me.cnr.it

Xavier Haulot Unité de Chimie des Interfaces, Université catholique de Louvain, Croix du Sud 2/18, 1348 Louvain-la-Neuve, Belgium;
e-mail: haulot@cifa.ucl.ac.be

Martin Hegner National Competence Center of Research in Nanoscale Science, Basel, Klingelbergstrasse 82, 4056 Basel, Switzerland;
e-mail: Martin.Hegner@unibas.ch

Peter Hinterdorfer Institute for Biophysics, Johannes Kepler University of Linz, Altenbergerstr. 69, 4040 Linz, Austria; e-mail: peter.hinterdorfer@jku.at

Andrea Ho Department of Mechanical Engineering, Northwestern University, 2145 Sheridan Road, Evanston, IL 60208-3111, USA

Sumio Hosaka Department of Production Science and Technology, Gunma University, 1-5-1 Tenjin-cho, Kiryu, Gunma, 376-8515, Japan

Yanxia Hou Ecole Centrale de Lyon, STMS/CEGELY,
36 Avenue Guy de Collongue, 69131 Ecully Cedex, France;
e-mail: yanxiahou24@yahoo.com

Atsushi Ikai Department of Life Science, Tokyo Institute of Technology, Nagatsuta, Midori-ku, Yokohama 226-8501, Japan; e-mail: aikai@bio.titech.ac.jp

Albena Ivanisevic Purdue University, Weldon School of Biomedical Engineering, 500 Central Drive, West Lafayette, IN 47907-2022, USA;
e-mail: albena@purdue.edu

Nicole Jaffrezic-Renault Ecole Centrale de Lyon, STMS/CEGELY,
36 Avenue Guy de Collongue, 69131 Ecully Cedex, France;
e-mail: Nicole.Jaffrezic@ec-lyon.fr

Ingomar L. Jäger Department of Materials Science, University of Leoben, Jahnstrasse 12, 8700 Leoben, Austria; e-mail: ingomar@unileoben.ac.at

K. S. Kanaga Karuppiah Department of Mechanical Engineering, 2025 H. M. Black Engineering Building, Iowa State University, Ames, IA 50011-2161, USA; e-mail: kskarup@iastate.edu

Kazushige Kawabata Division of Biological Sciences, Graduate School of Science, Hokkaido University, North 10 West 8, Kita-ku, Sapporo 060-0810, Japan; e-mail: kaw@sci.hokudai.ac.jp

Ferry Kienberger Institute for Biophysics, Johannes Kepler University of Linz, Altenbergerstr. 69, 4040 Linz, Austria; e-mail: ferry.kienberger@jku.at

Hyonchol Kim Division of Biosystems, Institute of Biomaterials and Bioengineering, Tokyo Medical and Dental University, 2-3-10, Kanda-surugadai, Chiyoda-ku, Tokyo, 101-0062, Japan; e-mail: kim.bmi@tmd.ac.jp

Joseph M. Kinsella Purdue University, Department of Biomedical Engineering, 500 Central Drive, West Lafayette, IN 47907-2022, USA; e-mail: jkinsel@purdue.edu

Andrzej J. Kulik Ecole Polytechnique Fédérale de Lausanne, EPFL – IPMC – NN, 1015 Lausanne, Switzerland; e-mail: Andrzej.Kulik@epfl.ch

Hans Peter Lang National Competence Center of Research in Nanoscale Science, Basel, Klingelbergstrasse 82, 4056 Basel, Switzerland and IBM Zurich Research Laboratory, Säumerstrasse 4, 8803 Rüschlikon, Switzerland; e-mail: Hans-Peter.Lang@unibas.ch

Tobias Lange Institute of Physiology II, Robert-Koch Str. 27b, 48149 Münster, Germany; e-mail: tolange@uni-muenster.de

Carmen LaTorre Owens Corning, Insulating Systems Business, 2790 Columbus Road, Route 16 (Bldg 20-1), Granville, OH 43023, USA e-mail: carmen.latorre@owenscorning.com

Małgorzata Lekka The Henryk Niewodniczański Institute of Nuclear Physics, Polish Academy of Sciences, Radzikowskiego 152, 31-342 Kraków, Poland; e-mail: Malgorzata.Lekka@ifj.edu.pl

Zoya Leonenko Department of Cell Biology and Anatomy, Faculty of Medicine; University of Calgary, Canada; e-mail: zleonenk@ucalgary.ca

Helga C. Lichtenegger Institute of Materials Science and Technology E308, Vienna University of Technology, Favoritenstrasse 9–11, 1040 Wien, Austria; e-mail: helga.lichtenegger@tuwien.ac.at

Claude Martelet Ecole Centrale de Lyon, STMS/CEGELY, 36 Avenue Guy de Collongue, 69131 Ecully Cedex, France; e-mail: Claude.Martelet@ec-lyon.fr

Vincent T. Moy Department of Physiology & Biophysics, University of Miami, 1600 NW 12th Ave., RMSB 5077, Miami, FL 33136, USA; e-mail: vmoy@miami.edu

Dessy Nikova Institute of Physiology II, Robert-Koch Str. 27b, 48149 Münster, Germany; e-mail: nikovad@uni-muenster.de

Michael Nosonovsky Nanomechanical Properties Group, Materials Science and Engineering Laboratory, National Institute of Standards and Technology, 100 Bureau Dr., Mail Stop 8520, Gaithersburg, MD 20899-8520, USA; e-mail: michael.nosonovsky@nist.gov

Hans Oberleithner University of Münster, Institute of Physiology II, Robert-Koch-Straße 27b, 48149 Münster, Germany;
e-mail: oberlei@uni-muenster.de

Toshiya Osada Department of Life Science, Tokyo Institute of Technology, Nagatsuta, Midori-ku, Yokohama 226-8501, Japan;
e-mail: tosada@bio.titech.ac.jp

Theeraporn Puntheeranurak Institute for Biophysics, Johannes Kepler University of Linz, Altenbergerstr. 69, 4040 Linz, Austria;
e-mail: theeraporn.puntheeranurak@jku.at

Félix Rico Department of Physiology & Biophysics, University of Miami, 1600 NW 12th Ave, RMSB 5077, Miami, FL 33136, USA;
e-mail: frico@med.miami.edu

Elisa Riedo Georgia Institute of Technology, School of Physics, 837 State Street, Atlanta, GA 30332-0430, USA; e-mail: elisa.riedo@physics.gatech.edu

Christoph Riethmüller University of Münster , Institute of Physiology II, Robert-Koch-Straße 27b, 48149 Münster, Germany; e-mail: chrth@uni-muenster.de

Robert A. Sayer Nanotribology Laboratory for Information Storage and MEMS/NEMS (NLIM), The Ohio State University, 201 West 19th Avenue, Columbus, OH 43210-1142, USA

Tilman E. Schäffer Center for Nanotechnology (CeNTech) and Institute of Physics, University of Münster, Gievenbecker Weg 11, 48149 Münster, Germany;
e-mail: tilman.schaeffer@uni-muenster.de

Hermann Schillers Institute of Physiology II, Robert-Koch Str. 27b, 48149 Münster, Germany; e-mail: schille@uni-muenster.de

Matthias F. Schneider University of Augsburg, Experimental Physics I, Universitätsstr. 1, 86159 Augsburg, Germany;
e-mail: matthias.schneider@physik.uni-augsburg.deF.

Thomas Schöberl Erich Schmid Institute of Materials Science of the Austrian Academy of Sciences, Jahnstrasse 12, 8700 Leoben, Austria;
e-mail: schoeber@unileoben.ac.at

Hiroshi Sekiguchi Department of Life Science, Tokyo Institute of Technology, Nagatsuta, Midori-ku, Yokohama 226-8501, Japan; e-mail: hsekiguc@bio.titech.ac.jp

Hayato Sone Department of Production Science and Technology, Gunma University, 1-5-1 Tenjin-cho, Kiryu, Gunma, 376-8515, Japan;
e-mail: sone@el.gunma-u.ac.jp

Daniel M. Steppich University of Augsburg, Experimental Physics I, Universitätsstr. 1, 86159 Augsburg, Germany;
e-mail: daniel.steppich@physik.uni-augsburg.de

Sriram Sundararajan Department of Mechanical Engineering, 2025 H. M. Black Engineering Building, Iowa State University, Ames, IA 50011-2161, USA, e-mail: srirams@iastate.edu

Robert Szoszkiewicz Georgia Institute of Technology, School of Physics, 837 State Street, Atlanta, GA 30332-0430, USA; e-mail: robert.szoszkiewicz@physics.gatech.edu

Stefan Thalhammer GSF-Institut für Strahlenschutz, Neuherberg, Germany; e-mail: stefan.thalhammer@gsf.de

Neil H. Thomson Molecular and Nanoscale Physics Group, University of Leeds, EC Stoner Building, Woodhouse Lane, Leeds, LS2 9JT, UK; e-mail: n.h.thomson@leeds.ac.uk

Hironori Uehara Department of Life Science, Tokyo Institute of Technology, Nagatsuta, Midori-ku, Yokohama 226-8501, Japan; e-mail: huehara@bio.titech.ac.jp

Tatsuo Ushiki Division of Microscopic Anatomy, Graduate School of Medical and Dental Sciences, Niigata University, 757 Asahimachi-dori, Chuo-ku, Niigata 951-8510, Japan; e-mail: t-ushiki@med.niigata-u.ac.jp

Claire Verbelen Unité de Chimie des Interfaces, Université catholique de Louvain, Croix du Sud 2/18, 1348 Louvain-la-Neuve, Belgium; e-mail: claire.verbelen@uclouvain.be

Hannspeter Winter Institut für Allgemeine Physik, Technische Universität Wien, Wiedner Hauptstraße 8-10/134, 1040 Wien, Austria; e-mail: winter@iap.tuwien.ac.at

Achim Wixforth University of Augsburg, Experimental Physics I, Universitätsstr. 1, 86159 Augsburg, Germany; e-mail: achim.wixforth@physik.uni-augsburg.de

Ewa P. Wojcikiewicz Department of Physiology & Biophysics, University of Miami, 1600 NW 12th Ave, RMSB 5077, Miami, FL 33136, USA; e-mail: e.wojcikiewicz@miami.edu

Part I
Instrumentation and Methods

35 Scanning Probe Microscopy in Biological Research

Tatsuo Ushiki · Kazushige Kawabata

From volume X

Abstract. Scanning probe microscopy (SPM) is especially attractive to biologists, because it has the advantage of obtaining three-dimensional images of sample surfaces at high resolution not only in a vacuum but also in a nonvacuum (i. e., air or liquid) environment. In addition to the visualization of biological structures, SPM has also been applied for measuring the physical properties (i. e., viscoelasticity) of living cells, and as a manipulation tool in biomaterials. The present review describes the application of SPM to biological fields for investigating the structure and function of biomaterials from biomolecules to living cells, and shows that SPM has great potential for broadening and opening up new fields in biomedical research.

Key words: Scanning probe microscopy, Atomic force microscopy, Near-field optic microscopy, Viscoelasticity, Nanomanipulation, Biomolecules, DNA collagen, Chromosomes, Living cells

Abbreviations

AFM	Atomic force microscopy
DNA	Deoxyribonucleic acid
GFP	Green fluorescent protein
HEPES	*N*-(2-Hydroxyethyl)piperazine-*N*′-ethanesulfonic acid
SEM	Scanning electron microscopy
SNOM	Scanning near-field optic microscopy
SPM	Scanning probe microscopy
STM	Scanning tunneling microscopy
TEM	Transmission electron microscopy

35.1 Introduction

Since the invention of the scanning tunneling microscope, various types of microscopes using a probing tip have been introduced, and are now referred to as a family of scanning probe microscopes. Among them, the scanning tunneling microscope is the first to be applied to biological fields, but its applications have been limited to observations of some biomolecules such as deoxyribonucleic acid (DNA) [1]. This is because the samples for scanning tunneling microscopy (STM) must be electrically conductive, indicating that STM is not well suited for analyzing biological samples, which are generally not conductive. In contrast with the scanning tunneling microscope, the atomic force microscope invented in 1986 has been widely

In: H. Fuchs, B. Bhushan (eds), *Biosystems - Investigated by Scanning Probe Microscopy*,
DOI 10.1007/978-3-540-74085-8_9, © Springer-Verlag Berlin Heidelberg 2010

applied in biological fields [2]. This microscope can create topographic images of samples three-dimensionally, which are similar to those obtained by scanning electron microscopy (SEM). In addition, atomic force microscopy (AFM) can be used for imaging both conductive and nonconductive samples at high resolution from the micrometer scale to the atomic scale. Thus, a number of researchers have been interested in studying the three-dimensional surface structure of biological samples from cells to biological molecules by AFM [3–7].

On the other hand, some other investigators have also been increasing their attention to the application of the atomic force microscope to the study of the physical properties of biological samples, because this microscope can measure the interaction forces between the tip and the sample [6–8]. Thus, AFM has been used for measuring the elasticity of single surface molecules in relation with function. There are also studies measuring the physical properties of the sample's surface, including surface changes, adhesion and elasticity of cells [8,9].

In addition to collecting surface information of biological structures, the atomic force microscope is expected to be used as a dissecting tool, because it can directly contact a probing tip with the sample [10]. Thus, dissections of biological samples such as chromosomes and DNA have been performed with an atomic force microscope.

In this chapter, we will firstly introduce the applications of scanning probe microscopy (SPM) for visualization of biological samples, secondly show the SPM applications for measuring the physical properties of biomaterials, and finally describe the possibility of SPM as a manipulation tool in biology.

35.2 SPM for Visualization of the Surface of Biomaterials

35.2.1 Advantages of AFM in Biological Studies

Among various types of SPM, AFM has been expected to be an essential tool for visualizing three-dimensionally the surface structure of biological samples. Although the AFM images are similar to those obtained by SEM, there are several advantages of AFM imaging, which are summarized as follows:

1. AFM provides three-dimensional information. Because the atomic force microscope traces the sample's surface by monitoring the interaction force between the tip and the sample, it can create the three-dimensional information on the sample topography. Thus, the AFM images are similar to SEM images.
2. The resolution of AFM is high. By AFM, individual atoms have been observed with a few hard crystalline samples. This means that AFM has the potential to provide three-dimensional information on the surface topography of samples at resolutions from the micrometer scale to the atomic scale, although AFM images of biological samples have not yet shown this level of resolution.
3. Samples do not need to have electron conductivity: Because SEM creates images by scanning the sample's surface with an electron beam, samples need to be conductively stained and/or metal-coated before SEM observation. This

implies that precise information on the surface topography is hidden owing to the presence of the coating and/or staining substances. In contrast, AFM enables the direct observation of nonconductive samples without a metal coating or any conductive treatment.

4. Samples can be observed in nonvacuum environments. In contrast with the scanning electron microscope operated in vacuum conditions, the atomic force microscope has the advantage of creating images not only in a vacuum but also in air or a liquid environment, indicating that functional information can be expected to be obtained directly from living organisms.
5. Quantitative height information can be provided. Usually, SEM images contain no quantitative information on the sample's height. However, AFM provides quantitative information in three dimensions, and the sample's profile containing height information can be easily obtained from AFM images.

35.2.2 AFM of Biomolecules

The structure of biomolecules has been studied by X-ray crystalline analysis. Recent investigators have reported that AFM has the potential for observing a two-dimensional crystalline sheet of molecules [7]. Some of the best examples to date have been obtained by Müller et al. [11,12]; they have succeeded in observing a crystalline array of bacteriorhodopsin on the cytoplasmic surface of the purple membrane. These studies imply that AFM has the potential to analyze the submolecular structure of certain samples. However, in general, AFM images of biomolecules have not shown this level of resolution. This is probably due to the unevenness of randomly distributed samples. On the other hand, the shape of macromolecules such as DNA has been visualized by AFM [13]. Similar images have been reported by transmission electron microscopy (TEM) of negatively stained or rotary shadowed samples. In contrast with the complicated and skillful preparation techniques for TEM, the techniques for AFM are rather simple and consistent, indicating that AFM is expected to become a powerful tool for visualizing macromolecules. Two examples are given in the following sections.

35.2.2.1 DNA Molecules

DNA is a polymer of nucleotides which forms a double helix. Because of the characteristic shape and dimensions of DNA, its visualization by AFM has attracted many investigators. The commonest method for DNA preparation for AFM is deposition of DNA on the substrate by dropping a DNA solution followed by evaporation of the solvent. For the preparation, the following two points are the most important: (1) the choice of substrate and (2) the method of DNA attachment to substrates. Although various materials such as sapphire and graphite have been reported, freshly cleaved mica is widely used as the substrate. Previous studies have been reported that DNA molecules do not bind strongly to mica and are usually moved or swept away by the scanning tip. This is probably because freshly cleaved mica and DNA are both negatively charged. Therefore, the mica surface was often treated with Mg^{2+} and a variety of other divalent and trivalent cations for AFM observation of DNA [13]. According to our experiments, DNA molecules dissolved in N-(2-hydroxyethyl)piperazine-N'-ethanesulfonic acid (HEPES)-Mg^{2+} buffer consistently binds to nontreated freshly cleaved mica, which is enough for AFM observation (Fig. 35.1). The operating mode suitable for DNA observation is a dynamic mode (or intermittent contact mode). With use of these techniques, clear images are consistently obtained in ambient conditions and even in liquid conditions. Our experiments have shown that DNA molecules dissolved in distilled water also consistently bind to freshly cleaved mica enough to image by dynamic mode AFM.

Although AFM images of DNA are comparable to those taken by a rotary shadowing technique for TEM, AFM has the advantage of observing directly the molecule without special staining or shadowing in ambient conditions and even in liquid conditions. AFM images also have the advantage of containing quantitative information on the sample's height: the DNA height in AFM images is usually about 1 nm or less. However, the image width typically varies from 8 to 20 nm. The reason why the width is much larger than the height of DNA is explained by the convolution effect of the shape of the probing tip: the radius of the probing tip is generally about 10 nm when the cantilevers are commercially purchased. To minimize this

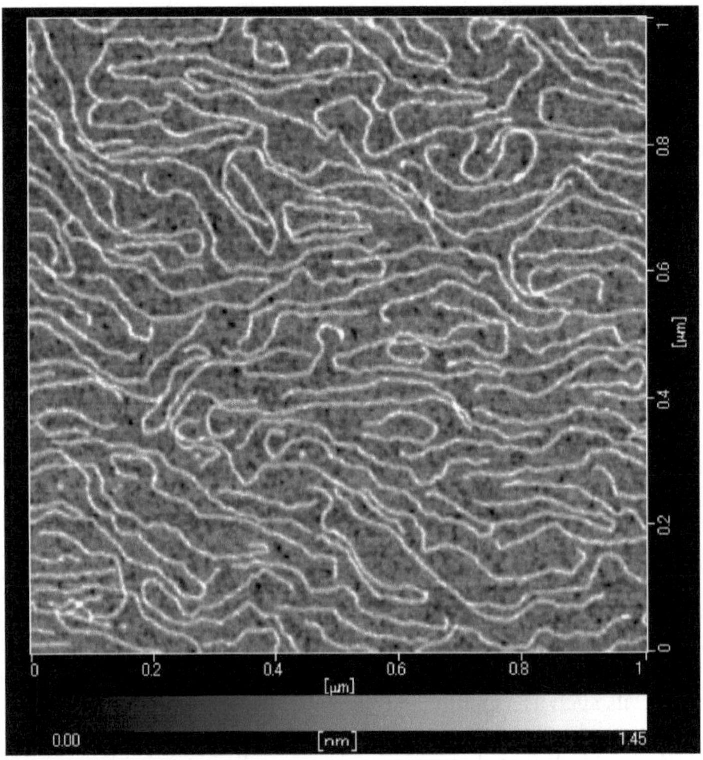

Fig. 35.1. Atomic force microscopy (AFM) image of DNA molecules absorbed on mica obtained in a dynamic mode in ambient conditions

artifact, the application of a carbon nanotube probe has been introduced by some researchers [14].

35.2.2.2
Collagen Type I Molecules

Collagen type I molecules are composed of three separate chains (α chains), each of which contains a characteristic sequence of amino acids twisted in the form of a left-handed helix. Biochemically, the length and the diameter of the molecules are estimated at about 300 and 1.5 nm, respectively. By AFM, the collagen molecules are observed as twisted threads about 280–310 nm long [15] (Fig. 35.2). The height and the width of the imaged molecules were 0.5–1 and 6–10 nm. At high magnification, somewhat globular bulges are usually present at both ends of the molecules. These terminal bulges are probably connected with the structure of the C- and N-terminals, which do not form the triple helix. A prominent depression, on the other hand, is often found at about 70 nm from one end of the molecules, suggesting that it might be a site of a relaxed triple helix which has been chemically studied. A characteristic bending of the molecules is sometimes found at one end of the molecules opposite the depression. Thus, because the surface morphology of collagen molecules is observed by dynamic mode AFM, it is expected to be used for future studies on the structure and function of various types of collagen molecules.

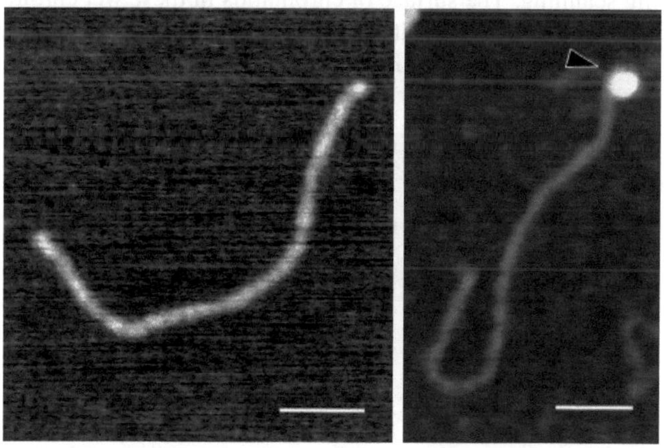

Fig. 35.2. AFM images of a collagen type I molecule (*left*) and a procollagen I molecule (*right*), obtained in a dynamic mode in ambient conditions. The collagen type I molecule is observed as a flexible thread 300 nm long. The procollagen I molecule is characterized by the presence of a globular C-terminal propeptide (*arrowhead*). *Bar* 50 nm. (Reproduced from [20])

35.2.3
AFM of Isolated Intracellular and Extracellular Structures

Various intracellular and extracellular structures have also been visualized by AFM. Among them, AFM imaging of chromosomes and collagen fibrils is introduced in the following sections.

35.2.3.1
Chromosomes

Chromosomes are produced by condensation of chromatin fibers, which consist of DNA and proteins. They appear in eukaryotic cells during cell division. For AFM imaging, the chromosomes are usually prepared in accordance with the standard method for light microscopy. Briefly, chromosome spreads are made by dropping a suspension of peripheral blood cultures (which are treated with a hypotonic solution of potassium chloride and fixed with a mixture of methanol and acetic acid). When these chromosomes are simply dried in air and observed by dynamic mode AFM, the samples are rather flattened apparently owing to the drying artifacts caused by the surface tension of water. On the other hand, the morphology of the critical-point-dried chromosomes was well preserved when observed by dynamic mode AFM in ambient conditions [16]; they were composed of the highly condensed chromatids of a mitotic pair, each of which was characterized by the presence of alternating ridges and grooves (Fig. 35.3). At high magnification, they are composed of tightly packed chromatin fibers about 50–60 nm thick.

AFM images of wet chromosomes can also be obtained using a dynamic mode in a liquid environment [17, 18]. Because of their extreme softness in liquid environments, the interaction force between the tip and the sample was especially carefully adjusted to the minimum degree, otherwise chromosomes were easily deformed during scanning. The surface of chromatids in these wet chromosomes was characterized by the presence of alternating ridges and grooves, as observed in the critical-point-dried chromosomes (Fig. 35.4).

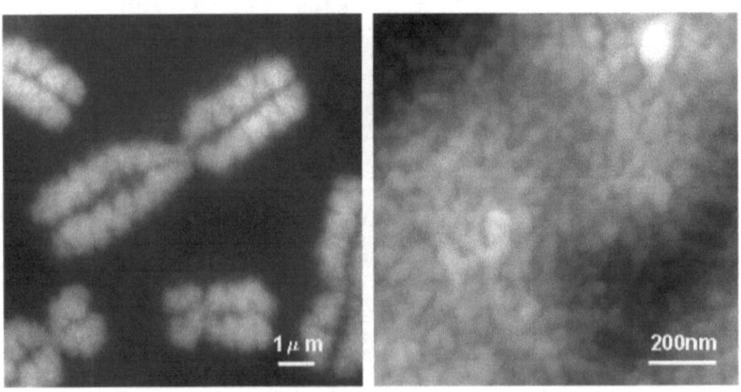

Fig. 35.3. AFM images of human metaphase chromosomes. At high magnification (*right*), the chromosomes were observed as an aggregation of strongly wound chromatin fibers

35.2.3.2
Collagen Fibrils

Collagen fibrils consist of collagen molecules which are assembled into a cylindrical fiber about 30–100 nm in diameter. When collagen fibrils with a diameter of about 40 nm are dried and observed by dynamic mode AFM, they appear to be semicylin-

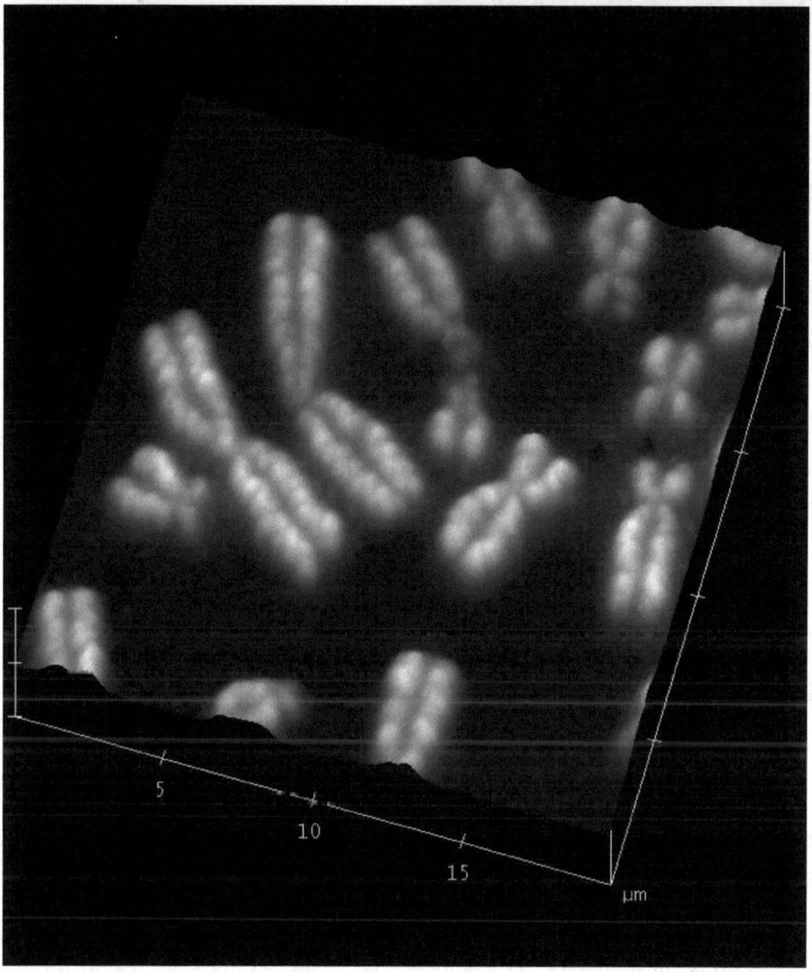

Fig. 35.4. AFM image of a chromosome spread in a phosphate-buffered saline solution. This image was obtained using dynamic mode AFM (Courtesy of Dr. Osamu Hoshi)

drical in profile, about 100 nm in width and 35–40 nm in height [19,20] (Fig. 35.5); the reason why the width of the fibrils is imaged about 3 times larger than their height is explained by the pyramidal shape of the AFM probe. Most fibrils examined by dynamic mode AFM showed a repeating pattern of grooves and ridges with a period of 60–70 nm, which is characteristic of collagen fibrils.

Very recently, high-speed AFM was introduced by Picco et al. [21]. With use of this technique, collagen fibrils were imaged both at high-speed video rate (30 fps) and at kiloheltz frame rates up to 1300 fps in ambient conditions. Thus, this high-speed AFM method is expected to become a powerful tool for investigating sequential processes of biological events occurring on a time scale of a few milliseconds.

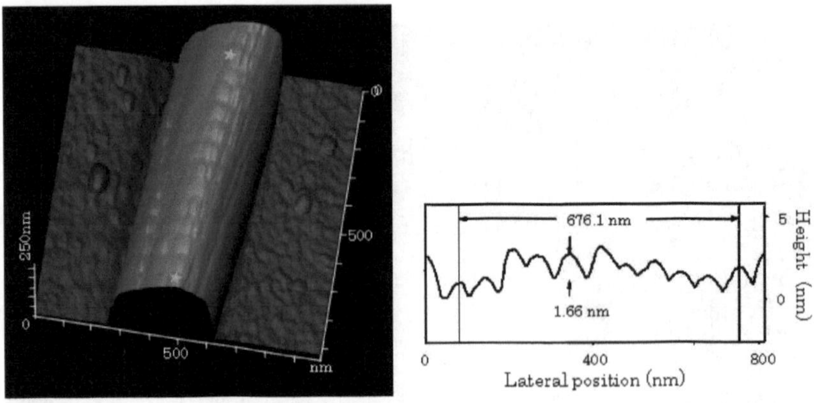

Fig. 35.5. AFM image of a bovine scleral collagen fibril. The fibril has periodic transverse grooves and ridges. Shallow longitudinal grooves are found on the surface of collagen fibrils. The longitudinal section profile of this fibril between the *two asterisks* is shown on the *left*. (Reproduced from [19])

35.2.4 AFM of Tissue Sections

In order to analyze internal structures of cells and tissues *in situ* with an atomic force microscope, relatively flat sectioned or fractured samples should be prepared because of the narrow *z*-range of the atomic force microscope scanner. Several investigators have attempted to observe the surface of ultrathin sections of LR white or epoxy resin with an atomic force microscope [22,23]. It is true that the contours of cellular structures such as unit membranes and chromatin blocks are traceable even in the AFM images of the resin sections. However, these cellular images are produced merely by depression or elevations of resin, probably owing to the uneven cleavage at the site of cellular structures during ultrathin sectioning. Thus, the possibility to obtain new information exceeding that from TEM observation is not expected with these sections. In this sense, embedment-free tissue sections are useful for AFM studies of cells and tissues because the internal structures can be observed directly at high resolution [24] (Fig. 35.6).

35.2.5 AFM of Living Cells and Their Movement

Because the atomic force microscope can be operated in liquid conditions, many researchers have been interested in visualizing the surface topography of cultured cells by AFM in liquid environments. For AFM imaging, flat cells should be attached firmly to the substrate (i. e., glass cover slip); spherical or domed cells are easily detached from the glass surface during scanning, probably because it is rather difficult to operate the tip accurately at the steep slope of the high sample. Adjustment of the imaging force to the weakest level is very important, otherwise cells are shrunk or damaged owing to the artificial forces overloaded by the scanning tip. When some kinds of cells of an epithelial nature are examined appropriately using contact mode

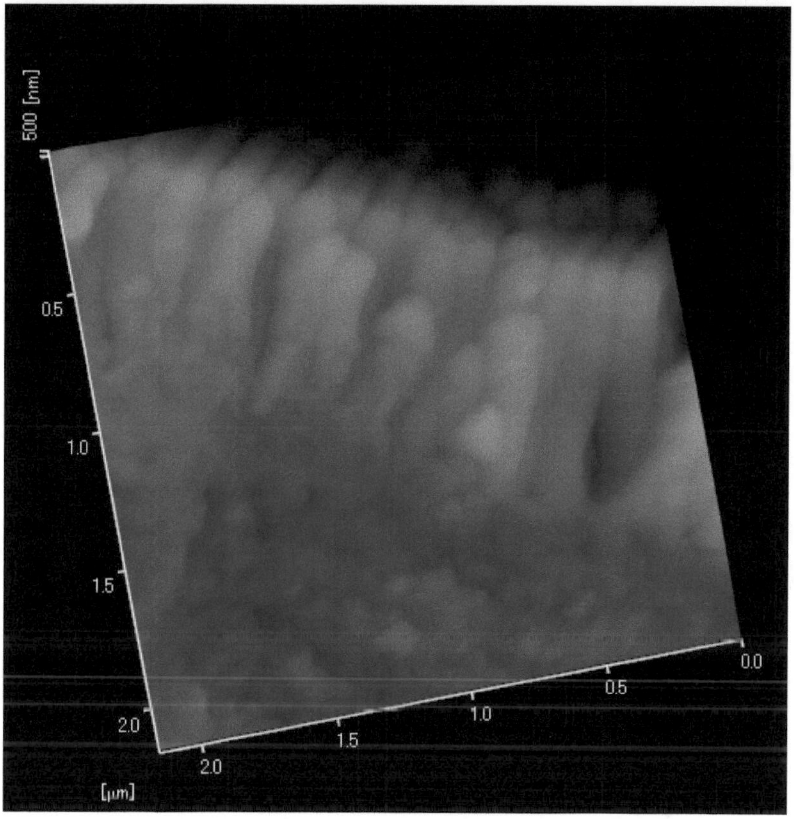

Fig. 35.6. AFM image of the surface of the embedment-free section of the mouse small intestine. Microvilli of the absorption cells are clearly three-dimensionally observed in this micrograph

AFM, cell processes such as lamellipodia are clearly observed at the cell margin of living cells as well as fixed ones [25, 26]. AFM also provides information on the contour of cytoskeletal elements just beneath the cell membrane, especially in living cells. In these studies, the height images are useful for measuring the thickness of these processes, but it may sometimes be difficult to visualize clearly the undulation of the cell surface. This is because the surface details may be hidden in the gradation of the wide range scale of the height images. Thus, variable deflection mode images (or error images) are often used in order to emphasize the contour of the cell surface as well as the shape of the cell processes (Fig. 35.7).

Because AFM can obtain images of living cells in liquid environments, there is the possibility that dynamic events of living cells can be analyzed in real time by AFM. The scan speed for AFM imaging of living cells depends on various factors, including the scan range, scanning mode and instrumentation. For example, in our previous studies of living epithelial cells, the practical scan speed was under 20 μm/s by contact mode AFM using our instrument [26]. This means that it takes 2 min or more to obtain a single image with a 40 μm × 40 μm scan area even when we use a 256 × 128 pixel size, and a series of AFM images with a 40 μm × 40 μm scan area

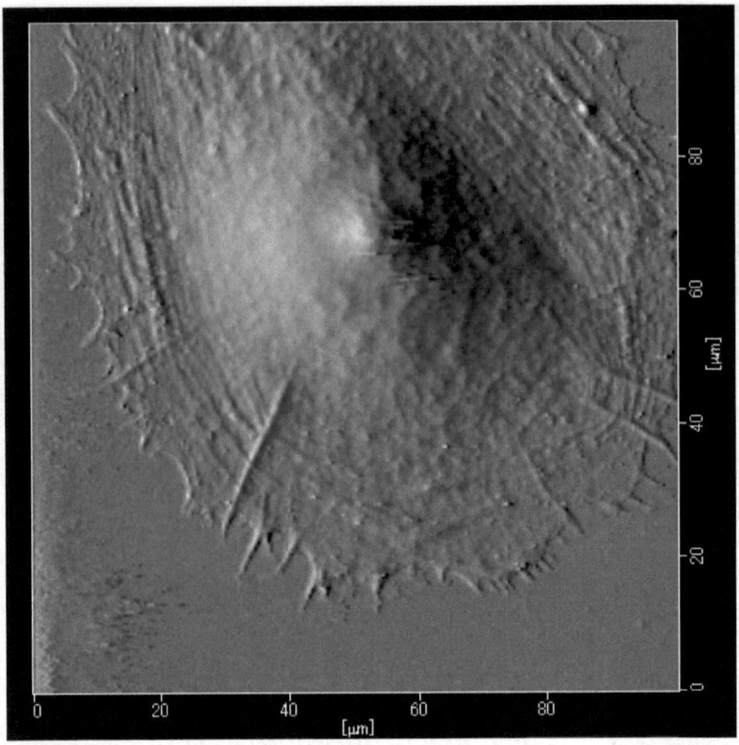

Fig. 35.7. AFM image of a living cultured cell (A549) in a liquid. This image was obtained using contact mode AFM

are acquired at time intervals of 2–8 min. Thus, at least, AFM is useful for examining dynamic events of living cells on a time scale of minutes. In order to observe living mammalian cultured cells by AFM under suitable culture conditions, it is necessary to maintain the pH of the culture medium either by exposing the chamber to 5% CO_2/95% air or by perfusing fresh culture medium into the chamber. Our studies showed that a fluid chamber system is useful for obtaining AFM images of living cells over the period of 1 h in a liquid environment. Several investigators have produced time-lapse photographic records from AFM images. These time-lapse movies are also very useful for realistically visualizing cellular dynamics. High-speed AFM is also expected to investigate cellular dynamics, including cell movements and secretory and excretory events, though these attempts are still awaited for with further studies [21].

35.2.6 Combination of AFM with Scanning Near-Field Optical Microscopy for Imaging Biomaterials

Recent advances in SPM have enabled the simultaneous collection of topographic and other physiological information of samples at the same portion of biological samples. AFM combined with scanning near-field optical microscopy (SNOM), or

SNOM/AFM, is one of the candidates for an attractive tool for the study of biological samples. This is because SNOM/AFM can collect both topographic and fluorescence images of the same portion of the samples simultaneously [27,28].

There are various types of SNOM/AFM instrumentation, but the SNOM/AFM instrumentation usually has an optical fiber probe or a special cantilever, either of which has a subwavelength (i.e., 50–100-nm) aperture at the tip. During the operation, excitation light is introduced to the aperture either through the optical fiber or from the rear side of the cantilever, resulting in the formation of the evanescent field between the tip and the sample (Fig. 35.8). When the fluorescent substance lies in the evanescent field, far-field light is emitted by the fluorescent substance, and is detected by light microscopy.

Using this instrument, one can observe a single DNA fiber at 100-nm resolution [29]. Several researchers including us have applied SNOM/AFM to chromosome research and have succeeded in simultaneously obtaining topographic and fluorescence images of such samples as fluorescence in situ hybridization treated chromosomes and immunostained chromosomes [28, 30]. For example, in chromosomes immunostained with an anti 5-bromo-2′-deoxyuridine (BrdU) antibody and Alexa Fluor 488 after incorporation of BrdU into DNA, the fluorescence images obtained by SNOM/AFM clearly showed the portion of sister-chromatid exchanges, and the corresponding topographic images provided precise information on the fine structure of those portions [30] (Fig. 35.9). Since the spatial resolution of the fluorescence images obtained is obviously higher than that of those obtained by conventional

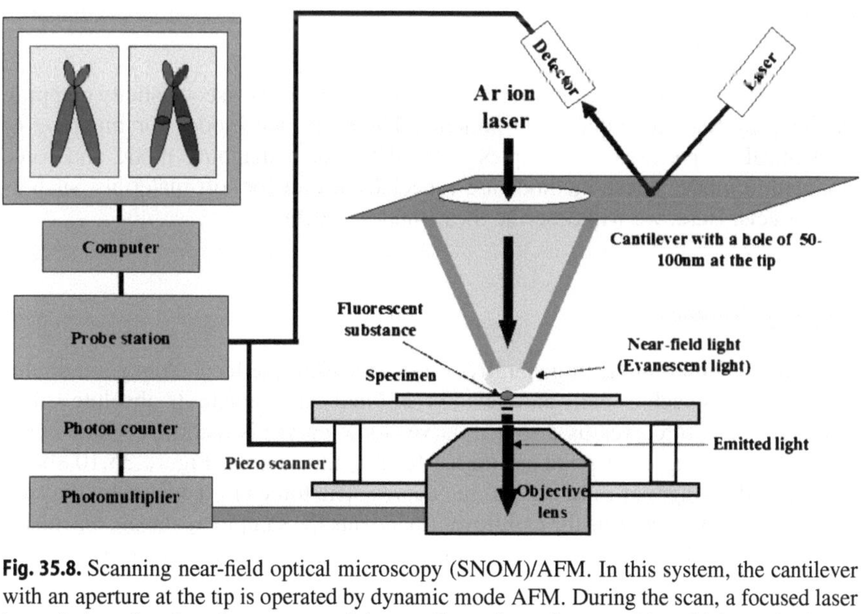

Fig. 35.8. Scanning near-field optical microscopy (SNOM)/AFM. In this system, the cantilever with an aperture at the tip is operated by dynamic mode AFM. During the scan, a focused laser beam is introduced to the aperture of the cantilever from the rear side, which results in the formation of an evanescent field between the tip and the sample. When a fluorescent substance is present in this field, light is emitted by the substance and is detected by a photomultiplier tube through the objective lens below the sample

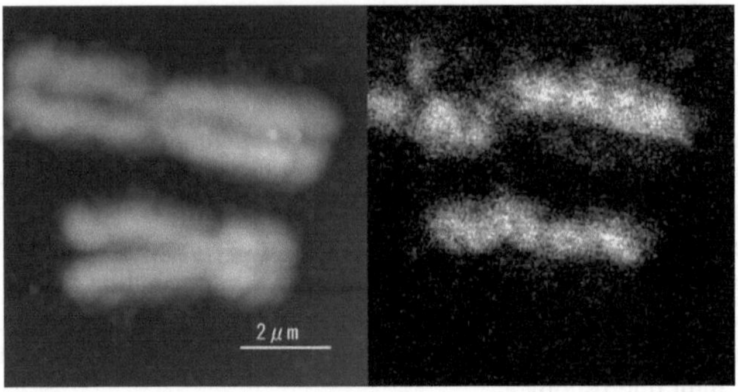

Fig. 35.9. SNOM/AFM images of 5-bromo-2′-deoxyuridine (BrdU) incorporated human chromosomes after differential staining with the anti-BrdU antibody. Topographic (*left*) and fluorescence (*right*) images of the same portion were simultaneously obtained by this technique. One of the paired sister chromatids was stained with fluorescence-labeled antibody. (Reproduced from [30])

fluorescence microscopy, SNOM/AFM is expected to serve as an important tool for future studies of the structure of biological samples in relation to the localization of their components.

35.3 SPM for Measuring Physical Properties of Biomaterials

35.3.1 Evaluation Methods of Viscoelasticity

Another important applications of SPM to biological fields is viscoelasticity mapping on living systems in liquid environments. There are two modes for mapping of mechanical properties over samples with SPM; force mapping mode and force modulation mode. These methods are especially useful for soft materials, such as cells or gels. Here, we will describe these modes in detail.

35.3.1.1 Force Mapping Mode

The force mapping mode is generally used to measure the local Young's modulus of soft samples such as cells and gels [31, 32] and can evaluate its absolute value. In this mode, a force versus distance curve (force curve) is recorded while a cantilever is indenting each point on the surface of the sample. Figure 35.10 shows schematic drawings for measuring force curves. The force (F) detected by the cantilever increases steeply, as the cantilever tip indents the sample by the piezoactuator translation (Z). The indentation depth of the sample δ and F are related by

$$Z - Z_0 = \frac{F}{k} + \delta ,$$

where k is the spring constant of the cantilever and Z_0 is the sample height under zero loading force.

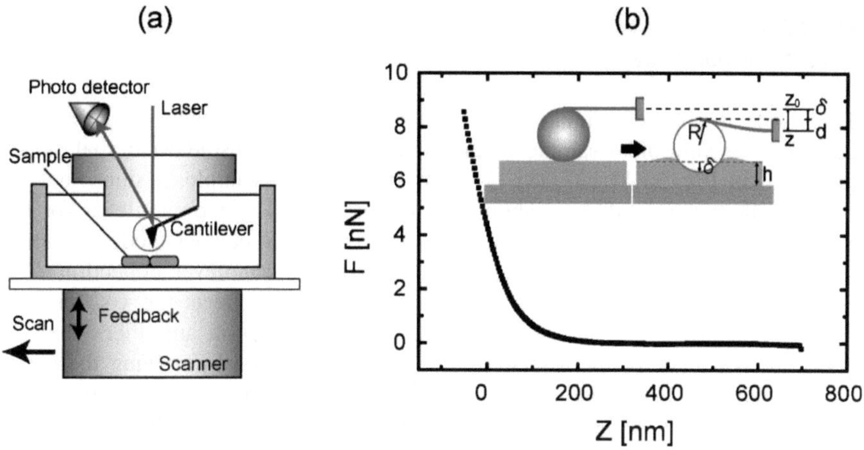

Fig. 35.10. Measuring the force curves in the force mapping mode

With use of nonlinear least squares fitting, the force curve obtained at one point of the sample's surface was fitted to Sneddon's model, a theoretical model in which the elastic contact theory of Hertz is applied to contact between an elastic plane and a rigid cone (the details are described in [33]). The applied loading force F is related to the indentation depth δ by

$$F = \frac{4\sqrt{R}}{3}\frac{E}{1-v^2}\delta^{2/3} ,$$

where E is the Young's modulus, v is the Poisson ratio of the sample, and R is the radius of the cantilever tip [34,35]. This fitting allowed the evaluation not only of the Young's modulus but also of the z-position, where the cantilever initially contacts the sample (i. e., the height of the sample at the point). For example, the force curve obtained from the cellular stiffness measurements was fitted in the loading force range 0.1–0.3 nN. The cellular Poisson ratio was assumed to be 0.5. Under the condition that a single force curve was taken for 100 ms, each image consisting of 64 × 64 pixels could be obtained in approximately 40 min. It should be noted that the spring constant of the cantilevers is chosen to be comparable to those of the samples, because this reduces sample damage and retains maximum sensitivity for elasticity measurements. This model is valid only when the indentation is much smaller than the thickness of the sample. For the case when the thickness of the sample is comparable to the indentation of the tip, Dimitriadis et al. [36] proposed an evaluation model that takes into account the finite thickness effect of samples on the estimation of the local elasticity. In this model, the relation between F and δ is modified as follows, where R is the radius of curvature of the cantilever tip and h is the thickness of the sample:

$$F = \frac{16E}{9}R^{\frac{1}{2}}\delta^{\frac{3}{2}}\left(1 + 1.133\frac{\sqrt{R\delta}}{h} + 1.283\frac{R\delta}{h^2} + \ldots\right) . \qquad (35.1)$$

35.3.1.2
Force Modulation Mode

This mode is able to estimate both local elasticity and viscosity of samples by detecting the sinusoidal sample indentation caused by sinusoidal external stress, as shown in Fig. 35.11 [37, 38]. To load stress on the sample, a sinusoidal vibration whose amplitude was about 10 nm was applied to the cantilever (or the piezo scanner) by using a function generator during the contact mode scanning. The periodic strain of the sample was detected as a cantilever deflection. The oscillating component of the cantilever deflection that reflects the viscoelasticity of the sample was extracted through a two-phase lock-in amplifier. The frequency dependence of the viscoelasticity can be measured by varying the oscillating frequency. In the liquid environment, however, the frequency should be set as low as possible, because the viscosity of the medium surrounding the sample cannot be neglected at higher frequencies. On the other hand, at lower frequencies, the temporal resolution becomes poorer. Then, the oscillating frequency was set at 100–500 Hz for viscoelastic imaging in a liquid. For example, a temporal resolution of 6 min for a 128 pixels × 64 lines image at 500 Hz. Since the feedback system for measuring topography cancels the periodic stress applied by the cantilever vibration, a band-elimination filter was used to remove the oscillating component from the cantilever deflection signal sent to the feedback system.

The viscoelasticity of the sample was calculated from amplitude and phase data by the method based on the linear viscoelasticity theory [38]. The output signals from the two-phase lock-in amplifier represent the amplitude ratio β and the phase lag θ_s of the cantilever deflection relative to the reference signals from the function generator. Using the experimental data β and θ_s, one can express Young's modulus E and the viscosity coefficient η of the sample as

$$E = C_1 \frac{\beta(\cos\theta_s - \beta)}{1 - \beta^2 - 2\beta\cos\theta_s}$$

and

$$\eta = C_2 \frac{\beta\sin\theta_s}{1 + \beta^2 - 2\beta\cos\theta_s},$$

where C_1 and C_2 are instrument constants including the spring constant of the cantilever, etc. For purpose of simplicity, we assumed that C_1 and C_2 were equal to

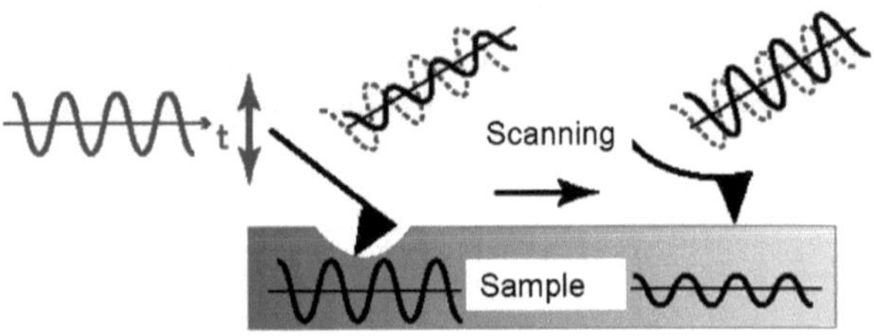

Fig. 35.11. The force modulation mode and a measurement system

1, so the values of E and η were relative quantities. In the case of the mechanical response originating only from the elastic property of samples, E simply represents Young's modulus of the materials.

35.3.2
Examples for Viscoelasticity Mapping Measurements

As examples for measuring physical properties of living systems by SPM, our results obtained from chromosomes, single cells and cell colonies are given in the following sections.

35.3.2.1
Chromosomes

The elasticity mapping of mitotic human chromosomes was carried out in phosphate-buffered saline (PBS) solution using the force mapping mode. The chromosomes were fixed with Carnor's fixative and kept in PBS solution to avoid drying. We used commercially available silicon nitride cantilevers with a length of 85 μm, a width of 20 μm, and a pyramidal tip with a typical radius of curvature of 50 nm. The spring constant of the cantilever was chosen to be 0.5 N/m to obtain the best sensitivity. The force curves obtained were evaluated by the Dimitriadis model taking account of the effect of the thickness of the sample [37]. In this study, we used only the first and second terms in (35.1) to compare the force curve with the model.

Figure 35.12 shows both topographic and elasticity images of the mitotic human chromosome in the PBS solution [39]. The height of the chromosome is about 150–360 nm. An imperceptible uneven surface structure of about 50–100 nm is visible in the morphology. This result agrees qualitatively with the spiral structure of the chromosome morphology previously reported in [16]. The elasticity distribution is not homogeneous over an entire chromosome, but a certain structure is visible. The values vary largely from 5 to 50 kPa.

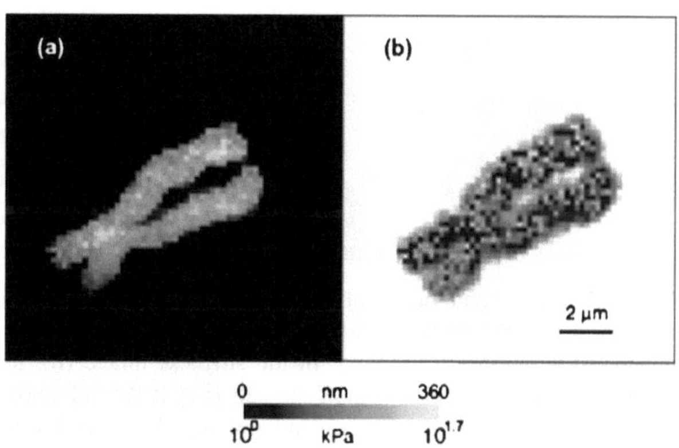

Fig. 35.12. **a** Topography and **b** elasticity images of a chromosome visualized by force mapping using a model that takes into account the effect of the thickness of the sample on the estimation of local elasticity

35.3.2.2
Living Single Cells

The viscoelastic measurements of living fibroblasts (NIH-3T3) were carried using the force modulation mode [40]. For the SPM imaging, the cell suspension was plated on a glass petri dish precoated with fibronectin, and then incubated for at least one night. To keep the pH constant during the SPM measurements, preheated HEPES buffer (pH 7.2–7.3) was substituted for the culture medium, and the samples were incubated for 1 h before the measurement.

Figure 35.13 shows typical topographic, local stiffness, and viscosity images of a living fibroblast. The cell extends thinly on the substrate and elongates toward the direction of crawling. Its height is about 3 μm. Many stress fibers appear on the cellular surface and align along the direction of movement. In the stiffness image, brighter areas represent greater hardness. We have confirmed that the distribution of the local stiffness measured using the force modulation mode is in good agreement with that obtained using the force mapping mode for an identical cell [38]. Therefore, the stiffness distributions range from several kilopascals to a few hundred kilopascals. It was found that local stiffness is not homogeneous on the cellular surface but varies largely from point to point. The stress fibers seen in the topographic image are harder than the surroundings, and this finding is consistent with the previous results [38, 41, 42]. In the viscosity image, brighter areas represent higher viscosity. The local viscosity of the area of the nucleus appeared to be lower than that of the periphery. However, further detailed, quantitative analysis will still be needed, because the values of the viscosity on the nanoscale have never been compared with those obtained by other methods. Sequential sets of topographic and local stiffness images of a typical moving cell are shown in Fig. 35.14. The topographic and stiffness images were taken simultaneously. The images were captured every 10 min. The SPM measurements were continued for about 60 min, which corresponds to the average time period of the intermittent cell crawling. The cell is changing its shape and moving downward in the image. A part of the cell is contracting, and the lamellipodium is extending gradually during the measurement. Accompanying the

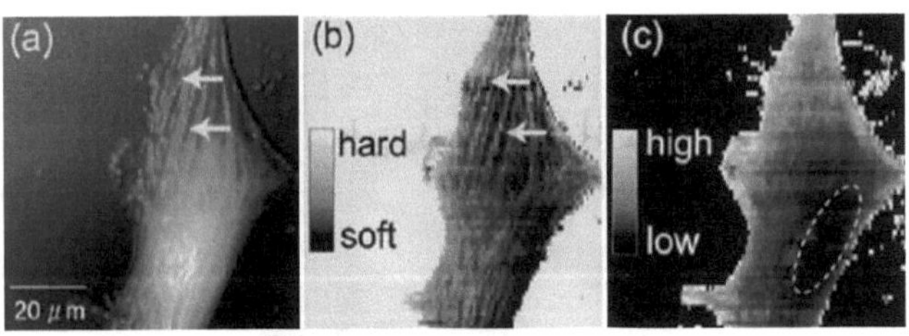

Fig. 35.13. Topographic (**a**), local stiffness (**b**), and viscosity (**c**) images of a fibroblast (NIH3T3) obtained using the force modulation mode. All images are 80 μm square. The topographic image (**a**) is graphically shaded. In the stiffness image (**b**), increased brightness represents greater hardness. The elasticity of the cell is spatially inhomogeneous. *Arrows* indicate stress fibers covering the cellular surface and aligning along the direction of the cell crawling. In the viscosity image (**c**), the brighter regions represent higher viscosity. The nuclear area within the *oval* appears to be less viscous than the periphery

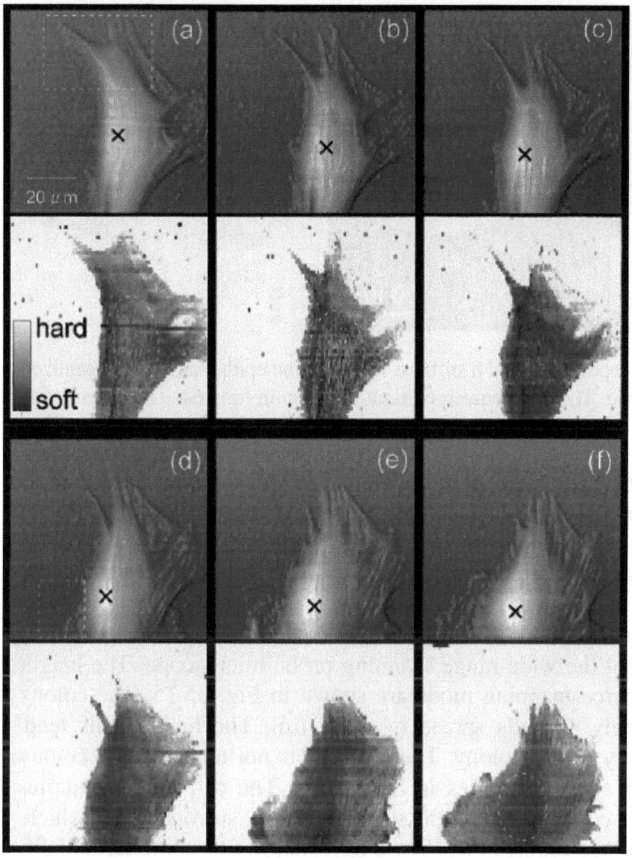

Fig. 35.14. Time-lapse images of topography and local stiffness of a moving cell. All images are 80 μm square. Topographic and stiffness images measured simultaneously are shown as coupled images. The coupled images were taken every 10 min. The posterior portion of the cell body (*dashed rectangle* in **a**) contracted, and a lamellipodiim (*dashed rectangle* in **d**) extended. The position of the nucleus (*cross*) moved downward. **a**, **b** The nucleus region (marked within an *oval*) is stiffer than the peripheral regions. Once the cell begins to move, the nucleus region becomes drastically softer

change in shape, the position of the nucleus is moving downward in the time course of the images.It should be noted that the local stiffness distribution also changes drastically with the cell migration.

35.3.2.3
Living Colony of Epithelial Cells

Epithelial cells (MDCK) were cultured in low-glucose Dulbecco's modified Eagle's medium containing 10% fetal bovine serum. The cells were grown in an incubator. The culture medium was substituted with HEPES buffer to avoid extreme pH de-

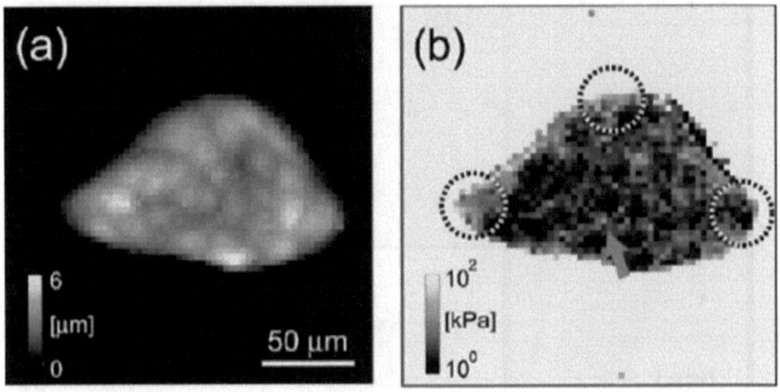

Fig. 35.15. **a** Topography and **b** stiffness images of an epithelial colony visualized using the force mapping mode. The data acquisition time was 30 min for a 64 × 64 pixel image

viation during the SPM measurements. A wide-range scanning probe microscope based on a conventional scanning probe microscope was developed [43]. In order to expand the measurement area, two conventional cylindrical piezotranslators were joined in tandem. We achieved the maximum scan ranges of 400 μm square in the xy-plane and 16 μm along the z-axis.

We measured the stiffness distribution and time-lapse images of a large epithelial colony using the wide-range scanning probe microscope. The height data imaged using the force mapping mode are shown in Fig. 35.15. The colony consisted of approximately 40 cells spread over 150 μm. The higher cells tend to locate on the boundary of the colony. The stiffness is not uniform in a colony; the average stiffness of each cell varies in the colony. The cell–cell boundaries (arrow) and the leading edges (circles) are stiffer than the surroundings, which is consistent with a previous report [44]. Figure 35.16 shows the time-lapse deflection images, which were measured in the constant force mode and captured every 1 h. The cells deformed cooperatively and induced migration of the whole colony. One portion of the colony (denoted by black circles in the figure) retracted inward, and then another portion (denoted by gray circles) extended outward. These results indicate that an epithelial colony moves collectively like single-cell migration. Recently we found the collective movements of thousands of epithelial cells on soft gels. [45] The present mechanical measurements are expected to provide the mechanism of the collective movements.

35.3.3 Combination of Viscoelasticity Measurement with Other Techniques

Combination of SPM with other techniques has the potential for investigating the physical properties of living systems in relation to their function. For example, in order to examine the response of living cells caused by external mechanical stimuli, stretching devices are very useful [46]. In our experimental systems, elastic chambers of silicone rubber are used to apply rapid, uniaxial deformation; a transparent bottom (20 mm × 20 mm) of 200-μm thickness and a wall of 5-mm height. These are deformable up to 20% along a single direction by the use of a handmade clamping de-

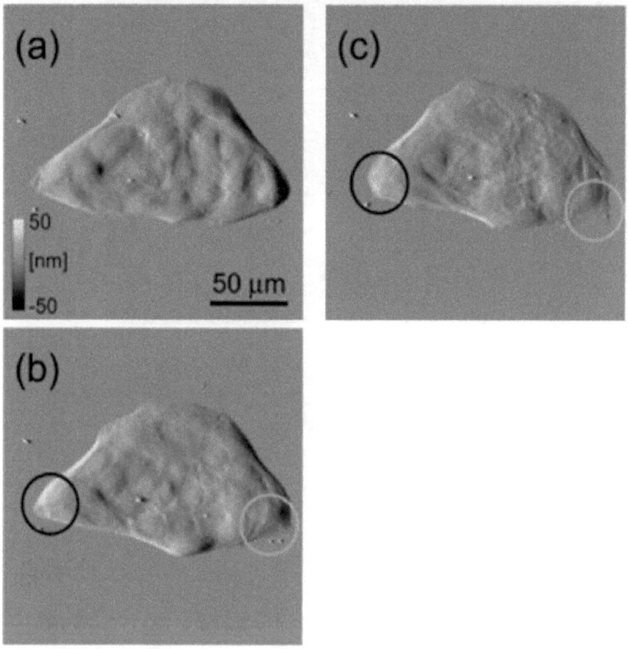

Fig. 35.16. Time-lapse deflection images of an epithelial colony taken every 1 h. The scanning frequency was 0.25 Hz. The posterior portion of the colony (*black circles*) contracted, and then the anterior portion (*gray circles*) elongated with a thin lammellipodium

vice. In order to fix living cells firmly, the silicone rubber substrate is usually coated with 20 μg/ml fibronectin in PBS for 30 min. During the experiment, trypsinized fibroblasts are plated onto either the prestretched or the unstretched chambers and incubated for 10 h. In order to observe a change in the cytoskeleton network, fibroblasts transfected with green fluorescent protein (GFP)–actin are stretched or compressed uniaxially on the silicon substrate, while the GFP fluorescence images are obtained by confocal microscopy.

A time-lapse series of topographic and stiffness images of a fibroblast, before and after the 8% elongation, is shown in Fig. 35.17. The images were captured every 30 min using the force mapping mode. The cell shape and stiffness were stable for 1 h before the elongation, but the height of the cell reduced and the stiffness of the cell increased immediately after the elongation. Many structures corresponding to the stress fibers became much stiffer after the elongation, and then softened gradually during the following 2 h. On the other hand, when the cells are contracted, the stiffness decreases instantly after the contraction and then increases gradually and stabilizes during the following 2 h. To examine whether or not the stress fibers are reconstructed after the deformation, fluorescence observation for GFP–actin was performed. Changes in the distribution of actin filaments beneath the cell membrane were observed in real time, while cells were deformed to a degree of 8%. Figure 35.18 shows a typical time-lapse series of fluorescence micrographs of GFP–actin, when cells were elongated or compressed. New thin stress fibers that did not significantly

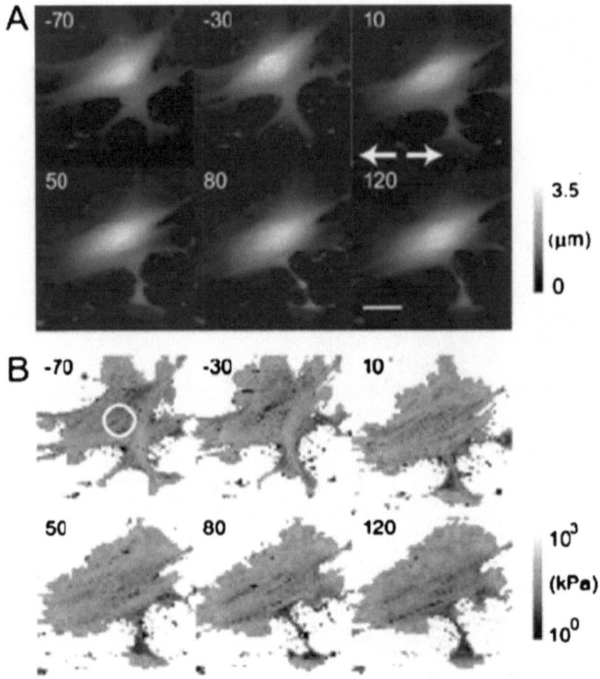

Fig. 35.17. Time-lapse SPM images of **a** topography and **b** stiffness distribution of a fibroblast before and after the application of force. Topographic and stiffness images measured simultaneously are shown as coupled images. *Numbers* represent the relative time (minutes) from the moment of substrate stretch. The *arrows* in the topography image represent the direction of the substrate stretch. The *circle* in the stiffness image represents the typical area used for the averaging of stiffness. *Bar* 20 μm

alter the cellular stiffness observed by SPM appeared slightly. These results indicate that the stiffness responses to the deformation originate from the contractile tension generated by the stress fibers, and the existence of tensional homeostasis at the cellular level.

35.4 SPM as a Manipulation Tool in Biology

In addition to imaging and/or measuring the physical properties of biological samples, the atomic force microscope has the potential to be used as a dissection tool for tiny substances at atomic-scale to submicron-scale levels. For example, with use of the contact mode, DNA molecules immobilized on mica can be dissected both in air and in a liquid environment by increasing the force applied to the AFM probe [47]. The size of these biomolecules is suitable for dissecting and manipulating by contact mode AFM using a standard probe. Dissection of the large biological structures such as a chromosome has also been performed by several investigators.

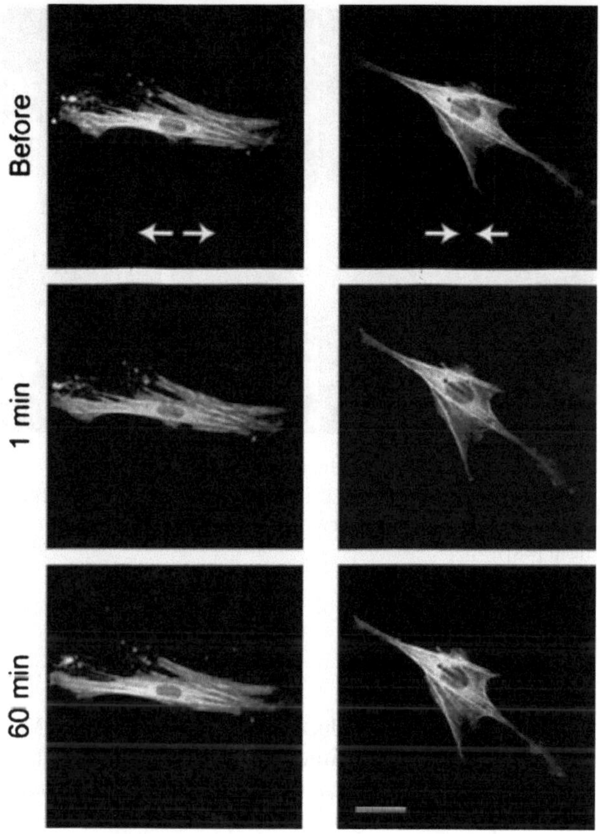

Fig. 35.18. Temporal behavior of green fluorescent protein–actin expressing fibroblasts, when subjected to elongation or 8% compression. The direction is shown by *arrows*. *Bar* 20 μm

However, it is often difficult to sharply dissect these samples using the contact mode, because samples are crushed or scratched by the scanning probe. Thus, dissection techniques using a dynamic force mode have been introduced for these thick and wide samples [48] (Fig. 35.19).

The development of a suitable and novel AFM probe is also an interesting subject for AFM manipulation of biological samples. For dissection of biomolecules at the nanoscale, a carbon nanotube probe has been applied. More recently, a knife-edged AFM probe was introduced for dissection of DNA and chromosomes [49]. With use of this probe, chromosomes can be cut in a dynamic mode more sharply and consistently than when using the conventional AFM probe. In addition, a tweezer-type AFM probe was also introduced for manupilation of nanomaterials [50]. This device has two thin probes, which are formed in the front of the silicon cantilever, with a vertical triangular shape. One functioned as an AFM imaging probe tip (sensing probe) for a dynamic mode, while the other can be made to work as a tweezer, by closing the vertical face of the probe (the movable probe) owing to the effect of the thermal expansion actuator. This tweezer-type AFM probe can be used for picking up small fragments after dissection of the samples.

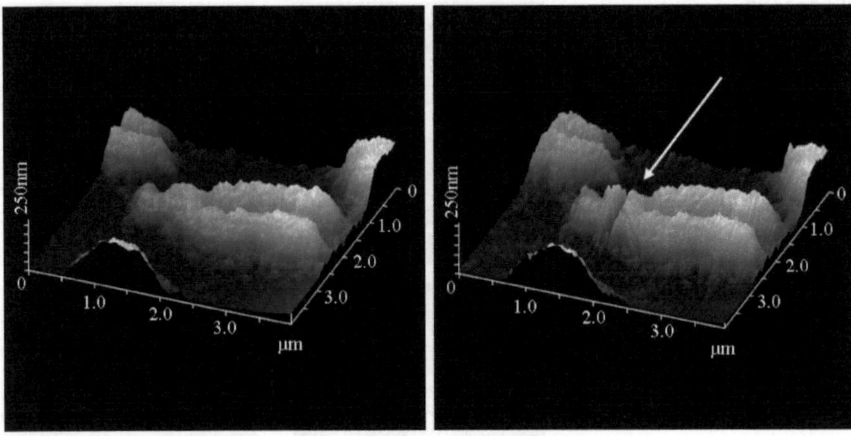

Fig. 35.19. AFM image of a human metaphase chromosome before (*left*) and after (*right*) dissection with an AFM probe using a dynamic mode. The *arrow* indicates the dissection place. A dissected fragment is observed on the left side of the cut part of the chromosome. (Reproduced from [48])

35.5 Conclusion

The present study introduced the application of SPM to biological fields for investigating various structures from biomolecules to living cells. As shown in this chapter, SPM provides much information which cannot obtained with other type of microscopes such as light microscopes and electron microscopes. In this sense, SPM has great potential for broadening and opening up new fields in biomedical research. SPM instrumentation and the technique of specimen preparation for biological studies are advancing, and we are just opening the door to the new microscopic world which will be made available by SPM for solving enigmas in biomedical fields.

References

1. Guckenberger R, Hartmann T, Wiegräbe W, Baumeister W (1992) In: Wiesendanger R, Güntherodt HJ (eds) Scanning tunneling microscopy II. Springer, Berlin, p 51
2. Binnig G, Quate CF, Gerber C (1986) Phys Rev Lett 56:930
3. Lal R, John SA (1994) Am J Physiol 266:C1
4. Ikai A (1996) Surf Sci Rep 26:261
5. Ushiki T, Hitomi J, Ogura S, Umemoto T, Shigeno M (1996) Arch Histol Cytol 59:421
6. Jena BP, Hörber JKH (2002) (eds) Atomic force microscopy in cell biology. Academic, San Diego
7. Hörber JKH, Miles MJ (2003) Science 302:1002
8. Jena BP, Hörber JKH (2006) (eds) Force microscopy. Applications in biology and medicine. Willy-Liss, Hoboken
9. Braga PC, Ricci D (2004) (eds) Atomic force microscopy. Biomedical methods and applications. Humana, Totowa
10. Liu Z, Li Z, Wei G, Song Y, Wang L, Sun, L (2006) Microsc Res Tech 69:998

11. Müller DJ, Heymann JB, Oesterhelt F, Moller C, Gaub H, Buldt G, Engel A (2000) Biochim Biophys Acta 1460:27
12. Müller DJ, Engel A (2002) In: Jena BP, Hörber JKH (eds) Atomic force microscopy in cell biology. Academic, San Diego
13. Zuccheri G, Samorì B (2002) In: Jena BP, Hörber JKH (eds) Atomic force microscopy in cell biology. Academic, San Diego
14. Hohmura KI, Itokazu Y, Yoshimura SH, Mizuguchi G, Masamura YS, Takeyasu K, Shiomi Y, Tsurimoto T, Nishijima H, Akita S, Nakayama Y (2000) J Electron Microsc 49:415
15. Yamamoto S, Nakamura F, Hitomi J, Shigeno M, Sawaguchi S, Abe H, Ushiki T (2000) J Electron Microsc 49:473
16. Ushiki T, Hoshi O, Iwai K, Kimura E, Shigeno M (2002) Arch Histol Cytol 65:377
17. Hoshi O, Owen R, Miles M, Ushiki T (2004) Cytogenet Genome Res 107:28
18. Hoshi O, Shigeno M, Ushiki T (2006) Arch Histol Cytol 69:73
19. Yamamoto S, Hitomi J, Shigeno M, Sawaguchi S, Abe H, Ushiki T (1997) Arch Histol Cytol 60:371
20. Ushiki T (2002) Arch Histol Cytol 65:109
21. Picco LM, Bozec L, Ulcinas A, Engledew DJ, Antognozzi M, Horton MA, Miles MJ (2007) Nanotechnology 18:1
22. Yamamoto A, Tashiro Y (1994) J Histochem Cytochem 42:1463
23. Yamashina S, Shigeno M (1995) J Electron Microsc 44:462
24. Ushiki T, Shigeno M, Abe K (1994) Arch Histol Cytol 57:427
25. Ushiki T, Yamamoto S, Hitomi J, Ogura S, Umemoto T, Shigeno M (2000) Jpn J Appl Phys 39:3761
26. Ushiki T, Hitomi J, Umemoto T, Yamamoto S, Kanazawa H, Shigeno M (1999) Arch Histol Cytol 62:47
27. Muramatsu H, Chiba N, Nakajima K, Ataka T, Fujihira M, Hitomi J, Ushiki T (1996) Scanning Microsc 10:975
28. Ohtani T, Shichiri M, Fukushi D, Sugiyama S, Yoshino T, Kobori T, Hagiwara S, Ushiki T (2002) Arch Histol Cytol 65:425
29. Yoshino T, Sugiyama S, Hagiwara S, Fukushi D, Shichiri M, Nakao H, Kim JM, Hirose T, Muramatsu H, Ohtani T (2003) Ultramicroscopy 97:81
30. Kimura E, Hitomi J, Ushiki T (2002) Arch Histol Cytol 65:435
31. Haga H, Sasaki S, Kawabata K, Ito E, Ushiki T, Sambongi T (2000) Ultramicroscopy 82:253
32. Nitta T, Haga H, Kawabata K, Abe K, Sambongi T (2000) Ultramicroscopy 82:223
33. Vinckier A, Semenza G (1998) FEBS Lett 430:12
34. Munevar S, Wang YL, Dembo M (2001) Mol Biol Cell 12:3947
35. Lauffenburger DA, Horwitz AF (1996) Cell 84:359
36. Dimitriadis EK, Horkay F, Maresca J, Kachar B, Chadwick RS (2002) Biophys J 82:2798
37. Radmacher M, Tillmann RW, Gaub HE (1993) Biophys J 64:735
38. Haga H, Nagayama M, Kawabata K, Ito E, Ushiki T, Sambongi T (2000) J Electron Microsc 49:473
39. Nomura K, Hoshi O, Fukushi D, Ushiki T, Haga H, Kawabata K (2005) Jpn J Appl Phys 44:5421
40. Nagayama M, Haga H, Kawabata K (2001) Cell Motil Cytoskeleton 50:173
41. Kawabata K, Nagayama M, Haga H, Sambongi T (2001) Curr Appl Phys 1:66
42. Rotsch C, Radmacher M (2000) Biophys J 78:520
43. Mizutani T, Haga H, Nemoto K, Kawabata K (2004) Jpn J Appl Phys 43(7B):4525
44. A-Hassan E, Heinz WF, Antonik MD, D'Costa NP, Nageswaran S, Schoenenberger CA, Hoh JH (1998) Biophys J 74:1564
45. Haga H, Irahara C, Kobayashi R, Nakagaki T, Kawabata K (2005) Biophys J 88:2250
46. Mizutani T, Haga H, Kawabata K (2004) Cell Motil Cytoskeleton 59:242

47. Hansma HG, Vesenka J, Siegerist C, Kelderman G, Morrett H, Sinsheimer RL, Elings V, Bustamante C, Hansma PK (1992) Science 256:1180
48. Iwabuchi S, Mori T, Ogawa K, Sato K, Saito M, Morita Y, Ushiki T, Tamiya E (2002) Arch Histol Cytol 65:473
49. Saito M, Nakagawa K, Yamanaka K, Takamura Y, Hashiguchi G, Tamiya E (2006) Sens Actuators A 130–131:616
50. Takekawa T, Nakagawa K, Hashiguchi G, Saito M, Yamanaka K, Tamiya E (2005) IEEJ Trans SM 125:448

5 Scanning Probes for the Life Sciences

Andrea M. Ho · Horacio D. Espinosa

From volume VIII

Abstract. Scanning probe based patterning techniques have the unique ability to deposit biological material into specific architectures on substrates and read and analyze the patterns using an atomic force microscope. Such devices are able to make much smaller biomolecule patterns, on the order of nanometers, than conventional techniques such as microcontact printing and optical lithography. A reduction in patterned feature size allows for greater sensitivity in biological studies and in life sciences applications such as drug screening and immunoassays. A variety of tools for the fabrication of nanoarrays are discussed. These include open- and closed-channel devices and pipette-based devices. Their potential for the integration of active components or augmentation to large-scale arrays for high-throughput deposition are examined. The mechanisms for deposition and biomolecule transport are also explained.

Key words: Nanoarrays, Patterning, Atomic force microscopy, Dip-pen nanolithography, Fountain probe, Protein array, Water meniscus

5.1 Introduction

In recent years novel techniques for depositing very small amounts of material, on the order of picoliters or less, onto surfaces at precise locations have been developed. While such efforts are clearly relevant in the fabrication of miniature electronics devices, it is now apparent that such techniques have many applications in biology and the life sciences as well. For example, the ability to spatially orient and immobilize biomolecules on a solid substrate is useful in the development of genomic or proteomic profiles of cells, drug screening, as well as biosensing, which requires precise and high-density arrays of biological material. Because cell functions are often mediated by the binding of a ligand to its membrane receptor, the study of cellular functions of living cells at the nanometer scale requires a device capable of delivering proteins to a particular location at a specific time [1]. This must be followed by an observation of the response.

In the design of a biomolecule-patterning device it is necessary to consider a multitude of factors. These include its resolution—the minimum feature size that can be patterned on the substrate; its reproducibility—small features must be repeatable and reliably patterned; and its ability for precise positioning. For use in high-throughput applications, a device should be easily scalable, i.e., its design should enable it to pattern arrays of biological material on the scale of a few centimeters. Devices that could deliver multiple solutions would have the advantage of depositing different

In: H. Fuchs, B. Bhushan (eds), *Biosystems - Investigated by Scanning Probe Microscopy*,
DOI 10.1007/978-3-540-74080-3_5,

molecules within one set of nanostructures spaced nanometers apart. For applications in the life sciences, a device must be able to function in ambient conditions. Finally, the ability to verify the successful deposition of biological material is also useful.

A key aspect in nanopatterning is the careful control of substrate surface chemistry. Nonspecific binding must be avoided otherwise intended signals will be masked in biosensing applications. Protein patterns must also be stable because in biosensors or bioanalytical devices immobilized proteins are often rinsed or washed with water, buffer solutions or surfactants [2]. This can be controlled by an appropriate choice of interactions used for immobilization. Immobilization via electrostatic interactions is normally reversible; proteins may be removed using certain buffers or surfactants. For long-term stability, covalent binding, which involves the formation of disulfide, imine or amide bonds, is often used.

In this chapter we review micro and nano patterning technologies with a particular emphasis on bioapplications. We begin by looking at microscale techniques for DNA and protein studies and then proceed to a review of nanoscale technologies, focusing on probe-based device designs and patterning results. The theoretical aspects behind nanoscale deposition is discussed followed by an overview of advances in the parallelization of the aforementioned devices.

5.2 Microarray Technology

Molecular spot arrays can be used for massively parallel determination and measurements of binding events and open up possibilities for automation. A major benefit of microarrays and nanoarrays is that they require only minute amounts of sample material. In essence, microarrays consist of many microscopic spots, each containing identical molecules, attached to a solid support. In a typical microarray experiment, or immunoassay, these spots contain one binding partner, such as a receptor or ligand. A sample containing targets to be investigated is then added to the array, and binding between the probe spots and the targets may occur. Fluorescent labels are often used to detect such binding events. DNA is likely the most interesting biomolecule because of its role in information storage [3]. Its stability also makes it useful in directing the immobilization and assembly of nanostructures. Patterned in arrays, DNA can be used as probes to analyze genetic defects and single nucleotide polymorphisms using lab-on-a-chip approaches.

There are two main methods for microarray production. The first involves synthesis on the chip to create libraries of short oligonucleotides (short sequences of nucleotides, the structural units of DNA and RNA) or peptides (compounds of two or more amino acids). This may be accomplished using techniques such as optical lithography. The second method involves separate synthesis with subsequent deposition on the chip, which is required for larger biomolecules such as polynucleotides or proteins. This type of deposition has been accomplished with restricted reproducibility via contact printing using pin tools dipped into a sample solution and brought into contact with the support material, thereby dispensing the solution. Optical lithography and microcontact printing will be discussed in greater detail in the following section. The experimental variability using microspotting techniques is expected to decrease with the recent advances being made [4]. Other devices are noncontact equipment such as inkjet printers which use the piezo effect to deposit nanoliter volumes of solution [3]. The distance between deposited drops has been minimized to 200 μm, resulting in arrays that are not very dense. Nevertheless, inkjet methods allow for in situ synthesis and can therefore deposit longer oligonucleotides than methods such as photolithography. Longer oligonucleotides offer sufficient specificity to detect genes using fewer probes [5].

5.2.1 Microcontact Printing

Microcontact printing is one method used to create microarrays. The technique covers an existing patterned surface, the master, with a liquid prepolymer. This is then cured to create an elastomeric stamp [6]. Typically, poly(dimethylsiloxane) (PDMS) is used. The hardened stamp is then peeled off the master. Ink solution is applied, the solvent is allowed to evaporate and the stamp is brought into contact with a substrate, at which point the ink is transferred. The ink forms a self-assembled monolayer (SAM) which is a replica of the master pattern.

Microcontact printing was first introduced by Whitesides and coworkers in 1993 as an alternative to photolithography [7]. The technique can be used to modify the adsorption properties of gold substrates for the attachment of proteins. Because proteins adsorb preferentially on some materials and are repelled by others, specific materials such as alkanethiols can be transferred via microcontact printing onto gold in order to functionalize it for protein patterning. However, proteins can also be directly transferred by the elastomeric stamp. Untreated PDMS provides a hydrophobic surface very much like the polystyrene used for adsorption of proteins in immunoassays. The process for protein patterning is as follows. A PDMS stamp is covered with protein solution for inking. This creates a monolayer of protein on the stamp surface. The stamp is rinsed and dried, and the pattern is then transferred onto a substrate. Contact need only be made for a few seconds and pattern transfer occurs only where the stamp contacts the substrate. Immunoglobulin G (IgG) protein was transferred onto a silicon wafer using a stamp made of the siloxane Sylgard 184. Features with dimensions as small as 500 nm were replicated. Atomic force microscopy (AFM) imaging verified that patterns had high contrast and resolution because the stamp was mechanically stable and proteins did not diffuse significantly across the surface. The features also retained their biological activity after printing.

For proteins that may not survive adsorption processes at surfaces, additional steps may be taken to immobilize them onto a substrate although this lengthens and complicates the process [8]. Aside from proteins, other patterned materials include lipid bilayers and poly(amino acids). In the case of lipid bilayers however, the printed lines were expected to be 9-μm wide, but were actually 19-μm wide. Because PDMS is deformable it would not be surprising if areas surrounding the intended pattern also made contact with the substrate.

Microcontact printing is suitable for printing over large areas at once. However, resolution is limited by the feature size of the master, mechanical deformation of the stamp and ink diffusion around the contact areas. The resolution may be improved by using stamps of higher stiffness. However, high-resolution masters require electron-beam patterning and reactive ion etching of a silicon-on-insulator (SOI) wafer. Michel et al. [9] used a high-stiffness PDMS stamp with 80-nm-diameter posts to transfer single antibody molecules. Although they claim 99% transfer efficiency, the resulting spots were not regular in size and shape. Finally, a major feature of microcontact printing is that the flexible silicone stamps require a casting mold, the master, with a predefined layout. This may be costly and may necessitate a series of photolithography processing steps [10].

5.2.2 Optical Lithography

The photolithographic technique can create DNA patterns with feature sizes as small as 18 μm [5]; however, it is expensive and places a limit on the allowable oligonucleotide length. For reliable detection of each gene, ten to 20 probes are needed, thereby limiting arrays to about 12000 genes per square centimeter. Smaller feature sizes are possible using polymeric photoresists used in the semiconductor industry which exhibit a nonlinear response to illumination intensity. Features as small as 8 μm have been constructed in this manner. UV lithography can be used to directly pattern alkanethiol SAMs with micron-scale resolution [11]. It is then possible to attach biomolecules to these patterns.

Lithographic patterning of DNA operates as follows (Fig. 5.1). Synthetic linkers modified with photochemically removable protecting groups are attached to a glass substrate [12]. The substrate is exposed to light directed through a mask to selectively deprotect and activate certain sites. Protected nucleotides can then attach to these

Table 5.1. Techniques suitable for use in the life sciences

Deposition method	Best resolution	Ink	Substrate	Reference
Microcontact printing	500 nm	IgG protein		[7]
	19 μm	Lipid bilayers		[8]
Optical lithography	8 μm	Alkanethiols		[11]
	500 nm	Streptavidin		[16]
Dip-pen nanolithography	30-nm lines	Collagen	Gold	[43]
	45-nm dots	IgG protein		[44]
Surface patterning tool	2–3-μm dots	Cy3-streptavidin		[51]
	150-nm lines	Quantum dots conjugated to streptavidin		
Microspotters	30-μm dots	IgG protein and oligonucleotides	Glass	[53]
Nanopipettes	440-nm dots	IgG protein	Glass	[61]
	510-nm dots	Biotinylated DNA		
Nanofountain probe	40-nm lines	MHA	Gold	[35]
	200–300-nm dots	DNA		[67]
	200–300-nm dots	IgG protein		Unpublished

IgG immunoglobulin G, *MHA* 16-mercaptohexadecanoic acid

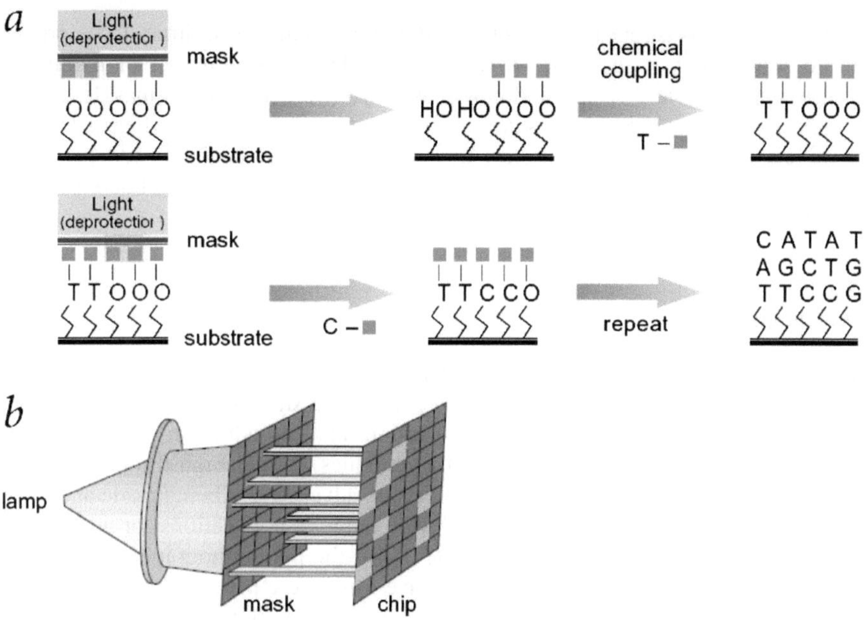

Fig. 5.1. a Photolithographic oligonucleotide synthesis. Light directed through a mask activates certain sites and protected nucleotides then couple to these sites. The process is repeated, with different sites being activated and different bases being coupled. **b** A lamp, mask and array [12]

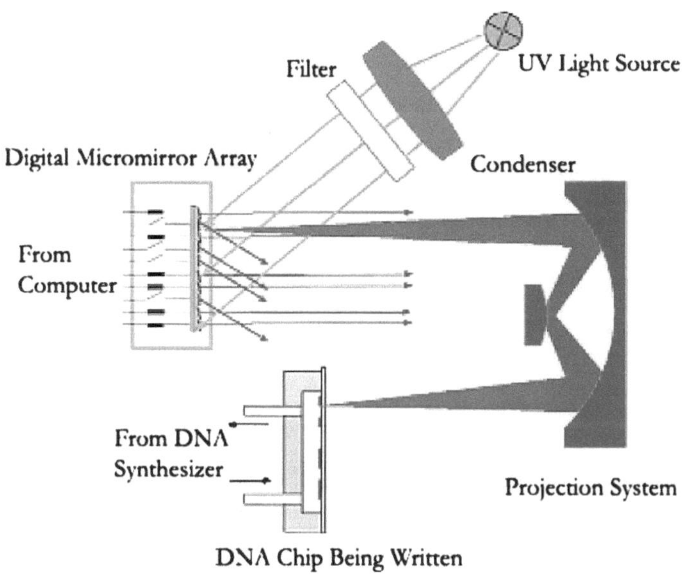

Fig. 5.2. The maskless array synthesizer [13]

sites. The process is repeated; coupling different bases allows arbitrary DNA probes to be constructed at each site. A drawback is that a changing set of monitored genes requires a new design and new masks. To overcome this, a maskless array synthesizer (MAS; Fig. 5.2) was developed [13]. Computer-generated virtual masks were relayed to a digital micromirror array which used 1:1 imaging to address pixels in a 10 mm × 14 mm area. The MAS system was used to print checkerboard patterns of 16-μm features (the size of each micromirror).

5.2.3 Protein Arrays

Protein patterning is useful for miniaturizing biochemical tests for high-throughput screening of new drug candidates and for performing studies in protein expression, protein–protein and protein–enzyme interactions and cell adhesion. Whereas DNA microarrays have been well established for genomics, no such high-throughput technologies yet exist for proteomics [14]. However, advances are being made and there are many protein array suppliers for the wide range of applications for which they are needed. (Listings are found in [14, 15]). For the photolithographic production of submicron protein patterns, pH-responsive films were patterned by exposure to 365-nm light. This allowed for the conjugation of aminooxy-modified biotin and the subsequent immobilization of streptavidin. Protein patterns as small as 500 nm were produced [16].

A conventional assay technique is the enzyme-linked immunosorbent assay (ELISA; Fig. 5.3), used to detect peptides, proteins, antibodies or hormones [17, 18].

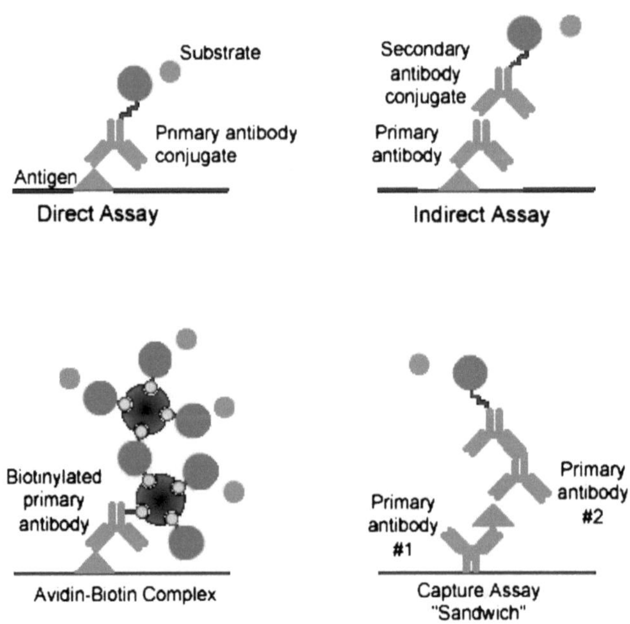

Fig. 5.3. ELISA immunoassay formats [17]

For antibody detection, an antigen is immobilized on a solid support. The sample for inquiry (any body fluid) is applied, and if the antibody in question is present it will bind to the antigen. Detection is accomplished via an enzyme directly linked to this primary antibody or a secondary antibody that recognizes the primary antibody. Or, if the primary antibody has been labeled with biotin, it can then be incubated with streptavidin. Incubation of this complex with an appropriate substrate produces a detectable product; typically a color change is imparted. The amount of color is measured proportional to the amount of antibody present in the sample. The most commonly used enzymes are horseradish peroxidase and alkaline phosphatase. ELISA is typically performed in 96-well or 384-well polystyrene plates coated with a ligand which will passively bind antibodies and protein; thus, unbound materials can be easily separated from the bound material during the assay. False positives are possible owing to the nonspecificity of protein–antibody interactions, and some analytes such as HIV require retesting using western blot, an electrophoretic technique where antibodies are directed against a number of viral proteins.

5.3 Nanoarray Technology

5.3.1 The Push for Nanoscale Detection

The delivery of biological material in increasingly smaller volumes has the potential to advance many applications. Studies of protein function are made possible, for instance, protein clustering in cell focal adhesion occurs at 5 200 nm length scales; delivery of material at this relevant length scale allows proteins of interest to be bound to a substrate [19]. Lateral control in creating specific adhesive and inert sites can create rigid ligand templates for cell binding. The specific interactions between binding pairs can be used for protein immobilization; these include affinity capture ligands (streptavidin–biotin binding) and antigen–antibody recognition for immunoassays [20]. Phospholipid deposition can be used to make model systems that can mimic the structural complexity of biological membranes [21]. It is also useful for studies investigating the binding of a protein and drug with supported lipid bilayers.

In the fabrication of arrays, it is important to reduce the sample volume, particularly in applications where limited sample amounts are available, such as in the analysis of multiple tumor markers from biopsy material [14]. Laser capture microdissection for cancer biomarker screening can obtain a few cells from a cancer tissue section, but it is very difficult to analyze the protein content of these cells using microarrays [22]. On the other hand, nanoarrays require only minute volumes of sample, approaching the volume of a single cell, so solid-phase testing on single cells can be performed.

A reduction in an array spot size suggests that statistically every molecule in an analyte has the opportunity to sample the entire capture surface in a reasonable amount of time [24]. In microarrays, not all of the sample will reach the capture surface unless energy is added to enhance the movement of the analyte, such as by agitation, mixing or electromotive force; thus, part of the analyte is effectively wasted.

Smaller arrays imply that biochemical reactions may not be diffusion-limited. In effect, nanoarrays offer the possibility of greater sensitivity in diagnostic tests, since disease progression is often correlated with protein levels [23]. Protein nanoarrays would be particularly useful, as protein signals cannot be amplified via methods such as polymerase chain reaction (PCR). Since many protein biomarkers are present in concentrations of only 10–100 pg/ml, improving assay sensitivity could lead to the discovery of new biomarkers that currently go undetected [24]. Essentially, the fabrication of nanoarrays will allow the screening of smaller volumes in shorter amounts of time. It will allow for higher-density arrays; one assay could then screen a greater number of targets.

Microarray analysis relies on optical readout methods, either by the observance of a color change as in ELISA, or the reading of fluoresence signals, which requires fluorescently labeled molecules for the detection of protein–protein interactions [25]. However, such labeling may cause deformational changes of the protein molecule, which may in turn affect the protein function. Radioisotopes are also sometimes used for labeling, but this requires the subsequent management of radioactive materials [26]. Labeling efficiency varies, so quantitation may not be reliable, and the labeling process is time-consuming and labor-intensive. For all these reasons, label-free detection is preferable. It also allows for real-time detection and in situ identification.

The atomic force microscope (Fig. 5.4) is an instrument capable of label-free sample analysis. It is a member of the family of scanning probe microscopes (SPM), which makes use of specialized probes to scan a sample surface to produce maps of topography, conductivity, binding force or friction among many others. These data sets can also be obtained all at once. The resolution of the technique is highly dependent on the probe quality and sharpness. Nanoarrays are well suited to atomic force microscope readout. For example, arrays for virus detection contain domains of

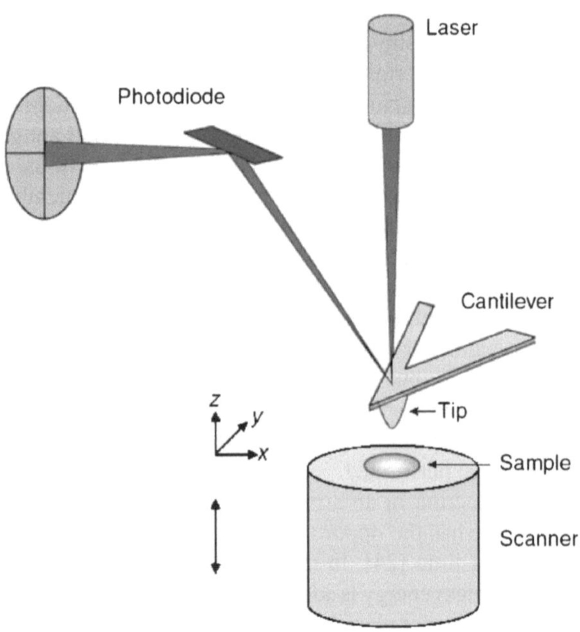

Fig. 5.4. Principle of atomic force microscopy (AFM). The sample is mounted on a piezoelectric scanner for three-dimensional positioning. The force between the tip and the surface is determined by measuring the cantilever deflection optically (with a laser and a photodiode) [28]

antibodies directed against specific viral species [22]. Any of these viruses present in a test solution will attach to certain domains and can be detected via AFM. A variety of solutions can be tested, including serum, sputum, sludge, coffee and urine, some of which inhibit other methods such as PCR. Perhaps most importantly, AFM is applicable in almost any environment, including liquids. This is required for studying cells and membranes in their native environment. Lastly, AFM-based methods are amenable to subsequent manipulation by the probe. For example, DNA can be induced to fold by controlled pushing by an atomic force microscope tip [27].

5.3.2 Probe-Based Patterning

The following section discusses a variety of scanning-probe-based lithography techniques suitable for use in the life sciences. They each have their own advantages and disadvantages. Their resolution and typical examples are given in Table 5.1. Many factors influence the ability to pattern very small features. These include environmental conditions, appropriate ink and substrate chemistry and the material and geometry of the probe tip.

5.3.2.1 Dip-Pen Nanolithography

Dip-pen nanolithography (DPN; Fig. 5.5) was first introduced to the research community in 1999 as a tool for patterning nanostructures, in which an atomic force microscope probe tip (approximately 20 nm radius) is dipped into a solution to coat it with the desired molecules [29, 30]. Patterning is typically accomplished with a commercial atomic force microscope which controls the movement of the probe. When the probe is brought into contact with a substrate, molecules diffuse from the probe to the substrate. In most cases, the ink molecules produce a local reactive functionalization of the surface, such that specific biochemical adhesion experiments can be conducted at these length scales. In general, the deposited material and the substrate must be paired, such that a chemical reaction occurs upon delivery, or a surface SAM is formed so that readout is possible using the atomic force microscope.

DPN is characterized by simplicity and high writing resolution (less than 100 nm). A large amount of literature is available on the use of single-probe DPN. Apertured probes [32, 33] can store and dispense larger numbers of ink molecules, but the writing speed is not increased substantially and the writing resolution is inferior to that of DPN. The resolution is affected by the geometry and aperture size of the probe tip and the wetting properties of the tip and the substrate. The smallest features produced were typically twice the diameter of the aperture; the smallest apertures produced were 35 nm in diameter [34]. Lastly, because such apertured probes are milled by a focused ion beam, they are not easily mass produced. Pulled-glass nanopipettes offer continuous ink delivery, but lower resolution (approximately 1 μm), and cannot be integrated by microfabrication into larger systems. Nanofountain probes (NFP) [35, 36] exhibit continuous ink delivery for a long-range writing capability, resolution close to that of DPN, and can be integrated in arrays and systems, although the probes and AFM systems become more complicated and costly.

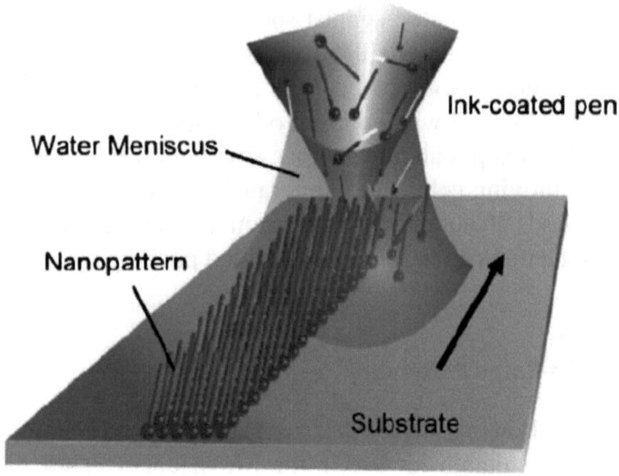

Fig. 5.5. Dip-pen nanolithography (DPN) patterning [31]

Indirect DPN

The most common demonstrations of DPN patterning have involved the deposition of n-alkanethiols onto gold surfaces because thiols self-assemble and strong thiol–gold bonds are formed [38]. The fabrication of patterns of biological materials began with indirect adsorption of the molecules of interest onto DPN-generated templates. 16-Mercaptohexadecanoic acid (MHA) could be deposited onto gold substrates via DPN; features as small as 100 nm were possible [39]. Nonpatterned areas were then passivated with 11-mercaptoundecyl tri(ethylene glycol) (PEG-SH). The desired proteins, lysozyme or rabbit IgG were attached by submersing the substrate into a 10 μg/ml protein solution for 1 h. No detectable nonspecific binding to the passivated areas was observed. It was also demonstrated that retronectin could adsorb specifically to the MHA-patterned dots 200 nm in diameter, spaced 700 nm apart. In turn, 3T3 Swiss fibroblast cells could be attached to these retronectin patterns, indicating that nanoarrays could be used for cell adhesion studies as well.

Metal ion-affinity templates can also be used to immobilize antibodies [40]. As metal ions are not susceptible to denaturation, they can be used as a linker to immobilize many unmodified polyclonal and monoclonal antibodies. First, DPN was used to pattern MHA spots. The surrounding areas were passivated with PEG-SH and the carboxylic acid groups of the MHA were coordinated to Zn^{II} ions. The substrates were then exposed to solutions of the desired antibody.

Indirect DPN techniques have the potential to be used for the creation of sandwich assays with the purpose of disease detection. For example, DPN can be used to deposit an array of MHA spots which then attract the p-24 antibody for HIV [41, 42]. The surrounding areas are passivated with bovine serum albumin. The array could then be used to detect HIV in patient samples, exceeding the detection limit of conventional ELISA-based immunoassays.

Direct DPN

Direct deposition of biological material eliminates the additional step of first patterning a linker molecule. It also eliminates cross-contamination of array features because the desired chemistry is carried out only at specific locations [29]. This is essential for the prevention of nonspecific adsorption, which becomes an increasing problem as arrays get smaller—a few nonspecifically bound molecules may overwhelm the entire intended signal [26]. Silicon atomic force microscope tips were used to deposit thiolated native collagen (1 mg/ml in 1 mM HCl) and collagen-like peptides (40 mg/ml) onto gold substrates via DPN [43]. Line widths down to 30 nm were achieved. Collagen arrays could conceivably be used to induce an assembly network of collagen scaffoldings that would direct cell attachment or as guest–host systems for other biological entities.

In most cases, however, direct deposition by DPN typically requires prior chemical modification of the atomic force microscope tip. For example, commercial atomic force microscope tips functionalized with 0.1 mM thiotic acid could be used to deposit lysozyme and rabbit IgG (Fig. 5.6) [44]. The modified atomic force microscope tip was immersed in protein solution (500 μg/ml) for 1 h and then used to pattern immediately. The patterned IgG dots were 45–200 nm in diameter, and the biomolecules maintained their biorecognition properties after being patterned. The study demonstrated that multicomponent arrays could be patterned without cross-contamination. However, high humidity (80–90%) was required for successful patterning; a relative humidity below 70% resulted in inconsistent ink transport properties. Furthermore, the transport rate depended on the composition of the protein.

2-[Methoxypoly(ethyleneoxy)propyl]trimethoxysilane can also be used to functionalize atomic force microscope tips. The material forms a biocompatible and hydrophilic surface layer on the tip, and prevents protein adsorption and denaturation on the tip surface. It also reduces the activation energy for transporting proteins from the tip to the substrate. Without it, the protein solutions (500 μg/ml) do not wet the silicon nitride (Si_3N_4) cantilevers and inconsistent or low-density protein patterns are produced. Lim et al. [45] immersed such an atomic force microscope tip in protein solution for 1 min and used it to pattern in 60–90% relative humidity. Rabbit IgG features ranging from 55 to 550 nm were fabricated. It was also demonstrated that human, goat and mouse IgG, and anti-human, anti-goat and anti-mouse IgG could be patterned in this manner.

Nickel-coated Si_3N_4 atomic force microscope tips were used to deposit histidine-tagged ubiquitin (300 μg/ml) and thioredoxin (250 μg/ml) proteins onto nickel oxide surfaces [46]. As in the previous cases, bare Si_3N_4 cantilevers could not be coated homogeneously with the proteins, resulting in inconsistent or low-density protein patterns.

The proteins angiogenin and integrin $\alpha_v\beta_3$ were deposited using gold-coated silicon cantilevers functionalized for 30 min in 1 mM mercaptoundecanoic acid [25]. The resulting SAM increased tip hydrophilicity and facilitated protein adsorption on the tip surface. The tip was then immersed in the desired protein solution for 1 h and the molecules were patterned via DPN. The smallest spots created were 120-nm wide and protein transfer was found to be affected not only by the tip–substrate contact time but also by the contact force. In order to produce regular patterns, high humidity was required, as in the work in [44].

Demers et al. [47] used DPN to directly pattern hexanethiol-modified oligonucleotides on polycrystalline gold, and oligonucleotides bearing 59-terminal acrylamide groups on derivatized silica. The atomic force microscope tip had to be

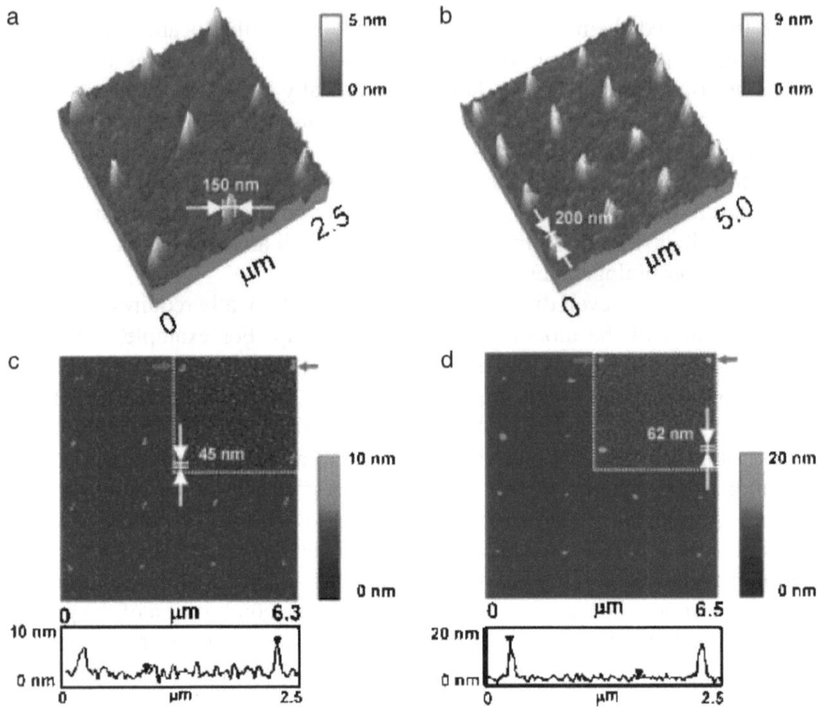

Fig. 5.6. AFM images of protein arrays generated via direct-write DPN. **a** Nanodot array of lysozyme; **b** array of immunoglobulin G (IgG); IgG dots before **c** and after **d** treatment with a solution of anti-IgG-coated nanoparticles [44]

first silanized by immersing it in a solution of 39-aminopropyltrimethoxysilane in toluene, to render it positively charged and hydrophilic.

Although protein delivery was not tested, PDMS-coated atomic force microscope tips were found to deliver MHA, octadecanethiol (ODT) and cystamine [48]. As cystamine is very volatile, its patterning via conventional DPN with Si_3N_4 tips is problematic. However, the PDMS-modified tips successfully patterned 200–300-nm-wide lines. The packing density of cystamine could be influenced by the writing speed. The PDMS acts as a reservoir that absorbs inks and allows for 1–2 h of continuous MHA patterning. The pattern resolution was not quite as good as that of regular DPN—the thinnest lines generated were 55-nm wide; however, these are much smaller features than can be obtained via methods such as microcontact printing.

It was demonstrated that a monolayer of succinic acid succinimidyl ester 5-thioyloxy-2-nitrobenzyl ester (SSTN) could be used as a photocleavable cross-linker to chemically link avidin to a gold-coated atomic force microscope tip [1]. When the tip was brought into contact with a biotinylated mica substrate, UV irradiation was applied to cleave the SSTN, releasing avidin onto the substrate. The protein-free atomic force microscope tip could then be used to image the deposited material. The delivered avidin was confined in a range of $100 \times 90\,\text{nm}^2$ and maintained its affinity to biotin.

Most DPN writing has been accomplished using the contact mode operation of the atomic force microscope. This is the most common method of AFM operation, in which the probe tip and the substrate remain in close contact while the substrate is being scanned. A drawback of this method is the exertion of large lateral forces on the substrate as the probe is dragged across it. In tapping mode AFM, a cantilever probe of lower stiffness is oscillated at its resonance frequency and taps the substrate for a small fraction of its oscillation period, thereby reducing the lateral force exerted on the substrate as well as the tip–substrate adhesion forces. Agarwal et al. [49] deposited a synthetic peptide MH_2 via tapping mode AFM, which allows for gentle imaging of deposited material as well as deposition on soft substrates. The drive amplitude was found to be a critical factor. When coated with peptide, the drive amplitude of the AFM probe decreased significantly, and as the peptide was being deposited, the drive amplitude required for imaging increased. In order to successfully deposit peptides in tapping mode, the drive amplitude had to be increased by a factor of 2–5 from its imaging value, causing the probe to exert a greater contact force on the substrate. Nevertheless this force is still less than that in contact mode DPN.

5.3.2.2 Open-Channel Pens

Reese et al. [5] micromachined stainless steel fountain pens (Fig. 5.7) with open trenches which narrowed to a width of 30 μm at the pen tip. The trench width dictated the patterned spot size; spots 10–30 μm wide and 20–140 μm long were deposited. The highest density arrays were 25,000 spots per square centimeter of fluorescent dye. With a single loading, a pen could print five to 20 spots. Printed oligonucleotides with a mean spot size around 3500 μm^2 were successfully hybridized. Nevertheless, such fountain pens are not capable of nanoscale patterning.

A device named the surface patterning tool (SPT) was developed by Henderson et al. [50]. An SPT consists of a cantilever with a split gap at the end, a reservoir on the handling chip, and a 1-μm-deep open transportation microchannel connecting the gap and the reservoir. Sample loading is carried out by filling the reservoir with sample solutions as well as by dipping the cantilever end into sample fluid. These designs, dedicated to biomolecular patterning, allowed reliable patterning of large molecular species and reduced reloading requirements. The length of the SPT cantilevers ranges between 200 and 300 μm, and the width ranges between 20 and 40 μm. The split gap is approximately 1 μm wide and approximately 40 μm long. At the fixed end of the cantilever, a 10-μm-deep rectangular reservoir is located on the handling substrate. The depth and the width of the microchannel are 1 μm and 1–10 μm, respectively.

Testing of the fabricated SPTs was performed using a dedicated commercial instrument called a NanoArrayer (BioForce Nanosciences, Ames, IA, USA). This instrument is equipped with a precision motion control system and an environmental chamber. Although this instrument uses the same optical lever deflection scheme employed in AFM, it does not scan or acquire images. SPTs are mounted to form a 12° angle with the deposition substrates such that only the tip end is in contact with the substrate. Patterning was demonstrated at a relative humidity of 35–50% using

Fig. 5.7. Stainless steel fountain pens [5]

Cy3–streptavidin [50]. A 10 × 10 array of spots with a diameter of 2–3 μm was routinely obtained; this value was limited by the width of the SPT's split-gap. A single loading of the tool printed at least 3000 spots in about 1 h. It was also demonstrated that quantum dots conjugated to streptavidin could be deposited in patterns of lines and spots using the SPT. These features had line widths of approximately 150 nm to 7 μm and spot diameters of 3–5 μm [51].

5.3.2.3
Microspotters

In microspotting technologies, a biochemical sample is loaded into a spotting pin by capillary action, and a small volume is transferred to a solid surface by physical contact between the pin and the solid substrate [52]. Other than microspotting pins, capillaries or tweezers can act as a printhead of biochemical samples. Printheads can be moved by the *XYZ* motion control system of an atomic force microscope and brought into a contact with a surface to transfer the sample.

Belaubre et al. [53, 54] fabricated microspotters (Fig. 5.8) on SOI substrates using conventional micromachining techniques.

Arrays of 2-mm-long, 210-μm-wide and 5-μm-thick cantilevers, spaced 450 μm apart, were microfabricated. These cantilevers were used to pattern a glass slide with 1-pl volumes of a solution containing cyanine3-labeled oligonucleotides (15-mers) [53]. Anti-goat IgG (rabbit), microarrays were also generated on a glass slide coated with dendrimer molecules as cross-linkers. In both cases, 30-μm-diameter spots were obtained. It was also demonstrated that no cross-contamination was observed for two different biological samples deposited with the same cantilevers if a cleaning procedure was used. Using a similar microspotter, Leichle et al. [55] deposited colloidal solutions containing poly(ethylene glycol) 600 and aminopropy-

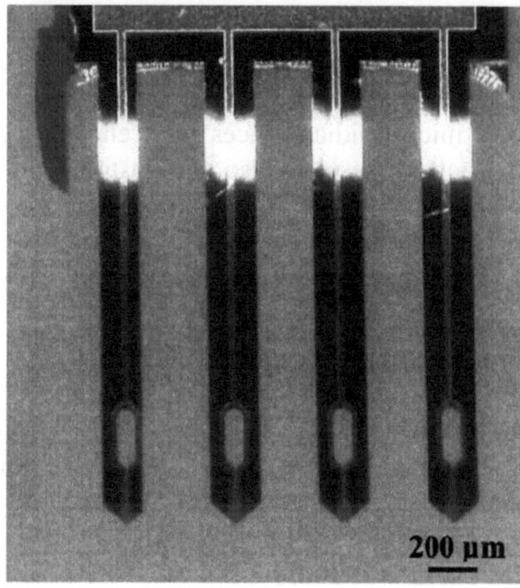

Fig. 5.8. Microspotter array [53]

ltriethoxysilane nanoparticles with diameters of approximately 300 and approximately 150 nm, respectively, on surfaces to form spots with diameters ranging from approximately 10 μm to more than 100 μm.

To further reduce droplet size, Saya et al. [56] introduced a technique to fabricate an in-plane sharp nanotip incorporated into the channel of a cantilever (Fig. 5.9). Tips with a curvature radius less than 100 nm were fabricated, and were connected to V-shaped microchannels 5 μm wide. The cantilevers were used to deliver water–glycerol droplets. The surface wettability of the nanotip and the substrate were important factors in determining the droplet size. For a hydrophilic tip contacting a hydrophobic substrate, droplets 3–4 μm in diameter were achieved, and the contact time did not significantly affect the drop size. Smaller droplets 2 μm in diameter could be achieved with high uniformity using a hydrophobic tip and a hydrophilic substrate; however, the contact force and time affected the spot size.

Pens with open microchannels integrated on cantilevers have the advantage of being clog-free and allow for easy cleaning and simple microfabrication. However, such open microfluidic elements such as microchannels and reservoirs are prone to cross-contamination via vapor from different types of samples, especially when loaded in arrays of cantilevers [57]. Evaporation may be critical in

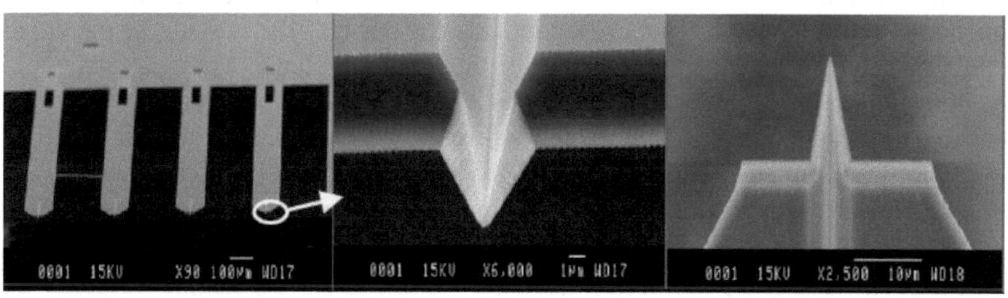

Fig. 5.9. Scanning electron microscopy (SEM) images of a cantilever array (*left*) with a nanotip (*center* and *right*) [56]

some applications although its rate can be reduced with environmental conditioning. Enclosed microchannels are beneficial in such cases, although they are more difficult to microfabricate and are subject to clogging. Pipettes are conventionally used microfluidic devices with enclosed channels. Microneedles with embedded microchannels were also demonstrated to deliver liquid materials [58]. In the following subsections, microcantilever devices with enclosed microchannels are described.

5.3.2.4
Pipettes

Bruckbauer et al. [59] developed a delivery system based on scanning ion-conductance microscopy. In this voltage-controlled nanopipette, an ion current flows between an electrode in the pipette and one in a bath; the pipette acts as the ink reservoir. Deposition of molecules to a surface occurs in the presence of aqueous buffer. Protein G was successfully delivered using the pipette. Biotin and single-standed DNA with spot sizes of 830–860 nm were also deposited.

Following this, goat anti-rabbit IgG antibody was delivered onto charged glass, with spot sizes of about 2 μm [60]. The flow of molecules began at an applied voltage of -0.5 V and increased linearly with applied voltage between -1.0 and -1.5 V. Although voltage control allows for the control of ink delivery at the single-molecule level, the diffusion of molecules in solution and on the substrate ultimately increases the patterned feature size. Because the nanopipette deposits in solution, protein denaturation may be prevented. However, the use of voltage subjects the sample to a high electric field. Nevertheless, owing to the small current, very little heating (less than 1 K) occurs. Indeed, the biomolecules were demonstrated to maintain their functionality after deposition. This was demonstrated via binding experiments with IgG and hybridization studies with DNA.

In order to reduce the patterned feature size, a double-barreled nanopipette (Fig. 5.10) operating in air was fabricated from 1.5-mm-diameter pulled-glass capillaries with a septum down the center [61]. The device was used to deliver fluorescently labeled rabbit IgG to a polyethyleneimine-coated glass surface, with the smallest spot produced being about 440 nm in diameter. Biotinylated DNA was delivered onto streptavidin-coated glass with an average spot size of 510 ± 40 nm for a 5-s deposition time. In both cases, molecules were delivered out of one barrel at any one time. Distinguishing itself from the original nanopipette, the double-barreled pipette could write with two different inks. Two different, noncomplementary sequences of biotinylated DNA were loaded into the barrels. The DNA to be patterned could be selected by changing the sign of the applied voltage. The two inks could be delivered to the same location on a substrate. The double-barreled pipette is also capable of topographical scanning but its resolution is not as good as that of AFM; it cannot track up steep slopes or tight grooves as effectively. Most recently, the double-barreled pipette was used to deposit water droplets, working under oil [62].

A major drawback of this device is that it is not easily scalable because the pipettes are individually fabricated by laser pulling. The smallest pipette that can be made using this approach has a radius of approximately 20 nm, limiting the pattern resolution [63].

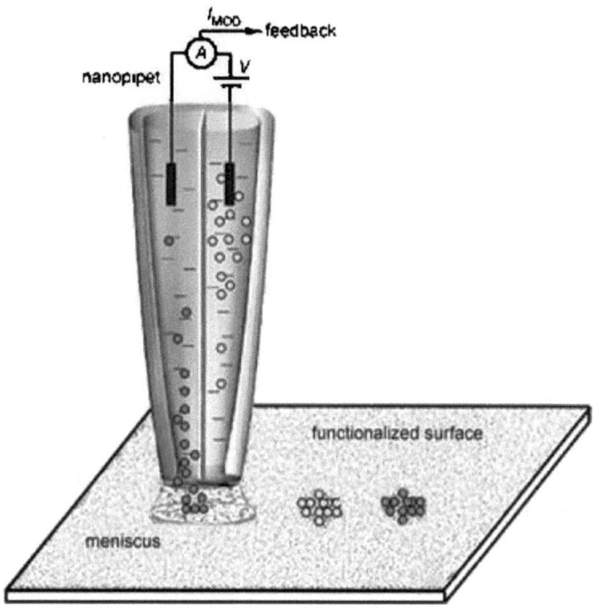

Fig. 5.10. Principle of double-barreled pipette deposition. The voltage between the two barrels controls the molecular delivery [61]

5.3.2.5
Closed-Channel Cantilevers

The concept of the nanofountain probe (NFP) was first introduced by Espinosa and coworkers in [64–66]. The device allows for both high-resolution patterning and continuous sample feeding through closed microchannels. The chip without its cantilevers has an overall size of 1.8 mm × 3.2 mm to fit easily into commercial AFM equipment. In the latest version of this chip, the two opposing sides of the chip each have 12 cantilevers; one set is 520 μm long and the other is 430 μm long. A volcano-like dispensing tip (Fig. 5.11) exists at the end of each cantilever and has a ring-shaped aperture and is able to generate sub-100-nm lines routinely.

For fast-evaporating solutions such as alkanethiols, the writing mode is similar to that of DPN, and pattern resolution is controlled by the radius of the core tip, not the aperture size. MHA (1 mM) in ethanol with line widths as small as 40 nm was patterned onto a gold substrate at a humidity of 60% using the NFP [35]. The NFP's imaging capability in topography was similar to results obtained using commercially available atomic force microscope tips, and its sensitivity in lateral-force imaging was better than with commercial tips, an advantage when imaging SAM patterns. This imaging capability is essential for immediate examination of deposited patterns and also serves as a means for realigning probes when writing multiple materials.

In typical operation, however, the NFP preserves the liquid state of the ink during the patterning process [36]. The liquid present near the writing tip provides a continuous source of solvent vapor, preserving a high local solvent vapor pressure, which may contribute to the formation of a more vigorous capillary condensation meniscus [65]. Solvent vapors may also condense on the substrate, leading to a solvent

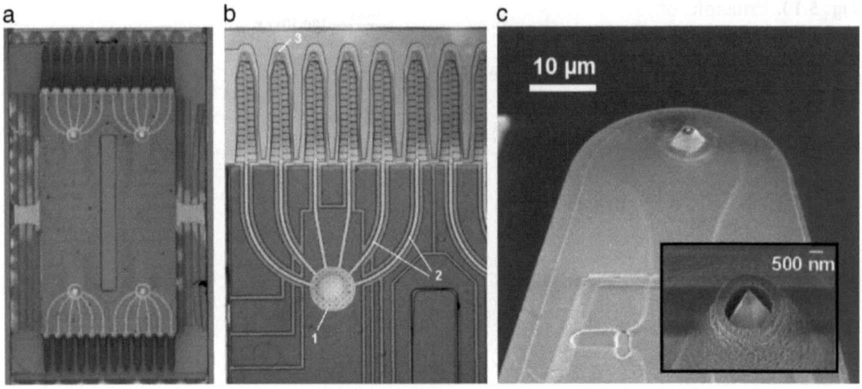

Fig. 5.11. Second-generation nanofountain probe (NFP) chip. **a** Optical image of the front view of an NFP chip **b** Closeup view of the chip showing *1* on-chip reservoir, *2* microchannels and *3* volcano tip **c** SEM image of a cantilever integrated with a volcano-like tip at the end; *inset* closeup of the volcano tip [36]

prewetting layer in the vicinity of the tip. The transport of ink molecules may also occur by coevaporation with the solvent and recondensation. This mechanism may lead to feature-growth dynamics different from that of DPN.

Batch fabrication allows for straightforward scaling to NFP arrays. Indeed, 12-cantilever NFP arrays have been fabricated, and their multiple on-chip reservoirs allow simultaneous patterning of two different solutions. Because the NFP can make use of two different reservoirs, it does not require a second tip to deposit an additional species. Simultaneous patterning with two different thiolate solutions, MHA and 1*H*,1*H*,2*H*,2*H*-perfluorododecane-1-thiol in acetonitrile was demonstrated with the NFP [36]. Also, use of the NFP does not require repeated reinking and realignment of the tip during patterning, the latter of which is time-consuming and difficult to accomplish precisely. Finally, deep channels (more than 500 nm) allow the delivery of larger particles, including nanoparticles [37]. High compliance of the core tip relative to the volcano shell structure was observed in the induced deformation during scanning electron microscopy (SEM) observation of dry tips and is helpful in preventing clogging of the probe orifice and mechanically aids the transfer of larger species from the volcano cavity to the substrate. However, this may negatively affect alignment accuracy.

What sets the NFP apart from other devices is that while nanopipettes, apertured pyramidal tips and quill-type SPTs function by the formation of an outer meniscus between the probe and the substrate, the NFP forms a meniscus between the ultra-sharp atomic force microscope-like tip and the substrate, as in DPN. However, the NFP does not require modification of the tip surface in order to deliver biomolecules, because solutions are directly delivered from the on-chip reservoir to the tip. The NFP was capable of *direct* deposition of hexanethiol-modified oligonucleotides on a gold substrate [58, 67].

As shown in Fig. 5.12, the NFP was used to pattern a gold substrate with alkanethiol-modified oligonucleotides. After passivation of the unpatterned areas, the DNA spots were hybridized with complementary DNA-functionalized gold

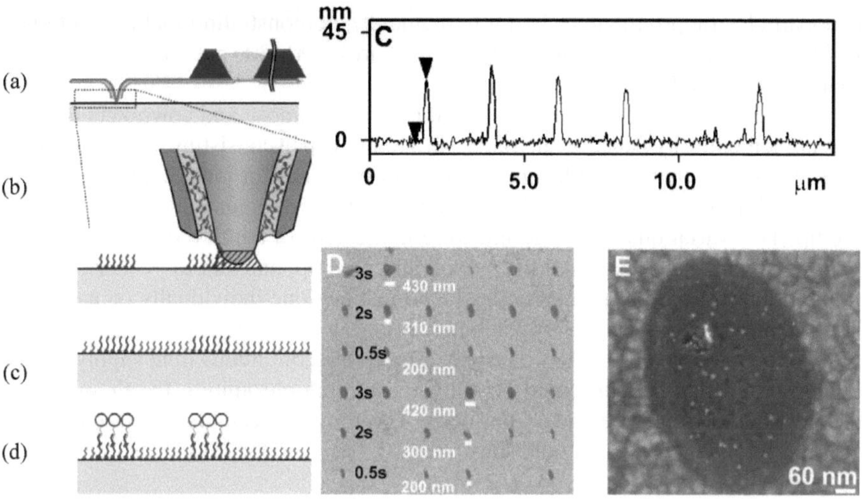

Fig. 5.12. Experimental procedure for DNA patterning: **a** molecular ink feeding, **b** direct patterning of a gold surface with alkanethiol-modified oligonucleotides, **c** passivation of the unpatterned areas with C6 thiol to avoid unspecific binding, **d** hybridization of the linker and probe DNA strands. **C** Height profile of the same array, acquired in tapping mode operation. **D**, **E** SEM image of a dot array and a single dot, respectively. Multiple gold nanoparticles are visible in **E** [67]

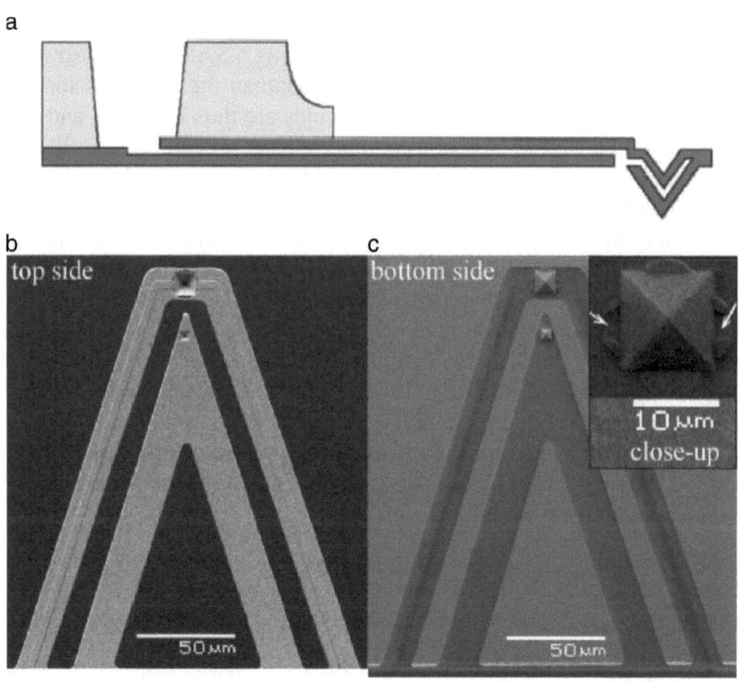

Fig. 5.13. **a** Cross section of the micromachined fountain pen; **b** *top view* and **c** *bottom view* of the fountain pen; *inset* closeup view of the pen tip, with outlet holes indicated by *arrows* [68]

nanoparticles (approximately 15 nm in diameter), demonstrating that the patterned DNA maintained its biological activity. Features 200–300 nm in diameter were routinely achieved.

A variation of the NFP concept introduced by Espinosa and coworkers [64, 65] was implemented by Deladi et al. [68,69] in a fountain pen consisting of two V-shaped cantilevers, one that is "multifunctional," and the other an inner cantilever solely for detection. A channel embedded in the first cantilever connected the reservoir to the tip. Two variations of this design were fabricated. In the first design, the outlet hole was located at the apex of the pyramidal probe tip, and had to be fabricated by focused ion beam milling, a process which must be done individually on each pen. The second design made use of batch processing, which dictated an opening at the base of the pyramid (Fig. 5.13). The pen allowed for continuous fluid supply. It was demonstrated that etchant could be delivered to etch chromium; the thinnest line etched was 350 nm wide and 14 nm deep. The thinnest features constructed were 500-nm-thick ODT lines on gold substrates.

5.3.3 Alternative Patterning Methods

5.3.3.1 Nanografting

Nanografting is a technique similar to DPN, but involves material removal as well as deposition. The basic procedure involves scanning a SAM on a substrate with an atomic force microscope tip using a force greater than the threshold force needed to displace SAM molecules [70]. SAM molecules are thus removed, and thiol inks previously adsorbed on the atomic force microscope tip can be adsorbed to these areas of the substrate. Care must be taken not to cause plastic deformation of the substrate beneath the SAM. Xu et al. [70] nanografted thiol solutions with concentrations ranging from 2 μM to 2 mM; the concentration did not appear to be a critical parameter in grafting. On the other hand, the scan rate was important. Slow scans often resulted in distorted patterns due to thermal drifts, but fast scans did not produce patterns with sufficient coverage. Liu et al. [71] demonstrated that alkanethiol features as small as $2 \times 4\,\mathrm{nm}^2$, and 10-nm-wide lines could be produced via nanografting. Wadu-Mesthrige et al. [2] used nanografting to prepattern SAMs with thiols terminated with protein-adhesive groups such as aldehyde and carboxylate, which would then dictate the subsequent adsorption of proteins. Nanopatterns of lysozyme, bovine serum albumin and rabbit IgG were fabricated, with lateral dimensions ranging from 10 nm to 1 μm.

5.3.3.2 Conductive DPN

In AFM charge writing, positive or negative surface charge patterns are created on an insulating substrate such as a polymer or silicon dioxide by applying voltage to an atomic force microscope tip. Mesquida et al. [10] created microarray charge patterns on poly(methyl methacrylate) substrates which were then immersed in water-in-perfluorinated oil emulsions where the water microdroplets were filled with $TTR_{105-115}$ peptide fibrils. The fibrils then attached to the charge patterns. The resolution was mainly limited by the size of the water droplets, which can typically be up to 5 μm in diameter. The use of water droplets as "containers" is advantageous in that the attachment process depends little on the content of the droplets, allowing the patterning of a variety of nano-sized biological materials.

Naujoks and Stemmer [72] used the same method to attach IgG–biotin stabilized emulsion droplets onto positively charged patterns. After drying, the protein molecules and salt from the buffer solution were left behind. The resulting spot sizes varied from 0.5 to 1.5 μm (Fig. 5.14). For these experiments, an atomic force microscope was operated in Kelvin probe force microscopy mode, where the lateral resolution was known to be in the range 50–100 nm, defined by the long range of electric forces and the geometry of the atomic force microscope tip. The pattern resolution might be enhanced by finding optimal emulsifying parameters and process conditions such as immersion time and drying ambiance.

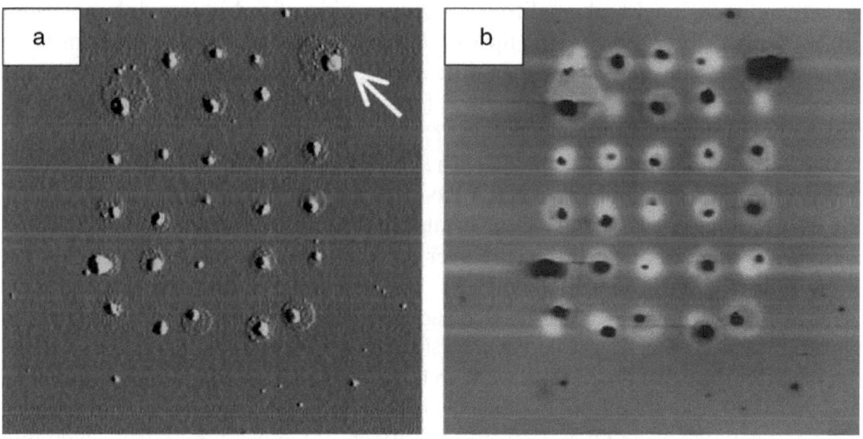

Fig. 5.14. **a** AFM (topography) image of IgG–biotin deposited on poly(methyl methacrylate). **b** Kelvin probe force microscopy (surface potential) image of positive charge pattern [72]

5.3.3.3
Electrochemical DPN

For the direct patterning of biological material, Agarwal et al. [73] used electrochemical DPN (e-DPN) to deposit a histidine-tagged peptide MH_2 and the histidine-tagged protein TlpA in tapping mode. An ionized nickel-coated substrate was held at ground potential and a negative bias was applied to the AFM probe tip. The method takes advantage of nickel–histidine binding and requires that a potential of about −2 V be applied. No functionalization of the silicon AFM probes was necessary. They were simply coated by immersing them in the desired solution for 15–30 s and then air-dried before patterning.

5.4 Nanoscale Deposition Mechanisms

Reproducibility is a concern for any fabrication technique. Nanopatterning results are not always repeatable and the resulting features are not necessarily well formed. An understanding of the ink deposition process is essential; however, in DPN, the mechanism for ink transfer is not completely understood and there is much controversy. It has been generally accepted that DPN patterning involves two processes: first, the transport of the molecular species from the atomic force microscope tip to the substrate through a water meniscus and, second, the adsorption of the ink onto the substrate. Control of these processes then would ultimately control the patterned feature resolution. Most studies on the patterning mechanism thus far have focused on the deposition of alkanethiols.

Theoretical studies on the dynamics of self-assembly have been performed in order to investigate the effects of deposition and atomic force microscope tip scanning rates on the resulting patterns generated via DPN [74]. The DPN process was treated as a two-dimensional Fickian diffusion process with a fixed source. Ink molecules on the atomic force microscope tip were driven by a concentration gradient to be deposited on a substrate. They then diffused over a monolayer of existing ink, and the diffusion was terminated by binding of the molecules to the bare substrate. In the patterning of dots by holding an atomic force microscope tip in contact with a substrate, the constant flux model predicted a $t^{1/2}$ dependence of the dot diameter, where t is the tip–substrate contact time. For the same number of molecules deposited, the final spot radius converged to the same value at long times, independent of the deposition rate. Increasing the diffusivity over the bare substrate resulted in circles that were fuzzier and not as well defined. For the generation of more complex patterns such as lines, a moving tip was needed and the tip scan rate had to be considered. In general, a fast scan or a slow deposition rate enhanced resolution, but as the scan rate increased further, patterned lines were no longer continuous. Such studies underline the necessity of determining optimal scan and deposition rates in order to produce coherent patterns using DPN. However, the effects of temperature, relative humidity and the presence of a water meniscus were not accounted for in these studies.

It is known that in ambient conditions, a water meniscus exists between an atomic force microscope tip and a substrate, and its volume increases with increasing relative humidity [75]. There have been suggestions that ink molecule transport occurs through this meniscus; however, Sheehan and Whitman [75] found that facile ODT deposition occurred even at 0% relative humidity in dry nitrogen. Since ODT is water-insoluble, its transport is not easily explained using a water meniscus model.

Cho et al. [76] proposed a "double-molecular layer" model in which the molecular ink on the atomic force microscope tip consists of a bulk solid region covered by a thin mobile layer on the surface (Fig. 5.15). A melting transition is suggested to explain the observed temperature dependence of the growth rate of patterns. The solid region serves as a reservoir of ink molecules which regulates the density of the surface layer, and the mobility and the number density of the mobile layer determine the molecular transport rate in DPN. It was proposed that three properties affect the mobility of the surface molecules—thermal energy, which increases the diffusion constant, residual solvent, which enhances diffusion of surface molecules, and ad-

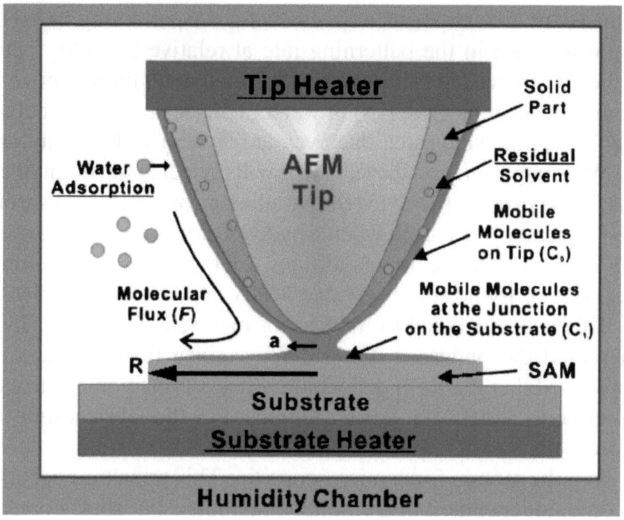

Fig. 5.15. Double molecular layer model for molecular ink transport [76]. *SAM* self-assembled monolayer

sorbed water, which enhances the mobility of hydrophilic molecular inks such as MHA and biomolecules.

In the general case where the diffusion and deposition rates are on the same order,

$$R \propto t^{\upsilon} \quad \text{and} \quad 1/3 \leq \nu \leq 1/2\,,$$

where R is the radius of a patterned dot, t is the tip–substrate contact time and ν is the scaling parameter. ν approaches $1/3$ as the deposition rate becomes comparable to the diffusion rate, and $\nu = 1/2$ in a constant deposition case. In the patterning of ODT, a hydrophobic ink, on gold substrates, there are unstable fluctuations in the scaling parameter. These experimental results were explained using the double-molecular layer model. Above the melting temperature of ODT, the solid region melts and a single molecular layer forms. There is no longer a solid "reservoir" that regulates the density of the surface layer, leading to the unstable fluctuations in ν. Essentially there is a phase transition of the molecular species moving through the nanoscale junction. In the patterning of MHA, the scaling parameter increases with increasing relative humidity. It is suggested that adsorbed water enhances the diffusion constant on the substrate more than it enhances the transport rate at the tip–substrate contact point, implying that water molecules are also adsorbed on the SAM regions, not just on the probe tip.

In any case, direct environmental SEM (ESEM) observations verified that the height of the water meniscus between a silicon nitride probe tip and a gold or silicon substrate increases exponentially with increasing relative humidity [77]. In addition, for gold substrates, there was no observable meniscus below a relative humidity of about 70%, although, since the ESEM resolution in the imaging conditions used was approximately 50 nm, it is possible there was still a small meniscus. Still, these results are consistent with experimental results for the patterning of MHA, in which there was little increase in the patterning rate at relative humidity between 0 and 50%, while the rate increased significantly at a relative humidity above 50%.

AFM studies by Rozhok et al. [78] of water meniscus formation between a probe tip and a NaCl substrate suggested that even at 0% relative humidity as defined by Sheehan and Whitman, water on the substrate will collect at the point of contact form to a meniscus. As it is very difficult to remove adlayers of water from surfaces, a small meniscus will typically form, even when a hydrophobic atomic force microscope tip is used. Only in ultra-high-vacuum conditions can the meniscus be eliminated.

Weeks et al. [79] found no observable MHA deposition at a relative humidity below 15%. This was attributed to the lack of meniscus formation. Increasing the humidity increased the patterned dot size for a given tip–substrate contact time. The group proposed that the DPN deposition process transitions from a dissolution-dominated regime to a diffusion-dominated one. For the fabrication of small alkanethiol features, the most important parameter is the activation energy of thiol detachment from the atomic force microscope tip. This transition occurs at a particular contact time, independent of the relative humidity. For short contact times, the transfer process is dominated by surface kinetics, whereas for long contact times, it is controlled by diffusion. Furthermore, no effect of contact force was observed; the water meniscus causes a capillary force between the tip and the substrate that is much greater, so capillarity dominates the total load force. This is in agreement with Zou et al. [80], who concluded that feature size is independent of contact force over a range from 0 to 10 nN.

It is thus fairly well agreed upon that the molecule transport rate is different in the fabrication of small and large patterns. It is slower for large features (longer tip dwell times) because the cantilever continues to deposit molecules on an area that has already been patterned. Ink depletion on the surface of the atomic force microscope tip also contributes to this phenomenon. Moreover, there is an equilibration time when patterning begins. This may imply that the solvent plays an important role in determining the mobility of the ink on the cantilever [81]. It may also be due to the fact that a water meniscus takes time to reach an equilibrium width when patterning begins, as observed via ESEM [82]. Moreover, the transport of thiol inks is affected by the state of the substrate, not only the ink already present on the substrate, but also adsorbates that may unintentionally be present. Uncontrolled surface chemistry may be one factor leading to differing results found in the DPN literature.

Schwartz [83] contends that the water meniscus is not universally responsible for molecular transport in DPN. In contrast to many other experimental results, both ODT and MHA were readily patterned on a gold substrate at 0.0% relative humidity. Two different but compatible models were proposed. For less polar molecules such as ODT, molecular transport occurs via molecular surface diffusion, known as "reactive spreading" in microcontact printing. This process depends on thermal energy. The second model is relevant to the transport of small ions such as DNA, as a solution in adsorbed water, and is compatible with the humidity dependence of the patterning rate of certain molecules. The degree of polarity of a molecule will determine its patterning behavior between the two extreme models. It has been suggested that MHA transport is due to two mechanisms. At low to moderate (50%) relative humidity, transport is caused by thermally activated surface diffusion [84]. As humidity is increased, dissolution and bulk transport enhance the transport rate; above 70% humidity, the water meniscus begins to have an effect as well. Furthermore, surface water is not responsible for molecular transport.

This is consistent with MHA dot patterns generated at high humidity [85]. At relative humidity less than 80%, filled MHA dot patterns were always observed, but increasing the humidity to approximately 84% resulted in the formation of hollow ring structures. A bulk meniscus transport model cannot explain such patterns because it predicts only filled circles. Rather, an annular diffusion model proposed that ink molecules are transported at the air–interface of the meniscus.

Whereas the models mentioned previously could describe the deposition of molecules which bind strongly to their substrates, Manandhar et al. [86] investigated the anomalous patterns formed by the weak binding of 1-dodecylamine (DDA) on mica, which cannot be explained by random walks. Anomalous patterns are sometimes formed when surface binding is weak and interactions between molecules dominates, determining the final pattern. For DDA, small patterns are isotropic, but become more anisotropic as the pattern size increases. Thus, to create large, regular patterns it may be necessary to move the atomic force microscope tip to fill the area, instead of relying on diffusion to produce the final pattern. Finally, a constant deposition rate was observed, implying that stable DPN writing is possible even for weak-binding cases. Lee and Hong [87] developed a theoretical model to explain these irregular patterns. The corresponding computer simulations showed that in the presence of strong intermolecular interactions and preferential substrate–molecule interactions while the diffusive motion of individual molecules is suppressed, irregular star patterns and fractal-like structures can be formed. On the other hand, if the intermolecular interactions between deposited molecules are weak relative to the thermal energy, diffusive motion becomes important and regular circular patterns are formed as predicted by random-walk models.

Beyond the physical capability of patterning dots with specific sizes, there is the question of the appropriate feature size for a particular application. In biodetection, there exists an optimal spot size for the attainment of a robust diagnostic readout. Individual molecules are dynamic in nature; in order to obtain a good signal there must be many sampling events or long sampling times for a single molecule. Array spots smaller than a certain size will not have enough active and correctly oriented capture molecules, leading to inaccurate quantitation and reduced dynamic range [24]. It was estimated that for typical IgG, the number of molecules in a 1-μm-diameter spot is on the order of 10^4 whereas a 250-nm-diameter spot will have on the order of 10^3 molecules. Spots below this 250-nm threshold may lack a sufficient number of molecules for reliable quantitative data to be obtained. Further studies must be conducted to determine the optimal feature size for particular bioassays.

5.5 AFM Parallelization

One requirement for a deposition technique is its capability for high-throughput patterning. A particular device design should be scalable, i.e., its ink-delivery probe should be arrayable and it should be possible to fabricate the entire device in mass quantities.

5.5.1 One-Dimensional Arrays

Xu et al. [88] reported a second version of the quill-type SPT described in Sect. 5.3.2.2, in which five cantilevers were arranged linearly (Fig. 5.16) [88]. The microcantilevers and corresponding microfluidic network were capable of transporting multiple fluid samples from macroscale reservoirs located on the SPT substrate through microscale channels to the distal end of the cantilevers. Five cantilevers and five reservoirs were arranged in a chip so that multiple biological samples could be transferred from the reservoirs to the SPT cantilever array. The overall size of an SPT chip was 3 mm × 6 mm. Each cantilever was 250 μm long, 30 μm wide and 2 μm thick, while consecutive cantilevers were separated by a spacing of 50 μm. The microchannel on each cantilever was 15 μm wide and 1 μm deep. Multiple ink loading and patterning were tested using two different fluorescent protein solutions, Cy2–donkey anti-goat IgG and Texas Red–donkey anti-rabbit IgG in phosphate-buffered saline, and alternatively loaded into the five reservoirs by hand-pipetting. The solutions transferred from the reservoirs to the distal end of each channel by capillary action and the fluids were confined inside the microchannels without observed cross-contamination. A dithiobis(succinimidyl undecanoate)/gold surface was patterned to generate 10 × 10 multiple-ink dot arrays, with the mean spot diameter being about 12 μm. The SPTs generated biological arrays with a routine spot size of 2–3 μm. Several thousand spots could be printed without reloading. The minimum spot size of the SPT was mainly limited by its gap width, which could be further reduced with a higher-resolution lithography technique.

Using single AFM probes in DPN limits a pattern to the 90 μm × 90 μm scan size of the AFM instrument. In order to create patterns that span macroscopic distances on the order of centimeters, it is necessary to use multiple-pen cantilever arrays. A 26-pen array of cantilevers was used to pattern a 10 mM $NHSC_{11}SH$ acetonitrile solution

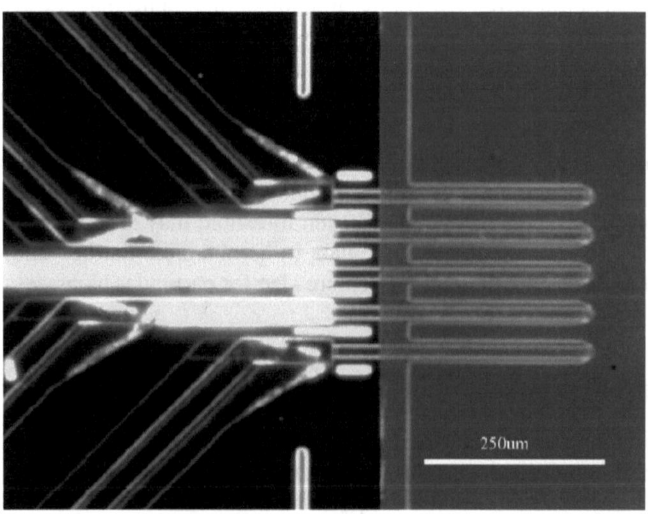

Fig. 5.16. Surface patterning tool cantilever array [88]

($NHSC_{11}SH$ is an amine-reactive alkyl thiol molecule) [89]. The surrounding areas on the substrate were passivated with PEG–SH. The dot array was then used as a template to immobilize protein A/G through covalent coupling. Highly dense arrays of 23,400 dots spaced 1 μm apart could be generated in this manner. The protein A/G templates could then be used for adsorption of human IgG, anti-β-galactosidase and anti-ubiquitin. It was demonstrated that the anti-β-galactosidase and anti-ubiquitin retained their biological activity after being deposited.

The second and third generations of the NFP introduced by Espinosa and coworkers allowed for multi-ink patterning using a linear array of 12 cantilevers [36, 66]. The device successfully deposited alkanethiols and oligonucelotides as described in Sect. 5.3.2.5, as well as bovine serum albumin and IgG proteins and gold nanoparticles [37].

5.5.2 Two-Dimensional Arrays

The IBM Millipede was likely the first example of a massively parallel cantilever array. The chip was an array of 32 × 32 cantilevers conceived for data storage using polymer thermal indentation [90, 91]. Feedback control in the z direction brought the entire array into contact with a substrate. This simplified the system but imposed a strict requirement on the uniformity of tip height and cantilever bending in order to minimize tip wear due to force variations over the array. Another two-dimensional cantilever array includes a 25 × 40 array of nickel probes fabricated by Zou et al. [80], with a tip-to-tip distance of 100 μm in the x and y directions.

In an initial attempt at a very large scale array for DPN, a 55,000-pen two-dimensional array occupying 1 cm^2 was fabricated using lithographic techniques [92]. The pens were spaced 90 and 20 μm apart in the x and y directions, respectively. As is the case for all arrays, the tips must all be aligned prior to patterning. Rather than implementing feedback for each individual cantilever, a gravity-driven alignment method was used for this large array. Complex patterns of 80-nm ODT dots on a gold substrate were successfully generated. One challenge that remains with DPN arrays is the difficulty in realigning the array after reinking or switching inks. Another is the variability in the inking process. In most cases, when the DPN tip is inked, excess ink is removed by a high-pressure blast of air. In an effort to increase reproducibility in inking the tip, Rosner et al. [31] developed arrays of microfabricated inkwells to supply ink to specific cantilevers in a probe array. A noninked probe could then be used for pattern inspection. A slower scan speed could be used to obtain better image quality while minimizing contamination onto the pattern. The inkwells employ open-channel meniscus-driven flow in microchannels to bring fluid stored in 1-mm-diameter reservoirs to a series of appropriately spaced inkwells. Different inks can be loaded onto different tips on the same cantilever array, allowing for multicomponent patterning.

The 55,000-pen array was used to deposit 1,2-dioleoyl-*sn*-glycero-3-phosphocholine (DOPC), a phospholipid, via DPN onto substrates of silicon wafers, glass slides and evaporated metal films [21]. Line features as thin as approximately 93 nm were patterned. The probe tips were coated with ink by immersing them into inkwells for at least 30 min. It was observed that in ambient humidity (30–50%), the DOPC did not flow onto the tips, but at a humidity greater than 70% the phospholipid ink became sufficiently fluid to easily coat the tips.

Although most cantilever probe arrays used in the experiments described are made of inorganic materials such as silicon, silicon nitride or other metals, polymer probes may be of interest because they offer different biochemical properties [80]. Furthermore, they have a smaller Young's modulus; thus, cantilevers can be made shorter while maintaining the same force constant. Processing of polymers does not require sophisticated equipment, so such probe arrays could be manufactured more cheaply. Zou et al. fabricated an array of more than 1,000,000 probes on a 7.5-cm glass slide. Tips and cantilevers made of photodefinable polyimide film were individually supported on SU-8 bonding structures; each of these structures had a footprint of 10 μm × 10 μm.

Passive pen arrays command that all pens in the array create the same pattern. Different inks must be amenable to the same environmental conditions in order to be deposited at the same time. Although they are more complicated to fabricate, active pens can create more complex structures, with multiple inks over large areas, and inks that do not necessarily deposit under the same environmental conditions can be deposited in series.

Rosner et al. [31] fabricated thermal bimorph actuators from evaporated Cr/Pt/Au on silicon nitride cantilevers. The gold thin-film resistor essentially acts as a heater and concentrates the heat away from the probe tip. When power is delivered, cantilevers are actuated in the direction normal to the heated surface. Bullen et al. [93] fabricated a similar array of ten thermal bimorph actuators, with cantilevers 300 μm long spaced 100 μm apart. Resistive heating actuates the cantilevers away from the substrate. Each cantilever was individually addressed by two lead wires. A typical actuation current of 10 mA resulted in an 8-μm deflection and an average probe temperature of 298 K above ambient. A ten-probe individually addressed, thermally actuated array of cantilevers 1400 μm long, 20 μm wide, 8 μm thick and spaced 100 μm apart was used to pattern ODT on gold substrates with sub-50-nm resolution [94]. Zou et al. [80] fabricated an array of 45 thermal bimorph probes in which cantilever alignment was achieved via contact sensing based on the detection of electrical continuity between the tips and the substrate. This required that both elements be partially or entirely conductive.

Limitations to thermal bimetallic actuation include the potential negative effects of the heat on inks such as biomolecules [95]. There may also be thermal crosstalk between probes—heat conduction and convection through the air between probes leading to unwanted tip deflection, effects which increase as the distance between probes decreases. Bullen and Liu [95] found that the smallest pitch in thermal bimetallic arrays where crosstalk was manageable was 100 μm. This led to the fabrication of an array of electrostatically actuated DPN probes with an array pitch of 30 μm (Fig. 5.17). The probes act as electrodes separated from the counter electrode, the array holder, by SU-8 photoepoxy acting as an insulator. The probes are grounded via conducting paste. Actuation occurs by applying a voltage (typically around 190 V) to the counter electrode; the probe tips are then pulled off the surface. Grounding the lithography surface ensures there is no electric field or actuator force between the tips and the substrate. While thermal crosstalk is avoided with electrostatically actuated probes, there may be crosstalk due to fringe electric fields.

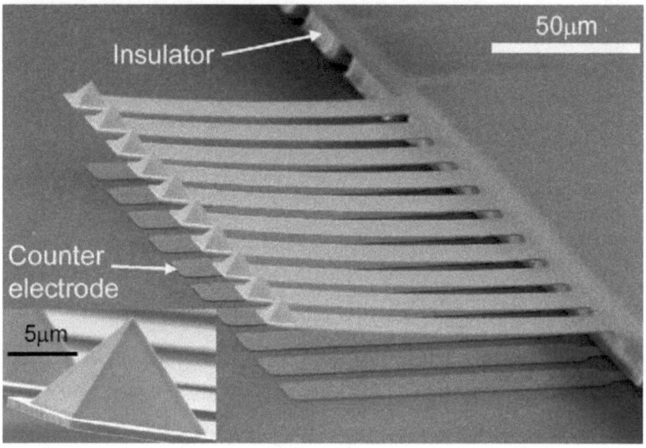

Fig. 5.17. SEM image of an electrostatically actuated DPN array [95]

Crosstalk deflection with these probes was similar to that in thermal bimetallic actuators, but at the substantially reduced array pitch of 30 μm.

A third method for probe actuation uses piezoelectric films. Lead zirconate titanate (PZT) has a high piezoelectric coefficient and can be integrated into cantilever probes. Zhu et al. [96] described a method to fabricate $50 \times 100\ \mu m^2$ microcantilevers with 1-μm-thick PZT membranes with almost 100% yield. Currently, the integration of piezoelectric bending actuators onto the NFP array is under development, and preliminary actuation results confirmed the feasibility of the operation with active probes [97]. With active probes capable of lifting the tips off the substrate, independent patterns can be made on individual writing sites, in one scanning stroke. Since scalability was proven with the one-dimensional array, the NFP is currently being augmented to a two-dimensional array with an integrated microfluidic network. This would bring the NFP one step closer to a production tool to generate, in a massively parallel fashion, templates that consist of broad range of analytes. While the augmentation to two-dimensional arrays is a matter of scalability and wafer-level-yields engineering, the operation of two-dimensional NFP arrays would require specially designed AFM systems rather than the presently available AFM instruments. With the adoption of piezoelectric precision stages or microelectromechanical system stages [98, 99], it is feasible that the two-dimensional NFP array could pattern large areas to produce multicomponent templates. In this case, leveling between the array and the substrate can be achieved with a three-probe feedback scheme, as reported for scanning probe data storage [100]. By avoiding the implementation of individual sensors on each probe, this approach can reduce the complexity of the system. With this scheme, the two-dimensional array can be maintained parallel to the substrate in the same way as an air table. Like one-dimensional arrays, passive probes are adequate if duplicate patterns, for example, nanoarrays, are to be fabricated. Alternatively, individual active probes allow the writing of different patterns with each probe.

For the operation of two-dimensional arrays of active probes with independent bending actuation, each cantilever requires one electrical connection for its electrode, assuming one set of electrodes shares the electrical path, while the others are indepen-

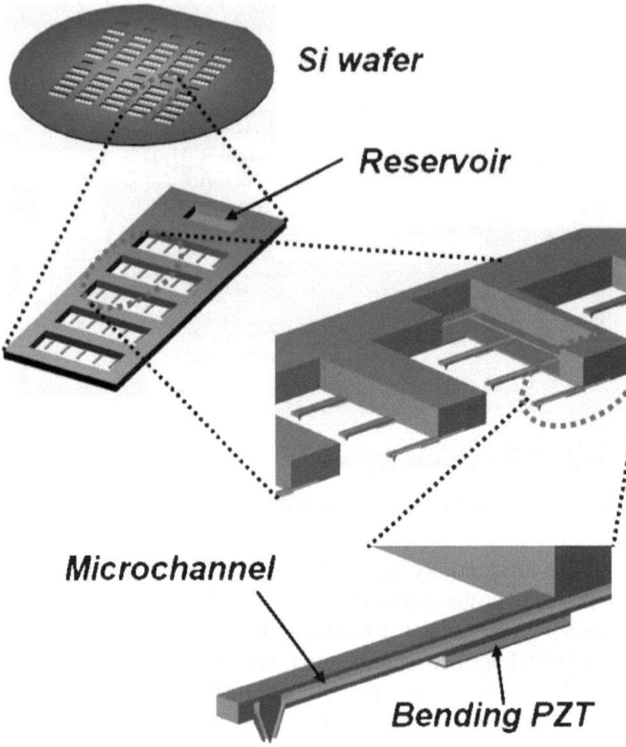

Fig. 5.18. Conceptual diagram of massively parallel NFP arrays for wafer-level patterning. *PZT* lead zirconate titanate

dently driven. Such direct drive is straightforward and simple when the size of the array is small. However, for a large array, such as a massively parallel two-dimensional array with a large number of cantilevers, a multiplexed addressing scheme needs to be employed to reduce the number of connections. This scheme is widely used in computer memories and liquid-crystal displays. Since the probe must be in continuous contact with the substrate between command signals, a bistable (memory) element must be integrated at the root of each cantilever. Furthermore, with multiple reservoirs on the wafer, individual reservoirs may contain one type of molecular ink to feed a cantilever or a group of cantilevers. Through this approach of multiple inks and massively parallel tips, NFP arrays have the potential to manufacture nanoscale features of chemical/biochemical materials at the wafer level (Fig. 5.18).

5.6 Future Prospects for Nanoprobes

Cantilevered nanoprobes offer the possibility of flexible protein deposition for diagnostic applications and drug screening. In addition to being compatible with SPM-readout methods, nanoprobe devices can be used in concert with classic protein analysis techniques such as ELISA. The probes can also deliver DNA for the purposes of genomic studies or nanoconstruction. The most recent advances in nanoprobes for biomolecule deposition involve the integration of microfluidics into cantilevers and the arraying of multiple probes in one and two dimensions. Higher-throughput patterning and a reduction in feature size have been achieved, and a better theoretical understanding of the patterning process has been acquired.

Aside from the direct and indirect patterning of biological materials, there are several other applications offered by cantilevered probes, such as in-depth studies of cell adhesion. Cells were able to adhere to DPN-generated patterns [39], although the pattern resolution was well below that of the length scale at which protein clustering in focal adhesion occurs [19]. Increasing the resolution attainable via probe-based patterning methods will improve the quality of rigid ligand templates for such studies. Controlled delivery can also be used for the study of cell communication: local perturbation of one cardiac myocyte in a small cluster of cells was achieved with the insertion of less than five α-toxin channels into the cell membrane using the nanopipette [101]. It could be observed that perturbation of one cell in a small cluster can affect the neighboring cells.

The atomic force microscope can also be exploited for its force sensitivity; it is able to measure piconewton forces associated with single molecules. In fact, AFM is currently the only force technique capable of mapping and analyzing single molecules with nanoscale lateral resolution [28]. AFM force spectroscopy involves measuring the interaction force between the AFM probe tip and the sample as the tip is pushed towards the sample and then retracted. The AFM probe deflection is measured as a function of the vertical displacement of the piezoelectric scanner. The technique can interrogate the forces and dynamics of interaction between individual ligands and receptors; this is essential to fundamental studies in molecular recognition, protein folding and unfolding, DNA mechanics and cell adhesion. The development of smaller cantilevers such as nanotube tips functionalized with single biomolecules should improve the force resolution, allowing for the measurement of smaller unbinding forces.

Finally, functionalized cantilevers may also operate as biosensors. For example, specific binding events cause changes in cantilever deflection [102]. Grogan et al. [103] demonstrated that antibody-coated cantilevers could be used to detect myoglobin concentrations in the range of normal physiological concentration in human serum. The change in frequency of a resonating cantilever can also be measured; this change is due to adsorption of biological material onto an appropriately functionalized tip. For example, Gupta et al. [104] fabricated arrays of silicon cantilever beams 20–30 nm thick to detect the mass of individual virus particles.

Cantilevered probes have the potential to enable new investigations in the life sciences, and improve diagnostic abilities in the health sciences. With continuing advancements, high-throughput assays performed rapidly and with high sensitivity are becoming a reality.

Acknowledgements. This work was supported primarily by the Nanoscale Science and Engineering Initiative of the National Science Foundation under NSF Award Number EEC–0647560. The authors acknowledge the Microfabrication Applications Laboratory of the University of Illinois at Chicago, the Cornell NanoScale Science & Technology Facility of Cornell University, the

Materials Processing and Crystal Growth Facility of Northwestern University, Northwestern University Atomic- and Nanoscale Characterization Experimental Center, and the Electron Microscopy Center of Argonne National Laboratory for the fabrication and characterization facilities.

References

1. Tang Q, Zhang Y, Chen L, Yan F, Wang R (2005) Nanotechnology 16:1062–1068
2. Wadu-Mesthrige K, Amro NA, Garno JC, Xu S, Liu G-Y (2001) Biophys J 80:1891–1899
3. Bier FF, Andresen D, Walter A (2006) In: Urban GA (ed) BioMEMS. Springer, Dordrecht, pp 167–198
4. Auburn RP, Kreil DP, Meadows LA, Fischer B, Matilla SS, Russell S (2005) Trends Biotechnol 23(7):374–379
5. Reese MO, van Dam RM et al (2003) Genome Res 13:2348–2352
6. Sotomayor Torres CM (ed) (2003) Alternative lithography: unleashing the potentials of nanotechnology. Kluwer/Plenum, Dordrecht
7. Bernard A, Renault JP, Michel B, Bosshard HR, Delamarche E (2000) Adv Mater 12(14):1067–1070
8. Hovis JS, Boxer SG (2000) Langmuir 16(3):894–897
9. Michel B, Bernard A, Bietsch A, Delamarche E, Geissler M, Juncker D, Kind H, Renault JP, Rothuizen H, Schmid H, Schmidt-Winkel P, Stutz R, Wolf H (2001) IBM J Res Dev 45(5):697–719
10. Mesquida P, Blanco EM, McKendry RA (2006) Langmuir 22(22):9089–9091
11. Tarlov MJ, Burgess DRF, Gillen G (1993) J Am Chem Soc 115(12):5305–5306
12. Lipshutz RJ, Fodot SPA, Gingeras TR, Lockhart DJ (1999) Nat Genet 21:20–24
13. Singh-Gasson S, Green RD, Yue Y, Nelson C, Blattner F, Sussman MR, Cerrina F (1999) Nat Biotechnol 17(10):974–978
14. Stoll D, Templin MF, Bachmann J, Joos TO (2006) In: Urban GA (ed) BioMEMS. Springer: Dortrecht, pp 245–268
15. Anonymous (2006) Nature 444(7121):963–964
16. Christman KL, Requa MV, Enriquez-Rios VD, Ward SC, Bradley KA, Turner KL, Maynard HD (2006) Langmuir 22(17):7444–7450
17. Pierce Biotechnology (2006) ELISA. Pierce Biotechnology, Rockford
18. Saliterman SS (2006) BioMEMS and medical microdevices. SPIE, Bellingham
19. Arnold M, Cavalcanti-Adam EA, Glass R, Blümmel J, Eck W, Kantlehner M, Kessler H, Spatz JP (2004) Chem Phys Chem 5(3):383–388
20. Ngunjiri JN, Li J-R, Garno JC (2006) In: Kumar CSSR (ed) Nanodevices for the life sciences, vol 4, 1st edn. Wiley-VCH, Weinheim pp 67–108
21. Lenhert S, Sun P, Wang Y, Fuchs H, Mirkin, Chad A (2007) Small 3(1):71–75
22. Henderson E, Lynch M, Nettikadan S, Johnson J, Huff J, Xu J, Mosher C, Vengasandra S, Radke K, Cao C, Ryckman K, Kristmundsdottir A (2004) Microsc Microanal 10:1432–1433
23. Christman KL, Enriquez-Rios VD et al (2006) Soft Matter 2:928–939
24. Lynch M, Mosher C, Huff J, Nettikadan S, Johnson J, Henderson E (2004) Proteomics 4:1695–1702
25. Lee M, Kang D-K, Yang H-K, Park K-H, Choe SY, Kang C, Chang S-I, Han MH, Kang I-C (2006) Proteomics 6(4):1094–1103
26. Yu X, Cheng DXQ (2006) Proteomics 6(20):5493–5503
27. Hu J, Zhang Y, Gao H, Li M, Hartmann U (2002) Nano Lett 2(1):55–57
28. Hinterdorfer P, Dufrene YF (2006) Nat Methods 3(5):347–355

29. Piner RD, Zhu J, Xu F, Hong S, Mirkin CA (1999) Science 283:661–663
30. Ginger DS, Zhang H, Mirkin CA (2004) Angew Chem Int Ed Engl 43(1):30–45
31. Rosner B, Duenas T, Banerjee D, Shile R, Amro N, Rendlen J (2006) Smart Mater Struct 15:S124–S130
32. Meister A, Jeney S, Liley M, Akiyama T, Staufer U, De Rooij NF, Heinzelmann H (2003) Microelectron Eng 644:67–68
33. Meister A, Liley M, Brugger J, Pugin R, Heinzelmann H (2004) Appl Phys Lett 85(25):6260–6262
34. Fang A, Dujardin E, Ondarcuhu T (2006) Nano Lett 6(10):2368–2374
35. Kim K-H, Moldovan N, Espinosa HD (2005) Small 1(6):632–635
36. Moldovan N, Kim K-H, Espinosa HD (2006) J Micromech Microeng 16:1935–1942
37. Wu B, Ho A, Moldoval N, Espinosa H (2007) Langmuir 23(17):9120–9123
38. Nafday OA, Weeks BL (2006) Langmuir 22:10912–10914
39. Lee K-B, Park S-J, Mirkin CA, Smith JC, Mrksich M (2002) Science 295:1702–1705
40. Vega RA, Maspoch D, Shen CKF, Kakkassery JJ, Chen BJ, Lamb RA, Mirkin CA (2006) Chembiochem 7(11):1653–1657
41. Lee KB, Kim EY, Mirkin CA, Wolinsky SM (2004) Nano Lett 4(10):1869–1872
42. NanoInk (2007) Protein arrays application note. NanoInk, Skokie. http://www.nanoink.net/d/appnote_protein_arrays.pdf
43. Wilson DL, Martin R, Hong S, Cronin-Golomb M, Mirkin CA, Kaplan DL (2001) Proc Natl Acad Sci USA 98(24):13660–13664
44. Lee K-B, Lim J-H, Mirkin CA (2003) J Am Chem Soc 125:5588–5589
45. Lim J-H, Ginger DS, Lee K-B, Heo J, Nam J-M, Mirkin CA (2003) Angew Chem Int Ed Engl 42(20):2309–2312
46. Nam J-M, Han SW, Lee K-B, Liu X, Ratner MA, Mirkin CA (2004) Angew Chem Int Ed Engl 43(10):1246–1249
47. Demers LM, Ginger DS, Park S-J, Li Z, Chung S-W, Mirkin CA (2002) Science 296:1836–1838
48. Zhang H, Elghanian R, Amro NA, Disawal S, Eby R (2004) Nano Lett 4(9):1649–1655
49. Agarwal G, Sowards LA, Naik RR, Stone MO (2002) J Am Chem Soc 125:580–583
50. Xu J, Lynch M, Huff JL, Mosher C, Vengasandra S, Ding G, Henderson E (2004) Biomed Microdevices 6(2):117–123
51. Vengasandra SG, Lynch M, Xu J, Henderson E (2005) Nanotechnology 16:2052–2055
52. Schena M, Heller RA, Theriault TP, Konrad K, Lachenmeier E, Davis RW (1998) Trends Biotechnol 16(7):301–306
53. Belaubre P, Guirardel M et al (2003) Appl Phys Lett 82(18):3122–3124
54. Belaubre P, Guirardel M et al (2004) Sens Actuators A 110:130–135
55. Leichle T, Silvan MM et al (2005) Nanotechnology 16(4):525
56. Saya D, Le T et al (2007) J Micromech Microeng 17(1):N1
57. Bietsch A, Zhang J, Hegner M, Lang HP, Gerber C (2004) Nanotechnology 15(8):873
58. Espinosa HD, Kim K-H, Moldovan N (2006) In: Kumar CSSR (ed) Nanodevices for the life sciences, vol 4, 1st edn. Wiley-VCH, Weinheim, pp 109–149
59. Bruckbauer A, Ying L, Rothery AM, Zhou D, Shevchuk AI, Abell C, Korchev YE, Klenerman D (2002) J Am Chem Soc 124(30):8810–8811
60. Bruckbauer A, Zhou D, Ying L, Korchev YE, Abell C, Klenerman D (2003) J Am Chem Soc 125(32):9834–9839
61. Rodolfa KT, Bruckbauer A, Zhou D, Korchev YE, Klenerman D (2005) Angew Chem Int Ed Engl 44(42):6854–6859
62. Rodolfa KT, Bruckbauer A, Zhou D, Schevchuk AI, Korchev YE, Klenerman D (2006) Nano Lett 6(2):252–257

63. Hu H, Xie S, Meng X, Jing P, Zhang M, Shen L, Zhu Z, Li M, Zhuang Q, Shao Y (2006) Anal Chem 78(19):7034–7039
64. Kim K-H, Ke C, Moldovan N, Espinosa HD (2003) In: Proceedings of the 4th international symposium on MEMS and nanotechnology. The 2003 SEM annual conference and exposition on experimental and applied mechanics, Charlotte, NC, 2–4 June, pp 235–238
65. In: Materials Research Society symposium proceedings, fall MRS meeting, 2004, p A5.56.1
66. Espinosa HD, Moldovan N, Kim K-H (2007) In: Bhushan B, Fuchs H (eds) Applied scanning probe methods, vol VII. Springer, Berlin, pp 77–134
67. Kim K-H, Sanedrin RG, Ho AM, Lee SW, Moldovan N, Mirkin CA, Espinosa HD (2007) Direct Delivery and Submicron Patterning of DNA by Nanofountain Probe. In press in Advanced Materials, 2007
68. Deladi S, Tas NR, Berenschot JW, Krijnen GJM, de Boer MJ, de Boer JH, Peter M, Elwenspoek MC (2004) Appl Phys Lett 85(22):5361–5363
69. Deladi S, Berenschot JW, Tas NR, Krijnen GJM, de Boer JH, de Boer MJ, Elwenspoek MC (2005) J Micromech Microeng 15:528–534
70. Xu S, Miller S, Laibinis PE, Liu G-Y (1999) Langmuir 15:7244–7251
71. Liu G-Y, Amro NA (2002) Proc Natl Acad Sci USA 99(8):5165–5170
72. Naujoks N, Stemmer A (2004) Colloids Surf A 249(1–3):69–72
73. Agarwal G, Naik RR, Stone MO (2003) J Am Chem Soc 125:7408–7412
74. Jang J, Hong S, Schatz GC, Ratner MA (2001) J Chem Phys 115(6):2721–2729
75. Sheehan PE, Whitman LJ (2002) Phys Rev Lett 88(15):156104
76. Cho N, Ryu S, Kim B, Schatz GC, Hong S (2006) J Chem Phys 124(2):024714
77. Weeks BL, Vaughn MW, DeYoreo JJ (2005) Langmuir 21:8096–8098
78. Rozhok S, Sun P, Piner R, Lieberman M, Mirkin CA (2004) J Phys Chem B 108(23):7814–7819
79. Weeks BL, Noy A, Miller AE, De Yoreo JJ (2002) Phys Rev Lett 88(25):255505
80. Zou J, Wang X, Bullen D, Liu C, Mirkin C (2006) Proc SPIE 6223:62230N-10
81. Hampton JR, Dameron AA, Weiss PS (2005) J Phys Chem B 109(49):23118–23120
82. Weeks BL, DeYoreo JJ (2006) J Phys Chem B 110(21):10231–10233
83. Schwartz PV (2002) Langmuir 18(10):4041–4046
84. Peterson EJ, Weeks BL, DeYoreo JJ, Schwartz PV (2004) J Phys Chem B 108(39):15206–15210
85. Nafday OA, Vaughn MW, Weeks BL (2006) J Chem Phys 125:144703
86. Manandhar P, Jang J, Schatz GC, Ratner MA, Hong S (2003) Phys Rev Lett 90(11):115505
87. Lee N-K, Hong S (2006) J Chem Phys 124(11):114711
88. Xu J, Lynch M, Nettikadan S, Mosher C, Vegasandra S, Henderson E (2006) Sens Actuators B 113(2):1034–1041
89. Lee SW, Oh BK, Sanedrin RG, Salaita K, Fujigaya T, Mirkin CA (2006) Adv Mater 18(9):1133–1136
90. Vettiger P, Despont M, Drechsler U, Durig U, Haberle W, Lutwyche MI, Rothuizen HE, Stutz R, Widmer R, Binnig GK (2000) IBM J Res Dev 44(3):323–340
91. Vettiger P, Cross G, Despont M, Drechsler U, Durig U, Gotsmann B, Haberle W, Lantz MA, Rothuizen HE, Stutz R, Binnig GK (2002) IEEE Trans Nanotechnol 1(1):39–55
92. Salaita K, Wang Y, Fragala J, Vega RA, Liu C, Mirkin CA (2006) Angew Chem Int Ed Engl 45(43):7220–7223
93. Bullen D, Chung S-W, Wang X, Zou J, Mirkin CA, Liu C (2004) Appl Phys Lett 84(5):789–791
94. Wang X, Bullen DA, Zou J, Liu C, Mirkin CA (2004) J Vac Sci Technol B 22(6):2563–2567
95. Bullen D, Liu C (2006) Sens Actuators A 125(2):504–511
96. Zhu H, Miao J, Chen B, Wang Z, Zhu W (2005) Microsyst Technol 11:1121–1126

97. Northwestern University (2007) Espinosa research group. http://clifton.mech.northwestern.edu/~nfp/
98. Rothuizen H, Drechsler U, Genolet G, Haberle W, Lutwyche M, Stutz R, Widmer R, Vettiger P (2000) Microelectron Eng 53(1–4):509–512
99. Kim CH, Jeong HM, Jeon JU, Kim YK (2003) J Microelectromech Syst 12(4):470–478
100. Lutwyche M, Andreoli C, Binnig G, Brugger J, Drechsler U, Haberle W, Rohrer H, Rothuizen H, Vettiger P, Yaralioglu G, Quate C (1999) Sens Actuators A 73(1–2):89–94
101. Ying L, Bruckbauer A, Zhou D, Gorelik J, Shevchuk A, Lab M, Korchev Y, Klenerman D (2005) Phys Chem Chem Phys 7:2859–2866
102. Ziegler C (2004) Anal Bioanal Chem 379:946–959
103. Grogan C, Raiteri R, O'Connor GM, Glynn TJ, Cunningham V, Kane M, Charlton M, Leech D (2002) Biosens Bioelectron 17(3):201–207
104. Gupta A, Akin D, Bashir R (2004) Appl Phys Lett 84(11):1976–1978

9 New AFM Developments to Study Elasticity and Adhesion at the Nanoscale

Robert Szoszkiewicz · Elisa Riedo

From volume V

Abbreviations and Symbols

α	Coefficient
δ	AFM tip–sample deformation depth
$\Delta\overline{F}$	Changes of the ultrasonic force
Λ	Dislocation density at the surface
a	Radius of a contact area between an AFM tip and a sample
A	Ultrasonic amplitude
AFAM	Acoustic force atomic microscopy
AFM	Atomic force microscopy or atomic force microscope
AH	Adhesion hysteresis
AH_{el}	Elastic part of adhesion hysteresis
AH^{rough}	Adhesion hysteresis in a contact between rough surfaces
BCP	Burnham–Colton–Pollock contact mechanics
CapH	Capillary forces hysteresis
d	AFM tip–sample distance
D	Fractal dimension
d_d	Dislocation length
d_o	Given indentation
DMT	Derjaguin–Muller–Toporov contact mechanics
d_s	Transducer's amplitude in SLAM
d_{tot}	Total normal or lateral displacement of an AFM tip (the sum of the contact deformation plus the cantilever bending)
$E_{1,2}$	Young's modulus of the AFM tip or the sample
E^*	Reduced Young's modulus between the AFM tip and the sample
F	Force
F_{all}	Sum of all interaction forces between an AFM tip and a sample
F_{adh}	AFM tip–sample adhesion force
F_o	Given force between an AFM tip and a sample
$F_{pull\text{-}off}$	Pull-off force (maximum tensile force at which an AFM tip and a sample are still in contact)
F_R	Frictional force
F_R^{rough}	Frictional force in a contact between rough surfaces
F_{SP}	AFM set point (load force set at tip–sample contact)
f-d curve	Force-distance curve
$G_{1,2}$	Shear modulus of the tip, or the sample

In: H. Fuchs, B. Bhushan (eds), *Biosystems - Investigated by Scanning Probe Microscopy*,
DOI 10.1007/978-3-540-37316-2_9, © Springer-Verlag Berlin Heidelberg 2010

G^*	reduced shear modulus between the tip and the sample
H	Hurst exponent
HF	High frequency
HFM	Heterodyne force microscopy
HOPG	Highly-oriented pyrolitic graphite
JKR	Johnson–Kendall–Roberts contact mechanics
k_{contact}	Normal or lateral stiffness of an AFM tip in contact with a surface
$k^{\text{l}}_{\text{contact}}$	Lateral stiffness of an AFM tip in contact with a surface
$k^{\text{n}}_{\text{contact}}$	Normal stiffness of an AFM tip in contact with a surface
k_{lever}	Normal or lateral stiffness of a free AFM tip
$k^{\text{n}}_{\text{level}}$	Normal stiffness of a free AFM tip
l_{c}	Contact length ($l_{\text{c}} = 2a$)
LF	Low frequency
l_{i}	AFM image size
NT	Nanotube
R	Reduced AFM tip–sample curvature radius
SLAM	Scanning local acceleration microscopy
SMM	Scanning microdeformation microscopy
S_{q}	RMS roughness at the acquired AFM images size l_{i}
UFM	Ultrasonic force microscopy
υ	Poisson ratio
$w_{\text{t–s}}$	AFM tip–sample adhesion energy
z_{c}	Cantilever's response amplitude in SLAM
z_{HF}	Cantilever response amplitude at the excitation frequency in AFAM

9.1 Introduction

The development of new materials with the size of a few nanometers has opened a new field of scientific and technological research. Nanomaterials such as carbon nanotubes, oxide nanobelts and semiconductor nanowires are promising building blocks in future integrated nanoelectronic and photonic circuits, nanosensors, interconnects and electromechanical nanodevices. The goal is to develop faster and better communication systems and transports, as well as smarter and smaller nanodevices for biomedical applications. To reach these objectives it is crucial to have knowledge and ability to control the mechanical behavior of these nanoobjects. In general, the mechanical properties of the materials at the nanoscale are not well-understood. The experimental challenge is to have an instrumentation that allows us to image and manipulate a nanoobject, characterize its atomic structure and measure forces of the order of nanonewtons. From the theoretical side, developing a theory of elasticity and friction at the nanoscale is an intriguing task that lies at the crossover between the atomic level and the continuum.

In this chapter we will describe recently developed methods to measure the elastic and adhesive properties of the materials at the nanoscale. Local elastic properties are usually studied by both static and dynamic AFM methods. The static methods include force-distance curves together with AFM-based nanoindentation tests as well as various local realizations of the classical triple point contact tests. The dynamic techniques utilize AFM cantilevers vibrating either at low frequencies, e.g., of a few kHz (modulated nanoindentation), or at ultrasonic frequencies. We will focus our description on two AFM-based dynamic methods: modulated nanoindentation and ultrasonic microscopy.

The chapter is divided in three parts: in Sect. 9.2 we describe continuum contact mechanics theories and their limitations. In the Sect. 9.3 we present the modulated nanoindentation method and we show how this technique is used to obtain high-resolution force versus indentation curves and the elastic modulus of nanomaterials. In the Sect. 9.4 we describe the use of ultrasonic AFM techniques to study local elasticity. We also highlight an application of ultrasonic AFM to investigate the local adhesive properties and their relation with friction at the tip–sample contact.

9.2 Contact Mechanics Theories and Their Limitations

The theories used to describe the contact between idealized macroscopic bodies are called continuum contact mechanics theories. These theories do not consider the local mechanical properties of the materials, but they treat each object in contact as a continuum. For this reason many questions can arise on the validity of these theories when we are dealing with nanocontacts. Here, we will present a brief compilation of classical continuum contact mechanics theories and we will discuss briefly the limitations of these theories in the last part of this section.

Hertz theory dates back to 1882 [1] This approach assumes that surface forces and adhesion can be neglected and that the elastic deformation due to the contact can be modeled taking into consideration only the geometry of the two contacting bodies and their elastic macroscopic properties. More advanced and general theories for bodies with axisymmetric shape versus a soft elastic surface have been developed by Sneddon [2]. When surface forces such as friction and adhesion are negligible Hertz and Sneddon models can be used, otherwise, other theories need to be considered.

The Derjaguin–Muller–Toporov (DMT) theory expands Hertz by including van der Waals forces between the bodies in the contact region [3]. DMT is applicable to AFM experiments where the reduced AFM tip is small, and the adhesion between the surfaces is small. JKR theory, named after Johnson, Kendall and Roberts, takes into account only short-range forces, such as adhesion, inside the contact region and neglects long-range forces outside of it [4]. For AFM experiments, this theory can be employed for highly adhesive systems when the tip has a low stiffness and large radius. A very complete contact mechanics theory was developed by Maugis and Pollock, and it deals with a wide range of behaviors [5]. By varying a single parameter this theory can go continuously from DMT to JKR, and can be applied to compliant, large and adhesive bodies as well as small rigid materials with low adhesion.

None of the afore-mentioned theories accounts for the AFM tip–sample forces outside the contact area. These forces are either semiempirically estimated (i.e., in the Burnham–Colton–Pollock theory [6, 7]), or calculated by first principles molecular dynamics simulations [8–10]. There is also an enormous bibliography on analytical and numerical studies about contact between varying geometries.

Contact mechanics theories assume no plastic deformation and no viscoelastic phenomena. These are accounted for through finite element studies for bodies of different geometries in contact. With typical AFM tip–sample contact radii of the order of a few nanometers and with loads of the order of tens of nanonewtons, the pressure into the tip–sample contact area is of the order of GPa. High pressure at the contact exceeds the bulk value of macroscopic tensile/compressive strengths (i.e., of about 250 MPa for mica), and the material can deform plastically. The compressive/tensile strengths depend, however, on the density of dislocations into the concerned volume. The average dislocation length d_d is about 100 nm, roughly 1000 times the interatomic distance. The dislocation density at the surface Λ reads: $\Lambda = d_\mathrm{d}^{-2}$. Thus, the dislocations are very rare in nanovolumes, and the values of the local tensile/compressive strengths approach their theoretical limits of about 10 percent of the corresponding Young's moduli E (e.g., $E = 70$ GPa for mica). Therefore, in typical AFM scanning conditions, the tip–sample interactions are mostly within the elastic limit. However, too high AFM tip–sample approaching speed and too high loads induce local plastic deformation at the surface [6].

Another important element not typically considered is surface roughness. The existence of roughness complicates tremendously the calculation of the real area of contact, because the contact occurs at several points where the asperities of the surfaces meet. How to incorporate roughness to contact mechanics has been studied for a long time. A recent review on the historical and modern understanding of the true contact area was written by Gao et al. [11]. There is extensive research on stochastic methods to treat contact among rough surfaces [12, 13]. There are also analytical models [14]. In particular, Greenwood and Williamson have assumed the surface as covered by hemispherical asperities with the same radii [15]. Whitehouse and Archard have extended that model by allowing random radii of curvature [16]. Nayak has introduced the random process theory with input parameters such as the distribution of asperities heights, the density of asperities, the mean surface gradient, and the mean curvature of asperities [17]. We will not treat this topic here, but we suggest some interesting reading, found in [18–26].

There is a lingering question as to whether continuum contact mechanics should be used at the nanoscale. At what point does the discreteness of matter start to play such an important role that continuum mechanics can no longer be used? There are some molecular dynamics simulations studying at what point continuum mechanics breaks down. Recently, Luan and Robbins [27] conducted molecular dynamics simulations to test whether the Hertz theory is still valid at the nanoscale. They found that the atomic discreteness at the bulk does not seem to affect the tip–sample contact area as much as roughness does. They concluded that continuum mechanics may underestimate the area of SPM contacts by up to 100% at small loads, though the error decreases with increasing load and radius. Miesbauer et al. [28] conducted molecular dynamics simulations of the contact between two NaCl nanocrystals. After comparing with the Hertz theory, they concluded that this theory can be employed to describe systems with sizes down to 5 nm. Furthermore, several friction force microscopy experiments support the validity of continuum contact mechanics at the nanoscale [29–31].

9.3 Modulated Nanoindentation

9.3.1 Force-Indentation Curves

The ideal and simplest way to measure the elasticity of nanoobjects is to indent an atomic force microscope tip in the nanoobject, while measuring force versus indentation curves. Measurements of force versus indentation curves are particularly challenging on nanoobjects because to insure the elastic regime, we usually have to measure forces of a few nanonewtons versus displacements of a few angstroms with subangstrom resolution.

An AFM force-distance (*f*-*d*) curve [32] is a measure of the normal force acting between an AFM tip and a sample surface as a function of the distance between the same tip and the undeformed sample surface. Thus, when the tip starts to be in contact with the sample, a *f*-*d* curve is a force versus indentation curve. During an AFM experiment the force is detected through the bending of the AFM cantilever, while the distance is given by the AFM scanner vertical position as the tip approaches the sample (Fig. 9.1). It is thus clear that *f*-*d* curves need to be calibrated [32] accounting on the cantilever's bending and the cantilever's spring constant. Recently, some convenient methods have been developed to calibrate the cantilever's spring constants based on thermal noise (see Sader et al. [33], Higgins et al. [34], Proksch et al. [35], and references therein, as well as the review written by Burnham [36] and Gibson [37]). The errors in *f*-*d* calibration come from many sources: detection of the cantilever's position and detection of the forces, coupling between torsion and vertical cantilever's bending, hysteresis and resonance of a piezoscanner, etc. Furthermore, Heim et al. [38] as well as Hutter [39] have recently reported on some systematic errors in *f*-*d* curves due to usual cantilever tilt with respect to the surface.

Some instrumental developments have pursued the improvement of the force-distance sensitivity through: (1) use of segmented scanners with lower hysteresis,

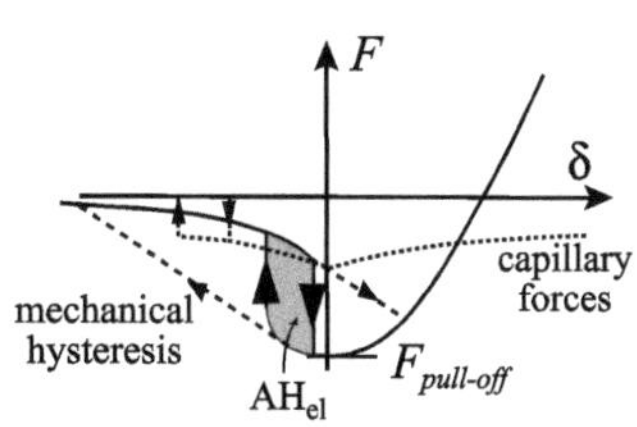

Fig. 9.1. An example of a force-distance curve obtained with an AFM. The *f*-*d* curves represent forces between an AFM tip and a sample as a function of the tip–sample indentation (positive value of δ) or separation (negative value of δ). The *arrows* mark that the curve is intrinsically hysteretic, e.g., different for approaching and retracting to/from the sample. The *shaded area* corresponds to the elastic part of adhesion hysteresis (AH$_{el}$), but there are other hysteretic effects present as well. Mechanical hysteresis results from a sudden cantilever snap-in and snap-out to/from the contact, when a gradient of existing forces exceeds a spring constant of a cantilever. Capillary forces hysteresis is associated with the presence of water on the sample. The pull-off force $F_{\text{pull-off}}$ is marked as well

(2) separation and minimization of X, Y, Z scanners to obtain better signal to noise ratio, and (3) use of low-coherence lasers to minimize any interference patterns on the f-d curves. Manyes et al. [40] with indentations smaller than one nanometer have obtained local hardness from the onset of a plastic deformation between diamond AFM tips and a Si surface. By improving the electronics to the limit and building scanners with very large resonance frequencies, Rost et al. [41] show that force-distance imaging can be obtained with video rates. Still however, the comparative force-distance measurements with the same AFM configuration are the most meaningful and trusted.

Recently several attempts have been pursued to attain more reliable methods to measure force versus indentation curves. Special attention has been dedicated in finding a technique to investigate the radial elasticity of carbon nanotubes (NTs). Some authors have proposed to vibrate an AFM cantilever in non-contact mode with small vibration amplitudes (a few angstroms) while it is scanned across the sample [42]. Because of van der Waals forces between the tip and the tube, the normal force can be considered in first approximation proportional to the amplitude of the cantilever vibrations. However, in this case it is extremely difficult to obtain quantitative results. In fact, this method requires a good description of tip–sample interactions and a good model of the cantilever vibrations taking into accounting the geometry of the problem and the different forces acting on the cantilever [43].

Another AFM-based method, modulated nanoindentation, has been demonstrated to be a powerful means to measure quantitatively the shear and radial elasticity of nanoobjects [44].

Modulated nanoindentation (Fig. 9.2) consists in indenting an AFM tip in a sample while laterally (parallel to the substrate) or normally (perpendicular to the substrate) oscillating the sample. The idea is to measure the slope of the f-d curves in each point (F_o, d_o). Here, we will describe a modulated nanoindentation experiment on a NT lying on a flat, hard substrate. The oscillations have to be very small, about one angstrom, so that the tip sticks on the nanoobject, i.e., on the NT. During the stick regime, the normal or lateral force necessary to move the NT vertically or laterally with respect to the cantilever support coincides with the force needed to elastically stretch two springs in series [45, 46]: the cantilever, with (normal or lateral) stiffness k_{lever}, and the tip–sample contact with (normal or lateral) stiffness k_{contact}. Mathematically we can write

$$\partial F/\partial d_{\text{tot}} = (1/k_{\text{lever}} + 1/k_{\text{contact}})^{-1} , \tag{9.1}$$

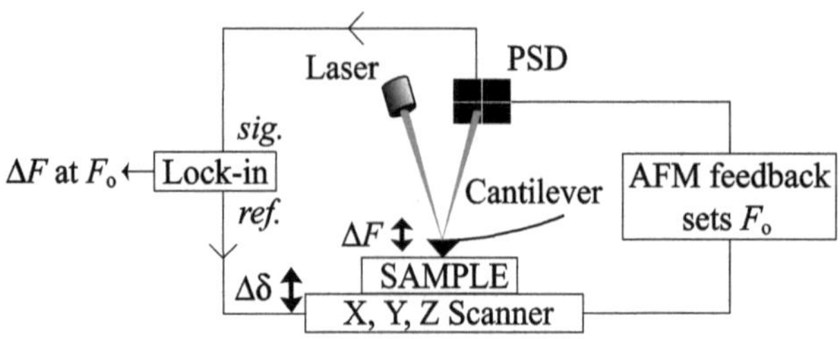

Fig. 9.2. Experimental setup for the modulated nanoindentation method

where F is the normal or lateral force and d_{tot} is the total normal or lateral displacement. Thus, the value of d_{tot} is the sum of the contact deformation plus the cantilever bending. Since k_{lever} is known, a measure of $\partial F/\partial d_{\mathrm{tot}}$ at different normal loads, F_{o}, gives allows for the obtainment of k_{contact} as a function of F_{o}.

The value of $\partial F/\partial d_{\mathrm{tot}}$ is experimentally measured by means of a lock-in amplifier. The lock-in amplifier modulates vertically or laterally the sample position Δd_{tot} by sending an appropriate signal to the piezotube, and it simultaneously measures the signal ΔF coming from the AFM (see Fig. 9.2). For the experiments on the NT the authors have chosen $\Delta d_{\mathrm{tot}} = 1.3$ Å, and they have worked at a frequency of 3.111 kHz. From normal modulation measurements they have obtained the normal contact stiffness $k^{\mathrm{n}}_{\mathrm{contact}}$ as a function of F_{o}, which contains all the information related to the radial elasticity of the NT. Besides, in lateral modulation experiments, we measure the lateral contact stiffness $k^{\mathrm{l}}_{\mathrm{contact}}$ versus F_{o}, which provides information on the shear elastic properties of the nanoobject.

In the following, first we describe how to derive force versus indentation curves from the measurements of $k^{\mathrm{n}}_{\mathrm{contact}}$. Second, we show how continuum contact mechanics is used to extract from $k^{\mathrm{n}}_{\mathrm{contact}}$ and $k^{\mathrm{l}}_{\mathrm{contact}}$ the radial Young's modulus and the axial shear modulus of the NT. In the analogy to Hook's law of elastic deformation, for normal modulation experiments we have $F = k^{\mathrm{n}}_{\mathrm{contact}}\delta$, where δ is the indentation of the tip in the NT. By integrating this equation we obtain

$$\int_{F(0)}^{F(d)} \left(\frac{1}{k^{\mathrm{n}}_{\mathrm{contact}}}\right) \mathrm{d}F = \int_{0}^{d} \mathrm{d}\delta \,. \tag{9.2}$$

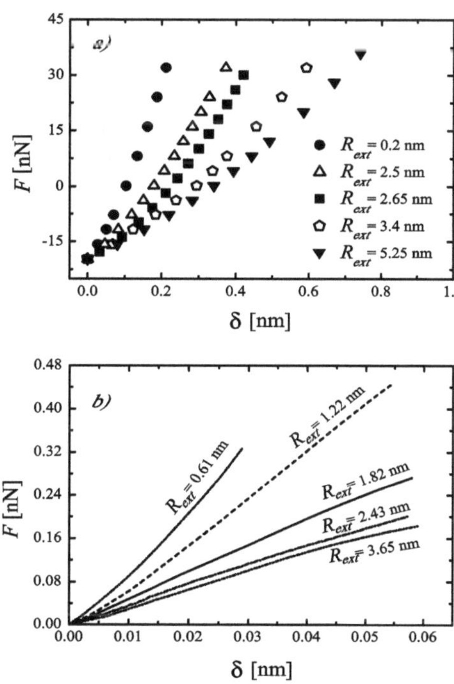

Fig. 9.3. Normal force as a function of indentation for NT of different radii, obtained by integration of experimental $1/k_{\mathrm{contact}}$ versus F curves using the trapeze method (**a**). Theoretical normal force as a function of indentation for NT of different radii (different number of layer N), obtained by simulating the indentation of a rigid sphere in a NT (**b**)

By substituting (9.1) into (9.2) we obtain from the experimentally measured $\partial F/\partial d_{\text{tot}}$ the F versus δ curves

$$\int_{F(0)}^{F(d)} \left(\frac{\Delta d_{\text{tot}}}{\Delta F} - \frac{1}{k^{\text{n}}_{\text{lever}}} \right) \mathrm{d}F = \int_{0}^{d} \mathrm{d}\delta \,, \tag{9.3}$$

where $k^{\text{n}}_{\text{level}}$ is the normal stiffness of a free AFM tip. In Fig. 9.3 we show the indentation curves obtained from experiments and by means of molecular dynamics simulations. Both experiments and theory show that the radial stiffness of carbon NT increases by decreasing the tube's external radius, e.g., by decreasing the internal radius. It is interesting to note that experiments and calculations agree in finding the same 3/2-power law dependence of F on indentation, $F \propto \delta^{3/2}$, which is expected in the framework of the Hertz contact mechanics [5].

9.3.2 Elastic Moduli

The elastic modulus of a material is usually extracted from force versus indentation curves by using the following relationship:

$$E^* = \frac{\sqrt{\pi}}{2} \frac{\partial F}{\partial \delta} \frac{1}{\sqrt{A}} \tag{9.4}$$

where E^* is the effective modulus of elasticity defined by $E^* = [(1 - \upsilon_1^2)/E_1 + (1 - \upsilon_2^2)/E_2]^{-1}$ with $\upsilon_{1,2}$ and $E_{1,2}$ the Poisson's ratio and Young's moduli of the indenter and sample and A is the contact area, respectively. In (9.4), $\frac{\partial F}{\partial \delta}$ is $k^{\text{n}}_{\text{contact}}$ and is measured during the experiments, whereas to determine A, we need to model the contact between the indenter, i.e., the AFM tip and the sample. The simplest way to do it is to use continuum contact mechanics models.

Molecular dynamics simulations show that continuum contact mechanics have some limits in the case of a nanometer size sphere in contact with a NT. Errors up to 30% have been calculated. However, the 3/2-power law dependence found in experimental and theoretical indentation curves is a strong indication of the reliability of the Hertz theory for our system. Following the continuum contact mechanics theories, for small normal and lateral forces, the contact area and the contact radius a [45, 46] for a sphere indenting a flat surface are given by

$$A = \pi a^2 \tag{9.5a}$$

$$a = R^{1/3}(F_{\text{SP}} + F_{\text{adh}})^{1/3}/(E^*)^{1/3} \,, \tag{9.5b}$$

where R is the sphere (tip) radius, F_{sp} is the load force set at the tip–sample contact, F_{adh} is the tip–sample adhesion force. Furthermore, we can show that the normal and lateral contact stiffness are directly proportional to the contact radius a [5]

$$k^{\text{n}}_{\text{contact}} = 2E^* a \tag{9.6b}$$

$$k^{\text{l}}_{\text{contact}} = 2G^* a \,, \tag{9.6b}$$

where G^* is the effective shear modulus defined by $G^* = [(2 - \upsilon_1)/G_1 + (2 - \upsilon_2)/G_2]^{-1}$ with $G_{1,2}$ being the shear moduli of the sphere and the flat surface. We underline that (9.5a) and (9.6b) are equivalent to (9.4), but now we also have access to the area of contact through (9.5b). Furthermore (9.5) and (9.6) can also be applied to the shear elasticity measurement. If the geometry in consideration is different from that one of sphere-plane, e.g., sphere-cylinder geometry as in the case of an AFM tip indenting a nanotube, relationships slightly different from those presented in (9.5) and (9.6) need to be considered. A detailed treatment of contact stiffness in different geometries can be found in [5].

In general, under the assumptions of continuum elasticity theories, the dependence of the contact stiffness versus the external force between two elastic solids in contact is an analytical function of the external force, the tip and the sample elastic moduli, and some geometrical parameters, e.g., the tip radius. By fitting the experimental k_{contact} versus F curves obtained by modulated nanoindentation with the appropriate analytical function, it is possible to obtain the normal and transversal modulus of the nanoobject.

Figure 9.4 shows the experimental values of $k^{\text{n}}_{\text{contact}}$ as a function of normal loads for a carbon NT. These curves have been fitted with the appropriate equations [44] for the sphere-cylinder geometry. We underline that in this fitting procedure the only free fitting parameter is the radial Young's modulus of the NT, which thus can be extracted without ambiguity. In Fig. 9.5 we show the radial modulus found by means

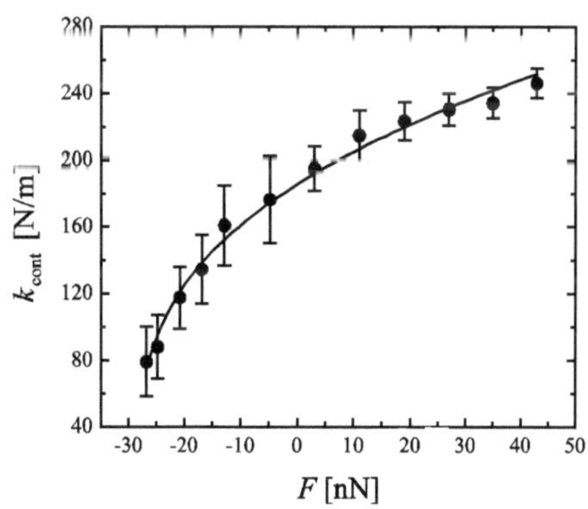

Fig. 9.4. Experimental normal contact stiffness versus normal indentation force for a 3 nm tube radius

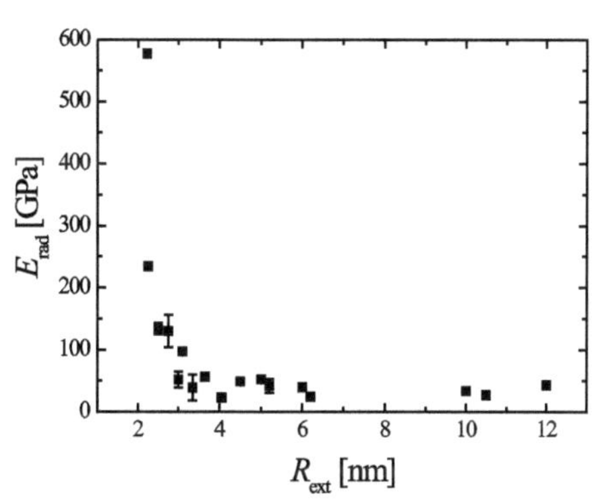

Fig. 9.5. Experimental values of the radial Young's modulus of carbon NTs as a function of external radius as obtained from normal modulation experiments. *Error bars* correspond to mean errors on tubes of the same diameter

of this method as a function of the NT radius for NTs with external to internal radius ratio equal to 2.

9.4 Ultrasonic Methods at Local Scales

Classical acoustic microscopy [47] is a very well-developed non-destructive method for evaluating the material's elasticity on a large scale because its resolution is limited, almost as in the case of an optical microscopy, up to about one micron. However, it has been reported that near-field techniques can greatly overcome physical barriers for microscope resolution [48–50]. Therefore, at local scales, the near-field combination of ultrasound with AFM has been widely applied. The experimental techniques include scanning microdeformation microscopy (SMM), acoustic force atomic microscopy (AFAM), scanning local acceleration microscopy (SLAM), heterodyne force microscopy (HFM), and ultrasonic force microscopy (UFM).

Also, an interesting recent application of the UFM technique has been proposed by Szoszkiewicz et al. [51]. With a modified UFM setup, they have been able to measure the elastic part of the adhesion hysteresis at micro- and nanoscales, and relate these measurements with corresponding friction in the case of fine-polished metallic samples as well as HOPG and mica [52–54]. This approach sets an ultimate way to measure local frictional forces, as will be described in the Sect. 9.4.3.

9.4.1 Brief Description of Ultrasonic Methods

The SMM method was developed by Cretin et al. [55, 56]. Here, a sapphire tip is mounted at the end of a silicon cantilever that is embedded on the piezoelectric transducer. Looped with a specific amplifier, this assembly constitutes an electromechanical oscillator oscillating on the first flexural mode of the cantilever and generating microdeformations at the sample surface. The strain field generated by the cantilever induces an acoustic wave in the sample. In the original setup [55] the acoustic wave was detected with a piezoelectric transducer placed underneath the sample. In the improved setup (Fig. 9.6), which works in reflection, the tip contacts the sample, and any variations in the local elastic properties of the sample induce

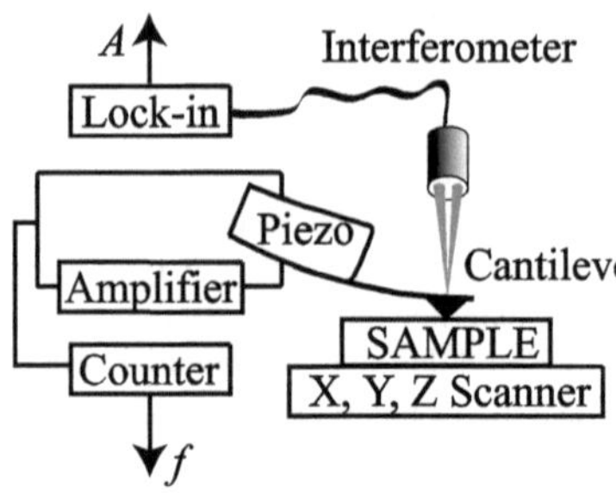

Fig. 9.6. The principles of the SMM technique. A sapphire tip is mounted at the end of a silicon cantilever embedded in the piezotransducer. Looped with an amplifier it is an oscillator at the first flexural mode of the cantilever. The strain field induces microdeformations of the sample and frequency change of the oscillator, which are detected by a frequency counter. Heterodyne interferometer measures the true amplitude of the cantilever's vibrations

variations of the resonance frequency of the oscillator. A counter detects these frequency variations, and a heterodyne interferometer detects the cantilever amplitude. The SMM technique produces an image of the subsurface defects or of the local elastic properties of the sample. Local elastic properties are inferred from a fit of the resonance spectrum (amplitude versus frequency) of the clamped (through contact stiffness) cantilever beam to the measured resonance spectrum of the cantilever. Due to high curvature radii of sapphire tips, the original resolution was limited to a few microns. Nowadays the tips can be crafted with finer precision, i.e., using the Focused Electron Beam (FEB) or the Focused Ion Beam (FIB) techniques [57, 58].

AFAM (Fig. 9.7) was originally developed by Rabe and Arnold [59]. The cantilever, in contact with the sample, is actuated by small amplitude vibrations of a few angstroms at one of its high-frequency (MHz) contact resonances (flexural or torsional). Unless a sample is extremely compliant (i.e., cells), high-frequency cantilever modes provide an enhanced sensitivity to contact stiffness [60, 61]. In AFAM, the cantilever response amplitude at the excitation frequency z_{HF} provides the local tip–sample elasticity, and the quasistatic cantilever's position changes provide topography of a sample. Insufficient AFM frequency bandwidth propelled the usage of a knife-edge detector [59] to measure the value of z_{HF} for higher ultrasonic modes. Later, it was realized that together with a reference sample of known elasticity both first and second modes of the cantilever's vibrations are sufficient to determine the tip–sample contact stiffness and an unknown elasticity of the sample. Nowadays, with a proper choice of AFM cantilevers, AFAM utilizes heterodyne down frequency converters or just commercial AFM systems without a need for knife-edge detectors [61–63]. Of the ultrasonic techniques, AFAM is probably the most intensively studied. The effects of microscope tip radius changes due to abrasion [62], humidity [64], and geometry of the tip [65] have been discussed. It has been also reported that attachment of a concentrated mass at the cantilever's extremity improves the AFAM's sensitivity [66].

The SLAM technique (Fig. 9.8) was developed by Burnham et al. [67]. Here, a transducer placed underneath the sample actuates it in the kHz range with amplitude d_s. The cantilever's response z_{HF} at the excitation frequency is detected by a standard beam-bounce technique. It has been shown in [68, 69] that at frequencies

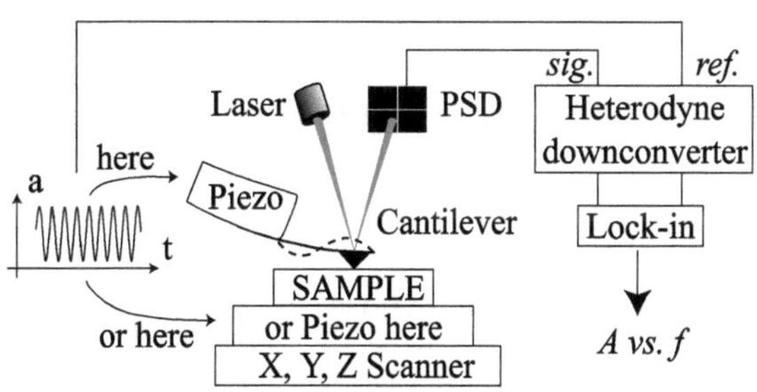

Fig. 9.7. The principles of the AFAM technique. The cantilever, in contact with a sample, is actuated by small amplitude vibrations of a few angstroms at its high-frequency contact resonance. Up to certain frequency (a few MHz) the cantilever vibrations at ultrasonic frequencies are detected with a typical photodiode and downshifted through a heterodyne downconverter

Fig. 9.8. The principles of the SLAM technique. A transducer placed below the sample actuates an AFM cantilever at the kHz range frequency. The lock-in amplifier insures precise detection of the cantilever's vibration at the excitation frequency

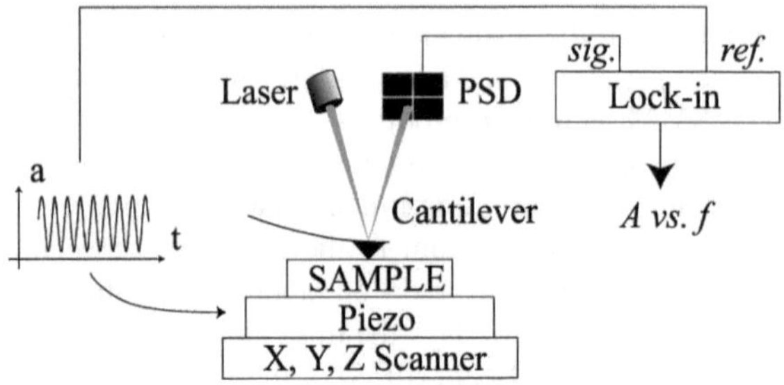

higher than the tip–sample contact resonance, and away from any spurious resonance, the ratio of d_s/z_{HF} is a sensitive measure of the contact stiffness. Therefore, subsequent SLAM imaging at different frequencies reveals the changes of contrast between areas of different elasticity, and allows calculations of the local Young's moduli.

Cuberes et al. [70] developed the HFM technique (Fig. 9.9). The HFM utilizes simultaneous excitation of both the tip and the sample. An ultrasonic transducer placed underneath actuates the sample and another ultrasonic transducer at the base of the cantilever actuates the tip. Both frequencies are in the range of MHz and differ only by a few kHz to produce beats. The cantilever's response in the beat frequency is detected by a conventional beam-bounce technique. The resulting HFM response (or image) depends on elasticity, but the quantitative interpretation remains unclear.

The UFM method was invented by Kolosov and Yamanaka [71]. Here (Fig. 9.10), an ultrasonic excitation of a few MHz is applied to a tip–sample contact via an actuator placed underneath the sample or above the AFM cantilever. The fast ultrasonic excitation is modulated by a slow frequency signal between 100 Hz up to a few kHz at which the cantilever can readily respond. The AFM feedback is usually set to be slow to maintain only the long time tip–sample position. Then, the cantilever reacts to the mean force between the tip and a sample (rectification) called the ultrasonic force and calculated over at least one period of vibrations. With increasing ultrasonic amplitudes, ultrasonic force increases only slightly up to the jump-up point when the tip–sample contact breaks. The ultrasonic force at the jump-up point

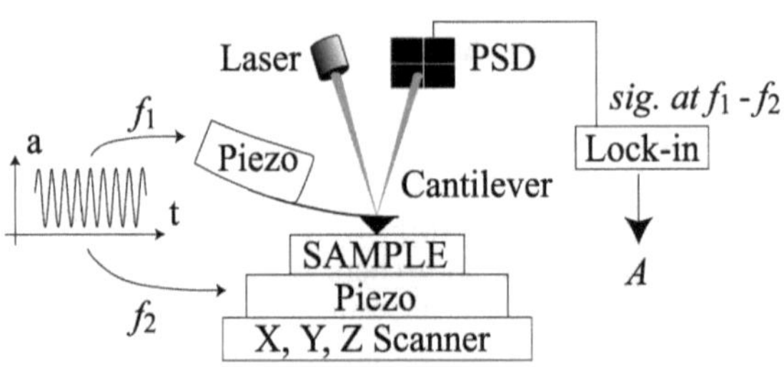

Fig. 9.9. The principles of the HFM technique. Two ultrasonic transducers (one below the sample, one above the tip) deliver ultrasonic vibrations differing by a few kHz. The lock-in amplifier insures precise detection of the cantilever's vibration at the beat frequency

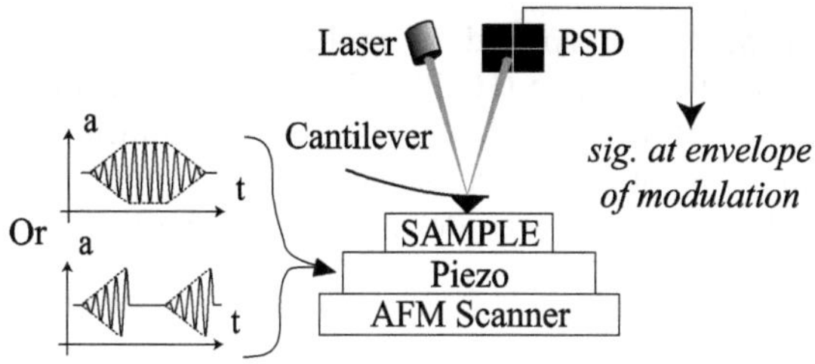

Fig. 9.10. The principles of the UFM technique. The cantilever is initially in contact with the sample, is actuated by a high-frequency ultrasonic signal at a few MHz, at which it responds only inertially. High-frequency signal is modulated by waveform of either a triangular or a trapezoidal shape. At sufficiently high ultrasonic amplitude the tip–sample contact breaks and the average cantilever's position at the modulating envelope jumps up

increases abruptly [52, 71, 72]. For larger ultrasonic amplitudes the tip stays outside the contact for a longer fraction of high-frequency cycles. Recently, it has been developed [52–54] that at least for metallic samples at ultrasonic frequencies avoiding any cantilever resonance and subharmonic [73–75], the cantilever response at the low frequency is a measure of local elasticity. In the absence of capillary forces between an AFM tip and a sample (either in dry environment or in liquids), and with the knowledge of R as well as the tip–sample adhesion energy $w_{\text{t–s}}$, the value of E^* is [53]

$$E^* \approx 0.2\pi^{5/2}[F_{\text{SP}}(A_2 - A_1) + \Delta\overline{F_2}A_2 - \Delta\overline{F_1}A_1]^{-3/2}R^2 w_{\text{t–s}}^{5/2} \tag{9.7}$$

where F_{SP} is the set point, and $\Delta\overline{F_1}$, $\Delta\overline{F_2}$ are changes of the ultrasonic force for ultrasonic amplitudes A_1, A_2 measured just before (index 1) and just after (index 2) the point of jump-up.

9.4.2 Applications of Ultrasonic Techniques in Elasticity Mapping

AFM-based ultrasonic techniques have been recently applied mostly to probe elasticity of thin films and embedded structures. As a result of the AFM tip blunting due to its high-frequency tapping [72], ultrasonic methods (SLAM, AFAM, UFM) achieve the lateral resolution of 20–100 nm. In the case of soft samples (i.e., polymers), usually only qualitative information is obtained owing to the complex relationship between the contact stiffness and the dynamic Young's modulus [68]. Moreover, any sharp variations in topography of the sample correspond to variations in the tip–sample contact area and produce some artifacts [72].

SLAM measurements have been performed on a wide range of engineering materials, from Al_2O_3/Al metal-matrix composites to PVC/PB polymer blends [68, 76, 77], and 100 thin silicon oxide coatings on polyamide [69].

AFAM has measured reduced Young's moduli for 50 nm thin films Ni deposited on Si, glass-fiber/polymer composites [78], and thin diamond-like carbon coatings [62].

The UFM method was applied to probe relative elasticity of defects in highly-oriented pyrolitic graphite (HOPG) [79], Ge dots on silicon substrate [80], and heterogeneous nanostructures [81]. Also polymeric materials embedded into stiff [82] or more compliant matrices have been measured. The UFM depth sensitivity for elastic imaging was determined to be several hundred nanometers [83].

9.4.3 UFM Measurements of Adhesion Hysteresis and Their Relations to Friction at the Tip-Sample Contact

The UFM method recently has been applied to measure adhesion hysteresis and local friction in fine-polished metallic samples, HOPG, and mica [52–54].

Adhesion hysteresis (AH) is the difference between the work needed to rupture and establish the contact between two bodies [84]. AH depends on elastic, viscoelastic, plastic, adhesive, geometric properties of contacting asperities [7,57], as well as on temperature, relative humidity [85], time of contact, and any chemical reactions at the contact [86,87]. In the realm of the nanoscale AFM measurements, the elastic part of adhesion hysteresis AH_{el} is usually entangled into simultaneously occurring mechanical hysteresis and capillary forces hysteresis (Fig. 9.1). Mechanical hysteresis stems from the snap-in and snap-out behavior of the AFM cantilevers when the gradient of the tip–sample forces exceeds the cantilever spring constant. Capillary force hysteresis is due to the presence of water at the surface.

A strong correlation between friction forces and adhesion hysteresis is confirmed by theoretical predictions and experiments at the macro- and microscale. Based on macroscale measurements, Briscoe et al. [88] reported that adhesion hysteresis and friction are interrelated, which has been supported by others [89, 90] through a systematic study in rolling as well as sliding motion using rubber and glass. Microscale measurements have been made with the surface force apparatus using molecularly thick polymeric surfactants sandwiched between curved mica surfaces [91].

Szoszkiewicz et al. [52–54] used a modified UFM setup to address measurements of micro- and nanoscale adhesion hysteresis and to correlate them with friction. They employed an AFM microscope as well as a nanoindenter. In their AFM-based setup, a heterodyne interferometer working in-situ is used to set the UFM high frequency (HF), which is above the first cantilever flexural resonance and does not coincide with any other cantilever's resonance or subharmonic. Trapezoidal shape of a modulating (LF) waveform was used, which allows for the tip–sample contact to be ruptured on a rising slope, and permanently reestablished on a declining slope. The mean LF cantilever deflection (UFM signal) is acquired by an AFM as a difference between instant cantilever deflection and the set point. In the nanoindenter-based setup, the nanoindenter resonance frequency (hundreds of Hz) is much lower than that of the AFM cantilever. Thus, HF has been downshifted by several decades, and an AFM scanner is used directly to vibrate a sample at high frequency with no additional transducer necessary between the AFM scanner and the tip. Besides, the nanoindenter diamond (Berkovitch) tip is more stable than any AFM tip.

Geometrical considerations of the tip–sample positions just before and just after the jump-up point on the hysteretic force-distance curves have yielded an elastic part of adhesion hysteresis in the following form [52]:

$$\mathrm{AH_{el}} = 4\left(A_2\overline{F}_2 - A_1\overline{F}_1\right) = 4\left(F_{\mathrm{SP}}(A_2 - A_1) + \Delta\overline{F}_2 A_2 - \Delta\overline{F}_1 A_1\right) . \quad (9.8)$$

Equation (9.8) can be applied as is to both AFM and nanoindenter-based measurements, because it is independent of the particular shape of $F(\delta)$ and geometry of the tip–sample contact area.

The values of $\mathrm{AH_{el}}$ have been theoretically and experimentally correlated with friction. Theoretical correlations were obtained based on the classical theory of adhesional friction, contact mechanics, and capillary hysteresis (CapH). It has been obtained that in case of such samples (e.g., with negligible viscoelastic and plastic contributions to adhesion hysteresis), adhesion hysteresis scales with friction at the tip–sample contact through the following relation:

$$\mathrm{AH_{el}} = 100\left(\frac{F_{\mathrm{SP}} + F_{\mathrm{cap}}}{3\pi R w_{\mathrm{t\text{-}s}}} + 1\right)^{-2/3} (w_{\mathrm{t\text{-}s}}/E^*) F_{\mathrm{R}} + \mathrm{CapH} , \quad (9.9)$$

where F_{cap} is the capillary force coming from the eventual presence of a capillary bridge between the tip and the sample, and F_{R} is the frictional force.

Considering that surfaces are often locally regarded as self-affine fractals [24,92], it has been found that adhesion hysteresis $\mathrm{AH^{rough}}$ and friction $F_{\mathrm{R}}^{\mathrm{rough}}$ on rough surfaces are reduced compared with their counterparts (AH, F_{R}) calculated on smooth surfaces

$$\mathrm{AH^{rough}}/\mathrm{AH} = 1 + 58.14 S_{\mathrm{q}} R (l_{\mathrm{c}}/l_{\mathrm{i}})^{H} l_{\mathrm{c}}^{\ 2} \quad (9.10)$$

$$F_{\mathrm{R}}^{\mathrm{rough}}/F_{\mathrm{R}} \approx 30^{1-H} , \quad (9.11)$$

where S_{q} is the RMS roughness at the acquired AFM images length l_{i}, l_{c} is the contact length calculated through the contact mechanics (no roughness assumed), and $H = 3 - D$ is the Hurst exponent, with D being the fractal dimension.

Correlations between adhesion hysteresis and friction have been measured using the ultrasonic force microscopy (UFM) on mica, calcite, and a few metallic samples (Pt, Au, Cu, Zn, Ti and Fe) both on nano- and microscale. Adhesion hysteresis measured at the nanoscale has ranged between 4×10^{-19} J and 10^{-18} J. Adhesion hysteresis measured at the microscale has ranged between 8×10^{-17} and 14×10^{-17} J. These results are similar to the results of other groups [93–96], which have measured adhesion hysteresis by probing energy losses with an AFM cantilever vibrating with ultrasmall amplitudes or with a surface force apparatus.

At the nanoscale, the experimental data quantitatively matched the calculated values when the surface roughness of the studied materials was included. In the microscale, the experimental data qualitatively matched the calculations, but were up to three times higher than the calculated results due to the shortcomings of the model. The model in the microscale has overestimated the calculated adhesion hysteresis, because it has implicitly considered only the single asperity tip–sample contact.

As a result the UFM method can be now used for reliable friction investigations with an unprecedented high lateral resolution, which is limited only by the tip–sample contact radius. AFM-based UFM measurements are also fast because they are limited only by a few periods of the modulating UFM frequency, which is roughly 100 Hz at a present time but can be extended into the kHz range as well. Then, a typical 256 × 256 pixel AFM friction image is recorded within one minute, which is a few times faster than acquiring a good quality image with currently available AFM systems.

Acknowledgements. We would like to thank B. Bhushan, G. Brune, A.J. Kulik and I. Palaci for stimulating discussions. We would like as well to acknowledge the financial support of the US NSF under the grant DMR-0405319, as well as US DOE under the grant number DE-FG02-06ER46293.

References

1. Hertz HJ (1882) Reine Angew Math 92:156
2. Sneddon IN (1965) Int J Eng Sci 3:47
3. Derjaguin BV, Muller VM, Toporov YPT (1975) J Colloid Interf Sci 53:314
4. Johnson KL, Kendall K, Roberts AD (1971) Proc R Soc Lon Ser A 324:301
5. Maugis D (2000) Contact, Adhesion, and Rupture of Elastic Solids. Springer, Berlin, Heidelberg, New York
6. Burnham NA, Colton RJ, Pollock HM (1991) J Vac Sci Technol A 9:2548
7. Bhushan B (1999) Handbook of Micro/NanoTribology, 2nd edn. CRC, Boca Raton
8. Schmauder S (2002) Annu Rev Mater Res 32:437
9. Barber JR, Ciavarella M (2000) Int J Solids Struct 37:29
10. Tsakmakis C (2004) Int J Plasticity 20:167
11. Gao JP, Luedtke WD, Gourdon D, Ruths M, Israelachvili JN, Landman U (2004) J Phys Chem B 108:3410
12. Sugimura J (1998) J Jpn Soc Tribol 43:933
13. Meakin P (1993) Phys Rep 235:189
14. Kragelsky IV, Dobychin MN, Kombalov VS (1982) Friction and Wear Calculation Methods. Pergamon, New York
15. Greenwood JA, Williamson JBP (1966) Proc R Soc Lon Ser A 295:300
16. Whitehouse DJ, Archard JF (1970) Proc R Soc Lon Ser A 316:97
17. Nayak PR (1971) J Lubr Technol T Asme 93:398
18. Majumdar A, Bhushan B (1991) J Tribol T Asme 113:1
19. Blackmore D, Zhou G (1998) Int J Mach Tool Manu 38:551
20. Zahouani H, Vargiolu R, Loubet JL (1998) Math Comput Model 28:517
21. Yan W, Komvopoulos K (1998) J Appl Phys 84:3617
22. Chung JC, Lin JF (2004) J Tribol T Asme 126:646
23. Persson BNJ (2000) Sliding Friction: Physical Principles and Applications, 2nd edn. Springer, Berlin, Heidelberg, New York
24. Persson BNJ (2001) Phys Rev Lett 87:116101
25. Persson BNJ, Tosatti E (2001) J Chem Phys 115:5597
26. Buzio R, Boragno C, Valbusa U (2003) Wear 254:917
27. Luan B, Robbins M (2005) Nature 435:929
28. Miesbauer O, Gotzinger M, Peukert W (2003) Nanotechnology 14:371

29. Enachescu M, van den Oetelaar RJA, Carpick RW, Ogletree DF, Flipse CFJ, Salmeron M (1998) Phys Rev Lett 81:1877
30. Schwarz UD, Zworner O, Koster P, Wiesendanger P (1997) Phys Rev B 56:6997
31. Schwarz UD, Zworner O, Koster P, Wiesendanger R (1997) Phys Rev B 56:6987
32. Cappella B, Dietler G (1999) Surf Sci Rep 34:1
33. Sader JE, Chon JWM, Mulvaney P (1999) Rev Sci Instrum 70:3967
34. Higgins MJ, Proksch R, Sader JE, Polcik M, Mc Endoo S, Cleveland JP, Jarvis SP (2006) Rev Sci Instrum 77:013701
35. Proksch R, Schaffer TE, Cleveland JP, Callahan RC, Viani MB (2004) Nanotechnology 15:1344
36. Burnham NA, Chen X, Hodges CS, Matei GA, Thoreson EJ, Roberts CJ, Davies MC, Tendler SJB (2003) Nanotechnology 14:1
37. Gibson CT, Smith DA, Roberts CJ (2005) Nanotechnology 16:234
38. Heim LO, Kappl M, Butt HJ (2004) Langmuir 20:2760
39. Hutter JL (2005) Langmuir 21:2630
40. Manyes SG, Guell AG, Gorostiza P, Sanz F (2005) J Chem Phys 123:114711
41. Rost MJ, Crama L, Schakel P, van Tol E, van Velzen-Williams GBEM, Overgauw CF, Horst H, Dekker H, Okhuijsen B, Seynen M, Vijftigschild A, Han P, Katan AJ, Schoots K, Schumm R, van Loo W, Oosterkamp TH, Frenken JWM (2005) Rev Sci Instrum 76:053710
42. Yu MF, Kowalewski T, Ruoff RS (2000) Phys Rev Lett 85:1456
43. Salvetat JP, Briggs GAD, Bonard JM, Bacsa RW, Kulik AJ, Stockli T, Burnham NA, Forro L (1999) Phys Rev Lett 82:944
44. Palaci I, Fedrigo S, Brune H, Klinke C, Chen M, Riedo E (2005) Phys Rev Lett 94:175502
45. Carpick RW, Ogletree DF, Salmeron M (1997) Appl Phys Lett 70:1548
46. Lantz MA, O'Shea SJ, Welland ME (1997) Phys Rev B 56:15345
47. Briggs GAD (1992) Acoustic Microscopy. Oxford University Press, Oxford
48. Takata K, Hasegawa T, Hosaka S, Hosoki S, Komoda T (1989) Appl Phys Lett 55:1718
49. Guthner P, Fisher C, Dransfeld K (1989) Appl Phys Lett 48:89
50. Khuri-Yakub BT, Akamine S, Hadimioglu B, Yamada H, Quate CF (1991) Proc SPIE 1556:30
51. Szoszkiewicz R, Huey BD, Kolosov OV, Briggs GAD, Gremaud G, Kulik AJ (2003) Appl Surf Sci 210:54
52. Szoszkiewicz R, Kulik AJ, Gremaud G (2005) J Chem Phys 122:134706
53. Szoszkiewicz R, Bhushan B, Huey BD, Kulik AJ, Gremaud G (2005) J Chem Phys 122:144708
54. Szoszkiewicz R, Bhushan B, Huey BD, Kulik AJ, Gremaud G (2006) J Appl Phys 99:014310
55. Cretin B, Sthal F (1993) Appl Phys Lett 62:829
56. Vairac P, Cretin B (1999) Surf Interface Anal 27:588
57. Bhushan B (2004) Springer Handbook of Nanotechnology, 1st edn. Springer, Berlin, Heidelberg, New York
58. Kueng A, Kranz C, Lugstein A, Bertagnolli E, Mizaikoff B (2005) Angew Chem Int Ed 44:3419
59. Rabe U, Arnold W (1994) Appl Phys Lett 64:1493
60. Dupas E (2000) PhD dissertation, Ecole Polytechnique Federale de Lausanne
61. Rabe U, Amelio S, Kester E, Scherer V, Hirsekorn S, Arnold W (2000) Ultrasonics 38:430
62. Amelio S, Goldade AV, Rabe U, Scherer V, Bhushan B, Arnold W (2001) Thin Solid Films 392:75
63. Hurley DC, Shen K, Jennett NM, Turner JA (2003) J Appl Phys 94:2347
64. Hurley DC, Turner JA (2004) J Appl Phys 95:2403

65. Passeri D, Bettucci A, Germano M, Rossi M, Alippi A, Orlanducci S, Terranova ML, Ciavarella M (2005) Rev Sci Instrum 76:093904
66. Muraoka M (2005) Nanotechnology 16:542
67. Burnham NA, Kulik AJ, Gremaud G, Gallo PJ, Oulevey F (1996) J Vac Sci Technol B 14:794
68. Oulevey F (1999) PhD dissertation, Ecole Polytechnique Federale de Lausanne
69. Rochat G, Leterrier Y, Plummer CJG, Manson JAE, Szoszkiewicz R, Kulik AJ, Fayet P (2004) J Appl Phys 95:5429
70. Cuberes TM, Briggs GAD, Kolosov OV (1998) AFM Modes for Non-Linear Detection of Ultrasonic Vibration. Oxford University Press, Oxford
71. Kolosov OV, Yamanaka K (1993) Jpn J Appl Phys 32:22
72. Dinelli F, Castell MR, Ritchie DA, Mason NJ, Briggs GAD, Kolosov OV (2000) Philos Mag A 80:2299
73. Hirsekorn S, Rabe U, Arnold W (1997) Nanotechnology 8:57
74. Turner JA, Hirsekorn S, Rabe U, Arnold W (1997) J Appl Phys 82:966
75. Stark RW, Drobek T, Heckl WM (2001) Ultramicroscopy 86:207
76. Burnham NA, Gremaud G, Kulik AJ, Gallo PJ, Oulevey F (1996) J Vac Sci Technol B 14:1308
77. Dinelli F, Burnham NA, Kulik AJ, Gallo PJ, Gremaud G, Benoit W (1996) J Phys IV 6:731
78. Hurley DC, Muller MK, Kos AB, Geiss RH (2005) Adv Eng Mater 7:713
79. Yamanaka K (1996) Thin Solid Films 273:116
80. Kolosov OV, Castell MR, Marsh CD, Briggs GAD (1998) Phys Rev Lett 81:1046
81. Dinelli F, Assender HE, Takeda N (1999) Surf Interface Anal 27:562
82. Porfyrakis K, Kolosov OV, Assender HE (2001) J Appl Polym Sci 82:2790
83. Geisler H, Hoehn M, Rambach M, Meyer MA, Zschech E, Mertig M, Romanov A, Bobeth M, Pompe W, Geer RE (2001) Proc Microscopy Semiconduct Materials 2001 169:527
84. Israelachvili J (1997) Intermolecular and Surface Forces, 3rd edn. Academic, San Diego
85. Szoszkiewicz R, Kulik AJ, Gremaud G, Lekka M (2005) Appl Phys Lett 86:123901
86. Chaudhury MK, Whitesides GM (1992) Science 256:1539
87. Luengo G, Heuberger M, Israelachvili J (2000) J Phys Chem B 104:7944
88. Briscoe BJ, Evans DCB, Tabor D (1977) J Colloid Interf Sci 61:9
89. Barquins M, Courtel R (1975) Wear 32:133
90. Kendall K (1975) Wear 33:351
91. Horn RG, Israelachvili JN, Pribac F (1987) J Colloid Interf Sci 115:480
92. Wang S (2004) J Tribol T Asme 126:1
93. Yoshizawa H, Chen YL, Israelachvili J (1993) J Phys Chem 97:4128
94. Yoshizawa H, Chen YL, Israelachvili J (1993) Wear 168:161
95. Hoffmann PM, Oral A, Grimble RA, Ozer HO, Jeffery S, Pethica JB (2001) Proc R Soc Lon Ser A 457:1161
96. Oral A, Grimble RA, Ozer HO, Hoffmann PM, Pethica JB (2001) Appl Phys Lett 79:1915

21 Application of SPM and Related Techniques to the Mechanical Properties of Biotool Materials

Thomas Schöberl · Ingomar L. Jäger · Helga C. Lichtenegger

From volume XIII

Abstract. Soon after the introduction of scanning probe microscopes (SPMs) originally developed for imaging purposes, their potential for mechanical analyses at the smallest scales was recognized and soon the method was applied to a variety of materials, including some of biological origin. Experimental techniques range from phase imaging to indentation and scratch testing. This chapter focuses on the use of instrumented indentation and related techniques such as scanning wear testing on biological tool tissues, materials often characterized by a high abrasion resistance. A brief overview of structure and composition of biological materials is given, since these factors are crucial in determining the mechanical properties and a basic understanding of such correlations is indispensable for the interpretation of the results. Furthermore, the influence of sample storage, preparation, and environmental conditions on mechanical tests is discussed, and relevant evaluation methods described. Finally, examples from the literature illustrating the successful application of SPM techniques on biotool tissues (mainly teeth) are presented.

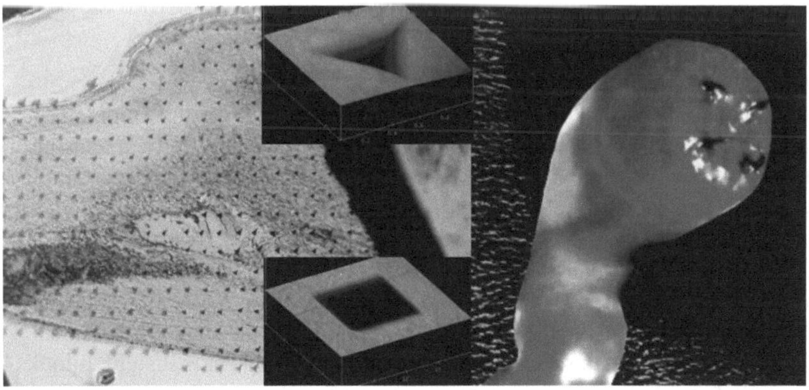

Key words: SPM, Indentation, Scratch test, Wear test, Biotool materials, Teeth, Jaws

Abbreviations

AFM	Atomic force microscope
O.P.-method	Oliver–Pharr method, a method for extracting hardness and elastic modulus from load-displacement curves

In: H. Fuchs, B. Bhushan (eds), *Biosystems - Investigated by Scanning Probe Microscopy*, DOI 10.1007/978-3-540-85049-6_4,

CSM	Continuous stiffness measurement. ... a dynamic indentation mode patented by MTS instruments, Oak Ridge, TN, USA
nanoDMA	A dynamic indentation mode patented by Hysitron Inc, Minneapolis, MN, USA
PTFE	Polytetrafluorethylen
SBR	Styrene butadiene rubber
HBSS	Hank's balanced salt solution, a salt solution containing phosphate and carbonate buffers
SEM	Secondary electron microscope
FTIR-RM	Fourier-transformed infrared spectroscopy in reflectance mode
SAXS	Small-angle X-ray scattering
qBEI	Quantitative backscattered electron imaging
DEJ	Dentin–enamel junction
TEM	Transmission electron microscopy
EDS	Energy-dispersive X-ray spectroscopy
EDTA	Ethylenediaminetetraacetic acid, used to sequester di- and trivalent metal ions
XPS	X-ray photoelectron spectroscopy
M	mol/l
His	Histidin

21.1 Introduction

Even small animals can cause huge detrimental effects: in the summer of 2004 coveys of desert locusts attacked eight countries in northern Africa, causing huge amounts of damage to the crop. Every day they ate 300 tons of forage. The largest covey observed had a number of 150 million distributed over an area of $15\,km^2$. Even though their number was huge, the eating activity of each single animal was enormous. For cutting and chopping up the forage, locusts possess mandibles which have to be stiff, hard and abrasion resistant, since their food often contains hard minerals. The material of the mandibles is, compared to common technical wear resistant materials, much softer and more compliant. Thus, at first sight, it looks like a mystery, how such a high abrasion resistance is achieved. Since many biological materials offer, apart from individual qualities, excellent and often specific mechanical properties, it is not surprising that they attract the attention of material scientists. One goal of this interest is to predict macroscopic material behaviour from the knowledge of chemistry, structures, and local mechanical properties on all smaller scales. As soon as at least part of this topic is understood the door is open to application of these concepts for the development of new synthetic materials. Even though the corresponding scientific field, biomimetics, is rather at its beginning, there already exist some successful applications, see [1–3].

The mechanical properties of biological materials span a wide range, depending not only on the specific demands, e.g. high elasticity, rigidity, or wear resistance, but also on the availability of chemical elements in the environment of the living organism.

A nice example is given by the formation of jaws and teeth of vertebrates: the chemical composition and the basic material structure on the nanometre level are more or less adopted from the skeleton (see below). For non-vertebrates such an already existing concept is at least not known, so that probably other strategies had to be found, taking into account the availability of chemical elements.

Before we delve into the details of biotools, let us shortly consider what we are talking about. In analogy to the Merriam-Webster Online Dictionary of a tool we define a biotool as "... something produced by an animal in a natural way to perform an operation otherwise impossible". In this strict sense the range of biotools is severely limited: only the tools for cutting and grinding food (the teeth of vertebrates and the mandibles of invertebrates) and the ovipostors of certain insects remain. But let us take a somewhat broader view; then the claws of a cat are biotools, as well as the hooves of ungulates, or the thick horny soles the camelids evolved for travelling over burning-hot sand or rugged terrain, because a cat without claws or a horse with a damaged hoof are severely handicapped. And since the competition for reproduction is a very important item for animals, maybe we should also include symbolic tools like, e.g. deer antler, which is regularly used for fighting for the dominance of a "harem", but seldom for real defence against predators. Or the bizarre "horns" some beetles carry (staghorn beetle!), which are quite useless for practical purposes, except, of course, to impress the females and for levering off rivals from a leaf. By the way, it is not that easy to draw the line because at least in some cases the fancy headgear is put to real use. The horns of a steer may easily be a symbol of male dominance, but a charging bull can use its horns with deadly precision, as any torero can tell you – if he survives. Finally, maybe a border case, the mollusc shells. They are certainly not "tools" to work something with, rather protective armour. But in a world full of hungry predators such armour can be life-saving for a stationary animal unable to flee or take cover. Therefore they should, at least, be mentioned.

For an insight into biological materials today a large number of experimental techniques exist. Apart from microscopy methods based on the interaction between electromagnetic waves or particles (visible light, X rays, electrons, ions, neutrons) with solid matter (optical methods in a more general meaning), scanning probe microscopy (SPM) and related techniques have become more and more useful and important. The former type of methods provides mainly structural and local chemical information, whereas the latter allows, apart from the imaging ability, the determination of mechanical properties on a very local scale, down to microns or even to the nanometre range. Apart from the classic SPM techniques, several refinements of mechanical measurements have been proposed, like dynamic mechanical analysis [4] or modulated lateral force microscopy [5]. However, the experimental limits of all devices can be clearly seen by the example of bone tissue. The basic building block is represented by mineral platelets embedded in a matrix of collagen [6]. The length and width of the platelets is some 10 nm, the thickness only a few nm. Most promising would be to measure directly on this scale, which is not practicable due to the small size. Systems on a larger, micron-scale, resembling the bone model do exist, e.g. the mineral rods in enamel or the structure of nacre. In both cases, however, the amount of the "hard" material is predominant. Even if a material with the basic bone structure would exist on larger scales, conclusions from such experiments on the material behaviour of bone tissue would be doubtful anyway, since its mechanical properties are intensely determined by the typical dimensions [7]. Since a straightforward experimental approach to "true" nanometre mechanical properties of solid-state matter does not yet exist (except tests on macromolecules), only sophisticated investigations on the submicron or micron scale can help to understand the effects which proceed on nanometre dimensions.

21.2 Typical Biotool Materials

21.2.1 Chemistry

In general, biological tool tissues consist of organic matter (mainly biopolymers), water and – as a further optional component – inorganic matter in the form of metal ions or mineral salt. Structural biopolymers can be grouped into proteins (chains of amino acids) and polysaccharides (chains of sugar molecules), both of which may occur in fibrous form or as a filler material. Fibrous biopolymers very commonly act as reinforcing elements in a fibre composite architecture, the main structural motif in biological tool tissues. In animal tissue, the fibrous component is usually proteinaceous (e.g. collagen in humans), an exception being the tissue of arthropods that consists of polysaccharide fibres (chitin) embedded in a protein matrix. Fibrillar structures with a moderate degree of cross-linking between the participating molecules can be very stiff in tension, but are notoriously weak in bending and compression. Therefore, they are used in nature to transmit tensile forces, if necessary around bends (Achilles tendon). Biotools such as teeth or claws, however, are usually stressed in various ways (compression, bending, shear), thus requiring additional stabilization strategies. One way towards greater hardness is to stiffen the organic matrix, by increasing the density of crosslinks. For example in keratin-containing biotools such as rhinoceros horn, bird beak, ungulate horn sheaths and hooves, finger nails and claws of predators [8], the hardness of the tissue is increased by disulphide bonds established between cysteines (sulphur-containing amino acid). Consequentially, high sulphur content is found predominantly in the hardest keratin-based tissues such as tortoise shell or biotools, in particular claws and beaks [9]. In chitinous arthropod tissue, the mechanical properties are controlled to a great extent by the degree of cross-linking of the protein matrix (sclerotization). Sclerotization is achieved by a process called tanning that usually also involves darker colouring and most importantly decreases the degree of hydration. The stiffness of fully tanned dry arthropod cuticle is about 20 GPa, while Young's moduli in the MPa range are found for wet and untanned cuticle [10]. Although chitin is mainly found in arthropods, it also occurs in the radula (a tongue-like rasping organ) of molluscs, the beak of squid and octopus, or in certain worms (e.g. grasping spines of arrow worms [11]). Alternatively or in addition to organic crosslinking, proteinaceous tissue may increase the density of crosslinks via coordination with metal ions. Such a mechanism was found for example in byssal threads of *Mytilus edulis* (holdfasts used by the marine mussel to stay attached to rocks), and also rather recently in the jaw of the marine worm *Nereis* that uses Zn ions to cross-link the protein matrix of its teeth [12]. A similar case may be present in *Glycera* worms (also belonging to the family of Annelida), albeit with Cu. Both mandibles contain considerable amounts of histidine, an amino acid well known for its affinity to transition metals. Remarkable amounts of transition metals (Zn and Mn) are also known to occur in arthropod tool tissue, in particular at the cutting edge where stresses are highest, e.g. locust and leaf cutter ant incisor, but the chemistry is not well known.

Very commonly, however, reinforcement of soft tissues in order to build structural elements and biotools is achieved by incorporation of mineral particles. In comparison to organic reinforcement strategies, mineral precipitation from solution is clearly a low-cost strategy: it saves metabolic energy needed to synthesize heavily cross-linked organic material and instead uses metals salts easily available from the respective environment as hard filler. From the chemical point of view, the assortment is rather limited. Vertebrates use bio-hydroxyapatite, a calcium phosphate mineral with the ideal stoichiometric formula $Ca_5(PO_4)_3OH$, but in vivo with up to 5% CO_3 groups substituting either the phosphate or the OH group. In vertebrate bone tissue, the mineral content is typically around 40–50% vol, embedded in a (hydrated) organic matrix consisting of mainly collagen (plus small amounts of proteoglycans and bone-specific proteins like osteopontin). In vertebrate tooth enamel, the mineral content reaches 90% mineral in an amelogenin matrix. By contrast, invertebrate animals favour a different calcium-containing mineral, namely calcium carbonate ($CaCO_3$). It is found in sea shells but also used by arthropods such as lobster (subphylum: Crustacea). In crustaceans the mineral is embedded in a protein–chitin matrix and used for the hard cuticle and their biotools, the claws [13]. Finally, there is a group of marine molluscs, the chitons, living in tidal waters and scraping algae from the rocks with a rasping tongue called radula. This radula carries many rows of tiny teeth and in order to prevent those teeth from being worn down too quickly by scraping calcite rocks they are quite literally armoured. Their cutting edge is impregnated with magnetite, whereas the rest of a tooth may contain Ca-minerals, silica (opal), and other Fe-minerals like limonite or lepidocrocite [14, 15]. Since iron minerals as structural elements are otherwise unknown in nature, this example is of special interest, the more so because for a long time it was believed that magnetite can only be synthesized using high temperatures.

Last, but not least: the marine worm *Glycera* makes use of both mechanisms to harden its teeth: reinforcement of a proteinaceous matrix by coordination with metal ions (Cu) and reinforcement by Cu mineral (atacamite) [16, 17], but the degree of mineralization ($\approx$10% vol) is much lower than that observed for other mineralized tissues. This rather unique combination is complemented by other peculiarities: *Glycera* worms are the only organism known to employ a Cu-based biomineral for structural purposes and their proteinaceous jaw matrix contains a considerable amount of melanin, not only providing the organic matrix with impressive hardness but also exceptional chemical stability.

21.2.2 Structures

As mentioned above, the most common structural concept for biotools is a fibre composite reinforced either by enhanced cross-linking or by incorporation of metal salts. In the following we shall concentrate on mineral particles as reinforcements and discuss a few principles of structuring biological materials using bone as an example. Bone is not only the most important structural material for vertebrates, but also a biotool: deer antler, for example, is slightly less mineralized bone, the sharp bill of sword fish and marlin consists of highly mineralized bone, and the horns of ungulates are bony protrusions, clad with a horn sheath. The structure of bone (like all biomaterials) is dominated by hierarchical levels. That is, certain principles and motifs can be found at various length scales, similar to a multi-fractal. Starting at the bottom, tropocollagen molecules self-assembly into subfibrils that are then mineralized. The mineral, bio-hydroxyapatite, usually forms platelets (although needle-like crystals are sometimes reported) within the fibrils as well as on their surface. Dimensions of the platelets vary considerably, as a rule of thumb they may be some tens of nanometres of length and width but, of course, not regularly shaped and only a few (3–5) nanometres thick. The volume fraction of the mineral component of bone and bone-like structures usually lies around 45%. The role of surface mineralization is not well understood, but within a fibril the platelets are arranged layer-by-layer, not in rank and file but in a staggered way, overlapping like the bricks in a well-laid wall. Jäger and Fratzl [6] succeeded in estimating the elastic modulus of such a structure using a simplified, two-dimensional model (Fig. 21.1), and were able to show that its longitudinal stiffness can be increased nearly arbitrarily simply by increasing the aspect ratio (the length-to-thickness ratio) of the platelets and therefore, implicitly, of the collagen layers in between. And – very important for biotools – this increase of stiffness does not cause a proportional decrease of fracture toughness. However, naturally, there is a price to pay. In this case it is a pronounced anisotropy. The model predicts very good mechanical properties along the direction of the longer axis of the platelets (which is in nature frequently oriented along the major stress direction), not quite as good along the minor axis and rather poor in the direction perpendicular to the plane of the particles. In certain cases, however, this is not necessarily a disadvantage. Trabeculae (the small struts forming the foam-like interior of bones, like the femur head and neck, but also vertebral discs and ribs) have been shown by SAXS (Small Angle X-Ray Scattering) [18] to have their mineral platelets oriented preferentially parallel to their long axes. It should be noted, however, that these trabeculae are arranged in such a way that they are almost exclusively stressed in tension or compression, but never in shear or bending. In cases where such a high degree of anisotropy would be less desirable the fibrils are arranged either in a crossed-plywood style, with the stiff axes alternating in their orientation with respect to the main stress, like in the secondary osteons in bone, or in an irregular, felt-like structure like in dentine.

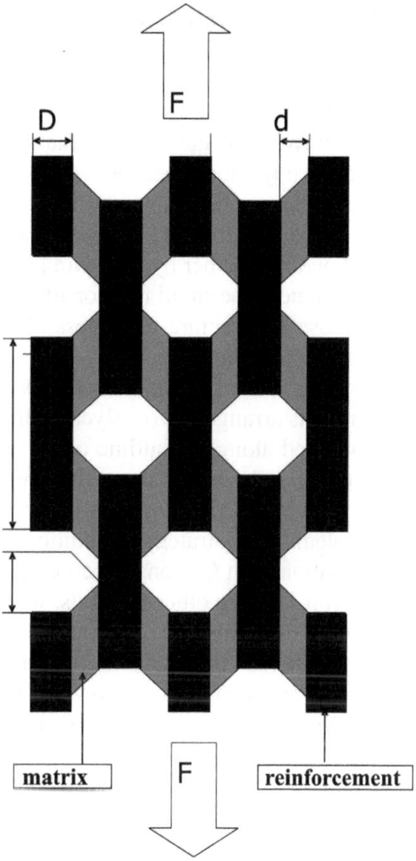

Fig. 21.1. Staggered model of bone: a tensile load *F* causes shear forces in the matrix (*light grey* regions) between the mineral platelets (*dark grey* regions). The lengths *D, L, d* and *δ* strongly influence the resulting mechanical properties

Another interesting aspect is the size of the particles. The staggered ("brick-and-mortar") model itself has no built-in length scale. But Gao et al. [7], using the Griffith criterion, showed that mineral particles below a certain critical length are insensitive to cracks and therefore display their (very high) theoretical strength, much in contrast to macroscopic ceramic bodies whose sensitivity to even small defects causes their brittleness. They estimated this critical length to be some tens of nanometres, which exactly corresponds to the lengths of the mineral platelets occurring in bone.

The existence of mineralized collagen fibrils, beautifully shown in SEM pictures of [19] and obviously connected by a very tough protein has led Gupta [20, 21] to propose a second hierarchical level of staggered arrangement consisting of stiff mineralized fibrils in a tough matrix of other proteins. But this idea remains controversial.

Other mineralized biomaterials follow similar concepts. The cuticle and claws of crustaceans, e.g. lobsters, are built from a chitin–protein fibre-composite, reinforced with calcite crystals and arranged in lamellae, in a crossed-plywood manner to avoid macroscopic anisotropy [13]. Enamel and nacre, the most heavily mineralized biomaterials known (above 90% vol mineral) strongly rely on very thin polymer layers between their mineral components in order to enhance fracture toughness. Enamel consists of "rods", bundles of thin but long hydroxyapatite crystals bound together by thin layers of the protein amelogenin and oriented more or less radially outward from the dentine–enamel junction perpendicular to the surface. Here the anisotropy mentioned above is fully exploited. Nevertheless enamel contains Hunter–Schreger bands, narrow regions of near-perfect crystal orientation that differ from one band to the next. Nacre is made from roughly hexagonal aragonite platelets, again bonded together by very thin layers of protein and arranged in a perfectly staggered manner. The small fraction of hydrated organic material ($\approx$5% vol) is reported to increase the fracture toughness of the composite by a factor of 3,000 [22] with respect to pure aragonite.

Finally, in the Cu-mineralized jaws of the marine worm *Glycera*, the Cu-mineral atacamite is arranged into polycrystalline fibres embedded in a proteinaceous matrix and oriented along the outline of the needle-like tip, an arrangement quite different from the ones described above. It should be kept in mind, however, that the degree of mineralization in *Glycera* jaws is very low and mineralization might be regarded as a complementary strategy in addition to the reinforcement of the organic matrix via coordination with Cu-ions. According to a recent investigation by Moses et al. [23] employing, among other methods, wear tests, the outer layer of the *Glycera* teeth is reinforced exclusively by Cu ions, resulting in a very hard, water and wear resistant surface, whereas the atacamite-mineralized fibres are restricted to the layer underneath and are thought to convey enhanced bending strength [24]. Most remarkably this layer is less wear resistant in spite of the mineral reinforcement [23].

21.2.3
Mechanical Properties

Most of the first micro- and nanomechanical tests have been performed on metals, glass, and ceramics. The "simplest" class of materials is represented by glasses, being homogeneous, isotropic and, at room temperature, almost non-viscous. The results of mechanical measurements should, thus, neither depend on the choice of the investigated area, nor on the orientation of the sample or the velocity of the test. Biomaterials, however, are usually highly anisotropic, inhomogeneous and their deformation behaviour depends on the deformation rate and on the degree of hydration [25–31]. In the literature several models describing the viscous deformation of polymers and/or biological materials can be found. They usually imply various combinations of springs (describing the response of an elastic solid) and dashpots (describing the response of a viscous fluid). The simplest version is the combination of one spring and one dashpot either in parallel (Voigt system) or in series (Maxwell system). A refinement was made by a three-element combined model [32], where a Voigt system is arranged in series with an additional spring. At least partly physical models for time-dependent deformation behaviour have been published by Sasaki and Enyo [33] and by Jäger [34,35]. Another, physics-based type of model has been proposed using poroelasticity [36–40], where the material is regarded as an elastic, porous body. With any deformation the fluid in the material must be "pressed" through the pores, causing a damped, viscous deformation response. Oyen et al. [40] performed a poroelastic analysis of nanoindentation data on bone with various fluid contents and compared the results with those already existing in the literature for poroelasticity at larger length scales. The resulting pore size was consistent with the scale of fundamental collagen–apatite interactions. However, such a model is somehow in contradiction with the fact that bone shows viscous behaviour even in a dry state. Models describing the mechanical behaviour of materials can never completely substitute experimental work. Therefore, for a full investigation of local mechanical properties it would be necessary to perform tests in various orientations, on a large number of spots in three dimensions ("mechanical tomography"), varying loading velocity and fluid content. In practice, however, such a vast amount of measurements is not feasible. It is, thus, necessary to reduce the total number of measurements by restricting the study to a few clearly defined questions. But even then, some fundamental constraints remain: the elastic modulus determined by indentation experiments is not identical with the classical, orientation-dependent Young's modulus [41,42], but reflects some average over the various orientations. Moreover, the choice of a certain position in the sample is often restricted by the available preparation methods. Under such conditions, thorough test protocols are highly recommended.

21.3 Experimental Methods and Setups

21.3.1 SPM and Indentation

The basis of all experiments described here is a probe tip interacting with a sample surface. A detailed description of these interactions can be found for example in [43]. For the purpose of mechanical tests the elastic–plastic interaction is of predominant interest, even though forces like adhesion may play a role, especially for very soft materials. The "classic" suspension of probe tips is that on a flexible "bending arm" (cantilever) which is the common setup for an SPM. The interaction between sample and tip leads to a bending of the cantilever acting as a force sensor [44,45]. Several techniques have been proposed for the spring force calibration of AFM cantilevers [46–60]. In addition to vertical probe tip–sample interactions, also lateral forces can be detected: by choosing the fast scan direction perpendicular to the cantilever axis, topographical extrusions as well as friction cause a torque on the cantilever with the consequence of tilting. The difference between the tilt angles from trace and retrace represents a measure of friction, which is usually only a semi-quantitative measure, mainly due to the fact that the resistance of a cantilever against torque is even harder to calibrate than the resistance against bending. Another, not fully quantitative mechanical measure is the damping of an oscillating cantilever hitting the sample surface, due to the interaction between tip and sample material, which affects the oscillation in frequency, amplitude and phase, where most commonly the resulting phase lag is utilized. The recorded "images" are maps of phase values with the high resolution of an AFM. However, the interpretation of such maps is not always straightforward. The most common application of this technique is

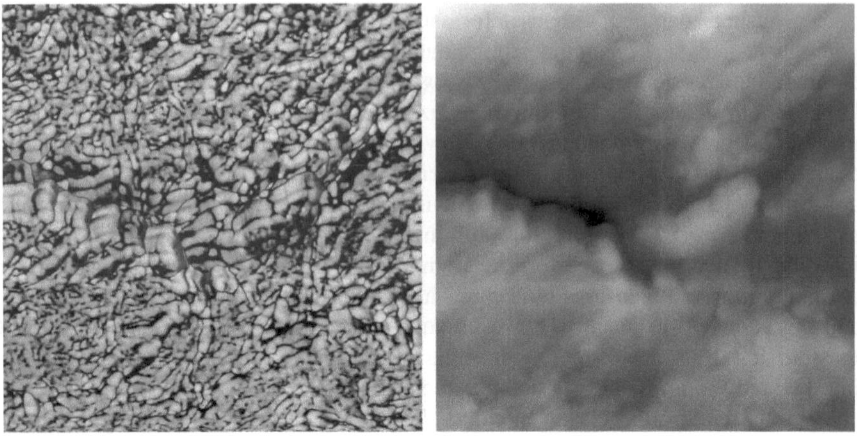

Fig. 21.2. AFM image of bone tissue. On the left the phase image, on the right the topographical image. Scan size 2 × 2 μm

imaging of components with highly different mechanical properties without the need of special preparation techniques. A nice example is given in Fig. 21.2 showing the phase image of bone in cross-section with the hard hydroxyapatite minerals in the soft collagen matrix, in comparison with the topographical image, containing only random surface features.

Another mapping technique is force volume imaging [61–66]. Here, force-distance curves are recorded at every point corresponding to a pixel of an AFM image. Since the tip is not only scanned over the surface, but travels also perpendicularly, the denomination "force volume" imaging has been established. By post-processing the force-distance curves, two dimensional maps of tip–sample interaction can be obtained, giving information on physico-chemical properties and mechanical material response. The main problem of the method is the long acquisition time which can give rise to sample drift. This fact and the memory usage for storage of all data points limits the resolution of force images, problems that can partly be overcome by the pulsed mode force technique [67, 68], where recording of force curves is possible at rather high frequencies. To avoid problems with data acquisition, only selected points of the force curves are captured. Apart from the mapping methods described above, the measurement of force-distance curves also allows one to obtain information about single molecules and molecule pairs [69–73]. In the first case, tensile tests are performed where a large molecule is bound to a substrate at one end, the other end is "picked up" and bound to the AFM tip. In the second case, specific interactions between biological pairs are investigated. In this chapter, however, these experiments are not discussed further.

In the contact regime of the tip–sample interaction, both in approach and in retraction, the elastic deformation of the sample can be related to the elastic modulus. There are several theories describing the sample deformation. In the Hertz model [74] the adhesion forces are neglected, whereas in the Johnson–Kendall–Roberts (JKR) model [75] and the Derjaguin–Müller–Toporov model [76–78] adhesion between tip and sample is taken into account. Since materials forming biotools are rather stiff, adhesion effects can usually be neglected. Probing the mechanical properties by driving and retracting a hard tip vertically into a sample and recording load-displacement curves is known as nanoindentation or instrumented indentation. One of the advantages over the conventional hardness test is that there is no need to measure the area of the residual imprint, which is a rather difficult endeavour and usually accompanied with large errors, especially for very small imprints. The most common data extracted from load-displacement curves are hardness and elastic modulus, or more precisely, the reduced elastic modulus which takes into account the deformation of the indenter tip and the lateral deformation of the sample material via its Poisson's ratio [79]. Since for many biological materials the latter is not or only poorly known, it is rather convenient to give the results of indentation experiments by the reduced modulus. As long as there is no plastic deformation of the sample, the modulus can be obtained simply by a Hertzian contact. In case of a spherical contact geometry a simple relationship between applied force, contact depth and the reduced elastic modulus can be found, i.e. the force is proportional to the reduced elastic modulus, to the square root of the (deformed) tip radius and to $h^{1.5}$, h being the penetration depth. The radius can be obtained from experiments on a calibration material with well-known reduced modulus. At the onset of plasticity, often a sudden displacement burst appears which allows us to calculate the yielding pressure at this point. In biological materials, however, such a distinct onset of plastic deformation is usually not found. Hardness and elastic modulus are, commonly, extracted from the unloading part of load-displacement curves, assuming that during unloading only elastic deformation takes place. The most popular method of this type was introduced by Oliver and Pharr in 1992 [79]. In contrast to earlier methods [80], it is suitable also for experiments with a curvature of the unloading curve, i.e. if the contact area reduces during unloading which is commonly observed in biomaterials. The method takes into account the elastic deflection of the surface immediately adjacent to the contact, based on the general relationships derived by Sneddon between load, displacement, and contact area [81].

In [79] it is shown that the load-displacement relationship for many simple punch geometries can be written as $P = \alpha\, h^m$, P being the load, h the displacement, and α and m constants. Values of the exponent m are, for example, $m = 1$ for flat cylinders, $m = 2$ for cones, and $m = 1.5$ for spherical indenters. Most of the commercially available instruments work with a software package based on the Oliver–Pharr (O.P.) method. The user should, however, be well aware of some shortcomings: indents are frequently surrounded by raised material, so-called pile-ups or by permanent sink-ins (not to be confused with the elastic, non-permanent sink-ins). Since the contact area is determined via the displacement, these effects are not taken into account by the Oliver–Pharr method. Special attention has to be focused on the calibration of the tip shape as well as on the machine compliance as will be described later.

21.3.2 Scratch and Wear Tests

Mechanical testing on small scales can be classified by dimensions of lateral tip movement: from this point of view indentation has the dimension zero (0D), followed by scratch tests (1D) and scanning wear experiments (2D). Besides the basic research on the origin of friction and wear, scratch and scanning wear tests turned out to be a valuable tool for the investigation of ultra-thin coatings, lubricants on hard discs and magnetic recording heads [82–88]. The physical mechanisms of deformation can be expected to be different for the various types of experiments: while with indentation elastic and plastic deformation should appear mainly in compression, in scratch and wear tests shear forces and the effect of ploughing and/or cutting play a role. In a scratch test, a sharp tip is moved over the surface of a test material. The scratch depth at a given load or the load at which the material fails serves as a measure of scratch and wear resistance. Scratch tests can be performed in various modes: with constant or increasing load, as a single scratch or oscillating scratch. Using scanner-based instrumentation, scratching and imaging can be performed with the same tip, so that immediately after scratching the scratch depth can be determined. In addition, pile-ups and debris caused by scratching can be imaged. However, depending on the tip shape, one must be aware that the result of imaging represents a convolution of topography and the tip contour which may lead in scanning wear experiments to entirely different values of volume of the worn crater and the pile ups, respectively. Depending mainly on the tip shape and radius, a minimum vertical load is necessary for the onset of measurable wear which can be estimated by simple geometrical considerations in connection with a Hertzian contact. In [89] it was shown for a sphero-conical indenter with a tip radius of $1.8\,\mu\text{m}$ that essentially two mechanisms occur. At loads below a threshold there is only ploughing in the material, no real material removal occurs. Eventually existing pile-ups are redistributed into the wear crater. At loads above the threshold material is removed from the beginning of the test. The transition from ploughing to immediate wear mainly depends on the lateral attack (critical contact) angle between tip and sample (including the pile ups in front of the tip). In [89] the attack angle was determined to be $19 \pm 2°$, independent of the material under investigation. This correlates to a transition from mainly compressive stresses to shear stresses. Similar thresholds have been observed for example by Degiampietro and Colaco on stainless steel [90] and Zhao and Bhushan on silicon [91].

21.3.3 Dynamic Modes

The techniques described involve "quasi-static" movement of the probe tip. An expansion of the utility of indentation devices has been achieved by introducing dynamical modes. The two most prominent variants from two different producers of indentation devices are described in the following, continuous stiffness (CSM) [79, 92, 93] measurements and nanoDMA. The CSM method is accomplished by superimposing a small, sinusoidally varying signal on top of the DC signal that drives the indenter. Data are obtained by means of a frequency specific amplifier (lock-in technique). At any point of the loading curve the contact stiffness can be measured, allowing the determination of the elastic behaviour at minimum penetration depths. Because of the short measurement time, due to frequencies above 40 Hz, this method is insensitive to thermal drifts and their fluctuations. CSM allows a fast examination of graded materials, layered structures as well as creep measurements at shallow penetration depths. Moreover, load cycles of sinusoidal shape allow the performance of fatigue tests on the nanoscale. Li and Bhushan described several applications of CSM [94]. Experiments on fused silica, PTFE and styrene butadiene rubber (SBR) revealed different results: for fused silica and PTFE, the contact stiffness increased linearly with the contact depth, elastic stiffness and hardness showed no change. For the SBR, however, the contact stiffness increased linearly with depth, while the measured elastic modulus decreased with depth, an effect not expected for this material. This result suggested that the contact depth calculated from the theory based on rigid plastic materials is not applicable to viscous materials, since time-dependent deformation is neglected. The dynamic method nanoDMA is also based on sinusoidal oscillation of the indenter tip. Various modes allow measurements with variable load, frequency or amplitude. Since this instrument is mounted on the scanner head of an AFM, it can be used, for example, for mapping the elastic modulus of a whole sample area within a rather short time.

21.3.4 Fracture Toughness Tests

In addition to hardness, elastic modulus, scratch and wear resistance fracture toughness is an important material parameter. Micro- and nanoindentation have been used to measure the fracture toughness of many brittle materials, mainly ceramics [95–108]. When brittle materials are indented with a sharp tip, radial cracks are generated during unloading. The length of these cracks can be correlated with the fracture toughness by various semi-empirical relationships. Rather simple relationships are given in [109–111], the fracture toughness being a function of elastic modulus, hardness and the crack length, all parameters which can be obtained immediately from indentation experiments. For details the reader is referred to [112]. However, a significant problem exists, namely that there are minimum loads for the occurrence of cracks, so-called cracking thresholds. Cracking thresholds have been measured, for example, by means of acoustic emission during indentation [113]. The cracking condition depends on the indenter geometry: the indentation thresholds can be reduced by using indenter tips with a small included tip angle, e.g. a Cube Corner tip instead of a Berkovic-type. Harding et al. showed [114] that the relationship between fracture toughness and crack length is the same for Berkovich and Cube Corner indenters, the only difference is a different geometric constant. The determination of fracture toughness is not totally restricted to pointed, pyramidal indenters on brittle materials: Yan et al. [115] reported on a simulation of internal cone cracks induced by conical indentation in brittle materials, and an attempt at estimating the fracture toughness of ductile materials by instrumented indentation was made by Lee and co-workers [116]. The applicability to biomaterials, however, has not yet been confirmed. Almost each of the experimental techniques above can be performed with a conventional AFM probe, however, with a few limitations, especially regarding indentation experiments and friction measurements. Calibrating the cantilever precisely is still not a simple task, especially a calibration of the torque resistance. Even the shortest and stiffest cantilevers have a spring constant too low to perform indentations in hard materials like enamel. Moreover, due to cantilever bending, the tip rotates during an indentation, thus the contact geometry is not known. From this point of view the introduction of a system where the cantilever plus its holder is replaced by a complete indentation setup (transducer) represents a considerable progress, combining indentation and scanning techniques [117].

21.4 Samples

21.4.1 Choice

Because of the inherent individual variations in biotool materials and the range of external factors (temperature, moisture, etc.) potentially influencing the mechanical properties, samples must be chosen with care and with regard to the questions to be answered. Bone, for example, occurs in a very large structural variety. Although composed of the same building blocks as bone on the nanometre scale, dentin exhibits a somewhat simpler structure. It contains rather densely packed channels (tubules) with a mutual distance on the order of 5–10 μm and a diameter of 1–3 μm, and consists of at least two types of material: intertubular dentin between the tubules and peritubular dentin surrounding the tubules, the two differing in the mineral content and, consequently, in hardness and stiffness. In order to elucidate the influence of the degree of mineralization, similar materials varying only in mineral content have to be found. Such an approach was undertaken, for example, by the investigation of turkey leg tendons [118], exhibiting large variations of mineral content with age: the mineral content starts from zero until an age of about 10 weeks and increases up to about 25% until an age of 23 weeks. In this study it could be shown that mechanical properties of bone-like material do not exclusively depend on the mineral content, they are influenced by variations of the collagen properties as well, a point of view important for clinic research on bone diseases. Clinical research usually requires statistical analysis making experiments on many samples from different individuals necessary. This approach, however, usually does not allow monitoring of the influence of external parameters and is therefore not feasible for basic research. In basic research usually a limited number of samples are thoroughly characterized under varying experimental and storage conditions, in order to extract valid property parameters. A study of the bone–cartilage interface of three human patellae [119] illustrates the latter strategy: the experimental approach was to perform indentation line scans across the interface and subsequently measure the local mineral content exactly at the locations of the indents. From these data a correlation between local stiffness and mineral content could be extracted, for bone and calcified cartilage, respectively. The final result could be obtained already from the first sample, the two additional samples served as (successful) confirmation. For a clinical study, on the other hand, three samples would be ridiculous.

21.4.2 Storage

Another important point when studying biological materials is that in vivo loading always takes place at physiological temperatures with a certain water or fluid content. Since in-vivo SPM measurements are hardly possible, the question arises, how the material properties change after isolation from the organism. For animal bone there is usually already a time span between slaughter and the deposition into a suitable storage fluid and/or freezing the samples. This delay can conveniently be avoided when working with tooth material by putting the teeth into a suitable fluid immediately after extraction. From this point of view dentin has an advantage as sample material over bone for basic research on bone-like material. But what are the "ideal" storage conditions, or which storage is the best compromise? Despite many investigations on biological materials, not much has been published on this issue. A few observations have been reported on teeth and bone: Moscovich et al. [120] studied the in-vitro Vickers hardness of dentin following gamma-irradiation and freezing. Freezing is one of the most widely used methods for storage of biomaterials, ionizing radiation can be employed as well but is more common for the sterilization of heat-sensitive medical devices, pharmaceutical packing and raw materials. The latter method has the advantage of excellent tissue penetration, even under wet conditions or refrigeration. The experiments revealed that the frozen samples showed no property changes, whereas the irradiated ones evidenced a slight increase in hardness. Hengsberger et al. [122, 123] reported mechanical degradation of bone samples on the micron scale after having performed indentation tests for 12 h under fluid. Habelitz et al. [124] compared the influence of various storage fluids on human dentin and enamel. The fluids were deionized water, normal salt solution and Hank's balanced salt solution (HBSS), respectively, the latter containing carbonate and phosphate buffer. For the first two fluids, a pronounced degradation of hardness and elastic modulus was observed, after one week of storage in these media, degradations up to 50% occurred. The most probable explanation for this effect is a demineralization especially in the near surface regions. Storage in HBSS for up to two weeks, however, caused no noticeable change in hardness and modulus. Obviously, the phosphate and carbonate buffers help to avoid demineralization. In another study, Guidoni et al. [125] investigated the influence of freezing tooth and bone samples on the mechanical properties, using HBSS as the storage medium. Freezing of bone did not remarkably alter hardness and modulus, in contrast to human dentin, where freezing led to a decrease in indentation hardness and modulus by about 20% after thawing the samples. This effect could be reduced by carefully removing the water content by replacing the water step-by-step with alcohol before freezing. From these findings the authors concluded that the freezing damage of dentin with some water content is mainly caused by the expansion of water in the tubules during freezing, since such an effect could not be observed with bone, where no such tubules exist. A second observation in this study strongly confirmed the results by Habelitz et al. [124]: using HBSS for the storage of a second premolar tooth conserved its hardness and modulus for even two years! It should be noted at this point that the sample handling in this study was performed with extreme care: beginning with the choice of the samples (totally healthy, "young" second premolars), they were deposited into HBSS (with a temperature of 4°C) within a few seconds after extraction, and, moreover, during all preparation steps the samples were rinsed with HBSS.

21.4.3 Preparation

Apart from possible influences by the storage of samples, their preparation may have a detrimental influence on the results of small-scale mechanical tests. The first step usually is cutting with a saw, where mainly two points have to be borne in mind: material adjacent to the cutting line should not be deformed too extensively, nor should the sample material be exposed to high temperatures. A low cutting speed combined with water cooling is usually satisfactory. In this context it should be mentioned, that suitable tooth samples can only be obtained from carefully extracted teeth without any cracks, otherwise damage by overheating cannot be excluded with certainty. The next usual step is embedding the samples in polymer resin, where the samples are brought into contact with the resin either by heating at elevated pressure, or just by being moulded in a two-component resin. Even in the latter case, curing of the polymer causes a temperature increase due to the exothermic character of the process. The most common procedure to obtain flat and smooth surfaces is adopted from the classical metallographic preparation techniques, i.e. grinding in steps with decreasing grit and then polishing, also usually in steps with decreasing grit. The polishing agent may be either a diamond paste or, more commonly for biological samples, an alumina-based polishing agent. The quality of the surface finish can considerably depend on the material properties. This can be seen, for example, from a polished section of a tooth: the (hard) enamel is usually much smoother compared to the (softer) dentin. In general, it appears that softer materials are harder to prepare which is often observed also for a (final) preparation with an ultra-microtome. In many cases an ultra-microtomed surface is smoother compared to the results from grinding and polishing. However, exceptions may confirm the rules, and, moreover, sensitive methods like cutting with an ultra-microtome require a skilled and experienced operator. In seeming contradiction to the demand for smooth surfaces it may be desirable to still be able to obtain some information on the microstructure from the topography on the surface after preparation. This is extremely valuable, if by a scanning probe the topography is imaged before a local mechanical test, therefore allowing one to choose the test position very precisely. Ho and co-workers investigated the effect of various sample preparation techniques on the determination of structures and nanomechanical properties of human cementum hard tissues [126]. The preparation methods included cryofracturing, ultrasectioning and polishing. The ultrasectioned sample revealed the lowest surface roughness, moreover the fibre structures in this material could be imaged with an AFM with a similar quality as obtained from the cryofractured samples. The 10–50-μm wide cementum dentin junctions could be clearly observed in the ultrasectioned specimens but not in the polished ones. The differences in hardness and modulus obtained when indenting along lines over the structures were more distinct with the ultrasectioned samples compared to the polished specimens, giving clear evidence that material was smeared and distributed laterally over the surface by polishing. All these findings suggest that for this type of material ultrasectioning is superior to polishing. In addition to the above considerations, it may be important to regard alterations by chemical sample treatment as well. Such treatment may be necessary to study the influence of a chemical component on mechanical properties. Such an approach has been reported by Broomell et al. [127], where the role of zinc in the tip region of jaws of the marine worm *Nereis* was investigated. The removal of zinc as well as its reconstitution was proven successfully, but, despite the well documented and carefully performed experiments, additional effects on the chemistry or nanometre structure could not be excluded.

Regarding all these considerations, we conclude that great care has to be taken with the choice, storage and preparation of samples for small-scale mechanical tests on biological materials.

21.5 Experimental Conditions

21.5.1 Moisture

The in vivo state of biomaterials is inherently linked to a certain fluid content, usually water or water-based fluids. Most commonly the exact fluid concentration in the material is not known. Thus, it is hard to decide, in which environment mechanical tests should be performed. In some cases at least reasonable assumptions can be made, e.g. for dentin, immersed in saliva from the environment in the oral cavity as well from the interior through the tubules, so that we may conclude that the material is saturated with fluid. The most straightforward experimental conditions are measurements either totally under fluid or doing the experiment in vacuum on completely dry samples. The second version has some convincing advantages: the sample condition is fairly well-defined; moreover certain types of in-situ experiments can be performed, observing the sample and its deformation in an SEM during a mechanical test. In-situ micro-bending and tensile tests on biological samples have been reported for example in [128]. Results from such experiments can be compared with parameters from indentation. In addition, as can be seen from the literature, the results of indentation tests on dry bone-like material and on enamel from different laboratories are very similar. For "wet" samples, however, the situation is a lot more complex: first, it is usually not well defined what is meant by "wet", the sample being either totally under water (including the upper surface), or only partly, the investigated surface being in air. Second, there is evidence that in a wet state the results of mechanical experiments depend remarkably on the indenter tip shape. Thus, it is not surprising that the results from different laboratories on similar materials in a "wet" state vary considerably. For example, the results from Hengsberger et al. [123] and Guidoni et al. [125] on bone and dentin agree quite well. All these experiments were performed with samples totally under fluid. However, if they are compared, e.g., with the results from Rho and Pharr and references therein

[129–132, 134, 135], large discrepancies are found between the first and the second group: Hengsberger et al. and Guidoni et al. observed a rather large difference between dry and wet state, modulus and hardness being lower in the wet state by a factor of about 1.5–1.7 and 2.5–3.3, respectively, whereas in [129–132, 134, 135] a factor of only 1.1–1.2 was reported. Schöberl and Jäger performed an experiment [136] where a tooth sample was stored under HBSS for several days. Then this sample was taken out of the fluid and immediately tested by a series of indentations. At the beginning the sample surface was still covered by a fluid film that evaporated within about 15 min. While at the beginning hardness and modulus agreed with the values reported in [123, 125], they increased continuously and approached the values reported in [129–132, 134, 135] after 10–15 min. Therefore, one may conclude that it makes a large difference if a wet material is investigated with the surface being covered with fluid or not. It should be noted, however, that such an experiment should be regarded with much caution. The reproducibility is rather poor mainly due to two effects: first, the evaporation of the fluid costs energy, usually leading to some decrease in temperature and, consequently, to thermal drift of the experimental setup. Second, the contact between probe tip and sample is surrounded by an amount of water causing adhesion forces which probably change during evaporation. In general it is found that at least most of the biological materials become softer and less stiff with increasing fluid uptake. In addition to hardness and stiffness, a high abrasion resistance is a very important property for many biotool materials, in particular those used for cutting and grinding. The results of abrasion tests on such materials strongly depend on the fluid content. Tests on dry samples often may lead to conclusions entirely different compared to those drawn from experiments under "wet" conditions (see the sections "Results" and "Examples from the Literature').

21.5.2 Temperature

Little or nothing has been reported on the influence of the temperature on the results of micron-scaled mechanical tests. Most of the investigations have been performed at the common laboratory temperatures (20°C or slightly more), whereas the in-vivo temperature of human bone is the well-known 37°C. Except for some influence on the viscous behaviour, no severe change of the mechanical properties is expected for a temperature difference of 15–17°C that can be considered negligible compared to the possible temperature rise of several tens of degrees during embedding and preparation.

21.5.3 Probe Tips

Now let us consider some basic aspects, concerning the determination of elastic modulus and hardness, the latter being no physical property on its own but related to the plastic flow stress in a complex way. It is well-known that even for standard materials like glass or simple metals, hardness and elastic modulus obtained from indentation experiments depend on the indenter tip shape. Similar observations have been reported for scratch, wear and friction tests. Since the mechanism of plastic deformation in biological tissues is not clear, the choice of a suitable indenter tip is more or less intuitive. The most common probe is a Berkovich-type with a large opening angle, causing mainly compressive stresses beneath the tip [137, 138]. This helps, to some extent, to avoid the formation of cracks, so that only "purely" plastic deformation occurs, whatever that may mean with the deformation of biological materials. Moreover, the results can easily be compared with data from the literature. The use of a Cube Corner tip makes sense, if either the lateral expansion of the plastic deformation shall be limited or if fracture toughness measurements are intended.

21.5.4 Test Velocity

Since almost all biological materials show viscous deformation behaviour it is obvious that the results of all types of mechanical tests depend on the loading rate. To gain information on the viscous behaviour, quasistatic indentation tests should be performed with a large variation of loading and unloading speed. However, the range of variations is limited: slow experiments are influenced mainly by long-term drift of the setup, fast experiments by the mass of the scanner-moved probe and the limited time resolution of data recording. If the quasistatic experiments are supplemented by dynamic measurements, a larger span of deformation velocities can be applied. Scanning wear tests are usually performed with scan speeds of some $\mu m/s$ to $100\,\mu m/s$. The velocities of practical abrasion applied on biotools in vivo, however, are higher by some orders of magnitude; thus, direct conclusions from a test in a laboratory on the true abrasion situation in nature are at least doubtful. From the literature, e.g. [139], it is known for viscous materials, that with increasing abrasion velocity the wear resistance increases. For comparative abrasion studies a fairly good compromise can be reached by varying the scan speed as much as possible (usually two orders of magnitude) and by looking for trends in friction and wear rate. If no remarkable dependence of friction and wear rate on the scan velocity occurs, or the dependence shows a continuous behaviour, one may extend the conclusions drawn from the comparatively slow tests to the natural, "fast" friction and wear situation [136].

21.6 Results

21.6.1 Sources of Error

The interpretation of results from scanning probe or indentation measurements is crucially dependent on possible errors occurring with such experiments. Even though appearing trivial, first of all we mention two very basic sources of error: if the sample is not fixed properly or the indenter tip is contaminated, there is no way to do any reliable correction, in contrast to the errors discussed below (e.g. wrong compliance or area function), where usually a correction can be made in retrospect, if the original curves are recorded properly. When testing biomaterials, the errors caused by the limits of instrumental resolution are usually small compared to all others.

The primary data derived from indentation tests are load and displacement with resolutions of clearly below 1 μN and 1 nm, respectively. In the literature, loads are typically at least 250 μN, and displacements at least 100 nm. Thus, the maximum error caused by the instrumental resolution is about or below 1%. The basic instrumental error in the displacement measurement is superimposed by additional influences, mainly by thermal drift of the whole setup. The drift can be minimized by enclosing the setup in a box protecting from air movements and acoustic (perhaps electrodynamic) influences and by waiting until potential temperature gradients have equilibrated. Moreover, stable long-term drifts can be easily taken into account by the software of most of the indentation devices. If, however, variations in the drift rate cannot be excluded for the whole measurement, an additional check may be performed, by introducing sections with constant load in the loading protocol. Especially for soft materials an additional problem occurs. Prior to an indentation experiment a preload of typically 1–10 μN is applied, causing a displacement of the indenter tip into the sample surface. This may lead to some zero-point shift. Solutions for this problem are usually already included in the software of the indentation device. Fortunately, among the large variety of biomaterials, the materials forming tools are rather stiff and hard, so that the zero-point shift usually does not introduce severe troubles. In order to derive mechanical properties from the load-displacement curves, the knowledge of the projected contact area is most crucial. Even though several methods for calibrating the tip area function exist, all methods suffer from certain shortcomings. One of the most common methods is to perform a series of indentations on a material with well-known mechanical properties, then "imposing" one known mechanical property (commonly the elastic modulus) while making a fit to the results [79]. The calibration material should be isotropic, homogeneous and its deformation behaviour independent of the deformation velocity (no viscous behaviour). Fused silica is usually considered "the standard calibration sample". Because of its high hardness and modulus, however, its application is limited to measurements to rather shallow indentation depths, mainly for two reasons: the maximum load of some instrumentation setups is limited and, especially for small open angle pointed indenter tips (Cube Corner geometry, sharp conical indenters), cracks are formed at high load indentations, making the use of results at least doubtful. Most of the indentation experiments on biomaterials, however, are performed with rather large probing depths. As "soft" calibration standards aluminium or polycarbonate have been proposed. Al is elastically isotropic with a well-known modulus of elasticity. However, Al is rather stiff which results in a steep unloading curve, which is very receptive to influences on the displacement measurement, mainly by thermal drift, or to uncertainties in the machine compliance (see below). Second, indents in Al are surrounded by pronounced pile-ups, causing uncertainties in the true actual contact area. Several attempts have been proposed, e.g. in [140] to account for the pile-ups, but due to the unknown pressure distribution below the indenter (does the pile up material take up the same pressure as the "bulk" material?), such corrections may lead to even an "overcorrection". Polycarbonate is isotropic and quite homogeneous. The main problem here is caused by its viscosity which makes hardness as well as modulus results dependent on the loading protocol. The problem of tip shape calibration is, fortunately, moderated at least for pyramidal indenter geometries: an indenter tip shape consists, roughly spoken, of the basic theoretical geometry and the tip rounding. The latter can be determined by shallow indents in fused silica. If the basic geometry can be precisely maintained by the manufacturer, the tip area function is completely defined. In contrast to some other tip geometries, the opening angle of three-sided pyramidal tips is maintained within 0.05° by several manufacturers. Thus, the use of three-sided pyramidal indenter tips is strongly recommended, unless other, specific needs suggest or even impose the use of a different tip shape. One basic question shall be briefly considered: why not use the hardness, instead of the elastic modulus, as a measure for tip shape calibration, since it reacts more sensitively on errors of the contact area? The main disadvantage utilizing the hardness is the fact that it is not a true physical parameter and, especially in metals, depends on the size of the plastic zone around the imprint (well known as indentation size effect, see e.g. [141–148]) and, in addition, on the indenter shape [149]. Another common method to determine the true tip shape is to scan the tip over a spike with extremely high aspect ratio and minimum tip radius, whereby the observed topography reflects the shape of the probe tip. If such a spike is not available, one may perform indents in a soft material with minimum elastic deformation, e.g. in silver. The residual imprint now reflects the "turned over" tip shape which can be imaged by AFM with a sharp, high aspect ratio tip. In addition to the tip shape, the compliance of the experimental setup must be known, i.e. the elastic response of the device plus tip holder. Even though, as for tip shape calibration, several methods for compliance measurements can be found in the literature [79, 149], it is still not a trivial task. An additional check can be performed by determining the elastic stiffness of various materials spanning a wide range of elastic moduli known from the literature [149]. If the results for all materials, especially the stiffest materials, are consistent with the literature values, the compliance is confirmed quite safely. Taking into account the fact that biological materials (except for highly mineralized tissues) are rather low modulus materials compared to most metals or ceramics, we conclude that an error in the machine compliance will rarely have a severe impact on the results.

All considerations given above are valid more or less for all kinds of materials. In the following, sources of errors typical for biological samples and the corresponding problems occurring with sample preparation shall be discussed. First of all, biological materials show viscous deformation behaviour. For quasistatic tests the method by Oliver and Pharr is the most common one, using the initial slope of the unloading curve as a key parameter. If the material creeps, there is still creep at least during the first part of the unloading cycle. This becomes obvious, when in a test on a viscous material without hold time at maximum load the first part of the unloading curve shows a pronounced bend with an even negative slope, which is in contradiction to the basic assumptions for the Oliver–Pharr method [79]. This problem can be overcome (at least partly) by introducing a long hold period at maximum load before unloading which minimizes the creep rate and/or by refined analysis of the

Fig. 21.3. In-situ indentation in a SEM. During unloading (from *top* to *bottom*) the contact area decreases remarkably within a very short time

load-displacement curve [150–152]. In addition to creep, biological materials exhibit viscoelastic recovery that becomes evident upon introducing a holding period at constant (low) load after partial unloading: the displacement decreases at constant load, converging to a constant value with time. This time-dependent elastic recovery can be nicely observed when performing experiments while imaging the indentation process in a scanning electron microscope (SEM) [149]. Such observations show, in addition, that during unloading the contact area decreases pronouncedly for such materials, see Fig. 21.3.

Another, notable shortcoming of the O.P.-method or other methods analyzing the unloading curves should be mentioned: after having deformed the sample plastically by loading, the structures below the indenter may have been damaged, thus changes in the elastic interaction between indenter and sample during unloading cannot be excluded. As a solution, an analysis of the very first part of the loading curve before the onset of plastic deformation may be proposed (Hertzian fit). Unfortunately, this procedure cannot be applied to most biomaterials, since there the plastic deformation usually starts at very low loads. Considering the long list of shortcomings discussed in the literature [153, 154], it may seem surprising that the O.P.-method is still quite commonly applied on viscous materials. It is impressive how modulus and hardness determined for many materials in different laboratories and on various indentation devices agree reproducibly. Some authors [153–155] propose to determine the hardness by measuring the residual imprint with an AFM or a scanning indenter tip. However, since the most common and reasonable definition of hardness is the maximum load divided by the projected contact area at this maximum load, this method is only valid, if the contact area does not change during unloading (see above), a prerequisite usually not fulfilled in polymers or biomaterials.

The possible precision of indentation, friction, scratch and wear experiments is, often to a large extent, limited by surface roughness. All the geometric considerations on indentation experiments are done on the assumption of an ideally flat and smooth sample surface which usually does not hold in practice. Biological samples in particular are mostly hard to prepare towards a perfect surface finish. As a consequence, most indentation experiments on biological materials are performed with rather large penetration depths (from this point of view the denomination nanoindentation is often not justified). If the indentation device allows scanning the tip over the surface before the mechanical test, it is possible to choose suitable regions for the experiment and image the area around the imprint after the test, allowing at least an estimate of the error resulting from sample topography including eventual pile ups.

The investigation of lamellar bone represents a good example for the possible influence of the topography on indentation results [123, 156]. After a standard sample preparation, usually classic metallographic grinding and polishing, the lamellar structure is clearly reflected by the topography with the lamellae protruding about 30 nm from the interlamellar regions. The interlamellar spacing is of the order of microns, thus the maximum indentation depth is severely limited. A comparison of mechanical data from the interlamellar region and the lamellae is superimposed by a systematic error due to the different contact situations: indentation in the interlamellar region represents an imprint "into a valley", whereas on the lamellae it is an imprint "on top of the hill". As a consequence modulus and hardness are overestimated for the former und underestimated for the latter. A method to minimize this systematic error has been proposed by Staedler et al. [156].

Measurements on biological materials are frequently conducted on areas with lateral dimensions of only microns, separated by pores or gaps etc. For handling and preparing (grinding, polishing) the samples they are usually embedded in polymer resin or similar material that partly fills any cavities. Another, similar situation occurs, if the experiments are performed on layered structures, where the thickness of individual layers is not drastically larger than the depth of the imprint. Several attempts have been made to include such a structural compliance into the analysis [157, 158, 169], a term that adds to the machine compliance, but unlike the latter the structural term can vary as a function of the position on the specimen [155]. For porous materials various attempts have been made to extract "true" material properties from indentation measurements [170–173], some of them applied for studies on bone.

21.6.2 Interpretation

Unfortunately, the modulus obtained from indentation measurements is not identical with the orientation dependent Young's modulus, measured anisotropies become "smeared" [41, 42, 174–177] due to the indentation geometry. Hardness, unfortunately, does not reflect a well-defined, physical property. Zhang and Sakai [178] investigated geometrical effects of pyramidal indenters with different face angles on the elasto-plastic contact behaviours of metals and ceramics. They found different hardness results for varying face angle. Moreover, they tried to derive a face angle dependent "true" hardness which was related to the flow stress by a constraint factor. Applying such considerations on biomaterials, however, is more or less hopeless, due to the lack of known constraint factors. Aside from hardness and stiffness, abrasion resistance is a very important property of biotools. Since hardness and modulus are the standard results from indentation measurements, it seems desirable to derive the abrasion behaviour from the former data. For metals, ceramics and coatings a high abrasion resistance is usually achieved by a combination of high hardness H at rather low elastic modulus E [179], the abrasion resistance frequently described as proportional to $H^{1.5}\,E^{-1}$. This proportionality factor has been taken into consideration in a number of studies on biotools, e.g. [12, 180–183]. Apart from copper, frequently Zn is found, for example in the tip region of the jaws of the marine worm *Nereis* [12] or the cutting edge of the desert locust mandible *Schistocerca gregaria* [136]. These mandibles represent an example for a high sensitivity of the mechanical behaviour on the water content as well as the fact that conclusions drawn from hardness and modulus results on the abrasion resistance can be erroneous. The mandibles have zinc-enriched (several at-%) regions at their cutting edge, i.e. in the material where high abrasion resistance is absolutely mandatory. Surprisingly, the abrasion rate depends strongly on the water content only in the cutting edge region. A similar observation was made for the abrasion behaviour of human teeth and the jaw of the marine worm *Nereis*: only for the material with the necessity of being highly abrasion resistant i.e. the tip of the jaw and the enamel, respectively, a high sensitivity on the water content was found [136]. The highest resistivity against abrasion of the locust mandibles at their zinc-enriched cutting edge was found for a certain water content. Hardness and modulus measurements showed that this is not at all the state of the largest value of $H\,E^{-1}$ or $H^{1.5}E^{-1}$. Scanning wear tests are basically not complicated experiments. However, in some cases they can be performed on mineralized materials only with some restrictions, e.g. on enamel which consists of mineral rods with protein in between. If the scanning area covers not only a single rod, but is larger, mineral particles may be pulled out which makes the definition of an abrasion rate almost impossible; results will often not be reproducible. A similar situation appears for the atacamite-enriched jaw regions of the marine worm *Glycera*, if a very sharp probe tip is used. In this case using a probe with larger tip radius and opening angle could help to achieve reproducible results [23].

21.7 Examples from the Literature

In the following some examples from the literature are given presented in order to elucidate the implications of the above considerations for scientific work on biotool materials. Although there is a vast variety of biotool tissues, here we shall focus on jaws and teeth and discuss the heavily mineralized version found in vertebrates as well as selected invertebrate tooth tissues containing little or no mineral.

A great number of studies have been performed on the mechanical properties of human teeth, e.g. [184–193]. In [193] area-scans of human coronal dentin were carried out in order to investigate correlations between local mechanical properties and the density, size and crystallinity of the mineral particles. The mineral content of dentin was found to decrease and the thickness of mineral crystals to increase towards the dentin–enamel junction (DEJ). Hardness and elastic modulus both decreased towards the DEJ. In a correlation analysis, the mineral content and, even more so, the thickness of mineral crystals were found to be crucial factors for the prediction of hardness. The dentin layer close to the DEJ exhibited a local minimum in hardness and elastic modulus, a configuration known to be an effective obstacle for crack propagation. It was concluded that the observed variations of mechanical and structural properties define crown dentin as a gradient material optimized for its mechanical function. Further studies on the dentin–enamel junction were performed for example in [194–197], where hardness, elastic modulus and fracture were studied as well as the functional width of the DEJ. Using AFM-based microscratching it could be shown that the width of the junction is only a few microns, much smaller than expected, being a nice example, how a complementary technique in addition to indentation allows deeper insight in such a system.

In stark contrast to vertebrate teeth based on large amounts of calcium mineral (enamel contains about 90% vol. hydroxyapatite), the jaws of the marine worm *Glycera* contain very low levels of a rare copper mineral (atacamite), the only one of its type known so far to be involved in biomineralization. The mineral is arranged in fibres running parallel to the outline of the jaw and was identified in 2002 by Lichtenegger et al. [180] who also performed instrumented indentation experiments in order to elucidate its function. Indentation was combined with local chemical analysis by microprobe measurements and SEM imaging on the same sample. A clear positive correlation between the local hardness and stiffness and the amount of atacamite was found, suggesting a structural function of the mineral. The performance index $H^{1.5} E^{-1}$ [198] was used as a rough estimate for abrasion resistance and found to correlate with the copper content. In a subsequent publication [182], H and E values of *Glycera* jaws and other materials were arranged into a double-logarithmic performance plot following the approach outlined by Ashby [198]. Data lying on a straight line with a slope of 2/3 should then represent materials with equivalent abrasion performance. From this plot it was, tentatively, suggested that compared to many materials, *Glycera* jaws have an excellent resistance to abrasion. Since *H* and *E* given in [180] were from dry samples embedded in polymer resin this estimate might not reflect the natural abrasion situation in sea water [153]. Indentation measurements performed by Moses et al. [182] on dry and wet *Glycera* jaw samples yielded similar spatial distributions of hardness and modulus in both cases, with an overall decrease of modulus by 25% and of hardness by 28% in wet samples. Furthermore, in this study samples were subjected to hydrolysis in boiling HCl for 48 h, which removed the proteinaceous portion of the jaw matrix as well as almost all Cu as evidenced by EDS (Energy-dispersive X-ray spectroscopy) measurements, but left the shape of the jaw intact. The residue, identified as melanin scaffold, exhibited a modulus reduced by 60% and a hardness reduced by 40% as compared to the untreated jaw.

Although it had been reported earlier that *Glycera* jaws also contain a certain amount of unmineralized Cu potentially coordinated with the histidin (His)-rich jaw matrix [199], it was only found in a recent study by Pontin et al. that the atacamite mineralized fibres are located underneath the jaw surface and are embedded in a layer rich in unmineralized copper [23]. In this work *Glycera* jaws were also subjected to wear tests. Selecting a tip with a comparatively large radius allowed the use of a relatively simple model for interpretation: a hard spherical indenter ploughing across a flat block of a perfectly plastic medium was assumed. The atacamite mineralized regions exhibited a wear depth about twice as large as the adjacent, copper-enriched but mineral-free matrix. It was also found that the wear rates increased by a factor of about three in all regions after hydration. Scratch tests revealed similar trends. Under both, wet and dry conditions, the regions with unmineralized copper emerged as most scratch resistant. The ratio between lateral and normal load (friction coefficient) was highest in the copper-free and the atacamite containing regions; hydration caused an increase by 30%, the increase of friction coefficient in copper-rich, unmineralized regions was about 22%.

These findings were explained with the coordination of Cu by the histidine-rich protein and melanin thus establishing inter-molecular crosslinks [23]. Such crosslinks are not only expected to enhance local hardness and stiffness, but may also reduce susceptibility to water, perhaps by occupation of intermolecular sites by Cu ions. The enhanced wear rate in atacamite-rich regions can be explained by chipping and cracking of the (brittle) mineral fibres, promoted by elastic/plastic mismatch between mineral and matrix. Even though the authors of this work presented some speculations, how the wear rate was increased by hydration, no real explanation for this behaviour could be given.

Similar studies were performed on the slightly different jaws of the marine worm *Nereis* [12] that contain Zn instead of Cu, albeit in entirely non-mineralized form and in lower amounts. Again, the jaw matrix is proteinaceous and rich in His, but in contrast to that of *Glycera* jaws it does not contain any melanin. The zinc is predominantly concentrated in the jaw tip, thus suggesting a mechanical function. The influence of Zn on stiffness and hardness was probed with a combination of indentation and microprobe analysis on dry jaws, yielding a positive correlation between local metal content and mechanical performance. The Zn content was shown to vary in parallel with the presence of Cl and local His levels. X-ray absorption experiments suggested a coordination of Zn by three His and one Cl in analogy to the arrangement found in Zn insulin [12]. In a recent study by Broomell et al. [127] *Nereis*

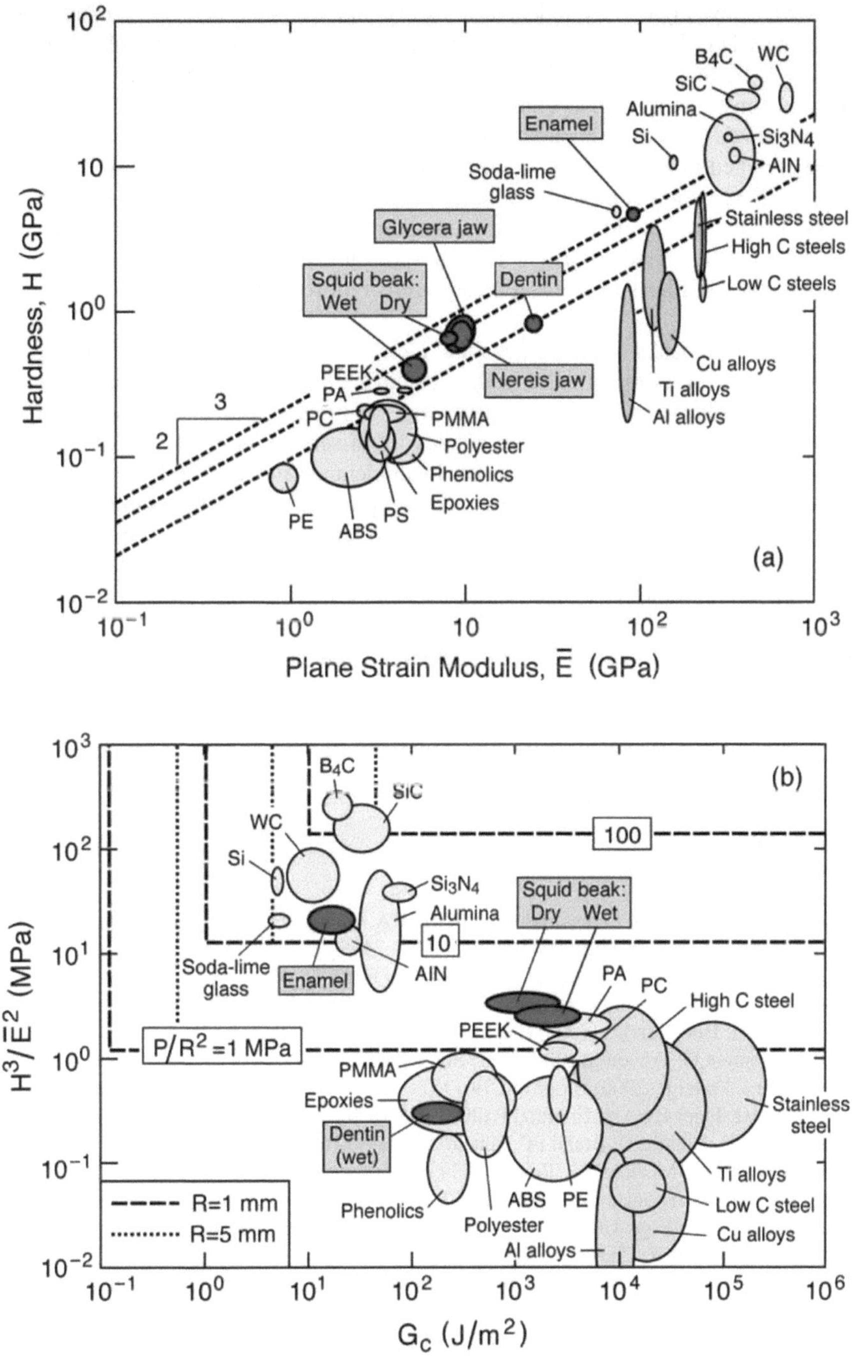

Fig. 21.4. Property maps showing (**a**) comparisons of hardness and stiffness of *Dosidicus* beaks with typical engineering materials and (**b**) critical loads for yielding and for cracking at contact with spherical rigid abrasives

jaws were investigated by indentation in dry and wet state and by element analysis using EDS. Samples were measured before and after treatment with EDTA for 96 h at 25°C that removed a large fraction of the Zn without causing changes in the organic constituents, as was confirmed by X-ray photoelectron spectroscopy (XPS) on pulverized jaws and secondary mass spectroscopy. In a further step, the samples were enriched in Zn again by incubation of pre-depleted jaw surfaces in $ZnCl_2$ for 72 h at 25°C. Not a complete, but partial reconstitution was evidenced by XPS and EDS. The authors could show that the measured hardness and modulus values correlated with the variations in Zn. The Zn removal resulted in over 65% reduction of hardness and stiffness in relevant areas and its reintroduction caused an appreciable recovery.

Finally, the beaks of the jumbo squid *Dosidicus gigas* represent a purely organic biotool material. Although the beaks contain neither metals nor biomineral, they exhibit a hardness and stiffness up to twice as high as that of the most competitive synthetic organic materials. In a study by Miserez et al. [181] indentation tests were carried out on squid beak samples in ambient air and under distilled water. The experiments were complemented by macroscopic fracture toughness measurements. Abrasion was assumed to be due to one of two failure mechanisms, either the onset of plasticity or the formation of cracks. The material was treated as flat plate and the abrading particle as a rigid solid of revolution, allowing treatment by classical contact mechanics. The normal load for yielding is then proportional to $H^3 E^{-2}$, whereas the critical load to develop a crack P_C was proportional to the shear modulus G. By plotting $H^3 E^{-2}$ vs. G, property maps were drawn (see Fig. 21.4), from which the deduced abrasion resistance was compared to the values of enamel, dentin, some polymers and metals, indicating a high abrasion resistance of the squid beaks. No direct measurements of the abrasion behaviour, however, were carried out.

References

1. Gorb SN, Sinha M, Peressadko A, Daltorio KA, Quinn RD (2007) Bioinsp Biomim 2:117
2. Otten A, Herminghaus S (2004) Langmuir 20:2405
3. Shan D, Wang S, He Y, Xue H (2008) Mat Sci Eng C28:218
4. Oulevey F, Burnham NA, Gremaud G, Kulik AJ, Pollock HM, Hammiche A, Reading M, Song M, Hourston DJ (2000) Polymer 41:3087
5. Pietrement O, Troyon M (2001) Surf Sci 490:L592
6. Jäger I, Fratzl P (2000) Biophys J 79:1737
7. Gao HJ, Jäger IL, Arzt E, Fratzl P (2003) PNAS 100:5597
8. Block RJ, Bolling D, Brand FC, Schein A (1939) J Biol Chem 128:181
9. Fraser RDB, Mc Rae TP, Rogers GE (1972) Keratins: their composition, structure and biosynthesis. Thomas CC, Springfield, IL
10. Vincent JFV, Wegst UGK (2004) Arthropod Struct Develop 33:187
11. Saito Y, Okano T, Chanzy H, Sugiyama J (1995) J Struct Biol 114:218
12. Lichtenegger HC, Schöberl T, Ruokalainen JT, Cross JO, Heald SM, Birkedal H, Waite JH, Stucky GD (2003) PNAS 100:9144
13. Sachs C, Fabritius H, Raabe D (2006) J Struct Biol 155:409
14. Evans LA, Alvarez R (1999) J Biol Inorgan Chem 4:166
15. Lee AP, Brooker LR, Macey DJ, Webb J, von Bronswijk W (2003) J Biol Inorgan Chem 8:256

16. Lichtenegger HC, Schöberl T, Bartl MH, Waite JH, Stucky GD (2002) Science 298:389
17. Lichtenegger HC, Schöberl T, Bartl MH, Waite JH, Stucky GD (2003) Science 301:5636
18. Rinnerthaler S, Roschger P, Jakob HF, Nader A, Klaushofer K, Fratzl P (1999) Calcif Tissue Int 64:422
19. Fantner GE, Rabinovych O, Schitter G, Thurner PH, Kindt JH, Finch MM, Weaver JC, Golde LS, Morse DE, Lipmann EA, Rangelow IW, Hansma PK (2006) Compos Sci Technol 66:1205
20. Gupta HS, Wagermaier W, Zickler GA, Aroush DR, Funari SS, Roschger P, Wagner HD, Fratzl P (2005) Nano Lett, 5: 2108
21. Gupta HS, Seto J, Wagermaier W, Zaslansky P, Boesecke P, Fratzl P (2006) PNAS 103:17741
22. Jackson AP, Vincent JFV, Turner RM (1988) Proc R Soc London Ser B 234:415
23. Moses DN, Pontin MG, Waite JH, Zok FW (2008) Biophys J doi:10.1529/ biophysj.107.120790
24. Lichtenegger HC, Birkedal H, Waite JH (2008) in: Met Ions Life Sci 4 (Biomineralization. From nature to application)
25. Lee EH, Radok JRM (1960) J Appl Mech 27:438
26. Ting TCT (1966) J Appl Mech 88:845
27. Johnson KL (1985) Contact mechanics. Cambridge University Press, Cambridge
28. Oyen ML (2006) Phil Mag 86:5625
29. Oyen ML (2005) J Mater Res 20:2094
30. Bembey AK, Oyen ML, Bushby AJ, Boyde A (2006) Phil Mag 86:5691
31. Bembey AK, Bushby AJ, Boyde A, Ferguson VL, Oyen ML (2006) J Mater Res 21:1962
32. Vandamme M, Ulm F-J (2006) Int J Solids Struct 43:3142
33. Sasaki N, Enyo A (1995) J Biomech 28:809
34. Jäger IL (2005) J Biomech 38:1451
35. Jäger IL (2005) J Biomech 38:1459
36. Cowin SC (1999) J Biomech 32:217
37. Wang HF (2000) Theory of linear poroelasticity with applications to geomechanics and hydrogeology. Princeton University Press, Princeton, NJ
38. Agbezuge LK, Deresiewicz H (1974) Israel J of Technol 12:322
39. Selvadurai APS (2004) Int J Solids Structures 41:2043
40. Oyen ML, Bembey AK, Bushby AJ (2007) Mater Res Soc Symp Proc 975
41. Vlassak JJ, Nix WD (1996) Phil Mag A67:1045
42. Swadener JG, Rho JY, Pharr GM (2001) J Biomed Res 51:108
43. Butt HJ, Capella B, Kappl M (2005) Surf Sci Rep 59:1
44. Meyer G, Amer NM (1988) Appl Phys Lett 53:2400
45. Alexander S, Hellemans L, Marti O, Schneir J, Elings V, Hansma PK, Longmire M, Gurley J (1989) J Appl Phys 65:164
46. Albrecht TR Akamine S, Carver TE, Quate CF (1990) J Vac Sci Technol A8:3386
47. Sader JE (1995) Rev Sci Instrum 66:4583
48. Neumeister JM, Ducker WA (1994) Rev Sci Instrum 65:2527
49. Hazel JL, Tsukruk VV (1999) Thin Solid Films 339:249
50. Butt H-J, Siedle P, Seifert K, Fendler K, Seeger T, Bamberg E, Weisenhorn AL, Goldie K, Engel A (1993) J Microsc 169:75
51. Sader JE, Larson I, Mulvaney P, White LR (1995) Rev Sci Instrum 66:3789
52. Senden TA, Ducker WA, (1994) Langmiur 10:1003
53. Maeda N, Senden TJ (2000) Langmuir 16:9282
54. Notley SM, Biggs S, Craig VSJ (2003) Rev Sci Instrum 74:4026
55. Degertekin FL, Hadimioglu B, Sulchek T, Quate CF (2001) Appl Phsy Lett 78:1628
56. Holbery JD, Eden VL, Sarikaya M, Fisher RM (2000) Rev Sci Instrum 71:3769

57. Cleveland JP, Manne S, Bocek D, Hnsma PK (1993) Rev Sci Instrum 64:403
58. Bonaccurso E, Butt H-J (2005) J Phys Chem B109:253
59. Hutter JL, Bechhoefer J (1993) Rev Sci Instrum 64:1868
60. Gibson CT, Watson GS, Myhra S (1996) Nanotechnology 7:259
61. Van der Werf KO, Putman CAJ, De Groth BG, Greve J (1994) Appl Phys Lett 65:1195
62. Cappella B, Paschieri P, Frdiani C, Miccoli P, Ascoli C (1997) Nanotechnology 8:82
63. Koleske DD, Lee GU, Gans BI, Lee KP, Dilella DP, Wahl KJ, Berger WR, Whitman LJ, Colton RJ (1995) Rev Sci Instrum 66:4566
64. Radmacher M, Fritz M, Cleveland JP, Walters DA, Hansma PK (1994) Langmuir 10:3809
65. Baselt DR, Baldeschwieler JD (1994) J Appl Phys 76:33
66. Heuberger M, Dietler G, Schlepbach L (1995) Nanotechnology 6:12
67. Rosa-Zeiser A, Weilandt E, Hild S, Marti O (1997) Meas Sci Technol 8:1333
68. Krotil HU, Stifter T, Waschipky H, Weishaupt K, Hild S, Marti O (1999) Surf Interf Anal 27:336
69. Janshoff A, Neitzert M, Oberdörfer Y, Fuchs H (2000) Angew Chem Int Ed 39:3212
70. Zhang W, Zhang X (2003) Prog Polym Sci 28:1271
71. Rief M, Oesterheld F, Heymann B, Gaub HE (1997) Science 275:1295
72. Fritz J, Anselmetti D, Jarchow J, Fernadez-Busquets X (1997) J Struct Biol 119:165
73. Kikuchi H, Yokoyama N, Kajiyama T (1997) Chem Lett 11:1107
74. Hertz H (1882) J Reine Angew Math 92:156
75. Johnson KL, Kendall K, Roberts AD (1971) Proc R Soc London A 324:301
76. Derjaguin BV, Muller VM, Toporov YP (1975) J Colloid Interf Sci 53:314
77. Müller VM, Yushchenko VS, Derjaguin BV (1980) J Colloid Interf Sci 77:91
78. Müller VM, Derjaguin BV, Toporov YP (1983) Colloids Surf 7:251
79. Oliver WC, Pharr GM (1992) J Mater Res 7:1564
80. Doerner MF, Nix WD (1986) J Mater Res 1:601
81. Sneddon IN (1965) Int J Engng Sci 3:47
82. Bhushan B, Koinkar VN (1995) Surf Coat Technol 76–77:655
83. Bhushan B, Lowry JA (1995) Wear 190:1
84. Koinkar VN, Bhushan B (1996) J Appl Phys 79:8071
85. Bhushan B (1999) Diam Rel Mater 8:1985
86. Miyake S, Kaneto R, Miyamoto T (1992) Diamond Films Technol 1:205
87. Jiang Z, Lu CJ, Bogy DB, Bhatia CS, Miyamoto T (1995) Thin Solid Films 258:75
88. Kaneko R, Miyamoto T, Yandoh Y, Hamada E (1996) Thin Solid Films 273:105
89. Schiffmann KI (2008) Wear, in press
90. Degiampietro K, Colaco R (2007) Wear 263:1579
91. Zhao X, Bhushan B (1998) Wear 223:66
92. Pethica JB, Oliver WC (1989) in: Bravman JC, Nix WD, Barnett DM, Smith DA (eds) Mater Res Soc Symp Proc 130:13
93. Syed Asif SA, Pethica JB (1997) in: Gerberich WW, Gao H, Sundgren JE, Baker SP (eds) Mater Res Soc Symp 436:201
94. Li X, Bhushan B (2002) Mater Charact 48:11
95. Lawn BR, Fuller ER (1975) J Mater Sci 10:2016
96. Ponton CB, Rawlings RD (1989) Mater Sci Technol 5:865
97. Cook RF, Pharr GM (1990) J Am Ceram Soc 73:787
98. Lawn BR, Wilshaw TR (1975) J Mater Sci 10:1049
99. Evans AG, Charles EA (1976) J Am Ceram Soc 59:371
100. Frank FC, Lawn BR (1967) Proc R Soc London A299:291
101. Lawn BR, Swain MV (1975) J Mater Sci 10:113
102. Marshall DB, Lawn BR (1979) J Mater Sci 14:2001

103. Marshall DB, Swain MV (1988) J Am Ceram Soc 71:399
104. Lathabai S, Rodel J, Dabbs T, Lawn BR (1991) J Mater Sci 26:2157
105. Burns SJ, Chia KY (1995) J Am Ceram Soc 78:2321
106. Burns SJ, Chia KY (1995) J Am Ceram Soc 78:2328
107. Sglavo VM, Green DJ (1995) J Am Ceram Soc 78:650
108. Fischer-Cripps AC, Lawn BR (1996) Acta Mater 44:519
109. Pharr GM (1998) Mater Sci Eng A253:151
110. Lawn BR, Evans AG, Marshall DB (1980) J Am Ceram Soc 63:574
111. Anstis GR, Chantikul P, Lawn BR, Marshall DB (1981) J Am Ceram Soc 64:533
112. Casellas D, Caro J, Molas S, Prado JM, Valls I (2007) Acta Mater 55:4277
113. Lankford J, Davidson DL (1979) J Mater Sci 14:1662
114. Harding DS, Oliver WC, Pharr GM (1995) Mater Res Soc Symp Proc 356:663
115. Yan J, Karlson AM, Chen X (2007) Eng Fract Mech 74:2535
116. Lee JS, Jang J, Lee BW, Choi Y, Lee SG, Kwon D (2006) Acta Mater 54:1101
117. Bhushan B, Kulkarni AV, Bonin W, Wyrobek JT (1996) Phil Mag A74:1117
118. Fratzl P, Gupta HS, Paschalis EP, Roschger P (2004) J Mater Chem 14:2115
119. Gupta HS, Schratter S, Tesch W, Roschger P, Berzlanovich A, Schöberl T, Klaushofer K, Fratzl P (2005) J Struct Biol 149:138
120. Moscovich H, Dreugers NHJ, Jansen JA, Wolke JGC (1999) J Dent 27:503
121. Halls N (1992) Med Device Technol August/September:37
122. Hengsberger S, Kulik A, Zysset P (2001) Eur Cells and Mater 1:12
123. Hengsberger S, Kulik A, Zysset P (2002) Bone 30:1
124. Habelitz S, Marshall G, Balooch M, Marshall SJ (2002) J Biomech 35:995
125. Guidoni G, Denkmayr J, Schöberl T, Jäger IL (2006) Phil Mag 86:5705
126. Ho SP, Goodis H, Balloch M, Nonomura G, Marshall SJ, Marshall G (2004), Biomaterials 25:4847
127. Broomell CC, Mattoni MA, Zok FW, Waite JH (2006) J Exp Biol 209:3219
128. Orso S, Wegst UGK, Eberl C, Arzt E (2006) Adv Mater 18:874
129. Rho JY, Pharr GM (1999) J Mater Sci Mater Med 10:485
130. Evans FG, Lebow M (1951) J Appl Physiol 3:563
131. Evans FG (1973) Mechanical properties of bone. Charles C Thomas, Springfield IL, p 43
132. Yamada H (1970) Strength of biological materials. Williams & Wilkins, Baltimore MD, p 74
133. Townsend PR, Rose RM (1975) J Biomech 8:199
134. Currey JD (1988) J Biomech 21:439
135. Jameson MW, Hood JAA, Tidmarsh BG (1993) J Biomech 26:1055
136. Schöberl T, Jäger IL (2006) Advanced Eng Mater 8:1164
137. Chu JL, Lin HY, Wu, Lee S (2000) Mat Sci Eng A282:23
138. Lichinchi M, Lenardi C, Haupt J, Vitali R (1998) Thin Solid Films 312:240
139. Schallamach A (1957) Wear 1:384
140. Lee YH, Baek U, Kim YI, Nahm SH (2007) Mater Lett 61:4039
141. Elmustafa AA, Stone DS (2002) Acta Mater 50:3641
142. Abu Al-Rub RK (2007) Mechan Mater 39:787
143. Abu Al-Rub RK, Voyiadis GZ (2004) Int J Multiscale Comput Eng 3:50
144. Chiu YL, Ngan AHW (2002) Acta Mater 50:2677
145. Durst K, Backes B, Franke O, Göken M (2006) Acta Mater 54:2547
146. Fleck NA, Muller GM, Ashby MF, Hutchinson JW (1994) Acta Metal Mater 42:475
147. Qu S, Huang Y, Nix WD, Jiang H, Zhang F, Hwang KC (2004) J Mater Res 19:3423
148. Chudoba T, Schwarzer N, Richer F, Beck U (2000) Thin Solid Films 377:366
149. Chudoba T, Schwaller P, Rabe R, Breguet JM, Michler J (2006) Phil Mag 86:5265
150. Feng G, Ngan AHW (2001) Mater Res Soc Symp Proc 649:7.1.1

151. Kermouche G, Loubet JL, Bergheau JM (2007) Mech Mater 39:24
152. Ngan AHW, Wang HT, Tang B, Sze KY (2005) Int J Solids Struct 42:1831
153. Schofield RMS, Nesson MH, Richardson KA (2002) Naturwissenschaften 89:579
154. Schofield RMS, Nesson MH (2003) Science 301:1049a
155. Jakes JE, Frihart CR, Beecher JF, Moon RJ, Stone DS (2008) J Mater Res 23:1113
156. Staedler T, Donnelly E, Van der Meulen MCH, Baker SP (2003) Mater Res Symp Proc 778:79
157. Stone DS (1998) J Mater Res 13:3207
158. Stone DS, Yoder KB, Sproul WD (1991) J Vac Sci Technol A9:2543
159. Yoder KB, Stone DS, Hoffman RA, Lin JC (1998) J Mater Res 13:3214
160. Soifer YM, Verdyan A, Kazakevich M, Rabkin E (2005) Mater Lett 59:1434
161. Hodzic A, Stachurski ZH, Kim JK (2000) Polymer 41:6895
162. Downing TD, Kumar R, Cross WM, Kjerengtroen L, Kellar JJ (2000) J Adhes Sci Technol 14:1801
163. Lee SH, Wang S, Pharr GM, Xu H (2007) Compos Part A-Appl Sci 38:1517
164. Gerber CE (1968) Contact problems for the elastic quarter-plane and for the quarter space. Stanford University, Palo Alto, CA, p 100.
165. Hetenyi M (1960) Trans ASME Series E, J Appl Mech 27:289
166. Hetenyi M (1970) Trans ASME Series E, J Appl Mec. 37:70
167. Keer LM, Lee JC, Mura T (1984) Intl J Solids Struct 20:513
168. Popov GY (2003) Mech Solids 38:23
169. Schwarzer N, Hermann I, Chudoba T, Richter F (2001) Contact modelling in the vicinity of an edge. Elsevier, San Diego, CA, p 371
170. Hengsberger S, Enstroem J, Peyrin F, Zysset P (2003) J Biomech 36:1503
171. Rice JC, Cowin SC, Bowman JA (1988) J Biomech 21:155
172. Sevostianov I, Kachanov M (2000) J Biomech 33:881
173. Hoffler CE, Moore KE, Kozloff K, Zysset PK, Brown MB (2000) Bone 26:603
174. Fratzl P, Gupta HS, Paschalis EP, Roschger P (2004) J Mater Chem 14:2115
175. Gindl W, Schöberl T (2004) Composites Part A35:1345
176. Gindl W, Reifferscheid M, Adusumalli RB, Weber H, Röder T, Sixta H, Schöberl T (2008) Polymer 49:792
177. Gindl W, Konnerth J, Schöberl T (2006) Cellulose 13:1
178. Zhang J, Sakai M (2004) Mater Sci Eng A 381:62
179. Bhushan B (1999) Principles and application of tribology. Wiley, New York
180. Lichtenegger HC, Schöberl T, Bartl MH, Waite H, Stucky GD (2002) Science 298:389
181. Miserez A, Li Y, Waite JH, Zok FW (2007) Acta Biomater 3:139
182. Moses DN, Harreld JH, Stucky GD, Waite JH (2006) J Biol Chem 281:3482
183. Moses DN, Mattoni MA, Slack NL, Waite JH, Zok FW (2006) Acta Biomater 2:521
184. Ho SP, Marshall SJ, Ryder MI, Marshall GW (2007) Biomaterials 28:5238
185. Zioupos P, Rogers KD (2006) J Bionic Eng 3:19
186. Nizam BRH, Lim CT, Cheng HK, Yap AUJ (2005) J Biomech 38:2204
187. Cheng HK, Ramli HN, Yap AUJ, Lim CT (2005) J Dentistry 33:363
188. Lippert F, David M, Parker DM, Jandt KD (2004) J Colloid and Interface Sci 280:442
189. Balooch G, Marshall GW, Marshall SJ, Warren OL, Asif SAS, Balooch M (2004) J Biomech 37:1223
190. Cuy JL, Mann AB, Livi KJ, Teaford MF, Weihs TP (2002) Archs Oral Biol 47:281
191. Finke M, Hughes JA, Parker DM, Jandt KD (2001) Surf Sci 491:456
192. Habelitz S, Marshall SJ, Marshall GW, Balooch M (2001) J Struct Biol 135:294
193. Tesch W, Eidelman N, Goldenberg F, Roschger P, Klaushofer K, Fratzl P (2001) Calcif Tissue Int 69:147

194. Nalla RK, Balooch M, Ager III JW, Kruzic JJ, Kinney JH, Ritchie RO (2005) Acta Biomater 1:31
195. Mahoney EK, Rohanizadeh R, Ismail FSM, Kilpatrick NM, Swain MV (2004) Biomaterials 25:5091
196. Kinney JH, Balloch M, Marshall SJ, Marshall GW, Weihs TP (1996) Archs Oral Biol 41:9
197. Angker L, Swain MV, Kilpatrick N (2003) J Dentistry 31:261
198. Ashby MF (1999) Materials selection in mechanical design, 2nd edn, Butterworth-Heinemann, Oxford
199. Waite JH, Lichtenegger HC, Stucky GD, Hansma PK (2004), Biochemistry 43:7653

9 Direct Force Measurements of Receptor–Ligand Interactions on Living Cells

Robert H. Eibl

From volume XII

Abstract. The characterization of cell adhesion between two living cells at the level of single receptor–ligand bonds is an experimental challenge. This chapter describes how the extremely sensitive method of atomic force microscopy (AFM) based force spectroscopy can be applied to living cells in order to probe for cell-to-cell or cell-to-substrate interactions mediated by single pairs of adhesion receptors. In addition, it is outlined how single-molecule AFM force spectroscopy can be used to detect physiologic changes of an adhesion receptor in a living cell. This force spectroscopy allows us to detect in living cells rapidly changing, chemokine SDF-1 triggered activation states of single VLA-4 receptors. This recently developed AFM application will allow for the detailed investigation of the integrin–chemokine crosstalk of integrin activation mechanisms and on how other adhesion receptors are modulated in health and disease. As adhesion molecules, living cells and even bacteria can be studied by single-molecule AFM force spectroscopy, this method is set to become a powerful tool that can not only be used in biophysics, but in cell biology as well as in immunology and cancer research.

Key words: Single-molecule force measurements, Living cells, Mouse, Human, Cell adhesion receptor, Cell adhesion molecule, Unbinding event, Rupture, AFM, Atomic force microscopy, Force spectroscopy, Lymphocyte homing, Integrins, Integrin activation, VLA-4, VCAM-1, LFA-1, ICAM-1, SDF-1, CD44, Chemokines

Abbreviations

$\alpha_4\beta_1$	Alpha(4)beta(1) (integrin VLA-4)
$\alpha_L\beta_2$	alpha(L)beta(2) (integrin LFA-1)
AFM	Atomic force microscope
Anti-VLA-4	Antibody against VLA-4
BSA	Bovine serum albumin
CD44	Cluster of differentiation 44
ConA	Concanavalin A
CXCL12	CXC chemokine receptor ligand 12 (SDF-1)
CXCR4	CXC chemokine receptor 4 (receptor for SDF-1)
Dynes/cm^2	Unit for shear forces
EDTA	Ethylene diamine tetraacetic acid
EGTA	Ethylene glycol tetraacetic acid
FACS	Fluorescence-activated cell sorter
F_c	Crystallizable fragment (of an antibody)

In: H. Fuchs, B. Bhushan (eds), *Biosystems - Investigated by Scanning Probe Microscopy*,
DOI 10.1007/978-3-540-85039-7_1,

GRGDSP	Amino-acid sequence, RGD-containing peptide (binding to, for example, integrin $\alpha_5\beta_1$)
ICAM-1	Intercellular cell adhesion antigen 1
IHC	Immunohistochemistry
HEPES	4-(2-hydroxyethyl)-1-piperazineethanesulfonic acid
HSA	Human serum albumin
LFA-1	Lymphocyte function-associated antigen 1 (integrin $\alpha_L\beta_2$)
MAb	Monoclonal antibody
μm	Micrometer
mN/m	Milli Newton per meter
min	Minutes
$NaHCO_3$	Sodium bicarbonate
ng/ml	Nanogram/milliliter
PBS	Phosphate-buffered saline
pN	Pico Newton
RGD	Arginine–glycine–aspartic acid (aminoacid sequence found in several integrin ligands)
RT	Room temperature
SDF-1	Stromal-derived factor 1 (CXCL12)
SFM	Scanning force microscope
SLeX ag	Sialylated Lewis X antigen
STM	Scanning tunnelling microscope
UV	Ultraviolet
VCAM-1	Vascular cell adhesion molecule 1
VLA-4	Very late antigen 4 (integrin $\alpha_4\beta_1$)

9.1 Introduction

In 1986 Binnig, Quate and Gerber introduced the first scanning force microscope (SFM) also known as an atomic force microscope (AFM) [1]. This AFM was derived from its precursor, the scanning tunnelling microscope (STM), developed by Binnig and Rohrer in 1981, for which they won the Nobel Prize in Physics in 1986. At the imaging level the AFM allows a resolution down to single atoms and subatomic features on a surface, i.e. an improvement of several orders of magnitude over the capacity of any conventional light microscopy. The AFM is a mechanical microscope the function of which is not affected by the optical diffraction limit, well known to be the limiting factor in classical light microscopy [2, 3]. Another application of AFM is the measurement of force-distance curves (Fig. 9.1). This force spectroscopic mode of operation has been routinely used to study even very weak forces in the range of pico-Newton (pN) and less. These are forces necessary to break atomic bonds, van der Waals forces, as well as forces needed to stretch and rupture biological molecules.

During the last two decades, the AFM has been applied in specialized fields linked to material sciences, applied physics and industrial processes. During the last

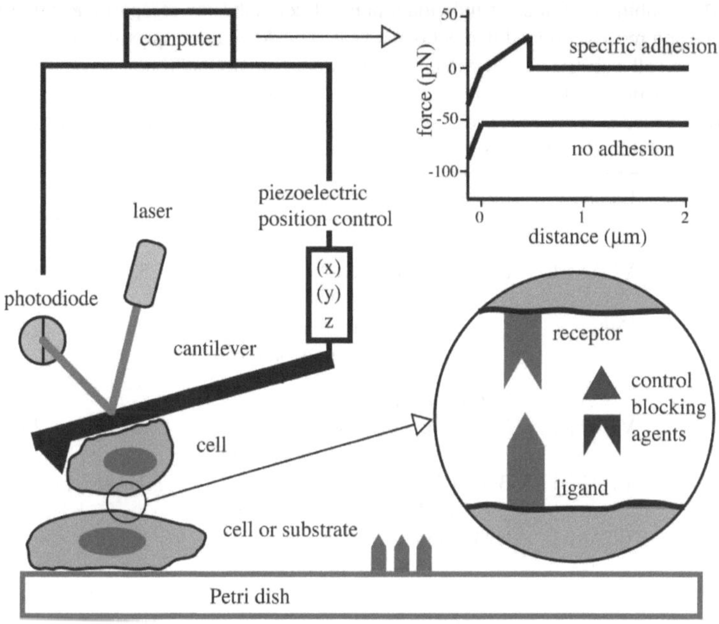

Fig. 9.1. Schematic setup of AFM measurements of a living cell

decade, however, the AFM has become a versatile tool in life sciences, where the measurements have to be adjusted for biological conditions. Applications vary from studies of unbinding forces of double-stranded DNA and stretching characteristics of isolated muscle titin proteins to measurements of viscoelastic properties of a living cell and the characterization of receptor-ligand bonds [4–15]. The specific interaction of biotin with streptavidin or avidin has been studied extensively and can serve as a reference model to guide AFM force measurements of isolated receptor–ligand interactions on a single-molecule level. The results obtained using other techniques estimating on and off-rates and energy landscapes of receptor ligand bonds can also be compared with AFM results as a reference [16–20] (Table 9.1).

Despite its enormous imaging resolution, the power of an AFM to directly measure cell adhesion forces brings the AFM force spectroscope into the focus of cell adhesion researchers, including immunologists and cell biologists, who are in general not very familiar with this technology. Many in the biophysical field still use the AFM on purified and often non-functional molecules which are studied isolated from the complexity of an intact living cell and not as highly motile members of an active cell membrane. For the critical reader it may appear rather challenging to measure the discrete unbinding force of a single pair of two distinct adhesion receptors, regardless whether these receptors were isolated and immobilized to a substrate or within a cell membrane of a large living cell. In 2000, Benoit and coworkers presented AFM measurements of the force needed to unbind the single adhesion bonds of homotypic receptors on a living amoeboid *Dictyostelium* cell with the identical, corresponding receptors on an adjacent amoeboid cell; this at a resolution of 23 pN

Table 9.1. Unbinding forces of individual pairs of cell adhesion receptors. Rupture forces of single adhesion receptor–ligand interactions revealed by AFM force spectroscopy on living cells (cell–cell, or cell–substrate interactions) and between isolated adhesion receptors (substrate–substrate interaction). Although it is not a pair of cell adhesion receptors the (strept)avidin–biotin system is included as a reference for one of the strongest single receptor–ligand interactions in biology, but not directly related to cellular adhesion

Receptor	Ligand	Interacting cell or substrate	Force (pN)	References
		Cell–cell		
VLA-4	VCAM-1	B16 melanoma cell –bEnd.3 endothelial cell	33	[9]
CsA	CsA	*Dictyostelium* amoeboid cell – *Dictyostelium* amoeboid cell	23	[30]
		Cell–substrate		
VLA-4	VCAM-1	RAW lymphoma cell – VCAM-1/F_C fusion protein	21–80	[4, 5, 31]
VLA-4	VCAM-1	B16 melanoma cell – VCAM-1/F_C fusion protein	21–45	Eibl (unpubl.)
VLA-4	VCAM-1	U937 monocytic cell – VCAM-1/F_C fusion protein	25–170	[24]
Integrin $\alpha_5\beta_1$	Fibronectin	K562 myeloid leukemic cell – protein	40–160	[26]
LFA-1	ICAM-1	3A9 lymphoma cell – ICAM-1/F_C fusion protein	100–300	[28]
Carbohydrates	ConA	NIH3T3 fibroblast–lectin	86	[32]
Integrin $\alpha_5\beta_1$	GRGDSP	Osteoclast–RGD peptide	32	[33]
Integrin $\alpha_V\beta_3$	GRGDSP	Osteoclast–RGD peptide	42	[33]
Integrin $\alpha_V\beta_3$	Osteopontin	Osteoclast–protein ligand	50	[33]
Integrin $\alpha_V\beta_3$	Echistatin	Osteoclast–protein ligand	97	[33]
		Substrate–substrate (related to cell adhesion)		
Integrin $\alpha_V\beta_3$	GRGDSP	Protein–RGD peptide	32	[34]
P-selectin	PSGL-1	Protein–protein	115–165	[35]
P-selectin	sLeX	F_C fusion protein – carbohydrate complex	40–200	[25]
VE-cadherin	VE-cadherin	F_C fusion protein – F_C fusion protein	15–150	[15]
Carbohydrate	ConA	Carbohydrate–lectin	35–65	[36]
Proteoglycan	Proteoglycan	Proteoglycan–Proteoglycan (from marine sponge)	400	[10]
		Substrate–substrate (unrelated to cell adhesion)		
HSA	Antibody	Protein–antibody	244	[37]
Avidin	Biotin	Protein–organic molecule	115–170	[17]

Table 9.1. (continued)

Receptor	Ligand	Interacting cell or substrate	Force (pN)	References
Avidin	Biotin	Protein–organic molecule	5–170	[20]
Avidin	Biotin	Protein–organic molecule	200	[38]
Avidin	Biotin	Protein–organic molecule	160	[19]
Avidin	Desthiobiotin	Protein–organic molecule	94	[18]
Avidin	Iminobiotin	Protein–organic molecule	85	[18]
Strepavidin	Biotin	Protein–organic molecule	257	[18]
Strepavidin	Iminobiotin	Protein–organic molecule	135	[18]

per average rupture. In their system, the specificity was obtained through the use of mutant cell lines that either could adhere to all other cells of the same sub line or did not adhere at all [21]. The non-adhering mutants did not show the characteristic rupture events. The authors then concluded that CsA-CsA adhesion receptors are indeed the interacting partners detected at the single-molecule level on the *Dictyostelium* surface. At that time, they could measure similar ruptures between mammalian cells, however, the identification of any specific pairs of mammalian receptors at the single-molecule level was not possible, because the cells studied were not well characterized and expressed more than one type of adhesion receptor on their surface [22]. Using well chosen cell lines and function blocking antibodies, Eibl and Benoit described in 2004 how to identify the unbinding forces of a specific pair of adhesion receptors between two living cells. Experiments described here, succeeded to directly measure—for the first time between two living mammalian cells—the rupture force of single bonds between the integrin very late antigen 4 (VLA-4) and its receptor, the vascular cell adhesion molecule 1 (VCAM-1), either protein expressed on B16 melanoma cells or on bEnd.3 endothelial cells, respectively [9].

It seems technically easier to detect unbinding forces of the interaction between a given pair of adhesion receptors fixed to a solid support, than those necessary to separate the same pair embedded in membranes of living cells. The potential difficulties to confirm the specificity of adhesion events when using two living cells could be reduced by using just one cell together with a corresponding and functional counter-receptor on an opposing surface. Meanwhile, integrins such as LFA-1, VLA-4, integrin $\alpha_5\beta_1$ and other cell adhesion receptors, for example, cadherins, have been studied using just one single cell expressing those receptors and the corresponding ligands either immobilized to the surface of a Petri dish or, alternatively, bound onto the AFM cantilever [15, 23–29] (Table 9.1).

Most of these reports include measurements performed under different conditions, as they are allowing variable off-rates by changing the velocity of the cantilever approaching the probe and retracting from it. As these reports describe, the experimental setup did allow for ambient temperatures and pH changes. Both parameters, temperature and pH, are known to be critical for the integrity of integrins and many other cell adhesion receptors. Such unphysiologic conditions did not prevent the cells adhering and reasonable single bond ruptures could be measured. One could conclude that these conditions may either not be too critical, or may even support the measurement at the single-molecule level, as the cell becomes less sticky with a reduced number of functional receptors on its surface. Furthermore, AFM measurements at room temperature may be much easier to perform than under conditions of higher temperatures, since the cantilever itself is sensitive to changes in temperature and to faster moving water molecules which can result in thermal drifts and higher background noise, respectively. Nevertheless, measuring the physiologic activation of integrin VLA-4 by the chemokine SDF-1 requires a higher temperature close to cell culture conditions of 37 °C as well as a constant and physiologic pH (Eibl, unpubl.) [4, 8]. An intact and viable cell allows for studies of single molecules in their native environment, in which different receptors embedded in the cell membrane can influence each other and be influenced by signalling molecules from outside and from inside the cell, for example, for inside-out and outside-in signalling or integrin–chemokine crosstalk. Purified full-length transmembrane molecules (what integrins are) are often difficult to study in the absence of membrane structures because hydrophobic transmembrane domains can then aggregate and the proteins may become insoluble.

More than often, the proteins loose their functional conformation. It is reasonable to experiment with purified deletion variants of these receptors, or to assemble a system with purified proteins and artificial and supported lipid bilayers. Although it has been shown that integrin receptors [39] as well as chemokine receptors [40] can be integrated in artificial lipid bilayers, it may take many more years to replace a living cell's membrane with an appropriate, integrin-coated artificial cell membrane, which will allow studies of the physiologic process of rapid SDF-1 triggered VLA-4 activation.

This activation step is part of the multi-step mechanism of lymphocyte homing, which has been studied since the 1980s and described independently by Butcher [41, 42] and Springer [43], who both were honoured with the Crafoord Prize of the Royal Swedish Academy of Sciences in 2004. Basically, lymphocytes and other white blood cells can use in health and disease a set of cell adhesion receptors for migrating to distant sites within the body, especially to attach to the blood vessel endothelium and resist shear forces imposed by the blood stream, this, just before they extravasate into the surrounding tissue of, for example, lymph nodes or Peyer's patches. Part of this homing process are discrete steps which comprise the initial tethering and cell rolling on the wall of a blood vessel, then the migrating cells arrest and stick to the endothelial monolayer, before they extravasate into the regional tissue (Fig. 9.2).

The initial tethering and rolling steps are mediated by selectins and a few integrins, such as VLA-4 or LFA-1. These integrins finally can mediate the firm adhesion

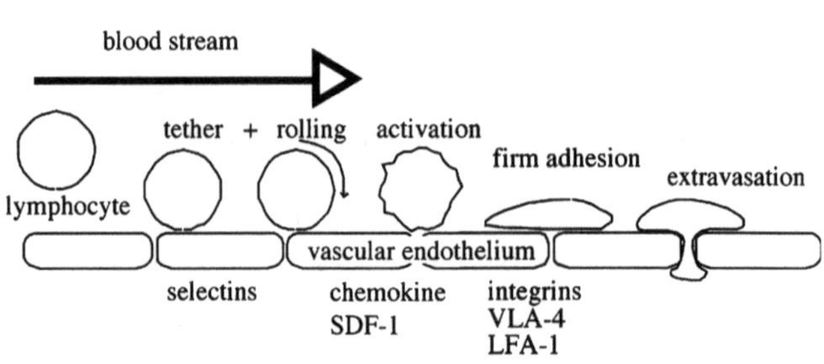

Fig. 9.2. Multi-step adhesion model of circulating and homing lymphocytes

or sticking onto the blood vessel wall. Interestingly, each of these few specialized integrin receptors, in particular VLA-4 and LFA-1 and their cognate ligands, VCAM-1 and ICAM-1, have been implicated in lymphocyte rolling as well as in the subsequent rapid arrest or sticking. It has been speculated that the contribution of integrins to this rolling phenomenon is that the integrin receptor functions as a fast low-affinity/high-affinity switch. A chemokine like SDF-1, usually presented by endothelial cells, can trigger the switch within milliseconds via its surface-bound chemokine receptor CXCR4, what then can lead to the rapid arrest of a rolling lymphocyte [44, 45]. More than a century after rolling cells were discovered in a fish fin, the knowledge about mechanisms and receptors involved in this phenomenon has become immense and has deepened significantly during the past three decades. Today's core methods include parallel-plate or capillary-based flow assays [46] and even intravital microscopy techniques used in mice and rats harbouring fluorescently labelled lymphocytes [47, 48]. The latter assays allow live imaging of rolling and sticking lymphocytes. Of course, many other methods have been developed during the past century to study the expression pattern and to characterize the functions of cell adhesion receptors. Receptor-specific antibodies or labelled ligands are routinely used in immunohistochemistry (IHC) and in cell sorting (e.g. fluorescence-activated cell sorting, FACS) [49, 50]. However, not every receptor specific antibody also blocks an adhesion receptor, therefore, the development of function blocking antibodies for each receptor is important for functional studies, especially on living cells. One functional assay, originally designed by Stamper and Woodruff [51], combines, for example, lymph-node tissue sections and lymphocytes as an in vitro model of lymphocyte homing. Lymphocytes which were layered over fixed sections of lymph nodes adhered selectively to specialized vascular endothelia. This assay can be applied to functionally characterize relevant pairs of homing receptors of circulating lymphocytes by using function-blocking antibodies. Such antibodies inhibit the adhesion of lymphocytes to specialized high endothelial cells on a lymph node section. The unbound or weakly adhering lymphocytes then can't resist applied centrifugation forces and are washed away. In another assay, the parallel-plate flow chamber assay, intact cells have to resist via their adhesion receptors the shear forces which are mainly dependent on the speed of the flow. This force imposed upon a cell is measured in $\mathrm{Dynes/cm^2}$ and is characteristic for a cell and the receptors studied. The migration properties of lymphocytes and other white blood cells can be studied in rolling chamber assays [46]. All red and white blood cells circulating within the blood stream are prevented from adhering to each other and to endothelial cells in order to avoid thrombus formation. However, a prerequisite to evade the blood stream is that actively migrating white blood cells need to use or activate adhesion receptors to initiate homing. An interesting note on the side is that metastatic cancer cells and adult stem cells may also share some of the lymphocyte adhesion receptors for their hematogenous spread to distant sites from a primary tumour or a metastatic lymph node [52, 53]. The sensitivity of all of these methods is far away from single-molecule techniques and usually depends on many functional receptor molecules per cell. To detect rarely expressed molecules like the chemokine receptor CXCR 4 on a cell's surface with immunohistochemistry, an additional signal-amplification process has to be included to increase the signal or signal-to-noise ratio. Although these methods are powerful in the investigation of cell-adhesion receptor function, none of these methods had allowed the direct measurement of cell adhesion forces on a single-molecule or single-receptor level and within the native environment of such a complex system as a living cell. Therefore, single-molecule AFM force spectroscopy is well positioned to become one method of choice when characterizing the regulation of cell adhesion receptors on living cells including migrating lymphocytes, adult stem cells and metastatic cancer cells.

The goal of this chapter is to bring the powerful method of single-molecule AFM force spectroscopy on living cells to a broader audience and to inform interested researchers and students, including cell biologists, cancer researchers and immunologists, in general not familiar with AFM-based force spectroscopy and its recent applications in life sciences. In addition, this chapter is of great interest for advanced AFM and biophysics experts, who have already gained experience in applying AFM to the measurement of single-molecules, but who are looking forward to using a living cell to study and manipulate single molecules within their native environment. Several supplementing step-by-step protocols will also guide and inspire fully advanced AFM and immunology enthusiasts. The protocol for activation measurements of integrin VLA-4 by the chemokine SDF-1 on a living cell and at the single-molecule level, as described in this chapter, can currently be considered as one of the most detailed descriptions of the method published in this field. The protocols are designed to help modify and adapt this method to similar questions for other cell types and adhesion receptors, as well as for many other physiologic or artificial activators or inhibitors. The principles described here need not just be applied to higher eukaryotic cells, but can easily be adapted for use with lower eukaryotic (e.g. yeast) and prokaryotic cells or combinations thereof, like for studies of the binding of bacteria to macrophages.

9.2 Procedure

When designing AFM experiments to measure cell–cell or cell–protein interactions, decisions have to be made about (a) which pair of adhesion receptors or set of molecules are of interest and fit, (b) which cells or cell lines express the relevant receptors on their surface, and (ideally) lack other, possibly interfering, adhesion receptors, (c) whether to use function-blocking antibodies, other inhibitors or activators in function control experiments, or agents in combination, (d) what the optimal experimental conditions (temperature, pH, growth medium) are for the selected cell culture, and (e) what the best AFM setup (including cantilever, its functionalization and retraction distance) is to comply with the cell culture conditions in order to obtain a good signal and signal-to-noise ratio. Note: I recommend refraining from (a) performing experiments with cells and biomolecules at suboptimal ambient temperatures and (b) using growth media with an unphysiologic or changing pH value. The pH of CO_2/HCO_3-buffered media can fluctuate dramatically if the medium is left unchecked for evaporation. Adding different amounts of Hepes buffer does not prevent CO_2 from leaving the media which appears to be critical for some adhesion measurements. Most cell lines can be adapted to more expensive, CO_2-independent medium which I recommend for AFM measurements of living cells.

9.2.1 Principle of AFM Force Spectroscopy

The AFM has the extraordinary capacity to measure single-molecule interaction forces (pico Newton) and displacements (nanometer) at the nanoscale level. The key part of the AFM is a microfabricated spring or cantilever. Usually, sample molecules are coated to the small pyramidal tip at the end of the cantilever and probed against other molecules on an opposing sample stage. Alternatively, a cell can be attached to one or both contacting parts of the AFM, i.e. to the end of the cantilever and/or to the bottom of a cell culture Petri dish (Fig. 9.1). By moving the cantilever (or the stage), both samples can be brought into contact. A piezoelectric element controls the movement of the cantilever in the direction of the z-axis (perpendicular to the sample stage) and is required to detect developing forces. By controlling the scanning in the x and y direction, additional piezoelectric elements provide data for the creation of three-dimensional AFM images. The combined data (position of the tip and measured force) allows pinpointing the location of different adhesion areas on a cell surface. AFMs used in cell adhesion experiments usually need a piezocontrol element that retracts the cantilever by up to 15 μm or more from the sample contact point. Manually controlled micrometer screws can be used for the spatial adjustment of the cantilever tip. A standard light microscope (not shown), to which a video camera is attached, is placed underneath the sample stage of the AFM to monitor the fishing of the cell onto the tip of the cantilever and to focus the laser beam onto the cantilever. The laser is deflected from the flip side of the sensor tip into the centre of a photo detector. The two segments of the photodiode are needed to adjust the laser and to measure the magnitude of deflection to one or the other half during the force scans.

The sample interaction force leads to bending of the cantilever and that alters the deflection angle of the laser beam and the location where the beam hits the photodiode. In order to translate the recorded deflection of the laser beam into a force scale, a proportionality constant—the spring constant of the cantilever—has to be determined. Although the manufacturers provide a nominal value of the spring constant, each cantilever should be calibrated since the spring constant may vary significantly. An estimation method based on the cantilever's thermal fluctuation has been proven useful [6,54]. For experiments with living cells soft and long cantilevers are required to achieve a spring constant between 10 and 15 mN/m. Lower constant values of up to 5 mN/m can be achieved by breaking away one of the two arms of a delta-shaped cantilever using a hand-held forceps under low magnificaton [9,55]. A series of force distance curves is recorded as a loop of forward and backward scans. Two examples of forward and backward scans are shown, one without any rupture event, the other with a single force step of 33 pN (Fig. 9.3). Specific adhesion events can be translated into rupture forces and further analyzed in a force spectrogram.

9.2.2 Cell–Cell Interactions

In order to measure cell–cell adhesion or deadhesion events, cells have to be immobilized on both sides of the probing instrument, for example, onto the tip of the cantilever and on the bottom of the Petri dish (Figs. 9.1 and 9.4). Although it is easy

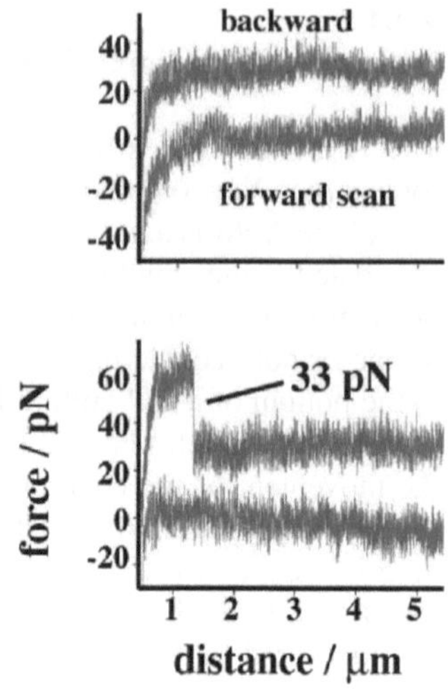

Fig. 9.3. AFM force scans of cell–cell interactions. Force-distance diagram of forward and backward scans. In the case of no adhesion after short contact of a B16 melanoma cell to a bEnd.3 endothelial cell, both, the backward and forward scans look similar and show the indentation force (40 pN) superimposed by the thermal noise, whereas an adhesion event of about 33 pN can be easily detected by the step in the backward force scan

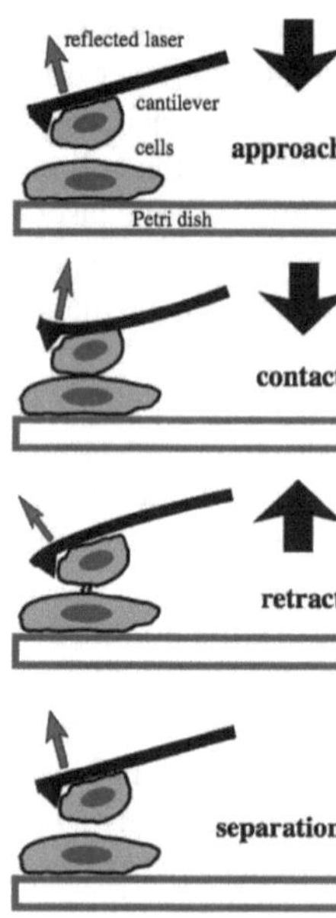

Fig. 9.4. Scanning steps in AFM-based force spectroscopy for cell–cell adhesion measurements

to grow most cell types on a Petri dish to which they adhere, it is more difficult to attach a living cell onto the tip of a cantilever and to maintain physiologic cell culture conditions for a cell that, perhaps, remains attached for up to several hours. In many cases, it has been useful to functionalize the cantilever with poly-D-lysine in order to catch a single cell onto the tip of the cantilever. A more detailed protocol for functionalizing the cantilever with concanavalin A (conA) is provided at the end of this chapter. In order to mimic the homing conditions of rolling and rapidly arresting lymphocytes, the ideal cell type for adhesion studies with AFM were several lymphoma cell lines expressing VLA-4 and an endothelial cell line expressing VCAM-1. The endothelial cells grow and attach strongly onto the plastic surface of a common Petri dish, whereas the lymphoma cell usually grows as a single cell or within a cluster of a few cells in suspension. Some lymphoma cells may grow weakly attached to a cell culture flask but can easily be separated by gentle shaking. Instead of placing a lymphoma cell onto the cantilever it is also possible to use a cell growing firmly attached to a cell culture flask. Then, mild trypsinization, Ca^{2+} ion removal by short exposure to a chelating agent, such as EDTA, or using both methods in combination, can separate even strongly attached cells. After separation of such cells, the attachment of one of these cells onto the tip of the cantilever should be done quickly, since the cell may bind firmly to the bottom of the Petri dish. A few separated cells are placed into the measuring Petri dish and the functionalized cantilever is placed above an unbound cell and smoothly lowered onto the top of the cell, monitored by an optical microscope placed underneath the Petri dish. After a few seconds of contact the cell adheres to the cantilever, which is then retracted and placed above firmly bound endothelial cells or a region of immobilized receptor proteins or a control region, respectively (Fig. 9.5).

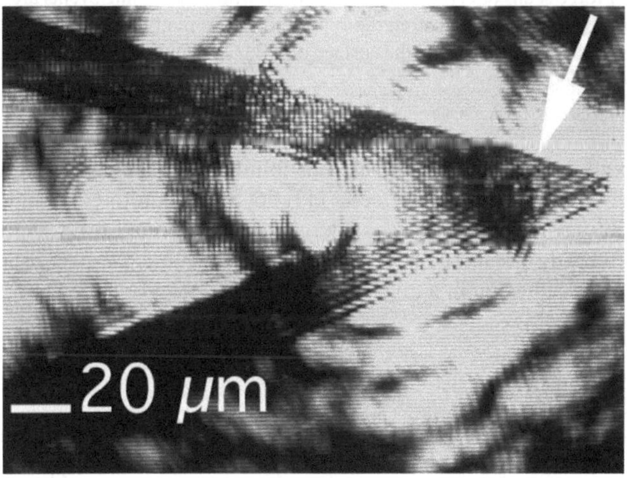

Fig. 9.5. Placing a cell on the cantilever. Photomicrograph of a living cell attached to the tip of an AFM cantilever previously functionalized with conA

9.2.2.1
Multiple Bond Ruptures

Different parameters can influence the quality of cell–cell adhesion contacts. Not only the cell type, the activation state (affinity) or the density (avidity) of the receptors involved, but also the approaching and retracting velocities, the contact time and force, and the detachment force may influence the adhesion strength. Choosing a longer contact time and higher contact forces before separating two cells usually leads to the detection of multiple bond ruptures (Fig. 9.6). This method can already

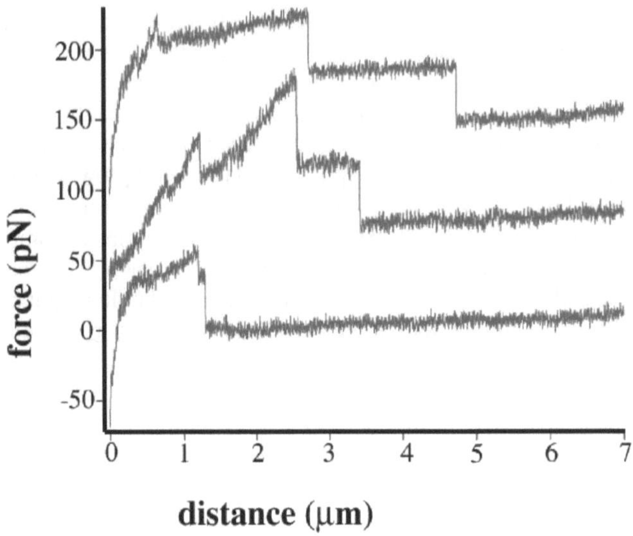

Fig. 9.6. AFM force scans of cell–cell interactions. Multiple unbinding events are shown in each of the three backward scans, indicating the rupture of more than one pair of VLA-4–VCAM-1 receptors, each of them expressed on the surface of a B16 melanoma cell or a bEnd.3 endothelial cell, respectively

be used to compare differently activated or inhibited cells or any drugs which may influence the adhesive strength of cells [11, 56, 57]. Therefore, the integrated area below the force-distance curve is considered as work force. Lower and higher work forces of differently treated cells allow conclusions about the overall adhesiveness and regulation of adhesion in a given cell.

9.2.2.2
Single Bond Ruptures

To directly measure single-receptor interactions between two living cells, several conditions have to be optimized, but they may vary between cell types and cell activation states, receptor types, receptor activation states and receptor densities on the cell surface. Adapting the contact force and duration of contact (e.g. 30 pN, 200 ms) to conditions which lead to adhesion events of 33% or less are considered to represent a probability of 83% or 91% to detect single-molecule rupture events [18, 21] (Fig. 9.7).

9.2.3
Cell-Substrate Measurements

For most cell adhesion studies, the measurement of receptor–ligand interaction between two living cells appears to be too complicated and not required, especially since a second cell may add a number of possible interaction partners not related

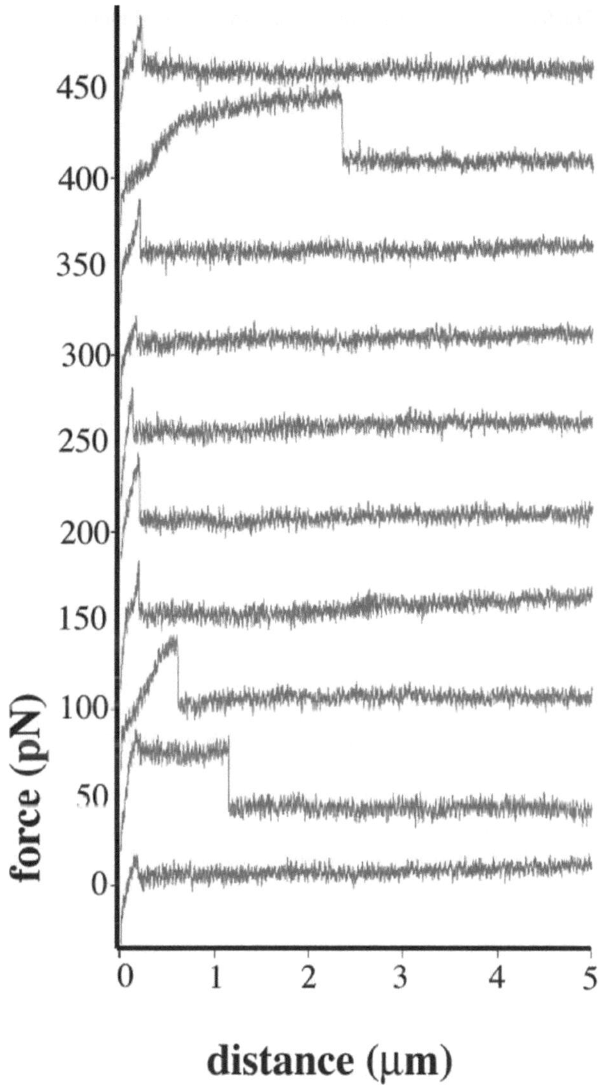

Fig. 9.7. AFM force-distance scans of cell–cell interactions optimized for single-molecule interactions. Ten consecutive adhesion events are shown with only a single rupture (no adhesions left out). The unbinding can occur at different distances reflecting the different length of tethers

to the receptor of interest on the first cell. Therefore, it is reasonable to study a receptor of interest in its native environment of a living cell and to probe it against a suitable counter-receptor immobilized on a surface (Fig. 9.8). This may help to increase the specificity and sensitivity, since a specific binding of additional ligands to the receptor of interest or binding of additional pairs of receptors is widely excluded, and non-specific binding, for example, of charged proteins to the probing surface is reduced. Nevertheless, it is still possible that a single type of immobilized ligand can recognize more than one or two types of counter-receptors on a cell. There appears to be no need for time-consuming control experiments with blocking antibodies. An experimental approach with just one living cell and immo-

Fig. 9.8. Scanning steps for cell–substrate measurements

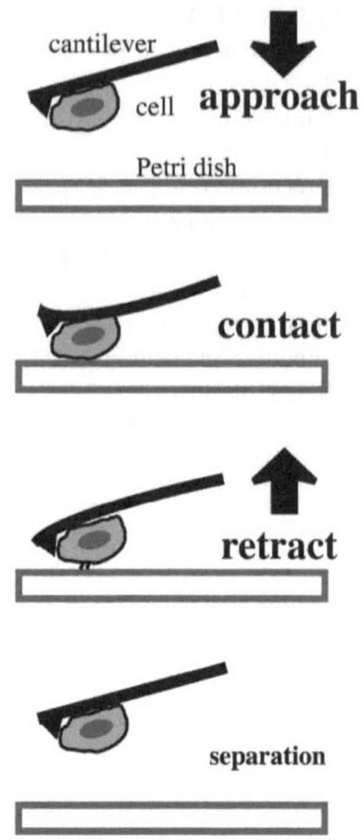

bilized counter-receptor appears to be ideal for most purposes addressing adhesion receptors and signalling cascades, such as the integrin–chemokine crosstalk, which hardly could be addressed in a cell-free system. In principle, the cell could be placed either on the cantilever or on the Petri dish, and vice versa, the isolated counter-receptor. Growing a cell firmly attached on the bottom of a Petri dish and functionalizing the probing cantilever with the suitable counter-receptor has the advantage of reducing cell culture procedures, but may result in otherwise stressed cells, since poking with a pyramidal tip of the cantilever onto the cell can damage the cell. On the other hand, such a procedure is used in studies of cell viscoelasticity, where the cells apparently survive the procedure. It often appears to be much easier and suitable to functionalize the bottom of a Petri dish with critical proteins and binding the cell to a cantilever rather than funtionalizing the pyramidal tip of a cantilever, since most of the protein receptors rapidly denature or are otherwise hard to bind to the cantilever.

9.2.4 Specificity and Blocking Antibodies

Highly specific function-blocking antibodies towards a number of cell adhesion receptors have been developed and are used to prevent cell adhesion to other cells or to the extracellular matrix. These antibodies, or any other inhibitors, such as small and fully synthetic function-blocking molecules, for example, with a sugar backbone, can bind to receptors, and thereby prevent its binding domain from recognizing and binding to its ligand. They exert their function either by binding directly to the binding domain, by masking the domain, or by inducing conformational changes in the receptor.

As shown with rolling chamber assays, lymphocytes, drawn through the chamber by otherwise cell-free medium, can roll and stick by adhering through their integrin VLA-4 receptors to endothelial VCAM-1 ligands; this can also be observed when analyzing metastasizing B16 melanoma cells (Eibl, unpublished results). In a rolling chamber assay, mimicking physiologic blood stream conditions, not only lymphocytes can roll and stick via their integrin VLA-4 receptors to endothelial VCAM-1 ligands, but also metastasizing B16 melanoma cells (Eibl unpubl.) (Fig. 9.9). Rolling and sticking of B16 melanoma cells was not observed when function-blocking antibodies against either VLA-4 or VCAM-1 were added, but did occur when control antibodies were used. In order to compare the results obtained from the rolling chamber assays, and to highlight the potential of single-molecule AFM force measurements, the complex system of B16 melanoma cells rolling on activated bEnd.3 endothelial cells was reconfigured for the analysis in an AFM.

The B16 melanoma cell can now be fixed onto the tip of an AFM cantilever and can be brought into contact with the endothelial cells on the sample stage for a set period of time. By manipulating cell–cell contact time and cantilever approaching and retracting velocities the touch-and-go rolling characteristic binding and unbinding events can now be re-examined, at least in part, in an AFM.

A typical force spectrum of single interactions between two living cells is shown (Fig. 9.10): a single B16 melanoma cell bound to the cantilever and a single bEnd.3 endothelial cell on the Petri dish. The adhesion frequency was adjusted to 33%, i.e. in 67% of all 960 cell–cell contacts binding and subsequent rupture events were not detected (Fig. 9.3). These 67% of force-distance curves (0–5 pN) showing no apparent rupture events were omitted from the histogram (0–5 pN). In this example the

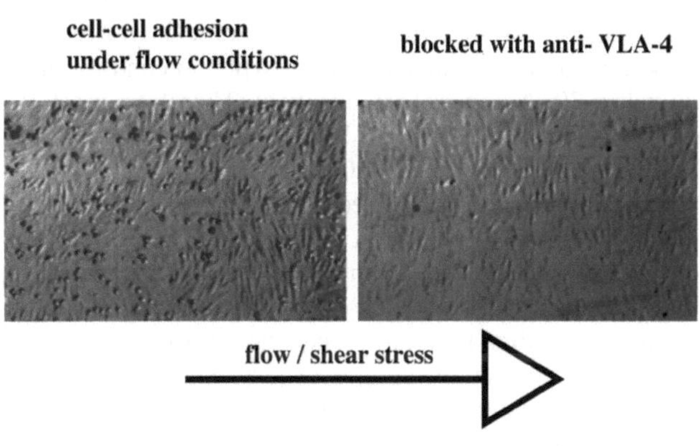

Fig. 9.9. VLA-4–VCAM-1 mediates cell–cell adhesion under flow conditions. Under physiologic shear stress in a parallel-plate flow chamber single B16 melanoma cells roll and stick on a monolayer of activated bEnd.3 endothelial cells (*left* photomicrograph). Function-blocking antibodies against VLA-4 can completely block this receptor–ligand interaction (*right*). Because of the speed of the flow, unbound melanoma cells are only visible as long and weak shadows across each picture frame

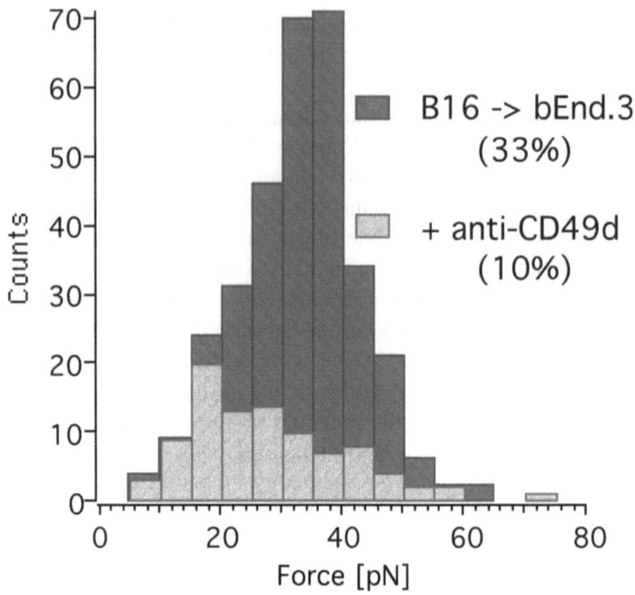

Fig. 9.10. Single-molecule AFM force spectrum of cell–cell interactions identifying VLA-4–VCAM-1 receptor-ligand bonds. Two typical histograms of AFM force measurements are shown, one of them including a function-blocking antibody against the VLA-4 adhesion receptor on the B16 melanoma cell (*lower* colloms) which leads to a significant reduction of binding events to the bEnd.3 endothelial cell

average unbinding force is 33 pN (standard deviation 12 pN). As expected results from blocking experiments performed in rolling assays show that antibodies, blocking either VLA-4 or VCAM-1, respectively, but not control antibodies, reduce the binding frequency and the average rupture force significantly. Blocking anti-VLA-4 antibodies block most of the interactions above 20 pN. In contrast, the average force and frequency below 20 pN appear to be identical. Therefore, most, if not all interactions around 33 pN can be attributed to VLA-4–VCAM-1 interactions. Therefore, as in the rolling chamber experiments, B16 melanoma cells can interact with bEnd.3 endothelial cells via one major pair of receptors, VLA-4 and VCAM-1, expressed on the surfaces of the interacting cells. In both experimental setups, this interaction can be blocked with inhibitors, such as function-blocking antibodies, supporting the ability of AFM force spectroscopy to measure very specific interactions between two living cells. Finally, according to the low frequency of rupture events resulting in a very high probability of detecting ruptures of just one pair of adhesion receptors, these rupture forces can be attributed specifically to VLA-4–VCAM-1 interactions. This view is supported by the lack of lower and higher peaks, which was observed under suboptimal conditions in amoeboid *Dictyostelium* [21].

The identical frequencies of single rupture events around 20 pN and below in both experiments can easily be attributed to non-specific binding of charged proteins on a cell's surface. On the other hand, it may point to another cell adhesion receptor that binds less frequently and at lower forces under the conditions optimized for

VLA-4–VCAM-1 interactions. Interestingly, under certain circumstances, for example, while testing cells with stronger and more rolling receptors such as VLA-4, other rolling mediating receptors known to be weak, such as CD44, may totally be masked in a rolling chamber assay or intravital microscopy experiment [58, 59]. Therefore, some other unknown rolling receptors may be masked in many of the standard experiments, even when blocking of a main receptor leads to a total inhibition of rolling and adhesion under flow conditions. AFM force spectroscopy may help to discover new adhesion receptors or new functions of adhesion receptors that currently escape other methods. Additional AFM force experiments may test some of the paradoxical functions of cell adhesion receptors such as VLA-4 and CD44 in normal development and disease [50,52,60–64]. Under some circumstances, VLA-4 mediates homotypic aggregation of cells which may reduce the migration potential of tumour cells to leave the primary tumour, which could result in a lower metastatic potential [62], but tumour cells entering the blood stream and expressing significant amounts of VLA-4 on their surface can use VLA-4 in at least one mechanism of metastasis (Eibl unpubl.) [64]. The AFM force measurements may help to identify a new, but weak receptor or a new and underestimated variation of a CD44 variant as a very common cell adhesion, lymphocyte homing and tumour metastasis receptor. This CD44 receptor is highly expressed on B16 melanoma cells but appears to be totally inactivated on all cells tested in standard rolling experiments (Eibl unpubl.).

9.2.5 Activation by SDF-1

AFM force measurements can be used to directly measure VLA-4–VCAM-1 interactions, either between two living cells [9,21] (Figs. 9.4, 9.6, 9.7, 9.10), or between one cell and immobilized VCAM-1 fusion proteins [8] (Figs. 9.11, 9.12, 9.13). Rolling experiments identified VLA-4 to mediate the rolling and sticking parts during the multistep mechanism of lymphocyte homing [41,43,65]. Usually, the integrin VLA-4 has to be triggered by a chemokine, which binds to its corresponding chemokine receptor, for example, SDF-1 (= CXCL12) and CXCR4, respectively [44,45]. This chemokine-mediated triggering is expected to change within milliseconds the activation state from a low-affinity to a high-affinity VLA-4 receptor, with both states recognizing their ligand VCAM-1, but with different characteristics. Direct AFM force measurements appear to have a unique potential to study such an immediate activation within the context of a living cell. In the following experiment, a VLA-4 expressing lymphoma cell is bound to the tip of the cantilever (Fig. 9.8) and different areas of the same Petri dish are functionalized with recombinant VCAM-1 fusion protein, as well as with controls, including LFA-1 ligand ICAM-1 and other control substances. Under conditions of higher adhesion frequencies single and multiple rupture bonds appear per backward scan (Fig. 9.11). Adding soluble SDF-1 many more rupture events appear per backward scan, as well as higher rupture forces at the last rupture of each scan (Fig. 9.11). This indicates that soluble SDF-1 can change the adhesive behaviour of a living cell in a setup of AFM force measurements without the need for intravital microscopy in a whole animal or in a parallel-plate chamber

Fig. 9.11. Effects of chemokine SDF-1 on AFM force scans of VLA-4–VCAM-1 cell–substrate interactions. Before activation (*above*), and after activation with the chemokine SDF-1

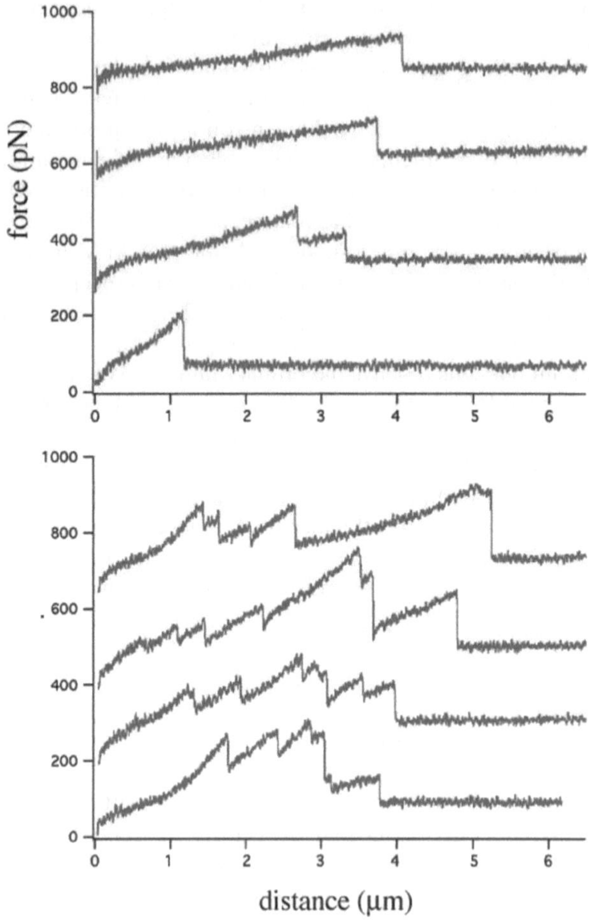

assay. Since the integrated area below the force-distance curve is larger after the addition of SDF-1 the mechanical energy required per detachment event is considerably increased. It is beyond the scope of this chapter to discuss more applications of easier to measure multiple rupture bonds and AFM applications such as biosensors in drug development. Such minimal changes in the workforce detected by AFM force measurements may be highly indicative for a good candidate in the screening of substances involved in inflammatory and autoimmune responses.

To preferentially measure the rupture of single VLA-4–VCAM-1 bonds instead of multiple bond ruptures, the critical parameters have to be adjusted to frequencies less than 33%. For pharmacological dose–response experiments it is reasonable to choose much lower frequencies of 2–10% as the starting frequency. Otherwise, SDF-1 may lead to multiple ruptures per bond. From many similar experiments, one representative experiment with increasing concentrations of SDF-1 is shown in eight typical force histograms (Fig. 9.12). The average forces were plotted in a dose–response diagram (Fig. 9.13). This experiment shows a limited number of rupture scans per histogram, since it is optimized for a higher number of increasing

Fig. 9.12. Affinity modulation of single VLA-4 receptors on a living cell by SDF-1. AFM force histograms of a lymphoma cell expressing integrin VLA-4 probed on immobilized VCAM-1 fusion protein. Increasing SDF-1 concentrations show a shift of the rupture forces to higher forces, then a shift back to normal

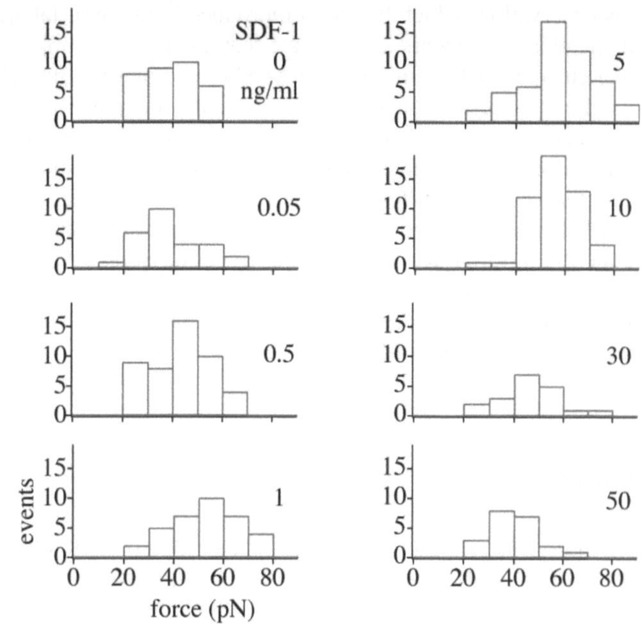

Fig. 9.13. Dose–response effects of chemokine SDF-1 revealed by AFM force spectroscopy. The dose–response curve shows the mean value of forces from the histograms shown in Fig. 9.12. The peak of unbinding single VLA-4–VCAM-1 adhesion bonds at physiologic conditions indicates a physiologic switch to a higher affinity state of the integrin VLA-4

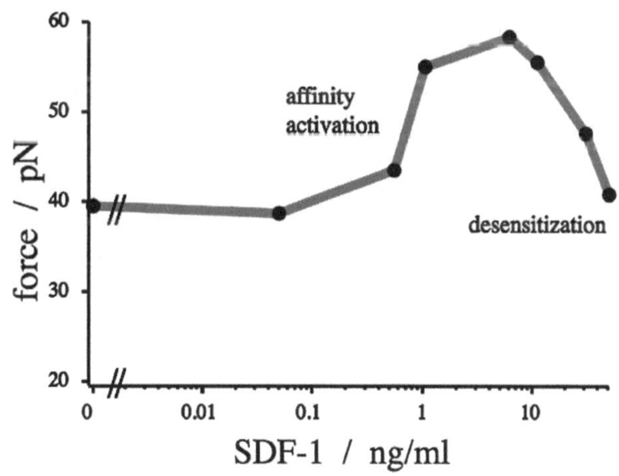

concentrations of SDF-1 to cover several orders of magnitude in the chemokine concentration, from no addition of SDF-1 up to physiologic levels, and above. Resting cells, i.e. cells with no addition of SDF-1, as well as cells at sub-physiologic concentration (0.05 ng/ml) show a force spectrum with rupture forces mainly between 20 and 60 pN. At physiologic concentrations of SDF-1 at 1–5 ng/ml, the rupture forces are shifted to higher levels of 40–80 pN. At higher concentrations of the chemokine, the average rupture forces shift back to 40 pN, i.e. to a level similar to that prior to the activation. Although there is some legitimate scepticism about the interpretation of this observation, which has been reproduced in several laboratories (personal communication), it appears safe to state that an arrest chemokine can induce quick cellular changes under conditions of AFM force spectroscopy. There is robust evidence that the force resolution at a single-molecule or single-receptor level is not only possible between isolated or "dead" receptors, but also between one living cell and an immobilized ligand and even between two living cells. Therefore, it may appear plausible that the force difference of low-affinity versus high-affinity integrin receptors could also be measured by AFM force spectroscopy. Future studies will have to correlate the different aspects of direct force measurements versus energy landscapes and off-rates of receptor-ligand binding, although for the immunologist such comparisons may not be necessary as long as a directly measured force can be translated into the function of an adhesion receptor.

9.3 Protocols

The following five step-by-step protocols should serve the interested student, biophysicist or cell biologist to choose the right experimental setup for their own needs. The protocols are designed to complement each other and should be easy to adapt to similar questions of cell adhesion and candidate drug screening. Starting with the functionalizing of the cantilever to attach a cell of interest onto its tip, the protocols continue with the description of using such an attached cell to probe it on another cell or on counter-receptors immobilized at different densities, and on suitable controls. A further protocol shows how to inhibit a specific cell adhesion binding by antibodies and other inhibitors which might be useful for the identification and prove the existence of specific interacting adhesion receptors. A state-of-the art protocol is included that allows us to measure the effect of potential adhesion activators such as a chemokine, a rare activating antibody or phorbol esters and other substances. Finally, a protocol for cell-free receptor ligand binding is provided.

9.3.1 Cantilever Functionalization

In order to probe a living cell with another cell or against an immobilized substrate the cell should be well attached to the tip of the cantilever (Fig. 9.14). The following protocol is more elaborate than the alternative protocol in which the cantilever is just incubated in poly-D-lysine for 30 min at RT and washed once with PBS. It has been shown that a very careful preparation of the cantilever pays off, and, therefore, this extended protocol is recommended for work on many cell types. However, the protocol includes the use of conA and if the protocol is being used on immune cells, one has to consider that conA has the potential to activate many different pathways in a living cell. Therefore, depending on the cell type and the underlying research question other procedures may need to be established first in the laboratory. This protocol can easily be adapted to functionalization procedures much more specific

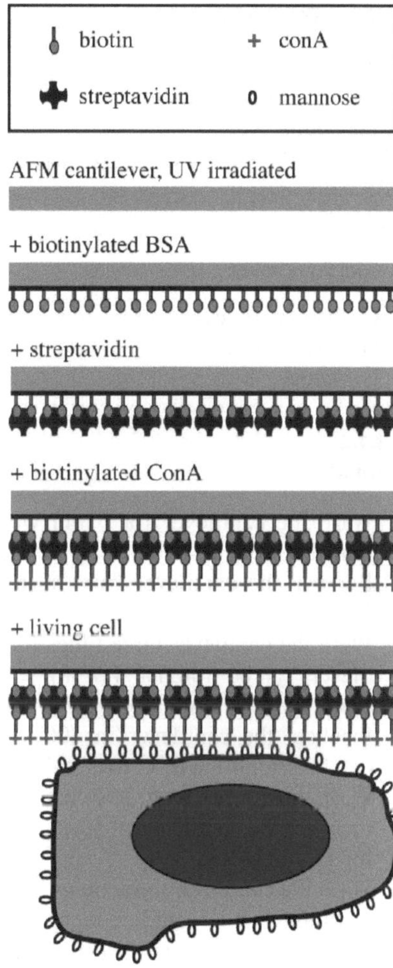

Fig. 9.14. Scheme of immobilizing an intact living cell to the AFM cantilever

for the cell to be attached. For example, biotinylated conA could be replaced by biotinylated antibodies recognizing and binding to the surface of the investigated cell ubiquitously expressed specific receptors. This may include antibodies against specific adhesion receptors, such as CD43, CD44, and similar receptors on white blood cells, but many of them are also involved in lymphocyte activation processes which may interfere with the original question [66, 67]. Therefore, several strategies should be applied when trying to answer specific questions about lymphocyte activation.

9.3.1.1
Materials

Caution: Refer to your laboratory guidelines for appropriate handling of material marked with <!>, i.e. flammable or hazardous material.

Buffers and Solutions

Acetone <!>
Biotinylated BSA (Sigma)
Streptavidin
Biotinylated concanavalin A (conA; Sigma) <!>
PBS
$NaHCO_3$

Special Equipment

- Unsharpened cantilevers (e.g. model MLCT-AUHW; Thermomicroscopes, Sunnyvale, CA)
- UV-lamp

9.3.1.2 Method

1. Wash the cantilever for 5 min with acetone in a glass (!) Petri dish under a fume hood.
2. Remove the cantilever from acetone bath and irradiate under a UV light for 15 min.
3. Wash the cantilever three times for 5 min in 0.01-M PBS, pH 7.4. Cleaned cantilevers can be stored at 4 °C for several weeks.
4. Wash once for 5 min in 0.1-M $NaHCO_3$, pH 8.6.
5. Incubate the cantilever in 100-μl biotinylated BSA (0.5 mg/ml) for 12–16 h (e.g. overnight) at 4 °C (or for 4 h at RT).
6. Wash three times with 0.01-M PBS, pH 7.4.
7. Incubate the cantilever in 50-μl streptavidin (0.5 mg/ml in 0.01-M PBS, pH 7.4) for 15 min at RT.
8. Wash the cantilever three times in 0.01-M PBS, pH 7.4 before use.
9. A biotinylated ligand recognizing receptors on the cell may be coupled to the streptavidin-functionalized tip (e.g. biotinylated concanavalin A; conA, 0.5 mg/ml, 15 min at RT).

9.3.2 AFM Measurement on Living Cells

This protocol describes in detail the preparations needed to succeed in extremely sensitive cell measurements with living cells. As an example a Petri dish is prepared with different densities of immobilized ligands including VCAM-1 fusion protein and useful controls [6]. Then, the AFM is assembled and the functionalized cantilever from the previous protocol used after its calibration to attach a cell onto the tip of the cantilever. Finally, the AFM can be used to directly detect rupture forces between the cell and the counter-receptors studied. Using physiologic conditions of temperature and pH, this protocol can be used to study chemokine-mediated triggering of integrin activation on lymphocytes or similar cells or cell lines, such as bone marrow stem or leukaemia-derived cells, and eventually metastatic cells. Additionally, this protocol can easily be adapted from cell–substrate to cell–cell adhesion measurements. In order to identify the interacting receptors, blocking agents such as function blocking antibodies against the receptors studied may be needed. The right choice of interacting cell lines is strongly suggested, for example, choosing cell lines known to bind just via one major pair of functional adhesion receptors and lacking all others.

9.3.2.1
Materials

Additional Buffers and Solutions

Biotinylated concanavalin A (ConA) (Sigma) <!>

Cells expressing cell adhesion receptors of interest, for example, integrin LFA-1 (or VLA-4)

Receptor protein(s), for example, recombinant Fc fusion protein with binding domains of receptor, ICAM-1 or VCAM-1, or native soluble receptor proteins, or other ligands

Cell culture medium

Note: I recommend refraining from (a) performing experiments with cells and biomolecules at suboptimal ambient temperatures and (b) using growth media with an unphysiologic or changing pH value. The pH of CO_2/HCO_3-buffered media can fluctuate dramatically if the medium is left unchecked for evaporation. Adding different amounts of HEPES (4-(2-hydroxyethyl)-1-piperazineethanesulfonic acid) buffer does not prevent CO_2 from leaving the media which appears to be critical for some adhesion measurements. Most cell lines can be adapted to more expensive, CO_2-independent medium which I recommend for AFM measurements of living cells.

Inhibitors of receptor-ligand interaction, for example, function-blocking antibodies or peptidomimetic compounds

Activators, for example, Mg^{2+}, Mn^{2+} plus EGTA, activating antibodies, phorbolesters (slow) or chemokines (fast)

Additional Special Equipment

An AFM apparatus with software is needed for the acquisition and analysis of force-distant curves along the *z*-axis. An AFM with an additional *x*-, *y*-axis is not really necessary for pure force spectroscopic measurements, but may help in combination with AFM imaging to locate the binding events on, for example, a cell surface (not shown here). Professional AFMs are equipped with software from the manufacturer. Custom AFMs may be used with Image SXM as a Macintosh application which is available based on NIH Image. We also use Igor Pro 4.0 for data analysis.

Standard cell culture lab for maintenance of cell lines or primary cells (with incubator, laminar flow hood, centrifuge).

Plastic tissue culture dishes (uncoated).

Heating device for physiologic temperature (some experiments may yield suitable results at RT). We developed our own temperature-controlled heating device based on several ceramic resistors and silicone forming a ring-like structure on which the Petri dish was placed.

9.3.2.2 Method

1. Prepare Petri dish with immobilized ligand adsorbed to the surface, for example, adsorb different concentrations of soluble ICAM-1—or VCAM-1— fusion proteins, 25 μl for each droplet in a humid chamber at 4 °C for 12–16 h, wash with PBS, block with 0.5-mg/ml BSA for 30 min, wash with PBS.
2. Functionalize the cantilever tip with conA (see previous Protocol 1).
3. Assemble quickly the AFM with the functionalized cantilever pointing into a cell culture dish with medium (Care: avoid dry-out of proteins on the cantilever).
4. Adjust focused laser path, mirror and photodetector to optimal signal (in our system we optimize the detected laser signal to a value of about 8.0 on a scale from 0 to 10.24 with an equal distribution on both parts of the segmented photodetector).
5. Calibrate cantilever and obtain spring constant (we usually can confirm the manufacturers' value for the spring constant of the largest cantilever in a range of 10–14 mN/m; too low spring constants are likely the result of a broken cantilever, too high could be the result of a wrong setup or wrong cantilever).
6. Position the cantilever over a region without ligand protein and at least 20-μm away from the bottom of the Petri dish. Add 2–10-μl of cells ($\sim 10^5/\text{ml}$) close to the cantilever; wait 1–2 min to allow the cells to settle down; locate a single cell with the light microscope, position the tip of the cantilever above the cell; with the positioning micrometer screw carefully lower the cantilever onto the cell; ideally the cell should rest directly behind the pyramid of the cantilever tip. After 1–5 s of smooth contact retract the cantilever slowly, the cell should be firmly attached to the cantilever. (*Note*: avoid pressing the cell too firm against the bottom of the Petri dish, several days/weeks of experience may help to improve this critical step considerably.)
7. Position the cantilever with the attached cell over a region of the Petri dish with the previously adsorbed ligand. Collect "test" force scans from areas with different densities of the ligand; choose a suitable area and density of ligand to adjust the frequency of rupture events to approximately 30% by then adjusting the contact force and contact time. Our typical settings include an approach rate of 1 μm/s, a contact duration of 10 ms and a compression force of 100 pN. (*Note*: for activation experiments the adhesion rate should be adjusted too much less than 30%, for example, 1–10%, for the non-activated cells, in order to remain under optimal conditions of single-rupture events after activation of the cell adhesion receptor.)
8. Collect force scans at the appropriate ligand density and from negative control regions. ICAM-1 – fusion protein serves as an ideal control for VCAM-1 – fusion protein, and vice versa. Additional controls include uncoated areas and areas

which were only blocked with 0.5 mg/ml BSA for 30 min and washed with PBS (see step 1).
9. Collect force scans at different loading rates to determine the force spectrum of the receptor–ligand interactions.

9.3.3 Inhibition with Blocking Antibodies, Peptidomimetic Inhibitors or EDTA

This method is useful to identify the nature of adhesion events. Function blocking antibodies are used in cell biology to inhibit specific interactions of one pair of adhesion receptors preventing a cell to adhere via these receptors. Any remaining strong binding should be due to other pairs of adhesion receptors. Weak additional binding could be due to either unspecific binding of charged protein–protein interactions, or could point to an additional, but otherwise masked specific pair of weak adhesion receptors (Eibl, unpubl.). The protocol could be adapted to combine two antibodies blocking two different pairs of adhesion receptors on one cell. If a combination of two antibodies can block most of the adhesion events, each of the blocking antibodies might be used to study the unblocked pair of adhesion receptors. This appears to be a crude experiment since the blocking antibodies may interfere in an unspecific way with the unblocked pair of adhesion receptors, but for some experiments it may be difficult to find the right set of cell lines and this might be an option, since the antibodies block very specifically one type of receptor without interfering significantly with the other receptors of interest (Eibl, unpubl.).

Of course, smaller peptidomimetic agents mimicking the blocking function [68, 69] can be tried instead of antibodies. The equivalent concentration of such smaller molecules should be translated accordingly. Chelating agents such as EDTA or EGTA can also be used to interfere with the cell adhesive function to be measured with AFM [70], but this appears to be in an unspecific way, since those chelating substances remove ions needed by many different cell adhesion molecules, including integrins, for their function. Usually EDTA is used to remove CA^{2+} ions from integrins to abolish their binding to their receptors.

9.3.3.1 Method

1. Optimize conditions as determined in Protocol 2 (Sect. 9.3.2) for measuring receptor–ligand interactions at an adhesion rate of approximately 30%.
2. Collect at least 100 force scans.
3. Add non-blocking control antibody (or control compound) to cell culture medium to a final concentration of 10–20 $\mu g/ml$.
4. After 30 min collect force scans to determine adhesion frequency.
5. Repeat experiment with function-blocking antibodies at a final concentration of 10–20 $\mu g/ml$ against the expressed receptor (or peptidomimetic compounds at suitable concentrations)—or add EDTA if analyzing a Ca^{2+}-sensitive adhesion receptor.
6. Determine changes in adhesion frequency and force value.

9.3.4 Activation with Mg^{2+}, Mn^{2+} Ions, Activating Antibodies, Phorbolester or Chemokines

This protocol should be the first protocol to measure on a single-receptor level the activation of any integrin by any chemokine on any living cell type. It may also be used for many different types of cells, several other integrins and chemokines, or any adhesion receptors which exist in different activation states like CD44 which has at least three different activation states including low- and high-binding affinity as well as a non-binding state to its major ligand hyaluronic acid [53, 71]. Other known activators, for example, of integrins include Mg^{2+} and Mn^{2+} ions, (very rare) activating antibodies and phorbolester.

9.3.4.1 *Method*

1. Optimize the conditions as described in Protocol 2 (Sect. 9.3.2), but to a much lower frequency of adhesion events, i.e. 1–15%, depending on the type of activators and cells, i.e. very potent activators need a lower adhesion frequency at the start of the experiment.
2. Collect at least 100 force scans, eventually with a suitable negative control (PBS, control antibody, or heat-inactivated chemokine, or inactive recombinant protein).
3. Add the activating substance. In the case of integrins, for example, Mg^{2+}, Mn^{2+} and EGTA, phorbol ester, or a chemokine-like SDF-1 (or a test compound as putative activator). For better dilution of the compounds and avoiding possible desensitization phenomena with high local concentrations of chemokines these biological compounds may be added by replacing part of the cell culture media with the new compound. (*Note*: avoid changes in temperature and pH.)
4. Wait 1–20 min to continue collecting force scans, depending on dilution and time needed for the activating mechanism of the substance.
5. Collect force scans, eventually collect series of force scans every 30 min during longer activation processes.
6. For pharmacological dose–response curves it is possible to add stepwise with 10–100-fold increments biologic active substances like chemokines such as SDF-1 starting from low concentrations below any detectable response up to concentrations covering the physiologic range and resulting in highest activity, and eventually up to doses which lead to desensitization of the chemokine receptor.

9.3.5 AFM Measurement—Cell Free

Although this chapter focuses on cell adhesion measurements of living cells the following short protocol is included for completion of the protocol section as an example that AFM force spectroscopy can also be used for isolated cell adhesion receptors without the need of a living cell [25].

9.3.5.1
Materials

Additional Buffers and Solutions

Receptor and ligand protein(s), one of them should be biotinylated, for example, P-selectin and sLeX-PAA-biotin (high molecular complex of sialyl Lewis X and Biotin)

Hanks buffer with 2-mM Ca^{2+}

Inhibitors of receptor–ligand interaction, for example, function-blocking antibodies (WAPS [72] against human P-selectin) and suitable control antibody.

9.3.5.2
Method

1. Prepare Petri dish with immobilized ligand, for example, adsorb 25 μl of P-selectin/Fc fusion protein in a humid chamber at 4 °C for 12–16 h, wash with PBS, block with 0.5-mg/ml BSA for 30 min, wash with PBS.
2. Functionalize cantilever tip with counter-receptor, for example, sLex-biotin (see Protocol 1 (Sect. 9.3.1), instead of biotinylated ConA use a suitable biotinylated ligand for the receptor chosen in the step before).
3. Assemble quickly the AFM with the functionalized cantilever pointing into the Petri dish filled with medium or Hanks' buffer including 2-mM $CaCl_2$. (*Note*: avoid dry-out of proteins on the cantilever.)
4. Adjust focused laser path, mirror and photodetector.
5. Calibrate cantilever and obtain spring constant.
6. Position the cantilever over a region with adsorbed ligand.
7. Collect force scans.
8. As a negative control position the cantilever over an uncoated region and/or over a region with a control protein. Collect force scans.
9. Collect force scans at different loading rates to determine the force spectrum of the receptor–ligand interactions.
10. Add inhibiting antibodies at 10–50 $\mu g/ml$ (e.g. WAPS against human P-selectin) to prove the specificity of the interaction. Control IgG antibodies should not interfere with the adhesion rate and force.
11. Instead of adding specific antibodies, removal of Ca^{2+} ions by adding 5–20-mM EDTA can also interfere with the folding and bond strength of Ca-sensitive receptors.

9.4
Conclusion and Future Developments

Interestingly, function-blocking antibodies have been used in many animal models to prevent a disease or reduce the outcome of a disease. In particular, antibodies against VLA-4 have been shown in mouse models to prevent lung metastasis [73] or multiple sclerosis [74], and therefore provide a new therapeutic approach for many different diseases in humans. In addition to the treatment of autoimmune or inflammatory diseases, the further study of cell adhesion receptors and their regulation and manipulation may lead to new therapies in medicine, including autoimmune diseases such as rheumatoid arthritis, multiple sclerosis, Sjögren's syndrome, type I diabetes [75], and perhaps metastasizing cancer. Combined analysis of cell adhesion functions with the genetic profile of a tumour may help to better understand and prevent the ability of tumour or tumour stem cells to migrate and form distant metastases [49, 76–87]. AFM force experiments on living cells are usually not designed for high-throughput analysis of many samples, but this methodology could still be used do characterize the action of a series of the most promising few candidate drugs and therefore may help to reduce the costs of an otherwise broad pharmaceutical screening of too many candidates.

Using sophisticated AFM force measurements should be useful to screen candidate substances for small changes in cell adhesion properties, perhaps just modulating cell trafficking and with the potential of reducing side effects of otherwise too potent inhibitors. As shown with a highly active substance like SDF-1, even smallest concentrations of 1 ng/ml can have very dramatic effects on a living cell within a short time interval. To measure direct forces in real-time, i.e. in milliseconds may add to the potential value of future experiments. The triggering of rapid arrest appears to be as fast as one can measure. More than a decade ago, it appeared to be very unreasonable to expect such a fast activation process, usually known only from signalling of neurons and cardiomyocytes. Originally, integrin activation was considered to be in a range of dozens of seconds, in the meantime it is very clear that the rapid activation of integrins is triggered by arrest chemokines and that it takes place in a range of a few milliseconds. AFM force measurement may also be helpful to study the kinetics of activation. Interestingly, SDF-1 is usually presented on the surface of endothelial cells, although it can come from other cells than the endothelial monolayer. There are conflicting speculations about the different activity potential of soluble SDF-1 versus SDF-1 molecules bound to and presented on the endothelial cell. It may be hard to fully exclude that even immobilized SDF-1 molecules separate from the endothelial cell and become locally soluble, or vice versa, soluble SDF-1 molecules could attach to the Petri dish or to endothelial cells within an experiment, and therefore could be presented as immobilized molecules. Future studies may address the question whether the postulated difference of soluble versus immobilized SDF-1 can be supported or not. Mouse and human lymphocytes appear to express homologous molecules with identical names and comparable functions: VLA-4, VCAM-1 and SDF-1 and CXCR4. In functional experiments these molecules of the two species are widely interchangeable. The only difference appears to be that human VCAM-1 has a second binding site for VLA-4. It may be interesting to find any differences in the use of these two binding sites in contrast to the single mouse binding site. For many experiments studying human VLA-4, a fusion protein with just one functional domain is used.

Two decades after its invention, AFM applications continue to broaden into very powerful applications in many different disciplines. Although more and more cell biologists appear to find interest in these powerful methods, most cell biologists remain unfamiliar with this methodology. This may be due to the different perceptions of physicists and biologists. Often, the physicist may want to invent a very new derivation or resolution of an AFM or a new combination of an AFM with another type of microscope or apparatus, whereas the biologist wants to use an existing method and can not find the appropriate support from the physicist who is not very interested in the cell biology question. In addition, the cell biologist may not have access to an existing AFM and does not want to invest in such a system without being used to the handling, controls and possible artefacts.

It may help if physicists could provide more access to perhaps older AFMs, or if the manufacturer would offer free trials for biology and cell adhesion labs. This may improve the existing AFMs for future cell adhesion experiments.

Acknowledgments. I gratefully acknowledge Heinz Höfler for his support on initiating the interdisciplinary project between nanotechnology and pathology, Irving Weissman, Eugene Butcher, James Campbell, Sara Michie and Peter Herrlich for helping me to greatly expand my knowledge in the field of lymphocyte rolling and cell adhesion, and Hermann Gaub, Vincent Moy, Xiaohui (Frank) Zhang, Olga Vinogradova and Martin Benoit for expert advice in the field of AFM. I thank Bernhard Holzmann, Horst Kessler and Ronen Alon for making available reagents and cell lines, Guttorm Haraldsen, Anna Müller, Albert Zlotnik, Markus Schneemann and Tim Springer for helpful discussions and Stephan Bärtsch and Matthias Maiwald for critical comments on the manuscript. This work was supported in part by the DFG (Ei378/1 + 2; Deutsche Forschungsgemeinschaft), a Deans fellowship from the Stanford University School of Medicine, a travel stipend of the GSO (German Scholar Organization) and prize money for travelling the author received from the German Cancer Research Centre (DKFZ, Heidelberg). For presenting part of this work the author received the first alumni prize of the German Cancer Research Centre, DKFZ Heidelberg.

References

1. Binnig G, Quate C, Gerber C (1986) Phys Rev Lett 56:930
2. Giessibl F, Hembacher S, Bielefeldt H, Mannhart J (2000) Science 289:422
3. Giessibl FJ (1995) Science 267:68
4. Eibl RH (2004) In: Immunology 2004: cytokine network, regulatory cells, signalling, and apoptosis. (Editor: Skamene E). Medimond, Milano, Italy, 1:115
5. Eibl RH, Benoit M, Weissman IL, Gaub HE, Moy VT (2004) Clinical and Investigative Medicine Suppl. 27:4
6. Eibl RH, Moy VT (2005) Methods Mol Biol 305:439.
7. Eibl RH, Moy VT (2004) In: Recent res. devel. biophys. (Editor: Pandalai, SG), TRN, Trivandrum, India, p 235
8. Eibl RH, Moy VT (2005) In: Protein-ligand interactions (Editor: Nienhaus GU), p 450
9. Eibl RH, Benoit M (2004) IEE Proc Nanobiotechnol 151:128
10. Dammer U, Popescu O, Wagner P, Anselmetti D, Güntherodt HJ, Misevic GN (1995) Science 267:1173
11. Hinterdorfer P, Dufrêne YF (2006) Nat Methods 3:347
12. Rief M, Clausen-Schaumann H, Gaub HE (1999) Nat Struct Biol 6:346
13. Rief M, Gautel M, Oesterhelt F, Fernandez JM, Gaub HE (1997) Science 276:1109
14. Wu HW, Kuhn T, Moy VT (1998) Scanning 20:389
15. Baumgartner W, Hinterdorfer P, Ness W, Raab A, Vestweber D, Schindler H, Drenckhahn D (2000) Proc Natl Acad Sci U S A 97:4005
16. Wong J, Chilkoti A, Moy VT (1999) Biomol Eng 16:45
17. Yuan C, Chen A, Kolb P, Moy VT (2000) Biochemistry 39:10219

18. Moy VT, Florin EL, Gaub HE (1994) Science 266:257
19. Florin EL, Moy VT, Gaub HE (1994) Science 264:415
20. Merkel R, Nassoy P, Leung A, Ritchie K, Evans E (1999) Nature 397:50
21. Benoit M, Gabriel D, Gerisch G, Gaub HE (2000) Nat Cell Biol 2:313
22. Thie M, Röspel R, Dettmann W, Benoit M, Ludwig M, Gaub HE, Denker HW (1998) Hum Reprod 13:3211
23. Zhang X, Wojcikiewicz EP, Moy VT (2006) Exp Biol Med (Maywood) 231:1306
24. Zhang X, Chen A, De Leon D, Li H, Noiri E, Moy VT, Goligorsky MS (2004) Am J Physiol Heart Circ Physiol 286:H359
25. Zhang X, Bogorin DF, Moy VT (2004) Chemphyschem 5:175
26. Li F, Redick SD, Erickson HP, Moy VT (2003) Biophys J 84:1252
27. Chen A, Moy VT (2000) Biophys J 78:2814
28. Zhang X, Wojcikiewicz E, Moy VT (2002) Biophys J 83:2270
29. Baumgartner W, Golenhofen N, Grundhöfer N, Wiegand J, Drenckhahn D (2003) J Neurosci 23:11008
30. Benoit M, Gaub HE (2002) Cells Tissues Organs 172:174
31. Eibl RH (2007) In: Hinterdorfer P, Schuetz G, Pohl P (eds) Advances in single molecule research for biology and nanoscience. Trauner, p 40
32. Chen A, Moy VT (2002) Methods Cell Biol 68:301
33. Lehenkari PP, Horton MA (1999) Biochem Biophys Res Commun 259:645
34. Kokkoli E, Ochsenhirt SE, Tirrell M (2004) Langmuir 20:2397
35. Fritz J, Katopodis AG, Kolbinger F, Anselmetti D (1998) Proc Natl Acad Sci U S A 95:12283
36. Dettmann W, Grandbois M, André S, Benoit M, Wehle AK, Kaltner H, Gabius HJ, Gaub HE (2000) Arch Biochem Biophys 383:157
37. Hinterdorfer P, Baumgartner W, Gruber HJ, Schilcher K, Schindler H (1996) Proc Natl Acad Sci U S A 93:3477
38. Wong SS, Joselevich E, Woolley AT, Cheung CL, Lieber CM (1998) Nature 394:52
39. Stenlund P, Babcock GJ, Sodroski J, Myszka DG (2003) Anal Biochem 316:243
40. Hussain MA, Agnihotri A, Siedlecki CA (2005) Langmuir 21:6979
41. Berlin C, Bargatze RF, Campbell JJ, von Andrian UH, Szabo MC, Hasslen SR, Nelson RD, Berg EL, Erlandsen SL, Butcher EC (1995) Cell 80:413
42. Butcher EC (1991) Cell 67:1033
43. Springer TA (1994) Cell 76:301
44. Campbell JJ, Hedrick J, Zlotnik A, Siani MA, Thompson DA, Butcher EC (1998) Science 279:381
45. Campbell JJ, Haraldsen G, Pan J, Rottman J, Qin S, Ponath P, Andrew DP, Warnke R, Ruffing N, Kassam N et al (1999) Nature 400:776
46. Altevogt P, Hubbe M, Ruppert M, Lohr J, von Hoegen P, Sammar M, Andrew DP, McEvoy L, Humphries MJ, Butcher EC (1995) J Exp Med 182:345
47. Vajkoczy P, Laschinger M, Engelhardt B (2001) J Clin Invest 108:557
48. Warnock RA, Askari S, Butcher EC, von Andrian UH (1998) J Exp Med 187:205
49. Eibl RH, Kleihues P, Jat PS, Wiestler OD (1994) Am J Pathol 144:556
50. Eibl RH, Pietsch T, Moll J, Skroch-Angel P, Heider KH, von Ammon K, Wiestler OD, Ponta H, Kleihues P, Herrlich P (1995) J Neurooncol 26:165
51. Stamper HBJ, Woodruff JJ (1976) J Exp Med 144:828
52. Gunthert U, Hofmann M, Rudy W, Reber S, Zoller M, Haussmann I, Matzku S, Wenzel A, Ponta H, Herrlich P (1991) Cell 65:13
53. Stamenkovic I, Amiot M, Pesando JM, Seed B (1989) Cell 56:1057
54. Hutter J, Bechhoefer J (1993) Rev Sci Instrum 64:1868
55. Benoit M (2002) Methods Cell Biol 68:91
56. Dufrêne YF, Hinterdorfer P (2007) Pflugers Arch 456:237

57. Wojcikiewicz EP, Zhang X, Chen A, Moy VT (2003) J Cell Sci 116:2531
58. DeGrendele HC, Estess P, Picker LJ, Siegelman MH (1996) J Exp Med 183:1119
59. DeGrendele HC, Estess P, Siegelman MH (1997) Science 278:672
60. Gallatin WM, Weissman IL, Butcher EC (1983) Nature 304:30
61. Weissman I (1967) Nature 215:315
62. Qian F, Vaux DL, Weissman IL (1994) Cell 77:335
63. Qian F, Hanahan D, Weissman IL (2001) Proc Natl Acad Sci U S A 98:3976
64. Gosslar U, Jonas P, Luz A, Lifka A, Naor D, Hamann A, Holzmann B (1996) Proc Natl Acad Sci U S A 93:4821
65. Butcher EC, Picker LJ (1996) Science 272:60
66. Mikulowska A, Johnson GG, Berberian JM, Butcher EC, McEvoy LM, Michie SA (1999) Cell Immunol 194:112
67. Johnson GG, Mikulowska A, Butcher EC, McEvoy LM, Michie SA (1999) J Immunol 163:5678
68. Gottschling D, Boer J, Schuster A, Holzmann B, Kessler H (2002) Angew Chem Int Ed 41:3007
69. Locardi E, Boer J, Modlinger A, Schuster A, Holzmann B, Kessler H (2003) J Med Chem 46:5752
70. Wojcikiewicz EP, Zhang X, Moy VT (2004) Biol Proced Online 6:1
71. Lesley J, He Q, Miyake K, Hamann A, Hyman R, Kincade PW (1992) J Exp Med 175:257
72. Picker LJ, Warnock RA, Burns AR, Doerschuk CM, Berg EL, Butcher EC (1991) Cell 66:921
73. Brocke S, Piercy C, Steinman L, Weissman IL, Veromaa T (1999) Proc Natl Acad Sci U S A 96:6896
74. Lauri D, De Giovanni C, Biondelli T, Lalli E, Landuzzi L, Facchini A, Nicoletti G, Nanni P, Dejana E, Lollini PL (1993) Br J Cancer 68:862
75. Michie SA, Sytwu HK, McDevitt JO, Yang XD (1998) Curr Top Microbiol Immunol 231:65
76. Clarke MF, Dick JE, Dirks PB, Eaves CJ, Jamieson CHM, Jones DL, Visvader J, Weissman IL, Wahl GM (2006) Cancer Res 66:9339
77. Reya T, Morrison SJ, Clarke MF, Weissman IL (2001) Nature 414:105
78. Ohgaki H, Eibl RH, Wiestler OD, Yasargil MG, Newcomb EW, Kleihues P (1991) Cancer Res 51:6202
79. Wiestler OD, Aguzzi A, Schneemann M, Eibl R, von Deimling A, Kleihues P (1992) Cancer Res 52:3760
80. von Deimling A, Eibl RH, Ohgaki H, Louis DN, von Ammon K, Petersen I, Kleihues P, Chung RY, Wiestler OD, Seizinger BR (1992) Cancer Res 52:2987
81. Wiestler OD, Brüstle O, Eibl RH, Radner H, Aguzzi A, Kleihues P (1992) Brain Pathol 2:47
82. Wiestler OD, Brüstle O, Eibl RH, Radner H, Von Deimling A, Plate K, Aguzzi A, Kleihues P (1992) Neuropathol Appl Neurobiol 18:443
83. Radner H, el-Shabrawi Y, Eibl RH, Brüstle O, Kenner L, Kleihues P, Wiestler OD (1993) Acta Neuropathol 86:456
84. Ohgaki H, Eibl RH, Schwab M, Reichel MB, Mariani L, Gehring M, Petersen I, Höll T, Wiestler OD, Kleihues P (1993) Mol Carcinog 8:74
85. Louis DN, von Deimling A, Chung RY, Rubio MP, Whaley JM, Eibl RH, Ohgaki H, Wiestler OD, Thor AD, Seizinger BR (1993) J Neuropathol Exp Neurol 52:31
86. Wiestler OD, Brüstle O, Eibl RH, Radner H, Aguzzi A, Kleihues P (1994) Recent Results Cancer Res 135:55
87. Kleihues P, Ohgaki H, Eibl RH, Reichel MB, Mariani L, Gehring M, Petersen I, Höll T, von Deimling A, Wiestler OD, et al. (1994) Recent Results Cancer Res 135:25

6 Self-Sensing Cantilever Sensor for Bioscience

Hayato Sone · Sumio Hosaka

From volume VIII

Abstract. A simple and high-sensitivity detection system is desired in the fields of biotechnology and medical science. In order to develop the system, one of the techniques is the use of a microcantilever mass sensor using a harmonic vibration with a resonance frequency. In this chapter, we describe a harmonic vibration-type self-sensing cantilever sensor in bioscience applications. Firstly, we introduce the cantilever mass sensor and its vibrations using theoretical analysis of cantilever motion and finite element method simulation. Then, we explain details of the self-sensing system using a piezoresistive cantilever. Finally, we demonstrate two application studies to achieve femtogram sensitivity, one for water molecule detection in air and the other for the biomolecular reaction between an antigen and an antibody in water.

Key words: Biosensor, Piezoresistive cantilever, Mass detection, Femtogram Sensitivity, Biomolecule

6.1 Introduction

As medical technologies advance, it is revealed that various kinds of chemical and biological materials affect our health. Of particular interest to the public are environmental hormone-disrupting chemicals, allergenic substances, and genetic materials. Analytical instrumentation techniques such as gas chromatography–mass spectrometry (GCMS) or liquid chromatography—mass spectrometry (LCMS) [1–3], immunoassay [4–6], the quartz crystal microbalance (QCM) method [7–9], and surface plasmon resonance (SPR) [10–12] have been developed to detect these materials. Although these techniques have some advantages, they also have a few disadvantages, as follows. GCMS and LCMS have a high mass resolution of about 1 pg, but the preparation and measurement of samples take a long time. QCM is a simplified measurement method, but it has a low sensitivity of about 30 pg/Hz. SPR has a high sensitivity equivalent to a few picograms, but the measurement instrument is complex and expensive. One of the solutions to these problems is a microcantilever-based mass sensor using the detection technique of vicinal deflection.

The cantilever-based mass sensor was introduced by Berger et al. [13, 14]. They proposed various physical and chemical sensor applications, including as a mass sensor, a force sensor [15], a temperature sensor [16], a calorimeter [17], and a thermogravimetric sensor [18]. Water adsorption in the nanogram range has been measured using a cantilever with a zeolite sensor [13]. Applications of biomolecular detection for bovine serum albumin and deoxyribonucleic acid hybridization using the

In: H. Fuchs, B. Bhushan (eds), *Biosystems - Investigated by Scanning Probe Microscopy*,
DOI 10.1007/978-3-540-74080-3_6, © Springer-Verlag Berlin Heidelberg 2010

nanomechanical responses of cantilever bending have been reported [19–21]. The binding of thiol molecules with a total mass of a few femtograms has been monitored using the resonance frequency shifts of cantilevers [22]. Nanometer-scale mass sensors with subattogram sensitivity have been fabricated using nanoelectromechanical oscillators [23]. Cantilever harmonic-vibration-type mass sensors with picogram mass sensitivity using a laser beam deflection detection system have been reported in experiments of molecular adsorption on the cantilever [24, 25]. These reported works, however, used a laser beam deflection detection system for the detection of the resonance frequency. This method involves technical difficulties such as heavy equipment and complex adjustment. Furthermore, in the case of biomolecule detection, the cantilever mass sensor has to operate in water. In order to solve these problems and to respond to bioscience needs more effectively, a self-sensing system has been proposed to replace the laser beam deflection detection system. As the self-sensing system, several researchers have used a piezoresistive strain sensor embedded in the cantilever [26–33]. The cantilever has been used to detect the deflection of the cantilever by measuring the piezoresistance in an atomic force microscope cantilever [34–36]. In thischapter, we will discuss the role of harmonic-vibration-type self-sensing piezoresistive sensors in bioscience applications.

6.2 Basics of the Cantilever Mass Sensor

When we vibrate a microcantilever using an exciting piezoactuator which supports the cantilever base, as shown in Fig. 6.1a, the cantilever equation of motion is given by

$$M\frac{\mathrm{d}^2 z}{\mathrm{d}t^2} + a\frac{\mathrm{d}z}{\mathrm{d}t} + kz = F_0 \sin(\omega t) , \tag{6.1}$$

where M is the effective mass of the cantilever, z is the deflection of the cantilever top, a is the viscous resistance modulus, k is the spring constant, and $F_0 \sin(\omega t)$ is the external vibration force from the exciting piezoactuator. The general solution is obtained using the differential equation of (6.1), as follows:

$$\begin{aligned} z(t) = {} & C_1 \exp\left[\left(-\frac{a}{2M} + \sqrt{\frac{a^2}{4M^2} - \omega_0^2}\right) t\right] \\ & + C_2 \exp\left[\left(-\frac{a}{2M} - \sqrt{\frac{a^2}{4M^2} - \omega_0^2}\right) t\right] \\ & + \frac{F_0}{M\left[\left(\omega_0^2 - \omega^2\right)^2 + \left(\frac{a\omega}{M}\right)^2\right]} \sin\left(\omega t + \delta\right) , \end{aligned} \tag{6.2}$$

where C_1 and C_2 are constants, $\omega_0^2 = k/M$ and $\delta = \tan^{-1}\left(-\frac{a\omega}{M\left(\omega_0^2 - \omega^2\right)}\right)$. When the cantilever is vibrated at the resonance frequency and given sufficient time, we only

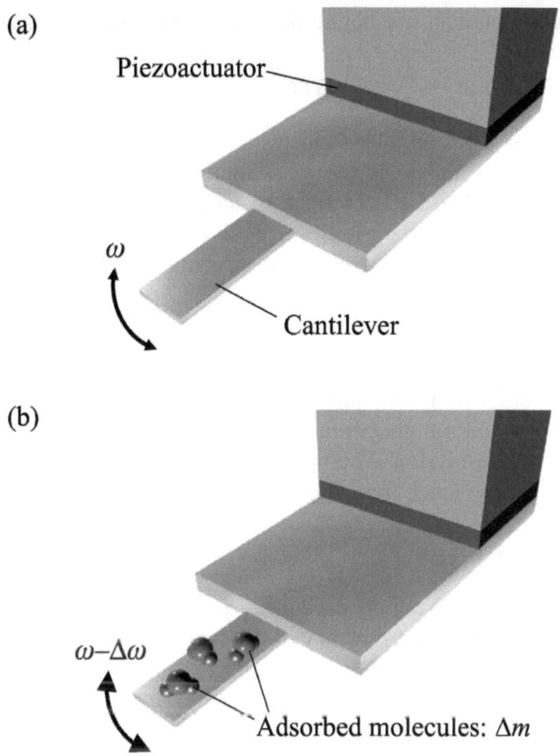

Fig. 6.1. The harmonic-vibration-type microcantilever sensor using a piezoactuator. **a** Resonance frequency, ω. **b** Resonance frequency with adsorbed molecules on the cantilever, $\omega - \Delta\omega$

need to consider the third term in (6.2). Since the vibration amplitude reaches its maximum value at the resonance frequency, ω is given by

$$\omega = \sqrt{\frac{k}{M} - \frac{a^2}{2M^2}} . \tag{6.3}$$

Additionally, the sensitivity of the sensor is derived from the total differential of (6.3), as follows:

$$\Delta\omega = \frac{1}{2}\left(\frac{\frac{a^2}{M^3} - \frac{k}{M^2}}{\sqrt{\frac{k}{M} - \frac{a^2}{2M^2}}}\Delta M + \frac{1}{M\sqrt{\frac{k}{M} - \frac{a^2}{2M^2}}}\Delta k \right) . \tag{6.4}$$

When the spring constant does not change, we can substitute the differential of the spring constant, $\Delta k = 0$:

$$\Delta\omega = \frac{1}{2}\frac{\frac{a^2}{M^3} - \frac{k}{M^2}}{\sqrt{\frac{k}{M} - \frac{a^2}{2M^2}}}\Delta M . \tag{6.5}$$

This result shows that the resonance frequency changes as the cantilever mass changes, and that the amount of the frequency change depends on the spring constant and the viscous resistance. When we vibrate the cantilever in a low-viscosity environment, a is negligibly small; therefore, $\Delta\omega$ is given by

$$\Delta\omega = -\frac{1}{2M}\sqrt{\frac{k}{M}}\Delta M = -\frac{\omega}{2M}\Delta M\,. \tag{6.6}$$

M of a rectangular-shaped cantilever is given by nm, where n is a proportional constant and m is the mass of the cantilever. Depending on the geometry, n is estimated to be 0.24 for a rectangular cantilever [37]. According to the material adsorption on the cantilever, as shown in Fig. 6.1b, the mass, m, changes to $m + \Delta m$ and the resonance frequency, ω, changes to $\omega - \Delta\omega$. From (6.6), the mass change, Δm, is given by

$$\Delta m = -2\frac{m}{\omega}\Delta\omega\,. \tag{6.7}$$

In order to detect small Δm values, it is necessary to use a cantilever with a small mass and a high resonance frequency and to detect small frequency changes with a fine resolution. The microcantilever satisfies the requirements of small mass and high resonance frequency. The spring constant of the cantilever is given by

$$k = \frac{Et^3 w}{4l^3}\,, \tag{6.8}$$

where E is Young's modulus, t is the thickness, w is the width, and l is the length. The resonance frequency of the cantilever is obtained from

$$\omega = \frac{1.02t}{l^2}\sqrt{\frac{E}{\rho}}\,, \tag{6.9}$$

where ρ is the density. Figure 6.2 shows the calculated results for k and the resonance frequency, f, of the rectangular silicon cantilever, where w is 10 μm. The calculated

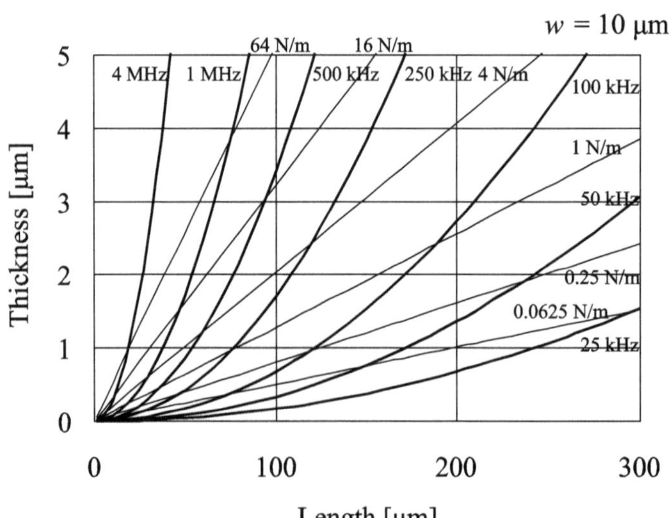

Fig. 6.2. Contour map of the resonance frequency and spring constant of the microcantilever at a cantilever width of $w = 10$ μm with changing length, l, and thickness, t

results show that k becomes larger with increasing t, and f becomes higher with increasing l. If we use a cantilever with a mass of 10 ng and a resonance frequency of 100 kHz, a mass sensitivity of $\Delta m/\Delta f = 200\,\text{fg/Hz}$ is obtained. This value is 100 times higher than that given by the QCM method.

6.3 Finite Element Method Simulation of the Cantilever Vibration

To realize highly sensitive mass detection using a piezoresistive cantilever, the resonance frequency and strain distribution were studied using a computer simulation and the commercially available software COMSOL Multiphysics (COMSOL). This software solves partial differential equations of physical parameters using finite-element modeling. In the simulation, the 3D structure of the cantilever is designed by the computer-aided design system, and the material properties, boundary conditions, and mesh size of the segment region are set in the designed 3D structures. Then, the software executes the calculations of the finite-element analysis together with adaptive meshing and error control using a variety of numerical solvers, and it can perform original frequency analysis or stationary-state analysis as well.

Figure 6.3a shows the strain distribution of the rectangular piezoresistive silicon cantilever. The length, width, and thickness of the cantilever are 110, 50, and 4 μm, respectively. The silicon base is fixed and the front end of the cantilever beam is pushed downward with an applied force of 40 nN. Consequently, the strain is widely distributed on the surface of the cantilever beam, and the maximum value of the strain, which concentrates in the base region of the cantilever beam, is approximately 0.17×10^{-6}. In addition, the original resonance frequency of the cantilever is about 473 kHz. In contrast, Fig. 6.3b shows the strain distribution on the cantilever base of the rectangular piezoresistive silicon cantilever with a narrow aperture 10 μm in length and 40 μm in width; i.e., the cantilever beam is connected with the silicon base via twin slim arms 10 μm in length and 5 μm in width. The length, width, and thickness of the cantilever and the applied force are same as for the previous rectangular cantilever. The results show that the strain is concentrated on the slim arms, and that the maximum value of the strain and the original resonance frequency are approximately 1.08×10^{-6} and 279 kHz, respectively. Therefore, it was found that the strain is concentrated in the slim arms of the cantilever base, and that it has the potential of generating a large resistance change, ΔR, unlike the cantilever without the narrow aperture.

The resonance frequency and strain of the rectangular piezoresistive cantilever with a narrow aperture for different cantilever lengths were calculated. Figure 6.4a shows the top view of the cantilever, and x_1 is the cantilever length. Figure 6.4b shows the calculated frequency and strain for cantilever lengths, x_1, of 20–110 μm. The frequency decreases from 5208 to 279 kHz with increasing length, while the strain increases linearly from 0.18×10^{-6} to 1.08×10^{-6} with increasing length. Although the higher resonance frequency and the smaller mass of the cantilever have the advantage of enabling the detection of small masses of adsorbed material, as shown by (6.7), the strain decreases with decreasing length. In addition, the detectable frequency is limited in the detection circuits. Under these conditions,

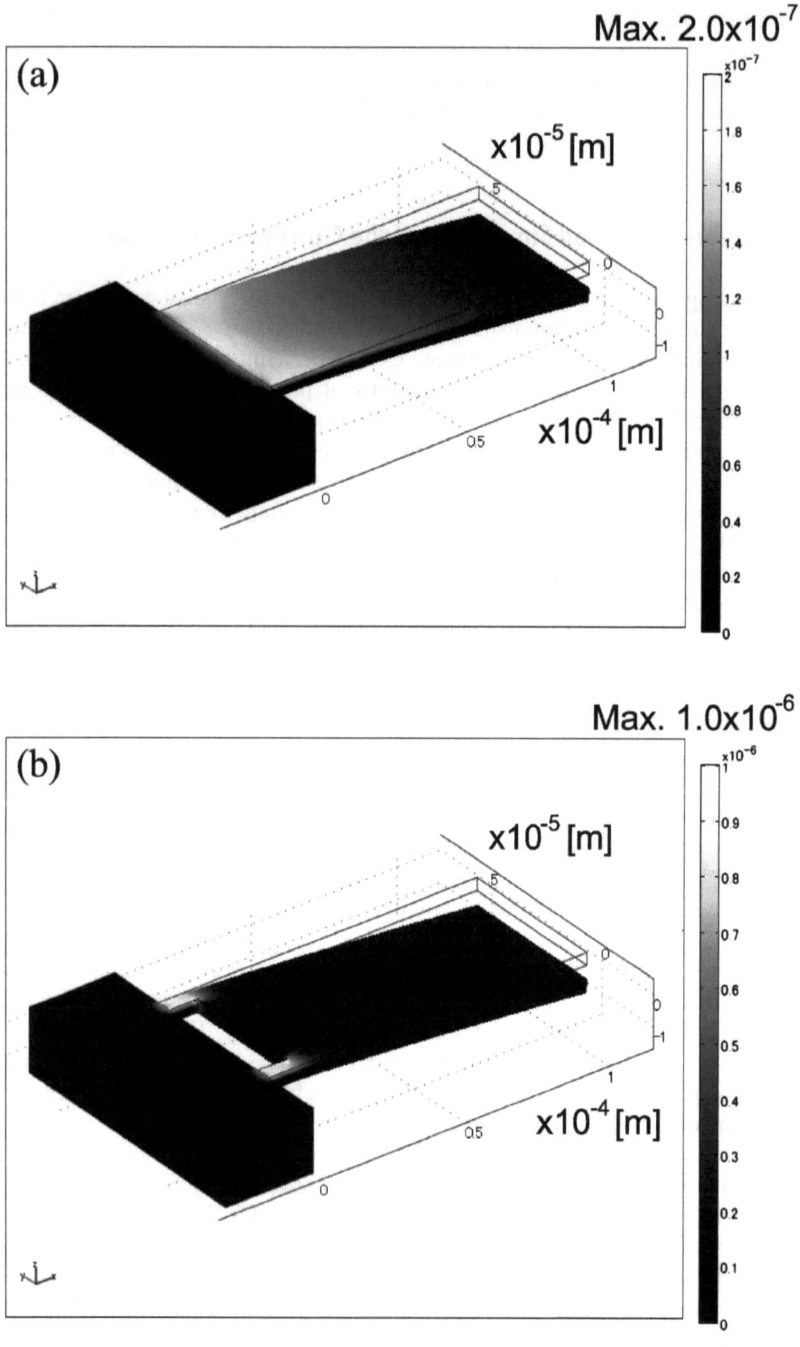

Fig. 6.3. Calculated results using finite-element modeling. **a** Strain distribution of the rectangular piezoresistive silicon cantilever. **b** Strain distribution of the rectangular piezoresistive silicon cantilever with a narrow aperture

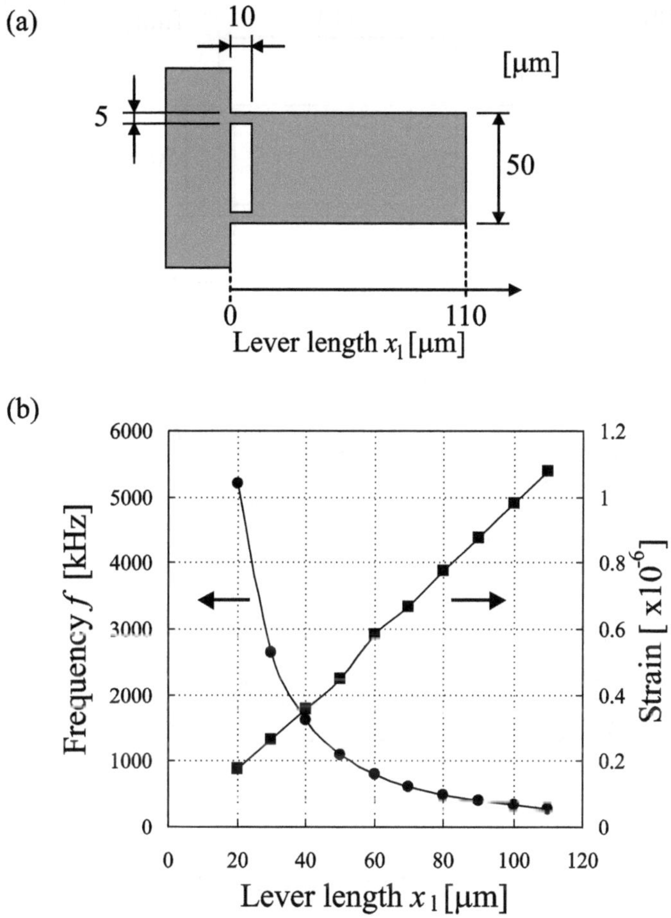

Fig. 6.4. **a** The piezoresistive cantilever with a narrow aperture. **b** Calculated results of the frequency and strain for cantilever lengths, x_l, from 20 to 110 μm

the optimum cantilever length (x_l) is considered to be about 80 μm. The optimum length changes depending on the cantilever width and the arm width.

Next, the resonance frequency and strain of the piezoresistive cantilever with an aperture for different aperture lengths were calculated. Figure 6.5a shows the top view of the cantilever, and x_a is the aperture length. Figure 6.5b shows the calculated results of the frequency and strain for aperture lengths, x_a, of 10–100 μm. The frequency decreases from 280 to 219 kHz in the x_a range from 10 to 50 μm, and increases from 219 to 312 kHz in the x_a range from 50 to 100 μm. The strain decreases rapidly from 1.08×10^{-6} to 0.93×10^{-6} in the x_a range from 10 to 30 μm, and increases gradually to 0.96×10^{-6} in the x_a range from 30 to 100 μm. As higher resonance frequency and higher strain have the advantage of enabling the detection of small mass changes, the optimum aperture length (x_a) is about 10 μm in the cantilever. This optimum length also changes depending on the cantilever width and the arm width.

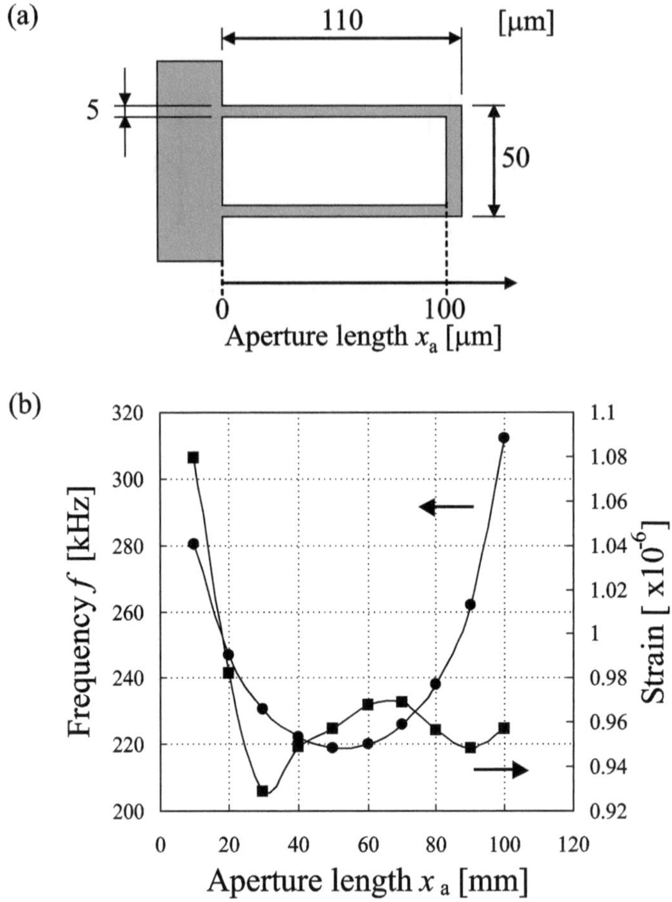

Fig. 6.5. a The piezoresistive cantilever with an aperture. **b** Calculated results of the frequency and strain for aperture lengths, x_a, from 10 to 100 μm

6.4 Detection of Cantilever Deflection

In order to evaluate the sensitivity of the cantilever sensor, we compared a position sensor using a laser beam deflection detection system with a piezoresistive sensor.

6.4.1 Using a Position Sensor

Laser beam deflection detection systems are widely used in atomic force microscope systems. A laser beam deflection detection system can magnify small position changes of the cantilever front end at a high optical magnification, from ×100 to ×1000. An incident laser beam irradiated from a laser diode is focused on the cantilever, and the position of the reflected laser beam is detected by a position-sensor

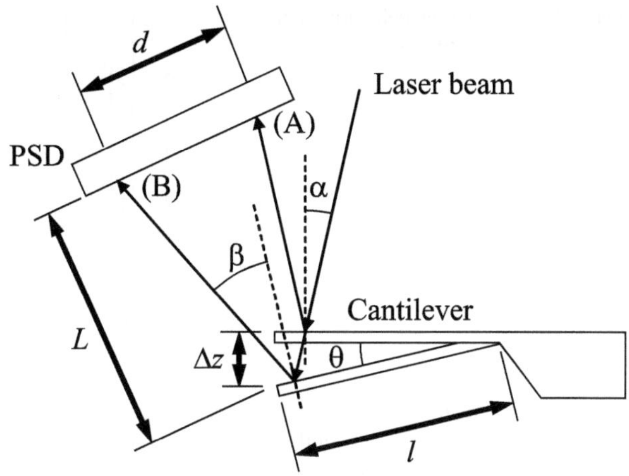

Fig. 6.6. The laser beam deflection detection system. *PSD* position-sensor diode

diode (PSD), as shown in Fig. 6.6. When the cantilever moves down with the angular change θ, the position of the laser beam reflected on the cantilever shifts from A to B on the PSD. The laser beams at A and B are reflected from the cantilever with the angles 2α and 2β, respectively. The position change of the cantilever top, Δz, is magnified to the distance d on the PSD. The magnification, $d/\Delta z$, of this system is given by

$$\frac{d}{\Delta z} = \frac{2L}{l}\,, \tag{6.10}$$

where l is the length of the cantilever and L is the distance between the cantilever and the PSD. A short cantilever should be used in order to obtain a large magnification; i. e., a system with a short cantilever has a high sensitivity for detecting the amplitude of the cantilever vibration. Here, we used a cantilever length, l, of 100 μm and an optical path length, L, of about 5 cm. Consequently, the magnification was 1000. As the PSD can detect a minimum beam position change of less than 0.1 μm, the minimum detectable amplitude of the cantilever vibration is about 0.1 nm; thus, this detection system can detect small cantilever vibrations.

6.4.2
Using a Piezoresistive Sensor

For biomolecule detection, the mass sensor must be operated in water. The reported detection method uses a laser beam deflection detection system for the detection of the resonance frequency, as discussed in Sect. 6.4.1. This method requires some technical adjustments of the laser path, focal point, etc. In order to respond to bioscience needs more effectively, a self-sensing system has been developed to replace the laser beam detection system. The self-sensing system is based on a piezoresistive sensor embedded in the cantilever [26–33].

The deflection of the piezoresistive cantilever is detected on the basis of the piezoresistance change generated by the bending strain. Figure 6.7 shows a cross-section diagram of a cantilever with thickness t and deflection length z. The length

Fig. 6.7. The bending behavior of the piezoresistive cantilever

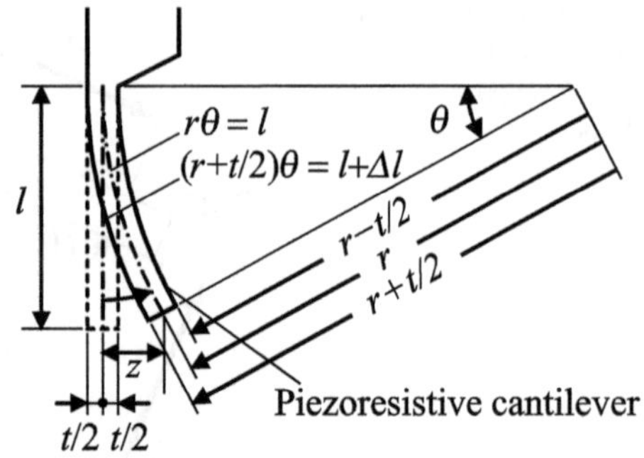

along the midline of the cantilever is l, which is the same as before bending. After bending the cantilever, when the angle of arc and the radius of the cantilever bending along the midline are θ (radians) and r, respectively, the length along the outer surface is given by

$$(r + \frac{t}{2})\theta = l + \Delta l \,, \tag{6.11}$$

where Δl is the stretch of the cantilever. From $l = r\theta$, Δl is given by

$$\Delta l = \frac{t}{2}\theta \,. \tag{6.12}$$

Therefore, the strain, ε, in the bending cantilever is defined by

$$\varepsilon \equiv \frac{\Delta l}{l} = \frac{t}{2r} \,. \tag{6.13}$$

The strain can be detected on the basis of the resistance change, ΔR, of the piezoresistor, as follows:

$$\varepsilon \equiv \frac{1}{K}\frac{\Delta R}{R} \,, \tag{6.14}$$

where K is the gage factor of the piezoresistor and R is the piezoresistance. Then, the deflection length, z, is given by

$$z = r(1 - \cos\theta) = r\left[1 - \cos\left(\frac{l}{r}\right)\right] \,. \tag{6.15}$$

From (6.14) and (6.15), we obtain

$$\begin{aligned} z &= \frac{t}{2\varepsilon}\left[1 - \cos\left(\frac{2\varepsilon l}{t}\right)\right] = \frac{tKR}{2\Delta R}\left[1 - \cos\left(\frac{2l\Delta R}{tKR}\right)\right] \\ &= \frac{tKR}{\Delta R}\sin^2\left(\frac{l\Delta R}{tKR}\right) \,. \end{aligned} \tag{6.16}$$

Since we use a piezoresistive cantilever, $l\Delta R$ is much smaller than tKR:

$$z \approx \frac{tKR}{\Delta R}\left(\frac{l\Delta R}{tKR}\right)^2 = \frac{l^2\Delta R}{tKR} \,. \tag{6.17}$$

Consequently, if we know the values of t, l, K, and R, the cantilever deflection can be obtained on the basis of the piezoresistance measurements. From (6.14), z is given by

$$z = \frac{l^2}{t}\varepsilon \,. \tag{6.18}$$

To detect the strain, ε, a strain meter with Wheatstone bridge circuits is used. The strain meter amplifies the strain signal generated by the cantilever deflection. For example, we obtain an output voltage from the strain meter of 1 V for a strain of 50×10^{-6}. As we can usually measure the voltage with a resolution of 10 mV, the minimum detectable strain is about 50×10^{-8}. If we use a piezoresistive cantilever length, l, of 100 μm and a thickness, t, of 4 μm, the minimum detectable deflection, z, is about 1 nm. This sensitivity is 10 times lower than that of the laser beam deflection detection system. But if we use small piezoresistive cantilevers or measure small voltage changes, it is possible to detect deflections about 1 nm.

Many kinds of piezoresistive cantilevers have been developed for atomic force microscope probes, and some of them are commercially available. We introduce here two types of piezoresistive cantilevers. One is the V-shaped piezoresistive silicon cantilever, shown in Fig. 6.8a, which is fabricated by a semiconductor microprocess. The cantilever has a length of 335 μm, a thickness of 3 μm, a surface area of about 5.8×10^{-4} cm^2 (total of both sides), a mass of about 200 ng, and a piezoresistance of about 1.3 kΩ. The piezoresistors R_1 and R_3 are formed by boron implantation on both beams of the cantilever, as shown in Fig. 6.9a. The other is a commercially available rectangular piezoresistive silicon cantilever (NPX1CTP004; SII Nanotechnology), shown in Fig. 6.8b. This cantilever has a length of 110 μm, a thickness of 4–5 μm, a surface area of about 9.8×10^{-5} cm^2 (total of both sides), a mass of about 46 ng, a spring constant of 40 N/m, and a piezoresistance of about 630 Ω.

The resistance change generated by the cantilever deflection was measured using a dynamic strain meter (e. g., DC-96A; Tokyo Sokki Kenkyujo) with Wheatstone bridge circuits, as shown in Fig. 6.9b. In the V-shaped piezoresistive cantilever shown in Fig. 6.8a, the bridge circuits comprise the piezoresistors R_1 and R_3 and the standard resistors R_2 and R_4. The voltage change, ΔV, is given by

$$\Delta V = \frac{R_1R_3 - R_2R_4}{(R_1 + R_2)(R_3 + R_4)}E \,, \tag{6.19}$$

where E is the voltage applied to the bridge circuits. In the case of $R_1 = R_3 = R + \Delta R$, $R_2 = R_4 = R$, ΔV is given by

$$\Delta V = \frac{2R\Delta R + \Delta R^2}{(2R + \Delta R)^2}E \,, \tag{6.20}$$

where ΔR is the piezoresistance change. As R is much larger than ΔR, ΔV is given by

$$\Delta V = \frac{\Delta R}{2R}E \,. \tag{6.21}$$

This equation substitutes for (6.17) and (6.18), as follows:

$$\varepsilon = \frac{t}{l^2}z = 2\frac{\Delta V}{KE} \,. \tag{6.22}$$

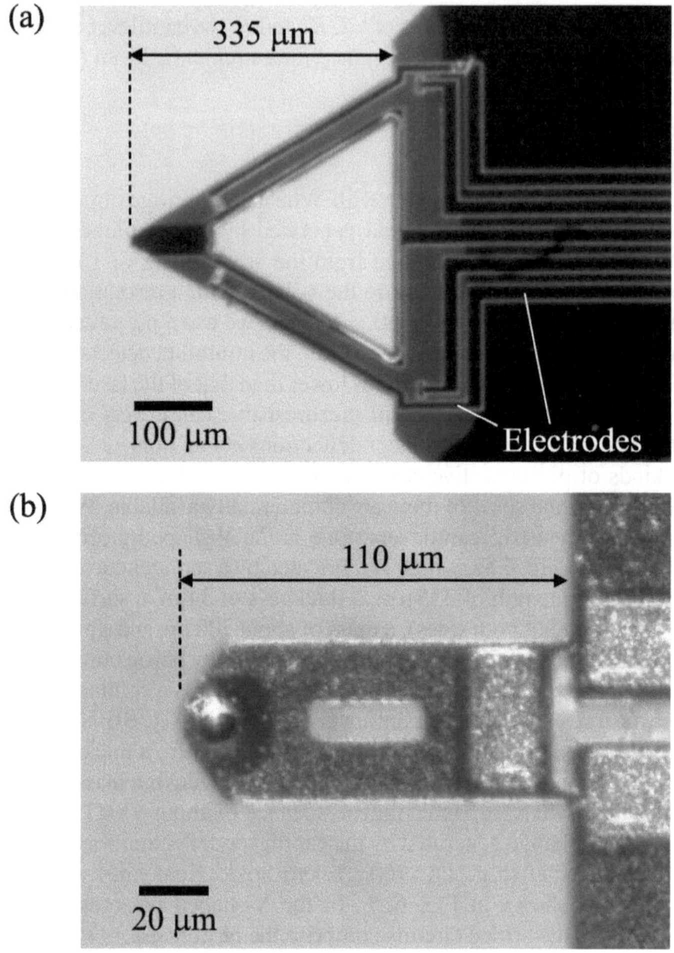

Fig. 6.8. Optical microscope image of the microcantilever. **a** V-shaped piezoresistive cantilever. **b** Commercially available rectangular piezoresistive cantilever

From this equation, the deflection of the cantilever can be detected on the basis of ΔV, and the resonance frequency of the cantilever is obtained from the change in the amplitude of the vibration signal.

The resistance, R, of the V-shaped piezoresistive cantilever is about 1.3 kΩ, and a voltage, E, of 2 V is applied to the bridge circuits. Consequently, the power induced in the piezoresistor is about 1.5 mW. In order to prevent the induced power from exerting any influence, we gave the bridge circuits ample time after supplying the power. Thus, we were able to measure the molecular adsorption on the cantilever under the thermal equilibrium condition.

The bridge circuits in Fig. 6.9b constitute a two-gage detection system that can calibrate the piezoresistance change with changing temperature. If we use a one-gage piezoresistive cantilever, as shown in Fig. 6.8b, which has the piezoresistance R_1,

ΔV is given by

$$\Delta V = \frac{\Delta R}{4R} E \, . \tag{6.23}$$

This equation substitutes for (6.17) and (6.18), as follows:

$$\varepsilon = \frac{t}{l^2} z = 4 \frac{\Delta V}{KE} \, . \tag{6.24}$$

The deflection of the cantilever can also be detected by a one-gage sensor on the basis of ΔV.

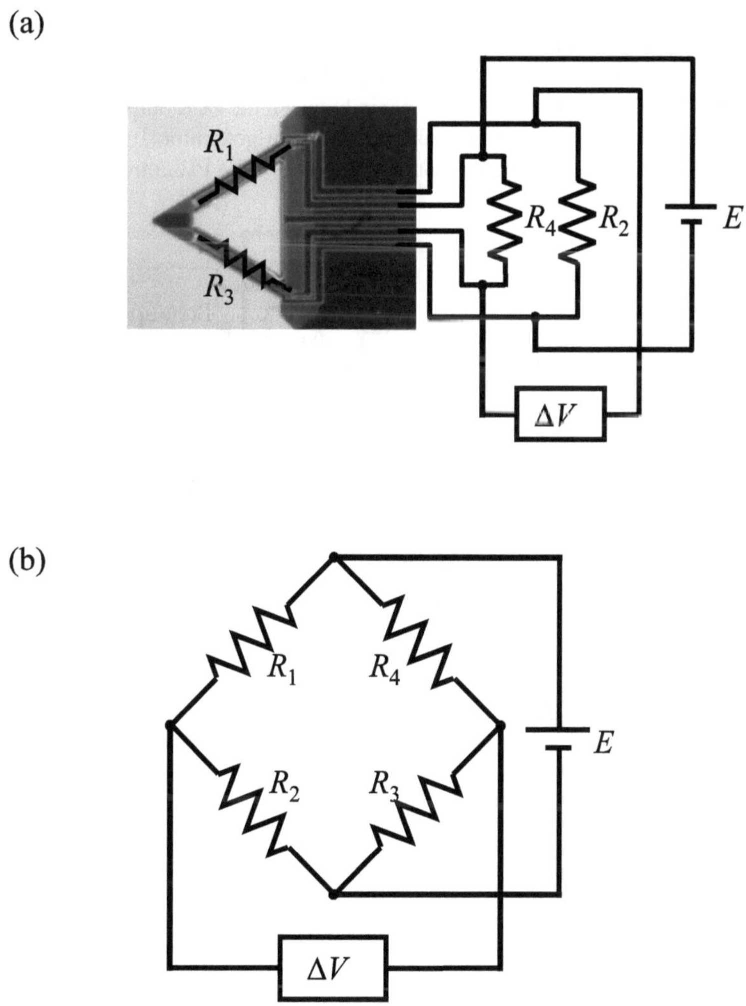

Fig. 6.9. **a** Piezoresistive cantilever and Wheatstone bridge circuits. **b** Equivalent circuits of **a**

6.5 Self-Sensing Systems

6.5.1 Vibration Systems

The cantilever mass sensor has the potential to achieve femtogram mass sensitivity using the harmonic-vibration method, as discussed previously. Figure 6.10 shows the setup of the piezoresistive cantilever mass sensor using the harmonic-vibration method. The mass sensor consists of a piezoresistive cantilever, a preamplifier, positive-feedback circuits, a piezoactuator, and phase-locked loop (PLL) circuits. To form an oscillator, the piezoresistive cantilever is set on a driver element using an exiting piezoactuator, and the output signal from the bridge circuits, ΔV, is input into the preamplifier and the positive-feedback circuits. Then, the output signal of the positive-feedback circuits is input into the piezoactuator to oscillate the cantilever. The oscillation signal output from the preamplifier is demodulated to the frequency signal by the PLL circuits. The relationship between the PLL output voltage and the frequency is calibrated in advance using a function generator. It is possible to use quadrature circuits or a frequency counter in place of the PLL circuits.

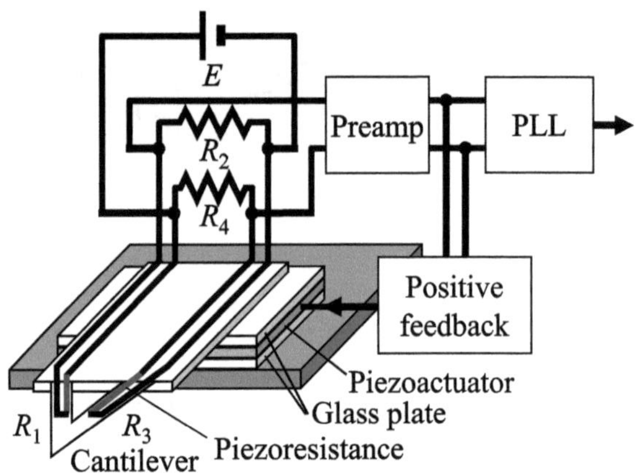

Fig. 6.10. Harmonic-vibration-type mass sensor system and detection circuits using a piezoresistive cantilever. *PLL* phase-locked loop

6.5.2 Vibration-Frequency Detection Systems

In order to detect the vibration frequency, the PLL circuits are usually used as the frequency modulation demodulator. Figure 6.11 shows a block diagram of the PLL system. The PLL circuits consist of a phase comparator, a loop filter, and a voltage-controlled oscillator (VCO). This system operates on the basis of the frequency and phase of the input signal, $\nu_i(t)$, consistent with those of the sine wave, $\nu_c(t)$, which is generated in the VCO. The output signal from the phase comparator, $\nu_e(t)$, corresponds to the phase difference between $\nu_i(t)$ and $\nu_c(t)$. The input signal, $\nu_i(t)$, is given by

$$\nu_i(t) = V_i \cos[\omega_i t + \theta_i(t)] . \quad (6.25)$$

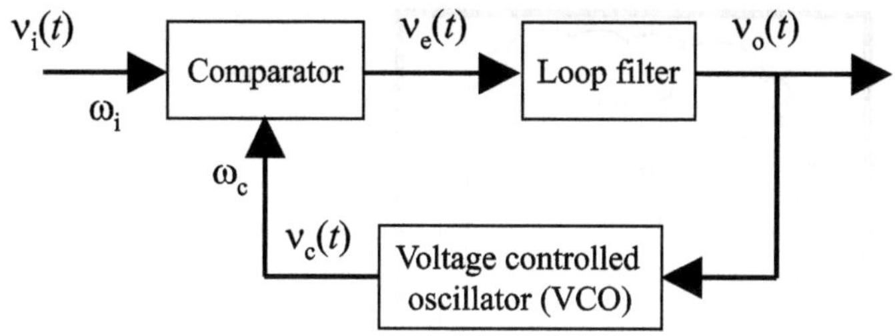

Fig. 6.11. The PLL circuits

The output signal from the VCO is given by

$$\nu_{\mathrm{c}}(t) = V_{\mathrm{c}} \sin[\omega_{\mathrm{c}} t + \theta_{\mathrm{c}}(t)] \,. \tag{6.26}$$

Since the phase comparator multiplies $\nu_{\mathrm{i}}(t)$ by $\nu_{\mathrm{c}}(t)$, the output signal from the comparator is given by

$$\begin{aligned} \nu_{\mathrm{e}}(t) &= V_{\mathrm{i}} V_{\mathrm{c}} \cos[\omega_{\mathrm{i}} t + \theta_{\mathrm{i}}(t)] \sin[\omega_{\mathrm{c}} t + \theta_{\mathrm{c}}(t)] \\ &= \frac{V_{\mathrm{i}} V_{\mathrm{c}}}{2} \{\sin[(\omega_{\mathrm{i}} + \omega_{\mathrm{c}})t + \theta_{\mathrm{i}}(t) + \theta_{\mathrm{c}}(t)] + \sin[(\omega_{\mathrm{i}} - \omega_{\mathrm{c}})t + \theta_{\mathrm{i}}(t) - \theta_{\mathrm{c}}(t)]\}. \end{aligned} \tag{6.27}$$

In the loop filter, the high-frequency component in the first term of $\nu_{\mathrm{e}}(t)$ is cut off, which means that the output signal from the loop filter, $\nu_{\mathrm{o}}(t)$, is given by

$$\nu_{\mathrm{o}}(t) = \frac{V_{\mathrm{i}} V_{\mathrm{c}}}{2} \sin[(\omega_{\mathrm{i}} - \omega_{\mathrm{c}})t + \theta_{\mathrm{i}}(t) - \theta_{\mathrm{c}}(t)] \,. \tag{6.28}$$

As this signal is fed back into the comparator through the VCO, the difference between ω_{i} and ω_{c} and the difference between $\theta_{\mathrm{i}}(t)$ and $\theta_{\mathrm{c}}(t)$ are gradually reduced. Finally, as the differences approach zero [$\omega_{\mathrm{i}} = \omega_{\mathrm{c}}$ and $\theta_{\mathrm{i}}(t) = \theta_{\mathrm{c}}(t)$], the $\nu_{\mathrm{o}}(t)$ that corresponds to ω_{i} is obtained.

6.6 Applications

Application studies for high-sensitivity sensors using a self-sensing system with a piezoresistive cantilever have been demonstrated. In this chapter, we describe two application studies, one for water molecule detection in air [30,31] and the other for the biomolecular reaction between antigen and antibody in water [32,33].

6.6.1 Water Molecule Detection in Air

Figure 6.12 shows a schematic diagram of the measurement system for determining the water molecule adsorption on a cantilever. A piezoresistive cantilever and

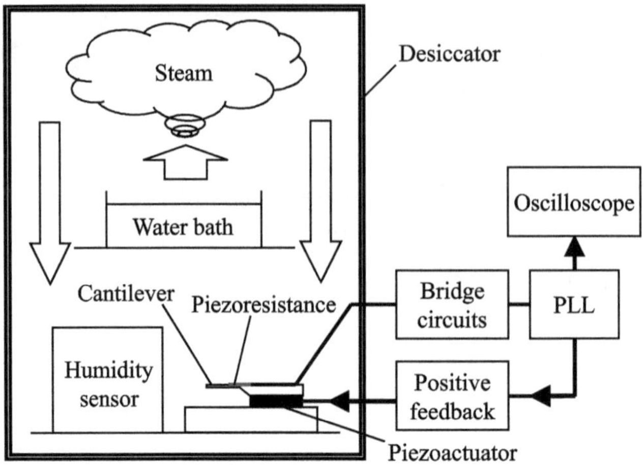

Fig. 6.12. The water adsorption measurement system using a piezoresistive cantilever sensor

a ceramic sensor (commercially obtained) were placed in a desiccator. The humidity inside the desiccator was reduced using silica gel before the experiment. First, we placed a water bath inside the desiccator. Then, we measured the PLL output voltage change with increasing humidity. Two types of piezoresistive cantilevers, shown in Fig. 6.8, were used in this experiment. The resonance frequency of the V-shaped piezoresistive cantilever is 174.2 kHz in air. Consequently, an oscillation frequency of about 174 kHz was obtained by the positive-feedback system, and the locked center frequency and PLL bandwidth were set at 173.5 and 1 kHz, respectively.

Figure 6.13a shows the resonance frequency, Δf, of the cantilever when the humidity increases from 19 to 57%. The humidity was measured by a ceramic humidity sensor. The resonance frequency decreases with increasing humidity. This fact shows that water molecules adsorb on the cantilever with increasing humidity.

Using (6.7), we can estimate the humidity dependence of the mass change due to the water molecule adsorption on the basis of the mass at the humidity of 19%, as shown in Fig. 6.13b. The total mass of the adsorbed water is about 550 pg for the humidity change from 19 to 57%. In this case, the resonance frequency shift is about 250 Hz, as shown in Fig. 6.13a. Consequently, the estimated mass detection sensitivity is about 2.2 pg/Hz, which is 10 times higher than the sensitivity of the quartz crystal oscillation method.

In Fig. 6.13b, the rate of mass change rapidly increases above the humidity of about 42%. The increased rates are about 10 pg/% and 21 pg/% in the humidity range under 42% and over 42%, respectively. It was considered that these differences are caused by the fact that the adsorbed water molecules covered the cantilever surface. The formation of water molecule islands on graphite, gold, and mica substrates has been observed by scanning force microscopy [38, 39]. The adsorbed water makes a number of water molecule islands on the cantilever surface at lower humidity. With increasing humidity, the height of each water island increases. Over the critical humidity, the islands join together, and the cantilever surface is covered by a continuous water film. This mechanism of water molecule adsorption on a cantilever was

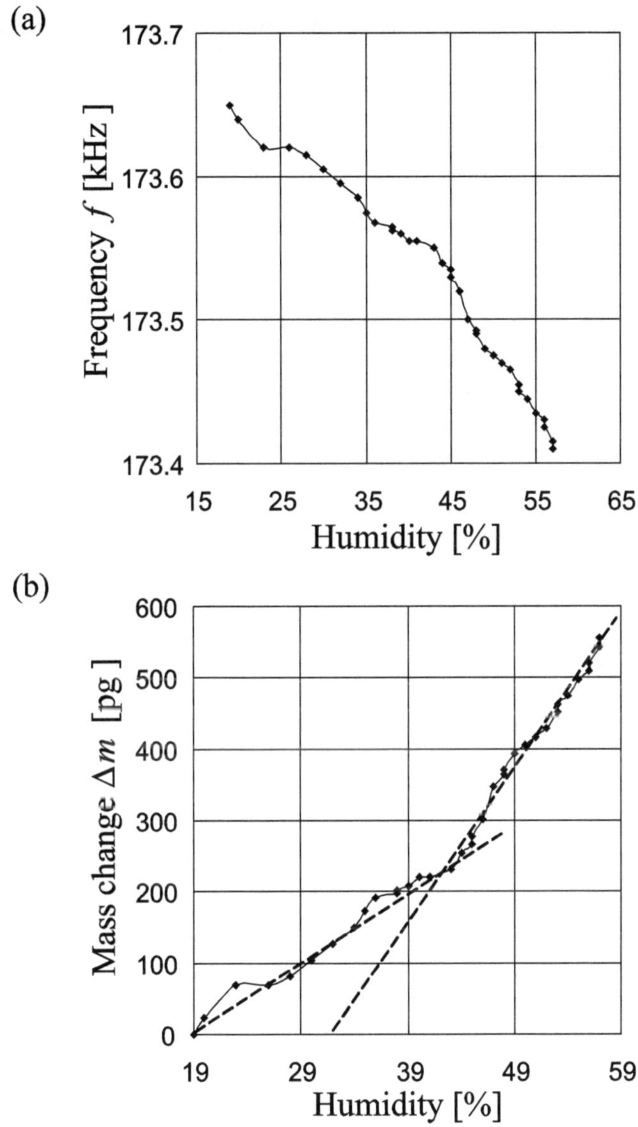

Fig. 6.13. Experimental results for the water molecule adsorption on the V-shaped piezoresistive cantilever. **a** Resonance frequency of the cantilever. **b** Humidity dependence of the mass change due to the water molecule adsorption

proposed in [25]. In our present experiment, it can be considered that water coverage expansion was achieved by an island growth mechanism in the humidity range under 42%. On the other hand, the coverage expansion shifted to a layer growth mechanism in the humidity range over 42%. In order to determine the amount of water adsorbed on the cantilever, we estimated the thickness of the water film at the critical humidity of 42%. Assuming that water adsorption occurs on the cantilever uniformly, the mass of a one-monolayer-thick water film, M_1, is given by

$$M_1 = \left(\frac{M_{H_2O}}{N_A}\right)^{1/3} S\,, \tag{6.29}$$

where M_{H_2O} is the molecular mass of water (18 g/mol), N_A is Avogadro's number (6.0×10^{23}), and S is the area of the cantilever surface (5.8×10^4 cm^2). In this case, M_1 is about 18 pg. Because the mass of the adsorbed water is 230 pg at a humidity of 42%, the number of water layers is about 13 monolayers. This value is almost consistent with the number of 12 monolayers found in the experiment for water molecule detection using a laser beam deflection detection system [25].

The critical humidity of the water adsorption is reported to be about 68% in [13] and about 60% in [25]. These values are different from the critical humidity of 42% found in our present experiments. We consider that the critical humidity changes depending on the condition of the cantilever surface, which was covered by zeolite and gold film in the studies described in [13] and [25], respectively.

In order to obtain a higher mass sensitivity, we used a rectangular piezoresistive cantilever, as shown in Fig. 6.8b. The measurement system in this experiment was the same as in the previous experiment. The resonance frequency obtained for the cantilever was 286.7 kHz in air. The locked center frequency and the PLL circuit bandwidth were set at 285.9 kHz and 300 Hz, respectively.

Figure 6.14a shows the resonance frequency shift, Δf, of the cantilever when the humidity increases from 19 to 26%. The humidity was measured by a ceramic humidity sensor with a humidity resolution of 1%. The resonance frequency shift is about 114 Hz for the humidity change from 19 to 26%. Figure 6.14b shows the humidity dependence of the mass change due to the water molecule adsorption on the cantilever surface, which is based on the mass of adsorbed water at a humidity of 19%. The total mass of the adsorbed water is about 36 pg in the humidity range from 19 to 26%. The resonance frequency shift is about 114 Hz, as shown in Fig. 6.14a. Consequently, the estimated mass detection sensitivity is about 320 fg/Hz, which is 100 times higher than the sensitivity of the quartz crystal oscillation method. This result demonstrates that this technique has the potential to enable the realization of a practical and highly sensitive molecular sensor, if we can detect frequency shifts with a resolution of less than 1 Hz. In this experiment, a minimum resonance frequency shift of 3 Hz, which corresponds to the detectable minimum mass change of about 1 pg, was detected. In order to estimate the minimum detectable frequency shift, we discuss thermal noise of the cantilever [40]. The mean-square frequency modulation due to thermal noise is given by

$$\langle \delta\omega \rangle = \sqrt{\frac{\omega_0 k_B T B}{k_L Q \langle z_{osc}^2 \rangle}}\,, \tag{6.30}$$

where ω_0 is the resonance frequency, k_B is the Boltzmann constant, T is the temperature, B is the bandwidth, k_L is the spring constant, Q is the quality factor, and $\langle z_{osc}^2 \rangle$ is the mean-square amplitude of the self-oscillating cantilever. In the case of the rectangular piezoresistive cantilever, ω_0 is 286.7 kHz, T is 300 K, B is 300 Hz, k_L is 40 N/m, $Q$is 350, and $\langle z_{osc}^2 \rangle$ is about 100 nm. This means that the mean-square frequency modulation is about 0.02 Hz. This result shows that we can detect frequency shifts of less than 1 Hz.

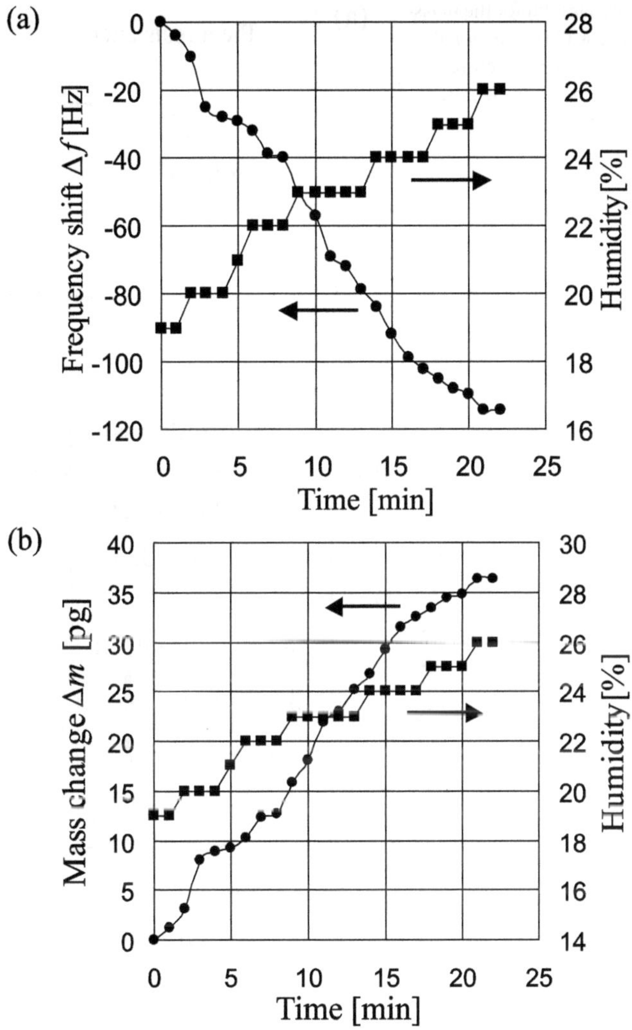

Fig. 6.14. Experimental results for the water molecule adsorption on the commercially available rectangular piezoresistive cantilever. **a** Resonance frequency shift (*circles*) of the cantilever with increasing humidity (*squares*). **b** Humidity (*squares*) dependence of the mass change (*circles*) due to the water molecule adsorption

To study the calibration of small-mass detection for this sensor, we prepared a cantilever on which a thin permalloy film was deposited by RF sputtering (MB96-1011, ULVAC), as shown in Fig. 6.15. Then, we measured the resonance frequency of the cantilever in air. The thickness of the deposited film was measured by atomic force microscopy (AFM). Figure 6.16 shows the frequency spectrums of the cantilever before and after the permalloy film deposition. The resonance frequency and quality (Q) factor were changed from 321.0 to 313.9 kHz and from 321 to 314 by the deposition, respectively. When a resonance frequency shift, Δf, occurs due to

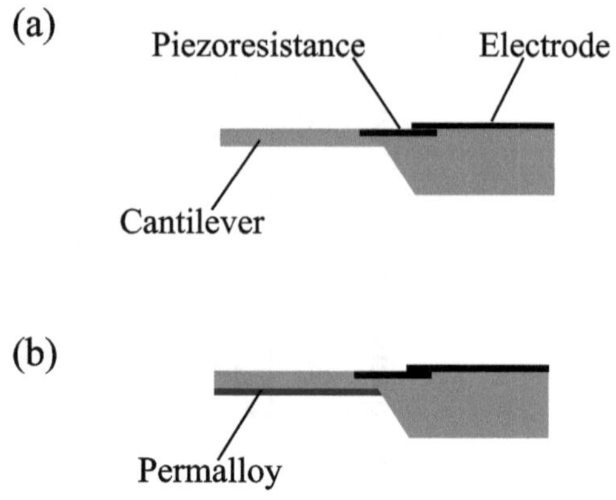

Fig. 6.15. The preparation for the mass-detection check. **a** Cross-sectional scheme of the cantilever before permalloy film deposition. **b** After permalloy film deposition

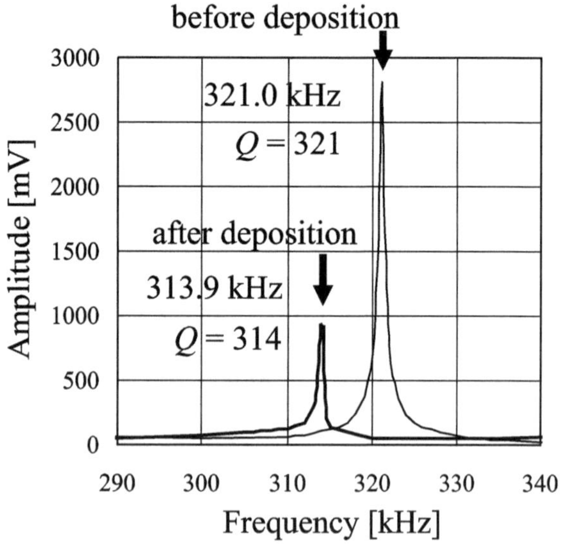

Fig. 6.16. Frequency spectrums of the cantilever before permalloy film deposition (*thin line*) and after deposition (*thick line*)

molecular adsorption, the mass change, Δm, can be obtained using (6.7). From the equation and the frequency shift of $\Delta f = 7.1$ kHz found in this experiment, we estimated the mass of the deposited film, Δm_e, to be about 2.02 ng. The thickness of the deposited film was about 90 nm, as measured by AFM. The mass, Δm_c, of the permalloy film, which had a thickness of 90 nm and a surface area of 4.9×10^{-5} cm^2 (single side of the cantilever), was about 3.83 ng. The Δm_c value almost agrees with the Δm_e value. The deviation between Δm_c and Δm_c can be considered by increasing the cantilever spring constant, which is suggested by decreasing the Q factor and decreasing the amplitude after deposition.

6.6.2 Antigen and Antibody Detection in Water

We measured the reaction between the antibody immunoglobulin G (22.5 μM) and the antigen egg albumen (440 μM). The molecular masses of the antibody and the antigen are 160 and 45 kDa, respectively. Figure 6.17 shows a schematic diagram of the measurement system for studying the reaction between the antigen and the antibody. The piezoresistive cantilever was submerged in 10 ml water. In the first experiment, we measured the resonance frequency shift of the cantilever upon supplying 10 μl (4.40 nmol) of the antigen using a micropipette, as shown in Fig. 6.18a. After the frequency shift saturated, we supplied 10 μl (0.225 nmol) of the antibody. In the second experiment, we supplied 10 μl of the antibody first and 10 μl of the antigen second, as the reverse of the first experiment, as shown in Fig. 6.18b.

Figure 6.19 shows optical images of the sensor system and the piezoresistive cantilever submerged in water, respectively. First, we measured the resonance frequency of the cantilever in air and in water, as shown in Fig. 6.20a and b, respectively. The resonance frequency and Q factor in air obtained were about 324.4 kHz and 368.6, respectively. Those in water were about 188.5 kHz and 11.8, respectively. In the water experiment, the resonance frequency shifted to a lower frequency, the Q factor decreased, and the amplitude decreased. These results suggest the influence of increased water viscosity. When we used a positive-feedback system, we obtained an oscillation frequency of about 190 kHz in water.

Figure 6.21a shows the experimental result for the frequency shift after supplying small volumes of the antigen and the antibody into the water bath. Seven minutes after supplying the antigen, the frequency shift became saturated. We assume that the cantilever surface was covered by the antigen. Next, we supplied a small volume of antibody into the water bath. After 18 min, the frequency shift became saturated again, which indicates that the reaction between the antigen and the antibody had finished. We estimated the mass change using (6.7), as shown in Fig. 6.21b. The mass change due to antigen adsorption was about 27.7 pg, and that due to antibody reaction was about 41.3 pg. Thus, a mass sensitivity of about 190 fg/Hz and a mass resolution of about 500 fg were obtained. This mass sensitivity is 100 times higher than that of

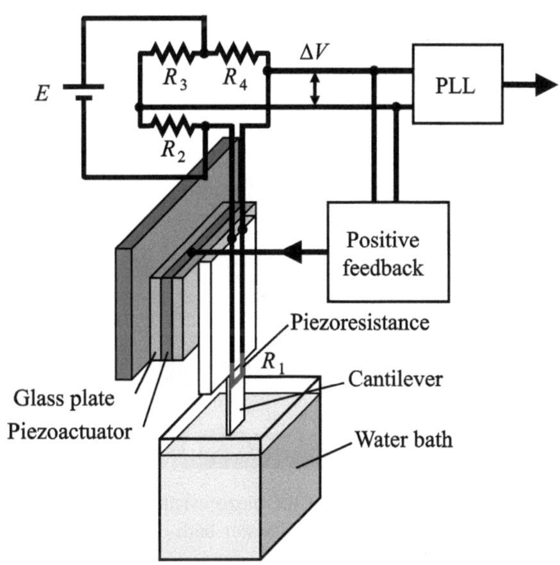

Fig. 6.17. Harmonic-vibration mass biosensor system using a piezoresistive cantilever in water

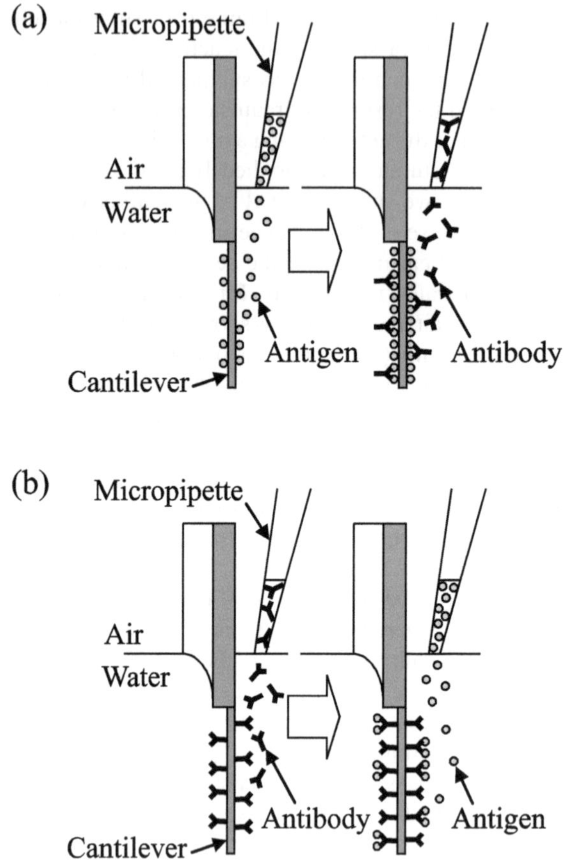

Fig. 6.18. The biochemical reactions for the adsorption of the antigen and the antibody. **a** Adsorption of the antigen on the cantilever followed by that of the antibody on the cantilever in water. **b** Adsorption of the antibody and antigen in the reverse process of that in **a**

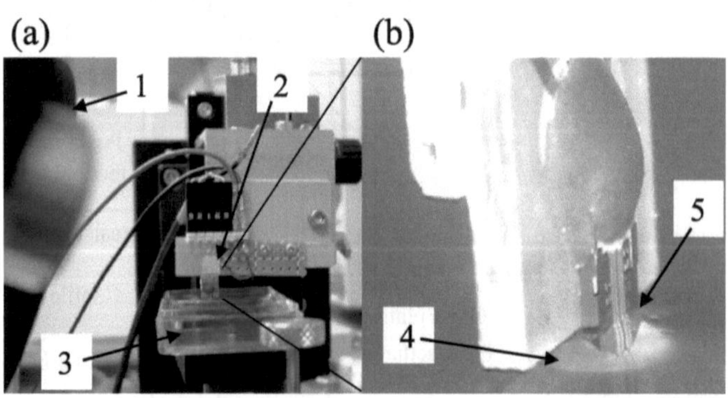

Fig. 6.19. Optical images of the piezoresistive cantilever biosensor. *a* Biosensor setup: *1* CCD camera, *2* cantilever sensor, *3* water bath. *b* Cantilever submerged in water: *4* water surface, *5* piezoresistive cantilever

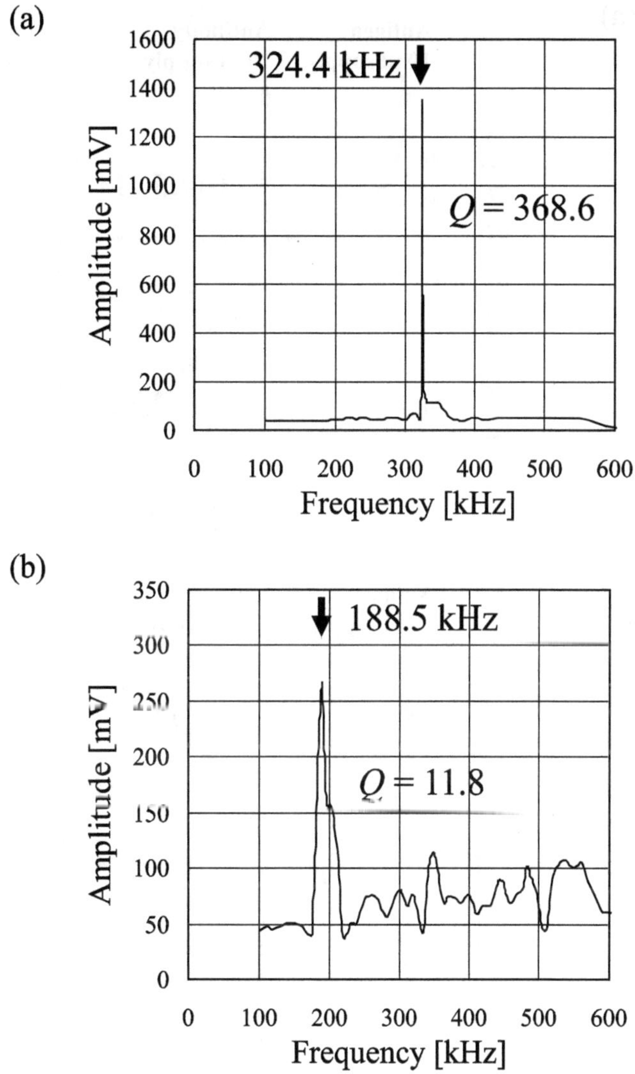

Fig. 6.20. Frequency spectrums of the cantilever **a** in air and **b** in water

the quartz crystal oscillation method. In this experiment, we used a commercially available piezoresistive cantilever designed for AFM measurement. If a cantilever with a small mass and a high resonance frequency is designed, a higher mass sensitivity of less than 190 fg/Hz can be achieved. In addition, we used a PLL with a locked center frequency and a detectable resonance frequency bandwidth of 190 and 1 kHz, respectively. We detected a minimum resonance frequency shift of about 2.5 Hz, which corresponds to the noise level of this sensor system. The mean-square frequency modulation due to thermal noise is less than 1 Hz [31,40]. Therefore, if we optimize the electric circuits of the detection system, we can detect frequency

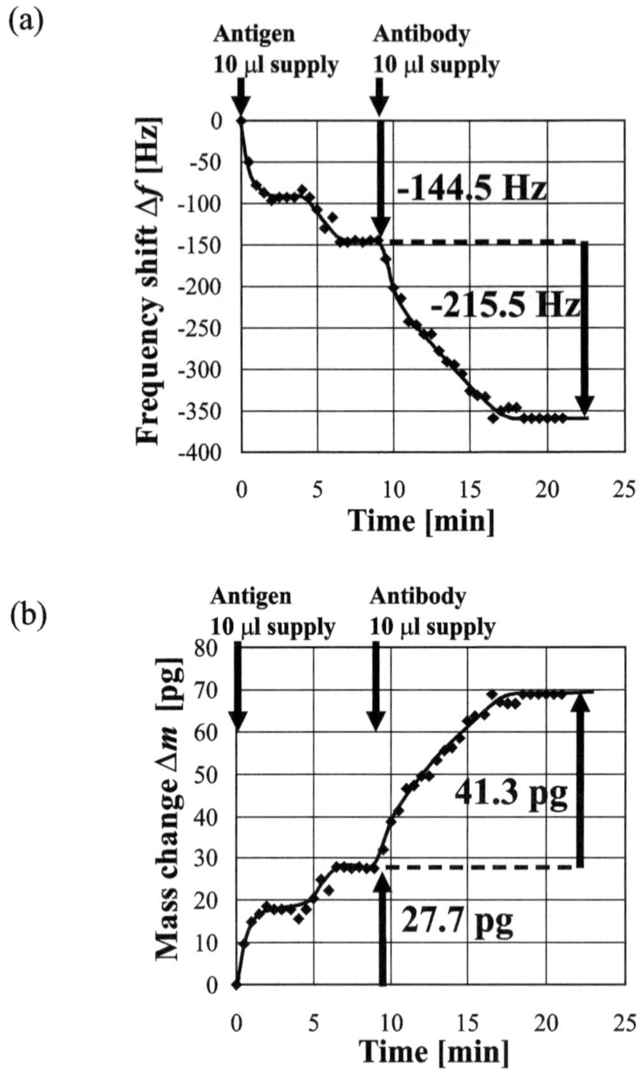

Fig. 6.21. Experimental results for the antigen–antibody reaction on the commercially available piezoresistive cantilever. **a** Frequency shift. **b** Mass change

shifts of less than 1 Hz. From the mass change, we were able to estimate the number of adsorbed molecules of the antigen and the antibody. The number of antigen molecules adsorbed on the cantilever was about 3.7×10^8, and that of the antibody reacted with the antigen was about 1.5×10^8. It has been reported in bioscience that an antibody has two binding sites for an antigen; therefore, perfect binding means that the ratio of antibody to antigen is 1:2. Our experimental results showed a ratio of 0.8:2.

Figure 6.22 shows the experimental results for the frequency shift and the mass change. The experiment was carried out using the reverse experimental procedure,

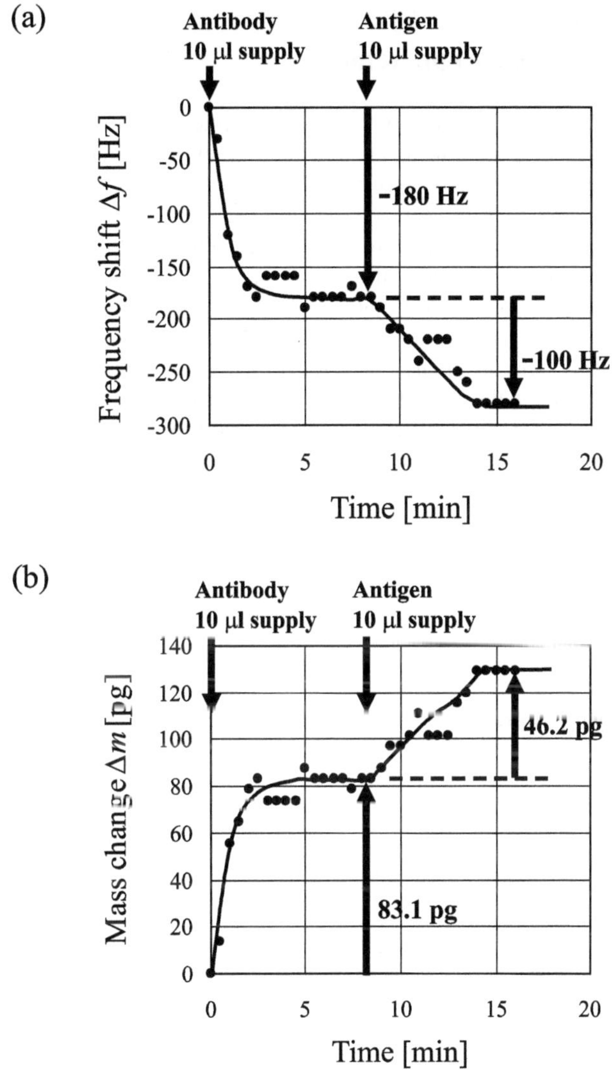

Fig. 6.22. Experimental results for the antibody–antigen reaction on the commercially available piezoresistive cantilever in the reverse process of that for Fig. 6.21. **a** Frequency shift. **b** Mass change

wherein we supplied the antibody first and the antigen second. We observed that the frequency shift saturated at 5 min owing to the adsorption of the antibody on the cantilever. At 13 min, the frequency shift saturated again, which indicates that the reaction between the antibody and the antigen had finished. The estimated mass changes of the adsorbed antibody and antigen were 83.1 and 46.2 pg, respectively. The estimated number of antibody molecules was about 3.1×10^8, and that of antigen molecules was about 6.2×10^8. These results show that the binding ratio for the antibody and the antigen is 1:2; therefore, we consider that the biosensor successfully monitored the reaction between the antibody and the antigen.

6.7 Prospective Applications

We conclude that the self-sensing mass sensor using a piezoresistive cantilever has the potential to enable femtogram-order mass detection in air as well as in water. Using this sensor system, we can monitor environmental changes in air and chemical or biological molecular reactions in water. It has a much broader range of application in the various fields of chemistry, biotechnology, and medical science, e. g., immunochemical assay, amino acid analysis, allergy testing, cancer testing, and genetic testing. Although some of the applications have already been studied, e. g., a cantilever array sensor system [14,41] and cantilever functionalization [19–21,42, 43], there are many obstacles for the commercialization of such products. In order to realize these applications, we must devise ways to modify the cantilever surface as well as time-efficient multiple-testing methods. Additionally, we must improve the sensitivity of the sensor to attain high reliability and high resolution.

Acknowledgements. The authors wish to express their special gratitude to H. Okano (Tokyo Sokki Kenkyujo Co.) and T. Izumi (Gunma University) for their valuable discussions and tremendous support. We would also like to thank Y. Fujinuma, T. Hieida, T. Chiyoma, and A. Ikeuchi for their great help and contribution. The authors gratefully acknowledge partial financial support from the Highland Kanto Liaison Organization.

References

1. James AT, Martin AJP (1952) Biochemistry 50:679
2. Liu R, Zhou JL, Wilding A (2004) J Chromatogr A 1022:179
3. Ma WT, Fu KK, Cai Z, Jiang GB (2003) Chemosphere 52:1627
4. Hamblin C, Barnett ITR, Hedger RS (1986) J Immunol Methods 93:115
5. Avremeas S (1969) Immunochemistry 6:43
6. Aozawa O, Ohta S, Nakano T, Miyata H, Nomura T (2001) Chemosphere 45:195
7. Janshooff A (2004) In: Baltes H (ed) Sensor update, vol 9. Wiley, New York, pp 313–354
8. Matsuno H, Niikura K, Okahata Y (2001) Chem Eur J 7:3305
9. Cooper MA (2004) J Mol Recognit 17:286
10. Liedberg B, Lundström I (1993) Sens Actuators B 11:63
11. Melendez J, Carr R, Bartholomew DU, Kukanskis K, Elkind J, Yee S, Furlong C, Woodbury R (1996) Sens Actuators B 35:212
12. Saenko E, Sarafanov A, Ananyeva N, Behre E, Shima M, Schwinn H, Josic D (2001) J Chromatogr A 921:49
13. Berger R, Gerber C, Lang HP, Gimzewski JK (1997) Microelectron Eng 35:373
14. Lang HP, Baller MK, Berger R, Gerber C, Gimzewski JK, Battiston FM, Fornaro P, Ramseyer JP, Meyer E, Güntherodt HJ (1999) Anal Chim Acta 393:59
15. Binnig G, Quate CF, Gerber C (1986) Phys Rev Lett 56:930
16. Gimzewski JK, Gerber C, Meyer E, Schlittler RR (1994) Chem Phys Lett 217:589
17. Berger R, Gerber C, Lang HP, Gimzewski JK, Meyer E, Güntherodt HJ (1996) Appl Phys Lett 69:40
18. Berger R, Lang HP, Gerber C, Gimzewski JK, Fabian JH, Scandella L, Meyer E, Güntherodt H-J (1998) Chem Phys Lett 294:363
19. Raiteri R, Nelles G, Butt H-J, Knoll W, Skladal P (1999) Sens Actuators B 61:213

20. Fritz J, Baller MK, Lang HP, Rothuizen H, Vettiger P, Meyer E, Güntherodt HJ, Gerber C, Gimzewski JK (2000) Science 288:316
21. Hansen KM, Ji H-F, Wu G, Datar R, Cote R, Majumdar A, Thundat T (2001) Anal Chem 73:1567
22. Lavrik NV, Datskos PG (2003) Appl Phys Lett 82:2697
23. Ilic B, Craighead HG, Krylov S, Senaratne W, Ober C, Neuzil P (2004) J Appl Phys 95:3694
24. Sone H, Fujinuma Y, Hieida T, Hosaka S (2003) Proc SICE Annu Conf 2985
25. Sone H, Fujinuma Y, Hosaka S (2004) Jpn J Appl Phys 43:3648
26. Boisen A, Thaysen J, Jensenius H, Hansen O (2000) Ultramicroscopy 82:11
27. Porter TL, Eastman MP, Pace DL, Bradley M (2001) Sens Actuators A 88:47
28. Zhou J, Li P, Zhang S, Huang Y, Yang P, Bao M, Ruan G (2003) Microelectron Eng 69:37
29. Tang Y, Fang J, Yan X, Ji H-F (2004) Sens Actuators B 97:109
30. Sone H, Fujinuma Y, Hieida T, Chiyoma T, Okano H, Hosaka S (2004) Proc SICE Annu Conf 1508
31. Sone H, Okano H, Hosaka S (2004) Jpn J Appl Phys 43:4663
32. Hosaka S, Chiyoma T, Ikeuchi A, Okano H, Sone H, Izumi T (2006) Curr Appl Phys 6:384
33. Sone H, Ikeuchi A, Izumi T, Okano H, Hosaka S (2006) Jpn J Appl Phys 45:2301
34. Tortonese M, Barrett RC, Quate CF (1993) Appl Phys Lett 62:834
35. Linnemann R, Gotszalk T, Hadjiiski L, Rangelow IW (1995) Thin Solid Films 264:159
36. Gotszalk T, Grabiecl P, Rangelow IW (2002) Ultramicroscopy 82:39
37. Chen GY, Warmack RJ, Thundat T, Allison DP, Huang A (1994) Rev Sci Instrum 65:2532
38. Luna M, Colchero J, Gil A, Gomez-Herrero J, Baro AM (2000) Appl Surf Sci 157:393
39. Gil A, Colchero J, Luna M, Gomez-Herrero J, Baro AM (2000) Langmuir 16:5086
40. Albrecht TR, Grütter P, Horne D, Rugar D (1991) J Appl Phys 69:668
41. Lang HP, Berger R, Andreoli C, Brugger J, Despont M, Vettiger P, Gerber C, Gimzewski JK, Ramseyer JP, Meyer E, Güntherodt H-J (1998) Appl Phys Lett 72:383
42. Wu G, Datar RH, Hansen KM, Thundat T, Cote RJ, Majumdar A (2001) Nat Biotechnol 19:856
43. Hyun S-J, Kim H-S, Kim Y-J, Jung H-I (2006) Sens Actuators B 117:415

28 Microfabricated Cantilever Array Sensors for (Bio-)Chemical Detection

Hans Peter Lang · Martin Hegner · Christoph Gerber

From volume IV

List of Abbreviations

AFM	atomic force microscopy
MW	molecular weight
PCA	principal component analysis
PEEK	poly-etheretherketone
PDMS	poly-dimethyl-siloxane
PSD	position sensitive detector
RIE	reactive ion etching
SFM	scanning force microscopy
VCSEL	vertical-cavity surface-emitting laser
μFN	microfluidic network

28.1 Introduction

28.1.1 Sensors

A sensor is a device that detects, or senses, a signal. A sensor is also a transducer, i.e. it transforms one form of energy into another or responds to a physical parameter. Transducers can be electrochemical (pH probe), electromechanical (piezoelectric actuator, quartz, strain gauge), electroacoustic (gramophone pick-up, microphone), photoelectric (photodiode, solar cell), electromagnetic (antenna, photocell, tape or harddisk head for storage applications), magnetic (Hall effect sensor), electrostatic (electrometer), thermoelectric (thermocouple, thermo-resistors), electrical (capacitor, resistor) or mechanical (deflection sensors). Sensors either directly indicate a state or a value (mercury thermometer), or use an indicator that indirectly displays the state or value by an analog or digital converter (display, computer).

(Bio-)chemical sensors convert changes of physical or chemical parameters due to the presence of molecules in the environment into a recordable signal (Fig. 28.1).

In: H. Fuchs, B. Bhushan (eds), *Biosystems - Investigated by Scanning Probe Microscopy*,
DOI 10.1007/3-540-26914-2_6, © Springer-Verlag Berlin Heidelberg 2010

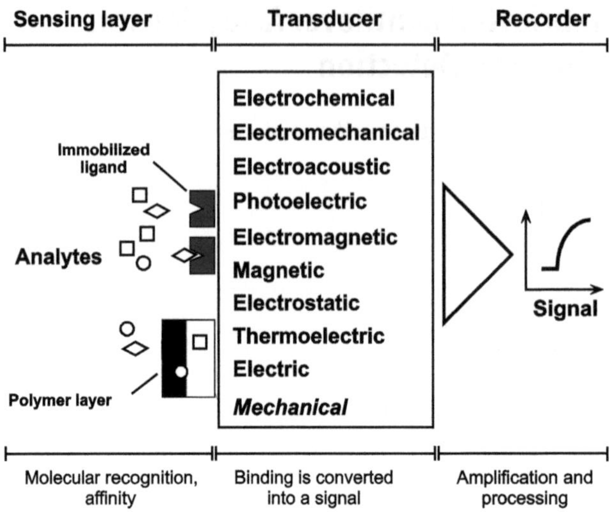

Fig. 28.1. Schematic of general sensor transduction. Analyte molecules present in the environment are recognized by the sensing layer of the sensor. Either recognition is very specific, i.e. recognition sites on immobilized ligand molecules recognize the analyte molecules, or recognition is partially specific, e.g. the analyte molecules diffuse into a polymer layer. The binding event is then transduced into a recordable signal via a variety of transduction mechanisms. The acquired signal is then further amplified and processed

28.1.2 Cantilevers

The term cantilever is understood here as a microfabricated rectangular bar-shaped structure that is longer than it is wide and has a thickness that is much smaller than its length or width. A cantilever is a horizontal structural element supported only at one end on a chip body; the other end is free. At the free end, a microfabricated sharp tip might be attached for use as a local probe to scan a surface. Cantilever beams have been applied since the mid 1980s as sensitive structures to measure interatomic forces in the piconewton range using a technique called scanning force microscopy (SFM) or atomic force microscopy (AFM) [1]. In this method, a cantilever with a sharp tip is scanned across a conductive or nonconductive surface by the use of an x-y-z actuator system (e.g. a piezoelectric scanner). Scanning is performed in a pattern of adjacent parallel lines to cover a rectangular or square area of the sample surface. Typical scanning speeds range from 1 nm/s to several micrometers per second, depending on the size of the area to be scanned. The tip can either be in direct contact with the surface (contact mode) or oscillated to interact with the surface only for a short time during the oscillation cycle (dynamic mode, noncontact mode). Common to SFM methods is the interaction of the cantilever tip with the surface. This interaction can be used to control a feedback loop intended to keep the force or force gradient between cantilever tip and sample surface constant. By recording the correction signal that has to be applied to the z-actuation drive to keep the interaction between tip and sample surface constant, a topography image of the sample surface can be obtained. SFM methods are nowadays well-established in scientific research, education and to a certain extent also in the industry. Beyond imaging of surfaces, other applications of cantilevers have been demonstrated.

The use of a sharp tip at the cantilever apex is not required for application as the (bio-)chemical sensor described here. Because of their flexibility, cantilevers can not only be used for probing the surface profile of a sample, but also to monitor processes taking place on the surface of the beam [2]. The use of cantilever beams as sensors, clamped at one end and freestanding at the other, allows the adsorption of molecules to be observed with unprecedented accuracy, because the formation of molecule layers on the cantilever surface will generate surface stress. This eventually results in a bending of the cantilever, provided the adsorption preferentially occurs on one surface of the cantilever. Adsorption is controlled by coating one surface (e.g. the upper surface) of a cantilever with a thin layer of a material that shows affinity to molecules in the environment (sensor surface). This surface of the cantilever is referred to as the "functionalized" surface. The other surface of the cantilever (e.g. the lower surface) may be left uncoated or be coated with a passivation layer, i.e. a chemical surface that does not exhibit significant affinity to the molecules in the environment to be detected (see Fig. 28.2). To facilitate the establishment of functionalized surfaces, a metal layer is often evaporated onto the surface designed as sensor surface. Metal surfaces, e.g. a gold surface, may be utilized to covalently bind a monolayer that represents the chemical surface sensitive to the molecules to be detected from environment. A typical example is a monolayer of thiol molecules covalently bound to a gold surface. The gold layer is also favorable for use as a reflection layer, if the bending of the cantilever is read out via an optical beam-deflection method.

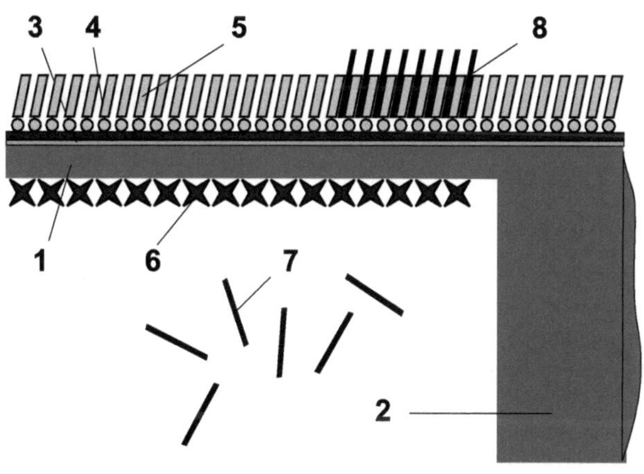

Fig. 28.2. Schematic sideview representation of a functionalized cantilever: (**1**) cantilever beam with a typical thickness of 1 μm, typically made from silicon, (*2*) chip body with a typical thickness of 500 μm, (**3**) 2 nm thick evaporated titanium layer required for adherence of (**4**) a 20 nm thick evaporated gold layer, (**5**) functional sensing layer, e.g. 1 nm thick self-assembled thiol monolayer, (**6**) passivation layer (several nanometers thick), e.g. a silanized polyethylene-glycol layer, (**7**) target molecules from the environment, (**8**) area in which target molecules have docked to the functionalized surface

28.1.3
Cantilever Operating Modes

28.1.3.1
Static Mode

The continuous bending of a cantilever as a function of molecular coverage with molecules is referred to as an operation in the "static mode", see Fig. 28.3. Adsorption of molecules onto the functional layer generates stress at the interface between the functional layer and the forming molecular layer. The stress is transduced towards the site at which the molecules of the functional layer are attached to the cantilever surface. Because the forces within the functional layer try to keep the distance between molecules constant, the cantilever beam responds with bending due to its extreme flexibility. The resulting surface stress change is calculated according to Stoney's formula [3]:

$$\Delta\sigma = Et^2/[4R(1-\nu)] \,, \tag{28.1}$$

where E is Young's modulus ($E_{\mathrm{Si}} = 1.3 \times 10^{11}\,\mathrm{N/m^2}$ for Si(100)), t the thickness of the cantilever, ν Poisson's ratio ($\nu_{\mathrm{Si}} = 0.25$), and R the bending radius of the cantilever.

The static-mode operation can be performed in various environments. In its simplest embodiment, molecules from gaseous environment adsorb on the functionalized sensing surface and form a molecular layer (Fig. 28.3a), provided

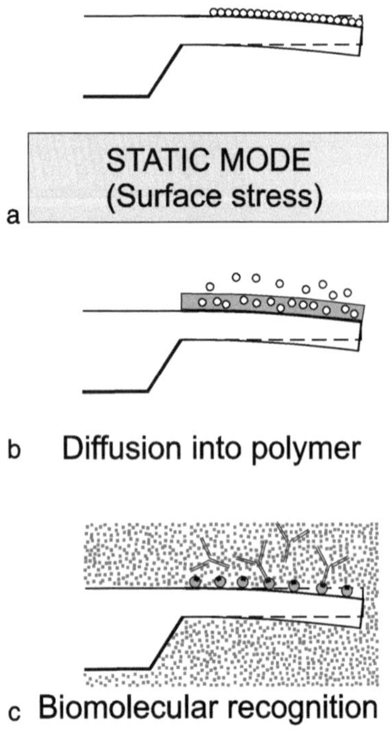

Fig. 28.3. (**a**) Static-mode operation. An adsorbing layer of molecules produces a change in surface stress on the upper functionalized surface of the cantilever, resulting in bending of the cantilever. (**b**) Diffusion of target molecules from the environment into a polymer layer induces swelling of the polymer layer and thereby a bending of the cantilever, because the swollen polymer layer expands more than the silicon cantilever. (**c**) Static-mode operation in liquids. Target molecules from the liquid environment are molecularly recognized by binding sites on the functionalized sensing layer on the upper surface of the cantilever

that the molecules exhibit some affinity to the surface. Polymer sensing layers show a partial sensitivity, because molecules from the environment diffuse into the polymer layer at different rates, mainly depending on the size and solubility of the molecules in the polymer layer (Fig. 28.3b). A wide range of hydrophilic/hydrophobic polymers can be selected, differing in their suitability for polar/unpolar molecules. Thus, the polymers can be chosen according to what the applications demand.

Static-mode operation in liquids, however, usually requires rather specific sensing layers, based on molecular recognition, such as DNA hybridization or antigen-antibody recognition (Fig. 28.3c).

28.1.3.2
Dynamic Mode

Because the bending of the cantilever is a direct result of the adsorbtion of molecules onto the cantilever surface, it is rather difficult to obtain reliable information on the amount of molecules adsorbed, as the surface coverage is basically not known. In addition, molecules on the surface might be exchanged with molecules from the environment in a dynamic equilibrium.

However, mass changes can be determined accurately by operation of a cantilever actuated at its eigenfrequency. The eigenfrequency is equal to the resonance frequency of an oscillating cantilever if the elastic properties of the cantilever remain unchanged during the molecule adsorption process and damping effects are negligible. This operation mode is called the dynamic mode (e.g., the use as a microbalance, Fig. 28.4a). Owing to mass addition on the cantilever surface, the cantilever's eigenfrequency will shift to a lower value. The mass change on a rectangular cantilever is calculated [4] according to

$$\Delta m = (k/4\pi^2) \times \left(1/f_1^2 - 1/f_0^2\right) , \tag{28.2}$$

where f_0 is the eigenfrequency before the mass change occurs, and f_1 the eigenfrequency after the mass change. The spring constant k of the cantilever is obtained using

$$k = Et^3 w/4l^3 , \tag{28.3}$$

where l, w, and t denote the length, width, and thickness of the cantilever, respectively.

Mass-change determination can be combined with varying environment temperature conditions (Fig. 28.4b) to obtain the method introduced in the literature as "micromechanical thermogravimetry" [5]. The sample to be investigated has to be mounted at the apex of the cantilever. Its mass should not exceed several hundred nanograms. In the case of adsorption/desorption/decomposition processes, mass changes in the picogram range can be observed in real time by tracking the resonance-frequency shift.

Dynamic mode operation in liquid environment poses problems, such as high damping of the cantilever oscillation due to high viscosity of the surrounding media.

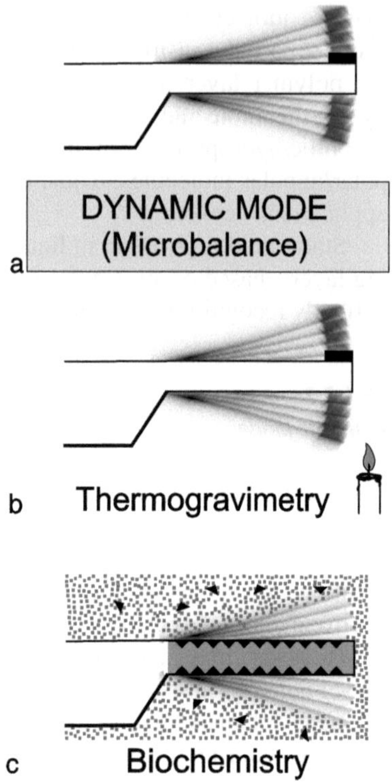

Fig. 28.4. (**a**) Cantilever operation in dynamic mode. The cantilever is excited at its resonance frequency. Tracking the resonance frequency allows one to determine mass changes. (**b**) Micromechanical thermogravimetry. Mass changes are measured in dependence of temperature as an external parameter. (**c**) Detection of biochemical processes involving mass changes due to adsorption and binding of molecules

This results in a low quality factor Q of the oscillation, and the resonance frequency shift is difficult to track with high resolution. The quality factor is defined as

$$Q = 2\Delta f / f_0 \, . \tag{28.4}$$

Whereas in air a frequency resolution of below 1 Hz is easily achieved, resolution values of about 20 Hz are already considered very good for measurements in a liquid environment.

28.1.3.2.1
Operation in a Large Damping Environment

In the case of damping or changes of the elastic properties of the cantilever during the experiment, e.g. a stiffening or softening of the spring constant by adsorption of a molecule layer, the measured resonance frequency will not be exactly the same as the eigenfrequency, and the mass derived from the frequency shift will be inaccurate. In a medium, the vibration of a cantilever is described by the model of a driven damped harmonic oscillator:

$$m^* \frac{\mathrm{d}^2 x}{\mathrm{d}t^2} + \gamma \frac{\mathrm{d}x}{\mathrm{d}t} + kx = F \cos(2\pi f t) \, , \tag{28.5}$$

where $m^* = \text{const.}(m_c + m_l)$ is the effective mass of the cantilever (for a rectangular cantilever the constant is 0.25). Especially in liquids, the mass of the co-moved liquid m_l adds significantly to the mass of the cantilever m_c. The term $\gamma \frac{dx}{dt}$ is the drag force due to damping, $F \cos(2\pi ft)$ is the driving force executed by the piezo-oscillator, and k is the spring constant of the cantilever.

If no damping is present, the eigenfrequencies of the various oscillation modes of a bar-shaped cantilever are calculated according to:

$$f_n = \frac{\alpha_n^2}{2\pi}\sqrt{\frac{k}{3(m_c + m_l)}}\,, \tag{28.6}$$

where f_n are the eigenfrequencies in the n-th mode, α_n are constants depending on the mode: $\alpha_1 = 1.8751$, $\alpha_2 = 4.6941$, $\alpha_n = \pi(n - 0.5)$; k is the spring constant of the cantilever, m_c the mass of the cantilever, and m_l the mass of the medium surrounding the cantilever, e.g. liquid [6, 7].

If mass is added to the cantilever due to adsorption, the effective mass is calculated according to

$$m^* = \text{const.}(m_c + m_l + \Delta m)\,, \tag{28.7}$$

where Δm is the additional mass adsorbed. Typically, the co-moved mass of the liquid is much larger than the adsorbed mass.

Figure 28.5 clearly shows that the resonance frequency is only equal to the eigenfrequency if no damping is present. With damping, the frequency at which the peak of the resonance curve occurs is no longer identical with that at which

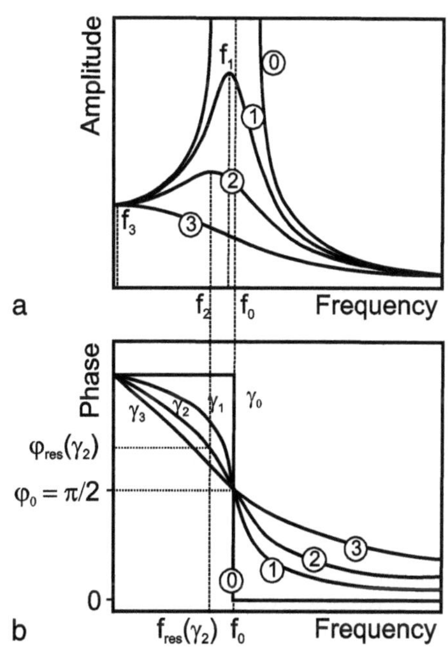

Fig. 28.5. (**a**) Resonance curve with no damping (**0**), and increasing damping (**1**)–(**3**). The undamped curve with resonance frequency f_0 shows a very high amplitude, whereas the resonance peak amplitude decreases with damping. This also involves a shift in resonance frequencies f_1 to f_3 to lower values. (**b**) Corresponding phase curves showing no damping (**0**), and increasing damping (**1**)–(**3**). The step-like phase jump at resonance of the undamped resonance gradually broadens with increasing damping

the turning point of the phase curve occurs. For example, resonance curve 2 with damping γ_2 shows its maximum amplitude at frequency f_2. The corresponding phase would be $\varphi_{\text{res}}(\gamma_2)$, which is not $\pi/2$. If direct resonance frequency tracking or a phase-locked loop is used to determine the frequency of the oscillating cantilever, then only its resonance frequency is detected, but not its eigenfrequency. Remember that the eigenfrequency, and not the resonance frequency, is required to determine mass changes.

28.1.3.3
Heat Mode

If a cantilever is coated with metal layers, thermal expansion differences in cantilever and coating layer will further influence cantilever bending as a function of temperature. This mode of operation is referred to as 'heat mode' and causes cantilever bending because of differing thermal expansion coefficients in the sensor layer and cantilever materials [2] (Fig. 28.6):

$$\Delta z = 1.25 \times (\alpha_1 - \alpha_2) \times (t_1 + t_2)/t_2^2\kappa \times l^3 P/(\alpha_1 t_1 + \alpha_2 t_2)w \,. \tag{28.8}$$

Here α_1, α_2 are the thermal expansion coefficients of the cantilever and coating materials, t_1, t_2 the material thicknesses, P is the total power generated on the cantilever, and κ a geometry parameter of the cantilever device.

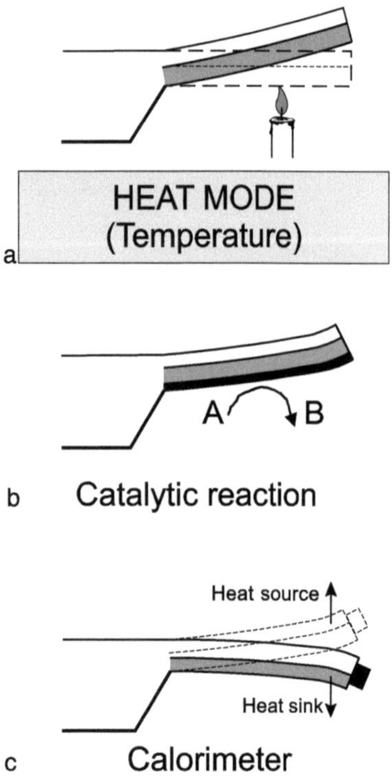

Fig. 28.6. (**a**) Cantilever heat mode. If a material with a different thermal expansion coefficient is evaporated as a thin layer onto the surface of a cantilever, then the cantilever bends when the external temperature is changed. (**b**) The heat change does not have to be generated by a change of temperature, but may also originate from heat production or drain during exothermal or endothermal reactions taking place on the cantilever surface. Exothermal processes are, for example, catalytic reactions. (**c**) If a sample is attached to the cantilever apex, a calorimetric experiment can be performed on the sample

Heat changes are either caused by external influences (change in temperature, Fig. 28.6a), occur directly on the surface by exothermal, e.g. catalytic, reactions (Fig. 28.6b), or are due to material properties of a sample attached to the apex of the cantilever (micromechanical calorimetry, Fig. 28.6c). The sensitivity of the cantilever heat mode is orders of magnitude higher than that of traditional calorimetric methods performed on milligram samples, as it only requires nanogram amounts of sample and achieves nanojoules [8] to picojoules [9, 10] of sensitivity.

These three measurement modes have established cantilevers as versatile tools to perform experiments in nanoscale science with very small amounts of material.

28.1.3.4 Other Measurement Modes

28.1.3.4.1 Photothermal Spectroscopy

When a material adsorbs photons, a fraction of energy is converted into heat. This photothermal heating can be measured as a function of the light wavelength to provide optical absorption data of the material (Fig. 28.7). The interaction of light with a bimetallic cantilever creates heat on the cantilever surface, resulting in a bending of the cantilever [11]. Such bimetallic-cantilever devices are capable of detecting heat flows due to an optical heating power of 100 pW, which is two orders of magnitude better than in conventional photothermal spectroscopy.

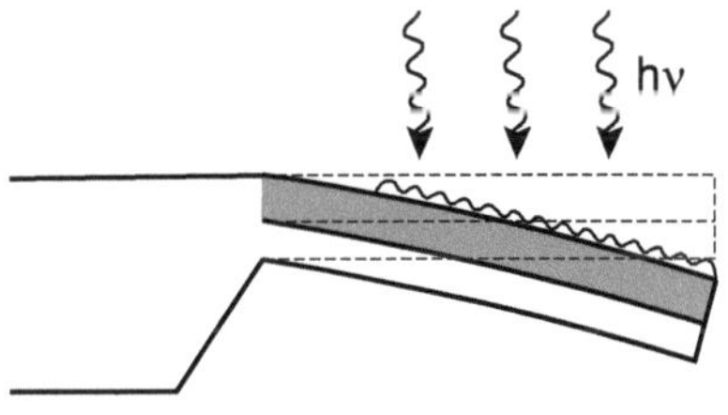

Fig. 28.7. Bending of a bimetallic cantilever due to heat production by incident photons

28.1.3.4.2 Electrochemistry

A cantilever coated with a metallic layer (measurement electrode) on one side is placed in an electrolytic medium, e.g. a salt solution, together with a metallic reference electrode, usually made of a noble metal (Fig. 28.8). Variations of the voltage between measurement and reference electrode induce electrochemical processes on the measurement electrode (cantilever), e.g. adsorption or desorption of ions from the electrolyte solution onto the measurement electrode. These processes lead to a bending of the cantilever due to changes in surface stress and in the electrostatic forces.

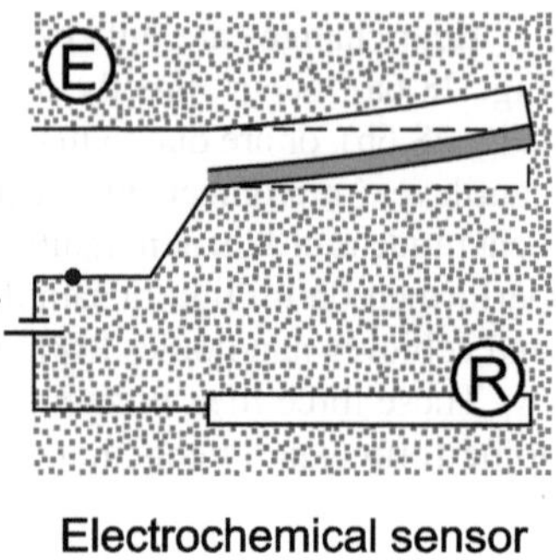

Fig. 28.8. A cantilever coated on one side with a metallic layer that is connected to a voltage source is immersed in an electrolytic fluid (**E**). The other pol e of the voltage source is connected to a reference electrode (**R**). Depending on the polarity and the magnitude of the voltage applied to the cantilever and reference electrode, ions from the electrolyte solution adsorb onto or desorb from the cantilever

28.1.3.4.3
Detection of Electrostatic and Magnetic Forces

The detection of electrostatic and magnetic forces is possible if charged or magnetic particles are deposited on the cantilever [12–14] (Fig. 28.9). If the cantilever is brought into the vicinity of electrostatic charges or magnetic particles, attractive or repulsion forces occur according to the polarity of the charges or magnetic particles present on the cantilever. These forces will result in an upwards or downwards bending of the cantilever. The magnitude of bending depends on the distribution of the charged or magnetic particles on both the cantilever and the surrounding environment according to the laws of electromagnetism.

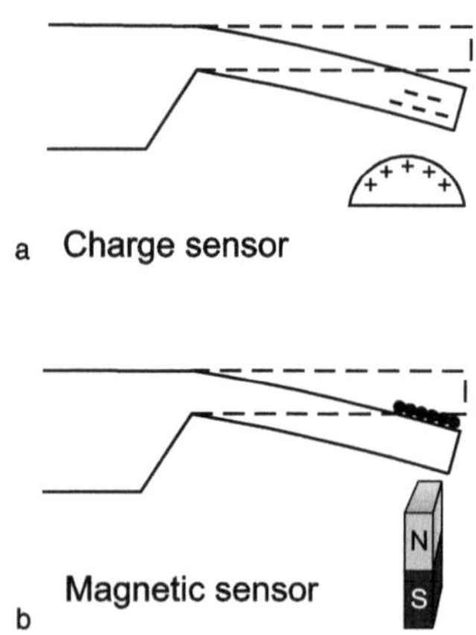

Fig. 28.9. (**a**) Cantilever bending due to electrostatic forces. (**b**) Cantilever bending due to magnetic forces

28.1.4
Cantilever Arrays

28.1.4.1
Disadvantages of Single-Cantilever Measurements

The bending response of a single cantilever is often influenced by various undesired effects, such as thermal drift and unspecific reactions taking place on the uncoated cantilever surface, resulting in additional cantilever bending. To avoid such effects, we have introduced measurements using reference cantilevers [15], i.e. cantilever sensors that will not react with the target analyte molecules (Fig. 28.10). As the difference in signals from the reference and sensor cantilevers shows the net cantilever response, even small sensor responses can be extracted from large cantilever deflections without being predominated by undesired effects.

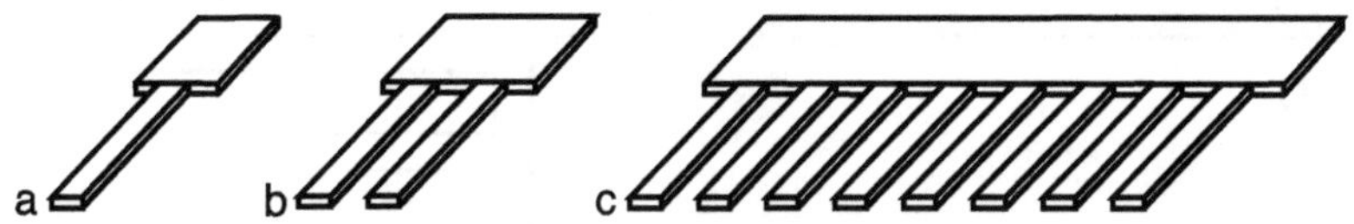

Fig. 28.10. (**a**) Single cantilever: no thermal-drift compensation is possible, both surfaces have to be chemically well-defined. One of the surfaces, e.g. the lower one, has to be passivated, otherwise the cantilever response will be convoluted with undesired effects originating from uncontrolled reactions taking place on the lower surface. (**b**) Dual cantilevers: one of them is used as the sensor cantilever (coated e.g. on the upper side with a molecule layer exhibiting affinity to the molecules to be detected), the other as the reference cantilever, coated with a passivation layer on the upper surface (exhibiting no affinity to the molecules to be detected). Thermal drifts are cancelled out if difference responses, i.e. differences in deflections of sensor and reference cantilevers are taken. Alternatively, both cantilevers are used as sensor cantilevers (sensor layer on the upper surfaces), and the lower surface is passivated. (**c**) Cantilever array: several cantilevers are used either as sensor cantilevers or as reference cantilevers so that multiple difference signals are possible simultaneously. Thermal drift is cancelled out as one surface of all cantilevers, e.g. the lower one, is left uncoated or coated with the same passivation layer

28.1.4.2
Cantilever Array Designs

Here, cantilever array designs are presented that have been used in the cantilever array sensor groups of IBM's Zurich Research Laboratory in Rüschlikon (Switzerland) and of the Institute of Physics, University of Basel (Switzerland). Accordingly, designs used by other research groups will not be discussed.

The first design in 1996 (Fig. 28.11a) consisted of a thick chip-body part (silicon, 500 μm thick, shown in grey in the figure), a doped epitaxial silicon layer for the cantilevers (1 μm thick, white), and a silicon top layer (2 μm thick, dark grey) for labeling purposes. The structures defining the chip body, the cantilevers and the top labeling layers are transferred to the silicon surfaces by means of photolithographic masks. The chips are thinned down from the lower side in an anisotropic wet-etching process using the different doping levels of the cantilevers as an etch stop. A second wet-etch step is applied from the top to release the cantilevers from the top and to etch the marks that are visible from the top. The uniformity of the cantilevers in one chip turned out to be within a few percent of the average resonance frequency. The wet-etch process steps exert considerable mechanical stress on the cantilever during production, so that defects or microcracks on some of the cantilever arrays in the wafer cannot be excluded. As all cantilevers are connected through the

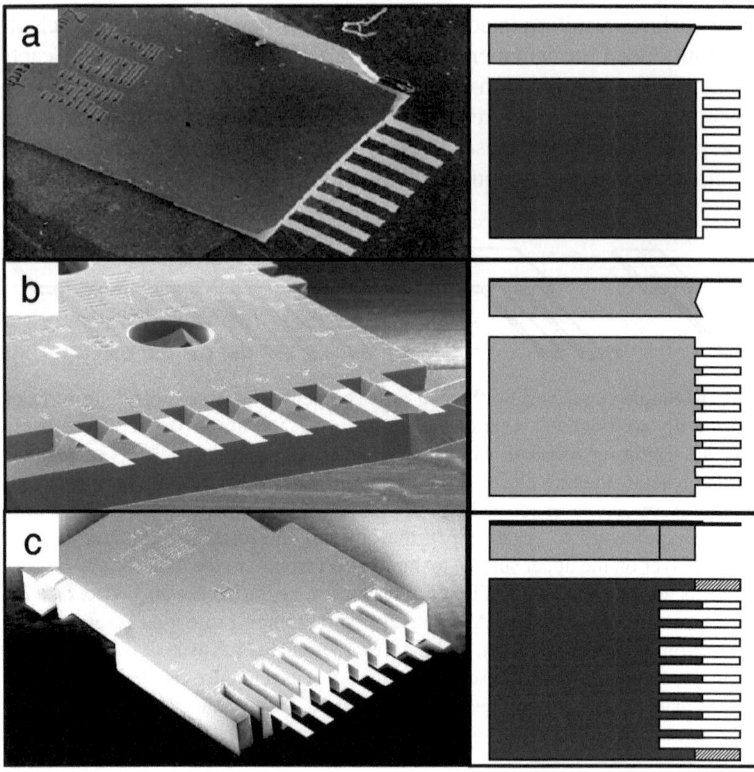

Fig. 28.11. (**a**) Cantilever array design, IBM 1996. (**b**) Cantilever array design, IBM 2000. (**c**) Cantilever array design, IBM 2003. Scanning electron micrographs on the left, side and top views on the right. The hatched area is a planned modification (not shown in the photo). Its purpose is to provide a rigid reference cantilever (full wafer thickness)

same membrane, crosstalk between responses from individual cantilevers cannot be completely excluded, although it was found to be only a few percent of the individual cantilever deflection signals.

To circumvent such crosstalk, the 2000 design was developed, which features special support structures for each cantilever, so that each cantilever is individually attached to the chip body (Fig. 28.11b). Whereas the chip body and the support structures are wet-etched from the lower side, the cantilevers are defined using dry-etching by reactive ion etching (RIE). This method produces high-aspect-ratio structures (grooves with sidewall angles larger than 85°).

To avoid time-critical wet-etching processes and the risk of damage to the cantilevers during the production process, the 2003 design relies entirely on dry-etching processes. The wall steepness has been increased to 89° by the use of improved RIE equipment, so that the etching from the lower side can also be performed by dry-etching, even if as much as 500 μm of silicon have to be etched off from the lower side. To further avoid crosstalk and to facilitate functionalization of the cantilevers, they are fixed to an extended (500-μm-long) bar-shaped structure (Fig. 28.11c).

28.1.4.3
Uniformity of Cantilever Arrays

Even if cantilever arrays consist of identical cantilevers by design, the properties might differ slightly from one cantilever to the next. For this reason, several test procedures are suggested to ascertain the uniformity of the cantilever arrays utilized in experiments.

28.1.4.3.1
Resonance-Frequency Investigation

The resonance frequency of a cantilever is basically determined by the material it consists of as well as its length, width and thickness. Measurement of the resonance frequencies in an uncoated cantilever array typically yields an absolute difference of less than 10 Hz at a typical resonance frequency of 4 kHz (1-μm-thick, 500-μm-long, 100-μm-wide cantilevers), which corresponds to a resonance frequency deviation of less than 0.25%.

28.1.4.3.2
Comparison of Responses to a Heat Pulse

One way to verify the homogeneity of gold-coated cantilevers is to observe the bending response of a bimetallic cantilever when a heat pulse is applied using a Peltier heater beneath the cantilever array. Figure 28.12 shows the response of eight cantilevers in an array to a heat pulse of 60 s duration (double arrow). The deviation in the deflection of the individual cantilevers is less than 9% of the maximum deflection observed. This gives a rough estimate of what is expected of cantilever responses in sensing experiments. However, the thermal (bimetallic) response of cantilevers may differ from the (bio-)chemical response, because the sensing layer might cause additional differences between cantilevers.

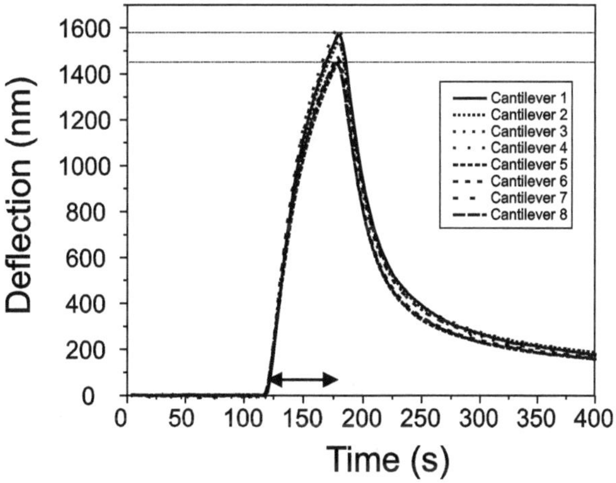

Fig. 28.12. Heat pulse of 0.71 V and 0.32 A applied for a duration of 60 s to an array of gold-coated cantilevers to test their uniformity. The *two horizontal lines* indicate smallest and largest cantilever deflections obtained from the heat pulse (data by J. Zhang, Univ. Basel, Switzerland)

28.2 Experimental Setup

28.2.1 Measurement Chamber

28.2.1.1 Designs for Gas and Liquid Environments

A measurement setup for cantilever arrays consists of four major parts: 1. the measurement chamber containing the cantilever array, 2. an optical or piezoresistive system to detect the cantilever deflection (e.g. laser sources, collimation lenses and a position-sensitive detector (PSD), 3. electronics to amplify, process and acquire

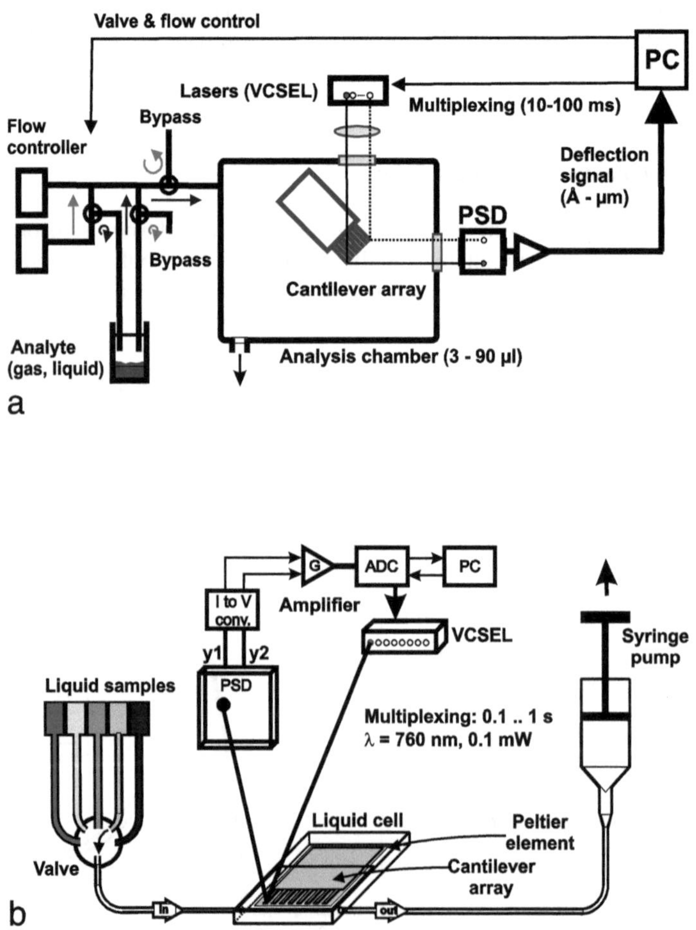

Fig. 28.13. Schematic of measurement setups for (**a**) a gaseous (artificial nose) and (**b**) a liquid environment (biochemical sensor)

the signals from the PSD, and 4. a gas- or liquid-handling system to reproducibly inject samples into the measurement chamber and purge the chamber.

Figure 28.13 shows the schematic setup of the experiments performed in (a) a gaseous and (b) a liquid (biochemical) environment. The cantilever sensor array is located in an analysis chamber of 3–90 μl in volume, which has inlet and outlet ports for gases or liquids. The cantilever deflection is determined by means of an array of eight vertical-cavity surface-emitting lasers (VCSELs) arranged at a linear pitch of 250 μm that emit at a wavelength of 760 nm into a narrow cone of 5 to 10°.

The light of each VCSEL is collimated and focused onto the apex of the corresponding cantilever by a pair of achromatic doublet lenses, 12.5 mm in diameter. This size was selected so that all eight laser beams pass through the lenses close to its center in order to minimize scattering, chromatic and spherical aberration artifacts. The light is then reflected off the gold-coated surface of the cantilever and hits the surface of a PSD. PSDs are light-sensitive photo-potentiometer-like devices that produce photocurrents at two opposing electrodes. The magnitude of the photocurrents linearly depends on the distance of the impinging light spot from the electrodes. Thus the position of an incident light beam can be determined with micrometer precision. The photocurrents are transformed into voltages and amplified in a preamplifier (see Fig. 28.14). As only one PSD is used, the eight lasers cannot be switched on simultaneously. Therefore, a time-multiplexing procedure is used to switch the lasers on and off sequentially at typical intervals of 10–100 ms (see Fig. 28.15). The resulting

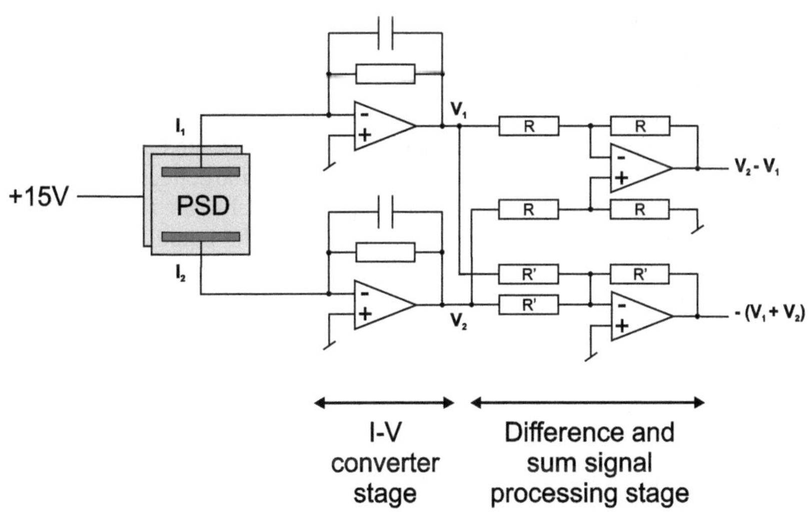

Fig. 28.14. Schematic circuit drawing of a current-to-voltage converter and preamplifier that produces difference and sum signals. The PSD is biased at +15 V to decrease its capacitance and enhance its response. The photocurrents I_1 and I_2 are fed into a current-to-voltage (I–V) converter stage to obtain voltages V_1 and V_2. These are further processed to yield a difference signal $V_2 - V_1$, which is used to measure cantilever deflection, and a sum signal $V_1 + V_2$, which measures the total intensity of light on the PSD and serves as a reference value to keep the intensity of the VCSELs constant (courtesy of A. Tonin, Univ. Basel, Switzerland)

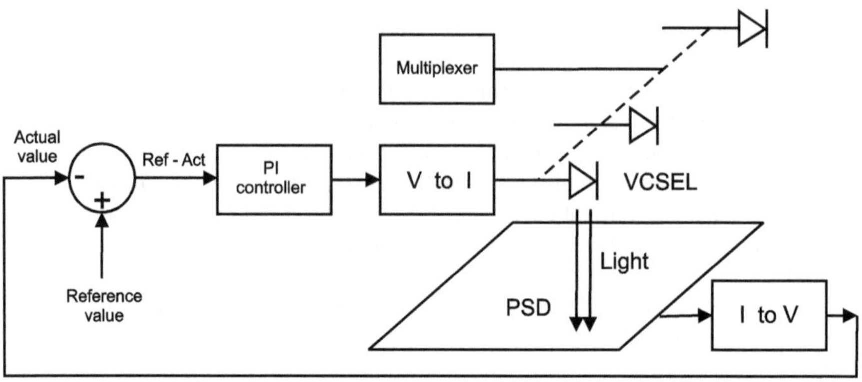

Fig. 28.15. Schematic circuit drawing of the VCSEL driver to switch on and off the eight VCSELs and to keep their light intensity constant. The light of a VCSEL impinges on the PSD and produces photocurrents $I_{1,2}$ that are converted into voltages $V_{1,2}$ in the preamplifier (I to V). The sum signal $V_1 + V_2$ is fed into a voltage comparator, which compares the actual value of the sum signal with a preset reference value. The difference between reference and actual value is then used as input to a PI controller that regulates the current to drive the VCSELs at constant intensity (V to I). A multiplexer switches all eight VCSELs on and off sequentially (courtesy of A. Tonin, Univ. Basel, Switzerland)

deflection signal is digitized and stored together with time information on a personal computer (PC), which also controls the multiplexing of the VCSELs as well as the switching of the valves and mass flow controllers used for setting the composition ratio of the analyte mixture.

The measurement setup for liquids (Fig. 28.13b) consists of a poly-etheretherketone (PEEK) liquid cell, which contains the cantilever array and is sealed by a viton O-ring and a glass plate. The VCSELs and the PSD are mounted on a metal frame around the liquid cell. After preprocessing the position of the deflected light beam in a current-to-voltage converter and amplifier stage (Fig. 28.14), the signal is digitized in an analog-to-digital converter and stored on a PC. The liquid cell is equipped with inlet and outlet ports for liquids. They are connected via 0.18 mm of i.d. teflon tubing to individual thermally-equilibrated glass containers, in which the biochemical liquids are stored. A six-position valve allows the inlet to the liquid chamber to be connected to each of the liquid-sample containers separately. The liquids are pulled (or pushed) through the liquid chamber by means of a syringe pump connected to the outlet of the chamber. A Peltier element is situated very close to the lumen of the chamber to allow temperature regulation within the chamber. The entire experimental setup is housed in a temperature-controlled box regulated with an accuracy of 0.01 K to the target temperature.

28.2.2 Cantilever Functionalization

To serve as sensors, cantilevers have to be coated with a sensor layer that is either highly specific, i.e. is able to recognize target molecules in a key-lock process, or partially specific, so that the sensor information from several cantilevers yields a pattern that is characteristic of the target molecules.

To provide a platform for specific functionalization, the upper surface of these cantilevers is generally coated with 2 nm of titanium and 20 nm of gold, which yields a reflective surface and an interface for attaching functional groups of probe molecules, e.g. for anchoring molecules with a thiol group to the gold surface of the cantilever. Such thin metal layers are believed not to contribute significantly to the bending due to surface-stress changes, because the temperature is kept constant. Many examples of molecular adsorption on cantilevers are described in the literature, for example, the adsorption of alkyl thiols on gold [16, 17], the detection of mercury vapor and relative humidity [18], dye molecules [19], monoclonal antibodies [20], sugar and proteins [21], solvent vapors [22–24], fragrance vapors [25], as well as the pH-dependent response of carboxy-terminated alkyl thiols [26], label-free DNA hybridization detection [27, 28], and biomolecular recognition of proteins relevant in cardiovascular diseases [29].

28.2.2.1
Functionalization Methods

There is a large variety of both simple and more advanced methods to coat a cantilever with material. The requirement is to develop a fast, reproducible and reliable method for the functionalization of one or both of the surfaces of a cantilever separately.

28.2.2.1.1
Simple Methods

Obvious methods to coat a cantilever include thermal or electron-beam-assisted evaporation of material, electro-spray or other deposition methods. The disadvantage of these methods is that inherently they do not provide sufficient resolution to coat individual cantilevers in an array, unless shadow masks are used. Such masks need to be accurately aligned with the cantilever structures, which is a time-consuming process.

Other methods to coat cantilevers use manual placement of particles onto the cantilever [2,5,8,19,30], which requires skillful handling of tiny samples. Cantilevers can also be coated by directly pipetting solutions of the probe molecules onto the cantilevers [22], or by means of air-brush spraying and shadow masks to coat the cantilevers separately [24].

All these methods only have limited reproducibility if a larger number of cantilever arrays has to be coated. As manual alignment is required, these methods are very time-consuming.

28.2.2.1.2
Microfluidic Networks

A step towards reliable and reproducible coating procedures is achieved by the use of microfluidic networks (μFN) [31]. μFN are structures of channels and wells, etched several tens to hundreds of micrometers deep into silicon-wafers. The wells can be filled easily using a laboratory pipette, so that the fluid with the probe molecules to be attached to the cantilever is guided through the channels towards openings at a pitch matched to the distance between individual cantilevers in the array (see Fig. 28.16).

The cantilever array is then approached to the opening by means of an *x*-*y*-*z* positioner, with an accuracy of better than 10 μm, and introduced into the open channels of the μFN that are filled with a solution of the probe molecules. The incubation of the cantilever array in the channels of the μFN takes from a few seconds (self-assembly of alkanethiol monolayers) to several tens of minutes (coating with protein solutions). To prevent evaporation of the solutions, the channels are covered by a slice of PDMS. In addition, the microfluidic network can be placed in a container filled with saturated vapor of the solvent used for the probe molecules. The obtained functional layers on the cantilevers are of good and reproducible quality, and the coating process is considerable faster than with the above-described simple techniques.

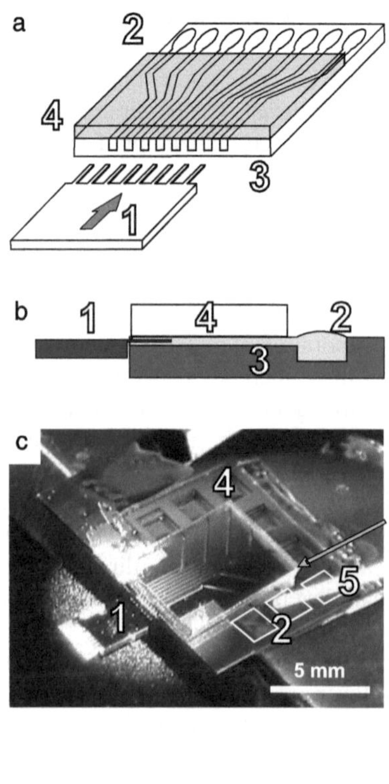

Fig. 28.16. Microfluidic network for functionalization of cantilever arrays. (**a**) Schematic, (**b**) side view, (**c**) photo of the actual device. (**1**) Cantilever array, (**2**) reservoir wells, (**3**) microfluidic network with channels, (**4**) PDMS cover to avoid evaporation of the liquid, (**5**) pipette to fill the reservoir well. Figure courtesy of A. Bietsch, Univ. Basel, Switzerland

28.2.2.1.3
Array of Dimension-Matched Glass Capillaries

As the positioning of the cantilever array relative to the μFN still requires manual alignment, we have developed a device in which the cantilever array is placed rigidly

Fig. 28.17. Setup for cantilever array functionalization using an array of dimension-matched microcapillaries. (a) Schematic, (b) side view, (c) photo. (1) Cantilever array, (2) glass capillaries, (3) tubing with liquid reservoir in a larger piece of tubing to compensate for liquid losses due to evaporation. Figure courtesy of A. Bietsch, Univ. Basel, Switzerland

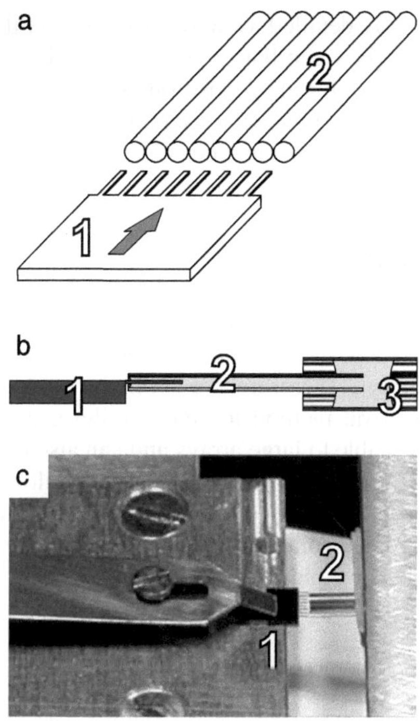

in a fit with an accuracy of better than 50 μm and fixed with a clamp (Fig. 28.17). Because the μFN needs to be cleaned thoroughly after each functionalization process, we replaced the μFN channels by an array of dimension-matched disposable glass capillaries. The outer diameter of the glass capillaries is 240 μm so that they can be placed neatly next to each other to accommodate the pitch of the cantilevers in the array (250 μm). Their inner diameter is 150 μm, providing sufficient room to insert the cantilevers (width: 100 μm) safely.

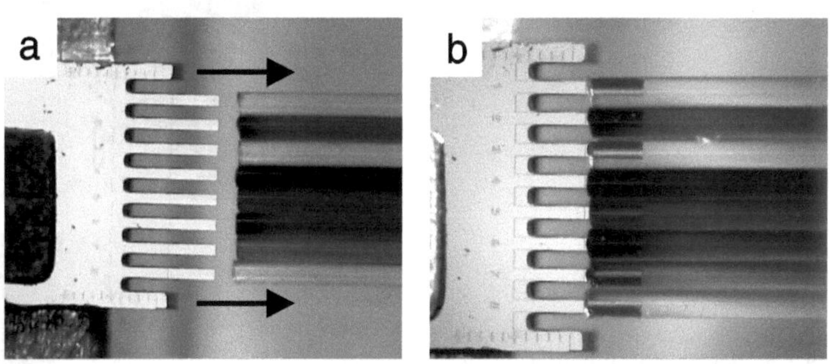

Fig. 28.18. Cantilever array (a) before and (b) after insertion into the open ends of an array of microcapillaries. The *different shades of grey* in the capillaries are due to the filling of the capillaries with various dyes. Note that the capillaries can only be moved until they touch the wafer-thick finger structures (see Fig. 28.11c). Courtesy of A. Bietsch, Univ. Basel, Switzerland

A detailed view of how the cantilevers in an array are inserted into an array of glass capillaries is shown in Fig. 28.18. This method has been successfully applied for the deposition of a variety of materials onto cantilevers, e.g. for polymer solutions [24], self-assembled monolayers [26], thiol-functionalized single-stranded DNA oligonucleotides [28], and protein solutions [29].

28.2.2.1.4
Inkjet Spotting

Even if the capillary-array technique yields reliable and reproducible layers that can be deposited within minutes, it is not ideal for coating a large number of cantilever arrays. Therefore, we applied the inkjet printing technique to establish a rapid and general method to coat cantilever arrays [32, 33] (Fig. 28.19). This technique is scalable to large arrays and can also coat arbitrary structures in noncontact manner.

An *x*-*y*-*z* positioning system allows a sharp nozzle (capillary diameter: 70 μm) to be positioned with an accuracy of approx. 10 μm over a cantilever. Individual droplets (diameter: 60–80 μm, volume 0.1–0.3 nl) can be dispensed individually by use of a piezo-driven ejection system in the inkjet nozzle. When the droplets are

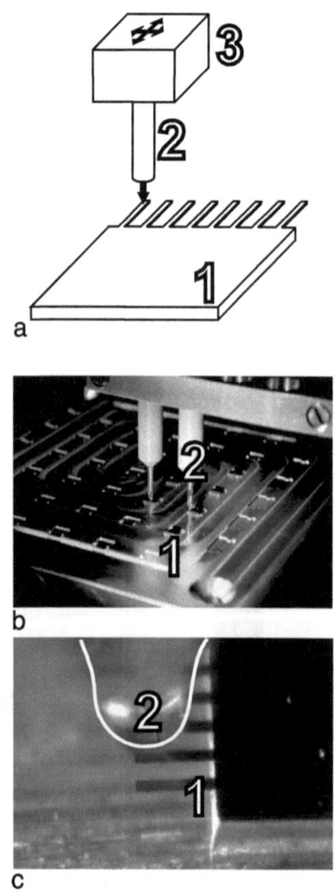

Fig. 28.19. Cantilever array functionalization by inkjet spotting. (**a**) Schematic, (**b**) photo of the actual device, (**c**) closeup. (**1**) Cantilever array, (**2**) inkjet nozzle, (**3**) *x*-*y*-*z* positioning unit. Courtesy of A. Bietsch, Univ. Basel, Switzerland

spotted with a pitch smaller than 0.1 mm, they merge and form continuous films. By adjusting the number of droplets deposited on cantilevers, the resulting film thickness can be controlled precisely.

The inkjet-spotting technique allows a cantilever to be coated within seconds and yields very homogeneous, reproducibly deposited layers of well-controlled thickness. Successful coating of self-assembled alkanethiol monolayers, polymer solutions, self-assembled DNA single-stranded oligonucleotides [33], and protein layers has been demonstrated.

In conclusion, inkjet spotting has turned out to be a very efficient method for functionalization that is not restricted to cantilevers, but can easily and rapidly coat even arbitrarily shaped sensors reproducibly and reliably [34, 35].

28.2.2.2
Functionalization Procedure

28.2.2.2.1
Polymer-Coated Cantilever Arrays

Polymer layers were deposited on a gold-coated cantilever array by inkjet spotting. Various commercial polymers were dissolved in solvents (5 mg/ml). Eight droplets of solution were dispensed onto the upper surface of one of the cantilevers of the array by inkjet spotting [32, 33], allowed to dry and form a homogeneous polymer layer of a few hundred nanometers in thickness. The following polymers were used: CMC: carboxymethylcellulose sodium salt in water (all concentrations 5 mg/ml); PEG: polyethylene glycol 6000 in water; PEI: polyethyleneimine in water; PSS: poly(sodium 4-styrene sulfonate), MW 70,000 in water; PAAM: poly(allylamine hydrochloride), MW 15,000 in water; PVP: poly(2 vinylpyridine) standard, MW 64,000 in ethanol; PVA: polyvinylalcohol 10-98 in dimethylsulfoxide; PMMA: polymethymethacrylate, MW 15,000 in methyl-isobutyl-ketone.

28.2.2.2.2
DNA-Coated Cantilever Arrays

For measurements in the liquid phase, functional monolayers were deposited by self-assembly of thiolated molecule layers, e.g. DNA oligonucleotide layers for hybridization experiments in liquid. Various oligonucleotide layers were applied in parallel and under identical conditions to the individual cantilever sensors by means of microcapillaries (see Fig. 28.17 and Fig. 28.18), filled with a 40-μM solution of 3′- or 5′-thiolated probe DNA in triethyl ammonium acetate buffer for 20 min. The coated arrays were rinsed in sodium saline citrate (ssc) 5× buffer and dried in nitrogen [28].

28.3
Measurements

Measurements using cantilever array sensors can either be highly specific to detect molecules that match a receptor site on the cantilever coating or partially specific, i.e. several cantilever sensors respond to the presence of a vapor, but only the pattern of responses of all eight cantilevers contains sufficient information to characterize a sample vapor. In the following, two experiments are presented. The first demonstrates the partial sensitivity of polymer-coated cantilevers as a kind of "artificial nose" [36] in a gaseous environment, the other shows the high specificity of a DNA hybridization recognition reaction in liquid.

28.3.1 Artificial Nose for Detection of Perfume Essences

A cantilever array consisting of eight differently coated cantilevers can be employed as an "artificial nose" to characterize vapors. The key elements of an artificial nose [36] are 1. chemical sensors composed of a physical transducer and a chemical interface layer or a receptor domain with partial specificity; 2. an appropriate pattern-recognition system to recognize simple or complex odors (pattern classifier), and 3. a sampling system to reproducibly perform measurements. The sensor array is exposed to the same sample, and produces individual responses, as well as a pattern of responses.

The main advantage of an artificial nose is that it is reproducible, does not wear out, and can be placed in environments that are harmful to humans.

Here we use polymer-coated cantilevers as chemical sensors prepared as described above. Detection of vapors proceeds via diffusion of the vapor molecules into the polymer, resulting in a swelling of the polymer and bending of the cantilever. The bending is specific to the interaction between the solvent vapor and polymer time- and magnitude-wise. To demonstrate the capability of the cantilever array artificial nose, we injected vapor (20 ml/min) from the headspace above 0.1 ml of ethanolic solution of perfume essence samples (lemon, wood, flower; main components: ethanol 35%, water 14%, dipropylene glycol 50%, fragrances $< 1\%$) in a stream of dry nitrogen. Figure 28.20 shows the deflection traces from eight polymer-coated cantilevers upon injection of perfume essence vapor during 60 s (starting at $t = 100$ s). The cantilever responses of three samples are shown (two wood essence and one lemon essence sample).

A "fingerprint" of the perfume essence sample is generated by reducing the deflection curves of all eight cantilevers to a discrete set of cantilever amplitudes at several points in time, e.g. at 110 s, 130 s, and 150 s. This yields in total $8 \times 3 = 24$ cantilever deflection amplitudes that account for a measurement data set.

This data set is then evaluated using principal-component-analysis (PCA) techniques [37], which extracts the most dominant deviations in the responses for the various perfume essence vapors. The largest differences in signal amplitudes of the fingerprint patterns are plotted in a two-dimensional graph, whereby an individual measurement (i.e. a set of 24 cantilever magnitudes) represents a single point in the PCA space. The axes refer to projections of the multidimensional datasets into two dimensions (principal components). This procedure is targeted at maximum distinction performance between analytes, i.e. several measurements of the same analyte should yield a cluster in principal-component space, whereas measurements of differing analytes should produce well-separated clusters of measurements.

Fig. 28.20. Bending patterns of polymer-coated cantilever sensors upon exposure to perfume essence vapor. (**a**) Wood flavor (first sample), (**b**) wood flavor (second sample), (**c**) lemon flavor. "Purge" means flushing the measurement chamber with dry nitrogen gas at a rate of 20 ml/min, "Inj." denotes the injection of dry nitrogen gas saturated with the vapor from the headspace of a vial filled with 100 μl of ethanolic solution of perfume essence. The cantilever deflection responses are offset for clarity reasons

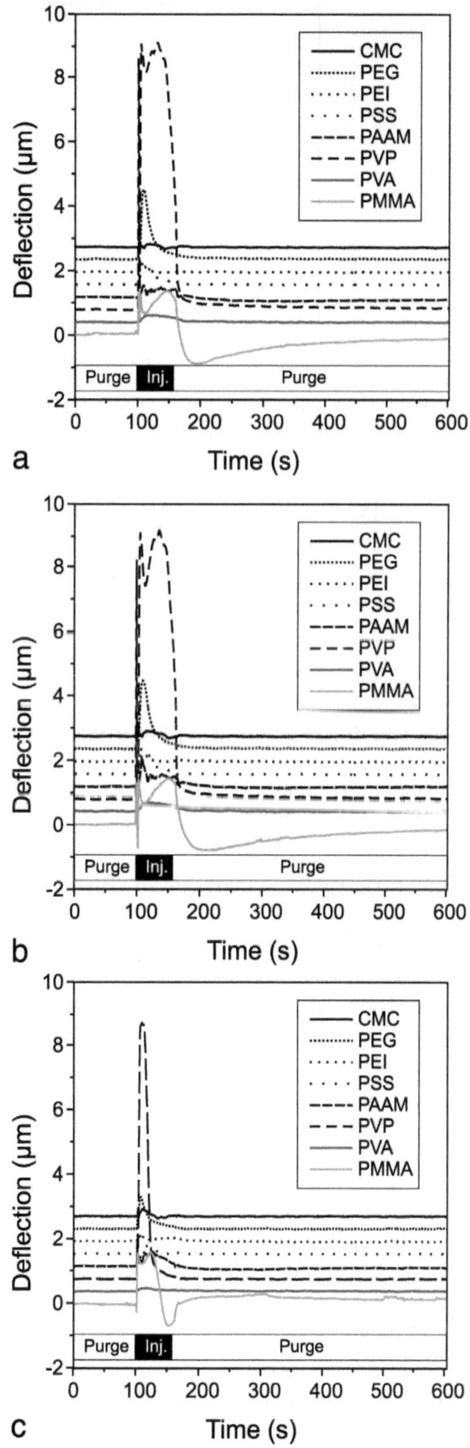

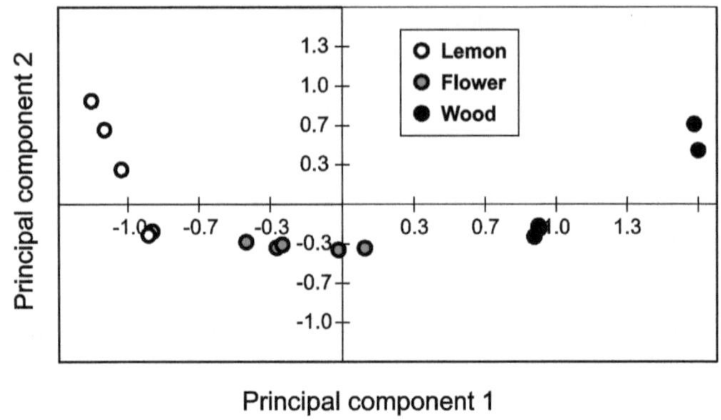

Fig. 28.21. PCA plot of perfume essence samples. Measurements of flower, wood and lemon perfume essences reveal three clearly distinguishable clusters

The PCA evaluation of the cantilever-sensor response curves is shown in Fig. 28.21. Clear clustering is observed for the perfume essences tested, demonstrating the successful recognition, as well as the selectivity of the method. However, the "artificial nose" can only recognize sample vapors that have been measured before. Therefore, it is a characterization rather than a chemical-analysis tool.

28.3.2 Label-Free DNA Hybridization Detection

The main advantage of cantilever array sensors is the possibility that measurements of differences in the responses of sensor and reference cantilevers can be evaluated. Measuring the deflection of only one cantilever will yield misleading results that might give rise to an incorrect interpretation of the cantilever-deflection trace [38].

Therefore, at least one of the cantilevers (the sensor cantilever) is coated with a sensitive layer that exhibits an affinity to the molecules to be detected, whereas other cantilevers are coated with a molecular layer that does not show an affinity to them (reference cantilevers). The biochemical system to be investigated here involves a DNA hybridization experiment in liquid using a thiolated 12-mer oligonucleotide sequence from the Bio B biotin synthetase gene (EMBL accession number: J04423). We selected three surface-bound probes, Bio B1 (5′-SH-C_6-ACA TTG TCG CAA-3′, C_6 is a spacer), Bio B2 (5′-SH-C_6-TGC TGT TTG AAG-3′) and Bio B6 (5′-SH-C_6-TCA GGA ACG CCT-3′), which are immobilized by thiol binding onto the gold-coated upper surface of a cantilever in an array (Fig. 28.22a).

The target complements are called Bio B1C, B2C and B6C are diluted in 5x ssc buffer at 100 pM concentration. Upon injection of the matching sequence to Bio B1, i.e. Bio B1C, the sensor cantilever coated with Bio B1 will bend, whereas the reference cantilever coated with Bio B2 will not bend (Fig. 28.22b). After thorough rinsing with an unbinding agent, the cantilever coated with Bio B1 will bend back to its initial position (Fig. 28.22c). The bending is due to the formation of surface stress during the hybridization process because of steric crowding (double-stranded DNA requires more space than single-stranded DNA).

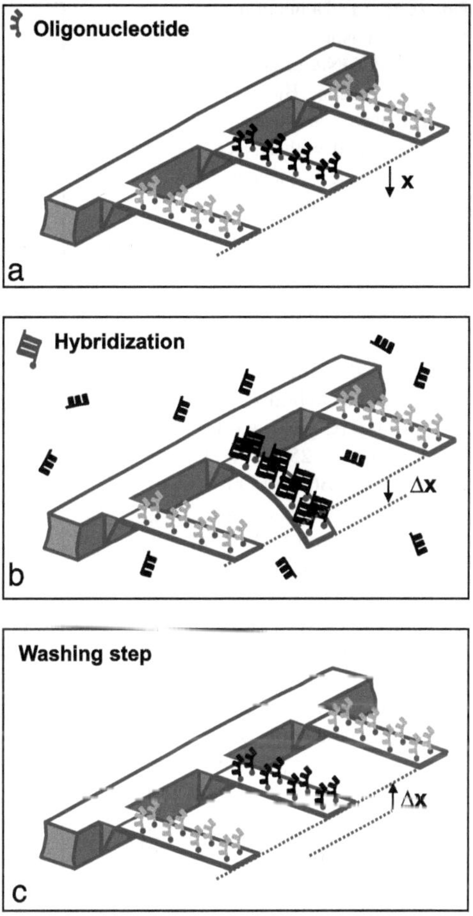

Fig. 28.22. Schematic of the cantilever array DNA hybridization sensor. (**a**) Each cantilever is functionalized with a self-assembled monolayer of thiolated oligonucleotides. (**b**) Upon injection of the complementary sequence to the oligonucleotide sequence shown in the middle, hybridization takes place and the cantilever bends downwards. (**c**) After thorough rinsing with an unbinding agent, the cantilever in the middle returns to its initial position. The schematics are courtesy of Jürgen Fritz, currently at the International University of Bremen, Germany

The actual experiment proceeds as follows (see Fig. 28.23a): First, the liquid cell with the functionalized cantilever array is filled with ssc buffer. After a stable base line has been achieved, ssc buffer is injected after 4 min for 3 min. The cantilevers deflect, but once the injection is over, a stable baseline is reached again. At 18 min, the target Bio B1C is injected, which is supposed to hybridize with the Bio B1 probe, but not with the Bio B2 or the Bio B6 probe. Both cantilevers deflect, but the deflection magnitude of the Bio B1-coated cantilever is larger than that of the Bio B2-coated cantilever. Finally, at 37 min, ssc buffer is injected again and a stable baseline is reached. From the deflection data shown in Fig. 28.23a, it is clear that no conclusive result can be obtained from individual cantilever responses only, as both the sensor and the reference cantilevers bend. However, a clear 20 nm deflection signal is observed when calculating the difference in deflection responses from probes Bio B1 (sensor) and reference Bio B2 (Fig. 28.23b), or the difference in deflection responses from probes Bio B1 and reference Bio B6. The differential deflection magnitudes obtained are 25 nm (B1–B2) or 30 nm (B1–B6), respectively.

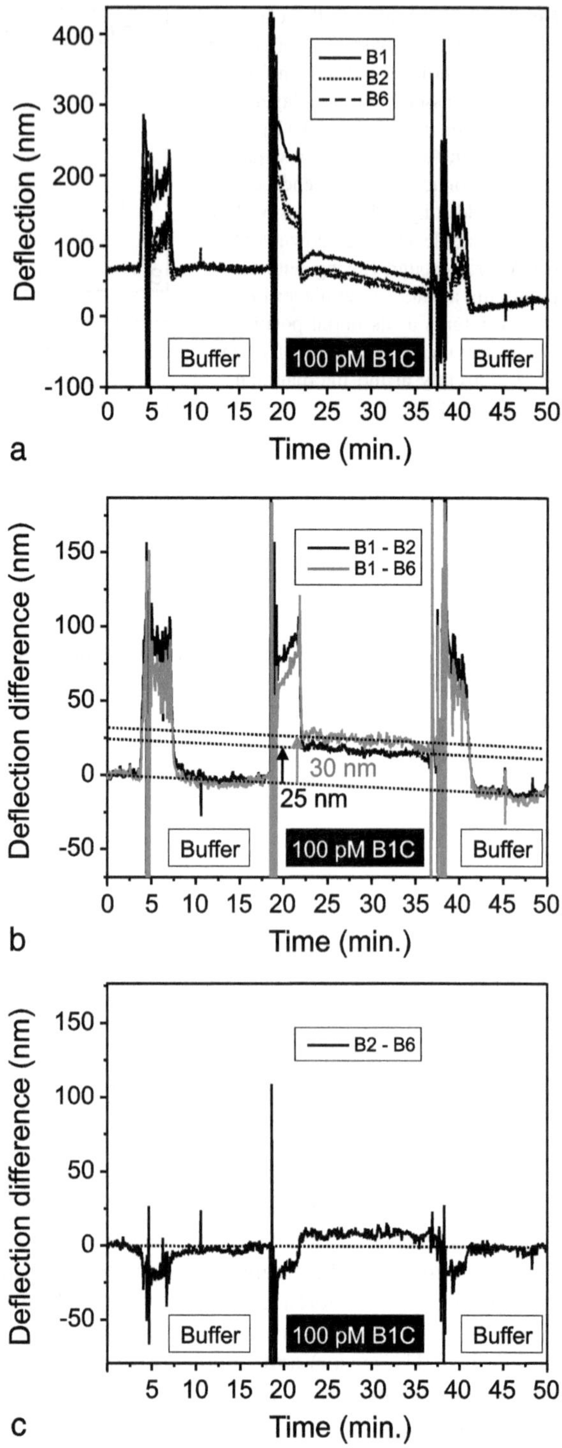

Fig. 28.23. (**a**) Deflection traces of sensor (functionalized with DNA oligonucleotide sequence Bio B1) and reference cantilevers (functionalized with DNA oligonucleotide sequences Bio B2 and Bio B6, respectively). (**b**) Differences B1 to B2 and B1 to B6 of the bending responses of the sensor cantilever B1 and the reference cantilevers B2 and B6. The *dotted lines* are guides to the eye. The hybridization of B1 with B1C yields a difference signal of 25 (B1–B2) or 30 nm (B1–B6). (**c**) Difference in responses of the two reference probes. B2–B6 only yields a small signal of 5 nm. "Buffer" indicates flushing the cell with 5x ssc buffer; "100 pM B1C" means injection of the complement to Bio B1 at a concentration of 100 pM in ssc buffer. Data courtesy of J. Zhang, University of Basel, Switzerland

The difference in deflection responses between two reference cantilevers yields no signal or only a very small signal that can be attributed to some unspecific binding of B1C to one of the reference probes, supposedly to B2, as the difference B2–B6 yields a positive signal of less than 5 nm, see Fig. 28.23c. We conclude that it is absolutely mandatory to use at least two cantilevers in an experiment, a reference cantilever and a sensor cantilever, to be able to cancel out undesired artifacts such as thermal drift or unspecific adsorption.

28.4 Applications and Outlook

The field of cantilever sensors has been very active in recent years [39–41]. Chief topics published 2003 and 2004 in the literature include the following studies: fabrication of silicon, piezoresistive [42,43] or polymer [44] cantilevers, detection of vapors and volatile compounds, e.g. mercury vapor [45], HF vapor [46,47], chemical vapors [48,49], as well as the development of gas sensors utilizing the piezoresistive method [50]. Pd based sensors for hydrogen [51], deuterium and tritium [52,53] have been reported, as well as sensors based on hydrogels [54–57] or zeolites [58]. A humidity sensor is suggested in [59]. Many articles investigate the topics of detection of explosives [60–65], pathogens [66], nerve agents [67], viruses [68], bacteria, e.g. *E. coli* [69–71], and pesticides like dichlorodiphenyltrichloroethane (DDT) [72]. Issues of detection of environmental pollutants are discussed in [73,74]. A chemical vapor sensor based on the bimetal technique is described in [75]. Several papers study the mechanism of static cantilever bending [76–78] and how the transduction can be improved by surface modification of the sensor layer on the cantilever [79,80], or how its properties can be improved [81–83]. In two publications, electrochemical redox reactions are measured with cantilevers [84, 85]. A large number of papers are focused on biochemical applications using cantilevers, e.g. to detect DNA [86–89], proteins [90,91], prostate-specific antigen (PSA) [92], peptides using antibodies [93,94] and living cells [95]. Most groups use the static deflection mode for measurements in liquids, whereas others use piezoresistive cantilevers [96] or the heat mode (calorimeter biosensor) [97]. The understanding of the chemistry [98] and the mechanism [99] of biosensing is of big importance, as is a strategy to improve the transduction mechanism [100]. Medical applications involve diagnostics [101], drug discovery [102] and detection of substances like glucose [103,104]. Mass measurement in dynamic mode is investigated in a variety of publications, e.g. down to the attogram regime in ultra-high vacuum [105] using Doppler vibrometer readout of the oscillation [106]. Several authors study the modeling of resonant sensors [107–114] and exploit higher modes [115], as well as torsional or lateral resonance modes for improved mass sensing in dynamic mode [116, 117]. To improve the oscillation [118], the quality factor can be enhanced electronically [119, 120]. Effects of temperature and pressure on the oscillation should also be taken into account [121]. Other researchers use piezoresonators integrated into the cantilever structure [122].

Ellipsometry combined with cantilever readouts has been suggested [123] as a combined technique. Cantilevers have been used to calibrate AFM cantilevers [124, 125]. Nanowire electrodes attached to cantilevers in an array can be used for local multiprobe resistivity measurements [126]. Two-dimensional microcantilever arrays have been proposed for multiplexed biomolecular analysis [127, 128].

We conclude that cantilever-sensor array techniques are very powerful and highly sensitive tools to study physisorption and chemisorption processes, as well as to determine material-specific properties such as heat transfer during phase transitions. Experiments in liquid environments have provided new insight on such complex biochemical reactions as the hybridization of DNA or molecular recognition in antibody/antigen systems or proteomics. Future developments point towards technological applications, e.g. to finding new ways to characterize real-world samples such as clinical samples. For example, the development of medical diagnosis tools requires the improvement of the sensitivity of a large number of genetic tests to be performed with small amounts of single donor-blood or body-fluid samples. On the other hand, from a scientific point of view, the challenge lies in optimizing cantilever sensors to improve their sensitivity to the ultimate limit: the detection of individual molecules.

Acknowledgements. We thank R. McKendry (University College London, London, U.K.), J. Zhang, A. Bietsch, V. Barwich, M. Ghatkesar, F. Huber, J.-P. Ramseyer, A. Tonin, H.R. Hidber, E. Meyer and H.-J. Güntherodt (University of Basel, Basel, Switzerland) for valuable contributions and discussions, as well as U. Drechsler, M. Despont, H. Schmid, E. Delamarche, H. Wolf, R. Stutz, R. Allenspach, and P.F. Seidler (IBM Research, Zurich Research Laboratory, Rüschlikon, Switzerland). This project was partially funded by the National Center of Competence in Research in Nanoscience (Basel, Switzerland), the Swiss National Science Foundation and the Commission for Technology and Innovation (Bern, Switzerland).

References

1. Binnig G, Quate CF, Gerber C (1986) Phys Rev Lett 56:930
2. Gimzewski JK, Gerber C, Meyer E, Schlittler RR (1994) Chem Phys Lett 217:589
3. Stoney GG (1909) Proc R Soc London Ser A 82:172
4. Thundat T, Warmack RJ, Chen GJ, Allison DP (1994) Appl Phys Lett 64:2894
5. Berger R, Lang HP, Gerber C, Gimzewski JK, Fabian JH, Scandella L, Meyer E, Güntherodt H-J (1998) Chem Phys Lett 294:363
6. Greenspoon JE (1961) J Acoust Soc America 33:1485
7. Zhang J, O'Shea S (2003) Sens Actuators B 94:65
8. Berger R, Gerber C, Gimzewski JK, Meyer E, Güntherodt H-J (1996) Appl Phys Lett 69:40
9. Bachels T, Schäfer R (1999) Chem Phys Lett 300:177
10. Bachels T, Tiefenbacher F, Schäfer R (1999) J Chem Phys 110:10008
11. Barnes JR, Stephenson RJ, Welland ME, Gerber C, Gimzewski JK (1994) Nature 372:79
12. Puers R, Lapadatu D (1996) Sens Actuators A 56:203
13. Huang JM, Liew KM, Wong CH, Rajendran S, Tan MJ, Liu AQ (2001) Sens Actuators A 93:273
14. Fricke J, Obermaier C (1993) J Micromech Microeng 3:100
15. Lang HP, Berger R, Andreoli C, Brugger J, Despont M, Vettiger P, Gerber C, Gimzewski JK, Ramseyer J-P, Meyer E, Güntherodt H-J (1998) Appl Phys Lett 72:383
16. Berger R, Delamarche E, Lang HP, Gerber C, Gimzewski JK, Meyer E, Güntherodt H-J (1997) Science 276:2021
17. Berger R, Delamarche E, Lang HP, Gerber C, Gimzewski JK, Meyer E, Güntherodt H-J (1998) Appl Phys A 66:S55

18. Wachter EA, Thundat T (1995) Rev Sci Instrum 66:3662
19. Scandella L, Binder G, Mezzacasa T, Gobrecht J, Berger R, Lang HP, Gerber C, Gimzewski JK, Koegler JH, Jansen JC (1998) Micropor Mesopor Mater 21:403
20. Raiteri R, Nelles G, Butt H-J, Knoll W, Skladal P (1999) Sens Actuators B 61:213
21. Moulin AM, O'Shea SJ, Welland ME (2000) Ultramicroscopy 82:23
22. Lang HP, Berger R, Battiston FM, Ramseyer J-P, Meyer E, Andreoli C, Brugger J, Vettiger P, Despont M, Mezzacasa T, Scandella L, Güntherodt H-J, Gerber C, Gimzewski JK (1998) Appl Phys A 66:S61
23. Lang HP, Baller MK, Berger R, Gerber C, Gimzewski JK, Battiston FM, Fornaro P, Ramseyer J-P, Meyer E, Güntherodt H-J (19999) Anal Chim Acta 393:59
24. Baller MK, Lang HP, Fritz J, Gerber C, Gimzewski JK, Drechsler U, Rothuizen H, Despont M, Vettiger P, Battiston FM, Ramseyer J-P, Fornaro P, Meyer E, Güntherodt H-J (2000) Ultramicroscopy 82:1
25. Battiston FM, Ramseyer J-P, Lang HP, Baller MK, Gerber C, Gimzewski JK, Meyer E, Güntherodt H-J (2001) Sens Actuators B 77:122
26. Fritz J, Baller MK, Lang HP, Strunz T, Meyer E, Güntherodt H-J, Delamarche E, Gerber C, Gimzewski JK (2000) Langmuir 16:9694
27. Fritz J, Baller MK, Lang HP, Rothuizen H, Vettiger P, Meyer E, Güntherodt H-J, Gerber C, Gimzewski JK (2000) Science 288:316
28. McKendry R, Zhang J, Arntz Y, Strunz T, Hegner M, Lang HP, Baller MK, Certa U, Meyer E, Güntherodt H-J, Gerber C (2002) Proc Nat Acad Sci USA 99:9783
29. Arntz Y, Seelig JD, Lang HP, Zhang J, Hunziker P, Ramseyer J-P, Meyer E, Hegner M, Gerber C (2003) Nanotechnology 14:86
30. Berger R, Gerber C, Lang HP, Gimzewski JK (1997) Microelectron Eng 35:373
31. Wolf H, Juncker, D, Michel B, Hunziker P, Delamarche E (2004) Biosens Bioelectron 10:1193
32. Bietsch A, Hegner M, Lang HP, Gerber C (2004) Langmuir 20:5119
33. Bietsch A, Zhang J, Hegner M, Lang, HP, Gerber C (2004) Nanotechnology 15:873
34. Lange D, Hagleitner C, Hierlemann A, Brand O, Baltes H (2002) Anal Chem 74:3084
35. Savran CA, Burg TP, Fritz J, Manalis SR (2003) Appl Phys Lett 83:1659
36. Gardner JW, Bartlett PN (1992) Sensors and Sensory Systems for an Electronic Nose. Kluwer, Dordrecht
37. Timm, NH (2002) Springer Texts in Statistics – Applied Multivariate Analysis. Springer, Berlin Heidelberg New York
38. Lang HP, Hegner M, Meyer E, Gerber C (2002) Nanotechnology 13:R29
39. Ziegler C (2004) Anal Bioanal Chem 379:946
40. Lavrik NV, Sepaniak MJ, Datskos PG (2004) Rev Sci Instrum 75:2229
41. Majumdar A (2002) Disease Markers 18:167
42. Tang YJ, Fang J, Yan XD, Ji HF (2004) Sens Actuators B 97:109
43. Forsen E, Nilsson SG, Carlberg P, Abadal G, Perez-Murano F, Esteve J, Montserrat J, Figueras E, Campabadal F, Verd J, Montelius L, Barniol N, Boisen A (2004) Nanotechnology 15:S628
44. McFarland AW, Poggi MA, Bottomley LA, Colton JS (2004) Rev Sci Instrum 75:2756
45. Rogers B, Manning L, Jones M, Sulchek T, Murray K, Beneschott B, Adams JD, Hu Z, Thundat T, Cavazos H, Minne SC (2003) Rev Sci Instrum 74:4899
46. Mertens J, Finot E, Nadal MH, Eyraud V, Heintz O, Bourillot E (2004) Sens Actuators B 99:58
47. Tang YJ, Fang J, Xu XH, Ji HF, Brown GM, Thundat T (2004) Anal Chem 76:2478
48. Yu XM, Zhang DC, Wang CS, Du XF, Wang XB, Ruan XB (2003) Chinese Phys Lett 20:1637
49. Abedinov N, Popov C, Yordanov Z, Ivanov T, Gotszalk T, Grabiec P, Kulisch W, Rangelow IW, Filenko D, Shirshov Y (2003) J Vac Sci Technol B 21:2931

50. Zhou J, Li P, Zhang S, Huang YP, Yang PY, Bao MH, Ruan G (2003) Microelectron Eng 69:37
51. Misna TE, McCorkle D, Warmack B (2003) Sens Actuators B 88:120
52. Fabre A, Finot E, Demoment J, Contreras S (2003) Ultramicroscopy 97:425
53. Fabre A, Finot E, Demoment M, Contreras S (2003) J Alloys Compd 356:376
54. Zhang YF, Ji HF, Brown GM, Thundat T (2003) Anal Chem 75:4773
55. Hilt JZ, Gupta AK, Bashir R, Peppas NA (2003) Biomed Microdevices 5:177
56. Liu K, Ji HF (200) Anal Sci 20:9
57. Zhang YF, Ji HF, Snow D, Sterling R, Brown GM (2004) Instrum Sci Technol 32:361
58. Zhou J, Li P, Zhang S, Long YC, Zhou F, Huang YP, Yang PY, Bao MH (2003) Sens Actuators B 94:337
59. Lee CY, Lee GB (2003) J Micromech Microeng 13:620
60. Muralidharan G, Wig A, Pinnaduwage LA, Hedden D, Thundat T, Lareau RT (2003) Ultramicroscopy 97:433
61. Pinnaduwage LA, Boiadjiev V, Hawk JE, Thundat T (2003) Appl Phys Lett 83:1471
62. Pinnaduwage LA, Thundat T, Gehl A, Wilson SD, Hedden DL, Lareau RT (2004) Ultramicroscopy 100:211
63. Pinnaduwage LA, Thundat T, Hawk JE, Hedden DL, Britt R, Houser EJ, Stepnowski S, McGill RA, Bubb D (2004) Sens Actuators B 99:223
64. Pinnaduwage LA, Wig A, Hedden DL, Gehl A, Yi D, Thundat T, Lareau RT (2004) J Appl Phys 95:5871
65. Pinnaduwage LA, Yi D, Tian F, Thundat T, Lareau RT (2004) Langmuir 20:2690
66. Weeks BL, Camarero J, Noy A, Miller AE, Stanker L, De Yoreo JJ (2003) Scanning 25:297
67. Yang YM, Ji HF, Thundat T (2003) J Am Chem Soc 125:1124
68. Gunter RL, Delinger WG, Manygoats K, Kooser A, Porter TL (2003) Sens Actuators A 107:219
69. Zhang J, Ji HF (2004) Anal Sci 20:585
70. Gfeller KY, Nugaeva N, Hegner M (2005) Biosens Bioelectron 21:528
71. Gfeller KY, Nugaeva N, Hegner M (2005) Appl Env Microbiol 71:2626
72. Alvarez M, Calle A, Tamayo J, Lechuga, LM, Abad A, Montoya A (2003) Biosens Bioelectron 18:649
73. Yan XD, Tang YJ, Ji HF, Lvov Y, Thundat T (2004) Instrum Sci Technol 32:175
74. Cherian S, Gupta RK, Mullin BC, Thundat T (2003) Biosens Bioelectron 19:411
75. Adams JD, Parrott G, Bauer C, Sant T, Manning L, Jones M, Rogers B, McCorkle D, Ferrell TL (2003) Appl Phys Lett 83:3428
76. Khaled ARA, Vafai K (2004) J Micromech Microeng 14:1220
77. Zhang Y, Ren Q, Zhao YP (2004) J Phys D 37:2140
78. Jeon S, Thundat T (2004) Appl Phys Lett 85:1083
79. Headrick JJ, Sepaniak MJ, Lavrik NV, Datskos PG (2003) Ultramicroscopy 97:417
80. Bottomley LA, Poggi MA, Shen SX (2004) Anal Chem 76:5685
81. Porter TL, Eastman MP, Macomber C, Delinger WG, Zhine R (2003) Ultramicroscopy 97:365
82. Kooser A, Gunter RL, Delinger WD, Porter TL, Eastman MP (2004) Sens Actuators B 99:474
83. Bumbu GG, Kircher G, Wolkenhauer M, Berger R, Gutmann JS (2004) Macromol Chem Phys 205:1713
84. Quist F, Tabard-Cossa V, Badia A (2003) J Phys Chem B 107:10691
85. Tian F, Pei JH, Hedden DL, Brown GM, Thundat T (2004) Ultramicroscopy 100:217
86. Su M, Li SU, Dravid, VP (2003) Appl Phys Lett 82:3562
87. Gunter RL, Zhine R, Delinger WG, Manygoats K, Kooser A, Porter TL (2004) IEEE Sensors J 4:430

88. Alvarez M, Carrascosa LG, Moreno M, Calle A, Zaballos A, Lechuga LM, Martinez-A C, Tamayo J (2004) Langmuir 20:9663
89. Gunter RL, Zhine R, Delinger WG, Manygoats K, Kooser A, Porter TL (2004) IEEE Sensors J 4:430
90. Lee JH, Yoon KH, Hwang KS, Park J, Ahn S, Kim TS (2004) Biosens Bioelectron 20:269
91. Lee JH, Kim TS, Yoon KH (2004) Appl Phys Lett 84:3187
92. Hwang KS, Lee JH, Park J, Yoon DS, Park JH, Kim TS (2004) Lab Chip 4:547
93. Kim BH, Mader O, Weimar U, Brock R, Kern DP (2003) J Vac Sci Technol B 21:1472
94. Kooser A, Manygoats K, Eastman MP, Porter TL (2003) Biosens Bioelectron 19:503
95. Saif MTA, Sager CR, Coyer S (2003) Annals Biomed Eng 31:950
96. Reasmussen PA, Thaysen J, Hansen O, Eriksen SC, Boisen A (2003) Ultramicroscopy 97:371
97. Arakawa ET, Lavrik NV, Rajic S, Datskos PG (2003) Ultramicroscopy 97:459
98. Yan XD, Lvov Y, Ji HF, Singh A, Thundat T (2003) Org Biomolec Chem 1:460
99. Liu F, Zhang Y, Ou-Yang ZC (2003) Biosens Bioelectron 18:655
100. Khaled ARA, Vafai K, Yang M, Zhang X, Orkan CS (2003) Sens Actuators B 94:103
101. Kumar S, Bajpai RP, Bharadwaj LM (2003) IETE Techn Rev 20:361
102. Zhang YF, Venkatachalan SP, Xu H, Xu XH, Joshi P, Ji HF, Schulte M (2004) Biosens Bioelectron 19:1473
103. Pei JH, Tian F, Thundat T (2004) Anal Chem 76:292
104. Yan XD, Ji HF, Lvov Y (2004) Chem Phys Lett 396:34
105. Ono T, Li XX, Miyashita H, Esashi M (2003) Rev Sci Instrum 74:1240
106. Ono T, Esashi M (2004) Meas Sci Technol 15:1977
107. Dufour I, Fadel L (2003) Sens Actuators B 91:353
108. Yum K, Wang ZY, Suryavanshi AP, Yu MF (2004) J Appl Phys 96:3933
109. Fadel L, Dufour I, Lochon F, Français O (2004) Sens Actuators B 102:73
110. Ryu WH, Chung YC, Choi DK, Yoon CS, Kim CK, Kim YH (2004) Sens Actuators B 97:98
111. Jeon S, Braiman Y, Thundat T (2004) Rev Sci Instrum 75:4841
112. Adams JD, York D, Whisman N (2004) Rev Sci Instrum 75:2903
113. Fadel L, Lochon F, Dufour I, Français O (2004) J Micromech Microeng 14:S23
114. Ren Q, Zhao YP (2004) In: Microsystem technologies, vol 10. Springer, Berlin Heidelberg New York, p 307
115. Kar AK, George MA (2003) J Appl Phys 94:4626
116. Sharos LB, Raman A, Crittenden S, Reifenberger R (2004) Appl Phys Lett 84:4638
117. Ilic B, Craighead HG, Krylov S, Senaratne W, Ober C, Neuzil P (2004) J Appl Phys 95:3694
118. Passian A, Muralidharan G, Mehta A, Simpson H, Ferell TL, Thundat T (2003) Ultramicroscopy 97:391
119. Tamayo J, Alvarez M, Lechuga LM (2003) Sens Actuators B 89:33
120. Tamayo J, Lechuga LM (2003) Appl Phys Lett 82:2919
121. Mertens J, Finot E, Thundat T, Falve A, Nadal MH, Eyraud V, Bourillot E (2003) Ultramicroscopy 97:119
122. Kim SJ, Hong KI, Choi DK (2003) J Appl Phys Pt 1 42:1475
123. Godin M, Laroche O, Tabard-Cossa V, Beaulieu LY, Grütter P, Williams PJ (2003) Rev Sci Instrum 74:4902
124. Gibson CT, Weeks BL, Abell C, Rayment T, Myhra S (2003) Ultramicroscopy 97:113
125. Hu ZY, Seeley T, Kossek S, Thundat T (2004) Rev Sci Instrum 75:400
126. Lin R, Boggild P, Hansen O (2004) J Appl Phys 96:2895
127. Khanafer K, Khaled ARA, Vafai K (2004) J Micromech Microeng 14:1328
128. Yue M, Lin H, Dedrick DE, Satyanarayana S, Majumdar A, Bedekar AS, Jenkins JW, Sundaram S (2004) J Microelectromech Syst 13:290

22 Nanomechanics and Microfluidics as a Tool for Unraveling Blood Clotting Disease

D.M. Steppich · S. Thalhammer · A. Wixforth · M.F. Schneider

From volume XIII

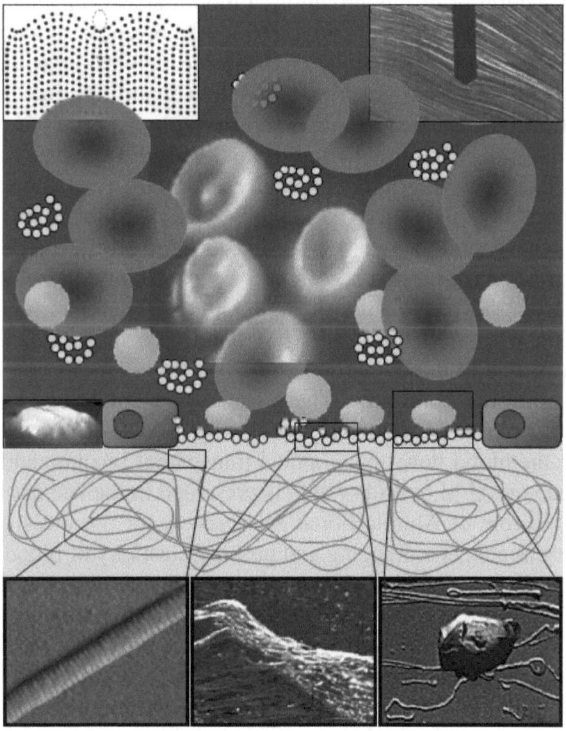

Key words: Blood Clotting, Von Willebrand Factor, Collagen Assembly, Nanomechanics, Erythrocyte Elasticity, AFM on Biomolecules, Lab on a Chip, Microfluidics

22.1 Introduction

One of the key advantages of Atomic Force Microscopy (AFM) over a variety of other techniques in life science is the freedom to operate in almost any medium.

In: H. Fuchs, B. Bhushan (eds), *Biosystems - Investigated by Scanning Probe Microscopy*,
DOI 10.1007/978-3-540-85049-6_5, © Springer-Verlag Berlin Heidelberg 2010

Especially for the investigation of cells, proteins, and membranes and their complex interplay, a prerequisite for reliable and biologically significant results is the choice of the right medium. Small changes from natural conditions in pH, temperature, salt concentration, etc., can make a huge difference for the interpretation of obtained data and for assigning them to a peculiar biological process in vivo. The use of an AFM is not limited to vacuum or a gaseous medium and further provides the possibility to operate in fluid media with variable buffer conditions.

In recent years, with a continuous increase in expertise for probing soft matter, AFM has developed into a decisive tool for approaching questions in life sciences. Scientists benefit from AFM in cell imaging, questions in regard to cell elasticity, the measurement of both specific lock-and-key and non-specific adhesion in a wide range of biological, physical, and also medical questions. On the following pages, we present a selection of results providing a more detailed understanding of blood clotting and hemostasis, which is obtained by a very inventive and non-conventional use of the AFM technique. A special challenge in this task was the consideration and integration of fluid flow, mimicking blood flow conditions, into the experimental setup. As mentioned above the medium is an extremely important factor in approaching biological questions. To further complicate things, the dynamic conditions inside a blood vessel like flow pattern, flow velocity, and velocity gradients influence and sometimes generate circumstances, which barely can be probed or reproduced in a static environment. Atomic Force Microscopy, as a rather direct technique of observation, further does have limited access into a sealed blood vessel or any common flow chamber trying to copy natural flow conditions. Therefore, unraveling secrets in blood clotting with the help of an AFM requires combination with a pumping system, which is able to mimic variable blood flow conditions and is accessible to the cantilever tip at the same time. These framework conditions led to the unique combination of a planar Surface Acoustic Wave (SAW)-driven microfluidic reactor, mimicking the blood flow on an optical transparent biochip, and an AFM. This microfluidic reactor–AFM hybrid system is presented in the last part of this chapter. To illuminate the role of Atomic Force Microscopy for the investigation of blood clotting we present results in imaging and probing elasticity of blood cells, platelets, and compounds of the subendothelial matrix exposed to different flow conditions. Subsequently, we introduce the SAW-driven microfluidic pumping system leading to a physicist's equivalent of a blood vessel. This approach is the direct precursor to "lab-on-a-chip" systems for probing blood clotting diseases. In the last section we present the microfluidic reactor–AFM hybrid combining the before mentioned techniques in a unique and valuable way. This hybrid system provided the means to unravel direct evidence for the formation and internal dynamics of protein networks responsible for blood clotting under elevated shear stress.

22.2 Topography

The invention of Atomic Force Microscopy opened the field for imaging surfaces with a lateral and vertical resolution in the range of Angstroms [1]. The resolution in imaging soft biological matter like membranes or cell surfaces often is far from that value and strongly depends on the kind of sample under investigation. For example, it may vary strongly according to wall and adhesion properties of different cells and the tip. The main contribution to the lack of resolution in cells is maybe the tip-induced deformation or even penetration of soft parts of proteins, membranes, and cells [2]. Nevertheless, the AFM technique proved to be a crucial tool in exploring biological questions. In the following sections, we will present a short scheme of the course of blood clotting and how AFM is applied to elucidate exciting features of the most important compounds involved in this process.

22.2.1 Little Story of Blood Clotting

Blood coagulation is a very complex process involving more than a dozen blood coagulation factors. We will sketch the course of wound healing in small arterioles as much as necessary for a proper understanding of the underlying basic processes while keeping it as simple as possible to motivate the basic questions under investigation in the AFM experiments. The focus will be set on vWF (von Willebrand Factor) network formation and platelet binding to the damaged vessel walls, which oppose high shear stress arising from the rapid flowing blood.

Wound healing is an extremely important adjustment of nature preventing continuous bleeding into the tissue after damage to the blood vessels. A normal human being is bleeding continuously in small arterioles and veins without even being aware of it. The first and an often neglected mechanism to prevent excessive blood loss is the contraction of the injured blood vessel, which can change the streaming properties of the blood and the forces on every single compound involved in blood clotting. Such bleeding, especially in these high shear regimes, is continuously sealed by an instantaneous repair mechanism (Fig. 22.1).

22.2.1.1 von Willebrand Factor Binding

An essential role for wound healing especially in small arterioles comprising high shear forces plays the biopolymer von Willebrand Factor. Its participation in wound healing has been known for a long time. The specific lock-and-key binding sites of vWF molecules to both extracellular matrix and damaged endothelial cells, exposed at sites of vascular injury, were characterized, but the mechanism of the activation of vWF's binding potential remained unclear. We could recently show directly that a conformational change in vWF molecules from a coiled to an activated unrolled state triggers its biological function [3]. In this activated conformation vWF's ability to bind to both exposed extracellular matrix at a site of vascular damage and to platelets out of the blood stream is rapidly increased. The extracellular matrix is mainly composed of different kinds of collagens (see also Sect. 22.2.2.1), where binding sites to vWF have already been identified [4]. This coiled–unrolled transition can be induced solely by mechanical stress on the molecule experienced due to shear stress of the blood stream. Comparable to the stream lines in a pipe, the velocity of blood in the blood vessel decreases from the center towards the endothelial cell surface. This velocity difference between contiguous layers of fluid results in

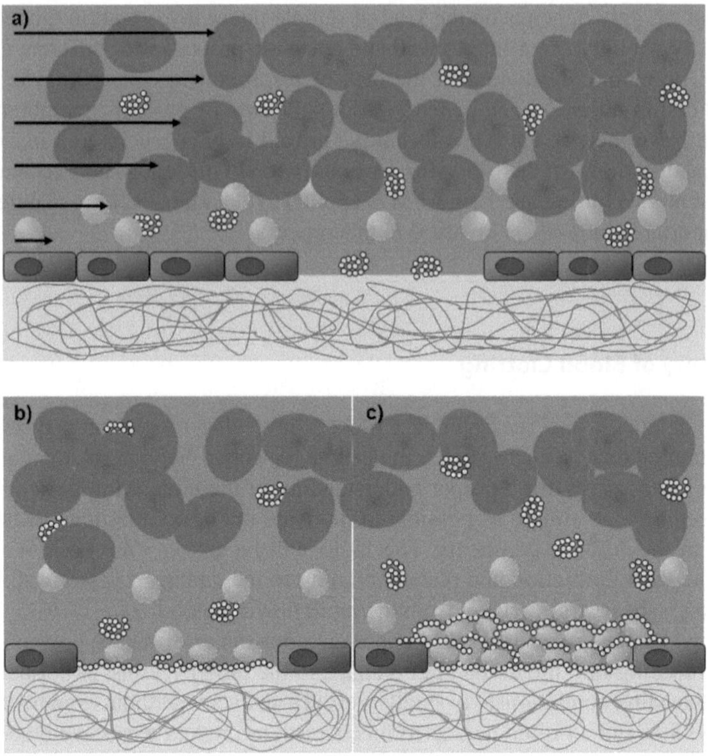

Fig. 22.1. (**a**) Erythrocytes (*red*), platelets (*yellow*), and von Willebrand Factor (vWF) (*blue*) facing a site of damage to the endothelial cell layer which seals the blood vessel against the extravascular tissue. vWF binding to collagens in the extravascular connective tissue is the initial step for wound healing in vessels with high fluid flow velocities. The *arrows* indicate the different flow velocities inside a blood vessel with a maximum speed in the center of the vessel and a minimum at the vessel wall. This velocity gradient induces shear stress on any compound in the blood stream. (**b**) Because of the increased shear stress at the vessel wall, immobilized vWF molecules are mechanically unrolled from their coiled into an activated stretched conformation building a sticky protein network. In this unrolled and activated conformation platelets can bind to vWF at the site of vascular damage, whereas there is no binding of vWF and platelets in the normal blood stream. (**c**) In the course of wound healing more and more platelets aggregate forming multiple layers stabilized by the vWF glue till a plaque has formed which is able to prevent further bleeding into the tissue

a shearing effect. The shear rate $\dot{\gamma} = \frac{\partial v}{\partial r}$ defined as the flow velocity gradient with respect to the vessel radius describes the acting shear field. The highest wall shear rates of approximately 470–4,700 1/s occur at the vessel walls of small arterioles (10–30-μm diameter) [5] and can induce vWF's activation.

From a physical perspective, in these high shear rate areas it seems extremely counterintuitive for any protein or cell to adhere to sites of vessel damage opposing the drag forces of the rapid surrounding flow. To solve this dilemma, evolution developed vWF, initializing its enhanced adhesion abilities exactly in this high shear environment, which makes commonly known adhesion mechanisms extremely unlikely or even impossible. The key in understanding vWF's binding potential for both compounds of the extracellular matrix and platelets is hidden at least in part in its enormous multimeric size. Large vWF molecules can be made up of up to a hundred identical subunits and its total mass can exceed 20,000 kDa [6]. In its coiled conformation a huge amount of vWF's binding sites are buried inside the protein whereas an unrolling of this biopolymer directly results in a drastic increase of accessible binding sites and therefore in the activation of the protein. The adhesive strength of vWF to exposed collagen at a vessel damage site or a damaged endothelial cell increases proportionally to its free binding sites.

22.2.1.2 Platelet Binding

Activated and bound to the extracellular matrix, vWF comprises binding sites for platelets too. Platelets are blood cells without a nucleus whose function is to prevent bleeding. While the vWF network represents the basis for wound healing, platelet aggregation to this network is the starting point for the growing plug, which finally seals the vessel damage. In this regard it is worth mentioning that platelets are enriched near the vessel wall while erythrocytes for example are found mainly in the center of the vessel. This condition enables a higher probability for platelets to adhere to sites of vascular damage and therefore seal lesions more effectively [7].

Platelets adhere to a vWF network at a site of injury in a two-step mechanism. First a weak bond with a high dissociation constant is formed. Because of the torque induced by the fluid streaming and the weak adhesion to the surface, platelets are decelerated and roll on the vWF network, giving a stronger and irreversible bond with a low dissociation rate enough time to form [8]. The process of rolling is conditioned by repeated forming and breaking of these weak bonds. Sometimes these weak adhesion contacts do not break while the platelet still translates over the surface. Because of the soft character of the platelets a membrane tube from the adhesion point on the vWF surface to the cell body along the direction of the fluid flow can be formed [9]. The morphology, formation, and detachment kinetics from the anchor point of these membrane tethers depend strongly on the acting shear field. At high shear rates the probability for tether formation and its growth rate are drastically increased while the lifetime is significantly lower. Dopheide et al. [10] speculated that the extension of membrane tethers may be a very important step in mediating strong adhesion by slowing platelets down and reducing the level of shear stress experienced by individual receptor–ligand interactions.

After a tight bond is formed, platelets tend to spread on the surface and therefore comprise even more binding sites for vWF while reducing the opposing drag forces at the same time. The stabilizing effect of platelet spreading is often accompanied with the formation of several filopodia. Although similar in their structure, filopodia are distinctly different from membrane tethers and protrude onto the surrounding surface without respect to the flow pattern.

As more and more platelets aggregate on the site of vascular damage, vWF serves also as a glue between multiple layers of platelets. This platelet plaque grows as long as it is sufficient to prevent further loss of blood. If the rent in the vessel wall is small,

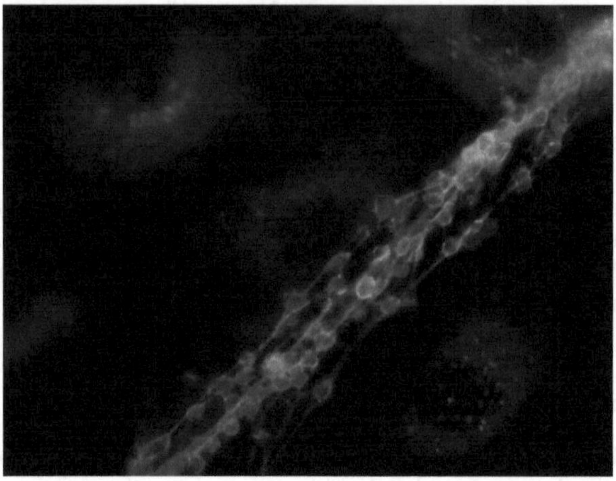

Fig. 22.2. The fluorescence image points out the functional activity of unrolled vWF (*green*) in platelet (*red*) binding. Platelets exhibit sizes in the order of 10 μm, which also highlights also the enormous size of the vWF network (Figure reprinted from George T, Kleinerüschkamp F, Barg A et al., Microfluidics reveals generation of platelet strings on tumor-activated endothelium, Thromb Haemost 2007; 98:285 with permission from Schattauer)

this platelet plug by itself can stop blood loss. A larger hole requires further stabilization of the platelet aggregate. In addition a meshwork of fibrin can interconnect platelets and blood cells forming a blood clot [11]. Figure 22.2 illustrates the platelet binding potential of a huge vWF network.

22.2.2 High-Resolution Imaging

22.2.2.1 Collagen Fibrils

Collagen molecules are a major compound of the extracellular matrix and are exposed to the blood stream at a site of vascular damage, building the first contact point for the initiation of blood clotting by network forming proteins. They consist of three polypeptide chains (α-chains), which form a unique triple-helical structure. More than 20 genetically distinct collagens exist in mammalian tissue, where collagen types I, II, III, V, and XI self-assemble into D-periodic cross-striated fibrils. Collagen molecules, forming the fibril, consist of an uninterrupted right-handed triple helix called tropocollagen [12], approximately 300 nm in length and 1.5 nm in diameter. The collagen self-assembles in cross-striated fibrils that normally occur in the extracellular matrix of connective tissues. These fibrils are stabilized by covalent cross linking of specific lysines and hydroxolysines of the collagen molecules, which are ordered parallel in a D-periodic pattern [13]. The stagger of molecules gives rise to a characteristic band pattern of light and dark regions when negatively stained and viewed using an electron microscope [14, 15]. Fundamental experiments for the understanding of the morphogenesis of collagen units and fibrils and influencing factors such as temperature, ionic strength, pH value and so on, were performed by Gross, Wood, Keech et al. [15–21] and further work was done by Bard and Chapman [22, 23].

Studies on collagen molecules by cryo-AFM [24] and normal AFM studies of segment-long spacing (SLS) crystals of collagen [25] have revealed variations in the structure. AFM investigations carried out by Paige et al. [26] show native fibrils and fibrous long spacing fibrils (FLS-fibrils) [27]. Cocoon-like fibrils, which are hundreds of nanometers in diameter and 10–20 mm in length, were found to coexist with mature FLS fibrils. On the basis of detailed AFM studies a stepwise process in the formation of FLS collagen was proposed. Different pH ranges were investigated to clarify the various stages in the self-assembly process of the fibrils. In early stages below pH 4 thin non-banded fibrils were formed. Filamentous structures showed protrusions at $\mathrm{pH}{\sim}4$ and at $\mathrm{pH}{>}4.6$ mature fibers emerged [28]. Gutsmann et al. observed that collagen fibrils from tendons behave mechanically like tubes [29]. They concluded that the collagen fibril is an inhomogeneous structure composed of a relatively hard shell and a softer, less dense core. For high-resolution AFM microscopy the collagen samples can be investigated immediately after self-assembly and drying on freshly cleaved mica. A novel collagen preparation technique [30], based on a dialysis system, allows the reproducible production of single collagen fibrils with different banding patterns (Fig. 22.3).

Automated electron tomography studies, performed on corneal collagen fibrils showed that collagen molecules are organized into microfibrils ($\approx$4 nm diameter) in a 36-nm diameter collagen fibril, which are tilted at $\approx$15° to the fibril long axis in a right-handed helix. Analysis of the lateral structure demonstrated that the microfibrils exhibit regions of order and disorder within the 67-nm axial repeat of the collagen fibrils [31].

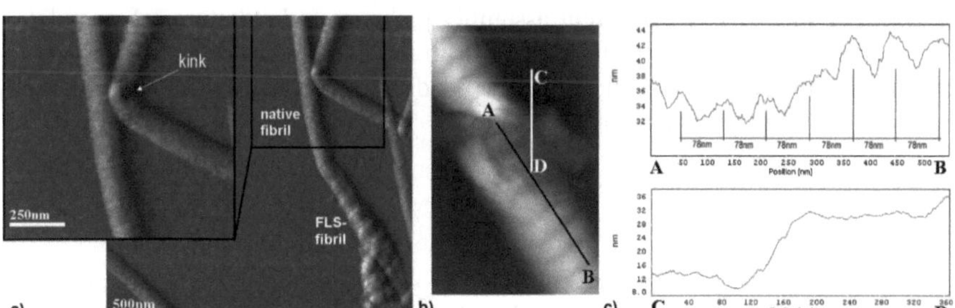

Fig. 22.3. (**a**) High-resolution AFM image of a single collagen fibril with a kink: due to kinks collagen are sometimes compared to tubes. The other fibril displays the polymorphism of collagen with native and FLS-parts in one fibril. (error signal, *scale bar* 500 nm). (**b**) AFM image of a microdissected collagen fibril: The core of the collagen fibril is revealed and the banding pattern can be recognized inside the fibril. The *arrows* indicate the overlap zones of the collagen molecules that arise during the self-assembly process of the collagen fibril (topography signal, *scale bar* 250 nm). (**c**) Line measurements on the microdissected collagen fibril **b**: Between point A and B the banding can be recognized on the shell as well as in the core. The banding pattern can be determined to be approximately 78 nm. Between point C and D the height difference of around 16 nm between the substrate and the level of the cut area can be seen

In contrast to other high-resolution imaging methods, like scanning electron microscopy (SEM) or transmission electron microscopy (TEM), which are operating in a vacuum with dried samples and conducting surfaces, the preparation for AFM investigations is very simple and does not alter the sample properties in most cases. For structural and mechanical investigations in the submicron range the AFM proved to be a suitable tool for manipulation and probing the mechanical properties of biological macromolecules in the nanometer range. Besides the manipulation of proteins like collagen it is also possible to probe and manipulate genetic samples. Thalhammer et al. demonstrated that it is possible to use the AFM tip as a new tool like a mechanical nanoscalpel and a nanoshuffle where minute amounts of material can be extracted and further processed [32]. The AFM-based microdissection on collagen fibers can be applied to reveal the inner structure of the specimen. During manipulation the shell and upper parts of the fibril are scratched away, to image the remaining core in a high-resolution mode (Fig. 22.3b, c). The dissection is carried out in a defined angle to the fibril axis in order to exclude artifacts originated by an orthogonal or parallel (to the fibril axis) scratching procedure. It could be shown that there are no major geometrical differences of the banding on the shell and in the core of native single collagen fibrils. The banding pattern inside the fibril fits to that on the shell in width and distance. The fibril has a banding pattern of 78 nm, a height of 30 nm, and a width of 270 nm. The cut area, shown in Fig. 22.3b, is located in the core of the fibril, which was confirmed by line measurements. Between the measured distance A and B the banding can be recognized on the shell as well as in the core. Among points C and D the height difference between the substrate and the level of the cut area can be seen. The height difference between the substrate and cut area was determined to be 16 nm [33]. The AFM could also be used to compose nanoscopic collagen matrices by orientating individual collagen fibers of self-assembled collagen layers [34]. AFM nanodissection of big FLS-fibrils with a width of about 1.7 μm and a banding pattern of 270 nm showed the FLS banding also in the core of the fibril [35]. For FLS-fibrils a different assembly pathway and structure are postulated. The characteristic banding mainly arises from the attachment of α1-acid glycoprotein in FLS-fibrils [28]. The characteristic banding of native fibrils is determined by the repetition of overlap and gap zones [36].

22.2.2.2
Elasticity

Several investigations using the AFM as a tool for measuring the tensile modulus of collagen fibrils and subunits revealed details of the protein assembly. Graham et al. calculated force elongation/relaxation profiles of single collagen fibrils using the AFM. The elongation profiles showed, that in vitro assembled human type I collagen fibrils are characterized by a large extensibility. It was shown that the fibrils are robust structures with a significant conservation of elastic properties [37]. Gutsmann et al. probed the crosslinks on a lower level of organization using an AFM cantilever to pull substructures out of the assembly. Two different rupture events were determined; the first with a strong bond and a periodicity of 78 nm (bonds between subunits) and a second weaker one with a periodicity of 22 nm (between molecules) [38]. Bozec and Horton [39] studied trimeric type I tropocollagen molecules by AFM, both topologically and by force spectroscopy, showing multiple stretching peaks on the molecular level similarly as shown by Gutsmann et al. [38]. Fratzl et al. and Puxkandl et al. investigated the fibrillar structure, viscoelastic and mechanical properties of collagen by recording stress/strain curves [40, 41]. The stress/strain curves can be divided into several regions [37]. At first crimps [42] and kinks [43] are removed, before a linear region is seen where the collagen triple helices are stretched, along with increase of the gap zones compared with the overlap zones. Slippage is first seen within fibrils at crosslink deficient collagen, and then higher strains lead to a disruption of the fibril. The mechanical behavior of native single fibrils was tested by recording force-distance curves on the shell and in the core of the fibrils to gain insights into the collagen assembly and mechanical properties [33]. The evaluation of the adhesion forces of the spectroscopy data indicates a higher adhesion in the core of the fibril (Fig. 22.4a). In Fig. 22.4b and c the height of the snap-out effect calculated from the retrace curves is shown. The average value was calculated for the shell to 5 nN and for the core to 6 nN, which points to a higher adhesion in the core of the fibril. Gutsmann et al. suggested the presence of more highly crosslinked collagen molecules near the fibril surface compared to the central region [29]. This could lead to a higher amount of binding capacity for the tip and cause higher adhesion forces during the measurement. However, the results could have been also influenced by the scratching process, which could have led to a rupture of molecules and destruction of crosslinks. The destroyed crosslinks and ruptured molecules could stick to the tip and increase the measured adhesion force. Moreover, the collagen fibrils were investigated in a dried state of preservation, which could have also an influence on the mechanical properties. The mechanical properties of collagen-rich tissues, e.g. tendons, are largely determined by the collagen structure [40]. An inhomogeneous assembly of collagen fibrils was published by several authors; Sarkar et al. [44] proposed that fluid domains in the collagen allow molecules to slip relative to one another, which was shown by Mosler et al. [45], in order to relieve applied stress. It has been proposed that, in case of high forces, the stiff outer shell of the collagen fibrils could break while the fluid core remains intact and might be used in the repair of the shell [29]. These morphological results as well as the statistical evaluation of the Young's modulus using the indentation measurements could not confirm a "fluid" core or different structures of core and shell. Solely the adhesion measurements show differences between core and shell.

22.2.2.3 von Willebrand Factor Network Formation and Platelet Binding

The multimeric von Willebrand Factor molecule is the key element in mediating adhesion of platelets to sites of injury at rapid flow conditions. Because of a conformational change from a globular state into an unrolled state, vWF is able to bind to various types of collagen from the extracellular matrix. Single vWF molecules and their adhesion and binding properties have been investigated by AFM technology. In early studies, Marchant et al. examined the adhesion and structural properties of vWF both on hydrated and dehydrated mica surfaces [46]. They experienced problems similar to those researchers are still facing in imaging soft matter under buffer conditions today. On hydrophilic mica, vWF is only interacting weakly with the

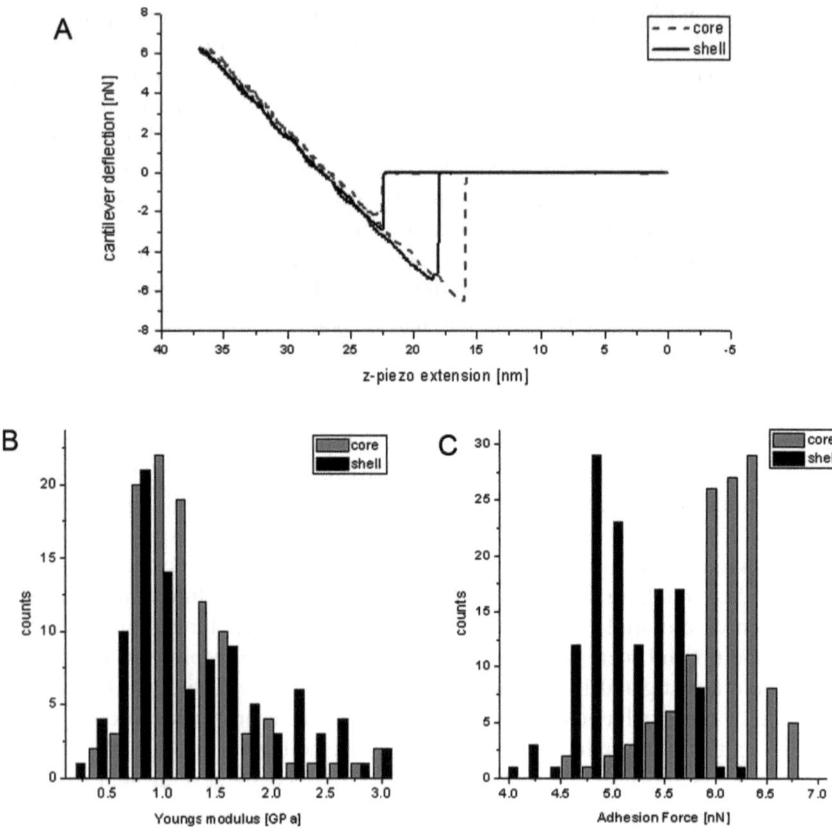

Fig. 22.4. (**a**) Typical sample of force-distance curves recorded on the shell and the core of a single collagen fibril. The *dark curve* shows a spectroscopy curve on the shell, whereas the *gray dashed curve* was recorded in the core. The slope of both curves is nearly the same and indicates identical elasticity. Measurements on the shell and in the core have to be performed in the same sample height to exclude "thin layer effects." The Hertzian model was fitted to the positive cantilever deflection range. (**b**) Elasticity measurement of a single collagen fibril. Force-distance curves were recorded on the shell and the exposed core. On both measuring points more than 100 curves were recorded. The diagram displays the Young's moduli of the force-distance curves versus the frequency. The nanoindentation experiment indicates no measureable difference between core and shell. (**c**) Adhesion measurement on a single collagen fibril. The data display the evaluation of the adhesion forces calculated from the height of the snap-out of the retrace curve. The values show a higher adhesion in the core of the microdissected fibril

surface and thus easily moved by the tip into a perpendicular orientation to the fast scanning direction. In contrast imaging of the protein in a dehydrated (but not very physiological) case is performed easily due to a tighter molecule immobilization to the surface (Fig. 22.5).

Marchant et al. speculated that the absence of water molecules in between the vWF protein and the mica substrate in the dehydrated sample accounts for the stronger protein–surface interaction. Raghavachari et al. extended this

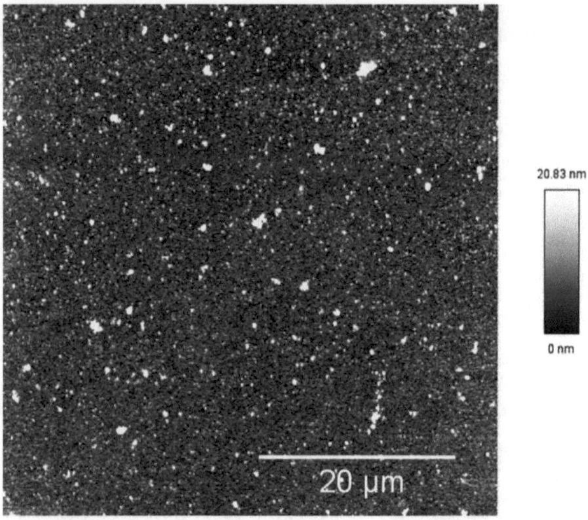

Fig. 22.5. Globular vWF molecules deposited on a thin collagen layer imaged in tapping mode in air. A broad distribution of vWF sizes is visible

surface-dependent adhesion study of vWF to hydrophobic substrates under physiological conditions [47]. Their data may provide a closer approach to the surface-dependent functional behavior of vWF molecules. On a hydrophobic OTS (octadecyltrichlorosilane) surface, vWF undergoes adsorption and molecular spreading. In contrast, vWF is loosely adsorbed and presents more extended structures on hydrophilic mica, which the authors explain as an optimization of hydration, electrostatic and van der Waals interactions. The first AFM study of unrolled vWF molecules was performed by Siedlecki et al. also on hydrophobic OTS [48]. The authors used a rotating disc system to unroll vWF on the substrate and imaged elongated vWF molecules using modified scanning probe tips and mathematical morphology modeling techniques. But also in this study, tip-induced irreversible deformation and elongation of globular vWF molecules occurred, which could be distinguished from unrolled vWF due to shear stress.

The AFM technique was proven to be a significant tool for investigating surface and conformation-dependent adhesion mechanisms and delivered valuable hints for both the driving forces for adhesion to different surfaces and simultaneously for the forces for the vWF monomer–monomer interaction, which keep the protein in its globular state under low shear conditions. In order to understand the blood clotting process it is important to extend the experiments presented above. Whereas the unrolling and activation of a vWF molecule on a site of vascular injury is the initial step for wound healing at high shear conditions, a single extended molecule would never have been able to diminish or even stop bleeding. Only the whole entity of a multitude of activated vWF molecules, self-assembling and forming huge networks, gains the adequate potential to bind enough platelets out of the rapid blood stream to seal a lesion. Imaging such networks is also a demanding task considering the elastic properties of protein assemblies and fibers within the network.

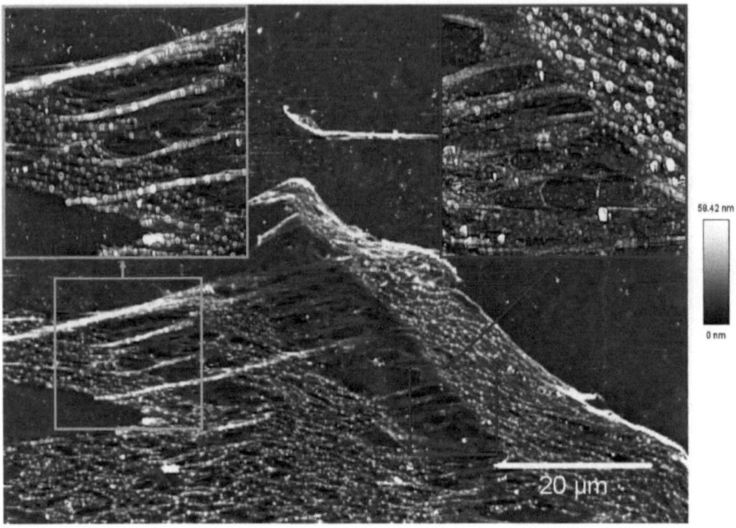

Fig. 22.6. A huge vWF network formed under shear conditions on a thin collagen layer. After dehydration the network was imaged in tapping mode in air. Both *insets* show different regions of the network exhibiting similar and also very different features. Both elongated thick vWF fibers (*left inset*) and globular vWF molecules (*right inset*) can be distinguished within the network

These networks can exhibit a broad variety of sizes dependent on the injury. In vitro networks with diameters in the range from 1 μm (and below) up to the order of mm were observed (Fig. 22.6). Both the very small and the almost macroscopic vWF networks proved to be hemostatically active and could mediate platelet binding.

Under high shear flow conditions platelet binding to a site of injury, or to the pre-formed vWF network to be precise, is accomplished in a two-step process. In a first initial tethering, platelets bind via their GP $Ib\alpha$ receptor to the A1 domain of vWF. This bond forms rapidly but has also a high dissociation rate and therefore a short half-life. This weak adhesion decelerates the platelet and keeps it near the surface, where a rolling motion is induced by the surrounding flow. The reduced velocity accompanied with an increased residence time on the surface enables a slower but tighter bond to be formed via the $\alpha_{IIb}\beta_3$ platelet binding site and the C1 domain of vWF in the second step [8]. The rolling and terminal stopping of the platelet is often accompanied by the formation of an extended cell membrane tether. These tethers are pulled from small localized contact points formed between the cell body and the vWF matrix via $\alpha_{IIb}\beta_3$ and the A1 domain. Dopheide et al. found also kinks in membrane tethers, indicating that multiple bond formation within a tether is possible [10]. Tether formation further increases the residence time of platelets on the vWF matrix compared to rolling platelets. Shao et al. addressed the physiologically important fact that compared to platelets without tethers, tether formation significantly reduces the force, experienced from the surrounding flow, which is working against the bond between platelet and vWF matrix [49]. Thus, a cell membrane tether further increases the residence time of platelets on the surface and increases the probability of tight bonding between platelets and the vWF surface. The observed tether formation is

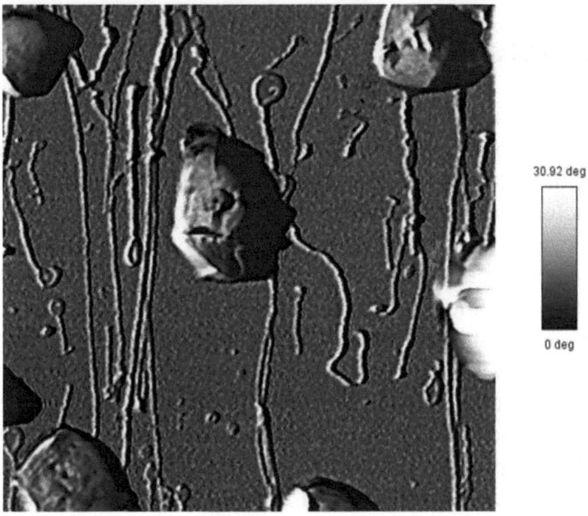

Fig. 22.7. Platelets were imaged in buffer conditions after a shear stress of 20,000 Hz was applied and subsequent formaldehyde fixation (image size $17 \times 17\,\mu m$). Membrane tethers were predominantly found in the direction of flow. But also kinked and broken tethers can be observed indicating multiple binding sites within the tether

not a biochemically driven rearrangement of the cytoskeleton but is a function of the shear rate of the surrounding liquid. Figure 22.7 illustrates immobilized platelets on a vWF-coated surface exhibiting various membrane tethers.

After a tight bond has formed platelets are activated and spread on the surface, exhibiting a smaller contact area for the surrounding flow and more binding sites for the next layer of platelets stabilized via vWF between adjacent platelets. Fritz et al. followed the activation process of platelets and were able to image the drastic reorganization of the cytoskeleton by tuning the scanning force of the tip to visualize internal structures of the cytoskeleton through the softer outer membrane [50].

22.3 Lab-on-a-Chip

In the previous section, the course of blood clotting was followed and investigated by AFM step-by-step from the exposure of subendothelial matrix at a site of vessel damage till platelet adhesion and thrombus formation occurred. On the following pages we will sketch how the AFM technique can be utilized in determining both distinct blood diseases and the age of bloodstains as an especially accurate measurement in forensic science. In the second part of this section we will highlight Surface Acoustic Wave technology as a relatively new and powerful tool for pumping small amounts of liquid. This technique is perfectly suited for mimicking blood flow conditions in vitro and can be combined to a variety of well-established methods for answering a broad variety of questions.

22.3.1 Nanomechanical Diagnostics

Atomic force microscopy has been widely applied in imaging biological samples under physiological conditions (for a review see [51]). The shape of red blood cells is characteristic of particular diseases and there is a growing interest in the identification of ultrastructural features associated with diseases. In early publications, images of fixed erythrocytes were presented [52, 53]. Easily recognizable erythrocytes were observed but little subcellular detail was seen. Häberle and coworkers immobilized the separated cells on a micropipette, which was mounted on the scanner. The imaging was performed on the protruding part of the cell and they achieved a resolution in the 10-nm range [54]. Several groups demonstrated the imaging of red blood cells in physiological media. At high resolution, the underlying cytoskeleton of the blood cell has been resolved and flaws in the cytoskeleton could be observed [2]. The membrane skeleton structures, without major distortions or deformations of the cell surface, were observed by Kamruzzahan and coworkers using tapping mode under near-physiological conditions [55]. With this approach significant differences in the morphology of red blood cells from healthy humans and patients with systemic lupus erythematosus were detected on topographical images. The surface of red blood cells from systemic lupus erythematosus patients showed characteristic circular-shaped holes approximately 200 nm in diameter under physiological conditions. Blood cells are subject to various kinds of stresses in flow fields. Hemolysis is the phenomenon in which a higher stress than normal damages the erythrocyte membrane and results in the leakage of its contents. Even if the stress is not strong enough to cause cell lysis, however, the cell membrane may sustain some damage. The fine surface structure of sheared erythrocytes was observed by Ohta and coworkers [56].

Blood analysis including blood cell measurement is one of the most basic items of hematological testing which is indispensable in health examination, disease diagnosis, and treatment. The common lab test is used to diagnose and monitor the body's response to diseases. Some tests measure the components of blood itself and others examine substances found in the blood to identify abnormal function of various organs. In forensic sciences the examination of bloodstains represents a major application during crime scene investigation. There exist a lot of reliable methods for the detection and identification of blood spots. For the evaluation of suspected bloodstains solutions such as phenolphthalein, tetramethylbenzidine can be used, as they change color when they come into contact with peroxidase or hemoglobin in the blood [57]. For the detection of even minute amounts of blood traces the presumptive luminol chemiluminescence test is widely used in forensic practical work [58]. It is further possible to unambiguously attribute the blood to a certain individual by using molecular biological techniques, such as genetic fingerprinting [59]. However, the determination of the age of a blood spot remains an unsolved problem in forensic routine work. For more than a hundred years forensic scientists have been engaged in finding a methodology, which allows determining the exact age of a dried bloodstain. Since then several approaches were proposed to have potentially solved this important problem during crime scene investigation, but none of these could ever be established in forensic practice. An attempt was undertaken by evaluating the time differences of the solubility of bloodstains of different ages [60]. In control experiments it turned out, that both methods considerably diverge from the real age of an investigated bloodspot. More recently, new methods were tested such as remission spectrophotometry [61] or electron-spin-resonance measurements [62], which can detect an age-dependent increase of signal intensity of meth-hemoglobin, non-hem-iron molecules, and organic radicals. Concordantly, these methods proved to have a high error rate, which only allows a rough estimation of the age of a bloodstain and does not excuse the high technical expense. In more recent times chromatographic methods were tested for use in dating blood spots. Inoue et al. used high-pressure liquid chromatography (HPLC) to measure the quantitative compound of the globin chains of the red blood dye hemoglobin [63]. They found a decrease of the α-chains related to the heme, the color defining prosthetic group of the red blood dye hemoglobin, with increasing age of a blood spot. The measurements of the standardized bloodstains revealed high deviations and revealed not to be practicable for routine forensic work. Moreover, a mobile application of this method for the use directly at a crime scene investigation seems not to be realizable. Taken together, all described methods proved to be not applicable in routine forensic work, as the results of the age measurement of the bloodstains provided too high deviations compared to the real age. Most of these attempts rely on advanced technical equipment and, therefore, do not allow application directly at a crime scene. Additionally, a part of the sometimes rare trace material has to be consumed for the use of these approaches, so that it will be lost for further important analysis, like molecular genetic investigations. A new AFM-based methodology, which could be used for the age determination of dried bloodspots during crime scene investigation, was introduced by Strasser and coworkers [64]. An AFM was used as a nanoindentor to monitor age-related changes of the elasticity of bloodstains under standardized conditions. Since the invention of the AFM, indentation experiments are generally applied in measuring elastic properties. Nanoscaled materials are probed with the AFM due to its high lateral and vertical resolution down to 0.01 nm. In contrast to other hardness testers the force resolution of the AFM can reach ranges of 10^{-4} nN [65]. The application of AFM-force spectroscopy is widespread and used for measuring polymer systems [66,67], bone elasticity [68], collagen [69,70], or cells [71–73]. For the analytic procedure a fresh blood spot was applied on a glass slide and the AFM detection started after drying of the blood drop. In a first step, an overview image was generated showing the presence of several red blood cells, which could easily be detected due to their typical "doughnut-like" appearance. The consecutive morphological investigations in a timeframe of 4 weeks could not show any alterations. Secondly, AFM was used to test the elasticity by recording force-distance curves. The measurements were performed immediately after drying, and at 1.5 h, 30 h, and 31 days. The conditions were kept constant at room temperature (20°C) and a humidity of 30%. The obtained elasticity parameters were plotted against a timeline and repeated several times. The elasticity pattern showed a decrease over time, most probably influenced by the alteration of the blood spot during the drying and coagulation process. The preliminary data demonstrates the capacity of this method for use in the development of calibration curves, which can be used for estimation of bloodstain ages during forensic investigations [64] (Fig. 22.8).

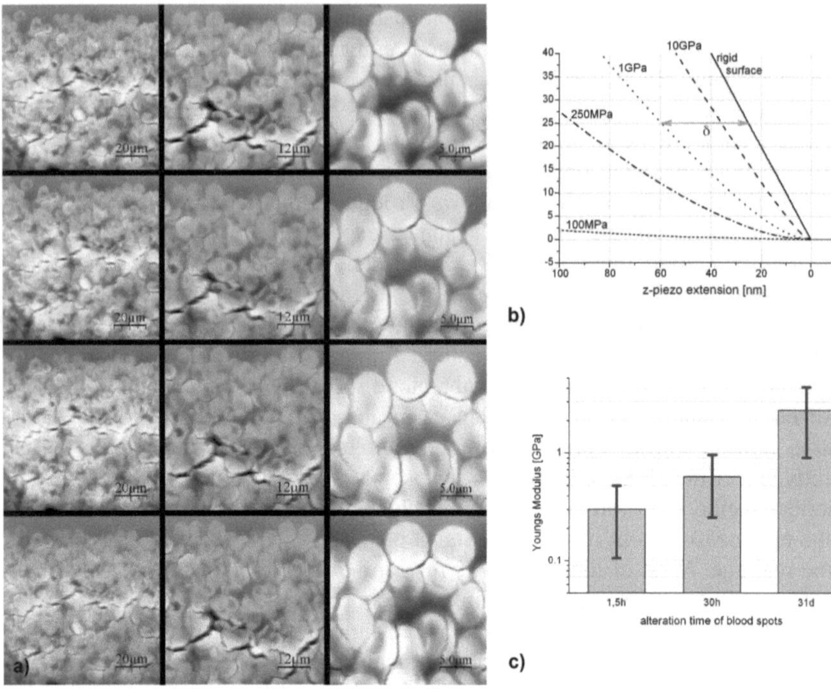

Fig. 22.8. (**a**) Results of the morphological investigations of a dried blood spot by AFM. The illustration is divided into rows and columns. The AFM images in the rows correspond to the individual measuring days. The columns display identical scan areas depending on the alteration time of the blood spot. The first series (first row) was scanned after 1.5 h, the second series after 1 week, the third series after 2 weeks, and the fourth series after 4 weeks. The morphology independent of the scan range (100, 60, and 25 μm) does not change with the alteration time; (**b**) Force-distance curves of various elastic samples calculated with the Hertzian theory. At rigid surfaces the AFM tip cannot penetrate into the sample, therefore the cantilever deflection corresponds to the extension of the z-piezo. The softer the sample the more penetrates the tip into the sample, therefore the slope is very smooth. To get the elasticity modulus (Young's modulus, E) the Hertzian or similar models have to be fitted to the measurement curves; (**c**) Elasticity change versus alteration time. The evaluation of the Young's moduli indicates an elasticity change depending on the alteration time. The bloodstain was measured after 1.5 h, after 30 h, and after 31 days. Every column was calculated from more than 100 individual force-distance curves. The *whisker* displays the standard deviation of the force curves

22.3.2 Mimicking Blood Flow Conditions on a Surface Acoustic Wave-Driven Biochip

So far, only static experiments on blood compounds after adhesion or fixation have been presented. The blood stream itself is a very dynamic process and caution has to be exercised in interpreting data of the functional behavior of its cells, proteins, etc. when achieved under static conditions. Apart from AFM there are a variety of

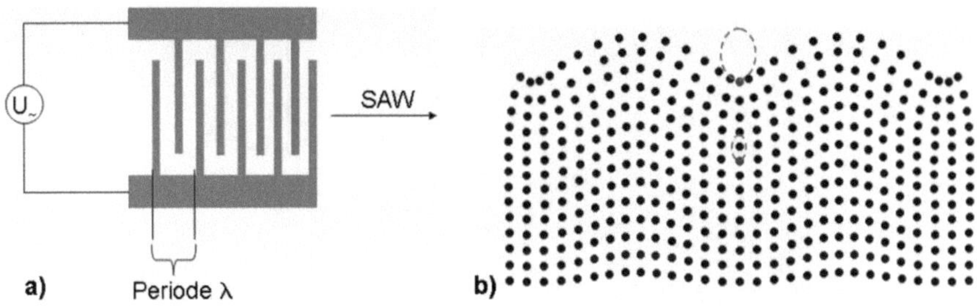

Fig. 22.9. (**a**) A voltage signal on a piezoelectric material results in a propagating mechanical deformation of the crystal. By adding a higher number of electrodes after one another and applying an alternating signal, matching the resonance frequency $f = \frac{v}{\lambda}$ (v: sound velocity of the piezoelectric material; λ: periodicity of the electrodes), the mechanical impulse is enhanced and a Surface Acoustic Wave is emitted. (**b**) Side view of a piezoelectric material excited via an IDT (not shown). The *red lines* display the elliptic course of the blue lattice points within one period of movement. The reason for the terminus Surface Acoustic Wave accounts for the exponential decrease into the substrate of the amplitude of an oscillating lattice point in the crystal

other techniques for investigation of blood clotting. SAWs offer an extremely potent method of driving small amounts of fluids and mimicking the blood vessel system on a planar substrate. The concept of Surface Acoustic Waves has been known since the late 19th century, when Lord Rayleigh first gave a theoretical description of earthquake waves [74]. Since that time SAWs have gone a long way. About 80 years later physicists started to apply this theoretical work to a scale Lord Rayleigh might not have thought about [75]. By lithographical methods Interdigital Transducers (IDTs) (Fig. 22.9) were deposited on piezoelectric substrates like $LiNbO_3$, where an alternating voltage signal excites a propagating mechanical high-frequency wave on the order of 100 MHz, an amplitude of 1 nm, and a wavelength of a few μm. The energy of the SAW is confined to the surface of the piezo and decreases almost totally within one wavelength into the material.

Dependent on the boundary conditions of the piezoelectric substrate, shear- or Rayleigh-waves, or a mixture of both wave modes, are excited. Only Rayleigh-waves exhibit an "out of plane" movement of the oscillating parts of the substrate, which can be applied for driving small amounts of fluids [76]. The surface oscillation dissipates energy into the liquid, which results in a pressure gradient within the fluid and hence induces fluid streaming. All further applications and experiments presented here are performed on $LiNbO_3$ chips and exploit the dissipated energy (leaky wave) of a Rayleigh-wave to induce acoustic streaming in a fluid [77].

There are a variety of advantages to using a planar SAW-driven microfluidic device compared to commonly used flow chambers. Because of the extremely small amount of sample volume (down to a few μl) even expensive or rare materials can be processed in contrast to conventional flow chambers with a typical working volume of some ml. By adjusting the voltage on the IDTs the flow velocity and therefore the shear rate of the fluid can be set very precisely. The 2D and optical transparent chip further offers the possibility to combine the setup with imaging techniques like fluorescence microscopy or AFM (see also Sect. 22.4). Furthermore, by

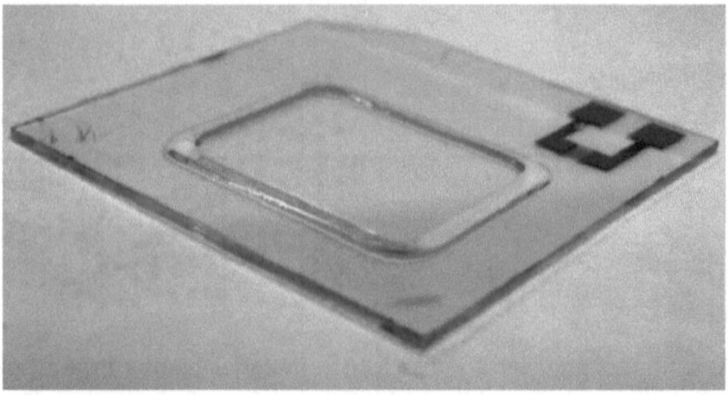

Fig. 22.10. Mimicking a blood vessel. The IDTs (*upper right*) on the 2D biochip (1×1 cm) drive the fluid clockwise within a hydrophobic-hydrophilic track. This track is deposited onto the chip by means of photolithography. A variety of different hydrophobic-hydrophilic structures can be produced to mimic even the smallest ramifications of blood vessels (Figure reprinted from Schneider MF, Guttenberg Z, Schneider SW et al., An Acoustically Driven Microliter Flow Chamber on a Chip for Cell-Cell and Cell-Surface Interaction Studies, ChemPhysChem 2008; 9: 641 with permission from Wiley-VCH)

means of standard optical lithography almost any flow geometry can be produced as a hydrophobic-hydrophilic track on a SAW-driven biochip. Using such a SAW-driven biochip, flow geometry and therefore the flow pattern can be fitted exactly to the physical, chemical, or medical problem under investigation. Figure 22.10 displays a hydrophobic-hydrophilic functionalized SAW-chip with a rectangular water channel representing a physicist's equivalent of a blood vessel. Full experimental freedom is achieved by the possibility to grow cells or coat the surface with a variety of different materials within the fluid track.

22.4 The Lab on a Chip – AFM – Hybrid

The SAW-driven artificial blood vessel provides the basis for testing variable blood flow conditions. In order to provide a first insight into the structural and dynamic properties of vWF-networks and -bundles we combined a microfluidic reactor, mimicking the hydrodynamic conditions in the microcirculatory system of our body [78] with an atomic force microscope for minute protein manipulation. After presenting the experimental setup and pointing out some basic advantages compared to commonly used AFM setups, we will present the hybrid setup in action, investigating the dynamic network properties of the blood clotting protein von Willebrand Factor in an amazing new way.

22.4.1 Experimental Setup

The completely planar reactor layout and fluid actuation system gives us the technological freedom to create a continuous flow in any desired geometry. No drain and hence no supply is necessary, allowing us to monitor the identical solution of

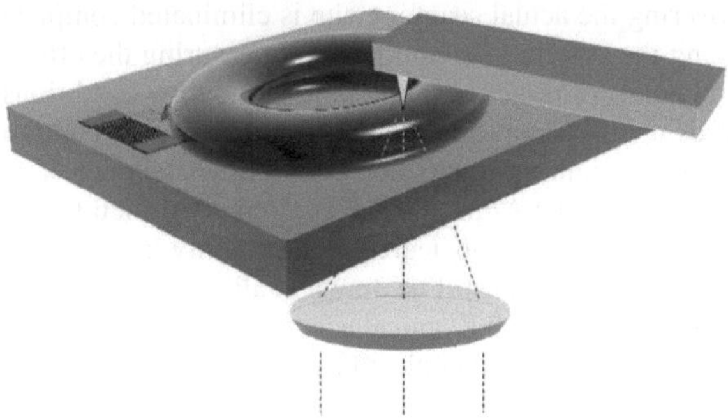

Fig. 22.11. The microfluidic set up. The unique combination of a microscope, an AFM tip, and a SAW-driven pumping system enables both the study of the formation of macromolecular VWF networks and conglomerates in real time and the mechanical manipulation of VWF aggregates and bundle pulling in particular (Figure reprinted from reference [79] with permission from Elsevier)

proteins, cells, vesicles, etc. without any loss of material or a change in any crucial parameter for long periods of time (~1 h). Although the fluid streaming induces a lot of vibration on the cantilever and hence too much noise for imaging, force spectroscopy under certain circumstances is still possible (Fig. 22.11).

Nevertheless, this setup provides distinct advantages in imaging too. Especially the measurement of the time course of sample alterations, induced by external stimuli, has always been a problem for AFM users. There are two significant deficits of normally operated devices in investigating samples before and after the application of either chemical or mechanical reagents or forces. One is the time scale till a reagent makes an impact on the sample and the second is the recovery of the region of interest (ROI) after this impact has taken place. Whereas chemical interactions such as changes in pH or salt concentration can be induced without moving either sample or tip, and thus do not affect the recovery of the ROI essentially, the equilibration process and therefore the reaction of the system is diffusion limited. Depending on the particle size of the added reagent the equilibration of the system may be extremely long. Without additional external mixing, equilibration of pH or salt concentration can be achieved within several minutes as a function of the sample volume. This may be sufficient for a variety of special applications, but is far from an ideal situation, where an immediate response is desirable to distinguish between a reagent effect and simple sample aging. An investigation of the influence of an enzyme of protein, for example a blood clotting protein, to a target sample would only be measurable as recently as several hours after addition depending on the protein size. Most biological samples would degrade within that time. The SAW pumping system provides mixing ability for a great variety of fluids with different viscosities, emulsions with particles of different sizes, and establishes equilibrium almost immediately. A tremendous advantage in doing so is that the sample does not have to be taken out of the AFM setup or moved at any time. Such actions increase the probability for sample contamination by other chemicals or dirt particles and open the door for other unpredictable interferences to the system in every experiment.

Furthermore, in the hybrid setup the cantilever can rest on the ROI and does not even have to be retracted during the few seconds of mixing. Thus, the fretful problem of recovering the actual scanning site is eliminated completely. Apart from mixing and driving the system into equilibrium, measuring the effect of shear stress on single cells, platelets of proteins, mimicking blood flow conditions or just adding mechanical forces on the sample is almost impossible in a common AFM-setup. Removal of the sample, putting it into an external flow chamber or any other device and mounting it back into the AFM, makes it extremely difficult to recover the ROI, and furthermore would cause a high degree of uncertainty in the system, if qualitative and quantitative conclusions can still be drawn at all.

The planar setup, providing full optical accessibility of the light path between the AFM tip and the fluidic track, is completed by an inverted microscope for recording the microfluidic channel and the AFM tip simultaneously [79]. This optical control system enables the online monitoring of any optically accessible event and thus provides a reliable means for reacting immediately with the AFM to every important change of the system. Furthermore, additional optical images of stained samples to a height profile for instance can often facilitate data interpretation a lot. In this way optical microscopy provides a direct reference system for AFM results. This hybrid system, combining the advantages of three different experimental approaches, does not simply serve as an inbuilt reference system. The assembled techniques supply information from different perspectives for the same question creating a unique and more complemented picture of the scientific problem. Thus, applying the AFM technique and optical microscopy simultaneously whilst maintaining fluid streaming to a particular sample, the acquired information is maybe more than just the sum of results of single experiments performed one after the other. The challenges of measuring a single sample in different devices after one another have already been discussed earlier.

This newly developed hybrid setup features high potential in every field where there is benefit or simply the necessity of the combination of the high-resolution imaging or the accuracy of force measurements of an Atomic Force Microscope, fluid mechanics, and optical accessibility.

In the next two sections we will proceed one more step and present the microfluidic reactor–AFM hybrid system illuminating new correlations in blood clotting by using the AFM cantilever not for measuring minute interaction forces or height profiles but as a hook for manipulating vWF networks.

22.4.2 Bundle Relaxation

vWF networks are a prerequisite for platelet aggregation at a site of vascular injury. The formation of vWF networks and conglomerates was shown to be shear-dependent and is thought to follow the same mechanism as the adhesion of vWF molecules to the extracellular matrix. Above a critical shear flow vWF molecules become activated, unroll and spontaneously self-aggregate with one another [3, 80]. On the other hand, there is only little knowledge about the actual binding properties and kinetics of vWF molecules inside such a network. Its reaction to an external induced force or the relaxation of single bundles, pulled out of such a network, can deliver valuable information about how such a network is made up on a molecular and energetic level.

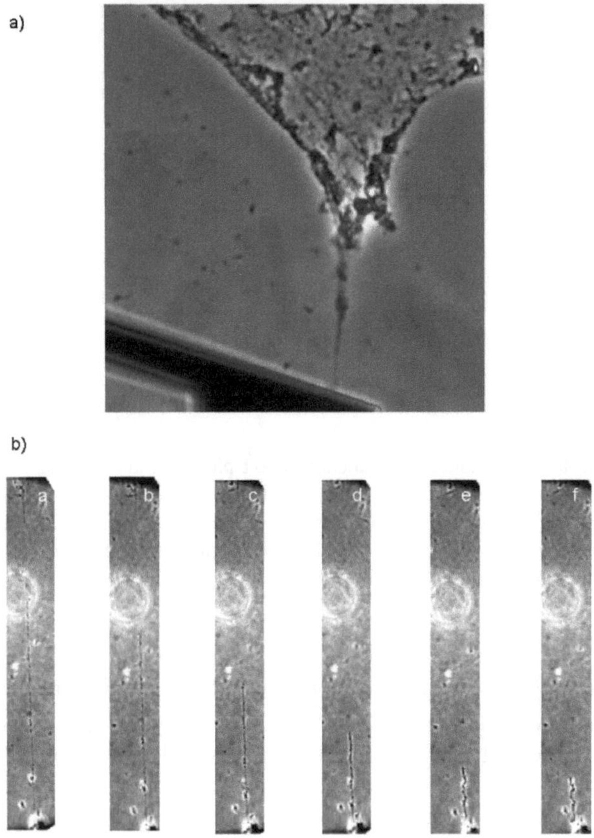

Fig. 22.12. (**a**) ULVWF immobilized on the AFM tip. (**b**) Relaxation after rupture (each image of the sequence is approximately 840 μm in height and 120 μm in width) (with courtesy of Elsevier) (Figure reprinted from reference [79] with permission from Elsevier)

Therefore, in the course of an experiment the AFM cantilever tip is approached to a protein network after the SAW-induced vWF conglomeration process is finished. vWF bundles are gently pulled by an untreated AFM tip until they finally rupture (Fig. 22.12). Acquired data of the time course of the relaxation process were analyzed using standard software. The hybrid setup enables us to study the relaxation of ultra-large vWF (ULVWF) bundles of different lengths pulled from vWF networks freshly formed by hydrodynamic stress in the microfluidic reactor while keeping all external parameters constant. In this set of experiments the AFM was solely used as a hook to pull vWF bundles until they finally rupture and not to quantify the required rupture-forces.

On a more quantitative basis, the dynamics of the system were monitored by measuring the end-to-end length of the bundle as a function of time. The maximum length L_0 was determined by the distance between the AFM tip and the point where rupture occurred. A typical trace of the end-to-end distance (i.e. the distance from

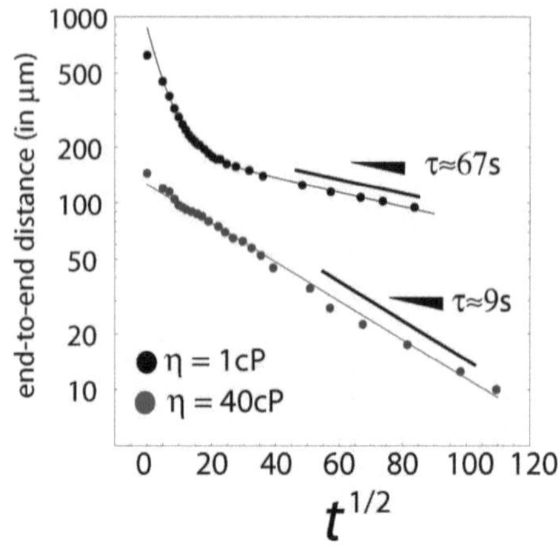

Fig. 22.13. End-to-end distance as a function of $\sqrt{t}$ plotted in a log-linear scale for different viscosities. The upper (*black*) curve corresponds to a viscosity of 1 cP, and the lower (*blue*) curve corresponds to a viscosity of 40 cP. In the low viscosity regime two distinct relaxation time scales could be observed (with courtesy of Elsevier) (Figure reprinted from reference [79] with permission from Elsevier)

the AFM tip to the free end of the ULVWF) as a function of time is presented in Fig. 22.13. We note that the relaxation has two characteristic time scales, and both of these periods are well described by stretched exponentials of the form: $L \propto e^{(t/\tau)^{\beta}}$ with $\beta \sim 0.5$. The fact that the relaxation process follows a stretched exponential implies that the underlying dynamics is governed by hopping events between random minima in a rough energy landscape. Equivalently, the relaxation spectrum is composed of a broad (random) distribution of exponential processes [81]. Such relaxation behavior is very common in nature, and has been found in different contexts. In particular, it has been observed in the electric birefringence relaxation of ensembles of synthetic polyelectrolytes [82], in the relaxation of DNA [83], and in proteins [84], to name a few. In these studies, the origin of the relaxation spectrum arises from the averaging over a multitude of parallel non-interacting (presumably exponential) relaxation processes. Theoretically, stretched exponential and power-law behavior has been extensively studied in the area of critical dynamics and glasses [85]. In the context of single polymer relaxation, Cherayil et al. have shown [86] that after inclusion of memory effects to the dynamics of the system, the relaxation is a stretched exponential with an exponent of $\beta \sim 0.5$. These memory effects included in this work accounted for the fact that a given conformation at earlier times affects the distribution of conformations at a later time along the relaxation process.

vWF bundles present new features compared to those described previously: (1) it has two clear relaxation time scales, and (2) the relaxation of the individual molecules is coupled to the one of all others in a non-trivial fashion. In this sense, we can say that vWF bundle relaxation is a more complex problem than those studied before. One of the striking features of our findings is that the longest relaxation rate k, corresponding to the inverse of the longest relaxation time τ, seems not to be dependent on the initial length of the bundle (data not shown). This implies that the dynamics of relaxation is dominated by conformational constraints, as well as effective internal friction limiting the rate of deformation of the object itself.

To further unravel the physical origin of the observed phenomenon the effect of viscosity of the solvent was studied on the long and short timescales of the bundle relaxation by adding a thickening agent, such as glycerol. In Fig. 22.13 we show two different traces of the end-to-end distance at different viscosities: 1 and 40 cP. As can be seen, the initial temporal decay is slowed down such that the full relaxation can be well described by a single stretched exponential. Interestingly, the slowest relaxation rate is now faster than the one for pure buffer solution (with lower viscosity). Further discussion on the origin of the double-time relaxation spectrum, and on the apparent enhancement of the long-time relaxation rate is given below.

In a high viscous environment, surprisingly, only a single relaxation time is observed. However, the relaxation spectrum found here is still different from the one of a single collapsing polymer. There, the relaxation of the end-to-end distance is supra-exponential, or in other words, the rate of compactification grows as the polymer relaxes [87]. If at all, we only find evidence of such a relaxation type during the first few 100 ms after rupture. This observation is not too surprising, as the ULVWF consists of many single fibers of different lengths relaxing at the same time in a complex fashion. Comparing it to the dynamics of a single polymer seems to at the least be doubtful or not valid at all. The complexity of our system arises because the relaxation of one fiber affects the relaxation of the other fibers, and vice versa.

For a better feeling of why the observed relaxation is a stretched exponential, it is probably better to describe the bundle as a single entity relaxing in a rough potential landscape determined by the configurations of the vWF single fiber constituents. The manifold of configurations is presumably strongly dependent on the number of self-association contacts, and hopping from one state to the other can only be done through thermally activated processes. The average barrier height will be reduced in the highly stretched state because of entropic forces, and thus the rate of relaxation should be faster at the beginning. Nonetheless, this argument would imply that one would see a continuous change in the relaxation of the bundle, and not two clear relaxation times as is found here. An alternative explanation is that there are two attractive interactions of different origins, which in this case would presumably correspond to self-association and hydrophobic attraction among monomers, driving the relaxation in each of these two regimes. To this date it is still unclear which of these interactions is the dominant one in the collapse of vWF, but we might argue that at long times, the hydrophobic interactions must become important since in principle the majority of the possible self-association contacts within the bundle should be present and only unfavorable surface interactions with the solvent and entropy can drive the further decrease in size.

With this unique combination of a Surface Acoustic Wave-driven planar microfluidic reactor, an optical microscope, and an AFM cantilever used in an unorthodox way, it was possible to unravel new aspects and concepts in protein network and bundle assembly. Beyond the scope of Imaging and Force Spectroscopy the AFM technique was successfully implanted into a fluidic system and functioned as a minute instrument in manipulating protein assemblies. Whereas bundle pulling was performed manually and no rupture forces were quantified, the sophisticated AFM control was a valuable tool for approaching and intruding protein aggregates and networks with distinct forces. In this case, the AFM technique, applied in a non-conventional manner, provided an ideal means to achieve more details on a very complex non-trivial question.

22.4.3 Stream Line Manipulation and Flow Sensoring

22.4.3.1 Stream Line Manipulation

The logical extension of the system presented above is to operate the AFM while exposed to fluidic streaming. Limitations in imaging due to an increased noise level can be compensated by benefits in other areas. This approach promises auspicious potential for real-time investigation of von Willebrand Factor, whose activity is triggered by the surrounding shear flow. With the help of the AFM cantilever, streamlines can be manipulated in a decisive way (Fig. 22.14), vWF network formation can be induced and moreover fluid streaming velocities can be measured with tip bound polymers.

Any stationary body inside a streaming fluid modifies the stream lines in a certain way. As far as the cantilever is thin and long enough, the overall flow pattern in the vicinity of the cantilever tip can be regarded as almost undisturbed. Around the tip the almost parallel flow lines are bent. Considering no slip boundary conditions at the tip, an extremely high shear rate can be induced dependent on the flow velocity of

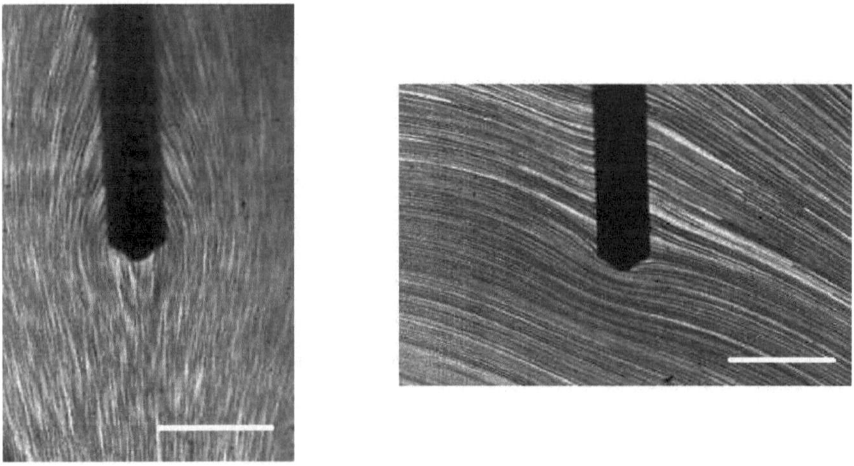

Fig. 22.14. An AFM cantilever tip (*scale bar* ~70 μm) is exposed to parallel and perpendicular fluid streaming. Because of the variable geometry of the planar SAW-driven microfluidic reactor and its adjustable pumping power, user-defined flow directions and velocities can be realized. Around the tip at the very end of the cantilever, the streamlines are softly deformed. The streamlines in the figures **a** and **b** are an overlay of a set of pictures from a high-speed camera, showing beads moving along the cantilever in a distinct way

the surrounding fluid. The shape, direction, and magnitude of the shear flow pattern including the shear rate acting on the tip can be varied both by an adjustment of the external voltage on the SAW-pumping system and by a change of the AFM cantilever position by the stepper control system in x-, y-, and z-direction relative to the variable stream lines.

Damage to a blood vessel also induces a mechanical obstacle and thus a disturbance of the flow pattern to the stream lines. Although a daring simplification from the point-of-view of a medical doctor, the effects of an injury to the vessel wall in the blood stream and of an AFM tip in the microfluidic reactor resemble each other in the basic aspects. Both create an increased shear rate in their close vicinity and hence are able to induce a conformational change of vWF molecules, if this shear rate is high enough. As discussed before, this conformational change is accompanied by an activation of vWF's binding potential resulting in a network formation both at the vascular injury and the AFM tip (Fig. 22.15).

The simulation of the actual flow conditions around the AFM tip derived from the experimentally detected stream lines for different flow conditions will soon make it possible to estimate the forces around the tip. Knowing the precise behavior of vWF under shear conditions will allow using vWF as a probe for sensing forces in a fluid with and without the AFM tip. This will not only enable us to compare field and point-like forces, but also to determine the forces on soft objects under flow, a field which is only very little explored due to the lack of experimental data.

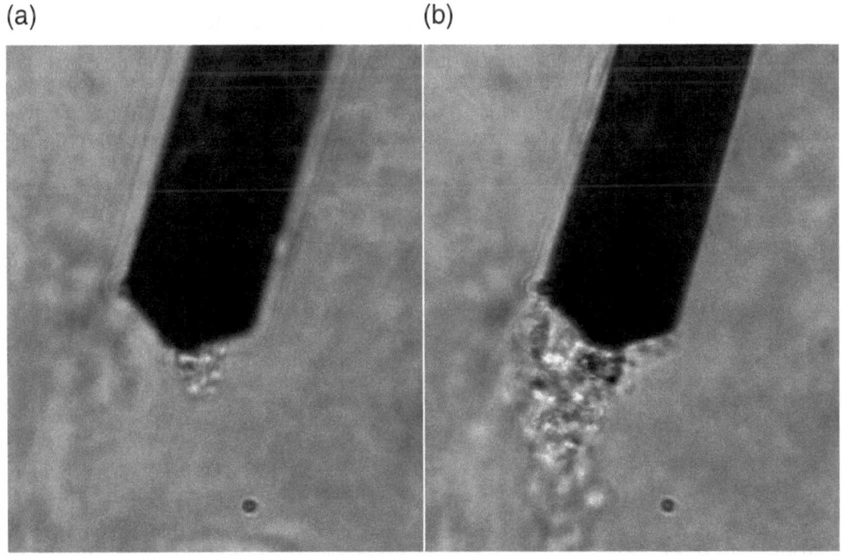

Fig. 22.15. The cantilevers were exposed to a vWF solution in a (**a**) low shear and (**b**) high shear regime for several minutes. Whereas in the low shear regime almost no or only minor protein adhesion could be observed, vWF self-associates continuously in the high shear regime forming a sticky three-dimensional conglomerate (cantilever width ~30 μm)

22.4.3.2
Flow Sensoring

Despite producing whole vWF conglomerates on the tip, it is much more difficult to bind single proteins or polymer fibers with one end onto the AFM tip, while keeping the rest of the protein unconstrained. This cannot be achieved simply by creating a high shear rate around the tip and waiting for spontaneous adhesion. Defined protein fibers are a prerequisite for investigating quantitative protein reactions to an external fluidic stream on one hand, and could also be a valuable tool for probing streaming properties by a protein "wind gauge" waving in the stream lines of a microfluidic reactor on the other hand.

Similar to AFM Force Spectroscopy measurements, where the tip scans for a protein on the surface first, protein fibers are "collected" from a vWF covered surface. The tip penetrates a protein conglomerate and a fiber or bundle of variable thickness is bound. Normally, a lot of such binding attempts have to be undertaken to pick up a suitable protein fiber. In contrast to the bundle relaxation experiments another prerequisite for a successful experiment is that the fiber does not rupture during the pulling sequence. Relaxation of a thin fiber to the cantilever tip will result most probably in an undefined and irreversible adhesion of the protein around the tip. In order to bind a loose and mobile vWF fiber with one end onto the AFM tip, it must not lose contact to another small and spatial unconfined body, acting as a contact point for external pulling forces. This approach is similar to the method of optical tweezers [88,89], where a protein is pinned to a solid surface (the AFM tip in this case) and experiences forces mediated by a bead tightly bound to the protein's second end. The role of the bead is taken over by a stiff conglomerate of vWF in the experiment shown in Fig. 22.16. In the course of the experiment, an extremely thin vWF fiber (below the optical resolution of the microscope) was linked to the AFM tip and remained attached to a freely moving vWF conglomerate (∼ vWF-"bead"). The force on the vWF assembly due to the fluid streaming elongated the vWF fiber. Throughout the experiment, the SAW-driven microfluidic system provided constant

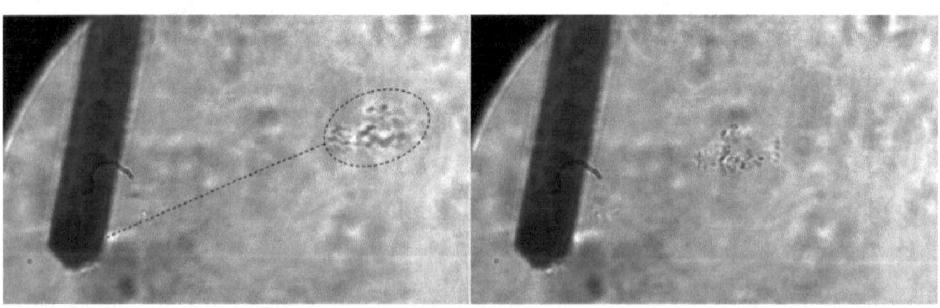

Fig. 22.16. A tip bound vWF fiber experiencing alternating forces by the fluid stream mediated by vWF assembly. The vWF assembly is highlighted by the spherical *inset* to guide the eye. The *dashed line* illustrates the optically not resolved extremely thin vWF fiber and furthermore points out the direction of the flow. Whereas on the picture on the left side the vWF fiber is stretched in a fast flowing regime, the fiber length drastically decreases under a critical flow velocity (cantilever width ∼30 μm)

power to pump the liquid. However, in particular the streaming velocity is not constant over the whole channel but is a function of the actual position inside the fluid. To check for the elongation of the vWF fiber as a function of the fluid streaming, the cantilever is retracted in 50-μm steps from its starting position far away from any wall. With increasing relative cantilever distance, the flow velocity and therefore the acting force on the fiber steadily decreased, resulting in a slight reduction of the fiber length. Above a critical height, or in other words falling below a critical force on the vWF-"bead," the protein fiber contracted in a non-linear fashion (Fig. 22.17). Note the continuous decrease of the flow velocity at the same time!

Similar experiments may help in understanding phase transitions in systems of fractal geometry. From a mathematical point-of-view, the protein fiber can only be considered as one-dimensional in a first approximation, but neither is it a two-dimensional object. The divergence from the one-dimensionality of the vWF fiber arises from multiple contact points of the linear molecule forming a protein ball of yarn. Therefore, probing the reversible elongation and contracting properties of such

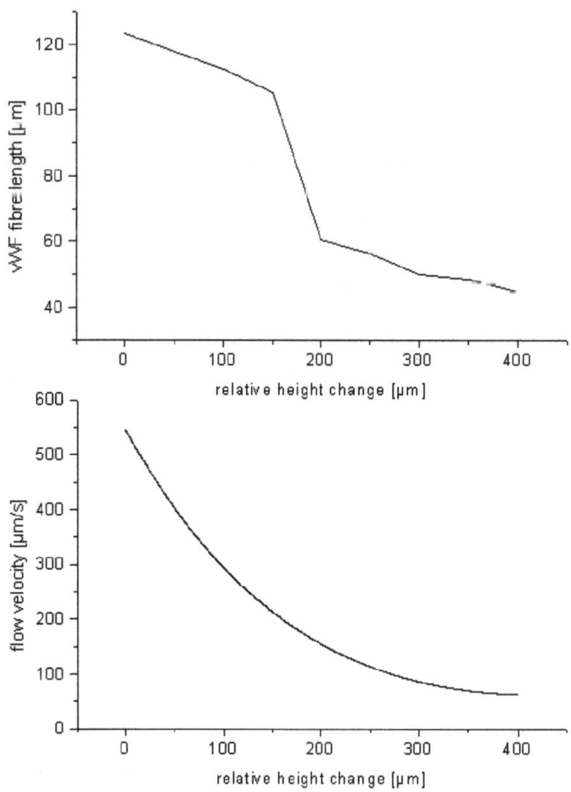

Fig. 22.17. While increasing the relative height the flow velocity steadily decreases whereas there is a significant discontinuous reduction of the vWF fiber length between 150 and 200 μm. Note that in first approximation the flow velocity is direct proportional to the acting force on the fiber

biopolymers may lead to a deeper insight into advanced mathematical concepts for a description of protein assemblies.

Taking advantage of almost unlimited possibilities in protein engineering for the replication of proteins [90–92], exhibiting similar dynamic conformational features as vWF, industry can apply these findings for the production of a new kind of flow sensor. Artificial proteins bound to an AFM tip on one end and to a bead on the other end can sense both the flow direction and the flow velocity, indicated by the polymer length. Such a sensor would be able to scan directly a three-dimensional vector field. Furthermore, a modified AFM tip comprising a distinct binding position for such a protein fiber would be of additional value. As this was not the case in the experiment presented above, the minute force detection of the AFM could not be taken into account. With an exactly known fiber contact point it would be possible to calculate acting forces from the bending of the cantilever, which could serve as an inbuilt reference measurement for the fiber elongation.

22.5 Summary and Outlook

In this chapter we presented AFM as a multi-functional tool for answering questions concerned with blood clotting. With the help of the AFM technique, phenomena like vWF network formation and its intrinsic dynamic structure, platelet adhesion to vessel lesions and pathological altered blood cells could be addressed and their underlying concepts elucidated. The combination of a Surface Acoustic Wave-driven two-dimensional microfluidic reactor with an AFM setup enabled the investigation of the dynamic properties of the mechanically activated blood clotting protein von Willebrand Factor not only under physiological buffer conditions but also under physiological mechanical flow conditions. From its start in 1986 AFM has gone a long way and its journey is far from over. Especially in combination with microfluidics the AFM technique is still at the very beginning and there are a lot of fields of activity left. Researchers will be able to probe dynamic correlations between components nobody has thought of yet. A demanding, but with current techniques possible task would be the investigation of blood clotting under atherosclerotic conditions. The properties of blood coagulation factors like vWF and also platelets depend strongly on the streaming pattern of the blood flow. Mimicking atherosclerotic conditions on a SAW-driven biochip would require only the production of a hydrophobic-hydrophilic track with a set of different "vessel" diameters which is easily achievable considering the recent state of the technology. Imaging of the resulting protein networks under these different physiological flow conditions can lead to a better understanding of thrombus formation, which may be a first step in preventing it. But this is just one example within the framework of blood flow. The essential concepts that form the basics of the results presented here will be extended to a far broader set of experiments and future applications.

Acknowledgments. The authors would like to thank Stefan Bössinger and Jennifer Angerer for laboratory work. Financial support by the Elite Netzwerk Bayern (CompInt) and the German Excellence Initiative via the "Nanosystems Initiative Munich" (NIM) is gratefully acknowledged.

References

1. Binnig G, Quate CF (1986) Phys Rev Lett 56:930
2. Nowakowski R, Luckham P, Winlove P (2001) BBA 1514:170
3. Schneider SW, Nuschele S, Wixforth A, Gorzelanny C, Alexander-Katz A, Netz RR, Schneider MF (2007) Proc Natl Acad Sci 104:7899
4. Sadler JE (1998) Annu Rev Biochem 67:395
5. Tangelder GJ, Slaaf DW, Arts T, Reneman RS (1988) Am J Physiol 254:1059
6. Sadler JE (1991) J. Biol. Chem. 266:22777
7. Perkkio J, Wurzinger LJ, Schmid-Schonbein H (1987) Thromb Res 45: 517
8. Ruggeri ZM (1997) J Clin Invest 99:559
9. Reininger AJ, Heijnen HFG, Schumann H, Specht HM, Schramm W, Ruggeri ZM (2006) Blood 107:3537
10. Dopheide SM, Maxwell MJ, Jackson SP (2002) Blood 99:159
11. Guyton AC, Hall JE (2000) Textbook of medical physiology, 10th edn. W.B. Saunders, Philadelphia
12. Ramachandra GN, Karthan G (1955) Nature 176: 593
13. Kadler KE, Holmes DF, Trotter JA, Chapman JA (1996) Biochem J 316:1
14. Williams BR, Gelman RA, Poppke DC Piez KA (1978) J Biol Chem 253:6578
15. Gross J, Highberger JH, Schmitt FO (1951) J Appl Phys 22:112
16. Gross J, Highberger JH, Schmitt FO (1952) P Soc Exp Biol Med 80:462
17. Gross J, Highberger JH, Schmitt FO (1954) Proc Natl Acad Sci 40:679
18. Gross J (1956) J Cell Biol 2:261
19. Gross J, Kirk D (1958) J Biol Chem 233:355
20. Wood GC, Keech MK (1960) Biochem J 75:588
21. Wood GC (1960) Biochem J 75:598
22. Bard JBL, Chapman JA (1968) Nature 219:1279
23. Bard JBL, Chapman JA (1973) Nature-New Biol 246:83
24. Shattuck MB, Gustafsson MGL, Fisher KA, Yanagimoto KC, Veis A, Bhatnagar RS, Clarke J (1994) J Microsc 174:Rp1
25. Fujita Y, Kobayashi K, Hoshino T (1997) J Electron Microsc 46:321
26. Paige MF, Rainey JK, Goh MC (1998) Biophys J 74:3211
27. Morris VJ, Kirby AR, Gunning AP (1999) Atomic force microscopy for biologists. Imperial College Press, London
28. Rainey JK, Wen CK, Goh MC (2002) Matrix Biol 21:647
29. Gutsmann T, Fantner GE, Venturoni M, Ekani-Nkodo A, Thompson JB, Kindt JH, Morse DE, Fygenson DK, Hansma P.K. (2003) Biophys J 84:2593
30. Strasser S, Zink A, Heckl WM, Thalhammer S (2006) J Biomech Eng 128:792
31. Holmes DF, Gilpin CJ, Baldock C, Ziese U, Koster AJ, Kadler KE (2001) Proc Natl Acad Sci 98:7307
32. Thalhammer S, Stark RW, Muller S, Wienberg J, Heckl WM (1997)J Struct Biol 119:232
33. Strasser S, Zink A, Janko M, Heckl WM, Thalhammer S (2007) Biochem Biophys Res Commun 354:27
34. Jiang FZ, Khairy K, Poole K, Howard J, Muller DJ (2004) Microsc Res Tech 64:435
35. Wen CK, Goh MC (2004) Nano Lett 4:129

36. Petruska JA, Hodge AJ (1964) Proc Natl Acad Sci 51:871
37. Graham JS, Vomund AN, Phillips CL, Grandbois M (2004) Exp Cell Res 299:335
38. Gutsmann T, Fantner GE, Kindt JH, Venturoni M, Danielsen S, Hansma PK (2004) Biophys J 86:3186
39. Bozec L, Horton M (2005) Biophys J 88:4223
40. Fratzl P, Misof K, Zizak., Rapp G, Amenitsch H, Bernstorff S (1998) J Struct Biol 122:119
41. Puxkandl R, Zizak I, Paris O, Keckes J, Tesch W, Bernstorff S, Purslow P, Fratzl P (2002) Philos T Roy Soc B 357:191
42. Diamant J, Arridge RGC, Baer E, Litt M, Keller A (1972) P Roy Soc Lond B Bio 180:293
43. Fratzl P, Fratzl-Zelman N, Klaushofer K (1993) Biophys J 64:260
44. Sarkar SK, Hiyama Y, Niu CH, Young PE, Gerig JT, Torchia DA (1987) Biochem 26: 6793
45. Mosler E, Folkhard W, Knorzer E, Nemetschekgansler H, Nemetschek T, Koch MHJ (1985) J Mol Biol 182:589
46. Marchant RE, Lea AS, Andrade JD, Bockenstedt P (1991) J Colloid Interface Sci 148:261
47. Raghavachari M, Tsai HM, Kottke-Marchant K, Marchant RE (2000) Colloids Surf, B 19:315
48. Siediecki CA, Lestini BJ, Kottke-Marchant KK, Eppell SJ, Wilson DL, Marchant RE (1996) Blood 88:2939
49. Shao JY, Ting-Beall HP, Hochmuth RM (1998) Proc Natl Acad Sci 95: 6797
50. Fritz M, Radmacher, Gaub HE (1994) Biophys J 66:1328
51. Hoerber JK, Miles MJ (2003) Science 302:1002
52. Butt HJ, Wolff EK, Gould SAC, Dixon Northern B, Peterson CM, Hansma PK (1990) J. Struct Biol 105:54
53. Gould SAC, Drake B, Prater CB, Weisenhorn AL, Manne S, Hansma HG, Hansma PK, Missie J, Longmire M, Elings V, Dixon Northern B, Mukergee B, Peterson CM, Stoeckenius W, Albrecht TR, Quate CF (1990) J. Vac. Sci. Technol. A8:369
54. Häberle W, Hoerber JK, Binnig G (1991) J Vac Sci Technol B9:1210
55. Kamruzzahan ASM, Kienberger F, Stroh CM, Berg J, Huss R, Ebner A, Zhu R, Rankl C, Gruber HJ, Hinterdorfer P (2004) Biol. Chem. 385:955
56. Ohta Y, Otsuka C, Okamoto H (2003) J Artif Organs 6:101
57. Cox M (1991) J Forensic Sci 36:1503
58. Quickenden TI, Cooper PD (2001) Luminescence 16:251
59. Gill P (2002) Biotechniques 32:366
60. Schwarzacher W (1930) Zeitschrift der Gesellschaft für Gerichtliche Medizin 15:119
61. Lins G, Blazek V (1982) Zeitschrift Fur Rechtsmedizin-J Legal Med 88:13
62. Miki T, Kai A, Ikeya M (1987) Forensic Sci Int 35:149
63. Inoue H, Takabe F, Iwasa M, Maeno Y and Seko Y (1992) Forensic Sci Int 57:17
64. Strasser S, Zink A, Kada G, Hinterdorfer P, Peschel O, Heckl WM, Nerlich AG, Thalhammer S (2007) Forensic Sci Int 170:8
65. Burnham NA, Colton RJ, Pollock HM (1991) J Vac Sci Technol A 9:2548
66. Tomasetti E, Legras R, Nysten B (1998) Nanotechnology 9:305
67. Vanlandingham MR, McKnight SH, Palmese GR, Elings JR, Huang X, Bogetti TA, Eduljee RF, Gillespie JW (1997) J Adhes 64:31
68. Thompson JB, Kindt JH, Drake B, Hansma HG, Morse DE, Hansma PK (2001) Nature 414:773
69. Bozec L, Horton M (2005) Biophys J 88:4223
70. Strasser S, Zink A, Janko M, Heckl WM, Thalhammer S (2007) Biochem Biophys Res Commun 354:27
71. A-Hassan E, Heinz WF, Antonik MD, D'Costa NP, Nageswaran S, Schoenenberger CA, Hoh JH (1998) Biophys J 74:1564
72. Goldmann WH (2000) Biotechnol Lett 2:431

73. Touhami A, Nysten B, Dufrene YF (2003) Langmuir 19:4539
74. Rayleigh LJWS (1885) Proc London Math Soc 17:4
75. White RM, Voltmer FW (1965) Appl Phys Lett 7:12
76. Wixforth A, Strobl C, Gauer C, Toegl A, Scriba J, von Guttenberg Z (2004) Anal Bioanal Chem 379:982
77. Sritharan K, Strobl CJ, Schneider MF, Wixforth A, Guttenberg Z (2006) Appl Phys Lett 88: 054102
78. Frisch T, Verga (2002) *Phys Rev E Stat Nonlin Soft Matter Phys* 66:041807
79. Steppich DM, Angerer JI, Sritharan K, Schneider SW, Thalhammer S, Wixforth A, Alexander-Katz A, Schneider MF (2008) Biochem Biophys Res Commun 369:507
80. Alexander-Katz A, Schneider MF, Schneider SW, Wixforth A, Netz RR (2006) PRL 2006.
81. Phillips JC (1996) Rep Prog Phys 59:1133
82. Degiorgio V, Bellini T, Piazza R, Mantegazza F, Goldstein RE (1990) Phys Rev Lett 64:1043
83. Hong MK, Narayan O, Goldstein RE, Shyamsunder E, Austin RH, Fisher DS, Hogan M (1992) Phys Rev Lett 68:1430.
84. Hagen SJ, Eaton WA (1996) J Chem Phys 104: 3395
85. Metzler R, Klafter J (2000) Phys Rep 339:1
86. Cherayil BJ (1992) J Chem Phys 97:2090
87. Frisch T, Verga (2002) PhylRev E 65:041801
88. Wang MD, Yin H, Landick R, Gelles J, Block SM (1997) *Biophys J* 72:1335
89. Arya M, Anvari B, Romo GM, Cruz MA, Dong JF,.McIntire LV, Moake JL, López JA (2002) Blood 99:3971
90. Huston JS, Levinson D, Mudgett-Hunter M, Tai MS, Novotný J, Margolies MN, Ridge RJ, Bruccoleri RE, Haber E, Crea R (1988) Proc Natl Acad Sci 85:5879
91. Itzhaki LS, Otzen DE, Fersht AR (1995) J Mol Biol 254:260
92. Matouschek A, Kellis JT, Serrano L, Fersht AR (1989) Nature 340:122

7 Quantitative Nanomechanical Measurements in Biology

Małgorzata Lekka · Andrzej J. Kulik

From volume II

Abbreviations

PBS	phosphate buffered saline
AFM	atomic force microscopy
FFM	friction force microscopy
TMAFM	tapping mode atomic force microscopy
FS	force spectroscopy
MT	microtubules
μF	microfilaments or actin filaments
IF	intermediate filaments
ECM	extracellular matrix
YM	Young's modulus
PSD	position sensitive detector

7.1 Stiffness of Biological Samples

7.1.1 Cell Structure

7.1.1.1 Cell

Cells are the basic units of all living organisms. Although we know the structure of cells, our knowledge of how cells work, their properties, and how these properties change due to the actions of various factors is still incomplete. Several structural elements are similar for all animal cells (see Fig. 7.1). All cells contain a plasma membrane separating the cell interior from the environment. The membrane regulates the movement of water, nutrients and wastes to and from the cell, and is composed of a lipid bilayer in which numerous proteins that participate in cell functions are embedded. At the center of the cell there is a cell nucleus that contains the cell's DNA, the genetic code that governs protein synthesis. In addition to the nucleus, inside the cell there are many other organelles – small structures that help carry out routine cellular operations. These include, for example, ribosomes (which participate in protein synthesis), lysosomes (which contain enzymes that help to digest nutrient molecules and other materials), and mitochondria (which are often referred to as the power-plants of the cell, as many of the reactions that produce energy take place in mitochondria) [1].

In: H. Fuchs, B. Bhushan (eds), *Biosystems - Investigated by Scanning Probe Microscopy*,
DOI 10.1007/3-540-27453-7_7,

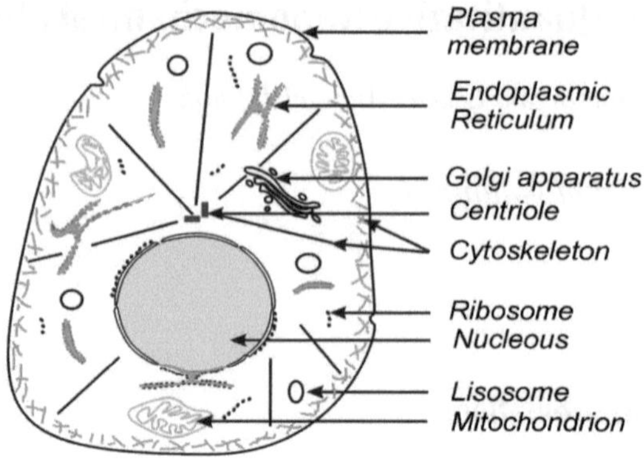

Fig. 7.1. Structure of a cell

7.1.1.2 Cytoskeleton

The shape of a cell is maintained by a three-dimensional protein structure called a cytoskeleton. The cytoskeleton is composed of three types of proteins forming different filaments (microfilaments, intermediate filaments and microtubules; see Fig. 7.2), organized into a network and bundled within the cytoplasm of eukaryotic cells. Although the term "cytoskeleton" gives an impression of a rather static structure, all of these three constituents are dynamic, constantly changing their shape through cycles of polymerization and depolymerization and through interactions with other proteins. As well as serving as the structural cell element, the cytoskeleton can participate in cell motility and it can regulate the activity of protein transport. In reaction to an extra- or intracellular stimulus, the cytoskeleton can self-assemble and adapt to a new situation within just a few seconds.

Microfilaments (μF, [2]) are linear filaments constructed from the globular protein actin (G-actin, molecular weight 43 kDa). These actin molecules polymerize

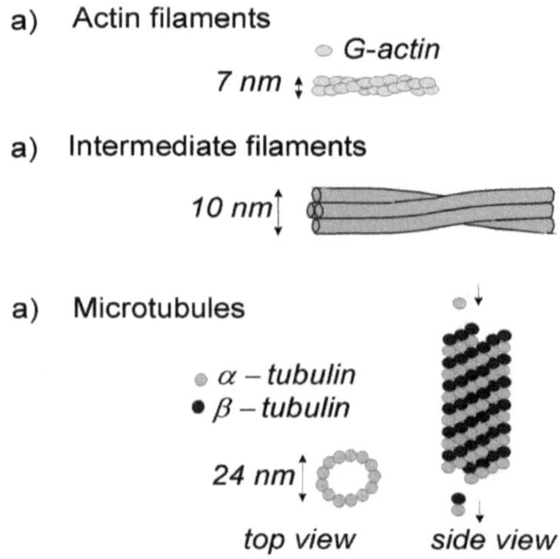

Fig. 7.2. The structural elements of the cell cytoskeleton: (**a**) actin filaments, (**b**) intermediate filaments, (**c**) microtubules

to form a long fiber. Two such fibers are twisted around each other in a helix to form a single actin filament with a diameter of about 7 nm, the thinnest cytoskeletal filament. Microfilaments, together with other types of filaments, are responsible for the mechanical strength of the cell and play a role in cell movement. They participate in cell division into two daughter cells, where their role is to anchor centrosomes at opposite poles of the cell during mitosis. The actin filaments also participate in the linkage of transmembrane (such as cell surface receptors) to cytoplasmic proteins.

Microtubules (MT, [2]) are small tubes constructed from tubulin dimer, composed of two subunits: α- and β-tubulin, both with a molecular weight of 55 kDa. They form so-called protofilaments. Each microtubule is typically composed of 13 protofilaments arranged around the circumference with diameters of about 24 nm and lengths in the range of 200 nm to 25 μm. A microtubule extends by the adding tubulin proteins to one of its ends. Microtubules are found in the cytoplasm of all eukaryotic cells, where they are often observed to spread out radially from a centrosome located near the nucleus. From this center, they provide a strong frame that supports the cell and determines its shape. They also serve as tracks along which cellular organelles can migrate. Microtubules can also form specialized structures such as centrioles, cilia, and flagella. Both cilia and flagella are cellular appendages consisting of a core of microtubules enclosed in an extension of the plasma membrane, and they play an important role in cellular locomotion.

Intermediate filaments (IFs, [2]) are the third type of fibrous cytoskeletal components. They are about 10 nm in diameter, so they are typically intermediate in size between microfilaments and microtubules. Unlike the microfilaments and microtubules, intermediate filaments are made of several different proteins. Therefore, intermediate filaments can be divided into five major types. Type I and II are composed of acidic and basic keratin, respectively. They are produced by different types of epithelial cells (bladder, skin). Type III encompasses intermediate filaments distributed in a number of cell types, including: vimentin in fibroblasts, endothelial cells and leukocytes; desmin in muscle; glial fibrillary acidic factor in astrocytes and other types of glia, and peripherin in peripheral nerve fibers. Type IV are neurofilaments and type V are made of laminin.

All three cytoskeletal components have distinct subcellular localizations. Actin filaments are usually present in a layer known as the "cell cortex", immediately beneath the plasma membrane, and in cell projections such as microvilli. In certain situations they are associated with a number of proteins and other structures, such as the contractile ring in dividing cells, or focal contacts linking the cytoskeleton with surface receptors. Microtubules extend from the nucleus region towards the cell periphery. Intermediate filaments are distributed in a similar pattern to microtubules. IFs and MTs are excluded from the actively expanding leading edge of the moving or "ruffling" cell [3].

7.1.1.3
Mechanical Properties of Cells

Over the past few decades, an increasing amount of evidence has suggested that many cellular processes depend on (or are regulated by) mechanical forces arising from either the cell alone or from its environment – an extracellular matrix (ECM) and/or neighboring cells [4]. It has been reported that these processes influence the ability of cells to either maintain or change their shapes under different types of external stresses. One property that decides whether the cell maintains or changes its shape is the cell deformability. Cell deformation depends on the overall viscoelastic mechanical properties of the cell, which are not static, as they change depending on the cell's state. The mechanical stability of the cell is maintained by its internal structure – the cytoskeleton [5–7]. Therefore, in order to resolve the cell mechanics at a molecular level, studies of the mechanical properties of the cell and its structural components are needed.

Alterations in cytoskeleton organization, which manifest themselves by changes in mechanical properties, can be induced by different factors. Depending upon the source of cytoskeleton rearrangement, the resulting changes can be beneficial or detrimental for the cell. They can reveal the adaptation of the cell to new circumstances but also they can indicate internal changes in cytoskeleton structure [1, 2]. Therefore, measurements of cell mechanical properties may help us to understand the causes of cytoskeleton reorganization.

There are different reasons for cytoskeleton reorganization. There may be a lack of a particular protein that is connected to the cytoskeleton and crucial to cell functioning (for example dystrophin, the absence of which is responsible for Duchenne muscular dystrophy, a serious muscle degenerative disease, [8]). As the cytoskeleton structure is not stable but changes according to cellular processes like cell division, cell locomotion, or transport of intracellular particles; many of these processes manifest themselves in altered cell deformability.

It is known that the cancerous transformation introduces significant changes in cell structure and behavior. Differences in cell growth, morphology, cell-to-cell interaction and also in the cell membrane and organization of the cytoskeleton have been reported [9, 10]. These differences can also cause changes in the mechanical properties of the cell, usually leading to higher cell deformability (lower Young's modulus). The changes are generally attributed to the altered cytoskeletal organization [5, 11]. The low stiffness of cancer cells may be caused by a partial loss of actin filaments and/or microtubules, and therefore by a lower density of the cellular scaffold. Poor differentiation of the cell and the reduced adhesive interactions characterize the vast majority of cancer cells [12]. A measurement of the Young's modulus of cells can probably help to determine the range of cytoskeleton changes and allow us to quantify them.

7.1.2
Determination of Young's Modulus

Many techniques have been applied to study the mechanical properties of biological specimens, at both the macro- and the microscale. However, considering the increased role of mechanical properties at the subcellular level, developing techniques that can deliver very local information from the cell surface is still a challenge. The most popular techniques used for the determining local mechanical properties are magnetic twisting cytometry [13], micropipette aspiration [14], microplate manipulation [5], and, recently, optical tweezers [15] and atomic force microscopy [16].

All of the above methods share the same idea of measurement: when a certain force is applied, the cell is deformed in some way and the deformation produced is recorded using various detection methods. *Magnetic twisting cytometry* applies a controlled mechanical stress to specific cell surface receptors using ligand-coated ferromagnetic beads, *micropipette aspiration* aspirates a single cell into a glass micropipette with a given suction pressure, *microplate manipulation* deforms a single cell situated between two microplates, *optical tweezers* use a strongly focused laser beam to manipulate cells, and *atomic force microscopy* produces deformation by sample indentation. The latter method, AFM, seems to be convenient method for studying the local mechanical properties of living cells.

AFM can measure the mechanical properties of many types of biological samples, ranging from biomolecules through to small cellular structures (such as microtubules) to very large components, like living cells [16]. The great advantage of AFM is the possibility of performing measurements in a nondestructive manner, with very good force and lateral resolution, and in a liquid environment, which is the natural milieu for most biological samples. Apart from surface topography, AFM can be used to measure the binding force between a molecule and its ligand [17], to determine the elastic properties of cells [18], to evaluate the friction coefficients of materials [19] and to perform high-resolution imaging of isolated proteins and structures [20]. Another advantage comes from its real-time or quasi-real-time measurements, which provide not only static but also dynamic information about the investigated specimens and allows the investigation of chemical modifications to surfaces and their influence on biological systems. In particular, control over biomaterial surface chemistry can result in a regulated cell response [21, 22].

Atomic force microscopy provides three-dimensional images of the topography of the investigated sample and information about its physical and mechanical properties. In AFM, a sharp small tip, which is attached to a compliant cantilever, is placed almost parallel to the surface and then moved over it, performing a raster scan. The force acting between the tip and the sample causes the cantilever to be deflected, and this deflection may be detected using, for example, the optical system shown in Fig. 7.3. The laser beam is focused at the end of cantilever and then, after reflection from the cantilever, it is detected by a position-sensitive detector (PSD). The position of the beam spot on a PSD surface provides information on cantilever displacement. Usually, commercially available instruments use a PSD photodiode that has an active area that is divided into four quadrants, which enables the recording of two orthogonal cantilever deflections: a normal one, (perpendicular to the investigated surface, corresponding to surface topography), and a lateral one (torsion, related to frictional forces). The great advantage of the method is that immersion of the sample in (transparent) liquid does not change this principle of operation [23].

During measurements of the sample stiffness, the AFM tip mounted at the end of the cantilever approaches the surface and indents it. The loading force is controlled by the amount of cantilever deflection and the spring constant. The measurement may be carried out in several different ways. The most popular method involves force spectroscopy, where so-called force curves are recorded. A force curve gives the relationship between the recorded cantilever deflection and the relative sample position. This dependence can be then converted into a force (load)–indentation curve, which is the basis for the Young's modulus determination [24].

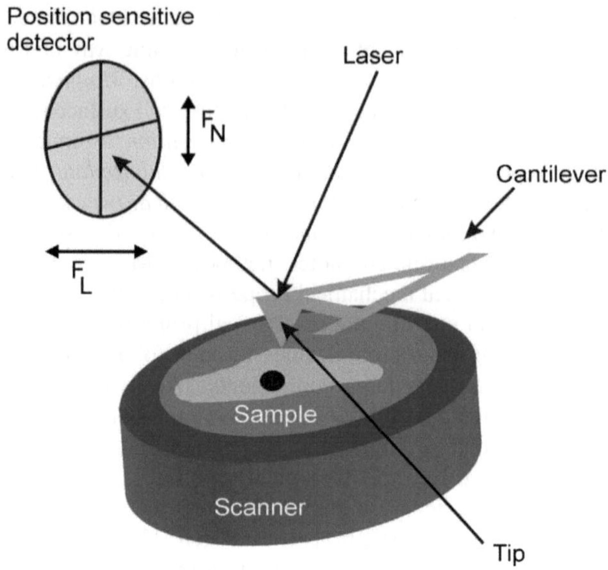

Fig. 7.3. Schematic view of an atomic force microscope

The other measuring technique uses AFM in a so-called *nano Swiss cheese* technique [25] that permits the axial and shear moduli to be determined. Although the latter technique was developed primarily for investigations of carbon nanotubes, it can be also be successfully applied to biological samples with similar fibrous structures (of course, providing that the experiment is performed in a liquid environment). The force modulation mode can also be applied to determine the mechanical properties of living cells, although this method gives only qualitative results [19, 26].

7.1.2.1 Force Spectroscopy

7.1.2.1.1 Force Curves

The most popular way of determining the stiffness of living cells uses so-called *force curves* to derive the Young's modulus of the sample [23]. In this approach, the AFM probe, mounted on a cantilever, is placed over the sample surface and then pushed against the sample; data acquisition is usually performed sequentially. The force–distance curve shows the relationship between the relative sample position (realized by the movement of the piezoelectric scanner) and the cantilever deflection (measured using a position-sensitive detector). Typical force curves taken for a compliant sample are presented in Fig. 7.4. These curves are then used to determine the local Young's modulus values or interaction forces. However, real experiments usually consist of measuring the voltage applied to the piezoelectric scanner (which is nonlinear and hysteretic) versus the relative cantilever deflection (which is uncalibrated). Therefore, a careful calibra-

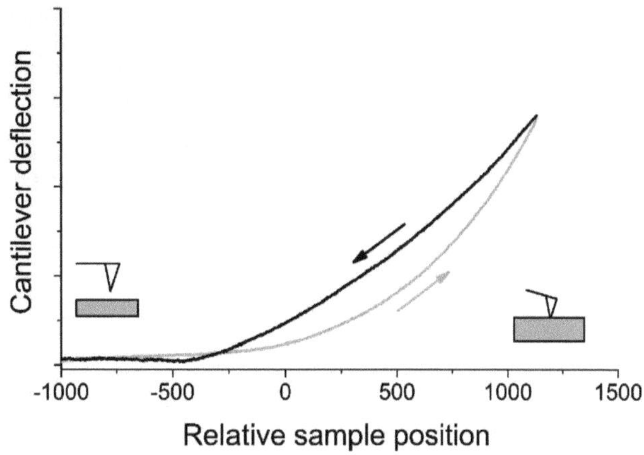

Fig. 7.4. Typical dependence between cantilever deflection and relative sample position (so-called "raw data") recorded for a bladder cell in culture conditions. The approach (*gray line*) and retraction (*black line*) traces do not overlap, indicating the hysteresis and the nonlinearity of the piezoelectric scanner

tion and linearization procedure should be applied in order to obtain quantitative results [27].

7.1.2.1.2
Scanner Linearization

Correcting for the nonlinearity and hysteresis of the piezoelectric scanner is crucial to gaining an accurate reading of a displacement. In commercially available AFMs, scanner linearization is frequently an integral hardware component, and thus scanner calibration is not needed. Real displacements of the sample in all three axes are measured using optical, capacitive or other methods. One simple way to verify the precision of the scanner is to measure a force–relative sample position curve on a stiff, nondeformable surface, and then to check whether the approach curve is identical (overlaps with) to the retraction curve and whether it is a straight line. If this is the case, the piezoelectric scanner is hardware-linearized. If not, the piezoelectric scanner should be linearized off-line. There are several methods of scanner linearization; out of these, the method of z-calibration proposed by Jaschke and Butt [28] is very straightforward and does not require any additional equipment. However, it can only be applied to AFMs with an optical detection system. In this method, a mirror is placed at the sample position in such a way that the inclination of the mirror is the same as the inclination of the cantilever. The slightly unfocused laser beam produces two reflected beams: one from the back of the cantilever and one from the surface of the mirror (Fig. 7.5a). When the mirror moves up and down, the optical path length between these beams changes and the resulting interference signal varies (Fig. 7.5b). The distance between subsequent maxima reflects the nonlinearity of the scanner, and the Bragg law can be used to calculate the true scanner extension. One can obtain parameters that describe the scanner nonlinearity from the above data by fitting a polynomial function (usually a parabolic function is sufficient, see Fig. 7.5c). The method is exact and does not require additional equipment, but it does require knowledge of the geometry of the system.

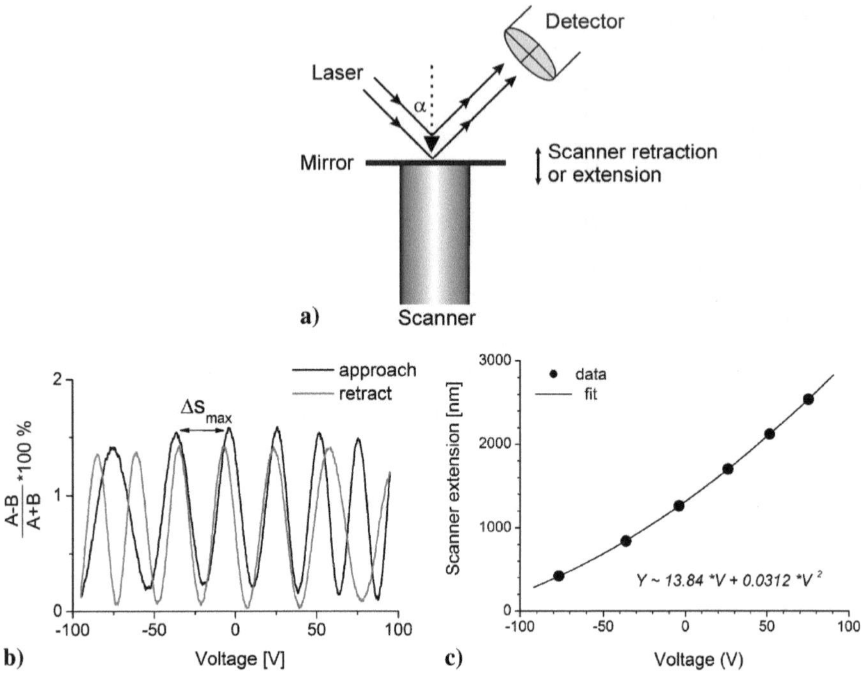

Fig. 7.5. (**a**) Schematic of the set-up used to linearize the scanner off-line. (**b**) Normalized photodiode signal of the extension and retraction of the scanner as a function of the applied voltage. (**c**) Scanner nonlinearity, fitted with a parabolic function (Reprinted with permission from [27], Copyright (1999) from Elsevier)

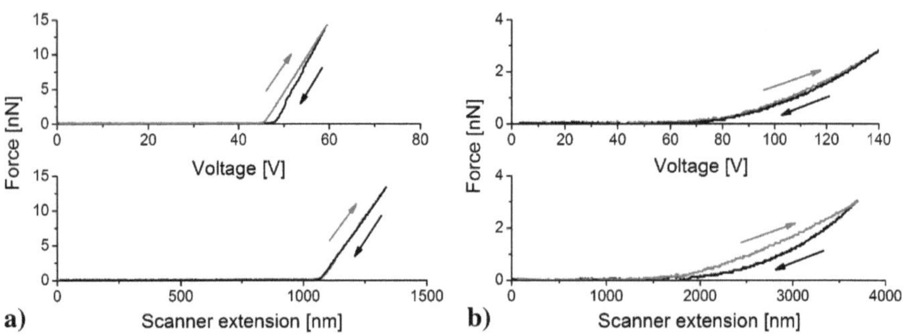

Fig. 7.6. Effect of scanner nonlinearity on force curves recorded on (**a**) a stiff (glass coverslip) and (**b**) compliant (cell) surfaces

The effect of scanner hysteresis and nonlinearity on force curves for stiff and compliant samples is shown in Fig. 7.6a,b. When a nonlinearized scanner is used on a stiff surface, the approaching trace does not overlap the retracting one. Both curves should overlap after being corrected for nonlinearity (Fig. 7.6a). Another effect is illustrated in Fig. 7.6b, where the scanner nonlinearity leads to an interpretation that is completely false for compliant samples. Both curves overlap in the upper plot of Fig. 7.6b, suggesting a purely elastic interaction. However, after correction, the true data exhibit an approach–retract hysteresis due to the contact having viscoelastic character.

7.1.2.1.3
PSD Signal Calibration

AFM measures the cantilever deflection by monitoring the position of the laser beam reflected from the cantilever. The active area of the measuring PSD photodiode is usually divided into four quadrants. The cantilever deflection that is perpendicular to the investigated surface is found by measuring the difference between the signals from the two upper and the two lower quadrants. In order to calibrate this cantilever deflection, the force curve should be recorded on a stiff, nondeformable surface (such as glass or mica), where the deflection directly reflects the position of the sample. This dependence is linear and is usually also employed as a reference. The slope determines the calibration coefficient between the PSD signal (in *volts)* and the real displacement (in *nanometers*).

7.1.2.1.4
Spring Constant Determination

Force calibration requires knowledge of the cantilever spring constant. The resonant frequency of a thermally excited cantilever is measured in the most widely applied method for spring constant calibration. The detailed procedure for a spring constant determination carried out in this way has been described in a publication from Sader [29]. Figure 7.7 presents noise spectra measured for two types of cantilevers ($k = 0.03$ N/m and 0.1 N/m). The spectra were obtained by applying a fast Fourier transform to an average of six noise spectra, each consisting of 4096 points [27]. The measured resonant frequencies are lower than the nominal values given by manufacturer: 11.8 kHz instead of 15 kHz and 27.7 kHz instead of 35 kHz. These

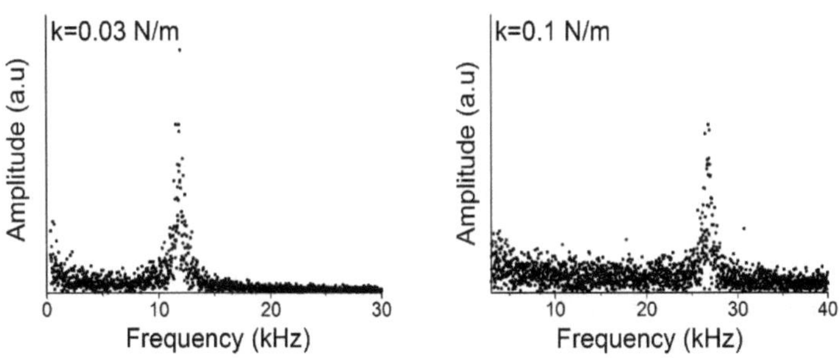

Fig. 7.7. Noise spectra for two types of cantilevers ($k = 0.03$ N/m and $k = 0.1$ N/m). The calculated resonant frequencies are lower than the nominal values: 11.8 kHz instead of 15 kHz and 27.7 kHz instead of 35 kHz (Reprinted with permission from [27], Copyright (1999) from Elsevier)

changes in resonant frequency are used to calculate the normal spring constant. It is worth noting that these properties differ even between tips carefully chosen from the same series of production.

7.1.2.1.5
Force–Indentation Curves

Force–indentation curves are obtained by measuring the deflections on stiff and compliant surfaces at a given relative sample position, and then noting the difference between these deflections (Fig 7.8). When a stiff material (not easily deformable like silicon or glass) is investigated, the deflection reflects the position of the sample. This is represented by a straight line and is usually employed as a reference line for the force calibration. For compliant samples, like cells, cantilever deflections are much smaller and the resulting force curve has nonlinear character. The difference between these curves determines the deformation of the sample surface [24].

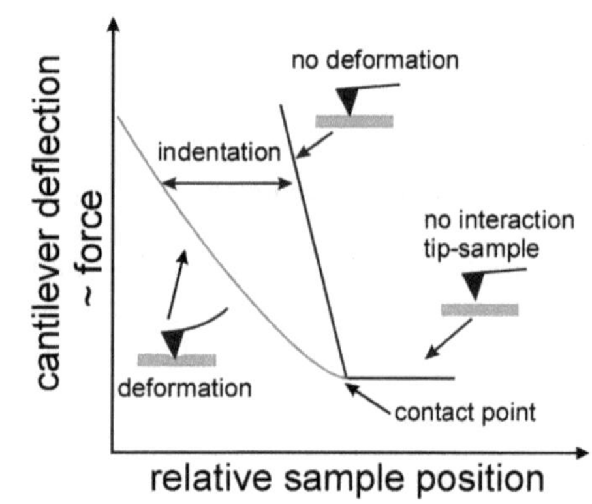

Fig. 7.8. Indentation determination from AFM force curve data

7.1.2.1.6
Young's Modulus Determination

The force versus indentation curve describes the mechanical response to an applied load, which is characteristic of a particular material. The Young's modulus, which characterizes the cell stiffness, can be evaluated using Hertz contact mechanics, which describes the deformation of two elastic spheres under an external load [30]. The Hertz model can also be extended [31] to account for an infinitely stiff indenter with a variety of AFM tip geometries (spherical, parabolic, conical or flat-ended) and a flat, deformable substrate. Usually, the AFM probe tip is a four-sided pyramid that can be modeled using either a cone or a paraboloid.

For a conical tip, the force–indentation curve is given by:

$$F(\delta) = \frac{2}{\pi \cdot \tan(\alpha)} \cdot E' \cdot (\delta)^2 \ , \tag{7.1}$$

and for a tip that is a paraboloid:

$$F(\delta) = \frac{4}{3} \cdot \sqrt{R} \cdot E' \cdot (\delta)^{\frac{3}{2}}, \tag{7.2}$$

where F is the loading force, α is the open angle, R is the radius of curvature of the AFM tip, E' is the reduced Young's modulus, while δ is the indentation depth.

The reduced Young's modulus is given by:

$$\frac{1}{E'} = \left(\frac{1 - \mu_{\text{tip}}^2}{E_{\text{tip}}} + \frac{1 - \mu_{\text{sample}}^2}{E_{\text{sample}}} \right) \tag{7.3}$$

When $E_{\text{sample}} \ll E_{\text{tip}}$ (which is the case for living cells), then

$$\frac{1}{E'} = \frac{1 - \mu_{\text{sample}}^2}{E_{\text{sample}}} \tag{7.4}$$

where μ is the Poisson ratio, which is related to the compressibility of the sample material.

The Poisson ratio for cells is difficult to determine. Therefore this value is usually set to be 0.5, since cells can be assumed to be incompressible material.

The final Young's modulus distribution is formed from values from single cell force curves (see Fig. 7.11).

7.1.2.2
The Nano Swiss-Cheese Method

Atomic force microscopy can be used to study the mechanical properties of not only cells but also their structural elements, like cytoskeletal components such as microtubules. The first application of AFM to investigating the mechanical properties of cell structural elements, by determining the Young's modulus, was performed by Vinckier et al. [32]. They studied glutaraldehyde-stabilized microtubules. The force curves were recorded for a mid section of a single microtubule lying on the stiff substrate. However, the main problem with this method was that it produced only a very small deformation of the microtubule, resulting in very small vertical displacements of the tip, which was difficult to measure reliably. Furthermore, an analytical description of the contact between a spherical tip and an anisotropic tube is difficult to establish. A more suitable method of studying the mechanical properties of fibrous structures was proposed by Salvetat et al. [25]. The method was initially developed for determining the Young's modulus of carbon nanotubes, but it was later shown to be useful for investigating the mechanical properties of microtubules. In this measurement method, single fibrils are deposited on the surface of a polished ultrafiltration membrane. The fibrils adhere to the surface, and some of them are deposited over holes in the surface spanning their edges (see Fig. 7.9, [33]). When a tube deposited over a hole in this way is found, a series of contact–mode AFM images are taken under increasing loads, and each image corresponds to the surface under a given load.

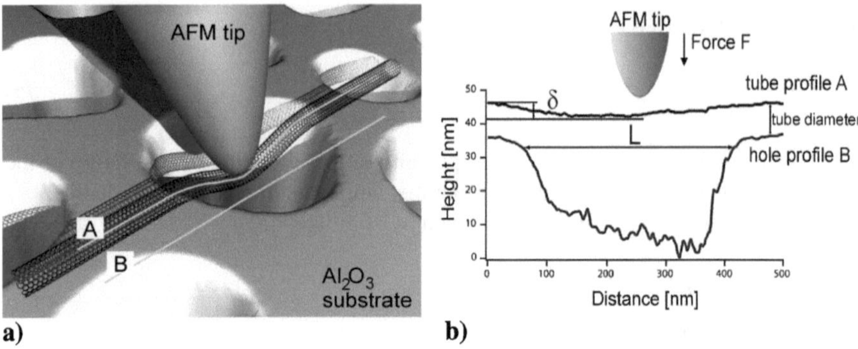

Fig. 7.9. Schematic drawing of a long tube with a relatively small diameter deposited on a porous substrate (**a**). During AFM imaging in contact mode, the tube deforms under the applied force (**b**), and this deformation is recorded as change in the local height in the corresponding AFM image (Courtesy of A.Kis, from [34])

For a range of applied normal loads, the midpoint deflection of a thin long tube under a nominal point load F can be found using the clamped beam formula:

$$\delta = \frac{FL^3}{192 \cdot E \cdot I} \tag{7.5}$$

where L is the suspended length, I is the second moment of the cross-section, which is $d^4/64$ for a solid cylinder of diameter D.

This variable load imaging technique is used to obtain quantitative information, as we can be sure that the AFM tip is in the desired location when the tube is deformed. The above equation is valid for small deformations if the tubes adhere well to the substrate, which is confirmed by images that reveal no displacement of the sections of the tube in contact with the membrane. Successive tube deflection measurements are usually performed for increasing and decreasing loads. Agreement between those measurements confirms the linearity and further establishes that the tube is firmly bound to the substrate.

7.1.2.3 Other AFM-based Methods

Biological samples are very delicate and are sometimes destroyed during scanning in contact mode. This lead to the introduction of dynamic force microscopy [35], in which the amount of force applied during measurements is minimized since it does not require continuous tip–sample contact. Among various types of dynamic force microscopy, tapping mode AFM (TMAFM) has become the most widely used [36]. In TMAFM, the cantilever is oscillated near its resonant frequency. During scanning, the oscillation amplitude is held constant using a feedback loop by adjusting the vertical position of the sample. Thus, the feedback signal does not reflect - as frequently stated - the surface topography [37], rather the resulting image shows the surface representing the constant amplitude of an oscillating cantilever.

Using this technique, the amplitude of oscillation and the phase can be recorded at the same time. The interaction between the tip and the surface depends not only on the topography of the sample but also on other material properties, such as stiffness, elasticity or adhesion. The phase shift recorded for a compliant sample can be used to detect, for example, differences in sample stiffness. However, in general it is difficult to measure the sample's physical properties using tapping mode images. Five physical properties of the sample are used in the model describing the vibrating cantilever tapping sample surface [38], and the same oscillating cantilever behavior versus average tip–sample distance can be obtained for many different combinations of all five physical properties. This makes it difficult to use dynamic force microscopy quantitatively, although the phase lag between cantilever detection and excitation can be used to measure energy losses during tapping [39]. The main advantage of dynamic force microscopy is that only small lateral forces are produced during scanning, which prevents sample displacement, modification or damage.

7.1.3 Brief Overview of the Application of AFM to Studies of Living Cells

Great efforts have been made to measure the stiffnesses of delicate biological samples, since this information can improve our understanding of cell mechanics. The examples described below represent only a small proportion of possible applications of the AFM technique to studying changes in the mechanical properties of living cells.

7.1.3.1 Living Cells

Cells are very compliant and complex objects, and so cell indentation generates responses from a variety of cell structural components, including glycocalix, the cell membrane and the interior of cell (the cytoskeleton and organelles immersed in cytosol). The responses from the glycocalix and the cell membrane are the most significant for small indentations (100 nm and less). However, these components are usually not visible in AFM measurements due to their very low compliance and the low accuracy of the contact point determination. The large deformation range is dominated by the responses from the internal cell structures such as cytoskeleton and cellular organelles. Comparing the dimensions of all interacting objects with the indentation depth suggests that the response of the cellular scaffold is the most important, and therefore the elastic properties measured reflect mainly the properties of the cytoskeleton [2,11]. In order to prove this hypothesis, a lot of effort has been made to investigate the role of cytoskeletal elements in the mechanical properties of the cell.

The contribution of the cell cytoskeleton to the mechanical properties determined by AFM has been discussed in several studies, where the importance of both the structural proteins of the cytoskeleton and other proteins associated with the cytoskeleton was demonstrated [40,41].

Since the cell cytoskeleton is composed of three main elements (actin filaments, intermediate filaments and microtubules), their individual roles have been investigated in order to find the main contributor to cell mechanics. The common method used for this is to apply so-called "cytoskeletal drugs" that promote or inhibit the formation of a given type of filament. Such measurements have been performed for various cell types [42–44], and complete and comparative studies on cytoskeleton integrity have been presented by Wu et al. [45] and Rotsch et al. [46], where the elastic properties of cells were measured after treatment with cytochalasin B, a protein that causes depolymerization of actin filaments (see Fig. 7.10), or nocodazole, which disassembles cell microtubules. The results have shown that the actin filaments are mostly responsible for the mechanical properties observed in AFM measurements, disrupting the microtubules produced no effect on the mechanical properties of cells as measured by AFM.

The role of cytoskeleton-associated proteins in measured cell elasticity has been demonstrated for cells possessing vinculin, a protein that is required for actin stress fiber formation. The loss of vinculin produces a noticeable reduction in cell adhesion and spreading and the presence of stress fibers. The comparison performed by Goldman et al. [39] in 1998 showed that vinculin-deficient F9 mouse embryonic carcinoma cells had a lower Young's modulus than wild-type cells. The authors attributed these changes to altered actin cytoskeletal organization [47], indicating that vinculin is an integral part of the cytoskeletal network.

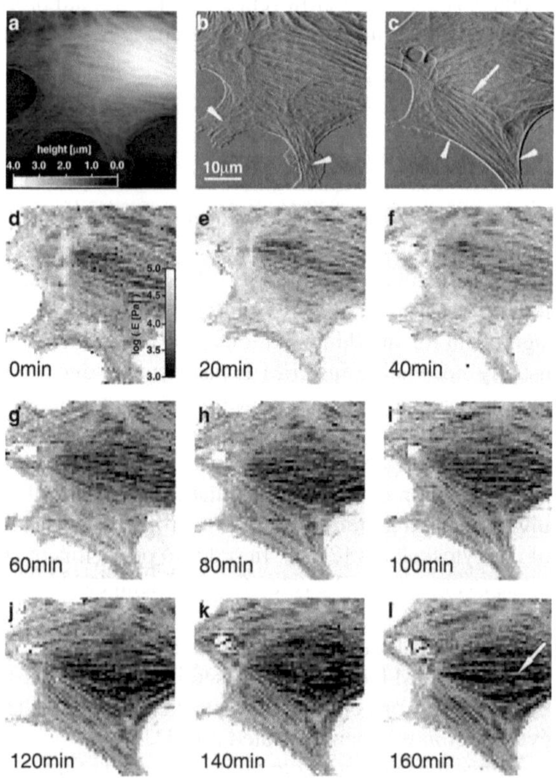

Fig. 7.10. Time series of elasticity maps of a 3T3 fibroblast showing disaggregation of the actin network by 10 μM cytochalasin B. (**a**) Reference map with marked stress fibers (*arrow*). (**b-i**) Successive degradation of the elasticity maps, indicating a decrease in the cell's Young's modulus. The stress fibers clearly visible in images (**a-f**) disappear completely after 120 min (**g-i**) (Reprinted with permission from [46], Biophysical Society 2000)

The cell elasticity determined via AFM is a very local feature; large discrepancies are seen between the Young's modulus measured for a population of cells and that for a single cell. It has been reported that cells in vitro have Young's modulus values in the range of 1–100 kPa [23, 48], which encompasses various types of investigated cells, including vascular smooth muscle cells, fibroblasts, bladder cells, red blood cells, platelets and other epithelial cells. Since different cell types are being measured, the large variation in Young's modulus is easy to understood. Despite large errors (defined as the half-width of the YM distribution peak at half maximum, the HWHM), a comparison of the cell elasticity clearly shows a much smaller Young's modulus (a larger deformability) for cancerous bladder cells than for reference cells [11]. The difference between the YM of these cells was almost one order of magnitude (Fig. 7.11).

The significant differences in the YM values found enabled alterations within cell populations arising from the actions of drugs to be traced. This was demonstrated in studies that showed a correlation between cell stiffness and glycolytic activity as a function of various degrees of chitosan deacetylation (defined as the ratio of the number of amino groups to the sum of amino and acetyl groups). An increase in Young's modulus was correlated with decreasing glycolytic activity (determined as the levels of ATP and lactate) in cancerous cells (Fig. 7.12). The YM values of the reference cells were unaffected, which highlights the cancer-specific action of chitosan [49].

It is known that the mechanical properties of cells depend strongly on the surface properties of the substrates used in cell cultures. There is also an increasing of evidence showing that substrates have a regulatory effect on cell growth, differentiation or proliferation. The surface properties of substrates are important, for example,

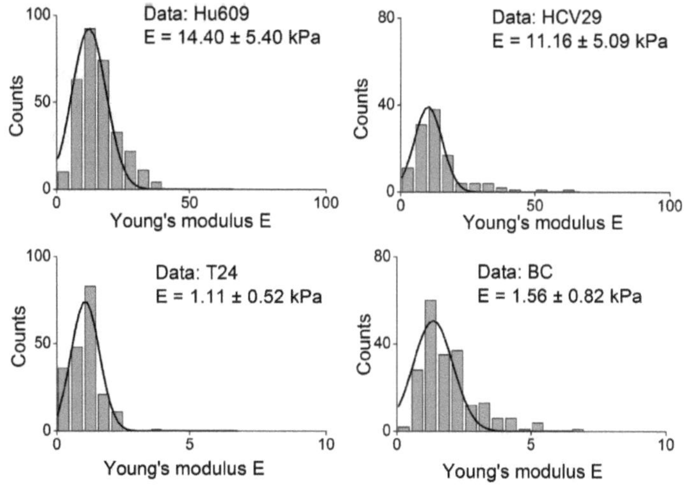

Fig. 7.11. Histograms showing the mean values and the distributions in Young's moduli measured for two reference cell lines: normal (Hu609 and HCV29) and cancerous (T24 and BC). In this evaluation it was assumed that the Poisson ratio is $\mu_{\text{cell}} = 0$ (Reprinted with permission from [11], Springer-Verlag, 1999)

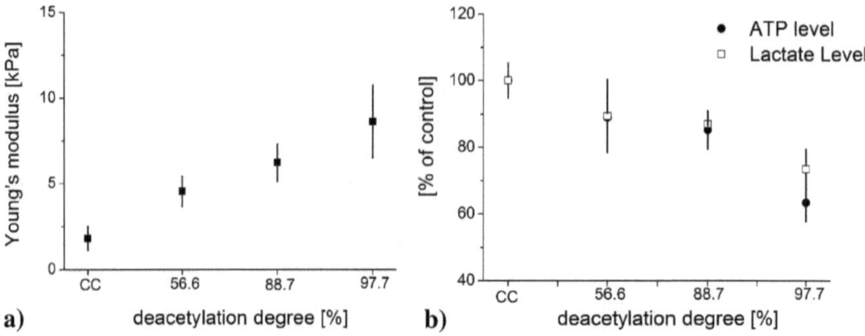

Fig. 7.12. Correlation between glycolytic activity and Young's modulus for bladder cells treated with various types of chitosan preparations (three degrees of deacetylation: 56.6%, 88.7%, 97.7%). *CC* indicates control cells. Chitosan with a degree of deacetylation of 97.7% showed both the largest increase in Young's modulus and the highest glycolytic inhibition

in implants, which need to be compatible with the living organism. The work performed by Domke et al. [22] investigated a whole range of parameters accessible by AFM, such as the cell morphology, the adhesion, and the cell elasticity of living osteoblasts. The properties of cells cultured on metallic surfaces were compared with those of cells growing on a standard glass surface and on a polystyrene surface in the tissue culture dish (Fig. 7.13). Figure 7.14 presents the effect of the substrate on cell elasticity. The Young's modulus changes significantly depending on substrate type.

The excellent lateral sensitivity and precision of AFM measurements allows the cell Young's modulus to be mapped, highlighting the heterogeneity of single cells. It has been reported that the Young's modulus of plasma membrane is 1–10 kPa, while the corresponding value for a membrane that is more strongly supported by the cytoskeleton is higher (15–50 kPa, [42]). This makes it possible to perform elasticity measurements within a single cell. The work described in [50] traces the alterations in elasticity during cell movement; the distribution of local fibroblast stiffness was found to be related to cell migration. When the cells were stationary, the

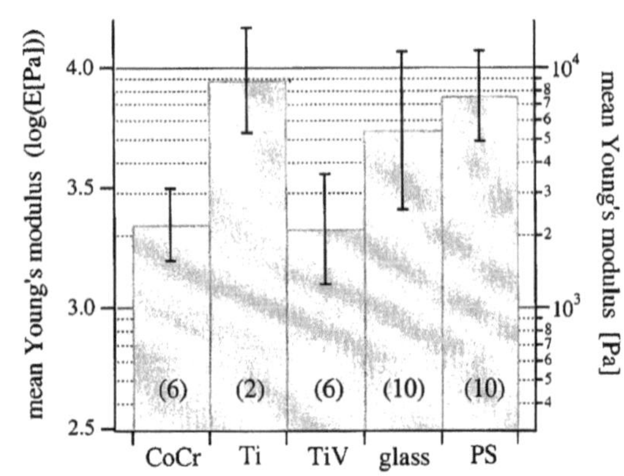

Fig. 7.13. The mean value of the Young's modulus for osteoblasts cultured on different types of substrates (CoCr, Ti, TiV, glass and PS tissue culture polystyrene surfaces) (Reprinted with permission from [22], Copyright (2000) from Elsevier)

distribution of stiffness on the cellular surface was more or less stable. Significant changes were observed during migration, where the stiffness in the fibroblasts' nuclear regions decreased. This observation was explained in terms of contractile force variations. The contractile force is generated by stress fibers made of actin filaments. When the cell starts to move, this force drags the cell body forward, causing nuclear displacement. Another example of the application of AFM to investigations of cellular processes is given by by Mantzke et al. [51], where cell division was studied via direct measurements of furrow stiffening.

7.1.3.2 Bacterial Cells

Another class of biological samples used in AFM studies of mechanical properties is bacterial cells [52]. However, while animal cells have been investigated extensively, bacterial cells have been investigated in only a limited number of AFM studies.

One of the first studies investigated magnetotactic bacteria by means of force modulation microscopy [53]. In this study, the strong contrast observed in the force modulation images was attributed to the presence of stiff magnetosomes inside the bacterial cells. Later on, the elastic properties of the proteinaceous sheath of the archeon *Methanospirillum hungatei* were probed via indentation experiments [54], giving Young's moduli of 20–40 GPa, indicating that the strength of the sheath could withstand an internal pressure of 400 atm. Another type of cell, gram-negative magnetic bacteria of the species *Magnetospirillum gryphiswaldense*, was investigated using AFM in buffer solution [55]. Multiple force–distance curves were obtained for the substratum and for the bacteria, which made it possible to determine the effective compressibility of the cell wall, which was about 42 mN/m.

AFM studies of bacteria have not only been used to determine the Young's modulus value of the bacterial wall. Changes during cellular processes, such as the germination of *Phanerochaete chrysosporium* spores [56], have also been studied. For dormant spores, no significant deviation from linearity was seen in the contact region, whereas a nonlinear curvature was observed for germinating spores, which was attributed primarily to sample compliance. This difference was consistent with the presence of a proteinaceous rodlet layer at the surface of the dormant spore and the absence of such layer on a germinating spore. Another example has highlighted the significant differences in mechanical properties between fibrillated *S. salivarius* HB and nonfibrillated *S. salivarius* HBC12 strains [57]. The cell surface appeared much more rigid due to the collapse of the fibrillar mass at various ionic strengths in 0.1 M KCl solution. The force–distance curves revealed that the nonfibrillated strain was much stiffer than the fibrillated one. Since elasticity maps can give information about the homogeneity of the investigated samples, a comparison of the maps for two types of bacteria (*Lactobacillus* family, [58]) has shown that bacteria *L. crispatus* expressing an S-layer have surfaces that are more homogenous than the surfaces of strains without such a layer. For another bacteria type, *L. johnsonii*, stiff and compliant regions have been found that have been explained by the composition of the surface.

7.1.3.3 Cellular Structures

Not only can the mechanical properties of whole cells be determined through AFM, but the high lateral resolution and precise control of the loading force afforded by AFM permit studies of the mechanics of cell organelles (like microtubules or stress fibers). Microtubules are the most commonly studied structural components. Their elastic properties were estimated by Vinckier et al. [32] in 1996 in buffer solution as a function of the concentration of glutaraldehyde (a cross-linking agent). The microtubules were more compliant when less cross-linking agent was present. Upon extrapolating to 0% glutaraldehyde concentration, the Young's modulus of MTs in buffer was found to be 3.1 ± 0.9 MPa.

Using the "nano Swiss-cheese" method, the bending modulus of the MTs was determined as a function of temperature (Fig. 7.14). The curve obtained showed a significant decrease in bending modulus upon a decrease in temperature from 37 °C to 5 °C. Results of bending experiments, using different suspended lengths of the microtubules, demonstrate that microtubules are highly anisotropic from a mechanical point of view. The tubulin subunits that form protofilaments are strongly bound along each single protofilament and weakly bound between neighboring protofilaments [33, 34].

7.1.4 Summary

AFM has established itself as a useful technique for determining the Young's moduli of a broad range of biological samples, in particular living cells. Determining the

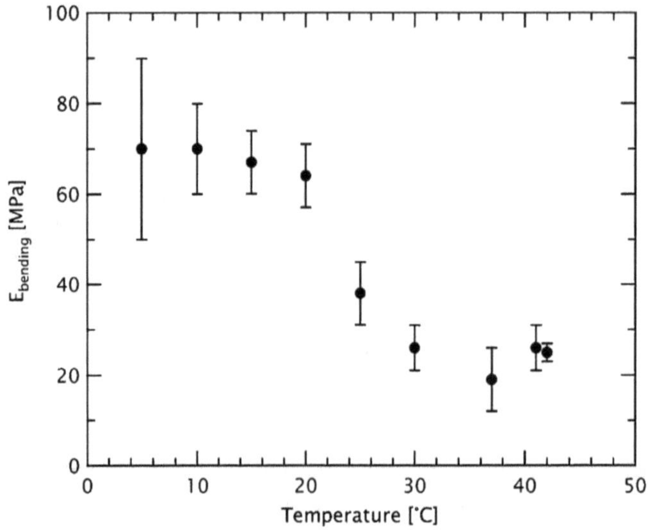

Fig. 7.14. Temperature dependence of the bending modulus for microtubules, showing the transition at 20–30 °C (Courtesy of A.Kis, from [34])

absolute value of Young's modulus value using AFM. Even if the calibration procedure is performed very carefully and with high accuracy, the calculated Young's modulus reflects only the elastic properties of the particular layer defined by the indentation depth (or layers that are deeper than the indentation depth). If we take into account the large heterogeneity of biological material too, the Young's modulus determined in this way cannot fully describe the overall mechanical properties of the cell(s) investigated.

Since one end of the cantilever is fixed, the tip does not move along a vertical line during the vertical movement of the sample but instead follows a circular trajectory. As a result, the tip does not remain at the same point on the sample surface, but rather slides laterally along the axis of the cantilever. This lateral movement can be estimated as being 10% of the vertical displacement. Some microscopes partially compensate for this tip displacement in the sample plane by coupling the movements along the z and y axes. Also noticing that that tip is usually not fully perpendicular to the sample plane, it is important to be very cautious during quantitative analyses of force-distance curves.

The most popular model, the Hertz model, is used to determine the mechanical properties of living cells under the assumption that the cells are purely elastic and isotropic. Actually, the measured materials are either purely elastic or they have a strong plastic component and so they usually behave like a combination of both. This is reflected in the shape of the force curve. The pure elastic character manifests itself in the exact overlapping of the approach and retract traces (assuming that the calibration procedure has removed any piezoelectric scanner nonlinearity). When the approach trace does not overlap the retract one, the plastic component starts to appear in the response of the investigated sample to the indenting force.

Another important source of errors originates from the physical models applied. Some approximation of the AFM tip shape must be assumed. Usually, the four-sided pyramid of the AFM tip is approximated by a cone or paraboloid. The cone fits the experimental data well for large deformations, but if the measured indentations are small, the paraboloidal approximation is used instead. For small indentations, the alternative "nano Swiss-cheese" method can be also applied. This method can be extended to measurements of other cytoskeletal filaments, like actin and intermediate filaments.

Together with the tip shape assumption, there are two other assumptions in the physical model. The first one states that the deformation of the AFM tip is negligible. The Hertz model delivers the reduced Young's modulus value, which is, in fact, a superposition of the Young's moduli of the AFM tip and that of the investigated sample. However, the living cells are much more compliant than the AFM tip, with stiffness values of 1–100 kPa. Thus the stiff AFM tip (modules: 150 GPa) deformation can be neglected and the Young's modulus determined reflects the Young's modulus of the investigated cell.

The second assumption is related to the Poisson ratio of the cell. The Poisson ratio, a characteristic material property that depends on the compressibility, ranges from 0 to 0.5, and is difficult to estimate for cells. Therefore, this factor is assumed to be constant for the same types of biological materials. It is usually set to 0.5, the value for materials that are incompressible.

AFM may also encounter problems from stiff substrate below the investigated cell. The influence of the stiff substrate on the indentation data leads to a dramatic increase in the estimated Young's modulus value. In the case of thin biological samples where the influence of underlying stiff substrate is not negligible, it is essential to take this effect into consideration. A strong correlation between the sample thickness and the Young's modulus value was demonstrated by Domke et al. [59] for gelatin films of different thickness.

The limitations of AFM when determining the Young's moduli of compliant, biological samples, especially at the microscopic scale, are being investigated and considered. These limitations should be addressed through both theoretical developments of the applied model and some practical solutions. For example, sharp cantilevers cause high local strains that are far from the linear material regime. This problem can be avoided by using microspheres as probes. Recently, corrections to the Hertz model that take into account the sample thickness were introduced by Dimitriadis et al. [60]. They present a simple theory that can be applied to determinations of the elastic moduli of compliant samples independent of the sample thickness.

These possible sources of error in quantitative estimations of Young's modulus using force curve spectroscopy may raise doubts about the usefulness of this method for measurements and analysis of the mechanical properties of living cells. However, we should note that exact knowledge of the absolute Young's modulus is not always needed. Very often relative changes or trends may provide valuable information about the state of a biological material. Also, living cells are highly heterogeneous and non-isotropic due to their internal structures, and this is reflected in different Young's moduli values when it is measured at different locations on a cell's surface. However, this diversity should not be considered to be a drawback; instead it should be seen as a spur to improve AFM in order to make it more suitable for measurements and analyses of the mechanical properties of living cells.

7.2 Friction Force Microscopy

Friction can be described as a resistance to motion that occurs when two solid objects that are in contact are moved with respect to each other. At the macroscopic scale, the main mechanisms of friction are adhesion and deformation. Adhesion is responsible for forming junctions between contacting bodies, the shearing of which requires work. Deformation dissipates energy through either plastic deformation or elastic hysteresis [61]. However, the above explanations do not describe friction on the microscopic scale, leaving the underlying physical mechanism poorly understood.

Contacts between sliding surfaces are of special interest to engineers designing mechanical devices. The lubrication of sliding contacts has always been a very important factor in machine efficiency. The studies of lubrication performed so far have targeted the minimization of friction forces, which has involved (among other things) investigating the origins of friction and searching for increasingly sophisticated lubricants based on organic molecules [62].

One way to study friction is through atomic force microscopy (AFM), which offers the unique ability to measure the normal (surface morphology) and lateral (friction) forces that act between the probe tip and the sample at the same time [63]. The development of the scanning force microscope has resulted in an increasing number of studies directly related to friction, due the ability of the technique to measure sliding contacts at the nanometer scale. This very high spatial resolution, when combined with quantitative measurements of interfacial properties, should lead to a better understanding of microscopic friction and other related phenomena at the molecular level. AFM-based techniques can provide an insight into chemical inhomogeneities at small length scales. Variations in surface frictional properties may correspond to variations in surface chemical composition and/or molecular organization. The simultaneous information about surface morphology and friction afforded by AFM may clarify effects that are difficult to investigate using other techniques. The well defined "single asperity" contact between an AFM tip and a sample surface provides the possibility of an analytical description of these forces [64].

7.2.1 Friction and Chemical Force Microscopy

7.2.1.1 Friction Force Microscopy

A friction force microscope (FFM) is an extension of a standard atomic force microscope working in the constant force mode (a constant force is maintained between the probe tip and the investigated surface) [62]. As in the standard AFM technique, the probe tip is mounted at the end of an elastic cantilever which is moved over a surface. Interaction forces cause a deflection of the cantilever which is recorded by an optical deflection system comprising a laser and a position-sensitive detector. The deflection and torsion of the cantilever are measured (the deflection in the normal direction corresponds to the surface morphology, while the torsion of the cantilever gives indirect information about the friction forces). The normal and lateral signals are separated as follows. The active area of the position-sensitive detector is divided into four quadrants. The difference in signal between the two upper quadrants and the two lower ones is a measure of the normal deflection, while the difference in signal between the two left quadrants and the two right ones gives the torsion of the cantilever, related to the friction forces.

However, it is important to note that the friction force is not measured directly. It can be only estimated or calculated from the cantilever torsion due to the friction force, and/or from the slope when the cantilever reaches the steep part of the slope (see Fig 7.15). To separate these two effects, the surface topography should be measured at the same time. Since the friction force is the lateral component of the total force acting on the moving tip, its value can be determined by subtracting the mean values (the average force values from a given region of the investigated surface) obtained from scans performed in two opposite directions (forward and backward scans) along the same path. The dependency observed upon performing such scans is called a friction loop, and the difference between two scan lines divided by two gives the value of the friction force (Fig 7.15, [65]).

7.2.1.2
Calibrating Friction Force

Quantitatively determining the friction force by AFM requires a lateral calibration that is similar to the calibration of the normal force. Two quantities are needed to calibrate the results: (1) the lateral photodiode signal (in order to convert the measured volts into the lateral displacement of the cantilever) and (2) the lateral (torsion) spring constant of the cantilever.

There are many ways to calibrate the photodiode's lateral sensitivity. A number of methods have been described for calibrating the photodiode's response to cantilever torsion. Perhaps the most direct one is the method of Meurk et al. [66], where the stepper motor is used to tilt a mirrored substrate, and the output voltage is measured as a function of angle. Alternatively, Liu et al. [67] proposed a calibration procedure based on the relationship between the torsional signal (arising due to friction forces) and the normal bending signal (originating from forces acting perpendicular to the investigated surface). The lateral sensitivity of the photodiode signal is proportional to the normal sensitivity (which is easy to measure from the slope of the force curve). This ratio is influenced by the shape and size of the laser spot and may be used for calibration only when the total photodiode signal is constant. Some workers [68] have proposed deriving the calibration factor from the initial slope of a friction loop. Assuming that the tip is pinned to the substrate, and that the lateral movement has been accurately calibrated, then the angle of twist can be obtained.

In contrast to the vertical spring constant measurement, the lateral spring constant is difficult to measure. Therefore, the lateral spring constant is usually calculated and not measured. For the most popular cantilevers, the lateral spring constant is calculated using finite element methods or by means of analytical formulae [69]. However, such calculations are only approximations, since they use a relatively

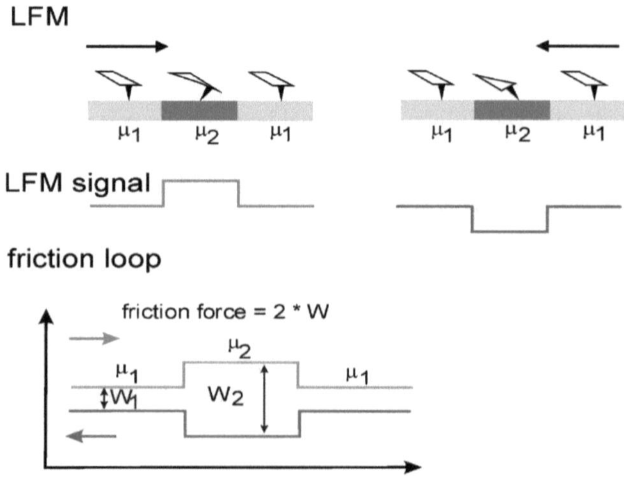

Fig. 7.15. The lateral force signal (LFM) in atomic force microscopy and the concept of the friction loop

simple cantilever model; in other words they assume an ideal geometry, a cantilever Young's modulus value for the bulk material neglecting coatings, a fixed thickness of cantilever, and so on. All of these factors may lead to a poor estimate of the lateral cantilever spring constant.

In addition to the two-step calibration procedures mentioned above, there are a few alternative methods of either calibration or direct measurement of the friction force. Ruan and Bhushan [70] suggested that the friction could be measured by moving the cantilever in the direction *parallel* to the long axis of the cantilever (the usual approach is to perform an AFM scan in a direction *perpendicular* to the long axis). In their method, the vertical spring constant multiplied by the vertical piezo movement necessary to hold the cantilever deflection constant provides an estimate for the friction force. The friction measured in this way was later used to calibrate conventional (carried out in the *perpendicular* mode) lateral friction measurements. Another method was proposed by Ogletree et al. [71], where friction loops were measured as a function of applied load on substrates with two well-defined slopes (Fig. 7.16).

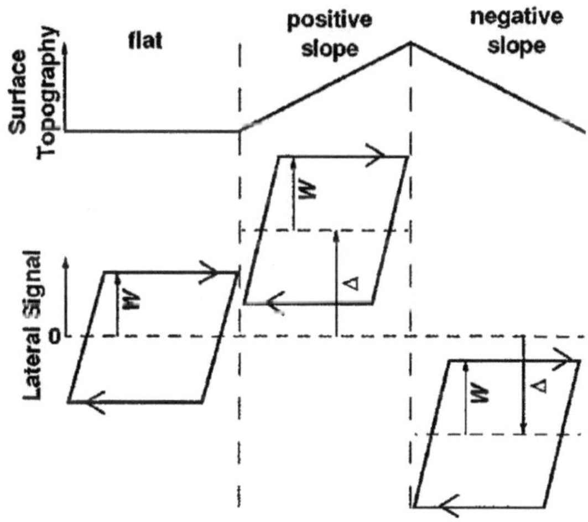

Fig. 7.16. The idea behind the wedge calibration method. Schematic of friction loops measured on flat, positively and negatively sloped surfaces for the same loading force. The friction loop half-width W is slightly different for the three cases, while the loop offset Δ is substantially different and is indicative of the overall tilt of each surface. The values of W and Δ are measured over a range of applied loads for known slopes and used to calibrate the lateral force response of the cantilever (Reprinted with permission from [71], Copyright (1996) American Institute of Physics)

7.2.1.3 Chemical Force Microscopy

Another variation of AFM, called *chemical force microscopy* (CFM, [72]), involves a probe tip that has been chemically modified, and it is used to investigate surfaces that

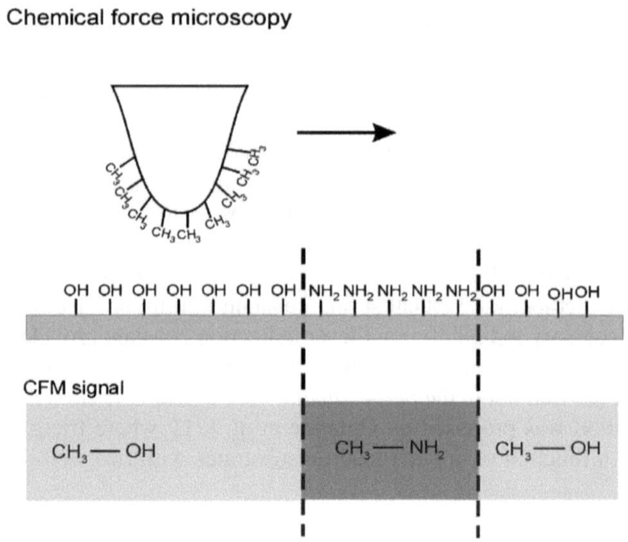

Fig. 7.17. The concept behind chemical force microscopy. The modified AFM probe is moved over a sample surface that is characterized by different chemical properties (such as different functional chemical groups). The different interaction forces between the probe and the surface produce different lateral signals

have particular chemical functional groups. In CFM the method of signal detection remains the same as in classical AFM, but the origin of the signal is different (Fig. 7.17). The chemical modification made to the probe tip makes it possible to image surfaces that are flat at the nanoscale but chemically different. This in turn makes it possible to map the chemical nature of surface in a variety of conditions, including different solvents, pH values, ionic strengths, temperatures, and different molecular structures on the surface. The experiments performed so far have largely focused on identifying specific functional groups on a surface constituting of organic layers, usually self-assembled monolayers (SAMs), as they form highly chemically defined surfaces [73].

To probe interactions between functional groups, the tip must be modified using well defined molecular layers. This modification can be accomplished by using monolayers of amphiphilic molecules adsorbed onto the surface of the AFM tip. Different types of interactions can be studied by varying the head groups of these molecules. This approach is somewhat limited because physisorbed layers are not very robust. Alternatively, stable molecular layers can be produced by covalently linking the molecular components to the tip and sample surfaces. Stable and robust monolayers of alkyl thiols or disulfides containing a variety of terminal groups can be readily prepared this way, enabling systematic studies of the interactions between basic chemical groups on the probe tip and similarly modified Au substrates [74].

7.2.2 Applications of FFM/CFM

7.2.2.1 Solid Surfaces

AFM/FFM has usually been used to study friction-related problems related to solid surfaces. AFM/FFM techniques are increasingly being used for tribological studies of engineering surfaces at scales ranging from atomic and molecular to the micron scale. The parameters studied include surface roughness, adhesion, friction, scratching, wear, indentation, detection of material transfer, and boundary lubrication.

The most commonly investigated phenomena are the nature of single asperity contact and frictional slip-stick phenomena on the atomic scale. Other studies of friction under ultrahigh vacuum conditions (that provide well-defined surface chemical conditions) are performed in order to test point contact theories relating to load, friction, adhesion and contact area.

At most solid–solid interfaces of technological relevance, contact occurs at numerous asperities of different shapes and sizes. The effect of the single asperity radius on the friction can be studied using AFM tips of different radii. A very sharp AFM/FFM tip provides a single asperity contact when sliding over a surface, which enables studies of friction force to be made at an atomic scale. Using the AFM/FFM technique, it has been proven that the friction laws for a single asperity are quite different from those known observed for macroscopic scale friction. The main result is that friction on the nanometer scale exhibits a sawtooth behavior: an atomic stick-slip [75]. An example of this kind of experiment is presented in Fig. 7.18. The surface of NaCl (100) was scanned with a silicon cantilever at an applied force of 0.65 nN and scan velocity of 25 nm/s. The friction loop was obtained by subtracting the two

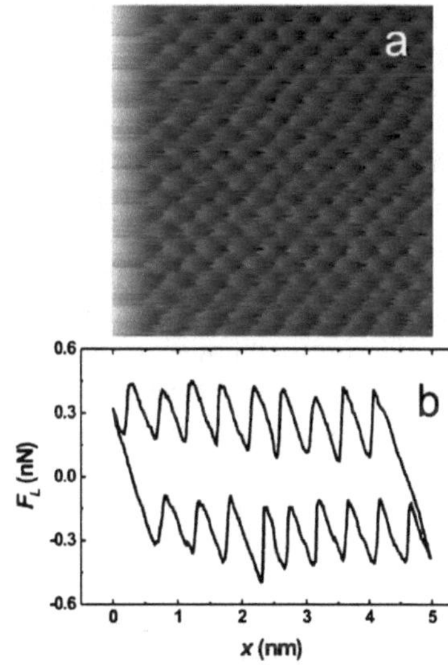

Fig. 7.18. (**a**) Lateral force image obtained for NaCl(100) for a loading force of 0.65 nN and a sliding velocity of 25 nm/s. (**b**) Friction loop obtained along the same scan lines recorded in forwards and backwards directions, showing the characteristic shape indicating stick-slip phenomena (Reprinted with permission from [75], Copyright (2000) by the American Physical Society)

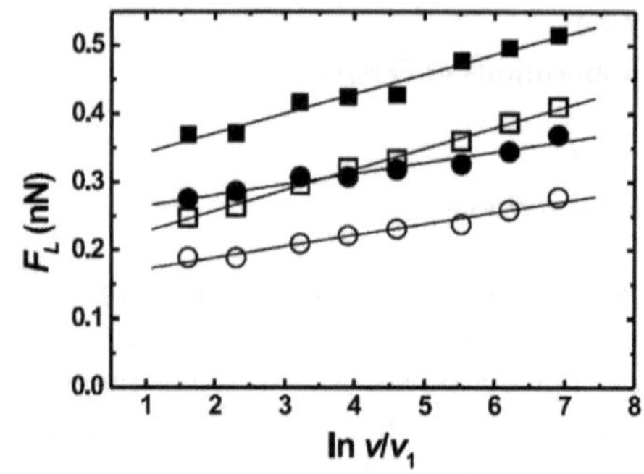

Fig. 7.19. Dependence of friction force on sliding velocity, which can be attributed to lateral loading rate. A gradual increase in friction force is visible as the velocity increases (Reprinted with permission from [75], Copyright (2000) by the American Physical Society)

lateral scan lines recorded in the forward and backward directions. The sawtooth character of the lateral signal highlights the slip-stick phenomenon (Fig. 7.18b).

In this experiment, it was reported that the friction depends on the sliding velocity. Further studies [76] have shown that the friction that occurs during periodic hopping may be strongly influenced by thermal activation. When the AFM tip moves on a surface, the atoms in contact must overcome an energy barrier. Thermal activation leads to a logarithmic increase in the friction force as the sliding velocity increases (Fig. 7.19).

Later on, it was shown that this increase in friction force stops at a certain critical velocity beyond which the friction remains constant. However, this effect was observed in measurements performed in vacuum or in air. In a humid environment, it has been reported that the friction force shows an exponential dependence on the sliding velocity when the surface is rough and hydrophilic (it is still logarithmic on a hydrophobic surface). The explanation for this involves spontaneous water adsorption [76, 77].

7.2.2.2
Organic Surfaces

Many papers [78–80] show good quality surface images that were obtained using chemical force microscopy, by recording lateral or friction forces. All of these studies used FFM/CFM to study interaction forces between organic surfaces and an AFM tip that had been previously modified with particular functional groups.

The functionalization of the AFM tip provides a way to probe the chemical diversity of a surface with a spatial resolution of nanometers. The use of self-assembled monolayers for this purpose enables frictional studies in conditions of well-controlled surface chemistry, which enables the characterization of complex heterogeneous materials. One of the deposition methods involves adsorption of a monolayer of a specific organic molecule on a particular surface. Nakagawa et al. [81] demonstrated that the alkylsilane monolayers can be deposited onto the surface of a silicon nitride cantilever. Cantilevers prepared in this way were used to determine the interaction forces with methyl-terminated and perfluorinated alkylsilanes deposited on the silicon surface. In the other method of adsorbing organic molecules, where organic molecules are extensively used, gold is first deposited on the AFM probe, and then it is immersed in a solution of organic molecules [82]. The efficiency of this method of functionalization was demonstrated for modified probes with either carboxylic acid (COOH) or methyl groups (CH_3) on their surfaces. These probes were used to image the surfaces with patterns composed of carboxylic- and methyl-terminated regions. The same pattern was observed in friction images for either tip functionality (COOH or CH_3), but the contrast was different for different functionalities, which corresponded to a particular chemical interaction. For example, when the surface was scanned with the carboxylic-terminated tip, the acid-terminated regions exhibited a brighter contrast (related to the higher friction force) than the methyl-terminated regions [82].

The sensitivity of friction force microscopy to changes in surface chemical properties was demonstrated by Wong et al. [83], who compared the frictional behavior of SAMs of alkyl chains of different (odd and even) lengths, which differ in the angle of orientation of the dipole moment of the terminal C–C bond. The results presented show that monolayers formed from odd-length alkyl chains have significantly smaller friction coefficients than those with even-numbered chains of the same length. The difference was large enough to enable the imaging of patterned monolayers formed by microcontact printing thiols with only one carbon atom difference in length.

The friction force is sensitive enough to image the molecular organization of organic molecules, which was demonstrated for thiolipid monolayers on mica [84]. These thiolipids form domains with different orientations of lipid molecules. The friction force measured on these domains showed unexpectedly strong friction depending on the angle of the tilt direction of the alkyl chains of the monolayer. The molecular tilt causing these frictional effects was less than 15°, demonstrating that even small molecular tilts can make a major contribution to the friction.

The FFM/CFM combination has been used not only for surface imaging with high chemical resolution, but also for investigating the molecular origin of friction forces. There are many papers that explore the relationships between the friction force (determined for a variety of organic molecules) and parameters that are essential to the frictional behavior of the studied surfaces, such as the load force, the velocity of the scanned AFM tip, the solution pH, the humidity, or the temperature [85]. In particular, alkylthiols and alkylsilanes that form self-assembled monolayers (SAMs) are ideal model systems that can be used for investigating the molecular origin of friction. Many authors have presented results on the relationship between SAM structure and frictional behavior. The friction in these systems has been studied as a function of structural properties such as alkyl chain length or terminal functionality, and also as a function of load and sliding velocity. The results presented in the work of Brewer et al. [86] showed different friction–load and friction–velocity correlations as a function of the type of terminal end functionality and the length of the alkyl chain, indicating the importance of the latter.

Friction forces are load-dependent, which means that the friction force increases with the load applied perpendicularly to the sample surface. For macroscopic friction, this dependence is linear and the coefficient of proportionality is called the coefficient of friction. However, in contrast to the macroscopic situation, the relationship between friction and loading forces is not expected to be linear for a single asperity contact. This conclusion was supported by experimental data obtained for some organic materials. Actually, a number of examples have shown a linear friction–

load relationship for organic monolayers, including for alkanethiol on gold [87] or phenyl-terminated SAMs [88], octadecylchlorodimethylsilane deposited on the oxide of silicon [89] or alkylsilanes on silica and Langmuir–Blodgett films of stearic acid [90]. However, there are also studies where nonlinear behavior was observed; for example complexes of naphthoylene benzimidazole and stearic acid [91]. More recently, Wei et al studied the frictional properties of monolayers of alkylsilanes on mica surfaces [92]. Their data showed a nonlinear relationship with the external load. This behavior can be explained by considering multiple asperity contacts and/or the plastic deformation of the samples. While the work of Xiao et al. [91] supports the hypothesis that plastic deformation may play a role, the other work by Kiely and Houston [93] suggests that monolayers may deform viscoelastically during contact with an AFM tip. This is an important suggestion, because many of the fundamental mechanistic studies of nanoscale friction behavior have focused on inorganic materials, characterized by high elastic moduli and similar bulk and surface chemical properties. The deformation of monolayers, consisting of relatively compliant organic compounds, on substrates that are often much stiffer, is significantly more complex from a theoretical perspective.

It is known that friction is independent of velocity at the macroscopic scale. At the microscopic scale, however, this is not true, and a dependence on velocity was observed for both solid [75, 76] and organic surfaces. Similar to the friction–load relationship, the character of the friction–velocity function depends on the type of organic molecules involved. As an example, the work performed by Brewer et al. [86] for a series of six self-assembled monolayers of alkanethiols with varying chain lengths and terminal group functionalities ($-COOH$, $-OH$, $-CH_3$) is presented. The friction–velocity dependencies were similar for all SAM monolayers when the methyl-terminated probe was in contact – the friction force gradually increased as the velocity increased. However, a different behavior was observed for carboxyl-terminated probes. Depending on the type of SAM monolayer deposited on the surface (polar or apolar), the friction force reached a plateau (Fig. 7.20). The effect was characteristic of polar-terminated SAMs. The authors explained the results

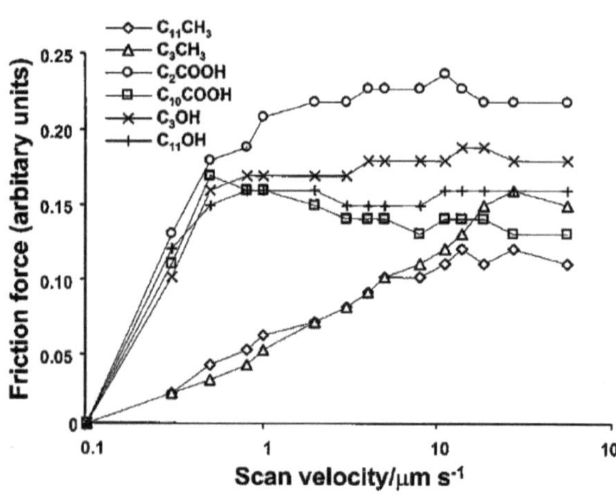

Fig. 7.20. Variation in friction force with scan speed for a carboxylic acid ($C_{10}COOH$)-functionalized tip in contact with six different SAMs (Reprinted with permission from [86], Copyright (2001) by the American Chemical Society)

by invoking cohesive forces between polar-terminated groups. According to them, cohesive forces are responsible for the breaking and forming of the hydrogen bonds that are important to SAM deformation when acid-terminated tips are used.

7.2.2.3 Biomolecular Surfaces

Proteins are molecules that participate in almost all biochemical reactions in cells. They are designed to bind to other molecules, from ions to large complex molecules such as other proteins or nucleic acid. The formation of enzyme–substrate complexes, recognition between antibodies and antigens, forces between complementary DNA strands, or the selectivity of lectins towards carbohydrates are just a few examples of this binding, which is involved in many biochemical phenomena, including molecular transport, signaling, and cellular adhesion.

The basic working principle of a molecular recognition event is the specific binding of a ligand molecule to a receptor molecule that is usually embedded in cell membrane. The binding interaction can be characterized by two features, the *specificity*, describing the preferential binding of one or a few closely related molecules, and the *affinity*, defining the strength of binding. For highly specific interactions, the structure of the binding site of the molecule should exactly fit its complementary protein. Protein binding (the recognition between biological molecules) is responsible for the majority of the interactions of cells with other cells and/or with various proteins in the external environment [94]. The loss or alterations of these interactions may affect cell functioning and lead to many pathological states. Numerous diseases are related to the malfunctioning of these molecular recognition processes. Investigations of the interaction mechanism are key to understanding a wide spectrum of biological processes, such as inflammation or cancer metastasis, which should lead to the development of highly specific drugs [95].

Many methods have been used to investigate unbinding processes between two molecules, especially those induced by the action of an external force that causes bond breaking. These methods encompass, for example, a biomembrane force probe with pipette suction [96], a hydrodynamic flow based method [97], optical tweezers [98], and also atomic force microscopy (AFM [99]). Over the last few years, the latter method has become an indispensable tool to biological studies of molecular interaction forces. Functionalization of the AFM probe significantly enhances the microscope's applicability, since molecular interactions can be studied with high chemical sensitivity. AFM offers a convenient way to measure forces involved in specific interactions by detecting unbinding events between the AFM tip, which is functionalized with one type of molecule, and the substrate, upon which its complementary partner is immobilized. The unbinding force, determined from the force required to separate the tip from the substrate, has been measured for single molecular complexes such as avidin–biotin [100], human serum albumin (HSA) and a polyclonal anti–HSA antibody [101], and even between complementary strands of DNA oligomers [102].

The chemical sensitivity of FFM suggests its possible applicability to measurements of interaction forces between biological molecules during recognition processes. Similarities in the molecular bond detection by means of friction and

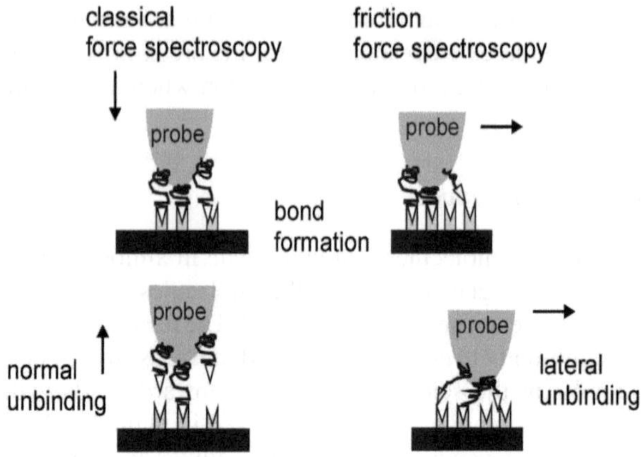

Fig. 7.21. Scheme of the molecular interaction between measurements of the pull-off (*left*, AFM) and friction force (*right*, FFM). The main difference is the direction of force separation(Reprinted with permission from [104], Copyright (2000) by the American Physical Society)

classical force spectroscopy in AFM are shown schematically in Fig. 7.21. Bonds are formed during contact between the tip and the surface. Tip–surface separation can be induced by the force applied to the cantilever, either by withdrawing it vertically (AFM) or by moving it over the surface (FFM).

There are few examples of the application of chemical force microscopy to detecting (bio)molecular interactions. The data presented in 1999 by Mazzola et al. [103] show the potential use of CFM to characterize the interactions between oligonucleotides, short segments of DNA. A patterned array of oligonucleotides was scanned with a DNA-modified AFM tip. The elements of the substrate arrays were relatively easily differentiated by means of friction force. The complementary nucleotides exhibited stronger friction, which was correlated with the strength of the interaction force between the tip and the functionalized surface. The noncomplementary interaction was shown to be weaker in terms of both friction and adhesion forces. The observed friction force was strongly dependent on the loading force.

In work by Lekka at al [104], both chemical (both the AFM tip and the surface were modified with a pair of proteins that recognize each other) and friction (the lateral signal was recorded in order to study molecular interactions) force microscopies were applied to study a typical carbohydrate–protein interaction. Measurements were carried out for two lectins: concanavalin A (from *Canavalia ensiformis*) and lentil lectin (from *Lens culinaris agglutinin*), which have the same type of specificity to mannose and glucose sugars. Carboxypeptidase Y was chosen as a complementary glycoprotein (a protein with a covalently attached carbohydrate chain), since its carbohydrate moiety is mainly composed of mannose. Proteins were deposited on the sample surface using by microprinting with PDMS stamps that had holes arranged periodically.

The sample surfaces with printed proteins were imaged with a ConA-functionalized AFM tip. The surface topography of the immobilized proteins (CaY or BSA) reflected the patterns of the PMDS stamp used. Figure 7.22 presents the surface topography of the CaY deposited on the modified glass surface. The friction force map obtained for the Con–CaY complex showed the pattern corresponding to the chemical diversity of the sample. The areas outside the circles were covered with CaY, and the interiors of the circles corresponded to the silanized and glutaraldehyde-

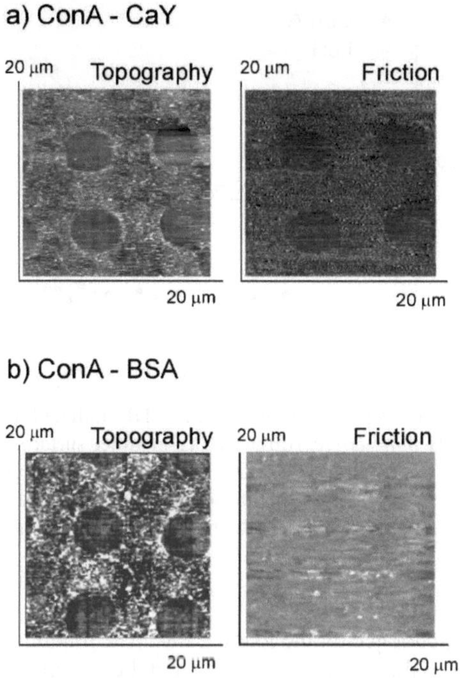

Fig. 7.22. Surface topography and friction force images recorded at the same lateral loading rate of 1.4×10^9 pN/s, measured for (**a**) ConA–CaY and (**b**) ConA–BSA molecular complexes in a buffer solution

treated glass surface. In order to verify the observed recognition interaction for the ConA–CaY complex, measurements were performed for the ConA-functionalized AFM tip and a BSA-printed surface. As was expected, the friction force did not show any pattern, while the surface topography reflected a similar pattern, just as for the patterned CaY surface.

When molecular forces are involved in interactions between (bio)molecules, it has been reported that the resulting unbinding force depends strongly on the applied loading rate (defining how rapidly the applied force breaks bonds [96]), which provides an insight into the energy landscape of the interacting molecules. The dependencies obtained so far show mostly linear region(s), indicating the presence of one or more energy barrier [96, 105, 106]. Therefore, assuming that the observed pattern for the ConA–CaY interaction is due to discrete dominating adhesion bonds, the dependence between the friction and the lateral loading rate was expected to be similar to that measured using classical force spectroscopy. Results have shown that such a dependence occurs only for interactions between ConA and CaY, where recognition events were expected. ConA–BSA does not show this complementariness, and their binding strength stems from forces acting outside the binding site, such as electrostatic and hydrophobic forces. For this type of interaction there was no dependence on scanning velocity (Fig. 7.23). It has already been demonstrated by classical force spectroscopy that the molecular unbinding process manifests itself

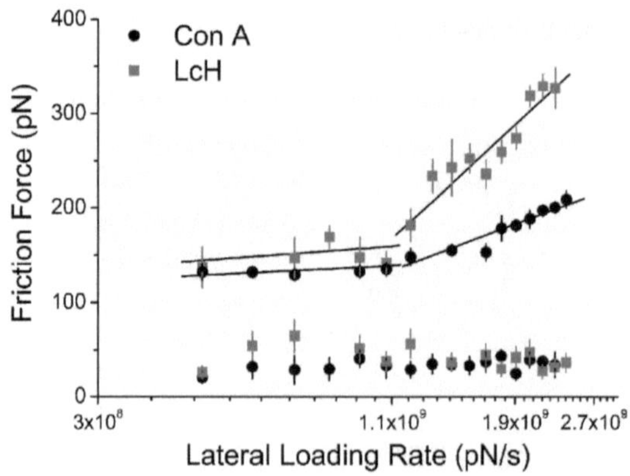

Fig. 7.23. Lateral loading rate dependence (measured in TBS buffer) for a lectin-functionalized (ConA, *gray squares*; LcH, *black circles*) tip and a CaY-modified surface. The specific interaction was inhibited by adding lectins (1 mg/ml) to the solution (*gray squares* and *black circles*, for ConA and LcH, respectively). After 30 minutes of incubation, the unbound or loosely bound lectins were removed by washing in buffer (Reprinted with permission from [104], Copyright (2005) American Institute of Physics)

in a characteristic dependence on the loading rate. The loading rate is usually determined as the product of the retraction velocity and the system spring constant (in the normal direction). For friction force, the loading rate can be defined as the product of the lateral spring constant and the scan velocity. On the other hand, the dominance of the adhesion component in the friction force was theoretically predicted by Filipov et al. [107]. The authors considered two plates connected by adhesive bonds, sliding against each other by the action of an external force. Their model describes the relationship between macroscopic frictional phenomena and the dynamics of formation and the rupture of microscopic bonds. The dynamic response of the system can be characterized by two constants that describe a competition between the rates of bond rupture and formation and the loading rate.

7.2.3 Summary

The combination of FFM and CFM offers high spatial chemical resolution that can be applied in order to study the properties of self-assembled monolayers and other well-ordered organic monolayers. It has yielded insights into the relationship between nanoscale friction, molecular structure and organization. FFM/CFM has proved to be exquisitely sensitive to changes in molecular organization and in chemical composition, and has several additional advantages: (i) the measurement is fast (information from one friction image is comparable to 10^3 friction–distance curves), (ii) it gives better control of normal force (as the measurement is carried out using feedback conditions), and (iii) the technique allows easy localization of binding sites (for example on cellular membrane). Its application to studies of interactions between biomolecules shows much promise, as the dependence between the friction force and the sliding velocity can provide information on the molecular energy landscape.

References

1. Lodish H, Berk A, Matsudaira P, Kaiser CA, Krieger M, Scott MP, Zipursky SL, Darnell J (2004) (eds) Molecular cell biology, 5th edn. WH Freeman, New York
2. Janmey P (1995) Cell membranes and the cytoskeleton. In: Lipowsky R, Sackmann E (eds) Structure and dynamics of membranes. Elsevier, Amsterdam, p 805
3. Janmey P (1998) Physiol Rev 78:763
4. Evans E (1995) Physical actions in biological adhesion. In: Lipowsky R, Sackmann E (eds) Structure and dynamics of membranes. Elsevier, Amsterdam, p 723
5. Thoumine O, Ott A (1997) Biorheology 34:309
6. McLean WHI, Lane EB (1995) Curr Opin Cell Biol 7:118
7. Sulton C, Stamenovic D, Ingber DE (2004) Ann Biomed Eng 32:520
8. Pasternak C, Wong S, Elson EL (1995) J Cell Biol 128:355
9. Ben-Ze'ev A (1997) Curr Opin Cell Biol 9:99
10. Ben-Ze'ev A (1985) Biochim Biophys Acta 780:197
11. Lekka M, Laidler P, Gil D, Lekki J, Stachura Z, Hrynkiewicz AZ (1999) Eur Biophys J 28:312
12. Mow VC, Guilak F, Tran-Son-Tray R, Hochmuth RM (1994) (eds) Cell mechanics and cellular engineering. Springer, Berlin Heidelberg New York
13. Trepat X, Grabulosa M, Puig F, Maksym GN, Navajos D, Farre R (2004) Am J Physiol – Lung C 284:L1025
14. Pierres A, Eymeric P, Baloche E, Touchard D, Bendiel AM, Bongrand P (2003) Biophys J 84:2058
15. Brandao MM, Fontes A, Barjas-Castro ML, Barbosa LC, Costa FF, Cesar CL, Saad ST (2003) 70:207
16. Vinckier A, Semenza G (1998) FEBS Letters 430:12
17. Almqvist N, Bhatia R, Primbs G, Desai N, Banerjee S, Lal R (2004) Biophys J 86:1753
18. Hassan EA, Heinz WF, Antonik MD, D'Costa NP, Nageswaran S, Schoenenberger CA, Hoh J (1998) Biophys J 74:1564
19. Colton RJ (2004) J Vac Sci Technol B 22:1609
20. Werten PJL, Rémigy HW, de Groot BL, Fodias D, Philipsen A, Stahlberg H, Grubmüller H, Engel A (2002) FEBS Lett 529:65
21. Sato H, Kataoka N, Kajiya F, Katano M, Takigawa T, Masuda T (2004) Colloids Surfaces B 34:141
22. Domke J, Dannohl S, Parak WJ, Mueller O, Aicher WK, Radmacher M (2000) Colloids Surfaces B 19:367
23. Cappella B, Dietler G (1999) Surf Sci Rep 34:1
24. Weisenhorn A, Khorsandi M, Kasas S, Gotzos V, Butt HJ (1993) Nanotechnology 4:106
25. Salvetat JP, Bonard JM, Thomson NH, Kulik AJ, Forró L, Benoit W, Zuppiroli L (1999) Appl Phys A 69:255
26. Li X, Bhushan B (2002) Mater Charact 48:11
27. Lekka M, Lekki J, Marszałek M, Golonka P, Stachura Z, Cleff B, Hrynkiewicz AZ (1999) Appl Surf Sci 141:345
28. Jaschke M, Butt HJ (1995) Rev Sci Instrum 68:1258
29. Sader JE, Larson I, Mulvaney P, White LR (1995) Rev Sci Instrum 66:3789
30. Hertz H (1882) J Reine Angew Mathematik 92:156
31. Sneddon IN (1965) Int J Eng Sci 3:47

32. Vinckier A, Dumortier C, Engelborghs Y, Hellemans L (1996) J Vac Sci Technol B 14:1427
33. Kis A, Kasas S, Babic B, Kulik AJ, Benoit W, Briggs GAB, Schonenberger CA, Catsicas S, Forro L (2002) Phys Rev Lett 89:248101-1
34. Kis A (2003) Mechanical properties of mesoscopic objects. Thesis No 2873, Ecole Polytechnique Federale de Lausanne (EPFL), Lausanne
35. Radmacher M, Tilmann RW, Gaub HE (1993) Biophys J 64:735
36. Putman CAJ, Van der Wetf KO, De Grooth BG, Van Hulst NF, Greve J (1994) Appl Phys Lett 64:2454
37. Nagao E, Dvorak JA (1999) Biophys J 76:3289
38. Burnham NA, Behrend OP, Oulevey F, Gremaud G, Gallo PJ, Gourdon D, Dupas E, Kulik AJ, Pollock HM, Briggs GAD (1997) Nanotechnology 8:67
39. Sahin O, Yaralioglu G, Grow R, Zappe SF, Atalar A, Quate C, Solgaard O (2004) Sensor Actuat A 114:183
40. Goldmann WH, Galneder R, Ludwig M, Kromm A, Ezzell RM (1998) FEBS Lett 424:139
41. Bhanu PJ (1997) Cell Biol Int 21:683
42. Bushell GR, Cahill C, Clark FM, Gibson CT, Myhra S, Watson SG (1999) Cytometry 36:254
43. Hoh J, Schoenenberger CA (1994) J Cell Sci 107:1105
44. Haga H, Sasaki S, Kawabata K, Ito E, Ushiki T, Sambongi T (2000) Ultramicroscopy 82:253
45. Wu HW, Kuhn T, Moy VT (1998) Scanning 20:389
46. Rotsch C, Radmacher M (2000) Biophys J. 78:520
47. Ezzell RM, Goldmann WH, Wang N, Parasharama N, Ingber DE (1997) Exp Cell Res 231:14
48. Radmacher M (1997) IEEE Eng Med Biol Mag 16:47
49. Lekka M, Laidler P, Ignacak J, Łabędź M, Lekki J, Struszczyk H, Stachura Z, Hrynkiewicz AZ (2001) Biochim Biophys Acta 1540:127
50. Rotsch C, Jacobson K, Radmacher M (1999) Proc Natl Acad Sci USA 96:921
51. Matzke R, Jacobson K, Radmacher M (2001) Nature (Cell Biology) 3:607
52. Dufrene YF (2004) Nat Rev Microbiol 2:451
53. Fritz M, Radmacher M, Petersen N, Gaub HE (1994) J Vac Sci Technol B 12:1526
54. Xu W, Mulhern PJ, Blackford BL, Jericho MH, Firtel M, Beveridge TJ (1996) J Bacteriol 178:3106
55. Arnoldi M, Kacher C, Bäuerlein E, Radmacher M, Fritz M (1997) Appl Phys A 66:S613
56. Dufrene YF (2001) Micron 32:153
57. van der Mei HC, Busscher HJ, Bos R, de Vries J, Boonaert CJP, Dufrene YF (2000) Biophys J 78:2668
58. Schaer-Zammaretti P, Ubbink J (2003) Ultramicroscopy 97:199
59. Domke J, Radmacher M (1998) Langmuir 14:3320
60. Dimitriadis EK, Horkay F, Maresca J, Kachar B, Chadwick RS (2002) Biophys J 82:2780
61. Bhushan B (1999) Handbook of micro/nanotribology, 2nd edn. CRC, Boca Raton, FL
62. Bhushan B (1996) Tribilogy and mechanics of magnetic storage devices, 2nd edn. Springer, Berlin Heidelberg New York
63. Marti O, Colchero J, Mlynek J (1990) Nanotechnology 1:141
64. Schwarz UD, Zwoerner O, Koester P, Wiesendanger R (1997) Phys Rev B 56:6987
65. Meyer E, Overney RM, Brodbeck D, Howald L, Luthi R, Frommer J, Guntherodt HJ (1993) Phys Rev Lett 69:1777
66. Meurk A, Larson I, Bergstrom L (1998) Mater Res Soc Symp Proc 522:427
67. Liu E, Blanpain B, Celis JP (1996) Wear 192:141
68. Toikka G, Hayes RA, Ralston J (1997) J Adhesion Sci Technol 11 :1479

69. Neumeister JM, Ducker WA (1994) Rev Sci Instrum 65:2527
70. Ruan J, Bhushan B (1994) ASME J Tribol 116:378
71. Ogletree DF, Carpick RW, Salmeron M (1996) Rev Sci Instrum 67:3298
72. Clear SC, Nealey PF (1999) J Colloid Interf Sci 213:238
73. Li J, Horton JH (2002) J Mater Chem 12:1268
74. Vezenow DV, Zhuk AV, Whitesides GM, Lieber CM (2002) J Am Chem Soc 124:10578
75. Gnecco E, Bennewitz R, Gyalog T, Loppacher C, Bammerlin M, Meyer E, Guntherodt H (2000) Phys Rev Lett 84:1172
76. Riedo E, Gnecco E (2004) Nanotechnology 15:S288
77. Opitz A, Ahmed SIU, Schaefer JA, Scherge M (2003) Wear 254:924
78. Kokkoli E, Zukoski CF (2001) Langmuir 17:369
79. He XH, Li QG, Zhou ZY, Zhang H, Li SFY, Liu ZF (2000) Langmuir 16:9683
80. Fujihira M, Tani Y, Furugori M, Akiba U, Okabe Y (2001) Ultramicroscopy 86:63
81. Nakagawa T, Ogawa K, Kurumizawa T (1994) J Vac Sci Technol B 12:2215
82. Noy A, Frisbie CD, Rozsnyai LF, Wrighton MS, Lieber CM (1995) J Am Chem Soc 117:7943
83. Wong SS, Takano H, Porter MD (1998) Anal Chem 70:5209
84. Liley M, Gourdon D, Stamou D, Meseth U, Fisher C, Lautz C, Stahlberg H, Vogel H, Burnham NA, Duschl C (1998) Science 280:273
85. Noy A, Vezenov DV, Lieber CM (1997) Ann Rev Mater Sci 27:381
86. Brewer NJ, Beake BD, Leggett GJ (2001) Langmuir 17:1970
87. Laibinis PE, Whitesides GM, Allara DL, Tao YT, Parikh AN, Nuzzo RG (1991) J Am Chem Soc 113:7152
88. Lee S, Puck A, Graupe M, Colorado R, Shon YS, Lee TR, Perry SS (2001) Langmuir 17:7364
89. Qian L, Tian F, Xiao X (2003) Tribol Lett 15:169
90. Bliznyuk VN, Everson MP, Tsukruk VV (1998) J Tribol 120:489
91. Xiao X, Hu J, Charych D.H, Salmeron M (1996) Langmuir 12:235
92. Wei ZQ, Wang C, Zhu CF, Zhou CQ, Xu B, Bai CL (2000) Surf Sci 459:401
93. Kiely JD, Houston JE (1999) Langmuir 15:4513
94. Lodish H, Berk A, Matsudaira P, Kaiser CA, Krieger M, Scott MP, Zipursky SL, Darnell J (2004) (eds) Molecular cell biology, 5th edn. WH Freeman, New York
95. Bongrand P (1999) Rep Prog Phys 62:921
96. Evans E (2001) Annu Rev Biophys Biomol Struct 30:105
97. Pierres A, Touchard D, Benoliel AM, Bongrand P (2002) Biophys J 82:3214
98. Kulin S, Kishore R, Hubbard J, Helmerson K (2002) Biophys J 83:1965
99. Allison DP, Hiterdorfer P, Han W (2002) Curr Opin Biotech 13:47
100. Florin EL, Moy VT, Gaub HE (1994) Science 264:415
101. Hinterdorfer P, Baumgartner W, Gruber HJ, Schilcher K, Schindler H (1996) Proc Natl Acad Sci USA 93:3477
102. Schumakovitch I, Grange W, Strunz T, Bertoncini P, Guntherodt HJ, Hegner M (2002) Biophys J 82:517
103. Mazzola LT, Frank CW, Fodor SPA, Mosher C, Lartius R, Henderson E (1999) Biophys J 76:2922
104. Lekka M, Kulik AJ, Jeney S, Raczkowska J, Lekki J, Budkowski A, Forró L (2005) J Chem Phys 123:014702
105. Merkel R, Nassoy P, Leung A, Ritchie K, Evans E (1999) Nature 397:50
106. Tees FJ, Waugh WE, Hammer DA (2001) Biophys J 80:668
107. Filipov AE, Klafter J, Urbakh M (2004) Phys Rev Lett 92:135503

11 Applications of Scanning Near-Field Optical Microscopy in Life Science

Pietro Giuseppe Gucciardi

From volume XII

Abstract. Scanning Near-Field Optical Microscopy (SNOM) is capable of attaining sub-diffraction resolution by exploiting nanoscopic light sources to illuminate the samples, such as nanoapertures or sharp metallic tips. In this chapter, after describing the basic principles of SNOM, we review the state-of-the-art in the experimental implementations and applications of fluorescence, infrared and Raman SNOM for the detection and imaging of DNA oligonucleotides, viruses and cellular membrane proteins.

Key words: Scanning near-field optical microscopy, Tip-enhanced Raman spectroscopy, Near-field fluorescence microscopy, Protein localization, Protein co-localization, DNA, Cells, Dendritic cells, Human skeletal cells, Malaria, Actin, Integrin, Sarcoglycan

Abbreviations

AFM	Atomic force microscopy
CCD	Charge coupled device
CLM	Confocal laser microscopy
DC	Dendritic cells
FITC	Fluorescein
GFP	Green fluorescent protein
HIV	Human immunodeficiency virus
imDC	Immature dendritic cells
IR	Infrared
MESA	Mature parasite-infected erythrocyte surface Antigen
NA	Numerical aperture
NIR	Near infrared
PIA	Polysaccharide intercellular adhesin
PBS	Phosphate buffered saline
PMT	Photomultiplier tube
PfHRP1	*Plasmodium falciparum* histidine rich protein
SEM	Scanning electron microscopy
SERS	Surface-enhanced Raman scattering
SM	Single molecule
SNOM	Scanning near-field optical microscopy
STM	Scanning tunneling microscopy

In: H. Fuchs, B. Bhushan (eds), *Biosystems - Investigated by Scanning Probe Microscopy*,
DOI 10.1007/978-3-540-85039-7_3, © Springer-Verlag Berlin Heidelberg 2010

TEM	Transmission electron microscopy
TERS	Tip-enhanced Raman scattering
TMV	Tobacco mosaic virus
TR	Tetramethylrhodamine
WD	Working distance

11.1 Introduction

Spectroscopic imaging techniques, among which are included absorption, fluorescence and Raman imaging, are non-invasive investigation means with a broad field of application in the life sciences. Fluorescence and Raman microscopy, in particular, have gained a firm position among the most important research tools in cell biology. They offer chemically specific contrast and allow for the study of living cells. To date fluorescence microscopy remains as the most widely used technique for cell imaging. Membrane components can be directly visualized via specific dye-conjugated antibodies or direct fusion with green fluorescent proteins (GFP) [1]. The employment of ultra-sensitive optical detectors, such as cooled charge coupled devices (CCD), photomultipliers (PMT) or avalanche photodiodes makes fluorescence imaging possible even at the single molecule (SM) level. SM fluorescence imaging [2, 3] offers a set of advantages compared to ensemble-averaged experiments, among which are (1) the study of the static and dynamic heterogeneity of a population of molecules and its relation to the specific environment; (2) the possibility to obtain information over the function of molecules on cell membranes; (3) the tracking of the conformational motion of biological macromolecules [4]; (4) the study of membrane transport mechanisms [5] as well as the structural change of proteins [6] and DNA [7].

Fluorescence microscopy at room temperature, however, suffers a principal drawback due to photobleaching. Photobleaching includes all those photochemical reactions occurring to the molecules while in the excited state which irreversibly alter the chemical structure of the molecules, causing the extinction of the photoluminescence after the emission of $\sim 10^6$ photons. The label-free, direct characterization of biomolecules is, therefore, of real importance since it avoids the use of an intermediary molecular species, the dye. Such a direct characterization is made possible by the use of vibrational spectroscopies, which are IR absorption and Raman scattering. Vibrational spectroscopies are powerful tools for label-free characterization of biological species since the vibrational modes are actual fingerprints of the whole molecular species or of the local binding between biomolecules. Moreover, they can be combined with standard microscopy techniques to obtain information on the local scale. Unfortunately, IR absorption and Raman scattering cross-sections are far too low ($\sim 10^{-24}$ and $\sim 10^{-30}$ cm^2, respectively) for fast imaging applications. Furthermore, the spatial resolution of these techniques, as well as for fluorescence microscopy, is limited by diffraction to about $\lambda/2NA$ [8], where λ is the wavelength of the radiation and *NA* the numerical aperture of the illumination/collection optics employed. Using far-field optics, such as microscope objectives in the visible and a Cassegrain objective or a parabolic mirror in the IR, the spatial resolution is always > 300 nm. Although new techniques, such as 4Pi microscopy and stimulated-emission depletion microscopy, have recently been introduced [9, 10] demonstrating an increased spatial and axial resolution, their applicability is limited to fluorescence microscopy and to the use of specific classes of dyes.

Scanning Near-Field Optical Microscopy (SNOM) [11–14] exploits a nanometric probe to confine the tip–sample optical interaction to scales between 10 and 100 nm, thus circumventing the diffraction limit (see [15, 16] for reviews). Probes can be either nanoapertures or sharp metallic tips [17]. Together with an unprecedented spatial resolution, Near-Field Microscopy offers several unique advantages: (1) similarly to other scanning probe techniques, it allows recovery of the surface morphology together with its optical properties, with a spatial resolution always in the 10-nm range; (2) being an optical microscopy technique, it allows implementation of all those contrast mechanisms available in confocal microscopy, such as elastic scattering, dichroism/birefringence, fluorescence and Raman; (3) owing to the exponential decay of the illumination field, SNOM is a surface-sensitive technique; (4) finally, due to the intriguing field enhancement effects provided by metallic tips, the Tip-Enhanced SNOM configuration is capable of locally amplifying the optical interaction by several orders of magnitude (2–6) making real absorption and Raman imaging with 10-nm spatial resolution possible (for reviews see [18, 19]).

Application possibilities of SNOM in life science are numerous. Fluorescence SNOM finds natural application for the identification of membrane lipid and proteins, for protein localization and co-localization on the cellular membranes and for the study of ion channel clusters. Tip-Enhanced Raman Spectroscopy (TERS) has recently demonstrated enormous potential applications for label-free detection of membrane proteins and DNA nucleobases.

The review of some of the most relevant experiments is the objective of this chapter. The chapter is organized as follows: in Sect. 11.2 we provide a general introduction to Near-Field experimental techniques; in Sect. 11.3 we give a review of the applications of SNOM in life-sciences; in Sect. 11.4 we draw conclusions on the state-of-art in bio-SNOM and propose some perspectives for the future.

11.2 Experimental Techniques in Near-Field Optical Microscopy

11.2.1 Principles of Near-Field Optical Microscopy

Optical microscopy has a prominent role in modern cell biology [20]. Confocal Laser Microscopy (CLM, Fig. 11.1a) is the most widely used technique, allowing reduction of the detection volume and reconstruction of three-dimensional images of the sample. In CLM a laser beam is focused by means of a high NA objective and scanned all over the sample surface. The back-reflected fluorescence is collected by the same objective and sent to the detector which reconstructs point-by-point the sample's image. Optical sectioning of the sample is achieved by using a spatial pinhole to

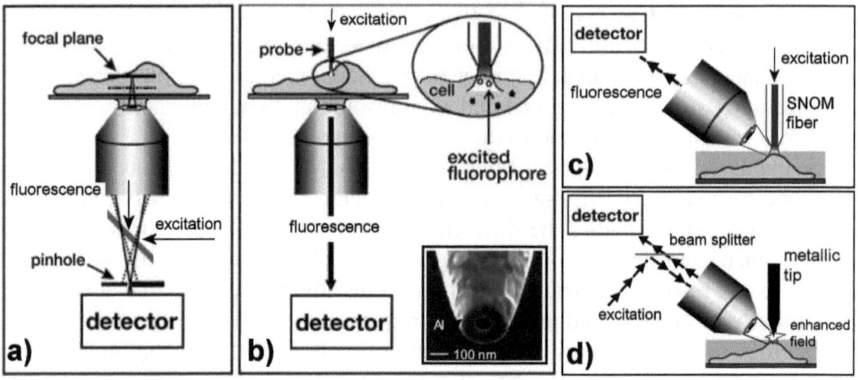

Fig. 11.1. Schematic diagrams of confocal (**a**) and of aperture SNOM microscopy working in transmission (**b**), and reflection (**c**) mode. The *inset* of (**b**) shows a SEM micrograph of the apical aperture of a SNOM fiber probe (reprinted from [25] with permission of the author). (**d**) Schematic of a TE-SNOM working in reflection mode

eliminate out-of-focus light in specimens that are thicker than the focal plane. CLM is, however, a far-field technique, therefore the lateral resolution is diffraction-limited to length scales larger than 250 nm. The microscope objective is located several wavelengths distant from the sample, in the so-called far-field. As a consequence, only propagative electromagnetic waves are detected in the light–matter interaction. In fact, due to the finite *NA* of the objective only waves with a transverse wavevector $\mathrm{k}_\perp$ whose components span in the interval $k_\perp = -NA\,(\omega/c)\ldots + NA\,(\omega/c)$ are collected. The spatial resolution, given by the Heisenberg principle as $|\Delta \mathrm{r}_\perp| \cdot |\Delta \mathrm{k}_\perp| \sim 2\pi$, therefore will be $|\Delta \mathrm{r}_\perp| \sim \lambda/2NA$. The enhanced spatial resolution of SNOM relies in the exploitation of evanescent, non-propagative waves in the tip–sample optical interaction [21]. Evanescent waves are characterized by having a transverse wavevector whose modulus $|\mathrm{k}_\perp| > \omega/c$. The angular spectrum of the radiation by an aperture of diameter $a \ll \lambda$, will contain components with a transverse wavevector roughly spanning between $k_\perp = -\pi/a \ldots + \pi/a$. The expected spatial resolution of aperture-SNOM will therefore be $|\Delta \mathrm{r}_\perp| \sim a \ll \lambda/2NA$. We note that evanescent waves, for which $|\mathrm{k}_\perp| > \omega/c$, have components of the wavevector parallel to the propagation direction $k_{//} = \sqrt{(\omega/c)^2 - k_\perp^2}$ that are purely imaginary. The corresponding electric field is exponentially attenuated during propagation. This makes SNOM a purely surface analysis technique. Moreover, due to the rapid decay length of the near-field (100 nm typically) it is imperative to keep the aperture in proximity of the sample surface ($< 10\,\mathrm{nm}$) during the scan. This is accomplished usually by means of an independent shear-force detection system [22, 23] controlling and stabilizing the tip–sample distance. Such a system, forcing the tip to follow the surface profile during the scan, allows recovery of the topography map of the sample simultaneously with its optical map.

Figure 11.1b shows the schematics of a so-called "aperture-SNOM" employing optical fiber probes. The sample excitation is accomplished through the nanometer-sized aperture (see inset of Fig. 11.1b) located at the apex of a tapered, metal-coated optical fiber [24, 25]. The light scattered by the sample is subsequently collected in the far-field. Depending on whether the sample is transparent or not, the transmitted (Fig. 11.1b) or the reflected (Fig. 11.1c) components of the scattered fields are collected. The transmission mode configuration takes advantage of the high light-gathering capabilities of oil-immersion objectives, while reflection mode SNOM is a more versatile configuration being independent of the opaqueness of the sample.

Different types of probes have been introduced in aperture-SNOM (see [17] for a review). Tapered, metal-coated optical fibers are still the most widely used sensors due to their versatility, especially in liquids, and good spatial resolution (50–100 nm). Nevertheless, they suffer several limitations among which are fragility, a very low light throughput (only a fraction between 10^{-3} and 10^{-6} of the light injected into the fiber is transmitted by the nanoaperture), and heating of the metallization due to absorption of the injected radiation [26–28] that limits to a few mW the maximum amount of light that can be coupled into the optical fiber before thermal damage of both the fiber and the sample under investigation can occur [29, 30]. The effectiveness of tapered uncoated fibers is still debated. Although it is recognized that elastic scattering images acquired with such probes are mainly dominated by artifacts, applications in photoluminescence detection of semiconductor heterostructures have demonstrated spatial resolution capabilities of the order of 150 nm [31,32]. More recently, cantilevered hollow tips [33] have demonstrated a higher robustness, a higher resistance to light injection, and a good polarization preservation [34]. Moreover, cantilevered hollow tips can be easily fitted to an Atomic Force Microscope (AFM) [35].

SNOM configurations using sharp metallic tips are referred to as "apertureless-SNOMs," "scattering-SNOM," (s-SNOM) or "Tip-Enhanced SNOM" (TE-SNOM) [16]. The Tip-Enhanced SNOM has attracted much attention due to the capability to push the resolution limit of aperture-SNOM even further, demonstrating elastic scattering, fluorescence and Raman imaging with 10-nm resolution [36]. In TE-SNOM the tip plays a dual role, confining the illumination field on length scales comparable to its radius of curvature and acting, at the same time, as a nanoantenna capable of collecting the near-fields scattered by the surface, radiating them in the far-field. As shown in Fig. 11.1d light is focused on a metallic tip which can be the metallic tip of a Scanning Tunneling Microscope (STM) [37] or the metalized cantilevered tip of an AFM. The backscattered light is collected, spectrally filtered and detected to reconstruct the object's image.

Among the TE-SNOM techniques, Tip-Enhanced Fluorescence [38] and Raman Spectroscopy [39, 40] promise the most interesting applications in terms of single molecule sensitivity.

11.2.2
Fluorescence Near-Field Optical Microscopy

Near-Field fluorescence imaging on cells is mainly carried out by using aperture-SNOM configurations. Here the tapered optical fiber acts as a light guide funneling the excitation to a spot whose width is given by the aperture diameter, well below the diffraction limit. SNOM configurations working in transmission-mode benefit from

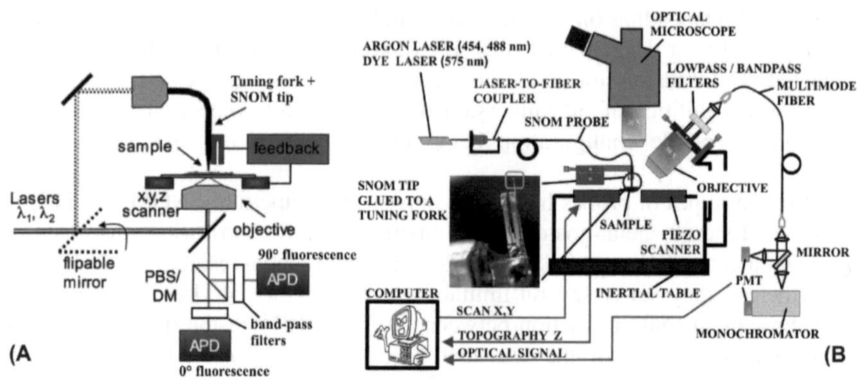

Fig. 11.2. (**a**) Layout of a confocal/SNOM apparatus with single-molecule fluorescence sensitivity working in transmission mode (reprinted from [85] with permission of the author). (**b**) Sketch of a SNOM apparatus for fluorescence imaging working in reflection mode

the large numerical aperture of immersion-objectives that allow the light radiated in the far-field to be gathered, thus featuring unique single-molecule sensitivity. A typical apparatus using optical fiber probes is depicted in Fig. 11.2a [41]. The system is integrated into an inverted microscope, and a flappable mirror enables switching between SNOM and confocal operation. One or more excitation laser beams (Ar^+/Kr^+, 457–647 nm) are used for localization or co-localization experiments. In the latter case both beams are controlled in terms of power level and polarization state, and then recombined by a beam splitter prior to injection into the Al-coated SNOM fiber probe. The tapered extremity of the probe is glued to a quartz tuning-fork [23], providing the shear-force signal that drives the tip–sample distance control feedback circuit. The shear-force detection works by dithering the tip parallel to the sample surface through a piezoelectric-slab and detecting the piezo-current generated by the tuning fork which is proportional to the tip oscillation amplitude. This oscillation amplitude decreases almost linearly when decreasing the tip–sample distance on sub-100-nm scales. Therefore, it can be used as the set parameter in a feedback loop to stabilize the tip–sample distance during the scan. The sample, typically fixed to 170-μm-thick glass cover slides, is scanned by a piezotable in all the three spatial dimensions. The emitted fluorescence is collected by an oil-immersion objective (1.4 NA) and spectrally selected by suitable band-pass filters. The fluorescence is then separated into two channels by means of a polarizing beam-splitter cube and detected by two single photon-counting avalanche photodetectors. Polarization-sensitive detection allows for simultaneous monitoring of the relative intensities of the polarized fluorescence signals, and is used to reconstruct the in-plane orientation of the emission dipole moment of the single molecules.

The reflection-mode SNOM configuration is essential when scattering or re-absorption of the radiation inside the sample are critical issues, preventing an efficient light collection "from below." Light is gathered on the same half-plane where excitation occurs through long working-distance objectives (WD $=$ 10–15 mm, NA $= 0.3$–0.55). We have developed a reflection-mode aperture-SNOM for biological purposes, whose design is inspired by some well assessed commercial models [42,43], implementing the electronic control and the software of a previous SNOM model employed for Raman spectroscopy and imaging [44,45]. The SNOM head is placed underneath an optical microscope (see Fig. 11.2b) working with long working-distance objectives (Olympus APOPLAN 20X, 50X). This allows for a first inspection of the sample and for the exact localization of the point we want to investigate by SNOM. We use commercial CrAl metal-coated fiber probes (Nanonics) with nominal apical apertures of 100 nm. The probe is glued to a quartz tuning-fork for non-optical shear-force detection, aimed at implementing the tip–sample distance control. The tuning fork is anchored, through a properly designed holder, to an XYZ micrometric translation stage used for coarse tip movement. The excitation radiation is provided by blue laser diodes (403 and 417 nm), Argon ion (454–515 nm) or HeNe (633 nm) lasers and is coupled to the SNOM fiber probe, used to illuminate the sample. The scattered light is collected by a long working-distance microscope objective (Olympus 50X, WD 10.6 mm, NA 0.52) rested at an angle of about 45°. The light is thus coupled to a multimode optical fiber and driven to a monochromator (Jobin-Yvon, Triax 190) for the spectral analysis of the fluorescence light. For imaging purposes, instead, light is spectrally filtered by means of appropriate band- or low-pass filters placed right after the collection objective. The fluorescence light is thus coupled to the multimode fiber and directly driven to the photomultiplier. The sample is rested on a 2.54-cm diameter piezo-tube capable of a $32 \times 32 \times 8$-μm^3 scan. A further piezo-scanner can be used for scans up to $80 \times 80 \times 10\,\mu m^3$. An aspheric lens ($d = 10\,mm$, $f = 8\,mm$, NA $= 0.5$) is mounted inside the 2.54-cm diameter piezo-tube (not shown in the picture), and can be used to collect the transmitted light. Maps of 128×128 and 256×256 pixels are typically carried out, with 10 or 20 ms integration time per point.

11.2.3 Near-Field Optical Microscopy in Liquid

Live cell imaging with SNOM requires the scanning tip to be immersed in liquid. Nevertheless, the shear-force feedback must still be able to detect pN interaction forces with soft cell membranes in order to control and stabilize the tip–sample distance during the scan, without damaging the sample. Several approaches have been proposed with optical shear-force detection [46], piezo-feedback mechanisms [47,48], or non-optical tuning-fork systems [49,50]. The main problem when working in liquid concerns the lowering of the oscillator Q-factor due to liquid viscosity, which decreases the sensitivity of the force feedback mechanism. An estimate of the interaction force is, in fact, given by $F = (kx_{res}/Q)\Delta\varphi$ where $\Delta\varphi$ is the phase shift caused to the force interaction, k is the spring constant of the tuning fork and x_{res} is the oscillation amplitude at resonance. With Q-factors of 700 and oscillation amplitudes of 0.1 nm, typical interaction forces in the 100-pN range can be detected, suitable for sensing soft samples such as cells membranes without damage. Figure 11.3a shows the "diving bell" concept introduced by Koopman et al. [50] in which the tuning fork and its holder are embedded into a glass tube. Through the air sealing of the holder, the glass tube acts as a diving bell for the tuning-fork, keeping the

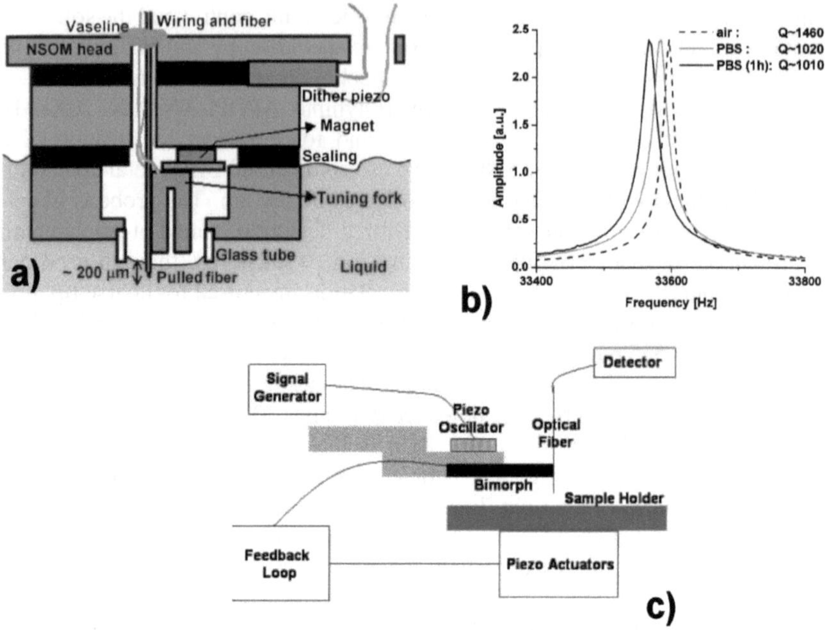

Fig. 11.3. SNOM configurations working in liquid. (**a**) Sketch of the "diving bell" concept. (**b**) Frequency resonance spectra of the tip in air (*dotted line*), in PBS (*solid gray line*), and in liquid after 1 h (*solid black line*), (reprinted from [50] with permission of the author). (**c**) Sketch of a tapping-mode SNOM configuration operating in liquid (reprinted from [48] with permission of the author)

air–liquid interface at a fixed level independently of the amount of liquid used. The optical fiber probe is glued to the tuning fork with the tip end protruding ~500 μm from the prong's end. The length of the glass tube is chosen such that a fiber length of ~200 μm protrudes outside the bottom plane of the glass tube. In such a way the tuning fork and its wiring are always dry, while only the tip apex penetrates the liquid surface. Figure 11.3b shows the tuning-fork resonance spectrum in air (dotted line). The resonance frequency is slightly lowered with respect to the free fork value (33.595 kHz against 32.768 kHz) due to the stiffening effect of the glass fiber on the prong. The Q-factor in air is 1,460, while immersion in phosphate-buffered saline solution (PBS) reduces the Q to 1,020 (solid gray line). Immersion for longer times (solid black line) only changes the Q by 1%.

Tapping-mode SNOM, in which the tip is vertically dithered, allows reduction of the tip–sample interaction forces to the pN range, and is therefore well suited for imaging soft samples. The schematics shown in Fig. 11.3c illustrate an apparatus operating in intermittent contact using straight optical fiber probes [48]. The tip is vertically vibrated by a piezoelectric tablet that produces the sinusoidal oscillation transmitted to the fiber probe through a long bimorph. The oscillation amplitude is detected with an external electronic circuit and is maintained constant by the tapping-AFM feedback loop.

11.2.4 Tip-Enhanced Near-Field Optical Microscopy

TE-SNOM is of particular interest due to possible implementation of IR and Raman contrast mechanisms which enables label-free imaging at the nanoscale. In TE-SNOM the use of metallic tips [51–53] allows one to take advantage of several physical processes such as resonant excitation of localized plasmons or antenna effects. A strong electric field is in fact created at the tip apex [54] and high field-enhancement factors are locally achieved [55]. The key to the success of TE-SNOM has been the discovery of different methodologies to extract the tiny near-field scattering signal from the huge far-field background source of artifacts. For IR absorption and scattering microscopy this is done through interferometric techniques [56–59]. Figure 11.4a shows a typical TE-SNOM apparatus for absorption and scattering measurements. The apparatus implements homodyne interferometric detection [60], and is mounted on a tapping-mode AFM framework. The microscope uses commercial Pt-covered, cantilevered Si tips with tapping frequency $\Omega = 33$ kHz and amplitude $= 25$ nm. The tip is illuminated by focused infrared radiation, typically from continuous wave CO laser beams spanning the spectral range around 9.6 and 10. 6 μm. The backscattered light is superimposed on a reference beam in a Michelson interferometer using a ZnSe partial reflector. The signal of a HgCdTe detector is demodulated at higher harmonics $n\,\Omega$ by a lock-in amplifier. Signal demodulation at a harmonic $n\,\Omega$ of the tapping frequency is required, along with interferometric detection, to eliminate the problem of background scattering artifacts [18, 61, 62]. The reference beam's phase is regularly shifted to separate and acquire simultaneously the amplitude and phase information.

TERS is an analytical technique for high-sensitivity Raman spectroscopy and nanometer-scale resolution imaging capabilities. Apparatus for TERS have been developed in both the transmission [39] and reflection configuration [40, 55, 63, 64]. The main difficulty in TERS is the extraction of the localized near-field tip–sample optical interaction from the huge far-field background coming from the sample's scattering from the diffraction-limited illumination spot. In TERS the Raman

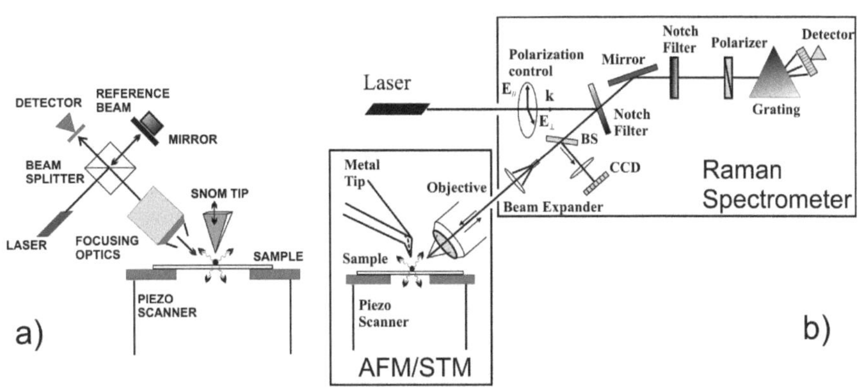

Fig. 11.4. Layout of TE-SNOM apparatus for (**a**) IR scattering and (**b**) Tip-Enhanced Raman scattering experiments

scattering intensity is approximately proportional to $\sim\xi^4\, a^2$ where ξ is the field enhancement factor provided by the tip, and a the radius of curvature of the tip. This value has to be compared to the far-field background proportional to $\sim(\lambda/NA)^2$. To allow for the near-field extraction, one can reduce the far field background by increasing the NA of the illumination objective and then limiting the size of the illumination spot area as much as possible. In the transmission configuration such a task is accomplished by using oil-immersion microscope objectives. In TERS setups with side illumination (reflection-mode configuration) oil immersion objectives cannot be used and the NA is limited to values close to 0.6. In such a case, first experiments suggest that a background reduction can be achieved by exploiting the local light depolarization induced by small particles [65] or metallic tips [66, 67]. Figure 11.4b shows a typical apparatus for TERS in reflection. It basically consists of a Raman spectrometer coupled to an AFM mounting a gold-coated cantilevered tip. Alternatively a shear-force microscope using metallic tips can replace the AFM. The system works in backscattering. Ar-ion, HeNe or diode lasers are used to cover the UV-NIS-NIR spectral range. Polarization control of the excitation is achieved through a set of quarter- or half waveplates. The beam is reflected by a notch filter centered at the laser wavelength and, after expansion, focused to the tip apex through long working-distance objectives. A proper imaging system is needed for precise focusing. The backscattered light is filtered by a double notch system and sent to the detector. An analyzer is typically used to single out the Raman scattering of a definite polarization state. Photomultipliers or CCD cameras are used to detect the radiation.

11.3 Applications of Near-Field Optical Microscopy in Life Science

11.3.1 Infrared Imaging of Tobacco Mosaic Virus with Nanoscale Resolution

Label-free compositional identification of cellular micro-organisms and proteins is a great challenge in biology. The highly confined field present at the apex of a metallic probe in s-SNOM, in particular, permits the mapping of the optical scattering of nanostructures, and to reconstruct the complex dielectric function of the material. This is of utmost interest in the infrared part of the electromagnetic spectrum, where spectroscopic s-SNOM can assess the optical dispersion and therefore one can recognize the infrared vibrational fingerprint of molecules, assigning its chemical composition. Brehm and co-workers [68] have applied IR s-SNOM to image individual tobacco mosaic viruses (TMV) demonstrating, at the same time, that its chemical fingerprint can be retrieved by this technique. For this experiment the s-SNOM (see Sect. 11.2.4 for details) has been interfaced to a CO laser tunable in the region around $\lambda = 6\,\mu\text{m}$ where TMV has distinct vibrational resonances. Figure 11.5 shows (a) the IR scattering amplitude and (b) the phase maps at different energies of an individual TMV cast on Si. It is observed that the contrast between virus and substrate varies with the energy. The scatter-

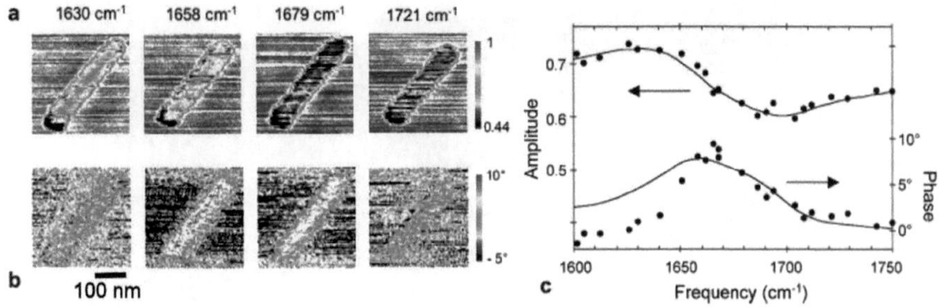

Fig. 11.5. IR scattering amplitude (**a**) and phase (**b**) maps of single TMV nanoparticles imaged at different energies. (**c**) Amplitude (*top plot*) and phase (*bottom plot*) spectra of a TMV particle showing typical dispersion behavior (reprinted from [68] with permission of the author)

ing amplitude shows a more pronounced absorption contrast at $1,679\,\text{cm}^{-1}$, while the phase map shows appreciable contrast only between 1,658 and $1,679\,\text{cm}^{-1}$. To confirm that the contrast observed in the images is indeed due to the amide I vibrational resonance of the peptide bonds of the virus shell proteins resonance spectra have been acquired on a single TMV. Figure 11.5c (black circles) shows the experimental data of the scattering amplitude and phase acquired as a function of the excitation energy in the $1,600$–$1,750\text{-cm}^{-1}$ range. The near-field spectra display a typical dispersion-like trend around the amide I vibrational resonance energy, and are well fitted (Fig. 11.5c, black lines) by a simple point-dipole model [16,69].

11.3.2 Co-Localization of Malarial and Host Skeletal Proteins in Infected Erythrocytes by Dual-Color Near-Field Fluorescence Microscopy

One of the most interesting aspects in biology relates to the organization of cells and how interactions between proteins determine important cellular processes such as signal transduction and receptor–ligand binding. Experiments to identify interactions in cellular systems are mostly performed through the co-localization imaging of two or more proteins, carried out by Fluorescence CLM using fluorescence antibody probes. Enderle and co-workers [70] have applied Fluorescence SNOM to study the co-localization of malaria (*Plasmodium falciparum* histidine rich protein, PfHRP1) and host membrane (protein 4.1, and mature parasite-infected erythrocyte surface antigen, MESA) proteins in infected red blood cells. Purified knob structures contain PfHRP1, erythrocyte skeletal proteins, spectrin, actin, dematin, and protein 4.1. The MESA has been localized on the cytoplasmic face of knobs by immune-electron microscopy, but analysis on purified knobs failed to identify MESA, suggesting the need of high spatial resolution immune-fluorescence investigations for a better understanding of these associations. Figure 11.6a shows the fluorescence image of an infected erythrocyte labeled with antibodies against PfHRP1 and immunolabeled with tetramethylrhodamine (TR) conjugated secondary antibody. Experiments

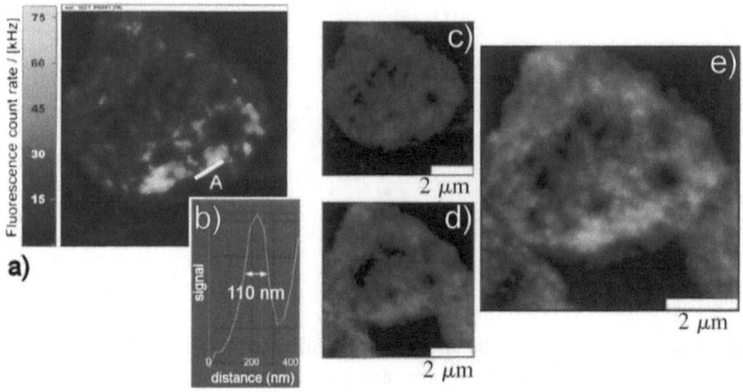

Fig. 11.6. (**a**) SNOM fluorescence map of an infected erythrocyte labeled with antibodies against PfHRP1. The cross-section in (**b**) demonstrates a resolution of the order of 100 nm. Co-localization of PfHRP1and protein 4.1 on a single knob: (**c**) SNOM fluorescence map of PfHRP1, (**d**) analogous for protein 4.1, (**e**) co-localization map of the PfHRP1/protein 4.1 pair (reproduced from [70] with permission of the authors)

were carried out with an aperture-SNOM working in transmission. The sample was excited with a SNOM fiber probe coupled to a 568-nm Ar/Kr laser, using a 100X 1.4 NA oil-immersion objective to gather the fluorescence which was finally detected by an avalanche photodiode. Fluorescence spots of 100–150 nm are typically resolved, as shown by the line profile in Fig. 11.6b. Their average density corresponds to the surface density of knobs deduced by electron microscopy, suggesting that the fluorescent spots originate from PfHRP1 in individual knobs on the infected erythrocyte membrane. Dual-color fluorescence SNOM was then used to study the co-localization of two host–parasite protein pairs. A thin blood smear was infected, fixed and reacted with either mouse monoclonal or rabbit polyclonal antibodies against the proteins under study. Secondary antibodies labeled with Fluorescein (FITC) and TR were used to stain the proteins. Two beams, the 488-nm line of an Ar laser and the 568-nm line of an Ar/Kr laser were collinearly combined and simultaneously injected into the SNOM fiber probe. Suitable band-pass filters were used to reject the excitation and to single out the two fluorescence signals, simultaneously acquired point-by-point. Figures 11.6c and d show the two pseudo-colored images belonging respectively to the TR and to the FITC channels for the host–parasite PfHRP1/protein 4.1 pair. Secondary antibodies conjugated with TRITC and TR were used to localize respectively protein 4.1 and PfHRP1. Figure 11.6e is the overlay of the two images obtained by mixing the two color channels. The map displays sparse green and red spots, arising from the separate emission of protein 4.1 and PfHRP1, and only few yellow zones indicating the simultaneous emission from the PfHRP1/protein 4.1 pair. The map provides clear evidence that the two proteins are only poorly co-localized and demonstrates that there is basically no association between PfHRP1 and protein 4.1 on the membrane of infected red blood cells. In particular this result suggests that, while both proteins are present on the knob structures, they do not directly interact as does MESA with protein 4.1.

11.3.3 Co-Localization of α-Sarcoglycan and β1D-Integrin in Human Muscle Cells by Near-Field Fluorescence Microscopy

The interaction between the extracellular matrix and the sarcolemma-associated cytoskeleton, leading to the dynamic of human muscle fibers, is mediated by the two protein complexes: the dystrophin-glycoprotein complex (DGC) and the vinculin-talin-integrin system [71–73]. The DGC contains, among other proteins, also dystrophin and the sarcoglycans sub-complex. These proteins play a key role in the pathogenesis of many muscular dystrophies and link the cytoplasmic myofibrillar contractile elements to the signal transducing molecules of the extracellular matrix, also providing structural support to the sarcolemma [74, 75]. The vinculin-talin-integrin system connects some components of the extracellular matrix with intermediate filaments of desmin, forming transverse bridges between Z and M lines [76]. Integrins are a family of transmembrane heterodimeric receptors which include at least 14 distinct α subunits and eight β subunits [77]. β1D-integrin, in particular, associates with at least ten α subunits to form distinct integrin dimers, capable of interaction with various extracellular matrix molecules as well as some cell adhesion molecules. The different spatial distribution of the DGC and vinculin-talin-integrin systems has been studied by immunofluorescence investigations using CLM, verifying the exact arrangement of the costameric bands of each protein, the co-localization of all proteins with each other, and, finally, the exact localization of the costameres [78]. The assessment of protein co-localization is, however, related to the spatial resolution capabilities of the microscope employed. We have studied the staining patterns of α-sarcoglycan and β1D-integrin in normal human skeletal muscle using Fluorescence-SNOM, demonstrating that SNOM can be fruitfully used to perform single and double localization of different protein complexes on human muscle cells with sub-diffraction resolution [79]. We have investigated samples of human skeletal muscle, obtained from vastus lateralis muscle biopsies and prepared following the protocol used to carry out indirect immune-fluorescence. In all reactions, TRITC-conjugated IgG anti-mouse in goat was used as the first fluorochrome. Sections were incubated with a second antibody, conjugated with FITC fluorochrome. Samples were then mounted on a reflection-mode SNOM (see. Sect. 11.2.2) and observed. Figure 11.7a shows the spectra of samples incubated either with TRITC (black circles) or with FITC (solid line) showing that the two red and green fluorescence emissions are well spectrally separated. The natural fluorescence of the tissues (gray line) is typically 10-times smaller than the FITC or the TRITC signals. Double localization of β1D-integrin with α-sarcoglycan was carried out by performing two subsequent scans of the same zone, looking each time for a single protein. Figures 11.7b and c display respectively the β1D-integrin and the α-sarcoglycan fluorescence maps, evidencing an alternation of green/red (β1D-integrin/α-sarcoglycan) and black bands (due to the absence of any reaction) at regular intervals. The co-localization of the two proteins was assessed by superimposing the two fluorescence maps, using green color tones to highlight the FITC fluorescence signal and red color tones for the TRITC one. The two topographies were exploited to compensate for lateral drifts between the two scans. The co-localization map (Fig. 11.7d) shows the alternation of yellow and black bands, indicating the spatial overlapping (co-localization) of

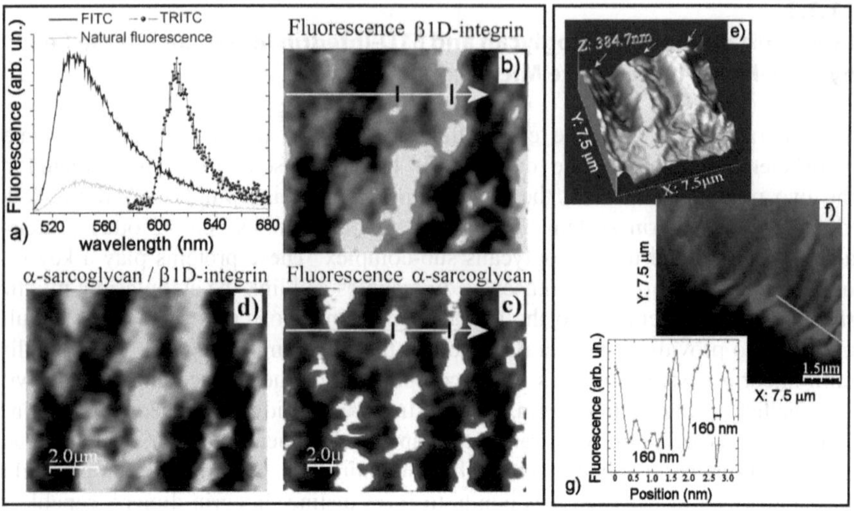

Fig. 11.7. (**a**) Fluorescence spectra of specifically bound FITC (*black line*) and TRITC (*black circles*) compared to natural tissues fluorescence (*gray line*) of human muscle cells immunolabeled against β1D-Integrin and α-sarcoglycan. SNOM fluorescence maps of (**b**) β1D-Integrin and (**c**) α-sarcoglycan. (**d**) Co-localization map obtained by superposing (**b**) and (**c**) using yellow color tones to code co-localization. (**e**) Topography of the costameric structure of a cell, acquired simultaneously to the fluorescence (**f**) after incubation with anti- α-sarcoglycan antibodies. The dye is concentrated into small channels separated by typically 150–200 nm as shown by the *line* profile in (**g**) (reprinted from [79] with permission of the author)

the two proteins. Figure 11.7e, f and g show the topography (e) and the fluorescence map (f), together with the fluorescence line profile (g) of a sample incubated with anti-α-sarcoglycan/TRITC. In the upper part of the 3D topography image we see costameric reliefs (yellow arrows) with an almost flat fluorescence counterpart, suggesting that the optical signal in this case is either due to natural tissue fluorescence, or to unspecific binding of the anti-α-sarcoglycan. We find, instead, a strongly modulated fluorescence in the central part of the map, showing strict channel-like structures, separated by narrower dark areas whose width is ∼160 nm. This value indicates the spatial resolution attainable in our measurements.

11.3.4 Single Molecule Near-Field Fluorescence Microscopy of Dendritic Cells

Dendritic cells (DCs) play a central role in both innate and adaptive immune response, since they are equipped with a variety of dynamically regulated pathogen-recognition receptors [80]. The surface receptor DC-SIGN is a C-type lectin exclusively expressed on DCs, playing an important role during the immune response and featuring extraordinary pathogen-recognition capabilities. In addition to a high-affinity binding to the human immunodeficiency virus (HIV-1) envelope glycoprotein gp120, DC-SIGN has been shown to mediate binding to several other viruses and microorganisms [81–83]. Still, it remains largely obscure how this receptor is capable to efficiently recognize such a large variety of pathogen agents. Immuno Electron Microscopy evidenced that DC-SIGN organizes in microdomains on the cell membrane of immature DCs (imDCs), and that thus this type of arrangement is crucial for binding HIV-1 particles in situ [84]. Garcia-Parajo and co-workers applied fluorescence-SNOM with single molecule sensitivity to investigate the organization of the pathogen receptor DC-SIGN on the membrane of intact DCs, with single molecule sensitivity [85, 86]. An apparatus working in liquid (see Sects. 11.2.2 and 11.2.3) has been developed in order to operate in physiological conditions and avoid potential drying artifacts. Standard immuno-fluorescence protocols were followed to prepare the specimens. In particular the cells were first incubated with primary monoclonal antibodies AZN-D1, and then a secondary incubation was performed with goat anti-mouse Cy5 to allow fluorescence detection of the anti-DC-SIGN antibodies. Cy5 was excited at 647 nm. Individual cells were first selected by bright field illumination and then analyzed by confocal and SNOM fluorescence microscopy. Figure 11.8a is the confocal fluorescence image of a 32-μm^2 area of a DC expressing DC-SIGN on the surface as round-shaped domains. Figures 11.8b and c are the simultaneous topography and SNOM images on the 16.5-μm^2 area highlighted in Fig. 11.8a. The topography image permits us to precisely identify the cell boundaries and is particularly useful to exclude fluorescence mis-assignments due to unspecific antibody bindings. The SNOM image and, in particular, the zoom shown in Fig. 11.8d (7 μm^2) evidence well-separated DC-SIGN domains with FWHM of 90 nm and 120 nm as measured from the line profiles (Fig. 11.8e and f)

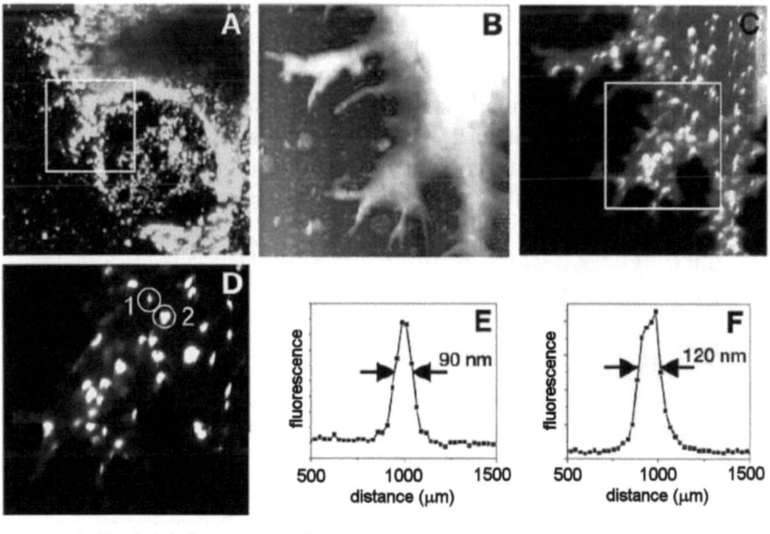

Fig. 11.8. (**a**) Confocal fluorescence image of an immature Dendritic cell in buffer solution. Simultaneous topography (**b**) and SNOM fluorescence map (**c**) of the zone *highlighted* in (**a**). (**d**) Close-up of the area *highlighted* in (**c**) clearly showing clusters of DC-SIGN with typical dimensions of 90 and 120 nm, as retrieved from the *line* profiles in (**e**) and (**f**) (reprinted from [86] with permission of the author)

relative to the domains highlighted in the Fig. 11.8d. The total photon counting rate on each spot is indeed proportional to the number of Cy5 molecules. Single-molecule sensitivity combined with quantitative analysis of the fluorescence intensity permits therefore to estimate the different number of DC-SIGN molecules within each domain. This gave an average of 5–10 molecules of DC-SIGN per cluster.

11.3.5 Chemical Information of Bacterial Surfaces and Detection of DNA Nucleobases by Tip-Enhanced Raman Spectroscopy

Label-free detection of membrane proteins by enhanced Raman spectroscopy is gaining considerable attention in biochemistry and biophysical sciences. Surface-Enhanced Raman Scattering (SERS) takes advantage of the optically resonant nanocavities formed in rough metallic substrates or colloids that are effective in providing Raman enhancement factors up to 10^{11}–10^{14} [87, 88]. Tip-Enhanced Raman Spectroscopy combines the extreme sensitivity of SERS with the possibility to confine the effective enhancement on length scales smaller than 50 nm. So far experiments have been carried out on specific probe molecules such as rhodamine 6G [89], benzenethiol [90], malachite green isothiocyanate [40], or on single-wall carbon nanotubes [39], demonstrating enhancement factors between 10 and 10^3. Deckert and co-workers reported the first TERS on bacterial cells [91]. Gram-positive *Staphylococcus epidermidis* bacteria were investigated, relevant as the cause of major infections associated with implanted medical devices. The cytoplasm membrane of Gram-positive bacteria is surrounded by a peptidoglycan layer, pervaded by other polysaccharides and a variety of surface proteins. Depending on the growth conditions, a biofilm consisting mainly of the two polysaccharides teichoic acid and PIA (polysaccharide intercellular adhesin) surrounds *Staphylococcus epidermidis* to a certain extent. TERS experiments were carried out on a system coupling an AFM with an inverted Raman microscope taking advantage of high numerical aperture objectives (NA 1.45) to maximize light collection efficiency. Figure 11.9a shows a

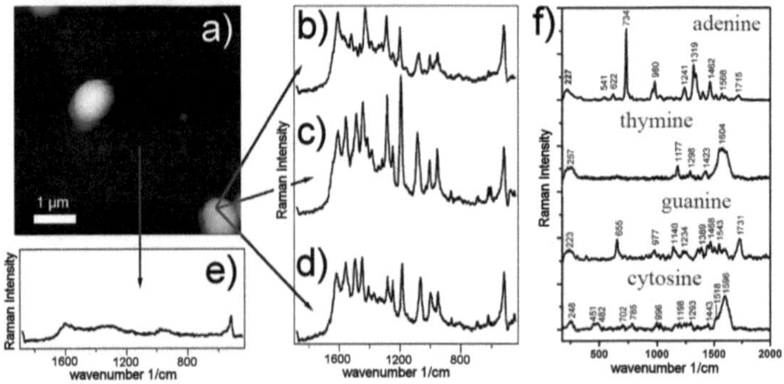

Fig. 11.9. (**a**) Topography of *Staphylococcus epidermidis* cells and corresponding TERS spectra of the membrane proteins (**b**, **c**, **d**). (**e**) Reference spectrum on glass (reprinted from [91] with permission of the author). (**f**) TERS spectra of the four DNA nucleobases deposited on Au(111) single crystals (reprinted from [92] with permission of the author)

topography map in which the *Staphylococcus epidermidis* cells show up as round-shaped features about 1 μm in diameter. Figures 11.9b, c, d and e show point TERS spectra recorded on the cell surface (b, c, d) and one (e) on the glass surface showing no signal. Most of the Raman bands are due to contributions from peptides and polysaccharides present on the cells surface. Sugar derivatives dominate with a peptidoglycan layer which is pervaded by teichoic acid and covered with PIA. Fluctuations observed in intensity and peak positions have been attributed to the persistent dynamics of the cell surface.

TERS allows as well for a highly sensitive vibrational analysis of molecules adsorbed in (sub)-monolayer quantities at atomically smooth surfaces. Pettinger and co-workers [92] have succeeded to obtain TERS spectra of (sub)-monolayers of the four DNA bases adenine, guanine, thymine, and cytosine adsorbed at Au(111) single crystals in picomole quantities. Experiments were carried out with a Raman/STM apparatus similar to the one described in Sect. 11.2.4, using gold tips as probes and exciting with the 633 nm of a HeNe laser (2 mW). The nucleobases were adsorbed at a freshly annealed Au(111) single crystal from 1-mM ethanolic solutions, rinsing ethanol to remove multilayers. According to the molecular dimensions the largest molecular density of adsorbed nucleobases can be estimated to around $2–2.5\ molecules/nm^2$ for closest packing in a monolayer, which corresponds to a density of only $420\,pmol/cm^2$ of target species present in the enhanced field region under investigation. Figure 11.9f shows the Raman vibrational fingerprints of monolayers of the four DNA bases after spectral background correction. These measurements, according to the authors, show that label-free sensoring of picomole quantities of small, non-resonant biomolecules is reproducible and straightforward with TERS.

11.4 Conclusions

In the last decade Near-Field Optical Microscopy has provided us with eyes to look into the NanoWorld. Different experimental concepts have been proposed and some specific configurations are now well assessed tools for analytical investigations in materials science, nanophotonics, and nanobiotechnology. In the biological field, aperture-SNOM turns out to be perfectly suited for fluorescence mapping of membrane proteins due to its high sensitivity (single molecule), high spatial resolution (100 nm), and small penetration depth ($< 100\,nm$). On the other hand the emerging Tip-Enhanced SNOM started providing convincing examples of the possibility to achieve label-free vibrational imaging of nano-organisms and important functional biomolecules. These experiments, in particular, show the great potential of tip-enhanced Raman spectroscopy for biological and biomedical applications. The technique permits us to improve the sensitivity of non-resonant Raman spectroscopy by several orders of magnitudes. As a non-invasive technique, TERS can characterize the chemistry at specific sites on biological samples with a spatial resolution down to a few tens of nm. This method might even be used to follow the dynamics on cell surfaces. For the future fluorescence-SNOM, IR-SNOM, and TERS, could represent key techniques for a detailed understanding of biological processes and mechanisms occurring on cell surfaces, and might give insights into the interaction of bacteria with their surroundings, including hosts and antibiotics with unprecedented sensitivity and spatial resolution.

Acknowledgments. P. Jones and M. Lamy de La Chapelle are acknowledged for critically reading the manuscript.

References

1. Stephens DJ, Allan VJ (2003) Science 300:82
2. Sako Y, Yanagida T (2003) Nature Cell Biol Suppl S, SS1.
3. Garcia-Parajo MF, Veerman JA, Bouwhuis R, Vallee R, van Hulst NF (2001) Chem Phys Chem 2:347.
4. Edman L, Mets U, Rigler R (1996) Proc Natl Acad Sci USA, 93:6710
5. Schmidt T, Schutz GJ, Baumgartner W, Gruber HJ, Schindler H (1995) J Phys Chem 99:17662
6. Lu HP, Xun L, Xie XS (1998) 282:1877.
7. Harada Y, Funatsu T, Murakami K, Nonoyama Y, Ishiama A, Yanagida T (1999) Biophys J 76:709
8. Abbe E (1873) Arch Mikroskop Anat 9:413
9. Hell SW, Stelzer EH (1992) J Opt Soc Am 9:2159
10. Klar TA, Kakobs S, Dyba M, Egner A, Hell SW (2000) Proc Natl Acad Sci USA 97:8206
11. Pohl DW, Denk W, Lanz M (1984) Appl Phys Lett 44:651.
12. Lewis A, Isaacson M, Harootunian A, Murray A (1984) Ultramicroscopy 13:227
13. Zenhausern F, O'Boyle MP, Wickramasinghe HK (1994) Appl Phys Lett 65:1623
14. Bachelot R, Gleyzes P, Boccara AC (1995) Opt Lett 20:1924
15. Labardi M, Gucciardi PG, Allegrini M (2001) La Rivista del Nuovo Cimento 23:1.
16. Patanè S, Gucciardi PG, Labardi M, Allegrini M (2004) La Rivista del Nuovo Cimento 27:1
17. Cefalì E, Patanè S, Spadaro S, Gardelli R, Albani M, Allegrini M (2007) In: Bushan B, Fuchs H, Tomitori M (eds) Applied scanning probe methods, vol VIII. Springer, Berlin, p78
18. Gucciardi PG, Bachelier G, Stranick S, Allegrini M (2008) In: Bushan B, Fuchs H, Tomitori M (eds) Applied scanning probe methods, vol VIII. Springer, Berlin, p 1
19. Gucciardi PG, Trusso S, Vasi C, Patanè S, Allegrini M (2007) In: Bhushan B, Fuchs H, Kawata (eds) Applied scanning probe methods, vol V. Springer, Berlin, p 287
20. Pawley JB (2006) Handbook of biological confocal microscopy, 3rd edn. Springer, Berlin
21. Novotny L, Hecht B (2006) Principles of nano-optics. Cambridge University Press, Cambridge
22. Betzig E, Finn P, Weiner JS (1992) Appl Phys Lett 60:2484
23. Karrai K, Grober RD (1995) Appl. Phys. Lett. 66:1842
24. Betzig E, Trautman JK, Harris TD, Weiner JS, Kostelak RL (1991) Science 251:1468
25. Veerman JA, Otter AM, Kuipers L, van Hulst NF (1998) Appl Phys Lett 72:3115
26. Gucciardi PG, Colocci M, Labardi M, Allegrini M (1999) Appl Phys Lett 75:3408
27. Ambrosio A, Allegrini M, Latini G, Cacialli F (2005) Appl Phys Lett 87:033109
28. Dickenson NE, Erickson ES, Mooren OL, Dunn RC (2007) Rev Sci Instrum 78:053712
29. Erickson ES, Dunn RC (2005) Appl Phys Lett 87:201102
30. Gucciardi PG, Patanè S, Ambrosio A, Allegrini M, Downes AD, Latini G, Fenwick O, Cacialli F (2005) Appl Phys Lett 86:203109
31. Intonti F, Emiliani V, Lienau Ch, Elsaesser Th, Noetzel R, Ploog KH (2001) Phys Rev B 63:075313

32. Intonti F, Emiliani V, Lienau Ch, Elsaesser Th, Noetzel R, Ploog KH (2001) Phys Rev B 64:155316
33. Mihalcea C, Scholz W, Werner S, Muenster S, Oesterschulze E, Kassing R (1996) Appl Phys Lett 68:3531
34. Biagioni P, Polli D, Pucci A, Ruggeri G, Labardi M, Cerullo G, Finazzi M, Duò L (2005) Appl Phys Lett 87:223112
35. Binnig G, Quate CF, Gerber C (1986) Phys Rev Lett 56:930
36. Novotny L, Stranick SJ (2006) Annu Rev Phys Chem 57:303
37. Binnig G, Rohrer H, Gerber C, Weibel E (1982) Phys Rev Lett 49:57
38. Ma Z, Gerton JM, Wade LA, Quake SR (2006) Phys Rev Lett 97:260801
39. Hartschuh A, Sanchez EJ, Xie S, Novotny L (2003) Phys Rev Lett 90:095503
40. Pettinger B, Ren B, Picardi G, Schuster R, Ertl G (2004) Phys Rev Lett 92:096101
41. De Bakker BI, de Lange F, Cambi A, Korterik JP, van Dijk EM, van Dijk EM, van Hulst NF, Figdor CG, Garcia-Parajo MF (2007) Chem Phys Chem 8:1473
42. Gucciardi PG, Vinattieri A, Colocci M, Damilano B, Grandjean N, Semond F, Massies J (2001) J Microscopy-Oxford 202:212
43. Gucciardi PG, Princi P, Pisani A, Favaloro A, Cutroneo G (2005) J Kor Phys Soc 47:S86
44. Gucciardi PG, Labardi M, Gennai S, Lazzeri F, Allegrini M (1997) Rev Sci Instrum 68:3088
45. Gucciardi PG, Trusso S, Vasi C, Patanè S, Allegrini M (2002) Phys Chem Chem Phys 4:2747
46. Lambelet P, Pfeffer A, Sayah A, Marquis-Weible F (1998) Ultramicroscopy 71:117
47. Brunner R, Hering O, Marti O, Hollricher O (1997) Appl Phys Lett 71:3628
48. Longo G, Girasole M, Cricenti A (2005) Phys Stat Sol B 15:3070
49. Höppener C, Molenda D, Fuchs H, Naber A (2003) J Microsc-Oxford 210:288
50. Koopman M, de Bakker BI, Garcia-Parajo MF, van Hulst NF (2003) Appl Phys Lett 83:5083
51. Ren B, Picardi G, Pettinger B (2004) Rev Sci Instrum 75:837
52. Billot L, Berguiga L, de la Chapelle ML, Gilbert Y, Bachelot R (2005) Eur Phys J-Appl Phys 31:139
53. Bonaccorso F, Calogero G, Di Marco G, Maragò OM, Gucciardi PG, Giorgianni U, Channon K, Sabatino G (2007) Rev Sci Instrum 78:103702
54. Novotny L, Bian RX, Xie XS (1997) Phys Rev Lett 79:645
55. Mehtani D, Lee N, Hartschuh RD, Kisliuk A, Foster MD, Sokolov AP, Caiko F, Tsukerman I (2006) J Opt A Pure Appl Opt 8:S183
56. Hillenbrand R, Keilmann F (2000) Phys Rev Lett 85:3029
57. Hillenbrand R, Keilmann F (2002) Appl Phys Lett 80:25
58. Labardi M, Patanè S, Allegrini M (2000) Appl Phys Lett 77:621
59. Taubner T, Keilmann F, Hillenbrand R (2004) Nanolett 4:1669
60. Taubner T, Hillenbrand R, Keilmann F (2004) Appl Phys Lett 85:5064
61. Gucciardi PG, Bachelier G, Allegrini M (2006) J Appl Phys 99:124309
62. Gucciardi PG, Bachelier G, Allegrini M, Ahn J, Hong M, Chang S, Jhe W, Hong SC, Baek SH (2007) J Appl Phys 101:064303
63. Ossikowski R, Nguyen Q, Picardi G (2007) Phys Rev B 75:045412
64. Gucciardi PG, Lopes M, Lamy de La Chapelle M (2008) Thin Sol Films 516:8064
65. Poborschii V, Tada T, Kanayama T (2005) Jpn J Appl Phys 44:L202
66. Mehtani D, Lee N, Hartschuh RD, Kisliuk A, Foster MD, Sokolov AP, Maguire JF (2005) J Raman Spectrosc 36:1068
67. Gucciardi PG, Lopes M, Deturche R, Julien C, Barchiesi D, Lamy de La Chapelle M (2008) Nanotechnol 19:215702
68. Brehm M, Taubner Th, Hillenbrand R, Keilmann F (2006) Nanolett 6:1307
69. Knoll B, Keilmann F (1999) Nature 399:134
70. Enderle Th, Ha T, Ogletree DF, Chemla DS, Magowan C, Weiss S (1997) Proc Natl Acad Sci USA 94:520

71. Ervasti JM, Ohlendieck K, Kahl SD, Gaver MG, Campbell KP (1990) Nature 345:315
72. Yoshida M, Ozawa E (1990) J Biochem 108:748
73. Mondello MR, Bramanti P, Cutroneo G, Santoro G, Di Mauro D, Anastasi G (1996) Anat Rec 245:481
74. Ohlendieck K (1996) Eur J Cell Biol 69:1
75. Yoshida M, Suzuki A, Yamamoto H, Noguchi S, Mizuno Y, Ozawa E (1994) Eur J Biochem 222:1055
76. Lazarides E (1980) Nature 283:249
77. Hynes RO (1992) Cell 69:11
78. Anastasi G, Amato A, Tarone G, Vita G, Monici MC, Magaudda L, Brancaccio M, Sidoti A, Trimarchi F, Favaloro A, Cutroneo G (2003) Cells Tissues Organs 175:151
79. Anastasi A, Cutroneo G, Pisani A, Bruschetta D, Milardi D, Princi P, Gucciardi PG, Bramanti P, Soscia L, Favaloro A (2007) J Microsc-Oxford 228:322
80. Akira S (2003) Curr Opin Immunol 15:5
81. Geijtenbeek TB, Torensma R, van Vliet SJ, van Duijnhoven GCF, Middel J, Cornelissen I, Nottet H, KewelRamani VN, Littman DR, Figdor CG, van Kooyk (2000) Cell 100:587
82. Alvarez CP, Lasal F, Carrillo J, Muniz O, Corbi AL, Delago R (2002) J Virol 76:6841
83. Pöhlmann S, Zhang J, Baribaud F, Chen Z, Leslie GJ, Lin G, Granelli-Piperno A, Doms RW, Rice CM, McKeating JA (2003) J Virol 77:4070
84. Cambi A, de Lange F, van Maarseveen NM, Nijhuis M, Joosten B, van Dijk EM, de Bakker BI, Fransen JAM, Bovee-Geurts PHM, van Leeuwen NF, van Hulst NF (2004) J Cell Biol 164:145
85. De Bekker BI, de Lange F, Cambi A, Korterik JP, van Dijk EM, van Hulst NF, Figdor CG, Garcia-Parajo MF (2007) Chem Phys Chem 8:1473
86. Garcia-Parajo MF, De Bekker BI, Koopman M, Cambi A, de Lange F, Figdor CG, van Hulst NF (2005) NanoBiotechnology 1:113
87. Kneipp K, Wang Y, Kneipp H, Perelman LT, Itzkan I, Dasari R, Feld MS (1998) Phys Rev Lett 78:1667
88. Nie SM, Emory SR (1997) Science 275:1102
89. Watanabe H, Hayazawa N, Inouye Y, Kawata S (2005) J Phys Chem B 109:5012
90. Ren B, Picardi G, Pettinger B, Schuster R, Ertl G (2005) Angew Chem 117:141
91. Neugebauer U, Rösch P, Schmitt M, Popp J, Julien C, Rasmussen A, Budich Ch, Deckert V (2006) Chem Phys Chem 7:1428
92. Domke KF, Zhang D, Pettinger B (2007) J Am Chem Soc 129:6708

3 Scanning Ion Conductance Microscopy

Tilman E. Schäffer · Boris Anczykowski · Harald Fuchs

From volume II

3.1 Introduction

The era of scanning probe microscopy was launched in 1981 with the construction of the first scanning tunneling microscope (STM) by Binnig, Rohrer and coworkers [1]. This microscope is based on the measurement of tiny tunneling currents between a sharp metal probe and a flat metal sample. For the first time, individual atoms could be laterally resolved on conducting surfaces. But more than that, the STM has triggered the development of a large family of new microscopes, the so-called scanning probe microscopes (SPM), that are all based on a small, locally confined probe which is sensitive to various kinds of physical quantities. In 1984, Pohl et al. invented the near-field scanning probe microscope (NSOM) that enables optical imaging of surfaces with sub-wavelength resolution [2]. In 1986, in the same year when they received the Nobel Price for their development of the STM, Binnig et al. invented the atomic force microscope (AFM) [3]. This microscope is based on the measurement of minute forces. One of its main applications is the creation of high-resolution topographic images of surfaces, where it has the advantage over STM of not requiring conducting probes and samples. The additional benefit of the AFM that it can image in different environmental conditions including aqueous buffer solution makes it well suited for biological applications [4]. Several related scanning probe microscopes have been developed so far, including the magnetic force microscope (MFM) [5, 6], the electric force microscope [7] and the scanning electrochemical microscope (SECM) [8].

In this chapter, we focus on the scanning ion conductance microscope (SICM), which was invented by Hansma et al. in 1989 [9, 10]. The SICM records the ionic current through a small aperture as this aperture is scanned over the surface of a sample immersed in an electrolyte solution. The aperture is usually formed by a nanopipette. Two electrodes are placed in the electrolyte: one inside the pipette (pipette electrode), and one outside the pipette in the bath over or under the sample (bath electrode). By applying a voltage between both electrodes and recording the ion current through the pipette, locally resolved images of the ion current over the sample surface can be generated. Despite the many possible applications of such a microscope, the SICM is one of the least developed scanning probe microscopy technique so far. Only few setups are described in literature [9–17].

Most researchers have used drawn-out glass capillaries as pipettes similar to those used in intracellular recording and patch clamp experiments [18, 19], but some also used microfabricated probes [20]. Typical opening diameters of the tips of the

In: H. Fuchs, B. Bhushan (eds), *Biosystems - Investigated by Scanning Probe Microscopy*,
DOI 10.1007/3-540-27453-7_7, © Springer-Verlag Berlin Heidelberg 2010

glass nanopipettes are 30–100 nm, and lateral scan resolutions in the same range are obtained. There is a strong dependence of the measured current on the tip-sample distance ("current squeezing effect"). The tip-sample distance can therefore be kept constant during scanning by using the current as an input to a feedback loop, thereby generating images of sample topography. This particularly works for non-porous samples where the bath electrode is positioned on the same side of the sample as the pipette. Pipette and sample do not come into mechanical contact with each other, and soft and delicate samples such as living cells can be imaged in this "non-contact" configuration [14]. To improve the imaging capabilities of the SICM, several methods have been devised that are mostly based on the principle of an ac measurement. By modulating either the pipette current or the tip-sample distance and by using lock-in amplifiers, these methods exhibit several advantages over the regular dc method [16, 17, 21].

All these methods use the measurement of the local ion conductance to generate images of sample topography. For many applications, however, it becomes important to measure complementary quantities, most prominently such that allow a determination of sample topography independent of the ion current. For this purpose, two different techniques have been developed so far: (a) SICM with a complementary AFM distance control [12] and (b) SICM with a complementary shear-force control [15]. The first technique uses a bent nanopipette that is coated with a reflective metal layer and that is scanned over the sample surface like an AFM tip. Different AFM imaging modes can be used this way, including contact mode and tapping mode imaging. The second technique uses a nanopipette that is positioned perpendicularly to the sample surface and that is set into transverse vibrations. Arising mechanical shear-forces between tip and sample provide an independent measure of sample height. Both techniques allow the simultaneous and complementary imaging of sample topography and local ion conductance.

We organize this chapter in the following way: in Sect. 3.2, we cover fundamental principles concerning the measurement of the local ion conductance. In Sect. 3.3, we present theoretical approaches and experimental implementations for measuring ion currents through nanopipettes close to surfaces. In Sect. 3.4, we move on to advanced techniques for the imaging of ion currents that are based on the local measurement of conductance only. In Sect. 3.5, we finally move our attention to the combination of SICM with other scanning techniques. Methods that provide an independent measurement of sample topography are especially important as they enable high-resolution measurements of the local permeability of a sample to ions.

3.2 Fundamental Principles

3.2.1 Basic Setup

The basic principle of SICM is shown in Fig. 3.1a. A nanopipette with a small opening diameter is positioned close to a sample surface. A voltage is applied accross the electrolyte via two electrodes, one inside the pipette and the other one in the bath

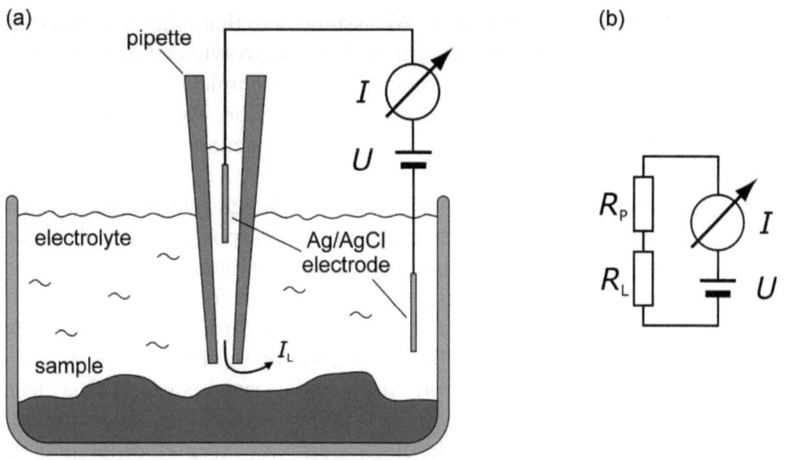

Fig. 3.1. (**a**) Schematic of a basic SICM setup for current measurements over a non-porous sample. The tip of a pulled nanopipette is brought into close proximity to the sample surface. A voltage, U, is applied between the electrode in the pipette and that in the bath, causing an ionic current to flow through the opening of the pipette tip. With decreasing tip-sample distance, this current decreases as a result of the "squeezing" of the current into a narrowing conductance path (gap) between pipette and sample. Since no mechanical contact occurs during this process until "touch-down", the measured leakage current, I_L, provides a contact-free measure of tip-sample distance. By scanning the sample with respect to the pipette tip, two-dimensional images of sample topography can be recorded. (**b**) Equivalent circuit for this setup. The resistance of the pipette, R_P, is in series with the resistance of the leakage path, R_L

solution. Usually, silver/silver chloride electrodes are used. The induced current flows from the pipette electrode through the pipette, leaves the pipette through the narrow gap between pipette and sample surface, and is collected by the bath electrode (the direction of the current can also be reversed). The closer the pipette tip is to the sample, the more this leakage current is restricted ("squeezed") by this narrow gap [9,22]. The detected current therefore directly gives a measure of the tip-sample distance. Especially important is the fact that this measurement is achieved without actual mechanical contact between tip and sample. It is therefore possible to record non-contact images of sample topography by scanning the sample relative to the tip and by using a feedback loop for controlling the tip-sample distance. We will see below that this property is especially valuable for the imaging of soft biological samples such as living cells.

To further characterize this basic SICM measurement setup, we consider the equivalent circuit (Fig. 3.1b). The applied voltage, U, drives a current between the electrodes. This current needs to pass through the narrow neck of the pipette tip that constitutes the resistance R_P, and, in series, through the leakage resistance of the thin gap between pipette and sample, R_L. Since it is this leakage resistance that is usually of interest as it provides a measure of the tip-sample distance, it must not be much smaller than the pipette resistance. Otherwise, the overall resistance and therefore the measured current, I, would be dominated by the pipette resistance. This makes it difficult to extract the leakage resistance which becomes increasingly larger for smaller tip-sample distances. We assume here that other resistances in the system such as resistances in the wires, in the bulk electrolyte and in the solid-liquid junctions are small. Furthermore, we neglect electrochemically induced potentials that can be compensated for by applying an offset voltage. We also note that, when fast changes of the current are to be measured, the capacitances in the system need to be considered.

In many cases it will be interesting to measure the current through conductive paths in the sample, such as pores or channels. For this purpose, we consider a slightly different setup (Fig. 3.2a). Characteristic for this type of measurement is that there is a separate compartment on each side of the sample. Each compartment is filled with electrolyte, and there are one or more conductive paths through the sample between the two compartments. Similarly to the setup described in Fig. 3.1, the thin end of a pulled pipette is brought into close proximity to the sample surface. One electrode is inside the pipette, and the other electrode is in the compartment below the sample. Again, a voltage is applied between the electrodes. Now there are different paths for the current. If the pipette would form an ideal, tight seal with the sample, then the only possibility for the current would be to travel through a conductive path (pore or channel) in the sample directly below the tip (I_C). In a realistic case, however, the seal is never perfect. Some current "escapes" through the thin gap between pipette and sample, and subsequently flows through the sample via other conductive paths, making up the leakage current, I_L, that acts as a background current.

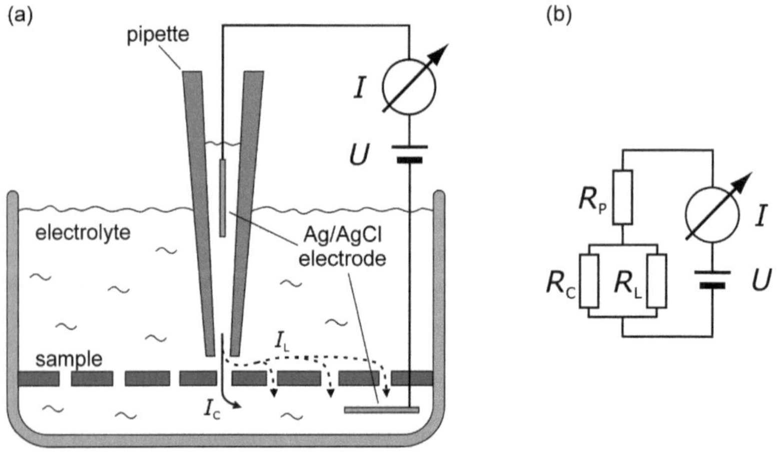

Fig. 3.2. (**a**) Schematic of a basic SICM setup for current measurements through a porous sample. There are two separate compartments with the sample at their interface. The pipette with the pipette electrode is placed above the sample, and the bath electrode is placed below the sample. The current can leave the pipette opening either through a conductive path in the sample directly below the tip (I_C) or can "escape" sideways through the thin gap between tip and sample as leakage current, I_L. The closer the tip is to the sample surface, the smaller this leakage current is (due to the current squeezing effect), and the better the conductance through the sample can be measured. (**b**) Equivalent circuit for this setup. The resistance of the conductive path through the pore directly below the tip, R_C, and the resistance of the leakage path, R_L, are in parallel with respect to each other, and in series to the pipette resistance, R_P

We again consider the equivalent circuit to further characterize this setup (Fig. 3.2b). The current first passes through the narrow neck of the pipette tip that constitutes the resistance R_P. The following two paths are symbolized by the channel resistance, R_C, and by the leakage resistance, R_L, in parallel to each other. Since we are usually interested in the current I_C that flows through the channel below the tip, we need to measure R_C. Therefore, the pipette resistance R_P should be as small as possible, and the leakage resistance R_L should be as large as possible, requiring the gap between pipette and sample to be small.

3.2.2 Nanopipettes

Pipettes for SICM face similar requirements as those used for intracellular recording. Most importantly, they need to be quite sharp at the tip, with opening diameters of a few tens of nanometers. This minimizes the risk of cell damage during cell puncturing in intracellular recording, and maximizes the resolution in SICM. Pipettes are usually drawn from glass (e.g., borosilicate) capillaries with an initial outer diameter of 1–2 mm. Pipette pullers based on a heated coil can be used, but laser-based pipette pullers are preferable. In these pipette pullers an infrared laser beam locally heats up the glass capillary which is subsequently drawn apart by force. This creates sharp tips with opening diameters down to a few tens of nanometers (Fig. 3.3).

A small opening diameter results in a relatively large pipette resistance R_P. To minimize this resistance, the pipette could be filled with a highly conductive solution (e.g., 2–3 M KCl). The diffusion of this solution into the bath solution

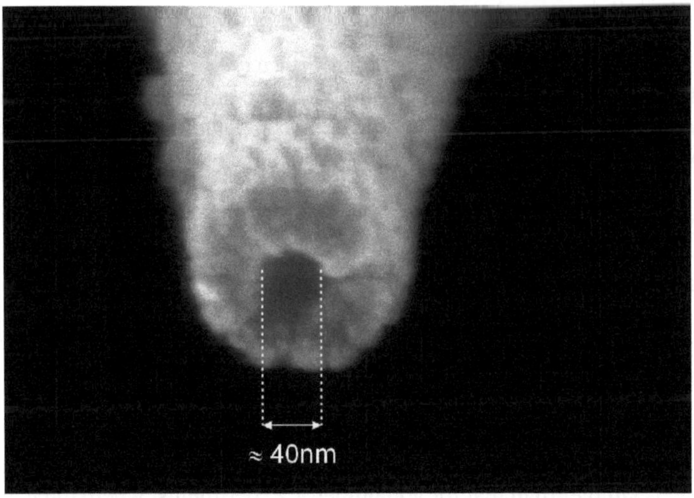

Fig. 3.3. Scanning electron microscope (SEM) image of the tapered end of a nanopipette as it is typically used as an SICM probe (sputter coated with a thin layer of platinum to improve SEM imaging). The size of the opening depends on the pulling parameters such as pulling force, velocity, and heat. Inner opening diameters in the range of 30–100 nm are routinely obtained

would be small due to the small tip diameter, and could be further reduced with the help of an agar salt bridge, for example. Physiological electrolyte solution of lower concentration, however, also works, resulting in typical pipette resistances on the order of 100 MΩ.

3.2.3 Electrodes

The electrodes that are typically used in SICM are silver/silver chloride electrodes. This type of electrode is also used as reference electrodes in potentiometry and voltammetry experiments. Silver/silver chloride electrodes have electrochemical properties that make them well suited for application in SICM. Furthermore, they are easily fabricated, for example by electrolytic deposition of silver chloride on a layer of metallic silver. The silver/silver chloride (Ag/AgCl) electrode consists of a silver (Ag) wire that is coated with solid silver chloride (AgCl) and immersed in an electrolyte containing chloride ions (e.g., a solution of KCl or NaCl). In thermodynamic equilibrium, the solution is saturated with silver chloride:

$$\mathrm{AgCl(solid)} \leftrightharpoons \mathrm{Ag^+(aq)} + \mathrm{Cl^-(aq)} \tag{3.1}$$

The equilibrium constant for this reaction is $K = 1.76 \times 10^{-10}$ ($T = 25\,°\mathrm{C}$), so the concentration of Ag^+ ions in solution is very small. The Ag/AgCl electrode is described by a reversible redox reaction, in which the chloride atoms in the solid silver chloride receive an electron and go into solution as chloride ions, leaving metallic silver, and vice versa:

$$\mathrm{AgCl(solid)} + \mathrm{e^-} \leftrightharpoons \mathrm{Ag(solid)} + \mathrm{Cl^-(aq)} \tag{3.2}$$

This reaction occurs close to the electrode surface (< 1 nm distance). The potential that will build up at the electrode in thermodynamic equilibrium is given by the Nernst equation. The Nernst equation relates the electrode potential, E, to the concentration (more precisely: the activity) of the chloride ions in solution [23].

$$\begin{aligned} E &= E_0 + \frac{RT}{F} \ln \frac{1}{[\mathrm{Cl^-}]} \\ &\cong +0.222\,\mathrm{V} - 0.0591\,\mathrm{V} \times \lg[\mathrm{Cl^-}] \end{aligned} \tag{3.3}$$

E_0 is the standard potential of the Ag/AgCl electrode (with respect to the Standard Hydrogen Electrode, at $T = 25\,°\mathrm{C}$), $R \cong 8.315\,\mathrm{J\,mol^{-1}\,K^{-1}}$ is the molar gas constant, T is the absolute temperature (here: $T = 298\,\mathrm{K}$), $F \cong 96{,}485\,\mathrm{C\,mol^{-1}}$ is the Faraday constant, and the square brackets denote the concentration.

In SICM, two Ag/AgCl electrodes are used as anode and cathode, respectively. This setup is simpler than that for voltammetric experiments, where three electrodes (working electrode, auxiliary electrode and reference electrode) are used and controlled by a potentiostat. For the SICM setup, simply an external voltage is applied between the two electrodes inducing a faradaic ionic current in the electrolyte which is measured with a low-current amplifier.

3.3 Ion Currents Through Nanopipettes

3.3.1 Background Theory

The measured current, I, that passes through the pipette depends on the effective ohmic resistance in the current path of the electrolyte, R_{eff}:

$$I = \frac{U}{R_{\mathrm{eff}}} . \tag{3.4}$$

U is the applied voltage. The resistance of the electrolyte is mainly determined by the narrow regions it passes through. Considering the equivalent circuit of Fig. 3.2b, the effective resistance becomes

$$R_{\mathrm{eff}} = R_{\mathrm{P}} + \frac{R_{\mathrm{C}} R_{\mathrm{L}}}{R_{\mathrm{C}} + R_{\mathrm{L}}} . \tag{3.5}$$

We are often interested in the channel resistance R_{C}. However, the channel resistance depends on the particular configuration in which it is measured. For configurations with simple boundary conditions, the channel resistance can easily be calculated. For example, for a cylindrical channel of length L and radius r that is capped by a (perfectly conducting) electrode at each opening, the ohmic channel resistance is given by

$$R_{\mathrm{C}}^{\mathrm{capped}} = \frac{L\rho}{\pi r^2} , \tag{3.6}$$

where ρ is the specific electrical resistivity of the electrolyte. Such an ideal configuration, however, never occurs in SICM. A more realistic configuration is when the electrodes are far removed from either end of the conductive channel. The resistance then increases, because the current additionally traverses the distance from the bulk medium to the disk-like mouths of the channel. This gives rise to an "access" resistance $R_{\mathrm{acc}}^{\infty}$ that acts in series with the channel resistance [18]:

$$R_{\mathrm{C}}^{\infty} \approx R_{\mathrm{C}}^{\mathrm{capped}} + R_{\mathrm{acc}}^{\infty} = \left(L + \frac{\pi r}{2}\right) \frac{\rho}{\pi r^2} . \tag{3.7}$$

But also this configuration does not precisely apply to SICM, mainly because one electrode is always confined to the interior of the pipette. The precise meaning of R_{P}, R_{C} and R_{L} in (3.5) therefore depends on the particular boundary conditions (for example, the tip-sample distance). Nevertheless, (3.5) is useful in many cases, since it provides a simple allocation of the resistances in the current path.

3.3.2 Simple Analytical Model

At the heart of SICM lies the dependence of the measured current on the tip-sample distance that comes about in so-called current-distance curves. Far away from the sample surface, the current through the pipette is determined by the pipette resistance R_P. Speaking in terms of (3.5), the leakage resistance is zero. In closer proximity to the sample surface, the leakage resistance increases due to the current squeezing effect. Therefore, the effective resistance increases and the measured current drops. In the ideal case of a perfect seal between pipette and surface at zero tip-sample distance, the current becomes zero at contact.

In the case of a conically tapered pipette tip with an opening angle that is not too large, the pipette resistance can be approximated by [15]

$$R_P \approx \frac{\rho L_P}{\pi r_0 r_i}\,, \tag{3.8}$$

where r_0 is the inner diameter of the thick end, r_i is the inner diameter of the thin end (at the pipette tip), and L_P is the length of the conical pipette (Fig. 3.4, inset). This formula can be derived by sectioning the cone into successive disks of infinitesimal thickness, and summing the resistances of all disks (this assumes that disk-shaped electrodes are placed in the openings of the thin and the thick end). Nitz et al. [15] constructed a simple analytical model for the distance-dependent leakage resistance R_L (in the absence of channels in the sample, i.e. $R_C \to \infty$). They obtained

$$R_L(z) \approx \frac{3\rho \ln(r_a/r_i)}{2\pi z}\,, \tag{3.9}$$

where r_a is the outer diameter of the thin end (pipette tip) and z is the tip-sample distance. The measured current thus becomes

$$I(z) \approx I_0 \left(1 + \frac{z_0}{z}\right)^{-1}\,, \tag{3.10}$$

where $I_0 = U/R_P$ is the measured current in the absence of the sample (or with a large tip-sample distance) and

$$z_0 = \frac{3 r_0 r_i}{2 L_P} \ln \frac{r_a}{r_i}\,. \tag{3.11}$$

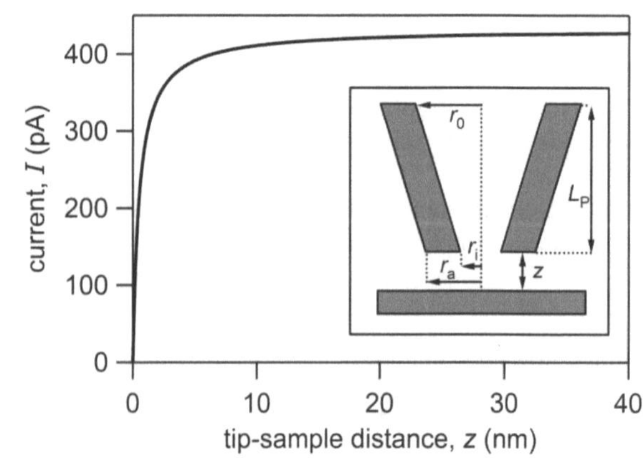

Fig. 3.4. Calculated current vs. distance (IZ) curve, based on a simple analytical model [*inset* and (3.10)]. The parameters used for calculating the curve are $r_0 = 0.3\,\mathrm{mm}$, $r_i = 50\,\mathrm{nm}$, $r_a = 85\,\mathrm{nm}$, $L_P = 10\,\mathrm{mm}$, $\rho = 1.09\,\Omega\mathrm{m}$ and $U = 100\,\mathrm{mV}$

Equation (3.10) describes a current that drops to zero for $z = 0$ and that asymptotically approaches I_0 for $z \to \infty$. At $z = z_0$, the current is half of that asymptotic value. For typical parameters ($r_0 = 0.3\,\mathrm{mm}$, $r_\mathrm{i} = 50\,\mathrm{nm}$, $r_\mathrm{a} = 85\,\mathrm{nm}$, $L_\mathrm{P} = 10\,\mathrm{mm}$, $\rho = 1.09\,\Omega\,\mathrm{m}$, $U = 100\,\mathrm{mV}$), we obtain $I_0 = 432\,\mathrm{pA}$ and $z_0 = 0.52\,\mathrm{nm}$. It becomes apparent that the current starts to significantly drop only at very small tip-sample distances (Fig. 3.4).

3.3.3 Finite Element Modeling

In order to further characterize and understand current flow when the pipette tip is close to a sample, we modeled the system using finite elements [24]. Due to the numerical nature of this modeling, more complex configurations can be analyzed. For example, it becomes straightforward to include a channel in the sample. We chose a conical pipette with an inner and outer tip opening diameter of 100 nm and 200 nm, respectively, and with an inner opening half-angle of 6°. We placed the pipette electrode in the plane 1000 nm into the pipette, and the bath electrode in the plane 320 nm below the tip opening. A cylindrical sample disk with a diameter and thickness of 700 nm and 30 nm, respectively, was placed concentrically below the pipette. Optionally, a thin concentric channel of 40 nm diameter was inserted in this sample disk. The electrolyte was assumed to be a 0.1 M NaCl solution, with a specific electrical resistivity of $\rho = 1.09\,\Omega\,\mathrm{m}$ [25]. The pipette walls and the sample were assumed electrical insulators. The governing differential equation of this system is the Poisson equation for the scalar electrical potential ϕ,

$$\Delta\phi = 0\,, \tag{3.12}$$

where Δ is the Laplace operator. The problem was then solved self-consistently with finite element modeling. Dirichlet boundary conditions apply at the pipette and bath electrodes, between which the voltage of 0.1 V was applied. Homogenous Neumann boundary conditions apply at the pipette walls and the sample, as well as at the system boundaries. The dimensionality of the system in the modeling process was reduced from three to two spatial dimensions by utilizing the cylindrical symmetry. The current density vector $\boldsymbol{j}$ directly derives from the electric field vector $\boldsymbol{E}$, which in turn directly derives from the scalar electric potential:

$$\boldsymbol{j} = \frac{1}{\rho}\boldsymbol{E} = -\frac{1}{\rho}\nabla\phi\,. \tag{3.13}$$

The current flowing through the pipette was then simply obtained by integrating the current density over an area A spanning fully across the pipette interior, where only components of the current density that are normal to the local area element were considered:

$$I = \int_A \boldsymbol{j}\cdot \mathbf{d}A\,. \tag{3.14}$$

The dot denotes the scalar product. We modeled three different experimental configurations. First, we removed the sample from the system, leaving only the pipette in the bath solution. The results of the finite element modeling process for this configuration are displayed in Fig. 3.5a. The local current density vectors in the axial symmetry plane are symbolized as arrows. Additionally, the absolute value of the current density is encoded in the grayscale, where darker areas denote larger current densities. It is apparent that the current density inside the pipette close to its tip is large ($\cong 1 - 2 \times 10^5\,\mathrm{A/m^2}$), and that it drops to almost zero soon after leaving the pipette opening (within a distance of about half of the tip opening diameter). The total current flowing through the pipette is $\cong 1.41$ nA (corresponding to $R_{\mathrm{eff}} \cong 70.9\,\mathrm{M\Omega}$). In Fig. 3.5b, we consider a different configuration by positioning an insulating sample surface 20 nm away from the pipette tip. This case demonstrates the current squeezing effect, as the current now has to escape sideways through the narrow gap between pipette tip and sample surface. But this relatively small tip-sample distance (20% of the tip diameter) does not lower the total current through the pipette ($\cong 1.29$ nA, corresponding to $R_{\mathrm{eff}} \cong 77.5\,\mathrm{M\Omega}$) by a large amount, compared to the case in Fig. 3.5a. This behavior will be discussed more below. In Fig. 3.5c, we add a small concentric channel with a diameter of 40 nm to the sample. Some of the current can now escape through this channel, thereby decreasing the effective resistance of the system. The total current through the pipette increases again slightly to $\cong 1.33$ nA (corresponding to $R_{\mathrm{eff}} \cong 75.2\,\mathrm{M\Omega}$).

It becomes apparent that even at these relatively small tip-sample distances, the influence of the sample, with or without channels, is relatively small. For larger effects to occur, the pipette tip needs to be positioned even closer to the sample, as we will see in the following paragraph.

We repeated these modeling steps for a series of tip-sample distances, and plotted the total current as a function of tip-sample distance (Fig. 3.6). In the absence of the conductive channel through the sample (dashed line), the current starts to drop significantly at tip-sample distances below 20 nm (about half of the tip radius). At zero distance, the current is zero. In the presence of the conductive channel (solid

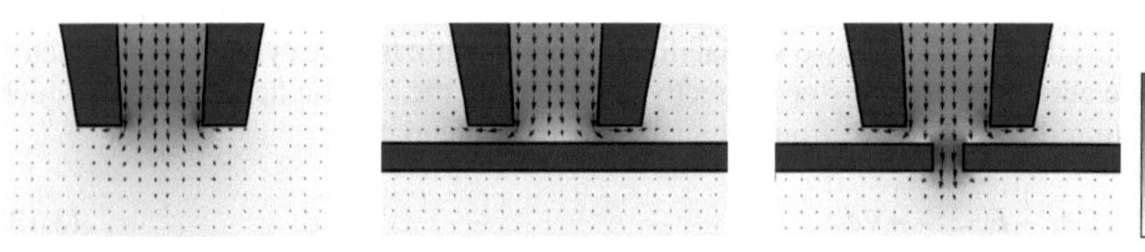

Fig. 3.5. Finite element analysis of the current squeezing effect at the tip of the pipette. (**a**) Pipette tip in the absence of a sample. The current density is large inside the pipette close to its end, and drops to almost zero soon below the pipette tip. (**b**) Pipette tip in the presence of a solid insulating sample. This added constraint causes the current to "leak" through the gap between tip and sample. (**c**) Pipette tip in the presence of a sample with a concentric channel below the tip. The current takes two alternative paths: it "leaks" through the gap between the tip and sample, and it flows through the channel in the sample. With a decreasing tip-sample distance, an increasing amount of current flows through the channel in the sample. For all three images, the *arrows* symbolize local current density vectors and the *grayscale* quantizes the absolute value of the current density. The parameters of the model are listed in the main text. *Grayscale range* (*white* to *black*): $0 - 2 \times 10^5\,\mathrm{A/m^2}$

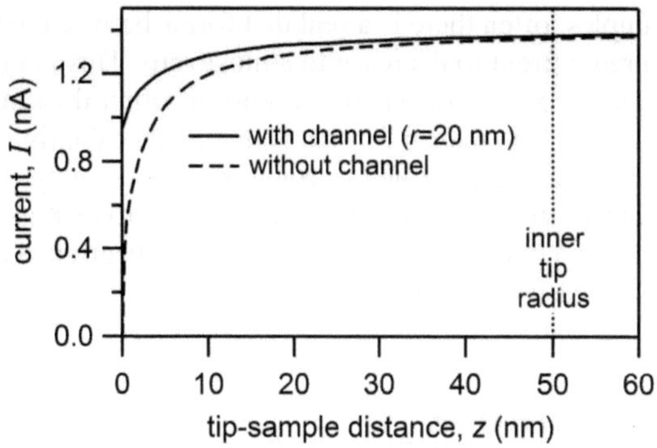

Fig. 3.6. Calculated current vs. distance (IZ) curve, based on a finite element analysis. Two cases are considered: a sample without a central conductive channel (*dashed line*) and a sample with a central conductive channel (*solid line*). In both cases, the current starts to drop significantly only at tip-sample distances below half of the pipette tip radius. In the presence of the channel, the current reaches a finite value at zero distance, while it drops to zero in the absence of the channel

line), the current behaves similarly, but does not drop below a minimum value (here: ≈ 0.95 nA). In this case, the leakage current is zero and all the current flows through the channel in the sample.

3.3.4 Experimental Current-Distance Curves

Current-distance curves manifest the basic principles of SICM, and provide useful information about the tip-sample interaction. In experimental current-distance curves, the pipette tip is ramped towards the sample surface with the help of a piezoelectric actuator (in z-direction). Simultaneously, the pipette current is recorded. A typical curve is shown in Fig. 3.7. When the pipette tip is far away from the sample surface, the current is about constant and independent of the z-piezo position (right hand side of graph). When the tip-sample distance gets in the range of the opening radius of the pipette tip, the cross section of the current path will decrease due to the thin gap forming. The resistance therefore increases and the measured current decreases. In

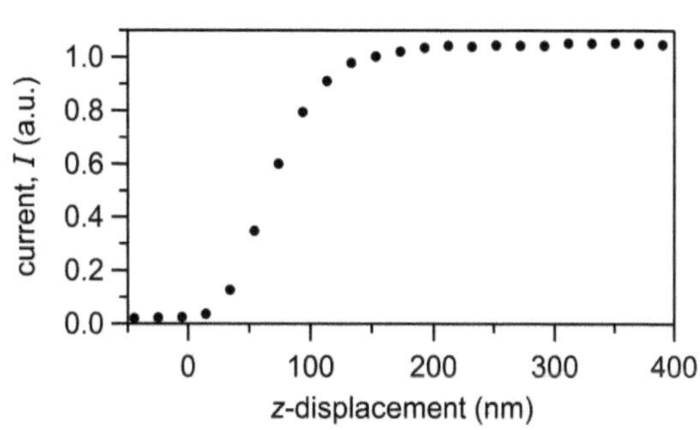

Fig. 3.7. Measured current vs. distance (IZ) curve on a glass surface. The current significantly decreases from a constant value to almost zero when approaching the surface. The inner tip radius of the pipette was 150 nm [15]

the case of soft samples, often there is a seal that forms between the tip and sample, causing the measured current to decrease to almost zero. This is in contrast to hard samples such as glass coverslips, where the current often does not reach zero at any tip-sample distance. The reason for this is that neither the tip nor the sample is deformable to even out a slightly tilted or rough pipette opening, so a gap of finite size will be maintained. In many cases, the current starts to increase again when the pipette is further lowered into the sample. This effect has been attributed to a bending of the pipette that increases the leakage current [15].

3.3.5 Imaging with Ion Current Feedback

The dependence of the measured ion current on the tip-sample distance can be utilized to control this distance. This allows scanning the tip laterally over a sample at a preset vertical distance. Generally, scanning is achieved with the help of piezoelectric actuators that move the tip relative to the sample in each of the three spatial dimensions. A feedback loop keeps the tip-sample distance constant by controlling the vertical piezoelectric actuator so that the measured ion current stays constant. Simultaneous scanning in lateral direction then generates a topographic image of the sample surface (Fig. 3.8). One characteristic of this ion current feedback is that topographic data is obtained without mechanical contact between tip and sample. Gentle imaging of very soft samples such as living cells is therefore possible. Korchev et al., for example, imaged the topography of living cells with SICM to monitor dynamic changes in cell volume at a resolution of 2.5×10^{-20} liter [26] and to localize single active ion channels on the cell surface [27]. The lateral resolution of SICM topography images, however, is typically limited to several dozen nanometers, set by the inner opening diameter of the pipette tip. In the next section, we will discuss improvements of the current feedback technique that allow an even better and more reliable tracking of the sample surface.

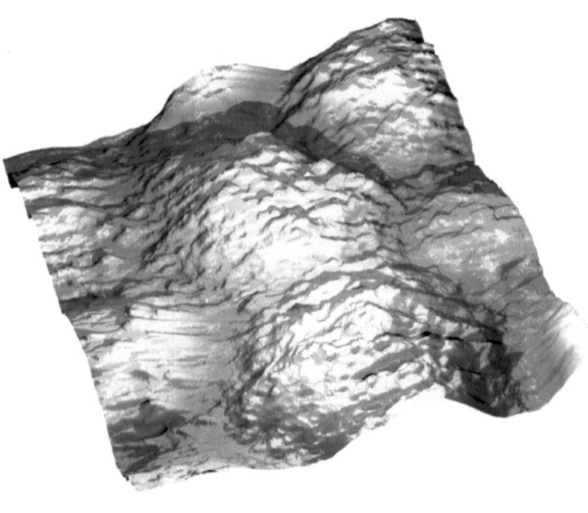

Fig. 3.8. Topography of a monolayer of MDCK cells on a glass support, imaged with SICM in ion current feedback. Vertical range: 8.2 μm. Image size: $26.4 \times 26.4\ \mu m^2$. Experimental data by P. Heidenreich, University of Münster

3.4 Advanced Techniques

The original SICM technique utilized the distance-dependence of the ion current to control the tip-sample distance with the help of a feedback loop. Provided that no local or temporal variations in the conductivity of the electrolyte occur, the feedback-controlled scanner will keep the distance between the probe and the sample constant while scanning the surface. Although this approach has successfully been applied to image the topography even of delicate biological samples, some inherent issues and weak points remain. Some of the problems of the original SICM setup will be discussed in the following paragraph.

Since the SICM relies on a measurement of the dc ion current flowing between the pipette and the bath electrode, its performance strongly depends on the long-time stability of the overall electrochemical conditions. Various factors can affect in particular the dc voltage and current between the electrodes. For instance, temperature drifts as they can easily occur during scanning for hours can cause significant dc offsets, which in turn have a direct effect on the current-based feedback system. Furthermore, drifts due to a gradual contamination of the electrodes over time can also degrade the performance of the scanning system, ultimately making it impossible to track the sample topography for a long time. If the SICM is to be combined with other electrophysiological investigations further problems may arise. For instance the application of drugs or additional voltages or currents can also have significant effects on the electrochemical potential between the SICM electrodes. Using the dc ion current as a feedback signal might therefore be impractical.

Another inherent weak point of a current-based feedback system results from the geometric shape of the pipette's tapered end. In order to obtain a significant current squeezing effect when the probe is brought into close proximity to the sample surface, the tapered end should be even and flat. Such a leveled probe typically shows a strong dependency of the ion current on the vertical tip-sample distance. At the same time, however, this geometry makes the probe practically insensitive to protrusions on the sample surface if they are approached from the side during lateral scanning. As a consequence, the probe runs into these protrusions and then bends due to the lateral movement of the scanner. Such a bending will result in a small tilt of the opening of the pipette with respect to the sample surface, leading to a reduction of the squeezing effect and thereby to an increase of the ion current. The feedback loop of the SICM will then force the probe even further toward the sample, ultimately causing a tip crash and damaging both sample and probe. For the same geometric reasons the probe can easily get trapped in pits or pores that might be present on the sample surface. In general, rough surface morphologies are therefore difficult to scan in a stable manner with a simple current-based feedback system.

As a first remedy the average distance between the probe and the sample could be kept rather large by simply adjusting the current setpoint of the feedback loop accordingly. Such a workaround, however, would degrade the sensitivity and the resolution of the SICM. Therefore, advanced techniques are required to overcome the inherent technical limitations of a simple current-based feedback system. A number of different approaches have been conceived to achieve this goal. They can be subdivided into two classes: SICM systems of the first class still rely on using the ion current as feedback signal for distance control, but apply special modulation techniques to minimize effects of the afore mentioned issues. We will discuss those techniques in the following paragraphs. In the second class of advanced SICM systems, the original concept of a current-based feedback system is abandoned. Instead, the SICM is provided with a current-independent feedback system to control the tip-sample distance. The local ion current is no longer used for distance control but recorded simultaneously with the surface topography as it is measured by a separate feedback system. Typical implementations are based on concepts well-known from other scanning probe techniques such as dynamic mode atomic force microscopy or shear force concepts. We will discuss them in detail in Sect. 3.5.

3.4.1 Modulation Methods

Different methods for improving the performance of an SICM while retaining the original idea of using the local ion current as feedback signal for controlling the tip-sample distance have been introduced. In particular, modulation techniques have proved to be effective for minimizing the inherent problems of a current-based feedback loop. They share the common idea that the risk of hitting a surface feature during lateral scanning can be effectively reduced by modulating the tip-sample distance. Furthermore, modulation techniques can be useful to overcome limitations of the original SICM concept regarding long-term stability as they are caused by dc offsets and drift effects.

Gitter et al. [28], Mann et al. [29] and Happel et al. [30] employ a type of point-spectroscopy technique as it is know from other scanning probe methods such as atomic force microscopy [31]. Here the probe is approached to and then retracted from the sample surface by typically several micrometers in each data point of the scanning area. Thus the lateral movement of the scanner can be performed while the probe is far away from the surface. Consequently, the risk of the probe laterally colliding with a protrusion on the surface or it getting trapped in a depression is reduced. However, this comes at the price of slow scanning speed, since the probe is not tracking the surface under continuous control of a feedback loop. Instead in each data point the probe is approached toward the sample until a predefined trigger point is reached, i.e. until a decrease of the ion current below a set threshold value signals that the probe is in close proximity to the surface. The z-position of this trigger point is stored and then the vertical movement is reversed and the probe retracted from the surface. This type of point-spectroscopy can practically be based on any type of trigger signal, for instance the "squeezing" effect of the dc ion current as it occurs typically a few tens of nanometers above the surface. While the point-spectroscopy approach provides an effective workaround for overcoming the inherently low lateral sensitivity of the probe, it does not address problems caused by time-varying dc offsets. To minimize such long term drift effects additional measures have to be taken.

One option is applying either short voltage or current pulses [29, 30] instead of applying a constant voltage between the pipette and bath electrodes. This reduces unwanted effects caused by slowly changing dc potentials at the electrodes. Another method to improve the performance of a SICM with a current-based feedback

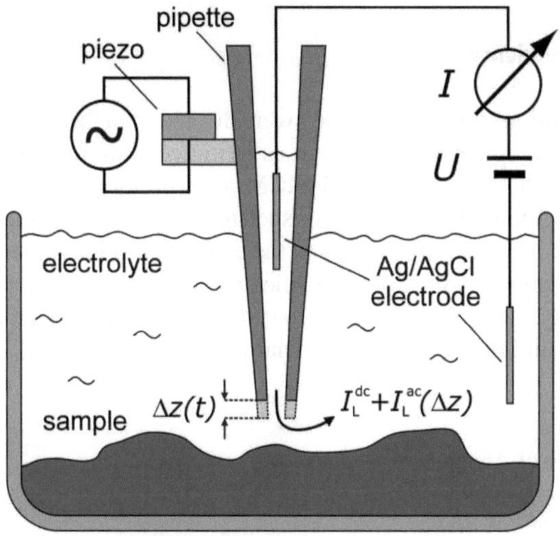

Fig. 3.9. Setup of a distance-modulated SICM. The tapered pipette is attached to a piezo-electric element which modulates the distance between the probe and the sample surface. Typically a modulation frequency in the range of a few hundred Hertz and a modulation amplitude of a few tens of nanometers is applied. If the tapered end of the probe is in close proximity to the surface the resulting leakage current I_L comprises a dc and an ac component. The amplitude of the latter depends on the average distance between the probe and the sample an serves as the input signal of a feedback loop

system has been proposed by the group of Shao [17, 21] and by the group of Korchev [16]. They keep the voltage between the two electrodes constant, but modulate the z-position of either the sample or the probe with an amplitude of a few tens of nanometers by means of a piezo-electric element (Fig. 3.9). If the probe is far away from the sample surface such a small distance modulation with a typical frequency of a few hundred Hertz does not significantly affect the ion current. If the tapered end of the pipette, however, is brought into close proximity to the surface so that the "squeezing effect" sets in, such a distance modulation leads to a modulation of the ion current. The amplitude of the resulting ac component in the ion current can then be detected with a lock-in amplifier. The latter allows recovering even noisy, low-level ion current signals by using the voltage signal which modulates the tip-sample distance as a reference signal. The detected amplitude of the modulated ion current is then used as feedback signal to control the average tip-sample distance, i.e. the feedback system tries to maintain a constant amplitude of the ion current while scanning the surface. This method of distance control has advantages over the conventional dc current-based approach because it makes the measurement less sensitive to changes in ionic strength or other dc drift effects.

3.4.2 Applications in Bioscience

Advanced modulation techniques as they were discussed in the previous section allow to scan delicate biological surfaces such as living cells more gently. The improved performance of a distance-modulated SICM allowed Gorelik et al. to obtain well resolved images of fine surface structures such as microvilli on living cells as depicted in Fig. 3.10 [32]. Since the introduction of the distance-modulation technique reduces the risk of uncontrolled contact between the probe and the sample and also improves long-term stability, it becomes possible to continuously image specific surface areas for several hours and thereby to study dynamic processes [33, 34].

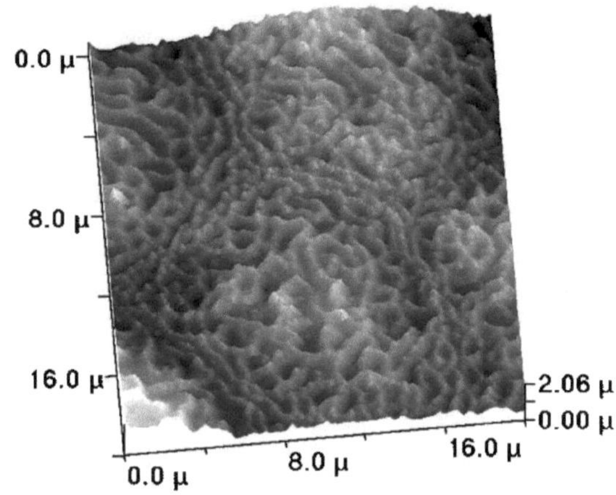

Fig. 3.10. Monolayer of A6 epithelial cells imaged by distance-modulated SICM. The fine structure on the cell surface is well resolved and shows numerous microvilli. Reprinted with permission from Gorelik et al. [32]

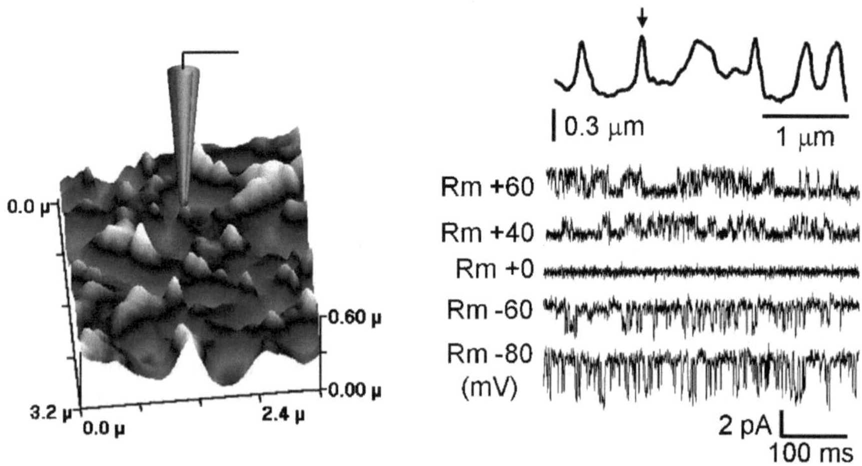

Fig. 3.11. SICM scan of an A6 cell surface (*left*) and ion channel recording on top of a single microvillus (*right*). The pipette's position is marked by an *arrow* in the line profile. The channel was identified as a K^+ channel on the basis of the reversal potential of the current-voltage curve. Reprinted with permission from Gorelik et al. [32]

Another type of application is to employ the SICM for performing patch-clamp experiments at specific surface sites [32, 34]. In such experiments the SICM is first used to scan the sample surface. The spatially resolved images of the surface topography allow identifying regions or structures of interest. The scan unit is then used to position the tapered SICM probe over a user-selected spot. Finally, the probe is engaged to the sample surface to form a gigaohm seal for subsequent patch-clamp recording. This combination of high-resolution SICM scanning and patch-clamping allows obtaining ion channel recordings on selective surface spots with lateral position control in the range of a few tens of nanometers. Such high precision is out of reach with conventional positioning methods such as those using light microscopy. With the SICM it becomes possible to investigate the activity of single ion-channels on nanometer-sized structures on native biological samples such as living cells. For instance, the tapered SICM pipette can be deliberately positioned on top of a single microvillus on an epithelial cell and used for subsequent patch-clamp recording (Fig. 3.11).

3.5 Combination with Other Scanning Techniques

The ability of the SICM to image biological samples in vivo with a resolution in the range of a few tens of nanometers makes it an interesting candidate for to combine with other scanning microscopy techniques. In doing so different strategies can be pursued: the first category of techniques sticks to the idea of using the ion current signal to maintain a constant distance between the probe and the sample while scanning. In addition a complementary data channel, for instance local optical data, is simultaneously recorded. Combinations of SICM with scanning nearfield optical microscopy (SNOM) or scanning confocal microscopy (SCM) fall in this category.

The groups of Korchev and Shao modified the original SICM setup in such a way that the end of the tapered pipette also serves as a near-field light source for SNOM [21, 35, 36]. This can be achieved by coupling laser light into the pipette via an optical fiber. Coating the outside of the pipette with a reflective metal layer helps to confine the laser light to the aperture, i.e. the tapered pipette's end. Provided that the sample and the substrate are transparent the SNOM signal can then be collected through an objective and detected by a photomultiplier located underneath the SICM head. Living cells were successfully investigated with such a combined SICM/SNOM setup by Korchev et al. An alternative way to utilize the SICM probe as a confined light source for SNOM was suggested by Bruckbauer et al. [37,38]. The method is based on fluorescence as it occurs when a calcium indicator, with which the pipette is filled, binds with calcium in the sample solution and is illuminated with a laser. The mixing zone where the fluorescent complex forms serves as a localized light source.

SICM has also been successfully combined with SCM [39]. The setup comprises an inverted light microscope fully configured for scanning confocal microscopy, on which the SICM head is placed. During lateral scanning the vertical position of the sample is controlled by a standard SICM, i.e. ion current based feedback loop. As a consequence the optical confocal volume, which is located just below the pipette's end, follows the topography of the sample. This allows capturing fluorescence images of a surface simultaneously with topographic data. Thereby interaction of fluorescent nanoparticles, for instance virus-like particles, with the surface of fixed or living cells can be studied.

An alternative strategy for combining SICM with other scanning microscopy techniques abandons the original idea of an ion current based feedback system. Instead, other distance control schemes are employed to keep the scanned probe at a fixed, predefined distance to the sample surface and then to record local variations in the ion current as they might be caused by variations in local ion-conductivity or sample-permeability for instance.

So far, we have considered only distance-control methods that are exclusively based on the measurement of the ion conductance. This works well for samples with a homogenous distribution of ion conductance. In many cases, however, it is of interest to study inhomogeneities in the local ion conductance, such as those that arise due to the presence of channels through a thin porous membrane, for example. In these cases, a mechanism independent from the ion conductance is required to keep the tip-sample distance constant. In the following sections we discuss different methods that solve this problem and present applications using these methods.

3.5.1 Combination with Atomic Force Microscopy

One possibility for a simultaneous measurement of ion current and sample topography is based on the use of a bent nanopipette as a cantilever [12]. With such a configuration, the sample topography can be imaged using a standard AFM measurement setup. Simultaneously, the ion current can be monitored and recorded as a complementary measurement signal. The invention of the tapping mode in liquid [40,41], which simplified imaging of soft samples in solution, initiated the development of a novel microscope: Tapping-mode AFM combined with a SICM [12]. In this design, a bent nanopipette [42,43] is used as both force sensor and ion conductance probe (Fig. 3.12). This bent nanopipette is vibrated perpendicular to the sample with the help of a piezo stack. Typical vibration amplitudes at the tip are ≈ 10 nm rms. The measured vibration amplitude of the pipette serves as input signal to a feedback loop that controls the pipette-sample distance, thereby generating the topography signal when scanning. Simultaneously, the ion current is recorded and used to generate a complementary image of the ion conductance. While such a microscope can also be operated in contact mode (using the dc deflection of the pipette), better resolution (both in topography and ion conductance) is generally obtained in tapping mode due to the absence of lateral imaging forces.

3.5.2 Application in Material Science

In this section, we will present an application of a tapping mode SICM in the field of biomineralization. This field deals with the processes by which living organisms synthesize organic-inorganic composite materials. These materials often exhibit elaborate micro- and nanostructures. They usually contain just a small

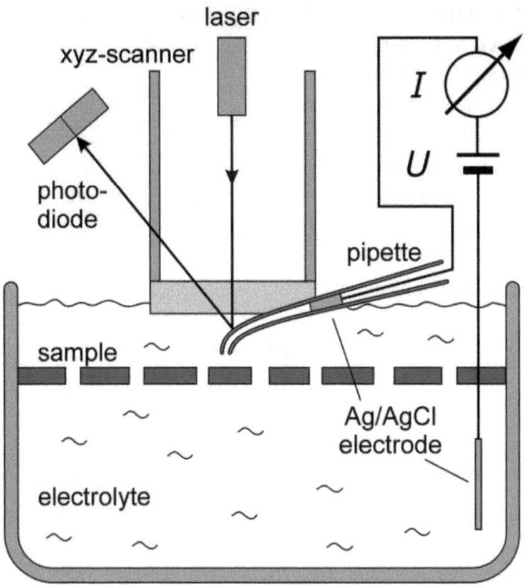

Fig. 3.12. Combined scanning ion conductance microscope and tapping-mode atomic force microscope. A bent nanopipette is used as both an ion conductance probe and an atomic force microscope cantilever. Tapping mode operation enables topographic imaging of soft samples. The tip-sample distance is thereby kept constant by mechanically vibrating the pipette and using feedback to keep the vibration amplitude constant. The vibration amplitude of the cantilever is detected with nanometer precision using the optical beam deflection method. The pipette is coated with a layer of gold (thickness ≈ 30 nm) to increase its reflectivity. Simultaneously to the topography, the ion current through the pipette is recorded, thereby generating complementary images of topography and local ion conductance

amount of biopolymers, but those biopolymers play a critical role in the assembly of the mineral. Furthermore, these organic-inorganic composite materials often exhibit extraordinary mechanical, electrical, magnetic and chemical properties. By improving the knowledge about the mechanisms of biomineralization, new models might be created that aid in the artificial synthesis of new, high-performance materials.

The example we will consider here is the formation of nacre ("mother of pearl") by the marine mollusc *Haliotis rufescens* (red abalone). There has been a great interest in this material, and significant progress in the elucidation of its structure and function has been made over the last years. Nacre is a micro-laminate composite of highly oriented calcium carbonate ($CaCO_3$) crystals in the aragonite polymorph and an organic matrix in the form of thin organic sheets, probably consisting of mostly proteins and polysaccharides. This composite material exhibits an intricate micro- and nanostructure of high regularity: it consists of an interdigitating brickwork array of crystal tablets and organic matrix. This composite structure is one of the reasons for the exceptionally large fracture toughness of nacre, which is ≈ 3000× larger than that of pure mineral.

Various models have been set up that explain the fabrication process of nacre. With the use of electron diffraction microscopy and atomic force microscopy, it has

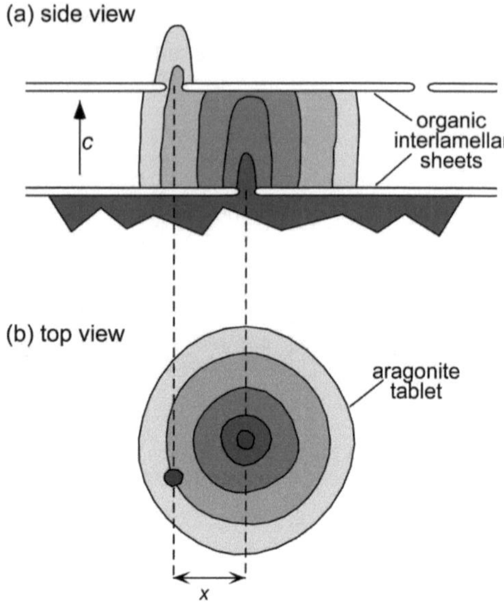

Fig. 3.13. Schematic of a simple model of nacre growth, in which aragonite crystal tablets grow through pores in organic interlamellar sheets. SICM was used to proof the presence of such pores in the organic sheets, and to image their distribution and conductivity. (**a**) Side view. (**b**) Top view. Reprinted with permission from Schäffer et al. [13]. Copyright (1997) American Chemical Society

been found that the tablets are single crystals, and that the crystallographic axes of these crystals are nearly identical within several layers of subsequent tablets. This observation has resulted in a prominent model that explains the formation of nacre by a process called "hetero-epitaxial nucleation". In this process, the growth of the crystal is periodically capped by an interlamellar organic layer, onto which a new crystal nucleates in a hetero-epitaxial fashion. Alternatively, abalone nacre may be formed by a continuous crystal growth of the mineral tablets through pores (channels) in the interlamellar organic sheets. Significant indirect and direct evidence has been favoring this alternate model that is based on growth through "mineral bridges" (Fig. 3.13).

SICM was one of the techniques that generated strong evidence for this alternate model. For that purpose, demineralized interlamellar organic sheets were prepared from abalone nacre by decalcification. These organic sheets were imaged with a SICM in tapping mode operation without drying, fixing or coating the sample. The tapping mode operation ensured that a constant tip-sample distance was maintained. There are areas of high and low ion conductance on the organic sheet (Fig. 3.14a). In some areas with a predominantly low ion conductance, there are small features of locally enhanced ion conductance (Fig. 3.14b). With the help of a statistical analysis, these features were attributed to individual pores in the organic sheet. Several observations support this conclusion. First, the average ion conductance of these features is 28 ± 11 nS (after subtracting a conductivity background of 50 nS that is caused by leakage). From this quantity, the diameter of the pores in the organic sheet with a thickness of ≈ 30 nm calculates to 15–40 nm, which is in agreement with observations using atomic force or scanning electron microscopy. Second, an estimate of the pore density in the organic sheet can be found from the integrated conductance over the scan area. This estimate, 58 ± 12 pores/μm^2, is in the same range as the pore density seen in AFM images. Third, calcified features

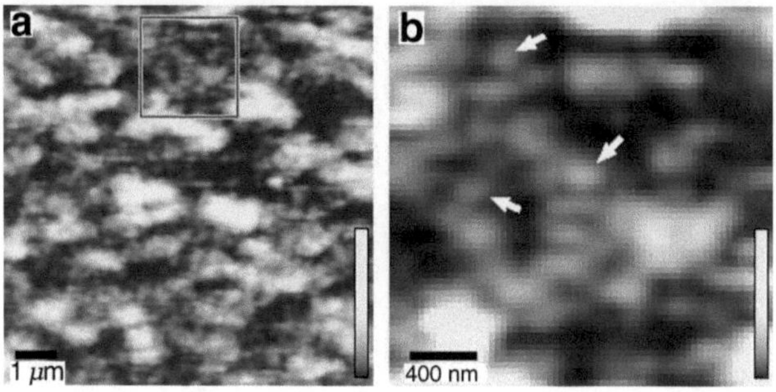

Fig. 3.14. SICM images of demineralized interlamellar organic sheets of abalone nacre. (**a**) Regions of small (*dark*) and large (*bright*) ion conductance can be seen, proving that the sheets are permeable for ions. Scalebar: 1 µm. *Grayscale range* (*black* to *white*): 50–200 nS. (**b**) Expansion of the box in (**a**). Some areas with predominantly low ion conductance show small, distinct features that might represent individual pores (*arrows*). Scalebar: 400 nm. *Grayscale range* (*black* to *white*): 50–200 nS. Reprinted with permission from Schäffer et al. [13]. Copyright (1997) American Chemical Society

that could be mineral bridges were observed using transmission electron microscopy (TEM).

The realization that there are pores in the interlamellar organic sheets has several important consequences. On one hand, these pores can transport ions across those sheets, and possibly even soluble proteins that are necessary for growth control of the crystal tablets. But most importantly, there is the possibility that the crystal itself grows through those pores, thereby forming mineral bridges through the interlamellar organic sheets. This alternate model of nacre growth readily explains several experimental observations, among those the mentioned coherence of the crystallographic axes between consecutive tablets.

3.5.3 Combination with Shear Force Microscopy

Another possibility for a simultaneous measurement of ion current and sample topography lies in the use of the nanopipette as shear force sensor. Shear force sensors are well known from SNOM setups, where they serve to keep an optical fiber at constant distance to the sample during scanning [44,45]. In a shear force configuration, the nanopipette is vertically oriented with respect to the sample surface (Fig. 3.15). A small piezo-electric stack that is attached to the nanopipette excites transverse mechanical vibrations in the drawn-out end of the nanopipette (parallel to the sample surface). The vibration amplitude strongly depends on the tip-sample distance due to shear forces. The shear forces increase sharply at small tip-sample distances, thereby decreasing the vibration amplitude. The vibration amplitude is therefore well suited as a measure for the sample topography. The technical challenge is finding a method of detecting this vibration amplitude on the nanometer scale. Several methods have been established so far, including optical detection [44] and detection using a piezo-electric tuning fork sensor [46]. Piezo-electric detection

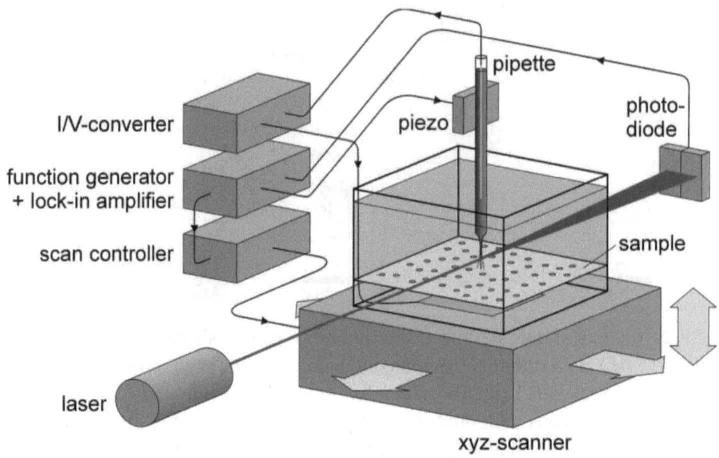

Fig. 3.15. Schematic illustration of the setup of a combined scanning ion conductance and shear force microscope. The lateral oscillatory motion of the probe's tapered end is induced by a piezo element to which the micropipette is attached. By focusing a laser beam on the tapered probe and detecting the diffracted light with a segmented photodiode the mechanical oscillation is converted into an electrical signal. This is analyzed by means of a lock-in amplifier whose reference channel is synchronized with the driving signal of the dither piezo. The amplitude of the probe's oscillation serves as the input signal of a feedback loop. The latter controls the vertical position of the scanner in such a way that a constant, predefined oscillation amplitude is maintained during *xy*-scanning. As a consequence the SICM probe follows precisely the sample topography. Simultaneously the local ion current flowing between the electrodes in the micropipette and under the porous sample is recorded. Thus possible variations in the local ion conductivity can be analyzed along with the surface topography

methods face some principal problems when they are used for imaging in liquids, since electrical shortcuts may occur in the electrolyte medium. Solutions include coating the tuning fork with an electrical insulating layer [47], applying a custom piezo-electrical detection [48], or using a diving bell concept, where a water-air interface is forced close to the sample surface [49]. On the other hand, optical detection of shear forces principally works in liquid just as well as in air [50]. In the optical detection method, a laser beam is focused onto the vibrating section of the nanopipette near its tip (Fig. 3.15). The incident laser beam is scattered by the vibrating tip, thereby modulating the intensity distribution of the beam. An optoelectronic detector, usually based on a split photodiode, detects these intensity modulations of the scattered beam and generates a signal that is proportional to the vibration amplitude of the nanopipette. This signal is then fed to the scan controller that attempts to keep the signal constant with the help of a feedback loop, thereby generating a topographic image of the sample surface when scanning. Simultaneously, the ion current from an electrode below the sample to the electrode in the nanopipette is recorded and used to generate a complementary image of the ion conductance through the sample.

One advantage of the shear force method over the tapping-mode method (Sect. 3.5.1) for measuring the tip-sample distance is that the pipettes need not go

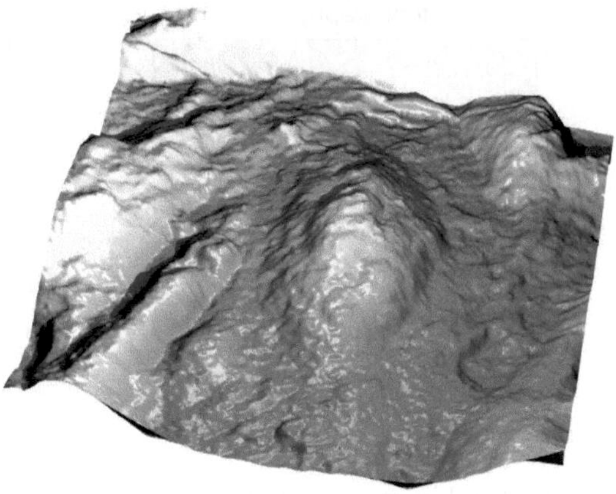

Fig. 3.16. Shear force topography image of a layer of fixed MDCK cells in buffer solution. Vertical range: 7.4 μm. Image size: 40 × 40 μm^2. Sample by S. Schrot, experimental data by M. Böcker, University of Münster

through the tricky process of being mechanically bent at their tip. One could think that the relatively large effective vertical spring constant of a vertically oriented pipette could impose a possible thread to soft samples. On the other hand, the non-contact nature of shear force imaging allows scanning of surfaces with small tip-sample interaction forces. It is possible to image soft samples as delicate as cells in liquid with a shear force setup (Fig. 3.16). The successful implementation of a combined scanning ion conductance and shear force microscope was demonstrated by Nitz et al. [15]. Using such a setup (Fig. 3.15), they acquired simultaneous images of topography and ion current on a polycarbonate membrane (Fig. 3.17). The locations of the pores can clearly be seen in the topography image as local depressions. The complementary ion current image shows an increased ion current at those locations. As further test samples, these researchers also used porous mica surfaces. Porous mica surfaces with defined pore sizes can be produced by a track-etching process [51]. In this process, a mica sample is irradiated with high energy ions, thereby inducing

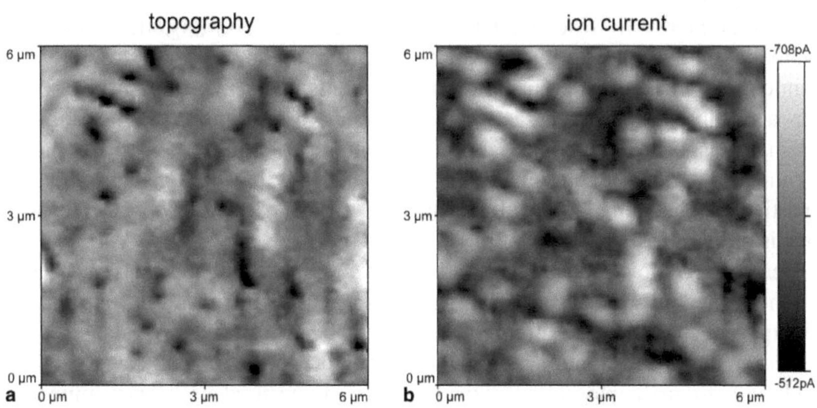

Fig. 3.17. Complementary (**a**) shear force topography and (**b**) ion current images of a polycarbonate membrane with pores of 0.4 μm diameter. Image size: 6 × 6 μm^2. Reprinted with permission from Nitz et al. [15]

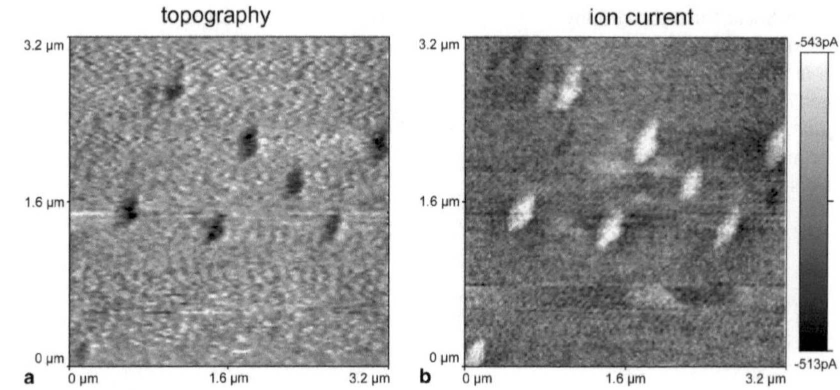

Fig. 3.18. Complementary (**a**) shear force topography and (**b**) ion current images of a Muscovite mica surface in which rhombic pores with an opening area of $200 \times 400\ \text{nm}^2$ were inserted by ion radiation and subsequent etching with hydrofluoric acid. Image size: $3.2 \times 3.2\ \mu\text{m}^2$. Reprinted with permission from Nitz et al. [15]

metastable states in the crystal lattice within a radius of 100 to 1000 nm from the ion trace. Subsequent etching with hydrofluoric acid produces rhombic pores with a diameter that can be influenced by the acid temperature and by the etching time. Simultaneous images of topography and ion current on such a porous mica surface show an increased ion current over the pores in the sample (Fig. 3.18), again demonstrating the feasibility of the shear force approach.

3.5.4 Application in Bioscience

The combined scanning ion conductance and shear force microscope is an ideal tool for investigating local variations in the ion conductance of biological specimen. This is of special interest for research in the field of the barrier-forming structures such as endothelial or epithelial cell layers. In multicellular organisms such structures form the interface between different fluid compartments and play an important role for inter- and transcellular processes [52, 53]. Gaining insight into the complex barrier-crossing transport mechanisms is a common interest of cell biology, medicine, and pharmacology, as malfunctioning of these barriers leads to pathological implications. In particular, knowledge about the permeability of barriers for substances such as drugs is of great relevance.

Hence special electrochemical and microscopic methods are required to study the ion-permeability of barrier-forming cell structures. For instance, experimental techniques such as the measurement of transepithelial electrical resistances (TER) provide valuable information about the barrier properties of cell layers [54, 55]. In addition to such integrating measurements of the total cell layer impedance, the combined scanning ion conductance and shear force microscope can provide further insight into cellular transport mechanisms, because it allows the simultaneous recording of topographic data and local ion-conductance with high lateral resolution.

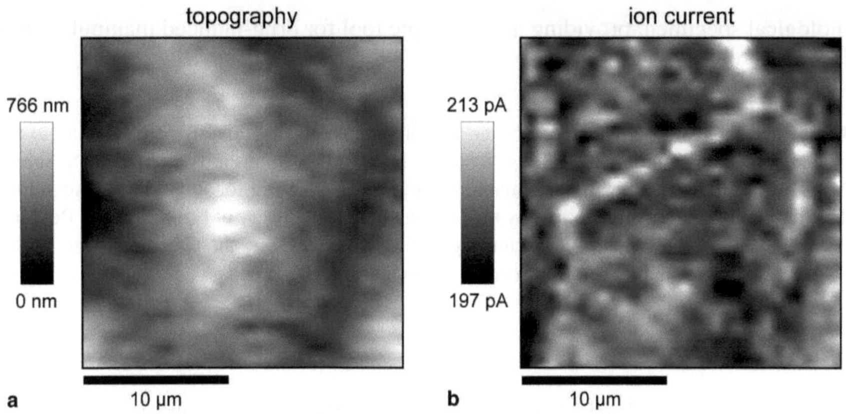

Fig. 3.19. Complementary (**a**) shear-force topography and (**b**) ion current images of a layer of live MDCK-II cells. Some but not all of the cell-cell contacts are visible in topography. The simultaneously recorded current image shows an increased ion-permeability over tight-junctions, independent of the local surface corrugations. Image size: $22 \times 22\,\mu m^2$. Sample provided by J. Seebach, experimental data by J. Kamp, University of Münster [66]

As depicted in Fig. 3.19, the shear-force-combined SICM has successfully been used to investigate intercellular structures, namely tight-junctions between living cells.

The results show the benefits of having two independent information channels: while the topographic shear-force data alone does not permit to differentiate between a permeable ion-channel and a closed cavity in the cell layer, the ion current image provides this important piece of information. The possibility of this type of combined microscope to recognize ionic transport channels in hard as well as soft samples opens perspectives for further applications in fields as diverse as biochemistry/pharmacology, corrosion research, quality control of coatings or artificial membranes.

3.6 Outlook

While the SICM itself already has a broad range of possible applications, further techniques which are based on similar components can be envisioned. For instance the tapered micropipette of the SICM can be used not only as a high-resolution scanning or patch-clamping probe for in vivo characterization of biological surfaces, but also as a source for localized material delivery. The material flux from the probe to the sample substrate can either be driven by electric fields, capillary forces, hydrostatic pressure, or by ultrasonic excitation. Such type of nanolithography has been successfully applied to deposit metals [56–59], chemicals [60, 61], and also biological material such as DNA, proteins or other biomolecules [62–65]. A possible field of application is the construction of nanostructured biosensors. Furthermore, having the ability to deliver material in a controlled manner through the micropipette of a SICM opens the perspective to study the effects of local drug delivery on biological specimen, providing a nanoscopic tool for drug-induced manipulation of individual cells.

Acknowledgements. We thank M. Böcker, P. Heidenreich and J. Kamp for contributing to this chapter with their experimental results. We thank H.-J. Galla, A. Janshoff, M. Madeja, S. Schrot, M. Schwarz, J. Seebach, E.-J. Speckmann, C. Steinem, J. Wegener for support and discussions. This research was supported in part by the German Federal Ministry of Education and Research (BMBF), and by the Deutsche Forschungsgemeinschaft (DFG) by the Gemeinnützige Hertie-Stiftung/Stifterverband für die Deutsche Wissenschaft (SCHA 1264).

References

1. Binnig G, Rohrer H (1982) Surface studies by scanning tunneling microscopy. Phys Rev Lett 49:57–61
2. Pohl DW, Denk W, Lanz M (1984) Optical stethoscopy: Image recording with resolution $\lambda/20$. Appl Phys Lett 44: 651–653
3. Binnig G, Quate CF, Gerber C (1986) Atomic force microscope. Phys Rev Lett 56:930–933
4. Drake B, Prater CB, Weisenhorn AL, Gould SA, Albrecht TR, Quate CF, Cannell DS, Hansma HG, Hansma PK (1989) Imaging crystals, polymers, and processes in water with the atomic force microscope. Science 243:1586–1589
5. Martin Y, Wickramasinge HK (1987) Magnetic imaging by "force microscopy" with 1000 Å resolution. Appl Phys Lett 50:1455–1457
6. Sáenz JJ, García N, Grütter P, Meyer E, Heinzelmann H, Wiesendanger R, Rosenthaler L, Hidber HR, Güntherodt HJ (1987) Observation of magnetic forces by the atomic force microscope. J Appl Phys 62:4293–4295
7. Martin Y, Abraham DW, Wickramasinghe HK (1988) High-resolution capacitance measurement and potentiometry by force microscopy. Appl Phys Lett 52:1103–1105
8. Bard AJ, Fan F-RF, Kwak J, Lev O (1989) Scanning electrochemical microscopy. Introduction and principles. Anal Chem 61:132–138
9. Hansma PK, Drake B, Marti O, Gould SAC, Prater CB (1989) The scanning ion-conductance microscope. Science 243:641–643
10. Prater CB, Drake B, Gould SAC, Hansma HG, Hansma PK (1990) Scanning ion-conductance microscope and atomic force microscope. Scanning 12:50–52
11. Olin H (1994) Design of a scanning probe microscope. Meas Sci Technol 5:976–984
12. Proksch R, Lal R, Hansma PK, Morse D, Stucky G (1996) Imaging the Internal and External Pore Structure of Membranes in Fluid–Tapping mode Scanning Ion Conductance Microscopy. Biophys J 71:2155–2157
13. Schäffer TE, IonescuZanetti C, Proksch R, Fritz M, Walters DA, Almqvist N, Zaremba CM, Belcher AM, Smith BL, Stucky GD, Morse DE, Hansma PK (1997) Does abalone nacre form by heteroepitaxial nucleation or by growth through mineral bridges? Chem Mater 9:1731–1740
14. Korchev YE, Bashford CL, Milovanovic M, Vodyanoy I, Lab MJ (1997) Scanning ion conductance microscopy of living cells. Biophys J 73:653–658
15. Nitz H, Kamp J, Fuchs H (1998) A combined scanning ion-conductance and shear-force microscope. Probe Microsc 1:187–200
16. Shevchuk AI, Gorelik J, Harding SE, Lab MJ, Klenerman D, Korchev YE (2001) Simultaneous Measurement of Ca^{2+} and Cellular Dynamics: Combined Scanning Ion Conductance and Optical Microscopy to Study Contracting Cardiac Myocytes. Biophysical journal 81:1759–1764

17. Pastré D, Iwamoto H, Liu J, Szabo G, Shao Z (2001) Characterization of AC mode scanning ion-conductance microscopy. Ultramicroscopy 90:13–19
18. Hille B (1992) Ionic Channels of Excitable Membranes, 2nd ed. Sinauer Associates, Sunderland, Mass.
19. Sakmann B, Neher E (1995) Single-Channel Recording, 2nd ed. Springer, Heidelberg
20. Prater CB, Hansma PK (1991) Improved scanning ion-conductance microscope using microfabricated probes. Rev Sci Instrum 62:2634–2638
21. Mannelquist A, Iwamoto H, Szabo G, Shao Z (2001) Near-field optical microscopy with a vibrating probe in aqueous solution. Appl Phys Lett 78:2076–2078
22. Bard AJ, Denuault G, Lee C, Mandler D, Wipf DO (1990) Scanning electrochemical microscopy: a new technique for the characterization and modification of surfaces. Acc Chem Res 23:357–363
23. Bockris JO, Reddy AKN (2000) Modern Electrochemistry: Electrodics in Chemistry, Engineering, Biology, and Environmental Science. Plenum Publishing Corporation, New York
24. QuickField SE. Tera Analysis Ltd., Svendborg, Denmark
25. Gehrtsen C, Kneser HO, Vogel H (1989) Physik, 16th ed. Springer, Berlin
26. Korchev YE, Gorelik J, Lab MJ, Sviderskaya EV, Johnston CL, Coombes CR, Vodyanoy I, Edwards CRW (2000) Cell Volume Measurement Using Scanning Ion Conductance Microscopy. Biophysical journal 78:451–457
27. Korchev YE, Negulyaev YA, Edwards CRW, Vodyanoy I, Lab MJ (2000) Functional localization of single active ion channels on the surface of a living cell. Nat Cell Biol 2:616–619
28. Gitter AH, Bertog M, J DS, Fromm M (1997) Measurement of paracellular epithelial conductivity by conductance scanning. Eur J Physiol 434:830–840
29. Mann SA, Hoffmann G, Hengstenberg A, Schuhmann W, Dietzel ID (2002) Pulse-mode scanning ion conductance microscopy: A method to investigate cultured hippocampal cells. J Neurosci Methods 116:113–117
30. Happel P, Hoffmann G, Mann SA, Dietzel ID (2003) Monitoring cell movements and volume changes with pulse-mode scanning ion conductance microscopy. J Microsc 212:144–151
31. Radmacher M, Cleveland JP, Fritz M, Hansma HG, Hansma PK (1994) Mapping interaction forces with the atomic force microscope. Biophysical Journal 66:2159–2165
32. Gorelik J, Yuchun, G., Spohr HA, Shevchuk AI, Lab MJ, Harding SE, Edwards CRW, Whitaker M, Moss GWJ, Benton DCH, Sanchez D, Darszon A, Vodyanoy I, Klenerman D, Korchev YE (2002) Ion channels in small cells and subcellular structures can be studied with a smart patch-clamp system. Biophys J 83:3296–3303
33. Gorelik J, Shevchuk AI, Frolenkov GI, Diakonov IA, Lab MJ, Kros CJ, Richardson GP, Vodyanoy I, Edwards CRW, Klenerman D, Korchev YE (2003) Dynamic assembly of surface structures in living cells. Proc Natl Acad Sci USA 100:5819–5822
34. Gorelik J, Zhang A, Shevchuk A, Frolenkov GI, Sanchez D, Lab MJ, Vodyanoy I, W ECR, Klenerman D, Korchev YE (2002) The use of scanning ion conductance microscopy to image A6 cells. Mol Cell Endocrinol 217:101–108
35. Korchev YE, Raval M, Lab MJ, Gorelik J, Edwards CRW, Rayment T, Klenerman D (2000) Hybrid Scanning Ion Conductance and Scanning Near-Field Optical Microscopy for the Study of Living Cells. Biophys J 78:2675–2679
36. Mannelquist A, Iwamoto H, Szabo G, Shao Z (2002) Near field optical microscopy in aqueous solution: implementation and characterization of a vibrating probe. J Microsc 205:53–60

37. Bruckbauer A, Ying L, Rothery AM, Korchev YE, Klenerman D (2002) Characterization of a novel light source for simultaneous optical and scanning ion conductance microscopy. Analytical Chemistry 2002,74:2612–2616
38. Rothery AM, Gorelik J, Bruckbauer A, Yu W, Korchev YE, Klenerman D (2003) A novel light source for SICM–SNOM of living cells. J Microsc 209:94–101
39. Gorelik J, Shevchuk A, Ramalho M, Elliott M, Lei C, Higgins CF, Lab MJ, Klenerman D, Krauzewicz N, Korchev Y (2002) Scanning surface confocal microscopy for simultaneous topographical and fluorescence imaging – Application to single virus-like particle entry into a cell. Proc Natl Acad Sci USA 99:16018–16023
40. Hansma PK, Cleveland JP, Radmacher M, Walters DA, Hillner PE, Bezanilla M, Fritz M, Vie D, Hansma HG, Prater CB, Massie J, Fukunaga L, Gurley J, Elings V (1994) Tapping mode atomic force microscopy in liquids. Appl Phys Lett 64:1738–1740
41. Putman CAJ, Werf KOVd, Grooth BGD, Hulst NFV, Greve J (1994) Tapping mode atomic force microscopy in liquid. Appl Phys Lett 64:2454–2456
42. Shalom S, Lieberman K, Lewis A, Cohen SR (1992) A Micropipette Force Probe Suitable For Near-Field Scanning Optical Microscopy. Rev Sci Instrum 63:4061–4065
43. Lewis A, Taha H, Strinkovski A, Manevitch A, Khatchatouriants A, Dekhter R, Amman E (2003) Near-field optics: from subwavelength illumination to nanometric shadowing. Nat Biotechnol 21:1378–1386
44. Betzig E, Trautman JK, Harris TD, Weiner JS, Kostelak RL (1991) Breaking the diffraction barrier: optical microscopy on a nanometric scale. Science 5000:1468–1470
45. Toledo-Crow R, Yang PC, Chen Y, Vaez-Iravani M (1992) Near-field differential scanning optical microscope with atomic force regulation. Appl Phys Lett 60:2957–2959
46. Karraï K, Grober RD (1995) Piezoelectric tip-sample distance control for near field optical microscopes. Appl Phys Lett 14:1842–1844
47. Rensen WHJ, van Hulst NF, Kämmer SB (2000) Imaging soft samples in liquid with tuning fork based shear force microscopy. Appl Phys Lett 77: 1557–1559
48. Brunner R, Hering O, Marti O, Hollricher O (1997) Piezoelectrical shear-force control on soft biological samples in aqueous solution. Appl Phys Lett 71:3628–3630
49. Koopman M, de Bakker BI, Garcia-Parajo MF, van Hulst NF (2003) Shear force imaging of soft samples in liquid using a diving bell concept. Appl Phys Lett 83:5083–5085
50. Lambelet P, Pfeffer M, Sayah A, Marquis-Weible F (1998) Reduction of tip-sample interaction forces for scanning near-field optical microscopy in a liquid environment. Ultramicroscopy 71:117–121
51. Fischer BE, Spohr R (1983) Production and use of nuclear tracks: imprinting structure on solids. Rev Mod Phys 55:907–948
52. Powell DW (1981) Barrier function of epithelia. Am J Physiol 241: G275-G288
53. Simionescu M, Simionescu N (1986) Functions of the endothelial cell surface. Ann Rev Physiol 48:279–293
54. Wegner J, Zink S, Rösen P, Galla H-J (1999) Use of electrochemical impedance measurements to monitor ß-adrenergic stimulation of bovine aortic endothelial cells. Eur J Physiol 437:925–934
55. Wegner J, Abrams D, Willenbrink W, Galla H-J, Janshoff A (2004) Automated multi-well device to measure transepithelial electrical resistances under physiological conditions. Biotechniques 37:590–597
56. Zhang H, Wu L, Huang F (1999) Electrochemical microprocess by scanning ion-conductance microscopy. J Vac Sci Technol B 17:269
57. Lewis A, Kheifetz Y, Shambrodt E, Radko A, Khatchatryan E, Sukenik C (1999) Fountain pen nanochemistry: Atomic force control of chrome etching. Appl Phys Lett 75:2689–2691
58. Müller A-D, Müller F, Hietschold M (1998) Electrochemical pattern formation in a scanning near-field optical microscope. Appl Phys A 66:S453–S456

59. Müller A-D, Müller F, Hietschold M (2000) Localized electrochemical deposition of metals using micropipettes. Thin Solid Films 366:32–36
60. Hong M-H, Kim KH, Bae J, Jhe W (2000) Scanning nanolithography using a material-filled nanopipette. Appl Phys Lett 77:2604–2606
61. Larson BJ, Gillmor SD, Lagally MG (2003) Controlled deposition of picoliter amounts of fluid using an ultrasonically driven micropipette. Rev Sci Instrum 75:832–836
62. Ying L, Bruckbauer A, Rothery AM, Korchev YE, Klenerman D (2002) Programmable Delivery of DNA through a Nanopipet. Anal Chem 74:1380–1385
63. Bruckbauer A, Ying L, Rothery AM, Zhou D, Shevchuk AI, Abell C, Korchev YE, Klenerman D (2002) Writing with DNA and protein using a nanopipet for controlled delivery. J Am Chem Soc 124:8810–8811
64. Bruckbauer A, Zhou D, Ying L, Korchev YE, Abell C, Klenerman D (2003) Multicomponent Submicron Features of Biomolecules Created by Voltage Controlled Deposition from a Nanopipet. J Am Chem Soc 125:9834–9839
65. Taha H, Marks RS, Gheber LA, Rousso I, Newman J, Sukenik C, Lewis A (2003) Protein printing with an atomic force sensing nanofountainpen. Appl Phys Lett 83:1041–1043
66. Seebach J (2001) Einsatz elektrisch leitender Kultursubstrate zur Charakterisierung von Zell-Zell- und Zell-Substrat-Interaktionen, Dissertation thesis, University of Münster

23 Scanning Probe Lithography for Chemical, Biological and Engineering Applications

Joseph M. Kinsella · Albena Ivanisevic

From volume IV

List of Abbreviations

AFM	atomic force microscope
SPL	scanning probe lithography
STM	scanning tunneling microscope
SAM	self-assembled monolayer
ODT	octadecanethiol
DPN	dip-pen nanolithography
UHV	ultrahigh vacuum
MHA	mercaptohexadecanoic acid
RH	relative humidity
MFN	meniscus force nanografting
ELP	elastin-like polypeptide
Trx-ELP	thioredoxin-ELP fusion protein
HIV-1	human immunodeficiency virus type 1
APTES	aminopropyltriethoxy silane
His-tagged	histidine-tagged
PAH	poly(allylamine hydrochloride)
Cys-CPMV	cow pea mosaic virus genetically modified with cysteine residues
LFM	lateral force microscopy
NPRW	nanopen reader and writer
HMDS	hexamethyldisilazane
MPTMS	3′-mercaptopropyltrimethoxysilane
OPA	octadecylphosphonic acid
FV	force volume
MEH-PPV	poly[2-methoxy-5-(2′-ethylhexyloxy)-1,4-phenylenevinylene]
SPAN	self-doped sulfonated polyaniline
PPy	polypyrrole
HRP	horseradish peroxidase
E-DPN	electrochemical DPN
EDOT	3,4-ethylenedioxythiophene
PAMAM	poly(amidoamine) dendrimer
NHS	*N*-hydroxysuccinimide
β-CD	β-cyclodextrin
pz	porphyrazines
BaFe	barium hexaferrite
MFM	magnetic force microscopy
μCP	microcontact printing
PDMS	polydimethylsiloxane

In: H. Fuchs, B. Bhushan (eds), *Biosystems - Investigated by Scanning Probe Microscopy*,
DOI 10.1007/3-540-26914-2_1,

23.1 Introduction

Taking a brief look back at lithographic and fabrication methods prior to 1959, it is interesting to note that the state of the art transistor in 1958 was 1 cm wide, while conventional photolithography by 2000 has reduced the size of similar structures to ~ 100 nm [1]. Conventional methods of fabrication are estimated to peak around 2010 when feature sizes will be less than 25 nm. To achieve devices with these dimensions alternative techniques will have to be employed, including E-beam and ion beam lithography, extreme ultraviolet photolithography, mask phasing and others. Many alternatives to these techniques have been developed based on scanning probe lithography (SPL) methods that may soon prove to be viable production methods.

This chapter will focus solely on such scanning probe techniques and applications thereof. In particular, we will explore the current state of research involved with employing these methods for biological and chemical modifications of surfaces on the nanoscale. The first reports of lithography using scanning probe methods date back to 1987 at the IBM Almaden labs, where researchers discovered the ability to position single atoms using STM [2–4]. These proof-of-concept experiments were only possible under ultrahigh vacuums, were done at very low temperatures and were painstakingly slow. Shortly after these experiments, researchers showed that STM could be used as a means to physically remove areas of octadecanthiolate SAMs from a gold surface by repeated scanning under ambient conditions [5]. Throughout the early to mid 1990s several groups explored techniques that involved the removal or modifications of SAMs using scanning probe tips [6–8]. These techniques will be expounded upon in subsequent sections of this chapter; it is also worth noting at this juncture that several thorough reviews of self-assembled monolayers have been written [9–11].

In 1995, the first experiments were performed showing that physisorbed organic molecules on an atomic force microscope probe could be transferred to an underlying substrate [12]. The report showed that ODT ($HS{-}(CH_2)_{17}{-}CH_3$) could be transferred to mica by scanning at very low forces and that the resulting patterns were stable and of a homogenous height profile. In 1999, the potential of this technique was stretched further than previously believed possible when controllable patterning of thiols on gold was achieved [13]. The importance of this report was in the control of ink deposition and the quality of the SAMs formed. The researchers showed that line widths as small as 30 nm and dots with diameters as small as 460 nm could be reproducibly formed. The group immediately noticed the parallels between this technique and the long known dip-pen writing technique and, therefore, appropriately named their method dip-pen nanolithography, Fig. 23.1.

DPN is an emerging technique that offers a means of fabrication in which molecular level control exists. One key feature that makes DPN such an attractive technique is its inherent versatility in selecting ink molecules. In principle, given the proper substrate, any chosen molecule can be either directly or indirectly bound to the surface in a site specific fashion. The present range of compatible inks varies from very large biomolecules, such as proteins, to small hydrocarbons. The range of substrates that have been shown to be compatible with this method to date includes metals, semiconductors, insulators and mica. The direct patterning of many

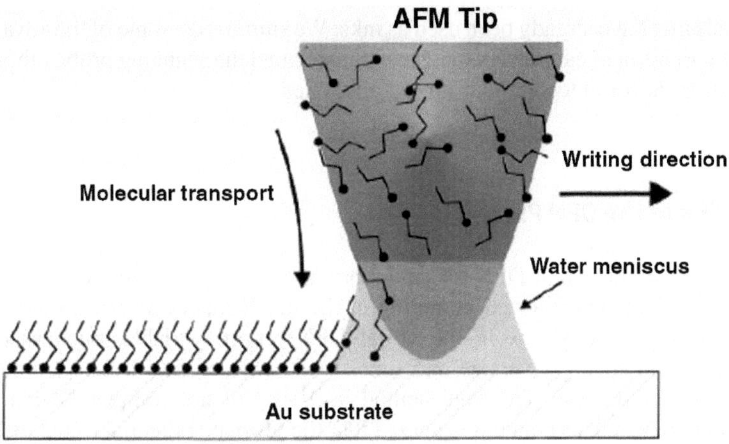

Fig. 23.1. A schematic representation of the DPN procedure. An AFM coated with physisorbed molecules is brought into contact with the substrate. Once in close proximity with the surface, a water meniscus is formed between the tip and the sample through capillary action. The water meniscus acts as a localized reaction vessel and is used to transport the physisorbed ink onto to the surface. The writing direction of the tip, as well as other environmental factors, is used to control the formation of the pattern generated. Figure reprinted with permission from [13]

of these materials was not previously possible using standard lithographic techniques. Photolithography, while of great importance in the semiconductor industry, is of little use for patterning of organic or biological molecules due to the destructive nature of the UV light used for exposure of resists. Soft lithography methods such as microcontact printing are also compatible with various molecular inks. Currently the resolution of soft lithography methods is limited due to their dependence on conventional lithography or micromachining to produce the master stamp.

This chapter highlights the importance of dip-pen nanolithography and related procedures for the fabrication of both traditional and non-traditional structures on the nanoscale. The chapter begins by presenting the fundamental mechanisms that control the transfer of ink. Theoretical studies describe the role of several parameters that are important for the patterning process: humidity, temperature, type of ink, and surface choice. The rest of this chapter is separated according to the types of inks used and the applications associated with them. Biological inks are currently being extensively studied in nanoscience applications and research, and are described first. These inks can be utilized in nanoscience as a method to fabricate templates for higher order bottom-up assemblies. Biology and medicine itself can greatly benefit from techniques such as DPN, because it provides a fabrication method capable of creating high-throughput nanoarrays and templates. Such patterned surfaces can be used to study fundamental problems associated with biological or biochemical interactions. In subsequent sections, the use of chemical inks is reviewed. We describe in detail the various types of chemicals that have already been patterned. Many of the inks that have been used to date are already used in commercial applications on the bulk scale. The engineering aspects of DPN are also presented in the concluding section. Many of the engineering applications described are possible by choosing from the materials that have already been used as inks. We summarize some of the advances in microfabrication of cantilevers that have transformed the scanning probe lithography methods from serial technologies to parallel ones.

23.2 Modeling of the DPN Process

The interest in the use of DPN for nanofabrication has spurred many researchers to determine a fundamental understanding of the mechanism involved in the transfer of ink from the AFM tip. In the initial DPN report, variables such as relative humidity, scan speed, scan size and time were considered to be the main factors in the transfer process. The true method itself is still not completely understood and highly debated, although it is agreed that the ultimate potential of the technique depends on a quantitative understanding of the ink transport mechanism. We will briefly attempt to objectively describe the work done by many researchers on this subject in an effort to describe the fundamental science and engineering that is involved in DPN, as well as to convey the complexity of the process from a scientific standpoint.

The first quantitative efforts made to understand the transport properties of DPN were reported in 2001 [14]. Their intent was to model and experimentally determine the diffusion and adsorption of SAM patterns using DPN. The model was based on the view that the molecules diffused down a concentration gradient from a high density region around the area of the tip to a lower density region that consists of the unoccupied binding sites of the substrate directly outside the patterned regions. This model assumes that the physisorbed ink molecules on the tip act as a diffusion point source when in static contact with a two-dimensional substrate. This model also assumes that the molecules diffuse out onto a region in which all of the available binding sites are occupied from previously deposited molecules. Furthermore, the molecules are assumed to continue to travel outward until they find an unoccupied site on the substrate and irreversibly bind to it. With these assumptions, the experiments model the dynamics of the self-assembly of the ink as a random walk on the substrate with the flow of ink as a constant flux. The results of the simulations showed that when the tip remained fixed the quality of the diffused SAM that formed was dependent on the number of molecules deposited. The rate of deposition during the simulation proved to have no effect when the same number of molecules was deposited. The circularity of the patterns formed however was found to increase when the number of deposited molecules was increased. When modeling patterns that were formed by a moving tip it was determined that there is a significant dependence between the tip speed and the deposition rate.

In a follow up study the role of the meniscus formation on the transport properties of the ink was examined [15]. Monte Carlo simulations were employed to examine the behavior of the meniscus for various geometries and energies. A variety of surfaces are assumed to all be completely wettable at any temperature. Also, the tip is assumed to be an ellipse and the water is modeled as a two-dimensional lattice gas that is confined between the tip and the substrate. Some of the variables that were examined include the wettability of the tip, which is highly dependent on the wettability of the ink physisorbed to it, the radius of curvature of the tip, the distance of the tip from the sample surface, as well as the interaction strength between both fluid and tip, and fluid and surface. The results of this simulation showed that the formation of the meniscus is thermodynamically controlled and the width of the meniscus can decrease in a number of situations, including increased tip curvature, increased separation of the tip-substrate and decreased wettability of the tip. The degree of saturation was also shown to be a critical component of the width of the meniscus and the simulation had shown that it can be a tunable property over a very large range. This is important because the limit of resolution of DPN patterns will depend on the minimum possible width for which a stable meniscus can be formed. The simulation showed that the smallest possible meniscus that could be formed would be ~ 2 nm, but that such a small meniscus would be prone to instabilities, thereby inferring that the actual limit of resolution for any reproducible large scale pattern would have to be larger than this.

Also in 2002, the deposition of ODT was reported in waterless environmental conditions using DPN [16]. The radius of the ODT pattern was experimentally measured as a function of contact time and relative humidity. It was also assumed in this work that the tip will act as an infinite reservoir of ink, which implies a constant concentration, not flux. Modeling the tip as a constant concentration allows for the extraction of kinetic data. It is also assumed that the diffusion rate is independent of the concentration and would be Fickian. The final assumption made was that, when the tip is removed from the surface, diffusion will cease at all areas except the periphery of the pattern. The researchers obtained data on deposition as a function of time. The data was fit to a radial diffusion model and allowed one to determine the surface diffusion coefficient of ODT onto clean gold to be $8400 \pm 2300\ \text{nm}^2\,\text{s}^{-1}$. Prior models based on constant fluxes indicated that the radius would increase with $t^{1/2}$. The data reported by Sheehan et al. did not fit these models. Humidity experiments were performed in air and under a nitrogen environment to precisely control the saturation of water on the substrate and, therefore, the ability of the meniscus to be formed during deposition. Repeated experimentation showed that the deposition of the ODT to the gold surface was independent of humidity and that the meniscus was likely to have no role in the ink transfer.

Experiments performed on atomically flat NaCl substrates have given further insight on the formation of the water meniscus between the AFM tip and the sample surface at different relative humidity conditions [17]. The meniscus formation was studied under humidity conditions ranging from 70% to 0%. An ultrahigh vacuum (UHV) chamber was used and the humidity dependence utilizing either hydrophobic or hydrophilic AFM contact mode tips was recorded. Tips were made hydrophilic by etching them with H_2SO_4 and a H_2O_2 solution. The hydrophobic tips were prepared by adsorbing ODT or dodecylamine on the tip surface. Theoretically, the hydrophilic tips should increase the meniscus size due to the large capillary forces between the tip and sample. The hydrophobic tips should minimize these forces, thus reducing or eliminating the meniscus formation. When a meniscus forms between the tip and NaCl substrate, the water meniscus causes the NaCl to dissolve and leaves a pit on the surface. Using the hydrophilic tips, it was noted that the NaCl substrate was etched by the formation of a water meniscus under all relative humidity conditions, excluding those under UHV. Under RH conditions exceeding 60%, it was noted that the substrate could not be etched due to the large water layer that extended across the entirety of the NaCl surface. The adhesion forces of the tip to the NaCl substrate were measured as a function of the atmospheric water conditions as well. This experiment was done to exclude the possibility that the pit formation is induced by a mechanical deformation of the sample during the experiments. It was noted that the adhesion force decreased with decreased humidity conditions. This result indicates that the capillary forces were negligible at low RH values. When the hydrophobic tips were studied, it was found that the substrate would still be dissolved under low RH values. In these cases the pits formed were not always circular. This observation was attributed to the variable meniscus shapes and sizes formed when coated tips were used. The study concluded that unless the atmospheric water conditions are controlled using a UHV chamber, a water meniscus will always form to some degree between the tip and the sample.

Similar experiments have been performed studying the transport phenomenon with respect to the relative humidity and temperature using ODT and MHA ($SH–(CH_2)_{15}–COOH$) [18]. The humidity dependence of the molecular transport of ODT to the surface corroborates the previous findings that there is no dependence on the transport for ODT and minor humidity dependence for MHA. The qualitative effects of a water vapor environment versus an ethanol environment were also studied to determine whether the ink-meniscus solubility would alter the deposition. It was observed that lines and dots of ODT in an ethanolic environment had poorer edge resolution compared with patterns formed in comparable water vapor environments. From the results of these experiments, it was concluded that the transport mechanism is heavily dependent on the solubility of the ink. With hydrophobic inks such as ODT and MHA, the dependence on humidity is negligible, but for ionic salts and highly charged ink there remains a strong dependency on the presence of water vapor to form a meniscus between the tip and substrate. In the transfer of the nonpolar inks it was determined that the transport was controlled by a simple surface diffusion mechanism in which a dependence on thermal energy exists.

A recent report highlights the temperature dependence found in patterning of ODT and MHA inks [19]. Patterns were formed nine times at nine different temperatures ranging from 22–33 °C. Holding all other variables constant and increasing the temperature systematically led to an increase in the dot diameter of the MHA and ODT patterns. It was established that an increase in temperature would affect a number of important variables, including the number of molecules solvated, the diffusion rate across the meniscus, and the adsorbate diffusion onto the surface. Similarly, the effect of the relative humidity (RH) was studied by varying the RH between 5–100%. Again nine humidity conditions were studied by depositing nine dot patterns of each molecule. The role that humidity played in the transfer of the two molecules differed. For ODT increasing the RH and maintaining a constant temperature caused a decrease in the dot sizes. The opposite effect was seen when MHA was patterned at increased humidity and fixed temperatures. From this data it was concluded that the formation of a larger meniscus at elevated humidity acts as a blocking layer for the insoluble ODT and a transport layer for the more soluble MHA. These experiments were in agreement with the previous work on the effects of humidity and meniscus formation for these two inks.

23.3 Patterning of Biological and Biologically Active Molecules

Several exciting discoveries have recently been made in the use of DPN for patterning biological molecules, viruses and bacteria. Research in the life sciences and medicine has already been revolutionized by the development of microarrays and assays for diagnostic screening. Microarrays are devices typically containing patterns of differing proteins or DNA strands that react specifically towards complementary oligonucleotide strands, short organic markers, ligands or specific antibodies in solution. These devices are widely used for research, clinical diagnostic testing, screening and profiling.

A typical microarray is formed by robotic spotting or photolithographic techniques, which limit their minimum feature size to $\sim 20\,\mu m$ [20]. In comparison, using DPN to pattern the molecule of interest on a surface with even poor resolution capabilities of hundreds of nanometers would increase the areal density of arrays by a 10,000 to 100,000 fold [21]. With these capabilities, the realization of whole genome single chip screening can not only be achieved, but can be done using a chip that is only centimeters in size. In comparison a present day whole genome chip hindered by current resolution limits would be on the orders of meters.

Another value of shrinking such devices lies in the reduced volume of analyte necessary for proper analysis. With a reduced volume of analyte the time of analysis is also decreased, which offers the possibility of on the spot detection for both clinical and research purposes. Time to purify and/or amplify the analyte would not be necessary in such a high-throughput device.

In addition to decreasing analyte concentration and increasing testing time, the value of shrinking the microarray to dimensions that are on the same order of magnitude as biological molecules lies in providing a way to study the interactions of several important biochemical reactions in situ. This includes the interactions of receptor-ligand and whole cell-ligand, DNA–DNA, DNA–RNA, RNA–RNA, peptide-RNA and several other important interactions. The study of these interactions on an artificial surface that properly mimics their in vitro state will give researchers much insight on the fundamental mechanisms that control such reactions. Knowledge of these mechanisms is not only important from a fundamental biological or biomedical standpoint, but also for a number of applications that rely on specific interactions.

Proof-of-concept experiments have shown that patterning of several biologically important molecules on a single chip is possible, although there is still much work to be done on detection methods. AFM based detection is a slow and tedious process at present, but it can be performed in situ allowing brilliant insight on the basic mechanisms of interactions. Integrated microelectronics are also being explored in a rapidly expanding field as a characterization method that offers very sensitive detection capabilities. Detection methods lie beyond the scope of this section. Here we will focus on the current state of patterning, surface chemistries and the possibilities offered by DPN with regard to patterning biomolecules.

Some examples of molecules that have already been patterned using DPN include: DNA oligonucleotides, proteins and peptides. These molecules have been written both directly and indirectly to a number of surfaces, including silicon oxide, nickel oxide, gold and mica. The success of direct writing approaches was demonstrated using thiolated DNA or cysteine residue containing proteins and peptides on gold surfaces. Indirect approaches can rely on surface modifications prior to writing and can be used very efficiently for the specific binding of the molecule to the modified surface. Indirect methods can be used on any properly modified surface and offer enormous potential in the modification of surfaces that prove difficult in direct writing. An example of the indirect patterning of proteins that is typical in standard protein arrays is the modification of glass or oxidized silicon with an aldehyde terminus and subsequent interaction of the amine terminus of the protein or a lysine side chain with the aldehyde via a hydrolysis reaction. One drawback to the indirect method of patterning is that typically only one type of biomolecule can be patterned. To overcome this, multiple steps involving patterning and adsorption can be employed, although the process is time-consuming.

23.3.1 DNA Patterning

It has been shown that thiolated oligonucleotides and 5′ terminal acrylamide functionalized oligonucleotides can be patterned via DPN onto gold and silica surfaces, respectively [22]. This report not only was groundbreaking in that it was the first to use DPN for the direct-write fabrication of biomolecular recognition arrays, but it was also one of the earliest reports of using DPN on a substrate other than gold. The report also notes that using standard silicon nitride tips results in poor patterning. To overcome this limitation, the tips were initially coated with 3-aminopropyltrimethoxy silane in order to make their surface more wettable. The wettable tip surface was more favorable for the physisorption of DNA ink molecules and contributed to more evenly distributed patterning on the surface. To test that the patterned oligonucleotide sequences retained their biological recognition, the patterned surfaces were treated with oligonucleotide modified gold nanoparticles. It was observed that complementary Watson–Crick base pairing bound the particles to sites were the oligonucleotides were patterned exclusively. Multiple DNA types were also patterned and incubated with their complementary strands, which were modified with fluorophores for detection purposes. The results of these experiments showed that the binding of the complementary strands was specific and that even after rinsing the fluorophore modified complementary strand off of the substrate with water, the patterned spots remained and further testing revealed that they had retained their activity.

Oligonucleotides immobilized onto surfaces can also serve as templates to organize the assembly of multidimensional and multicomponent structures using solution phase bottom-up assembly [23]. This was shown by patterning DNA indirectly through the use of an MHA ink and subsequent coupling of the DNA to the MHA pattern using alkylamine-modified oligonucleotide strands. To generate patterns of different sequences of DNA on the surface a multistep fabrication scheme is necessary due to the indirect nature of this approach. To do so, patterning of MHA and subsequent adsorption of the DNA was performed using different DNA strands each time and repeated as many times as necessary. Once the desired numbers of different DNA strands were immobilized, they were then used as templates to bind materials that contain the complementary sequences in a specific manner. In this experiment, two different DNA strands were chosen and their complementary strands were immobilized onto either 13 nm or 30 nm gold nanoparticles. When the nanoparticles were allowed to interact with the surface, it was noted that the interaction of the surface immobilized strands and the particle bound complementary strands would guide the assembly of the structures in a controlled fashion.

Other scanning probe lithography techniques have been used to immobilize DNA onto surfaces. MFN is a method that employs the removal of a layer of molecular resist on a surface through the shear force provided by the surface tension of a thin film of the solution present during scanning that is to be patterned [24]. This technique is similar to nanografting, where a high contact force is applied to remove a monolayer from a substrate [25]. Since the force applied to remove the resist in MNF is the surface tension supplied by the inking solution, the need for AFM feedback control is unnecessary and the cantilever can scan large areas very rapidly while removing the resist. It was also noted that the surface tension created by an aqueous thin film on the surface significantly exceeded the force that is typically applied by AFM feedback controls. Depending on the tip radius, the wettability of the surface and the volume of fluid in the film, the applied force due to the surface tension of the thin film could approach micronewton ranges. Typical conditions for the applied force in contact imaging, DPN or traditional nanografting are only in the nanonewton range. These large forces provided pressures that exceeded the yield strength of the underlying gold substrate and, therefore, would mechanically flatten the gold area that was patterned by MFN. This technique was successfully employed to pattern single strands of DNA onto gold substrates. Line patterns were formed with scan rates ranging from 20–160 μm/s. The patterning of the DNA occurred simultaneously with the grafting due to the presence of DNA in the solution used to provide the surface tension. The lithographically defined DNA on the surface was incubated with different sizes of nanoparticles containing complementary strands to those patterned to verify its viability.

Another process that closely resembles both DPN and robotic stamping methods has been developed using an array of silicon based microcantilevers to mechanically spot DNA and proteins on glass surfaces [26]. The theory for the technique relies on the same principles as DPN, although the array of cantilevers is designed with microreservoirs, microchannels and aluminum electrodes that can be used to actively control the height of the liquid rise when the arrays are dipped into an inking solution. This increases the liquid loading efficiency of the tips and allows for the tips to be cleanable unlike conventional coated tips in DPN experiments. To create a microarray, the liquids to be spotted are placed on the surface and the cantilever array is brought into proximity. The cantilevers are then dipped into the solutions and a small electric field is applied for a short time. The applied voltages induce electrowetting on the surface of the cantilevers due to the changes in surface charge distribution, thereby forcing the liquid to spread across the cantilever and into the embedded microreservoirs. Not only were these cantilever arrays successful in the patterning of DNA, but they also were shown to be able to fabricate protein microarrays. The size of the features spotted was in the micron range. With optimization of the cantilever design, contact force and contact time this technique may contend with DPN in terms of feature size.

23.3.2
Protein Patterning

Proteins offer a challenge to patterning by DPN based on their relatively large size, high molecular weights and intricate three-dimensional structure. The typical protein patterned for diagnostic microarrays is the antibody. Antibodies are Y shaped proteins that have average dimensions of $\sim$5–15 nm. The conformation of the antibody on the surface is also critical because the top two chains of the molecule (light chains) contain the variable regions and determine the specific biorecognition properties. If these regions of the protein are not exposed to the analyte, then subsequent interaction is severely hindered or rendered impossible. Because of the importance of the higher order structure of the protein with respect to its recognition abilities, the transfer process of the molecule to the surface must be done in a way that will not cause the protein to denature. Intact cytoplasmic proteins at physiological pH's are often highly charged molecules with hydrophilic amino acids exposed towards the outside to solvate the molecule and make it thermodynamically stable. Adsorption can, therefore, occur through electrostatic interactions between the substrate and the protein or through covalent interactions of a modified substrate and an exposed amino acid of the protein. Other proteins may contain a combination of both hydrophilic and hydrophobic residues on their surfaces. Such proteins may diffuse at different rates or interact differently with surfaces.

In the theory section, we addressed the concept of simple surface diffusion of the inking molecules from the tip onto the underlying substrate. The models generated were fit to data generated by inks that were short organic molecules such as ODT, MHA or ionic salts. The composition of proteins and their interactions with surfaces differ greatly when compared to these inks. Several considerations regarding these limitations need to be taken into account when designing DPN experiments with protein inks. In the simple diffusion model, the molecules that were modeled diffused outward from the tip until they bound onto the next available binding site. The diffusion is impeded in the patterning of proteins due to their high molecular weights according to classic Fickian diffusion. Steric factors and/or electrostatic repulsions between adsorbed proteins and those being deposited may create difficulty in the ability to generate controlled patterns.

The physisorption of the protein onto the tip is also a problem that may follow a mechanism very different to that of smaller hydrophobic molecules. Proteins are susceptible to denaturation upon physisorption onto standard AFM cantilevers by the surface forces present. The minimization of denaturation, however, is imperative when attempting to fabricate an operable protein array. To overcome the possibility of denaturing of the protein, modifications of the tip itself have been employed. Coating of the tip with a low surface energy SAM prior to inking has proven to be successful. Several different types of SAMs can be employed to coat the tip as long as they are both hydrophilic and biocompatible. Formation of the meniscus can also be increased when a more wettable tip is used, resulting in more reliable patterning. Typical protein patterning experiments are also performed at a heightened humidity to further enhance formation of the water meniscus.

The initial work indicating that DPN could be a viable method to generate protein patterns was performed by writing thiolated collagen and collagen-like peptides on a gold surface [27]. The spatial control over the collagen patterns generated ranged from ~800 nm line widths to 30–40 nm line widths. At smaller line widths the level of organization of the patterned collagen was not observed to match that typically found in native collagen structures. The helical architecture of native collagen was easily observed for larger pattern sizes. Bulk deposition of the inking solution onto the surface failed to generate the higher order structures, which led researchers to conclude that the actual mechanism of formation of the triple helices found in native collagen was facilitated by the writing method. Reactions with collagen-specific primary antibodies were employed to test the biological activity of the patterned collagen. The patterned structures interacted with the antibody, indicating their retention of biological function.

Experimental work has also been done to show that antibodies containing a cysteine residue, which contains a thiolated side chain, can be patterned onto gold under high humidity conditions [28]. Other work has been done patterning proteins onto glass, metals and oxidized silicon. In the case of the gold substrate, the patterned antibodies retained biological activity and interacted strongly with their appropriate antigens. Patterned antibodies on aldehyde modified silicon oxide and on negatively charged oxidized silicon were also shown to retain their activity even when the patterned area was as small as 55 nm diameter dots [29], Fig. 23.2. Experiments on silicon are important because this is the surface of choice in both conventional array technologies and in microelectronics.

Nickel surfaces are also often used in biological arrays due to the strong affinity for histidine residues. This would imply that in principle DPN can be used to directly write proteins or polypeptides that contain histidine residues in their amino acid sequences onto nickel surfaces. N-terminal histidine-tagged ubiquitin and thioredoxin protein arrays have been fabricated on nickel oxide surfaces using this interaction [30]. Feature sizes as small as 80 nm and as large as 500 nm were constructed. The activity of the patterned proteins was shown to be maintained after writing by staining against fluorescent antibodies.

Protein nanoarrays have also been formed via the indirect assembly of proteins using DPN patterning of templates to guide the assembly of the proteins onto the substrate [31]. Many proteins will have a high affinity for carboxylic acids at physiological pH levels. Patterning of MHA onto gold using DPN has been used as an indirect method to immobilize a variety of proteins under these conditions. The patterning of MHA is simple and well-defined in comparison to the direct patterning of proteins. The inability to immobilize a number of differing proteins using an indirect method, without several cycles of subsequent patterning followed by immobilization, is a severe limitation to the effectiveness of this method.

Examples of indirect patterning using MHA include immobilizing IgG [31,32], Lysozyme [31], thioredoxin-ELP fusion proteins [33] and modified monoclonal antibodies against HIV-1 [34] onto gold substrates. By passivating the bare gold surface after MHA patterning, the non-specific binding of these proteins to the unpatterned areas of the surface can be minimized. The height profiles of MHA patterns after adsorption of the proteins showed that one or two protein molecules would adsorb to each individual MHA spot. Importantly, the bound proteins retained their activity once adsorbed. Cell adhesion was also studied by adsorbing retronectin onto the MHA patterns. Retronectin is a protein that can cause certain cell types to adhere via

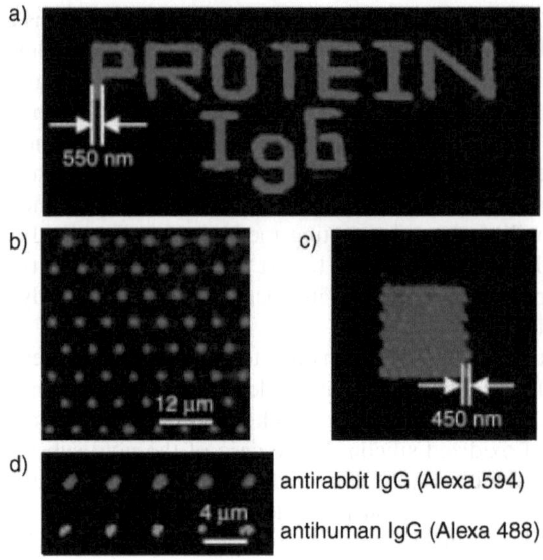

Fig. 23.2. (**a–c**) Antirabbit IgG proteins patterned on negatively charged silicon oxide surfaces using DPN imaged by fluorescent microscopy. The antirabbit IgG proteins were labeled using Alexa 594 as a fluorescent marker. (**d**) Protein patterns formed on aldehyde modified surfaces of antirabbit IgG. The *upper line* consists of dot patterns of the antirabbit IgG stained with Alexa 594. The *lower line* is dot patterns of antihuman IgG stained by Alexa 488. This demonstrates the ability of DPN to generate patterns of multiple inks on the same surface. The functionality of the deposited proteins is tested against specific probes in this reference demonstrating the potential of DPN in the functionalization of surfaces for protein nanoarray manufacturing. Figure reprinted with permission from [29]

the specific interaction of its RGD sequence with cytoplasmic integrins on the cellular membrane. It was found that the cells would attach to the patterns of retronectin almost exclusively in comparison to the bare gold surface. The morphology of the cells attached to retronectin differed from those that were attached to the gold as well. Such findings indicate that the use of DPN patterned proteins can be useful not only for diagnostic and biosensing applications, but also in cellular studies and biomaterials.

The indirect patterning method used to capture the thioredoxin-ELP fusion protein (trx-ELP) relied on the patterning of MHA using DPN and the subsequent immobilization of elastin-like polypeptide (ELP) sequences onto the patterned MHA [33]. The surface bound ELP could be used to capture or release the trx-ELP depending on the environmental conditions, such as the ionic strength of the buffer or temperature. This use of indirect patterning and biorecognition between the ELP polypeptide and the trx-ELP is an example of higher order patterning in the nanoscale using biologically inspired bottom-up assembly procedures.

Indirect capture of proteins has also been achieved using the patterning of MHA followed by the binding of biotin, streptavidin and then biotinylated proteins [35]. This recognition-based assembly of proteins onto gold relies on the small-molecule ligand biotin that shows a very high affinity for the streptavidin molecule. The biotin-streptavidin assembly procedure is important due to the ubiquity of existing biotin-tagged molecules. The high affinity of the molecules permits selectivity of the patterning that results in almost no non-specific binding when non-fouling SAM is used to passivate the underlying substrate.

23.3.3 Peptide Patterning

Peptides have been of much interest to nanoscience research recently. Solid state synthesis techniques have made the production of peptides cheap and efficient. As proteomics and enzymology uncover the active sites of large proteins, it is typically discovered that the function relies on only a fraction of the amino acids that make up the entire protein. These amino acid sequences can be reproduced in the form of peptides and used for catalytic, recognition or sensing purposes. Like for larger proteins the conformation of peptides is critical in their function and its integrity must be preserved. Many researchers have modified bulk surfaces using peptides and have shown that upon monolayer formation the peptides remain active [36–38]. An example of this would be the modification of gold using an RGD peptide [37]. The RGD sequence is one of many that have been shown to actively attach integrin proteins found on cell membranes to promote adhesion of specific cell types. Such work is important in biomaterial science or tissue engineering applications were implants may be engineered to actively attach cells onto their surface in order to depress immune responses and/or integration of the implant with surrounding tissues.

A class of TAT peptides derived from the TAT protein has been patterned onto SAMs modified SiO_x surfaces using DPN [39], Fig. 23.3. The TAT protein has been isolated from the human immunodeficiency virus type 1 (HIV-1) and shown to function in cell permeation and viral replication. The TAT peptide that was patterned

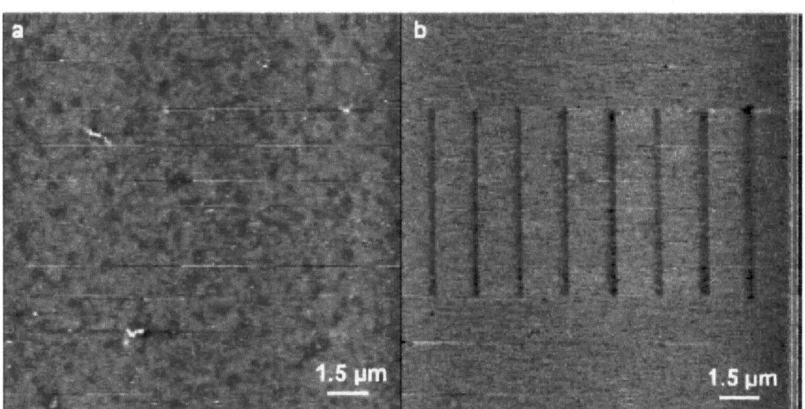

Fig. 23.3. AFM images (**a**) height and (**b**) lateral force microscopy (LFM) of DPN generated lines of TAT peptide on SMPB modified silicon oxide surfaces. Figure reproduced with permission from [39]

has been shown to interact with specific sequences of RNA. The modification of the bulk SiO_x surface prior to patterning included silanization of the surface using aminopropyltriethoxy silane (APTES) and subsequent coupling of a heterobifunctional cross-linker SMPB. The TAT peptide covalently binds to the cross-linker through a terminal cysteine residue that contains a thiol group. To successfully write the peptide onto the modified surface it was important to maintain high humidity conditions and apply minimum force with the tip so that the underlying bulk surface modifications would not be affected. The covalently bound peptide was shown to retain its recognition properties when incubated with a specific RNA sequence.

Directed immobilization of peptides via the thiol bond of a cysteine residue onto gold surfaces has also been achieved [40]. Contrary to most studies using DPN, the deposition of these peptides was performed using tapping-mode AFM. Tapping-mode AFM is typically used in the imaging of biological or "soft" surfaces in order to minimize damage to the adsorbed molecules. One of the key factors to depositing the peptides on the surface using tapping-mode DPN was to increase the drive amplitude of the probe to 2–5 times the value typically used for imaging purposes, scanning at slower speeds (0.2–1 Hz). Even at the elevated drive amplitudes used for the patterning, the force between the tip and the sample is still considerably less than the tip–sample forces found in contact mode imaging or deposition. The amplitude of the coated tapping-mode cantilever was found to be dampened by the presence of the physisorbed peptides when compared to the amplitude of the bare tapping-mode cantilever. The verification of the transfer was shown using line scans acquired through tapping-mode images to determine if the height of the generated pattern was higher than the surrounding non-patterned areas. The immunolabeling of the peptides after deposition failed to show any fluorescent signal indicating an inactivity of the adsorbed peptide. This may be attributed to the quenching of the fluorescence by the gold background or the conformation of the patterned peptides.

A follow-up study using an electrochemical variation on the DPN method (E-DPN) was performed in tapping-mode to pattern both His-tagged proteins and peptides onto a nickel surface [41]. The electrochemical dip-pen nanolithography resembles the typical DPN technique although the transfer of the ink is directed by the presence of an externally applied electric field. A negative bias was applied at the tip during patterning and the nickel surface was held at ground potential. This causes the nickel surface to ionize once the meniscus is formed between the tip and the sample. The choice of the histidine-tagged proteins and peptide was made due to their strong affinity towards Ni^{2+}, which was formed at the surface during the ionization process. The potential was held between -10 and $+10$ volts, the drive amplitude was increased to 5–10 times the value that would be used for imaging and slow scan speeds ($\sim$0.5 Hz) were employed. When no voltage was applied it was noted that no deposition had occurred. This was due to the lack of nickel ions on the surface for the deposited peptides to bind to. The optimal tip voltage for the controlled transfer of the peptides was determined to be in the range of -2 to -3 V. A low positive bias did not induce the ionization of the nickel, therefore, no deposition was observed. The optimal transfer of the proteins also occurred when the tip voltage was held at -2 V. The patterned proteins displayed fluorescence once tagged, indicating their viability once immobilized. Height scans of the peptides indicated that the peptide was in an upright position with the terminal histidine residue bound to the nickel surface.

23.3.4 Patterning of Templates for Biological Bottom-Up Assembly

Bottom-up assembly is a highly researched area of nanotechnology and science. Many of the roots of self-assembly mechanisms used in nanoscience have origins or inspirations based on mechanisms found in natural in biological processes. Biological self-assembly often relies on non-covalent interactions such as hydrogen bonding, hydrophobic/hydrophilic interactions, van der Waals forces and electrostatic interactions in the building of three-dimensional structures. Understanding these mechanisms has allowed the patterning of materials that interact using these forces to selectively guide the assembly or the placement of biological molecules. A basic example of this was seen in the development of arrays using both DNA and proteins. In the case of DNA the H-bonding base pairs of the complementary oligonucleotides in a solution will interact specifically with the strands patterned on the surface. For proteins, the antibodies interact through these secondary forces to tightly bind specific antigens. Although these structures are complex, their interactions can be exploited on surfaces or in solution to fabricate three-dimensional architectures.

Biological macromolecules and biological entities such as bacteria and virions can also be indirectly patterned using non-covalent interactions. Patterning of polyelectrolytes, which will be explained in further detail later, can create specific areas on the substrate that are highly charged. Charged areas on the surface can be used in the electrostatic assembly of oppositely charged biological molecules. As an example, PAH can be utilized to impart a pattern of positive charge. DNA bears a net negative charge at physiological pH due to the phosphate backbone. When stretched onto a negatively charged hydrophilic surface that contains positive PAH patterns, the placement of the DNA will be guided by electrostatic interactions [42], Fig. 23.4. Much work has also been done using DNA itself as a template for the bottom up assembly of metals, polyelectrolytes and nanoparticles using electrostatic interactions.

Another important biological entity that has gained much interest in nanoscience is the virus. Typical virus capsids are ~10 nm in size and can be considered macromolecules in the sense that they are composed entirely of protein molecules. Virions have also been the subject of studies in genetic engineering and controlling features such as capsid protein composition is easily accomplished. The genetically modified cow pea mosaic virus that contains a cysteine residue (Cys-CPMV) at specific spots on the capsid has been used to specifically adsorb onto patterns of covalently modified gold surfaces produced via DPN [43], Fig. 23.5. To this end, an amine terminated thiol was patterned directly onto the gold surface. The remaining unpatterned areas were then passivated using a protein resistant SAM. The reactive terminal amine was then used to interact with a maleimide cross-linker, while the protein resistant SAM was not. The cross-linker is highly specific in its interactions and reacts with the thiol of the cysteine residue on the Cys-CPMV. This reaction formed a covalent

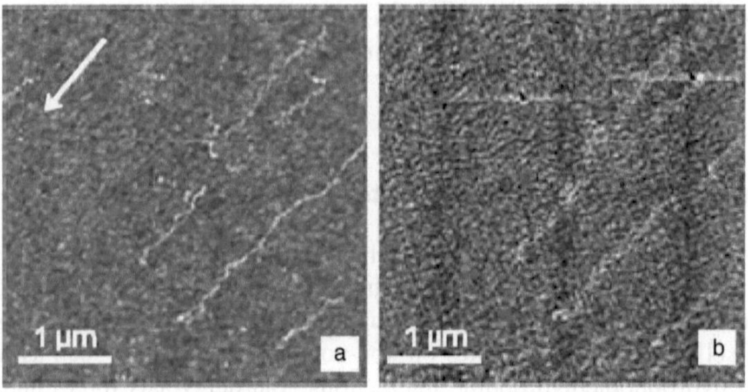

Fig. 23.4. AFM images of DNA stretched onto DPN generated patterns of the polyelectrolyte poly(allylammonium) hydrochloride (PAH). (**a**) Height image. (**b**) Phase image. The DPN generated patterns of PAH are clearly visible as *vertical lines* in the phase image. They do not appear in the height image due to the small height of the deposited polyelectrolyte when patterned using DPN. Figure reprinted with permission from [42]

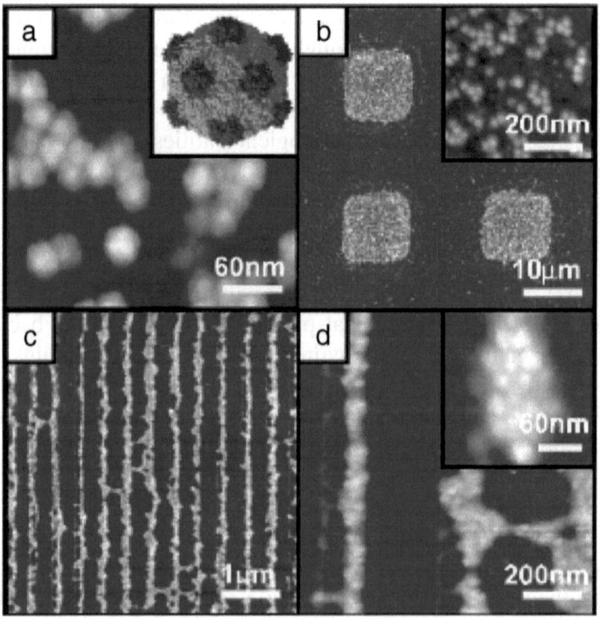

Fig. 23.5. AFM height image of genetically modified cowpeamosaic virus (CPMV). *Inset* of figure (**a**) shows a model of the virus surface. The surface of the virus particle is specific modified, through genetic engineering, to express cysteine labels. These cysteine residues are present at each of the five corners of the pentagon structures on the capsid of the CPMV. (**b**) AFM height images of the virus assembled onto templates. *Inset* shows high resolution of the virus coated templates. (**c**) Lines of virus assemblies formed through nanografting of the surface and subsequent chemical template assembly. (**d**) High resolution AFM imaging of the virions assembled on the chemoselective template lines of (**c**). Figure reprinted with permission from [43]

link between the virus and the maleimide modified DPN pattern. It was also reported that, due to the large attractive interaction that the virions displayed for one another, stacking occurred on the patterned areas.

Similar work has been done to capture cysteine modified virus particles that employed the writing of a mixed monolayer onto gold [44]. The mixed monolayer consisted of two chemically distinct dialkyl disulfide molecules. The first contained a maleimide group that would interact specifically with the virus while the second contained a penta(ethylene glycol) group that worked to inhibit the non-specific adsorption of proteins onto the patterned surface. The density of available maleimide groups on the surface was determined by controlling the fraction of the maleimide molecule in the inking solution.

23.4 Chemical Patterning

Decades of research in surface science have given nanoscience a predetermined beginning to take advantage of. The original DPN work involving patterning of ODT and MHA soon escalated into the patterning of various other small organic molecules, organics, nanoparticles, metal ions, polyelectrolytes, polymers, salts, supramolecular complexes and sol-gels. An equal number of surfaces have been used in these experiments ranging from metals to insulators and semiconductors.

23.4.1 Thiols

The original work of DPN focused on the patterning of ODT and MHA on gold, which was not surprising since these thiolated molecules have been well-studied as bulk SAMs on gold surfaces [11]. On a gold surface assumed to be atomically flat, the formation of a monolayer follows a stepwise mechanism. The initial step is the adsorption of the molecules to the surface. During this step, the molecules typically lie flat across the surface. When the coverage of the surface is nearing capacity, the flat lying molecules are forced to reorient into positions that allow the remaining surface vacancies to be occupied; this step is referred to as nucleation. Nucleation does not occur simultaneously across the surface, but instead typically occurs in dense patches. The orientation of the molecules in these patches is one where the carbon chains align at a specific angle relative to the surface. This angle between the adsorbed molecules and the surface is typically close to but not exactly normal. This specific orientation allows for the dense formation of the monolayer. This non-perpendicular arrangement also infers that the height of the monolayer on the surface is not the same as the length of the molecule adsorbed but will be slightly less. DPN patterns of thiolated molecules on gold have been proven to follow a similar adsorption mechanism as that which occurs during bulk adsorption of the same molecules on gold [45].

The morphology of the patterns generated with alkylthiols on gold can be controlled by tuning the writing speed [46]. This evidence provides further proof that the mechanism of formation of the adsorbed monolayer patterned by DPN is similar to that of the bulk SAM formation mechanism. The control over the morphology can be explained by the influence of the time that the adsorbed molecules are exposed to the meniscus used for writing. It was found that the relative height of the patterned monolayers would decrease as the writing speed was increased. This implies that the initial deposition of molecules occurs and the orientation of the molecules is flat on the surface. The quick writing speed would move the meniscus away from the deposited molecules too quickly and will not be allowed to induce their reorientation. The lack of reorientation of the deposited molecules is due to the lack of molecules competing for the same surface sites and a shortage of solvent due to the absence of the meniscus. Under these quick writing conditions an incomplete monolayer is formed.

23.4.2 ω-Substituted Thiols

The interfacial properties of the DPN patterned monolayer of a thiolated molecule can be controlled by employing a molecule with a specific terminal functionality opposite the thiol end of the molecule (ω-terminated thiols). Alternatively, a mixed monolayer can also be formed through chemical modification of the monolayer after patterning [47]. A number of ω-terminated thiols exist and those typically used for SAM formation are the amino, carboxylic acid and hydroxyl terminated varieties. These molecules provide versatile platforms for further bottom-up directed nanofabrication schemes.

Such control over the interfacial properties of the surface can be seen by comparing the properties of ODT and MHA. Octadecanethiol is a long thiolated hydrocarbon that terminates with a methyl group. Mercaptohexanoic acid is a thiolated hydrocarbon that is of similar length as ODT but contains a terminal carboxylic acid. When a full monolayer of either molecule is allowed to adsorb on the surface the wetting properties of the differing surfaces should be different due to the difference in the hydrophobicity between the two molecules. This difference can be examined using lateral force microscopy (LFM) to image patterns formed by the different molecules [48]. The force between the hydrophobic ODT monolayer and the tip will be much less than the force between the more hydrophilic MHA monolayer and the tip when compared to the surrounding unpatterned gold surface. The process depends on the quality of the monolayer formed during patterning. A pattern of MHA written with a faster speed will appear to have the same attraction to the tip as a completely formed monolayer pattern of ODT. Such an observation is thought to be due to the orientation of the molecules and the nature of the terminal chemical groups.

We previously stated that thiolated molecules that are typically used in DPN patterning are the amino, carboxylic acid and hydroxyl terminated thiols. Once patterned, the terminus of the molecule can be used as a template for higher order nanofabrication schemes. In earlier sections, we discussed the fabrication of a virus specific pattern using DPN fabricated templates of amine terminated thiols [44]. Additionally, we also discussed how the carboxylic acid terminated MHA has been used as a template for the indirect patterning of proteins [31].

The orthogonal assembly of charged nanostructures has been formed using redox-active ferrocenylalkylthiol inks and polyanionic oligonucleotide-modified gold particles [49]. Two different ferrocenylalkylthiols were chosen, ($Fc(CH_2)_{11}SH$ and $Fc(C{=}O)(CH_2)_{11}SH$), due to their non-overlapping redox processes. The inks were patterned on gold with line widths as small as 55 nm. An electrochemical cell was then employed using the patterned substrate as the working electrode to apply potentials that would oxidize the generated patterns. The oxidized patterns could be used for the concomitant assembly of the polyanionic particles. Due to the large difference in the redox potentials, the guided assembly of the particles could be controlled using linear sweep voltammetry.

Thiols have also been used to form mixed monolayers on a surface using traditional SAM formation procedures. The properties of a mixed monolayer can be controlled by the types of thiolated molecules (different ω-terminated thiols, different number of methylene units) that are used and the stoichiometric concentrations of each type of thiolated molecule present. In classic formation of SAMs, the molecules were allowed to compete with each other for surface binding sites until complete coverage was attained. A method of directed formation of mixed monolayer patterns was discovered using a technique that employs a scanning probe lithography method called the nanopen reader and writer (NPRW) [50]. This method is similar to DPN in that a coated tip is used to transfer molecules. What is different from DPN is that typically the surface is already modified with one type of SAM. The coated tip is used to do two tasks simultaneously. The first task is to remove the SAM below the area of scanning by using a high contact force. The high contact force shears off the adsorbed monolayer. The second task is to fill-in the newly exposed bare area of the substrate with the molecule physisorbed onto the tip. The resulting substrate will have a monolayer of patterns with resolutions comparable to DPN.

23.4.3 Silanes and Silazanes

Silanes and silazanes are important in the field of solid state microfabrication. They are often used as adhesion layers to promote the film formation of photoresists onto surfaces. Neither compound reacts with gold surfaces, but the silane and silazane ends of the molecules can interact with surface oxide groups to become immobilized. Several of the commonly used trichlorosilanes and trialkoxysilanes undergo polymerization reactions when exposed to water.

Hexamethyldisilazane (HMDS) was used to pattern both silicon oxide and gallium arsenide surfaces by DPN [51]. This silazane is a common adhesion layer promoter film and is not likely to polymerize when in contact with the meniscus during the transfer procedure. DPN generated patterns smaller than 100 nm were achieved on both types of semiconductor surfaces. It was reported that the transfer of the silazane ink was much slower than the reported transfer rates of thiols on gold. The rate of transfer was altered by the increase in the local temperature of the tip and substrate. This data indicates that temperature may be an important variable in the transfer of certain inks in the DPN process.

Despite the difficulties associated with patterning of alkoxysilanes and trichlorosilanes, it has been shown that under a controlled environment that impedes the polymerization of these molecules patterning is possible [52,53]. These are important molecules because standard microarray technology employs them as the starting

SAM molecules on glass and silicon oxide surfaces. Successfully patterning these molecules on substrates is an important advancement for the fabrication of nanoarrays and biosensors. The molecule patterned was 3′-mercaptopropyltriethoxysilane (MPTMS) on glass [52]. The functional groups of other silanes tend to interact with oxide surfaces and, therefore, interfere with the patterning, while the thiol group is relatively inert towards glass surfaces. The patterning was performed under humidity conditions from $\sim 0\%$ to ambient humidity in order to limit the polymerization of the silanes. It was found that at low relative humidity values MPTMS could be controllably patterned without significant polymerization. The proposed mechanism of reaction for silanes and silazanes on glass and silicon oxide involves the presence of adsorbed water in order to covalently bind the molecule to the surface. The silanes were also shown to remain active after patterning by coupling the thiol group of the adsorbed silane with biotin and Cy3-streptavidin. The coupled complex displayed fluorescence on the areas that were specifically patterned.

23.4.4 Deposition of Solid Organic Inks

AFM cantilevers fabricated with embedded resistive heating elements have been employed to deposit organic inks. The organic inks were chosen to exist as solids at room temperatures [54]. A resistive heater was used to increase the local temperature

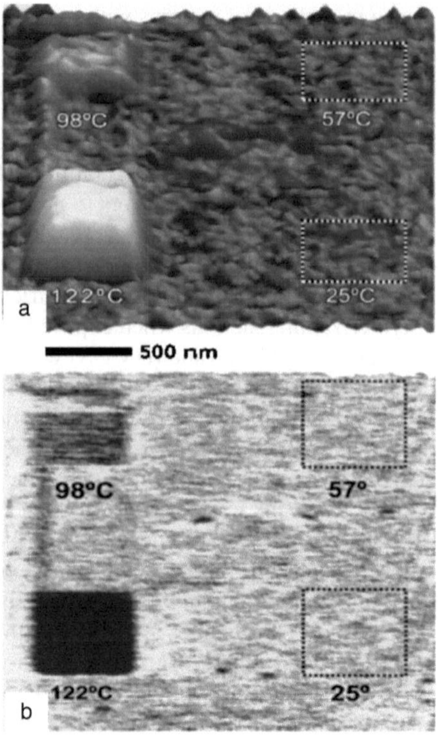

Fig. 23.6. (**a**) AFM height image of OPA patterns generated using thermal dip-pen nanolithography (tDPN) on a mica surface by scanning for 256 s on a 500 nm × 500 nm area. The melting point of OPA is 99 °C. Patterning below this temperature did not deposit any ink on the surface. As the temperature approached and exceeded the melting point of the ink deposition occurred. (**b**) AFM friction image of the same surface. Image reprinted with permission from [54]

of the tip to the melting point of the solid ink physisorbed onto it. Upon heating the ink melts and deposition onto the surface occurs. The use of the heating element allows for controlled deposition of the ink that can be stopped without withdrawal of the tip from the substrate. The tip can also be used in its "cold" state to image the patterned area without risk of further ink deposition. The ink used in the proof-of-concept study was OPA, which has a melting temperature of 99 °C, Fig. 23.6. This ink can form SAMs on mica, stainless steel, aluminum and metal oxides in its liquid state. Deposition of the OPA onto mica was studied at several different localized tip temperatures. Controllable patterning of lines that were 100 nm in width was routinely accomplished.

23.4.5 Polymers

Advances in polymer synthesis have accelerated greatly due to the increase in applications of polymers in nanoscience [55]. The electrical, optical, catalytic and mechanical properties of polymers are dictated by their design and synthesis, which opens up a host of possible applications and opportunities. Polymers may be synthesized to interact with biological cells, tissues or organs in predetermined ways and are currently being explored for uses such as drug delivery, material coatings and implants.

Polymers can either be directly immobilized onto surfaces using covalent bonding, electrostatic forces or van der Waals interactions, or they can be deposited in their monomer state and polymerized in situ. By employing the direct writing of polymers on surfaces, one can create controlled feature sizes and placement. In principle, due to the high molecular weights and low vapor pressures, the resolution of polymer inks should be lower than that encountered with thiol inks.

Using DPN to pattern luminescent conductive polymers into nanowires onto glass has been demonstrated using MEH-PPV as ink [56], Fig. 23.7. This polymer displays luminescent properties that make it an attractive candidate for the fabrication of nano-LEDs. Nanowires of conductive polymers such as MEH-PPV can also be

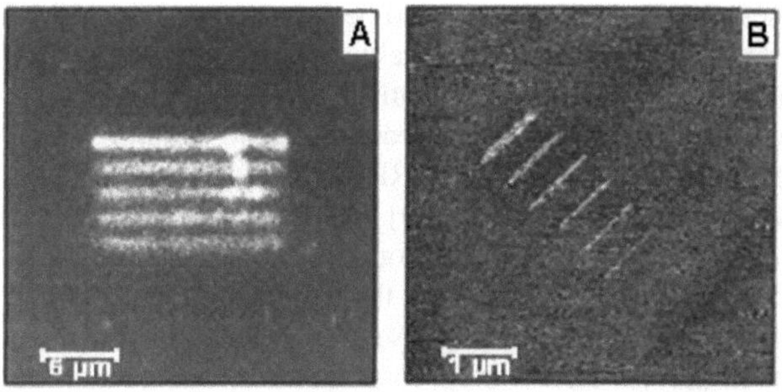

Fig. 23.7. (**a**) Scanning confocal microscopy images of a series of MEH-PPV lines generated using DPN on glass. The scanning speed used to generate the *lines* decrease from the top line down. With the decrease in scanning speed, the fluorescence intensity is also seen to decrease. (**b**) LFM of MHA lines patterned on gold. Reprinted with permission from [56]

used to study the basic principles of photoluminescence and electrical transport in an idealized one-dimensional system. The polymer wires that were deposited had line widths as small as 30 nm. Like the patterning of more traditional chemical inks, such as ODT, it was determined that the line width was inversely proportional to the scanning speed.

The electrostatic attraction between conductive polymers and surfaces has itself been employed as the driving force to fabricate assemblies of nanoscale features using DPN [57]. To this end, silicon substrates were appropriately modified to impart either a net positive charge or a net negative charge onto the surface. The oppositely charged polymers were used as inks and the electrostatic interactions between the ink and the substrate were used to immobilize the polymer patterns. SPAN, a negatively charged polymer, and doped PPy, a positively charged polymer, were used as the polymer inks. It was noted that the SPAN polymer was not significantly deposited onto the negative substrate in control experiments and the PPy polymer was not significantly deposited onto the positively charged substrates. These experiments concluded that the mechanism causing the transfer was the long-range electrostatic attraction between the surface charges and the charge of the physisorbed ink on the tip as patterning was occurring. Electrochemical analysis of the deposited polymer patterns confirmed that the transfer occurred successfully.

The previous applications involved the transfer of the polymer to the surface after polymerization occurred. The ability to transfer monomers onto the surface and to polymerize them in situ is also possible. This technique can be advantageous because many of the polymers are insoluble in water while their monomers are soluble. Therefore, controlling the deposition of the polymers may be challenging using a water meniscus. A drawback to the in situ polymerization is that controlled doping of the polymers may be made difficult, which limits their functionality in applications. Proof-of-concept experiments have demonstrated that a monomer ink can be patterned onto SiO_2 substrates [58]. The chosen ink was required for the acid-promoted polymerization of pyrrole. The deposition was dependent on the degree of polymerization that the ink had undergone prior to writing. The monomer was also deposited into an electrode gap so that I–V measurements could be performed to characterize the conductance of the patterned material. A dependence of the resistance on the relative light intensity was reported, which indicates the presence of photoluminescent properties that would be expected in polymerized pyrrole.

In an effort to reduce the environmentally damaging products used in the synthesis of polyaniline polymers and its derivatives, HRP was introduced as a catalyst to induce the free-radical polymerization reaction. Deposition of 4-aminothiolphenol using DPN on gold and subsequent HRP catalyzed reaction to form polymeric conducting wires was accomplished [59]. 4-aminothiolphenol is a monomer that forms the polymer structure poly(4-aminothiolphenol). The line widths and heights changed upon polymerization, indicating the patterned monomers had polymerized upon exposure to HRP. This and spectroscopic evidence of the polymerization led to the conclusion that the HRP catalyzed polymerization reaction occurred on the monomer patterns.

E-DPN has been employed to write poly(thiophene) polymer nanowires on semiconducting and insulating surfaces [60]. This method employs an electrochemical in situ polymerization reaction between the tip and sample during patterning. The monomer of interest EDOT was employed as the ink. During patterning, a negative bias voltage is applied to the tip resulting in polymerization. The polymer was then directly deposited as in typical DPN experiments. When no voltage was applied to the system, no polymer was deposited. The control over the morphology of the patterns was determined to be dictated by the standard conditions used to control deposition via DPN, as well as the applied voltages. The ability to pattern these polymers on both semiconducting and insulating surfaces allows the possibility of integration into novel device structures.

23.4.6 Polyelectrolytes

Polyelectrolytes are a class of macromolecules that have been extensively researched in nanoscience due to their ability to self-assemble when deposited onto charged surfaces. The self-assembly of these molecules has been used in a method called layer by layer assembly. Polyelectrolytes do not covalently link as polymers do, but instead rely on electrostatic interactions to form ordered complexes. Layer by layer assembly exploits the electrostatic interactions of polyelectrolytes by forming layer upon layer of oppositely charged thin films on surfaces. Since several polyelectrolytes are commercially available, it is possible to alternate not only the charge, but also the composition of several different layers during assembly. Polyelectrolytes have been used on several types of substrates and on particle surfaces. We have discussed the use of polyelectrolytes that have been patterned onto SiO_x as templates for the guided assembly of stretched long DNA molecules [42].

An interesting study comparing the mechanical properties of polyelectrolyte patterns created using both DPN and microcontact printing has been carried out [61]. The polyelectrolyte PAH was patterned using DPN onto SiO_x. Microcontact printing was then performed to generate micron sized lines on top of the DPN patterns. By direct comparison of tapping-mode images, it was noted that the DPN patterns are visible only by phase contrast imaging due to their relatively small heights, while the microcontact printed patterns can be discerned both in phase and height mode images. FV imaging was performed to discern whether the mechanical properties of the two patterns generated varied. From the FV experiments, it was deduced that the relative hydrophilicity of the DPN generated PAH patterns was greater than that of the microcontact printing PAH patterns.

23.4.7 Dendrimers

Dendrimers are an important class of macromolecules that offer researchers the flexibility of polymers but are often easier to control. Dendrimers are branched three-dimensional molecules that can be designed with specific surface functionalities through chemical modifications. These molecules have attracted extensive research interest in sensing, electronics, luminescent and drug delivery applications. Starburst PAMAM has shown promise in the transport of DNA across cell membranes [62]. The high degree of flexibility in the surface functionality makes dendrimers the perfect candidates for both direct and indirect writing experiments on several surfaces.

Direct writing has been performed with amino terminated dendrimers onto silicon oxide surfaces using DPN [63]. These experiments were already successful when attempted by microcontact printing and the resulting patterns were determined to be stable [64]. It should be noted that dendrimers are macromolecules that have high molecular weights that, in principle, impede transport during patterning. To further understand the transfer kinetics, the patterning of five generations of starburst PAMAM was performed, each of differing molecular weights. It was noted that the deposition of the higher molecular weight generations of PAMAM was slower than the lower molecular weight generations, as expected from diffusion models.

PAMAM has also been patterned onto NHS ester SAMs modified gold [65]. The esters exposed by SAMs allow for covalent attachment to the PAMAM by amide bond formation. Direct DPN patterning onto glass and mica was shown to be difficult due to a lack of stabilizing interactions between the surface and ink. When PAMAM ink was covalently coupled to NHS, stable patterns below 100 nm could be routinely fabricated.

23.4.8 Deposition of Supramolecular Materials

Supramolecular compounds have been of recent interest due to their flexibility in chemical, electronic and recognition functionalities. The recognition functionality is displayed by the specificity of host-guest interactions between calixarenes host

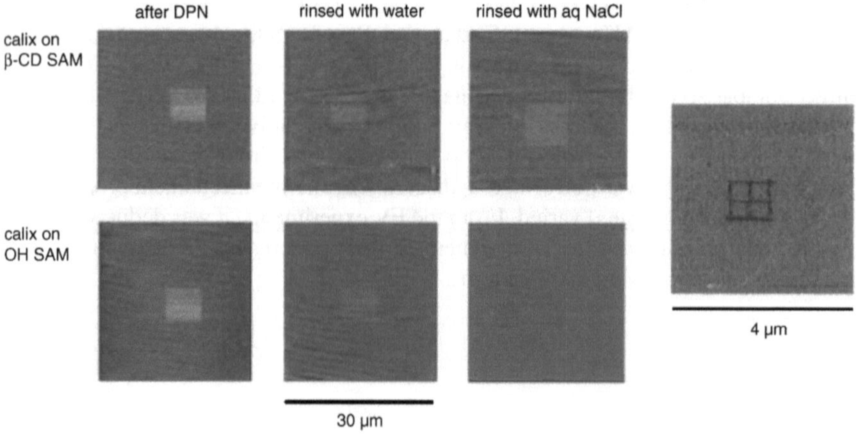

Fig. 23.8. LFM images (*left*) of calixarene patterns generated using DPN. The *first row* shows images of calixarenes patterned onto β-cyclodextrin (β-CD) SAMs on gold. The *second row* shows images of calixarenes patterned onto 11-sulfanyl-1-undecanol (OH) SAMs on gold. The patterns generated on β-CD terminated SAMs were shown to remain intact even after washing in both water and NaCl solutions. The patterns generated on OH SAMs rinsed off when washed in water and NaCl solutions. The image on the *right* shows an LFM image of 60 ± 20 nm lines of calixarene molecules generated on gold. Reprinted with permission from [66]

molecules and univalent guest molecules in solution. These interactions can be used for sensing, electronic and optical purposes through the introduction of specific guest molecules. As a proof-of-concept experiment calixarene inks were patterned on gold surfaces modified with β-cyclodextrin (β-CD) or onto 11-sulfanyl-1-undecanol SAMs, [66] Fig. 23.8. Patterns of the calixarene were generated on the β-CD SAM which remained stable even after washing. The patterns generated on the 11-sulfanyl-1-undecanol SAMs were easily removed from the surface by rinsing. The resolution of the patterns was sub 100 nm on both surfaces.

Heteroatom-functionalized porphyrazines synthesized to be terminated on mercaptides have also been patterned on gold surfaces using DPN [67]. These experiments employed two types of pz molecules designed to adsorb either perpendicular to the surface or lie flat on top of the surface. To this end, the pz molecule was either synthesized with disulfide linkers on the same side (perpendicular arrangement) or with disulfide linkers on opposite sides (flat lying). Electrochemical measurements confirmed that in solution the redox potential of the molecules was identical. The redox potential varied based on orientation when SAMs of each molecule were formed. The change in the redox potential induced by the orientation of the molecules may be useful in the design of molecular devices.

23.4.9 Deposition of Metals

Inorganic materials are important inks for DPN based on their many important applications, including electronics, optics and catalysis. Electronic solid state fabrication procedures typically involve thermal evaporation, e-beam evaporation or sputtering of a desired metal or alloy to form a thin film of the metal on the surface [1]. This is typically followed by an etching step to remove the metal from unwanted areas of the substrate. The power of DPN for the deposition of metals is that this technique can be used in either bottom-up assembly or top-down fabrication.

The assembly of inorganic nanostructures using the bottom-up approach relies on techniques that are similar to those we have already seen, although incorporation of metal or nanoparticle inks instead of thiols, silanes or biomolecules is employed. The top-down method can be achieved by depositing molecules as a resist onto a thin film of metal and subsequently etching away the unpatterned areas. Both approaches have been demonstrated and the top-down approach for fabricating metal nanostructures will be covered in greater detail in the section on engineering applications of DPN.

An important metal in modern IC device structures is gold. Gold is often used for interconnects and contacts on microelectronic devices [1]. DPN has been used to deposit gold onto silicon substrates with sub 100-nm resolution [68]. The mechanism of the feature formation was dependent on the use of an $HAuCl_4$ ink deposited onto an HF etched Si substrate. The Au(III) metal is reduced to an insoluble form, Au(0), by the silicon surface. Once patterned, the reduced metal nanostructures are stable to washing with both aqueous and organic solvents, heating to below 500 °C, and HF etching. It was found that the patterns underwent a phase transition above 500 °C and formed aggregates of gold on the surface above this temperature. These findings indicate the utility of the generated Au patterns and their ability to withstand some abrasive conditions typical of solid state fabrication procedures.

Using E-DPN, several other metal patterns have been generated on silicon surfaces [69]. The underlying mechanism was the reduction of the metal salts using a voltage applied to the tip during transport to create an electrochemical cell out of the meniscus. Platinum, gold, silver, copper, palladium and germanium have all been patterned using this technique. The compositions of these features was also tested by HF etching and thermally induced phase transitions to ensure that the applied voltage was depositing the metal and not oxidizing the substrate. The platinum features displayed catalytic activity when reacted with ethylene, which was characteristic of bulk platinum surfaces.

The use of DPN to deposit metal clusters or nanoparticles onto surfaces has also been of much recent interest. Metal nanoparticles exhibit optical, electronic and catalytic properties which may be size dependent and differ from their complementary bulk materials. For electronic applications, the positioning of metal nanoparticles with high registration is of great importance in fabricating working structures. Early work prior to the advent of DPN technology showed that the tip of the atomic force microscope could be used to controllably manipulate the positions of nanoparticles on surfaces [70–73]. To this end, particles on the surface were pushed with the tip in contact mode, resulting in the movement of the particle as directed by the motion of the cantilever. This technique remains serial and the movement of large numbers of particles is time-consuming because only single particles or small aggregates of clusters are moved. DPN can be used to generate patterns of particle inks on a number of different surfaces through both direct writing and indirect templating procedures [74].

The NPRW system employed to fabricate precise mixed monolayers has also been used to successfully generate 3D patterns of thiol passivated gold nanoparticles [75]. The physisorbed gold particles on the tip are transferred to the surface during scanning while SAM resist on the surface is simultaneously removed due to the shear forces applied. Single monolayer deposition of the particles was achieved when a slow scan rate was employed.

23.4.10 Deposition of Solid-State Materials

In typical top-down techniques, the underlying semiconductor material is etched away in a step-wise process to fabricate devices of specific geometries, orientations and chemical compositions. The procedures to fabricate these devices are parallel processes that generate structures on a wafer scale level. It is imperative that patterns of materials typically used by the semiconductor industry be successfully patterned in order to successfully achieve bottom-up approaches to nanofabrication of electrical circuitry.

To meet this goal, direct patterning of solid state materials using inorganic precursor salts with amphiphilic block copolymer surfactants has been employed to generate patterns of Al_2O_3, SiO_2 and SnO_2 on silicon and silicon oxide substrates [76, 77]. Bulk surface depositions of these materials have been performed using sol-gel chemistry on semiconductor-oxide surfaces. The patterned composite hydrolyzes on the underlying substrate forming the desired inorganic nanostructure. To remove the organic copolymer, the surfaces were heated to 400 °C, causing the organic residues to combust. It was noted in these experiments that the diffusion of the ink qualitatively acted in a similar way to thiol inks.

23.4.11 Deposition of Magnetic Materials

Magnetic materials are often found in applications such as high density recording media, magnetoelectronics and biotechnology. In the case of high density recording media, shrinking these features to the nanoscale would increase the information density by orders of magnitude over the current limits. Conventional fabrication of these devices is challenging due to current etching procedures. Because these devices shrink, using top-down methods becomes ever more difficult. Patterning by DPN is an alternative that can allow nanoscale control over feature size and interfeature distances, which are both important for the standard operation of such devices.

The magnetic material BaFe is considered a "hard" magnet and is useful for high density recording applications. Patterns of BaFe have been generated on silicon oxide substrates with sub 100 nm resolution using a sol-based ink containing a mixture of precursors necessary for the synthesis of BaFe [78]. After patterning, the substrates were thermally treated to yield polygonal BaFe particles with a mean diameter of $\sim$35 nm. The arrays patterned displayed characteristics of single domains or at most just a few domains. The elemental composition was characterized using various spectroscopic analysis techniques and the magnetic properties were displayed using MFM.

The direct writing and indirect templating of magnetic nanoparticles with nanoscale precision has been demonstrated using DPN. In direct writing experiments, Fe_2O_3 nanoparticles with an 11 nm diameter were deposited onto both silicon and mica [79], Fig. 23.9 The generated patterns exhibited sharp edges, which indicated little lateral diffusion and a high control over the positioning of the particles. Indirect templating experiments relied on the DPN patterning of MHA and electrostatic interaction as the driving force to generate arrays of iron oxide nanoparticles [80]. Iron oxide nanoparticles capped with tetramethylammonium hydroxide were synthesized with a $\sim$ 10 nm diameter. Features were routinely generated with less than 100 nm diameters on gold. The cross-sectional topography of the templated structures indicated that magnetic particles formed single layers on the MHA patterns.

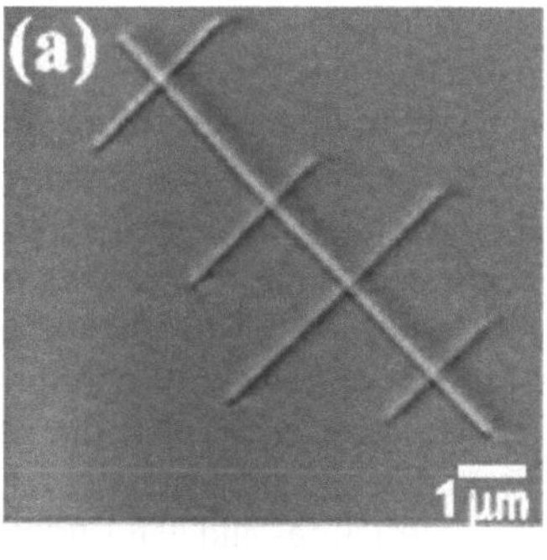

Fig. 23.9. Contact mode AFM height image of γ-Fe_2O_3 nanoparticle patterns generated by DPN on a mica surface. Control over the position of the particles during patterning without substantial lateral diffusion can be seen. Figure reprinted with permission from [79]

23.5 Engineering Applications of DPN

Several of the inks described in the previous sections can be used in a wide range of engineering applications. Biomolecules are commonplace in catalysis and fermentations. Biomolecules also provide a potential source as components for the fabrication of 3D electrical or mechanical architectures. Many of the chemicals patterned also have inherent properties that make them useful in applied technologies. The interfacial control achieved over surfaces through the patterning of thiols, the photochemical properties of patterned silanes, silazanes and other polymers, as well as the electrical properties of deposited metals and sol-based inks are all useful for engineering purposes. The deposition of these important materials on the nanoscale also allows researchers to study the fundamental properties of the materials and gain further insight regarding their potential.

Integration of DPN with existing methods for creating microstructures is an area of research that is actively being explored and may function to increase both the recognition of DPN and to concurrently improve these existing technologies. Typical photolithography approaches depend on the use of a photomask to fabricate structures from the top down. The photomask can be fabricated on various types of fused silica, while the opaque layer is typically chromium [1]. Any defect in the chromium structure of the mask will be translated directly on the fabricated features resulting in defects. Typically, these defects are repaired by additional deposition of the metal when found. As device size shrinks to sub 100 nm resolution, it is extremely important that the opaque layer contains zero defects. DPN can be employed in the repair of such small mask defects through directed site specific deposition of chromium onto defective sites.

Other non-conventional soft lithography microfabrication methods can also benefit from the integration of DPN. One such method is microcontact printing. μCP is a soft lithography method that involves the transfer of a molecular ink to a substrate using an elastomeric stamp. The process has the same versatility of inks and substrates as DPN. The resolution of μCP and its derivatives is typically on the microscale, although by using edge-transfer methods or controlling the diffusion of the ink, one can decrease the resolution to the nanoscale [81]. One advantage of the technique is that the physical size of the stamp is not limited and patterns are formed in parallel. However, one of the problems typically encountered when μCP is used is the inability to pattern multiple types of inks with high registration, since the technique relies on the physical stamping of the ink to the surface to define the patterns. Once the pattern is generated on the substrate, traditional SAM techniques can be used to passivate bare areas on the surface, although this typically only allows the patterning of two distinct regions on the surface: the μCP generated pattern and the area passivated. To overcome this limitation, a procedure similar to DPN has been used to ink the features of the elastomeric stamp used for μCP with different molecules [82]. This method of integrating scanning probe methods and μCP has allowed the transfer of multiple ink types using a single stamp.

We have also noted the generation of polyelectrolyte layers on silicon using first DPN and subsequently μCP [61]. This technique can be used to generate both microstructures and nanostructures on the same surface. The alignment and registration between the deposited using the two different lithographic techniques remains challenging. The direct comparison of the chemical, mechanical and electrical properties of materials deposited through these two techniques can be used to answer fundamental questions regarding the differing nature of the inks when existing in the microscale versus the nanoscale.

Researchers have also combined the two techniques of DPN and μCP by fabricating dip-pen nanolithography stamp tips. To this end, standard AFM tips were coated with the elastomer PDMS and allowed to cure [83], Fig. 23.10. Patterning of MHA, ODT and cystamine on gold was displayed, as well as G6-OH dendrimer patterning on gold and SiO_x. It was noted that the PDMS coated tip absorbs the ink and the solvent used during tip coating. Structures of MHA and ODT were patterned on sub 100 nm scales almost approaching the limits of DPN generated patterns and exceeding the limits of those patterns generated by μCP. Efforts have also been made to generate cantilevers from PDMS elastomers [84]. Such cantilevers have been used to generate sub-500 nm patterns of ODT on gold, although the lack of reflective metal coatings on the cantilever make control over the AFM feedback difficult.

The deposition of biomolecules, nanoparticles, organic molecules and magnetic materials on microfabricated devices using DPN has also been used to study the nanoscopic properties of these materials. The registration and resolution of the DPN technique has allowed researchers to deposit inks in electrode gap junctions, allowing one to directly probe the electrical properties of the ink. Electrical characterization of these materials on the nanoscale is important on a fundamental level when considering their uses for future nanodevices, (bio)arrays, sensors or single molecule electronics.

DPN has also been used to pattern catalysts in desired locations in order to grow nanowires [85]. Specifically, $Ni(NO_3)_2$ patterns were generated directly onto SiO_2 substrates. After patterning, the substrates were exposed to elemental gallium and ammonia to induce the growth of gallium nitride (GaN) nanowires through a vapor–liquid–solid (VLS) mechanism. Nanowires several microns long were grown from

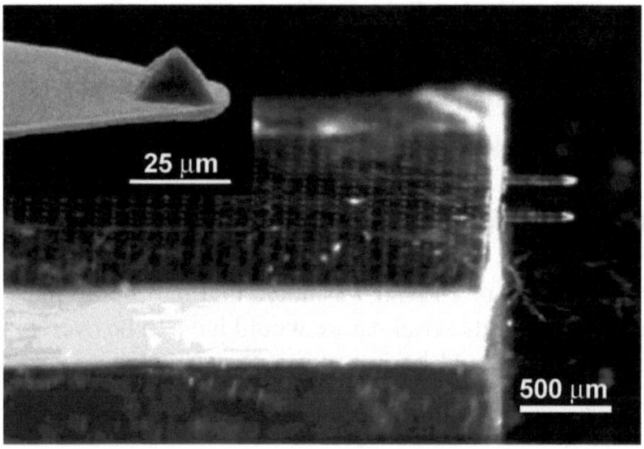

Fig. 23.10. Optical micrograph of a chip containing two scanning probe contact printing tips. *Inset*: Scanning electron micrograph of scanning probe contact printing tip composed of a PDMS tip and a polyimide cantilever. Image reprinted with permission from [84]

the catalyst islands. Metal electrodes were evaporated onto the wires after growth and the electrical characteristics of the wires were determined to match their expected values.

Existing top-down technologies relying on wet chemical etchants and wet chemical oxidants may also be interfaced with DPN to control the depletion or oxidation of site-specific areas of a substrate. Wet chemical oxidants have been employed as inks and used to controllably fabricate heterostructure nanowire materials. Further control over the degree of oxidation of the native gallium nitride (GaN) nanowires has been demonstrated by employing E-DPN to tune the deposition of the oxidant [86]. Specifically, KOH ink was employed to site-specifically oxidize native GaN nanowires forming GaN/gallium oxide heterostructures. Once the E-DPN induced oxidation occurred, the resulting gallium oxide regions of the nanowires could be etched using an acidic etchant solution. Devices containing the GaN nanowires were fabricated to test the electronic properties of the native wires, the heterostructured wires and the acid etched wires. It was noted that the heterostructured wire was far less conductive that the native GaN wire. This was attributed to the behavior of the gallium oxide as a barrier to electron transport through the wire. After etching of the gallium oxide, a significant decrease in the current was always observed. Exploitation of the bias voltage applied during patterning, the dwell time of the tip and the relative humidity can all be used to form oxide structures that are of different diameters on the GaN wires. By controlling these parameters, one can control the degree of heterostructures and thereby control electronic properties of the individual wires.

Several reports of using DPN to pattern resist layers on a number of surfaces have been published recently [87]. ODT patterns have been used as resists on gold and palladium surfaces to protect areas of the substrate from wet chemical etching. MHA has also been successful as a resist on gold and silver substrates. Using these methods, arrays of gold structures functionalized with oligonucleotides have been formed. MHA ink was deposited as the etch resist onto a gold surface to form the arrays. The Au was etched resulting in nanostructures protected by the MHA patterns. The samples were irradiated with UV light for 10 hrs and rinsed with water to remove the MHA. The result was bare gold nanostructures that could be functionalized with disulfide modified oligonucleotides. The controlled formation of nanogaps between gold electrodes has also been demonstrated using MHA inks. Nanogaps with distances less than 50 nm can be routinely fabricated using this method.

23.6 Future Challenges and Applications

Shortly after the discovery of DPN, it was noted that in order to make the technique a production scale method, its serial nature would have to be overcome. To this end, arrays of cantilevers that could be used to generate patterns in parallel had to be fabricated. The proof-of-concept experiments employed standard contact mode tips that contained multiple cantilevers adjacent to one another as the multi-tip array. This array of tips was shown to be able to write both single ink types and multiple ink types in parallel with the alignment that was determined by the dimensions of the cantilevers. Since then, several groups have fabricated tip arrays of different tip geometries and writing capabilities. To further enhance the parallel development of DPN, cantilevers containing integrated microfluidic inking channels have also been developed [88].

These initial experiments were soon followed by the fabrication of probe arrays to directly address the parallel writing potential of DPN. Prior to the advent of DPN researchers had developed methods to fabricate arrays of multiple tips in an effort to increase the scan size and speed of imaging for AFM [89]. The probe arrays fabricated for the purpose of DPN differed from the arrays meant for imaging. Factors in fabrication of the DPN arrays included the force constants, length of the probe, sharpness of the tips and the tip to tip spacing. Two types of arrays were developed in these experiments: the first was a 32 pen array that could be used to fabricate structures of ODT with line widths averaging 260 nm. The second array was fabricated with sharper tips and 8 pens and achieved line widths of ODT ranging from 80–100 nm.

Recent experiments in this area have focused on fabricating arrays of individually addressable cantilevers. Thermal bimorph actuated arrays were fabricated where each individual cantilever contained its own actuator [90]. When a cantilever was heated using an embedded electrical heater, the thermal mismatch strain between the chromium/gold and silicon nitride layers of the cantilever would cause the probe to deflect away from the surface. When the heating element was switched off, the cantilever would approach the surface and begin writing. The vapor-deposited ODT ink arrays of 10 individually addressable cantilevers enabled the simultaneous generation of the numerals 0–9 on a gold surface with sub 100 nm resolutions.

23.7 Conclusions

In the few short years that the DPN technique for nanofabrication has been utilized, a number of different platforms with different applications have been actively explored. Many of these applications existed before the use of DPN, although no reproducible methods to segue these technologies into the nanoscale were available. The potential of DPN to revolutionize fields ranging from basic biology, chemistry, materials science, physics and engineering has yet to be fully reached. Advances in the parallel patterning and in atomic force microscopy in general will help this technique to achieve its full potential.

References

1. Campbell S (2001) Science and Engineering of Microfabrication. 3rd edn. Oxford University Press, New York
2. Becker RS, Golovchenko JA, Swartzentruber BS (1987) Nature 325:419
3. Eigler DM, Schweizer EK (1990) Nature 344:524
4. Crommie MF, Lutz CP, Eigler DM (1993) Science 262:218
5. Kim YT, Bard AJ (1992) Langmuir 8:1096
6. Liu G-Y, Salmeron MB (1994) Langmuir 10:367
7. Xiao X-D, Liu G-Y, Charych DH, Salmeron M (1995) Langmuir 11:1600

8. Kelley SO, Barton JK, Jackson NM, McPherson LD, Potter AB, Spain EM, Allen MJ, Hill MG (1998) Langmuir 14:6781
9. Whitesides GM, Mathias JP, Seto CT (1991) Science 254:1312
10. Dubois LH, Nuzzo RG (1992) Annu Rev Phys Chem 43:437
11. Ulman A (1996) Chem Rev 96:1533
12. Jaschke M, Butt H (1995) Langmuir 11:1061
13. Piner RD, Zhu J, Xu F, Hong S, Mirkin CA (1999) Science 283:661
14. Jang J, Hong S, Schatz GC, Ratner MA (2001) J Chem Phys 115:2721
15. Jang J, Schatz GC, Ratner MA (2002) J Phys Chem 116:3875
16. Sheehan PE, Whitman LJ (2002) Phys Rev Lett 88:156104
17. Rozhok S, Sun P, Piner R, Lieberman M, Mirkin CA (2004) J Phys Chem B 108:7814
18. Schwartz PV (2002) Langmuir 18:4041
19. Rozhok S, Piner R, Mirkin CA (2003) J Phys Chem B 107:751
20. Fodor SPA (1997) Science 277:393
21. Ginger DS, Zhang H, Mirkin CA (2004) Angew Chem Int Ed 43:30
22. Demers LM, Ginger DS, Park SJ, Li Z, Chung SW, Mirkin CA (2002) Science 296:1836
23. Demers LM, Park S-J, Taton TA, Li Z, Mirkin CA (2001) Angew Chem Int Ed 40:3071
24. Schwartz PV (2001) Langmuir 17:5971
25. Xu S, Liu GY (1997) Langmuir 13:127
26. Belaubre P, Guirardel M, Garcia G, Pourciel JB, Leberre V, Dagkessamanskaia A, Trevisiol E, Francois JM, Bergaud C (2003) Appl Phys Lett 82:3122
27. Wilson DL, Martin R, Hong S, Golomb MC, Mirkin CA, Kaplan DL (2001) Proc Natl Acad Sci USA 98:13660
28. Lee KB, Lim JH, Mirkin CA (2003) J Am Chem Soc 125:5588
29. Lim JH, Ginger DS, Lee KB, Heo J, Nam JM, Mirkin CA (2003) Angew Chem Int Ed 42:2309
30. Nam J, Han S, Lee K, Liu X, Ratner M, Mirkin CA (2004) Angew Chem Int Ed 43:1246
31. Lee KB, Park SJ, Mirkin CA, Smith JC, Mrksich M (2002) Science 295:1702
32. Zhang H, Lee K-B, Li Z, Mirkin CA (2003) Nanotechnology 14:1113
33. Hyun J, Lee W, Nath N, Chilkoti A, Zauscher S (2004) J Am Chem Soc 126:7330
34. Lee K-B, Kim E-Y, Mirkin CA, Wolinsky SM (2004) Nano Lett 4:1869
35. Hyun J, Ahn SJ, Lee WK, Chilkoti A, Zauscher S (2002) Nano Lett 2:1203
36. Boateng S, Lateef SS, Crot C, Motlagh D, Desai T, Samarel AM, Russell B, Hanley L (2002) Adv Mater 14:461
37. Xiao S-J, Textor M, Spencer ND (1998) Langmuir 14:5507
38. Kinsella JM, Ivanisevic A (2004) Appl Surf Sci 243:7
39. Cho Y, Ivanisevic A (2004) J Phys Chem B 108:15223
40. Agarwal G, Sowards LA, Naik RR, Stone MO (2003) J Am Chem Soc 125:580
41. Agarwal G, Naik RR, Stone MO (2003) J Am Chem Soc 125:7408
42. Nyamjav D, Ivanisevic A (2003) Adv Mater 15:1805
43. Cheung CL, Camarero JA, Woods BW, Lin T, Johnson JE, Yoreo JJD (2003) J Am Chem Soc 125:6848
44. Smith JC, Lee KB, Wang Q, Finn MG, Johnson JE, Mrksich M, Mirkin CA (2003) Nano Lett 3:883
45. Hong S, Zhu J, Mirkin CA (1999) Langmuir 15:7897
46. Barsotti RJ, O'Connell MS, Stellacci F (2004) Langmuir 20:4795
47. Laibinis PE, Whitesides GM, Allara DL, Tao YT, Parikh AN, Nuzzo RG (1991) J Am Chem Soc 113:7152
48. Zhou Y, Fan H, Fong T, Lopez GP (1998) Langmuir 14:660
49. Ivanisevic A, Im J-H, Lee K-B, Park S-J, Demers LM, Watson KJ, Mirkin CA (2001) J Am Chem Soc 123:12424

50. Amro NA, Xu S, Liu G-Y (2000) Langmuir 16:3006
51. Ivanisevic A, Mirkin CA (2001) J Am Chem Soc 123:7887
52. Jung H, Kulkarni R, Collier CP (2003) J Am Chem Soc 125:12096
53. Pena DJ, Raphael MP, Byers JM (2003) Langmuir 19:9028
54. Sheehan PE, Whitman LJ, King WP, Nelson BA (2004) Appl Phys Lett 85:1589
55. Loiacono MJ, Granstrom EL, Frisbie CD (1998) J Phys Chem B 102:1679
56. Noy A, Miller AE, Klare JE, Weeks BL, Woods BW, DeYoreo JJ (2002) Nano Lett 2:109
57. Lim JH, Mirkin CA (2002) Adv Mater 14:1474
58. Ming S, Aslam M, Fu L, Wu N, Dravid V (2004) Appl Phys Lett 84:4200
59. Xu P, Kaplan DL (2004) Adv Mater 16:628
60. Maynor BW, Filocamo SF, Grinstaff MW, Liu J (2002) J Am Chem Soc 124:522
61. Nyamjav D, Ivanisevic A (2004) Chem Mater 16:5216
62. Bielinska AU, Kukowska-Latallo JF, Johnson J, Tomalia DA, Jr JRB (1996) Nuc Acids Res 24:2176
63. McKendry R, Huck WTS, Weeks B, Fiorini M, Abell C, Rayment T (2002) Nano Lett 2:713
64. Li H, Kang D-J, Blamire MG, Huck WTS (2002) Nano Lett 2:347
65. Degenhart GH, Dordi B, Schonherr H, Vansco GJ (2004) Langmuir 20:6216
66. Auletta T, Dordi B, Mulder A, Sartori A, Onclin S, Bruinink C, Peter M, Nijhuis C, Beijleveld H, Schonherr H, Vansco GJ, Casnati A, Ungaro R, Ravoo BJ, Huskens J, Reinhoudt D (2004) Angew Chem Int Ed 43:369
67. Vesper BJ, Salaita K, Zong H, Mirkin CA, Barrett AGM, Hoffman BM (2004) J Am Chem Soc 126:16653
68. Maynor BW, Li Y, Liu J (2001) Langmuir 17:2575
69. Li Y, Maynor BW, Liu J (2001) J Am Chem Soc 123:2105
70. Schaefer DM, Reifenberger R, Patil A, Andres RP (1995) Appl Phys Lett 66:1012
71. Junno T, Deppert K, Montelius L, Samuelson L (1995) Appl Phys Lett 66:3627
72. Resch R, Baur C, Bugacov A, Koel BE, Madhukar A, Requicha AAG, Will P (1998) Langmuir 14:6613
73. Resch R, Lewis D, Meltzer S, Montoya N, Koel BE, Madhukar A, Requicha AAG, Will P (2000) Ultramicroscopy 82:135
74. Liu S, Maoz R, Sagiv J (2004) Nano Lett 4:845
75. Garno JC, Yang Y, Amro NA, Cruchon-Dupeyrat S, Chen S, Liu G-Y (2003) Nano Lett 3:389
76. Su M, Liu X, Li SY, Dravid VP, Mirkin CA (2002) J Am Chem Soc 124:1560
77. Su M, Li S, Dravid VP (2003) J Am Chem Soc 125:9930
78. Fu L, Liu X, Zhang Y, Dravid V, Mirkin CA (2003) Nano Lett 3:757
79. Gundiah G, John NS, Thomas PJ, Kulkarni GU, Rao CNR (2004) Appl Phys Lett 84:5341
80. Liu X, Fu L, Hong S, Dravid VP, Mirkin CA (2002) Adv Mater 14:231
81. Cherniavskaya O, Adzic A, Knutson C, Gross BJ, Zang L, Liu R, Adams DM (2002) Langmuir 18:7029
82. Gerding JD, Willard DM, Orden AV (2005) J Am Chem Soc 127:1106
83. Zhang H, Elghanian R, Amro NA, Disawal S, Eby R (2004) Nano Lett 4:1649
84. Wang X, Ryu KS, Bullen DA, Zou J, Zhang H, Mirkin CA, Liu C (2003) Langmuir 19:8951
85. Li J, Lu C, Maynor B, Huang S, Liu J (2004) Chem Mater 16:1633
86. Maynor BW, Li J, Lu C, Liu J (2004) J Am Chem Soc 126:6409
87. Zhang H, Li Z, Mirkin CA (2002) Adv Mater 14:1472
88. Ryu K, Wang X, Shaikh K, Bullen D, Goluch E, Zou J, Liu C, Mirkin C (2004) Appl Phys Lett 85:136
89. Bullen D, Chung S-W, Wang X, Zou J, Mirkin CA, Liu C (2004) Appl Phys Lett 84:789
90. Wang X, Bullen DA, Zou J, Liu C (2004) J Vac Sci Technol B 22:2563

[illegible]

13 Scanning Probe Microscopy: From Living Cells to the Subatomic Range

Ille C. Gebeshuber · Manfred Drack · Friedrich Aumayr ·
Hannspeter Winter · Friedrich Franek

From volume III

Abbreviations

AFM	atomic force microscopy
ATP	adenosine triphosphate
BDPA	a, g-bisdiphenylene b-phenylallyl
CDOS	charge density-of-states
ESD	electron stimulated desorption
HOPG	highly oriented pyrolytic graphite
ID	interstitial defect
MCI	multiply charged ion
MEMS	microelectromechanical system
MR	magnetic resonance
MRFM	magnetic resonance force microscopy
MRI	magnetic resonance imaging
NEMS	nanoelectromechanical systems
PS	potential sputtering
PSD	photon stimulated desorption
rms	root mean square
SEM	scanning electron microscopy
SPM	scanning probe microscopy
UHV	ultra-high vacuum
VD	vacancy defect

13.1 Introduction

In this chapter the reader will be introduced to scanning probe microscopy of samples varying by seven orders of magnitude in size (Fig. 13.1). The largest samples presented are living cells, measuring some hundreds of micrometers. Small units of life, biomolecules with only some tens of nanometers, are the next sample. They are investigated while interacting with each other in real-time. One more step down in size, small ion-induced defects on atomically flat crystals represent structures in the nanometer regime. New data storage devices might result from such investigations. Finally, single electron spin detection (dozens of atomic layers beneath the surface) and the imaging of atom orbitals extend scanning probe microscopy to the subatomic regime. Gathering of 3D atomic-level information of (bio)molecules embedded in

In: H. Fuchs, B. Bhushan (eds), *Biosystems - Investigated by Scanning Probe Microscopy*,
DOI 10.1007/3-540-26910-X_2, © Springer-Verlag Berlin Heidelberg 2010

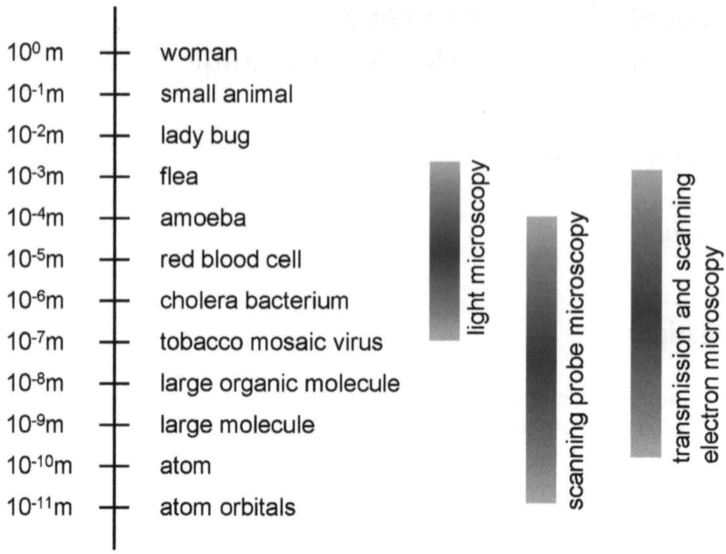

Fig. 13.1. The major types of microscopy cover at least eight orders of magnitude in size. Common examples for every scale are given. Note that scanning probe microscopy covers seven orders of magnitude

their natural environment or single defect imaging in bulk silicon might be possible with these new techniques in the near future.

These versatile applications demand methods such as scanning tunneling microscopy at ultra-low temperatures (1.6 K) or atomic force microscopy in ultra-high vacuum (10^{-11} mbar). Furthermore, in many cases, specially engineered and/or functionalized scanning probe tips are needed.

13.2 Cells *In Vivo* as Exemplified by Diatoms

13.2.1 Introduction to Diatoms

Diatoms [1] are unicellular microalgae with a cell wall consisting of a siliceous skeleton enveloped by an organic case essentially composed of polysaccharides and proteins [2].

Diatoms are small, mostly easy to cultivate, highly reproductive and, since many of them are transparent, they are accessible by different kinds of optical microscopy methods.

The cell walls form a pillbox-like shell (siliceous exoskeleton). This shell consists of two valves and a series of girdle bands. Diatoms vary greatly in shape, ranging from box-shaped to cylindrical; they can be symmetrical as well as asymmetrical and exhibit an amazing diversity of nanostructured frameworks (Fig. 13.2).

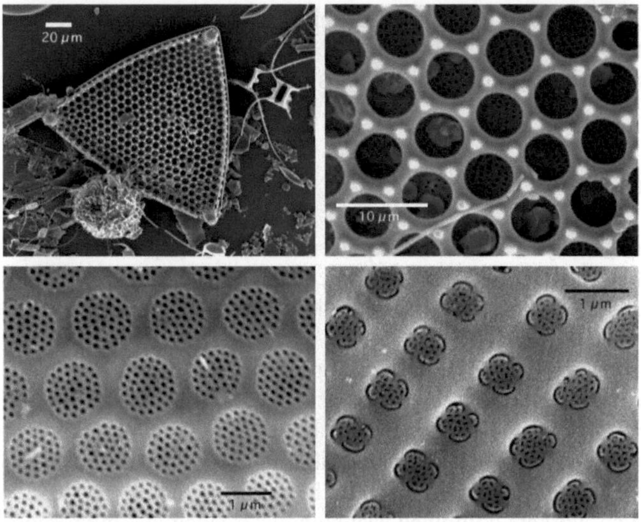

Fig. 13.2. Siliceous exoskeletons of three diatom species imaged with scanning electron microscopy. *Top*: *Tricaeratium favus*, whole cell (*left*), detail (*right*). *Bottom*: *Roperia tessellata* (*left*) and *Achnathes brevipes* (*right*). Reprinted with permission from Gebeshuber IC, Thompson JB, Del Amo Y, Stachelberger H, Kindt JH (2002) Mat Sci Technol 18:763 [4] © 2002 IoM Communications Ltd.

These naturally nanostructured surfaces gained the attention of nanoscientists, and diatom nanotechnology developed as a new interdisciplinary field of research [3].

Diatoms are found in freshwater, brackish and marine environments, as well as in moist soils, and on other regularly moist surfaces. They are either freely floating (planktonic forms) or attached to a substratum (benthic forms), and some species may form colonies in the form of chains of cells of varying length. Individual diatoms range from two micrometers up to several millimeters in size, although only few species are larger than 200 micrometers. Diatoms as a group are very diverse with 12,000 to 60,000 species reported [5,6].

These unicellular organisms are interesting from the point of view of materials science and biomimetic studies, since they master challenges as diverse as building nanostructured glass-like shells with high load capacity (a problem interesting for lightweight structures architecture) and engineering strong and robust adhesives that are stable in wet environments (most man-made adhesives fail to bond in wet conditions, owing to chemical modification of the adhesive or its substrate). Furthermore, diatoms excel at preventing dissolution of their silica shells in water owing to a covering layer (up-to-date technology is currently facing the problem that man made glass fiber reinforced polymers show rapid deterioration when used in water).

Currently, human chemical synthesis cannot produce siliceous structures with the hierarchical structural detail of the diatom frustules nor can ordered siliceous structures be produced synthetically under the benign conditions of diatom biomineralization. Biosilicification occurs at ambient temperatures and pressures, whereas artificial chemical synthesis of silica-based materials (e.g. resins, molecular sieves and catalysts) requires extreme conditions of temperature, pressure and pH.

13.2.2
SPM of Diatoms

The first AFM study of diatoms was presented in 1992 [7]. In this study, the surface structure of six different diatom species collected from a mud sample was imaged after the cells had been briefly rinsed with ethanol to kill, clean and immobilize them.

Topography and micromechanical properties like elasticity and hardness of dead diatom cells were reported by Almquist et al. in 2001 [8].

In contrast to these AFM images of dead cells, topography and micromechanical properties (such as viscoeleastic properties, adhesion forces and hardness) of the surface of the living diatom cell has been investigated [e.g. 5, 10–13].

Lee and co-workers combined scanning electrochemical microscopy and scanning optical microscopy to obtain simultaneous electrochemical and optical images of living diatoms in a constant-current mode [13]. This kind of microscopy might prove useful in mapping the biochemical activity of a living cell.

The defense potential of the diatom shell was investigated by Hamm and co-workers by measuring its strength [14]. It was found that diatoms are remarkably strong by virtue of their architecture and the material properties of the diatom silica.

In 2004 Arce and co-workers used the AFM to compare the adhesion of diatoms to several surfaces. Tipless AFM cantilevers were functionalized with living diatom cells, and the surfaces investigated were tested with the same diatom bioprobe [15].

13.2.2.1
Diatom Topography as Investigated with AFM

Owing to the poor adhesion to the substrate, it is impossible to obtain stable images of most benthic diatom species with the AFM.

AFM-compatible diatom species can be selected from a large sample by following a simple and effective strategy: Freshwater aquarium plants covered with benthic diatoms are placed in a jar filled with water, as well as two left-handed European freshwater snail species, *Physa fontinalis* and *Planorbarius corneus*, and some glass slides. In the following weeks, the diatoms will colonize the jar and the glass slides. The snails will feed on the diatoms, predominantly leaving the species behind, which obviously strongly attach to the substrate.

By this strategy, Gebeshuber and co-workers [10] selected three different diatom species: *Eunotia sudetica, Navicula seminulum* and a yet unidentified species, and subsequently imaged them in contact mode AFM (Fig. 13.3).

The natural adhesives of these diatoms, which attach them to the substrate as well as to each other (all of them are colonial forms), prove to be sufficiently strong that stable AFM imaging conditions are achieved without further sample preparation.

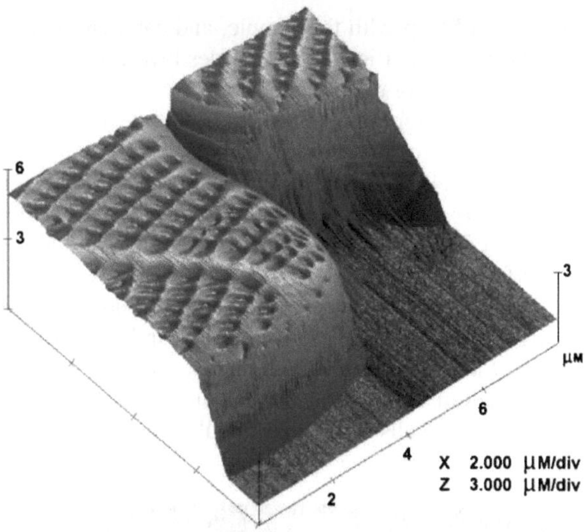

Fig. 13.3. AFM image of parts of two living diatom cells of the species *Navicula seminulum* growing on a glass slide. Note that the flat area does not correspond to the surface of the glass slide, but is determined by the maximum possible extension of the z-piezo of the microscope. Image acquired using AFM contact-mode imaging in water, imaging parameter topography, scan size 8 × 8 μm^2, scanning frequency 1 Hz. Reprinted with permission from Gebeshuber IC, Kindt JH, Thompson JB, Del Amo Y, Stachelberger H, Brzezinski M, Stucky, GD, Morse DE, Hansma PK (2003) J Microsc 212:292 [10] © 2003, The Royal Microscopical Society

The cells are imaged in their culture medium or in tap water while they are still growing on the glass slides. Tapping-mode as well as contact mode imaging is easy to achieve as long as engaging the cantilever takes place on the cell surface.

Navicula seminulum grows in stacks of cells pointing out from the glass slide. These chains of cells can be about 10 cells high, as investigated by SEM (data not shown). Figure 13.3 reveals detailed surface patterning of the top valve faces of two adjacent cells of *Navicula seminulum.*

The chains of *Eunotia sudetica* and of the yet unidentified species grow with the valve faces perpendicular to the surface of the glass slide, allowing for AFM investigation of the girdle bands.

The cells are alive and continue to divide after imaging.

13.2.2.2
Diatom Adhesives Investigated by SPM

Most man-made adhesives fail to bond in wet conditions, owing to chemical modification of the adhesive or its substrate. Engineering strong and robust underwater adhesives that are stable in wet environments is a challenge to current technology. Diatoms produce excellent adhesives that are stable and robust in wet environments.

Phase images depict the phase delay between the drive and response of the cantilever. These images contain information about the energy dissipated during the interaction of the AFM tip with the sample, and can help us to understand the viscoelastic and adhesion properties of the surfaces investigated, specifically of the organic material responsible for diatom adhesion.

Because phase imaging highlights edges and is not affected by large-scale height differences, it provides clearer observation of fine features that can be hidden by rough topography (Fig. 13.4). To investigate the natural adhesives utilized to attach cells to each other and to the substratum, it was tried to probe the cleft between two connected diatom cells with the AFM. In the yet unidentified species, the cleft at the cell–cell interface proved too deep. In this region, even the use of electron-beam-deposited AFM tips with high aspect ratio merely results in tip imaging. Phase imaging reveals slight differences (2°) in viscoelastic and adhesion properties of the two adjacent valves. *Eunotia sudetica*, by contrast, is very convenient for *in situ* investigation of the diatom adhesive at the cell interface, because there is barely any cleft between adjacent cells and valve undulations are less pronounced than

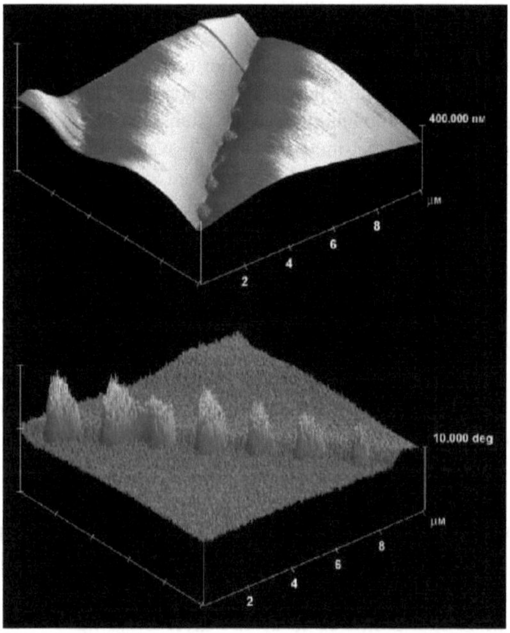

Fig. 13.4. (**a**) The adhesives in the contact region of two cells of *Eunotia sudetica* are apparent as small topographic features on the slightly undulated cell interface. The corrugation of the bead-like structures is between 10 and 20 nm, and their lateral dimension and spacing is about 1 mm. (**b**) In the phase image these features are far more striking. The diatom adhesive causes a phase lag of about 10° compared with the rest of the frustule surfaces, where on a single frustule it is within 1°. Note the 2° interfrustule phase step, which reveals slightly different viscoelastic properties of the two neighboring valves. Tapping mode, topography and phase, scan size 10 × 10 mm^2, scan rate 5 Hz. Note that for better view (**b**) is rotated clockwise by 90° as compared with (**a**). Reprinted with permission from Gebeshuber IC, Kindt JH, Thompson JB, Del Amo Y, Stachelberger H, Brzezinski M, Stucky, GD, Morse DE, Hansma PK (2003) J Microsc 212:292 [10] © 2003, The Royal Microscopical Society

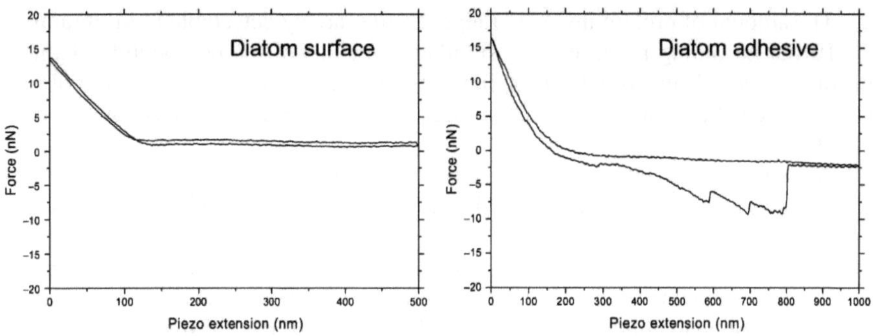

Fig. 13.5. Force–distance curves. *Left*: no adhesion can be recognized on the diatom surface. *Right*: representative data for the diatom adhesive that attaches *Eunotia sudetica* to the substrate. Several debonding events occur. Reprinted with permission from Gebeshuber IC, Kindt JH, Thompson JB, Del Amo Y, Stachelberger H, Brzezinski M, Stucky, GD, Morse DE, Hansma PK (2003) J Microsc 212:292 [10] © 2003, The Royal Microscopical Society

in the other species investigated (Fig. 13.4). The diatom adhesive is apparent as small topographic features at the cell interface. The bead-like structures are 10–20 nm high, have lateral dimensions of about 1 μm and are about 1 μm apart. The phase image clearly depicts the altered viscoelastic properties of these structures: the diatom adhesive causes a phase difference of up to 10° compared with the phase difference on the rest of each of the two frustules, where it is within 1° on each, apart from a 2° phase difference between the two adjacent valves, a feature which also appears in the other species, where the adhesives are not accessible because of deep clefts between the single organisms.

Force–distance curves on the surface and on the adhesive of *Eunotia sudetica* reveal basic differences in adhesion properties (Fig. 13.5). On the diatom surface, no adhesion force can be detected (Fig. 13.5 left). The diatom adhesive, by contrast, is strong and robust in the wet environment. To gain reproducible access to this natural adhesive, a chain of *Eunotia sudetica* that was embedded in a densely packed field of *Navicula seminulum* was scraped away from the glass slide with an STM-tip mounted on a three-dimensional micromanipulator. Over a period of several hours, force–distance curves were taken on the adhesive molecules that were used to attach the diatom cells to the glass slide (Fig. 13.5, right). No change in the basic shape of the force–distance curves can be detected within hours of repetitive pulling in the area where the colony was located. Typically, several debonding events occur until the natural adhesive molecules finally debond at a tip–surface separation of about 600 nm. For a detailed description of this study, see [4, 10].

13.3 Interaction of Large Organic Molecules

Conformational diseases such as Parkinson's disease, Alzheimer's disease, kuru, scrapie, BSE and vCJD (variant Creutzfeldt-Jakob Disease) result from misfolded proteins aggregating into detrimental structures like amyloid fibers [16–18].

The amount of protein involved ranges from scarcely detectable to kilograms.

Partial unfolding might expose significant regions of the polypeptide chain to the outside world, allowing the protein to aggregate and convert into amyloid fibrils. Once formed, the strong hydrogen bonding between molecules can make this process effectively irreversible.

As with crystallization, the formation of amyloid fibrils is "seeded" by preformed aggregates, a phenomenon that might also be responsible for the rapid progression of sporadic diseases such as Alzheimer's once the symptoms become evident. BSE, for example, has almost undoubtedly resulted from the highly unnatural practice of feeding young cows on the remains of old ones, with the disease then being transmitted to humans as vCJD. Both kuru and BSE have virtually disappeared as a result of effective action taken once their origins were understood.

The proteins that have emerged under evolutionary pressure are normally robust enough to resist reversion to aggregated states. Evolutionary processes have selected sequences of amino acids with the remarkable ability to form monomeric structures in which the main chain is folded in a unique way within the mass of close-packed side chains, preventing it from interacting with other molecules.

Furthermore, "chaperone" proteins help to protect against such changes.

Chaperones are proteins whose function is to assist other proteins in achieving proper folding: They prevent protein aggregation by providing encapsulated hydrophobic environments that allow the protein to fold properly.

Many chaperones are heat or cold shock proteins, that is, proteins expressed in heat or cold shock conditions. The reason for this behavior is that protein folding is severely affected by extreme temperatures. Chaperones act to counteract the potential damage. Although most proteins can fold in the absence of chaperones, a minority strictly requires them.

A large number of chaperones need adenosine triphosphate (ATP) to function properly. Chaperones recognize unfolded proteins by the hydrophobic residues these expose to the solvent. Exposed hydrophobic residues are unusual for properly folded proteins. Since the environment of the cell is characterized by hydrophilic groups (mostly water), incompletely folded or misfolded proteins with exposed hydrophobic groups have a tendency to aggregate to larger structures, where again, the hydrophobic residues would be hidden from the surrounding.

Chaperonins are a subset of chaperone proteins found in prokaryotes, mitochondria and plastids.

The AFM has proven to be a useful tool for studying proteins at the single molecule level. For a review on single molecule techniques in biomedicine and pharmacology, see [19].

Many of the single molecule studies with the AFM have been restricted by noise and speed limitations.

The first protein–protein interactions on the single molecule level imaged in real time were presented in 2000 [20]. This study demonstrated the enormous contributions AFM can make to molecular biology. Bulk results are interesting, but there are many valuable properties that can only be investigated on the single molecule level.

This work was enabled by the development of small cantilevers [21–23]. Small cantilevers allow for faster imaging and faster force spectroscopy of single biopolymers, because they have higher resonant frequencies and lower coefficients of viscous damping.

A new generation of AFMs using small cantilevers will enable the study of biological processes with greater time resolution, possibly at video refresh rates. Furthermore, small cantilever AFMs allow to narrow the gap in time between results from force spectroscopy experiments and molecular dynamics calculations.

The small cantilevers are fabricated out of low stress silicon nitride. They are ten micrometers long, have widths of 3–5 mm, and their thickness is about 75 nm. These cantilevers can measure smaller forces than larger cantilevers with the same spring constant because they have lower coefficients of viscous damping.

The prototype small AFM detects the motion of small cantilevers by using high numerical aperture optics to focus a laser beam onto the cantilever and then measuring angular changes in the reflected light beam.

This microscopy was used to observe, in real time, the interactions between individual molecules of the *Escherichia coli* chaperonin protein GroES binding to and then dissociating from individual *E. coli* GroEL proteins, which were immobilized on a mica support.

Both X-ray crystallography and cryoelectron microscopy studies have been used to resolve the structures of GroEL and the GroEL–GroES complex in different stages of the folding cycle (Fig. 13.6, e.g. [24–29]).

A prototype small cantilever AFM [23] image of both GroEL deposited on mica and the GroEL–GroES complex repeatedly without the aid of fixing agents (Fig. 13.7).

GroEL adsorbs to mica in end-up orientation. The average diameter of the molecules in this image agrees with the X-ray and cryoelectron microscopy data.

Upon the addition of GroES and ATP into the buffer solution, GroES molecules were observed as features that extend 3.6 ± 1 nm higher than the GroEL film (Fig. 13.8).

The height of these features is also consistent with X-ray crystallography and cryoelectron microscopy data. The same sample region can be scanned repeatedly without excessively disturbing the GroEL–GroES complexes (for details, see [20]).

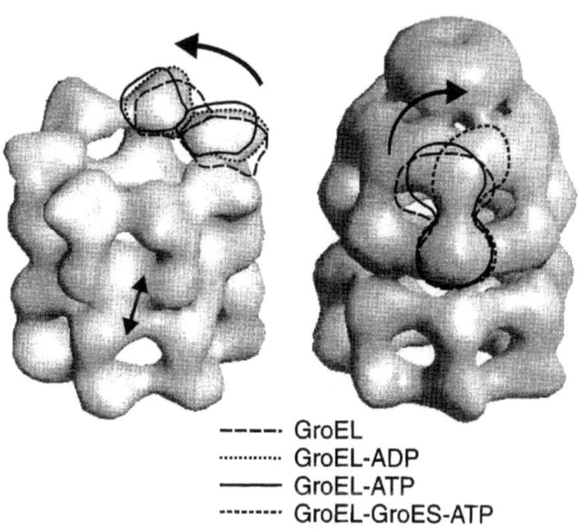

Fig. 13.6. Cryoelectron microscopy images of GroEL (*left*) and the GroEL–GroES complex. The height of the GroEL molecule is about 15.1 nm, the height of the GroEL–ES complex is about 18.4 nm. Upon interaction with ADP or ATP, domain movements occur, as indicated. Reprinted with permission from Roseman AM, Chen S, White H, Braig K, Saibil HR (1996) Cell 87:241 [29] © 1996, Elsevier

Fig. 13.7. GroEL film deposited on mica scanned in two dimensions (*left*) and in one dimension (*right*). In this image of GroEL, the movement along the slow scan axis was disabled half way through the scan. From then on the AFM tip repeatedly scanned the same line of proteins. Each *horizontal line*, therefore, shows changes in time of an individual molecule. Reprinted with permission from Viani MB, Pietrasanta LI, Thompson JB, Chand A, Gebeshuber IC, Kindt JH, Richter M, Hansma HG and Hansma PK (2000) Nature Struct Biol 7:644 [20] © 2000, Nature Publishing Group

Therefore, in order to obtain the temporal resolution required for observing the formation and dissociation of the GroEL–GroES complexes in the presence of Mg-ATP, the sample was scanned in one dimension rather than two (Fig. 13.8).

The time/height diagram of the protein lines displays repetitive well-defined step-like variations in height (Fig. 13.8). The magnitude of these steps is 3.6 ± 1 nm. The observed height variations result from GroES molecules attaching to and then separating from the respective GroEL molecules.

Without GroES and Mg-ATP no such steps can be observed.

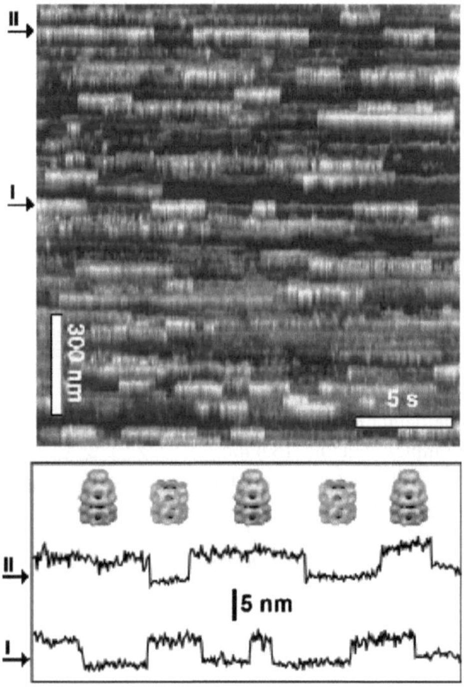

Fig. 13.8. Tapping mode AFM in liquid. *Top*: after the addition of GroES and Mg-ATP into the buffer solution, variations in height along the lengths occur in the single protein lines, as exemplified by *arrows I* and *II*. *Bottom*: time/height diagram of the protein lines indicated with the *arrows* in the top image. The height changes between two values that differ by 3.6 ± 1 nm. This indicates the binding and unbinding of GroES. The cryoelectron microscopy images of GroEL and the GroEL–GroES complex are from Roseman et al., 1996 (reprinted with permission, © 1996, Elsevier). Adapted with permission from Viani MB, Pietrasanta LI, Thompson JB, Chand A, Gebeshuber IC, Kindt JH, Richter M, Hansma HG, Hansma PK [20] (2000) Nature Struct Biol 7:644 [20] © 2000, Nature Publishing Group

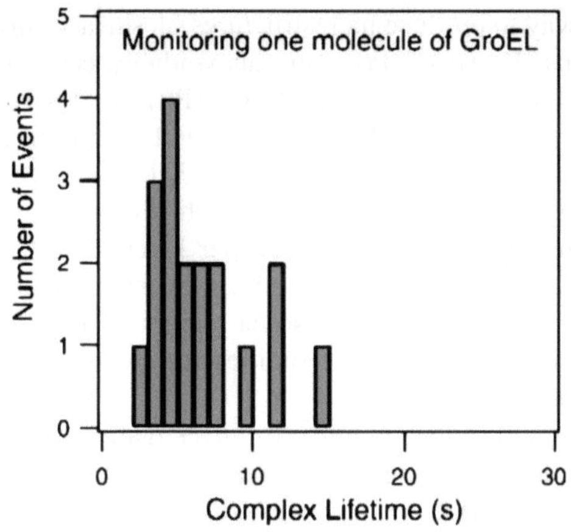

Fig. 13.9. Histogram of measured GroEL–GroES complex lifetime in the presence of Mg-ATP. Individual GroES molecules attach to and then separate from the same GroEL molecule 18 times during an observation period of about 120 seconds. Note the absence of events with lifetimes < 2 seconds. This is interesting in itself, telling us about the GroEL-GroES complex lifetime on the single molecule level, and furthermore indicates gentle measuring, since strong disturbance of complex formation by the cantilever would also lead to subsecond complex lifetimes. Reprinted with permission from Viani MB, Pietrasanta LI, Thompson JB, Chand A, Gebeshuber IC, Kindt JH, Richter M, Hansma HG, Hansma PK [20] (2000) Nature Struct Biol 7.644 © 2000, Nature Publishing Group

A histogram of the complex lifetime for a single GroEL molecule that was investigated for about 120 seconds is shown in Fig. 13.9. During this time interval, 18 times a complex with GroES has formed. The distribution of complex lifetime peaks near five seconds and the average lifetime is $\sim (7 \pm 1)$ s ($n = 18$).

In future application of this kind of single molecule studies with the AFM, misfolded proteins could well be involved and, e.g. the effect of various pharmaceuticals on folding efficiency could be tested.

13.4 Nanodefects on Atomically Flat Surfaces

Most of the small structures currently used in technology are in the micrometer range. One reason for this is silicon micromachining technology, which works fast and at low cost in this regime.

However, needs for increased data-storage density and smaller devices call for nanometer-sized structures.

Nanofabrication techniques comprise techniques such as electron beam and nano-imprint fabrication, epitaxy and strain engineering, scanning probe techniques, as well as self-assembly and template manufacturing [30].

Nanotransfer printing is a more recent high-resolution printing technique, which uses surface chemistries as interfacial "glues" and "release" layers to control the transfer of solid material layers from stamp relief features to a substrate [31].

One important way to produce nanostructures on surfaces involves kinetic sputtering by "fast" ions. However, fast ions unavoidably cause unwanted radiation damage. As opposed to this, potential sputtering (PS), i.e. desorption induced by the potential energy of slow multiply charged ions (MCI), holds great promise for more gentle nanostructuring of insulating surfaces [32, 33].

It can cause high sputter yields even at such low ion impact energies where kinetic sputtering and defect creation in deeper layers is not possible. While the physical mechanisms of PS have been the subject of extensive investigation [34–38], technical applications of slow MCI have so far remained largely unexplored, despite the fact that they provide unique opportunities for etching, ultra-thin film growth and nanostructure fabrication.

The AFM is the microscope of choice for investigating ion induced nanodefects on flat crystals, because of its unprecedented resolution and of the fact that it can also image insulating materials.

13.4.1 Ion Bombardment of Highly Oriented Pyrolytic Graphite (HOPG)

HOPG is used as a diffracting element in monochromators for X-ray and neutron scattering and as a calibration standard for STM and AFM. The graphite surface is easily prepared as a clean atomically flat surface by cleavage with an adhesive tape. HOPG is, therefore, used in many laboratories as the surface of choice for "seeing atoms".

Surface defects in HOPG produced by the impact of individual (singly charged) ions have already been investigated via STM/AFM by a number of groups [40–49, and further references therein].

However, first results were reported only recently for impact of slow multiply charged ions and the effect of the projectile charge state (or potential energy) on the size of the produced nanodefects [49–52].

Moreover, in most previous studies, either STM in air was used or the irradiated samples were transported in air towards STM inspection after ion bombardment. If, e.g., chemical bonds at the surface are broken due to the ion impact, impurities could preferentially adsorb at these sites and thus change the topography of the surface (and the resulting STM image) during contact with air. Therefore, here MCI bombardment has been followed by STM/AFM investigations without breaking the ultra-high vacuum. In this way, possible influences from target surface exposure to air can be ruled out.

Figure 13.10 shows STM and AFM scans of the HOPG surface before bombardment.

The STM image of HOPG bombarded with 800 eV Ar^+ ions reveals a large number of individual nanosized defects as a result of ion bombardment (Fig. 13.11). In AFM scans of the same surface, no significant topographic changes could be detected [53].

For very highly charged projectile ions, surface defects have recently also been observed in AFM studies [51, 52].

Meguro and co-workers found that HCI impact and subsequent treatment either by electron injection from an STM tip or by He-Cd laser irradiation induce a localized

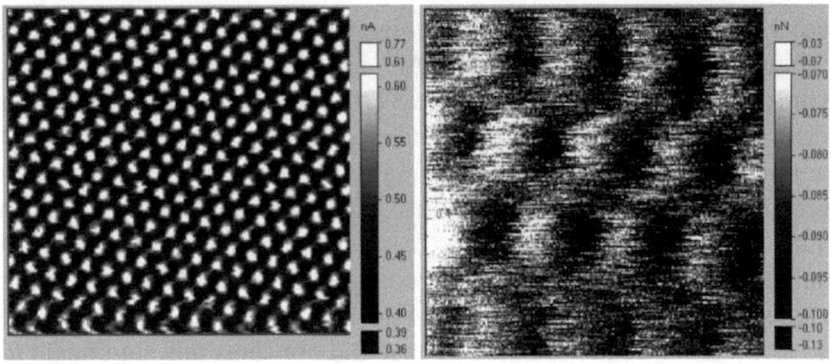

Fig. 13.10. Highly oriented pyrolytic graphite imaged in ultrahigh vacuum with atomic resolution. *Left*: scanning tunneling microscopy image, image size 4×4 nm^2. *Right*: atomic force microscopy image, image size 1×1 nm^2

transition from sp^2 to sp^3 hybridization in graphite, resulting in the formation of nanoscale diamond-like structures (nanodiamonds) at the impact region [54].

In an investigation of HOPG bombarded with 400 eV Ar^+ and Ar^{8+} ions involving Raman spectroscopy, Hida and co-workers found that the charge state of the ions as well as their mass have an influence on the disordering of HOPG and that the defects introduced by Ar^{8+} are not simple vacancies, but assumed to be vacancy clusters in contrast to their results for Ar^+ irradiation [55].

Several hundred defects from different sample positions have been statistically analyzed for each projectile type (Ar^+, Ar^{8+}, Ar^{9+}).

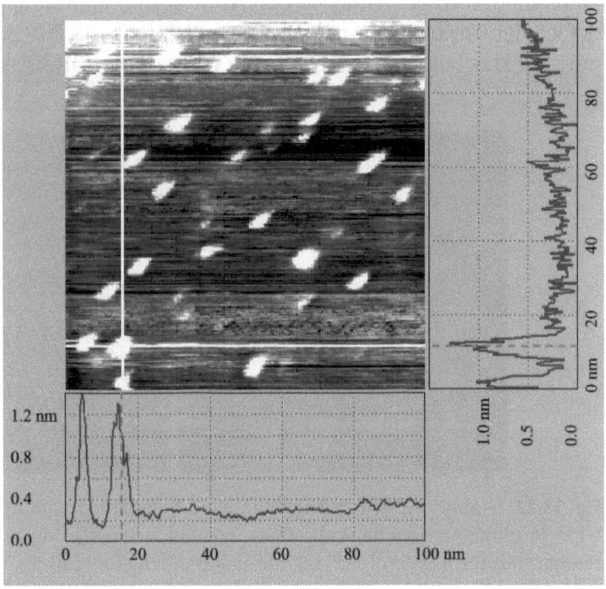

Fig. 13.11. Highly oriented pyrolytic graphite bombarded with 800 eV Ar^+ ions imaged with STM in ultrahigh vacuum. Image size 100×100 nm^2. The ion induced nanodefects are clearly visible

Figure 13.12 (right trace) shows the enlarged STM image of a typical defect on HOPG created by the impact of a single Ar^+ ion of 800 eV kinetic energy

The only surface defects found in the STM images are "protrusions" (hillocks) with a mean lateral size of 0.8–1.25 nm and an average equivalent height of 0.22 nm. They are randomly dispersed on the originally flat surface. Their area density is in good agreement with the applied ion dose, implying that nearly every single ion impact has caused one protrusion. A $\sqrt{3} \times \sqrt{3}$R30° surface, as characteristic for interstitial defects in HOPG [56–58], surrounded by undisturbed surface parts is observed in the vicinity of most defects (see Fig. 13.12). Scanning with the AFM down to atomic resolution on the irradiated surface did not show any significant topological changes due to ion bombardment. Therefore, we conclude that the nanodefects produced by slow ion impact are of electronic rather than of topographic nature.

For impact of singly charged ions, our findings are in good agreement with previous observations [43, 57].

As a remarkable result, however, it was found that the measured mean diameter of the "hillocks", and to a somewhat lesser extent their "height", increase with the projectile charge state [53]. In a careful STM study, Hahn and Kang [57] showed that generally two kinds of defects in HOPG are created by low energy (100 eV) Ar^+ bombardment, namely carbon vacancy defects (VDs) and interstitial defects (IDs) formed by trapping the projectile beneath the first carbon plane.

Both types of defects are detected as protrusions in the STM topographic image. The dangling bonds at the VD site cause an enhancement of the local charge density-of-states (CDOS) near the Fermi energy, seen as a protrusion in the STM image [57].

The protrusion observed in the STM image at ID sites results from a small geometric deformation of the graphite basal plane due to the trapped projectile (not large enough to be visible in our AFM scans) and an apparently larger electronic defect due to an increased CDOS. A $\sqrt{3} \times \sqrt{3}$R30° surface was reported [57] only for IDs but not for VDs. From this $\sqrt{3} \times \sqrt{3}$R30° superlattice structure also

Fig. 13.12. Highly oriented pyrolytic graphite bombarded with 800 eV Ar^+ ions imaged with STM in ultrahigh vacuum with atomic resolution (*right*). Image size $10 \times 10\,nm^2$. The fast Fourier transform (*left*) of the Ar^+ ion-induced defect reveals a $\sqrt{3} \times \sqrt{3}$R30° surface: the ion induced features are larger than the features from the HOPG lattice and they are rotated with respect to them by 30 degrees

observed in our experiments (see Fig. 13.12), we therefore conclude that the majority of the "hillocks" observed are due to IDs, or VDs created along with IDs. The strong increase of the lateral protrusion size with increasing charge state of the projectile ion is interpreted as a "pre-equilibrium" effect of the stopping of slow multiply charged ions in HOPG, as has so far only been observed for higher charge states [44].

Although MCI are converted already into neutral hollow atoms (i.e., an atom whose inner shells remain essentially unoccupied) during their approach towards the surface, their captured electrons remain in highly exited states until surface impact, where they are gradually peeled off and replaced by conduction band electrons forming a partial screening cloud around the MCI [59].

Before final deexcitation of the hollow atom can take place within the solid, reduced screening should result in a strongly increased energy loss of the projectiles. According to SRIM-2000 (© IBM) calculations [60], the mean range of 150 eV Ar projectiles in HOPG is about two monolayers. An increased stopping and straggling of the higher charged Ar projectiles would lead to IDs located closer to the surface, as well as to more VDs due to a higher momentum transfer to the carbon atoms of the first plane. Because of the extreme surface sensitivity of STM, this pre-equilibrium effect in the stopping power is not masked by (equilibrium) bulk effects and is apparently observable with unprecedented clearness.

Extending pertinent work by other groups with singly charged ions only, our combined STM/AFM studies revealed nanodefects that comprise a disturbance of the electronic density-of-states of the surface rather than its topography.

Whereas the size of these defects increases with the ion charge (here up to $q = 9$), as expected for any conducting target surface they showed no evidence for potential sputtering.

For more detailed information on these studies, see [50, 53].

13.4.1.1 Revealing the Hidden Atom in Graphite by Low-Temperature AFM

Despite the proverbial ease of imaging graphite by STM with atomic resolution, every second atom in the hexagonal surface unit cell remains hidden, and STM images show only a single atom in the unit cell.

The reason for this is that the tunneling current is not a function of the surface topography, but of the local electronic structure. On the graphite surface, there are two different types of carbon atoms in the basal plane, as distinguished by the presence (α) or absence (β) of a carbon atom in the plane immediately below the surface. The α atoms are located directly above another α atom, in the layer directly underneath, the β atoms are located above a hollow site. These local electronic structure variations imply that the STM can only detect every other atom on the graphite surface. Consequently, an alternative imaging method is required to detect the "hidden" α atoms on the graphite surface [61].

Also in contact-mode AFM images of graphite the quasi-atomic resolution images show only one protrusion per unit cell [62].

Recent progress in dynamic AFM allows researchers to routinely achieve true atomic resolution on conductors and insulators [63, 64], but once again only one maximum within a hexagonal unit cell of the graphite surface was obtained in the attractive noncontact mode [65].

In 2003, Hembacher and co-workers presented measurements with a low-temperature atomic force microscope with pico-Newton force sensitivity that reveal the hidden surface atom [66].

The instrument used in this investigation is a combined ultrahigh vacuum (UHV) STM/AFM to simultaneously probe the charge density at the Fermi level and the total charge density of graphite by recording tunneling currents and forces, respectively. The instrument is immersed in a liquid He bath cryostat, yielding a sample temperature of 4.89 K and exceptionally low thermal drifts of about 0.02 nm/h (at room temperature, even with drift correction, currently 2–10 nm/h are achieved). To protect the microscopy from external vibrations, the setup is built on a foundation with a mass of 30,000 kg.

In their dynamic AFM images of graphite, recorded at small oscillation amplitudes and with weak repulsive forces, both the α atoms and the β atoms are detected. The reason for this is that the repulsive forces utilized in AFM involve different electrons in the tungsten tip than in the STM mode.

Revealing the hidden atoms in graphite by means of room-temperature AFM might become possible with miniaturized AFMs based on nano- or microelectromechanical systems (NEMS/MEMS) technology (since they show small drift). In such an instrument, the operating frequency could be commensurately increased and there would be no need for a 30,000 kg fundament [61].

13.4.2 Bombardment of Single Crystal Insulators with Multicharged Ions

Systematic STM/AFM investigations on nanoscopic defect production at atomically clean insulator surfaces of Al_2O_3 after bombardment by slow (impact energy $\leq$ 1.2 keV) singly and multiply charged ions under strict UHV conditions is the topic of this section.

It will be demonstrated that on monocrystalline insulator surfaces, well-defined topographic features of typically nm extensions are produced ("potential sputtering"). For Al_2O_3, there exists a clear dependence of the defect size on the projectile ion charge. These results are discussed in view of possible new nanoscopic surface structuring and modification methods for which the kinetic projectile energy plays only a minor role.

Impact of slow ions on solid surfaces can give rise to inelastic processes that modify the geometric and electronic structure at and below the surface, cause emission of electrons and photons as well as neutral and ionized target particles (atoms, molecules, clusters), remove surface-adsorbed material and lead to projectile neutralization. The transfer of electrons between surface and projectile possibly acts as precursor for the above-mentioned processes and makes them proceed irrespective of the kinetic projectile energy.

The importance of such "electronic" processes increases with multicharged projectile ions and their role is elucidated when slow ions of the same kinetic energy, but with different charge states are applied as projectiles.

For certain insulator surfaces, the impact of slow multicharged ions (MCIs) Z^{q+} gives rise to considerably stronger ablation than the well-established kinetic sputtering by neutral or ionized projectiles. First experimental evidence for this PS was reported for alkali-halide surfaces and explained by "Coulomb explosion" [67], i.e. the creation of small positively charged surface spots from the rapid electron capture by impinging MCI, and the subsequent ablation because of strong mutual target ion repulsion.

"Coulomb explosion" was also invoked in order to explain AFM observations of blister-like defects on mica samples produced by highly charged ions Z^{q+} (kinetic energy 1–3 keV/atomic mass unit) [36, 68].

However, studies for impact of slow ($\leq$ 1 keV) MCI on thin polycrystalline films of alkali-halides (LiF, NaCl) and Al_2O_3 deposited on quartz microbalance crystals [69] suggested a different explanation for PS, namely defect-stimulated desorption induced by very efficient electron capture [35].

It has been established that such desorption processes are induced by electrons (electron stimulated desorption, ESD) or photons (photon stimulated desorption, PSD) on such materials where self-trapping of specific crystal defects proceeds via electron–phonon coupling in the crystal lattice [70].

However, such defect trapping as the prerequisite for PS may also be caused or at least supported by the kinetic projectile energy ("kinetically assisted PS" [37]), which could also explain some PS-like effects reported for target species where no electron–phonon coupling can take place, i.e. for semiconductors like Si and GaAs [68]. In any case, for slow ion impact, the self-trapping mechanism is most relevant for PS. Consequently, for metal and semiconductor surfaces no slow MCI-induced PS can be observed, so far [71].

As the surface region from which a slow MCI does capture electrons should be rather small (nm extensions), it is probable that the surface defects caused by PS are of similar size. In order to study such defect structures, we applied AFM in UHV on monocrystalline target surfaces of insulator species for which PS by slow MCI impact has already been demonstrated on polycrystalline thin films [37, 69, 71].

The results for Al_2O_3 presented below are of possible interest for nanostructuring these surfaces.

Observations of slow ion-induced nanodefects on different atomically clean target surfaces were performed under strict UHV conditions with a combined AFM/STM instrument (UHV-AFM/STM, OMICRON Nanotechnology GmbH, Germany). Nanodefects were looked for on freshly prepared surfaces of sapphire *c*-plane $Al_2O_3(0001)$ after irradiation with low doses of slow singly and multiply charged ions. In order to avoid disturbing noise from an ion irradiation chamber directly attached to the AFM/STM instrument, a transportable UHV vault for target transfer, which was alternatingly coupled via UHV locks to the target ion irradiation chamber and the AFM/STM was used. This procedure kept the target surfaces under permanent UHV conditions after initial cleaning, thermal annealing, and during subsequent slow ion irradiation until completion of the AFM/STM inspection. Ion irradiation of the insulator surfaces was accompanied by low-energy ($\leq$ 4 eV) electron flooding to compensate for surface charge-up, which otherwise strongly inhibits AFM observation or makes it even impossible. The electron gun was arranged at 2 cm distance to the sample. All AFM observations were made in the contact mode, with the base pressure in the AFM/STM chamber kept at about 10^{-10} mbar during measurements.

The singly and multiply charged ions for target irradiation have been extracted from a 5 GHz electron cyclotron resonance ion source [72], magnetically analyzed and guided via electrostatic lenses to the UHV irradiation chamber. The ions were decelerated in front of the target surface to their desired impact energy ($\leq$ 1.2 keV). Uniform irradiation was assured by rapidly scanning the ion beam across the target surface by means of deflection plates.

13.4.2.1
Production of Slow Ion-Induced Surface Defects on Al_2O_3 Insulator Targets

Polished Al_2O_3(0001) *c*-plane single crystals (TBL Kelpin, Neuhausen, Germany) were CO_2 snow cleaned (to remove micrometer and submicrometer particles and hydrocarbon-based contamination) and then annealed in UHV for 3 h at 400 °C. This preparation technique yields very flat crystal surfaces.

AFM contact mode studies on 14 samples prepared by the standard preparation technique revealed a root mean square (rms) roughness of 0.093 ± 0.06 nm rms. Bombardment with Ar ions of different charge states and kinetic energies (500 eV Ar^+ and Ar^{7+}, 1.2 keV Ar^+, Ar^{4+} and Ar^{7+}) results – as seen in AFM contact mode – in hillock-like nanodefects (see Fig. 13.13).

The ion-induced defects on the sapphire single crystal surface can be removed by annealing at 450 °C for 5 h. The density of nanodefects does not directly correspond with the applied ion dose: an ion dose of 5×10^{12} ions/cm^2, which is equivalent to five ions per 10 nm $\times$ 10 nm, leads to a rather small, however reproducible, density of defects on the sapphire surface: about 10 nanodefects per 1000 nm $\times$ 1000 nm can be observed after bombardment in the energy range reported here. This is equivalent

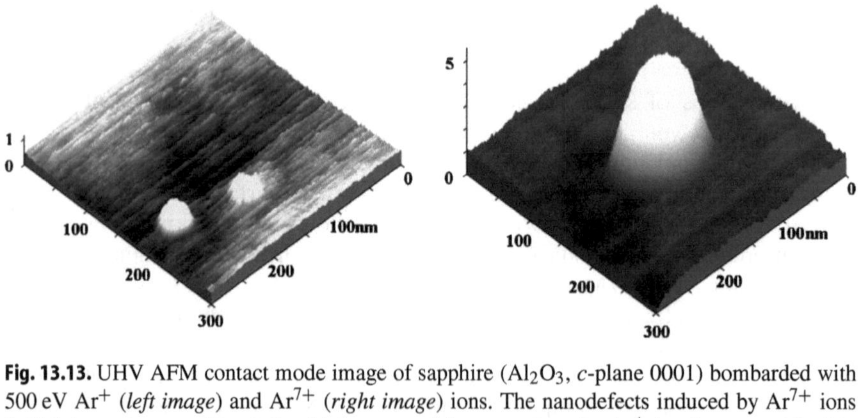

Fig. 13.13. UHV AFM contact mode image of sapphire (Al_2O_3, *c*-plane 0001) bombarded with 500 eV Ar^+ (*left image*) and Ar^{7+} (*right image*) ions. The nanodefects induced by Ar^{7+} ions (which have the same kinetic but higher potential energy than the Ar^+ ions) are considerably higher and wider than those caused by singly charged ions. The defects are real topographic features; the units on the three axes are nanometers. Reprinted with permission from Gebeshuber IC, Cernusca S, Aumayr F, Winter HP (2003) Int J Mass Spectrom 229:27 [53] © 2003, Elsevier Science B.V.

to a dose to defect ratio of 5000. More detailed experiments with different ion doses are needed. Analysis of the statistics of random impacts will clarify how many individual ion impacts are needed to form a visible nanodefect on the insulator surface.

A possibly similar migration and subsequent recombination of point defects at the surface has previously been reported for silicon bombarded by 5 keV He ions above 160 K [73].

In fact, the only case where the number of defects corresponded fairly well to the applied ion dose was for the conducting HOPG samples (see Sect. 13.4.1).

The Al_2O_3 *c*-plane proved to be the insulator surface showing most clearly a dependence of the ion bombardment induced defects with the kinetic energy and charge states of the projectiles. 500 eV Ar^+ ions produce defects that are about 1 nm high (Fig. 13.13) and have lateral dimensions of some tens of nanometers (one should keep in mind that the height is more accurately measurable with the AFM than lateral dimensions), whereas the defects produced by 500 eV Ar^{7+} ions are several nanometers high (Fig. 13.13) and show lateral dimensions of about 100 (!) nanometers. At higher kinetic energy the differences in the slow ion-induced nanodefects on the sapphire c-plane became even more distinct.

1.2 keV Ar^+-induced defects are up to about 8 nm high and their width is some 10 nm. For a higher charge state such as Ar^{4+}, two different kinds of defects occurred on the sapphire surface.

They have about the same height, but their lateral dimensions vary considerably: some are nearly 200 nm wide, whereas the smaller defects are only about 50 nm wide.

The height of both kinds of defects is about 2 nm. For Ar^{7+}, only one kind of defect was visible in the AFM images, with about 50 nm diameter and about 2 nm height (for a more detailed description of these results and for similar investigation on SiO_2 surfaces, see [53]).

Al_2O_3 is, therefore, a good candidate for PS-induced nanostructuring. This material is relevant for applications in microelectronics and nanotechnology.

13.5 Subatomic Features

In this section, the detection of atomic orbitals and single electron spins by means of SPM is described. In many cases, sophisticated signal acquisition techniques have to be applied, and the instruments have to be operated at very low temperature, since extremely small drift is required.

13.5.1 Atom Orbitals

Silicon and tungsten are the two chemical elements that already have been investigated with SPM regarding their atomic orbitals.

13.5.1.1 Silicon (111)-(7 × 7) Surface

Publications concerning the imaging of subatomic features with the AFM started in the year 2000, when Giessibl and co-workers published their paper on imaging of subatomic features on the reconstructed silicon (111)-(7 × 7) surface [74]. For a review on semiconductor surface reconstruction, see [75].

A scientific discussion, in which Hug and co-workers questioned this result by proposing that the subatomic features are caused by a feedback artifact, followed this publication [76].

In the course of this argument, Giessibl and co-workers presented refined calculations, showing striking similarities to the experimental images (see Fig. 13.14).

In 2003, Huang and co-workers presented a theoretical work demonstrating the feasibility of seeing atomic orbitals on the Si(111)-(7 × 7) surface with AFM [77].

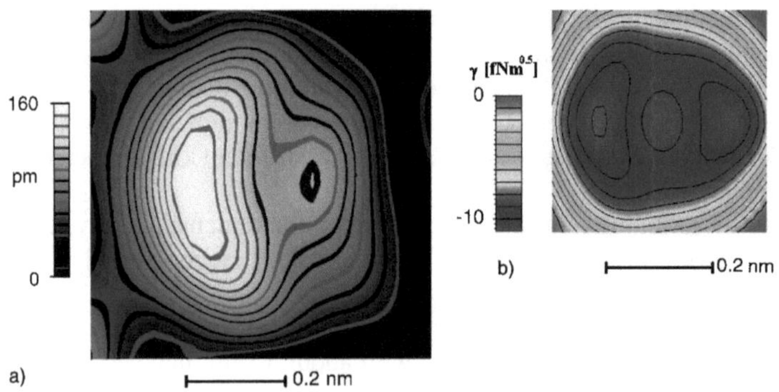

Fig. 13.14. Refined calculations of the normalized frequency shift of a single adatom on the reconstructed silicon (111)-(7 × 7) surface (*right*) performed by the Giessibl group, showing striking similarities with the experimental images (*left*). Reprinted with permission from Hug HJ, Lantz MA, Abdurixit A, van Schendel PJA, Hoffmann R, Kappenberger P, Baratoff A, Giessibl FJ, Hembacher S, Bielefeldt H, Mannhart J (2001) Science 291:2509 [76] © 2001, AAAS

13.5.1.2 Tungsten

In 2004, Giessibl and co-workers finally ended this discussion by presenting images of an individual tungsten atom by AFM at a resolution of 77 pm [78]. The diameter of a tungsten atom is 274 pm. Four distinct peaks that are attributed to highly localized electron clouds can be identified (Fig. 13.15).

The experiment was performed in UHV at a temperature of about five Kelvin. The microscope was isolated from vibrations by a 30 t foundation and from sound and electromagnetic stray fields by a metal chamber.

In contrast to STM (which only probes the most loosely bound electrons with energies at the Fermi level) AFM can resolve the charge density variations within a single atom, because the forces between the AFM tip and the sample are of electrostatic nature.

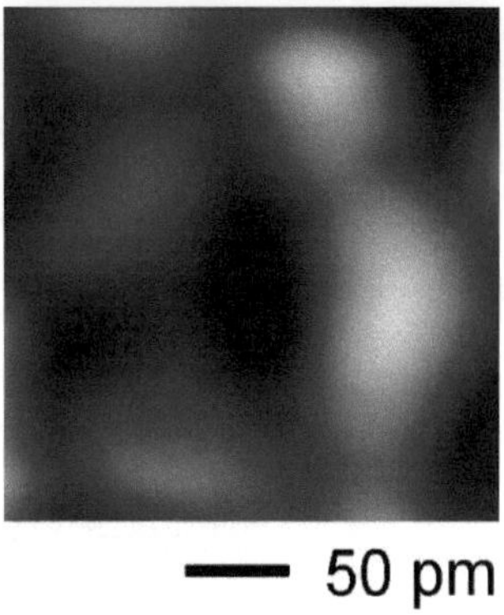

Fig. 13.15. UHV low temperature (5 K) AFM constant-height mode image reveals four-fold symmetry in the amplitudes of the higher harmonics signal (centered close to the maximum of the tunneling current, data not shown). Reprinted with permission from Hembacher S, Giessibl FJ and Mannhart J (2004) Science 305:380 [78] © 2004, AAAS

The electron structure originates from the quantum-mechanical nature of tungsten bonding: tungsten develops a body centered cubic crystal structure such that every tungsten atom is surrounded by eight nearest neighbor atoms, causing "arms" of increased charge density which point to the next neighbors. Four of these precisely localized electron clouds are visible on surface atoms.

The role of tip and sample was switched in the experiment: the front atom in a sharp tungsten tip was imaged by a light carbon atom of a graphite surface. The reason a light atom was used for probing was to minimize image blurring, since the mapping of one atom with another atom always involves a convolution of the electronic states. The tunneling current is confined to the top atom because of the sharp increase of tunneling probability with decreasing distance.

Instead of measuring static deflections or frequency changes, higher harmonics triggered by forces between the tip and the sample are recorded in this technique. These higher harmonics are much more sensitive to short-range interactions than static deflections or frequency changes.

13.5.2 Single Electron Spin Detection with AFM and STM

Single-spin detection is a vital goal for read-out in quantum computing, and single nuclear spin detection could solve the problem of how to distinguish between materials at the atomic level.

Several research groups have reported various single spin-detection methods [79–86].

In 1989, Manassen and co-workers presented the first direct observation of the precession of individual paramagnetic spins on partially oxidized silicon (111) surfaces [79].

They used an STM to detect the modulation in the tunneling current at the Larmor frequency. The Larmor frequency is the frequency at which magnetic resonance can be excited. It is given by the Larmor equation, which states that the resonant frequency is proportional to the overall (macroscopic and microscopic) magnetic field. Balatsky and Martin presented the theoretical explanation of this result [87] in 2001.

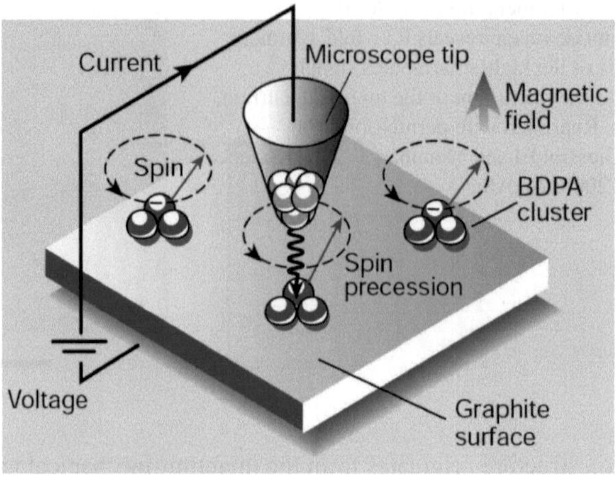

Fig. 13.16. Spin detection through the union of high-resolution microscopy and resonance techniques. The sample is a HOPG surface coated with clusters of organic BDPA molecules. In the applied magnetic field, the electron-spin vectors associated with free radicals in the molecules precess at a certain frequency. The STM tunneling current is modulated at the precession frequency. Detecting the modulation effectively measures electronic spin in the molecule. Reprinted with permission from Manoharan HC (2002) Nature 416:24 [88] © 2002, Nature Publishing Group

In 2002, Durkan and Welland published an article that essentially reproduced this experimental result with a different sample: BDPA (a, g-bisdiphenylene b-phenylallyl) on HOPG [80].

The idea of combining magnetic resonance (MR) with force microscopy in magnetic resonance force microscopy (MRFM) was published as a concept in 1991 [89].

Sidles settled on force microscopy because the performance of induction coils, the detectors in conventional MR, scales unfavorably with size. Shrinking the coil to detect a single spin reduces the signal irretrievably below noise. A force microscope, on the other hand, becomes more sensitive the smaller it gets.

In 1992, Rugar and co-workers demonstrated that the force exerted by 10^{12} electron spins could be detected at room temperature with a conventional cantilever for AFM [90].

Since then they have improved their spin detection limit by 12 orders of magnitude:

MRFM was proposed as a means to improve detection sensitivity to the single-spin level, and thus enable 3D imaging of (bio)molecules with atomic resolution [91, 92].

MRFM is essentially a combination of 3D magnetic resonance imaging (MRI) with the unprecedented resolution of AFM.

For an overview on MRFM, see Hammel and co-workers, 2003 [93] and for the theory of spin relaxation in MRFM, see Mozyrsky et al. 2003 [94].

In the year 2004, the force exerted by a single electron spin was measured by MRFM ([86], Fig. 13.17).

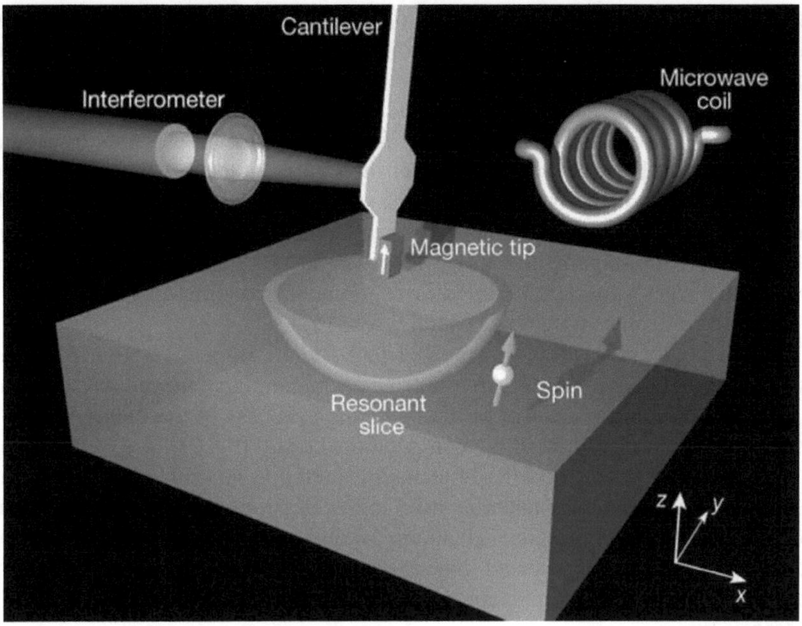

Fig. 13.17. A wiggling cantilever with a tiny CoSm magnet is the key element of a magnetic resonance force microscope. Elaborated signal acquisition makes it possible to detect a single electron spin dozens of atomic layers beneath the surface. In this way, scanning probe microscopy left the surface regime. Perhaps even atomic resolution images of molecules beneath the surface might be possible in the near future. Reprinted with permission from Rugar D, Budakian R, Mamin HJ, Chui BW (2004) Nature 430:329 [86] © 2004, Nature Publishing Group

The force detected in the Rugar 2004 experiment is a million times smaller than the forces usually encountered in AFM (van der Waals forces, electrostatic forces). The single electron spin was buried 250 nm below the surface of an irradiated vitreous silicon sample and exerted a force of 2 aN (2×10^{-18} N).

The sample had been prepared by irradiation with a 2 Gy dose of ^{60}Co gamma rays, producing a low concentration of Si dangling bonds containing unpaired electron spins known as E' centers. Unpaired electrons and many atomic nuclei behave like tiny bar magnets. Estimated spin concentration was between 10^{13} and 10^{14} cm^{-3}. The experiment was performed at 1.6 K in a small vacuum chamber that fits within the bore of a superconducting magnet. The low operating temperature minimizes thermal noise and reduces the relaxation rate of the spins.

The force exerted by a single electron spin is the smallest ever detected. Currently, the smallest volume elements in an image must contain at least 10^{12} nuclear spins for MRI-based microscopy [95], or 10^7 electron spins for electron spin resonance microscopy [96].

The cantilever used in the experiment is only 100 nm thick and had to be aligned vertically to the surface. In the conventional AFM configuration with the cantilever parallel to the surface, van der Waals forces and electrostatic forces would make it stick on the surface. Directly on the cantilever a strong 150 nm wide CoSm magnet is attached. It generates a field gradient of 200,000 T/m. The cantilever is slowly scanned over the surface. A laser interferometer records the cantilever deflections and sophisticated measurement signal acquisition techniques are needed for successful single spin detection. The strong magnetic field gradient allows for distinguishing magnetic resonance signals arising from different spatial locations, enabling accurate spin localization. By scanning the tip over the sample, a local magnetic resonance force is detected, which corresponds with a spatial resolution of about 25 nm. This spatial isolation of the signal is also the main argument that a single spin is being detected. Currently this method is very slow. As Stokstad mentions in his "Science News of the Week" article on the Rugar experiment, scanning a 170 nm stretch of the irradiated silicon sample took several weeks [97].

MRFM could serve as an invaluable tool for the implementation of a spin-based solid state quantum computer. It provides an attractive means for addressing the characterization and control of the fabrication process of the device during its construction and the readout of the computational result [98, 99].

If developed further, the MRFM technique could prove useful for investigating the atomic structure inside materials used in the electronics industry and to image biomolecules – such as proteins – at atomic resolution. However, to reach this goal, nuclear spins have to be detected. Nuclear spins are harder to detect than electron spins, because a proton's magnetic moment is 658 times smaller than that of an electron.

13.6 Conclusions and Outlook

In this review, we have presented scanning probe microscopy across dimensions from large samples like single cells, via single biomolecules and nanometer small ion induced defects on crystal surfaces to subatomic features like electronic orbitals and single electron spins.

Scanning probe microscopy is on its way to a standard laboratory method: subatomic features can be imaged, and with magnetic resonance force microscopy it has even left the two-dimensional surface regime. Perhaps in the not too distant future 3D-imaging of (complex) molecules, at surfaces or in the bulk state, with atomic resolution might become possible with these powerful techniques. The 3-D MRFM would also deliver chemical specific information because each magnetic nucleus has a unique gyromagnetic ratio.

Acknowledgements. Part of this work was supported by the Austrian Science Fund (FWF) and was carried out at Vienna University of Technology.

Part of this work was funded by the "Austrian Kplus-Program" and was done at the "Austrian Center of Competence for Tribology", AC^{2}T research GmbH.

References

1. Round FE, Crawford RM, Mann DG (1990) Diatoms: biology and morphology of the genera, Cambridge University Press
2. Hecky RE, Mopper K, Kilham P, Degens ET (1973) Marine Biol 19:323
3. Gordon R, Sterrenburg F, Sandhage K (guest eds) (2005) Diatom nanotechnology, special issue J Nanosci Nanotechnol 5:1
4. Gebeshuber IC, Thompson JB, Del Amo Y, Stachelberger H, Kindt JH (2002) Mat Sci Technol 18:763
5. Werner D (1977) The biology of diatoms, University of California Press
6. Gordon R, Drum RW (1994) Int Rev Cytol 150:243
7. Linder A, Colchero J, Apell H-J, Marti O, Mlynek J (1992) Ultramicrosc 42-44:329
8. Almqvist N, Delamo Y, Smith BL, Thomson NH, Bartholdson A, Lal R, Brzezinski M, Hansma PK (2001) J Microsc 202:518
9. Higgins MJ, Crawford SA, Mulvaney P, Wetherbee R (2002) Protist 153:25
10. Gebeshuber IC, Kindt JH, Thompson JB, Del Amo Y, Stachelberger H, Brzezinski M, Stucky GD, Morse DE, Hansma PK (2003a) J Microsc 212:292
11. Higgins MJ, Sader JE, Mulvaney P, Wetherbee R (2003a) J Phycol 39:722
12. Higgins MJ, Molino P, Mulvaney P, Wetherbee R (2003b) J Phycol 39:1181
13. Lee Y, Ding Z, Bard AJ (2002) Anal Chem 74:3634
14. Hamm CE, Merkel R, Springer O, Jurkojc P, Maier C, Prechtel K, Smetacek V (2003) Nature 421:841
15. Arce FT, Avci R, Beech IB, Cooksey KE, Wigglesworth-Cooksey B (2004) Biophys J 87(6):4284
16. Solomon B, Taraboulos A, Katchalski-Katzir E (2002) Conformational diseases: a compendium based on the 1st international workshop on conformational diseases, S Karger Pub, Tunbridge Wells, UK
17. Dobson CM (2002) Nature 418:729
18. Dobson CM (2003) Nature 426:884
19. Greulic KO (2004) Curr Pharmac Biotech 5:243
20. Viani MB, Pietrasanta LI, Thompson JB, Chand A, Gebeshuber IC, Kindt JH, Richter M, Hansma HG, Hansma PK (2000) Nature Struct Biol 7:644
21. Walters DA, Cleveland JP, Thomson NH, Hansma PK, Wendman MA, Gurley G, Elings V (1996) Rev Sci Instrum 67:3583
22. Viani MB, Schäffer TE, Chand A, Rief M, Gaub HE, Hansma PK (1999a) J Appl Phys 86:2258
23. Viani MB, Schäffer TE, Paloczi GT, Pietrasanta LI, Smith BL, Thompson JB, Richter M, Rief M, Gaub HE, Plaxco KW, Cleland AN, Hansma HG, Hansma PK (1999b) Rev Sci Instrum 70:4300
24. Braig K, Otwinowski Z, Hegde R, Boisvert DC, Joachimiak A, Horwich AL, Sigler PB (1994) Nature 371:578
25. Hunt JF, Weaver AJ, Landry SJ, Gierasch L, Eisenhofer J (1996) Nature 379:37
26. Boisvert DC, Wang J, Otiwonowski Z, Horwich AL, Sigler PB (1996) Nature Struct Biol 3:170
27. Xu Z, Horwich AL, Sigler PB (1997) Nature 388:741
28. Chen S, Roseman AM, Hunter AS, Wood SP, Burston SG, Ranson NA, Clarke AR, Saibil HR (1994) Nature 371:261
29. Roseman AM, Chen S, White H, Braig K, Saibil HR (1996) Cell 87:241
30. Ziaie B, Baldi A, Atashbar MZ (2004) Introduction to micro/nanofabrication In: Bhushan B (ed) Springer handbook of nanotechnology. Springer, Berlin Heidelberg New York, p 147

31. Rogers JA (2004) Stamping techniques for micro and nanofabrication: methods and applications. In: Bhushan B (ed) Springer handbook of nanotechnology. Springer, Berlin Heidelberg New York, p 185
32. Arnau A, Aumayr F, Echenique PM, Grether M, Heiland W, Limburg J, Morgenstern R, Roncin P, Schippers S, Schuch R, Stolterfoht N, Varga P, Zouros TJM, Winter HP (1997) Surf Sci Rep 27:113
33. Winter HP, Aumayr F (2002) Europhys News 6:215
34. Sporn M, Libiseller G, Neidhart T, Schmid M, Aumayr F, Winter HP, Varga P, Grether M, Niemann D, Stolterfoht N (1997) Phys Rev Lett 79:945
35. Aumayr F, Burgdörfer J, Varga P, Winter HP (1999) Comm Atom Molecul Phys 34:201
36. Schenkel T, Hamza AV, Barnes AV, DH Schneider (1999) Progr Surf Sci 61:23
37. Hayderer G, Cernusca S, Schmid M, Varga P, Winter HP, Aumayr F, Niemann D, Hoffmann V, Stolterfoht N, Lemell C, Wirtz L, Burgdörfer J (2001) Phys Rev Lett 86:3530
38. Hayderer G, Schmid M, Varga P, Winter HP, Aumayr F, Wirtz L, Lemell C, Burgdörfer J, Hägg L, Reinhold CO (1999) Phys Rev Lett 83:3948
39. Porte L, de Villeneuve CH, Phaner M (1991) J Vac Sci Technol B 9, 1064
40. Coregater R, Claverie A, Chahboun A, Landry V, Ajustron F, Beauvillain J (1992) Surf Sci 262:208
41. You HX, Brown NMD, Al-Assadi KF (1992) Surf Sci 279:189
42. Mazukawa T, Suzuki S, Fukai T, Tanaka T and Ohdomari I (1996) Appl Surf Sci 107:227
43. Mochiji K, Yamamoto S, Shimizu H, Ohtani S, Seguchi T, Kobayashi N (1997) J Appl Phys 82:6037
44. Reimann KP, Bolse W, Geyer U, Lieb KP (1995) Europhys Lett 30:463
45. Habenicht S, Bolse W, Feldermann H, Geyer U, Hofsäss H, Lieb KP, Roccaforte F (2000) Europhys Lett 50:209
46. Neumann R (1999) Nucl Instrum Meth B 151:42
47. Hahn R, Kang K, Song S, J Jeon (1996) Phys Rev B 53:1725
48. Hahn R, Kang K (1999) Phys Rev B 60:600
49. Minniti R, Ratliff LP, Gillaspy JD (2001) Phys Scr T92:22
50. Hayderer G, Cernusca S, Schmid M, Varga P, Winter HP, Aumayr F (2001) Phys Scr T92:156
51. Terada M, Nakamura N, Nakai Y, Kanai Y, Ohtani S, Komaki K, Yamazaki Y (2004) Observation of an HCI-induced nano-dot on an HOPG surface with STM and AFM In: Rudzikas Z (ed) Abstracts HCI-2004 12th international conference on the physics of highly charged ions, European Physical Society, p 208
52. Terada M, Nakamura N, Nakai Y, Kanai Y, Ohtani S, Komaki K, Yamazaki Y (2005) Nucl Instrum Meth Phys Res B 235:452
53. Gebeshuber IC, Cernusca S, Aumayr F and Winter HP (2003) Int J Mass Spectrom 229:27
54. Meguro T, Hida A, Koguchi Y, Miyamoto S, Yamamoto Y, Takai H, Maeda K, Aoyagi Y (2003) Nuc Instrum Meth B 209:170
55. Hida A, Meguro T, Maeda K, Aoyagi Y (2003) Nuc Instrum Meth B 205:736
56. Hahn R, Kang K, Song S, Jeon J (1996) Phys Rev B 53:1725
57. Hahn R, Kang K (1999) Phys Rev B 60:6007
58. Krasheninnikov AV, Elsin F (2000) Surf Sci 519:454
59. Winter HP, Aumayr F (1999) J Phys B: At Mol Opt Phys 32: R39
60. Ziegler JF, Biersack JP, Littmark U (1985) The stopping and range of ions in matter 1. Pergamon, New York
61. Hersam MC and Chung Y-W (2003) Proc Natl Acad Sci 100:12531
62. Albrecht TR and Quate CF (1988) J Vac Sci Technol A 6:271
63. Morita S, Wiesendanger R, Meyer E (2002) (eds) Noncontact atomic force microscopy. Springer, New York

64. Giessibl FJ (2003) Rev Mod Phys 75:949
65. Hölscher H, Allers W, Schwarz UD, Schwarz A, Wiesendanger R (2000) Phys Rev B 62:6967
66. Hembacher S, Giessibl FJ, Mannhart J, Quate CF (2003) Proc Nat Acad Sci 100:12539
67. Bitensky IS, Murakhmetov MN, Parilis ES (1979) Sov Phys Tech Phys 25:618
68. Schneider DH, Briere MA, McDonald J, Biersack J (1993) Radiat Eff Def Sol 127:113
69. Hayderer G, Schmid M, Varga P, Winter HP, Aumayr F (1999) Rev Sci Instrum 70:3696
70. Townsend P (1983) In: Behrisch R (ed) Sputtering by particle bombardment II. Berlin, Heidelberg, New York, p 147
71. Varga P, Neidhard T, Sporn M, Libiseller G, Schmid M, Aumayr F, Winter HP (1997) Phys Scr T73:307
72. Leitner M, Wutte D, Brandstötter J, Aumayr F, Winter HP (1994) Rev Sci Instrum 65:1091
73. Bedrossian PJ, de la Rubia TD (1998) J Vac Sci Technol A 16:1043
74. Giessibl FJ, Hembacher S, Bielefeldt H, Mannhart J (2000) Science 289:422
75. Duke CB (1996) Chem Rev 96:1237
76. Hug HJ, Lantz MA, Abdurixit A, van Schendel PJA, Hoffmann R, Kappenberger P, Baratoff A, Giessibl FJ, Hembacher S, Bielefeldt H, Mannhart J (2001) Science 291:2509
77. Huang M, Cuma M, Liu F (2003) Phys Rev Lett 90:256101
78. Hembacher S, Giessibl FJ, Mannhart J (2004) Science 305:380
79. Manassen Y, Hamers RJ, Demuth JE, Castellano AJ jun (1989) Phys Rev Lett 62:2531
80. Durkan C, Welland ME (2002) Appl Phys Lett 80:458
81. Wrachtrup J, von Borczyskowski C, Bernard J, Orritt M, Brown R (1993) Nature 363:244
82. Köhler J, Disselhorst JAJM, Donckers MCJM, Groenen EJJ, Schmidt J, Moerner WE (1993) Nature 363:242
83. Jelezko F, Popa I, Gruber A, Tietz C, Wrachtrup J, Nizovtsev A, Kilin S (2002) Appl Phys Lett 81:2160
84. Elzerman JM, Hanson R, Willems van Beveren LH, Witkamp B, Vandersypen LMK, Kouwenhoven LP (2004) Nature 430:431
85. Xiao M, Martin I, Yablonovitch E, Jiang HW (2004) Nature 430:435
86. Rugar D, Budakian R, Mamin HJ, Chui BW (2004) Nature 430:329
87. Balatsky AV, Martin I (2002) Quant Inf Proc 1:355
88. Manoharan HC (2002) Nature 416:24
89. Sidles JA (1991) Appl Phys Lett 58:2854
90. Rugar D, Yannoni CS, Sidles JA (1992) Nature 360:563
91. Sidles JA (1992) Phys Rev Lett 68:1124
92. Sidles JA, Garbini JL, Bruland KJ, Rugar D, Züger O, Hoen S, Yannoni CS (1995) Rev Mod Phys 67:249
93. Hammel PC, Pelekhov DV, Wigen PE, Gosnell TR, Midzor MM, Roukes ML (2003) Proc IEEE 91:789
94. Mozyrsky D, Martina I, Pelekhov D, PC Hammel (2003) Appl Phys Lett 82:1278
95. Ciobanu L, Seeber DA, Pennington CH (2002) J Magn Reson 158:178
96. Blank A, Dunnam CR, Borbat PP, Freed JH (2003) J Magn Reson 165:11
97. Stokstad E (2004) Science 305:322
98. Berman GP, Doolen GD, Hammel PC, Tsifrinovich VI (2000) Phys Rev B 61:14694
99. Pelekhov DV, Martin I, Suter A, Reagor, DW, Hammel PC (2002) Proc SPIE 4656:1

Part II
AFM of Biomolecules

15 Atomic Force Microscopy of DNA Structure and Interactions

Neil H. Thomson

From volume VI

Abbreviations

SPM	scanning probe microscope/microscopy
STM	scanning tunneling microscope/microscopy
AFM	atomic force microscope/microscopy
FM	frequency modulation
AM	amplitude modulation
ShFM/TDFM	shear force microscopy/tranverse dynamic force microscopy
SEM	scanning electron microscopy
UHV	ultrahigh vacuum
TIRFM	total internal reflection fluorescence microsopy
DNA	deoxyribose nucleic acid
RNA	ribonucleic acid
bp	base-pairs
dsDNA	double-stranded DNA
ssDNA	single-stranded DNA
APTES	aminopropyltriethoxysilane
APS	aminopropylsilatrane
ATP	Adenosine triphosphate
ADPNP	5′-adenylyl β,γ-imidodiphosphate
E. coli	*Escherichia coli*

15.1 Introduction: The Single-Molecule, Bottom-Up Approach

Scanning probe microscopes (SPMs) are perceived as tools central to the goals of the burgeoning field known as nanotechnology. In the late 1950s, the physicist Richard Feynman, envisaged and encouraged a new field of manipulating matter down to the atomic scale [1]. In his visionary talk "There's plenty of room at the bottom" he suggested that electron microscopes needed to be made more powerful and introduced the idea of top-down fabrication to reach out to the nanoworld. One of the many ideas that arose naturally from his presentation, is that we should only have one machine or device interfacing between us and the nanoworld, because errors in scaling down will be too large, i.e. in building a machine to make a smaller machine that makes a smaller machine and so on, unacceptable tolerances would be needed. He used the example of smaller and smaller pantographs joined together,

In: H. Fuchs, B. Bhushan (eds), *Biosystems - Investigated by Scanning Probe Microscopy*,
DOI 10.1007/978-3-540-37319-3_5, © Springer-Verlag Berlin Heidelberg 2010

but perhaps did not envisage just one device that could not only image directly in the nanoworld but manipulate it also. Indeed, he said "Perhaps you can figure a better way than that to get down to small scale more rapidly." This way has now been achieved (at least in one sense) and the family of scanning probe microscopes that measure and manipulate matter at the nanoscale were possibly of what he was dreaming. Electron microscopes, of course, were well on their way already, and since that time they have been made more powerful and are now important imaging devices for nanotechnology, while electron beams are essential tools for top-down nanofabrication. In contrast, SPMs offer flexibility in addressing a probe directly to a surface, particularly the versatile atomic force microscope (AFM), which can be a nanopen, nanoscribe or nanofinger, allowing the human user to write with molecules [2], scratch or draw lines on surfaces smaller than the wavelength of light [3] or simply feel atoms and molecules. While SPMs can also be used as devices for top-down fabrication in lithographic modes of operation, they are also suited to bottom-up fabrication because they can, in principle, pick up and move atoms and molecules and construct new devices atom-by-atom or molecule-by-molecule.

Another important aspect of bottom-up fabrication in nanotechnology is the property of certain molecules to self-assemble in a spontaneous fashion [4]. Evolutionary theory tells us that biological organisms themselves have been self-assembled through myriad generations of Earth's history and incorporated ever-increasing levels of complexity. To understand biomolecular interactions and molecular mechanisms on a fundamental level, one major approach is to isolate pure samples of the molecules of interest so that only one or two, or a handful of species are interacting within an in vitro model environment. By adding more and more species sequentially as understanding of the fundamental system is gained, the complexity of biology may be grasped in a bottom-up approach. The counterpoint to this approach is also analogous to the top-down approach mentioned in nanofabrication, whereby markers (e.g. fluorescent probes) and imaging devices (e.g. confocal microscopy) are used on cells in an attempt to understand biological processes by seeing what happens in vivo, whilst all the complexity is present.

While the AFM is capable of imaging atomic periodicity, and in certain cases true atomic resolution [5] and even subatomic resolution [6], it has a large dynamic range that enables it to image molecules up to larger structures, such as cells [7]. The resolution on soft, large structures such as cells is compromised to a certain degree because the tip indents the surface, increasing its effective size. Additionally, as a surface-profiling technique, AFM is unable to "look" or should one say feel inside a cell without causing irreversible damage. For these reasons, the AFM is perhaps less suited to the in vivo approach for biology and could be claimed to be better suited to studying in vitro systems. Having said that there are many interesting questions that can be asked about cells using AFM because much of the communication in biology occurs at the cell interfaces, however, in this chapter we will focus on the use of AFM to study biomolecular systems in vitro, from the single-species case (e.g. DNA alone), increasing the complexity molecule-by-molecule and species-by-species (e.g. DNA plus one protein type) in a bottom-up approach to understanding molecular biology, which may eventually lead to spin-off applications in nanotechnology. This review of AFM research in the last twenty years is centered on DNA: the molecule central to the propagation of living organisms, the molecule that preserves the genetic code.

We will see that we have not come far on this bottom-up journey into biological complexity, and that introducing more complexity than DNA alone has only fairly recently been realized. However, it is hoped that this review of the field today will demonstrate the increasing importance of this approach and illustrate that new gains in the knowledge of DNA and its interactions with other molecules will continue to be made using the AFM in this fashion for many years to come.

15.2
DNA Structure and Function

Deoxyribonucleic acid: the name does not exactly trip off the tongue, but its abbreviation DNA does and it is arguably the most well-known of biomolecules outside scientific circles. Its key functions are storage and retrieval of genetic information. It acts both as the blue-print for maintaining the survival and propagation of an organism: it is the "library" of the cell allowing portions of the genetic information to be used, for example, for the synthesis of enzymes and other proteins. This chapter will outline the basic structure and interactions of double-stranded DNA (dsDNA) in the context of AFM studies but for more details on DNA structure and function see for example [8, 9].

Its double-stranded structure is the key to its function, as first recognized by Watson and Crick [10]. The hydrogen bonding between the base-pairs makes the structure very stable allowing the propagation of genetic information through generations upon generations of organisms with low error rates and little genetic drift. The base-pairs (bp) that encode the information are hidden "inside" the helix, protected by the sugar phosphate backbone (with negatively charged phosphate groups) meaning that the structure has to be actively distorted for the genetic code to be read or changed.

The base-pairing in DNA occurs between the two classes of chemical bases, known as purines and pyrimidines. Both are based on carbon-based rings with nitrogen atom substitutions, the pyrimidines being a single ring and the purines being a double ring of 5 and 6 atoms fused, where two carbon atoms are shared between the rings (see Fig. 15.1). The most stable base-pairings are always between purines and pyrimidines giving rise to the standard Watson–Crick pairings: adenine and thymine (A–T) and cytosine and guanine (C–G). There are two hydrogen bonds between A–T base-pairs and three between C–G, giving rise to a binding energy difference

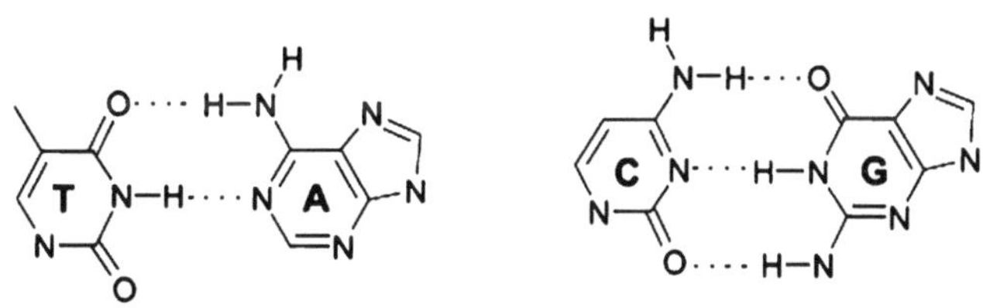

Fig. 15.1. The chemical structures of base-pairing between adenine and thymine (A–T) and between cytosine and guanine (C–G) through 2 and 3 hydrogen bonds respectively. A and G are purine bases while T and C are pyrimidine bases. Courtesy of Neal Crampton

between the two sets of genetic letters of around $2kT$ [11], allowing discrimination in energetic terms as well as structural between the letters in the genetic code. Other base-pairings can occur, i.e. base mismatches, where two purines or two pyrimidines are paired together, but these situations cause distortions in the overall DNA structure because the pyrimidines cannot bridge the gap between strands very easily and purines are too large to be accommodated by the usual helix structure.

The bases of DNA are arranged into a helical structure through the presence of the scaffold of the sugar moieties, deoxyribose, linked together to by phosphate groups via phosphodiester bonds. The chemical moiety consisting of the base and sugar is known as a nucleoside. The bases are attached to the C1′ carbon of the sugar ring, while the alternating phosphates between these nucleosides are attached at the hydroxyl groups on the C3′ and C5′ positions. (Note: a prime (′) is used to distinguish the carbons on the sugar from those in the base.) This staggering of the phosphodiester bonds gives rise to a polarity in each single-strand of DNA such that one is running in the 3′ to 5′ direction, while the other runs from the 5′ to 3′ end in an antiparallel arrangement. The two strands are intertwined into a double-helix that takes two basic structural forms: known as A and B. Both these structures were originally identified from X-ray diffraction of DNA fibers, in low (75%) and high (92%) humidities respectively. The "fully" hydrated B-form is thought to be the most common conformation found in vivo [8]. In B-DNA, the base-pairs lie essentially perpendicular to the helix axis, while in A-DNA the base-pairs are tilted with respect to the helix axis and lie a longer way off axis (Fig. 15.2). While the base-pairs in B-DNA span the helix, the helix in A-DNA opens up to form a channel down the center. The consequences of this structural transition are that the number of bp per helix turn goes from 10.5 to 11.0, while the helix diameter goes from about 2 to 2.6 nm, as the DNA is dehydrated. In essence, the pitch of the helix becomes less steep, which causes the molecule to become broader, and shortens the length of a linear dsDNA molecule. In B-DNA the axial rise per base-pair is 0.33 nm whereas in A-DNA this falls to only 0.23 nm. It is unclear as yet whether A-DNA even exists in vivo but has been proposed to occur at specific sites for protein binding [8]. There are also more unusual DNA structures, such as Z-DNA, where the helix turns inside out and becomes left-handed. Again, the prevalence of such structures within organisms is still under debate [8].

The genomes of organisms are organized into compressed structural architectures, chromosomes, which serve the dual purposes of conserving space and regulating gene expression. In cells containing a nucleus, the DNA is wound around histone protein complexes, like cotton on reels, to form chromatin, which is then further hierarchically condensed into chromosomes. For the survival and maintenance of cells a large number of genetic processes are required, which fall into the categories: RNA and protein synthesis; DNA repair; DNA replication; genetic recombination. All these processes are mediated by enzymes (proteins) that carry out work on the DNA templates made accessible as a result of decondensing the chromatin through the action of other proteins. When genes are expressed, a fundamental molecular process, known as DNA transcription is required to copy the genetic code into messenger RNA, which is subsequently used by the cell as a code to manufacture proteins in the process known as translation. Transcription is mediated (or catalyzed) by the enzyme RNA polymerase and later, we will see that AFM has been used to visualize

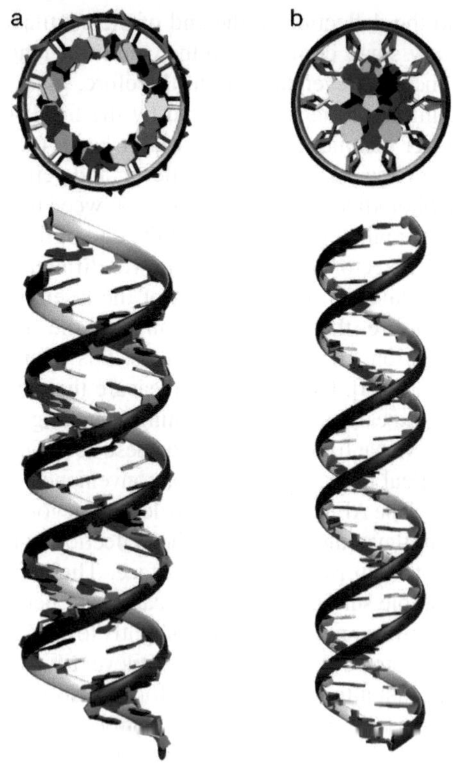

Fig. 15.2. The structure of double-stranded DNA in the A-form (*left*) and the B-form (*right*). The *upper panels* show top down views of the two structures demonstrating that dehydration of the helix leads to a opening up of the two strands to form a channel down the middle in the A-form. These model helices are formed from repeating sequences of GC bases for 30 bp. Note that the top view of the B-form is slightly magnified compared to the other images for clarity. Courtesy of Sarah A. Harris

the process of transcription in situ at single-molecule resolution [12]. This was the first visualization of an enzyme in action by AFM and heralded the beginning of AFM single-molecule enzymology. Since that time, still only a handful of real-time visualizations by AFM have been achieved (see Sect. 15.5.3.5). This highlights the difficulty of the technique and this chapter attempts to explain the reasons behind this slow progress. For more details on the molecular mechanisms of genetic expression and maintenance, see for example [13].

15.3 The Atomic Force Microscope

The atomic force microscope (AFM) [14] can be considered as the most versatile of all scanning probe microscopes (SPM). Since forces arise from the properties of matter, they are ubiquitous, and therefore the AFM can in principle be applied to any solid-state material. As with all SPMs, the roughness of the sample is a limiting factor, and curvature and topographical gradients on the sample should ideally not be comparable to or larger than the radius of curvature and opening angle of the AFM tip respectively. The "atomic forces" that arise between the apex of the AFM tip and the sample are detected because the tip is on the end of a flexible cantilever spring. The cantilever obeys Hooke's law ($F = kx$), so that the force detected is proportional to the deflection of the end of the cantilever beam (Fig. 15.3). The deflections are very small compared to the cantilever length and hence the deformation is elastic. The cantilever deflection, therefore, gives a measurement of the resistance of the sample to the force exerted by the tip. It is commonly measured using an optical lever technique whereby a laser beam (e.g. 5 mW red solid-state laser) is focused onto the back of the cantilever and reflected onto a split photodetector (usually a photodiode). The path length between the cantilever end and the detector amplifies the small angular motion of the end of the cantilever into large linear motion across the detector interface. The power difference on the two halves of the photodetector is related to the movement of the laser spot. In this sense, this technique actually measures the change in the angle of the end of the cantilever rather than its absolute position. For small deflections, this approximation of the cantilever deflection is valid [15]. Other ways to measure the cantilever deflection are interferometry and in the earliest cases a scanning tunneling microscope (STM), piggy-backing on the AFM cantilever [14]. Both these techniques are more difficult to set up and the optical lever is preferred for convenience.

The AFM can be used to determine forces by calibrating the spring constant of the cantilever and measuring the deflection of the cantilever as a function of the position of the tip relative to the surface. These forces are of course acting normal to the sample surface because the cantilever bending occurs along this axis. In reality, the cantilever must be inclined slightly toward the surface so that the cantilever itself does not catch the surface. This cantilever tilt angle can be 10 to 15°, which slightly affects the magnitude of the normal forces resolved. This tilt also means that the tip will travel across the surface during indentation experiments. Lateral forces can be inferred by measuring torsional motion of the cantilever using a quadrant photodetector and taking the power difference between the left and right sides of the detector.

As well as measuring forces along a single axis, the AFM can be used as an imaging tool to gain three-dimensional topographical information about samples by using a scanning system with a feedback loop to manipulate the tip across the sample. Figure 15.4 shows a schematic representation of the AFM and its major components. The scanning is usually generated through the use of piezoceramic materials that change

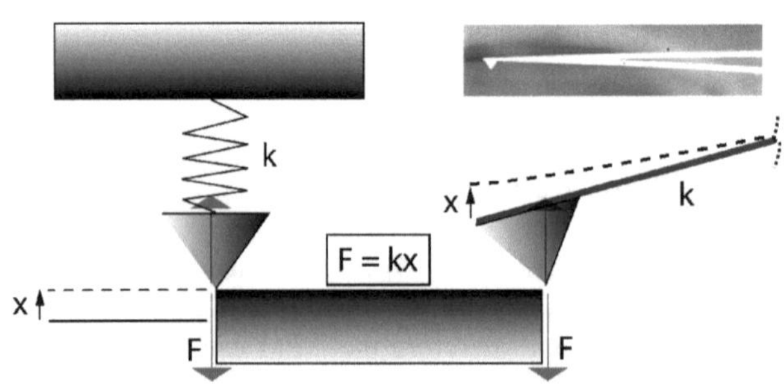

Fig. 15.3. RHS: The AFM cantilever (RHS) obeys Hooke's law for the small nanometer scale deflections it makes compared to the length of the cantilever, typically 100 or 200 μm. LHS: The AFM tip can represented as being on the end of a spring with stiffness k, known as the spring constant of the cantilever. The *top right panel* shows an SEM micrograph of a 100 μm-long silicon nitride AFM cantilever, with an integral tip, which has an open triangular shape

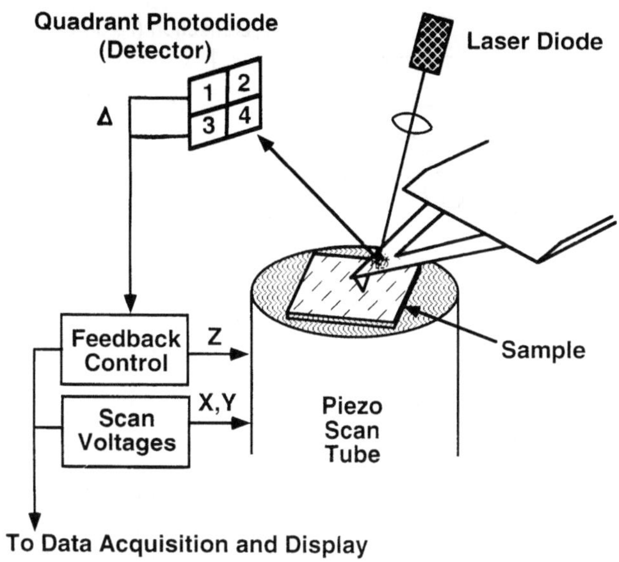

Fig. 15.4. Schematic representation of the major components of an atomic force microscope. Either the sample or tip assembly is scanned on a piezoceramic transducer that can create motion in all three axes. The microfabricated cantilever is addressed to the surface and the deflection of the end of the cantilever measured using an optical lever technique. The relative size of the cantilever in this schematic is greatly exaggerated. The deflection signal of the cantilever is used to determine how the feedback circuit and scanners drive the tip across the three-dimensional surface

shape on application of high voltages, such as lead zirconium titanate (PZT). Minute changes in the equilibrium position of the atoms in the crystals allow subnanometer positioning accuracy, although they do exhibit nonlinear behavior and hysteresis. Newer generations of commercial AFMs overcome these limitations by accurately measuring the position of the scanner and using an additional feedback loop to hold the tip in the required position. These are known as closed loop scan systems and eliminate image distortion from piezo nonlinearity and thermomechanical drift. In principal, any scan pattern can be generated by the scanners, but typical images are square and built up in a raster pattern in a series of lines, so that there are fast and slow scan axes. Typical scan rates of current commercial AFMs are between 1 and 4 Hz (i.e. scan lines per second), although the tip velocity across the surface is the limit for imaging speed, and this is typically up to 10 μm/s. Images are typically sampled at 512 × 512 pixels, which means that a 1 μm^2 scan will take nearly one minute. AFM scan technologies are developing all the time, however, and some groups are working on making systems that perform at video rates or better (see Sect. 15.6).

There are two imaging strategies for AFM, (or indeed any SPM): to have the feedback enabled or disabled. Figure 15.5 shows schematically how these strategies work and what information is obtained. Without feedback, the tip is simply scanned across the surface and the deflection measured to build up an image of the surface. In this case, the force applied to the sample is proportional to the topographical changes across the sample, since this is related to the cantilever deflection. This approach

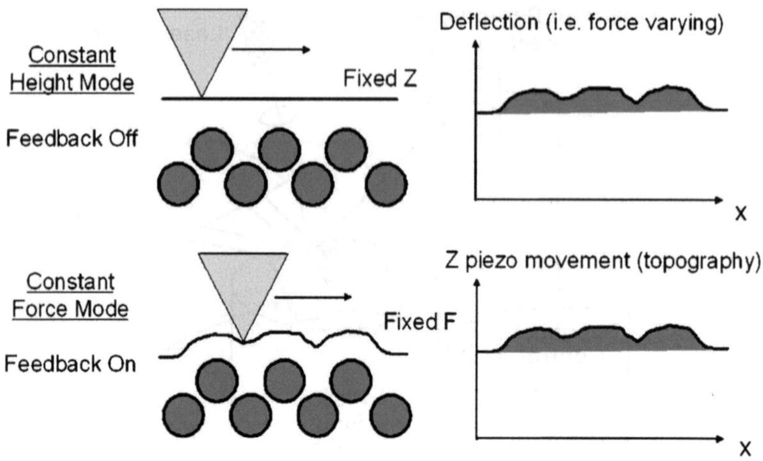

Fig. 15.5. Schematic representation of the two fundamental ways of operating a scanning probe microscope. For the atomic force microscope, constant-height mode implies information on the surface topography resides in the cantilever deflection (**a**), whereas in constant-force mode with feedback on the movement of the vertical Z-piezo gives topographical information also (**b**)

is almost always unsuitable for soft samples, such as polymers and biomaterials because the forces that build up as the deflection increases when a sample feature is encountered cause sample damage. To eliminate this problem a feedback loop can be used to sample the deflection as a function of time and move the scanner up or down to keep the cantilever deflection and therefore the contact force constant and ideally minimized. With feedback on, the imaging method is known as constant force, while with feedback off the imaging method is known as constant height. In practice, there are user-controlled gains for the amplifiers in the feedback loop that allow essentially a continuous spectrum between the feedback off and the fully on situations. The two fundamental signals that are typically monitored are the cantilever deflection and the voltage sent to the vertical scanner element, denoted as the Z-piezo. The distribution of information between these two images depends on the response time of the microscope. If the feedback loop could be driven infinitely quickly then the deflection signal would be perfectly flat. In practice, the cantilever usually has the quickest response time in the whole system (being a small mechanical device) and therefore there is always residual information in the deflection signal. The inertial response of the piezos means that the topographical signal (i.e. Z-piezo position) contains the lowest frequency information about the surface, while the deflection signal contains the higher-frequency information up to the frequency of the cantilever. Altering the gains in the feedback loop essentially allows the user to partition information between these two signals, but for biomolecular samples such as DNA, gains are almost always kept at the maximum possible (above which scanner resonance would occur).

Whether operation is performed in constant height or constant force imaging, there are two basic imaging modes that the AFM can be operated in. In analogy with electronics, these are termed DC or AC modes, which relate to whether the cantilever is used statically or dynamically (Fig. 15.6). In the typical and traditional DC mode, known as contact mode, the AFM tip is addressed mechanically towards the sample surface and the quasi-static deflection of the cantilever is monitored as the tip is scanned back and forth across the sample surface [14]. This mode creates large shear forces between tip and sample because the tip is in constant contact with the surface. Contact mode is generally unsuitable for soft materials such as polymers and biomaterials because they are distorted and/or damaged quite readily. To overcome these limitations, AC techniques in which the cantilever is vibrated around its resonant frequency were developed. In these methods, the tip is intermittently addressed to the sample surface, once per cycle of cantilever vibration and importantly contacts the surface in essentially the surface normal direction (i.e. vertically). It appears that viscoelastic material can support compressive forces more readily than the shear forces present in contact mode. AC imaging techniques can be divided into those that use the cantilever amplitude as the detection signal and those that use the cantilever frequency. In analogy with radio electronics, these are known as amplitude modulation (AM) and frequency modulation (FM) techniques, respectively [16]. Historically, AM AFM has been used in ambient and liquid environments on soft samples (e.g. polymers and biomaterials) while FM AFM is used under UHV conditions, usually to try and achieve atomic resolution on solid-state systems. The divergence in the use of AM or FM detection in AFM AC techniques for different applications arose because FM detection usually requires cantilevers with a very high quality factor (Q) where environmental damping in UHV is negligible. By contrast, in AM AFM a significant change in the cantilever amplitude is required when the tip interacts with the sample, therefore cantilevers with a lower Q and higher damping environments (air and liquids) are acceptable. The AM imaging mode is often known as tapping mode, since the AFM tip repeatedly taps the sample surface as it is scanned [17].

Having established that constant-height imaging is not suitable for imaging biomolecular samples and realizing that the feedback is almost always set to the quickest response time possible, of the two imaging modes the AC (AM) approach is almost always more successful for biological specimens. Having said this, as we will learn later, the imaging environment plays an influential role on the ability of the AFM tip to image molecular structure reproducibly. The AFM can be operated under the full range of environments, from vacuum through gaseous or ambient

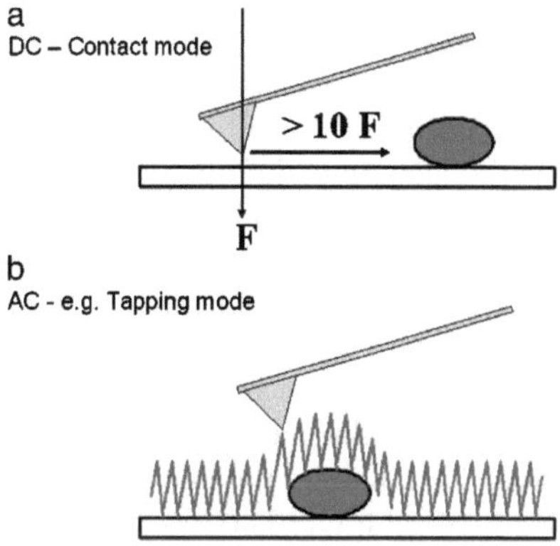

Fig. 15.6. Schematic representation of the two fundamental imaging modes of an atomic force microscope. (**a**) In contact mode the static DC deflection is used as the detection of the surface, but large lateral shear forces can build up between tip and sample damaging or sweeping away weakly bound molecules. (**b**) In amplitude-modulation techniques, such as the tapping mode, the cantilever amplitude is the detected signal. In this mode the tip contacts the surface in the normal direction eliminating damaging shear forces

(air) to liquid, including aqueous media. The environment affects the magnitude of different forces operating between the tip and the sample. For instance, in ambient air conditions there is usually a capillary neck between the tip and surface formed through condensation of water vapor on the contacting surfaces. This produces a large attractive force that sucks the tip onto the surface and limits the lowest tracking force that can be achieved. This effect can be eliminated by going to vacuum, using dry environments or submersing the tip under liquid. The cantilever is chosen to have a suitable stiffness (spring constant) depending upon the imaging environment and mode. For example, tapping mode in air requires relatively stiff cantilevers (e.g. $k > 10\,\mathrm{N/m}$), to prevent the tip from adhering irreversibly to the sample surface through the capillary interaction. Contact mode requires relatively soft cantilevers (e.g. $k < 1\,\mathrm{N/m}$) to give a measurable deflection at loading forces that are sufficiently low as to not cause sample (or tip) damage. Tapping mode under liquid also uses these softer levers to prevent sample damage during mechanical drifting of the microscope.

The ultimate resolution in the AFM is determined by the size and shape of the apex of the scanning tip. The cantilevers are microfabricated from silicon and silicon nitride and usually include integrated tips that have slightly different shapes and are often sharpened through oxidation to produce the finest apexes. Typically, the radius of curvature of the tips is around 10 nm, which means that in AFM images, DNA will also appear broadened by approximately twice the radius of the tip [18], limiting the lateral resolution obtained. Resolution along the backbone could be higher than this because the contact area between tip and sample is at its lowest when the sample is locally convex.

It should be realized that the AFM only produces a true topographical image of a surface if the mechanical response under the AFM tip is homogeneous across the surface. For soft biological molecules adhered to a relatively hard support surfaces, it is unlikely that the AFM will produce an exact topograph. For instance, biological cells are readily indented by the AFM tip and the image produced is usually representative of the internal cytoskeleton, i.e. the stiffest components of the cell. For single biomolecules, the situation is more complex but it is interesting to note that the heights of DNA, particularly when imaged in air are almost always lower than would be expected. Typically, the maximum height of the DNA backbone above the support surface obtained by AFM is around 1 nm or so, sometimes less, whereas we might expect 2 to 2.6 nm depending on the local conformation of the DNA. One explanation for this effect has been to say that the DNA is compressed under the large local pressures exerted by the AFM tip, however, it is difficult to verify this experimentally. In tapping mode at least, the image-contrast-generation mechanism is not well understood. Other factors may also influence the measured height, including: deformation of the DNA molecule due to surface-binding forces; erroneous background support levels due to surface deposition of salt layers; complex dynamics of the vibrating cantilever in AC imaging modes; influence of longer range forces (e.g. electrostatic) on topographical imaging. The full picture of the image-contrast-generation mechanisms for AFM will require much more work. For imaging samples in air the capillary neck present between tip and sample is likely to have a big influence on the imaging mechanism, both in contact mode and tapping mode [19]. On 2D protein crystals under aqueous fluid it has been shown that the electrostatic forces between tip and sample also play an important role [20].

15.4 Binding of DNA to Support Surfaces

15.4.1 Properties of Support Surfaces for Biological AFM

Since the AFM, like all SPMs, is a surface-profiling instrument, study of molecular systems requires that the molecules of interest are deposited irreversibly onto a suitable support surface (sometimes known as the substrate[1]) that can be scanned by the AFM tip. This support surface should fulfil a number of criteria.

First, the roughness of the surface should be small compared to the height of the molecules in question so that the AFM tip can distinguish them from the background in the topographical (Z-piezo) signal. Secondly, the interaction between the molecules and the surface should be strong enough such that the influence of the scanning tip does not distort, move or damage the molecules during repeated scanning. Finally, the surfaces must be easy to prepare, reproducible and free of contamination. These different aspects of the supporting surface will now be discussed in more detail.

The criterion that the surface should be extremely flat, at least on local scales (nm to μm), tends to be more important for globular biomolecules, such as proteins, compared to long polymer molecules, such as DNA. Even if the resolution is not high enough to pick up a globular feature in a rough background the correlation of many points along an elongated structure (even if it is only one pixel wide in an image) allows the structure to be discerned from the noise (background surface features). Development of high-resolution chemical force microscopy (CFM) [21] or molecular recognition force microscopy (MRFM) [22], where the tip is functionalized with a chemical species or biological receptor respectively, would generate better material contrast and may, to some extent, obviate the need for very flat surfaces (see Sect. 15.6 for a brief discussion of CFM and MRFM techniques).

There are a number of ways that molecules can be adhered to a surface such that they are immobile under the scanning AFM tip. Essentially we can think of these as either a chemical strategy where molecules are covalently linked to the surface, or a physisorption process, where the natural forces involved attract and immobilize the molecules. The most straightforward sample preparation uses physisorption processes, whereby the chosen surface and the environment (air, vacuum or liquid) determine if and how molecules bind. Chemical immobilization requires either a derivatization of the surface or a chemical crosslinker to be introduced between the molecule and surface or a combination of both approaches. Molecular immobilization strategies, with particular reference to DNA, are discussed in more detail in Sect. 15.4.2 below.

Ideally, the support surface should be very easy to prepare, having as few preparation steps as possible to prevent introduction of any contamination. This means that physisorption approaches are preferred over chemical strategies if and only

[1] The term substrate is also used in biochemistry to mean the entity, e.g. molecular species, upon which an enzyme acts during its catalytic cycle. Consequently, we will refer to the underlying surface in the AFM sample as the support surface.

if they achieve the required molecular surface binding. Introduction of chemical functionalization may also make reproducible formation of surfaces more difficult to achieve. The preferred support surface for many biomolecular AFM studies that traditionally has been used is the mineral, mica. It is a layered crystal in which the silicate layers are ionically bonded together comparatively weakly. The anisotropy in the crystal, means it forms flat flake-like structures, and allows the top surface of the flakes to be simply cleaved away using sticky tape (the best kinds to use are those with low tackiness, usually semitransparent and white in appearance). This produces a clean surface every time molecules are required for deposition. The layers are atomically flat over many hundreds of micrometers to millimeters, allowing easy distinction of biomolecules such as DNA. There are other such layered materials available to use in SPM studies of molecular systems, such as highly oriented pyrolytic graphite (HOPG), crystalline graphite in which the planes are regularly ordered over large dimensions. They are usually a couple of centimeters across and can also be cleaved with sticky tape because of the weak bonding between the 2D covalently bonded carbon planes. It is, however, a very hydrophobic surface (due to its non-polar nature) to which hydrophilic molecules, such as DNA, do not bind very strongly and is therefore inappropriate for most AFM studies. Mica, by contrast, is a hydrophilic surface that is well-suited to biomolecular deposition for all the reasons given above.

15.4.2 DNA Binding to Surfaces

As we have discussed, mica is the most common surface on which to deposit DNA and its complexes for study by AFM. Mica is in fact a group of minerals with over 30 members belonging to the silicates class. Silicates are thought to make up 90% of the Earth's crust and are based on tetrahedral SiO_4 motifs. These tetrahedra can be arranged singularly or doubly and in chains, sheets, rings and other frameworks. Mica is a phyllosilicate with a sheet-like structure and is related to clay minerals, all of which have a plate-like or layered structure. Clays hold water in between the silicate sheets, whereas micas are anhydrous with no water present. The mica used in AFM is one of the most common members of the group, called muscovite mica, being potassium aluminum silicate hydroxide $KAl_2(AlSi_3O_{10})(OH)_2$. It consists of silica tetrahedra and alumina octahedra covalently bonded together into sheets 1 nm thick that are weakly bonded together ionically with potassium ions (K^+) (Fig. 15.7). This gives rise to its good basal cleavage, revealing a surface terminated with oxygen atoms. Potassium ions sit in a regular array of binding sites at the surface. When immersed in aqueous solution the K^+ ions diffuse away from the surface to be replaced by a counterion cloud. This renders the mica surface negatively charged with a diffuse double layer of cations above it.

Since DNA is a negatively charged semiflexible rod [23] it will not bind to the mica unless the electrostatic repulsion between itself and mica is overcome. This is usually achieved through the introduction of divalent metal cations into the aqueous medium. These replace the K^+ ions in the mica binding sites and effectively render the mica positively charged. Different cations have a different efficacy of inducing DNA binding to mica, which appears to be related to their

Fig. 15.7. The atomic structure of mica. Reused with permission from Shojiro Miyake, Applied Physics Letters, 67:2925 (1995). Copyright 1995, American Institute of Physics

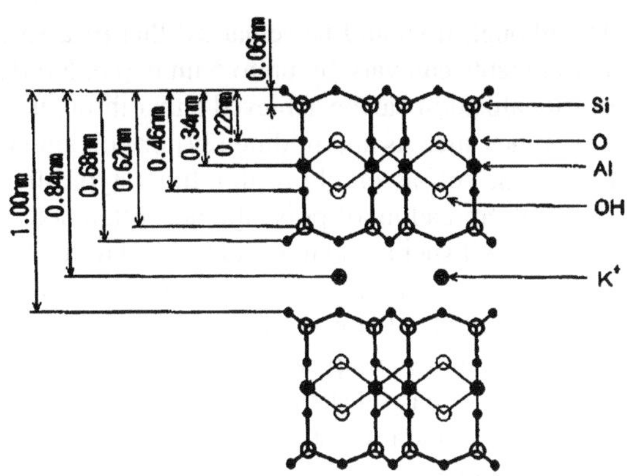

enthalpy of hydration [24]. The transition-metal ions, Ni(II), Co(II) and Zn(II) have been shown to be the most effective. A simplistic picture is the idea that smaller ions are the only ones that physically fit into the binding sites, however, along with these ions, Mg(II), despite being a group II element, has a similar ionic radius and yet does not bind DNA well enough for it to be imaged in tapping mode, at least not under aqueous fluid. Larger transition-metal cations, like Mn(II) bind DNA weakly, while Cd(II) and Hg(II) do not appear to bind DNA at all to mica. Those cations, however, that have a larger enthalpy of hydration bind DNA more efficiently (e.g. Ni(II)), while Mg(II) and Mn(II) having intermediate values. The enthalpy of hydration of an ion is related to the strength of the coordination bonds it makes with ligands. Those ions that make strong bonds with water can also make strong bonds with the recessed hydroxyl groups in the mica surface once the coordinated water molecules are removed. It appears that the immobilization of the divalent cations into the mica binding sites is the most important factor for immobilization of DNA. The real situation is somewhat more complex, however, since in Mg(II) buffers DNA binds sufficiently strongly to mica that it can be imaged using tapping-mode AFM in air, after the sample has been rinsed with excess water and dried with a stream of nitrogen gas. There are two ideas used to explain the mechanism of binding of DNA to the mica; one is a static model based on salt bridges and the other is a dynamic picture invoking counterion correlation forces. In the static picture, the divalent cation simply acts as an electrostatic bridge between the negatively charged DNA and mica [25]. In the dynamic model, the ions are in constant motion, and an overall attractive force arises through sharing of divalent counterions in a correlated way [26]. This picture seems realistic since the attraction is ion-concentration dependent, where higher concentrations of divalent cations electrostatically screen these correlated interactions. Pretreating the mica surface with transition-metal ions that bind irreversibly into the K^+ sites on the mica (e.g. Ni(II)) increases the correlation force and lowers the net negative surface charge, encouraging DNA to bind more strongly.

In essence, immobilization of the DNA requires a positively charged surface since the DNA is highly negatively charged. Another way to create positively charged mica surfaces is to chemically functionalize the surface with positively charged groups, such as amines (NH_2). These can accept a proton to give a $+1$ charge around neutral pH, although it should be remarked that pKa values between solution and surface environments can vary by up to 5 units (i.e. 5 orders of proton concentration). This functionalization can be achieved either through the spreading of polycations, using polyamines such as spermidine [27] or polyaminoacids such as polylysine [28], or through the self-assembly of thin films, typically a monolayer, using silane chemistry [29]. Spreading of polycations is often not very reproducible because they are usually polydisperse and may not be ideal for DNA applications because the polymer widths are comparable to DNA widths and linear surface features arising from the polymer could be confused with DNA molecules.

Lyubchenko's group has pioneered the formation and use of aminosilanes to create positively charged surfaces [29]. Silanes are chemically unstable molecules that crosslink with themselves and/or surface hydroxyl groups, using a condensation reaction, eliminating water to form Si–O bonds. They react strongly with the hydroxyl groups on silicon surfaces and much of the characterization of the resultant layers has been performed on silicon. The preferred silane for mica functionalization is 3-aminopropyltriethoxysilane (APTES) partly because of its commercial availability. A trifunctional silane such as this has the ability to crosslink to a surface through one alkoxyl group, while the other two can be involved in lateral crosslinking of the molecules into a monolayer. The amine group on the end of the alkane is then expected to be pointing upwards on the surface creating an array of positively charged sites under aqueous fluid. Figure 15.8 shows the chemical structure of APTES and how it self-assembles into monolayers on surfaces. There is increasing evidence that the stability of APTES layers on mica comes solely from lateral crosslinking rather than any direct surface crosslinking, which is unlikely because the hydroxyl groups on mica are recessed 0.22 nm below the surface, whereas the Si–O bond length is 0.16 nm [30]. It is possible that plasma treatment of mica disrupts the surface enough to allow covalent interactions between silane and the surface [31]. The formation of these layers is strongly affected by the formation conditions, e.g. from vapor versus liquid phase, and the instability of the silane can easily lead to 3D polymerization in the bottle before the surface is made. If the silane "goes

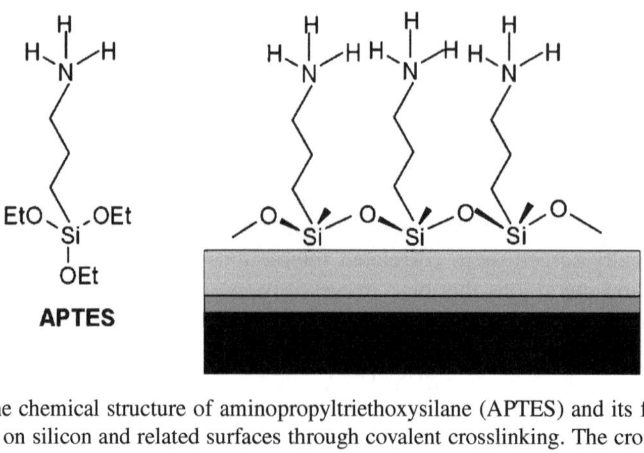

Fig. 15.8. The chemical structure of aminopropyltriethoxysilane (APTES) and its formation of monolayers on silicon and related surfaces through covalent crosslinking. The crosslinks form through a condensation reaction eliminating alcohol. Courtesy of Neal Crampton

off' in this way, reproducible surfaces cannot be made. Liquid-phase preparation also tends to produce uncontrolled polymerization at the surface into 3D globular structures also unsuitable for molecular imaging. Lyubchenko and coworkers [32] have got round these problems by making a cyclic derivative of APTES, aminopropylsilatrane (APS), that is stable in aqueous solution and can only crosslink in the presence of a surface. Figure 15.9 shows schematically how the interactions of DNA in solution are mediated with mica and aminosilane treated mica to promote surface binding.

Vapor-phase deposition is the preferred method of formation of APTES-mica, although the humidity of the surrounding environment can strongly affect the structure of the silane layers formed [30]. The water layers that are present on hydrophilic mica under ambient conditions influence the extent and structure of the layer formation. Below 25% relative humidity (RH) a monolayer of APTES is formed that can bind DNA without the need for divalent cations in the buffer. Above 25% RH a bulk water layer starts forming on the mica and the APTES forms a bilayer with mobile silanes present that induce DNA condensation. This condensation effect can be removed upon baking the surfaces to drive off the mobile silanes, but DNA does then not bind without the presence of divalent cations, suggesting a gross structural rearrangement of the structure. Generally, APTES-mica for DNA applications is therefore produced under environments as dry as possible [32].

DNA that is bound to surfaces adopts conformations that can be influenced by the binding forces present. The two extremes of behavior are known as equilibrated and kinetically trapped [33]. Figure 15.10 shows examples of AFM images taken in these two conditions. In the first situation, the DNA is bound to the surface but free to diffuse sideways in two-dimensions and therefore adopts a minimum energy conformation in the 2D plane of the surface (Fig. 15.10a). In the second case, as part of the DNA molecule touches the surface it is irreversibly pinned at that point and the DNA molecule effectively collapses on top of itself and can no longer move sideways (Fig. 15.10b), although local motions of regions of the DNA can be allowed [34]. The equilibration of DNA occurs on mica surfaces where the interaction is mediated via the Mg(II) cations. This is allowed because the Mg(II) ions are mobile and not bound into the cavities on the mica surfaces, enabling the DNA to move in the plane of the surface. For the divalent cations that are irreversibly bound into the cavities such

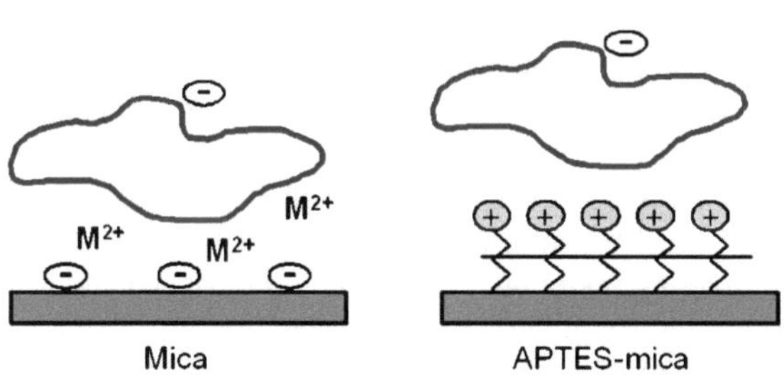

Fig. 15.9. Schematic representation of the binding mechanisms of dsDNA to mica and APTES-mica. The binding to mica is mediated by divalent metal cations, while the negatively charged phosphate backbone of the DNA interacts directly with positively charged amino groups on the APTES surface

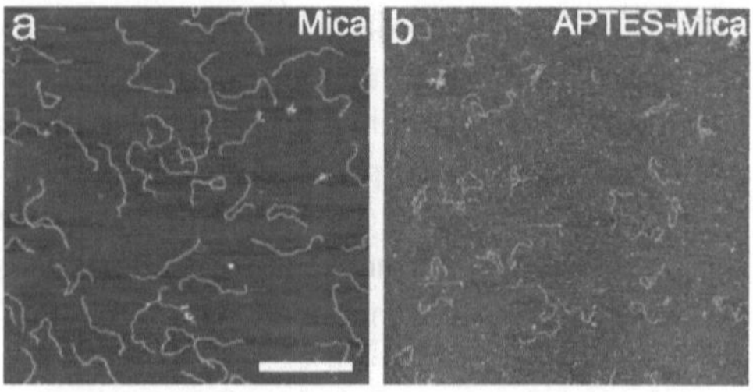

Fig. 15.10. Tapping-mode AFM images of a 1074 bp linear dsDNA fragment on dried surfaces imaged in ambient laboratory conditions. (**a**) The mica surface confers open conformations in 2D equilibrated state. (**b**) APTES-mica binds the DNA molecules in more compact conformations. *Scale bar* = 500 nm. Courtesy of Neal Crampton

as Ni(II) the DNA becomes trapped [35]. Generally, for an electrostatic interaction with the DNA where the positive charges are immobile the DNA will be kinetically trapped at least to some extent. Increasing the ionic strength of the buffer above the Ni(II)-treated mica can alleviate the trapped state and leads to molecules closer to the 2D surface equilibrated conformation [26]. Surfaces that induce trapping include amine-functionalized and plasma-treated mica, essentially those that retain a net positive charge.

Measurement of average end-to-end distance (by the shortest path) of a distribution of imaged DNA molecules indicates whether the DNA is equilibrated or trapped. The expected distance can be calculated using polymer-chain statistics models; the equilibrated case being equivalent to a random walk along the DNA backbone. For strong trapping of molecules, the average end-to-end distance tends towards a limit that demonstrates that the conformation of the DNA on the surface is a projection of the 3D solution conformation (random coil) onto the 2D surface [33]. It is important to know in which conformation the DNA molecules have bound to the surface when considering structural interactions between proteins and DNA, particularly measurement of DNA wrapping and DNA bend angles (see Sect. 15.5.3.4). As a full understanding of the mechanisms by which DNA binds to mica it then becomes possible to control the DNA binding through altering the buffer conditions [35, 36] and/or surface chemistry for real-time in situ experiments (see Sect. 15.5.3.5).

15.4.3
DNA Transport to Surfaces

Before DNA binds to the surface it has to arrive at the surface from the bulk solution. Counting the number of molecules present on the surface as time progresses indicates that for very dilute DNA solutions at least the transport of DNA to the surface in sessile-drop preparation is governed by diffusion [33]. Practically, to make a sample for AFM imaging, a solution of DNA molecules is pipetted onto the surface and left a given time (incubation) to bind. The sample is then rinsed and kept wet for imaging under bulk aqueous fluid or rinsed and dried for structural studies. The surface density of DNA molecules per unit time that are binding through a diffusion-controlled mechanism follows a $t^{1/2}$ power law (Fig. 15.11). Recently, we have demonstrated that there can be some competition in binding from heterogeneous solutions (Crampton, Bourbigot, Thomson; unpublished). This illustrates that for complex multicomponent systems, what is on the surface may not be a true representation of the distribution of structures in solution.

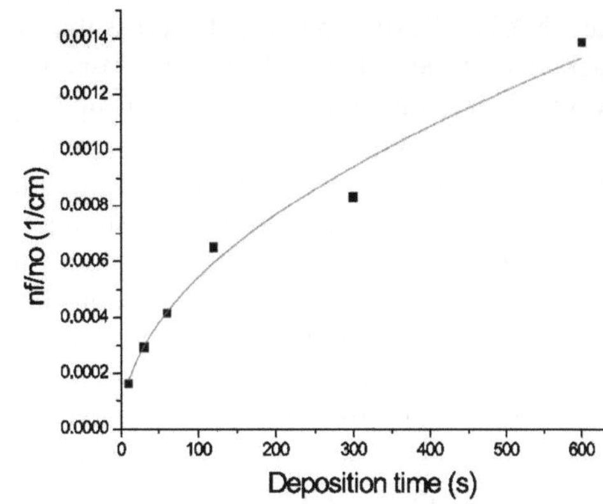

Fig. 15.11. Surface density of 1074 bp linear dsDNA molecules on mica deposited from a Mg(II) containing buffer as a function of time. The fit is to a $t^{1/2}$ dependence. Courtesy of Neal Crampton

15.5 AFM of DNA Systems

This chapter concentrates on the use of AFM imaging to study DNA structure, interactions and dynamics. While the AFM has the ability to mechanically stretch DNA and thereby study DNA interactions through force measurements, the force-range capabilities of AFM are not always low enough. This is particularly true for protein molecular motors that operate using forces of about 50 pN or less, where optical or magnetic tweezers are usually the preferred biophysical tool. Many more imaging studies have been carried out to date with the AFM and this next section reviews current progress. We begin by discussing the relative merits of undertaking structural or dynamics investigations. Then we relay the history of the efforts made to make DNA imaging by AFM reproducible and to push the boundaries of resolution. In the final sections, we review the types of interactions with DNA that have been studied, from small organic molecules, such as drugs, to larger polymer systems that invoke DNA condensation, to biomolecular protein interactions. We conclude by highlighting the advances made in studying the dynamics of DNA systems in situ, using what is known as real-time or time-lapse imaging.

15.5.1 Static Imaging versus Dynamic Studies

Since the AFM can be operated in air and liquid there are two basic ways that biomolecular systems can be studied: imaging of static structures in air or imaging of dynamic processes under aqueous fluids. In practice, of course, there is nothing to stop one performing studies of static systems under bulk liquid, but biomolecules in buffers will be moving unless rigidly attached to the support surface. We have already discussed protocols for immobilization of molecules with varying degrees of complexity involved. The most straightforward way to immobilize DNA on mica is to deposit it from a Mg(II)-containing buffer, but imaging under Mg(II) buffer with typical conventional scanners (where imaging speeds are restricted to below about 4 Hz line rates) DNA molecules cannot be resolved. The interpretation is that the DNA is diffusing too quickly across the surface for the tip to be able to trace the backbone of the polymer. However, if one rinses the sample with excess water (to avoid salt precipitation) and dries it (using N_2 gas, for example) the DNA molecules are clearly resolved. The relative ease of preparing dried samples means that imaging of biomolecular systems in air make up a significant proportion of AFM experiments, and is a useful starting point for investigations of new systems, e.g. DNA-protein. More importantly, since mica is hydrophilic and a very thin (1 or 2 nm) water layer exists on the surface at typical laboratory humidities (RH – 30 to 40%) [37–41], biomolecules such as DNA appear to remain hydrated [42] and it is not clear that going to bulk liquid for pure structural work is necessarily always an advantage. Some water, however, needs to be present for dsDNA to maintain its correct structure. For dynamic processes, where one would like to study the interactions of molecules in situ, imaging under liquid is required, allowing the imaging buffer to be exchanged online to introduce other components, such as chemical fuel (e.g. ATP) or proteins or other cofactors.

15.5.2
The Race for Reproducible Imaging of Static DNA

Applications of AFM to biology began a few years after its conception and DNA was a particular favorite, no doubt because of its central importance to life processes. It can also be regarded as a model system for the following reasons. Large enough DNA fragments are thermodynamically stable, giving it a well-defined unchanging structure (in the same environment) allowing the morphology (i.e. semiflexible polymer) to be easily distinguished in topographical images. Also, as discussed already it has well-defined physical properties that allow a good understanding of how it binds to model surfaces and what influence those surfaces exhibit on its structure. As mentioned in Sect. 15.2, genetic maintenance, expression and regulation are controlled by DNA-dependent proteins, and proteins generally have globular morphology that is easily recognizable from DNA in micrographs, therefore standard AFM techniques are well suited to studying DNA-protein systems (see Sect. 15.5.3.4 below).

The first images of DNA alone using an SPM were obtained using the scanning tunneling microscope (STM) [43], in essence because the STM was developed before the AFM. It soon became obvious however, that the STM was not, in general, a particularly convenient tool to image DNA. STM is the original SPM developed that uses the quantum-mechanical tunneling current between a conducting tip and sample to map surface electronic states. The DNA, therefore, had to be deposited on a conducting substrate, often chosen to be HOPG because of its atomically flat terraces. However, as mentioned already, DNA does not bind very strongly to such a hydrophobic surface and early claims that atomic resolution of the double-helix was achieved under ultrahigh vacuum conditions [44] were hard to justify in the longer term, especially after steps on the HOPG surface were shown to be realistic mimics of a helical structure of similar dimensions [45]. The low hit rate of finding DNA on these surfaces, indicated that it either did not bind well enough and/or was swept away by the scanning tip without being imaged, supporting the idea that most, if not all, images were artefacts. Generally, STM has not been particularly successful at biological imaging, also because interpretation of the image contrast that is generated is difficult to relate to molecular structure. A novel method by Guckenburger's group demonstrated that STM can be performed on insulating mica surfaces that are sufficiently hydrated [46], where the best imaging conditions are 60% RH with bias voltages of 5 V or more [47]. This method was coined hydration STM and the conduction mechanism is thought to occur via protons in the water layer and the image contrast of DNA on the surface is generated through path-length differences from the STM tip to the sample electrode. Hydration STM has also not been taken up by the wider scientific community. This development came somewhat later, when tapping-mode AFM was already becoming an established technique, and early STM studies of biomolecules often still required metal coatings, which obviated the potential advantage of SPM over scanning electron microscopy (SEM) [48].

The first comparative study between the performance of AFM and STM to image uncoated DNA under water came from Lindsay et al. [49] in 1989. Lindsay's group preferred gold as the support surface and used an electrochemical deposition method to hold the DNA firmly in place. This improved the reproducibility of the STM method and allowed evaluation of the relatively crude first AFM images. The early AFM studies up to about 1993, all used contact-mode imaging since AC techniques such as tapping mode were yet to be developed. The shear forces generated in contact mode required good immobilization of DNA and new methods were developed, such as covalent immobilization to a fatty-acid monolayer, to try and image single bases in ssDNA under water [50]. Single-stranded DNA is particularly difficult to image because of its tendency, in common with RNA, to fold up into structures with more complex 3D topology. The attempted high-resolution studies have not really stood the test of time and most studies today involve imaging to obtain resolution of simply the dsDNA backbone alone, usually using tapping mode. This approach is well suited to studying the influence of proteins on gross DNA structure (see Sect. 15.5.3.4 below). The best resolution reliably achieved to date on DNA alone used contact-mode imaging of a continuous monolayer of DNA molecules on a cationic lipid bilayer support surface under aqueous fluid, where a regular repeating helical pitch of 3.4 nm was detected [51]. Despite the advances in sample preparation and DNA immobilization for contact-mode techniques [52], once tapping mode was demonstrated to be effective in most sample preparation cases and in particular in the study of proteins bound to DNA, contact mode fell out of favor, principally because of the complexity of sample preparation.

However, before we review progress from 1993 onwards (post development of tapping mode so to speak), it is worth taking an aside to talk more generally about imaging of DNA in contact mode. Particularly in light of recent developments in fast scan systems development for AFM, one of which is based on contact-mode principles with static deflection sensing [53] (see Outlook Sect. 15.6 below). Bustamante's group discovered that DNA could be routinely imaged using contact mode on mica provided that the sample was dried [18]. They appreciated that under these ambient conditions the capillary neck of water that forms between the tip and the sample is responsible for creating a large proportion of the shearing force that creates sample distortion and damage, and ultimately prevents imaging [54]. They also realized that treating the mica with Mg(II) or making it positively charged through glow-discharge would assist DNA binding [55]. Along with other groups they found that the imaging humidity had to be kept below a critical value to avoid high forces, caused by the capillary neck sucking the AFM tip down onto the surface. On mica, the humidity had to be below 35% RH [56,57] whereas on glow-discharged mica the humidity could be as high as 45% RH [18]. Recent work in our laboratory suggests that the structure of surface-bound DNA is not strongly compromised when imaged at RH as low as 7% (Crampton and Thomson; unpublished) leading to the possibility that useful structural studies could be carried out using contact-mode imaging in dry conditions or possibly even under moderate vacuums.

Imaging DNA using contact mode under aqueous fluid is difficult because although the capillary forces are no longer present the DNA must be bound strongly enough not to diffuse across the surface. Collaboration between the Hansma and Bustamante groups discovered that imaging under alcohols, such as propanol, gave reproducible imaging, because alcohols precipitate DNA [58]. The DNA was sufficiently strongly bound to the mica that plasmid DNA could be incised by the AFM tip by increasing the contact force in a small scan area. However, the DNA was previously dried onto Mg(II)-treated mica before imaging under propanol and therefore this approach did not hold much promise for studying dynamic processes by contact-mode AFM. This appeared to be the limit in the usefulness of contact-mode AFM to image DNA structure because AM imaging soon took over.

Contact-mode AFM of DNA under water could be achieved by first drying the sample and rehydrating it [59], which was also helped by functionalizing the surface with aminosilane [60]. The major breakthrough came, however, with the introduction of tapping mode that first allowed reproducible imaging of DNA in air [59] and then ultimately under bulk aqueous liquid [61]. This advance opened up the possibility to study a wide range of structural and dynamic problems where other molecules interacted with the DNA.

15.5.3 Applications of Tapping-Mode AFM to DNA Systems

This section reviews the applications that tapping-mode AFM has made to date on various DNA systems: DNA alone; DNA after interactions with small organic molecules, such as drugs; DNA condensation in the presence of multivalent cations and on surfaces; and DNA-protein systems. It finishes by highlighting achievements of using AFM to study DNA dynamical processes in situ.

15.5.3.1 DNA Structure

As mentioned already, typical AFM resolution limits imaging to the mapping of the overall morphology of the dsDNA backbone. Contour-length measurements from the images along the DNA backbone allow one to infer something of the internal structure of the DNA. For instance, careful measurements of several hundred molecules on air-dried samples, allowed Rivetti and Codeluppi [62] to infer that 15 to 30% of the length of linear DNA fragments was in the A-form (compared to the expected hydrated B-form) when it is bound to the mica surface and imaged under typical ambient laboratory conditions (e.g. 30 to 40% relative humidity). Other work indicates that the drying is at least partly responsible for the shortening, since imaging of a 1036 bp linear dsDNA fragment under buffer and air led to apparent contractions of 2% and 6%, respectively [63]. Lindsay's group have demonstrated that a full B to A transition can occur under aqueous solutions if the DNA is bound to a mica surface cooled to 2 °C [64]. This is not a sequence specific event and requires both the presense of the surface and the lower temperature. Introduction of ethanol into bulk aqueous fluid can also induce full contraction to the A-form, when the ethanol concentration goes over 25% [65]. Again the presence of the mica surface helps to induce this transition, since in bulk liquid it occurs at $>$ 70 to 85% ethanol. As we will see later, accurate contour-length measurements are important in understanding structural interactions between proteins and DNA.

Different DNA structures can be distinguished in the AFM on the basis of morphology alone. To date, ssDNA, dsDNA and triple-stranded DNA [66] and quadraplex DNA [67] have been unambiguously identified on air-dried samples. Triple-stranded DNA arises through Hoogsteen base-pairing in DNA strands each containing only purines or pyrimidines that are present in a ratio of 2:1 or 1:2 [8], whereas quadruplex can only occur in long repeat sequences of guanine bases, where four poly G strands hydrogen bond to form a structure known as a G-wire. These have a channel running down the middle and are found in the telomeres of chromosomes [67]. Telomeres are the structures at the end of linear chromosomes and in higher eukaryotes are made up of multiple repeats of the sequence 5′-TTAGGG-3′. These are disposable and act as a buffer zone during DNA replication as they get shortened through repeated cell division. They are therefore important structures to study in relation to the aging process.

As described before, ssDNA folds up to form compact structures with globular submorphology [68], which makes interpretation of how other molecules interact with it difficult from AFM topographical images. Triple-stranded DNA is morphologically similar to dsDNA but appears twice as high [66]. Quadraplexes appear more rod-like due to their increased stiffness but the apparent heights varied depending on the formation conditions of the buffer, ion concentrations etc. [67].

For dsDNA, lengths of linear fragments that have been imaged vary from the very short (100 or 200 bp) [66] up to extremely long λ DNA from bacteriophages (48.5 kbp) [29]. The kinetics of binding will vary with length since transport to the surface is often diffusion limited (see Sect. 15.4.3), therefore longer fragments will require longer surface incubation times. Longer fragments may also be affected more by rinsing procedures. Indeed, there is a technique known as molecular combing in which long DNA fragments, bound at one end to hydrophobic silane surfaces, are overstretched on the surface due to the passing of the water meniscus [69,70]. To gain statistics on linear DNA samples, model templates around 1 kbp are used to optimize the efficiency of imaging enough molecules. Typically, this would yield around 50 equilibrated molecules per square micrometer. As well as linear DNA templates, circular DNA templates such as plasmids can also be readily imaged using AFM [18]. Comparison of DNA in various states of supercoiling can be also be studied using air-dried samples [71] or under aqueous fluid using aminosilane treated mica [72], while the dynamics of supercoiled molecules can also be probed [73,74] (see Sect. 15.5.3.4 below). A combined AFM and EM study showed that sample preparation, i.e. adsorption onto a surface, may affect the appearance of supercoiled DNA [75], particularly in AFM studies where the DNA is adsorbed on mica and prefers to relax into the 2D plane, interchanging writhe for helical twist. The three-dimensional information in AFM images allows one to assign the chirality of supercoiling [76], which may be advantageous when studying topoisomerase proteins. More unusual structures have also been studied mostly by Lyubchenko's group and these include hairpins [77], loops and hemiknots [78], three-way junctions [34], and cruciforms (4-way junctions) [79, 80], which have structural similarities to Holliday junctions. The three-way junctions and cruciforms were seen to be non-planar structures that exhibit dynamical movement [79]. It has been shown that the structural transition between different cruciform conformations can influence communication at a distant point on plasmid DNA mediated by superhelical stress [79]. AFM has also shown that p53 protein, a transcription factor involved in tumor supression, binds exclusively at a cruciform on supercoiled DNA templates [81]. Holliday junctions are formed in the cross-strand exchange process of genetic recombination, where two dsDNA molecules with homologous sequences exchange one of their strands. AFM has the ability to investigate the dynamics of these structures (see Sect. 15.5.3.5).

The tertiary structures of DNA described above are consequences of sequence-dependent effects, where competition of internal base-pairing between the single strands affects the structural outcome. More subtle sequence-dependent effects are also possible to pick up, such as flexibility differences and intrinsic curvature of the DNA double helix. AFM has determined the average bend angle of an A-tract sequence, which contains six A–T bps in a row, to be 13.5° [82]. Linear DNA molecules with palindromic symmetry have been used to demonstrate that the apparent DNA conformation arises from intrinsic curvature and local flexibility and such a symmetric construct allows the separation of these two effects [83]. Sequence-dependent flexibility was also confirmed to be strongly influenced by A–T rich sequences [84]. Further, polymer-chain statistical models have been developed for DNA sequences containing intrinsic curvature, discrete bends or different regions of persistence length (e.g. dsDNA and ssDNA) [82]. Minicircles of dsDNA as small as 168 and 126 bps have been imaged and these have been shown to contain kinked structures in the presence of Zn(II) ions [85]. The kinks are thought to arise because of the highly strained conformations of these minicircles.

The highest resolution on DNA that has been achieved in both contact mode and tapping mode is to resolve the helical pitch of the DNA (around 3.4 nm). Tapping mode achieved this infrequently on single isolated molecules under propanol [68], while contact mode on an ordered 2D layer of DNA bound to the surface of positively charged lipid bilayers allowed detection of the helical pitch over the whole molecular array [51]. Non-contact AFM has also been claimed to resolve the helical pitch in UHV conditions [86], but the importance of this result for understanding biology may be limited, since the DNA should be dehydrated.

It should be noted here that RNA is more difficult to image than DNA using the AFM and very few such studies are published. RNA is analogous to ssDNA, it is single-stranded and base-pairs with itself to fold up into complex 3D morphologies that are difficult to adsorb to surfaces reliably and are also inaccessible to the AFM tip. Some palindromic RNA sequences that form recognizable structures, such as stems and loops, have been successfully studied [87], but in any case RNA lies outside the scope of this chapter.

15.5.3.2
Small Organic Molecule Interactions

The binding of small organic ligands, such as drugs or intercalators, to DNA can be monitored indirectly through observation of DNA contour length changes. Molecules that intercalate between the base-pairs will increase the length of linear DNA, allowing discrimination between those that intercalate and those that bind by other means, e.g. minor groove binders [88]. Accurate length determination can give information on site occupancy, binding affinity, exclusion numbers and cooperativity. Distinctions between intercalators and minor groove binders can also be made using AFM in force-spectroscopy mode, on the basis of differing mechanical responses [89]. Intercalators all contain closed-ring structures that are planar and hydrophobic and interact well with the planar hydrophobic faces of the DNA nucleotides. These molecules insert between base-pairs, some specific to sequence [90], and some non-specifically, e.g. ethidium bromide that is used to fluorescently stain DNA for detection in agarose gel electrophoresis. As well as causing length increases on linear DNA molecules, intercalators can cause topological changes in closed circular molecules, due to build up of stress. For instance, AFM has been used to show that anthracycline intercalators, chemotherapeutic antitumor agents, induce positive superhelical coiling, such that negatively supercoiled plasmid DNA becomes relaxed and relaxed plasmid DNA becomes positively supercoiled [91]. The effect on DNA structure of a number of other intercalators, used as anticancer drugs, has been studied using AFM [92–94].

The tertiary structure of plasmid DNA (how the DNA helical axis is arranged spatially), i.e. its supercoiling, has been studied in detail for various stoichiometries of binding of ethidium bromide [71]. It was found that increasing amounts of the intercalator first induced toroidal loops in the DNA before plectonemically supercoiled structures, where the DNA helix is tightly twisted around itself, were nucleated. At the highest concentrations of ethidium bromide studied, where 8 intercalator molecules were available per bp, the plasmid DNA was in a fully plectonemic structure.

Some intercalators, such as psoralen, have been used in a novel way to immobilize DNA to surfaces. Psoralen will covalently crosslink to pyrimidine bases upon radiation with soft UV light (320–400 nm). Shlyakhtenko et al. [95] used a derivatized psoralen that could then crosslink to the amine groups on an APTES mica surface. This was seen as a way of periodically attaching the DNA to a surface covalently such that dynamics of segments between crosslinks could be investigated by AFM.

Higher concentrations of intercalators, such as YOYO dyes can cause more severe tertiary and quaternary effects, such as the formation of toroids under acidic pH conditions [96]. YOYO carries a +4 charge that presumably initiates this condensation process despite YOYO containing two sets of ring structures that intercalate between DNA bases. We will move on in the next section to consider what structural changes occur when DNA interacts with larger polymeric molecules (polyelectrolytes) and more highly charged small-molecule species. Relatively low concentrations of highly positively charged ionic species will cause various degrees of DNA condensation.

15.5.3.3
DNA Condensation

Condensation is the process whereby either individual DNA molecules self-aggregate or a number of DNA molecules self-assemble to form structures with well-defined morphologies and sizes, such as globular aggregates, rods and toroids. As we have said, DNA is a negatively charged polyelectrolyte that will only interact with itself if this charge is compensated for or neutralized. The structure that the condensed DNA adopts is determined through interplay between the DNA mechanical properties and the forces involved in the condensation; stiffer polyelectrolytes will form toroids, whereas more flexible ones form disordered globules [97]. The forces of attraction between two neighboring DNA segments arise due to correlation of fluctuations of condensed counterions on the DNA surfaces [98–100]. The helical structure of the DNA plays an important role in determining the outcome of the condensation process [100], while the size limits for the toroids are determined by both kinetic and thermodynamic factors within a nucleation and growth framework [101].

To induce condensation in solution the DNA must be incubated with multivalent cations with a charge of greater than +2. Relatively high concentrations of divalent metal cations, such as Mg(II), can begin the process of condensation. Increasing the valency of the cation will cause condensation at lower co-ion concentrations. The evidence from many different AFM studies seems to imply that this process occurs initially through DNA looping, where two distant points on the DNA helix contact each other. As many of these contacts are made, the aggregates take on a flower-like appearance and at higher counterion concentrations eventually toroidal structures of well-defined size are formed [102]. Examples of multivalent cations that can induce DNA condensation include: metal cations [103] and their complexes [104]; positively charged proteins, such as protamine [105]; polyamines, such as spermine and spermidine [106–108]; positively charged polypeptides such as poly L-lysine [109, 110]; polysaccharides [111]; cationic silanes [112, 113]; or positively charged synthetic polymers [114]. For a review of recent studies on the formation of DNA toroidal condensates see [101]. AFM is particularly suited to studying DNA condensation since it involves larger hierarchical structure where the limits of the microscope resolution are not being tested so strongly.

There is a great practical interest in DNA condensation techniques for gene-therapy applications. To replace a gene within a cell, the therapeutic DNA has to get through the outer cell membrane, across the cell, through the nuclear envelope and inside the nucleus. Notwithstanding the practical issues of delivering the DNA to a target tissue within the body, the DNA must in any case be condensed to prevent it being degraded enzymatically or chemically by the body before it reaches the target. AFM has been used to characterize a variety of putative gene-therapy vectors, where the DNA has been condensed with polyethylimines [115], methacrylates [116], cationic liposomes [117], poly L-lysine [118] or a protein/lipid system that can condense bacterial artificial chromosomes about 150 kbp large enabling them to transfect cells [119]. There are a whole variety of positively charged polymeric systems that can be used, one of the most popular being lipid-based systems, because of their biocompatibility. A recent paper reviews the use of SPM and particularly AFM for studying their structure [120].

We turn now to surface-directed condensation of DNA. On surfaces where there are mobile multivalent cations it is possible for DNA to condense on the surface. This has been a prevalent feature of surfaces modified with aminosilanes used to try and bind DNA strongly. Fang and Hoh [121] discovered that DNA toroids and rods formed on silicon surfaces treated with aminosilanes and attributed this to non-polymerized silane molecules remaining on the surface. The surface induced condensation was eliminated after treatment of the silane surfaces with vacuum storage or baking at 100 °C for 2 h. The majority of studies that use aminosilanes are made on mica from vapor phase following the protocol of Lyubchenko and coworkers [29, 60, 122]. Recent work has shown that the relative humidity in the deposition chamber must be kept below 25% RH to prevent surface-induced DNA condensation by APTES [30]. In this study, baking at 150 °C for 2 h was ineffectual at eliminating the condensation on surfaces made in RH $> 25\%$ because the silane layer underwent a dramatic structural reorganization.

Interestingly, in the study of imaging DNA under various concentrations of ethanol to induce a B-to-A transition, it was also discovered that high concentrations of ethanol ($> 20\%$) in aqueous media induce DNA condensation, including toroid formation [65]. This observation is consistent with recent theoretical work that describes DNA condensation in terms of poor solvents rather than counterion correlations [123]. As we learned earlier, other alcohols such as propanol were used to increase the adhesion of DNA to mica surfaces to obtain reproducible contact-mode imaging [58]. DNA is not soluble in alcohols and they are used routinely to precipitate it from solution.

Binding of DNA onto the surface of positively charged supported lipid bilayers has also been described as a 2D condensation process [124, 125]. In this description, however, the DNA strands are packed closely together (the interhelical distance being 4 to 6 nm), but aligned in a 2D monolayer array across the top of the bilayer. As we have seen, these close-packed arrays have allowed high-resolution imaging of DNA using contact mode, detecting the helical repeat [51], in an analogous way that high-resolution images of 2D protein crystals can be made.

As we shall see in the next section, DNA is usually condensed inside the nucleus of the cell in conjunction with a range of proteins, as a way of storing and protecting the genome of the organism.

15.5.3.4
DNA-Protein Interactions

The AFM is particularly adept at studying DNA-protein systems, since proteins that interact with DNA are globular and compact, and therefore can be distinguished from the DNA polymer on the basis of morphology alone. Figure 15.12 shows an image of a single DNA gyrase protein molecule bound to a 1070 bp linear dsDNA molecule. The images can be used to visualize how a single protein modifies local DNA structure, either through a change in bend angle [126] or through wrapping of the DNA around the protein. The ability of the AFM to accurately measure the DNA contour length can give information about wrapping of DNA around bacterial molecular motors such as *E. coli* RNA polymerase [126] and *E. coli* DNA gyrase that is a structural requisite for these enzymes' work [42]. Where two proteins bind

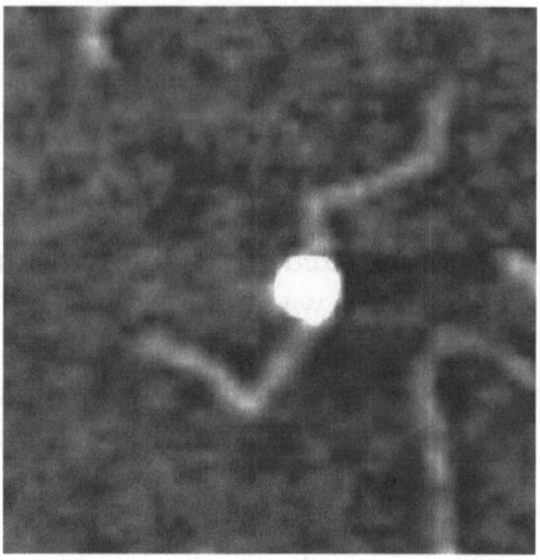

Fig. 15.12. Tapping-mode AFM image of a complex of a single *E. coli* DNA gyrase molecule bound to a linear dsDNA template 1070 bp long. The sample has been dried and imaged under ambient laboratory conditions. The DNA gyrase protein is the globular feature bound to the center of the DNA molecule. Image size = 250 × 250 nm. (Data: Thomson NH, Heddle JG, Maxwell A)

distantly on a single DNA template their interactions can often be visualized through the formation of loops in the DNA. This situation may occur, for example, during the regulation of DNA transcription through the binding of transcription factors to RNA polymerase, when the polymerase is at the promoter region [127, 128].

Complexes of RNA polymerase on linear DNA templates have demonstrated that this protein can induce a bend in the DNA and requires wrapping of the DNA around itself, to a greater or lesser extent, at various stages in the transcription cycle [129]. Bend-angle distributions tend to be very broad, and therefore are not all that informative about the local DNA deformation, mainly because the finite size of the AFM tip means that there is a "shadow" around the protein where the tip cannot contact the surface to profile the path of the DNA backbone close to the protein. However, it has been shown that the DNA bending of two different types of methyltransferases can be discriminated and is associated with their two different modes of action [130]. Wrapping of DNA around DNA gyrase, a topoisomerase that is unique in being able to introduce supercoils into dsDNA, depends on the conformational state the protein is in during its enzymatic cycle. Addition of a nonhydrolyzable ATP analog, ADPNP, was shown to remove DNA wrapping consistent with the enzyme performing one cycle of work [42]. The sizes of DNA that wraps around *E. coli* RNA polymerase and *E. coli* DNA gyrase are comparable, about 100 bp, but slightly larger wraps have been identified, such as the 160 bp wrap around histones in reconstitued chromatin [131]. In this case, the wrap is thought to involve two whole turns, whereas for RNA polymerase and DNA gyrase it is about one. Chromatin, the basic DNA-protein complex that makes up chromosomes, consists of DNA wound round a series of histone protein complexes. Histones are octameric-containing dimers of four different subclasses of histone proteins. The basic structural unit of DNA wrapped around the histone is known as a nucleosome, which can be studied in its native form by extraction from cells or can be reconstituted in vitro. Nucleosome structure has been shown to resemble "beads on a string" [132]. AFM has been used quite extensively to study chromatin to investigate how and what controls nucleosome positioning on the DNA [133–136].

There are numerous other investigations of DNA-protein systems in the literature, so many, that they cannot all be reviewed here. This chapter aims to give a flavor of what can be achieved with the AFM in this research area. We have seen, therefore, that the AFM has a role in performing single-molecule experiments to determine structural and mechanistic aspects of DNA-dependent molecular motors that cannot be accessed by ensemble techniques. To date it has been used mostly to study interactions between DNA and single proteins, but eventually it should be possible to apply AFM in a bottom-up manner to study more complex systems, such as transcription roadblocks, replication forks, convergent transcription and other important aspects of genetic regulation. It is very good at visualizing center-of-mass motions of proteins relative to linear DNA templates and as such the next generation of experiments are likely to feature systems with two or more proteins acting on a single DNA template.

15.5.3.5 Studies of DNA Dynamic Systems

As already alluded to, the study of DNA processes in situ within the AFM are the most difficult of all DNA systems to study. The main reasons for this are the difficulties of achieving suitable surface binding coupled with the relatively slow scan speeds currently available in most AFM technology. To study DNA dynamics there is a choice between adjusting buffer conditions on mica to have the DNA transiently bound or using a positively charged surfaces at neutral pH, such as aminosilanes, so that the imaging buffer can be varied to some extent without affecting the DNA surface-binding characteristics. AFM studies of this kind are usually called "real-time" or "time-lapse" imaging. Time-lapse is perhaps the most realistic description at present because the images are built up line-by-line in a raster scan pattern, and with the typical scan rates available, the time between the same point in successive images can be up to a minute or more (depending on the image size). As faster scan technologies are developed (see Sect. 15.6) then the term "real-time" may become more appropriate.

After the realization that divalent metal cations could enhance DNA adhesion to mica to the extent that samples could be dried for air imaging [24], this procedure was taken forward for imaging under bulk aqueous fluid [36]. Transition-metal cations had been shown to be most effective for increasing DNA-mica adhesion, therefore, Zn(II) was chosen in a gravity-fed flowthrough system where the buffer is continuously flowing through the AFM imaging cell and can be switched between buffers with and without Zn(II). In this manner, the DNA could be reversibly bound and released from the mica surface on the timescale of tens of seconds. In hindsight, this was probably not the best divalent cation to choose, since Zn(II) precipitates as zinc hydroxide around pH 8 and it was clear that the DNA adhesion was in part due to a precipitate salt layer forming on the surface. More recently, Ni(II) has become the preferred co-ion for DNA adsorption under liquid and it has been demonstrated that mica can be pretreated with Ni(II) solutions to enhance DNA binding [137]. This is because the Ni(II) ions are irreversibly bound into the surface sites on mica that allows experiments to be performed without the Ni(II) present in the imaging solution. This is important for biological processes, since transition-metal cations can poison the action of many enzymes. For example, RNA polymerase contains two Mg(II) ions that are required for its mechanistic action, which can become displaced by transition-metal cations.

Figure 15.13 shows a sequence of tapping-mode AFM images taken under a Mg(II)-containing buffer on mica pretreated with nickel chloride. One can see that there are portions of the molecules that are not firmly bound to the mica surface and move underneath the scanning tip. The molecules remain in the same approximate conformations for timescales of minutes but there are local motions of the DNA that occur from one frame to the next, as highlighted with the arrows. In one instance, there is a concerted motion where the movement of one DNA molecule dramatically affects the conformation of a neighboring one (see white arrows, Figs. 15.13b,h). It is apparent that the motions of the DNA molecules are not strongly influenced by the scanning tip. The center-of-mass of DNA molecules can be tracked under similar conditions where they exhibit diffusion in the 2D plane of the surface to show that the AFM tip in tapping mode does not influence the motion [138]. As we will see for the enzyme studies below, is it crucial that the DNA can equilibrate into the 2D surface plane of mica but still retain some motion. Recent theoretical and experimental work that has already been discussed has gone the furthest in understanding the amount of control an experimenter can impose of the dynamics of the DNA on mica surfaces [26, 35].

We have already seen that while positively charged surfaces can trap DNA, motions of local DNA structures such as three- and four-way junctions can be observed on the aminosilane surfaces [34, 79]. Importantly, whole-molecule center-of-mass motions can occur if the buffer conditions are suitable. Despite the previous implication that these surfaces always trap DNA molecules, recent work in the author's laboratory shows that conditions that look like 2D surface equilibration can occur. Figure 15.14 shows a sample of linear dsDNA 1070 bp long incubated with the protein DNA gyrase, which is a bacterial enzyme that controls DNA topology. Both images show the same area of the sample before and after addition of a nonhydrolyzable analog of ATP, known as ADPNP. One can see that during the time interval between the two images (a few minutes) that the gyrase molecules (globular features) have stayed stationary, whilst the DNA molecules have all moved across the surface, showing that even on surfaces where trapping might be expected, under these

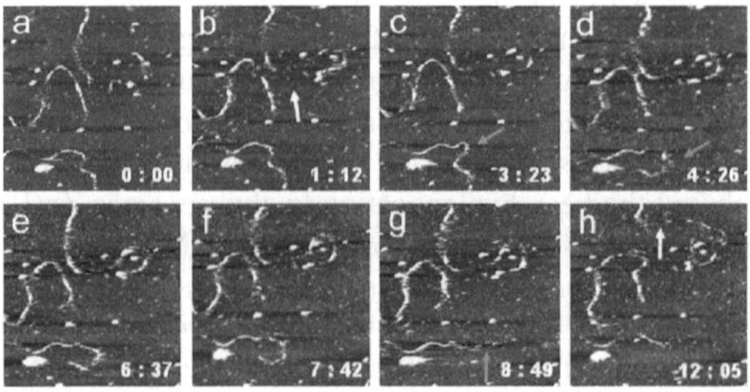

Fig. 15.13. Time-lapse AFM movie of 1074 bp DNA moving on Ni(II)-treated mica under a Mg(II) containing buffer. The *gray* and *white arrows* highlight regions of the DNA molecules that move position over time. The *gray arrows* (**c**,**d**,**g**) indicate a molecule that aligns along the fast scan direction and the *white arrows* (**b**,**h**) indicate a correlated motion between two neighboring DNA molecules. The time of each frame is in minutes; seconds is shown in the *bottom right corner*. The scan size = 690 × 690 nm. Courtesy of Fiona Hurrell

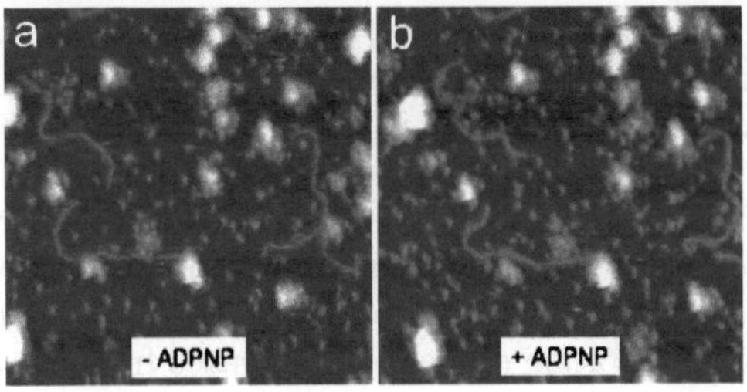

Fig. 15.14. Tapping-mode AFM images under buffer of a sample of DNA gyrase and linear dsDNA 1070 bp long. The same area of the sample was imaged before (**a**) and after (**b**) addition of ADPNP through direct exchange of buffers during imaging. The offset between the images is due to instrumental drift. While the proteins appear unchanged during the experiment the DNA molecules have moved, either as local movements or center-of-mass movement. (Data: Thomson NH, Heddle JG, Maxwell A)

buffer conditions DNA can still diffuse in the surface plane. This result illustrates the complexity of understanding DNA binding and dynamics on surfaces such as mica and aminosilanes. Aminosilane surfaces are useful for tightly binding the DNA in a kinetically trapped configuration and observing local chain dynamics of the DNA. For example, APTES-mica was used on which to observe the mobility of three-way junctions [34], while branch migration behavior of Holliday junctions was observed on APS-mica [139].

The first AFM images of motion of linear dsDNA were obtained on Ni(II)-pretreated mica and were combined with studies of the effect of nuclease DNase I, where even DNA that was well bound to the surface experienced degradation [137]. This was seen as a loss of regions of the linear DNA from the surface while the rest of the molecule remained bound. Unambiguous identification of DNase I was difficult due to its motion in solution. Recent time-lapse studies have followed the effects of DNase I on the degradation of DNA complexes condensed with polyamidoamino dendrimers [140]. These complexes are potential candidates for gene therapy and it is therefore interesting to investigate how the condensation affords protection against DNA degradation by nucleases. The image sequences revealed that the degree of protection is correlated to the amount of condensation. In some cases, the DNase I proteins can be observed diffusing across the surface encountering a DNA toroid and causing a large structural change, uncondensing the DNA.

One of the most studied proteins using these techniques is RNA polymerase (RNAP), that mediates the transcription of the genetic information from dsDNA into RNA. This is a central molecular process to gene expression and therefore generates great research efforts to understand mechanistically how this enzyme works. Early AFM studies under bulk aqueous solutions followed directly the assembly of *E. coli* RNAP onto DNA templates [141]. A few years later with great collaborative efforts from the Bustamante and Hansma laboratories, the first AFM "movies" of transcription at the single-molecule level were captured [12, 138]. In these experiments, the complexes of RNAP and DNA were formed in vitro with the RNAP in a stalled state on the DNA. These complexes were deposited in the AFM sample chamber under continuously flowing buffer that was driven by gravity to avoid vibrational interference to the AFM [36]. Switching the buffer to one containing the nucleoside triphosphates that make up the RNA transcript reinitiates transcription during continuous image acquisition. In these time-lapse "movies" the RNA transcript cannot be imaged, presumably because it is too small and too mobile, but rather one observes the relative motion between the protein and the DNA. In these cases, the RNAP is non-specifically bound to the mica surface and the DNA is translocated by the enzyme (see Fig. 15.15). Many experiments are performed to capture a few transcription events, because the interactions of the enzymes with the support surface can inactivate a large proportion of them. One can see from these experiments that the AFM is adept at measuring the center-of-mass motions of the protein relative to DNA but internal protein motions are difficult to detect with current technologies. Further experiments showed how the RNAP finds its promoter sequence, the portion of the DNA from where it starts transcribing, through a 1D diffusional search mechanism [138].

A handful of studies using similar approaches have been published on other DNA enzymes. In a dedicated home-built AFM the association, dissociation and movement of photolyase over linear dsDNA molecules was observed [142]. In another study, snap shots of the translocation and cleavage of DNA by EcoK1 endonuclease were obtained [143]. The interactions of p53 have also been studied by time-lapse AFM and indicated that the p53 could directly bind to a target sequence or bind non-specifically and search by 1D diffusional search along the DNA [144]. Most of these studies are carried out on linear fragments about 1 kbp long, where the motion of the protein relative to the DNA can be readily observed, but some have been able

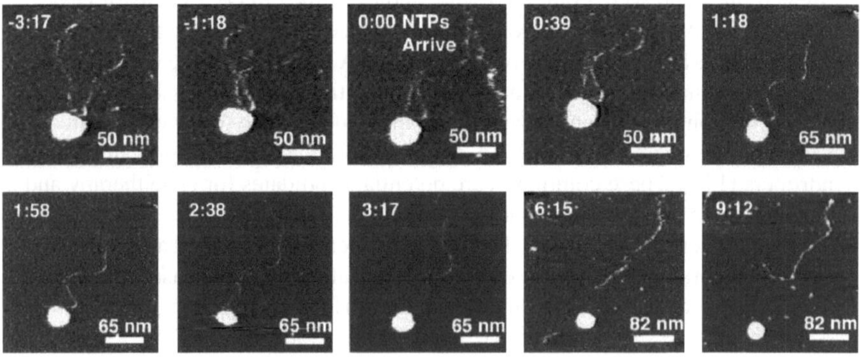

Fig. 15.15. Time-lapse AFM movie of DNA transcription by *E. coli* RNA polymerase (RNAP) at single-molecule resolution. The time of each frame is shown in minutes: seconds in the *top left corner*. The RNAP is bound to the center of a linear dsDNA template 1047 bp long. At time 0:00 the nucleoside triphosphates (NTPs) that fuel the enzyme and are building blocks for the RNA polymer that the RNAP fabricates arrive into the AFM sample chamber. As time progresses the right-hand arm of the DNA gets longer and the left arm concomitantly gets shorter, as the RNAP translocates the DNA. Eventually the RNAP releases the DNA as it transcribes off the end of the template, and the buffer is changed to a Zn(II)-containing buffer that firmly adheres the DNA to the mica. The penultimate image shows a small feature on the opposite side of the RNAP to the DNA ejection from the enzyme. This could be the RNA transcript that is formed, but better time resolution is required. Reproduced with permission from *Biochemistry* (1997) 36(3):461–468 [12]. Copyright 1997 American Chemical Society

to use supercoiled plasmids [143]. The few studies published to date highlight the difficult nature of these experiments and a lack of control of the DNA adsorption to the surface. They all require the DNA to be transiently bound to the surface, sufficiently immobilized to allow the AFM tip to track the DNA position but not fixed, at least in the surface plane in order to allow enzymes to operate on the DNA. The last section covers developments in faster scan technologies that should allow these type of experiments to become more widespread and the data gained from them to be more informative.

15.6 Outlook

It is hoped that this chapter demonstrates that the AFM has an important role to play in the research of DNA structure and interactions. In particular, its ability to study single-molecule complexes allows it to assess the heterogeneity of systems, albeit at a resolution much lower than atomic. It is expected that the bottom-up approach to increasing complexity will enable the AFM to play an even bigger role as a biophysical tool for DNA research in the future. As the macromolecular systems studied become larger and more complex, ensemble techniques such as NMR and X-ray crystallography may become increasingly challenging. AFM technology itself is still immature and there are many technical developments that can and need to be made before its full promise can be fulfilled.

Far and away the most important development necessary is to increase scan speeds of the tip to levels where image-acquisition times become comparable with enzyme turnover rates. While it has been demonstrated that AFM is a useful tool for structural studies of DNA systems, especially DNA-protein complexes, its particular advantage in terms of a high-resolution microscopy is its ability to image under aqueous solutions. Progress on increased scan speeds is already under way. There are two basic ways that this problem has been approached, although both rely on miniaturizing further one or more of the AFM components. Since the AFM is essentially a mechanical microscope, instead of a microscope that relies on the focusing of radiation (either optical or electron waves), miniaturizing components will allow them to be driven at higher speeds, since their resonant frequencies increase. Miniaturizing cantilevers has allowed images of DNA under aqueous buffer at about 8 nm pixel size to be taken every 1.7 s [145], while miniaturizing the scanners also has allowed image rates of 80 ms at a pixel size of 2.4 nm allowing the conformational switching of myosin V molecular motors to be followed in situ [146]. These approaches continue to use an amplitude-modulation imaging mode (e.g. tapping mode) as the mechanism of interacting the tip with the surface. Another approach has gone back to a type of contact mode where the deflection of the cantilever is monitored while the tip is scanned over the surface. It implements a passive mechanical feedback loop enabling it to collect images of continuous polymer films in under 20 ms [53]. This technique commercialized as videoAFM^{TM} has achieved much greater scan rates to date, but reliable operation under liquid has yet to be demonstrated. This method requires a direct force to be applied through the tip onto the surface so that the tip reliably tracks the surface.

It is still unclear how reliable this approach will be to biological systems, particularly single molecules, because the shearing interactions are not well understood at these scan rates. It has a potential advantage over the small cantilevers approach because it can image larger areas. Since AFM cantilevers need to be tilted for the tip to address the surface without the cantilever snagging, smaller cantilevers require flatter samples or are restricted to smaller areas on samples with larger topographical variations.

If a kind of real-time single-molecule enzymology is to be realized with the AFM, as well as the fast-scan approaches to the instrumentation, greater understanding and control of DNA adsorption and behavior at surfaces will be required. The recent work on counterion correlation forces between DNA and mica has allowed a greater degree of control of DNA conformations under bulk aqueous fluid. More work needs to be done on control of DNA orientation and positioning on surfaces and may benefit from techniques such as dip pen nanolithography (DPN) [147] and dielectrophoresis [148]. DPN can chemically pattern surfaces locally on nanometer scales to spatically control DNA adsorption, while microcontact printing can chemically pattern surfaces over larger areas. Molecular combing can align DNA on surfaces, however, it requires a meniscus being drawn across the surface, effectively drying it and causing irreversible binding of the DNA. This approach is unlikely to enable dynamic studies to be performed upon sample rehydration.

Gentler imaging modes in force microscopy will also assist in the goals of imaging the delicate interactions between proteins and DNA. For example, phase imaging with tapping-mode AFM under buffer can be used to follow processes at low interaction forces [149]. Using a very high setpoint where the tip is a long way from the surface, the feedback loop cannot follow DNA molecules on the mica to produce a faithful topographical image, but the phase signal is still sensitive to the motion of the molecules. Other force microscopy setups such as shear force microscopy (ShFM), also known as tranverse dynamic force microscopy (TDFM), might also be an interesting avenue to explore [150]. These instruments are interesting propects because they avoid the jump-to-contact problem of cantilevers in AFM, since they use a very stiff vertical probe and monitor the tranverse motions of the probe.

As more complex DNA-protein interactions are studied, i.e. where two or more proteins interact at the same binding site on a DNA molecule, the ability of the AFM to distinguish different proteins on a morphological basis alone will become compromised. The AFM will require molecular-recognition capability or at the least a chemical recognition capability. Chemical force microscopy (CFM), through controlled functionalization of the AFM tip, has been demonstrated on crystal surfaces [21], where chirality of surface groups can be determined, and on more complex protein–biomineral systems [151]. If the chemistry on the tip is defined and known, then knowledge of the operating environment (e.g. pH and ionic strength) allows the user to infer the chemistry on the surface. The signal for imaging can be either an adhesion event in an AC imaging technique or change in the lateral force signal (i.e. cantilever torsion behavior). The functionalization of AFM probes has been taken further, to link biological molecules such as antibodies to the tip. This approach has been called molecular recognition force microscopy (MRFM), and allows mapping of the interaction sites between the antibody on the tip and its antigen sites on the sample [22, 152]. Since antibodies can be produced against a wide variety different biological targets this approach shows a lot of promise. Generally, the resolution of CFM and particularly MRFM will be lower than AFM since the tip has molecules bound to it. However, Hinterdorfer's group have devised a method to simultaneously extract a topographic and recognition image using an AC imaging mode that can discriminate individual binding sites [152]. Development of CFM and MRFM for DNA systems will allow complexes on rougher support surfaces and with more complex topology and larger topographical changes to be investigated. Alternatively or additionally, one might also envisage combining single-molecule optical techniques with AFM to understand DNA-protein interactions in more detail. A useful surface technique is total internal reflection fluorescence microsopy (TIRFM) that allows fluorescently labeled biomolecules at surfaces to be identified and tracked.

As well as imaging, the AFM can manipulate DNA molecules, stretch them and measure inter- and intramolecular forces [153]. The limited amount of work that has been undertaken to date lies outside the scope of this chapter, but force measurements and sensor technology based on AFM cantilevers [154] has the ability to probe DNA interactions. These approaches should allow binding assays and some sequencing applications to be developed.

In conclusion, the AFM has already proved to be a versatile tool for studying the structure of DNA and its interactions with other molecules, while in situ dynamic experiments are beginning to gain more insight into the local behavior of DNA systems through direct imaging under bulk aqueous fluid. As AFM technology develops further one can see from this review of work to date that there is great potential for the range of applications to DNA research to broaden further and it is expected that the AFM will play a more important role in the areas of structural molecular biology and bionanotechnology in the future.

Acknowledgements. The author wishes to thank former researchers from his group, in particular Neal Crampton and Fiona Hurrell for their contributions to the work. I also extend thanks to current collaborators Jonathan Heddle and Tony Maxwell on the DNA gyrase project. I appreciate the critical evaluation and review of the chapter by Bill Bonass and Sarah Harris, who both stimulated insightful discussions. I also thank Sarah for providing molecular models of DNA structures in Fig. 15.2. Work from my group has been supported by the EPRSC, BBSRC and the University of Leeds.

References

1. Feynman RP (1959) There's plenty of room at the bottom. Annual Meeting of the American Physical Society, http://www.zyvex.com/nanotech/feynman.html
2. Piner RD, Zhu J, Xu F, Hong SH, Mirkin CA (1999) Science 283:661
3. Garcia R, Martinez RV, Martinez J (2006) Chem Soc Rev 35:29
4. Ulman A (1996) Chem Rev 96:1533
5. Giessibl FJ (2003) Rev Mod Phys 75:949
6. Hembacher S, Giessibl FJ, Mannhart J (2004) Science 305:380
7. Dvorak JA (2003) Methods. Academic Press, Elsevier, Science 29:86

8. Bates AD, Maxwell A (2005) DNA Topology, 2nd edn. Oxford University Press, Oxford
9. Neidle S (2002) Nucleic Acid Structure and Recognition, Oxford University Press, Oxford
10. Watson JD, Crick FHC (1953) Nature 171:737
11. Harris SA (2004) Contemp Phys 45:11
12. Kasas S, Thomson NH, Smith BL, Hansma HG, Zhu XS, Guthold M, Bustamante C, Kool ET, Kashlev M, Hansma PK (1997) Biochemistry 36:461
13. Alberts B, Bray D, Lewis J, Raff M, Roberts K, Watson JD (1994) Molecular Biology of the Cell, 3rd edn. Garland Publishing Inc., New York
14. Binnig G, Quate CF, Gerber C (1986) Phys Rev Lett 56:930
15. Gould SAC, Drake B, Prater CB, Weisenhorn AL, Manne S, Kelderman GL, Butt H-J, Hansma H, Hansma PK, Magonov S, Cantow HJ (1990) Ultramicroscopy 33:93
16. Garcia R, Perez R (2002) Surf Sci Rep 47:197
17. Zhong Q, Inniss D, Kjoller K, Elings VB (1993) Surf Sci Lett 290:L888
18. Bustamante C, Vesenka J, Tang CL, Rees W, Guthold M, Keller R (1992) Biochemistry 31:22
19. Zitzler L, Herminghaus S, Mugele F (2002) Phys Rev B 66:155436
20. Muller DJ, Fotiadis D, Scheuring S, Muller SA, Engel A (1999) Biophys J 76:1101
21. McKendry R, Theoclitou ME, Rayment T, Abell C (1998) Nature 391:566
22. Riener CK, Stroh CM, Ebner A, Klampfl C, Gall AA, Romanin C, Lyubchenko YL, Hinterdorfer P, Gruber HJ (2003) Anal Chim Acta 479:59
23. Frank-Kamenetskii MD (1997) Phys Rep 288:13
24. Hansma HG, Laney DE (1996) Biophys J 70:1933
25. Shao ZF, Mou J, Czajkowsky DM, Yang J, Yuan JY (1996) Adv Phys 45:1
26. Pastre D, Pietrement O, Fusil P, Landousy F, Jeusset J, David MO, Hamon C, Le Cam E, Zozime A (2003) Biophys J 85:2507
27. Tanigawa M, Okada T (1998) Anal Chim Acta 365:19
28. Bussiek M, Mucke N, Langowski J (2003) Nucleic Acids Res 31:e137
29. Lyubchenko Y, Shlyakhtenko L, Harrington R, Oden P, Lindsay S (1993) Proc Natl Acad Sci USA 90:2137
30. Crampton N, Bonass WA, Kirkham J, Thomson NH (2005) Langmuir 21:7884
31. Kim S, Christenson HK, Curry JE (2002) Langmuir 18:2125
32. Shlyakhtenko LS, Gall AA, Filonov A, Cerovac Z, Lushnikov A, Lyubchenko YL (2003) Ultramicroscopy 97:279
33. Rivetti C, Guthold M, Bustamante C (1996) J Mol Biol 264:919
34. Shlyakhtenko LS, Potaman VN, Sinden RR, Gall AA, Lyubchenko YL (2000) Nucleic Acids Res 28:3472
35. Pietrement O, Pastre D, Fusil S, Jeusset J, David MO, Landousy F, Hamon L, Zozime A, Le Cam E (2003) Langmuir 19:2536
36. Thomson NH, Kasas S, Smith B, Hansma HG, Hansma PK (1996) Langmuir 12:5905
37. Beaglehole D, Radlinska EZ, Ninham BW, Christenson HK (1991) Phys Rev Lett 66:2084
38. Beaglehole D, Christenson HK (1992) J Phys Chem 96:3395
39. Kohonen MM, Christenson HK (2000) Langmuir 16:7285
40. Xu L, Lio A, Hu J, Ogletree DF, Salmeron M (1998) J Phys Chem B 102:540
41. Hu J, Xiao XD, Ogletree DF, Salmeron M (1995) Surf Sci 344:221
42. Heddle JG, Mitelheiser S, Maxwell A, Thomson NH (2004) J Mol Biol 337:597
43. Binnig G, Rohrer H (1982) Helv Phys Acta 55:726
44. Driscoll RJ, Youngquist MG, Baldeschwieler JD (1990) Nature 346:294
45. Clemmer CR, Beebe TP (1991) Science 251:640
46. Guckenberger R, Heim M, Cevc G, Knapp HF, Wiegrabe W, Hillebrand A (1994) Science 266:1538
47. Heim M, Steigerwald R, Guckenberger R (1997) J Struct Biol 119:212

48. Keller RW, Dunlap DD, Bustamante C, Keller DJ, Garcia RG, Gray C, Maestre MF (1990) J Vac Sci Technol A 8:706
49. Lindsay SM, Nagahara LA, Thundat T, Knipping U, Rill RL, Drake B, Prater CB, Weisenhorn AL, Gould SA, Hansma PK (1989) J Biomol Struct Dyn 7:279
50. Hansma HG, Weisenhorn AL, Gould SAC, Sinsheimer RL, Gaub HE, Stucky GD, Zaremba CM, Hansma PK (1991) J Vac Sci Technol B 9:1282
51. Mou JX, Czajkowsky DM, Zhang YY, Shao ZF (1995) FEBS Lett 371:279
52. Allison DP, Bottomley LA, Thundat T, Brown GM, Woychik RP, Schrick JJ, Jacobson KB, Warmack RJ (1992) Proc Natl Acad Sci USA 89:10129
53. Humphris ADL, Miles MJ, Hobbs JK (2005) Appl Phys Lett 86:034106
54. Yang G, Vesenka J, Bustamante CJ (1996) Scanning 18:344
55. Vesenka J, Guthold M, Tang CL, Keller D, Delaine E, Bustamante C (1992) Ultramicroscopy 42:1243
56. Thundat T, Warmack RJ, Allison DP, Bottomley LA, Lourenco AJ, Ferrell TL (1992) J Vac Sci Technol A 10:630
57. Yang J, Mou J, Yuan JY, Shao ZF (1996) J Microsc (Oxford) 182:106
58. Hansma HG, Vesenka J, Siegerist C, Kelderman G, Morrett H, Sinsheimer RL, Elings V, Bustamante C, Hansma PK (1992) Science 256:1180
59. Hansma HG, Sinsheimer RL, Groppe J, Bruice TC, Elings V, Gurley G, Bezanilla M, Mastrangelo IA, Hough PVC, Hansma PK (1993) Scanning 15:296
60. Lyubchenko YL, Gall AA, Shlyakhtenko LS, Harrington RE, Jacobs BL, Oden PI, Lindsay SM (1992) J Biomol Struct Dyn 10:589
61. Hansma PK, Cleveland JP, Radmacher M, Walters DA, Hillner PE, Bezanilla M, Fritz M, Vie D, Hansma HG, Prater CB, Massie J, Fukunaga L, Gurley G, Elings V (1994) Appl Phys Lett 64:1738
62. Rivetti C, Codeluppi S (2001) Ultramicroscopy 87:55
63. Schulz A, Mucke N, Langowski J, Rippe K (1998) J Mol Biol 283:821
64. Feng XZ, Bash R, Balagurumoorthy P, Lohr D, Harrington RE, Lindsay SM (2000) Nucleic Acids Res 28:593
65. Fang Y, Spisz TS, Hoh JH (1999) Nucleic Acids Res 27:1943
66. Hansma HG, Revenko I, Kim K, Laney DE (1996) Nucleic Acids Res 24:713
67. Marsh TC, Vesenka J, Henderson E (1995) Nucleic Acids Res 23:696
68. Hansma HG, Laney DE, Bezanilla M, Sinsheimer RL, Hansma PK (1995) Biophys J 68:1672
69. Li J, Bai C, Wang C, Zhu C, Lin Z, Li Q, Cao E (1998) Nucleic Acids Res 26:4785
70. Bensimon A, Simon A, Chiffaudel A, Croquette V, Heslot F, Bensimon D (1994) Science 265:2096
71. Pope LH, Davies MC, Laughton CA, Roberts CJ, Tendler SJB, Williams PM (2000) J Microsc (Oxford) 199:68
72. Lyubchenko YL, Shlyakhtenko LS (1997) Proc Natl Acad Sci USA 94:496
73. Zuccheri G, Dame RT, Aquila M, Muzzalupo I, Samori B (1998) Appl Phys A 66:S585
74. Nagami F, Zuccheri G, Samori B, Kuroda R (2002) Anal Biochem 300:170
75. Cherny DI, Jovin TM (2001) J Mol Biol 313:295
76. Samori B, Siligardi G, Quagliariello C, Weisenhorn AL, Vesenka J, Bustamante CJ (1993) Proc Natl Acad Sci USA 90:3598
77. Oussatcheva EA, Shlyakhtenko LS, Glass R, Sinden RR, Lyubchenko YL, Potaman VN (1999) J Mol Biol 292:75
78. Lyubchenko YL, Shlyakhtenko LS, Binus M, Gaillard C, Strauss F (2002) Nucleic Acids Res 30:4902
79. Shlyakhtenko LS, Hsieh P, Grigoriev M, Potaman VN, Sinden RR, Lyubchenko YL (2000) J Mol Biol 296:1169

80. Shlyakhtenko LS, Potaman VN, Sinden RR, Lyubchenko YL (1998) J Mol Biol 280:61
81. Jett SD, Cherny DI, Subramaniam V, Jovin TM (2000) J Mol Biol 299:585
82. Rivetti C, Walker C, Bustamante C (1998) J Mol Biol 280:41
83. Zuccheri G, Scipioni A, Cavaliere V, Gargiulo G, De Santis P, Samori B (2001) Proc Natl Acad Sci USA 98:3074
84. Scipioni A, Anselmi C, Zuccheri G, Samori B, De Santis P (2002) Biophys J 83:2408
85. Han WH, Dlakic M, Zhu YWJ, Lindsay SM, Harrington RE (1997) Proc Natl Acad Sci USA 94:10565
86. Maeda Y, Matsumoto T, Tanaka H, Kawai T (1999) Jpn J Appl Phys 38:L1211
87. Andersen ES, Contera SA, Knudsen B, Damgaard CK, Besenbacher F, Kjems J (2004) J Biol Chem 279:22243
88. Coury JE, McFaillsom L, Williams LD, Bottomley LA (1996) Proc Natl Acad Sci USA 93:12283
89. Krautbauer R, Fischerlander S, Allen S, Gaub HE (2002) Single Molecules 3:97
90. Sun J-S, Francois J-C, Montenay-Garestier T, Saison-Behmoaras T, Roig V, Thuong NT, Helene C (1989) Proc Natl Acad Sci USA 86:9198
91. Viglasky V, Valle F, Adamcik J, Joab I, Podhradsky D, Dietler G (2003) Electrophoresis 24:1703
92. Mukhopadhyay R, Dubey P, Sarkar S (2005) J Struct Biol 150:277
93. Berge T, Jenkins NS, Hopkirk RB, Waring MJ, Edwardson JM, Henderson RM (2002) Nucleic Acids Res 30:2980
94. Saito M, Kobayashi M, Iwabuchi SI, Morita Y, Takamura Y, Tamiya E (2004) J Biochem136:813
95. Shlyakhtenko LS, Gall AA, Weimer JJ, Hawn DD, Lyubchenko YL (1999) Biophys J 77:568
96. Kuyper CL, Brewood GP, Chiu DT (2003) Nano Lett 3:1387
97. Ou ZY, Muthukumar M (2005) J Chem Phys 123
98. Borukhov I, Bruinsma RJ, Gelbart WM, Liu AJ (2001) Phys Rev Lett 86:2182
99. Grosberg AY, Nguyen TT, Shklovskii BI (2002) Rev Mod Phys 74:329
100. Cherstvy AG (2005) J Phys Condens Matter 17:1363
101. Hud NV, Vilfan ID (2005) Ann Rev Biophys Biomol Struct 34:295
102. Golan R, Pietrasanta LI, Hsieh W, Hansma HG (1999) Biochemistry 38:14069
103. Moreno-Herrero F, Herrero P, Moreno F, Colchero J, Gomez-Navarro C, Gomez-Herrero J, Baro AM (2003) Nanotechnology 14:128
104. Liu D, Wang C, Li JW, Lin Z, Tan ZK, Bai CL (2000) J Biomol Struct Dyn 18:1
105. Allen MJ, Bradbury EM, Balhorn R (1997) Nucleic Acids Res 25:2221
106. Dunlap DD, Maggi A, Soria MR, Monaco L (1997) Nucleic Acids Res 25:3095
107. Lin Z, Wang C, Feng XZ, Liu MZ, Li JW, Bai CL (1998) Nucleic Acids Res 26:3228
108. Fang Y, Hoh JH (1998) J Am Chem Soc 120:8903
109. Wolfert MA, Seymour LW (1996) Gene Therapy 3:269
110. Hansma HG, Golan R, Hsieh W, Lollo CP, Mullen-Ley P, Kwoh D (1998) Nucleic Acids Res 26:2481
111. Strand SP, Danielsen S, Christensen BE, Varum KM (2005) Biomacromolecules 6:3357
112. Fang Y, Hoh JH (1999) FEBS Lett 459:173
113. Liu Z, Li Z, Zhou H, Wei G, Song Y, Wang L (2005) J Microscop (Oxford) 218:233
114. Rackstraw BJ, Martin AL, Stolnik S, Roberts CJ, Garnett MC, Davies MC, Tendler SJB (2001) Langmuir 17:3185
115. Kleemann E, Dailey LA, Abdelhady HG, Gessler T, Schmehl T, Roberts CJ, Davies MC, Seeger W, Kissel T (2004) J Control Rel 100:437

116. Chim YTA, Lam JKW, Ma Y, Armes SP, Lewis AL, Roberts CJ, Stolnik S, Tendler SJB, Davies MC (2005) Langmuir 21:3591
117. Wangerek LA, Dahl HHM, Senden TJ, Carlin JB, Jans DA, Dunstan DE, Ioannou PA, Williamson R, Forrest SM (2001) J Gene Med 3:72
118. Chan CK, Senden T, Jans DA (2000) Gene Therapy 7:1690
119. Montigny WJ, Houchens CR, Illenye S, Gilbert J, Coonrod E, Chang YC, Heintz NH (2001) Nucleic Acids Res 29:1982
120. Mozafari MR, Reed CJ, Rostron C, Hasirci V (2005) J Liposome Res 15:93
121. Fang Y, Hoh JH (1998) Nucleic Acids Res 26:588
122. Bezanilla M, Manne S, Laney DE, Lyubchenko YL, Hansma HG (1995) Langmuir 11:655
123. Pereira GG, Williams DRM (2001) Biophys J 80:161
124. Fang Y, Yang J (1997) J Phys Chem B 101:441
125. Fang Y, Yang J (1997) J Phys Chem B 101:3453
126. Rivetti C, Guthold M, Bustamante C (1999) EMBO J 18:4464
127. Rippe K, Guthold M, von Hippel PH, Bustamante C (1997) J Mol Biol 270:125
128. Wyman C, Grotkopp E, Bustamante C, Nelson HCM (1995) EMBO J 14:117
129. Rivetti C, Codeluppi S, Dieci G, Bustamante C (2003) J Mol Biol 326:1413
130. Garcia RA, Bustamante CJ, Reich NO (1996) Proc Natl Acad Sci USA 93:7618
131. Sato MH, Ura K, Hohmura KI, Tokumasu F, Yoshimura SH, Hanaoka F, Takeyasu K (1999) FEBS Letters 452:267
132. Leuba SH, Yang GL, Robert C, Samori B, Vanholde K, Zlatanova J, Bustamante C (1994) Proc Natl Acad Sci USA 91:11621
133. Yodh JG, Lyubchenko YL, Shlyakhtenko LS, Woodbury N, Lohr D (1999) Biochemistry 38:15756
134. Yodh JG, Woodbury N, Shlyakhtenko LS, Lyubchenko YL, Lohr D (2002) Biochemistry 41:3565
135. Leuba SH, Bustamante C, van Holde K, Zlatanova J (1998) Biophys J 74:2830
136. Leuba SH, Bustamante C, Zlatanova J, van Holde K (1998) Biophys J 74:2823
137. Bezanilla M, Drake B, Nudler E, Kashlev M, Hansma PK, Hansma HG (1994) Biophys J 67:2151
138. Guthold M, Zhu X, Rivetti C, Yang G, Thomson NH, Kasas S, Hansma HG, Smith B, Hansma PK, Bustamante C (1999) Biophys J 77:2284
139. Lushnikov AY, Bogdanov A, Lyubchenko YL (2003) J Biol Chem 278:43130
140. Abdelhady HG, Allen S, Davies MC, Roberts CJ, Tendler SJB, Williams PM (2003) Nucleic Acids Res 31:4001
141. Guthold M, Bezanilla M, Erie DA, Jenkins B, Hansma HG, Bustamante C (1994) Proc Natl Acad Sci USA 91:12927
142. van Noort SJT, van der Werf KO, Eker APM, Wyman C, de Grooth BG, van Hulst NF, Greve J (1998) Biophys J 74:2840
143. Ellis DJ, Dryden DTF, Berge T, Edwardson JM, Henderson RM (1999) Nat Struct Biol 6:15
144. Jiao YK, Cherny DI, Heim G, Jovin TM, Schaffer TE (2001) J Mol Biol 314:233
145. Viani MB, Schaffer TE, Paloczi GT, Pietrasanta LI, Smith BL, Thompson JB, Richter M, Rief M, Gaub HE, Plaxco KW, Cleland AN, Hansma HG, Hansma PK (1999) Rev Sci Instrum 70:4300
146. Ando T, Kodera N, Takai E, Maruyama D, Saito K, Toda A (2001) Proc Natl Acad Sci USA 98:12468
147. Piner RD, Zhu J, Xu F, Hong SH, Mirkin CA (1999) Science 283:661
148. Germishuizen WA, Tosch P, Middelberg APJ, Walti C, Davies AG, Wirtz R, Pepper M (2005) J Appl Phys, p 97
149. Argaman M, Golan R, Thomson NH, Hansma HG (1997) Nucleic Acids Res 25:4379

150. Antognozzi M, Szczelkun MD, Round AN, Miles MJ (2002) Single Molecules 3:105
151. Smith DA, Connell SD, Robinson C, Kirkham J (2003) Anal Chim Acta 479:39
152. Stroh CM, Ebner A, Geretschlager M, Freudenthaler G, Kienberger F, Kamruzzahan ASM, Smith-Gil SJ, Gruber HJ, Hinterdorfer P (2004) Biophys J 87:1981
153. Zlatanova J, Lindsay SM, Leuba SH (2000) Prog Biophys Mol Biol 74:37
154. McKendry R, Zhang JY, Arntz Y, Strunz T, Hegner M, Lang HP, Baller MK, Certa U, Meyer E, Guntherodt HJ, Gerber C (2002) Proc Natl Acad Sci USA 99:9783

26 Nanostructuration and Nanoimaging of Biomolecules for Biosensors

Claude Martelet · Nicole Jaffrezic-Renault ·
Yanxia Hou · Abdelhamid Errachid · François Bessueille

From volume VII

Abbreviations

AFM	atomic force microscopy
LB	Langmuir–Blodgett
SAM	self-assembled monolayer
ODT	octadecanethiol
OBP	odorant-binding protein
MHDA	16-mercaptohexadecanoic acid
EDC	*N*-ethyl-*N*-[dimethylaminopropyl] carbodiimide
NHS	*N*-hydroxysuccinimide

26.1 Introduction and Definition of Biosensors

Recognition between a selective molecule and an analyte constitutes the primary event of mechanisms involved in biosensing. Thus the engineering of the biospecies in the recognizing layer is critical for the design of biosensors. For a few years now, such devices tended to be a reliable alternative to classical bioanalytical methods in numerous fields such as diagnostics, health care, pharmaceuticals, food and process control, environmental monitoring, defense and security, etc.

26.1.1 Definition

Biosensors are analytical devices containing immobilized biological sensitive materials in contact with or integrated within a transducer, which ultimately converts a biological signal into a quantitatively measurable electrical signal. From the concept we can see that a biosensor is composed of two essential components: a bioreceptor and a transducer. Figure 26.1 shows a schematic view of a biosensor.

26.1.2 Biosensor Components

A bioreceptor is the most critical component of a biosensor. It is responsible for the selective recognition of the analytes, generating the physicochemical signal monitored by the transducer. Ultimately, it determines the sensitivity of the device.

In: H. Fuchs, B. Bhushan (eds), *Biosystems - Investigated by Scanning Probe Microscopy*,
DOI 10.1007/978-3-540-37321-6_6,

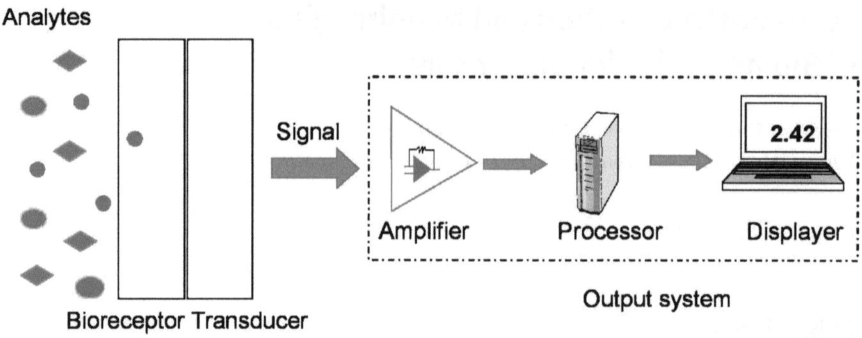

Fig. 26.1. Schematic view of a biosensor

Table 26.1. Fundamental components used for biosensor construction

Receptors	Transducers
Enzymes	Optical
Antibodies	Electrochemical
Organelles	Thermometric
Cell receptors	Piezoelectric
Tissues	Magnetic
Microorganisms	
Nucleic acids	
Synthetic receptors	

There is a wide range of naturally produced molecules from plants, animals, and microorganisms that can be used as the receptor in biosensor elaboration.

A transducer is the another crucial component for the biosensor, which converts a specific biological response to a useful electronic output, and must also be suitable for the receptor immobilization on, or close to, its surface. The output system involves amplifying, processing and displaying the signal in an appropriate format. In Table 26.1, the most frequently used sensing elements and transducers in the construction of biosensors are listed.

26.1.3 Immobilization of the Bioreceptor

For designing a biosensor, the biological component has to be properly attached to the transducer. This process is known as immobilization. One key point in biosensor fabrication is the development of immobilization technologies for stabilizing biomolecules and tethering them to surfaces. During the immobilization of the bioreceptor the recognizing properties have to be kept. Two immobilization techniques LB and SAMs, allowing the build up of the sensing layer at a molecular level, have been proven to be quite efficient, used alone or in combination. In this chapter these strategies will be detailed and illustrated with specific cases, AFM techniques appearing as a valuable tool to understand the mechanisms involved at a molecular level.

26.2
Langmuir–Blodgett and Self-Assembled Monolayers as Immobilization Techniques

26.2.1
Langmuir–Blodgett Technique

26.2.1.1
History

When describing the history of Langmuir or Langmuir–Blodgett films one is bound to start with an American statesman, Benjamin Franklin, who showed the first interest in monomolecular layers and reported the following to the British Royal Society in 1774 [1]:

> *At length at Clapman where there is, on the common, a large pond, which I observed to be one day very rough with the wind, I fetched out a cruet of oil, and dropped a little of it on the water. I saw it spread itself with surprising swiftness upon the surface; but the effect of smoothing the waves was not produced; or I had applied it first on the leeward fide of the pond, where the waves were largest, and the wind drove my oil back upon the shore. I then went to the windward fide, where they began to form; and there the oil, though not more than a teaspoonful, produced an instant calm over a space several yards square, which spread amazingly and extended itself gradually until it reached the leeside, making all that quarter of the pond, perhaps half an acre, as smooth as a looking glass.*

Over one hundred years later Lord Rayleigh suspected that the maximum extension of an oil film on water represents a layer with the thickness of one molecule. At the same time the foundation for our ability to characterize monolayers at an air/water interface was set by Agnes Pockels. She developed a rudimentary surface balance in her kitchen sink, which she used to determine (water) surface contamination as a function of area of the surface for different oils. Publication of Pockels's work in 1891 in Nature set the stage for Langmuir's quantitative work on fatty acid, ester and alcohol monolayers [2, 3].

Irwing Langmuir was the first one to perform systematic studies on floating monolayers on water in the late 1910s and early 1920s. He investigated the pressure-area relationship of molecules on an aqueous surface. The areas occupied by molecules such as acids, alcohols and esters were found to be independent of the hydrocarbon chain length, thus showing that only the hydrophilic head groups were immersed in the subphase. Langmuir was awarded the Nobel Prize for Chemistry in 1932 for his studies of surface chemistry, using floating monolayers to learn about the nature of intermolecular forces.

Katherine Blodgett, who worked with Langmuir on the properties of floating monolayers, developed the technique of transferring the films onto solid substrates and hence building up multilayer films. These built-up monolayer assemblies are therefore referred to as Langmuir–Blodgett films. The term "Langmuir film" is normally reserved for a floating monolayer at the air/water interface.

Interest in Langmuir–Blodgett films subsided with the outbreak of the Second World War, and remained low until the 1960s when Kuhn and Möbius showed how monolayers could be used to construct precise supermolecular structures. They used the Langmuir–Blodgett technique to demonstrate the fluorescence and quenching of dye molecules attached to fully saturated fatty acids under ultraviolet light [4]. This work and the publication of Gaines' "Insoluble Monolayers at Liquid Gas Interfaces" in 1966 initiated a revival of interest in the field. The first international conference on LB films was held in 1979 and since then the use of this technique has been increasing widely among scientists working on various different fields of research.

26.2.1.2
Langmuir Films

The formation of Langmuir films consists of two fundamental steps:

(i) Spread

To form a Langmuir film, the surface-active molecule, normally amphiphile, is dissolved in a volatile organic solvent (frequently chloroform or hexane) that will not react with or dissolve into the subphase. A very small quantity of this solution is spread on the surface of the subphase with a micropipette, and as the solvent evaporates, the surfactant molecules spread.

(ii) Compress

Sweeping the barrier over the subphase surface causes the molecules to come closer together and eventually to form a compressed, ordered monolayer.

26.2.1.2.1
Amphiphilic Molecules

An amphiphile is a molecule that is insoluble in water, with one hydrophilic end, which is preferentially immersed in the water; and the other hydrophobic part, which preferentially resides in the air (or in the nonpolar solvent) [5]. With correct "amphiphatic balance", namely the balance between the hydrophilic and hydrophobic constituents within the same molecules, such amphiphilic molecules can orient themselves at an air/water interface with their hydrophobic chains upwards in the air and their hydrophilic head groups downwards into the aqueous phase.

The hydrophobic part of amphiphile molecule usually consists of hydrocarbon or fluorocarbon chains, while the hydrophilic part consists of a polar group ($-OH$, $-COOH$, $-NH_3^+$, $-SO_3^-$, etc.). The hydrocarbon chain of the substance used for monolayer studies has to be long enough for formation of an insoluble monolayer. A rule of thumb is that there should be more than 12 hydrocarbons or groups in the chain ($(CH_2)_n$, $n > 12$). If the chain is shorter, though still insoluble in water, the amphiphile on the water surface tends to form micelles. On the other hand, if the chain is too long the amphiphile is prone to crystallize on the water surface, not to form a monolayer at the air/water interface. It should be noted that the solubility of an amphiphile in water depends on the balance between the alkyl chain length

Table 26.2. Effect of different functional groups on Langmuir films formation with C16-compounds

No film	Unstable film	Stable film	Dissolve in water
Hydrocarbon	$-CH_2OCH_3$	$-CH_2OH$	$-SO_3^-$
$-CH_2I$	$-C_6H_4OCH_3$	$-COOOH$	$-OSO_3^-$
$-CH_2Br$	$-COOCH_3$	$-CN$	$-C_6H_4SO_4^-$
$-CH_2Cl$		$-CONH_2$	$-NR_4^+$
$-NO_2$		$-CH=NOH$	
		$-C_6H_4OH$	
		$-CH_2COCH_3$	
		$-NHCONH_2$	
		$-NHCONH_3$	

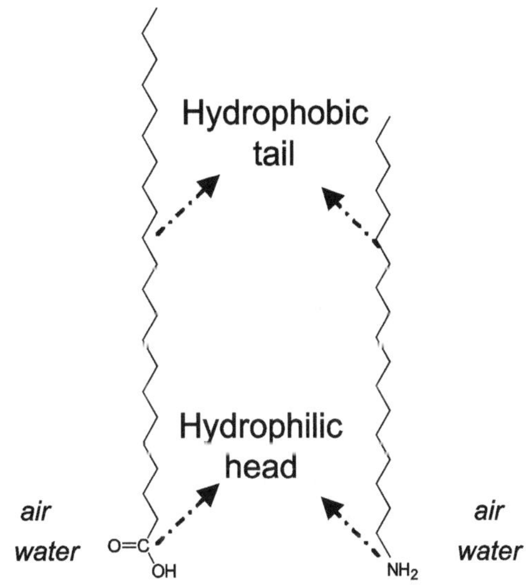

Fig. 26.2. Chemical structures of BA and ODA, and their spreading at the air/water interface

and the strength of its hydrophilic part. Abraham Ulman presented the effect of different functional groups on monolayer formation of C16-compounds as shown in Table 26.2.

The amphiphilic molecules used in our study are behenic acid (BA) ($C_{21}H_{43}$ COOH) and octadecylamine (ODA) ($C_{18}H_{37}NH_2$). Their chemical structures are shown in Fig. 26.2.

26.2.1.2.2
LB Instrument

The formation of Langmuir and Langmuir–Blodgett films is performed in a Langmuir trough, which is equipped with a Wilhelmy balance in order to measure surface pressure of the monolayer, one or two moveable barriers to compress the monolayer at the air/water interface, and a dipper mechanism permitting transfer of the monolayer

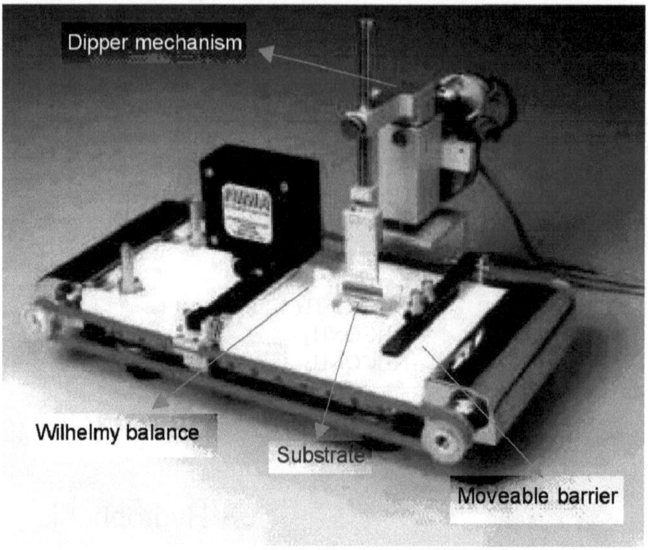

Fig. 26.3. Langmuir trough produced by NIMA

onto solid substrate, as shown in Fig. 26.3. Usually the trough is placed in a chamber in order to prevent external contamination.

26.2.1.2.3
Surface Pressure–Molecular Area Isotherm

The most important indicator of the monolayer properties is given by measuring the surface pressure as a function of the area of each molecule. This is carried out at a constant temperature and is known as the surface pressure–molecular area isotherm (referred to as Π-A). Usually an isotherm is recorded by compressing the film at a constant rate while continuously monitoring the surface pressure. A schematic Π-A isotherm is shown in Fig. 26.4.

The surface pressure–molecular area isotherm is rich in information on stability of the monolayer at the air/water interface, the reorientation of molecules in the two-dimensional system, phase transitions, and conformational transformations.

A simple terminology used to classify different monolayer phases of fatty acids has been proposed by Harkins as early as 1952 [6]. As the monolayer is compressed, it shows three phases: gas, liquid or a solid one.

In the gas state, the area per molecule is large, and ideally there should be no interaction (lateral adhesion) between the molecules, and therefore, the surface pressure is low.

As the barriers move, the monolayers are compressed, accompanyied by an intermolecular distance decrease and a surface pressure increase. A phase transition (from the gas to the liquid state) takes place. In the liquid phase, the monolayer is coherent, and the molecules have more degrees of freedom and gauche conformations can be found in the alkyl chains.

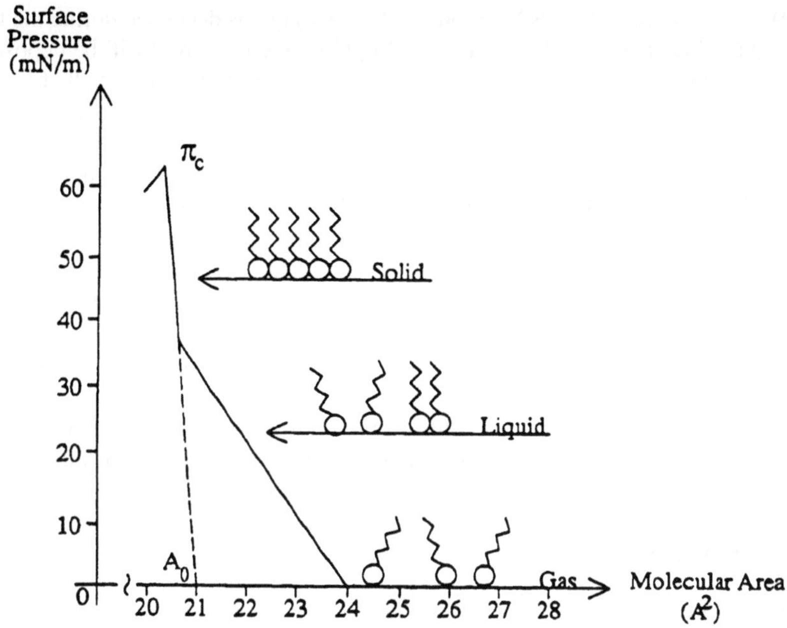

Fig. 26.4. Schematic Π-A isotherm

When the barriers compress the film further, a second phase transition can be observed, from the liquid to the solid state. In the condensed solid phase, the molecules are closely packed and uniformly oriented. If further pressure is applied on the monolayer, it collapses due to mechanical instability, and consequently a sharp decrease in the surface pressure is observed.

26.2.1.3 Langmuir–Blodgett Films

26.2.1.3.1 Principle

The term "Langmuir–Blodgett film" traditionally refers to monolayers that have been vertically transferred off the water subphase and onto a solid support. Vertical deposition is the most common method of LB film transfer, called the Langmuir–Blodgett technique. An alternative way to deposit the monolayer is the Langmuir–Schaeffer (LS) technique, which differs from the LB technique only in the sense that the solid substrate is horizontally lowered in contact with the monolayer for transferring the monolayer.

The LB deposition is traditionally carried out in the "solid" phase with surface pressure kept constant. In this case the surface pressure is high enough to ensure sufficient cohesion in the monolayer, so that the monolayer will not fall apart during transfer to the solid support. This also ensures building up of homogeneous multilayers.

When the solid surface is hydrophobic the first layer is deposited during the first down pass. When the solid support is hydrophilic, the first layer will be deposited during the first up-stroke pass, because the hydrophobic tails are repelled by the hydrophilic support as it is immersed in the water. Once the first layer is deposited, more layers will be deposited on each subsequent pass of the support through the air/water interface. Multilayers can therefore be deposited onto the support.

There are several parameters that can affect quality of LB films:

- Quality of the monolayer at the air/water interface, particularly the stability;
- Properties of the subphase (temperature, ingredient, pH);
- Surface pressure for transferring LB films, namely the target pressure;
- Transfer rate;
- Surface properties of support (cleanliness, surface free energy, chemical composition, etc).

In order successfully to deposit LB films onto the support, usually the substrate is previously functionalized to provide a favorable platform for the deposition.

26.2.1.3.2 Types of Deposition

Different film architectures can be obtained upon deposition by the Langmuir–Blodgett technique (Fig. 26.5). Y-type multilayers are most common and can be prepared on either hydrophilic or hydrophobic substrates with the reversed head–tail arrangement. They are the most stable architectures because of strong head–head and tail–tail interactions. X-type and Z-type films are reported less often. In fact, with some amphiphiles, even when films have been deliberately transferred by an X- or Z-method, the spacing between the hydrophilic headgroups shows a packing similar to that of Y-type films, implying that some molecular reorganization could occur during or shortly after deposition [7].

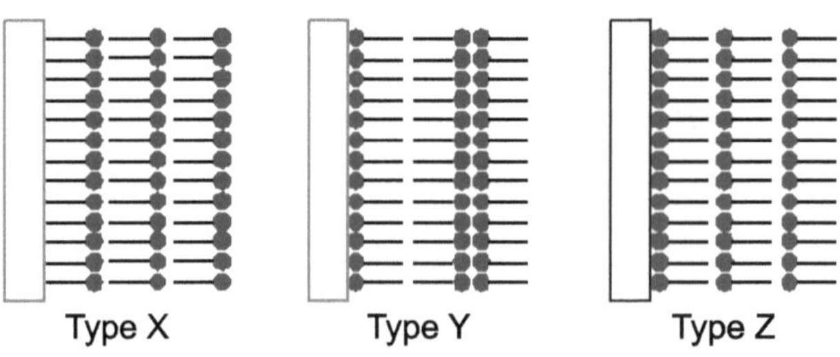

Fig. 26.5. Possible architectures for Langmuir–Blodgett films

26.2.1.3.3 Transfer Ratio

The quantity and the quality of the deposited monolayer on the solid support can be measured by transfer ratio (TR), given by the equation:

$$TR = A_l / A_s \ ,$$

where A_l is the area of monolayer removed from subphase at a constant pressure, A_s is the area of substrate immersed in water. Such a ratio is usually measured for each substrate pass through the air/water interface. For an ideal transfer the TR is equal to 1.

26.2.1.4 Mixed Amphiphile/Protein LB Films

Langmuir–Blodgett films have been studied for application in a great variety of domains, such as nonlinear optics, molecular electronics, sensors, etc. Since the 1980s, extensive studies have been carried out to immobilize biomolecules by the LB technique for elaboration of biosensors. Proteins were introduced in the amphiphile monolayer to form mixed amphiphile/protein LB films. There are several different ways to prepare mixed amphiphile/protein LB films.

26.2.1.4.1 Methods for Preparation of the Mixed Amphiphile/Protein LB Films

Adsorption of Proteins Under the Compressed Amphiphile Monolayer In 1971, Fromherz used this method to make the mixed layer with a multicompartment Langmuir trough [8]. As shown in Fig. 26.6, in the first compartment, the amphiphile was injected on the subphase of pure water and the monolayer was compressed till the surface pressure in the range of 5 to 30 mN m^{-1}. After that, under a constant surface pressure, the layer was moved to the second compartment, in which the proteins were present in the subphase. The proteins were adsorbed onto the head groups of amphiphile molecules by electrostatic interaction and the mixed monolayer formed. Finally, the mixed layer was moved to the third compartment with pure water as the subphase, where the mixed layer was transferred onto solid supports. Arisawa and Yamamoto [9] developed a urea sensor by immobilizing urease onto an ISFET by using this LB technique.

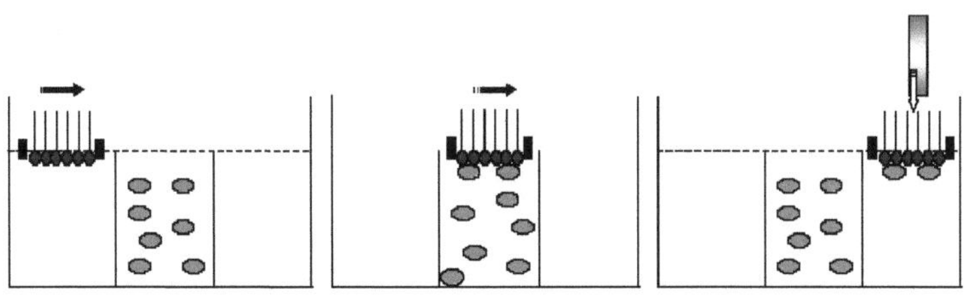

(1) (2) (3)

Fig. 26.6. Formation of the mixed LB films by adsorption of proteins under the compressed amphiphile monolayer

Injection of Proteins Under the Compressed Amphiphile Monolayer This method was first developed by Zaitsev et al. [10, 11], they injected a solution of protein in

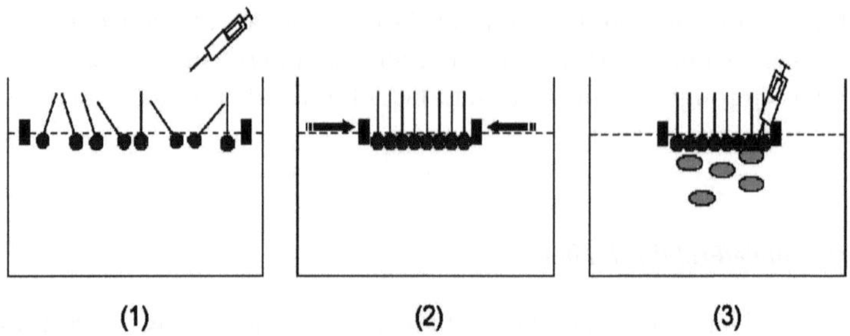

Fig. 26.7. Formation of the mixed LB films by injection of proteins under the compressed amphiphile monolayer

the water subphase under the amphiphile layer that was previously formed at the air/water interface, as shown in Fig. 26.7. However, these two methods show poor reproducibility for the formation of mixed LB films.

Adsorption of Proteins Under Noncompressed Amphiphile Monolayer The most commonly used method is based on adsorption of proteins under a noncompressed amphiphile monolayer. As shown in Fig. 26.8, first, the amphiphile solution is injected on the subphase of protein solution, proteins are gradually adsorbed onto the amphiphile layer by the electrostatic interaction. When the adsorption reaches equilibrium, the mixed monolayer is compressed to the target pressure. And finally at this surface pressure the stable mixed layer is transferred onto solid substrates.

Such mixed LB films have been investigated by the group of Valleton [12–20] and in our laboratory [21–23]. Compared to the first two methods mentioned above, this method has several advantages: simplicity for obtaining well-ordered multilayers

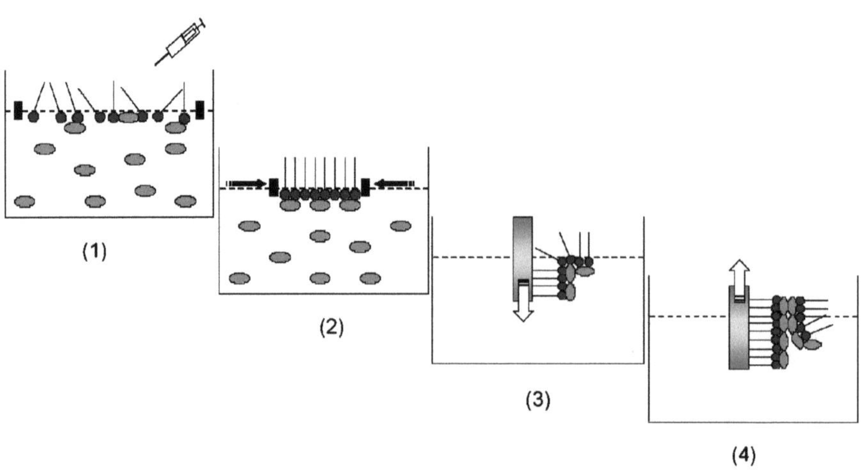

Fig. 26.8. Formation of the mixed LB films by adsorption of protein under the noncompressed amphiphile monolayer

of LB films, high reproducibility for depositing of LB films, and highly homogenous structures of LB films. There are still some other LB methods, which are not commonly used, to elaborate the mixed amphiphile/protein LB films. For example, Girard-Egrot et al. [24,25] first deposited amphiphile LB films on the substrates, and then immersed the coated substrates in a solution of enzyme glutamate dehydrogenase for adsorption onto an amphiphile multilayer. Okahata et al. [26] prepared an enzyme–lipid complex in the first step, and then transferred the monolayer of the complex onto a Pt substrate as the sensor membrane.

26.2.1.4.2 Structure of the Mixed Monolayer at the Air/Water Interface

Protein can be adsorbed onto amphiphile monolayers spread at the air/water interface by electrostatic interaction between the protein and the polar heads of amphiphile molecules. Zaitsev [27] investigated the influence of the electrostatic effect on the adsorption of protein by studying the mixed monolayer of glucose oxidase with different kinds of amphiphiles: phosphatidylcholine (neutrally charged), cetyltrimethylammonium bromide (positively charged) and stearic acid (negatively charged). The results showed that the enzyme GOD, which is negatively charged in buffer solution at pH 7.0, is more strongly adsorbed on positively charged lipid monolayers than on zwitterionic ones, and only slightly adsorbed on negatively charged monolayers of stearic acid.

The dynamic formation and structure of mixed monolayer with BA/GOD at the air/water interface proposed by the group of Valleton is shown in Fig. 26.9 [18].

The results showed that before spreading the amphiphile behenic acid, the surface pressure was about $2\,mN\,m^{-1}$ showing that some of enzyme molecules were adsorbed at the air/water interface. Enzyme GOD was allowed to adsorb under the

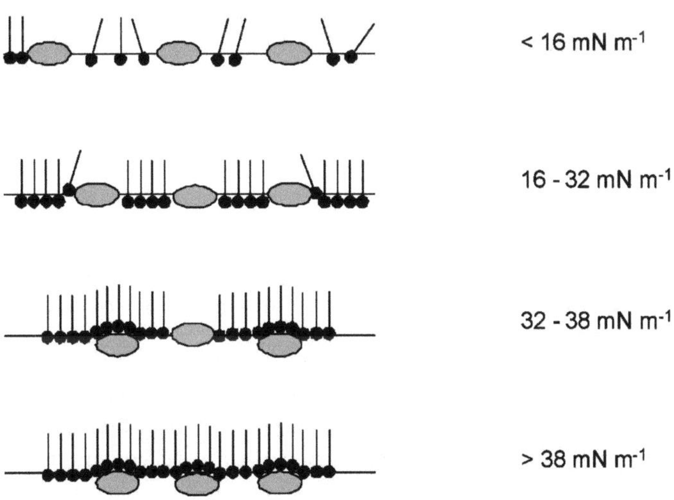

Fig. 26.9. Dynamic form ation and structure of mixed monolayer with BA/GOD at the air/water interface

behenic acid film at the air/water interface. It took about 45 min to reach adsorption equilibrium and the surface pressure reached $16\,\mathrm{mN\,m^{-1}}$. Then the monolayer was compressed at a constant rate, during the compression the surface pressure increased. In the range of surface pressure from $16\,\mathrm{mN\,m^{-1}}$ to $32\,\mathrm{mN\,m^{-1}}$, the mixed monolayer was thought to be in a liquid phase. With further compression, the enzyme molecules were gradually expelled from the air/water interface and pushed down under the polar heads of BA molecules, the monolayer is considered to reach gradually the solid phase. At a high surface pressure (above $38\,\mathrm{mN\,m^{-1}}$), the mixed film was constituted of a behenic acid monolayer in a solid phase, under which a GOD sublayer was adsorbed by the electrostatic interaction as shown in Fig. 26.9.

However, the phase change above $40\,\mathrm{mN\,m^{-1}}$ was not mentioned. If further compression was applied on the mixed monolayer, there would be a risk of overlapping of the layer and even a complete collapse of the monolayer.

26.2.1.4.3 Application of LB Technique for Biosensor Elaboration

The quest for biosensors has brought the need for highly selective and sensitive organic layers, with tailored biological properties that can be incorporated into electronic, optical, or electrochemical devices. The molecular dimension of LB films makes the use of even highly expensive molecules economically attractive, therefore, the development of sophisticated detection systems becomes possible.

Up to now, the LB technique has been widely used to immobilize enzymes [28–36], antibodies [37–44] and DNA [45], etc. on the substrate for the elaboration of various biosensors.

26.2.2 Self-Assembled Monolayers

26.2.2.1 Introduction

26.2.2.1.1 Definition of Self-Assembled Monolayers

Self-assembled monolayers are molecular assemblies that are formed spontaneously by the immersion of an appropriate substrate into a solution of a reaction-active surfactant in a suitable solvent.

From the energetics point of view, a self-assembling surfactant molecule can be divided into three parts, as shown in Fig. 26.10. The first part is the head-anchoring group that provides the most exothermic process while anchoring on the substrate surface by the chemisorption. The energy associated with the chemisorption is of the order of tens of kcal/mol (e.g., $\sim 40-45\,\mathrm{kcal\,mol^{-1}}$ for thiolate on gold). The second part is the alkyl chain, and the energy associated with its interchain van der Waals interaction is of the order of a few (< 10)

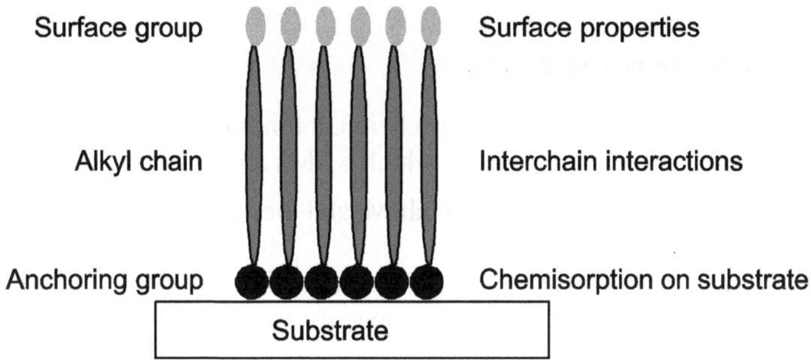

Fig. 26.10. A schematic view of the self-assembling surfactant molecule in SAMs

kcal mol^{-1} (exothermic). The third part in the surfactant is the terminal surface group, which determines the surface properties (e.g. wetting, friction, adhesion, etc.) of the SAMs functionalized substrate surface. For example, in the case of a simple alkyl chain, the terminal methyl (CH_3) group provides a hydrophobic surface. The ordering of alkanethiol is driven by the strong affinity between the head groups and the substrates, the lateral van der Waals interactions between the tethered alkyl chains, and the dipole interactions between the polar end groups.

26.2.2.1.2
Types of Self-Assembled Monolayers on Electrode

SAMs can form on many materials that possess well-defined surface chemistries, including metals, semiconductors, oxides and so on [16]. In Table 26.3, surfactants and the related substrates, on which they can form SAMs are listed.

Table 26.3. Chemical systems of substrates and the corresponding surfactants that form SAMs

Surface	Substrates	Surfactants
Metal	Au	R–SH, R–SS–R, R–S–R, R–NH_2, R–NC, R–Se, R–Te
	Ag	R–COOH, R–SH
	Pt	R–SH, R–NC
	Pd	R–SH
	Cu	R–SH
	Hg	R–SH
Semiconductor	GaAs	R–SH
	InP	R–SH
	CdSe	R–SH
	ZnSe	R–SH
Oxide	Al_2O_3	R–COOH
	TiO_2	R–COOH, R–PO_3H
	ITO	R–COOH, R–SH, R–Si$(x)_3$
	SiO_2	R-Si$(x)_3$

26.2.2.1.3
Advantages of Self-Assembled Monolayers

The advantages of SAMs arise from the chemical binding of the anchoring group at the substrate surface and the ease with which SAMs are formed.

- SAMs are very stable as mentioned above and they are able to survive after long exposure to vacuum.
- Different surface properties of the substrates can be provided by terminal groups of the surfactants without disrupting the self-assembly process or destabilizing the SAMs.
- SAMs provide a means of attaching the biomolecules to the electrode with a diverse set of structures ranging from modified monolayers to multilayers.
- In electrochemistry, the thiol-based SAMs survive the electrochemical treatment and experiments. The anchored sulfur atoms are believed to resist oxidation, reduction, and desorption in a range of potentials.
- A monolayer is deposited on the metal in a manner of seconds to minutes.
- Usually, this method does not require anaerobic or anhydrous conditions; nor does it require vacuum.
- Weakly adsorbed impurities have a slight effect on the formation of SAMs, since high affinity interaction between the anchoring group and the substrate enables the assembling layer to displace more weakly adsorbed impurities.
- Curvature or accessibility of the substrate surface is not a serious obstacle factor; because substrates can range from the macro- to the nanoscale, from smooth to porous, as well as from plane to spherical (nanoparticles).

Another advantage is the amenability to application in controlling interfacial (physical, chemical, electrochemical) properties.

Compared to LB films SAMs are more stable. Since LB films are deposited onto the substrates by physisorption, mainly by hydrophilic and/or hydrophobic interaction, they are more sensitive to the external environment, such as temperature, pH, solvent, etc.

26.2.2.2
Preparation of Self-Assembled Monolayers

In further examples shown, SAMs of alkanethiol on gold and silver were performed. Therefore, only preparations of SAMs on gold and silver substrates will be discussed here.

Usually, the gold and silver substrates are obtained by evaporating or sputtering gold and silver films (typically 50–200 nm thick) on glass, silicon, or cleaved mica. To improve the adhesion of the gold and silver film to the oxide surface, a thin layer of chromium or titanium is often deposited first.

26.2.2.2.1
Preparation of SAMs on Gold Substrates

Pretreatment of Gold Substrates Gold is the most popular substrate for thiol SAMs. Owing to its noble character, gold substrate can be handled in air without the formation of an oxide surface layer, and can survive harsh chemical treatments for the removal of organic contaminants. The formation of a monolayer on gold is strongly dependent on the cleanliness and structure of gold prior to modification [47], since it is well known that gold is a soft metal and is readily contaminated by organic and inorganic species during manual handling [48].

Before use, gold substrates need to be cleaned well in order to remove the contaminants on the surface that may affect the integrality of the SAMs formation. The importance of the cleanliness of the gold surface on formation and reproducibility of SAMs has been recognized by many groups working in the field of SAMs. In the literature numerous different cleaning procedures have been proposed. Most of them involve gold surface oxidation to remove organic and inorganic contaminants on the gold surface: electrochemical oxidation [49], immersion into strong oxidizing solutions [50], a combination of both [51], or exposure to reactive oxygen species formed by either UV irradiation or oxygen plasma [52, 53].

In most examples discussed hereafter, a strong oxidizing solution, H_2SO_4:H_2O_2 (7:3 v/v) was used. First, the gold substrates were cleaned with acetone in an ultrasonic bath for 10 min in order to remove the protection layer of polymer and dried under nitrogen flow, after they were immersed into the hot cleaning solution for 1 min. Subsequently, the gold substrates were rinsed thoroughly in absolute ethanol, and dried under nitrogen flow. The immersion time in acetone and piranha was optimized.

The wetting properties of gold surface before and after pretreatment were studied with a contact-angle meter. It is widely accepted that clean gold is hydrophilic. In this thesis, the contact angle of gold substrate as received is about 90°, which is due to the presence of a carbon- and oxygen-containing contaminant layer on the gold surface [54]. After cleaning, the contact angle is $< 10°$, which proves that the gold surface is free of contaminants. The contact-angle images are shown in Table 26.4.

Table 26.4. Contact angle values and images of the gold substrates before and after different treatments

Samples	as received	after cleaning	with SAMs of ODT	with mixed SAMs
Contact angle	$90° \pm 3°$	$< 10°$	$110° \pm 3°$	$9° \pm 3°$
Contact angle image				

Formation of the SAMs on Gold Substrate The most common approach for formation of SAMs on solid substrate is by immersing the substrate into a homogeneous solution of the self-assembling molecule at room temperature. There are a number of experimental factors that can affect the structure of the resulting SAMs: solvent, temperature, concentration of the adsorbate, immersion time, purity of the adsorbate, concentration of oxygen in solution, cleanliness of the substrate, and chain length or more generally, structure of the adsorbate.

Usually, any solvent capable of dissolving the adsorbate is suitable. Ethanol is the most popular solvent because some plausible gold oxide will be reduced in ethanol, but sometimes other solvents are preferred. The thiol concentration can be varied from micromolar levels to the net liquid thiol. However, very low concentrations of thiols are favorable for obtaining large crystalline domains of alkanethiols via slow self-assembly. Immersion time varies from several minutes to several hours for alkanethiol, while for sulfides and disulfides, immersion times of several days may be necessary.

In the examples presented, gold substrates were functionalized with two kinds of SAMS. In the first case, as soon as the cleaning procedure is finished, the gold substrates were put into a 1 mM 1-octadecanethiol (ODT) ethanol solution for 21 h at room temperature to render their surfaces hydrophobic. Since such a hydrophobic surface is favorable to the deposition of LB films. Then the substrates were thoroughly cleaned with ethanol to remove physically adsorbed thiol on the surface and dried under a N_2 flow. The contact-angle image of the gold substrate modified with SAMs of ODT is given in Table 26.4.

In the second case, after cleaning, gold substrates were immersed immediately into a mixed solution of 16-mercaptohexadecanoic acid (MHDA) and 1,2-Dioleoyl-*sn*-Glycero-3-Phosphoethanolamine-*N*-(Biotinyl) sodium salt (biotinyl-PE) in absolute ethanol at the concentration of 1 mM and 0.1 mM, respectively, for 21 h in order to obtain mixed self-assembled monolayers. Then the substrates were thoroughly cleaned with ethanol to remove physically adsorbed thiol and biotinyl-PE on the surface, and finally dried under a N_2 flow. The contact-angle image of the gold substrate with the mixed SAMs is also shown in Table 26.4. Figure 26.11 shows the procedures for the functionalization of gold substrate by thiol SAMs.

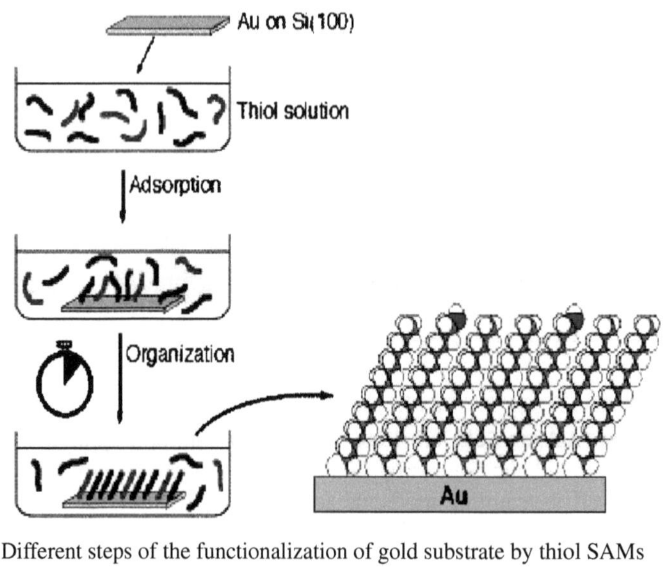

Fig. 26.11. Different steps of the functionalization of gold substrate by thiol SAMs

There are some other deposition methods for SAMs. When the thiol is sufficiently volatile, SAMs deposition can be performed in the vapor phase of the thiol, both in a vacuum [56] and at ambient pressures [57]. SAMs deposition can also be performed in a solid phase, Xia and Whitesides [58] developed the technique of microcontact printing μCP: It uses the relief pattern on the surface of a PDMS stamp to form patterns of self-assembled monolayers on the substrate surface by contact. In μCP of alkanethiols on gold, for example, the PDMS stamp is wetted with an "ink" (typically, a 2 mM solution of thiol in ethanol) and is brought into contact with the gold surface for 20 s. The thiol is transferred from the stamp to the gold surface upon contact, SAMs with predesigned patterns are stamped on the gold surface.

26.2.2.2.2
Pretreatment and Formation of SAMs on Silver Substrates

For silver substrates, mild cleaning procedures are necessary due to their high reactivity to the oxygen in air [55]. Samples were only cleaned with hot ethanol for 5 min, and then in order to minimize the effects of oxidation and contamination, they were immediately immersed into thiol solution of ODT with a concentration of 1 mM for 24 h. Since they are easily oxidized, the formation of SAMs was performed in the presence of N_2 flow, as shown in Fig. 26.12. After the substrates were thoroughly cleaned with ethanol to remove physically adsorbed thiol and dried under a N_2 flow. The contact angle of ODT-functionalized silver surface is 115 ± 3°.

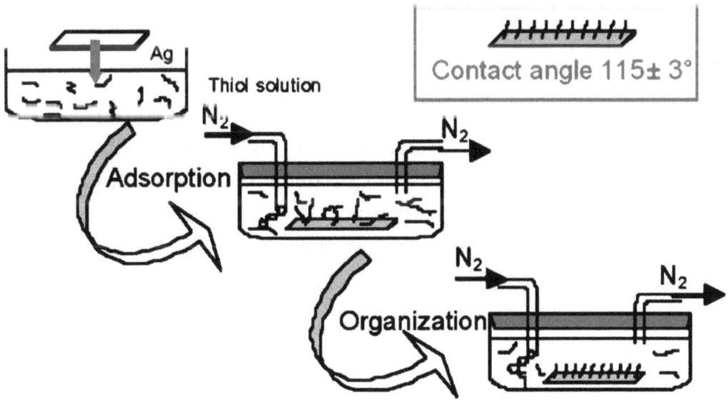

Fig. 26.12. SAMs functionalization of silver substrate with 1 mM ODT

26.2.2.3
Kinetics and Mechanism of Formation of SAMs

There are substantial differences among reports in the literature regarding the kinetics of formation of the SAMs on the solid substrates from liquid solution. The most important factors affecting the kinetics features appear to be the presence of preadsorbed contaminants, the solution phase hydrodynamics, and solvent effects [59].

The general kinetic of alkanethiol adsorption onto Au(111) in relatively dilute solution (10^{-3} M) is believed to have two distinct adsorption kinetics: a very fast step, which takes a few tens of seconds to minutes, by the end of which the contact angles are close to their desired values and the thickness are about 80–90% of their maximum; a slow step, which may last several hours, at the end the thickness and contact angles will reach their final values [60].

The initial step is a diffusion-controlled Langmuir adsorption, its kinetics is governed by the surface–head group reaction. The second step can be described as a surface reorganization process, during which the alkyl chains undergo a phase change from the disordered state to an ordered state and form a two-dimensional quasicrystal. Its kinetics is related to the chain disorder, different components of chain–chain interaction, and surface mobility of chains.

Extensive X-ray photoelectron spectroscopy (XPS) experiments suggest that chemisorption of alkanethiols on a gold surface yields the gold thiolate species. In the case of alkanethiols, the mechanism of alkanethiols binding to gold substrates is considered to be an oxidative addition of the S–H bond to gold substrate followed by reductive elimination of the hydrogen, thus resulting in the formation of a thiolate species, as shown in reaction (I):

$$\text{R–SH} + \text{Au}_n^0 \rightarrow \text{RS}^-\text{–Au}^+ + (1/2)\,\text{H}_2 + \text{Au}_{n-1}^0\,. \tag{I}$$

In the case of disulfides, the mechanism is thought to be:

$$\text{R–S–S–R} + \text{Au}_n^0 \rightarrow 2\text{RS}^-\text{–Au}^+ + \text{Au}_{n-2}^0\,. \tag{II}$$

Another common SAMs system is silanization of silicon oxide. For the formation of the SAMs on the silicon oxide, for example, with octadecyltricholorosilane (OTS), two possible monolayer growth modes were proposed by Dutta and coworkers: island growth and uniform growth [61, 62]. In the island mode, the molecules form dense islands of (nearly) vertical molecules. Additional molecules attach at island edges or form new islands as the coverage increases. In the latter mode, molecules deposit randomly on the substrate, falling over at low coverage. As more molecules adsorb on the substrate, the molecules begin to stand up.

26.2.2.4
Structure of SAMs on Gold and Silver Substrates

Alkanethiols on Au ⟨111⟩ form a $(\sqrt{3} \times \sqrt{3})$ R30° (R = rotated) structure. Sulfur atoms bond to the three-fold hollow sites on the Au surface and the extremely strong surface bond (44 kcal mol^{-1}) contributes to the stability of the SAMs. Alkanethiols on Ag ⟨111⟩ form a $(\sqrt{7} \times \sqrt{7})$ R10.9° structure with an S–S distance of 0.441 nm. Because of this shorter S–S distance, alkanethiols on Ag are more densely packed, which results in near-perpendicular orientation of molecules on the Ag surfaces. However, in the case of Au ⟨111⟩, the alkyl chain is titled 30° from the surface normal to decrease the lattice energy. The difference in orientation is mainly due to the interchain van der Waals interaction [63].

26.2.2.5 Protein Immobilization Using SAMs

As referred to above, SAMs are well-ordered, once formed they are quite stable due to their high affinity to the substrate. Moreover, the variety of terminal functionality of SAMs can provide the needed design flexibility, both at the individual molecular and at the material levels, and offer a vehicle for investigation of specific interactions at interfaces. Such well-ordered monolayers are suitable for protein immobilization, close to the electrode surface with a high degree of control over the molecular architecture of the recognition interface. In practice, SAMs have been used for fundamental investigations of the interactions of proteins with surfaces and particularly for the fabrication of a variety of biosensors including enzyme biosensors [64–66], immunosensors [67], and DNA hybridization biosensors [68], etc.

There are two basic approaches for proteins immobilization onto gold substrates surface by SAMs. Proteins can be attached to thiol-containing moieties and then bind to the gold surfaces by SAMs, or they can be bound to previously SAMs-functionalized gold substrates.

26.2.2.5.1 Protein Modification

In this case, proteins are first tagged in the solution with an appropriate sulfur-containing molecule via reaction with specific reactive groups present on the biomolecule surface, e.g. the amino group. And then the modified proteins are self-assembled onto the gold substrate via the sulfur moiety.

It has been reported that proteins can be chemically modified by *N*-succinimidyl 3-(2-pyridyldithiol)propionate (SPDP) (Reaction A) and 2-iminothiolane (Reaction B). During such modification, a disulfide moiety is connected to the primary amine group of the protein via an amido linkage [69, 70]. Then modified proteins were immobilized onto the gold surface by SAMs [71], as shown in Fig. 26.13.

Such a method allows a monolayer of protein to be attached to the electrode. In addition, single-point labeling of the protein allows control over the orientation of the protein. However, Dong et al. [72] showed that the immobilization of thiol-labeled glucose oxidase produced a less stable sensing interface than enzyme immobilized onto a preformed SAM.

26.2.2.5.2 Substrate Modification

Through such a method the substrate is functionalized by a SAM monolayer of a sulfur-containing surfactant. Then proteins are bound to the SAMs via reaction using specific moieties present on the biomolecule structure.

Noncovalent Binding One common noncovalent approach for immobilizing protein to the surface of SAMs is by electrostatic binding. This method is simple and

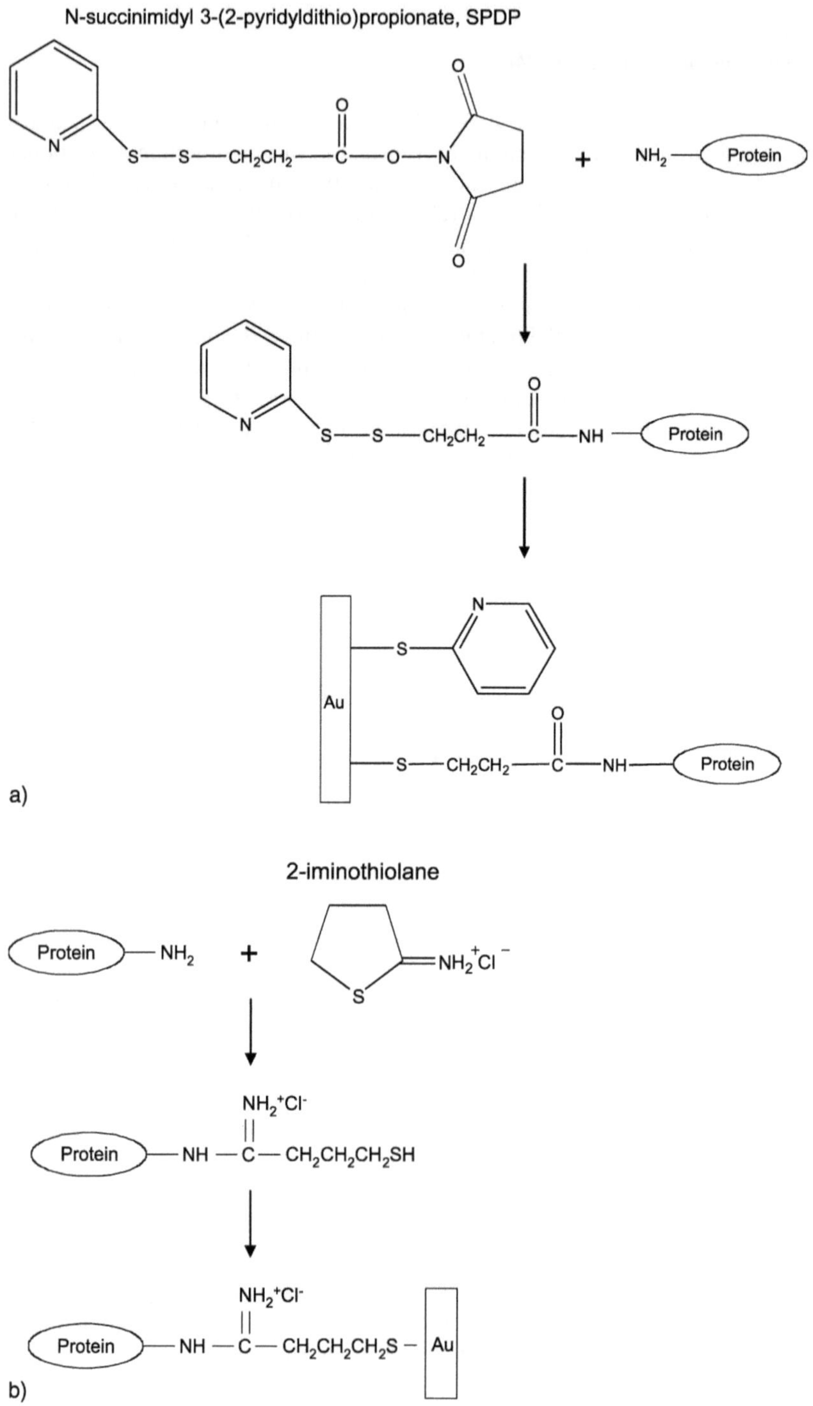

Fig. 26.13. Previously modified protein immobilization by SAMs

gentle, and it can provide a potential control over the orientation of the immobilized protein molecules depending on the charge distribution on the shell of the protein.

Mixed protein/amphiphile LB films have been transferred onto SAMs prefunctionalized gold and silver substrates by noncovalent binding and used for designing impedimetric immunosensors [85, 86].

The major drawback of the electrostatic binding is that the strength of the bond is dependent on the solution conditions. The change in ionic strength and pH can lead to removal of proteins from the surfaces of the SAMs.

Crosslinking Crosslinking proteins to the surfaces of SAMs with bifunctional reagent glutaraldehyde is a relatively uncontrolled method. It suffers from poor reproducibility and protein leaching. Usually multilayers of protein are produced on the substrate surface.

Direct Covalent Attachment Direct covalent attachment is the most potential method for elaboration of biosensor due to the high stability of the resulting bond. The immobilization is conducted in two stages: activation of the carrier, and coupling of the biomolecule.

Usually proteins are covalently bound to COOH-terminated SAMs by using carbodiimide coupling that allows amines to be coupled to carboxylic acids. The most often used carbodiimide coupling are *N*-ethyl-*N*-[dimethylaminopropyl] carbodiimide (EDC) and *N*-hydroxysuccinimide (NHS). In the reaction EDC converts the carboxylic acids into a reactive intermediate which is susceptible to be attacked by amines. In some cases, EDC and NHS or *N*-hydroxysulfosuccinimide (NHSS) are used together as they produce a more stable reactive intermediate that has been shown to give a greater reaction yield.

For covalently coupling proteins onto amine-terminated SAMs, a bifunctional linker glutaraldehyde is often used. GA can react with amino tails of SAMs to create –C–N=C– bond, after that the proteins with the amine group are coupled with the other aldehyde group.

Mirsky et al. [67] reviewed surface-activation methods for immobilization of proteins on the monolayers of 16-mercaptohexadecanoic and 16-mercaptohexadecylamine on gold electrodes. Here, we only list the most commonly used activation methods of carboxy and amino groups for the elaboration of biosensors, as shown in Fig. 26.14.

Direct Affinity Attachment More recently a new strategy for more site-specific binding of biomolecules based on a streptavidin/biotin pair has been extensively investigated. Streptavidin is a tetrameric protein, approximately 4.5 × 4.5 × 5.8 nm, and with a molecular weight of 60 kDa. It can bind biotin at four binding sites, two binding sites on each of two opposite faces, with an extremely high binding constant ($K_d = 10^{-15}$ M), which makes the binding process effectively irreversible [73]. Moreover, diverse biomolecules can be easily functionalized by biotin with no disturbing effect on their specific sites. All these properties make the avidin/biotin binding pair highly suitable for immobilizing proteins on the substrates and for the construction of biosensors.

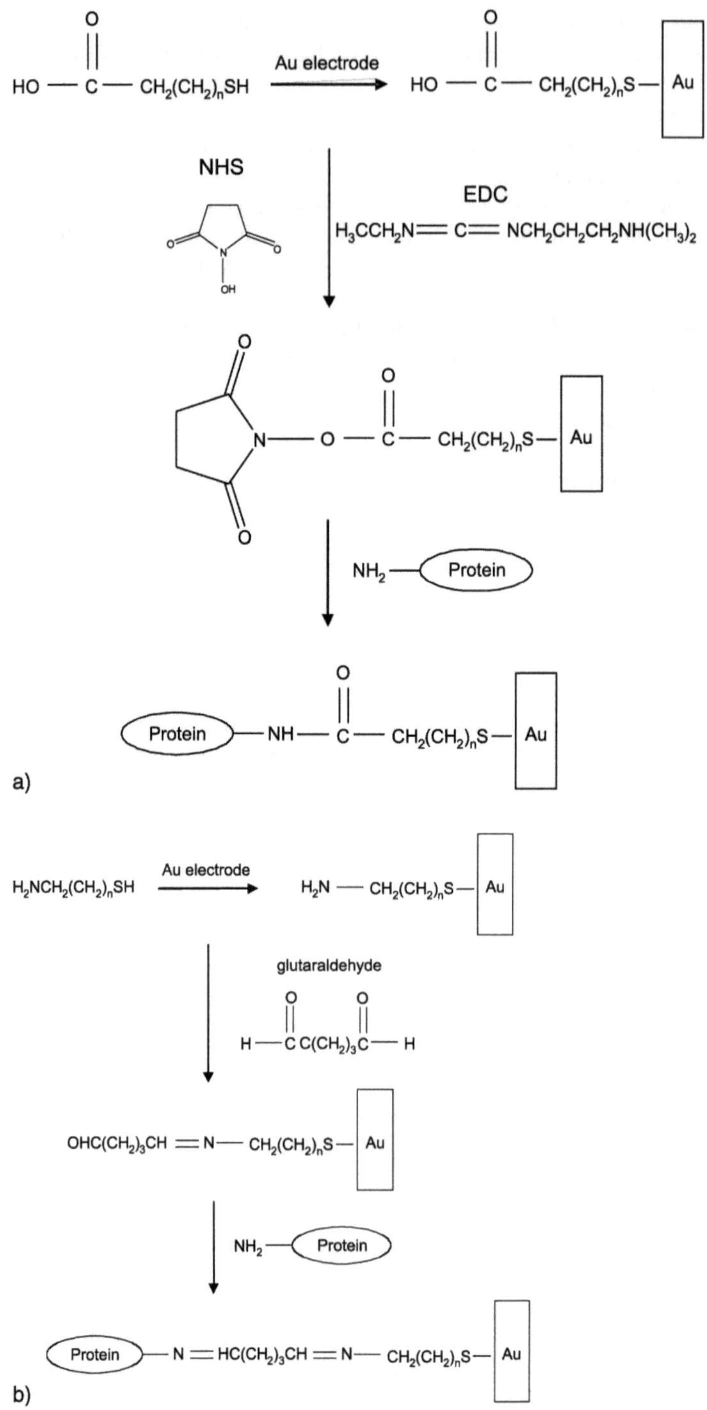

Fig. 26.14. Protein immobilization on previously SAMs-functionalized substrates by covalent attachment

The formation of SAMs with biotin molecules on gold substrate can be realized either by covalently coupling biotin to previously chemisorbed ω-functionalized thiolates [74–77], or by directly self-assembling of biotinylated thiols or disulfides [78, 79]. Usually, in order to optimize the binding properties of the monolayer, accordingly to reduce the nonspecific interaction between streptavidin and the surface, SAMs of biotinylated thiol are diluted by hydroxyalkanethiol with a spacer segment [73, 80–83].

However, these methods suffer from the complicated chemical treatment or organic synthesis of biotinylated thiol, which is almost not commercially available. As a basis of such molecular building a mixed SAMs of MHDA and biotinyl-PE, commercially available, can be chosen. The mixed SAMs, in which the biotinyl-PE is inserted and bound to MHDA by the electrostatic interaction, were simply prepared by immersing the pretreated gold substrates into the mixed solution of MHDA and Biotinyl-PE in ethanol. The ratio between two components determines the density of specific biotinyl-anchoring sites.

The biotin-containing SAMs are a desirable platform for construction of self-assembled multilayer of protein on biosensors, in particular, immunosensors. The structure of the system is shown in Fig. 26.15: the mixed SAMs containing biotinyl sites, streptavidin, biotinylated antibody and antigen in the way of a stepwise assembled system.

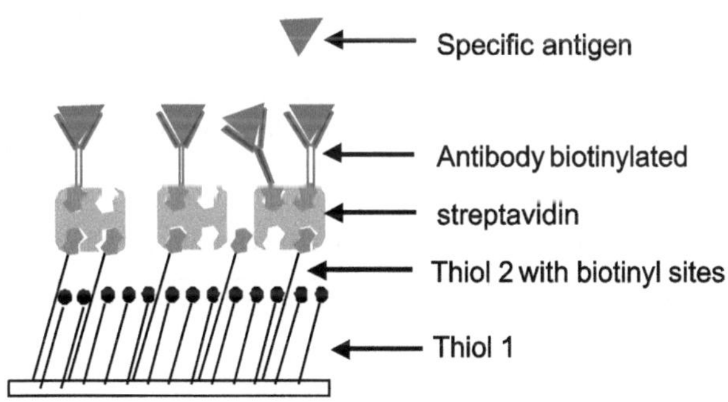

Fig. 26.15. Self-assembled multilayer structure based on biotin/avdin affinity interaction

26.2.2.6 Application of SAMs to Biosensor Elaboration

As discussed above, SAMs are effective methods for protein immobilization and they have many important advantages for the construction of biosensors, as listed in the following:

- Easy formation of ordered, pinhole-free and stable monolayers;
- Flexibility to design the head group of SAMs with various functional groups in order to obtain hydrophobic or hydrophilic surfaces as the requirement;
- Only a minimum amount of biomolecule is needed for immobilization on theSAMs;

– Reasonable stability for extended periods, allowing several reliable measurements and reproducible immobilization of the recognition molecule in a favorable microenvironment;
– Molecular level control of the interfacial chemistry can be exploited to enable ready approach of the target analyte to the interface with the limitation of the access of interfering substances.

Up to now, SAMs have been used as a major immobilization technique for elaboration of all kinds of biosensors, such as enzyme biosensors, immunosensors, DNA-based biosensors and so on. With the development of electronics and nanotechnology, the size of electrodes is gradually being minimized to the micro- and even the nanometer scale. SAMs can be expected to become an increasingly prevalent and critical tool due to their clear advantages.

26.2.3 Characterization of SAMs and LB Films

To obtain accurate information on the structure of thin organic films such as SAMs and LB films, it is essential to use a variety of characterization techniques that are able to determine certain properties of the layer. The most frequently used techniques are divided into five groups and listed in Table 26.5.

Contact angle can be used to examine the general hydrophobicity or hydrophilicity of a surface, it is a simple and effective method to obtain a first impression about the structure and composition of the layer.

A quartz crystal microbalance can detect mass change on a scale of several nanograms. It can be used to study the kinetics of layer adsorption of SAMs and LB films.

Table 26.5. Analytical techniques for characterizing the ultrathin films [84]

Analytical techniques	Structural information
General	
Contact angle	Hydrophobicity, order
QCM	Change in mass
Optical	
IR spectroscopy/FTIR	Functional groups, molecular orientation
Fluorescence spectroscopy	Density of adsorbates
Ellipsometry	Layer thickness
SPR	Layer thickness
Vacuum	
XPS	Elemental composition
Microscopy	
AFM	Molecular packing
STM	Molecular packing
Electrochemical	
Cyclic voltammetry	Thickness, order/defects, electrical properties
Impedance spectroscopy	Thickness, order/defects, electrical properties

Infrared (IR) spectroscopy is an important method for characterization of SAMs and LB films. The spectra can provide not only the usual structural data, but also information about the average orientation of the molecules in the thin layer.

Fluorescence spectroscopy can be used to observe the packing of SAMs and LB films on the substrates and it can provide information on the density of adsorbates.

Ellipsometry is one of the most common optical techniques to estimate the thickness of the thin organic films. It is based on a plane-polarized laser beam that is reflected by the substrate, which results in a change of the phase and amplitude of the light. Then, by comparing these two sets of parameters for an uncovered and covered substrate, the thickness of the layer can be calculated.

Surface plasmon resonance is based on the angle-dependent reflection of a p-polarized laser beam at thin metal layers and it can give important information on thickness of the layer. It is sensitive to the variation of the mean thickness and refractive index of the SAMs layer.

X-ray photoelectron spectroscopy is a useful technique for quantitative analysis of surface composition. It provides information on the chemical state of the constituents, but gives very little, or no, lateral resolution. XPS has been used to probe the chemical nature of the SAMs. Briefly, incident X-rays bombard the sample, and electrons are ejected from the core shells of the atoms within the SAMs. These electrons are collected in an analyzer; by measuring the energies of the electrons entering the analyzer, the binding energies are calculated. These are specific to each element and give indications of the oxidation states of the elements.

Atomic force microscopy and scanning tunneling microscopy (STM) allow characterization of the organic thin films at a molecular resolution. The generated images are the result of the forces acting between the tip and the surface or the surface electron density for AFM and STM, respectively.

The electrochemical characterization of SAMs on gold is usually performed by cyclic voltammetry (CV) and electrochemical impedance spectroscopy. For CV, large potential sweeps are applied to the gold electrode and the resulting current is recorded. However, for EIS only very small sinusoidal potential sweeps with frequencies varying from 100 mHz to 10 kHz are applied to the gold electrode, which makes it a less disturbing technique than CV. Associated with the use of a redox probe CV curves allow evaluation of the electrical blocking character of the layer. Furthermore impedance spectroscopy allows probing of the biomolecular interactions occurring on the bioengineered layers and can serve as a basis to develop impedimetric biosensors.

26.2.3.1
AFM Technique: A Tool for Characterization of Biofunctionalized Surfaces

The AFM technique has become increasingly a reference technique for the characterization of biofunctionalized surfaces at the nanometer level. In order to increase the resolution obtained with AFM measurements when imaging biological samples, the knowledge and control of the interactions between the tip and the sample is a crucial factor. The tip surface can be functionalized with chemicals in order to lead to lower adhesion force interactions. Silicon tip functionalization with a highly hydrophobic compound (organosilane: perfluorosilane) allows the imaging of cells (Fig. 26.16) and proteins (Fig. 26.17), in contact mode, in water, with quite a good resolution.

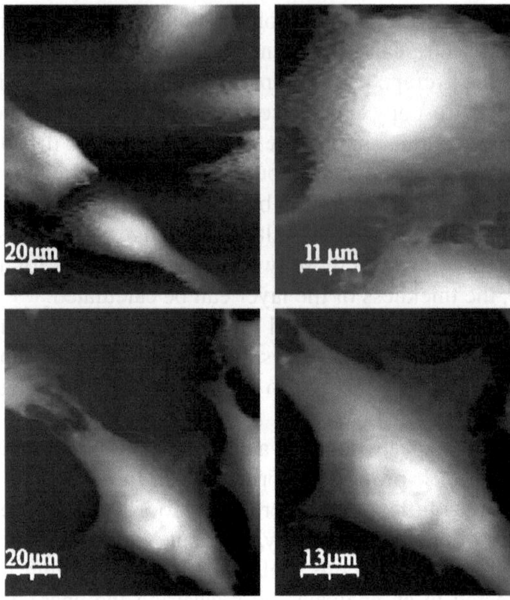

Fig. 26.16. Cell images obtained with Pico Plus II (MI) AFM system equipped with a silicon tip (0.9 N/m) in contact mode in water. At the *top*: Standard Si tip, high friction forces, unstable image. At the *bottom*: Silicon tip fucntionalized with perfluorosilane, low friction forces, stable and well-defined image

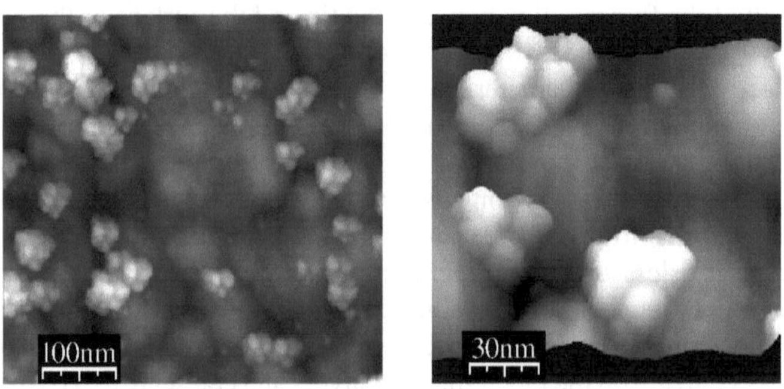

Fig. 26.17. AFM images of a gold surface functionalized with a mixed biotinylated self-assembled monolayer and an antihaemoglobin antibody, after haemoglobin recognition. Pico Plus II (MI) AFM system with a silicon tip functionalized with perfluorosilane, in contact mode (avoid any protein motion in contact mode)

26.2.3.2 Examples of Characterization of Nanostructured Biomolecules by AFM

26.2.3.2.1 Characterization of LB Films for Odorant Biosensors Applications

This example deals with the preparation of ultrathin films in which odorant-binding proteins (Rat OBP-1) were incorporated by the combination of LB films and SAMs, with the aim of drawing up a preliminary design for an artificial olfactory system to detect specific odorant molecules. Under optimized experimental conditions (phosphate buffer solution, pH 7.5, OBP-1F concentration of 4 $mg L^{-1}$, target pressure 35 $mN m^{-1}$), the mixed monolayer (OBP-1F/ODA LB film) at the air/water interface is very stable and has been efficiently transferred onto gold supports, which were previously functionalized by self-assembled monolayers with 1-octadecanethiol [87].

Topographic images (Fig. 26.18a,b) of the mixed OBP-1F/ODA LB films remain almost unchanged within the resolution of the imaging conditions of the experiments after exposure to a small, hydrophobic odorant molecule, isoamyl acetate. The white spots on the topographic images are probably due to dust or salts remaining after the rinsing process.

However, a strong change in the AFM phase images was induced by the exposure to the odorant (Fig. 26.18a′,b′), the sample surface is almost uniform (variations of $\sim 1°$ maximum are observed), there is no phase contrast between the oval domains and the rest. This proves that the LB films obtained are well ordered and uniform, and OBP-1F aggregates are well covered with ODA molecules. After exposure to the odorant, a sharp contrast was observed between the aggregates and the other sample constituents, with phase variations of up to $\sim 9°$. Because phase atomic force microscopy images by recording the phase lag between the signal that drives the cantilever and the cantilever oscillation output signal, it provides information about

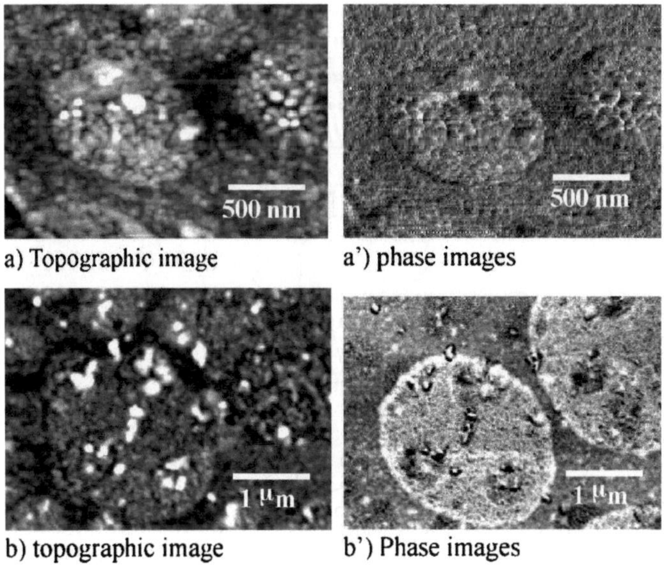

Fig. 26.18. AFM topographic (**a**) and phase (**b**) images of two layers of mixed OBP-1F/ODA LB film transferred onto functionalized gold substrate at 35 $mN m^{-1}$ before exposure to odorant. Scan area $2.1 \times 2.1\ \mu m^2$. Vertical scale (**a**) 10 nm, (**b**) 1.3°. AFM topographic (**c**) and phase (**d**) images of a mixed OBP-1F/ODA LB film transferred onto functionalized gold substrate at 35 $mN m^{-1}$ after exposure to odorant. Scan area $4 \times 4\ \mu m^2$. Vertical scale (**c**) 10 nm, (**d**) 9.4°

surface properties such as elasticity, adhesion and friction, and it can be concluded that one or all of these mechanical properties in the aggregates were modified.

Small odorant molecules can be assumed to penetrate the outer lipid layer to reach the OBP-1F aggregates. Interaction between the odorant molecules and OBP-1F can induce a conformational change of the protein, as observed by Nespoulous et al. [88] and consequently a stronger interaction between the hydrophilic part of OBP-1F and ODA polar heads, leading to a partial breakdown of LB films.

26.2.3.2.2
Characterization of Rhodopsin Immobilized on a Self-Assembled Multilayer

Rhodopsin, the G protein-coupled receptor that mediates the sense of vision, was immobilized onto gold electrode by two different techniques: Langmuir–Blodgett and a strategy based on a self-assembled multilayer. It was observed that LB films of rhodopsin are not stable on gold electrode. Thus a new protein multilayer was prepared on gold electrode by building up layer-by-layer a self-assembled multilayer. It is composed of a mixed self-assembled monolayer, as described in Sect. 26.2.5.2.4, formed by MHDA and biotinyl-PE, followed by a biotin-avidin system that allows binding of biotinylated antibody specific to rhodopsin. Figure 26.19 shows the schematic diagram for self-assembled multilayer formation on gold electrode.

In order to characterize the formation of the multilayer film and to obtain information on its architecture, AFM measurements were done in tapping mode. After the

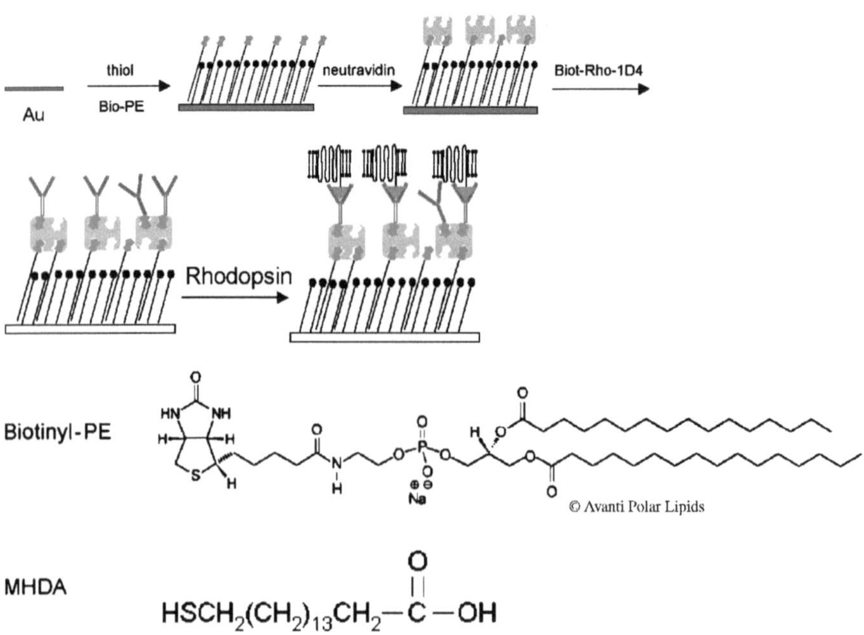

Fig. 26.19. Schematic diagram for immobilization of rhodopsin on gold, using self-assembled multilayer formation via biotin/avidin pairs

rhodopsin membrane fraction was immobilized on the self-assembled multilayers, an AFM image was taken, as presented in the image in Fig. 26.20, which reveals randomly located grainy features that range in size from 30–100 nm and are about 5 nm high. These sizes are consistent with those observed by negative staining of the rhodopsin membrane fraction preparation using electron microscopy [89]. In addition, the thickness of the rhodopsin-containing layer is approximately equivalent to the thickness of the cell membrane, indicating that membrane vesicles open upon immobilization. Further, during the scanning, we found that on some areas of fraction membrane, the tip interaction with the surface increased, which confirms that the rhodopsin is present as a membrane fraction.

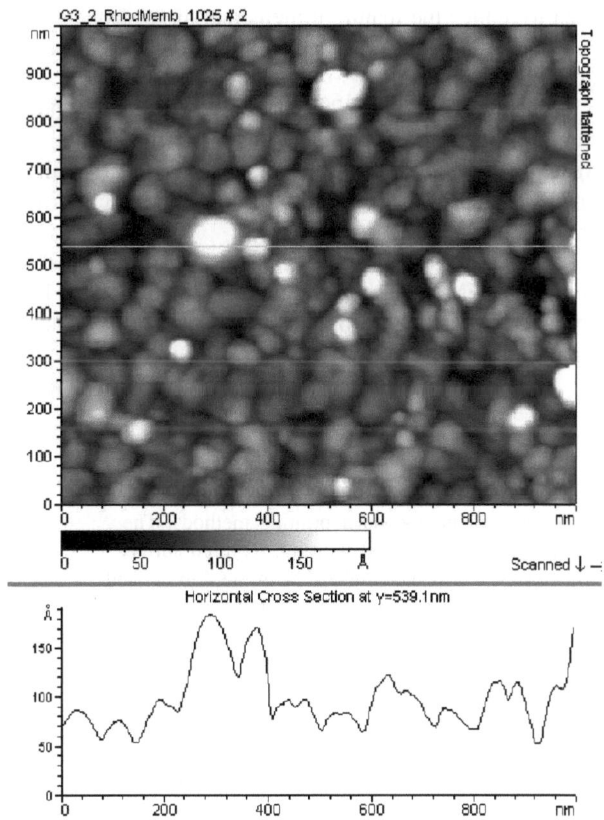

Fig. 26.20. Topographic AFM image (1 μm × 1 μm) obtained in tapping mode: after immobilization of rhodopsin membrane fraction on the self-assembled multilayers

26.3 Prospects and Conclusion

The fast development of microarrays, labs on a chip and other multimicrosensing devices need practical processes for a local molecular engineering of functional layers. Among these techniques microcontact printing appears as a promising and simple way for rapid patterning of molecules. Microcontact printing (μCP) using self-assembled monolayer (SAM) "inks" has been developed for rapid patterning of molecules over large surface areas. μCP is used to pattern gold surfaces using thiols, glass using silanes and so forth [90]. Unfortunately, conventional μCP suffers from problems limiting pattern resolution. These include limits on the structural aspect ratio (defined as the ratio of structure width vs. structure height) and the printing conditions dictated by the elastomeric stamp. The problem of structural collapse could be alleviated if the μCP is carried out in a liquid medium that supports the stamp during printing and coincidentally confines the inking molecules to the area of mutual contact. Xia and Whitesides [91, 92] have published results to show that the spreading of thiols microcontact printed under water can be used to reduce the feature dimensions of subsequently etched gold substrates. The schematic diagram of submersed microcontact printing method (SμCP) is presented in Fig. 26.21.

As shown in Fig. 26.22, it is only possible to pattern gold surface with PDMS stamp (aspect ratio 46:1) (inked using 10 mM alkylthiol solution in ethanol) if the SμCP method is used and not the classical μCP method in air. The SμCP method greatly increases the feasibility of designing microarrays of SAMs on large surface areas.

In conclusion, molecular engineering of functional layers, joined to the imaging power of AFM, has opened new research directions in the field of supramolecular engineering leading to practical applications for chemical and biosensing at the nanolevel.

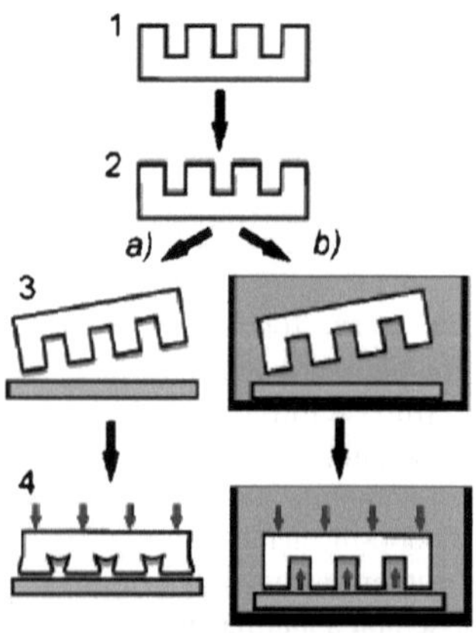

Fig. 26.21. Schematic diagram of (**a**) conventional and (**b**) submersed microcontact printing methods. The PDMS stamp is produced with the required surface structures (*1*). It is then "inked" with the thio (*2*) and placed in contact with the gold substrate (*3*). A small amount of pressure is applied manually (*4*) before the stamp is removed, leaving the patterned SAM on the substrate. In air, the applied pressure causes the stamp to deform; when submersed, the stamp is supported by the water. (From Bessueille F, Pla-Roca M, Mills C, Martinez E, Samitier J, Errachid A (2005) Langmuir 21:12060)

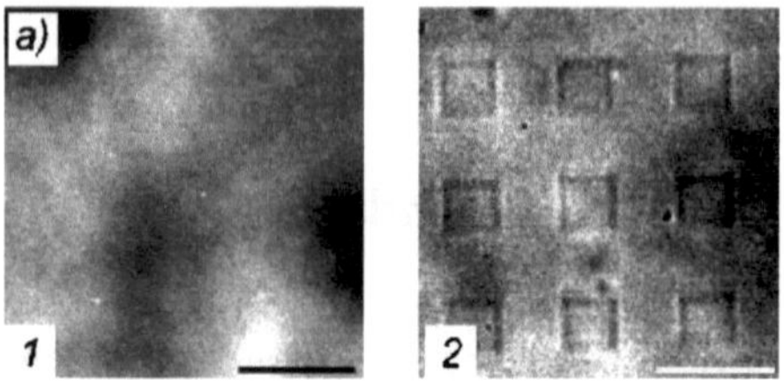

Fig. 26.22. Optical images μ contact printing obtained with stamps with ratio 46:1 (*Bar* = 75 μm) *1*) shows the collapse of the stamp using conventional μCP, 2) shows the stamp under SμCP conditions. (From Bessueille F, Pla-Roca M, Mills C, Martinez E, Samitier J, Errachid A (2005) Langmuir 21:12060)

References

1. Roberts G (1990) Langmuir–Blodgett films, Plenum Press, New York
2. Pockels A (1891) Nature 43:437
3. Pockels A (1892) Nature 46:418
4. Langmuir–Blodgett Troughs-operating manual, 6th Edition, Nima Technology Ltd.:21
5. Ulman A (1991) An Introduction to Ultrathin Organic Films: From Langmuir–Blodgett to Self-Assembly, Academic Press, Boston
6. Harkins WD (1952) The Physical Chemistry of Surface Films, Reinhold, New York
7. Talham DR (2004) Chem Rev 104:5479
8. Fromherz P (1971) Biochim Biophys Acta 225:382
9. Arisawa S, Yamamoto R (1992) Thin Solid Films 210/211:443
10. Zaitsev SY (1993) Colloids Surf A 75:211
11. Zaitsev SY (1995) Sens Actuators B 24–25:177
12. Fiol C, Alexandre S, Delpire N, Valleton JM (1992) Thin Solid Films 215:88
13. Dubreuil N, Alexandre S, Fiol C, Sommer F, Valleton JM (1995) Langmuir 11:2098
14. Fiol C, Alexandre S, Dubreuil N, Valleton JM (1995) Thin Solid Films 261:287–295
15. Dubreuil N, Alexandre S, Fiol C, Valleton JM (1996) J Colloid Interface Sci 181:393
16. Alexandre S, Dubreuil N, Fiol C, Lair D, Duc TM, Valleton JM (1997) Thin Solid Films 293:295
17. Dubreuil N, Alexandre S, Lair D, Sommer F, Duc TM, Valleton JM (1998) Colloids Surf B 11:95
18. Chovelon JM, Provence M, Jafrrezic-Renault N, Derue V, Lair D, Alexandre S, Valleton JM (2001) J Biol Phys Chem 1:68
19. Alexandre S, Lafontaine C, Valleton JM (2001) J Biol Phys Chem 1:21
20. Alexandre S, Derue V, Valleton JM, Sommer F, Duc TM (2002) Colloids Surf B 23:183
21. Zhang A, Jaffrezic-Renault N, Wan K, Hou Y, Chovelon JM (2002) Mater Sci Eng C 21:91
22. Chovelon JM, Wan K, Jaffrezic-Renault N (2000) Langmuir 16:6223
23. Chovelon JM, Gaillard F, Wan K, Jaffrezic-Renault N (2000) Langmuir 16:6228
24. Girard-Egrot AP, Morelis RM, Coulet PR (1997) Thin Solid Films 292:282
25. Girard-Egrot AP, Morelis RM, Coulet PR (1997) Langmuir 13:6540
26. Okahata Y, Tsuruta T, Ijiro K, Ariga K (1998) Langmuir 4:1373
27. Zaitsev SY (1993) Colloids Surf A 75:211
28. Fiol C, Chovelon JM, Dubreuil N, Barbey G, Barraud A, Ruaudel-Teixier A (1992) Thin Solid Films 210/211:489
29. Tatsuma T, Tsuzuki H, Okawa Y, Yoshida S, Watanabe T (1991) Thin Solid Films 202:145
30. Wan K, Chovelon JM, Jaffrezic-Renault N (2000) Talanta 52:663
31. Choi JW, Kim YK, Lee IH, Min J, Lee WH (2001) Biosens Bioelectron 16:937
32. Zhang A, Hou Y, Jaffrezic-Renault N, Wan J, Soldatkin AP, Chovelon JM (2002) Bioelectrochemistry 56:157–158
33. Singhal R, Gambhir A, Pandey MK, Annapoorni S, Malhotra BD (2002) Biosens Bioelectron 17:697
34. Hou Y, Jaffrezic-Renault N, Zhang A, Wan J, Errachid A, Chovelon JM (2002) Sens Actuators B 86:143
35. Sharma SK, Singhal R, Malhotra BD, Sehgal N, Kumar A (2004) Electrochim Acta 49:2479
36. Sharma SK, Singhal R, Malhotra BD, Sehgal N, Kumar A (2004) Biosens Bioelectron 20:651
37. Vikholm I, Teleman O (1994) J Colloid Interface Sci 168:125
38. Barraud A, Perrot H, Billard V, Martelet C, Therasse J (1993) Biosens Bioelectron 8:39

39. Katsube T, Yaji T, Suzuki K, Kobayashi S, Kawaguchi T, Shiro T (1998) Appl Surf Sci 33–34:1332
40. Dubrovsky TB, Demcheva MV, Savitsky AP, Yu. Mantrova E, Yaropolov AI, Savransky VV, Belovolova LV (1993) Biosens Bioelectron 8:377
41. Chudinova GK, Chudinov AV, Savransky VV, Prokhorov AM (1997) Thin Solid Films 307:294
42. Lepesheva GI, Azeva TN, Knyukshto VN, Chashchin VL, Usanov SA (2000) Sens Actuators B 68:27
43. Oh BK, Chun BS, Park KW, Lee W, Lee WH, Choi JW (2004) Mater Sci Eng 24:65
44. Nicolini C, Adami M, Dubrovsky MT, Erokhin, Facci VP, Paschkevitsch P, Sartore M (1995) Sens Actuators B 24:121
45. Sukhorukov GB, Montrel MM, Petrov AI, Shabarchina LI, Sukhorukov BI (1996) Biosens Bioelectron 11:913
46. Smith RK, Lewis PA, Weiss PS (2004) Prog Surf Sci 75:1
47. Bain CD, Troughton EB, Tao YT, Evall J, Whitesides GM, Nuzzo RG (1989) J Am Chem Soc 111:321
48. Yang Z, Gonzalez-Cortes A, Jourquin G, Vire JC, Kauffmann JM (1005) Biosens Bioelectron 10:789
49. Strong L, Whitesides GM (1998) Langmuir 4:546
50. Evans SD, Sharma R, Ulman A (1991) Langmuir 7:156
51. Finklea HO, Snider DA, Fedyak J (1990) Langmuir 6:371
52. Samant MG, Brown CA, Gordon JG (1991) Langmuir 7:437
53. Ron H, Rubinstein I (1994) Langmuir 10:4566
54. Ron H, Matlis S, Rubinstein I (1998) Langmuir 14:1116
55. Majid A, Bensebaa F, L'Ecuyer P, Pleizier G, Deslandes Y (2003) Rev Adv Mater Sci 4:25
56. Poirier GE, Pylant ED (1996) Science 272:1145
57. Chailapakul O, Sun L, Xu C, Crooks RM (1993) J Am Chem Soc 115:12459
58. Xia Y, Whitesides GM (1998) Annu Rev Mater Sci 28:153
59. Pan W, Durning CJ, Turro NJ (1996) Langmuir 12:4469
60. Ulman A (1996) Chem Rev 96:1533
61. Richter AG, Yu CJ, Datta A, Kmetko J, Dutta P (2000) Phys Rev E 61:607
62. Richter AG, Yu CJ, Datta A, Kmetko J, Dutta P (2000) Materials Res CAT Sector 10:136–139
63. Sandhyarani N, Pradeep T (2003) Int Rev Phys Chem 22:221
64. Gooding JJ, Hibbert DB (1999) Trends Anal Chem 18:523
65. Creager SE, Olsen KG (1995) Anal Chim Acta 307:277
66. Chaki NK, Vijayamohanan K (2002) Biosens Bioelectron 17:1
67. Mirsky VM, Riepl M, Wolfbeis OS (1997) Biosens Bioelectron 12:977
68. Gooding JJ, Mearns F, Yang W, Liu J (2003) Electroanalysis 15:81
69. Kenny JW, Fanning TG, Lambert JM, Traut RR (1979) J Mol Biol 135:151
70. Carlsson J, Drevin H, Axen R (1978) Biochem J 173:23
71. Ferretti S, Paynter S, Russell DA, Sapsford KE, Richardson DJ (2000) Trends Anal Chem 19:530
72. Dong X, Lu J, Cha C (1997) Bioelectrochem Bioenerg 42:63
73. Spinke J, Liley M, Schmitt FJ, Guder HJ, Angermaier L, Knoll W (1993) J Chem Phys 99:7012
74. Yam C, Pradier CM, Salmain M, Marcus P, Jaouen G (2001) J Colloid Interface Sci 235:183
75. Yam C, Pradier CM, Salmain M, Fischer-Durand N, Jaouen G (2002) J Colloid Interface Sci 245:204
76. Lasseter TL, Cai W, Hamers RJ (2004) Analyst 129:3

77. Cui X, Pei R, Wang Z, Yang F, Ma Y, Dong S, Yang X (2003) Biosens Bioelectron 18:59
78. Häussling L, Ringsdorf LH, Schmitt FJ, Knoll W (1991) Langmuir 7:1837
79. Herron J, Müller W, Paudler M, Riegler H, Ringsdorf H, Suci PA (1992) Langmuir 8:1413
80. Pérez-Luna VH, O' Brien MJ, Opperman KA, Hampton PD, Lopez GP, Klumb LA, Stayton PS (1999) J Am Chem Soc 121:6469
81. Jung LS, Nelson KE, Stayton PS, Campbell CT (2000) Langmuir 16:9421
82. Nelson KE, Gamble L, Jung LS, Boeckl MS, Castner DG, Campbell CT, Stayton PS (2001) Langmuir 17:2807
83. Spinke J, Liley M, Guder HJ, Angermaier L, Knoll W (1993) Langmuir 9:1821
84. Flink S, Frank C, Van Veggel JM, Reinhoudt DN (2000) Adv Mater 12:1315
85. Hou Y, Tlili C, Jaffrezic-Renault N, Zhang A, Martelet C, Ponsonnet L, Errachid A, Samitier J, Baussels J (2004) Biosens Bioelectron 20:1126
86. Tlili C, Hou Y, Hafsa KY, Ponsonnet L, Martelet C, Errachid A, Jaffrezic-Renault N (2005) Sens Lett 2:1
87. Hou Y, Jaffrezic-Renault N, Martelet C, Tlili C, Zhang A, Pernollet JC, Briand L, Gomila G, Errachid A, Samitier J, Salvagnac L, Torbiero B, Temple-Boyer P (2005) Langmuir 21:4058
88. Nespoulous C, Briand L, Delage MM, Tran V, Pernollet JC (2004) Chemical Senses 29:189
89. Minic J, Grosclaude J, Aioun J, Persuy MA, Gorojankina T, Salesse R, Pajot-Augy E, Hou Y, Helali S, Jaffrezic N, Bessueille F, Errachid A, Gomila G, Ruiz O, Samitier J (2005) Biochim Biophys Acta 1724:324
90. Xia Y, Whitesides GM (1998) Angew Chem Int Ed 37:550
91. Xia Y, Whitesides GM (1997) Langmuir 13:2059
92. Xia Y, Whitesides GM (1995) J Am Chem Soc 117:3274

Part III
AFM of Biological Membranes, Cells and Tissue

10 Imaging Chemical Groups and Molecular Recognition Sites on Live Cells Using AFM

David Alsteens · Vincent Dupres · Etienne Dague · Claire Verbelen · Guillaume André · Grégory Francius · Yves F. Dufrêne

From volume XII

Abstract. Imaging the nanoscale distribution of specific chemical and biological sites on live cells is an important challenge in current life science research. In addition to imaging the surface topography of live cells, atomic force microscopy (AFM) is increasingly used to probe their chemical groups and biological receptors. In chemical force microscopy, AFM tips are modified with specific functional groups, thereby allowing investigators to probe chemical sites and their interactions on a scale of only ~25 functional groups. In molecular recognition imaging, tips are functionalized with specific biomolecules, or samples labeled with immunogold particles, enabling researchers to localize specific receptors. Clearly, these nanoscale investigations provide new avenues in cellular biology and microbiology for elucidating the structure–function relationships of cell surfaces. In this chapter, we discuss the principles of these AFM modalities and their applications in life science research.

Key words: Atomic force microscopy, Cells, Chemical force microscopy, Molecular recognition studies, Single molecules

10.1 Introduction

Specific molecular recognition interactions between receptors and cognate ligands are ubiquitous in life sciences and increasingly used in nanobiotechnology [1–3]. In addition, nonspecific intermolecular interactions, such as hydrophobic and electrostatic forces, also play essential roles in biology since they promote crucial events like protein folding and cell adhesion [4, 5]. Despite the importance of these noncovalent intermolecular forces, their quantification has long been challenging. Force measuring techniques like the surface forces apparatus (SFA) [6] and the optical and magnetic tweezers [7, 8] have been increasingly used in that respect; yet these approaches are limited by their poor lateral resolution which does not provide access to the spatial distribution of (bio)chemical interactions. In this context, atomic force microscopy (AFM) has recently opened remarkable opportunities. On the one hand, chemical force microscopy (CFM), involving modification of AFM tips with specific functional groups, has enabled researchers to map the spatial arrangement of chemical groups and their interactions on organic surfaces and on cell surfaces. This nanoscale, chemically sensitive imaging tool offers two major advantages over classical probing methods: (1) hydrophobic and charged groups, and their interactions,

In: H. Fuchs, B. Bhushan (eds), *Biosystems - Investigated by Scanning Probe Microscopy*,
DOI 10.1007/978-3-540-85039-7_2,

are measured directly and quantitatively on live cells, and (2) nanoscale variations of hydrophobicity and charge can be resolved. On the other hand, single-molecule force spectroscopy with biologically modified tips provides a means to map individual receptors on cellular surfaces and to measure their molecular recognition forces. As a complement, topographic imaging combined with immunogold labeling has also proved useful for localizing molecular recognition sites. In this chapter, we describe the principle and methodology of chemical force microscopy and molecular recognition imaging, and highlight some of their applications.

10.2 Chemical Force Microscopy

10.2.1 Methods

The general idea of CFM is to use AFM tips with well-defined chemistry for measuring adhesion and friction, and/or for imaging surfaces (for recent reviews [6, 9]). AFM cantilevers and tips are usually made of silicon (Si) or silicon nitride (Si_3N_4) using microfabrication techniques. Because the surface chemistry of such commercial tips is poorly controlled and often contaminated with gold and other materials, reliable CFM measurements require functionalizing the tip surface with organic monolayers terminated by specific functional groups (e.g., OH or CH_3). A common method to achieve this is based on the formation of self-assembled monolayers (SAMs) of alkanethiols on gold surfaces. The procedure involves coating, by thermal evaporation, microfabricated cantilevers with a thin adhesive layer (Cr or Ti), followed by a 15–100-nm thick Au layer, immersing the coated cantilevers in dilute (0.1–1 mM) ethanol solutions of the selected alkanethiol, rinsing with ethanol and drying, using a gentle nitrogen flow. Although the protocol is fairly simple, it is important to validate the quality of the surface modification, which can be done by treating model supports (silicon) in the same way as the tips and characterizing them by means of surface analysis techniques (e.g., contact angle measurements or X-ray photoelectron spectroscopy). Another important point is to use the functionalized tips immediately after they are prepared in order to minimize surface contamination and alteration.

The most common application of CFM is the measurement of the adhesion strength between chemical groups via force spectroscopy. Here, the cantilever deflection is recorded as a function of the vertical displacement of the piezoelectric scanner, i.e. as the sample is pushed towards the tip and retracted. This yields a raw "voltage-displacement" curve which can be converted into a "force-distance" curve as follows. Using the slope of the curves in the region where tip and sample are in contact, the voltage can be converted into a cantilever deflection. In order to minimize the possible effects of repulsive surface forces and/or sample deformation, it is recommended to consider the slope of the retraction curve. The cantilever deflection is then converted into a force (F) using Hooke's law: $F = -k \times d$, where k is the cantilever spring constant. The curve can be corrected by plotting F as a function of $(z - d)$. The zero separation distance is then determined as the position of the vertical linear parts of the curve in the contact region. Note that force mapping can be performed by recording spatially resolved force-distance curves in the (x, y) plane. In adhesion measurements, the hysteresis or "pull-off" force observed during retraction is used to estimate the adhesion (unbinding) force between tip and sample. To get an accurate knowledge of the measured forces, it is important to determine actual spring constants experimentally since they may substantially differ from values quoted by the manufacturer [10].

10.2.2 Probing Hydrophobic Forces

A variety of chemical groups (hydrophobic/hydrophilic; charged/uncharged) and their interactions in different solvents have been investigated using CFM [9, 11–16]. Recently, Alsteens et al. [17–19] validated the use of the technique for probing hydrophobic groups and for quantifying short-range hydrophobic forces. Although hydrophobic forces have been known for 70 years and are of prime importance in biology (protein folding and aggregation, membrane fusion and cell adhesion), their detailed mechanisms are not fully understood. Various effects have been proposed to explain these forces, such as entropic effects due to disruption of the arrangement of water near hydrophobic surfaces (solvent organization), separation-induced phase transition (cavitation), hydrodynamic fluctuating correlation, bridging of submicron bubbles, electrostatic effects, correlated charge fluctuations or correlated dipole interactions.

As a proof of concept, SAMs of CH3- and OH-terminated alkanethiols mixed in different proportions were probed using water contact angle measurements and CFM with hydrophobic, CH3-terminated tips (Fig. 10.1). Consistent with the expectation (work of adhesion, Young equation), the contact angle and adhesion force values measured on mixed SAMs increased gradually with the molar fraction of CH3-terminated alkanethiols (Fig. 10.1a, b), yielding a linear relationship between the adhesion force and the cosine of the contact angle (Fig. 10.1c). This excellent agreement demonstrates that the measured adhesion forces reflect surface hydrophobicity, thus hydrophobic forces.

Notably, as we shall see below, interpretation of the data in terms of interfacial thermodynamics reveals that the measured adhesion forces do not originate from true, direct tip–sample interactions, but may rather reflect entropy changes associated with the restructuration of water near hydrophobic surfaces. We start by considering the Johnson–Kendall–Roberts (JKR) model which links the adhesion force (F_{adh}) to the work of adhesion (W_{adh}) [12]:

$$F_{adh} = 1.5\pi R W_{adh} \tag{10.1}$$

where R is the radius of curvature of the AFM tip. The work of adhesion in water is given by [20]:

$$\begin{aligned} W_{adh} &= \gamma_{sample,water} + \gamma_{tip,water} - \gamma_{tip,sample} \\ &= W_{tip,sample} - W_{sample,water} - W_{tip,water} + 2\gamma_{water} \end{aligned} \tag{10.2}$$

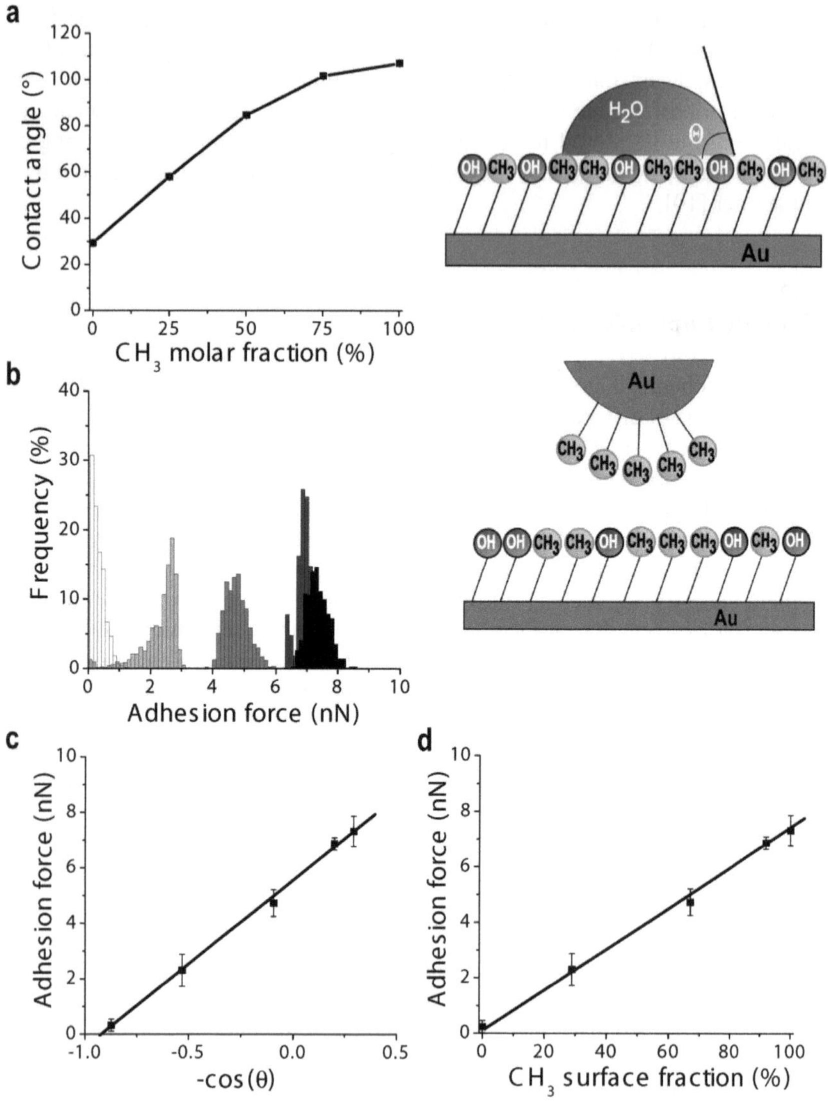

Fig. 10.1. Chemical force microscopy (CFM): principle and application to the probing of hydrophobic forces. (**a**) Water contact angle (θ) values measured for mixed self-assembled monolayers of CH3- and OH-terminated alkanethiols as a function of the molar fraction of CH3-terminated alkanethiols. (**b**) Histograms of adhesion forces measured on the mixed SAMs using CFM with hydrophobic CH3-tips. (**c**) Variation of adhesion forces as a function of the cosine of the water contact angle. (**d**) Adhesion force as a function of the surface fraction of CH3-terminated alkanethiols computed using Cassie's law. Reprinted with permission from [17]

where $\gamma_{X,Y}$ is the interfacial energy, $W_{X,Y}$ is the work of adhesion in a vacuum, and γ_X is the surface energy. The term $W_{sample,water}$ can be deduced from the water contact angle θ, using the Young equation:

$$\begin{aligned} W_{sample,water} &= \gamma_{sample} + \gamma_{water} - \gamma_{sample,water} \\ &= \gamma_{water}(1 + \cos\theta) \end{aligned} \tag{10.3}$$

Combining Eqs. (10.1), (10.2), and (10.3), yields the following expression:

$$\frac{F_{adh}}{1.5\pi R} = W_{tip,sample} - W_{tip,water} - \gamma_{water}\cos\theta + \gamma_{water} \tag{10.4}$$

Assuming that the $W_{tip,sample}$ and $W_{tip,water}$ values are close, which appears reasonable since the CH_3-modified tip is involved only in dispersion interactions (London forces) (10.4) becomes:

$$\frac{F_{adh}}{1.5\pi R} = \gamma_{water} - \gamma_{water}\cos\theta \tag{10.5}$$

Considering the surface tension of water (72.6 mN/m) and the tip radius (~20 nm) then leads to the following linear equation $F_{adh} = 6.84–6.84\cos\theta$, which agrees remarkably well with the regression equation of Fig. 10.1c, i.e.: $F_{adh} = 5.60–6.08\cos\theta$, where the uncertainties on slope and intercept are 0.14 and 0.10 nN, respectively. The agreement between experimental and theoretical adhesion forces supports the hypothesis that $W_{tip,sample}$ and $W_{tip,water}$ are similar, thus that only dispersion forces are responsible for the measured forces.

Accordingly, this reasoning indicates that the measured adhesion forces do not originate from true, direct tip–sample interactions, but may rather reflect entropy changes associated with the restructuration of water near hydrophobic surfaces. This work is also of practical interest since it shows that CFM can be used for quality control of chemically modified tips, which is not feasible by common analytical techniques. As we shall discuss below, further interpretation of cellular data implied expressing the measured adhesion force as a function of the surface fraction of CH_3-terminated alkanethiols, determined using Cassie's law and contact angle values (Fig. 10.1d).

In earlier work, AFM was also used to probe charged groups and their electrostatic forces [12, 13, 21]. For instance, force-distance curves were recorded at different pH values between tips and surfaces modified with carboxylic groups [22]. The curves showed differences according to pH (Fig. 10.2): while no adhesion was observed at pH 10, significant adhesion forces, of 1.4-nN magnitude, were measured at pH 3. Curves recorded in the 3–10 pH range showed adhesion forces for pH values smaller than 5. In addition, at pH 10 a long range repulsion force was seen starting at 40 nm, while a jump to contact was noted at pH 3. The repulsion force at high pH was attributed to electrostatic double layer forces between negatively charged COO^-/COO^- surfaces. The jump to contact at low pH was due to attractive van der Waals forces and indicated a lack of repulsion between uncharged COOH/COOH surfaces. In the same way, the strong adhesion observed at low pH values could be attributed to hydrogen bonding between COOH/COOH groups while the lack of adhesion at high pH was suggested to reflect the electrostatic repulsion

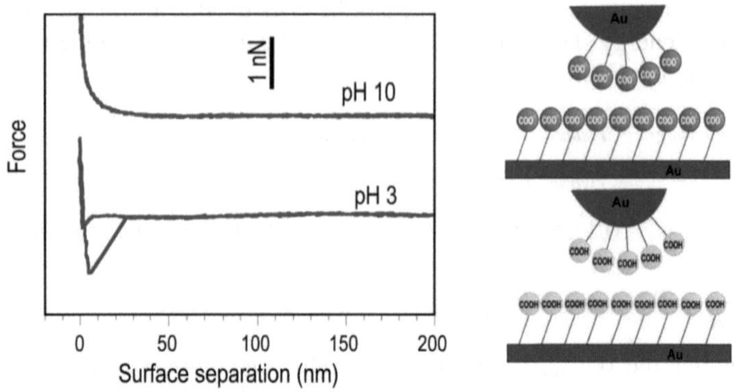

Fig. 10.2. AFM force-distance curves recorded at pH 10 and pH 3 between modified tips and model surfaces terminated with COOH groups. As shown in the drawing the differences observed according to pH can be related to a change of ionization state of the surfaces. Adapted with permission from [22]

between the negatively charged COO^-/COO^- surfaces. Hence, the above studies show that functionalized tips are very sensitive to hydrophobic and charged groups.

10.2.3 Chemical Force Microscopy of Live Cells

For the first time, we showed that the CFM technique can probe chemical groups and their interactions on live cells on a scale of only ~25 functional groups [17–19, 23]. Dague et al. [19] used CFM with methyl-terminated tips to measure the hydrophobicity of the human opportunistic pathogen *Aspergillus fumigatus*. Topographic images revealed the presence of regularly arranged rodlets on *A. fumigatus* conidia (Fig. 10.3a). These structures are composed of hydrophobins, a family of small, moderately hydrophobic proteins that favor spore dispersion by air currents and mediate adherence to host cells. Force curves recorded across these surfaces with a hydrophobic tip showed large adhesion forces, of ~3,000 pN magnitude (Fig. 10.3b, c). Comparison with the data obtained on reference surfaces (Fig. 10.1d) indicated that the conidial surface has a marked hydrophobic character, corresponding to a surface composed of ~10 CH_3 and ~15 OH groups, which is fully consistent with the presence of an outermost surface layer of hydrophobins and provides direct indications as to their putative functions as dispersion and adherence structures. In agreement with the uniform surface structure, adhesion maps were rather homogeneous (Fig. 10.3b), supporting the idea that the *A. fumigatus* conidial surface is homogeneously hydrophobic.

Two hydrophilic controls confirmed that the measured hydrophobic properties are associated with hydrophobins [19]. First, topographic images recorded on conidia treated with NaOH, a procedure that removes all cell wall proteins including rodlet proteins, revealed a surface lacking any rodlet, but exposing underlying cell wall polysaccharides (Fig. 10.3d, e, f). These structural changes were correlated with profound decrease of cell surface hydrophobicity, the adhesion force towards the hydrophobic tip being only ~300 pN. Second, similar results were obtained on the

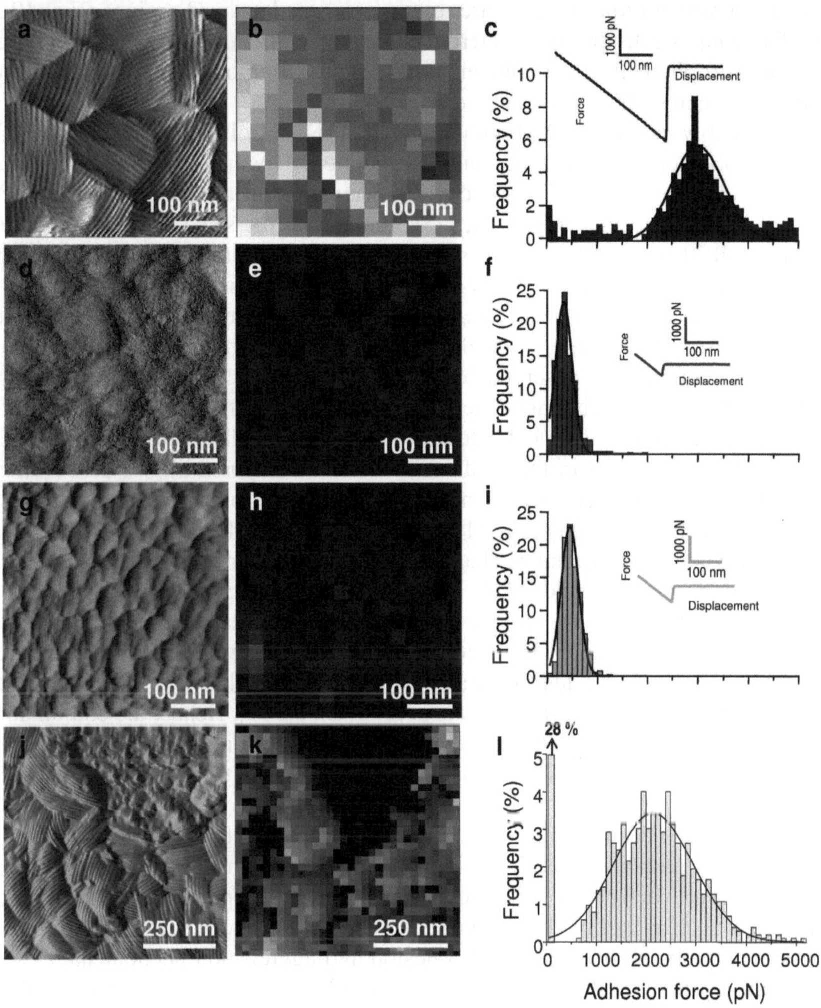

Fig. 10.3. CFM of *Aspergillus fumigatus*. (**a**) High-resolution image of a wild-type conidial surface in aqueous solution revealing rodlets. (**b**) Adhesion force map (z-range: 6 nN) and (**c**) adhesion force histogram ($n = 512$) recorded with a hydrophobic tip, indicating that the rodlet surface is uniformly hydrophobic. (**d, e, f, g, h, i**) High-resolution images, adhesion force maps and histograms ($n = 512$) obtained on NaOH-treated conidia (**d, e, f**) and on the ΔrodA ΔrodB double mutant (**g, h, i**). The purely hydrophilic surface is attributed to cell wall polysaccharides. (**j, k, l**) High-resolution image, adhesion force map and histogram obtained on SDS-treated conidia, revealing highly correlated structural and hydrophobic heterogeneities. Reprinted with permission from [19]

ΔrodA ΔrodB double mutant which does not produce rodlets (Fig. 10.3g, h, i). Considering the above reference SAMs surfaces (Fig. 10.1d), it was concluded that the two conidial surfaces are purely hydrophilic, corresponding to purely OH-terminated surfaces, which agrees well with the exposure of cell wall carbohydrates. Notably, nanoscale variations of hydrophobic properties were also resolved on SDS-treated conidia, for which rodlet patches are missing in very localized regions (Fig. 10.3j, k, l). These nanoscale structural heterogeneities were directly correlated with differences in hydrophobicity, the rodlet and polysaccharide regions displaying contrasted hydrophobic and hydrophilic characters.

Remarkably, real-time CFM with a temperature-controlled stage enables researchers to probe not only structural, but also chemical dynamics on cells. In one such study, the changes of *A. fumigatus* conidia occurring during germination at 37 °C were tracked in real time (Fig. 10.4, [23]). Images of the same spore after 20 min, 60 min, and 120 min clearly revealed significant structural alterations, the rodlet layer changing into a layer of amorphous material, presumably reflecting the underlying polysaccharides. Consistent with this, adhesion maps with hydrophobic tips revealed a dramatic loss of hydrophobicity with time. After 2-h of germination, heterogeneous hydrophobic contrast was observed, reflecting the coexistence of hydrophobic rodlets and hydrophilic polysaccharides. These data nicely demonstrate that CFM is capable of resolving submicron chemical heterogeneities on live cells as they grow.

Of particular interest in the medical context, is the possibility to probe differences of chemical properties and interactions on cells following treatments with drugs. For instance, Alsteens et al. [18] showed that the surface of mycobacteria has a remarkably strong hydrophobic character due to the presence of an outermost layer of hydrophobic mycolic acids. This finding is of biomedical relevance since these hydrophobic constituents are thought to represent an important permeation barrier to common antibacterial agents. By contrast, treatment of the cells with two antibiotics, isoniazid and ethambutol, led to a dramatic decrease of cell surface hydrophobicity, attributed to the removal of the mycolic acid layer [18, 19]. Hence, the combination of topographic imaging with CFM provides unique opportunities to gain insight into the action modes of antibiotics as well as into the nanoscale organization of bacterial cell walls.

CFM has also been used to probe charged groups and their interactions on live cells, which is particularly relevant since most microorganisms possess a negative surface charge under physiological conditions due to the presence of anionic surface groups such as carboxyl and phosphate. The resulting cell surface charge plays an important role in controlling cell adhesion and aggregation phenomena, as well as antigen–antibody, cell–virus, cell–drug and cell–ions interactions. An example of this is provided by the work of Ahimou et al. [22] who used AFM tips functionalized with ionizable carboxyl groups to probe the electrostatic properties of *S. cerevisiae*. In Fig. 10.5, one can see that force-distance curves recorded in these conditions were strongly influenced by pH: while no adhesion was measured at neutral/alcaline pH,

Fig. 10.4. Tracking the structural and chemical dynamics of germinating *A. fumigatus* cells. Series of high-resolution deflection images (*left*) and adhesion force maps (*right*) recorded on a single spore during germination. Within less than 3 h, the crystalline rodlet layer changed into a layer of amorphous material, presumably reflecting inner cell wall polysaccharides. After 2 h, both rodlet and amorphous regions were found to coexist (separated by *dashed line*). Consistent with this structural dynamics, substantial reduction of adhesion contrast was noted with time (*right* images), reflecting a dramatic decrease of hydrophobicity. After 2 h, heterogeneous contrast was observed in the form of hydrophobic patches (*dashed line*), surrounded by a hydrophilic sea. Reprinted with permission from [23]

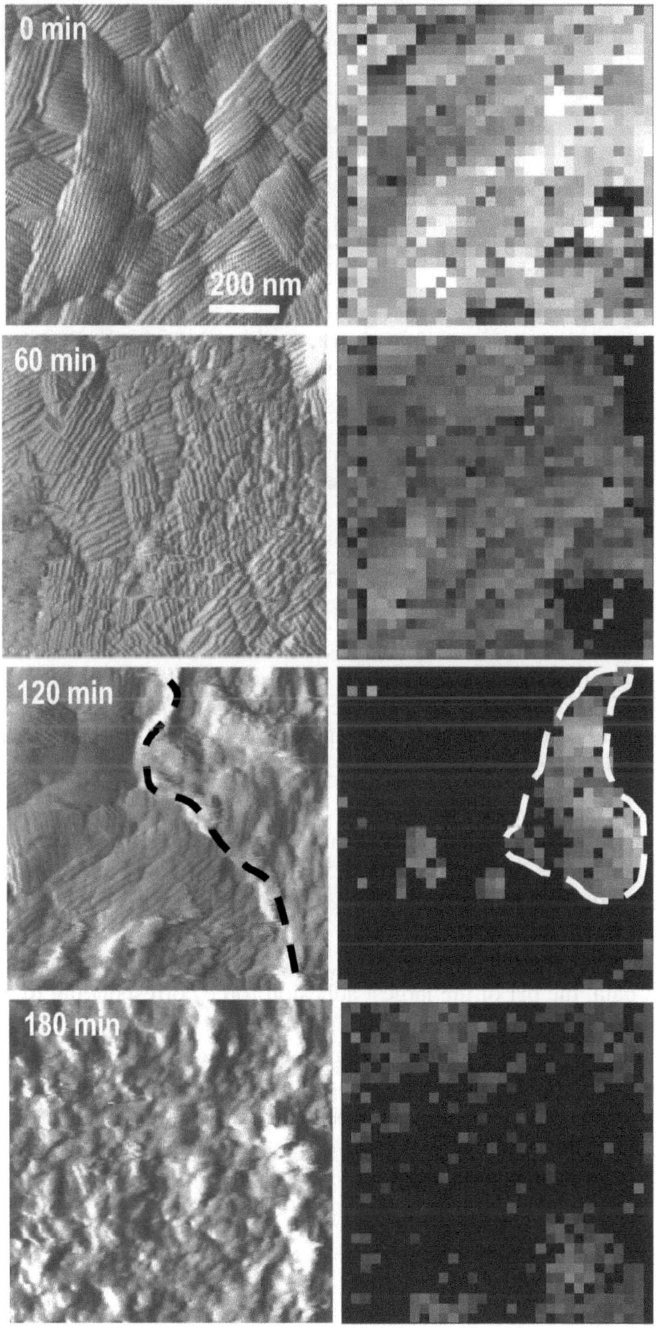

Fig. 10.4.

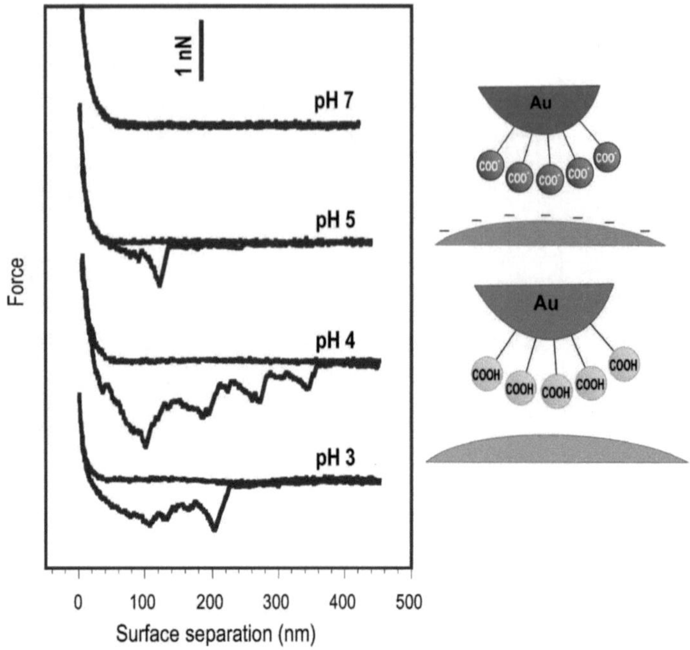

Fig. 10.5. Force-distance curves recorded in solutions of varying pH between the surface of *Saccharomyces cerevisiae* and an AFM tip functionalized with carboxyl groups. The differences in adhesion forces observed with pH were related to a change of ionization state of the cell surface. Adapted with permission from [22]

reflecting electrostatic repulsion between the negatively charged surfaces, multiple adhesion forces were recorded at pH ≤ 5 which we attribute to hydrogen bonding between the protonated tip surface and cell surface macromolecules. These changes were shown to be related to differences in the ionization state of the cell surface functional groups: the adhesion force *vs.* pH curve was correlated with microelectrophoresis data, the pH of the largest adhesion force corresponding to the cell isoelectric point (pH 4).

To summarize, the above data demonstrate that the CFM method allows investigators to detect specific chemical groups and measure their interaction forces on live cells, thereby complementing methods currently available for assessing surface properties. Also, the technique can resolve nanoscale chemical heterogeneities while the cells grow or interact with drugs.

10.3 Molecular Recognition Imaging

Single-molecule force spectroscopy (SMFS) with biologically modified tips enables us to measure molecular recognition interactions at the level of single molecules, providing valuable information on the molecular dynamics within the complexes, and to map individual receptors on surfaces. In parallel, molecular recognition sites may also be detected using topographic imaging combined with immunogold labeling.

10.3.1 Spatially Resolved Force Spectroscopy

So-called "affinity imaging" using spatially resolved SMFS provides unique possibilities for the localization of specific receptors on cell surfaces. The method implies recording multiple force curves between the modified tip and sample, assessing the unbinding force between complementary receptor and ligand molecules from the adhesion "pull-off" force observed upon retraction, and displaying the values as gray pixels [24]. The measured unbinding forces are typically in the 50–400-pN range, depending on the experimental conditions, i.e. number of interacting molecules and the loading rate. The method has been applied to different cell types, including red blood cells [25], osteoclasts [26], and endothelial cells [27].

The power of the approach in microbiology is illustrated in Fig. 10.6 with three key examples. Flocculation (i.e. aggregation) of yeast cells in fermentation technology, such as brewing and wine-making, is mediated by specific lectin–carbohydrate interactions. To measure these interactions on *S. carlsbergensis*, AFM tips were functionalized with the plant lectin concanavalin A (Con A). As can be seen in Fig. 10.6b, the force curves obtained on top of a single yeast cell showed single or multiple unbinding forces of respectively 53 ± 6 pN, 94 ± 12 pN and 145 ± 17 pN magnitude reflecting the specific interaction between one, two, and three lectins and cell surface mannose or glucose residues. In addition, the homogeneous adhesion maps suggested that the distribution of carbohydrate residues was homogeneous.

In the medical context, molecular recognition events mediate the interaction between adhesins on bacterial pathogens and host receptors. A prominent example is *M. tuberculosis* which adheres to heparan sulfates on epithelial cells via the heparin-binding hemagglutinin (HBHA). Spatially resolved SMFS was used to map the distribution of single HBHA on live mycobacteria (Fig. 10.6c, d; [28]). While high-resolution topographic images of the cells revealed a smooth and homogeneous surface, affinity maps recorded with a heparin-modified tip were highly contrasted (Fig. 10.6d), adhesion events (bright pixels) being observed in about half

Fig. 10.6. Molecular recognition imaging of live cells using spatially resolved SMFS. (**a**) AFM image of a *Saccharomyces carlsbergensis* cell showing a homogeneous, smooth morphology. (**b**) Adhesion force map (*gray scale*: 200 pN) recorded with an AFM tip functionalized with the plant lectin concanavalin A, revealing that mannose/glucose residues are homogeneously distributed on the surface. (**c**) AFM topographic image recorded in PBS showing two *M. bovis* BCG cells on a polymer substrate. (**d**) Representative adhesion force map (*gray scale*: 100 pN) recorded with a heparin-modified AFM tip. Adhesion events (*bright pixels*) reflect the detection of single cell adhesion proteins (HBHA), which apparently are concentrated into nanodomains. (**e**) AFM image of *Lactococcus lactis* cells during the course of the division process, showing a well-defined division septum rich in nascent peptidoglycan. (**f**) Adhesion force map (gray scale: 100 pN) recorded with a vancomycin tip on the septum region. Adhesion events were essentially located in the septum region (*dashed line*), suggesting newly formed peptidoglycan is inserted there. Reprinted with permission from [28] and from [29]

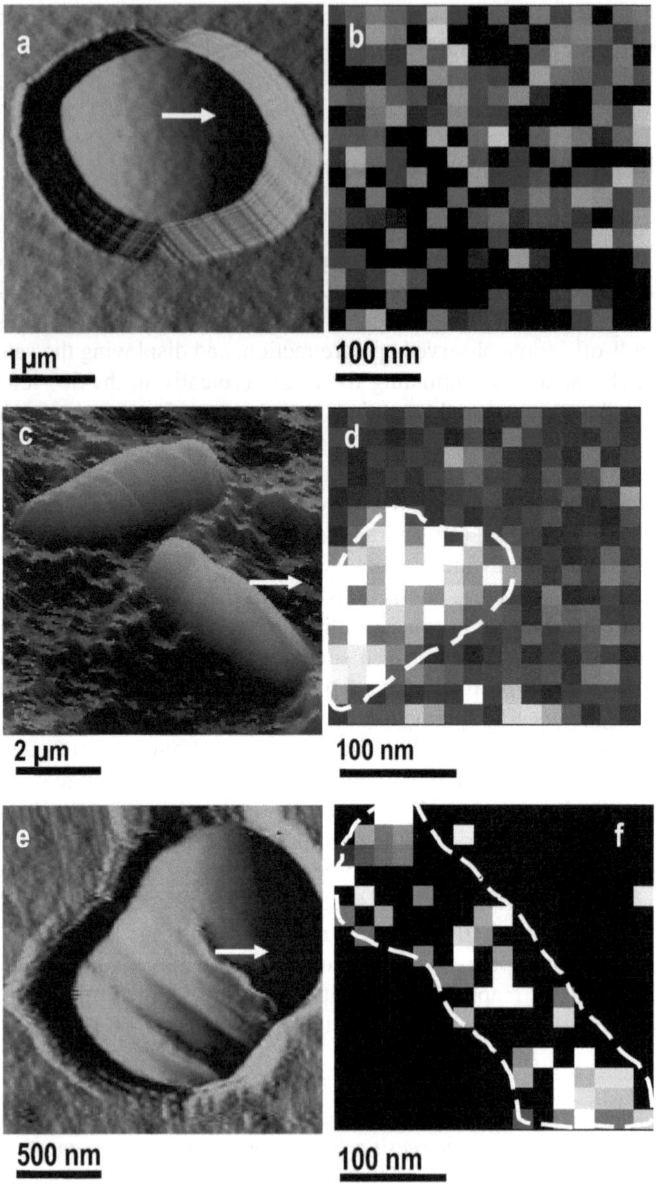

Fig. 10.6.

of the locations. The adhesion force magnitude was very close to the value of single HBHA–heparin interactions, supporting the notion that single adhesins were detected. This was further confirmed by showing that a mutant strain lacking HBHA did not bind the heparin tip. Interestingly, the HBHA distribution was not homogeneous, but apparently concentrated into nanodomains which may promote adhesion to target cells by inducing the recruitment of receptors within membrane rafts. In the future, these molecular recognition studies may help in the development of new drugs capable to block bacterial adhesion.

More recently, spatially resolved SMFS with antibiotic-modified tips was used to map individual binding sites on live bacteria (Fig. 10.6e, f; [29]). Fluorescence microscopy with a fluorescent vancomycin probe was used to visualize D-Ala-D-Ala sites of nascent peptidoglycan in the cell wall of dividing *Lactococcus lactis* cells. Fluorescence staining of the wild-type strain was found around the septum, while no fluorescent labeling was detected for a mutant strain producing peptidoglycan precursors ending by D-Ala-D-Lac instead of D-Ala-D-Ala. AFM topographic images of *L. lactis* cells revealed a smooth and elongated cell morphology as well as a well-defined division septum (Fig. 10.6e). Ring-like structures were seen at a certain distance from the septum, presumably formed by an outgrowth of the cell wall. Notably, adhesion force maps demonstrated that binding sites were essentially located in the septum region, and more specifically on the equatorial rings (Fig. 10.6f), suggesting that newly formed peptidoglycan was inserted in these regions. This study shows that AFM with vancomycin tips is a complementary approach to fluorescent vancomycin to explore the architecture and assembly process of peptidoglycan during the cell cycle of Gram-positive bacteria. While fluorescence microscopy generates microscale images allowing the localization of peptidoglycan in the entire cell wall, AFM adhesion force mapping reveals the distribution of single peptidoglycan molecules on the outermost cell surface.

While spatially resolved SMFS provides a quantitative analysis of unbinding forces, it is limited by its time resolution. The time currently required to record a map is on the order of 2–15 min depending on the acquisition parameters, which is much greater than the time scale at which dynamic processes usually occur in biology. An exciting alternative in this context is dynamic recognition force mapping (TREC) [30] which records topography and recognition images at the same speed as that used for conventional topographic imaging, typically 1–2 images per minute. Recently, Chtcheglova et al. [31] applied TREC imaging to microvascular endothelial cells to demonstrate that cadherins, involved in homophilic cell-to-cell adhesion, are organized into nanodomains ranging from 10 to 100 nm in diameter. Clearly, a challenging issue for future research would be to apply TREC imaging in the microbiological context.

10.3.2 Immunogold Imaging

A different way of exploiting molecular recognition for AFM imaging is to use immunogold labels as cell-surface markers, as is traditionally used in electron microscopy. Here, cells are first incubated with monoclonal antibodies directed against specific cell wall constituents, then further incubated with the corresponding gold-conjugated secondary antibodies, and finally imaged using topographic imaging. In pioneering work [32], application of the approach to dried immunogold-labeled human lymphocytes enabled researchers to resolve the location of antigens on the cell surface. Similarly, types I and II collagen fibers were revealed on dried rat fibroblasts and human chondrosarcoma cells [33]. For the first time, immunogold

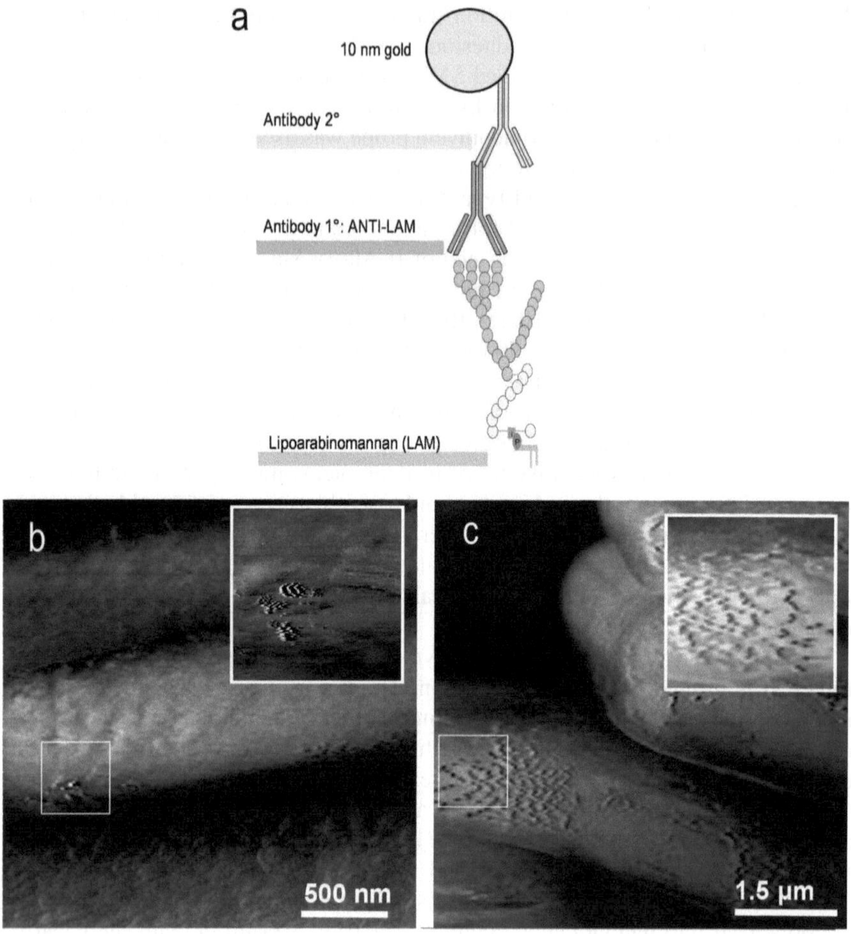

Fig. 10.7. Molecular recognition imaging of live cells using immunogold labeling. (**a**) *M. bovis* BCG cells were incubated with monoclonal anti-LAM antibodies, followed by another incubation with the corresponding gold-conjugated secondary antibodies. (**b**, **c**) AFM tapping mode (phase) images of immunogold-labeled cells prior (**b**) and after (**c**) treatment for 24 h with isoniazid, revealing that the drug induces the massive exposure of LAM at the surface. Reprinted with permission from [18]

AFM imaging was used to detect and localize lipoarabinomannan (LAM) on the surface of live, hydrated mycobacteria (Fig. 10.7) [18]. Contact mode and tapping mode images were obtained prior and after treatment with two antibiotics, INH and EMB. Gold particles were never seen in contact mode, emphasizing the need to use tapping (phase) mode for such in situ immunogold studies. Using tapping mode, the surface of native cells showed essentially no labeling, suggesting that LAM is not exposed at the surface. This finding was consistent with the uniform distribution of cell surface hydrophobicity measured on native cells. By contrast, INH and EMB-treated cells revealed a large coverage of gold particles, indicating that LAM was exposed. This observation, which correlates with topographic and CFM data, provides direct evidence that the two drugs lead to the massive exposure of LAM at the cell surface. In summary, this study showed that combining immunogold detection with topographic imaging and CFM allows us to shed new light into the 3-D organization of bacterial cell walls.

10.4 Conclusions

Owing to its ability to observe and manipulate biosystems under physiological conditions, AFM is revolutionizing the way in which today researchers explore live cells. Here, we have shown that two AFM modalities, CFM and molecular recognition imaging, offer fascinating prospects for mapping the chemical and biochemical properties of live cells. These nanoscale analyses should have an important impact on future life science and biomedical research, particularly to understand the molecular bases of cell–drug and host–pathogen interactions. Yet, it must be realized that procedures for attaching chemical groups, biomolecules and cells to AFM cantilevers remain labor intensive and require specific expertise that is usually not found in biology laboratories. In the future, defining simple standard protocols for tip functionalization and making them readily available to the biological community should contribute to spread the use of force spectroscopy in the various life science disciplines.

Acknowledgments. This work was supported by the National Foundation for Scientific Research (FNRS), the Région wallonne, the Foundation for Training in Industrial and Agricultural Research (FRIA), the Université Catholique de Louvain (Fonds Spéciaux de Recherche), the Federal Office for Scientific, Technical and Cultural Affairs (Interuniversity Poles of Attraction Programme), and the Research Department of Communauté Française de Belgique (Concerted Research Action). Y.F.D. and D.A. are Research Associate and Research Fellow of the FRS-FNRS, respectively.

References

1. Fritz J, Baller MK, Lang HP, Rothuizen H, Vettiger P, Meyer E, Guntherodt HJ, Gerber C, Gimzewski JK (2000) Science 288:316
2. Turner APF (2000) Science 290:1315
3. Niemeyer CM, Mirkin CA (2004) Nanobiotechnology: Concepts, applications and persepectives. VCH (ed). Wiley, Weinheim, p 469.
4. Jahn TR, Radford SE (2005) FEBS J 272:5962
5. Doyle RJ (2000) Microb Infect 2:391
6. Noy A (2006) Surf. Interface Anal 38:1429
7. Smith SB, Finzi L, Bustamante C (1992) Science 258:1122
8. Ashkin A, Schutze K, Dziedzic JM, Euteneuer U, Schliwa M (1990) Nature 348:346
9. Vezenov DV, Noy A, Ashby P (2005) J Adhes Sci Technol 19:313

10. Burnham NA, Chen X, Hodges CS, Matei GA, Thoreson EJ, Roberts CJ, Davies MC, Tendler SJB (2003) Nanotechnology 14:1
11. Noy A, Frisbie CD, Rozsnyai LF, Wrighton MS, Lieber CM (1995) J Am Chem Soc 117:7943
12. vanderVegte EW, Hadziioannou G (1997) Langmuir 13:4357
13. Vezenov DV, Noy A, Rozsnyai LF, Lieber CM (1997) J Am Chem Soc 119:2006
14. Frisbie CD, Rozsnyai LF, Noy A, Wrighton MS, Lieber CM (1994) Science 265:2071
15. Sinniah SK, Steel AB, Miller CJ, ReuttRobey JE (1996) J Am Chem Soc 118:8925
16. Cappella B, Dietler G (1999) Surf Sci Rep 34:1
17. Alsteens D, Dague E, Rouxhet PG, Baulard AR, Dufrêne YF (2007) Langmuir 23:11977
18. Alsteens D, Verbelen C, Dague E, Raze D, Baulard AR, Dufrêne YF (2008) Eur J Physiol 456:117
19. Dague E, Alsteens D, Latge JP, Verbelen C, Raze D, Baulard AR, Dufrêne YF (2007) Nano Lett 7:3026
20. Dupont-Gillain CC, Nysten B, Hlady V, Rouxhet PG (1999) J Colloid Interface Sci 220:163
21. Ducker WA, Senden TJ, Pashley RM (1991) Nature 353:239
22. Ahimou F, Denis FA, Touhami A, Dufrene YF (2002) Langmuir 18:9937
23. Dague E, Alsteens D, Latgé JP, Dufrêne YF (2008) Biophys J 94:656
24. Ludwig M, Dettmann W, Gaub HE (1997) Biophys J 72:445
25. Grandbois M, Dettmann W, Benoit M, Gaub HE (2000) J Histochem Cytochem 48:719
26. Lehenkari PP, Charras GT, Nykänen A, Horton MA (2000) Ultramicroscopy 82:289
27. Almqvist N, Bhatia R, Primbs G, Desai N, Banerjee S, Lal R (2004) Biophys J 86:1753
28. Dupres V, Menozzi FD, Locht C, Clare BH, Abbott NL, Cuenot S, Bompard C, Raze D, Dufrêne YF (2005) Nat Methods 2:515
29. Gilbert Y, Deghorain M, Wang L, Xu B, Pollheimer PD, Gruber HJ, Errington J, Hallet B, Haulot X, Verbelen C, Hols P, Dufrene YF (2007) Nano Lett 7:796
30. Raab A, Han WH, Badt D, Smith-Gill SJ, Lindsay SM, Schindler H, Hinterdorfer P (1999) Nat Biotechnol 17:902
31. Chtcheglova LA, Waschke J, Wildling L, Drenckhahn D, Hinterdorfer P (2007) Biophys J 93:L11
32. Putman CAJ, Degrooth BG, Hansma PK, Vanhulst NF, Greve J (1993) Ultramicroscopy 48:177
33. Arntz Y, Jourdainne L, Greiner-Wacker G, Rinckenbach S, Ogier J, Voegel JC, Lavalle P, Vautier D (2006) Microsc Res Tech 69:283

14 Single-Molecule Studies on Cells and Membranes Using the Atomic Force Microscope

Ferry Kienberger · Lilia A. Chtcheglova · Andreas Ebner · Theeraporn Puntheeranurak · Hermann J. Gruber · Peter Hinterdorfer

From volume VI

Abbreviations

AFM	atomic force microscope
DFM	dynamic force microscopy
PEG	poly(ethylene glycol)
BR	bacteriorhodopsin
SPM	scanning probe microscopy
NHS	*N*-hydroxysuccinimide
PSGL-1	P-selectin glycoprotein ligand-1
EM	electron microscopy
TEM	transmission electron microscopy
NPC	nuclear pore complex
NE	nuclear envelope
APTES	aminopropyltriethoxysilane
HSP	heat shock protein
S-layer	surface layer
HRV	human rhinoviruses
HIV	human immunodeficiency virus
LFA-1	leukocyte function-associated antigen-1
ICAM-1	intercellular adhesion molecule-1
WLC	worm-like chain
OspA	outer surface protein of Borrelia burgdorferi
NhaA	sodium-proton antiporters
TREC	simultaneous topography and recognition imaging
MyEnd cells	microvascular endothelial cell line from mouse myocardium
VE-cadherin	vascular endothelial cadherin
EDTA	ethylenediaminetetraacetic acid

14.1 Abstract

Atomic force microscopy (AFM) has been developed into an imaging method that yields fine structural details on biological samples in their physiological environment. Topographic AFM imaging enables a broad range of biomolecular complexes that are bound to the cells and membranes to be investigated. Dynamic processes can be visualized, single biomolecules can be observed at work, and the formation

In: H. Fuchs, B. Bhushan (eds), *Biosystems - Investigated by Scanning Probe Microscopy*, DOI 10.1007/978-3-540-37319-3_4, of biomolecular assemblies can be studied. In addition to high-resolution imaging, the measurement of mechanical forces at the molecular level has provided detailed insights into structure–function relationships of many biological systems. The detection of inter- and intramolecular forces yields information of structural parameters of the binding pocket, on the molecular dynamics of the recognition process, and on the energy landscape of the interaction. Single-molecule recognition measurements have also been demonstrated on cells and membranes and open new perspectives in exploring the regulation of cellular processes. By combining topographic imaging with force measurements, receptor binding sites can be localized with nanometer accuracy rendering possible the identification of specific components on the cellular surfaces.

14.2 Introduction

At present, atomic force microscopy (AFM) is extensively used in a wide range of disciplines such as molecular biology, solid-state physics and material science [1]. The major application is imaging of surfaces on scales from micro- to nanometers with the objective to visualize and properly characterize surface textures and shapes. For instance, it has been applied to characterize various biological samples (e.g. proteins, DNA, membranes, cells) at high lateral resolution [2]. Moreover, it is the unique technique to provide subnanometer resolution at a reasonable signal-to-noise ratio under physiological conditions. Due to continuous developments of sample preparation, imaging techniques and instrumentation, AFM has evolved into a companion technique of X-ray crystallography and electron microscopy (EM) for the determination of protein structures [3]. Regarding membranes and cells, AFM has successfully complemented EM studies. Recently, AFM images of native membranes have been obtained at subnanometer resolution for the first time [1–4]. AFM allows visualization of the samples in buffers that preserve their structure over extended periods of time. Most importantly, AFM does not rely on symmetry averaging and crystallization. Thus, AFM can reveal defects and structural anomalies that are not observable in classical ensemble measurements [5]. Unlike EM, AFM yields three-dimensional maps with an exceptionally good vertical resolution (less than a nanometer). When applied to living cells, new cellular surface structures and their physiological functions have been identified [6–8]. For instance, topographical images of living pancreatic acinar cells revealed the presence of fusion pores at the apical plasma membrane and their structure and dynamics [6].

In addition to high-resolution imaging of proteins, nucleotides, membranes and living cells [9], the measurement of mechanical forces at the molecular level has provided detailed insights into the function and structure of biomolecular systems [10]. Inter- and intramolecular interactions can be studied directly at the molecular level, as exemplified by the analysis of polysaccharide elasticity [11, 12], DNA mechanics [13, 14], the function of molecular motors [15, 16], and the binding potentials of receptor–ligand pairs involved in cell adhesion [17, 18]. In the latter case, defined forces are exerted on a receptor–ligand complex and the dissociation process is followed over time. Dynamic aspects of molecular recognition are addressed in force spectroscopy experiments, where distinct force–time profiles are applied to monitor the changes of conformational and states during receptor–ligand dissociation. Consequently, dynamic force spectroscopy allows detection of energy barriers not detectable by conventional near-equilibrium assays and to probe the free-energy surface of proteins and molecular complexes [19]. The capability of AFM to resolve nm-sized details, together with its force-detection sensitivity, has led to the development of molecular recognition imaging [20]. By combining topographical imaging with force measurements, receptor sites are localized with nanometer accuracy. Topography and recognition of target molecules are thereby simultaneously mapped. Thus, the AFM can identify specific components in a complex biological sample and retain its high resolution in imaging. In summary, several examples demonstrate the capability of AFM to resolve characteristic substructures of single individual proteins, to study conformational changes, and to analyze molecular interactions and the assembly of molecular complexes.

14.3 Principles of Atomic Force Microscopy

The AFM, a member of the scanning probe microscopy family, operates by moving a sharp tip on a microcantilever horizontally across a surface, thus rendering it possible to record the topographical map of the scanned surface [21] (Fig. 14.1a). Deflections of the cantilever caused by the tip–sample interaction are usually detected by a four-segment split photodiode. The magnitude of the deflection is registered by the changes in the angle of a laser beam reflected at the end of the cantilever (Fig. 14.1b). The deflection signal of the cantilever is used to keep the applied force constant by moving either the sample or the cantilever up and down via a feedback loop. The surface topography is then reconstructed from the vertical movements of the sample or the cantilever, respectively. The AFM can be operated in various modes. In the contact-force imaging mode, images are created by bringing the tip and sample into contact and scanning the tip across the surface. In dynamic force microscopy (DFM) imaging, the AFM tip is oscillated near its resonant frequency while it scans over the surface, such that the tip does not continuously touch a sample [22–24]. The amplitude reduction upon intermittent tip–surface contact is held constant through the feedback loop. As the tip touches the sample surface at the very end of its downward movement only, the lateral forces are greatly reduced during imaging. This is of advantage when imaging soft biological samples [25].

Both contact and dynamic force microscopy have been employed for high-resolution imaging of biomolecules. In fact, some of the highest resolution images reported in the literature were obtained using the contact mode. In these studies, mostly two-dimensional arrays of membrane proteins were imaged at subnanometer resolution, using conventional AFM cantilevers in aqueous buffered conditions [26]. The high image resolution was made possible by the close packing of the biomolecules, which makes them more resistant to the deformations possibly caused by the forces of the cantilever during imaging. In contrast, DFM imaging has mainly been applied to single, isolated biomolecules on surfaces and to supramolecular complexes [27] because it overcomes the strong lateral imaging forces associated with contact-mode AFM [28]. Moeller et al. [29] have recently demonstrated that DFM allows for imag-

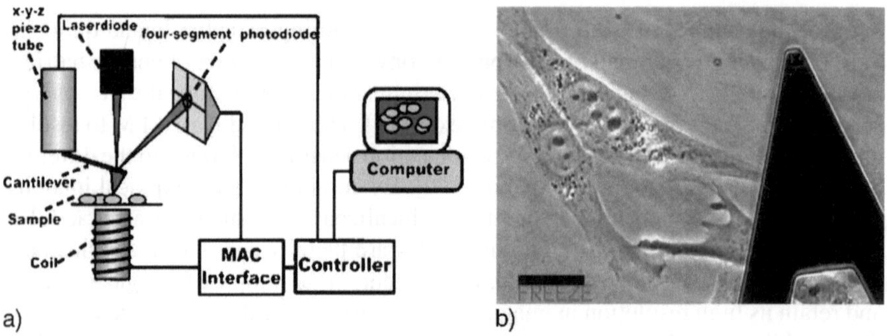

Fig. 14.1. Principles of scanning probe microscopy. (**a**) Schematic diagram of an atomic force microscope. A sample is probed by an ultrasharp stylus mounted on a cantilever, which moves in a raster manner over the surface. A reflected laser beam amplifies and reports deflections of the cantilever to a split photodiode. Photoelectric circuitry then converts the deflections into height information recorded as a digital image. Picture taken from [27]. (**b**) Phase-contrast micrograph of a cantilever and living cells. For cell measurements, the can tilever tip is carefully positioned on top of a cell. *Scale bar* = 25 μm. Image taken from [63]

ing of native protein surfaces with similar resolution as in contact-mode AFM. For contact-mode and DFM imaging the samples need to be tightly adhered to the support. In recent years, the art of sample preparations has been optimized in many laboratories with respect to stable sample supports, buffer composition to control the substrate–sample, and tip–sample interactions [30]. The challenge of imaging soft biological material has been met by employing force–distance curves to adjust the forces, the scanning speed and feedback parameters.

14.4 Imaging of Membrane–Protein Complexes

In the following, AFM imaging of biological membranes with a lateral resolution of ∼ 1 nm is discussed and the new insights into biological processes are summarized. In topographical imaging, the heights of membrane-protruding structures can accurately be measured with a vertical resolution of about 1 Å [4]. The high signal-to-noise ratio of the instrument allows for assessing the oligomeric state and the sidedness of membrane proteins directly in raw-data images. Even poorly ordered single particles can be recognized and imaged with high resolution [31]. AFM has developed to a point that now permits detection of flexible and stable extrinsic domains of membrane proteins and the ability to perform time-lapse imaging, thereby monitoring structural changes of molecules and molecular assemblies as a function of time [32].

14.4.1 Membranes of Photosynthetic Bacteria and Bacterial S-Layers

The first application of the AFM in the field of photosynthetic bacteria was an investigation of the peripheral light harvesting complex LH2 that is embedded in a highly organized assembly of transmembrane proteins that are responsible for absorption of light and its transformation into chemical energy [33]. High-resolution images of LH2 (Fig. 14.2a) revealed a nanomeric (i.e. a nine-fold symmetry) organization of the regularly packed cylindrical complexes which were incorporated into the membrane in both orientations. Topographs of LH2 complexes had a lateral resolution of ∼8 Å and a vertical resolution of ∼1 Å confirming the nanomeric organization of the heterodimers determined by EM. The cylindrical complexes were found to strongly protrude (by ∼14 Å) from one side (periplasmic side) and weakly (by ∼6 Å) from the other (cytoplasmic) side of the membrane. The heights of the protruding rings provide information about the position of the LH2 cylinder with respect to the lipid bilayer. Such data exemplify the additional benefit of AFM imaging in complementing EM studies, as well as the precision of AFM when analyzing oligomeric states of membrane proteins.

Other native membranes, which have been used for AFM investigations, are bacterial cell surface layers (S-layers) [34]. S-layers were identified as the outermost cell-envelope component in many different species and form highly ordered arrays with various symmetries [35]. S-layer proteins can be functionalized with capture molecules by means of genetic engineering. Such sensing layers have been already exploited for label-free detection methods, as microadsorbents in blood purification, and for the development of novel types of antiallergic vaccines [36]. S-layer

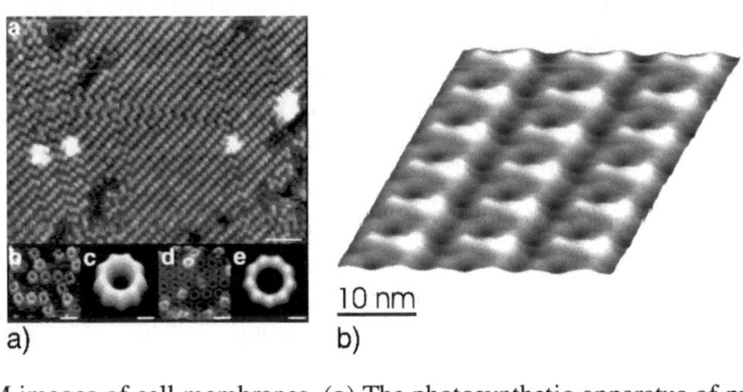

Fig. 14.2. AFM images of cell membranes. (**a**) The photosynthetic apparatus of purple bacteria. Medium-resolution AFM topograph of a reconstituted light harvesting complex LH2 crystal showing areas of crystallinity interspaced by more randomly packed domains. To produce 2D crystals of these complexes, the general method of reconstitution started with the purified proteins and a suitable combination of lipids, both solubilized in detergent. Next, the detergent was removed by dialysis resulting in the formation of lipid bilayers in which the proteins were reconstituted as 2D crystals or densely packed proteins. *Scale bar*: 50 nm. The *lower panel* shows high-resolution topographical analyzes of the periplasmic (*left two images*) and cytoplasmic surface (*right two images*) of the LH2 complex. The *scale bars* are 10 nm and 2 nm for raw-data topographs and for average images, respectively. The protrusion height is 1.6 nm and 0.5 nm for the periplasmic and cytoplasmic complex, respectively. Image taken from [33]. (**b**) Recrystallized bacterial cell surface layer (S-layer) fused with streptavidin. Functional heterotetramers consisting of one molecule fusion protein and three molecules streptavidin were obtained and recrystallized on gold substrates. The two-dimensional protein crystal displays streptavidin in defined repetitive spacing. In the lattice of the fusion protein, streptavidin shows up with periodical heights of 0.5 nm. The unit cell dimensions of the S-layer crystal are 10.2 nm and 8.0 nm with a base angle of 76°. Image taken from [37]

fusion proteins with an easily addressable binding site, such as fused streptavidin, offer the opportunity to specifically bind other proteins tagged with specific ligands. Figure 14.2b shows a high-resolution AFM image of an S-layer functionalized with streptavidin tetramers [37]. Using gentle DFM imaging, details of the S-layer protein lattice with bound streptavidin have been obtained in buffer solution. Streptavidin molecules are evident with periodical heights of 0.5 nm, with unit cell dimensions of 10.2 nm and 8.0 nm, and a base angle $\gamma = 76°$. These values are in good agreement with data from transmission electron microscopy (TEM) of the same system. The streptavidin moiety was observed in the topographical image at a high signal-to-noise ratio (Fig. 14.2b). In contrast to the TEM data that were derived from negatively stained preparations, the AFM images were obtained under physiological conditions.

14.4.2
Nuclear Pore Complexes

Protein transport into the cell nucleus is mediated by nuclear pore complexes (NPCs), supramolecular structures that are embedded in the nuclear envelope (NE) [38]. Usually, AFM studies of NPC are performed on NE spread on mica or glass. Figure 14.3 shows typical topographical images of NPCs in NE, either at high density (Fig. 14.3a) or at low density (Fig. 14.3b) [39]. Images of the cytoplasmic side of NE show NPCs as smooth rings with diameters of about 100 nm and heights of about 10 nm. Time-resolved images of aldosterone-induced conformational changes of individual NPCs are shown in Fig. 14.3c [40]. Before aldosterone injection, NPCs appear as smooth rings at the NE surface (Fig. 14.3c, first panel). Two minutes after hormone injection, the ring periphery of NPCs are decorated with macromolecules, termed flags (second panel). Eight minutes after injection, NPCs again appear as smooth rings (third panel). This is caused by the lack of flags at this stage of hormone stimulation. Nineteen minutes after aldosterone injection, rather large masses (termed plugs) in the central channels have been detected (fourth panel). The plugs have been identified as ribonucleoproteins exported by the NPCs in response to aldosterone stimulation. The flags were detectable only within the first few minutes after aldosterone injection. They are likely to be mineral corticoid receptors, because (i) a competitive receptor antagonist inhibited NPC flagging (not shown) and (ii) the molecular weight of the flags estimated by AFM matches the molecular mass of the mineral corticoid receptor. In summary, hormone-induced protein import across nuclear pores and ribonucleoprotein export from the nucleus to the cytoplasm have been visualized at the molecular level.

14.4.3
Cell Membranes with Attached Viral Particles

Until recently, transmission electron microscopy (TEM) has been virtually the only source of detailed images of virions and virion-infected cells [41]. Cross sections of cells that had been embedded, stained, and visualized with TEM showed the presence of virions near and at cell surfaces in the processes of maturation and

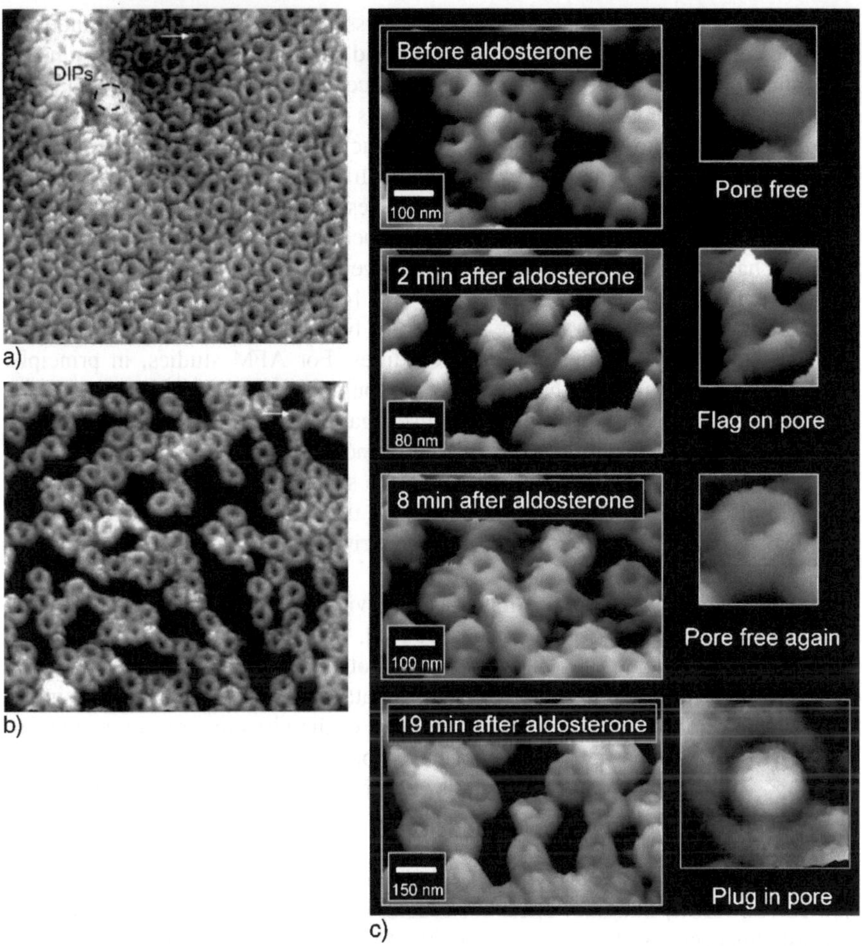

Fig. 14.3. AFM images of nuclear membranes. (**a**) The nuclear envelope of the cell nucleus is perforated by nuclear pore complexes (NPCs). The image shows the cytoplasmic side of the nuclear envelope in Xenopus laevis oocytes. Densely packed NPCs with ~ 100 nm in diameter are observed. Image size is 1.6 μm. The *gray scale* ranges from 0 to 40 nm. Image taken from [39]. (**b**) Topographical image of the cytoplasmic side of the nuclear envelope with a low density of NPCs. Same image size and *gray scale* as before. Image taken from [39]. (**c**) Aldosterone-induced conformational changes of individual NPCs. Images of nuclear envelope cytoplasmic surfaces are shown before and after aldosterone injection. Before aldosterone injection, the NPCs are clearly visible (pore free). Two minutes after injection, most of the NPCs are flagged (flag on pore). Eight minutes after injection, the NPCs are aga in free of flags (pore-free again). Nineteen minutes after injection, the NPCs are plugged (plug in pore). The z-scale ranges from 0 to 20 nm. Image taken from [40]

budding. Although these images are quite impressive, they have certain limitations. First, the cells had been embedded in plastic and subsequently exposed to a heavy metal stain, either or both of which might introduce distortions into the cell and virion structures. While the lateral resolution of TEM is usually high, the determination of height, i.e. of the third dimension is problematic for nonsymmetric objects. More recently, cryo-EM has been used to visualize both immature and mature retroviruses isolated from culture medium [41]. These studies are technically more demanding but they are free of many of the artefacts introduced by staining and dehydration.

AFM has its virtues and its drawbacks as well, but fortunately they tend to be complementary to those of TEM and cryo-EM [42]. With AFM, samples may be imaged in fluids, including culture medium or buffer, in situ, or after processing according to established histological procedures. For AFM studies, in principle, a single cell or virion is all that is required, though several hundred of either may be present on the substrate. In previous investigations of viruses with AFM it has been shown that the technique is sufficiently incisive and reproducible that even individual capsomeres can be visualized on the surfaces of both plant and animal viruses. Furthermore, it was demonstrated that the structures of viruses observed by AFM are entirely consistent with models derived by X-ray crystallography and cryo-EM [42].

Figure 14.4a shows an image of human rhinoviruses (HRV) on an artificial membrane interface that mimics the cell surface [43]. The specific and site-directed immobilization of viral particles via cell receptor proteins resulted in a two-dimensional crystal with hexagonal symmetry; the average lattice constant obtained was similar to the diameter of the virus, i.e. 35 nm. Binding of the virus to a receptor-modified bilayer closely mimics the binding to its receptor on the host-cell plasma membrane and is thus ideally suited for studies of viral dynamics and recognition under near physiological conditions. In addition to the hexagonal arrangement, a regularly spaced pattern was observed on almost all viral particles (Fig. 14.4b). This pattern reflects the protrusions on the surface of the viral capsid, which is well known from cryo-EM and X-ray imaging. These substructures are more clearly discernable upon contrast-enhancement (Fig. 14.4c). Roughly 20 blobs or protrusions with diameters of 2–4 nm and heights of ~ 0.3 nm were counted on each viral particle; they were arranged in lines with ~ 5 protrusions per line (see labeling in Fig. 14.4c). Virus capsids did not appear strictly globular in shape, but rather the projection of the particles exhibited some straight borders akin to their polygonal surface features, which were also observed by cryo-EM. The center-to-center distance of single protrusions is about 7 nm (Fig. 14.4c). This is in the same range as the distance between the plateau at the three-fold axis and the plateau at the five-fold axis as seen in the structures resolved by cryo-EM. In summary, high-resolution imaging of the rhinovirus capsid revealed a regular arrangement of 3-nm-sized protrusions. The achieved resolution was comparable to the resolution of EM, but the latter method required the sample to be frozen in amorphous ice, whereas AFM measurements were carried out under near physiological conditions.

In another study, human immunodeficiency virus (HIV)-infected lymphocytes have been imaged in buffer solution [44]. Figure 14.4d–f show images of HIV virions as they emerge from, or are attached to, the surfaces of infected cells. Presumably, they represent a mixture of immature and mature particles and possibly

Fig. 14.4. AFM images of virus–membrane complexes. (**a**) Crystalline arrangement of human rhinovirus (HRV) on a cell mimicking membrane. Specific and site-directed immobilization of HRV on a receptor modified bilayer results in a hexagonal symmetry and 35 nm spacing. *Inset (upper right)*: The Fourier spectrum exhibits hexagonally arranged spots (*arrows*). *Inset (lower right)*: Averaged three-dimensional representation of the virus arrangement. Image taken from [43]. (**b**) Three-dimensional representation of rhinovirus particles with 30 nm diameters. Roughly 20 protrusions are counted on each single virus particle. Image size is 80 nm. The *z*-scale ranges from 0 to 10 nm. Image taken from [43]. (**c**) Contrast-enhanced image for a better visibility of the protrusions. Singly resolved structures (labeled one to five) with diameters of ~3 nm and heights of ~0.3 nm can be clearly obtained. Image size is 80 nm. *Inset*: External surface of HRV as derived from X-ray coordinates. Image taken from [43]. (**d**) Human immunodeficiency virus (HIV)-infected lymphocytes imaged under physiological conditions. On the surface of cultured lymphocytes, emerging HIV particles can be seen. The particles are not present on uninfected cells (not shown). The image size is 5 × 5 µm. Image taken from [44]. (**e**) Small scan size image of the surface of an infected human lymphocyte. The image size is 1.5 × 1.5 µm. Image taken from [44]. (**f**) The average diameter of single HIV particles emerging from or attached to the cells is ~127 nm. The image size is 500 × 500 nm. Image taken from [44]

virions in intermediate states. The particles have an average diameter of 127 nm, with a variation of 30 nm. A similar range of size diversity of particle diameters has been observed by cryo-EM as well. Some viral particles can be observed that are considerably smaller than 127 nm that had the gross appearance of virions, indicating that these particles might lack their cores. Also frequently present were some anomalously large particles that otherwise had the appearance of virions. They had diameters in the range of 160 to 240 nm and probably represent virions that contained multiple cores. The external appearances of the virions having diameters of about 127 nm, the vast majority, are similar to one another and are the same as those of the free virions. In either case, the surfaces of HIV particles exhibit protrusions, presumably gp120, which do not physically resemble spikes. The protrusions, which number about 100 per virus particle, have average diameters of about 20 nm, but with a large variance. They likely consist of arbitrary associations of small numbers of gp120 monomers on the surface of HIV particles.

14.5 Single-Molecule Recognition on Cells and Membranes

A number of techniques are presently available to investigate intermolecular forces acting between single biomolecules and cellular surfaces. The most prominent tools are the atomic force microscope, optical tweezers [45], and the biomembrane force probe [46]. These techniques span a measurable force window ranging from entropic forces at several femto-Newtons (1 fN $= 10^{-15}$ N) up to the rupture of covalent bonds at several nano-Newtons (1 nN $= 10^{-9}$ N) [47]. Using AFM, many different types of interactions have been studied either on isolated proteins in vitro or on cellular surfaces in vivo. For studying specific binding between biomolecules, an increasing force is exerted onto the molecular complex, and the dissociation process is followed over time. Dynamic aspects of recognition are addressed in force spectroscopy experiments, in which the timescale is systematically varied in order to evaluate the changes in conformations and states during receptor–ligand dissociation [48]. Such experiments allow the estimation of affinity, rate constants, and structural data of the binding pocket.

14.5.1 Principles of Recognition Force Measurements

In molecular recognition force spectroscopy experiments, the binding of ligands immobilized on AFM tips towards surface-bound receptors (or vice versa) is studied by applying a force to the receptor–ligand complex until the bond breaks at a measurable unbinding force [49]. Such experiments require that one or several ligand molecules are permanently tethered to the apex of the AFM tip, usually by covalent bonding via a flexible linker molecule [50]. As an example, the covalent coupling of ligands to gold-coated tip surfaces via a free SH-group of the ligand guarantees a sufficiently stable attachment, because these bonds are about ten times stronger than typical ligand–receptor interactions [51]. This chemistry has been used for various types of receptor–ligand studies, but it requires a prior deposition of a gold layer onto the silicon nitride or silicon oxide tip surface. Another frequently used method of functionalization is the direct tethering of ligand molecules to the surface of a silicon or silicon nitride tip (Fig. 14.5a). In the first step, amino groups are generated on the tip surface, either with ethanolamine hydrochloride or with silane derivatives such as aminopropyltriethoxysilane (APTES) [52,53]. These two procedures do not cause stickiness and yield a rather low number of amino groups on the apex of the tip [54], as is desired for the single-molecule experiments. In the second step of the anchoring protocol, a distensible and flexible linker is often used to space the ligand molecule from the amino-functionalized tip surface by several nanometers. The ligand on the spacer molecule can freely orient and diffuse within a certain volume

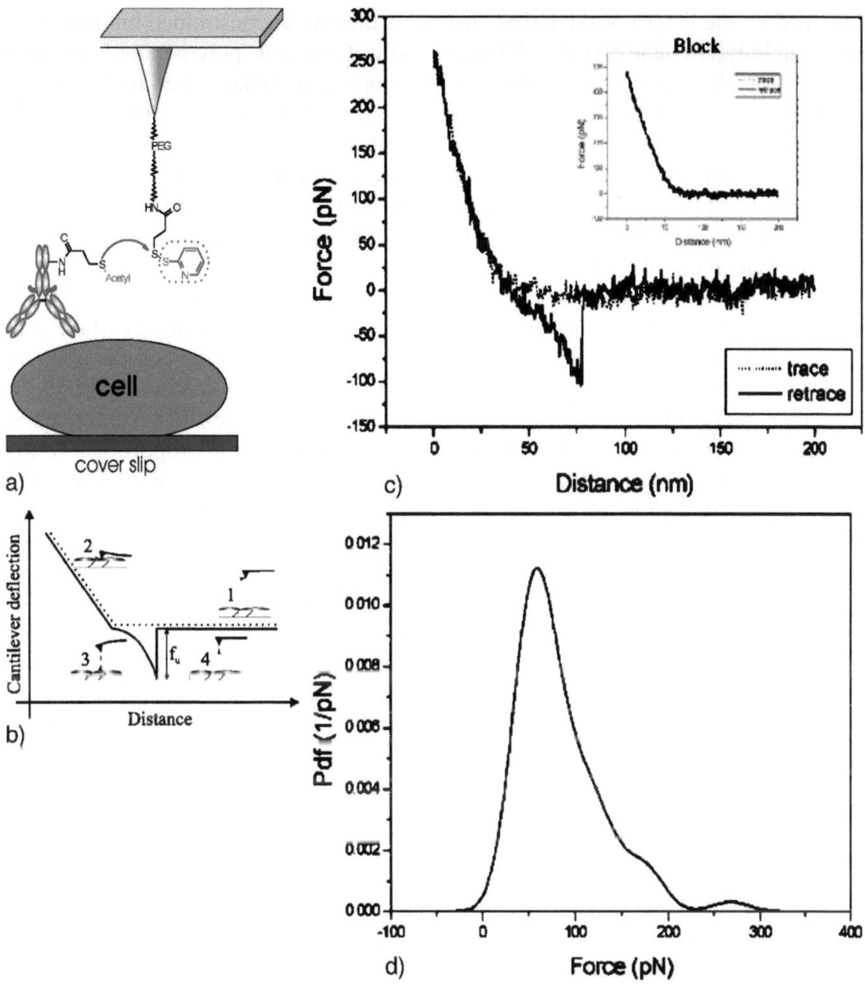

Fig. 14.5. Specific immobilization of biomolecules to the AFM tip for studying single-molecular interaction forces on cells. (**a**) A distensible and flexible polymer linker (polyethylene-glycol, PEG) was used to bind a single antibody covalently to the tip, enabling the ligand to freely reorient and to achieve unconstrained binding to its cognate receptor at the cell surface. (**b**) In force–distance cycles, an AFM tip with ligands is brought into contact with a cell expressing the corresponding receptor at high density, so that a ligand–receptor bond is formed. During a subsequent retraction of the tip, an increasing force is applied to the ligand–receptor bond until the interaction breaks at a critical force, the unbinding force f_{u}. (**c**) A force–distance cycle on the surface of a living cell with an antibody-coated AFM-tip shows specific interaction in the retrace (jump at 75 nm). The interaction is blocked using free antibody in solution (*inset*), proving the specificity of the detected force jump. (**d**) The probability density function (pdf) is constructed from an ensemble of forces and gives the distribution of unbinding forces f_{u}, similar to a histogram. The maximum of the pdf depicts the most probable unbinding force at a given loading rate. Data taken from [58]

provided by the length of the tether, thereby achieving unconstrained binding to its receptor. Poly(ethylene glycol) (PEG), an inert water soluble polymer has been often used as flexible linker [55]. The heterobifunctional crosslinkers used for tip–ligand coupling carry two different reactive groups at their ends [56]. One of these is usually a carboxyl group, activated in the form of an N-hydroxysuccinimide (NHS) ester. This group couples to the amino group on the tip, yielding a stable amide bond. In the third step, the outer end of the crosslinker (e.g. a thiol-reactive group) reacts with a suitable function on the ligand molecule (e.g. a thiol group), resulting in the flexible attachment of the ligand to the tip (Fig. 14.5a).

Interaction forces of single ligand–receptor pairs are measured in force–distance cycles using a ligand-carrying tip mounted on a cantilever and a target surface with firmly attached receptor molecules (e.g. receptors anchored in the cell membrane). A sketch showing the principles of a force–distance cycle is given in Fig. 14.5b. At a fixed lateral position, the tip vertically approaches the surface and is subsequently retracted. During this cycle, the cantilever deflection (which is proportional to the force) is continuously measured and plotted versus tip–surface separation (i.e. distance). At the beginning of the tip–surface approach (Fig. 14.5b, dotted curve), the cantilever deflection remains zero. Upon tip–surface contact, the cantilever bends upward, consistent with a repulsive force that linearly increases with the distance. Subsequent tip–surface retraction (Fig. 14.5b, solid line) first leads to relaxation of the cantilever bending until the repulsive force drops to zero. Upon further retraction, the cantilever progressively bends downwards, reflecting an attractive force that increases with increasing tip–surface separation. The shape of this nonlinear force–distance profile is determined by the entropic properties of the flexible PEG crosslinker and shows parabolic-like characteristics, which mirrors the increase of the spring constant of the polymer chain during extension [57]. The physical connection between tip and surface exerts the increasing force until the ligand–receptor complex finally dissociates at a certain critical force, termed the unbinding force f_u, whereupon the cantilever jumps back to the resting position. A typical force–distance cycle with an antibody modified AFM-tip and the surface of living cells is shown in Fig. 14.5c [58]. If the ligand on the tip does not form a specific bond with the receptor on the cell surface, the recognition event (i.e. the parabolic-shaped curve) is missing and the retrace looks like the trace (inset in Fig. 14.5c). In addition, the specificity of ligand–receptor binding is usually demonstrated by blocking experiments with free ligands, which are injected into the solution in order to block the receptor sites on the cell surface. As a consequence, almost all specific recognition signals completely disappear and only occasional adhesion events are observed.

After acquiring a few hundred force–distance cycles, empirical probability density functions (pdf) from the detected unbinding forces f_u can be constructed (Fig. 14.5d) [58]. The maximum of the distribution (Fig. 14.5d) reflects the most probable force upon which a single antibody-cell surface bond dissociates under the force ramp used. An overall binding probability, which is the probability to record an unbinding event in a force–distance cycle of 15.4% was obtained. Blocking experiments, performed by injection of free antibody in solution reduced this probability to 5.8% [58]. No binding was found at all when a bare tip was used instead of an antibody-coated tip. These results strongly support the specificity of the binding events observed.

Viewing ligand–receptor binding on the single-molecule level, the average lifetime of a ligand–receptor bond, $\tau(0)$, is given by the inverse of the kinetic off-rate constant, $\tau(0) = 1/k_{\text{off}}$. In the thermal activation model, the lifetime of a complex in solution is described by a Boltzmann equation, $\tau(0) = \tau_{\text{osc}} \exp(E_{\text{b}}/k_{\text{B}}T)$ [59], where τ_{osc} is the inverse of the natural oscillation frequency and E_{b} the energy barrier for dissociation. Hence, due to the thermal energy, there is a finite probability of overcoming the energy barrier E_{b}, which leads to the separation of the ligand–receptor complex. The force acting on a binding complex deforms the interaction energy landscape and lowers the activation energy barrier. The lifetime $\tau(f)$ of a bond loaded with a constant force f is given by $\tau(f) = \tau_{\text{osc}} \exp(E_{\text{b}} - fx/k_{\text{B}}T)$ [60], x being interpreted as the distance of the energy barrier from the energy minimum. Using AFM, an effective force increase or loading rate r can be deduced from $r = \text{d}f/\text{d}t$, equal to pulling velocity times effective spring constant [61]. The combination of the Boltzmann equation with the stochastic description of the unbinding process predicts different unbinding force distributions at different loading rates r [62]. The maximum of each force distribution, $f^*(r)$, reflects the most probable unbinding force for the respective loading rate r. f^* is related to r through $f^*(r) = k_{\text{B}}T/x \ln(r.x/k_{\text{B}}T.k_{\text{off}})$. Apparently, the unbinding force f^* scales linearly with the logarithm of the loading rate. For a single barrier, this would give rise to a simple, linear dependence of the force on the logarithm of the loading rate. In cases where more barriers are involved along the escape path, the curve will follow a sequence of linear regimes, each of which marks a particular barrier. The predicted logarithmic dependence of the unbinding force on the loading rate in the thermally activated regime was confirmed by a large number of unbinding experiments. A beautiful demonstration revealing hidden barriers between biotin and avidin was provided by Merkel et al. [63]. Most notably, the unbinding kinetics was dominated by different activation energy barriers positioned along the force-driven unbinding pathway and were consistent with the location of prominent transition states revealed by molecular dynamics simulations.

14.5.2 Force-Spectroscopy Measurements on Living Cells

Recently, recognition force spectroscopy has also been applied to cells. In one of the early studies, Lehenkari and Horton [17] measured the unbinding forces between integrin receptors present on the surface of intact cells and various ligands bound to the tip. The unbinding forces measured were found to be cell- and amino acid sequence-specific and sensitive to the pH and divalent cationic composition of the cellular culture medium. In another study, Chen and Moy [64] used AFM to measure the adhesive strength between concanavalin coupled to the tip and receptors on the surface of fibroblasts. Crosslinking of receptors on the cell surface led to an increase in adhesion that was attributed to enhanced cooperativity among adhesion complexes. Pfister et al. [58] investigated the surface localization of heat shock proteins (HSP) on stressed and unstressed endothelial cells. By detecting specific single-molecule binding events between the monoclonal antibody tethered to the tip and HSPs on cells, clear evidence was found for the occurrence of HSPs on the surface of stressed cells, but not on unstressed cells.

Figure 14.6 shows force-spectroscopy measurements on living cells, where the interactions between leukocyte function-associated antigen-1 (LFA-1) with its cognate ligand, intercellular adhesion molecule-1 (ICAM-1) was probed [65]. Force–distance cycles between cells expressing LFA-1 and ICAM-1 have been carried out under different buffer conditions (Fig. 14.6a). A typical measurement involved the formation of multiple LFA-1/ICAM-1 complexes that did not necessarily rupture simultaneously during retraction. The "saw-tooth" profile observed in the force–distance cycles suggests that these complexes often ruptured sequentially before the final separation, with each sharp transition in the retraction trace being attributed to a breakage of one or more LFA-1/ICAM-1 complexes (Fig. 14.6a). The unbinding forces required to detach the cell are quantitative measurements of cell adhesion. In terms of the unbinding force, cells stimulated with Mg^{2+} (i.e. activated cells) adhered more tightly to ICAM-1 than resting cells. As shown in Fig. 14.6b, the average unbinding force of the LFA-1/ICAM-1 complex increases over three orders of magnitude change in loading rate. Moreover, two loading regimes in the LFA-1/ICAM-1 interactions were evident in the force-spectroscopy plot (Fig. 14.6b). There was a gradual increase in unbinding force with increasing loading rate up to 10,000 pN/s. Beyond this point, there was a second loading regime that exhibited a faster increase in unbinding force. Activated cells resulted in higher unbinding forces that were pronounced in the slow-loading regime (Fig. 14.6b). Interestingly, there was no significant difference in the unbinding forces of the low and high-affinity complexes in the fast loading regime (at loading rates > 10,000 pN/s). The force-spectroscopy measurements revealed details of the complex process of ligand–receptor unbinding in cell adhesion. Two activation barriers were found in the dissociation of the LFA-1/ICAM-1 complex (Fig. 14.6c), as deduced from the two slopes in the force-spectroscopy plot (Fig. 14.6b). Each of the activation barriers (TS_1 and TS_2) is characterized by two parameters: a dissociation rate constant, k_i, and the position of the transition state, x_i, whereby $i = 1$ refers to the inner activation barrier and $i = 2$ to the outer activation barrier of the complex. Both k_i and x_i can be extracted from the force-spectroscopy plot (Fig. 14.6b). The values obtained are comparable to the dissociation rate constants measured by conventional methods. Furthermore, the dissociation rate constants can be used to estimate the energy differences between transition-state energies of high- and low-affinity complexes (Fig. 14.6c), showing that the outer activation barrier of the high-affinity complex is $3.2k_{\mathrm{B}}T$ higher than that of the low-affinity complex (Fig. 14.6c).

In another study, single-molecule dynamic force spectroscopy has been applied to study the interaction of antibodies and different sugars with the Na^+-glucose cotransporter SGLT1 in living cells (Fig. 14.7) [66]. The transporter drives the translocation of glucose into cells by use of the electrochemical gradient for Na^+ across the membrane and it is a prime example of the large family of biologically important ion-gradient-driven cotransport systems. Several methods have been employed to investigate the topology of the SGLT1, indicating that the SGLT1 contains 14 transmembrane helices [67]. The *N*-terminal half of the protein contains the Na^+-binding sites, whereas the sugar pathway is located in the C-terminal domain. Phlorizin, an aromatic glucoside, is known to be a potent competitive inhibitor of the SGLT1 transporter. It is postulated that

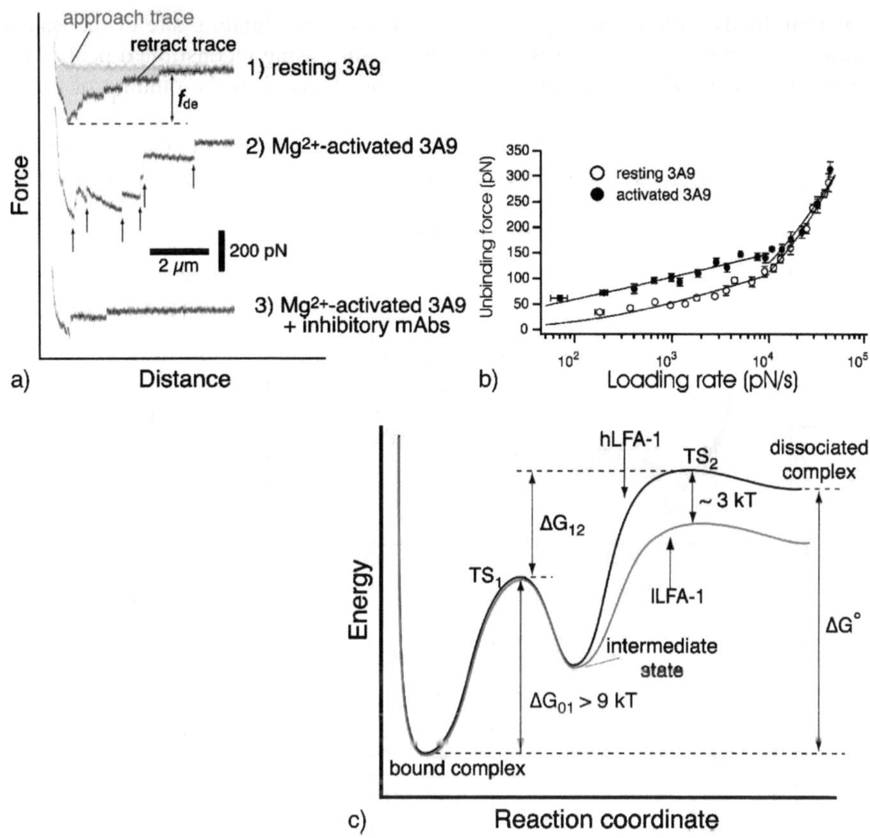

Fig. 14.6. AFM force measurements of cell adhesion. (**a**) A series of force–distance cycles between a living cell and intercellular adhesion molecule-1 (ICAM-1) acquired under different conditions. For traces 2 and 3, the cell was treated with Mg^{++}. F_{de} is the detachment force supported by the adhesive bonds formed between the cell and the substrate. The *arrows* in trace 2 point to positions where a cell/ICAM-1 complex ruptured. (**b**) Measurement of cell/ICAM-1 unbinding forces as a function of the loading rate. Plotting the most probable unbinding force against the loading rate results in the force-spectroscopy plot for ICAM-1 and resting cells (*open circles*) or activated cells (*filled circles*). The slopes of the linear regimes in the force-spectroscopy plot mark distinct activation barriers along the direction of force and yield the energy landscape along the reaction coordinate. The fitted curves (*solid lines*) were derived from fitting Eq. 2 to the data. (**c**) Intermolecular potentials of the cell/ICAM-1 interaction calculated from the force-spectroscopy plot. The dissociation of ICAM-1 from the cell surface occurs via two transition states, TS_1 and TS_2. The inner activation barriers of the high-affinity (hLFA-1) and the low-affinity (lLFA-1) complexes are $> 9k_BT$. Differences between the high- and low-affinity complexes ($\sim 3k_BT$) stemmed from the energetics of the outer barriers. Estimates of the energies of transition states were obtained from the dissociation rate constants. Data taken from [65]

phlorizin binds with a two-step mechanism to the sugar binding site of the transporter. Site-directed mutagenesis studies and studies using reconstituted peptide in vitro have shown that the C-terminal loop 13 is involved in the binding of phlorizin [68].

To study the structural and functional dynamics of plasma membrane transport proteins in living cells in finer detail, dynamic force spectroscopy was applied at

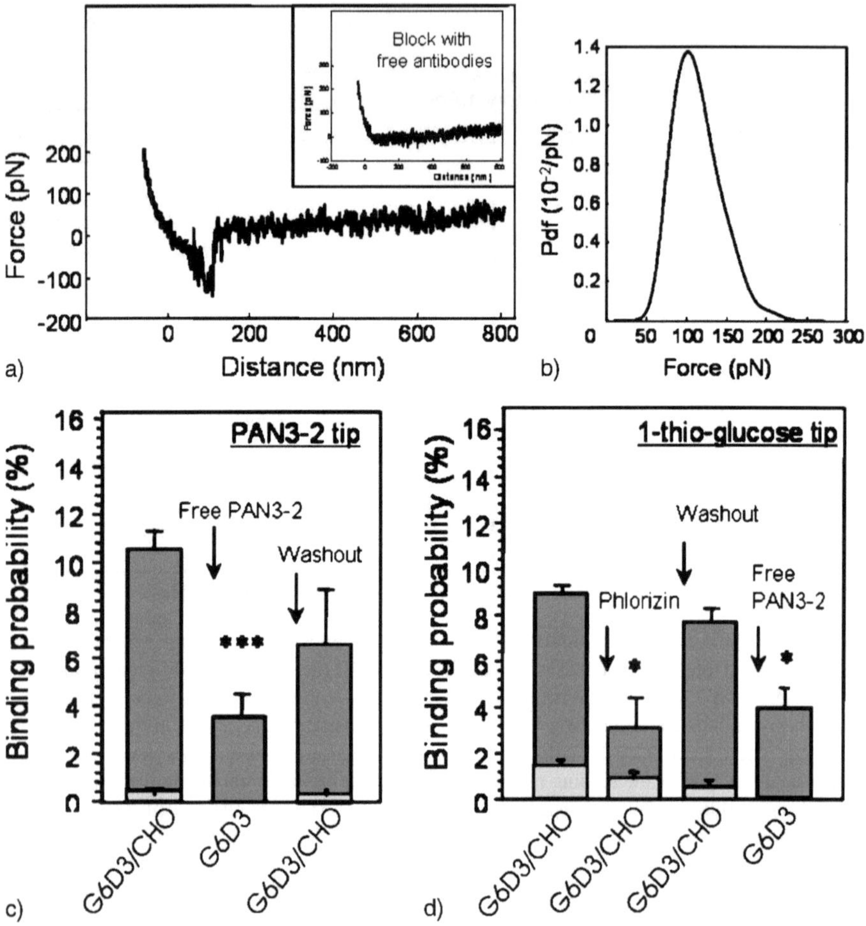

Fig. 14.7. Force measurements of antibodies and substrates with the Na^+-glucose cotransporter SGLT1 in living cells. (**a**) Recognition of SGLT1 on the surface of intact cells by an AFM tip carrying an epitope specific antibody (PAN3-2). Force curve showing specific interaction between the antibody and SGLT1 upon tip–surface retraction. The specific interaction is blocked by injecting free PAN3-2 antibodies in the solution (*inset*). (**b**) Probability density function giving the distribution of the unbinding force (f_u) of PAN3-2 to SGLT1. (**c**) Quantitative comparison of binding probabilities of PAN3-2-coated tips on G6D3 cells (*dark gray*) and CHO cells (*light gray*) in the absence or presence of free PAN 3–2 in the medium. (**d**) Recognition of SGLT1 on the surface of intact cells by a D-glucose primed AFM tip and the effect of phlorizin and the antibody PAN3-2 on the binding probabilities on G6D3 and CHO cells. Data taken from [66]

the single-molecule level [66]. The molecular recognition of SGLT1 on the cell membrane surfaces of living cells in different buffer conditions was probed with single antibody and with single D-glucose molecules that had been coupled to AFM tips. Scanning SGLT1-transfected cells with AFM tips carrying an epitope-specific antibody directed against the C-terminal loop 13 resulted in specific recognition events detected in force–distance cycles. Figure 14.7a shows a typical force curve (retraction) for a single-molecule antibody–antigen recognition event. For confirmation of the specificity of recognition events, blocking experiments were performed by injecting free antibodies. As illustrated in Fig. 14.7a (inset), no binding events were observed in the presence of free antibodies since no free epitopes were present on the cell surface. By constructing an empirical probability density function (pdf) of the unbinding forces (Fig. 14.7b), the maximum of the distribution was found to be 100 ± 30 pN. In addition to unbinding forces, binding probabilities from several experiments were quantified (Fig. 14.7c). For G6D3 cells, a binding probability (the probability to find an unbinding event in one force distance cycle) of 10.6% was observed. Another indication for specificity was that a very low binding probability was found for cells that contained no functional SGLT1. When free PAN3-2 antibodies were present in solution, the binding probability on G6D3 cells was significantly decreased to 3.6% (Fig. 14.7c). After subsequent washout of free PAN3-2 antibodies a high binding probability was recovered. These results demonstrate that SGLT1 on the surface of living cells can be specifically detected at the molecular level by using epitope specific antibodies. In addition, it was concluded that the epitope that is part of the loop 13 of SGLT1 is localized on the extracellular part of the cell membrane. In order to study the initial event of the glucose transport cycle, i.e. the binding of the sugar to SGLT1 on the surface of intact cells, glucose was coupled to the AFM tip. With this tip, force–distance cycles were performed on live cells, either on G6D3 cells expressing SGLT1 or on CHO control cells lacking SGLT1. Distinct glucose-recognition events were only observed in G6D3 cells, whereas in CHO cells the recognition signals were almost absent (Fig. 14.7d). These results demonstrate that the glucose binding site of SGLT1 is accessible on living cells for AFM force spectroscopy investigations.

14.6 Unfolding and Refolding of Single-Membrane Proteins

Dynamic force spectroscopy not only allows investigation of the interaction between receptors and ligands but also measurement of intramolecular force profiles of single molecules. The application of mechanical force to biological polymers like proteins, polysaccharides and DNA produces conformations that are different from those previously investigated by chemical or thermal denaturation [69]. The force-induced domain unfolding in proteins [70], length transitions caused by conformational changes in the sugar rings [71] or in the secondary structure of polysaccharides [72], and alterations of the secondary structure of DNA molecules were studied in great detail [73], elucidating the molecular determinants of mechanical stability and the role of force-induced conformational changes in the regulation of physiological function. In these experiments, the molecule is held between the tip and the support and its viscoelastic properties are studied in force–distance cycles (Fig. 14.8). Similarly as in molecular recognition force-spectroscopy experiments, a detailed picture of the complex mechanical unfolding pathway through a rough energy landscape can be gained by varying the dynamics of pulling.

As a recent example, OspA, an outer-surface protein of Borrelia burgdorferi containing a unique single-layer β-sheet, was unfolded with AFM (Fig. 14.8a) [74]. In this study, a monomeric unit was stretched by pulling it from its N- and C-terminal residues. Two peaks in the force–extension cycle were detected before the final rupture of a fully extended polypeptide occurred (Fig. 14.8b). The peaks in the unfolding curves are consistent with the results of previous thermodynamic studies showing two cooperative units in OspA. Mutations near the boundary of the cooperative units reduced the height of the first unfolding peak to undetectable levels

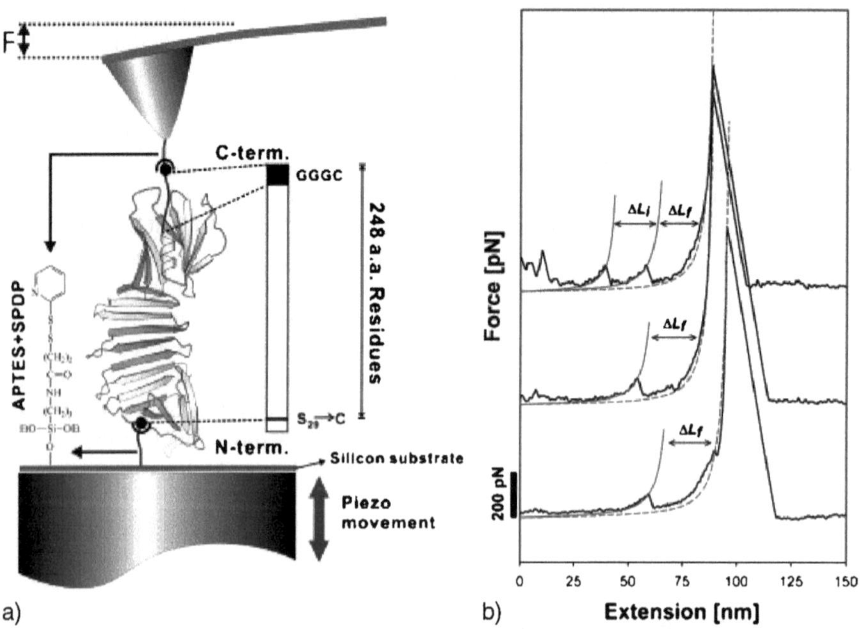

Fig. 14.8. Unfolding mechanics of an isolated membrane protein using AFM. (**a**) Experimental scheme for mechanical unfolding of a single, monomeric form of the outer surface protein A (OspA), a Lyme disease antigen containing a unique single-layer β-sheet. The engineered protein was covalently sandwiched through the formation of disulfide bonds between the substrate and the AFM tip. The sandwiched protein was then mechanically stretched by lowering the sample stage, leading to a characteristic and reproducible unfolding force pattern. (**b**) Force–extension curves of wild-type OspA (*upper panel*) and modified OspA (*lower panels*). Two peaks are detected in the force-extension profile before the final rupture of the fully extended protein occurs, which are interpreted as unfolding of multiple substructures in OspA. The contour length difference between two intermediate force peaks (ΔL_{i}) and that between the rupture peak and its adjacent peak (ΔL_{f}) are given. The persistence length of 0.4 nm was used to fit the worm-like-chain (WLC) model to all force peaks (*dashed lines*). Data taken from [74]

and marginally affected the second one (Fig. 14.8b). Based on a worm-like chain (WLC) analysis of the data, a model for mechanical unfolding of OspA was proposed. The analysis of these data revealed both similarities and differences between the mechanical unfolding process in the AFM and the thermal unfolding process at elevated temperature. This work demonstrates that the dynamic force spectroscopy studies of monomeric proteins can elucidate details of the intramolecular mechanics of protein substructures.

In another study, controlled unfolding and refolding of a sodium-proton antiporter has been analyzed with AFM (Fig. 14.9) [75, 76]. Single-molecule force spectroscopy was employed to unfold and refold single sodium-proton antiporters (NhaA) from membrane patches. For this purpose, the AFM tip was pressed onto the membrane surface with a contact force of about 1 nN for 1 s, and then withdrawn while recording the cantilever deflection versus tip–sample distance. In about 3% of the cases, one terminal domain of the protein actually adhered to the AFM tip, which resulted in a force spectrum over a distance of more than 95 nm (Fig. 14.9a, upper panel). Such force curves reflect the stretching and unfolding of the entire polypeptide AFM stylus (Fig. 14.9a, lower panel). The force spectra contained detailed information on the unfolding process, each peak representing an internal potential barrier that was built up by molecular interactions within the protein. Many of the potential barriers were detected either from the C-terminal or the N-terminal end. In this way, it was found that some molecular interactions stabilizing secondary structural elements were directional while others were not.

The above-described unfolding experiments of membrane proteins were further refined to allow for controlled refolding of individual secondary structures (Fig. 14.9c) [76]. Once distinct helices and loops had been unfolded, the whole system was relaxed for a given period of time, allowing the polypeptide to refold (Fig. 14.9b). Repeated recording of a force spectrum detected whether the polypeptide refolded to its native secondary structure. An example of such experiments is shown in Fig. 14.9b. First, NhaA was unfolded until only two helices remained anchored in the membrane (Fig. 14.9b, upper panel). The force peaks of the unfolding spectra indicated that the protein was unfolded from its C-terminal end. The AFM stylus was then lowered towards the membrane (Fig. 14.9b, lower panels). After the system had been allowed to relax for 5 s, the AFM stylus was separated from the membrane to restretch the polypeptide chain. The force spectrum then indeed contained the record of the repeated unfolding process (Fig. 14.9b, lower panels). All major peaks (potential barriers) observed during the initial unfolding were detected. These data indicate that several secondary structures observed for native NhaA had actually been refolded, and they support the hypothesis that unfolding and refolding of transmembrane helices may be fully reversible, including reinsertion of the transmembrane segments into the lipid bilayer.

14.7 Simultaneous Topography and Recognition Imaging on Cells (TREC)

The combination of high-resolution topography imaging with single-molecule force spectroscopy provides unique possibilities for the detection of specific molecular-

Fig. 14.9. Unfolding and refolding of single-membrane proteins. (**a**) Mechanical unfolding pathways of single sodium-proton antiporters (NhaA) from membrane patches. A characteristic pattern of force peaks corresponding to the unfolding of single NhaA molecules was observed. *Upper panel*: Superimposition of dozens of force curves highlights common features of the unfolding events and reduces deviations. The *gray scale* corresponds to the number of force curves superimposed. *Red lines* represent WLC fits with their numbers indicating the contour length of the stretched polypeptides. *Medium panel*: Detailed analysis of the positions and the probability of main peaks and side peaks. *Lower panel*: To compare the polypeptide length derived from the WLC fits with the NhaA structure, the secondary structure model has been chosen. Data taken from [76]. (**b**) Single-molecule unfolding and refolding of NhaA. *Upper panel*: Mechanical unfolding of NhaA yields force–distance curves containing characteristic peaks (*upper red trace*). The scheme shows the major unfolding pathway of NhaA. WLC fits of major force peaks and side peaks are shown by *continuous* and *broken black lines*, respectively. *Lower panel*: Scheme of a refolding experiment. Initially, the intact NhaA molecule was unfolded. The tip was then retracted from the membrane up to about 80 nm, so that the last helical pair remained anchored in the membrane. Then the stretched polypeptide was relaxed as the tip was brought into proximity (about 10 nm) to the surface (*upper panel, blue trace*). After a specified delay time (ranging from 0.5 s to 15 s; *upper panel*), the polypeptide was pulled repeatedly while recording a force–distance trace (*red traces* in the *upper panel*). After certain refolding times, the force peaks, which were detected of the refolded peptide, occur at positions identical with those measured upon initial unfolding of NhaA. Since this spectrum exhibits the characteristic unfolding peaks of native NhaA, it follows that the peptide folded into the native structure. Data taken from [75]

recognition events [77, 78]. The identification and localization of specific receptor binding sites on biosurfaces such as cells and membranes are of particular interest in this context. For this purpose, high-resolution imaging should be combined with the detection of specific ligand–receptor binding in the way that binding sites can be easily assigned to biomolecular structures. By combining dynamic force microscopy imaging with molecular recognition force spectroscopy, an imaging method for the detection and localization of binding sites has recently been developed [79–81]. This new imaging mode allows separation of the topography signal from the simultaneously recorded recognition signal. Topography and recognition images are thereby acquired using tips, which are magnetically oscillated during scanning and contain ligand molecules directed against the cognate receptor on the surface. For cantilevers driven at frequencies below resonance, the surface contact only affects the downward deflections (i.e. the minima) of the oscillations, whereas binding of the ligand on the tip to the receptor molecule on the surface affects only the upward deflections (i.e. the maxima) of the oscillations. Accordingly, the oscillation minima are used to record the topography image, and the oscillation maxima provide the data for the recognition image.

Figure 14.10 shows the application of the simultaneous imaging of topography and recognition sites on cells [82]. Real-time visualization and quantification of receptor binding sites on cell surfaces is a fundamental challenging task in cell biology. This is normally achieved by using immunofluorescence techniques such as immunostaining. However, the observed lateral resolution is not better than 200 nm. Recently, the simultaneous imaging of topography and recognition sites has been applied to vascular endothelial (VE) cells in order to localize VE-cadherin bind-

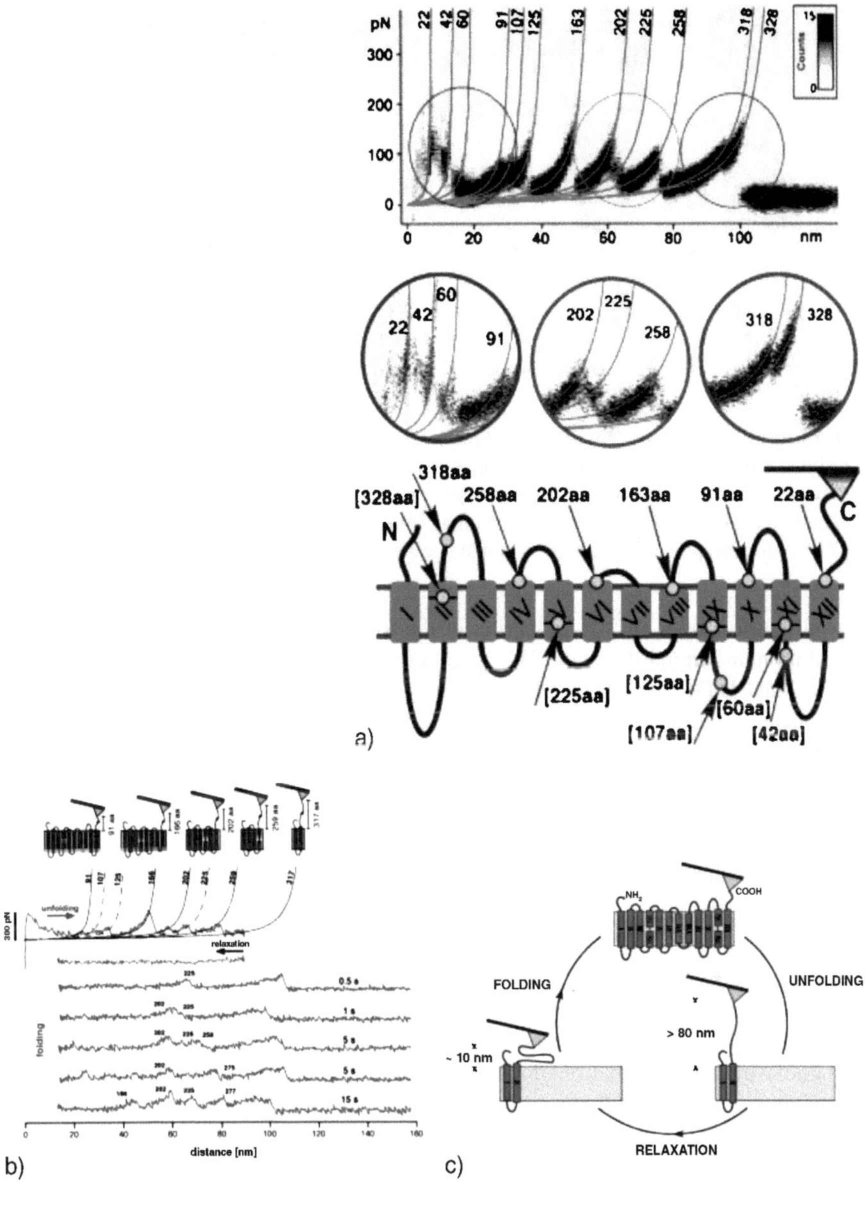

ing sites on the cell membrane [82]. In this study, topography (Fig. 14.10a) and recognition (Fig. 14.10b) images were recorded on gently fixed cells using AFM tips, which were chemically functionalized with fibrinogen. Besides the prominent role of fibrinogen in haemostasis, fibrinogen molecules can bind specifically to VE-cadherins on the cell membrane. Many specific interactions were observed in the recognition image (Fig. 14.10b). The recognition dark spots in the recognition image

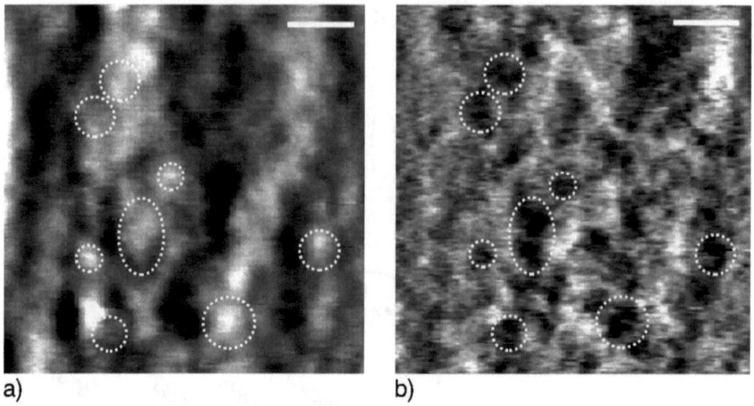

Fig. 14.10. Recognition imaging on cells. Simultaneous recorded topography (**a**) and recognition (**b**) images on gently fixed MyEnd cells acquired with a fibrinogen-coated AFM tip. The specificity of the detected interactions was tested by the addition of EDTA in the solution. *Scale bars* on both images are 200 nm. The z-scales range from 0 to 25 nm (from *black* to *white*) and 0 to 0.7 V for topography and recognition images, respectively. Data taken from [82]

reveal the locations of VE-cadherins in the topographical image with high lateral resolution and high efficiency. The specificity of the recognition process was tested by adding EDTA in solution. After injection, the dark spots in the recognition image disappeared, while the topographical image did not change (not shown). Thus, the simultaneous investigation of both topography and recognition open up a wide field of applications for investigating biological structure–function relationships in native environments on the nanometer scale, because the technique can map composition on top of a topographical image and can detect compositional changes occurring during biological processes.

14.8 Concluding Remarks

By using atomic force microscopy as an imaging technique one can at present gain the fine structural information of native membranes and protein organizations such as membranes of photosynthetic bacteria, bacterial S-layers, nuclear pore complexes, or even of viral arrangement on an artificial cell membrane under near physiological conditions. Due to continuous progress in the technical aspects of the AFM and in tip functionalization procedures, the investigation of receptor–ligand interactions on living cell has become possible. Moreover, the AFM as a force tool enables the dynamic pathways of the mechanical unfolding and subsequent refolding of single proteins in cell membranes to be followed. Thus, details in the intramolecular mechanics of protein substructures and of molecular interactions that apparently stabilize secondary structural elements are elucidated. The recent development of the TREC technique demonstrates a way, alternative to fluorescence studies, to visualize, identify, and quantify local receptor binding sites and to assign their locations to the topographical features of the cell surface with nm lateral resolution.

Thus, with improvements of the sensitivity and acquisition speed the AFM will open the way to fundamental challenging tasks in cell molecular biology.

Acknowledgements. This work was supported by Austrian Science Foundation projects P14549/P15295, the GEN-AU initiative of the Austrian Ministry of Education, Science and Culture, and EU projects Tips4Cells, NASSAP, HFSP RGP 0053/2004C and Force Tools. We thank Dr R. Zhu for the careful reading of the manuscript.

References

1. Binnig G, Quate CF, Gerber C (1986) Phys Rev Lett 56:930
2. Horber JK, Miles MJ (2003) Science 302:1002
3. Mueller DJ, Janovjak H, Lehto T, Kuerschner L, Anderson K (2002) Prog Biophys Mol Biol 79:1
4. Scheuring S, Levy D, Rigaud JL (2005) Biochim Biophys Acta 1712:109
5. Malkin A, Plomp J, McPherson A (2005) Meth Mol Biol 292:85
6. Jeremic A, Kelly M, Cho SJ, Stromer MH, Jena BP (2003) Biophys J 85:2035
7. Franz CM, Muller DJ (2005) J Cell Sci 118:5315
8. Zaman MS, Goyal A, Dubey GP, Gupta PK, Chandra SL, Das TK, Ganguli M, Singh Y (2005) Microsc Res Tech 66:307
9. Engel A, Muller DJ (2000) Nat Struct Biol 7:715
10. Allison DP, Hinterdorfer P, Han W (2002) Curr Opin Biotechnol 13:47
11. Rief M, Oesterhelt F, Heymann B, Gaub HE (1997) Science 275:1295
12. Marzsalek PE, Oberhauser AF, Pang YP, Fernandez JM (1998) Nature 396:661
13. Smith S, Cui Y, Bustamante C (1996) Science 271:795
14. Strick TR, Allemend JF, Bensimon D, Bensimon A, Croquette V (1996) Biophys J 271:1835
15. Veigel C, Coluccio LM, Jontes JD, Sparrow JC, Milligan RA, Molloy JE (1999) Nature 398:530
16. Tanaka H, Homma K, Iwane AH, Katayama E, Ikebe R, Saito T, Yanagida T, Ikebe M (2002) Nature 415:129
17. Lehenkari PP, Horton MA (1999) Biochem Biophys Res Com 259:645
18. Benoit M, Gabriel D, Gerisch G, Gaub HE (2000) Nat Cell Biol 2:313
19. Evans E (2001) Annu Rev Biophys Biomol Struct 30:105
20. Kienberger F, Ebner A, Gruber HJ, Hinterdorfer P (2006) Acc Chem Res 39:29
21. Sarid D (1991) Scanning Force Microscopy, Oxford University Press, New York
22. Putman CAJ, Vanderwerf KO, de Grooth BG, Vanhulst NF, Greve J (1994) Appl Phys Lett 64:2454
23. Hansma PK, Cleveland JP, Radmacher M, Walters DA, Hillner PE, Bezanilla M, Fritz M, Vie D, Hansma HG, Prater CB, Massie J, Fukunaga L, Gurley J, Elings V (1994) Appl Phys Lett 64:1738
24. Han W, Lindsay SM, Jing T (1996) Appl Phys Lett 69:1
25. Kienberger F, Stroh C, Kada G, Moser R, Baumgartner W, Pastushenko V, Rankl C, Schmidt U, Muller H, Orlova E, LeGrimellec C, Drenckhahn D, Blaas D, Hinterdorfer P (2003) Ultramicroscopy 97:229
26. Muller DJ, Schabert FA, Buldt G, Engel A (1995) Biophys J 68:1681
27. Kienberger F, Zhu R, Moser R, Blaas D, Hinterdorfer P (2004) J Virol 78:3203
28. Lantz M, Liu YZ, Cui XD, Tokumoto H, Lindsay SM (1999) Interface Anal 27:354
29. Moeller C, Allen M, Elings V, Engel A, Muller DJ (1999) Biophys J 77:1050
30. Muller DJ, Amrein M, Engel A (1997) J Struct Biol 119:172
31. Scheuring S, Sturgis JN (2005) Science 309:484

32. Ando T, Kodera N, Takai E, Maruyama D, Saito K, Toda A (2001) Proc Natl Acad Sci USA 98:12468
33. Scheuring S, Reiss-Husson F, Engel A, Rigaud JL, Ranck JL (2001) The EMBO Journal 20:3029–3035
34. Pum D, Neubauer A, Gyoervary E, Sara M, Sleytr UB (2000) Nanotechnol 11:100
35. Sara M, Sleytr UB (2000) J Bacteriol 182:859
36. Sara M, Pum D, Schuster B, Sleytr UB (2005) J Nanosci Nanotechnol 5:1939
37. Ebner A, Kienberger F, Huber C, Kamruzzahan ASM, Pastushenko V, Tang J, Kada G, Gruber HJ, Sleytr UB, Sara M, Hinterdorfer P (2006) Chem Bio Chem 7:588
38. Pante N, Aebi U (1993) J Cell Biol 122:977
39. Shahin V, Albermann L, Schillers H, Kastrup L, Schaefer C, Ludwig Y, Stock C, Oberleithner H (2005) J Cell Phys 202:591
40. Schaefer C, Shahin V, Albermann L, Hug MJ, Reinhardt J, Schillers H, Schneider SW, Oberleithner H (2002) Proc Natl Acad Sci USA 99:7154
41. Chiu W, Baker ML, Jiang W, Dougherty M, Schmid MF (2005) Structure 13:363
42. Kuznetsov YG, Malkin AJ, Lucas RW, Plomp M, McPherson A (2001) J Gen Virol 82:2025
43. Kienberger F, Rankl C, Pastushenko V, Zhu R, Blaas D, Hinterdorfer P (2005) Structure 13:1247
44. Kuznetsov YG, Victoria JG, Robinson W, McPherson A (2003) J. Virol. 77:11896
45. Svoboda K, Schmidt CF, Schnapp BJ, Block SM (1993) Nature 365:721
46. Evans E, Ritchie K, Merkel R (1995) Biophys J 68:2580
47. Clausen-Schaumann H, Seitz M, Krautbauer R, Gaub HE (2000) Curr Opin Chem Biol 4:524
48. Hinterdorfer P, Baumgartner W, Gruber HJ, Schilcher K, Schindler H (1996) Proc Natl Acad Sci USA 93:3477
49. Hinterdorfer P (2002) Meth Cell Biol 68:115
50. Hinterdorfer P, Gruber HJ, Kienberger F, Kada G, Riener C, Borken C, Schindler H (2002) Coll Surf B 23:115
51. Grandbois M, Dettmann W, Benoit M, Gaub HE (1999) Science 283:1727
52. Hinterdorfer P, Schilcher K, Baumgartner W, Gruber HJ, Schindler H (1998) Nanobiology 4:177
53. Lyubchenko YL, Gall AA, Shlyakhtenko LS (2001) Meth Mol Biol 148:569
54. Riener CK, Stroh CM, Ebner A, Klampfl C, Gall AA, Romanin C, Lyubchenko YL, Hinterdorfer P, Gruber HJ (2003) Anal Chim Acta 479:59
55. Hinterdorfer P, Kienberger F, Raab A, Gruber HJ, Baumgartner W, Kada G, Riener C, Wielert-Badt S, Borken C, Schindler H (2000) Single Mol 1:99
56. Riener CK, Kienberger F, Hahn CD, Buchinger GM, Egwim IOC, Haselgruebler T, Ebner A, Romanin C, Klampfl C, Lackner B, Prinz H, Blaas D, Hinterdorfer P, Gruber HJ (2003) Anal Chim Acta 497:101
57. Kienberger F, Pastushenko V, Kada G, Gruber HJ, Riener C, Schindler H, Hinterdorfer P (2000) Single Mol 1:123
58. Pfister G, Stroh CM, Perschinka H, Kind M, Knoflach M, Hinterdorfer P, Wick G (2005) J Cell Sci 118:1587
59. Bell GI (1978) Science 200:618
60. Evans E, Ritchie K (1997) Biophys J 72:1541
61. Evans E, Ritchie K (1999) Biophys J 76:2439
62. Strunz T, Oroszlan K, Schumakovitch I, Guntherodt H, Hegner M (2000) Biophys J 79:1206
63. Merkel R, Nassoy P, Leung A, Ritchie K, Evans E (1999) Nature 397:50
64. Chen A, Moy VT (2000) Biophys J 78:2814
65. Zhang X, Wojcikiewicz E, Moy VT (2002) Biophys J 83:2270
66. Puntheeranurak T, Wildling L, Gruber HJ, Kinne KH, Hinterdorfer P (2006) submitted

67. Xia X, Lin JT, Kinne RK (2003) Biochemistry 42:6115
68. Raja MM, Tyagi NK, Kinne RK (2003) J Biol Chem 278:49154
69. Carrion-Vazquez M, Oberhauser AF, Fowler SB, Marszalek PE, Broedel SE, Clarke J, Fernandez JM (1999) Proc Natl Acad Sci USA 96:3694
70. Rief M, Gautel M, Oesterhelt F, Fernandez JM, Gaub HE (1997) Science 276:1109
71. Marszalek PE, Pang YP, Li H, El Yazal J, Oberhauser AF, Fernandez JM (1999) Proc Natl Acad Sci USA 96:7894
72. Li H, Rief M, Oesterhelt F, Gaub HE (1998) Adv Mater 3:316
73. Rief M, Clausen-Schaumann H, Gaub HE (1999) Nat Struct Biol 6:346
74. Hertadi R, Gruswitz F, Silver L, Koide A, Koide S, Arakawe H, Ikai A (2003) J Mol Biol 333:993
75. Kedrov A, Ziegler C, Janovjak H, Kuehlbrandt W, Muller DJ (2004) J Mol Biol 340:1143
76. Kedrov A, Janovjak H, Ziegler C, Kuehlbrandt W, Muller DJ (2006) J Mol Biol 355:2
77. Willemsen OH, Snel MME, van der Werf KO, de Grooth BG, Greve J, Hinterdorfer P, Gruber HJ, Schindler H, van Kyook Y, Figdor CG (1998) Biophys J 57:2220
78. Raab A, Han W, Badt D, Smith-Gill SJ, Lindsay SM, Schindler H, Hinterdorfer P (1999) Nat Biotech 17:1999
79. Stroh CM, Ebner A, Geretschlaeger M, Freudenthaler G, Kienberger F, Kamruzzahan ASM, Smith-Gill SJ, Gruber HJ, Hinterdorfer P (2004) Biophys J 87:1981
80. Stroh CM, Wang H, Bash R, Ashcroft B, Nelson J, Gruber H, Lohr D, Lindsay SM, Hinterdorfer P (2004) Proc Natl Acad Sci USA 101:12503
81. Ebner A, Kienberger F, Kada G, Stroh CM, Geretschlager M, Kamruzzahan ASM, Wildling L, Johnson WT, Ashcroft B, Nelson J, Lindsay SM, Gruber HJ, Hinterdorfer P (2005) Chem Phys Chem 6:897
82. Chtcheglova LA, Waschke J, Kada G, Drenckhahn D, Hinterdorfer P (2006) (manuscript in preparation)

21 Atomic Force Microscopy: Interaction Forces Measured in Phospholipid Monolayers, Bilayers, and Cell Membranes

Zoya Leonenko · David Cramb · Matthias Amrein · Eric Finot

From volume IX

Abstract. Atomic force microscopy (AFM) is a powerful technique which is commonly used to image surfaces at the nanoscale and single-molecule level, as well as to investigate physical properties of the sample surface using a technique known as force spectroscopy. In this chapter, we review our recent research where we used AFM to investigate physical properties of phospholipid monolayers, bilayers, and cell membranes. We describe the experimental procedures for AFM imaging, force measurements, and theoretical models to analyze force spectroscopy data. The data obtained allowed correlations between AFM topography and local adhesion and mechanoelastic properties of supported lipid bilayers in water, supported pulmonary surfactant films in air, and the plasma membrane of epithelial type II cells. Finally, AFM was applied to help elucidate the effect of anesthetics and cholesterol present in the lipid films.

Key words: Phospholipid, Monolayer, Bilayer, Pulmonary surfactant, Bovine lipid extract surfactant, Epithelial type II cell, Atomic force microscopy, Adhesion, Cholesterol, Halothane

21.1 Introduction

Lipid membranes are integral components of the cellular organization in all living organisms [1]. As the major constituents of the biological membranes that form the outer boundary of cells and organelles they are involved in a vast number of biological processes. Their functions are to compartmentalize the cells, sort, regulate, and mediate biomolecular interactions. These complex functions are possible owing to the diverse physical-chemical properties of these membranes, which include two-dimensional fluidity, material elasticity, chemical diversity, and rich phase behavior [2]. Understanding the physical and chemical properties of lipid membranes [3,4] and how they control biological processes is critical to our understanding of the molecular mechanisms of many diseases and finding a cure for them. The variety in structure and physical properties of cell membranes governs their biomolecular assembly and plays an important role in many fundamental biological processes [5]. For example, small, condensed rafts of spingolipids, cholesterol, and raft-associated protein play a crucial role in signaling, intracellular trafficking of lipids, platelet activation, membrane fusion, and protein binding [6,7]. Lipid rafts present in a cell membrane serve as entry and exit sites for microbial pathogens, such as influenza virus, measles virus, and HIV. Lipid–protein mixtures not only form membranes. Pulmonary surfactant [8,9], which is a specific lipid–protein complex,

In: H. Fuchs, B. Bhushan (eds), *Biosystems - Investigated by Scanning Probe Microscopy*,
DOI 10.1007/978-3-540-74083-4_9, © Springer-Verlag Berlin Heidelberg 2010

forms a molecular film that covers the lung's interface with the air [10]. The major functions of pulmonary surfactant are to reduce the surface tension of the airspace–liquid interface, provide stability to the alveolar structure, and reduce the work of breathing [11]. Understanding the lateral distribution and interactions among the lipids and proteins in pulmonary surfactant monolayers is extremely important in order to develop methods of preventing alveolar collapse in cases of surfactant deficiency such as in premature infants or dysfunctional surfactant such as in adult respiratory distress syndrome (ARDS) [12].

Supported phospholipid bilayers (SPBs) [13] and monolayers are accepted and convenient model systems which mimic many fundamental properties of biological membranes and pulmonary surfactant. SPBs are composed of phospholipids adsorbed to a planar solid support. They also are of importance for biotechnology applications, as they form membrane-mimetic materials useful for creation of new classes of biosensors, diagnostic tools, and high-throughput characterization platforms for rapid and early detection of interactions between cells and their environment [14, 15]. Such lipid films can serve as templates into which functional biomolecules can be incorporated, creating a powerful platform for the development of nanosensor and nanodevices. For biomedical and nanotechnology applications it is important to control processes involving the interaction of lipid membranes and monolayers with DNA and proteins, as well as polymers and inorganic nanoparticles. Adhesion of particles to the supported lipid films is important in biotechnology and medicine applications, for example, to study toxicity of nanoscale air pollutants and also for the development of effective pulmonary drug delivery systems [16].

Atomic force microscopy (AFM) has proven to be a valuable tool for imaging "soft" biological samples such as lipid films [17–21] especially in intermittent contact mode [22–24]. With AFM, one can not only image membranes in their native hydrated state. One can also measure the forces of interaction between the AFM probe and the sample, obtaining information about the physical properties of biological systems [25–29].

Measuring the physical and chemical properties of lipid membranes and monolayers involves measuring a multitude forces of weak noncovalent bonds (e. g., electrostatic, van der Waals, and/or hydrogen bonds) or hydrophobic interactions between geometrically complementary surfaces. To understand a response of a biological system to its environment one needs to measure the physical and chemical properties of the macromolecular assembly, such as membranes and monolayers, and how these properties are affected by mechanical stress, thermal change, and/or molecular adsorption. It concerns the mechanical properties such as two-dimensional fluidity and anisotropic elasticity of the membrane as well as thermodynamic properties underlying its rich phase behavior. Those bulk properties can also be investigated by a thermodynamic approach such as calorimetry, densitometry, contact angle measurement, or analysis of adsorption kinetics using surface plasmon resonance or ellipsometry. On a molecular level, molecular dynamics simulation has been employed, providing an insight into the organization of lipid structures and the stability. AFM is an ideal tool to measure directly the molecular force from 0.1 pN up to micronewtons between two surfaces or even single molecules. On approach of the AFM probe tip to the sample, the force–distance curve can be used to characterize surface properties, such as van der Waals and electrostatic forces, solvation, hydration, and compression-related steric forces [30]. The retraction force curves often show a hysteresis referred to as an adhesion "pull-off" event, which can be used to estimate the adhesion forces. Many experimental force data are now available in the literature, and theoretical models have been developed for the analysis of forces acting between two solid surfaces [31–36]. In contrast, only a few publications have addressed force measurements of soft thin layers [37] such as thiol self-assembled monolayers [38–41] and phospholipid bilayers [42–44].

Here we review our recent research on interaction forces measured with AFM on SPBs in water and supported phospolipid monolayers in air, as models for biological membrane and pulmonary lung surfactant films, and on epithelial type II cells which cover the lung's interface with the air.

21.2 Phase Transitions of Lipid Bilayers in Water

The complex structural dynamics of phospholipid bilayers are governed by temperature-dependent parameters such as average interfacial area per lipid, bilayer thickness, and disorder of the hydrophobic tails. Although SPBs differ from free-standing membranes such as those found in vesicles or liposomes, they serve as good models for biological membranes. The SPB can exist in several lamellar phases: gel phase, liquid-crystalline or fluid phase, subgel phase, and ripple phase [45]. Often, phase behavior is dominated by the main L_{β}–L_{α} (gel–fluid) transition. A large number of stable, metastable, and transient lamellar gel structures can be adopted by different lipids. For saturated phosphatidylcholines, such as 1,2-dipalmitoyl phosphatidylcholine (DPPC), there are four recognized lamellar phases, namely, a liquid-crystalline phase (L_{α}) or phases with ordered hydrocarbon chain arrangements, a ripple phase (P_{β}), a gel phase (L_{β}), and "subgel" or "crystal" phase (L_{c}) [46]. Phase-related bilayer properties have been examined using a variety of experimental techniques [47–50]. The phase transition from the gel phase to the liquid-crystalline phase is shown schematically in Fig. 21.1. Physical properties of lipids such as density and thickness have been studied by differential scanning calorimetry, pressure calorimetry or dilatometry, and densitometry and other methods [51, 52]. Temperature-dependent AFM studies allow for visualizing these phase transitions and provide valuable information about changes in structural and physical properties of bilayers during phase transitions. Temperature-dependent AFM studies involving imaging bilayers have been reported [53–57]. In addition, we have examined the force measurements as a function of temperature [58].

In this section we review our results on force measurements in SPBs during phase transitions associated with melting or anesthetic incorporation. Dioleoyl phosphatidylcholine (DOPC; lyophilized or in chloroform solution) (Sigma, Oakville, ON, USA) and DPPC (Avanti Polar Lipids, Alabaster, ME, USA) were used without further purification. Phosphate buffer (pH 6.8) and distilled, nanopure water were used in the preparation of all vesicles. Freshly cleaved ASTMV-2 quality, scratch-free ruby mica (Asheville-Schoonmaker Mica) was used as the solid support. Supported planar bilayers were prepared for AFM imaging by the method of vesicle fusion [59, 60]. All vesicles were prepared as previously reported [59]. Aliquots of

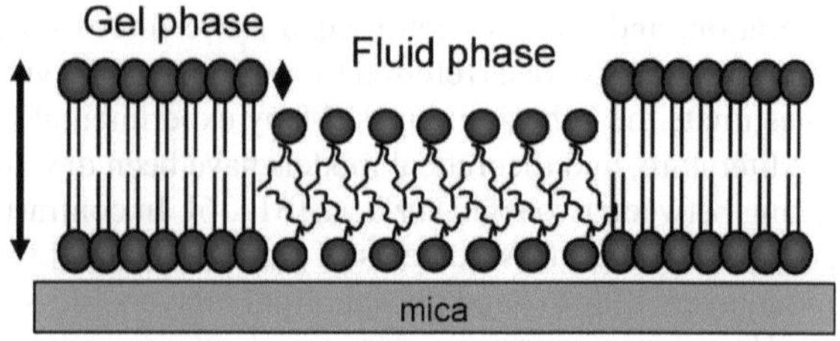

Fig. 21.1. Domain formation in a phospholipid bilayer supported on mica. The fluid phase domains correspond to a lower thickness than the gel-phase domains

the vesicle solution were deposited on freshly cleaved pure mica. After a controlled period of time the mica was gently rinsed with nanopure water. Surfaces were imaged with an atomic force microscope (Pico SPM, Molecular Imaging) equipped with an AFMS-165 scanner. The nominal spring constants of Au–Cr-coated MAClevers used were approximately 0.6 N/m. The tip radius of curvature was typically 25 nm. The scan rate was 20 μm/s. All imaging and force measurements were performed in a liquid cell under nanopure water. For temperature-dependence experiments the temperature was varied from 22 to 65 °C, within 0.1 °C accuracy with a 1 °C/min ramp.

21.2.1 Morphology Change During Lamellar Phase Transition

21.2.1.1 Temperature-Dependent Phase Transition

A DPPC phospholipid bilayer normally exists in the gel phase at room temperature. When heated above the melting transition temperature it transforms into the fluid phase. Domains of lower thickness are readily formed during this process. Such domains are shown schematically in Fig. 21.1 and are easily observed by AFM imaging (Fig. 21.2).

A DPPC supported bilayer was imaged while it was heated from room temperature to 65 °C and also while it was cooled back to room temperature. The topography images (Fig. 21.2) show the typical formation of lipid domains during the phase transition. In Fig. 21.2, panel A, a DPPC bilayer is in the gel phase, L_β, at 22 °C. When sample was heated, several phase transitions were observed as a function of increasing temperature: a broad main transition was observed at 42–52 °C, and another transition at 53–60 °C within the fluid phase. The first transition, at 42–52 °C, is attributed to the main L_β–L_α transition, and the second transition, between 53–60 °C, to the formation of a fluid disordered phase, perhaps with interdigitated or partly interdigitated lipid chains. Slowly cooling the system back showed that the two higher-temperature transitions at 42–52 and 53–60 °C are reversible and that the bilayer can be restored to its initial thickness [61]. A DOPC bilayer, which exists in a fluid phase at room temperature, when heated shows thickness reduction as well as an increase in surface coverage [62], but no domain formation. Previous studies have established the L_β–L_α phase transition temperature for DPPC at 41 °C [63]. The presence of two broad transitions that we observed on supported bilayers can be attributed to the effect of the mica support and to separating the phase transitions

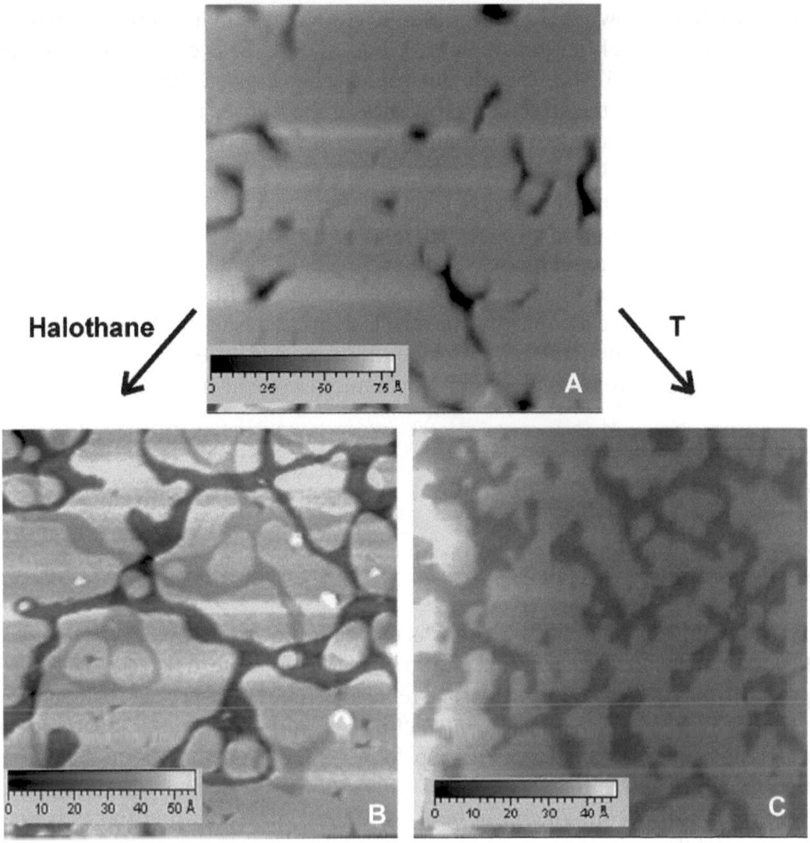

Fig. 21.2. Atomic force microscopy (AFM) topography images (scan size 1.5 μm by 1.5 μm) showing a phase transition in a 1,2-dipalmitoyl phosphatidylcholine (DPPC) bilayer. **A** DPPC bilayer at room temperature, 22 °C. **B** DPPC bilayer with excess of halothane (60 μl) [57]. **C** 50 °C—melting transition [61]. The *scale bars* show the deviation in height

in the bilayer's upper and lower leaflets. A similar effect was reported for supported bilayer transitions observed using differential scanning calorimetry by Yang and Appleyard [64].

21.2.1.2
Phase Transition Due to Incorporation of Halothane

Understanding the changes in physical and chemical properties of biological membranes owing to the incorporation of anesthetics is of great interest for understanding the mechanism of anesthetic action. Currently, it is understood that anesthetic directly affects membrane proteins; however, because they are lipophilic, they partition into the membrane and may interact with membrane proteins via the lipid bilayer.

AFM imaging of a DPPC bilayer (Fig. 21.2, panel B) shows domain formation in the presence of general anesthetic halothane similar to what we observed during heating. We have shown that changes in bilayer thickness induced by halothane are similar to the first melting transition [58]. For a DPPC SPB containing halothane, only one transition at 52–60 °C was evident during heating and cooling. The presence of halothane changes the structural properties of the bilayer in such a way that the other transitions observed for the pure bilayer are not possible. We assume that in the presence of halothane, the water interface layer is disturbed by halothane molecules, thus eliminating the low-temperature transition. In spite of the visual similarity between an anesthetic-induced domain formation (Fig. 21.2, panel B) and the heat-induced gel–liquid phase transition (Fig. 21.2, panel C), observed by AFM imaging, the mechanism of anesthetic action is likely to be different from the effect of membrane melting. We assume that the physical properties of the domains produced in lipid bilayers by halothane and by temperature are also different. To address these physical properties we used the atomic force microscope also as a force apparatus.

21.2.2 Change in Forces During Phase Transition

21.2.2.1 Force Measurements

In AFM force measurements, the probe is moved towards the sample and cantilever deflection is measured as a function of the extension of the piezoelectric tube. During approach to the surface, the attractive and repulsive forces measured are characterized by van der Waals and electrostatic interactions as well as solvation, hydration, and compression-related steric forces [26, 36, 52, 65]. The probe is then retracted from the surface. The retraction force curves often show a hysteresis referred to as an adhesion pull-off event, which can be used to determine adhesion forces between the probe and the sample.

In our experiments, we imaged the supported bilayer first in tapping MAC mode. After the image was complete and had been saved we switched the regime into the contact mode and then using the same cantilever, we measured forces of interaction by positioning the atomic force microscope tip on different locations using the image obtained in tapping mode.

It is a well-known fact that the force interactions depend on the velocity of the surface approach [26], especially for soft, viscoelastic materials. We varied the frequency sweep over about four decades, from 0.01 to 10 Hz. The Z-range was 50 nm. Note that varying the scan frequency affects simultaneously the velocity at which the tip is withdrawn from the bilayer and the contact time between the tip and the sample. At a fixed applied load, increasing the frequency sweep rate increases the tip velocity, but decreases the contact time. Data were collected over a time period of 2 h at ten locations on the bilayer and a total of 100 force curves were analyzed for each sample. Force curves were saved as plots and then converted into text files for analysis. After the raw data had been collected they were converted into the "force versus separation plots", averaged, and analyzed.

Here we review our results on force measurements in DOPC and DPPC SPBs during heating above T_M from 22 to 65 °C. We also compare them with results obtained on bilayers saturated with halothane at 20 °C.

21.2.2.2 AFM Force Analysis

Force measurements using the atomic force microscope (described in Sect. 21.2.2.1) give us raw data of cantilever deflection δV versus piezo movement z_{piezo}. Raw data were transformed numerically using the MatLab environment. Raw data (Z_c versus Z_p) were then converted into force F versus surface–tip separation D using Hooke's law: $F = kZ_c$, where k is the spring constant of the cantilever, and the geometric relationship $D = Z_c - Z_p$ for incremental changes.

Statistical data were extracted from a large set of measurements. The bin size of the histogram was determined statistically by dividing the range of measurements by the square root of the number of measurements. The average of the adhesion force and the standard deviation were obtained from the analysis of the whole set of force curves, which were represented in histogram form, $P(x)$, and then adjusted using Gaussian and Poisson distributions.

21.2.2.3 Adhesion Forces

To extract information about adhesion forces we analyzed the retrace part of the force curve. Raw data of the cantilever deflection corresponding to the retraction of the tip from the surface, after contact, are shown in Fig. 21.3.

Force data were collected as a function of temperature for a DPPC SPB. When a pure DPPC bilayer was heated, the temperature increase resulted in a gradual change of the force profile. The force required to detach the tip from the bilayer increased from 3 nN at 36 °C to 10 nN at 50 °C. In addition, the jump out of the contact was not abrupt, compared with that for a "solid-like" surface (i.e., mica). The distance the tip travels before it separates from the surface increased with increasing temperature. As van der Waals forces are known to vary only slightly with temperature, the change in the profile must be due to tip–bilayer interactions that depend on the bilayer phase. Incorporation of halothane into DPPC also leads

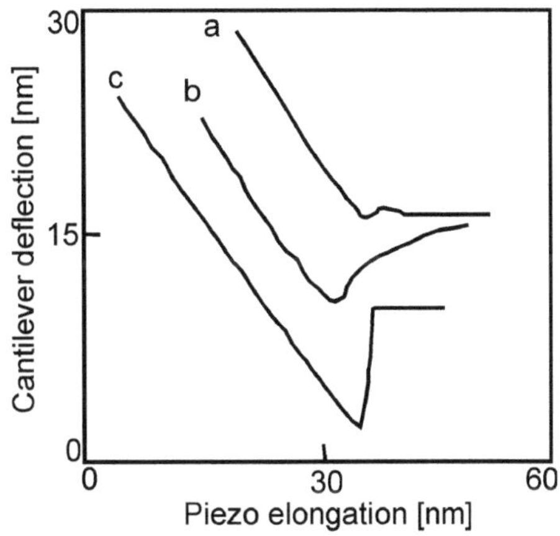

Fig. 21.3. Cantilever deflection dZ_c as a function of the piezotube elongation when the tip is retracted from the sample. DPPC bilayers deposited on mica were measured under three conditions: **a** the DPPC bilayer at 22 °C; **b** the DPPC bilayer at 50 °C; **c** the DPPC bilayer with halothane incorporated at 22 °C

to a significant change in the force profile. As a bilayer adsorbed on mica is known to screen the van der Waals forces [66], the huge adhesion force observed cannot originate solely from the van der Waals forces. This adhesion behavior can be compared with that observed in the phase transition on DPPC from the gel to the liquid crystal observed on increasing the temperature (Fig. 21.3, plots b and c). Halothane induces then a transition of the bilayer into the fluid phase in a similar manner as melting, and induces the increase in adhesion, although the adhesion with excess halothane was greater than the adhesion observed on pure DPPC at elevated temperature.

To analyze the force histogram (Fig. 21.4), we employed Poisson statistics. It has been demonstrated that the force of a unit interaction between an atomic force microscope tip and a surface can be determined from a statistical analysis of a series of detachment force measurements. For a statistical analysis based on adhesion force originating from a discrete number n of individual interactions or bonds, F_s was used. The total force distribution follows Poisson statistics, where both the adhesion force F_{adh} and the variance σ originate from a number of individual tip–surface bonds, n:

$$F_{adh} = nF_s \quad \text{and} \quad \sigma^2 = nF_s^2 \,. \tag{21.1}$$

The force of one bond, F_s, is therefore given by the square of the variance of the force divided by the mean adhesion force F_{adh}. The value n is the ratio between F_{adh} and F_s. This analysis has the advantage that knowledge of the mean radius of curvature is not required [67] and gives information on the nature of the bilayer. This analysis allows one to correlate microscale changes in the bilayer physical properties with

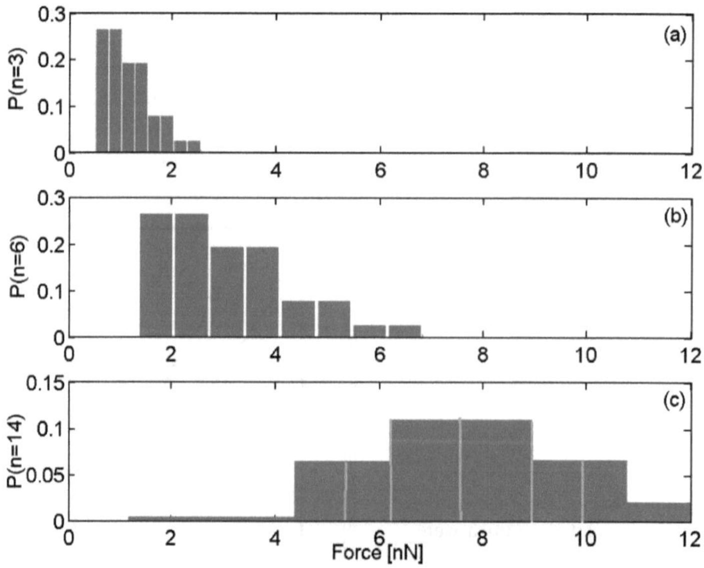

Fig. 21.4. Distribution of adhesion force F_{adh}, measured in phospholipid bilayers in water: **a** DPPC $F_s = 0.3$ nN, $n = 3$; **b** DPPC $F_s = 0.5$ nN, $n = 6$ [58]; **c** DPPC + halothane $F_s = 0.5$ nN, $n = 14$ [58]

Table 21.1. Experimental adhesion force F_{adh} and its standard error σ on a 1,2-dipalmitoyl phosphatidylcholine (DPPC) bilayer at room and elevated temperature and DPPC with halothane, measured at room temperature; calculation of the number of bonds n and the mean force of a single bond F_s

	DPPC[a] 22 °C	DPPC[b] 22 °C	DOPC[a] 22 °C	DPPC[a] 60 °C	DPPC + halothane[b] 22 °C
F_{adh} (nN)	1	1.5	3	5	7
σ (nN)	0.6	0.8	1.2	2	2
F_s (nN)	0.3	0.4	0.5	0.8	0.5
n	3	4	6	6	14

DOPC dioleoyl phosphatidylcholine
[a] From [61]
[b] From [58]

the molecular structure and the mobility of individual lipid molecules. The lipids in biological membranes have a high lateral mobility since they can easily exchange their positions. The fluidity of the membrane can be then thought of as the lateral motion of the constituents of the membranes. This fluidity should depend on the chain length and the composition of the membranes. The mobility of tails can be seen as the number of molecules n in contact with the AFM probe. The larger n is, the higher the mobility. The change in mobility increases the number of bonds n and therefore F_{adh} and σ.

The statistical distribution of the adhesion forces (Table 21.1) clearly demonstrates that adhesion forces observed for a pure DPPC bilayer increase with heating and increase even more with halothane incorporation.

A fluid-phase DPPC bilayer at high temperature is typically ascribed to a higher number of tip–sample bonds [61]. The force of a single bond is increased in a fluid phase. In addition, in a fluid phase, n increases on decreasing the speed of the tip motion v. The number of bonds, n, involved is higher in the fluid phase. It has been shown previously that the force required to remove a single lipid from a bilayer is of the order 30–140 pN [68,69], depending on the lipid and on the technique employed. The lowest values that we measure are at least twice this range, suggesting that we do not observe single lipid binding events. Instead, our unit "bonds" must represent patches containing more than three lipids. The detachment force between the patch and the tip must be smaller than the extraction force of the entire patch otherwise we would see evidence of lipids coating the tip. This is consistent with our previous work where continuous imaging of bilayers displayed no resolution degradation or changes in force curves, which would result from uncontrollable lipid coating the atomic force microscope tip.

The effect of halothane is similar to the effect of melting. A DPPC bilayer with halothane corresponds to higher adhesion force, higher standard error, and a larger number of single binding events between the atomic force microscope tip and the sample. This number, $n = 14$, is more than 2 times larger than for a fluid phase produced by melting. The single force F_s was not altered significantly by the presence of halothane. The presence of halothane in the bilayer increases, therefore, the fluidity of the DPPC molecules in the bilayer. The number of binding events, n, could also be higher owing to the contribution of halothane molecules themselves, interacting with the atomic force microscope tip. The unit force F_s for fluid-phase DPPC is 0.8 compared with 0.5 for the DOPC bilayer (Table 21.1); the umber of unit bonds is 6 for both fluid-phase DPPC and DOPC bilayers. This correlation permits evaluation of a bilayer fluidity and is in a good agreement with imaging data. Adhesion forces depend, therefore, on the physical properties of the bilayer and the force analysis gives valuable information on the nature of the bilayer.

21.2.2.4 Repulsive Forces

Repulsive forces were extracted by analyzing the approach part of experimental force curves. Figure 21.5 shows plots of the repulsive forces measured while the atomic force microscope tip was approaching the sample surface of pure mica for a DPPC bilayer at room temperature, a DPPC bilayer at 50 °C [61], and a DPPC bilayer with an excess of halothane [58].

For pure DPPC, three exponential regimes clearly exist, one at the long range from 4 to 7 nm, another at short range between 0.3 and 4 nm, and the third, steep one in close contact with mica, which we will not consider.

The *long-range* interactions can be explained by *electrostatics*. Electrostatic repulsion likely originates from an effective charge density at the bilayer–water interface, where zwitterionic headgroups interact with water molecules. The approximate expression of the electrostatic force (also called double-layer force) is given by [70]

$$F = 4\pi\sigma R\psi_T \exp(-K_D D) , \quad (21.2)$$

where R is the tip radius, σ is the surface charge, ψ_T is the tip potential, and K_D is the inverse of the Debye length. K_D can be estimated to be 0.25/nm.

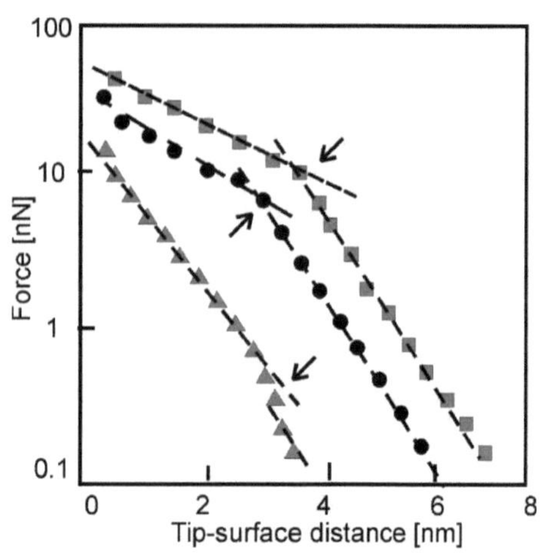

Fig. 21.5. Logarithmic representation of the force acting on the DPPC layer at 22 °C (*squares*), on the DPPC layer at 50 °C (*circles*), and on the DPPC layer with halothane at 22 °C (*triangles*)

The long-range force does not seem to be affected by temperature, as the two curves in the log plot are almost parallel (Fig. 21.5). In the presence of halothane, (Fig. 21.5), the long-range electrostatic forces appear to be reduced, in the range above 4 nm. Therefore, halothane clearly changes the surface charge density of the DPPC bilayer by replacing water molecules and reducing water shielding of the lipid bilayer, thereby decreasing the magnitude of the electrostatic forces.

The second, *short-range* regime, between 0.3 and 4 nm, is attributed to the *deformation of the bilayer*. Most studies on the bilayer report a jump corresponding to the piercing of the bilayer for forces ranging from approximately 3 to 25 nN [71–73]. This jump into contact is observed when the force gradient exceeds the spring constant of the cantilever k, which is the case when $k < 1\,\mathrm{N/m}$. Although the force sensitivity can be altered, a high cantilever spring constant ($k = 30\,\mathrm{N/m}$) was chosen for monitoring all the data during the penetration. The force gradient attributed to the layer indentation was measured to be around $10\,\mathrm{N/m}$ and the force needed to puncture the lipid bilayer is in good agreement with that found in other studies.

The repulsive force in the short-range regime at the contact of the bilayer cannot be easily treated using common elastic mechanical models such as Hertz's theory, or any other models based on continuum mechanics. The heads of the phospholipids are responsible for a crystalline elasticity, i. e., a change of the head-to-head distance (high elasticity but low rupture). The lipid chain behaves like rubber (low elasticity), and this behavior is related to the change of entropy (increase of order when one presses the layer). It results in the property that the cell membranes can be easily deformed in the plane but do not resist the expansion (distance between the headgroup cannot vary more than 2%). With our AFM experiments we are examining very small, discrete areas where nanoindentation induces not only macroscopic compression of the bilayer, but also nanoscale structural changes in the bilayer. Therefore, a molecular approach, considering steric forces [42], is preferable.

The exponential increase of the repulsive force likely originates from tail steric repulsion induced by the atomic force microscope tip compressing the bilayer. The tails in the bilayer will resist compression, because compression will increase the local concentration of the alkyl segments, increasing the free energy of the system. The steric forces of compression act on the bilayer until the loading force is close to the yield force, when the tip penetrates through the bilayer. The breakthrough force decreases as the temperature increases, and a DPPC bilayer at elevated temperatures shows similar instability in the force plot as a fluid-phase DOPC bilayer and a fluid-phase 1,2-dioleoyl-3-trimethylammonium propane (DOTAP) bilayer [44].

The steric force per unit area that the bilayer exerts on the tip can be modeled using the formalism developed in the work of de Gennes [74]. This model was further developed by Cappella and Dietler [26], where interaction between two polymer surfaces was considered. We modified this formula, considering the interaction of the AFM probe and the bilayer, which gives the formula for the steric repulsion force:

$$F_{\mathrm{ste}} = 100kTRL\gamma^{3/2}\exp(-2\pi D/L)\ , \tag{21.3}$$

where k is the Boltzmann constant, R is the tip radius of curvature, T is the temperature, γ is the tail density (inverse of the area per lipid), D is the distance between the atomic force microscope tip and the mica surface, and L is the uncompressed bilayer thickness. Similar to the equation for the two interacting polymer surfaces, this formula is valid in the range of distances where $0.2 < D/L < 0.9$. Since the tip radius is not well defined, it is difficult to estimate the tail density from the fit, but the transition between the two exponential regimes can be a good way to measure the thickness of the bilayer L.

The steric forces can, however, be used to determine the thickness of the layer adsorbed at the surface in good agreement with the image cross-section analysis. With the temperature increase, the steric force departs from a pure exponential law owing to the increase of the lateral mobility of the phospholipid tails. The transition between the electrostatic regime and the steric regime where the tip is at the top of the bilayer can be an effective way to estimate the bilayer thickness. At 22 °C, a transition occurs at $D = 4$ nm, corresponding to a thickness $L = D/0.9 = 4.5$ nm, which agrees with what is known from the literature (4.7 nm). With increasing temperature, the magnitude and the range of the steric force decrease contrary to the change in the electrostatic force. The steric force in a fluid phase, at elevated temperatures, is reduced by 50% compared with that a gel phase, as in Fig. 21.5. This can be explained by the reduction in the tail density with temperature. The tail density is reduced from $1/50$ to $1/67.1$ (considering an area per lipid of 50 Å^2 at room temperature and 67.1 Å^2 at 65 °C [75–77]), which makes 25% of difference and therefore a change of 40% in the force. In a fluid regime, i. e., at $T > 50$ °C, the shape of the steric part of the force–distance curve starts to depart from an exponential dependence and is similar to the linear abrupt jump observed for a fluid-phase DOTAP bilayer at room temperature. It appears that (21.3) is only valid for a gel-phase DPPC bilayer, where lateral motion of the lipids is lower.

With halothane incorporation (Fig. 21.5, triangles) the force is mostly governed by the bilayer deformation, described by steric forces. The induced phase transition of the DPPC has already been characterized by a decrease in the "effective thickness". The major difference with the temperature-induced phase transition comes from the fact that the incorporation of halothane changes the electrostatic interaction, which was not the case during the melting transition. Our force measurements confirmed that assumption and clearly showed the differences in the physical properties of the bilayer in thin domains, produced by the melting transition and partitioning of halothane.

When the adhesion forces are compared for DPPC and DOPC bilayers, they are similar for the fluid-phase DPPC bilayer at elevated temperatures and the fluid-phase DOPC bilayer at room temperature. The unit force F_s for fluid-phase the DPPC bilayer is 0.8 compared with 0.5 for the DOPC bilayer (Table 21.1); the number of unit bonds is 6 for both fluid-phase DPPC and DOPC bilayers. This correlation permits evaluation of a bilayer fluidity and is in a good agreement with imaging data. Therefore, adhesion forces depend on the physical properties of the bilayer, and force analysis gives important information on the nature of the bilayer. In the repulsive part of the force plot using a steric force model also provides an important insight into the properties of the bilayer and ability of lipids to resist compression. Steric forces reduce with the temperature increase for the DPPC bilayer and correlate well with low steric forces for the fluid-phase DOPC bilayer at room temperature. In general, a fluid-phase bilayer can be characterized by the reduced repulsive forces, compared with the gel-phase bilayer. The major contribution to this change is provided by changes in steric forces. Lipid compression is much easier in the fluid phase owing to the higher disorder of tails and mobility of lipids, and lower tail density.

Individual bond forces can be estimated for lipid bilayers and changes in binding number can be used as a distinctive feature of the increase of bilayer fluidity. A force analysis revealed that the interaction between the tip and the surface was governed by both the electrostatic forces, for longer range, and the steric forces, at shorter range. Electrostatic repulsion likely originates from an effective charge density at the bilayer–water interface, where zwitterionic headgroups interact with water molecules. The steric forces can be used to determine the thickness of the layer adsorbed at the surface in good agreement with the image cross-section analysis. With the temperature increase, the steric force departs from a pure exponential law owing to the increase of the lateral mobility of the phospholipid tails. The transition between the electrostatic regime and the steric regime where the tip is at the top of the bilayer can be an effective way to estimate the bilayer thickness.

The temperature-induced phase transition of DPPC has been characterized by a decrease in the, "effective thickness" and we also observed similar behavior for bilayers containing halothane [58]. The major difference comes from the fact that the incorporation of halothane changes the electrostatic interaction, which was not the case during the melting transition. Halothane is a small polar molecule and is assumed to increase the polarity of the bilayer at the interface, and has been shown to replace water molecules bound to lipids at the interface. Our force measurements confirmed that assumption and clearly showed the differences in the physical properties of the bilayer in thin domains, produced by the melting transition and partitioning of halothane.

21.3 Force Measurements on Pulmonary Surfactant Monolayers in Air

Pulmonary surfactant is a specific mixture of phospholipids and surfactant-specific proteins. It forms a molecular film at the interface of the hydrated lung epithelium with the air and thereby reduces the surface tension of the interface to near zero. This is required for normal respiration and structural stability of the lung. In adult respiratory distress syndrome (ARDS), surfactant function fails. As a result, the lung is less compliant, the gas exchange area is reduced, and blood oxygenation is strongly decreased. The surfactant film is also the first barrier of airborne particles in polluted air. When inhaled, ultrafine particles with an aerodynamic diameter of 100 nm or less produce a major health threat. They cross the surfactant film, cross the lung epithelium, and enter the bloodstream. We describe force spectroscopy experiments aimed at a better understanding of surfactant failure in ARDS. Force spectroscopy was used to study the interactions between airborne particles and the surfactant. Particles (modeled by an atomic force microscope tip) interact with the surfactant film before they get into close contact because of the electrical surface potential of the film. This interaction depends on surface potential, which is locally strongly variable, depending on the film composition.

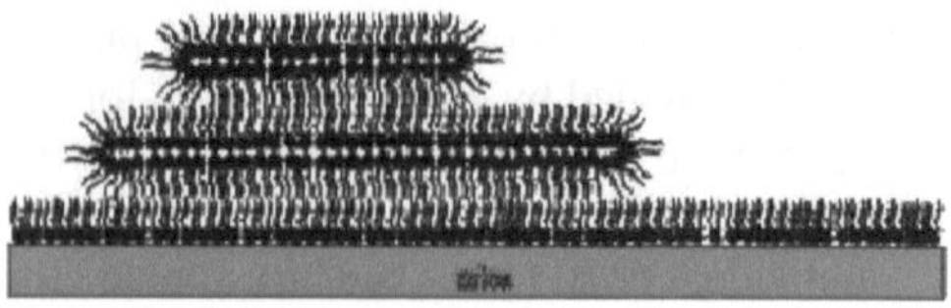

Fig. 21.6. Supported lipid monolayer and multilayers on mica, transferred from the liquid–air interface by Langmuir–Blodgett deposition; these multilayers are shown on the AFM topography image in Fig. 21.7a

We have shown that the ability of surfactant to reduce surface tension is associated with a distinct molecular architecture of the surfactant film. The functional pulmonary surfactant forms a pattern of molecular monolayer areas and areas of lamellar stacks, cross-linked to the monolayer at the air–aqueous interface (Fig. 21.7a). We recently found that cholesterol in excess of a physiological proportion abolishes surfactant function, and surfactant fails to lower the surface tension upon compression [78]. This failure of function is associated with a change in the molecular architecture [79]. The formation of multilayer stacks, which is a characteristic feature of functional surfactant, does not occur in the presence of excess cholesterol.

We performed atomic force measurements on supported pulmonary surfactant films in air to address the effect of cholesterol on the physical properties of lung surfactant films. We used bovine lipid extract surfactant (BLES). BLES is a hydrophobic extract of bovine lung lavage that differs from natural surfactant in the lack of surfactant-specific proteins SP-A and SP–D and cholesterol. Phosphatidylcholines represent 80% of its mass with half of the phosphatidylcholines being the disaturated DPPC. Between 5 and 10 mass % is the negatively charged phosphatidylglycerol, and two hydrophobic surfactant-associated proteins (SP-B, SP-C). BLES in nonbuffered normal saline (pH 5–6) with a phospholipid concentration of 27 mg/ml was a kind gift from the manufacturer (BLES Biochemical, London, ON, Canada). Cholesterol was purchased from Sigma Chemicals. A solution of 1:1:1 methanol, chloroform, and BLES by volume was first vortexed and then spun at 100*g* for 5 min. The methanol–water phase was discarded and the BLES in chloroform was retained and either 5 or 20% of cholesterol (by mass) with respect to phospholipids in chloroform was added. Each solution was then dried under N_2 and resuspended with Goerke's buffer [140 mM NaCl, 10 mM *N*-(2-hydroxyethyl)piperazine-N'-ethanesulfonic acid and 2.5 mM $CaCl_2$; pH 6.9] to obtain an aqueous suspension of BLES and cholesterol at a concentration of 27 mg/ml phospholipids. BLES solutions were spread at the air–liquid interface. Supported planar monolayers on mica were prepared using Langmuir–Blodgett technique and transferred on mica support when films were compressed to a surface tension of 25 mN/m. Supported films were imaged in air using a NanoWizard atomic force microscope from JPK Instruments, Germany. All measurements were performed at 25 °C, over five different velocities, collecting each time ten force curves. Silicon cantilevers from Micromash, Spain, were used with cantilever spring constants of 0.6 and 0.7 N/m as determined by a thermofluctuation method using JPK SPM software. Forces were measured on several different structural areas of the film, by positioning the atomic force microscope tip after the image had been collected.

21.3.1
Adhesion Measurements: Monolayer Stiffness and Function

Atomic force measurements helped us to elucidate how the loss of mechanical stability is related to the local mechanical properties of the film. Films containing 5 and 20% w/w cholesterol were compared. The presence of 20% cholesterol in BLES resulted in a decrease of the observed adhesive interaction, and an increase in the rigidity of the film. The force measurements were performed both on the monolayer area and on a first bilayer adjacent to the monolayer. The typical height of the bilayer area is 4.5 nm, and that of the monolayer is 2.5 nm. Typical areas where forces were measured are shown in Fig. 21.7a. For BLES with 20% cholesterol, only a monolayer was present and was used for force measurements (Fig. 21.7a, point 1).

BLES with 5% cholesterol at the bilayer area has an adhesion of 60 nN, which is higher than that of the monolayer. BLES with 20% cholesterol has a lower adhesion than the BLES monolayer with 5% cholesterol. It is interesting that the "single force" F_s was slightly lower in the presence of 20% cholesterol, 3 nN compared with 4 nN with 5% cholesterol, but the number of bonds decreased significantly from 15 and 10 to 7 (Table 21.1 [79]). Therefore, the presence of 20% cholesterol decreases the fluidity of the monolayer.

A force histogram (Fig. 21.8) was used to estimate the single force F_s and the number of bonds n. It is interesting that F_s was slightly lower in the presence of 20% cholesterol, 3 nN compared with 4 nN with 5% cholesterol, but n decreased significantly from 15 and 10 to 7 (Table 21.2). Therefore, the presence of 20%cholesterol decreases the fluidity of the monolayer.

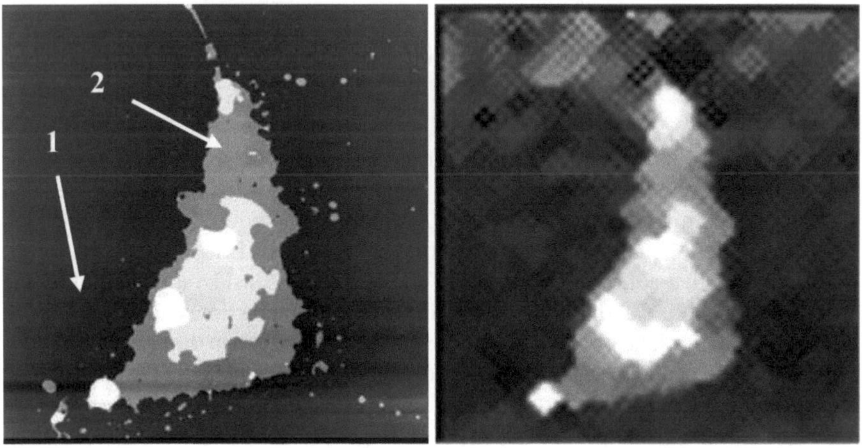

A B

Fig. 21.7. A AFM topography image of supported bovine lipid extract surfactant (BLES) film, scan area 15 μm by 15 μm. **B** Adhesion force map image 64 by 64, collected on the same area. The adhesion force map was collected while forces of interaction at each of 64 by 64 points were measured. The adhesion peak of the retrace plot was used to generate the adhesion force map image. *1* area of a typical monolayer; *2* area of a typical bilayer, adjacent to a monolayer

Table 21.2. Experimental adhesion force F_{adh} and its standard error σ on a bovine lipid extract surfactant (*BLES*) bilayer with with 5% cholesterol, a BLES monolayer with 5% cholesterol, and a BLES monolayer with 20% cholesterol, measured at room temperature; calculation of the number of bonds n and the mean force of a single bond F_s. Rate 200 nm/s, load 20 nN. Adhesion force on pure mica does not exceed 1.5 nN

	BLES 20% cholesterol monolayer	BLES 5% cholesterol monolayer	BLES 5% cholesterol bilayer
F_{adh} (nN)	20	40	60
σ (nN)	8	12	15
F_s (nN)	3	4	4
n	7	10	15

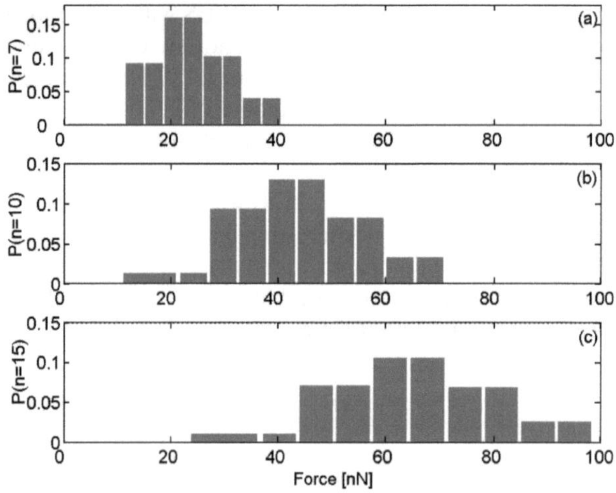

Fig. 21.8. Distribution of adhesion force, F_{adh}, in air: **a** BLES monolayer + 20% cholesterol; **b** BLES monolayer + 5% cholesterol; **c** BLES bilayer + 5% cholesterol

The increase in the adhesion force for a bilayer as compared with a monolayer may be understood by an increase in the contact area between the AFM probe and the sample at a given load. BLES with 20% cholesterol shows the lowest adhesion force, compared with monolayer and multilayer areas of BLES with 5% cholesterol. This indicates that adhesion properties of the film were considerably altered by the incorporation of 20% cholesterol, decreasing adhesive interaction, and increasing the rigidity of the film. The lipid molecules are not mobile enough in the presence of 20% cholesterol and cannot be easily rearranged around the tip to increase the contact area and adhesion force. The increased rigidity of the film and decreased adhesion that we observed may play an important role in preventing the monolayer–bilayer conversion.

21.3.2 Repulsive Forces: The Interaction of Charged Airborne Particles with Surfactant

The repulsive forces (Fig. 21.9) observed in air between the lipids film and the AFM probe were analyzed using two theoretical models for electrostatic interaction. The atomic force microscope tip was modeled once as a point charge and once as

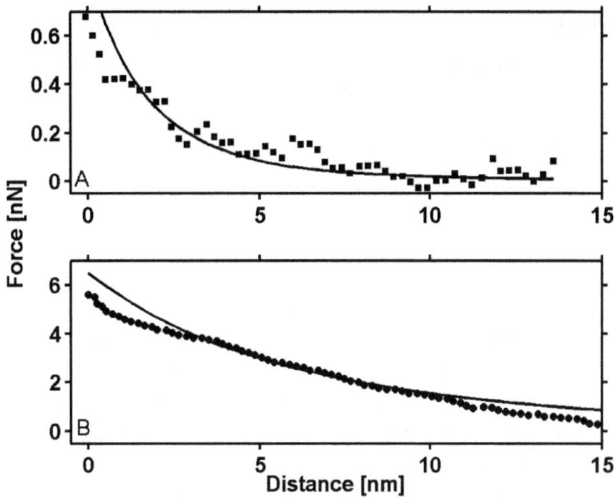

Fig. 21.9. Electrostatic force measured in air on BLES film with 0% cholesterol (*top*) and with 20% cholesterol (*bottom*). Experimental data are shown by *dots* and theoretical fits with the point-charge model are drawn in *solid lines*

a charged semisphere. Here q is a charge of the atomic force microscope tip, k is a constant, ε is the dielectric permittivity of the medium, and r is the separation distance between the tip and the medium. The supported BLES surfactant monolayer is composed of lipid molecules, which are organized and oriented in such a way that the dipole moment is directed along the surface normal. Therefore, the patch of the monolayer can be treated as two charged planes, each having charge Q_+ and Q_- and surface charge density σ_+ and σ_-, with the separation distance δ_o, which corresponds to the thickness of the monolayer and is equal to 2.5 nm. The surface charge density of the sample, σ, is related to the surface potential V, which we have determined experimentally using Kelvin probe force microscopy [80, 81] and used here as a parameter.

In the case of a point charge, the total electrostatic force acting on q is given by

$$F = \frac{V_q}{2\delta_0 A_0}\left(\frac{\beta}{\sqrt{1+\beta^2}} - \frac{\alpha}{\sqrt{1+\alpha^2}}\right) , \tag{21.4}$$

where

$$\frac{D}{R_0} = \alpha ,$$
$$\frac{D+\delta_0}{R_0} = \beta .$$

For the semisphere model, the force acting on the semisphere tip with a radius R_{so} will be

$$F_s = \frac{V_q}{2\delta_0 A_0}\left(\sqrt{1+\beta_1^2} - \sqrt{1+\alpha_1^2} - \sqrt{1+(\beta_1-\gamma)^2} + \sqrt{1+(\alpha_1-\gamma)^2}\right) , \tag{21.5}$$

where

$$\frac{D + R_{S0}}{R_0} = \alpha_1 ,$$
$$\frac{D + R_{S0} + \delta_0}{R_0} = \beta_1 ,$$
$$\frac{R_{S0}}{R_0} = \gamma .$$

Kelvin probe microscopy was used [81] to determine the surface potential of the lipid layer in air. Cholesterol induces a nonuniform distribution of the electric surface potential even at small physiological concentrations, showing nanoscale domains in the Kelvin image, while the AFM topography image remains uniform. Such domains vary in surface potential from −0.6 V, measured for 0% cholesterol, to −0.4 V, at 20% cholesterol.

Both models we used fit the experimental data well, showing the presence of electrostatic interactions in the net force measured between the tip and the sample. To better fit the experimental data we varied the charge on the atomic force microscope tip and the size of the charged disks, R_0, which experimentally corresponds to the size of domains already observed with both AFM and Kelvin probe force microscopy imaging. We modeled the AFM probe also with a charged sphere; in this model the effective charge depends on the radius of the tip. Samples of BLES with 20% cholesterol exhibit stronger electrostatic repulsion at longer distances, for all models used.

Interestingly, the electrostatic force was found to be more sensitive to the size of these domains than to the changes in the surface potential observed owing to the effect of cholesterol. The experimental data and the theoretical fit correlate well and both revealed that the size of the domains is approximately 5 times larger for the film containing 20% cholesterol than that for the film with no cholesterol present. This confirms the important role of cholesterol—it produced small domains which differ by electrostatic potential, initially invisible with AFM, which causes the increase in the total electrostatic force. These findings correlate with the results of molecular dynamics simulations [82], where cholesterol was shown to order individual lipids adjacent to cholesterol molecules, but abolished the long–range cooperativity between lipids, characteristic for the gel phase with true latticelike spatial order, and altered the lipid electrostatic potential [83]. In the presence of cholesterol [84] it was shown that the membrane electrostatic potential has a much larger variance depending on the distance along the bilayer normal, compared with that for a pure DPPC bilayer.

21.4 Interaction Forces Measured on Lung Epithelial Cells in Buffer

Elevated levels of ultrafine particles in air pollution are associated with increased morbidity and mortality [85]. Unlike larger particles, ultrafine particles reach the peripheral lung and are able to cross the various barriers of the lung, including the lipid–protein layer of pulmonary surfactant at the air–alveolar interface and the cell membranes of the epithelial and endothelial cells underneath and reach the bloodstream [86–89]. The toxicity of inhaled nanoparticles entering the body through the lung is thought to be initially defined by the electrostatic and adhesive interaction of the particles with the wall of the lung, and lung epithelial cells. Because of their substantial uptake, ultrafine particles are also of great interest for delivery of drugs to the peripheral lung and the body. Despite the importance of the early penetration steps in the biological effects of inhaled particles, the nature of the interactions with the various barriers of the lung remain poorly understood. A detailed study, therefore, promises substantial progress in the understanding of the health threat posed by fine and ultrafine air pollutants and the effectiveness of drug delivery by aerosols. The adhesive interaction includes electrostatic, van der Waals, steric, and hydrophobic forces as well as the line tension between particle, cell, and aqueous medium and depends on the elastic and plastic properties of the cell. The free energy due to these interactions will be minimized upon particle wetting by the lining layer. The lower the surface free energy of the particle, the less it will be wetted by the cell. A thermodynamic model using the "wettability criterion" was indeed successful in predicting passive particle uptake by cells [90]. Another thermodynamic analysis combined with a molecular dynamics simulation found negative line tension values for nanometer-sized particles, but positive values for those an order of magnitude larger [91]. The engulfment of particles in the nanometer range by unbalanced capillary forces was first suggested by Shanahan [92].

We demonstrated that the initial thermodynamic aspects and the time course of the uptake of nanoparticles by lung epithelial cells can be studied using AFM [93] in the force measurement mode. We investigated the first steps of the interaction of nanoparticles with lung epithelial cells using the atomic force microscope as a force apparatus. The apex of theatomic force microscope tip can be used as a model of the nanoparticles, thereby enabling the monitoring of the interaction forces between the nanoparticles and the cell over time. The adhesion force and the work of adhesion have been investigated and correlated with the mechanical properties of the cell.

21.4.1 Cell Culture/Force Measurement Setup

Lung epithelial cells, type II, were isolated from male, pathogen-free Sprague-Dawley rats weighing 150–250 g. The cells were cultured at 6×10^8 cells per milliliter in Dulbecco's modified Eagle's medium supplemented with 1% L-glutamine, containing 10% heat-inactivated fetal bovine serum, and 300 μl of a penicillin–gentamycin solution in γ-irradiated culture dishes. The cells were grown on light microscope glass cover slips. The cells were kept in an incubator under 5% CO_2 and *at*37 °C until they created a monolayer with approximately 80% of confluence (4–6 days). At this stage, the phenotype of the lung epithelial cells had changed from type II to type I in that the cells were spread out flat on the interface and ceased to produce surfactant.

Lung epithelial cells were first imaged using a light microscope (Zeiss Axiovert 200). For force spectroscopy, a medium-covered cover slip with the cells was mounted in the liquid cell of the atomic force microscope. The setup used allows for the observation of the sample and the atomic force microscope cantilever across the cover slip from below by an inverted light microscope at the same time as force spectroscopy (or AFM imaging) is performed. The tip of the atomic force microscope was thus placed over a suitable cell and cell location was under light-microscope control (Fig. 21.10a). Figure 21.10b shows a typical force scan where the cantilever deflection was monitored as a function of the sample position.

At first, the probe approached the cell with the lever undeflected (horizontal line of the force trace), indicating no interaction. The tip then made contact with the sample and became deformed. Forces of interaction between the probe and the cell were then measured in medium at 35 °C, by approach of the tip to the cell interface until a preset load was reached (trace of force curve). The force was acquired as a function of the tip–sample separation h. After the tip has come into contact with the cell, h becomes the penetration depth for the tip into the cell. Separate force curves were acquired for a preset load of 100, 200, and 300 pN. The tip was now kept in contact with the cell for a predetermined time (delay time), while the preset load was kept constant under feedback control (i. e., when the load decreased, the tip was moved forward). We varied the delay time from 0 to 1800 s. Thereafter, the tip was retracted and the adhesion observed. Cantilevers (Micromash), with a spring constant of 57 mN/m, and pyramidal silicon nitride tips were used (tip height 2.9 nm, tip radius less than 20 nm (typical 10 nm), tip angle (face to face) 25 – 45° (top to around 300 nm down). Cells were kept in the medium at all times. The spring constant for each cantilever was measured before and after the experiment using the procedures implemented by the manufacturer of the atomic force microscope

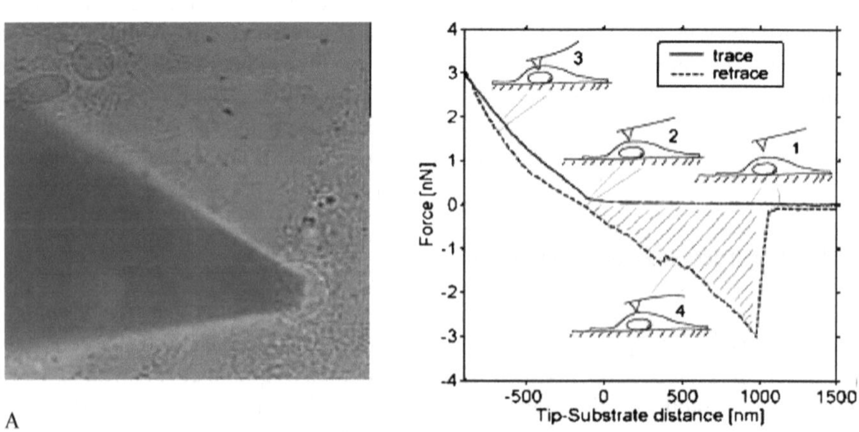

Fig. 21.10. Optical microscopy image of epithelial type II cells and atomic force microscope cantilever. For force spectroscopy, an atomic force microscope tip was placed over the central region of a cell. Trace and retrace force curves were measured on a cell. Force versus tip–sample separation is shown in **b**. The approach part (trace) reveals how the cell deforms upon contact by the tip. The retrace or withdraw part of the curve reveals a large adhesion peak which is often split in multiple peaks. The sketches explain the physical situation for each part of the curve. The *hatched area* denotes the work of adhesion (adhesion energy) between the tip and the cell. The atomic force microscope tip was allowed to stay in contact for various times (delay time) in position \$3

(thermal noise method). The sensitivity and the spring constant were found to be not different before and after the experiment.

Forces of interaction between the probe and the cell were then measured in the medium at 35 °C.

21.4.2 Mechanical Properties

The approach part of the force curve was used to determine the penetration depth of the tip into the cell as well as the mechanical properties of the sample. Reviews on this topic are given in [94, 95].The measurement of the penetration depth will depend on the accurate determination of the contact point between the tip and the cell. A cantilever spring constant lower than or similar to that of the cell is preferable. In this case, we can assume that the contact point corresponds in the trace force curve to the first upward deflection. After the tip has come into the contact with the cell, D becomes the penetration depth h of the tip into the cell. To evaluate the mechanical properties of the sample from the approach force profile, the Sneddon model was employed [26, 96]. Note that the use of macroscopic physical law should be reserved for homogeneous bulk materials. Young's modulus should be viewed as an effective parameter to analyze the trend of the in-plane organization of supramolecular structures. The effective parameter can differ from one experiment to another. A more sophisticated approach was to model the mechanical response of the cell membrane explicitly taking into account the mechanics of the cell membrane and that of the underlying cytoskeleton [97]. As the penetration depth h can reach 1 μm, the tip shape must be considered as a function of h. The effective Young's modulus, E', was estimated point by point for h considering the three regions determined by the geometry of the tip [98].

For h between 0 and 10 nm, the AFM probe was considered as a semisphere with a radius R of 10 nm :

$$E_{\text{sphere}} = \frac{3}{4\sqrt{R}} \frac{F}{h^{3/2}} . \tag{21.6}$$

For further penetration, the tip profile was viewed as a cone whose half angle varies linearly from $\alpha_1 = 25°$ at $h = 10$ nm to $\alpha_2 = 45°$ when h reaches 300 nm:

$$E_{\text{cone}} = \frac{\sqrt{2}}{\tan\alpha} \frac{F}{h^2} . \tag{21.7}$$

For an indentation deeper than 300 nm, α is considered constant to α_2.

Figure 21.11 shows the change in the Young's modulus when increasing the penetration depth h. At first, the tip experienced a layer of increasing stiffness to a penetration depth of about 100 nm, at which point the resistance of the cell to penetration reached a maximum. Thereafter the cell became more compliant again. The stiff layer may represent the outer layer of the cytoskeleton anchored to the cell membrane. The Young's modulus of the membrane can be thought of as the surface tension of the membrane (0.5 N/m) divided by the thickness of the membrane (around 10 nm), i. e., a pressure of around 50 MPa. The lipid head is responsible

Fig. 21.11. When analyzing the approach curve, one can determine the Young's modulus of the sample surface. Young's modulus, which gives a measure of cell elasticity in our case, was found to depend on indentation depth, showing two distinctive parts of the plot: when the peripheral part of the cell was penetrated, up to about 100 nm into the cell, the cell became increasingly stiffer. The stiffness then dropped sharply upon further penetration until it leveled out after penetration to about 200 nm deep

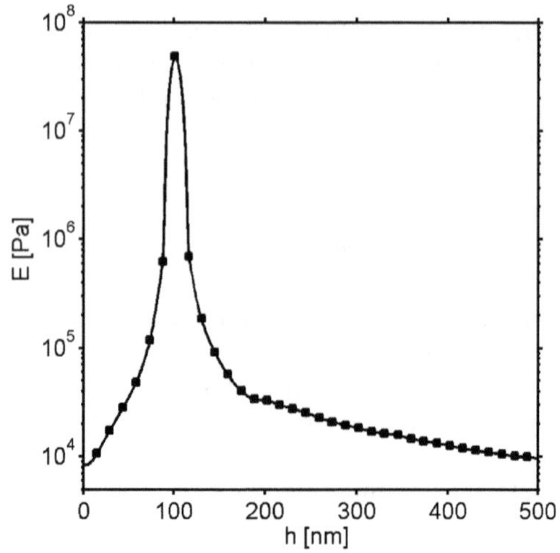

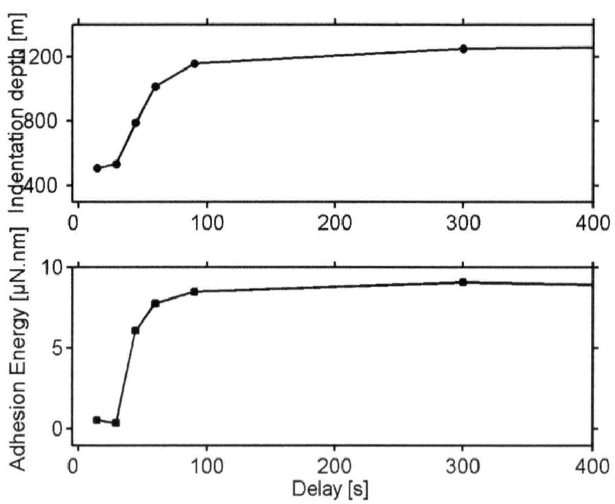

Fig. 21.12. *Top*: Indentation depth as a function of time. The load was kept constant at 200 pN. At time zero, the indentation was about 500 nm. Over time, the tip first penetrated rapidly and then slowly from 500 to 1300 nm. This is evidence of a time-dependent response of the cell to the tip. *Bottom*: Adhesion energy as a function of time. The tip was kept in contact with the cell at a constant load of 200 pN. The longer the tip stays in contact with the cell, the larger the adhesion that is observed. Statistical analysis shows that considerable changes occur during the first 100 s, after which the mean value of the adhesion energy no longer changes

for crystalline elasticity, i.e., the change in the head-to-head distance can change (high elasticity but low rupture). The lipid chain behaves with rubber elasticity (low elasticity), which is related to the change of entropy (increase of order upon pressure). It results in the properties that the cell membranes can be easily deformed in the plane but do not resist the expansion (increase between the headgroup cannot vary more than 2%).

Further measurements of the Young's modulus of cells have been performed with human platelet cells ($E = 1-50\,\text{kPa}$) [99], cardiac cells ($E = 100\,\text{kPa}$), skeletal muscle cells ($E = 25\,\text{kPa}$), and endothelial cells (1–7 kPa) [100].

The analysis of the forced indentation of the epithelial cells by the tip reveals the fate of a nanoparticle coming into contact with the lung's epithelium and then being actively loaded onto a cell. This situation will indeed occur in the lung. The lung's epithelium towards the air is covered by a thin aqueous layer and a molecular film of pulmonary surfactant at the air–water interface. Transmission electron microscopy of lung thin sections as well as scanning electron microscopy of the lung have indeed shown that particles trapped between the surfactant layer and the epithelial cells strongly deform the cells [101, 102].

The penetration depth was also investigated as a function of the contact time between the tip and the cell at a given load. Figure 21.12 shows that the tip first penetrated the cell down to 500 nm in less than 100 s. Equilibrium was reached for approximately 1-μm depth (around half of the cell thickness). We assume that the tip is either actively taken up or the cell rearranges its plasma membrane and the cytoskeleton elements to accommodate the tip.

21.4.2.1 Adhesion Work

After the initial indentation, the tip was kept loaded onto the cell with 1 nN for preset times ranging from 0 to 900 s under feedback control. Thereafter, the tip was retracted and the tip adhesion was acquired.

The retrace curve was used to calculate the adhesion energy Γ between the tip and the cell:

$$\Gamma = \sum_{h_1}^{h_2} F_{\text{retrace}}(h_i)(h_{i+1} - h_i)\,. \tag{21.8}$$

$F_{\text{retrace}}(h_i)$ is the adhesion force at h_i. The point of the initial contact between the tip and the cell surface, h_1, was determined from the approach part of the force curve; h_2 corresponds to the last jump out of contact.

Upon the penetration into the cell, the adhesion energy also increases. This indicates that the tip becomes engulfed by the membrane and the area of interaction increases (Fig. 21.12). Our results suggest that the process of particle uptake by lung epithelial cells occurs over approximately 100 s. This appears to be the time necessary for the nanoparticle to increase its surface area in contact with the cell and, hence, its adhesive interaction. Other authors have also shown [103] that the increase in adhesion between a polystyrene sphere to mica with increasing load or contact time is due to the plastic deformation. Load dependence and contact time dependence also indicate plastic and viscoelastic deformation [104]. A linear dependence of adhesion force on the reduced radius $R_{\text{eff}} = R_1R_2/(R_1 + R_2)$, where R_1 and R_2 are the radii of the two particles, was found by Heim et al. [105].

Particles of 50–100 nm in diameter are not phagocytosed [106, 107]. They enter cells in the absence of clathrin and caveolin, associated with active uptake. Even

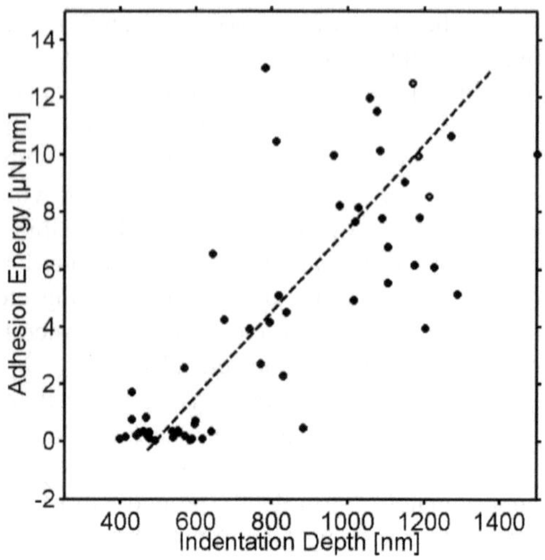

Fig. 21.13. Correlation between the indentation depth (Fig. 21.12, top) and the adhesion energy (Fig. 21.12, bottom) for all the measurements (around 100 points). The deeper the indentation, the higher is the adhesion energy

when an actin-based mechanism has been ruled out [107], nanoparticles inside a cell are not surrounded by membrane and, hence, are not taken up through vesicle formation.

The AFM investigation suggests the adhesive interaction in particle uptake can now be measured by our approach and used as a significant parameter test. Particles with different surface properties and different geometries can be mimicked by modified atomic force microscope tips. The adhesive interaction of a particle with the plasma membrane in the lung depends also on the particle history. Airborne particles, after crossing a surfactant layer at the air–water interface of a cell culture dish, have been shown to be more readily taken up than particles added to the media [108]. They may have become coated with a film of pulmonary surfactant and, as a consequence, have penetrated the epithelial cells differently. Whether this is related to a change in adhesive interaction can be tested by coating atomic force microscope tips with surfactant.

21.5 Conclusions

Measurements of forces with the atomic force microscope can determine the local changes in the mechanical and electrical properties induced by the absorption of proteins within lipid monolayers or bilayers. Such studies have offered new perspectives in the identification and characterization of the pulmonary surfactant films in air, model membranes in liquid media, as well as cell membranes in media.

In air, the adhesion forces of particles are known to be 10 times higher than in aqueous media. For a given applied load, the differences in the magnitude of the force do not depend on the contact area, which is similar in air or in a liquid. In fact, the van der Waals forces are screened under a liquid, thereby reducing the total adhesion forces. However, the adhesion forces can be modulated by varying the contact area between the particle and the layer. We have shown that the temperature or the anesthetic incorporation can largely modify the fluidity of the membrane.

Using the AFM probe as a nanoparticle model is an idea to evaluate the ability of the particle to interact with the cell. Although many kinds of force curve cannot yet be interpreted, for instance, the role of the hydrophobic attraction between tails, our work is a starting point to understand how the particle can penetrate through the lipid layer. For example, the nonnegligible role of the electrical forces in air has been proved and the electrical potential of the layer is of importance in the adsorption process involving the electrical barrier.

The mechanical properties of the cell, and in general the plastic deformation, have been investigated. The role of the membrane as a mechanical barrier is clearly observed by measuring the Young's modulus as a function of the penetration depth. Those properties depend on many parameters: the particle speed, the particle size, the applied load. Also, the duration of the contact is crucial in the incorporation ability of the particles. The complexity of the phenomenon cannot of course be satisfactorily and fully described, but some empirical trend can be locally and statistically defined.

Acknowledgements. The authors gratefully acknowledge financial support from the National Sciences and Engineering Research Council of Canada, the Canadian Institute for Health Research, the University of Calgary, and the University of Burgundy and also thank C. Le Grimellec and M. Moskovits for critical reading of the manuscript and helpful discussion.

References

1. Cullis PR, Dekruijf B (1979) Biochim Biophys Acta 559:399–420
2. Seelig J, Seelig A (1980) Q Rev Biophys 13:19–61
3. Brown DA, London E (1998) J Membr Biol 164:103–114
4. Nagle JF, Tristram-Nagle S (2000) Biochim Biophys Acta Rev Biomembr 1469:159–195
5. Bloom M, Evans E, Mouritsen OG (1991) Q Rev Biophys 24:293–397
6. Brown DA, London E (2000) J Biol Chem 275:17221–17224
7. Brown DA, London E (1998) Annu Rev Cell Devel Biol 14:111–136
8. Vangolde LMG, Batenburg JJ, Robertson B (1988) Physiol Rev 68:374–455
9. Lewis JF, Jobe AH (1993) Am Rev Respir Dis 147:218–233
10. Possmayer F (1988) Am Rev Respir Dis 138:990–998
11. Vedhuizen R, Nag K, Orgeig S, Possmayer F (1998) Biochim Biophys Acta Mol Basis Dis 140:890–108
12. Rooney SA, Young SL, Mendelson CR (1994) FASEB J 8:957–967
13. Tamm LK, McConnell HM (1985) Biophys J 47:105–113
14. Tien HT, Ottova AL (1998) Electrochim Acta 43:3587–3610
15. Heyse S, Ernst OP, Dienes Z (1998) Biochemistry 37:507–522
16. Muller RH, Mader K, Gohla S (2000) Eur J Pharma Biopharma 50:161–177
17. Leonenko ZV, Carnini A, Cramb DT (2000) Biochim Biophys Acta 1509:134–140
18. Zasadzinski J, Helm CA, Longo ML, Weisenhorn AL, Gould SAC, Hansma PK (1991) Biophys J 59:755–760
19. Hui SW (1995) Mol Membr Biol 12:45–50
20. Singh S, Keller DJ (1991) Biophys J 60:1401–1410
21. Beckmann M, Nollert P, Kolb HA (1998) J Membr Biol 161:227–233

22. Han W, Lindsay SM, Jing T (1996) Appl Phys Lett 69:4111–4113
23. Han W, Lindsay SM (1998) Appl Phys Lett 72:1656–1658
24. Fasolka MJ, Mayes AM (2001) Annu Rev Mater Res 31:323–355
25. Lee GU, Kidwell DA, Colton RJ (1994) Langmuir 10:354–357
26. Cappella B, Dietler G (1999) Surf Sci Rep 34:1–104
27. Senden TJ, Drummond CJ, Ke'kicheff P (1994) Langmuir 10:358–362
28. Cleveland JP, Schäffer TE, Hansma PK (1995) Phys Rev B 52:8692–8695
29. Israelachvili J (2005) Q Rev Biophys 38:331–337
30. Livadaru L, Kovalenko A (2006) Nano Lett 6:78
31. Biggs S (1995) Langmuir 11:156–162
32. Dufrene YF, Boland T, Schneider JW, Barger WR, Lee GU (1998) Faraday Discuss 111:79
33. Finot E, Lesniewska E, Mutin J-C, Goudonnet J-P (1999) J Chem Phys 111:6590–6598
34. Finot E, Lesniewska E, Mutin J-C, Goudonnet J-P (2000) Langmuir 16:4237–4244
35. Finot E, Lesniewska E, Mutin J-C, Goudonnet J-P (2000) Appl Surf Science 161:316–322
36. Butt HJ, Cappella B, Kappl M (2005) Surf Sci Rep 59:1–152
37. Finot E, S Roualdes, Kirchner M, Rouessac V, Berjoan R, Durand J, Goudonnet J-P, Cot L (2002) Appl Surf Sci 187:326–338
38. Mollica V, Relini A, Rolandi R, Bolognesi M, Gliozzi A (2000) Eur Phys J E 3:315
39. Van EW der Vegte, Hadziioannou G (1997) Langmuir 13:4357–4368
40. Dimitriadis EK, Horkay F, Maresca J, Kachar B, Chadwick RS (2002) Biophys J 82:2798–2810
41. Legay G, Finot E, Meunier-Prest R, Cherkaoui-Malki M, Latruffe N, Dereux A (2005) Biosens Bioelectron 21:627–636
42. Franz V, Loi S, Muller H, Bamberg E, Butt H-J (2002) Colloids Surf B 23:191–200
43. Butt HJ, Franz V (2002) Phys Rev E 66:1601
44. Loi S, Sun G, Franz V, Butt HJ (2002) Phys Rev E 66:1602
45. Tenchov B, Koynova R, Rapp G (2001) Biophys J 80:1873–1890
46. Meyer HW, Semmler K, Rettig W, Pohle W, Ulrich AS, Grage S, Selle C, Quinn PJ (2000) Chem Phys Lipids 105:149–166
47. Bizzotto D, Nelson A (1998) Langmuir 14:6269–6273
48. Tamm LK (1985) Biol Chem Hoppe-Seyler 366:859–859
49. Kalb E, Frey S, Tamm LK (1992) Biochim Biophys Acta 1103:307–316
50. Nollert P, Kiefer H, Jahnig F (1995) Biophys J 69:1447–1455
51. Exerowa D (2002) Adv Colloid Interface Sci 96:75–100
52. Ebel H, Grabitz P, Heimburg T (2001) J Phys Chem B 105:7353–7360
53. Giocondi MC, Vie V, Lesniewska E, Milhiet PE, Zinke-Allmang M, Le C Grimellec (2001) Langmuir 17:1653–1659
54. Leonenko Z, Cramb D (2002) Nano Lett 2:305–309
55. Leonenko ZV, Merkle DD, Lees-Miller SP (2002) Langmuir 18:4873–4884
56. Tokumasu F, Jin AJ, Dvorak JA (2002) J Electron Microsc 51:1–9
57. Leonenko Z, Cramb D (2004) Can J Chem 82:1–11
58. Leonenko Z, Finot E, Cramb D (2006) Biochim Biophys Acta Biomembr 1758:487–492
59. Leonenko ZV, Carnini A, Cramb DT (2000) Biochim Biophys Acta 1509:131–147
60. Brian AA, McConnell HM (1984) Proc Natl Acad Sci USA 81:6159–6163
61. Leonenko ZV, Finot E, Ma H, Dahms T, Cramb DT (2004) Biophys J 86:3783–3793
62. Shamrakov LG, Leonenko Z, Finot E, Cramb DT (2006) Aust J Chem 59:386–389
63. Biltonen RL, Lichtenberg D (1993) Chem Phys Lipids 64:129–142
64. Yang J, Appleyard J (2000) J of Phys Chem B 104:8097–8100
65. Binnig G, Quate CF, Gerber C (1986) Phys Rev Lett 56:930–933
66. Israelachvili J (1992) Intermolecular and surface forces. Academic, New York
67. Williams JM, Han T, Beebe TP (1996) Langmuir 12:1291–1295

68. Marrink SJ, Berger O, Tieleman P (1998) Biophys J 74:931–943
69. Evans K, Evans MF, Laker MF (1995) Ann Clin Biochem 32:261–280
70. Butt HJ (1991) Biophys J 60:1438–1444
71. Butt HJ, Franz V (2002) Phys Rev E 66:1601
72. Carnini A, Phillips HA, Shamrakov LG, Cramb DT (2004) Can J Chem 82:1139–1149
73. Garcia-Manyes S, Oncins G, Sanz F (2005) Biophys J 89:4261–4274
74. de Gennes PG (1987) Adv Colloid Interface Sci 27:189–209
75. Petrache HI, Dodd SW, Brown MF (2000) Biophys J 79:3172–3192
76. Nagle JF, Tristram-Nagle S (2000) Biochim Biophys Acta Rev Biomembr 1469:159–195
77. Janiak MJ, Small DM, Shipley GG (1979) J Lipid Res 20:183–199
78. Gunasekara L, Schurch S, Schoel WM (2005) Biochimi Biophys Acta Mol Cell Biol Lipids 1737:27–35
79. Leonenko Z, Finot E, Amrein M (2007) Ultramicroscopy 107:948–953
80. Leonenko Z, Rodenstein M, Dohner J, Eng LM, Amrein M (2006) Langmuir 22:10135–10139
81. Leonenko Z, Gill S, Baoukina S, Monticelli L, Doehner J, Gunasekara L, Felderer F, Rodenstein M, Eng LM, Amrein M (2007) Biophys J 93:674–683
82. Chiu SW, Jakobsson E, Scott HL (2001) J Chem Phys 114:5435–5443
83. Smondyrev AM, Berkowitz ML (1999) J Chem Phys 111:9864–9870
84. Smondyrev AM, Berkowitz ML (1999) J Chem Phys 110:3981–3985
85. Atkinson RW, Anderson HR, Strachan DP, Bland JM, Bremner SA, de Leon AP (1999) Eur Respir J 13:257–265
86. Oberdorster G, Sharp Z, Atudorei V, Elder A, Gelein R, Lunts A, Kreyling W, Cox C (2002) J Toxicol Environ Health 65:1531–1543
87. Geiser M, Rothen-Rutishauser B, Kapp N, Schurch S, Kreyling W, Schulz H, Semmler M, Hof VI, Heyder J, Gehr P (2005) Environ Health Perspect 113:1555–1560
88. Nemmar A, Hoet PHM, Vanquickenborne B, Dinsdale D, Thomeer M, Hoylaerts MF, Vanbilloen H, Mortelmans L, Nemery B (2002) Circulation 105:411–414
89. Takenaka S, Karg E, Roth C, Schulz H, Ziesenis A, Heinzmann U, Schramel P, Heyder J (2001) Environ Health Perspect 109:547–551
90. Chen H, Langer R, Edwards DA (1997) J Colloid Interface Sci 190:118–133
91. Bresme F, Quirke N (1999) J Chem Phys 110:3536–3547
92. Shanahan MER (1990) J Phys D Appl Phys 23:321–327
93. Le C Grimellec, Lesniewska E, Giocondi MC et al (1998) Biophys J 75:695–703
94. Costa KD (2003) Dis Markers 19:139–154
95. Radmacher M (2002) Methods Cell Biol 68:67–90
96. Sneddon JN (1965) Int J Eng Sci 3:47
97. Mc MElfresh, Baesu E, Balhorn R, Belak J, Allen MJ, Rudd RE (2002) Proc Natl Acad Sci USA 99:6493–6497
98. Stark RW, Drobek T, Weth M, Frickef J, Heckel WM (1998) Ultramicroscopy 75:161
99. Radmacher M, Fritz M, Kacher CM, Cleveland JP, Hansma PK (1996) Biophys J 70:556–567
100. Mathur AB, Collinsworth AM, Reichert WM, Kraus WE, Truskey GA (2001) J Biomech 34:1545–1553
101. Bachofen H, Schurch S, Possmayer F (1994) J Appl Physiol 76:1983–1992
102. Gehr P, Geiser M, Im V Hof, Schurch S, Waber U, Baumann M (1993) Microsc Res Tech 26:423–436
103. Biggs S, Spinks G (1998) J Adhes Sci Technol 12:461–478
104. Reitsma M, Craig VSJ, Biggs S (2000) J Adhes 74:125–142
105. Heim LO, Blum J, Preuss M, Butt H-J (1999) Phys Rev Lett 83:3328–3331

106. Geiser M, Rothen-Rutishauser B, Kapp N, Schurch S, Kreyling W, Schulz H, Semmler M, Hof VI, Heyder J, Gehr P (2005) Environ Health Perspect 113:1555–1560
107. Rothen-Rutishauser BM, Schurch S, Haenni B, Kapp N, Gehr P (2006) Environ Sci Technol 40:4353–4359
108. Blank F, Rothen-Rutishauser BM, Schurch S, Gehr P (2006) J Aerosol Med 19 (in press)

16 Atomic Force Microscopy Studies of the Mechanical Properties of Living Cells

Félix Rico · Ewa P. Wojcikiewicz · Vincent T. Moy

From volume IX

Abstract. The study of viscoelastic and adhesive properties of cells falls under the discipline of cellular mechanics. Our knowledge of cell mechanics has been limited until the development of tools that allow us to measure minute forces (piconewton) and small dimensions (nanometer) at the cellular level. The atomic force microscope (AFM) is a very useful tool for determining the mechanical properties of cells as it allows experiments to be conducted under physiological conditions. The focus of this review is to provide an overview of the AFM as a tool for measuring the mechanical (viscoelastic and adhesive) properties of living cells. First, the different modes of operation of the AFM are briefly introduced, drawing special attention to the force mode. Then, the methods used to determine the viscoelasticity of cells are described, followed by a review of the available methods for determining cell adhesion at both single-molecule and whole-cell levels.

Key words: Atomic force microscopy, Atomic force microscope, Cell adhesion, Cell mechanics, Viscoelasticity, Young's modulus, Dynamic force spectroscopy, Scanning probe microscopy, Cantilever, Tips, Indentation, Force, Shear modulus, Elasticity, Single molecules

16.1 Introduction

Mechanics is the branch of physics that studies forces. As such, mechanics describes the relations between the strain and the applied stress to obtain the elastic and viscous properties of bodies. But mechanics also describes the forces required to maintain two bodies in contact when pulled apart. Thus, mechanics involves both viscoelasticity and adhesion, two properties that are very difficult to study independently, as we will see in the following sections [1].

Cells are living systems that are constantly subjected to or exerting forces. That is the reason why the mechanics of living cells is proving to be as important to cells' functions as their biochemical counterparts [2–5]. The structure, shape and function of living cells are regulated in part by their mechanical properties, and vice versa [6–13]. Until recently, little was known about single cell mechanical properties owing to the lack of adequate tools to control displacements and forces at the cellular level. Our understanding of the mechanics of the cell has increased thanks to the emergence of nanotechnology. The currently available nanotools enable us to manipulate biomolecules and cells with subnanometer resolution and simultaneously measure forces at the piconewton scale. One of the most versatile nanotools available today is the atomic force microscope (AFM). The AFM is capable of acquiring both topographical images and mechanical measurements under physiological conditions,

In: H. Fuchs, B. Bhushan (eds), *Biosystems - Investigated by Scanning Probe Microscopy*,
DOI 10.1007/978-3-540-74083-4_4,

making it a very powerful tool for studying living systems at the piconewton and nanometer scales.

The AFM was first introduced in 1986 as a tool to investigate the surfaces of insulators in air [8, 10–12, 14, 15]. It was then rapidly applied to measure forces and nanomechanical properties of different material surfaces [16, 17]. Its application for imaging biological systems came about once measurements could be conducted in liquids [18–20]. In 1992, the AFM was first used to probe the mechanics of biological samples. Tao et al. [21] applied the AFM to measure the elasticity of sectioned biological material, and Hoh and Schoenenberger [22] used the AFM to measure the mechanical properties of living cells. Following these pioneering studies, there has been a continuous increase in studies assessing the mechanical properties of living cells under different physiological conditions as well as other novel methods to improve AFM measurements of biological samples, especially of living cells [23–46]. The focus of this review is to provide an overview of the AFM as a tool for measuring the mechanical (viscoelastic and adhesive) properties of living cells. The first part of the review will focus on the basic principles of atomic force microscopy operation, and will be followed by descriptions of the methods used to determine the viscoelastic and adhesive properties of cells.

16.2 Principle of Operation

The AFM and its precursor, the scanning tunneling microscope, were first applied to obtain topographical information of hard surfaces in air [14, 47–49]. For that purpose a sharp tip was used to probe the sample while scanning its surface in the horizontal plane. Atomic force microscopy is still mostly used as an imaging technique, especially in the field of material sciences. However, its capability of imaging nonconducting surfaces in fluids led to its immediate application to biological specimens [50–54].

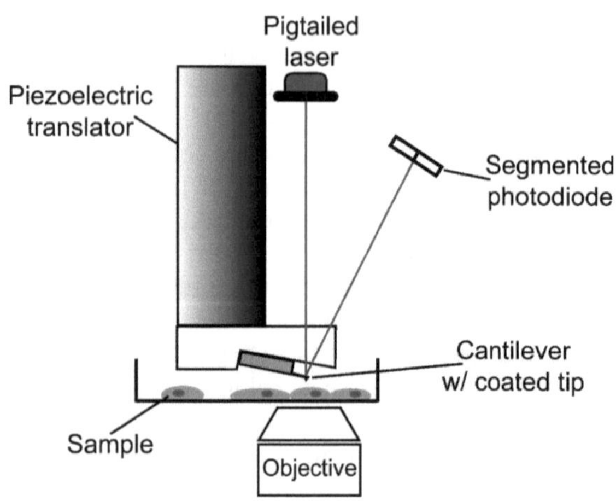

Fig. 16.1. A typical atomic force microscope (AFM)

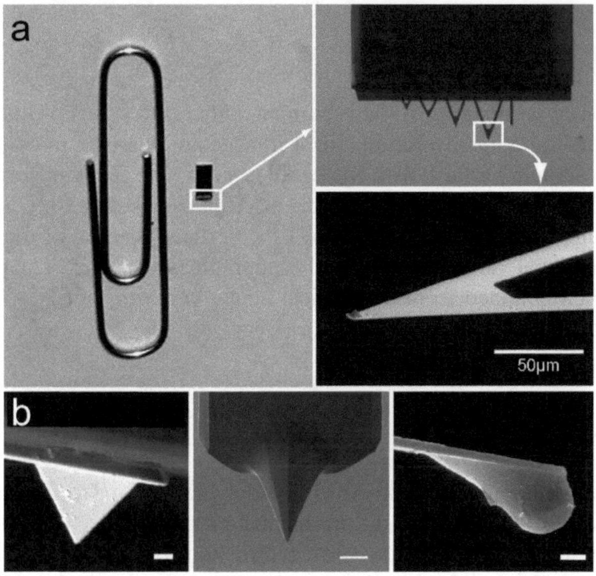

Fig. 16.2. **a** Au-coated Si_3N_4 cantilever chip compared with a regular paper clip. The *upper-right image* shows a magnification of the front of the chip where five different cantilevers can be observed, each one has a different spring constant (the longest cantilever is 320 μm long). The *lower-right image* is a scanning electron microscopy lateral view of the end of one of the cantilevers. **b** Commercially available AFM cantilever tips include, e. g., four- and eight-sided pyramidal tips, and spherical tips. *Bars* stand for 1, 5 and 2 μm, respectively

The AFM makes use of a silicon or silicone nitride flexible cantilever with a tip at its end for probing the sample surface (Fig. 16.1). The cantilever is the most important part of the AFM, since it determines the applied and measured forces and the lateral resolution. There are different tips and cantilevers commercially available, some of which are shown in Fig. 16.2. The probe is moved relative to the sample by means of piezoelectric elements, ceramic materials that expand or contract when different voltages are applied. Owing to the inherent nonlinearity of piezoelectric materials, newer AFM models use position sensors to detect the actual extension, and correct the nonlinearities using electrical feedback circuits. When the probe tip makes contact with the surface, the cantilever bends or deflects. This deflection is usually measured using the optical lever method [55, 56]. A laser beam is focused on the cantilever surface and the position of the reflected beam is detected using a segmented photodiode. The difference in the photocurrents of the photodiode segments provides a measure of the deflection change. AFM cantilevers are very flexible and act as springs with a characteristic constant, k. Thus, the applied force of the cantilever (F) can be obtained using Hooke's law, $F = -kd$, where d is the cantilever's deflection. An illustrative sketch of an AFM is shown in Fig. 16.1.

16.2.1 AFM Imaging

The operation of an imaging AFM is similar to that of a stylus profiler or a record player but with superior resolution. AFM images are obtained in contact mode by scanning the sample surface with the cantilever tip at a constant force. The force level is kept constant by using feedback electronics (usually digital proportional integral circuits) that continuously correct the vertical position of the cantilever as the deflection signal varies as a result of topographical changes. The topographical images are then obtained from the vertical height corrections used to maintain the force constant.

16.2.2 Force Measurements

Apart from imaging purposes, the AFM is widely used to apply and measure forces on materials and to determine their mechanical properties [57, 58]. To achieve this, the AFM cantilever is moved in the vertical direction at a constant speed until the tip contacts the sample, and the cantilever undergoes deflection before being retracted away from the sample. The result of this complete cycle is termed a force–distance curve (F–z) or, simply, a force curve. The approach or the loading trace of the force–distance curve is usually analyzed to obtain the elastic properties of the sample, while the retract or the unloading trace of the curve is mainly used to measure adhesion. Figure 16.3 shows an example of a force curve obtained with a pyramidal tip on a living cell, together with an illustration of the position of the cantilever at each stage of the cycle. When the cantilever tip makes contact with a soft material, such as a living cell, the resulting deformation or indentation of the sample is recorded in the loading curve. If, in addition, the sample adheres to the cantilever, the retraction curve may present negative force values, as shown in the figure.

As we mentioned before, cantilever deflection is measured using the laser beam that is reflected onto the photodiode. To obtain a quantitative measure of such a deflection, it is necessary to calibrate the photodiode signal before any measurement. To do so, a force curve is obtained on a hard surface where the cantilever will deflect as much as the piezoelement displaces it. Thus, knowing the applied displacement and recording the photodiode signal, one obtains the optical lever sensitivity (usually reported in units of volts per nanometer), enabling the conversion of the deflection from voltage to distance. In practice, this is obtained by computing the slope of the F–z curve obtained on a hard surface. In addition, to obtain mechanical measurements, it is necessary to convert the deflection of the cantilever into force. To do so, the spring constant of the cantilever has to be determined. Several methods have been reported to obtain the spring constant. These range from calculated values derived from the cantilever's geometry and material properties to measured values using a reference lever [59–71]. There are a number of reviews which provide an excellent overview of these methods [72, 73]. However, the most commonly used method to calibrate the spring constant of the cantilevers is the thermal fluctuation method introduced by Hutter and Bechhoefer [65]. It consists of modeling the behavior of the cantilever to a simple harmonic oscillator. Then, by measuring the thermal fluctuations ($\langle d^2 \rangle$) of the cantilever away from the surface and applying the equipartition theorem, one can relate $\langle d^2 \rangle$ to the thermal energy ($k_B T$, k_B being the Boltzmann constant and T the absolute temperature) to obtain the cantilever spring constant. As suggested by Butt and Jaschke, the method may require correcting for the secondary modes of oscillation which are not considered in the Hutter and Bechhoefer method but are measured in practice [66, 69]. Typically, there are several cantilevers with different spring constants on an AFM cantilever chip (Fig. 16.2a). They are intended for different applications. Hard samples will require stiffer cantilevers, with spring constants of a few thousand piconewtons per nanometer, while more delicate and malleable samples, such as living cells, will call for softer cantilevers. The softer cantilevers commercially available have a spring constant of approximately 6 pN/nm (Biolever, Olympus, Tokyo, Japan). Two of the most widely used chips of cantilevers for cellular use are the Microlever series commercialized by Veeco (Santa Barbara, CA, USA) and the OMCL-TR series by Olympus (Tokyo, Japan). All these cantilevers are made of silicon nitride since this material allows for the fabrication of thinner cantilevers, which results in lower spring constants.

16.3 Cell Viscoelasticity

The approach trace of the force curve shown in Fig. 16.3 (dotted line) has a zone of relatively constant force when the tip is away from the cell surface, followed by a nonlinear increase when the tip makes contact with and indents the surface. In this case, the curve was obtained using a pyramidal tip, and the nonlinear response does not come from the nonlinear behavior of the cell, but from the increasing contact area between the tip and the sample. The shape of the curve will thus depend on the

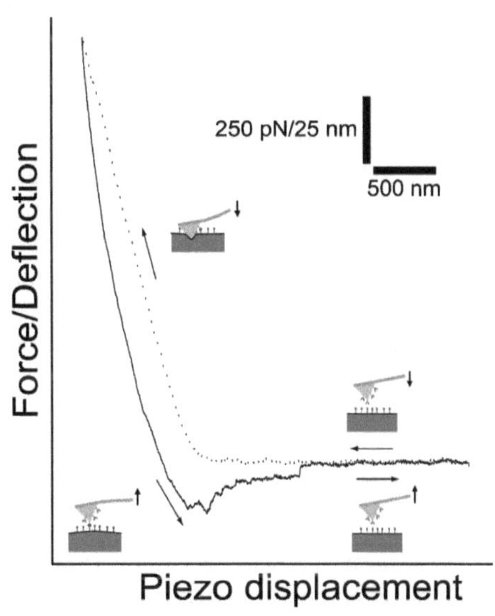

Fig. 16.3. Force–displacement (deflection–displacement) curve acquired at approximately 1.8 μm/s on a living cell using a pyramidal tip. Approach (*dotted line*) and retract (*solid line*) traces are plotted. The various steps of the tip–sample interaction are depicted

tip geometry and on the elastic properties of the sample, its stiffness. On the other hand, the retract trace of the curve follows a similar shape, although slightly shifted, i. e., the complete cycle has hysteresis. This hysteresis is mainly due to the viscosity of the sample. Taken together, these behaviors reflect the viscoelastic nature of living cells. This section will review the most-used methods to quantitatively determine the elasticity and viscoelasticity of cells with atomic force microscopy.

16.3.1 AFM Tip Geometries

As mentioned in Sect. 16.1, cantilever tips can have a variety of geometries, and the choice of tip will depend on its final application. The cantilever tips most widely used in the biological field are silicon nitride four-sided pyramidal tips (Fig. 16.2a). These cantilever tips are very suitable since they allow both mechanical measurements and imaging with high lateral resolution [74]. Sharper tips are also being used for imaging and probing single molecules, where the interaction is limited to the very end of the tip, typically with a curvature radii of approximately 10 nm (Fig. 16.2b) [75–77]. In contrast, when a smooth contact with the sample surface is required, a tip with dull geometry is a better choice. Thus, some authors have suggested the use of spherical beads of a few microns in diameter attached to the end of conventional cantilevers (Fig. 16.2b) [30, 39, 78]. This procedure will provide a more global estimate of the mechanical properties and may render the simultaneous imaging mode inoperative. As we will see in the following section, the geometry of the tip greatly affects force measurement analysis.

16.3.2 Elasticity: Young's Modulus

All commercially available tip geometries (e. g., conical, pyramidal and spherical[1]) lead to nonlinear force–indentation responses. The force increases as the tip indents the sample as a result of the increasing area of contact during the indentation process. The force-indentation relationship ($F(\delta)$) of two elastic bodies in contact was first described by Hertz [79] in 1881 for two spheres. Later on, Sneddon [80] solved the problem for various tip geometries. These so-called elastic models relate the force as a function of the indentation and the sample elastic modulus, taking tip geometry into consideration. Because of his pioneering work, elastic models are also commonly referred to as Hertz models. The most frequently used elastic models are summarized in Table 16.1. Note that the only model that leads to a linear force response is that of the flat-ended cylinder, the only geometry which maintains a constant area of contact independently of the indentation depth.

The most common approach used to obtain the elastic or Young's modulus (E) of living cells involves acquiring force curves (i. e., indenting the cell surface) and then fitting the suitable elastic model to the loading trace [32, 37, 81–84]. In principle, the unloading trace could also be used; however, using the loading trace minimizes

[1] Novascan Technologies (Ames, IA, USA) commercializes cantilevers with spherical beads of various materials attached at their end.

Table 16.1. Contact elastic models for common rigid-tip geometries indenting an elastic half space. E and ν are Young's modulus and the Poisson ratio of the sample, respectively

Tip geometry	Contact models, $F(\delta)$
Sphere of radius R	$F = \frac{4E\sqrt{R}}{3(1-\nu^2)}\delta^{3/2}$ [79, 85]
Cone of semi-included angle θ	$F = \frac{2E\tan\theta}{\pi(1-\nu^2)}\delta^2$ [80]
Flat-ended cylinder of radius a	$F = \frac{2Ea}{(1-\nu^2)}\delta$ [80]
Four-sided regular pyramid of semi-included angle θ	$F = \frac{0.7453E\tan\theta}{(1-\nu^2)}\delta^2$ [168]

artifacts due to adhesion between the tip and the sample. The elastic modulus of living cells has been measured using different techniques, revealing a wide range of values from a few pascals to hundreds of kilopascals, depending on cell type and the region of the cell being probed. Compared with everyday material such as rubber and jelly, cells are relatively soft[2] and a very soft cantilever is required to determine their Young's modulus. A reliable estimation of a cell's elastic modulus requires acquisition of experimental force curves obtained at an adequate indentation range. It has been shown that the elastic modulus is difficult to determine from curves with indentations lower than approximately 200 nm [74, 85]. The heterogeneity of the cell's glycocalix and membrane as well as the poorly defined geometry of the tip at such low indentations make it difficult to obtain a good estimation of E. In addition, the use of very high indentation forces which result in the probing of the hard substrate on which cells are plated is also not recommended for obtaining accurate Young's modulus values [83, 86]. It should also be noted that under very high indentation forces cells can behave nonlinearly, i. e., cells can have a nonelastic response [87–94]. Thus, it is important to limit the indentation in order to work within the linear regime of cellular elasticity. The actual indentation range may vary between cell types and regions of the cell being probed. However, as a rule of thumb, the hard surface will not affect the E values as long as the maximum indentation does not exceed 10–20% of the cell's thickness [83]. Secondly, two main issues are crucial for a reliable estimation of E during data analysis: accurate determination of the point of initial contact and suitability of the elastic model to the tip geometry being used [26, 39, 43, 74, 95–98]. The point of contact coordinates are defined by the force (or deflection) offset (F_{off} or d_{off}) and the vertical position point at which the cantilever starts to deflect (z_{c}) and are used to determine the absolute indentation [$\delta = z - z_{\mathrm{c}} - (d - d_{\mathrm{off}})$]. As noted in Fig. 16.4, it is difficult to visually determine the point of contact for living cells since the deflection increases very smoothly following contact owing to the cells' softness. Thus, an objective method to determine F_{off} and z_{c} should be used. The inclusion of the noncontact region of the loading trace in the fitting process allows for the simultaneous determination of the point of contact coordinates and the Young's modulus [26, 74]. Figure 16.4 shows a representative force curve overlaid with the fit of the appropriate contact model. In addition, the

[2] Rubber has a Young's modulus in the megapascal range, while that of jelly falls in the kilopascal range.

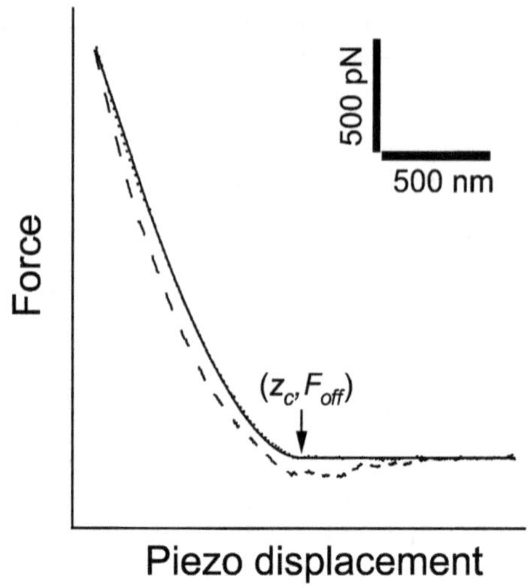

Fig. 16.4. Example of a representative approach trace of a force curve (*dotted line*) used to determine the cell elasticity. The fit of the contact model (Table 16.1) for a four-sided pyramid with a 35° semiopen angle is also shown (*solid line*). The Young's modulus obtained was 4.5 kPa. The *arrow* indicates the estimated point of contact. The retract trace (*dashed line*) presents hysteresis owing to the viscoelastic nature of living cells. The curve was acquired at a speed of approximately 1.8 μm/s by indenting an epithelial cell using a pyramidal tip

suitability of the contact elastic model depends on the similarity of the tip geometry to the idealized geometry and on the accuracy of the actual geometry parameters (radii, semiopen angles, etc.). As an example, it has been shown that using the common pyramidal tips (Fig. 16.2b) leads to a spherical-like F–δ response at low indentation, followed by a conical-like response at higher indentation[3] [81, 99]. This behavior was attributed to the actual bluntness of pyramidal tips, which lead to a region in the approach trace of tens of nanometers with a spherical-like geometry. There are a number of publications that take these more realistic blunted geometries of actual tips into account [74, 97]. In summary, a reliable estimation of the Young's modulus of living cells requires appropriate experimental conditions (indentation range), an accurate and objective determination of the point of contact, and the application of an elastic model that describes the actual tip geometry.

16.3.3 Viscoelasticity: Complex Shear Modulus

Even if we obtain a reliable estimation of cellular elasticity, a great number of studies have shown that cells are not only elastic but also viscous: they have a viscoelastic nature [18, 27, 81, 89, 100–106]. Viscoelasticity is roughly deduced from the hysteresis between the approach and retract traces of the curves shown in Figs. 16.3

[3] Pyramidal tips were first modeled as cones. Indeed, the indentation dependence of the force is the same for a pyramid and for a cone, apart from a scaling factor (Table 16.1).

and 16.4. However, the viscoelastic dynamic response of living cells has been shown to be more complex [104, 107–112]. Several authors have measured the viscoelasticity of living cells by applying a force or deformation step and measuring the resulting response [27, 101, 113, 114]. The principal limitation of these approaches is that the estimated viscous parameters are limited by the time window used during the measurements [85]. A more elegant method to probe dynamic viscoelastic cell behavior involves the application of low-amplitude oscillations at a wide frequency range and subsequent derivation of the complex shear modulus $G^*(\omega)$. The complex shear modulus is divided into a real or in-phase component, which stands for the elasticity or the stored energy, and an imaginary or out-of-phase component, which stands for the viscosity or lost energy. The real part is known as the elastic modulus (G'), while the imaginary part is called the loss modulus (G''). This technique has been successfully applied to a wide variety of living cells studied under different conditions [39, 86, 88, 103, 104, 108, 110, 111, 114].

In AFM viscoelastic measurements conducted on living cells, the oscillatory technique involves applying low-amplitude oscillations (50–100 nm) at an operating indentation (δ_0). Under these conditions, a contact elastic model (Table 16.1) described by $F(\delta)$ can be approximated by taking the first two terms of its Taylor expansion around the fixed point (δ_0)

$$F \cong F(\delta_0) + \frac{\partial F}{\partial \delta}(\delta - \delta_0) + \dots \,. \tag{16.1}$$

For the sake of simplicity, we will follow the derivation using the elastic model for a flat-ended cylinder of radius a, whose expansion leads to

$$F = F_0 + \frac{2Ea}{1-\nu^2}(\delta - \delta_0) \,. \tag{16.2}$$

Expressing (16.2) in terms of the shear modulus, using $E = 2(1+\nu)G$, and solving for G, we obtain

$$G = \frac{1-\nu}{4a}\frac{F - F_0}{\delta - \delta_0} \,. \tag{16.3}$$

By applying the Parseval theorem, we can transform (16.3) into the frequency domain, resulting in an expression for the complex shear modulus

$$G^*(\omega) = \frac{1-\nu}{4a}\frac{F(\omega)}{\delta(\omega)} \,, \tag{16.4}$$

where ω is the angular frequency ($\omega = 2\pi f$) and $F(\omega)$ and $\delta(\omega)$ are the Fourier transforms of force and indentation, respectively.

When samples are probed in liquid conditions, (16.4) has to be corrected for the hydrodynamic drag force exerted on the cantilever by the surrounding liquid. It has been determined that the drag contribution is not negligible for oscillatory frequencies above 0.3 Hz [115]. The drag force (F_d) depends on the geometry and dimensions of the cantilever, which give us the drag factor b, and on the velocity of the cantilever's free end $v = \mathrm{d}\delta/\mathrm{d}t$ [i.e., $F_\mathrm{d} = b(0)v$]. As described by Alcaraz et al. [115], the drag factor increases as the tip approaches the surface. Thus, it is important to measure $b(h)$ experimentally by oscillating the cantilever at various tip–surface distances (h) and extrapolate its value to zero separation [84, 110, 115]. The hydrodynamic drag force contribution in the frequency domain has the form $F_{\mathrm{d}}(\omega)/\delta(\omega) = \mathrm{i}\omega b(0)$ (i being the imaginary unit). To correct for the drag force, the last expression should be included in (16.4), resulting in the final form of the complex shear modulus

$$G^*(\omega) = \frac{1-\nu}{4a}\left(\frac{F(\omega)}{\delta(\omega)} - \mathrm{i}\omega b(0)\right) . \tag{16.5}$$

Notice, again, that the cylindrical punch has no dependency on the absolute indentation (the operating indentation, δ_0), since its force response is linear. That feature makes it ideal to probe cell mechanics at different indentation depths. The corresponding shear modulus for spherical or pyramidal models does depend on the absolute indentation [74]. A recent work proposed another viscoelastic approach that took into account the possible contribution of the hard substrate in oscillatory measurements. The authors used this procedure to determine the viscoelastic properties of thin regions of living cells [86]. Even if its application is quite a bit more complicated than the one presented here, it might be useful when probing very thin samples.

Experimentally, the procedure to carry out oscillatory measurements would first involve the acquisition of a high-amplitude force curve (a few microns) to determine the point of contact. Then the probe would be moved to an operating indentation in the linear regime (approximately 0.5 μm), and low-amplitude oscillations approximately (50 nm) would be applied at a given frequency or at a multifrequency, recording a few tens of cycles. The resulting data would be analyzed in the frequency domain according to (16.5). The oscillatory method is well suited for experiments conducted on living cells since it enables us to measure their dynamic viscoelastic response with a single measurement by applying a multifrequency wave as the oscillatory signal [85]. This procedure provides faster and less invasive measurements, convenient for living systems.

16.3.4 Cell Adhesion

Cell adhesion plays a major role in many biological processes, including developmental and immune system functions. It involves many cell types, from nonadherent cells, such as platelets or leukocytes, to adherent cells that comprise tissues, such as endothelial or epithelial cells [116–119]. Cell adhesion is mediated by cell adhesion receptors that adhere specifically to ligands on other cells or to the extracellular matrix [120, 121]. The AFM can be used to study cell adhesion at the single-molecule level as well as at the whole-cell level, where the contribution of multiple receptor–ligand interactions can be examined. The principle of operation of adhesion measurements by atomic force microscopy and other techniques is the same for both types of experiments. The method involves measuring the forces necessary to rupture the ligand–receptor bonds. We have seen that the approach trace of a force curve can provide us with an estimate of the cellular elasticity. In adhesion measurements, the unloading curve, or the retract trace, is usually used to acquire adhesive information at both molecular and cellular levels (Fig. 16.3). Different approaches have been used for that purpose. One method involves the functionalization or coating of the cantilever tip with receptors and acquiring force measurements by bringing the cantilever tip into contact with a cell or ligand-coated surface below. This is the most common approach when measuring adhesion between cell ligands that specifically attach to extracellular matrix proteins [122, 123]. A second technique involves attaching a cell to the end of a cantilever and bringing it into contact with a surface functionalized with the ligand of interest or another cell [40, 45, 124]. These two approaches are equivalent and their use is more or less appropriate depending on the system being studied. In both types of experiments, the first step is the coating of the cantilever tip with the desired molecules. The most widely used procedures for studying cell adhesion are explained in this section.

16.3.4.1 Functionalization of AFM Tips

There are many different protocols for coating surfaces, since surface chemistry modification is an intensive field of study [125]. We will describe two commonly used protocols, one involves covalently attaching receptors to AFM cantilevers, and the second is used to immobilize living cells on the cantilevers.

16.3.4.2 Attachment of Protein

- Clean cantilevers in acetone and irradiate them with UV light for 10 min.
- Immerse cantilevers in a glass dish containing 2% 3-aminopropyltriethoxysilane in acetone and shake them for 10 min at room temperature.
- Rinse extensively with nanopure water.
- Incubate for 30 min in 0.1% glutaraldehyde in phosphate-buffered saline (PBS) at room temperature.
- Rinse extensively with PBS.
- Air-dry cantilevers.
- Cover dried cantilevers with the protein solution and incubate for at least 1 h at room temperature or overnight at 4 °C.
- Rinse with tris(hydroxymethyl)aminomethane (Tris) buffered saline (TBS).
- Store coated cantilevers in TBS at 4 °C until use.

16.3.4.3 Attachment of Living Cells

- Clean cantilevers in acetone and irradiate them with UV light for 10 min.
- Incubate cantilevers in biotinylated bovine serum albumin (0.5 mg/ml in $NaCO_3$, overnight at 4 °C).
- Rinse cantilevers with PBS.
- Incubate cantilevers with streptavidin (0.5 mg/ml) for 10 min.
- Rinse again with PBS.
- Incubate cantilevers in a solution of biotinylated concanavalin-A (0.5 mg/ml).
- Rinse with PBS.
- Store coated cantilevers in PBS at 4 °C until use.

16.3.4.4
Dynamic Force Spectroscopy

The study of single-molecule interactions shed light on the ultimate biophysical determinants of cell adhesion. Measurements can be carried out on isolated receptor–ligand complexes attached to synthetic surfaces or directly on living cells [23, 75, 126–134]. An updated review on single-molecule measurements using atomic force microscopy has recently been published [77]. Single-molecule measurements are performed by allowing the cantilever tip coated with receptors or a living cell to make contact with a ligand-coated substrate or another cell. The resulting force curve is used to measure the rupture forces of the adhesion events. The bottom curve in Fig. 16.5 shows the rupture force (f_u) of a single bond. The forces necessary to break single bonds range from tens to hundreds of piconewtons. All force curves in Fig. 16.5 reveal fluctuations, especially visible in the noncontact region. This noise is mainly due to thermal fluctuations, although mechanical perturbations of the system and turbulence of the liquid also have an effect. The noise level resulting from thermal fluctuations is on the order of tens of piconewtons; thus, rupture events involving forces below 10–20 pN are difficult to resolve and are usually rejected [135, 136]. In order to obtain single-molecule measurements on living cells, force curves are obtained at low indentation forces to minimize the contact area and with a brief contact time (on the order of fractions of a second), thus reducing the binding probability. Under these conditions, most of the force curves will show no adhesion, as illustrated in Fig. 16.5. This effect is desirable, since, assuming Poisson statistics, an adhesion frequency of less than 30% in the force measurements ensures that more than 83% of the events are caused by single bonds [137, 138]; therefore, thousands of measurements need to be acquired, from which only curves presenting rupture events will be analyzed. Automatic procedures for detecting and

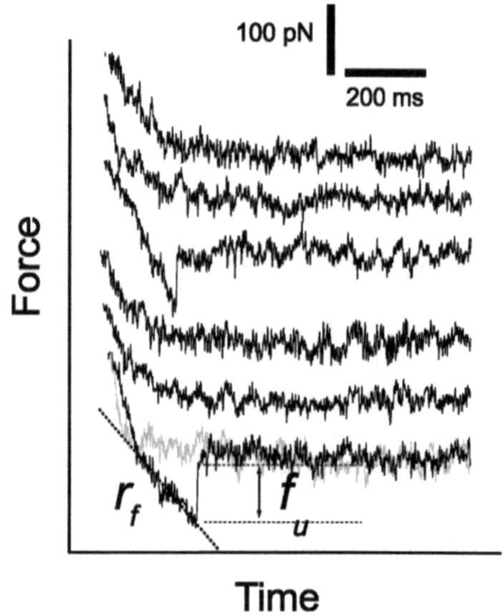

Fig. 16.5. Unbinding curves obtained from single-molecule measurements. Measurements were acquired at speeds of 5 μm/s and conditions that minimized contact (approximately 50 ms contact time and approximately 100 pN indentation force) ensuring an adhesion frequency of less than 30%. Adhesion was observed in the third and sixth traces. f_u is the unbinding force of the interaction, and r_f is the actual loading rate derived from the slope of the force–time graph

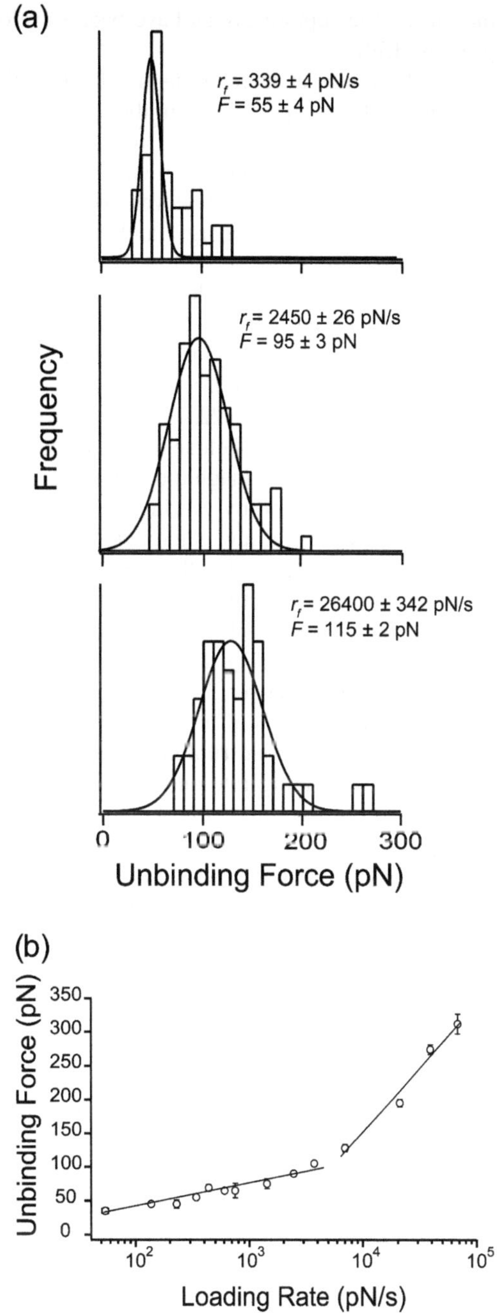

Fig. 16.6. **a** Rupture force histograms obtained from measurements at different loading rates of the interaction between a T-cell integrin and one of its ligands. Rupture events were grouped according to loading rates. The most probable unbinding forces were taken from the Gaussian fit of each histogram (*solid lines*). **b** Dynamic force spectrum resulting from the histograms in **a**. Measurements were acquired at a loading rate range from 50 to 60,000 pN/s, which was achieved by varying the retraction rate of the cantilever from 0.1 to 26 μm/s. Forces acquired at cantilever retraction speeds exceeding 1 μm/s were corrected for hydrodynamic drag [150]. The fits were obtained using (16.6) on each linear regime (*solid lines*). The *error bars* represent the standard error of the mean. Two linear regimes can be differentiated corresponding to inner (fast loading rates) and outer (slow loading rates) barriers

measuring the rupture events have been developed to accelerate the data analysis process [139].

Weak molecular interactions involved in biology can be characterized by their free-energy landscape, which, in the simplest case, can be represented by a potential well with a single barrier. This barrier has to be overcome in order to break the molecular interaction. Thermal motion determines the kinetics of unbinding when no force is applied. However, AFM measurements involve the application of a force that distorts the energy landscape by lowering its barrier height. This theory was first proposed by Bell [120] in 1978 for a constant applied force. Evans and Ritchie [140, 141] developed it for an increasing ramp of force, which better describes common force measurements. According to this theory, the rupture force will depend on the logarithm of the loading rate, i. e., the rate at which force is applied ($\mathrm{d}F/\mathrm{d}t$ or r_f). A large number of studies have tested this prediction experimentally [24, 25, 41, 122, 132, 141–146]. Thus, to completely characterize the dynamic strength of molecular bonds, rupture forces have to be measured ranging over several orders of magnitude of the loading rate. In AFM experiments, this parameter is controlled by varying the spring constant of the cantilever and the retraction speed, from a few namometers per second to tens of microns per second. Owing to the elasticity of cells and adhesion molecules, the actual loading rate is determined by the effective spring constant of the cantilever–sample system in series. The loading rate is then obtained from the slope immediately before rupture in a force–time curve (Fig. 16.5). The loading rates measured in experiments using living cells usually cover 3 or 4 orders of magnitude, from tens of piconewtons per second to hundreds of nanonewtons per second.Owing to the stochastic nature of single-molecule bond rupture, hundreds of rupture events have to be grouped for each fixed loading rate and organized into histograms. A Gaussian fit is applied to each histogram and provides the mean rupture force for each loading rate (Fig. 16.6a) [122, 127, 142, 147, 148]. The resulting force versus loading rate plot is called the dynamic force spectrum of the interaction (Fig. 16.6b). According to Bell's theory, each linear regime in the semilogarithmic plot corresponds to an energy barrier whose width (x_β) and dissociation rate (k^o) under no force can be obtained from the following equation:

$$F = \frac{k_\mathrm{B}T}{x_\beta} \ln\left(\frac{x_\beta}{k^\mathrm{o} k_\mathrm{B} T}\right) + \frac{k_\mathrm{B}T}{x_\beta} \ln r_\mathrm{f} \,. \tag{16.6}$$

Fitting the above equation for each linear regime to the experimental data will provide us with the unknown variables that characterize the molecular interaction. More elaborate models have been proposed, which may also allow us to determine the potential barrier height, although their practical application appears to be more complex [149].

16.3.4.5 Whole-Cell Adhesion Measurements

Even if single-molecule interactions are the ultimate determinants of cell adhesion, cells usually attach to each other and to the extracellular matrix by forming clusters of receptors which may cooperate and break in a more intricate manner [151–156]. To explain the rupture of multiple-bond attachments, the three simplest configurations are N identical bonds loaded in series, zipperlike attachment of N identical bonds, and N identical bonds loaded in parallel. The first configuration is common in protein unfolding measurements, in which each bond (or protein) experiences the same force and ruptures (or unfolds) in random order [157]. In a zipper, the same force propagates from one bond to the next after rupture of the lead bond has taken place. Bond rupture occurs in sequence from first to last. Finally, in the parallel-bonds arrangement, the force is shared by all the bonds that are involved in the attachment and rupture may occur cooperatively. The last of these is probably the configuration more physiologically relevant and the one which better describes multiple adhesion measurements obtained with atomic force microscopy on living cells.

Whole-cell adhesion is usually measured by obtaining high-amplitude curves (tens of microns) and keeping the cell in contact with the ligand-coated surface for a longer time (a few seconds) than in the single-molecule measurements. The retract trace is then analyzed by measuring the maximum force and the total work required to completely detach the two surfaces [40,41,132]. Owing to the large area of contact and the long contact time, multiple bonds are typically formed, leading to multiple rupture events [124]. Retract traces of the force–distance curves usually present a peak of force followed by a cascade of rupture events until complete detachment is achieved (Fig. 16.7). In living cells, long tethers may also form. Curves then present a long and constant offset of force, whose value determines the force needed to extract the tether from the cell membrane [144]. The maximum peak of force in the retract trace of the curve determines the detachment force, which is usually hundreds of piconewtons or a few nanonewtons. The work of adhesion (or deadhesion) is computed from the area under the unloading curve on a force–distance plot (shaded region in Fig. 16.7). The dependence of the single-bond rupture force on the loading rate may have an influence on the measured adhesion strength; thus, dynamic measurements at different loading rates might be required. The slope just before initial rupture in a force–time plot provides an estimate of the loading force. This loading rate, however, is not usually determined since the force increase before

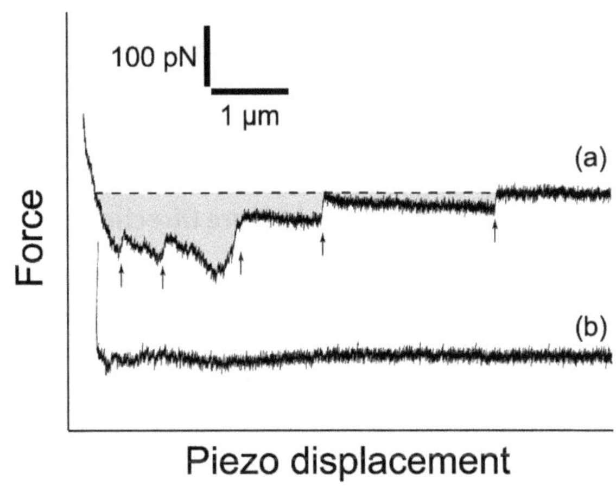

Fig. 16.7. Unbinding curves involving the rupture of multiple bonds. **a** A T cell, which expresses integrins on its surface, was attached to the cantilever and brought into contact with a surface coated with integrin ligands. **b** The same cell in the presence of ligand-blocking antibody. The retraction curves were obtained at a cantilever retraction speed of 5 μm/s

rupture is usually nonlinear; therefore, the cantilever retraction speed is the parameter used. The forces involved in whole-cell adhesion depend on the contact area, the time of contact, the loading rate, the density of receptors and the strength of the specific interaction [124]. As an illustrative example, Fig. 16.7 shows the retraction curves obtained with living cells under different conditions. The decrease in both force and work of de-adhesion is noticeable when blocking the ligand with antibody and reflects the suitability of these measurements to quantify cell adhesion strength. Both force and work of deadhesion are good estimates of cell adhesion strength, but they depend on the specific experimental conditions. Therefore, in order to be able to compare data from different cell systems and methods, such as the biomembrane force probe, it is necessary to estimate a parameter which describes intrinsic cell adhesion properties such as force or energy density. This may require knowledge of the contact area, which is not always possible.

In both single-bond and multiple-bond measurements, the hydrodynamic drag force introduces a systematic bias in the force that has to be corrected. As mentioned before, the drag force is proportional to the tip velocity and to the drag coefficient b. Both rupture forces are corrected by adding the product of $vb(h)$ to the measured peak force. Since the drag coefficient depends on the tip–sample distance (h), a previous calibration may be obtained to determine the correction factor taking h into account. Typically a constant drag coefficient is estimated and used to correct for the drag force. At high retraction speeds (tens of microns per second) and when the interaction lasts several hundreds of nanometers, this simplified approach may introduce important errors and should be avoided, especially in single-molecule measurements where the drag force is of the same order of magnitude as the actual rupture forces [158]. Therefore, measurements at high retraction speeds may be inaccurate and should be avoided. High loading rates can be achieved by using stiffer and smaller cantilevers and lower retraction speeds, thus minimizing the drag forces.

Finally, an important issue at either single-molecule or multiple-molecule level concerns the specificity of the interaction. Negative controls are required to ensure that the probed ligand–receptor interaction is specific. There are different procedures available. Similar measurements using uncoated surfaces are the simplest. More sophisticated and effective measures may require cell mutagenesis to obtain cells without the specific protein, or the use of specific antibodies directed against the molecules involved to block the interaction. In practice, it is recommended to carry out complementary negative controls.

16.4 Concluding Remarks and Future Directions

We have seen that the AFM is a very versatile tool to measure cellular mechanics in living cells, a feature that complements its ability to image cells. Recent articles also show the ability of the AFM to map the viscoelastic and adhesive properties of the whole cell surface with nanometer resolution and single-molecule recognition [77, 95, 123, 134]. The development of models describing the behavior of living cell mechanics is the next area of research to advance. There has been intense progress in this area in recent years [9, 15, 92, 93, 104, 107, 149, 151, 153, 154, 156, 159–167].

The next technological step would be the miniaturization of the AFM to enable us to measure the mechanical properties of living cells in their native environment, the human body.

References

1. Johnson KL (1985) Contact mechanics. Cambridge University Press, Cambridge
2. Zhu C, Bao G, Wang N (2000) Annu Rev Biomed Eng 2:189
3. Heidemann SR, Wirtz D (2004) Trends Cell Biol 14:160
4. Janmey PA, Weitz DA (2004) Trends Biochem Sci 29:364
5. Sheetz MP, Sable JE, Dobereiner H-G (2006) Annu Rev Biophys Biomol Struct 35:417
6. Chicurel ME, Chen CS, Ingber DE (1998) Curr Opin Cell Biol 10:232
7. Goldmann WH, Ingber DE (2002) Biochem Biophys Res Commun 290:749
8. Huang S, Ingber DE (1999) Nat Cell Biol 1:E131
9. Ingber DE (1993) J Cell Sci 104:613
10. Ingber DE, Prusty D, Zhengqi S, Betensky H, Ning W (1995) J Biomech 28:1471
11. Maniotis AJ, Chen CS, Ingber DE (1997) Proc Natl Acad Sci USA 94:849
12. Pourati J, Maniotis A, Spiegel D, Schaffer JL, Butler JP, Fredberg JJ, Ingber DE, Stamenovic D, Wang N (1998) Am J Physiol 274(5 Pt 1):C1283
13. Wang N, Butler JP, Ingber DE (1993) Science 260:1124
14. Binnig G, Quate CF, Gerber C (1986) Phys Rev Lett 56:930
15. Ingber DE, Dike L, Hansen L, Karp S, Liley H, Maniotis A, McNamee H, Mooney D, Plopper G, Sims E (1994) Int Rev Cytol 150:173
16. Burnham NA, Colton RJ (1989) J Vac Sci Technol A 7:2906
17. Weisenhorn AL, Hansma PK, Albrecht TR, Quate CF (1989) Appl Phys Lett 54:2651
18. Butt HJ, Wolff EK, Gould SA, Dixon NB, Peterson CM, Hansma PK (1990) J Struct Biol 105:54
19. Haberle W, Horber JKH, Binnig G (1991) J Vac Sci Technol B 9:1210
20. Henderson E, Haydon PG, Sakaguchi DS (1992) Science 257:1944
21. Tao NJ, Lindsay SM, Lees S (1992) Biophys J63:1165
22. Hoh JH, Schoenenberger CA (1994) J Cell Sci 107:1105
23. Florin EL, Rief M, Lehmann H, Ludwig M, Dornmair C, Moy VT, Gaub HE (1995) Biosensors Bioelectron 10:895
24. Baumgartner W, Hinterdorfer P, Ness W, Raab A, Vestweber D, Schindler H, Drenckhahn D (2000) Proc Natl Acad Sci USA 97:4005
25. Baumgartner W, Hinterdorfer P, Schindler H, Drenckhahn D (2000) Biophys J 78:446A
26. Crick SL, Yin FC (2007) Biomech Model Mechanobiol 6:199
27. Wu HW, Kuhn T, Moy VT (1998) Scanning 20:389
28. Goldmann WH, Galneder R, Ludwig M, Kromm A, Ezzell RM (1998) FEBS Letters 424:139
29. Goldmann WH, Galneder R, Ludwig M, Xu W, Adamson ED, Wang N, Ezzell RM (1998) Exp Cell Res 239:235
30. Dimitriadis EK, Horkay F, Maresca J, Kachar B, Chadwick RS (2002) Biophys J 82:2798
31. Dvorak JA (2003) Methods 29:86
32. Domke J, Dannohl S, Parak WJ, Muller O, Aicher WK, Radmacher M (2000) Colloids Surf B 19:367
33. Charras GT, Horton MA (2002) Biophys J 83:858
34. Horton MA, Charras GT, Ballestrem G, Lehenkari PP (2000) Single Mol 1:135
35. Lehenkari PP, Charras GT, Nykanen A, Horton MA (2000) Ultramicroscopy 82:289

36. Mathur AB, Collinsworth AM, Reichert WM, Kraus WE, Truskey GA (2001) J Biomech 34:1545
37. Rotsch C, Radmacher M (2000) Biophys J 78:520
38. Obataya I, Nakamura C, Han S, Nakamura N, Miyake J (2005) Nano Lett 5:27
39. Mahaffy RE, Shih CK, MacKintosh FC, Kas J (2000) Phys Rev Lett 85:880
40. Wojcikiewicz EP, Zhang X, Chen A, Moy VT (2003) J Cell Sci 116:2531
41. Zhang XH, Chen A, De Leon D, Li H, Noiri E, Moy VT, Goligorsky MS (2004) Am J Physiology Heart Circ Physiol 286:H359
42. Lee I, Marchant RE (2000) Colloids Surf B 19:357
43. A-Hassan E, Heinz W, Antonik MD, D'Costa NP, Nageswaran S, Schoenenberger CA, Hoh JH (1998) Biophys J 74:1564
44. Han SW, Nakamura C, Obataya I, Nakamura N, Miyake J (2005) Biosensors Bioelectron 20:2120
45. Benoit M, Gaub HE (2002) Cells Tissues Organs 172:174
46. You HX, Lau JM, Zhang S, Yu L (2000) Ultramicroscopy 82:297
47. Binnig G, Rohrer H (1982) Helv Phys Acta 55:726
48. Bustamante C, Dunlap D (1991) Semin Cell Biol 2:179
49. Dunlap DD, Bustamante C (1989) Nature 342:204
50. Muller DJ, Sapra KT, Scheuring S, Kedrov A, Frederix PL, Fotiadis D, Engel A (2006) Curr Opin Struct Biol 16:489
51. Hansma H, Pietrasanta L (1998) Curr Opin Chem Biol 2:767
52. Hansma HG, Bezanilla M, Zenhausern F, Adrian M, Sinsheimer RL (1993) Nucleic Acids Res 21:505
53. Poggi MA, Gadsby ED, Bottomley LA, King WP, Oroudjev E, Hansma H (2004) Anal Chem 76:3429
54. Simon A, Durrieu M-C (2006) Micron 37:13
55. Hansma PK, Drake B, Grigg D, Prater CB, Yashar F, Gurley G, Eling SV, Feinstein S, Lal R (1994) J Appl Phys 76:796
56. Putman CAJ, Degrooth BG, Vanhulst NF, Greve J (1992) J Appl Phys 72:6
57. Butt HJ, Cappella B, Kappl M (2005) Surf Sci Rep 59:1
58. Cappella B, Dietler G (1999) Surf Sci Rep 34:1
59. Gibson CT, Watson GS, Myhra S (1996) Nanotechnology 7:259
60. Chon JWM, Mulvaney P, Sader JE (2000) J Appl Phys 87:3978
61. Sader JE (1995) Rev Sci Instrum 66:4583
62. Sader JE (1998) J Appl Phys 84:64
63. Sader JE, Larson I, Mulvaney P, White L (1995) Rev Sci Instrum 66:3789
64. Sader JE, White L (1993) J Appl Phys 74:1
65. Hutter JL, Bechhoefer J (1993) Rev Sci Instrum 64:1868
66. Stark RW, Drobek T, Heckl WM (2001) Ultramicroscopy 86:207
67. Cleveland JP, Manne S, Bocek D, Hansma PK (1993) Rev Sci Instrum 64:403
68. Walters DA, Cleveland JP, Thomson NH, Hansma PK, Wendman MA, Gurley G, Elings V (1996) Rev Sci Instrum 67:3583
69. Butt HJ, Jaschke M (1995) Nanotechnology 6:1
70. Gibson CT, Watson GS, Myhra S (1997) Scanning 19:564
71. Gibson CT, Weeks BL, Abell C, Rayment T, Myhra S (2003) Ultramicroscopy 97:113
72. Burnham NA, Chen X, Hodges CS, Matei GA, Thoreson EJ, Roberts CJ, Davis MC, Tendler SJB (2003) Nanotechnology 14:1
73. Gibson CT, Smith DA, Roberts CJ (2005) Nanotechnology 16:234
74. Rico F, Roca-Cusachs P, Gavara N, Farre R, Rotger M, Navajas D (2005) Phys Rev e72
75. Rief M, Grubmuller H (2002) Chemphyschem 3:255

76. Frederix P, Akiyama T, Staufer U, Gerber C, Fotiadis D, Muller DJ, Engel A (2003) Cur Opin Chem Biol 7:647
77. Hinterdorfer P, Dufrene YF (2006) Nat Methods 3:347
78. Stolz M, Raiteri R, Daniels AU, VanLandingham MR, Baschong W, Aebi U (2004) Biophys J 86:3269
79. Hertz H (1881) Hertz's miscellaneous papers (miscellaneous papers). Macmillan, London, p 146
80. Sneddon IN (1965) Int J Eng Sci 3:47
81. Radmacher M, Fritz M, Kacher CM, Cleveland JP, Hansma PK (1996) Biophys J 70:556
82. Radmacher M (1997) IEEE Eng Med Biol Mag 16:47
83. Domke J, Radmacher M (1998) Langmuir 14:3320
84. Rico F, Alcaraz J, Fredberg JJ, Navajas D (2005) Int J Nanotechnol 2:180
85. Roca-Cusachs P, Almendros I, Sunyer R, Gavara N, Farre R, Navajas D (2006) Biophys J 91:3508
86. Mahaffy RE, Park S, Gerde E, Kas J, Shih CK (2004) Biophys J 86:1777
87. Coughlin MF, Stamenovic D (2003) Biophys J 84:1328
88. Trepat X, Grabulosa M, Puig F, Maksym GN, Navajas D, Farre R (2004) Am J Physiol Lung Cell Mol Physiol 31:L1025
89. Fernandez P, Pullarkat PA, Ott A (2006) Biophys J 90:3796
90. Stamenovic D, Coughlin MF (1999) J Theor Biol 201:63
91. Stamenovic D, Coughlin MF (2000) J Biomech Eng 122:39
92. Stamenovic D, Ingber DE (2002) Biomech Model Mechanobiol 1:95
93. Wang N, Naruse K, Stamenovic D, Fredberg JJ, Mijailovich SM, Tolic-Norrelykke IM, Polte T, Mannix R, Ingber DE (2001) Proc Natl Acad Sci USA 98:7765
94. Wang N, Tolic-Norrelykke IM, Chen J, Mijailovich SM, Butler JP, Fredberg JJ, Stamenovic D (2002) Am J Physiol Cell Physiol 282:C606
95. Touhami A, Nysten B, Dufrene YF (2003) Langmuir 19:4539
96. Costa KD, Yin FCP (1999) J Biomech Eng 121:462
97. Briscoe BJ, Sebastian KS, Adams MJ (1994) J Phys D Appl Phys 27:1156
98. Costa KD, Sim AJ, Yin FC (2006) J Biomech Eng 128:176
99. Radmacher M, Fritz M, Hansma PK (1995) Biophys J 69:264
100. Hansma PK, Cleveland JP, Radmacher M, Walters DA, Hillner PE, Bezanilla M, Fritz M, Vie D, Hansma HG, Prater CB, Massie J, Fukunaga L, Gurley J, Elings V (1994) Appl Phys Lett 64:1738
101. Thoumine O, Ott A (1997) J Cell Sci 110:2109
102. Thoumine O, Ott A, Cardoso O, Meister JJ (1999) J Biochem Biophys Methods 39:47
103. Maksym GN, Fabry B, Butler JP, Navajas D, Tschumperlin DJ, Laporte JD, Fredberg JJ (2000) J Appl Physiol 89:1619
104. Fabry B, Maksym GN, Butler J, Glogauer M, Navajas D, Fredberg JJ (2001) Phys Rev Lett 87:148102
105. Puig-de-Morales M, Grabulosa M, Alcaraz J, Mullol J, Maksym GN, Fredberg JJ, Navajas D (2001) J Appl Physiol 91:1152
106. Puig-de-Morales M, Millet E, Fabry B, Navajas D, Wang N, Butler JP, Fredberg JJ (2004) Am J Physiol Cell Physiol 287:C643
107. Fabry B, Maksym GN, Butler JP, Glogauer M, Navajas D, Taback NA, Millet EJ, Fredberg JJ (2003) Phys Rev e68
108. Alcaraz J, Buscemi L, Grabulosa M, Trepat X, Fabry B, Farre R, Navajas D (2003) Biophys J 84:2071
109. Stamenovic D, Suki B, Fabry B, Wang N, Fredberg JJ (2004) J Appl Physiol 96:1600
110. Smith BA, Tolloczko B, Martin JG, Grutter P (2005) Biophys J 88:2994

111. Trepat X, Grabulosa M, Buscemi L, Rico F, Farre R, Navajas D (2005) J Appl Physiol 98:1567
112. Bursac P, Lenormand G, Fabry B, Oliver M, Weitz DA, Viasnoff V, Butler JP, Fredberg JJ (2005) Nat Mater 4:557
113. Bausch AR, Hellerer U, Essler M, Aepfelbacher M, Sackmann E (2001) Biophys J 80:2649
114. Desprat N, Richert A, Simeon J, Asnacios A (2005) Biophys J 88:2224
115. Alcaraz J, Buscemi L, Puig-de-Morales M, Colchero J, Baro A, Navajas D (2002) Langmuir 18:716
116. Burridge K, Chrzanowska-Wodnicka M (1996) Annu Rev Cell Dev Biol 12:463
117. Burridge K, Tsukita S (2001) Curr Opin Cell Biol 13:528
118. Vogel V, Sheetz M (2006) Nat Rev Mol Cell Biol 7:265
119. Chen CS, Tan J, Tien J (2004) Annu Rev Biomed Eng 6:275
120. Bell GI (1978) Science 200:618
121. Springer TA (1990) Nature 346:425
122. Lehenkari PP, Horton MA (1999) Biochem Biophys Res Commun 259:645
123. Hyonchol K, Arakawa H, Osada T, Ikai A (2002) Colloids Surf B 25:33
124. Zhang X, Wojcikiewicz EP, Moy VT (2006) Exp Biol Med (Maywood) 231:1306
125. Wagner P (1998) FEBS Lett 430:112
126. Florin EL, Moy VT, Gaub HE (1994) Science 264:415
127. Moy VT, Florin EL, Gaub HE (1994) Biophys J 66:A340
128. Hinterdorfer P, Baumgartner W, Gruber HJ, Schilcher K, Schindler H (1996) Proc Natl Acad Sci USA 93:3477
129. Schutz GJ, Sonnleitner M, Hinterdorfer P, Schindler H (2000) Mol Membr Biol 17:17
130. Moy VT, Chen A (2000) Biophys J 78:447A
131. Hinterdorfer P, Schutz G, Kienberger F, Schindler H (2001) J Biotechnol 82:25
132. Zhang XH, Wojcikiewicz E, Moy VT (2002) Biophys J 83:2270
133. Zhang XH, Bogorin DF, Moy VT (2004) Chemphyschem 5:175
134. Stroh CM, Ebner A, Geretschlager M, Freudenthaler G, Kienberger F, Kamruzzahan AS, Smith-Gill SJ, Gruber HJ, Hinterdorfer P (2004) Biophys J 87:1981
135. Rief M, Junker JP, Schlierf M, Hell K, Neupert W (2006) Biophys. J 91:2011
136. Ainavarapu SRK, Li L, Fernandez JM (2006) Biophys J 91:2009
137. Tees DFJ, Waugh RE, Hammer DA (2001) Biophys J 80:668
138. Evans E (1998) Faraday Discus 111:1
139. Baumgartner W, Hinterdorfer P, Schindler H (2000) Ultramicroscopy 82:85
140. Evans E, Ritchie K (1997) Biophys J 72:1541
141. Merkel R, Nassoy P, Leung A, Ritchie K, Evans E (1999) Nature 397:50
142. Benoit M, Gabriel D, Gerisch G, Gaub HE (2000) Nat Cell Biol 2:313
143. Zhang XH, Craig SE, Kirby H, Humphries MJ, Moy VT (2004) Biophys J 87:3470
144. Sun M, Graham JS, Hegedus B, Marga F, Zhang Y, Forgacs G, Grandbois M (2005) Biophys J 89:4320
145. Li FY, Redick SD, Erickson HP, Moy VT (2003) Biophys J 84:1252
146. Wojcikiewicz EP, Abdulreda MH, Zhang X, Moy VT (2006) Biomacromolecules 7:3188
147. Li YQ, Tao NJ, Pan J, Garcia AA, Lindsay SM (1993) Langmuir 9:637
148. Kokkoli E, Ochsenhirt SE, Tirrell M (2004) Langmuir 20:2397
149. Dudko OK, Hummer G, Szabo A (2006) Phys Rev Lett 96
150. Evans E, Leung A, Hammer D, Simon S (2001) Proc Natl Acad Sci USA 98:3784
151. Seifert U (2000) Phys Rev Lett 84:2750
152. van Kooyk Y, Figdor CG (2000) Curr Opin Cell Biol 12:542
153. Evans E (2001) Annu Rev Biophys Biomol Struct 30:105
154. Seifert U (2002) Europhys Lett 58:792

155. Prechtel K, Bausch AR, Marchi-Artzner V, Kantlehner M, Kessler H, Merkel R (2002) Phys Rev Lett 89:028101
156. Williams PM (2003) Anal Chim Acta 479:107
157. Rief M, Gautel M, Oesterhelt F, Fernandez JM, Gaub HE (1997) Science 276:1109
158. Janovjak HJ, Struckmeier J, Muller DJ (2005) Eur Biophys J 34:91
159. Stamenovic D, Wang N (2000) J Appl Physiol 89:2085
160. Hummer G, Szabo A (2005) Acc Chem Res 38:504
161. Hummer G, Szabo A (2003) Biophys J 85:5
162. Hummer G, Szabo A (2001) Proc Natl Acad Sci USA 98:3658
163. Dudko OK, Filippov AE, Klafter J, Urbakh M (2003) Proc Natl Acad Sci USA 100:11378
164. Stamenovic D, Fredberg JJ, Wang N, Butler JP, Ingber DE (1996) J Theor Biol 181:125
165. Erdmann T, Schwarz US (2004) J Chem Phys 121:8997
166. Erdmann T, Schwarz US (2004) Europhys Lett 66:603
167. Erdmann T, Schwarz US (2004) Phys Rev Lett 92:108102
168. Bilodeau G (1992) J Appl Mech Trans ASME 59:519

[illegible]

19 Application of Atomic Force Microscopy to the Study of Expressed Molecules in or on a Single Living Cell

Hyonchol Kim · Hironori Uehara · Rehana Afrin · Hiroshi Sekiguchi · Hideo Arakawa · Toshiya Osada · Atsushi Ikai

From volume IX

Abstract. One of the most useful ways to study the physical and biochemical properties of expressed molecules in or on living cells is to directly manipulate the molecules in question. In this chapter, we describe how the atomic force microscope (AFM) was used as a nanomanipulator, and membrane receptors or expressed messenger RNA (mRNAs) were touched, pulled and picked up to study the expressed information and biophysical properties of these molecules. First, the chemical and physical modifications of the AFM tip are introduced. For the study of membrane receptors over a large area of cell surface using an AFM, the colloidal probe method is useful. Details of probe preparation, chemical modification of colloids and data analysis are introduced. Furthermore, a useful method to measure the interaction force between a single molecular pair on a soft material is also introduced. Second, some examples of expressed receptor mapping are given. Ligand molecules were immobilized on a microbead attached to the AFM tip and force measurements were performed one by one over a large area of living cell surface. Distributions and amounts of expressed receptors on the cell surface were studied by measuring ligand–receptor interactions. Third, extractions of both membrane receptors and mRNAs are introduced. For the extraction of membrane receptors, the AFM tip was treated with cross-linkers. The tip was moved close to the cell surface and covalent bonds were formed between the tip and the receptors through the cross-linkers. Extraction of membrane proteins was confirmed by monitoring the force in each assay. For the extraction of mRNAs, the AFM tip was inserted into the cell body, and incubated for a few minutes. After incubation, the tip was withdrawn and inserted into a PCR tube. Collection of mRNAs was then confirmed with PCR analysis. By application of this method, the time course expression of specific mRNAs was successfully measured. The methods introduced in this chapter represent new and useful applications of the AFM to the study of the properties of single cells.

Key words: Single cell, Nanomanipulation, Membrane receptor, Messenger RNA, Colloidal probe

Abbreviations

AFM	Atomic force microscope
CHO	Chinese hamster ovary
CHX	Cycloheximide
ECM	Extracellular matrix
FBS	Fetal bovine serum
FN	Fibronectin
GFP	Green fluorescent protein
mRNA	Messenger RNA
PCR	Polymerase chain reaction

In: H. Fuchs, B. Bhushan (eds), *Biosystems - Investigated by Scanning Probe Microscopy*,
DOI 10.1007/978-3-540-74083-4_7,

PG	Prostaglandin
SD	Standard deviation
VN	Vitronectin

19.1 Introduction

A recent trend in drug discovery is to find molecular markers which determine cellular characteristics and functions. For example, finding specific molecular markers for the determination of stem cell characteristics has critical importance not only for the development of regenerative medicine but also for the understanding of the occurrence and metastasis of cancers [1–3]. Recently, several methods have been developed to identify expressed molecules in or on cells. One of the promising methods is the application of array-based technologies, such as DNA microarray [4–8]. This technology is suitable for the simultaneous detection of a large number of expressed molecules; however, it is difficult to detect the expression of molecules in a single cell without amplification such as by polymerase chain reaction (PCR) and hence the result generally gives the average value of the properties of many cells. Another suitable method is the application of in situ hybridization technology [9–12]. By application of this method, expressed molecules in each cell can be visualized; however, as this method is usually based on fluorescent dye, spatial resolution of target detection is restricted by optical limits, and quantification of the target number is difficult because of the photobleaching of fluorescent dyes. Thus, to study the cellular properties and characteristics of each cell at the molecular level, a new technology should be developed.

As examples of molecular markers representing cellular characteristics and hence good candidates for detection, two types should be mentioned: receptors on a cell surface, and messenger RNAs (mRNAs) in a cell. Cellular condition is usually controlled by outer environmental stimulation, and those signals are recognized at receptors on the cell surface. The cellular condition is also usually related to the amount of expressed mRNAs in a cell. Taking these points into account, direct observation and quantification of both cellular receptors and mRNAs are key factors to understand each cellular characteristic and to discover key biomarkers.

In this chapter, methods to identify target receptors and mRNAs using atomic force microscopy force measurement mode are introduced. To determine the cellular conditions and responses of cells to environmental stimulations, the amount and the distribution of expressed receptors are good parameters, because the expression of receptors is usually controlled by the environment. To study the distribution of receptors using an atomic force microscope (AFM), it is useful to employ a microbead as a probe to increase the area of contact with the cell. With use of this colloidal probe tip, the distribution of molecules such as integrins has been determined on a living cell [13, 14]. Membrane receptors were also harvested using an AFM tip modified with bifunctional cross-linkers [15–17]. This technology is especially remarkable for the establishment of a method to identify expressed characteristic marker proteins on the cell surface at the single-molecule level. Moreover, we can harvest expressed mRNAs from a cell using an AFM tip as a nanomanipulator without serious damage to the cell, and time-course developments of mRNA expressions have been studied using this method [18–20]. One of the advantages of using the AFM is that observation and manipulation can be made simultaneously in a liquid environment. From this benefit, all of the abovementioned experiments could be performed using living cells under physiological conditions. Another advantage is that quantitative force measurement is possible. Both observation of the distribution of receptors and harvest of mRNA were carried out by monitoring the interaction force between the AFM tip and the sample. By using force as a parameter, we can carry out quantitative nanomanipulations for a single living cell.

19.2 Methods of Manipulation To Study Molecules in or on a Living Cell Using an AFM

19.2.1 AFM Tip Preparation To Manipulate Receptors on a Cell Surface

For the observation and manipulation of receptors on a cell surface, several kinds of chemical treatments or physical modifications have been applied to the AFM tip. In comparison with the size of a cell, the usual AFM tip is too sharp and small; thus, the colloidal probe method is useful for observation of the distribution of receptors. The colloidal probe was first introduced by Ducker et al. [21, 22]. Silica particles are a popular material for colloidal probes, because the chemistry of silica is well developed [23, 24]. By changing the material and the size of the colloidal particles, one can easily control the chemical properties or the contact area of the colloid. As the contact area of the colloid is larger than that of the usual AFM tip, many molecular pairs make bonds during an assay. When ligand molecules are immobilized on a colloidal particle and force measurements are performed on a living cell, the strength of interaction forces depends on the number density of receptors under the contact area at the position of each measurement. With use of this method, the distribution of several specific kinds of receptors over a large area of a cell surface has been studied [13, 14, 25].

The method of preparing the colloidal AFM probe is as follows: a suitable size and material for microbeads should be chosen to increase the contact area between the sample and the tip of the AFM. A suitable colloid for the modification of protein is carboxylated polystyrene microbeads, which are ideal for the immobilization of proteins or as a size reference, being a series of monodispersed polystyrene particles, and carboxyl groups are exposed on the bead surface. We can wash and gather dispersed colloids in solution by centrifugation. After they have been washed, beads are suspended in suitable solvent such as ethanol.

The procedure to mount a microbead onto the AFM cantilever is as follows (Fig. 19.1). First, an AFM cantilever, epoxy resin, and a drop of microbead solution are put onto the warmed hot plate, respectively. Ethanol in the bead suspension immediately evaporates and the epoxy resin melts. A small amount of melted epoxy resin is picked up by the glass pipette of the micromanipulator, and placed on the end of the cantilever. A microbead is also picked up using another glass rod by physisorption and is placed on the epoxy resin coated spot on the cantilever, and the temperature is allowed to fall to room temperature. Figure 19.2 shows scanning electron microscopy images of an AFM cantilever to which a microbead has been fixed with epoxy resin. The microbead is tightly attached to the cantilever, while its surface facing the sample is not covered with epoxy resin. Figure 19.3 shows a schematic image of the experiment to mechanically locate receptors on

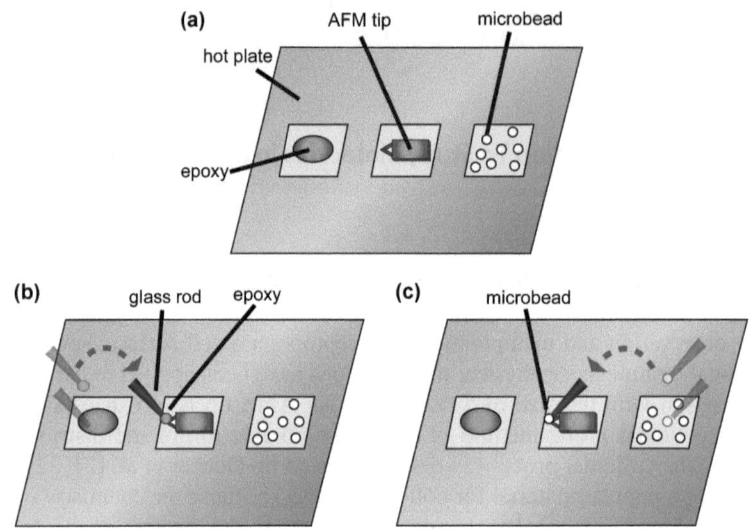

Fig. 19.1. A method to make a colloidal probe using a manipulator. **a** An atomic force microscope (*AFM*) cantilever, epoxy resin, and a drop of microbead solution were put onto warmed glass plates, respectively. Ethanol in the bead solution immediately evaporated and epoxy resin melted. **b** A small amount of molten epoxy resin was picked up by immersing a glass pipette in melted epoxy resin, and was placed on the end of the cantilever. **c** A microbead was picked up using another glass rod and placed on the epoxy resin coated spot on the cantilever

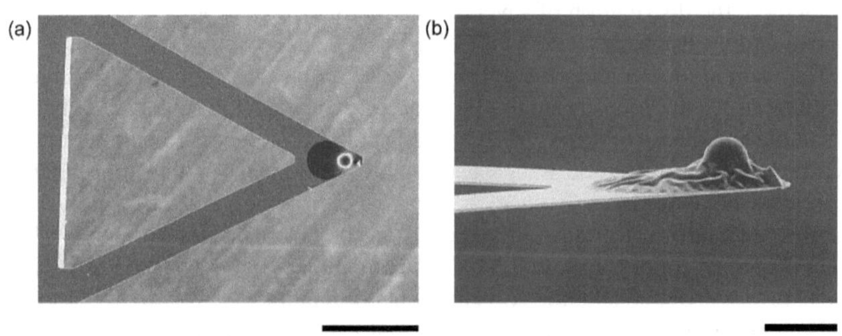

Fig. 19.2. Scanning electron microscopy images of an AFM cantilever with a polystyrene microbead. A microbead was placed at the end of the cantilever using a micromanipulator. **a** *Overhead view* of the cantilever. The length of the cantilever is 200 μm and the diameter of the microbead is about 10 μm. *Scale bar* 50 μm. **b** *Side view* of the cantilever. *Scale bar* 10 μm

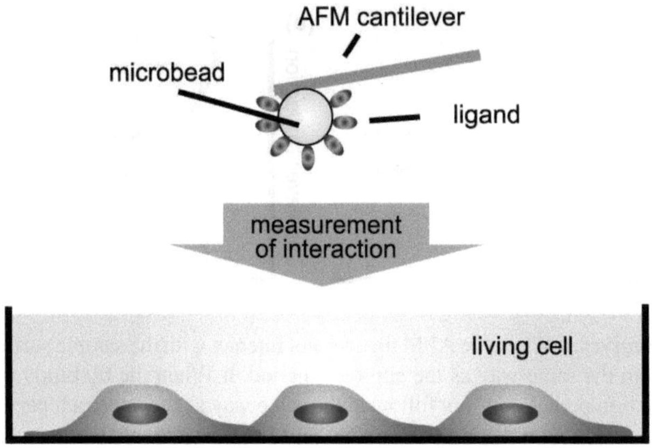

Fig. 19.3. The experimental system. A microbead was attached to an AFM cantilever and several kinds of cell-adhesive molecules or antibodies were covalently immobilized on the bead surface. Interactions between these molecules and target receptors were measured in a liquid. Distributions of receptors were obtained from measurement results

the cell surface. As shown in this figure, ligand molecules are covalently immobilized to the bead surface and AFM force measurements are carried out. When using carboxylated polystyrene beads, ligand molecules are covalently immobilized to microbeads with 1-ethyl-3-(3-dimethylaminopropyl)carbodiimide hydrochloride and *N*-hydroxysuccinimide, which are well-known reagents to connect the amino group of the former with the carboxyl group of the latter. Measurements of interaction between ligand molecules and target receptors are then performed. The distribution of receptors on a cell surface is obtained from these measurements.

19.2.2 Analysis of Molecular Interactions Where Multiple Bonds Formed

The interaction force is usually analyzed using a "force curve" (also usually referred to as a "force–distance curve"). Figure 19.4 shows a schematic image of the force curve. The abscissa indicates the vertical piezo movement relative to the sample surface, while the ordinate indicates the deflection of the cantilever. Cantilever deflection can be converted to applied force by multiplying it with the cantilever spring constant based on Hooke's law. The downward deflection of the cantilever on the retraction sequence reflects complex adhesive interactions between the tip and the sample. Such adhesive interactions are terminated with a rupture force indicated by a sudden jump of the cantilever to the free level.

Figure 19.5 shows a schematic image of the force curve, where a colloidal probe was used. Because of the large contact area, multiple point adhesions and detachments were involved in the measurement and the shape of the force curve was complicated. To evaluate the number of receptors under the contact area, the forces of all peaks must be added and calculated; however, it was almost impossible to distinguish between small force peaks and noise. Thus, force curves were converted to force–extension curves (Fig. 19.5b), where the abscissa indicates the

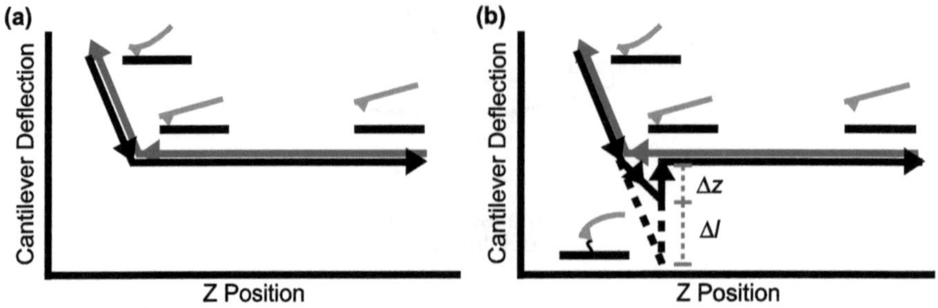

Fig. 19.4. Force curves. **a** When the AFM tip does not interact with the sample surface, a retraction period follows in the same way as the approach period. **b** When the tip binds with the sample surface, a retraction period does not follow in the same way as the approach period, because the tip bends in the direction of the sample surface. This bend is suddenly recovered by detachment of the tip from the surface, and the binding force is calculated from the cantilever deflection, Δz. Extension of the sample is also calculated from the difference between cantilever deflection and piezo movement, which corresponds to Δl

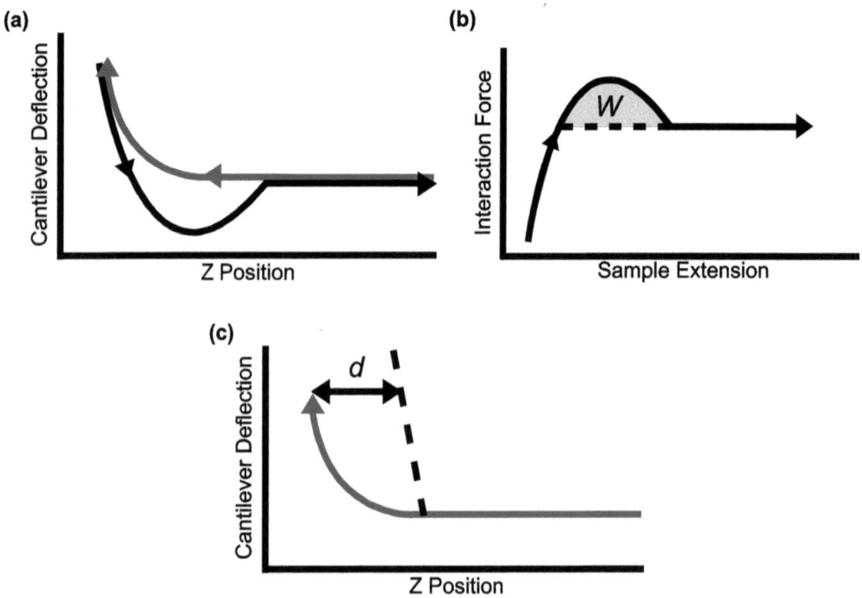

Fig. 19.5. The method used to calculate both the separation work, W, and the indentation depth. **a** The force curve at which a microbead was used as a probe. **b** Conversion of the force curve to a force–extension curve. The retraction period of the force curve was transformed to a force–extension curve, and W, which corresponds to the area as indicated in the image, was calculated. **c** Method to calculate indentation depth. The *broken line* shows linear tip deflection calculated from piezo movement. The depth of indentation of the cell surface, d, corresponds to the length of the *arrow*, and was calculated from the compression part of the force curve

elongation of molecules on a sample surface and the ordinate indicates the interaction force calculated from the cantilever deflection. It is expected that the area under the force–extension curve is a better measure of interaction [26], and the force–extension curve was integrated to obtain the quantity with the dimension of energy (joules or piconewton nanometers) referred to as "separation work" (denoted as W, hereafter) [13, 14, 25, 27, 28].

To quantify the number of receptors, the contact area of the colloidal probe on the cell surface should be calculated. The contact area between a microbead and the cell is estimated according to the Hertzian contact model [29–34] as explained below. Figure 19.5c shows a schematic image of force curves obtained during separation of the bead from the cell surface. The broken line in the figure corresponds to tip deflection, which is proportional to piezo movement. The depth of indentation of the cell surface corresponds to the length of the arrow in Fig. 19.5c, and is calculated from the extension of the force curve. According to the Hertzian contact model for a spherical indenter, the depth of indentation, d, is given as

$$d = \{[3F(1-\nu^2)]/4ER^{1/2}\}^{2/3}\,, \tag{19.1}$$

with the tip radius, R, loading force, F, Young's modulus, E, and Poisson ratio, ν, for the cell. The contact radius, a, is given according to the Hertzian contact model as

$$a = [3FR(1-\nu^2)/4E]^{1/3}\,. \tag{19.2}$$

From (19.1) and (19.2), the relation between d and a is obtained as

$$a = (Rd)^{1/2}\,. \tag{19.3}$$

The contact area, S, corresponds to the spherical cup of the radius R and height d, and is given as

$$S = 2\pi R^2 \int_0^{\theta} \sin\theta\,\mathrm{d}\theta\,, \tag{19.4}$$

where θ is defined as

$$\sin\theta = a/R = (d/R)^{1/2}\,. \tag{19.5}$$

We can therefore calculate the contact area S based on d obtained from a force curve. In their interesting trial, Charras et al. [34] estimated the contact area for living cells from the analysis of fluorescence images, but optical errors in their analysis were large. The calculation of the exact contact area remains as an important problem in the studies of cellular properties with an AFM, and a theoretical approach using the Hertzian contact model may be one of the useful approaches, as indicated by the result of Charras et al.

19.2.3 Measurement of Single-Molecule Interaction Strength on Soft Materials

For the measurement of interaction between a single ligand molecule and its receptor on the cell surface, the usual sharp AFM tip was used. In particular, a gold-coated AFM tip is convenient for chemical modification. One of the procedures of tip

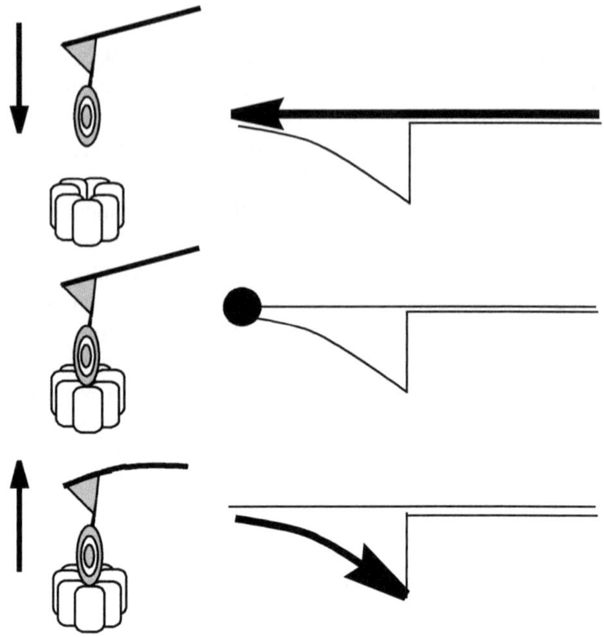

Fig. 19.6. "Compression-free force spectroscopy." An AFM tip on which a ligand molecule was immobilized is false-engaged, gradually approaches the cell surface, and is retracted before pushing into the cell

modification is as follows [25]. The AFM tip is cleaned with a UV–ozone cleaner and reacted with 10-carboxy-1-decanethiol. After the washing and activation procedure, the amino group of the ligand molecule is covalently cross-linked to the carboxylated tip in the same way as microbeads.

To measure the interaction of a single receptor–ligand on a cell surface, "compression-free force spectroscopy", where piezo movement is reversed from approaching the tip to retraction immediately before the start of the upward deflection of the cantilever, is used to reduce the area of contact (Fig. 19.6) [35, 36]. An AFM tip on which a ligand molecule has been immobilized is false-engaged and set into the force curve mode. The parameters of the force curve are the same as for microbeads. The tip gradually approaches the cell surface using a step motor. When the tip approaches the cell surface and retracts, we can observe interactions without pushing the cell surface.

19.3 Observation of the Distribution of Specific Receptors on a Living Cell Surface

19.3.1 Distribution of Fibronectin Receptors on a Living Fibroblast Cell

As the first example of receptor mapping, fibronectin (FN) receptor on a living cell surface was observed using an AFM and its colloidal tip [13]. FN is a large, extracellular matrix (ECM) glycoprotein that exists in two major states, as a soluble dimeric form in plasma and as an insoluble component within ECMs [37]. FN is composed of three types of repeating modules, termed type I, II, and III repeats, which are organized into functional domains [38, 39]. FN mediates its biological effects through binding to the heterodimeric transmembrane glycoproteins, integrins, which physically couple the cytoskeleton to the ECM [40–43]. The majority of integrin-mediated interactions of FN with cells occur through the Arg–Gly–Asp (RGD) cell binding sequence. Through these interactions, integrins play a critical role in the regulation of a number of cellular functions, including adhesion and migration, cytoskeletal reorganizations, and the cell cycle progression. Disruption of the FN gene results in an embryonic lethal phenotype, confirming the importance of FN in cellular development [43]. By taking this clinical importance into account, the distribution of FN receptors has been studied on the single-cell level. In detail, FN molecules were immobilized onto the surface of the AFM colloidal tip, and force measurements were performed at several places on a single living cell.

Figure 19.7 shows a histogram of *W* obtained between FN-coated beads and living cells. Because of the softness of the cell surface, the contact area changed depending on the position of the cell, so the contact area was calculated from the extension realm of the force curve according to the Hertzian contact model, and *W* was normalized with this value. The mean and the standard deviation (SD) of normalized *W* between FN-coated beads and living cells was

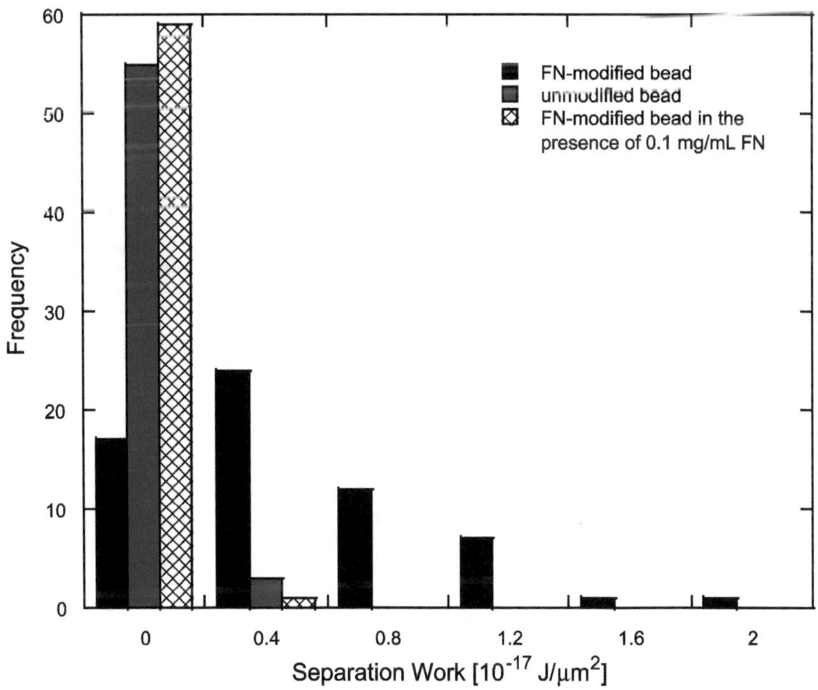

Fig. 19.7. Histogram of *W* obtained between a fibronectin (*FN*) coated bead and a living cell, between an unmodified bead and a cell, and between a FN-coated bead and a cell in the presence of 0.1 mg/mL FN

$0.69 \pm 0.45 \times 10^{-17}$ J/μm². In the presence of 0.1 mg/mL FN, the values decreased to $0.10 \pm 0.08 \times 10^{-17}$ J/μm². Unmodified microbeads gave a mean and a SD of $0.15 \pm 0.13 \times 10^{-17}$ J/μm². Because of the observed difference of cell adhesion strength from place to place on a cell surface, the data distribution was far from the normal distribution and the SD was large for the data given in Fig. 19.7. However, the mean and the SD were calculated to emphasize local differences of cell adhesion strength on a cell surface and to compare cell adhesion strength numerically with control experiments. From these results, it was concluded that W represented specific interactions between FN and the cell surface.

Figure 19.8 shows the results of measurement of W in 16 subdivided areas on a living cell, a process we call W-mapping. All W-mappings were done in the same region. It took about 1 min for one cycle of mapping. The pattern of W-mapping was reasonably reproducible for six consecutive runs. To check the reproducibility of interaction between FN and a living cell, Student's t test was carried out. The average P value of results during each pattern was 0.45, indicating that there were no clear differences between all possible pairs of the patterned data. Since the diffusion coefficient of the integrin β_1 subunit in the membrane is reported to be in the range 0.01–0.1 μm²/s [44–46], the receptor could migrate a distance covering a few subdivisions within a map between two consecutive measurements. The observed reproducibility of this results can be interpreted as follows: first, there was no large-

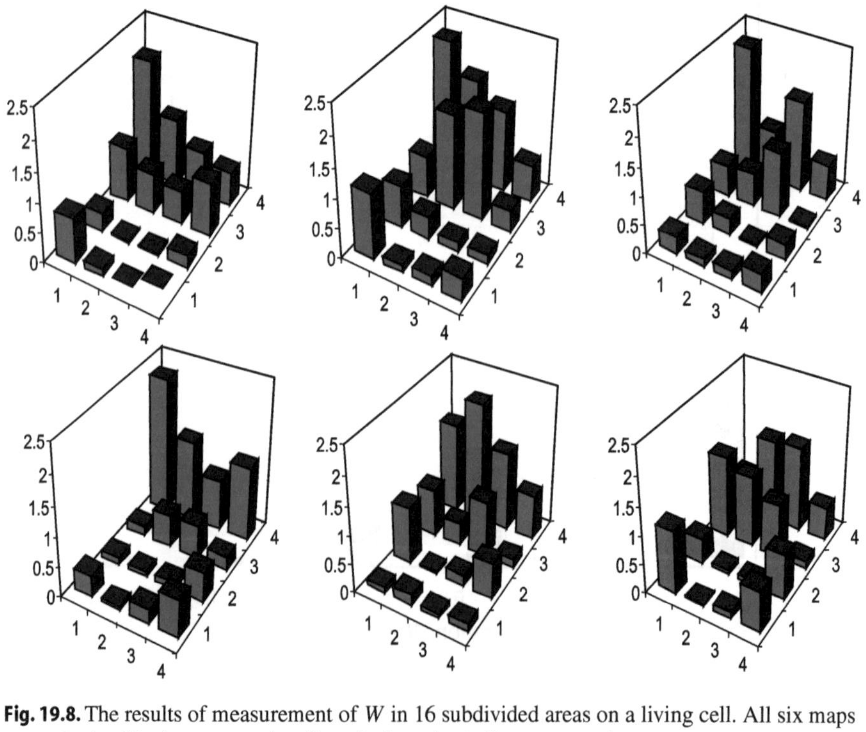

Fig. 19.8. The results of measurement of W in 16 subdivided areas on a living cell. All six maps were obtained in the same region. It took about 1 min for one mapping

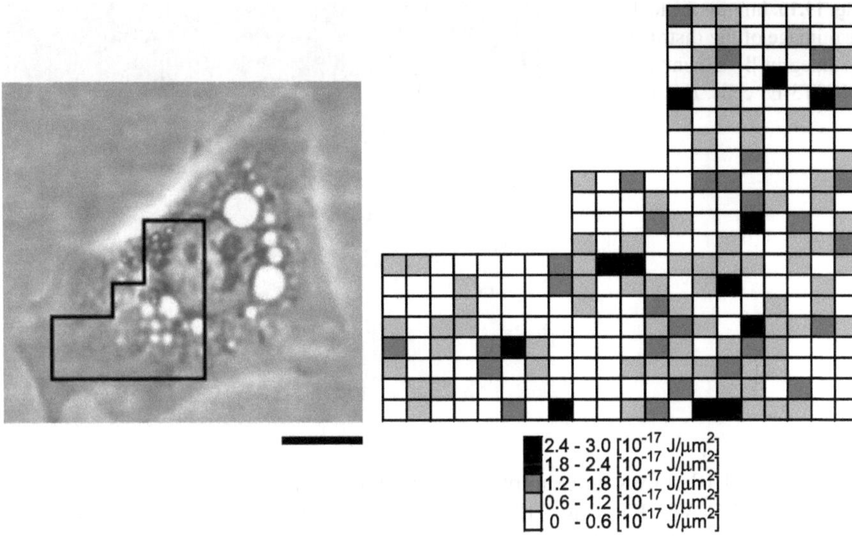

Fig. 19.9. W-mapping over a large area of the cell surface. The area surrounded with a *solid line* was subdivided into 1-μm^2 units, and measurements were carried out at each point. *Bar* 10 μm

scale migration of receptors on the time scale of this measurement, and, second, it is unlikely that the receptors were extracted from the cell membrane during the experiments. If the receptor proteins were extracted, the maps would not have been as reproducible as observed with respect to their local patterns. This conclusion does not exclude the possibility of small-scale migrations of receptors.

Figure 19.9 shows a W-mapping over a large area of a living cell. The position at which FN was strongly adhered to the cell was scattered across the surface without recognizable patterns. This result was easily obtained by increasing the contact area, indicating that the use of increased contact area is useful for the quantitative analysis of cell-adhesive interactions and for studying the membrane at the level of the entire cell.

19.3.2
Distribution of Vitronectin Receptors on a Living Osteoblast Cell

As the next example, the distribution of vitronectin (VN) receptors was studied [14]. One of the major VN receptor is integrin $\alpha_v\beta_5$ receptor, which only binds to VN with a high specificity [41,47]. VN belongs to a group of adhesive glycoproteins that play key roles in the adhesion of cells to their surrounding matrix, and its involvement in various kinds of diseases has been reported [48]. For example, the locomotion of keratinocytes which plays an important role in the reepithelialization of cutaneous wounds is promoted by VN, and the promotion is inhibited by antibodies raised against the integrin $\alpha_v\beta_5$ receptor [49], which indicates medical importance of VN, especially its interaction with integrin $\alpha_v\beta_5$ receptor.

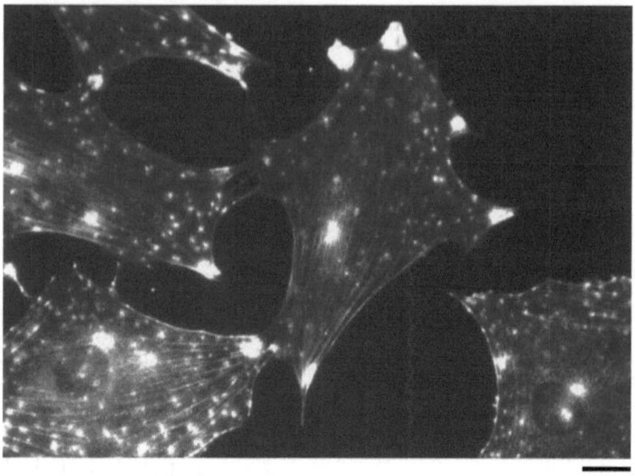

Fig. 19.10. Immunohistochemical image of the distribution of integrin β_5 subunits on MC3T3-E1 cells. *Bar* 10 μm

Figure 19.10 shows an example of immunofluorescence images of integrin β_5 subunits on MC3T3-E1 cells. This integrin subunit only makes a heterodimer with the integrin α_v subunit; thus, the image in Fig. 19.10 was almost equivalent with a distribution of integrin $\alpha_v\beta_5$ receptors, which are one of the major VN receptors. There have been several reports on the localization of VN receptors, including integrin $\alpha_v\beta_5$, around focal contacts in cultured cells [50–52]. In this figure, the integrin β_5 subunit seemed to be localized on the cell membrane, in agreement with previous findings. Localization was particularly conspicuous at the tiplike end of cells.

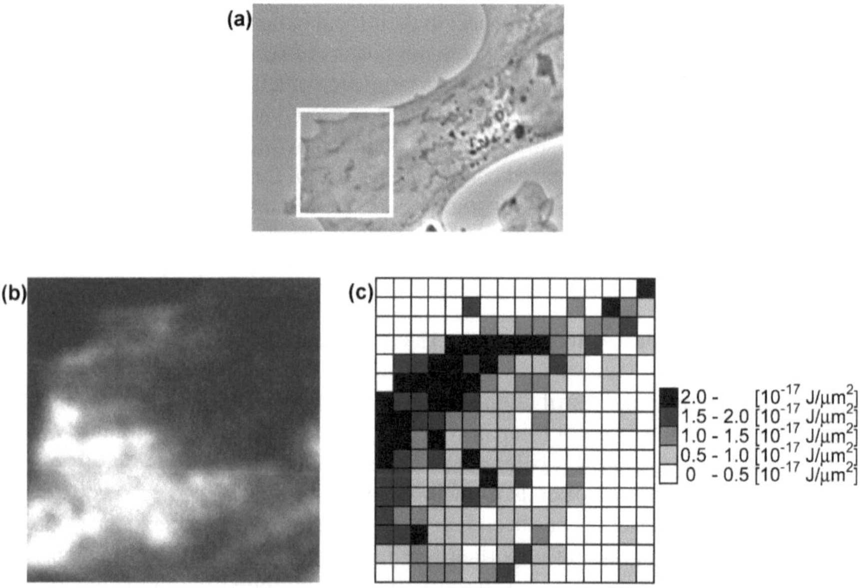

Fig. 19.11. Images of a murine osteoblastic cell taken with **a** an optical microscope and **b** a fluorescence microscope, and **c** mapping of *W* using an AFM. The *square* in **a** corresponds to the area of 32 × 32 μm^2 in which the measurement using the AFM was performed

To study the distribution of binding sites for VN, a specific ligand of integrin $\alpha_v\beta_5$, on the cell membrane, and to compare the result with that for the integrin β_5 subunit, measurements of cell-adhesive strength between VN and the cell were carried out using an AFM. Figure 19.11 shows a *W*-mapping between VN and the receptors on the cell surface, and a comparison of it with the immunohistochemical findings. Results of AFM measurements indicated that VN receptors, including integrin $\alpha_v\beta_5$, were localized in a similar manner as obtained by immunohistochemistry. The distribution of high values of *W* was had a similar pattern to that of the fluorescence image, but was not completely the same. This might be because, first, all of the integrin β_5 subunits on the top and bottom sides of the cell were stained in immunohistochemistry, but only those on the dorsal side of the cell surface were probed in the AFM experiment, and, second, there are VN receptors other than integrin $\alpha_v\beta_5$. Though there are other kinds of VN-binding molecules, a comparison with the distribution of integrin $\alpha_v\beta_5$ has special importance, because integrin $\alpha_v\beta_5$ interacts only with VN, and their biological role might be controlled by VN. From the abovementioned results, it was concluded that VN receptors are localized and some of the peripheries of the cell were covered with receptors. These results indicated that this AFM-based method is suitable for studying the membrane receptors at the level of a single cell, and a comparison of the results of AFM measurements with immunohistochemistry is useful to study distributions of receptors on the cell surface.

19.3.3 Quantification of the Number of Prostaglandin Receptors on a Chinese Hamster Ovary Cell Surface

Cell-adhesive interactions have been studied at the single-cell level using the AFM and notable results were obtained. This AFM-based method does not require artificial cellular treatment, such as cell fixation, and it is easy to make a quantitative evaluation of receptor expression, because the work to separate a molecular pair, which is extensively variable, has been referred to as a parameter. In these previous studies, cell-adhesive strength was successfully measured; however, the numbers of receptors could not be quantified. For quantitative evaluation, values of cell-adhesive strengths, which can be measured using a microbead probe, should be normalized with that of single molecular interaction. To quantify the numbers of receptors, many single ligand–receptor interactions should be measured one by one, but it is not so easy and a general method is desired. For such a purpose, it is useful to genetically introduce a green fluorescent protein (GFP) molecule to the receptors [53–55]. The advantage is, firstly, many kinds of receptors, the distributions of which have been observed with a microbead, will be quantified by normalizing the value with that of the single GFP:anti-GFP antibody interaction when the GFP is set as the target of the measurement of the distribution by immobilizing the anti-GFP antibody to the microbead, and, secondly, it is also useful to compare the result of AFM-based receptor mapping with that of fluorescence observation. Though we can also identify the distribution of the receptor by the fluorescence of GFP molecules, there are advantages with the use of the AFM. Firstly, it is not necessary to consider the photobleaching problem; hence, it has the potential to detect a low number of target molecules with high reliability. Secondly, quantitative evaluation is easy in the case of the AFM-based method.

A prostaglandin (PG) receptor, EP3, which is known to be a target of many kinds of medicines was investigated [25]. PGs exert a variety of actions in various types of tissues and cells. Milton and Wendlandt [56] were the first to suggest PGE_2 as a central mediator of fever. Indeed, E type PGs are powerful fever inducers when injected into the brain, and the level of PGE_2 increased in the preoptic area during lipopolysaccharide-induced fever, and indomethacin completely abolished both lipopolysaccharide-induced fever and increased the levels of PGE_2 in the preoptic area [56–60]. The actions of PGE_2 are mediated through rhodopsin-type cell surface receptors [61]. These receptors are four subtypes of the PGE_2 receptor: EP1, EP2, EP3, and EP4 [61]. Ushikubi et al. [62] used mice lacking each subtype of the PGE receptor to address PGE_2-mediated fever generation. They found that only the mice lacking the EP3 receptor failed to show febrile responses. This study thus clearly demonstrated that PGE_2-mediated fever generation by acting on the EP3 receptor.

By taking such clinical importance into account, the distribution of EP3 receptors was studied at the single-cell level using an AFM. For the identification, GFP, which was genetically introduced to the extracellular region of EP3 receptor on a chinese hamster ovary (CHO) cell (the cell is denoted as GFP+/CHO cell hereafter), was used as a target of interaction measurement, and an AFM-based measurement of the interaction between the GFP and its antibody, which was immobilized on a microbead tip, was carried out (Fig. 19.12). The force curves obtained were transformed to force–extension curves, the separation work, W, was calculated, and a W-mapping between the anti-GFP antibody and GFP on a GFP+/CHO cell surface was obtained (Fig. 19.13). The result was also compared with the immunohistochemical findings at the same area. The results of both AFM measurement and immunohistochemistry indicated that EP3 receptors were distributed nonuniformly on the GFP+/CHO cell surface in small patches. To compare the AFM finding with that of fluorescence, the former was averaged and the latter was subdivided into 4 × 4 regions (Fig. 19.13d,e) and the correlation of each data point

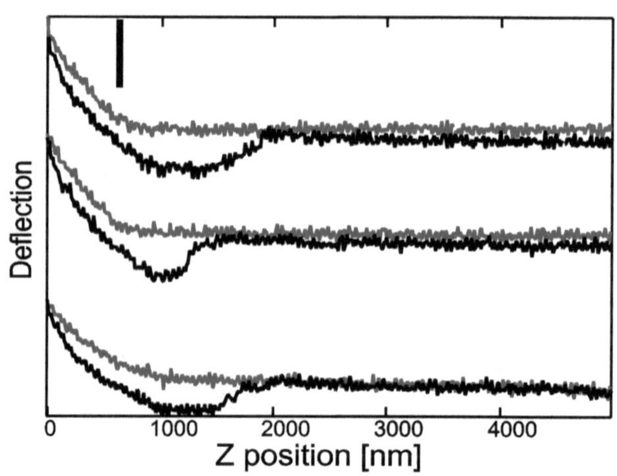

Fig. 19.12. Force curves obtained between the GFP+/CHO cell and anti-GFP antibody where a microbead was used as a probe. *GFP* green fluorescent protein, *CHO* Chinese hamster ovary. *Bar* = 5 nm

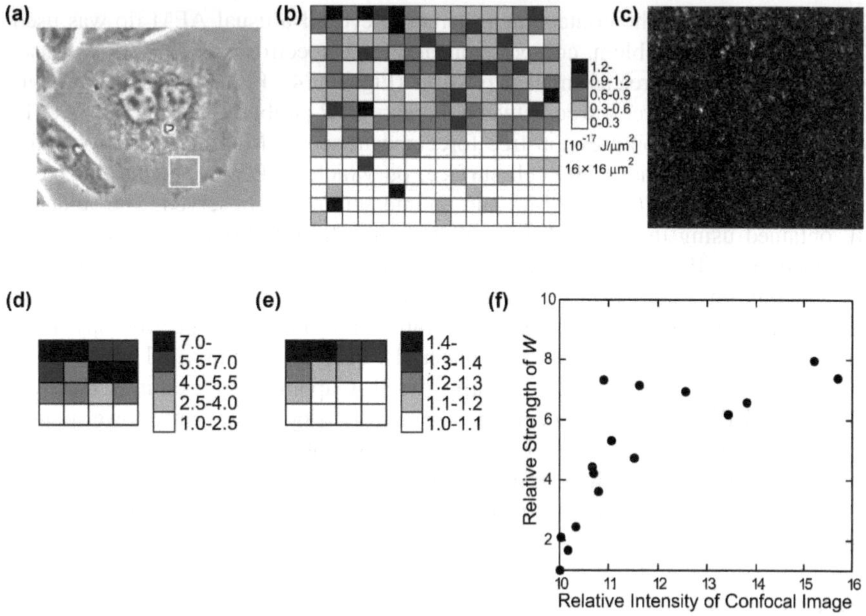

Fig. 19.13. Distribution of EP3 receptors on the GFP+/CHO cell surface. **a** Optical microscope image of a living GFP+/CHO cell on which the measurement of W was performed. The area surrounded by a *solid line* was subdivided into 1-μm^2 units, and measurements of W were carried out at each point using a microbead probe. **b** A map of W between the anti-GFP antibody and the GFP+/CHO cell surface. **c** Immunohistochemical image obtained using a confocal microscope. **d** The W map obtained with 16×16 data points was averaged to a 4×4 map to compare with it the fluorescence image. Values were normalized using the weakest value. **e** The fluorescence image was subdivided into 4×4 data points, and the intensity of each point was calculated. Values were also normalized using the weakest value. **f** Correlation between the intensity of the confocal image and the result for W, obtained by comparing each data point between **d** and **e**

was studied (Fig. 19.13f). There was a good correlation between the intensity of the confocal image and the value of W, which confirmed the reliability of the result of W-mapping. The advantages of the AFM measurement are, firstly, that the AFM can detect small amounts of receptors, which are difficult to visualize by immunohistochemistry (in this case, the fluorescence of GFP was enhanced with the immunohistochemical method because of too weak fluorescence), without any chemical treatment, such as cell fixation, and, secondly, it is very easy to compare the relative numbers of receptors at each position of the cell membrane individually with a high signal-to-noise ratio by comparing the values of W. These results indicated that the AFM-based method is suitable for studying the membrane receptors at the single-cell level, and a comparison of the results of AFM measurements with immunohistochemistry is useful to study the distributions of receptors on the cell surface.

To quantify the number of EP3 receptors, a single molecular interaction between GFP and the anti-GFP antibody was measured on the GFP+/CHO cell surface. Because the cell surface is too soft, the cells were easily deformed by applying a loading force and the contact area increased even if a usual AFM tip was used. To overcome this problem, compression-free force spectroscopy [35, 36] was used to make the contact area as small as possible (Fig. 19.14). Figure 19.14b shows typical images of the force curve between the anti-GFP antibody and the GFP+/CHO cell obtained with compression-free force spectroscopy. Measurements were also performed on a normal CHO cell surface, on which GFP was not introduced, in a control experiment (Fig. 19.14c) and no interaction was observed. The values of W obtained using the method were plotted, and discrete populations of W were obtained (Fig. 19.15). The value of W to separate a single GFP molecule from its antibody on a cell surface was then estimated to be about 1.5×10^{-18} J. This result was firstly obtained by applying compression-free force spectroscopy to the measurement of interaction on soft materials, such as a living cell surface. Using this value, we can roughly quantify the W-mapping. For example, there are about 880 EP3 receptor molecules in the area of Fig. 19.13, and if we assume the area of the cell surface to be $5000\,\mu\text{m}^2$ [20], about 1.7×10^4 receptors exist on the cell surface. These results indicate that the AFM-based method is one of the more powerful ways to quantify the number of receptors on a living cell surface.

19.4 Further Application of the AFM to the Study of Single-Cell Biology

19.4.1 Manipulation of Expressed mRNAs in a Living Cell Using an AFM

The AFM was used not only for the nanoprobe to measure interaction strength, but also as a mechanical nanomanipulator to pick up biomolecules from a living cell without serious damage. In this section, examples of a method to study expressed mRNAs in a single living cell are introduced [18, 19].

In this method, the AFM tip is inserted into a living cell body, and expressed mRNA is harvested by the AFM tip. Details of the procedure are as follows. Living cell lines were grown in culture dishes and cells were washed with culture medium devoid of serum components and used for AFM experiments. The AFM tip was positioned above a target cell under an AFM combined with an optical microscope. The AFM tip was approached using a step motor and the contact of the AFM tip with the cell surface was detected by the deflection signal of the AFM cantilever. Higher loading forces were applied to insert the AFM tip into the cell membrane. The AFM tip was inserted into the cytoplasm region around the nucleus to collect mRNA. After the AFM tip had been inserted into a single cell, it was placed into a PCR tube and PCR was performed.

For the first example, the β-actin mRNA expression of individual living cells was examined using rat fibroblast-like VNOf90 cells and mouse osteoblast-like MC3T3-E1 cells (Fig. 19.16a). Although β-actin mRNAs are usually distributed throughout the cytoplasm uniformly, they are localized to the leading edge of the cells when the cells start to migrate [63, 64]. Thus, the single cells, which were surrounded by other cells that inhibit the migration of the target cells, were chosen as targets to study

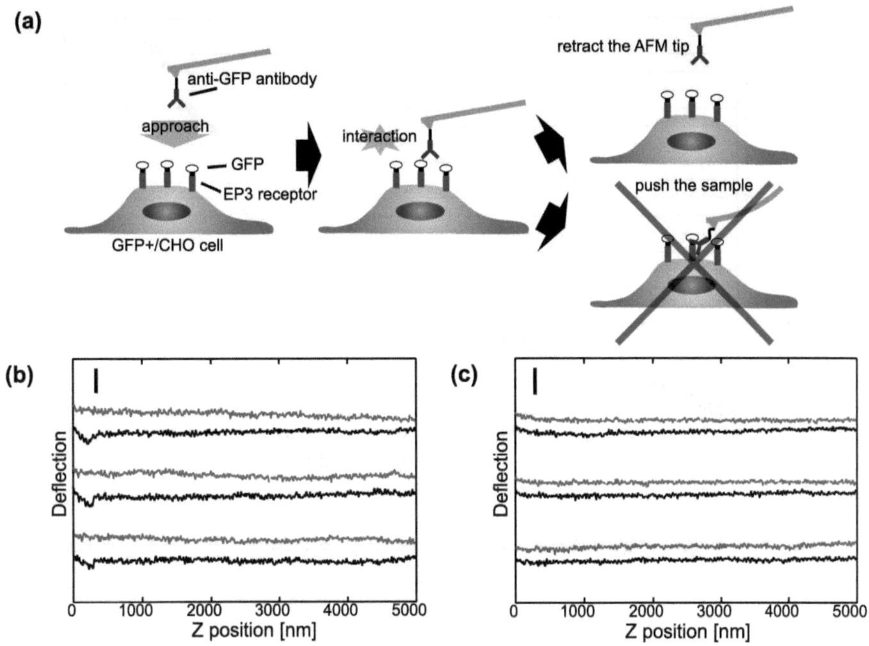

Fig. 19.14. Observation of single molecular interactions on a cell surface. **a** Compression-free force spectroscopy. Piezo movement was reversed from approaching the tip to retraction immediately before the start of the upward deflection of the cantilever. **b** Typical images of the force curve between anti-GFP antibody and the GFP+/CHO cell using compression-free force spectroscopy. **c** Typical images of the force curve between the anti-GFP antibody and the normal CHO cell in the control experiment. Approach of the AFM tip to the cell surface was confirmed by a small deflection of the cantilever. *Bars* 2 nm

β-actin mRNA expression. PCR products for rat and mouse β-actin mRNAs were detected as shown in the even-numbered lanes of Fig. 19.16b and c. In the negative control, PCR products were not detected without the insertion of the tip into the cell (odd-numbered lanes in Fig. 19.16b, c). Experiments for the detection of β-actin mRNA and the negative control were performed alternately.

In Table 19.1, results of the detection of β-actin mRNA from single VNOf90 cells are presented. The AFM-based mRNA extraction was performed on 102 single living cells. The number of assays against single cells ranged from one to six. The time interval between one assay and the next one against the same cell ranged from 5 to 60 min. When six sets of assays had performed against three single cells, the following results were obtained. In cases of two single cells, PCR products were detected in all six assays. In the case of another cell, PCR products were detected five times out of six assays. Seventy-two positive results were obtained when one assay was tried for 73 single cells. In total, 189 assays were performed on 102 single living cells, and 182 positive results were obtained (96.3%). The other seven negative results may have resulted for the following three reasons: (1) the procedure for inserting the tip into the cell was not successful; (2) the PCR was not successful;

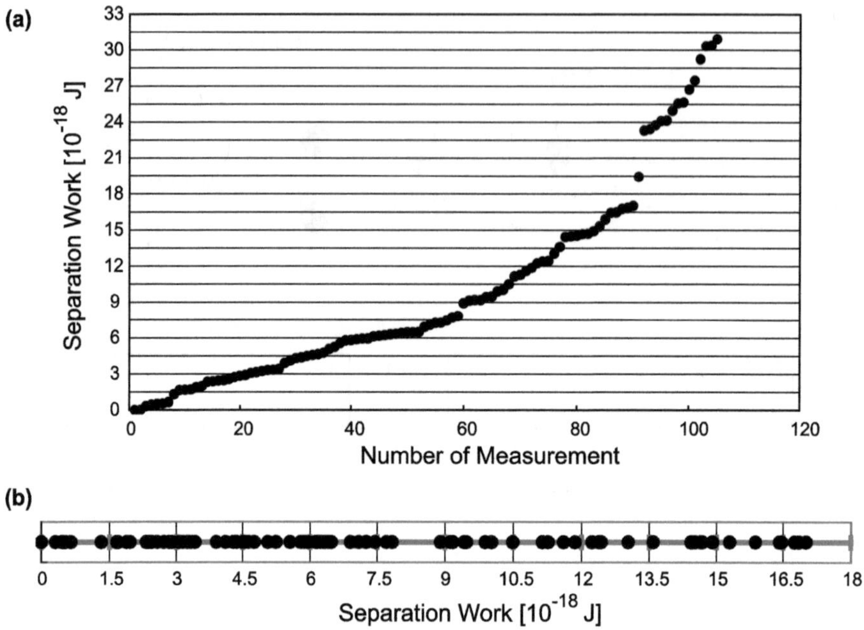

Fig. 19.15. Determination of *W* which corresponds to the strength of a single molecular interaction. **a** *W* obtained using compression-free force spectroscopy. **b** Population of the values of *W* obtained. All data points were plotted on the axis of *W*

(3) β-actin mRNA did not exist in the cell where the tip was inserted. The first two reasons are considered to be experimental errors, and the last one is considered to result from the probability of the existence of mRNA. Nevertheless, the high probability of detecting β-actin mRNA indicated that it could be used as a positive marker for examining the expression of other genes which would exclude the first two reasons from consideration. A population of mammalian cells grown in culture medium contains about one million mRNAs per cell. Fibroblast cells are reported to contain 2000–3000 copies of β-actin mRNA per cell [65]. This method can be applied to detect other mRNAs whose copies number more than several thousand per cell.

In the next trial, it was examined whether this AFM-based method can be used to determine more than one kind of gene at the same time or not. As shown in Fig. 19.17a, VNOf90 cells expressed FN protein as well as β-actin. AFM tips inserted into the single living cells were placed into PCR solution that contained primers for both β-actin and FN. In total, ten single living cells were assayed and the results of five single cells are shown in Fig. 19.17b. PCR products for both β-actin and FN mRNAs were detected in all of the ten different single cells.

In order to examine time-dependent gene expression of single living cells, the response of rat VNOf90 cells to serum has been used as a model for studying the changes in c-fos gene expression (Fig. 19.18). The cells were induced to enter a quiescent state with culture medium containing 0% fetal bovine serum (FBS) for

Table 19.1. Detection of β-actin messenger RNA from single VNOf90 cells

Number of assays against single cells	Number of β-actin positive cases	Number of cells	Total number of β-actin positive cases	Total number of assays in each case
1	1	72	72	72
1	0	1	0	1
2	2	4	8	8
2	1	1	1	2
2	0	0	0	0
3	3	1	3	3
3	2,1 or 0	0	0	0
4	4	12	48	48
4	3	3	9	12
4	2,1 or 0	0	0	0
5	5	4	20	20
5	4	1	4	5
5	3,2,1 or 0	0	0	0
6	6	2	12	12
6	5	1	5	6
6	4,3,2,1 or 0	0	0	0
		102	182	189

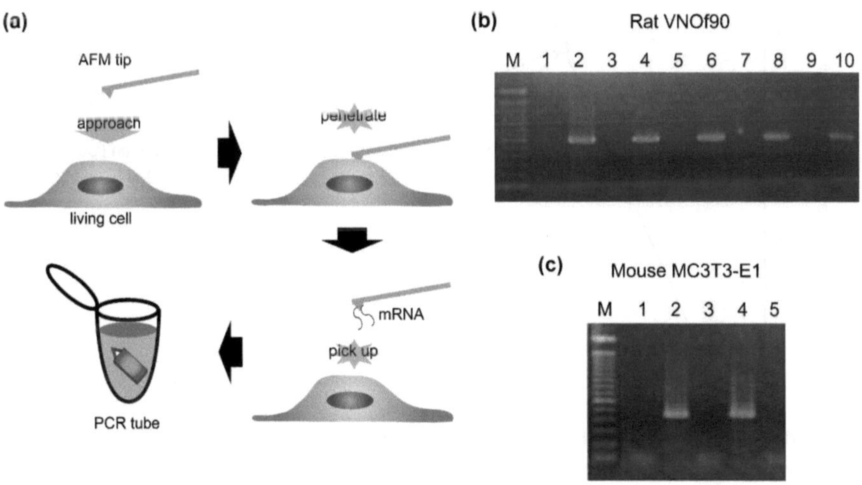

Fig. 19.16. Extraction of messenger RNAs (*mRNAs*) from living cells. **a** The experimental procedure. A target cell was positioned underneath the AFM tip through the observation of an inverted optical microscope combined with AFM. The AFM tip was then lowered onto the cell and inserted, and held for approximately 45 s to allow the tip to bind the cell ingredient containing mRNA with physical absorption. The tip was then lifted off the cell and placed into a polymerase chain reaction (*PCR*) tube. **b, c** β-Actin mRNA expression of five rat VNOf90 cells (**b**) and two mouse MC3T3-E1 cells (**c**) was examined as shown in the *even-numbered lanes*. In the negative control (*odd-numbered lanes*), the tip contacted the medium without its insertion into the cell

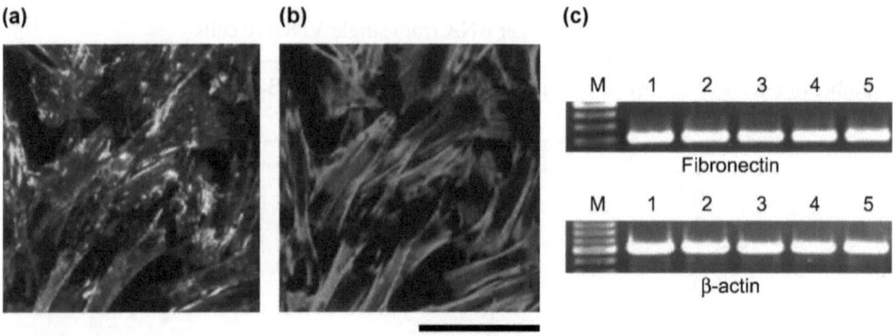

Fig. 19.17. Detection of both β-actin and FN mRNAs from single living cells. Immunocytochemistry demonstrated VNOf90 cells expressing both FN (**a**) and β-actin (**b**). *Bar* 100 μm. Both β-actin and FN mRNAs were detected in all five different single cells (**c**)

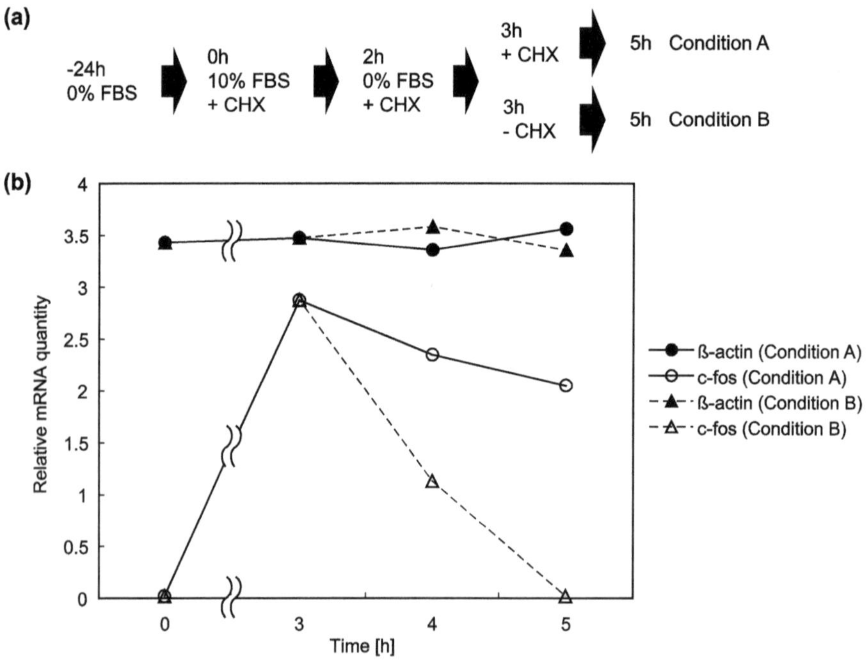

Fig. 19.18. Cell population analysis. **a** Response of VNOf90 cells to serum and cycloheximide (*CHX*) was used as a model for studying time-dependent gene expression. VNOf90 cells were induced to enter a quiescent state by serum deprivation for 24 h and then stimulated by the addition of culture medium containing 10% fetal bovine serum (*FBS*) and 10 μg/mL CHX. The addition of serum and CHX induced changes in gene expression, including c-fos mRNA. **b** Relative mRNA levels of the cell population were measured with quantitative PCR. The ordinate represents the natural logarithm of the relative initial amount of mRNA

24 h and then stimulated by addition of 10% FBS and 10 μg/mL cycloheximide (CHX) for 2 h. CHX is well known to induce increases in c-fos mRNA expression level to prevent degradation of c-fos mRNA. The solution was then changed to culture medium containing no FBS and 10 μg/mL CHX for 1 h. Finally, the solution was changed to culture medium containing 10 μg/mL CHX (condition A) or no CHX (condition B). Total RNA from confluent cells on 60-mm dishes at different times was extracted and quantification of mRNA was carried out with quantitative PCR. The β-actin mRNA levels were more constant than the c-fos mRNA levels in both conditions A and B (Fig. 19.18b). Although the c-fos mRNA levels decreased rapidly in the absence of CHX from 3 to 5 h following stimulation, they decreased gradually in the presence of CHX from the cell population analysis. The detection of c-fos and β-actin mRNAs from single cells, termed A1, A2, and A3 cells (from condition A), was examined with an AFM-based manipulation method (Fig. 19.19). The change

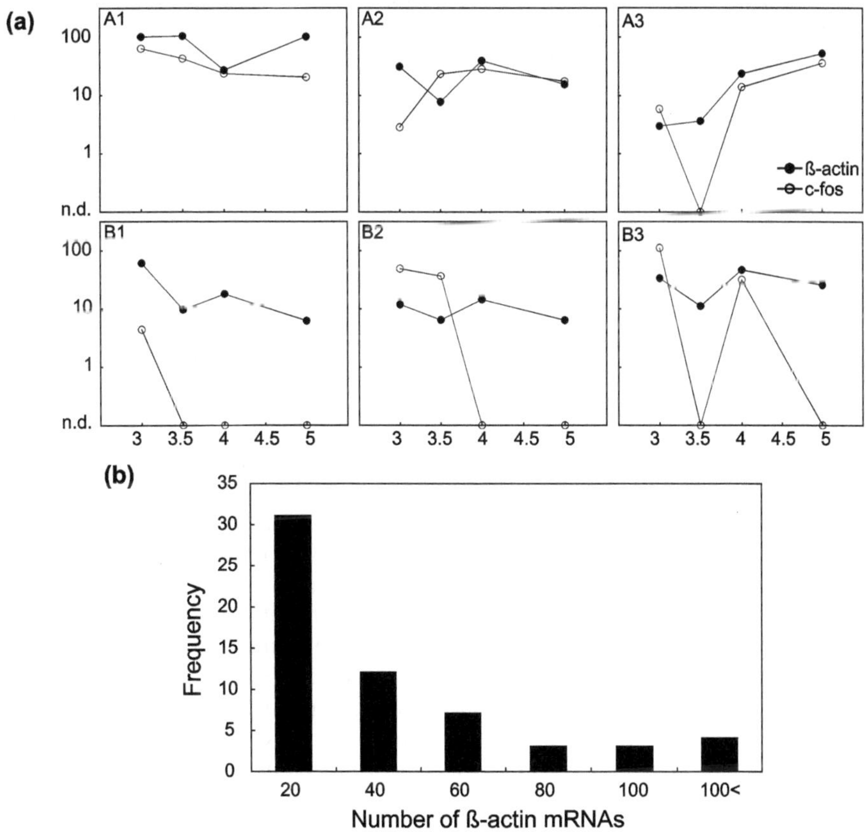

Fig. 19.19. Single-cell analysis. **a** Time-dependent mRNA expression of single living cells was examined in both condition A (*A1–A3*) and condition B (*B1–B3*). The ordinate represents the natural logarithm of the number of initial mRNA copies, and the abscissa represents time in hours. **b** Number of β-actin mRNAs detected with single-cell assay. *n.d.* product was not detected with quantitative PCR

of β-actin mRNA levels was also small compared with that of c-fos mRNA levels in single-cell analysis as well as in the cell population analysis. Although the expression profile of c-fos mRNA in the A1 cell was very similar to the result obtained from the cell population analysis, the A2 and A3 cells showed different expression profiles. These results suggest that c-fos mRNA expression levels are not uniform among individual cells. Although c-fos mRNA was not detected in the A3 cell at 3.5 h, β-actin mRNA was detected, indicating that the experimental errors were able to be eliminated. In condition B (B1, B2, and B3 cells), c-fos mRNA expression was not observed in any of the cases at 5 h following stimulation. This result corresponded to the result from the cell population analysis. The expression of c-fos mRNA in the B1 cell was observed only at 3 h. In the B2 cell, c-fos mRNA expression was detected at 3 h and at 3.5 h and not after 4 h. In the B3 cell, c-fos mRNA expression was detected at 3 h and not at 3.5 h, and was then detected again at 4 h. This type of pattern was also observed in the A3 cell. These results indicate the possibility that the fluctuation of c-fos mRNA expression level occurs at the single-cell level. Figure 19.19b shows the histogram of the number of β-actin mRNAs bound to AFM tips. In most cases, their number was equal to or less than 20.

In conclusion, a powerful new technique to study gene expression in mammalian single cells was provided. The results of these studies demonstrate that time-dependent gene expression patterns in single cells are not always similar to those in large cell populations. At the single-cell level, gene expression patterns sometimes fluctuate up and down. These phenomena will aid in the understanding of the properties of individual cells.

19.4.2 Manipulation of Membrane Receptors on a Living Cell Surface Using an AFM

The extraction of membrane protein molecules has been examined using an AFM for establishment of a new technique to harvest and identify critical molecular markers on the cell surface as shown in Fig. 19.20 [15–17]. In this study, the AFM tip was modified with bifunctional covalent cross-linkers, and membrane proteins were indiscriminately picked up using AFM force measurement mode. In Fig. 19.21, typical force curves, which were obtained on a glass surface (Fig. 19.21a) and on the surface of the live fibroblast cell (Fig. 19.21b) using an unmodified silicon nitride tip, are presented. In the case of Fig. 19.21a, there was no indentation on either the tip or the sample, whereas in the case of Fig. 19.21b, the sample was deformed under the vertical stress inflicted by the tip. The curve in Fig. 19.21c represents a typical force curve obtained on a living cell surface using a tip modified with a bifunctional covalent cross-linker. Although the approach part of the curve is similar to that of the curve in Fig. 19.21b, its return part is characterized by a prolonged downward deflection which was abruptly ended after a sample extension of approximately 115 nm. The extension of the sample was calculated as the difference between the cantilever deflection (ordinate) and the distance of the piezo movement (abscissa) from the sample surface. The curve shows that an adhesive interaction was established between the tip and the sample surface during their contact of approximately 300–400 ms and that, during the retraction period of the curve, the cell surface was lifted by a vertical distance of approximately 115 nm, whereas the cantilever was

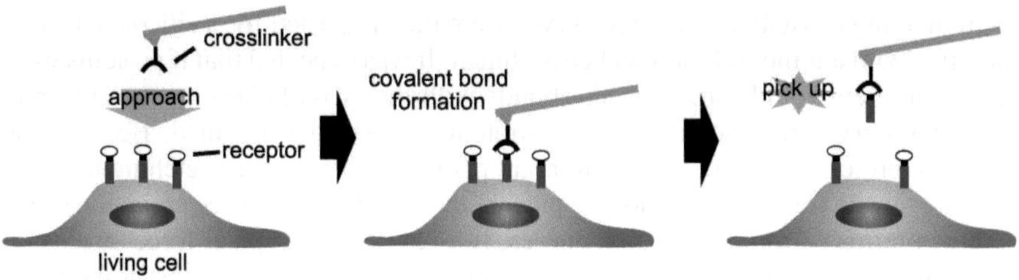

Fig. 19.20. Extraction of receptors. An AFM tip was modified with a cross-linker and brought to the living cell surface. A covalent bond was formed between the cross-linker on the tip and receptors. Receptors were then picked up with the retraction of the AFM tip

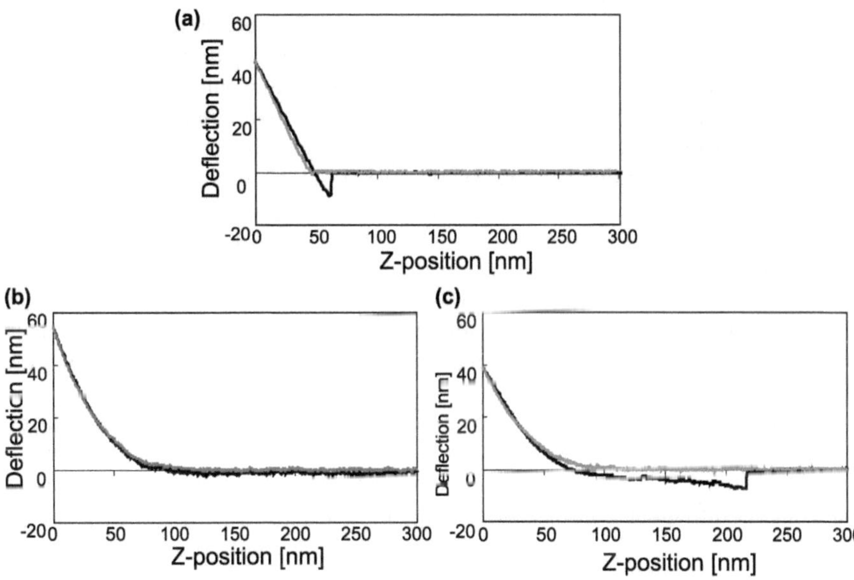

Fig. 19.21. Force curves obtained under three different conditions: **a** against a bare glass surface using an unmodified tip, **b** against a cell surface using an unmodified tip, and **c** against a cell surface using a tip modified with a cross-linker. The *gray line* corresponds to approach and the *black line* corresponds to retraction for all the force curves

pulled down by approximately 7 nm. The tensile force at the final separation of the tip from the cell surface was calculated as approximately 0.42 nN in this case by knowing the force constant of the cantilever to be approximately 0.06 nN/nm. When unmodified silicon nitride tips were used on live cells, 90% of the force curves ($N = 50$) showed no positive indication of adhesive interactions, and the remaining 10% exhibited weak interaction forces of less than approximately 50 pN with the cell surface.

To verify that the relatively strong separation force observed in Fig. 19.21c corresponded to the extraction of membrane proteins through covalent bond formation, the modified cantilevers used on cell surface to obtain more than 50 force curves were remounted on the AFM after treating them with glycine/glycinamide to deactivate the remaining cross-linkers. Force curves were then recorded for a silicon substrate modified with another bifunctional cross-linker. It was expected that if proteins were on the tip they would form covalent bonds with the cross-linkers on the substrate and force curves representing protein stretching would be obtained. Because the length of protein extension is an important parameter in protein stretching together with the tensile force, it is suitable to present the results of the force curve measurements in terms of force–extension curves. In Fig. 19.22, a typical force–extension curve, which was obtained using a used tip on a cell surface against a modified silicon surface, is presented together with a histogram of the extended length before the final separation of the tip from the substrate. For the force–extension curve in Fig. 19.22a, the force of the final separation was approximately 1.3 nN, proving the establishment of a covalently bonded system between the tip and the substrate. Proof for this comes from a report on the strength of covalent bonds under mechanical stress. Grandbois et al. [66] reported their theoretical prediction that the weakest bond in a similar cross-linking system to that in this study would be the Si–C bond, and concluded from their experimental results that covalent bond rupture occurs at 2.0 ± 0.3 nN (with a distribution ranging mainly between 1.0 and 2.0 nN) at a loading rate of 10 nN/s. In a similar loading rate range, 1.0–2.0 nN of the final rupture force is also most often obtained for carbonic anhydrase using a covalently immobilized tip or substrate under the AFM [67, 68]. The histogram in Fig. 19.22b shows that the tensile material covalently sandwiched between the used tip and the modified silicon substrate was stretched up to 200–250 nm with a mean extension length of approximately 85 nm, strongly suggesting that the extracted macromolecules, most probably proteins, were transferred from the cell surface to the AFM tip during the abovementioned operation.

Figure 19.23 shows a histogram of rupture forces obtained on a cell surface using a modified AFM tip. In this histogram, the force values mainly clustered between 0.4 and 0.8 nN, with a mean and deviation of 0.45 ± 0.22 nN. Because a force larger than 1 nN is infrequent and may involve multiple adsorption events, a force range

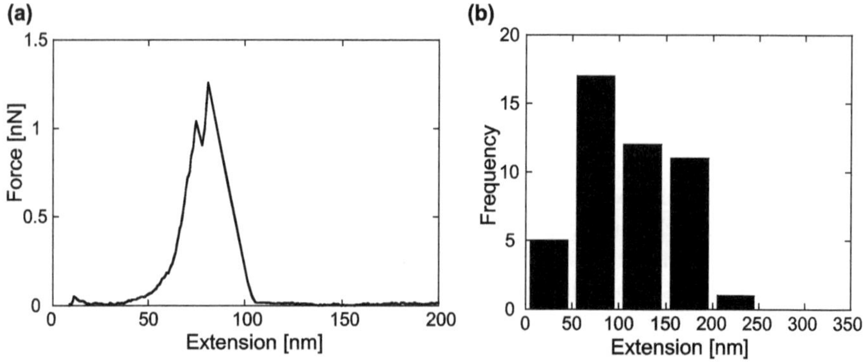

Fig. 19.22. Confirmation of the extraction of membrane receptors using an AFM. **a** F–E curve obtained when a modified tip used on the cell surface was tested on a modified silicon surface with a crosslinker, as characterized by a total extension of approximately 85 nm and a final rupture force of 1.3 nN. **b** The maximal extension length of 44 curves (final rupture force >1.0 nN) is collected in a histogram

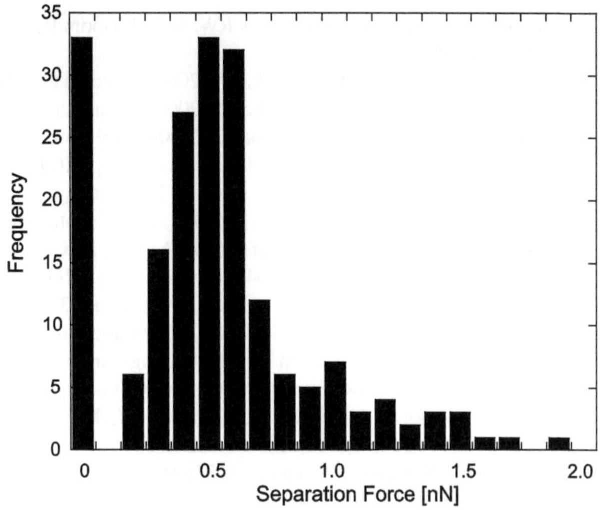

Fig. 19.23. Rupture forces obtained on the cell surface with cross-linker-modified tips

of 0.4–0.6 nN was considered as the most likely force range required to extract or, according to Bell, "uproot" intrinsic membrane proteins, including glycoproteins, from the cell membrane [69].

In conclusion, a method to pick up the membrane receptors is suggested using an AFM tip on which reactive cross-linkers are modified. This technology is especially remarkable for the establishment of a method to identify expressed characteristic marker proteins on a cell surface at the single-molecule level.

References

1. Reya T, Morrison SJ, Clarke MF, Weissman IL (2001) Nature 414:105
2. Pardal R, Clarke MF, Morrison SJ (2003) Nat Rev Cancer 3:895
3. Krivtsov AV, Twomey D, Feng Z, Stubbs MC, Wang Y, Faber J, Levine JE, Wang J, Hahn WC, Gilliland DG, Golub TR, Armstrong SA (2006) Nature 442:818
4. Schena M, Shalon D, Davis RW, Brown PO (1995) Science 270:467
5. Lockhart DJ, Dong H, Byrne MC, Follettie MT, Gallo MV, Chee MS, Mittmann M, Wang C, Kobayashi M, Horton H, Brown EL (1996) Nat Biotechnol 14:1675
6. Cawley S, Bekiranov S, Ng HH, Kapranov P, Sekinger EA, Kampa D, Piccolboni A, Sementchenko V, Cheng J, Williams AJ, Wheeler R, Wong B, Drenkow J, Yamanaka M, Patel S, Brubaker S, Tammana H, Helt G, Struhl K, Gingeras TR (2004) Cell 116:499
7. Bertone P, Stolc V, Royce TE, Rozowsky JS, Urban AE, Zhu X, Rinn JL, Tongprasit W, Samanta M, Weissman S, Gerstein M, Snyder M (2004) Science 306:2242
8. Cheng J, Kapranov P, Drenkow J, Dike S, Brubaker S, Patel S, Long J, Stern D, Tammana H, Helt G, Sementchenko V, Piccolboni A, Bekiranov S, Bailey DK, Ganesh M, Ghosh S, Bell I, Gerhard DS, Gingeras TR (2005) Science 308:1149
9. Rudkin GT, Stollar BD (1977) Nature 265:472
10. Bauman JG, Wiegant J, Borst P, van Duijn P (1980) Exp Cell Res 128:485
11. Nederlof PM, Robinson D, Abuknesha R, Wiegant J, Hopman AH, Tanke HJ, Raap AK (1989) Cytometry 10:20

12. Raap AK, van de Rijke FM, Dirks RW, Sol CJ, Boom R, van der Ploeg M (1991) Exp Cell Res 197:319
13. Kim H, Arakawa H, Osada T, Ikai A (2002) Colloids Surf B 25:33
14. Kim H, Arakawa H, Osada T, Ikai A (2003) Ultramicroscopy 97:359
15. Afrin R, Arakawa H, Osada T, Ikai A (2003) Cell Biochem Biophys 39:101
16. Afrin R, Yamada T, Ikai A (2004) Ultramicroscopy 100:187
17. Afrin R, Ikai A (2006) Biochem Biophys Res Commun 348:238
18. Osada T, Uehara H, Kim H, Ikai A (2003) J Nanobiotechnol 1:2
19. Osada T, Uehara H, Kim H, Ikai A (2004) Adv Clin Chem 38:239
20. Uehara H, Osada T, Ikai A (2004) Ultramicroscopy 100:197
21. Ducker WA, Senden TJ, Pashley RM (1991) Nature 353:239
22. Ducker WA, Senden TJ, Pashley RM (1992) Langmuir 8:1831
23. Lee GU, Kidwell DA, Colton RJ (1994) Langmuir 10:354
24. Takano H, Kenseth JR, Wong SS, O'Brien JC, Porter MD (1999) Chem Rev 99:2845
25. Kim H, Arakawa H, Hatae N, Sugimoto Y, Matsumoto O, Osada T, Ichikawa A, Ikai A (2006) Ultramicroscopy 106:652
26. Smith BL, Schaffer TE, Viani M, Thompson JB, Frederick NA, Kindt J, Belcher A, Stucky GD, Morse DE, Hansma PK (1999) Nature 399:761
27. Kim H, Arakawa H, Osada T, Ikai A (2002) Appl Surf Sci 188:493
28. Kim H, Tsuruta S, Arakawa H, Osada T, Ikai A (2004) Ultramicroscopy 100:203
29. Sneddon IN (1965) Int J Eng Sci 3:47
30. Weihs TP, Nawaz Z, Jarvis SP, Pethica JB (1991) Appl Phys Lett 59:3536
31. Ikai A (1996) Surf Sci Rep 26:263
32. Haga H, Sasaki S, Morimoto M, Kawabata K, Ito E, Abe K, Sambongi T (1998) Jpn J Appl Phys Part 1 37:3860
33. Haga H, Sasaki S, Kawabata K, Ito E, Ushiki T, Sambongi T (2000) Ultramicroscopy 82:253
34. Charras GT, Lehenkari PP, Horton MA (2001) Ultramicroscopy 86:85
35. Sekiguchi H, Arakawa H, Okajima T, Ikai A (2002) Appl Surf Sci 188:489
36. Sekiguchi H, Arakawa H, Taguchi H, Ito T, Kokawa R, Ikai A (2003) Biophys J 85:484
37. Romberger DJ (1997) Int J Biochem Cell Biol 29:939
38. Sekiguchi K, Hakomori S (1980) Proc Natl Acad Sci USA 77:2661
39. Brumfeld V, Werber MM (1993) Arch Biochem Biophys 302:134
40. Hynes RO (1987) Cell 48:549
41. Ruoslahti E, Pierschbacher MD (1987) Science 238:491
42. Hynes RO (1992) Cell 69:11
43. George EL, Georges-Labouesse EN, Patel-King RS, Rayburn H, Hynes RO (1993) Development 119:1079
44. Duband JL, Nuckolls GH, Ishihara A, Hasegawa T, Yamada KM, Thiery JP, Jacobson K (1988) J Cell Biol 107:1385
45. Felsenfeld DP, Choquet D, Sheetz MP (1996) Nature 383:438
46. Choquet D, Felsenfeld DP, Sheetz MP (1997) Cell 88:39
47. Smith JW, Vestal DJ, Irwin SV, Burke TA, Cheresh DA (1990) J Biol Chem 265:11008
48. Preissner KT, Seiffert D (1998) Thromb Res 89:1
49. Kim JP, Zhang K, Chen JD, Kramer RH, Woodley DT (1994) J Biol Chem 269:26926
50. Wayner EA, Orlando RA, Cheresh DA (1991) J Cell Biol 113:919
51. Stuiver I, Ruggeri Z, Smith JW (1996) J Cell Physiol 168:521
52. DePasquale JA (1998) Histochem Cell Biol 110:485
53. Rocheville M, Lange DC, Kumar U, Patel SC, Patel RC, Patel YC (2000) Science 288:154
54. Bouvier M (2001) Nat Rev Neurosci 2:274
55. Zhang J, Campbell RE, Ting AY, Tsien RY (2002) Nat Rev Mol Cell Biol 3:906

56. Milton AS, Wendlandt S (1970) J Physiol 207:76P
57. Stitt JT (1986) Yale J Biol Med 59:137
58. Sirko S, Bishai I, Coceani F (1989) Am J Physiol 256:R616
59. Kluger MJ (1991) Physiol Rev 71:93
60. Sehic E, Szekely M, Ungar AL, Oladehin A, Blatteis CM (1996) Brain Res Bull 39:391
61. Narumiya S, Sugimoto Y, Ushikubi F (1999) Physiol Rev 79:1193
62. Ushikubi F, Segi E, Sugimoto Y, Murata T, Matsuoka T, Kobayashi T, Hizaki H, Tuboi K, Katsuyama M, Ichikawa A, Tanaka T, Yoshida N, Narumiya S (1998) Nature 395:281
63. Lawrence JB, Singer RH (1986) Cell 45:407
64. Kislauskis EH, Zhu X, Singer RH (1997) J Cell Biol 136:1263
65. Latham VM Jr, Kislauskis EH, Singer RH, Ross AF (1994) J Cell Biol 126:1211
66. Grandbois M, Beyer M, Rief M, Clausen-Schaumann H, Gaub HE (1999) Science 283:1727
67. Wang T, Ikai A (1999) Jpn J Appl Phys Part 1 38:3912
68. Alam MT, Yamada T, Carlsson U, Ikai A (2002) FEBS Lett 519:35
69. Bell GI (1978) Science 200:618

17 Towards a Nanoscale View of Microbial Surfaces Using the Atomic Force Microscope

Claire Verbelen · Guillaume Andre · Xavier Haulot · Yann Gilbert · David Alsteens · Etienne Dague · Yves F. Dufrêne

From volume IX

Abstract. In recent years, the atomic force microscope (AFM) has greatly improved our understanding of microbial surfaces. AFM imaging has proved to be a powerful tool for visualizing membrane proteins and live cells at high resolution and in physiological conditions. In addition, AFM force spectroscopy has enabled us to probe and map a variety of properties, including the elasticity of cell walls and cell surface molecules, and the unfolding forces of single proteins, and has allowed the detection and functional analysis of molecular recognition sites. These unique capabilities allow researchers to answer a number of questions that were inaccessible before, such as how does the surface architecture of microbes change as they grow or as they interact with an antibiotic, what are the conformational changes in single membrane proteins, and what are the molecular forces responsible for the interaction between host cells or specific molecules?

Key words: Atomic force microscopy, Cell surfaces, Living cells, Imaging, Force spectroscopy, Ultrastructure, Molecular recognition, Single molecule, Surface properties, Membrane proteins

17.1 Introduction

Characterization of microbial surfaces is a key to gain insight into their structure and function and is particularly useful in the biomedical context for understanding microbial infection processes such as biofilm formation on medical implants and pathogen–host interactions. Traditionally, elucidation of the cell surface architecture has relied on transmission electron microscopy (TEM) and scanning electron microscopy techniques [1–5]. Remarkably, the development of cryo-TEM of frozen-hydrated thin sections has allowed researchers to get more natural views of bacterial cell envelopes. Unfortunately, these cryogenic methods are very demanding in terms of sample preparation and analysis and are only applied in a few laboratories worldwide. Valuable information on the composition, properties and interactions of cell surfaces can also be gained using electron microscopy approaches, biochemical analysis, biophysical techniques and surface analysis methods [4, 5]. Although very useful, these techniques generally require cell manipulation prior to examination and often provide averaged information obtained on large ensembles of cells. Consequently, there is clearly a need to develop ultrasensitive, high-resolution probes for mapping the structure, properties and interactions of microbial surfaces.

Recent studies have revealed that atomic force microscopy (AFM) can provide information on microbial surfaces that cannot be obtained with classical

In: H. Fuchs, B. Bhushan (eds), *Biosystems - Investigated by Scanning Probe Microscopy*,
DOI 10.1007/978-3-540-74083-4_5, © Springer-Verlag Berlin Heidelberg 2010

methods [5–9]. Basically, AFM offers two unique advantages over other microscopies. First, it provides surface images of the sample with nanometer (even subnanometer) resolution, in real time and under physiological conditions. Second, the force sensitivity of the instrument can be exploited to measure physical properties and interactions (e.g., elasticity, adhesion) on a local scale, and to detect and analyze single molecular recognition events. The aim of this chapter is to provide a flavor of the different possibilities offered by AFM in microbiology.

17.2 Imaging

17.2.1 Sample Preparation

An important prerequisite for successful AFM experiments is sample preparation, i.e., the specimen must be firmly attached to a solid support in order to withstand the lateral forces exerted by the scanning tip. While isolated membranes are generally well immobilized by simple adsorption on mica [10], stronger attachment is often needed for whole cells. This can be achieved by drying the sample or by bonding the cells covalently to the support. However, these treatments may cause significant rearrangement, denaturation or contamination of the cell walls, yielding information that may no longer be representative of the native state. More appropriate approaches include treating the support with a polycation to strengthen cell adhesion or immobilizing the cells mechanically in an agar gel or in a porous membrane. In the agar method, the gel is used as a soft, deformable immobilization matrix, thereby allowing direct visualization of growth processes [11]. This enabled in situ investigation of the growth process of *Saccharomyces cerevisiae* cells and direct visualization of micrometer-sized bud scars. In the porous membrane method, spherical cells are trapped in a polymer membrane with a pore size comparable to the dimensions of the cell, allowing repeated imaging without cell detachment or cell damage [12, 13].

17.2.2 Visualizing Membrane Proteins at Subnanometer Resolution

During the past decade, the use of AFM has afforded researchers a glimpse of many different membrane proteins, of which *Bacillus* S-layers [14], the hexagonally packed intermediate (HPI) layer of *Deinococcus radiodurans* [15], purple membrane from the archeon *Halobacterium* [16] and porin 2D crystals of *Escherichia coli* [17, 18]. AFM yields structural information to a resolution of 0.5–1 nm under physiological conditions, which makes it a complementary tool to X-ray and electron crystallography. Remarkably, function-related conformational changes can be monitored on single proteins, as nicely demonstrated for porin OmpF in which voltage-induced and pH-induced channel closure was observed [19]. In addition, there is clear evidence that the instrument is evolving from an imaging tool to a “lab

Fig. 17.1. High-resolution imaging of native membrane proteins. High-resolution atomic force microscopy (AFM) image of a high-light adapted native photosynthetic membrane of *Rhodospirillum photometricum* revealing the complex mixture of LH2 (*small rings*) and core complexes (*large rings*). (Reproduced with permission from S. Scheuring)

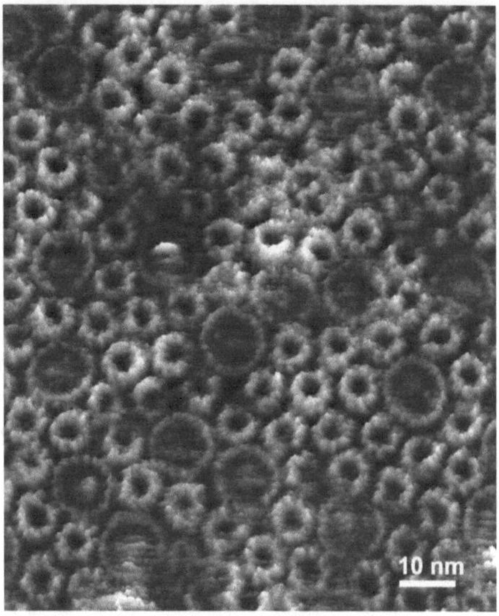

on a tip" multifunctional tool [8, 20], enabling the study of the unfolding pathways, the assembly, oligomeric states and molecular interactions of membrane proteins. In the first such study, Müller et al. [21] used the atomic force microscope tip as a molecular tweezer to reversibly push away single loops of bacteriorhodopsin proteins, thereby accessing structures otherwise hidden. More recent work has shown that the combined use of AFM imaging and single-molecule force spectroscopy makes it possible to pull on single membrane proteins, thereby providing novel molecular insight into their unfolding pathways and assembly forces [22–26].

Recently, a number of studies have focused on noncrystalline native membranes of different photosynthetic bacteria, including *Rhodopseudomonas viridis* [27], *Rhodospirillum photometricum* [28, 29], *Rhodobacter sphaeroides* [30], *Rhodobacter blasticus* [31] and *Rhodopseudomonas palustris* [32], providing new insight into the supramolecular organization of membrane protein complexes. In an elegant study, AFM was used to investigate how the composition and architecture of photosynthetic membranes of *R. photometricum* change in response to light [29] (Fig. 17.1). Despite large modifications in the membrane composition, the local environment of core complexes remained unaltered, whereas specialized paracrystalline light-harvesting antenna domains grew under low-light conditions. It was concluded that such structural adaptation ensures efficient photon capture under low-light conditions and prevents photodamage under high-light conditions.

17.2.3 Live-Cell Imaging

AFM is being used increasingly for imaging living cells, revealing structural details of hydrated cell walls with unprecedented resolution. Examples of microbial species that have been explored by AFM are *S. cerevisiae* [33], *Phanerochaete chrysosporium* [13], *Aspergillus oryzae* [34], *Aspergillus nidulans* [35], *Pinnularia viridis* [36], lactic acid bacteria [37,38], *Bacillus* spores [39,40], *Staphylococcus aureus* [41,42], *E. coli* [41], *Mycobacterium bovis* [43] and vaccinia virus [44].

Figure 17.2 illustrates the power of high-resolution AFM imaging in mycological research [34]. As can be seen, the surface of *A. oryzae* spores is covered with a crystalline-like array of rodlets that are 10 nm in diameter. These "nanobio-arrays" are known to cover the surface of many fungal spores and are thought to have different biological functions, such as spore dissemination and protection. Dramatic changes of cell surface structure are observed upon spore germination, the rodlet layer changing into a layer of soft material within a few hours. On close examination, this material shows streaks oriented in the scanning direction, suggesting that soft, loosely bound material is interacting strongly with the scanning tip. These direct observations are in good agreement with previous structural and chemical studies which showed that germination of *Aspergillus* spores results in the disruption of the proteinaceous rodlet layer and reveals inner spore walls which are essentially composed of polysaccharides.

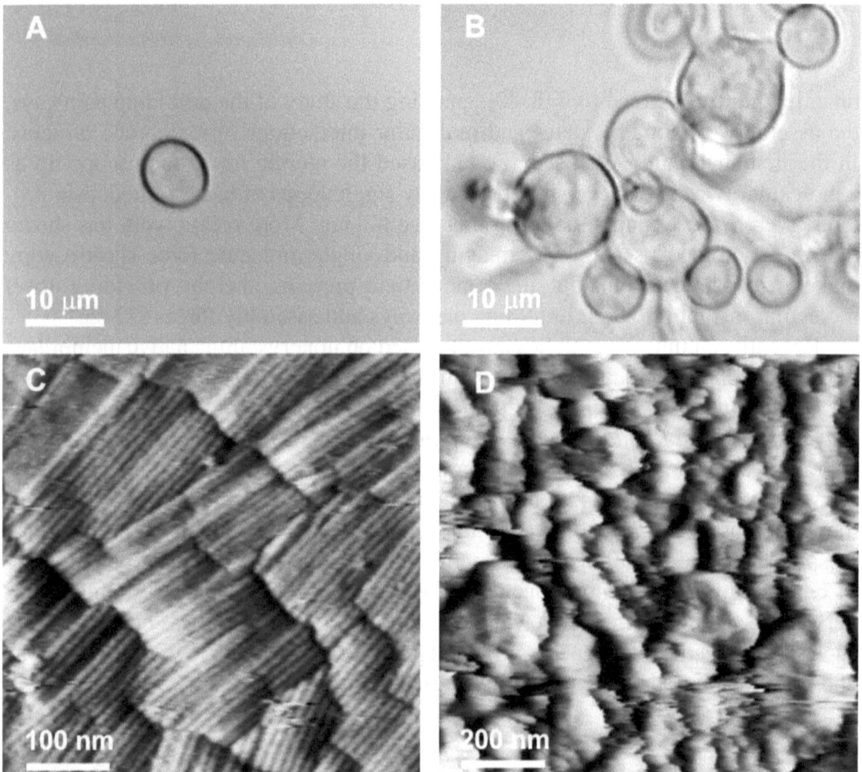

Fig. 17.2. Live-cell imaging. Optical micrographs of fungal spores from *Aspergillus oryzae* in the dormant state (**a**) and after germination for a few hours in a liquid medium (**b**). AFM deflection images recorded in aqueous solution for the surface of dormant (**c**) and germinating (**d**) spores revealing that, upon germination, the crystalline rodlet layer changes into a layer of soft material. (Reproduced with permission from [34]. Copyright 2001 American Chemical Society)

AFM can be combined with thin-section TEM to monitor cell growth and division events, as nicely demonstrated for *S. aureus* [42]. Nanoscale holes were seen around the septal annulus at the onset of division and were attributed to cell wall structures possessing high autolytic activity. After cell separation, concentric rings were observed on the surface of the new cell wall and were suggested to reflect newly formed peptidoglycan.

Remarkably, real-time imaging enables researchers to follow cell surfaces as they interact with enzymes and antibiotics. For example, the effect of protease on the cell surface architecture of *S. cerevisiae* cells was monitored in real time [33]. Progressive

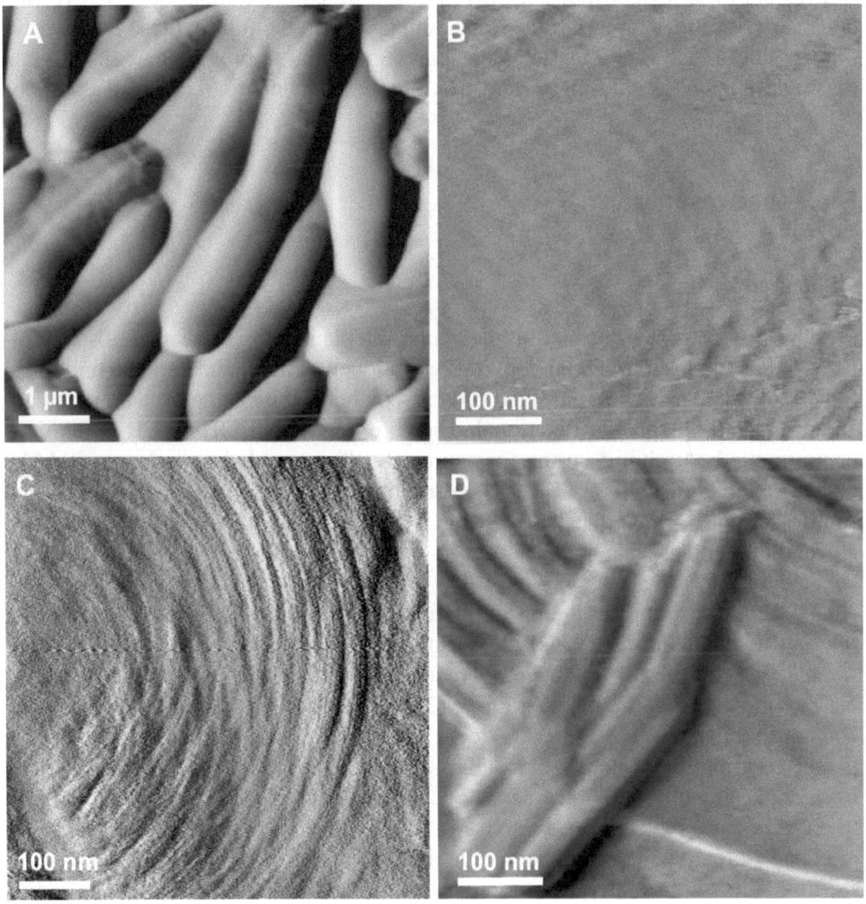

Fig. 17.3. Nanoscale imaging of drug-induced alterations. **a** Deflection image recorded in aqueous solution showing native mycobacteria immobilized on a polymer surface. **b** High-resolution image recorded on top of a native cell revealing a smooth, homogeneous surface. **c** High-resolution image of the cell surface following incubation for 24 h with 4 µg/ml ethambutol showing that the drug created concentric striations. **d** High-resolution image of the cell surface following incubation for 24 h with 10 µg/ml ethambutol revealing alteration of cell wall layers of 8- and 12-nm thicknesses, attributed to the outer electron-opaque and thick electron-transparent layers usually seen by electron microscopy. (Reproduced with permission from [43])

changes of the cell surface topography were clearly observed after protease addition. With time, the surface became eroded and showed large depressions, about 500 nm in diameter, surrounded by protruding edges, about 50 nm in height. By contrast, no modification of the cell surface was noted upon addition of amyloglucosidase, which was consistent with the cell wall biochemical composition. In a biomedical context, in situ AFM was used to investigate the effect of the drug ethambutol on the cell wall architecture of mycobacteria [43] (Fig. 17.3). Consistent with the mode of action of ethambutol, which primarily inhibits arabinan polymerization, the drug was found to dramatically alter the fine-surface architecture of the cells, by removing outer cell wall layers to an extent that depends on the drug concentration and by revealing concentric striations on top of the remaining layers. In the future, these noninvasive ultrastructural investigations will have a great potential to provide new insight into the mode of action of antibiotics as well as into the mechanisms of cell wall assembly.

17.3 Force Spectroscopy

17.3.1 Customized Tips

Probing surface properties and biomolecular interactions by force spectroscopy often requires the use of customized tips bearing well-defined chemical or biological moieties [45]. Chemically modified tips can be obtained by using organic monolayers terminated by specific functional groups (e.g., OH, COOH or CH_3), the most common method relying on the formation of self-assembled monolayers (SAMs) of alkanethiols on gold surfaces. The procedure involves coating microfabricated cantilevers with a thin adhesive layer of Cr or Ti, followed by another thin layer of Au, and immersing the coated cantilevers in dilute ethanol solutions of the selected alkanethiol [46,47].

Molecular recognition studies imply functionalizing the tips (and supports) with relevant ligands (and receptors) [45]. Here, several important issues must be considered: (1) the forces which immobilize the molecules should be stronger than the intermolecular force being studied; (2) the attached biomolecules should have enough mobility so that they can freely interact with complementary molecules; (3) the contribution of nonspecific adhesion to the measured forces should be minimized; (4) attaching biomolecules at low surface density is recommended in order to ensure single-molecule detection; (5) site-directed coupling may be desired to orientate all the interacting molecules in the same way.

Immobilization strategies commonly used for making biofunctionalized tips are illustrated in Fig. 17.4 [45]. A first approach consists of modifying tips by adsorption of biotinylated bovine serum albumin and further reaction with avidin (or streptavidin) to attach biotinylated molecules [48–50]. However, this approach is not ideally suited for single molecular recognition studies because (1) parallel breakage of multiple bonds is often observed and (2) the biotin–avidin system has a fairly low binding strength compared with that of covalent bonds.

To solve this problem, biomolecules can be firmly attached on atomic force microscope tips using either thiol or silane chemistry. In the first approach, biomolecules

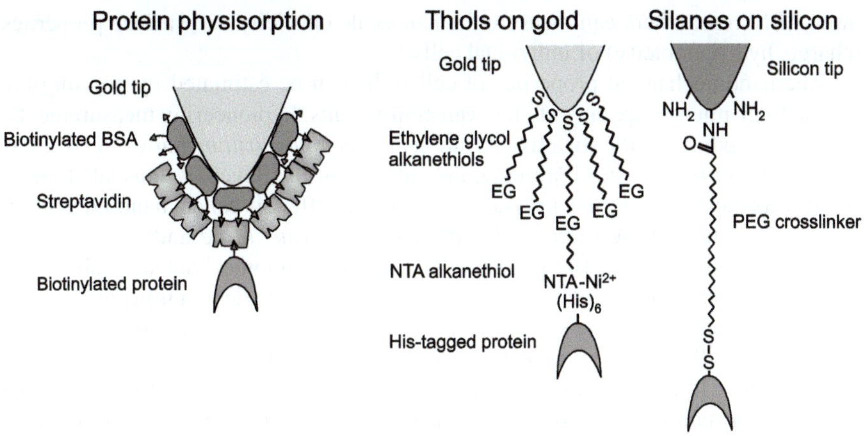

Fig. 17.4. Strategies commonly used for modifying atomic force microscope tips for single-molecule recognition studies: physisorption of proteins such as biotinylated bovine serum albumin (*BSA*), chemisorption of alkanethiols on gold and covalent coupling of silanes on silicon oxide. *EG* ethylene glycol, *NTA* nitrilotriacetate, *PEG* poly(ethylene glycol) (Adapted with permission from [45])

are covalently attached onto SAMs of functionalized alkanethiols on gold tips, using 1-ethyl-3-(3-dimethylaminopropyl)carbodiimide and *N*-hydroxysuccinimide (NHS) [51]. Mixing long-chain alkanethiols with COOH terminal functions in a matrix of shorter OH-terminated alkanethiols ensures a certain mobility of the attached biomolecules and minimizes nonspecific adsorption [52]. In the second approach, biomolecules are covalently attached onto silicon tips using various amine-functionalization procedures [53]. The amino terminated surfaces are reacted with a cross-linker (spacer) which provides the ligands with motional freedom and prevents their denaturation. The use of long spacer molecules is also advantageous for data treatment, for separating specific from nonspecific adhesion forces. Cross-linkers typically carry two different functional ends, e.g., an amine reactive NHS group on one end and 2-pyridyldithiopropionyl or vinyl sulfone groups, which can be covalently bound to thiols, on the other. Importantly, in the abovementioned approaches it is possible to orientate all biomolecules in the same way, thereby allowing an optimal exposure of the C-terminal or N-terminal domains of proteins, by attaching recombinant histidine-tagged proteins onto an atomic force microscope tip terminated with nitrilotriacetate groups [54,55].

Finally, it is worth mentioning that for probing cellular interactions, cell probes can be created by attaching microbial cells onto atomic force microscope cantilevers, using glutaraldehyde [56], glue [57], amino groups [58] or lectins [59].

17.3.2 Probing Nanoscale Elasticity and Surface Properties

Because of the small size of microbes, measuring the physical properties of the cell wall at the subcellular level has always been very challenging. For the first time, AFM researchers can measure the nanoscale elasticity and surface properties (charge, hydrophobicity) of individual cells.

The nanomechanical properties of cell walls can be estimated by pressing the atomic force microscope tip onto the wall components. In pioneering measurements, the elastic modulus of the sheath of the archeon *Methanospirillum hungatei* GP1 was found to be in the 20–40-GPa range, indicating that this layer of unusual strength could withstand an internal pressure of 400 atm [60]. So-called nanoindentation experiments showed the turgor pressure of *Enterococcus hirae* and *Pseudomonas aeruginosae* to be in the 10^5-Pa range [61]. Similar nanomechanical analyses revealed that the elasticity of the *S. cerevisiae* cell wall varies significantly across a single cell, i. e., the Young's modulus is 6 MPa on the bud scar and 0.6 MPa on the surrounding cell surface, in agreement with the accumulation of chitin in the bud scar area [62]. In this nanomechanical context, a crucial issue is to distinguish true sample deformation from repulsive surface forces, which can be difficult, particularly when working in aqueous solution. One way to clarify these two contributions is to vary the ionic strength or pH, as done by Gaboriaud et al. [63], who quantified the nanomechanical properties (bacterial spring constant, Young's modulus) of *Shewanella putrefaciens* at different pH values. The data demonstrated a swelling effect induced by raising the pH that was attributed to a change of water mobility due to intermolecular repulsive forces inside the polymeric fringe. Beside force spectroscopy, it is worth emphasizing that atomic force microscope cantilevers have also been used to measure local nanomechanical motion of *S. cerevisiae* cell walls, in relation to metabolic processes within the cell [64].

Hydrophobicity and charge are two crucial cell surface properties that are involved in a variety of interfacial events, including cell–support, cell–cell, antigen–antibody, cell–drug and cell–ions interactions. Various experimental methods have been developed to assess these properties, among which are microelectrophoresis, colloid titration, isoelectric focusing of cells, ion-exchange chromatography and surface conductivity, water contact-angle measurements on cell lawns, adhesion to hydrocarbons, partitioning in aqueous two-phase systems and hydrophobic interaction chromatography [5]. Because these assays only provide averaged information obtained on a large ensemble of cells, developing complementary tools for probing properties at the subcellular level is an important challenge. For the first time, functionalized atomic force microscope tips enable researchers to measure local surface charge and surface hydrophobicity. Ahimou et al. [47] used tips functionalized with ionizable carboxyl groups (COO–/COOH) to probe the surface charges of *S. cerevisiae* at the nanometer level. Force–distance curves were strongly influenced by pH: while no adhesion was measured at neutral/alkaline pH, multiple adhesion forces were recorded at acidic pH. Adhesion maps always showed fairly homogeneous contrast, indicating that the cell surface properties were homogeneous. The change of adhesion forces as a function of pH was interpreted as resulting from a change of cell surface electrostatic properties, a claim which was supported by the correlation obtained between the AFM titration curve constructed by plotting the adhesion force versus pH and the electrophoretic mobility versus pH curve obtained by classical microelectrophoresis. With use of a similar approach, it has been shown that hydrophilic (OH) and hydrophobic (CH_3) tips can be used to map cell surface hydrophobicity [46].

17.3.3
Stretching Cell Surface Polysaccharides and Proteins

By pulling on single proteins and polysaccharides, researchers can probe their intramolecular and intermolecular forces, thereby providing new insight into their folding/unfolding forces, stability and elasticity [23,65]. In particular, single-molecule AFM has allowed researchers to visualize single subunits of S-layer proteins and to probe their molecular interactions. In the first such study, force–extension curves recorded for *D. radiodurans* S-layers showed sawtooth patterns with six force peaks attributed to the sequential pulling out of multiple protomers of the S-layer. After the force curve had been recorded, a molecular defect the size of a hexameric complex was visualized by means of high-resolution imaging [22]. A similar approach was

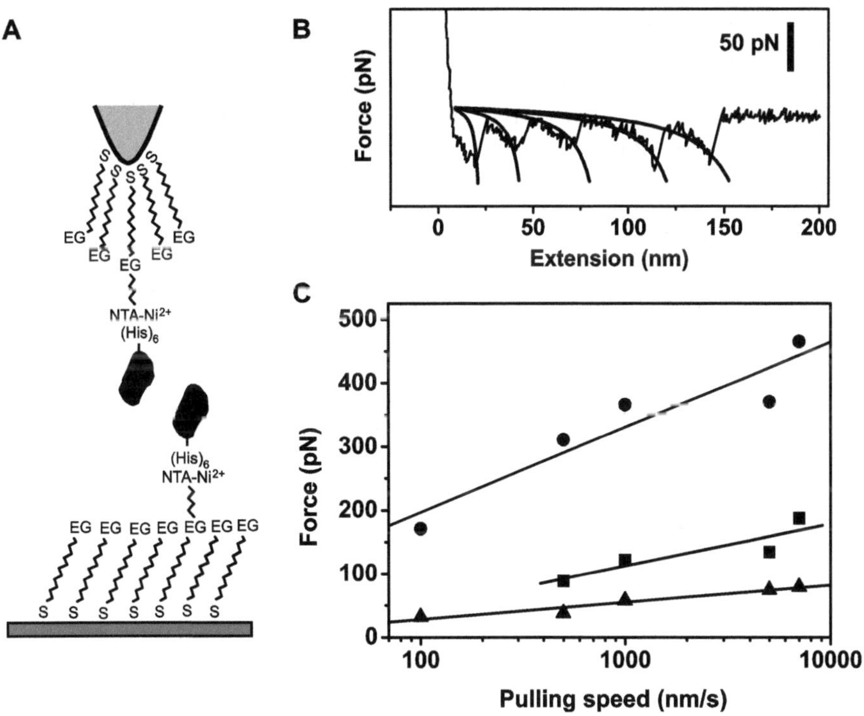

Fig. 17.5. Measuring the molecular forces driving the folding and assembly of S-layer proteins. **A** The surface chemistry used to functionalize atomic force microscope tips and supports with CbsA proteins at low surface density. The entire mature CbsA protein as well as CbsA peptides truncated in their N-terminal and C-terminal regions were engineered with a His-tag for site-directed immobilization onto NTA/EG terminated surfaces. **B** Typical force–extension curve obtained in phosphate-buffered saline (PBS) between tips and supports modified with CbsA proteins. The well-pronounced force peaks forming a sawtooth pattern are well-described using the wormlike chain model and are attributed to the unfolding of individual α-helical structures. **C** Interaction forces of the CbsA protein (*triangles*) and of CbsA peptides truncated in their N-terminal (*circles*) and C-terminal (*squares*) regions as a function of the pulling speed

Fig. 17.6. Detection and mapping of individual adhesins. **a** Representative force curves and adhesion force histogram obtained in PBS between a heparin-binding hemagglutinin adhesin (HBHA) tip and a heparin surface. The adhesion force histogram revealed a bimodal distribution reflecting the binding strength of one and two adhesins. **b** Mapping cell-adhesion nanodomains on living bacteria. Adhesion force maps (*gray scale* 100 pN) recorded in PBS with heparin-modified atomic force microscope tips over *Mycobacterium bovis* BCG bacteria expressing HBHA adhesins. Adhesion events (*clear pixels*) were attributed to the detection of single HBHA molecules. Notably, the HBHA distribution was not homogeneous, but apparently concentrated into nanodomains which may play important biological functions. *Scale bars* 100 nm. (Reprinted with permission from [55])

used to unzip the S-layer of *Corynebacterium glutamicum*, providing new insight into the stability of this protective bacterial surface coat [66].

Recently, single-molecule force spectroscopy was used to gain insight into the molecular forces driving the folding and assembly of the CbsA S-layer protein from *Lactobacillus crispatus*, which is known to have important functions, including promoting binding to collagen, triggering self-assembly of the protein subunits and mediating attachment to teichoic acids of the cell wall [67] (Fig. 17.5). Force curves recorded between tips and supports modified with CbsA proteins (Fig. 17.5a) showed sawtooth patterns with multiple force peaks of 58 ± 26 pN that were attributed to the unfolding of α-helices, in agreement with earlier secondary structure predictions (Fig. 17.5b). The average unfolding force increased with the pulling speed but was independent of the interaction time (Fig. 17.5c). Force curves obtained for CbsA peptides truncated in their C-terminal region showed similar periodic features, except that fewer force peaks were seen. Furthermore, the average unfolding force was 83 ± 45 pN, suggesting the domains were more stable. By contrast, cationic peptides truncated in their N-terminal region showed single force peaks of 366 ± 149 pN, presumably reflecting intermolecular electrostatic bridges rather than unfolding events. Interestingly, these large intermolecular forces increased not only with pulling speed but also with interaction time. It was concluded that the measured intramolecular and intermolecular forces may play a significant role in controlling the stability and assembly of the CbsA protein.

AFM has also been used to pull on microbial cell surface polysaccharides. For instance, force curves recorded on germinating *A. oryzae* spores showed large attractive forces along with elongation forces that were attributed to the stretching of single cell surface polysaccharides [34]. Indeed, fitting the curves with a theoretical model from statistical mechanics yielded parameters similar to those reported for individual amylose and dextran molecules [65]. It was concluded that the stickiness and flexibility of cell surface polysaccharide chains may promote spore aggregation via macromolecular bridging interactions between opposing cells. In another work, mannan polymers were stretched at the surface of *S. cerevisiae* cells using tips that were functionalized with the lectin concanavalin A [68].

17.3.4 Nanoscale Mapping and Functional Analysis of Molecular Recognition Sites

Functionalization of atomic force microscope tips with ligands (Sect. 17.3.1) enables researchers to detect complementary receptor molecules either on model surfaces or

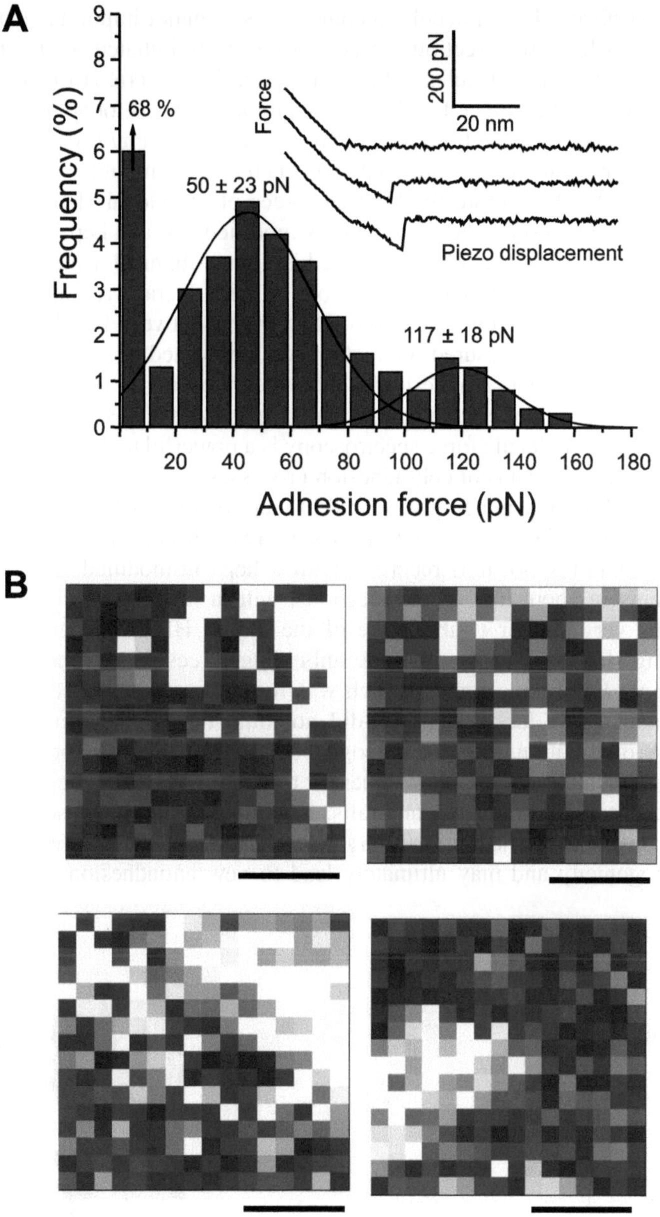

A
Frequency (%)
68 %
50 ± 23 pN
117 ± 18 pN
Force
200 pN
20 nm
Piezo displacement
Adhesion force (pN)
B

on cellular surfaces [45]. Molecular recognition is of major importance in bacterial pathogenesis, where the infectious process is generally initiated by the interaction between cell-adhesion molecules on the bacterial surface, referred to as adhesins, and specific ligands on the host cell surface. An example of this is found in tuberculosis, which is caused by the infectious agent *Mycobacterium tuberculosis*. Mycobacteria adhere to heparan sulfates on epithelial cells via the heparin-binding hemagglutinin adhesin (HBHA). To explore the binding forces of individual HBHA molecules, force curves were recorded between HBHA-modified tips and heparin-coated supports [55] (Fig. 17.6a). The adhesion force histogram obtained using a loading rate of 10,000 pN/s revealed a bimodal distribution with average rupture forces of 50 and 117 pN, attributed to one and two binding events between HBHA and heparin. The specificity of the measured interaction was confirmed by showing a dramatic reduction in the number of adhesion events when working in the presence of free heparin or with a bovine serum albumin coated tip instead of a HBHA tip. This work shows that single-molecule force spectroscopy is a powerful tool to gain insight into the molecular mechanisms of cell-adhesion processes.

Next, AFM affinity imaging was used to map the distribution of single HBHA molecules on bacterial surfaces with nanoscale resolution [55] (Fig. 17.6b). Affinity maps (300 nm × 300 nm) recorded with a heparin-modified tip revealed adhesion events in about half of the locations, with a magnitude of approximately 50 pN, thus very similar to the value of the model HBHA–heparin interaction. This finding strongly suggests that the unbinding forces correspond to the detection of single HBHA molecules, which was further supported by the observation that a mutant strain lacking HBHA did not bind the heparin tip. Interestingly, the HBHA distribution was not homogeneous, but apparently concentrated into nanodomains which may promote adhesion to target cells by inducing the recruitment of receptors within membrane rafts. In summary, these nanoscale functional analyses contribute to shedding new light onto the molecular mechanisms of bacterial pathogenicity and may ultimately lead to new antiadhesion approaches for therapy.

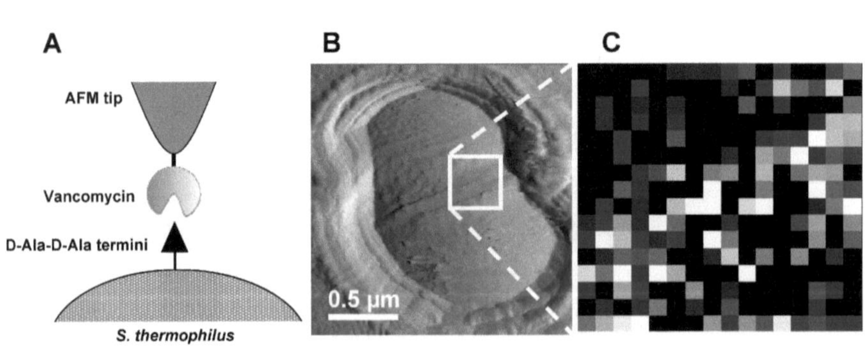

Fig. 17.7. Imaging vancomycin binding sites on living *Streptococcus thermophilus*. **a** The experimental setup. **b** AFM image showing a single cell during the course of the division process. **c** Affinity map recorded with a vancomycin tip on the septum region highlighted by the *white box* in **b**. Single adhesion events are detected in the septum region and are attributed to the specific recognition of individual terminal D-Ala–D-Ala of nascent peptidoglycan

Molecular recognition is also involved in the action mode of important antibiotics such as vancomycin, which binds with high affinity and specificity to the terminal D-Ala–D-Ala of peptidoglycan precursors in Gram-positive bacteria, thereby preventing their incorporation into the bacterial cell wall. The use of vancomycin-terminated atomic force microscope tips to detect single D-Ala–D-Ala sites on living bacteria was recently established (Fig. 17.7). Topographic images of living *Streptococcus thermophilus* cells revealed a smooth morphology as well as a well-defined division septum. Affinity maps recorded with vancomycin tips revealed that D-Ala–D-Ala sites were essentially located in the septum region, suggesting that newly formed peptidoglycan is inserted in these regions. Accordingly, these novel antibiotic nanoprobes offer exciting prospects for understanding the assembly of bacterial cell walls and for elucidating the mode of action of antibiotics that target cell wall constituents.

17.4 Conclusions

AFM has recently opened a variety of novel possibilities for imaging and manipulating microbial surfaces in their native environment. While AFM imaging is useful for visualizing surface structures at high resolution and in physiological conditions, force spectroscopy enables researchers to probe a variety of cell surface properties, including the elasticity of cell walls and cell surface molecules, the unfolding forces of single proteins, and the detection and functional analysis of molecular recognition sites. These nanoscale analyses allow researchers to answer a number of questions that were difficult to address before, such as how does the surface architecture of microbes change as they grow or interact with antibiotics; what is the force required to unfold a membrane protein; and what are the molecular forces driving the interaction between a pathogen and a host surface or between an antibiotic and a bacterial surface?

There are several technological issues to address in the future. First, the relatively slow imaging speed of commercially available atomic force microscopes (30–60 s per image) represents a crucial bottleneck for exploring cell surface dynamic processes like conformational changes. Fortunately, fast instruments have been introduced in the few last years, which enable recording up to 200 frames per second and open up fascinating new perspectives to explore molecular and cellular dynamics [69, 70]. Another key issue when dealing with living cells is that the image resolution is often limited because the extremely soft cell surface is deformed by the scanning tip. The use of novel dynamic imaging modes which minimize the applied force [71], combined with improved sample preparation methods, may help solve this problem, thereby providing reliable images at much higher resolution than currently available. With such further technological developments (faster imaging, lower interaction forces), it should become possible to observe, for example, the binding of individual antibodies or drugs to distinct locations of the cell surface, function-related conformational changes of surface proteins, or local alterations of the cell surface topography associated with cellular processes such as cell growth and division.

Advances in single-molecule force spectroscopy will allow elucidation of the elasticity, conformational changes and unfolding forces of an increasing number of membrane proteins and other cell surface constituents. Progress in developing reliable protocols for attaching biomolecules and cells to atomic force microscope cantilevers will enable the measurement of molecular interactions on microbial surfaces in the daily routine, providing novel insight into the molecular mechanisms of microbial infection processes (biofilms on medical implants, pathogen–host interactions) and new ideas for antiadhesion strategies.

Acknowledgements. This work was supported by the National Foundation for Scientific Research (FNRS), the Foundation for Training in Industrial and Agricultural Research (FRIA), the Université Catholique de Louvain (Fonds Spéciaux de Recherche) and the Federal Office for Scientific, Technical and Cultural Affairs (Interuniversity Poles of Attraction Programme). Y.F.D. is a Research Associate of the FNRS.

References

1. Beveridge TJ (1999) J Bacteriol 181:4725
2. Matias VRF, Beveridge TJ (2005) Mol Microbiol 56:240
3. Matias VRF, Beveridge TJ (2006) J Bacteriol 188:1011
4. Beveridge TJ, Graham LL (1991) Microbiol Rev 55:684
5. Ubbink J, Schar-Zammaretti P (2005) Micron 36:293
6. Engel A, Müller DJ (2000) Nat Struct Biol 7:715
7. Dufrêne YF (2004) Nat Rev Microbiol 2:451
8. Müller DJ, Sapra KT, Scheuring S, Kedrov A, Frederix PL, Fotiadis D, Engel A (2006) Curr Opin Struct Biol 16:489
9. Gaboriaud F, Dufrêne YF (2007) Colloids Surf B 54:10
10. Müller DJ, Amrein M, Engel A (1997) J Struct Biol 119:172
11. Gad M, Ikai A (1995) Biophys J 69:2226
12. Kasas S, Ikai A (1995) Biophys J 68:1678
13. Dufrêne YF, Boonaert CJP, Gerin PA, Asther M, Rouxhet PG (1999) J Bacteriol 181:5350
14. Györvary ES, Stein O, Pum D, Sleytr UB (2003) J Microsc 212:300
15. Müller DJ, Baumeister W, Engel A (1996) J Bacteriol 178:3025
16. Müller DJ, Schabert FA, Büldt G, Engel A (1995) Biophys J 68:1681
17. Schabert FA, Henn C, Engel A (1995) Science 268:92
18. Scheuring S, Ringler P, Borgnia M, Stahlberg H, Müller DJ, Agre P, Engel A (1999) EMBO J 18:4981
19. Müller DJ, Engel A (1999) J Mol Biol 285:1347
20. Janovjak H, Kedrov A, Cisneros DA, Sapra KT, Struckmeier J, Müller DJ (2006) Neurobiol Aging 27:546
21. Müller DJ, Büldt G, Engel A (1995) J Mol Biol 249:239
22. Müller DJ, Baumeister W, Engel A (1999) Proc Natl Acad Sci USA 96:13170
23. Oesterhelt F, Oesterhelt D, Pfeiffer M, Engel A, Gaub HE, Müller DJ (2000) Science 288:143
24. Janovjak H, Kessler M, Oesterhelt D, Gaub H, Müller DJ (2003) EMBO J 22:5220
25. Kedrov A, Ziegler C, Janovjak H, Kühlbrandt W, Müller DJ (2004) J Mol Biol 340:1143
26. Sapra KT, Besir H, Oesterhelt D, Müller DJ (2006) J Mol Biol 355:640
27. Scheuring S, Seguin J, Marco S, Lévy D, Robert B, Rigaud JL (2003) Proc Natl Acad Sci USA 100:1690

28. Scheuring S, Rigaud JL, Sturgis JN (2004) EMBO J 23:4127
29. Scheuring S, Sturgis JN (2005) Science 309:484
30. Bahatyrova S, Frese RN, Siebert CA, Olsen JD, van der Werf KO, van Grondelle R, Niederman RA, Bullough PA, Otto C, Hunter CN (2004) Nature 430:1058
31. Scheuring S, Busselez J, Lévy D (2005) J Biol Chem 280:1426
32. Scheuring S, Gonçalves RP, Prima V, Sturgis JN (2006) J Mol Biol 358:83
33. Ahimou F, Touhami A, Dufrêne YF (2003) Yeast 20:25
34. van der Aa BC, Michel RM, Asther M, Zamora MT, Rouxhet PG, Dufrêne YF (2001) Langmuir 17:3116
35. Ma H, Snook LA, Kaminskyj SGW, Dahms TES (2005) Microbiol SGM 151:3679
36. Crawford SA, Higgins MJ, Mulvaney P, Wetherbee R (2001) J Phycol 37:543
37. Schaer-Zammaretti P, Ubbink J (2003) Ultramicroscopy 97:199
38. Schaer-Zammaretti P, Ubbink J (2003) Biophys J 85:4076
39. Chada VGR, Sanstad EA, Wang R, Driks A (2003) J Bacteriol 185:6255
40. Plomp M, Leighton TJ, Wheeler KE, Malkin AJ (2005) Biophys J 88:603
41. Doktycz MJ, Sullivan CJ, Hoyt PR, Pelletier DA, Wu S, Allison DP (2003) Ultramicroscopy 97:209
42. Touhami A, Jericho M, Beveridge TJ (2004) J Bacteriol 186:3286
43. Verbelen C, Dupres V, Menozzi FD, Raze D, Baulard AR, Hols P, Dufrêne YF (2006) FEMS Microbiol Lett 264:192
44. Malkin AJ, McPherson A, Gershon PD (2003) J Virol 77:6332
45. Hinterdorfer P, Dufrêne YF (2006) Nat Methods 3:347
46. Dufrêne YF (2000) Biophys J 78:3286
47. Ahimou F, Denis FA, Touhami A, Dufrêne YF (2002) Langmuir 18:9937
48. Lee GU, Kidwell DA, Colton RJ (1994) Langmuir 10:354
49. Florin EL, Moy VT, Gaub HE (1994) Science 264:415
50. Fritz J, Katopodis AG, Kolbinger F, Anselmetti D (1998) Proc Natl Acad Sci USA 95:12283
51. Dammer U, Popescu O, Wagner P, Anselmetti D, Güntherodt HJ, Misevic GN (1995) Science 267:1173
52. Berquand A, Xia N, Castner DG, Clare BH, Abbott NL, Dupres V, Adriaensen Y, Dufrêne YF (2005) Langmuir 21:5517
53. Hinterdorfer P, Baumgartner W, Gruber HJ, Schilcher K, Schindler H (1996) Proc Natl Acad Sci USA 93:3477
54. Kienberger F, Kada G, Gruber HJ, Pastushenko VPh, Riener C, Trieb M, Knaus HG, Schindler H, Hinterdorfer P (2000) Single Mol 1:59
55. Dupres V, Menozzi FD, Locht C, Clare BH, Abbott NL, Cuenot S, Bompard C, Raze D, Dufrêne YF (2005) Nat Methods 2:515
56. Ong YL, Razatos A, Georgiou G, Sharma MM (1999) Langmuir 15:2719
57. Bowen WR, Hilal N, Lovitt RW, Wright CJ (1998) Colloid Surf A 136:231
58. Lower SK, Hochella MF, Beveridge TJ (2001) Science 292:1360
59. Benoit M, Gabriel D, Gerisch G, Gaub HE (2000) Nat Cell Biol 2:313
60. Xu W, Mulhern PJ, Blackford BL, Jericho MH, Firtel M, Beveridge TJ (1996) J Bacteriol 178:3106
61. Yao X, Walter J, Burke S, Stewart S, Jericho MH, Pink D, Hunter R, Beveridge TJ (2002) Colloids Surf B 23:213
62. Touhami A, Nysten B, Dufrêne YF (2003) Langmuir 19:4539
63. Gaboriaud F, Bailet S, Dague E, Jorand F (2005) J Bacteriol 187:3864
64. Pelling AE, Sehati S, Gralla EB, Valentine JS, Gimzewski JK (2004) Science 305:1147
65. Rief M, Oesterhelt F, Heymann B, Gaub HE (1997) Science 275:1295
66. Scheuring S, Stahlberg H, Chami M, Houssin C, Rigaud J-L, Engel A (2002) Mol Microbiol 44:675

67. Verbelen C, Antikainen J, Korhonen TK, Dufrêne YF (2007) Ultramicroscopy 107:1004
68. Gad M, Itoh A, Ikai A (1997) Cell Biol Int 21:697
69. Ando T, Kodera N, Takai E, Maruyama D, Saito K, Toda A (2001) Proc Natl Acad Sci USA 98:12468
70. Humphris ADL, Hobbs JK, Miles MJ (2003) Appl Phys Lett 83:6
71. Humphris ADL, Miles MJ (2002) Methods Cell Biol 68:337

18 Cellular Physiology of Epithelium and Endothelium

Christoph Riethmüller · Hans Oberleithner

From volume IX

Abstract. Epithelial and endothelial cell layers have many essential functions for keeping higher organisms alive. One main task which they both share is the regulation of water balance in the body. Regulation of water permeability can be performed by altering the conductivity between the cells (paracellularly) or through the cells (transcellularly). Here, four examples are shown of how atomic force microscopy (AFM) can indirectly assess the effects of these physiological functions on primary cell preparations under buffer conditions. (1) assessing the tightness of paracellular junctions via their height profile, (2) softening of the cytoskeleton in response to thirst hormone, (3) probing intracellular vacuoles and (4) exploring the invagination formed by a transmigrating immune cell after nanomanipulation. It can be concluded that AFM can help us to examine mechanisms regulating the permeability of cell layers.

Key words: Physiology, Epithelium, Endothelium, Cell Junctions, Permeability, Cell Elasticity

18.1 Introduction

Physiology is a discipline of biology that investigates phenomena of organisms with respect to their function in keeping the individual alive. The spectrum of organisms covers plants, animals and humans. Not only whole organisms, but also parts of them, like isolated organs, cell ensembles, single cells or even organelles, are used to study their respective role in enabling a normal life. An important aim of physiological investigation of isolated organs or at the cellular level is to discover regulatory cycles that allow the organism adapt to the ever-changing environmental conditions. Classic examples of physiological models are isolated beating hearts kept in buffer solution to study the effects of flow resistance or prepared single nephrons to investigate fluid absorption. The search for mechanisms can make it necessary to dissect a regulatory cycle down to the molecular level, i. e., single metabolic steps. Especially in human physiology, out-of-tune states can be of considerable interest for understanding the development of diseases, the knowledge of which is comprised in the subdiscipline pathophysiology. The latter is intimately intertwined with pharmacology, because the impact of chemical compounds on cellular and organ function forms the basis for exploring their therapeutic potential.

Major categories of physiological interest are the uptake and transport of nutrients, circulation, secretion, hormonal regulation, muscle contraction and bone mechanics as well as biological data acquisition (eye, ear, skin) or information processing through neurons. Since a physiological measurement focuses on entities like pressure, flow, concentration gradients and electric potential or current, it

In: H. Fuchs, B. Bhushan (eds), *Biosystems - Investigated by Scanning Probe Microscopy*,
DOI 10.1007/978-3-540-74083-4_6,

mainly relies on biophysical and biochemical methods. Reversely, biological questions sometimes drive the technical improvement of measuring devices, for instance, the patch-clamp technique promoted the design of ultra-low-noise amplifiers [65,89]. For the atomic force microscope (AFM) as a rather young "tactile" instrument, new aspects have been gained in the past decade when it was applied to old questions and more are expected in the years to come.

The fact that the AFM can not only be run in a high vacuum like its predecessor the scanning tunneling microscope but it can also be run under atmospheric pressure made it feasible for many more applications. And after it had surprisingly been proven to work even under fluid buffer conditions, the biosciences started to recognize its huge potential. Since then, the AFM has experienced a continuously growing acceptance in the life sciences [15,17,28,32,38,44,52,63,67,75,99,111]. New insights have been obtained into lipid biochemistry [4, 36, 49, 61], structural biology [24, 31, 43, 51, 81, 91], cell biology [14, 19, 41, 44, 67], and pharmacology [14,22,79,88] to name just a few fields.

In cellular physiology of higher animals, two closely related cell types, namely epithelium and endothelium, have diverse functions of outstanding importance. They form cellular monolayers separating different body compartments and thus build pseudo-two-dimensional organs with a maximal surface. As such they are ideally suited specimens for investigation by the surface probing tool the AFM. The results promise to be of physiological relevance, because the surface maximization follows their function [34,35,52,69]. Here, we focus on these two types of cells in mammals. This chapter will spotlight that specific area of AFM application of the past decade and provide some examples of ongoing experiments with a clear perspective for physiological implications in the future.

18.2 Epithelium

18.2.1 Transport Through a Septum

The mammalian body is separated from its environment by at least one epithelial cell layer. This layer primarily forms a barrier between inside and outside to control any kind of transport in both directions. A regulation is very important, because literally all of the substances like water, oxygen, salt and nutrients or waste products going either into or out of the body have to cross epithelial layers. The asymmetry between up and down faces of the cells is termed "polarization". Basically, the outside (with respect to the organism) is called "apical" and the inside is called "basolateral." For in vitro experiments, isolated cells are usually cultivated in a petri dish coated with basal membrane proteins for better growth. For this reason, the cells interpret the interface to the culture dish as "inside" the organism (basolateral) face. Hence, only the apical surface of polarized cells is accessible to the AFM tip, unless a special membrane preparation is used [3,23,46].

A well-differentiated layer develops cell-to-cell borders consisting of adherent transmembrane proteins interacting with their counterparts in the neighboring cell to form a zipper-like structure. Fibrillary actin is found in a higher density in this

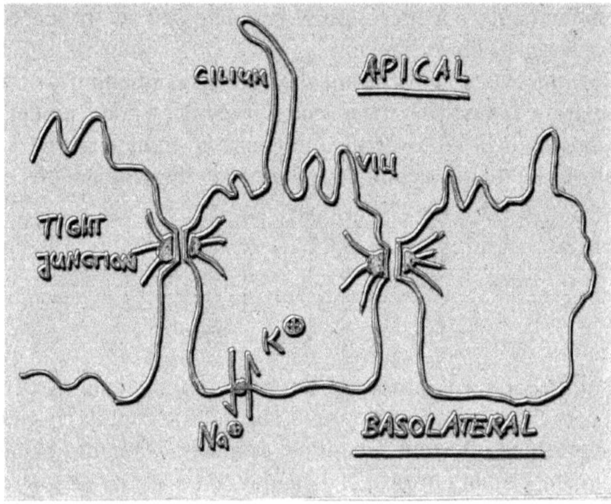

Fig. 18.1. Epithelial cells. The environment of an epithelial cell is asymmetric: the apical side is oriented to the outside of the organism, the basolateral side to the inside. The tight junctions accomplish and regulate the barrier function of an intact epithelium. In the basolateral plasma membrane, the Na/K exchanger, a transmembrane protein, builds up concentration gradients and membrane potential. Microvilli and cilia on the apical membrane increase the interaction area available for the transport of substances

belt, offering mechanical support from inside the cell as well as a better coupling of barrier properties to an intracellular regulation. The junctions are tight enough to build up an electrical resistance in the range of kiloohm square centimeters. Accordingly, the conductivity for solutes between the cells (paracellularly) is reduced down to transport rates comparable to the minute amounts passing through the cells (transcellularly). The belt of tight-junction proteins effectively hinders the diffusion of membrane proteins within the lipid bilayer from the basolateral to the apical side or in the reverse direction. As a consequence, transport proteins, ion channels and receptors are differently partitioned between apical and basolateral faces, reflecting the asymmetric nature of epithelial function [21, 104, 108]. For the physiologist, only a well-polarized epithelium is a good one, because it can be expected to represent the in vivo situation.

In functionally different regions, the epithelial layers are specialized according to their respective tasks. The skin barrier protects against entrance of particles and bacteria as well as against the loss of water and solutes. Alveolar epithelium of the lung enables an effective exchange of gaseous oxygen and carbon dioxide. Epithelium of the guts absorbs sugar and peptides, in the stomach it secretes protons, and the epithelia of secretory glands produce saliva, sweat or tears. The process of epithelial secretion including the formation of pores has been described in pancreatic acinar cells [44, 92]. Another secretion process of high physiological importance is that of the proteoglycan "surfactant" of alveolar type II cells in the lung, which reduces surface tension at the gas–fluid interface. Although the protein has been well studied [56, 100] and alveolar cells can be subjected to AFM imaging [1], the process of secretion could not successfully be addressed yet, probably owing to its tenside-like properties of the surfactant.

In the following, we will focus on the kidney epithelium as an example. It separates urine from interstitial space and serves as a good model for different types of transport. Typical AFM applications can be transferred to other epithelia. Technical details of methods beyond normal contact imaging are given at the end of the chapter.

18.2.2
In the Kidney

The main task of the kidney is to control the water and salt balance of the organism. It consists of a complex system with around one million capillaries called nephrons. A nephron is segmented into four functional domains: (1) in the glomerulus, blood is filtrated to give the primary urine; (2) then, the vast majority of nutrients and water are reabsorbed by the proximal tubulus epithelium; (3) salt is actively pumped out of the loop of Henle into the interstitial space, setting up an osmotic gradient; (4) the fine-tuning of water reabsorption is performed via regulating the water permeability of the cell membranes.

The first reports on kidney cells viewed with the AFM date back to the early 1990s [40, 50, 68]. The studies used immmortalized epithelial cells from canine kidney collecting duct (MDCK), which has established as the standard cell line for AFM studies on epithelium. Since then, numerous reports have extended the spectrum of cells to various types of freshly isolated specimens, so-called primary culture, which can be regarded as better representing the in vivo characteristics [20,75,77,109]. Following the functional segmentation of nephrons, glomerular epithelial cells have very recently been imaged [103]. They are termed "podocytes" owing to their feet(toe)like interdigitations that have a role in the filtration process. However, efforts to get such a morphologically differentiated cell culture exhibiting this special feature have not been successful yet and are still ongoing. Cells from the proximal tubule have been immortalized to give the LLCPK-1 line, which has been demonstrated to offer the functional characteristics and is convenient to work with [82]. While cells from the loop of Henle have not been addressed by AFM studies, cells of the collecting duct, especially the MDCK line, have been serving as a model for a vast number physiological and biophysical studies since the mid-1990s [68,95,96].

18.2.2.1
Protrusions

The luminal surface of epithelia is covered by (micro-)villi, which are small protrusions of the plasma membrane (in the micron range). Many membrane proteins, like transporters, receptors and ion channels, are preferentially localized in the villi. The AFM is a good tool to measure the height of microvilli accurately. This is of significance, since villi are the physiological means to increase the interfacial area between cells and the environment. This area can be a limiting factor, as seen from highly active transporting epithelia like in the ileum of the intestine or in proximal tubules of the kidney, which exhibit a higher density and length of villi. In other words, the transport capacity correlates to the density and height of villi. The number and correct size of these microstructures is obtained rather accurately and easily with atomic force microscopy as compared to transmission electron microscopy. In Fig. 18.2, typical examples from freshly isolated nasal and kidney cells are given.

The data from atomic force microscopy are in good accordance with those from scanning ion conductance microscopy [29], a method that uses the resistance of a patch pipette as a feedback signal instead of the deflection of the cantilever in the AFM [33, 47]. Combination of atomic force microscopy with fluorescence

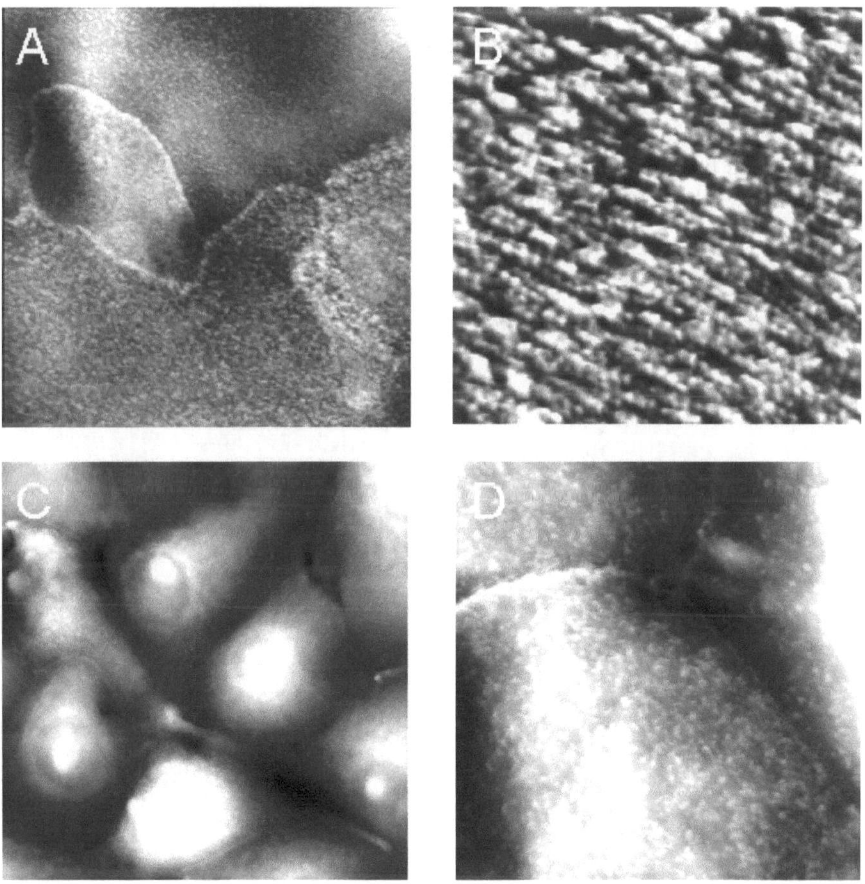

Fig. 18.2. Fixed epithelial cells. Epithelial cells freshly isolated from **a** and **b** human nasal biopsies or **c** and **d** rat inner medulla collecting duct (IMCD) were fixed with glutardialdehyde and scanned in physiological buffer in contact mode. **a** At 70-μm^2 scan area, the cell borders can be seen. **b** A zoom into 10 μm resolves the microvilli protruding into the lumen by around 0.5 μm. **c** An area of $(100\ \mu m)^2$ covers around six cell bodies, four of which exhibit a ciliary structure in their center. **d** Zooming to 25 μm resolves the microvillar pattern and shows the seam of a cell border (in the *upper third* of the image)

microscopy demonstrated that these villi contain a high density of fibrillary actin, probably in order to keep them in an upright position to avoid a loss of interaction surface [80]. Occasionally, the villi do not only protrude, but they also seem to build long tubes lying flat on the surface [30].

Moreover, well-differentiated epithelial cells show a central cilium on their apical side. This prominent structural feature carries many transmembrane proteins and signaling molecules [101]. Since bending of this cilium activates intracellular messenger cascades, it has been interpreted as a sensor antenna, which reports via its bending state on the velocity of urine flow in the nephron. Its loss is associated with a severe illness, polycystic kidney disease [53, 76]. This prominent structure is known primarily from ex vivo preparations for electron microscopy, but could recently be imaged by atomic force microscopy in a cell-culture model of inner medulla epithelial cells from rat collecting duct [9, 57] (Fig. 18.3).

Another functionally important element is the cell–cell junctions. Epithelial junctions vary in terms of their tightness. Apart from measuring transepithelial electrical resistance or solute permeabilities, it might be useful to align morphological data to these functional characteristics. In contrast to the sketches in textbooks on cell biology, AFM contact imaging clearly demonstrates the cell-to-cell border to

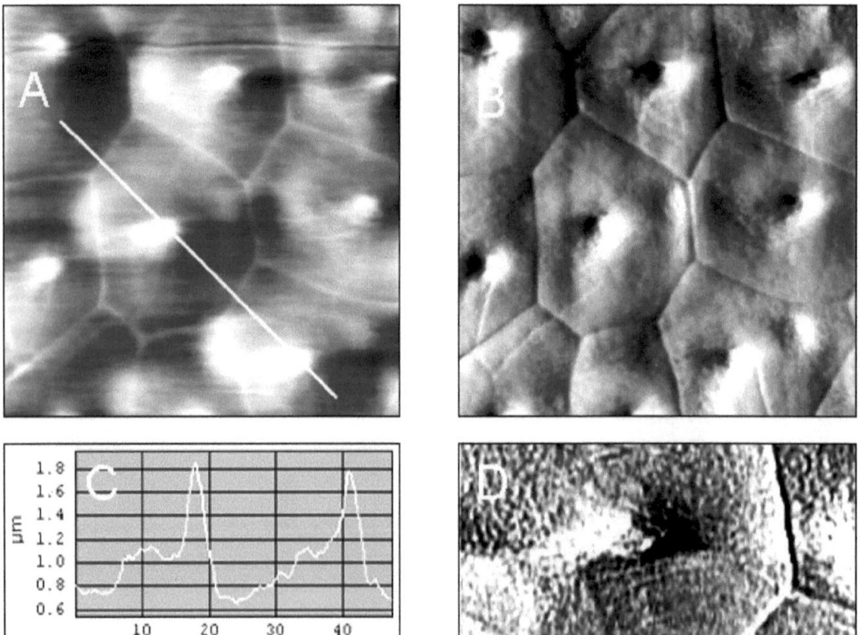

Fig. 18.3. Living epithelial cells. Cells isolated from rat collecting duct were imaged in physiological buffer at 37 °C. **a** In a 50-µm-height image, central cilia show up in the middle of each cell body. **b** The corresponding deflection image better depicts the hexagonal "lattice" of this well differentiated cell layer. **c** From a section along the *white line* in **a** the height of the central cilia can be taken (1 µm). **d** The application of an image-filtering routine can resolve the microvillar pattern even in living cells

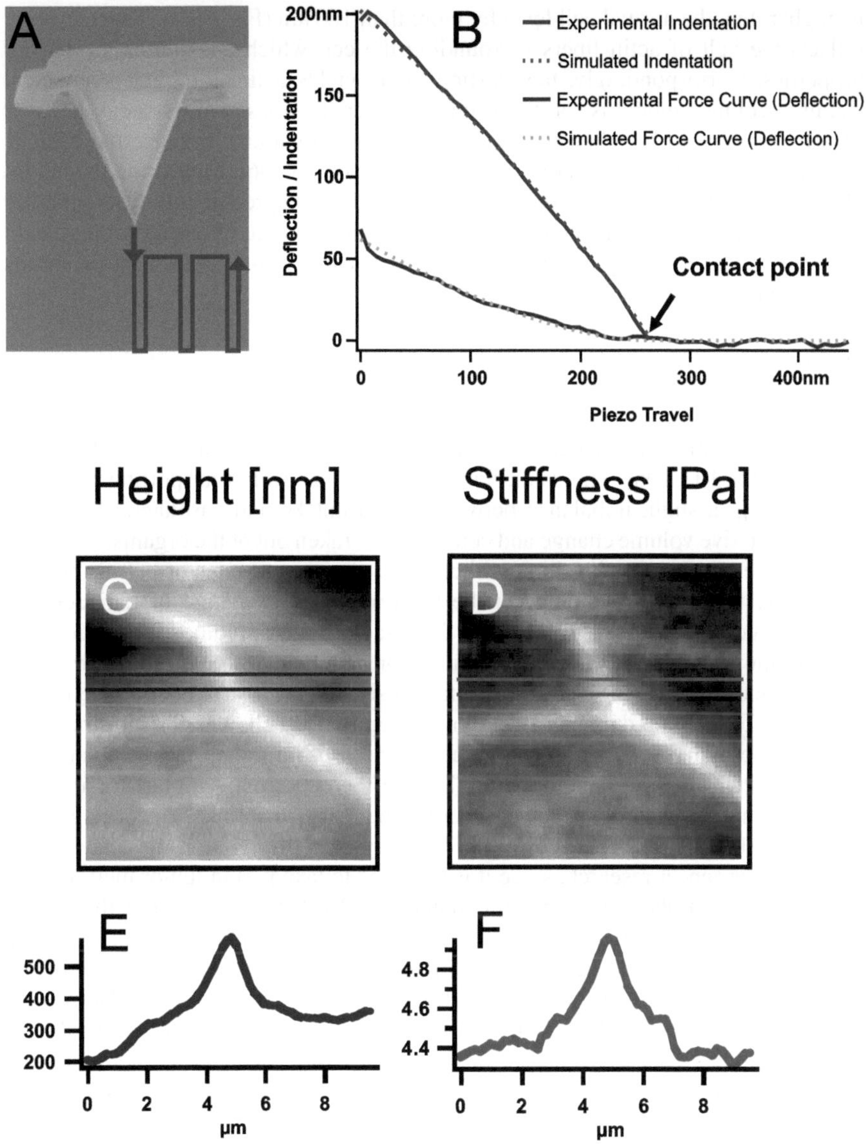

Fig. 18.4. Stiffness map of living epithelial cells. **a** In the so-called force volume mode, the tip is run like the needle in a sewing machine, thus recording an array of force–distance curves **b**. From the raw data, maps of **c** true height and **d** stiffness can be obtained. The 10-μm maps on a border area of four IMCD cells show analogously elevated values (represented by *light gray*) for height and stiffness along the tight junctions. **e** ,**f** From sections along the *horizontal lines* in **c** and **d**, the values can be deduced to resemble **e** 150-nm elevation and **f** a factor of 3 in stiffness (log scale)

be higher than the normal cell level by more than 100 nm (Fig. 18.4). This elevation reflects the belt of actin fibers surrounding the cell which constitutes the barrier properties. Correspondingly, the elasticity map yields 3 times higher mechanical rigidity (Young's modulus) on these junctional areas. These values speak in favor of primary cells to be taken for physiological investigations, because the MDCK cell line does not form comparably prominent junctions [96]. Measuring the height elevation and the stiffness increase might indicate the degree of cell differentiation and thus could replace an electrical or functional assessment of barrier properties in cases when cells have to grow on optically impermeable substrates, e. g., biomaterials for implants.

18.2.2.2 Volume

Accurate regulation of its volume is very important for a cell to survive [13,25,30,37, 45,66,71,93,102,106]. This is difficult to fulfill by actively transporting epithelia, because only a slight imbalance between apical uptake and basolateral loss rates leads to massive volume change and cell damage. Taken out of the organism context, cell volume changes in vitro hint at alterations of transport rates or permeabilities. Measuring the cellular volume can be performed via exchange of radioactively labeled water, for instance. Apart from the disadvantage of radioactive hazard, this method allows only for end-point determinations, without any possibility to monitor volume changes in living cells. Advantageously, the AFM has been made use of in a study on volume dynamics in migrating epithelial cells [72,94]. It turned out that volume dynamics are aligned to the direction of migration in a population of MDCK-F cells. At the lamellipodium, where the cell front searches for its way, a volume increase can be detected, whereas the rear end shrinks. Accordingly, the stiffness exhibits a gradient too [48]. The study of cell migration may have some relevance for cancer research, since it is regarded to take part in tumor metastasis.

An important player in terms of cell and blood volume regulation is the steroid hormone aldosterone. Being secreted by the suprarenal glands, it increases blood volume and pressure. The mechanism involves cell swelling via activation of the epithelial sodium channel, which is highly expressed in epithelia [74]. Narrowing of the nephron lumen with concomitant increase of flow resistance is among the responsible effects discussed, because this would lead to subsequent counterregulation via angiotensin [54,55,85]. This mechanism would complete the "vicious cycle" of raising blood pressure. Endothelial aspects of this hormone action are discussed later.

18.2.2.3 Invaginations

Tightly connected with the volume regulation is a cell's ability to take up or secrete fluid. Secretion is the main physiological function in glandular epithelia that produce saliva, sweat or pancreatic enzymes. Secretion of enzymes is usually performed via an exocytotic process. Vesicles store the protein product of the cell and upon appropriate signals fuse with the plasma membrane. The secretion process has been addressed by various AFM studies on pancreatic cells [44,92].

A process closely related to secretion is the translocation of membrane proteins which are stored in the vesicle membrane and are inserted into the plasma membrane upon the respective stimulus. The process itself is more or less the same, but the focus is on the membrane protein instead of the vesicle containment. A well-described model of membrane insertion is the translocation of aquaporin-2 to the apical side of collecting duct cells. When cells which have been stimulated with arginine vaso-

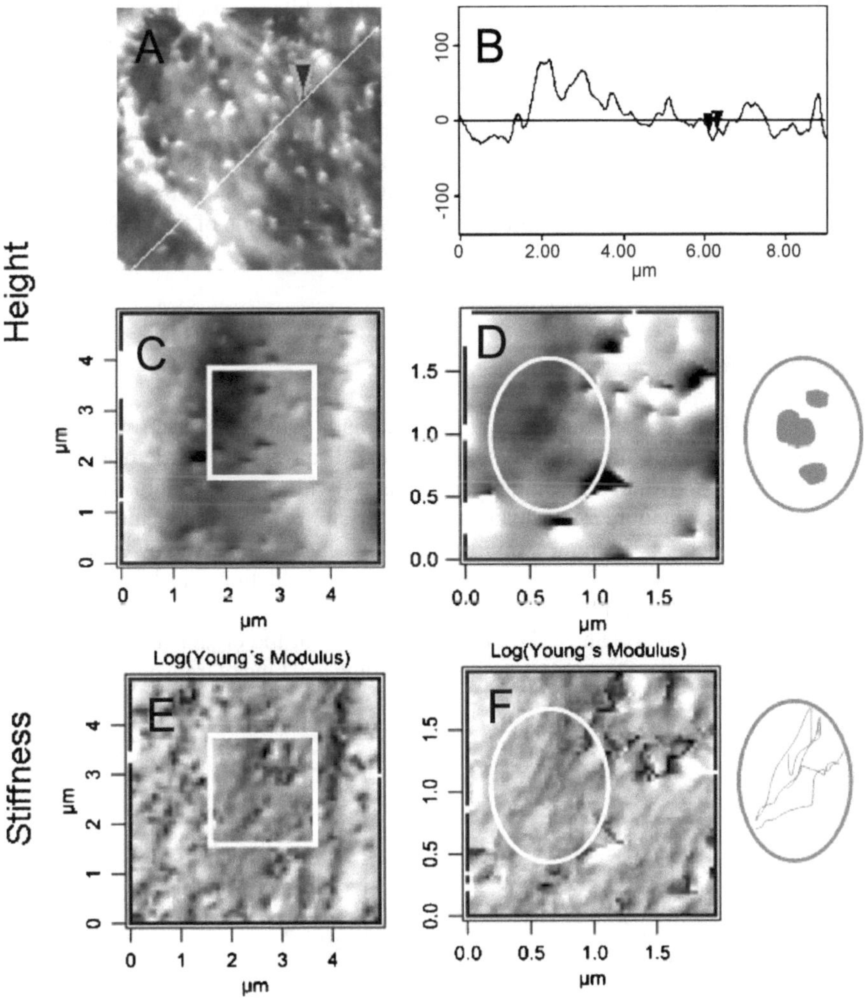

Fig. 18.5. Stiffness map of fixed epithelial cells. IMCD cells were stimulated with arginine vasopressin (AVP), fixed with glutardialdehyde and imaged in contact mode (**A**). **B** From a section along the *white line* in **A**, shallow circular depressions of the plasma membrane can be found to measure around 200 nm in diameter. **C, E** A 10-µm force volume map was recorded for this sample, showing deviants from mean values in **C** height and **E** stiffness mainly at the microvilli. **D, F** Performing a 2-µm force volume exhibits **D** shallow (5–10-nm) depressions between microvilli, which **F** do not correspond to soft spots

pressin (AVP) are fixed with glutardialdehyde, some of the vesicles should be caught in the process of fusion. Dimples of the expected diameter (200 nm) could be found in such a preparation, which might represent such cases (Fig. 18.5).

From a geometric point of view, any intense interaction of a vesicle with the plasma membrane, be fusion or fission, must produce alterations of the actual cell height during the process. The AFM is ideally suited to monitor height changes with nanometer accuracy on living cells. Such height changes have been reported for bacteria as well as for single proteins [39, 70]. While yeast cells seem to prefer a certain frequency in the kilohertz region [78], this has not been reported for other cells. We have tested the inner medulla collecting duct (IMCD) cell model, which was proven to translocate vesicles upon a stimulation with AVP. Indeed, the height changes determined on a single spot as a function of time gave less than 1 nm mean square z-displacement (R_{ms}) for the control and 1.3 nm after addition of AVP. It remains to be clarified how much actomyosin interaction contributes to the effect. However, no possible result would contradict an interpretation of the z-vibrations as a sign of membrane traffic.

18.3 Endothelium

The endothelium is a specialized sort of epithelium, which is inwardly bound and does not have any contact with the environment. It covers the inside of blood and lymph vessels and shares the main characteristics with the epithelium. Apart from strictly forming not more than one layer, it has the ability to communicate in a highly sensitive manner. One extreme example in a long row may be given by the diatomic radical of nitric oxide, which mediates blood vessel dilation via diffusion in the nanomolar concentration range.

Endothelial cells organize the internal workflow of the organism on a humoral (non-neuronal) basis. They are central in controlling any kind of transport from blood

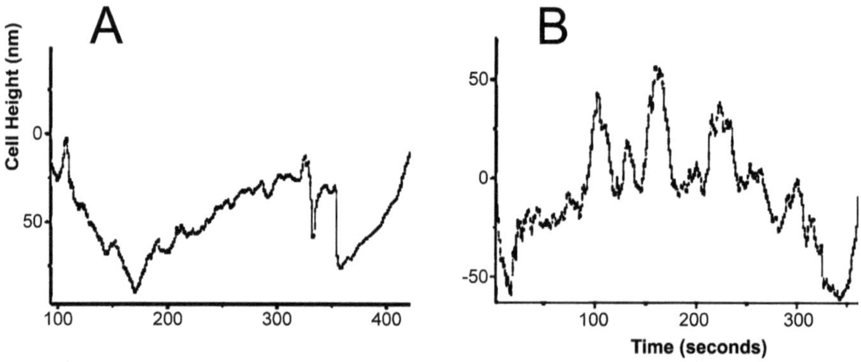

Fig. 18.6. Height fluctuations of living epithelial cells. The height of living IMCD cells at one spot was monitored in physiological buffer at 37 °C **A** before and **B** after stimulation with AVP. AVP elevates intracellular cyclic AMP and leads to membrane translocation of aquaporin-containing vesicles. The height fluctuations correlate to the intracellular activities

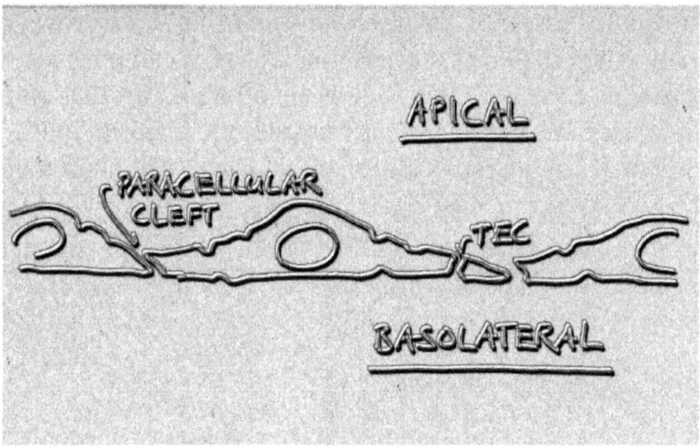

Fig. 18.7. Endothelial cells. Analogous to epithelial cells, endothelial cells are asymmetrically constructed. The apical side is oriented to the bloodstream and the basolateral side to the interstitial space (tissue). Most of the fluid that crosses the cell layer passes through the paracellular clefts between the cells, but in special cases also transendothelial channels (*TEC*) form a shortcut between the compartments

to intercellular space and vice versa. They regulate the bloodstream and the healing of wounds and they modulate the immune response. Endothelial dysfunction leads to disorders as severe as bleeding, dementia, hypertension and arteriosclerosis—most of which develop slowly, but tend to be irreversible and lethal. Therefore, improved knowledge of endothelial cell biology deserves the interest of the physician.

The first studies on endothelium date from a little later than those on on epithelium: images of fixed cells were reported in 1995, when a comparison with electron microscopy was performed [10, 11]. (Atomic force microscopy has been applied to endothelial cells of vascular origin only; lymphatic endothelial cells are virtually nonexistent in cell-culture laboratories.) Not only cell morphology has been addressed, but also cytoskeletal mechanics [58, 62, 90]. Moreover, atomic force microscopy has demonstrated a shape-dependence of cells from the biological coating of the culture support. This effect is caused by different degrees of adhesion forces between the cell and the support [8].

18.3.1 Paracellular Gaps

One of the permanent tasks of endothelia is to let pass a suitable amount of liquid and solutes from the blood plasma to the diverse tissues and backwards. The endothelium in different regions is morphologically adapted to the requirements of the respective tissues. Its shape ranges from a discontinuous appearance in liver sinusoids, which is highly permeable, to a tight layer in the brain with characteristics close to those of epithelium. Rather than transporting small inorganic ions via membrane transporters through apical and basolateral membranes like epithelia do, the typical passage across endothelial layers is a rather the unspecific flow of fluid between the cells, i.e., paracellularly. An imbalance in regulating fluid balance may lead to tissue swelling (edema), as happens in the course of an inflammation. Transport of fluid is experimentally assessed by physiologists using cells grown on permeable filter membranes. When the layer reaches confluence, a cargo is added to the apical side and its appearance on the other side is monitored. Not that atomic force microscopy could replace this measurement, but since the majority of the passage takes a paracellular route, it can inspect cell-to-cell contacts. Although a direct access to

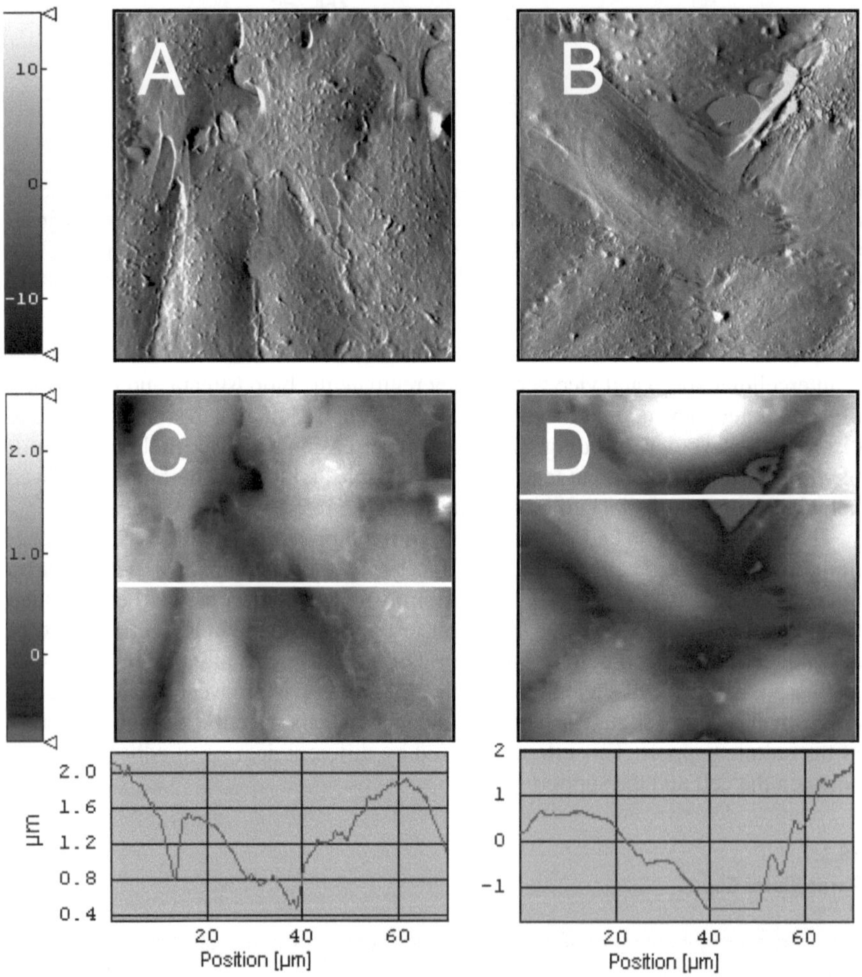

Fig. 18.8. Paracellular gap formation. Endothelial cells of human umbilical vein are isolated, fixed with glutardialdehyde and imaged in contact mode (70 µm). **A, C** Deflection and height images of a control sample. **B, D** Deflection and height images of a sample having been treated with ionomycin (1 µM) 5 min before fixation. Ionomycin elevates intracellular calcium concentration and induces cell contraction. This leads to the formation of paracellular gaps (*red area* in **D**), that perforate the layer. From the sections *below C and D*, it can be seen that the bottom of the culture dish is exposed in **D** the stimulated sample (straight horizontal part of the line between 40 and 50 µm), as opposed to **C** the control

the intercellular cleft (normally only few nanometers wide) is not possible owing to steric hindrance, the formation of paracellular gaps can be observed by the AFM. An interplay of two counteracting factors balances the opening of junctions. On the one hand, transmembrane proteins interact with their counterparts in the neighboring cells to define a certain adherent force. On the other hand, intracellular contractile force is generated from actomyosin interaction. Increasing tension as a consequence of elevated intracellular calcium concentrations leads to a local detachment of cells at sites of weakest adherence. Eventually, these "pores"—when big enough—allow the AFM tip to access the culture dish (Fig. 18.8). This state of cell activation reflects pathophysiological conditions in vivo. The paracellular gaps demonstrate a clear contribution of intracellular tension in the induction of endothelial hyperpermeability on a time scale of minutes, which is definitely too short for downregulation of transmembrane proteins like cadherins [5,98,105]. These proteins have been shown to regulate the barrier properties of endothelial cells over long range.

It is of considerable interest to the biologist that filter membranes can be subjected to AFM imaging as seen from Fig. 18.8. There is not a big difference in appearance compared with cells grown on glass, but the development of internal signaling pathways can be expected to better represent the in vivo situation. The various degree of barrier function can be inspected from cells of different origin. The most intimate cell—cell contacts are built up by endothelial cells in brain capillaries (blood–brain barrier). A morphological correlate to this function is called "marginal folds", which are cell lobes that overlap neighboring cells and are easily measured by the AFM technique as heightened lines [97, 107].

18.3.2 Cellular Drinking

In special cases of proinflammatory activation, a contribution of transcellular transport could be indirectly shown [86]. The study compared the diffusion of a fluid marker and the paracellular electrical resistance. Upon the proinflammatory stimulus bradykinin, which is known to cause edema in vivo, both detection channels should alter in parallel, but they surprisingly pointed in different directions, i.e., closing of the paracellular gaps coincided with increased dextran diffusion. This paradox could only be solved by assuming transcellular transport of fluid [86].

Transcellular transport of fluid was hypothesized earlier, not at least owing to diverging results between in vitro and in vivo experiments. Transendothelial channels have been found which resemble a tunnel made out of plasma membrane spanning from the apical side directly to the basolateral side [59,60]. These have been reported to increase in number as determined by transmission electron microscopy, when intracellular calcium concentration is raised [64]. However, the mouth of such a channel has not been observed by atomic force microscopy yet, although observation should be possible from a geometric point of view. One reason might be that it is inaccessible to the AFM tip owing to coverage by the glycocalix. The term describes a layer of long, branched and hydrophilic proteoglycans anchored at membrane proteins. Proteoglycans can be imagined as a kind of microscopic seaweed that dampens turbulence in plasma flow and mediates the adhesion of immune cells. Their role in AFM image convolution has not been clarified yet.

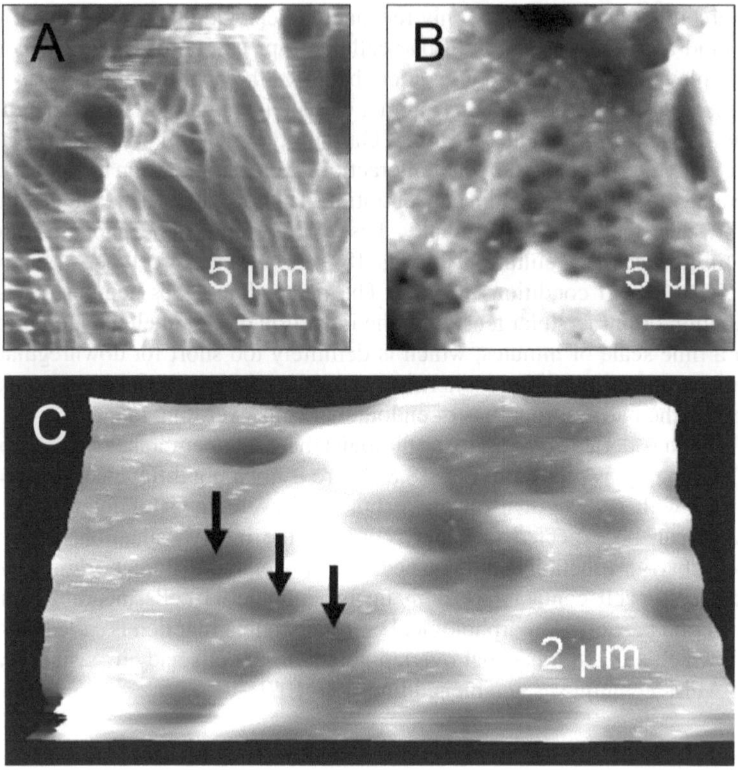

Fig. 18.9. Comparison of living and fixed endothelial cells. Human endothelial cells (HUVEC) were imaged in contact mode either **A** before or **B** after fixation with glutardialdehyde. **A** In living cells, the cytoskeletal fibers are predominant. **B** After fixation, they almost disappear, generating dimples as a new structural feature. **C** Three-dimensional view with three of the characteristic depressions marked by *arrows*

An unspecific uptake of fluid by cells is termed pinocytosis ("cellular drinking"). From electron-microscopic studies it is known that their size ranges from 200 to more than 1000 nm. The lack of specificity does not only concern the cargo, but also reflects the lack of specific markers for these organelles. Fluorescence microscopy can easily identify those, that were internalized while being exposed to a solution containing a fluorescently labeled and biologically inert compound like fluorescein isothiocyanate–dextran [42]. But the fraction of pinocytotic organelles that were formed before exposition to fluid marker remain dark. Atomic force microscopy can add a completely different approach to this question. The technical "trick" is that the cross-linking of intracellular proteins elevates the cellular stiffness by roughly an order of magnitude [88]. This leads to an embedding of the actin fibers that have dominated the picture of living cells. These prominent structures are mechanically buried in the polymeric biocement made from glutardialdehyde and protein-bound amino groups. But organelles devoid of (cytosolic) proteins like pinosomes or vacuoles remain as soft as they were before; hence, in an elasticity map they appear as soft spots like bubbles of air in a pudding [87].

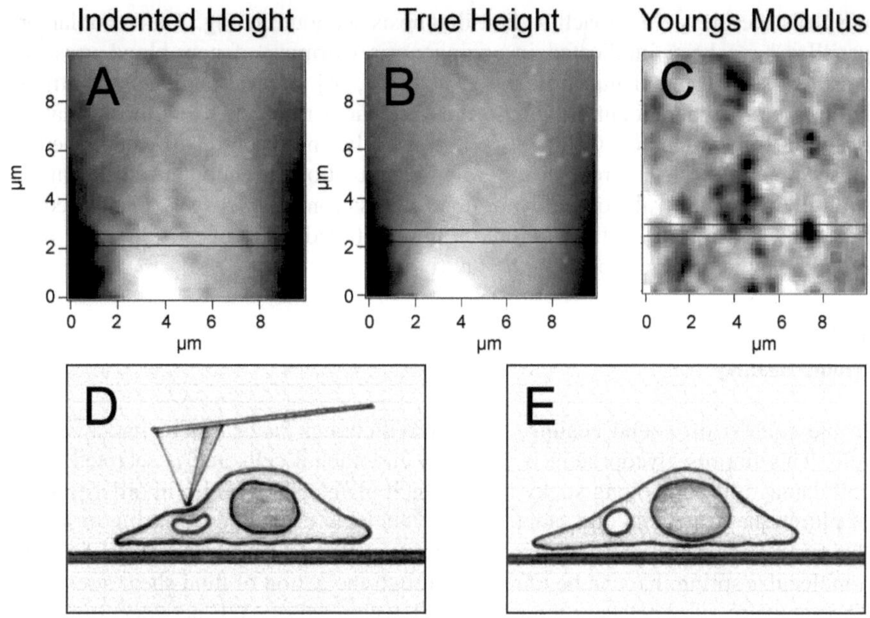

Fig. 18.10. Stiffness map of fixed endothelial cells. Isolated human endothelial cells were fixed with glutardialdehyde and subjected to force volume recording. By calculation from the raw data, maps of **A** indented height, **B** true height and **C** Young's modulus are obtained. The indented height map is reconstructed from all points at a force load of 0.5 nN and the true height is at zero force, as illustrated by the sketches in **D** and **E**, respectively. The stiffness map (**C**) shows a colocalization of soft spots with the depressions of **A**. Consistently, they disappear (**B**) concomitant to vanishing force load

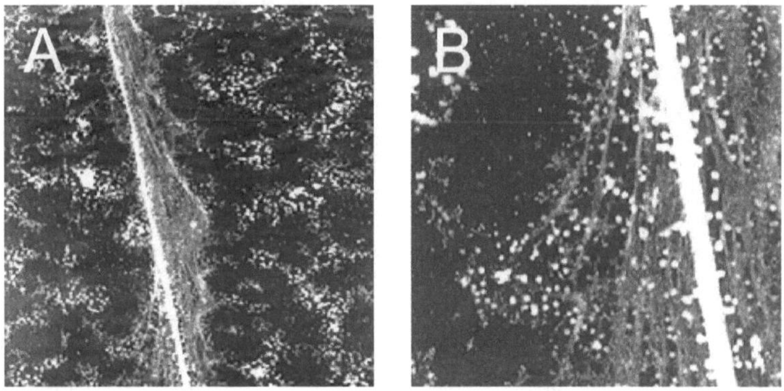

Fig. 18.11. von Willebrandt fibers. Purified glycoprotein of von Willebrandt factor are deposited on mica, dried and imaged in contact mode under air. **A** Long fibers can be found in samples of this adherent protein. **B** A zoom reveals a treelike structure

Mapping the elasticity overcomes the limitation of the AFM to the surface of a sample and enables intracellular organelles to be detected via their lack of mechanical support. It compares to the palpation of intraabdominal organs by a physician, which is the classic approach to first diagnosis without cutting. The vacuolar organelles have been implicated in angiogenesis (forming of new blood vessels), a critical step in tumor nutrition and growth [18, 26]. Typical angiogenesis stimulating factors concomitantly increase the vascular permeability and the formation of intracellular vacuoles [6, 7, 16]. Up to now, it has not been investigated,whether these organelles have a role in the volume regulation of endothelial cells in response to the mineralocorticoid hormone aldosterone, which was mentioned in Sect. 18.2.2.2 [67, 72, 73]. Therefore, a more detailed study of vacuolization may offer some therapeutic aspects in the future.

18.3.3 Wound Healing

In the context of wound healing, the von Willebrand factor (vWF) has a central role. This fibrous glycoprotein is stored in endothelial cells and is secreted upon cell damage. It builds long sticky fibers which platelets adhere to, in order to kind of plumb the lesion site of a blood vessel. Technical difficulties in trying to image the purified protein by atomic force microscopy led to the notion of vWF being a molecular spring that can be unrolled through the action of fluid shear stress [2]. The activation of platelets as the next step downstream of vWF-network formation has also been imaged by atomic force microscopy [27, 84].Owing to the small size of around 1 μm, classic assays for determining the aggregation state have to rely on

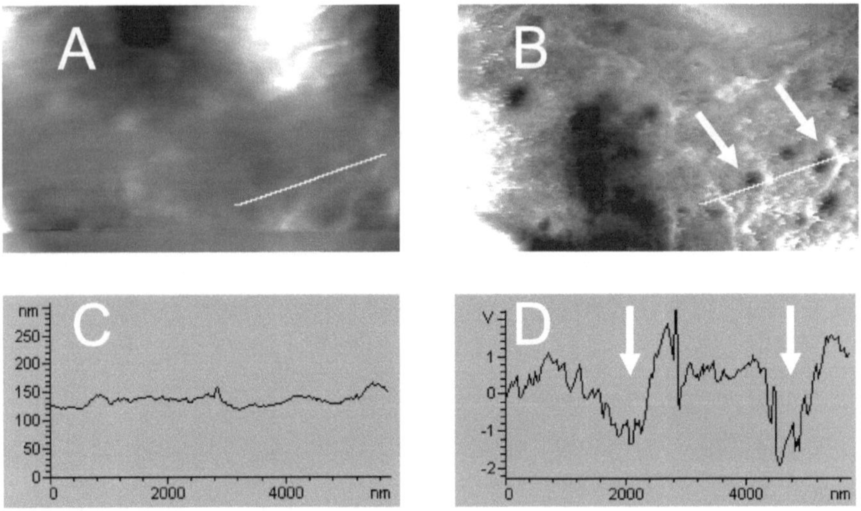

Fig. 18.12. TREC™ imaging of living endothelial cells. TREC™ mode has been developed by P. Hinterdorfer, University of Linz, Austria. A cantilever is functionalized with an antibody (here: anti-β1 integrin) and with use of a modified AFM controller, the simultaneous recording of **A** topography and **B** molecular recognition is possible. Integrins are membrane proteins that are important for cell adhesion. The recognition events are *black spots* marked by *arrows*. From **C** and **D**, it can be seen that there are **C** no depressions of the surface height that correlate to **D** the recognition events

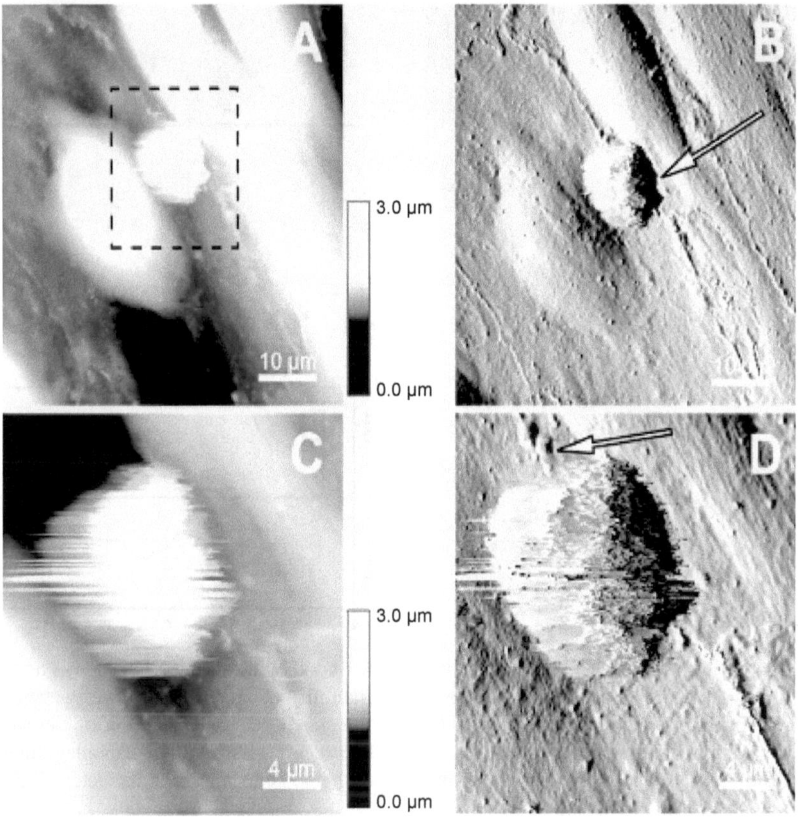

Fig. 18.13. Immune cells on endothelium. Endothelial cells were freshly isolated from porcine brain and treated with a proinflammatory stimulus. Isolated polymorphonuclear neutrophiles (PMN), a sort of leukocyte, are added and the sample is fixed. **a** Height and **b** deflection images of one PMN (*arrow*) in close vicinity of an endothelial cell nucleus. The zoom (**c**, **d**) reveals the interendothelial cell border (*arrow*) lying underneath the PMN. (The image was gratefully obtained from H.-J. Galla, University of Münster, Germany; S. Schrot, K. Psathaki and H.-J. Galla, unpublished results)

macroscopic turbidity measurement via light scattering methods which need a high number of platelets; to get an image of single platelets, electron microscopy was necessary. Nowadays, atomic force microscopy has taken advantage of working in buffer solution to investigate a single platelet's activation state via its morphology [84].

18.3.4 Transmigration of Leukocytes

The concept of transcellular transport in a proinflammatory context has very recently been shown to apply even to intact cells of the immune system. For some decades the notion of paracellular pathways exclusively dominated the notion of leukocyte transmigration. The term describes the puzzling fact that blood-born white immune

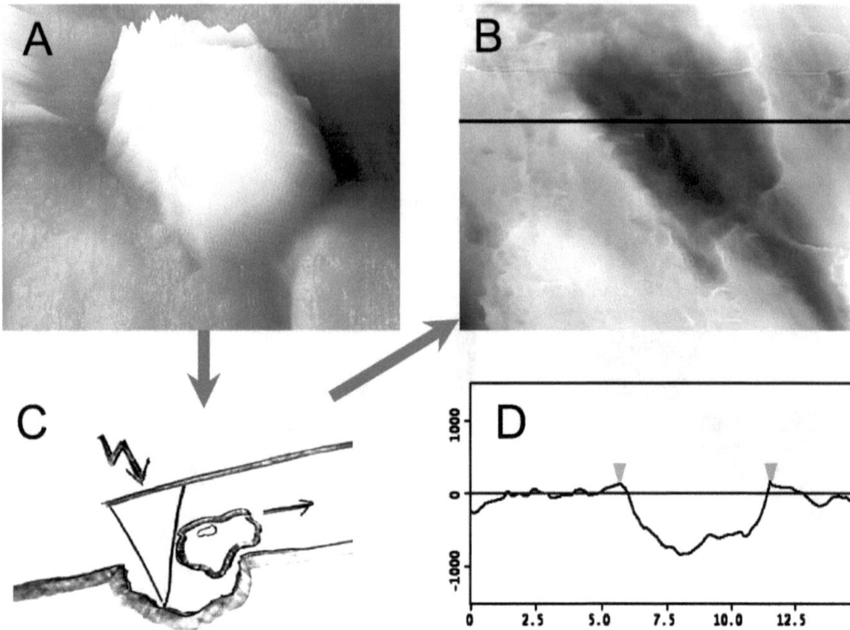

Fig. 18.14. Footprint of a leukocyte in endothelium. Endothelial cell line bEND5 was added to freshly isolated PMNs and fixed (samples were gratefully obtained from D. Vestweber, Max Planck Institute, Münster, Germany). **A** Contact image. **B** Nanomanipulation of the sample by application of high force. **C** Image of the same region as in **A** after nanomanipulation; a transmigratory cup remains imprinted in the endothelial cell. **D** It has a diameter of around 10 μm and a depth of 800 nm

cells of 10 μm in size get across the endothelial barrier into the tissues, while the plasma fluid is mainly kept in the blood vessel. This paradox has been solved by the (experimentally proven) assumption that leukocytes squeeze through the zipper of interendothelial clefts. But recently, it was shown by video microscopy that leukocytes can even invade and cross endothelial cells without even touching a cell border [110]. In the course of this transmigration, the endothelial cell forms a complementary entry channel, named a "transmigratory cup" [12]. Atomic force microscopy is not only able to image and characterize leukocytes adhering to endothelium in both fixed and unfixed states, but also can be used as a nanomanipulator to scratch off the adhering leukocyte, uncovering the transmigratory cup lying underneath (Fig. 18.12). Atomic force microscopy uniquely offers the possibility to image and manipulate cells. The high-resolution capabilities render it a valuable tool for endothelial cell biology and physiology.

18.4 Technical Remarks

The data shown here were acquired under physiological buffer conditions (except those in Fig. 18.11) using scanning microscopes built by Digital Instruments

(Santa Barbara, USA). Mostly, standard contact imaging in the low-nanometer range was applied using DNP or MLCT cantilever substrates (purchased through Veeco, Mannheim, Germany). The force–volume recordings were performed in 64 × 64 matrices with the softest cantilevers available (0.01 N/m) at very low indentation (100–200 nm) and piezo travel (less than 3 μm/s). The data were processed by a software routine for IGOR Pro™ based on the Hertz model for elasticity [83].

18.5 Summary

Atomic force microscopy throws light from a different angle onto the fascinating scenery of cells whose physiological architecture is nearly two-dimensional. Regulatory cycles in physiological contexts can uniquely be addressed by this technique on living specimens with unprecedented resolution. Nanometer movements of cells can be monitored to get insight into their intracellular machinery. The future will see the AFM providing the basis of physiological assays for pharmacological testing on single cells.

Acknowledgements. We would like to thank Rainer Matzke (Munich), Philippe Carl (Mannheim) and Tilman Schaeffer (Münster) for advice with the force curve analysis. Primary cultures were gratefully obtained from Kristina Kusche (human nasal airway), Bayram Edemir (rat collecting duct) and Joachim Wegener (porcine brain endothelial cells) (all University of Münster).

References

1. Alcaraz J, Buscemi L, Grabulosa M, Trepat X, Fabry B, Farre R, Navajas D (2003) Biophys J 84:2071
2. Alexander-Katz A, Schneider MF, Schneider SW, Wixforth A, Netz RR (2006) Phys Rev Lett 97:138101
3. Allen S, Rigby-Singleton SM, Harris H, Davies MC, O'Shea P (2003) Biochem Soc Trans 31:1052
4. Almofti MR, Harashima H, Shinohara Y, Almofti A, Baba Y, Kiwada H (2003) Arch Biochem Biophys 410:246
5. Baumgartner W, Hinterdorfer P, Ness W, Raab A, Vestweber D, Schindler H, Drenckhahn D (2000) Proc Natl Acad Sci USA 97:4005
6. Bayless KJ, Davis GE (2002) J Cell Sci 115:1123
7. Berlin RD, Oliver JM (1980) J Cell Biol 85:660
8. Bhadriraju K, Hansen LK (2002) Exp Cell Res 278:92
9. Birnbaumer M, Seibold A, Gilbert S, Ishido M, Barberis C, Antaramian A, Brabet P, Rosenthal W (1992) Nature 357:333
10. Braet F, de Zanger R, Kalle W, Raap A, Tanke H, Wisse E (1996) Scanning Microsc Suppl 10:225
11. Braet F, de Zanger R, Seynaeve C, Baekeland M, Wisse E (2001) J Electron Microsc (Tokyo) 50:283
12. Carman CV, Springer TA (2004) J Cell Biol 167:377
13. Charles CJ, Nicholls MG, Rademaker MT, Richards AM (2002) J Endocrinol 173:123
14. Costa KD (2003) Dis Markers 19:139

15. Costa KD (2006) Methods Mol Biol 319:331
16. Davis GE, Bayless KJ (2003) Microcirculation 10:27
17. Dufrene YF (2003) Curr Opin Microbiol 6:317
18. Dvorak AM, Feng D (2001) J Histochem Cytochem 49:419
19. Dvorak JA (2003) Methods 29:86
20. Dvorak JA, Nagao E (1998) Exp Cell Res 242:69
21. Ebnet K, Suzuki A, Ohno S, Vestweber D (2004) J Cell Sci 117:19
22. Edwardson JM, Henderson RM (2004) Drug Discov Today 9:64
23. Ehrenhofer U, Rakowska A, Schneider SW, Schwab A, Oberleithner H (1997) Cell Biol Int 21:737
24. Engel A, Gaub HE, Muller DJ (1999) Curr Biol 9:R133
25. Evans EA (1989) Methods Enzymol 173:3
26. Feng D, Nagy JA, Pyne K, Hammel I, Dvorak HF, Dvorak AM (1999) Microcirculation 6:23
27. Fritz M, Radmacher M, Gaub HE (1993) Exp Cell Res 205:187
28. Fritz J, Katopodis AG, Kolbinger F, Anselmetti D (1998) Proc Natl Acad Sci USA 95:12283
29. Gorelik J, Zhang Y, Shevchuk AI, Frolenkov GI, Sanchez D, Lab MJ, Vodyanoy I, Edwards CR, Klenerman D, Korchev YE (2004) Mol Cell Endocrinol 217:101
30. Gorelik J, Zhang Y, Sanchez D, Shevchuk A, Frolenkov G, Lab M, Klenerman D, Edwards C, Korchev Y (2005) Proc Natl Acad Sci USA 102:15000
31. Graham JS, Vomund AN, Phillips CL, Grandbois M (2004) Exp Cell Res 299:335
32. Hansma PK (2006) Structure 14:390
33. Hansma PK, Drake B, Marti O, Gould SA, Prater CB (1989) Science 243:641
34. Henderson RM (2003) Methods Mol Med 86:139
35. Henderson RM, Oberleithner H (2000) Am J Physiol Renal Physiol 278:F689
36. Henderson RM, Edwardson JM, Geisse NA, Saslowsky DE (2004) News Physiol Sci 19:39
37. Hillebrand U, Hausberg M, Stock C, Shahin V, Nikova D, Riethmuller C, Kliche K, Ludwig T, Schillers H, Schneider SW, Oberleithner H (2006) Cardiovasc Res 69:916
38. Hinterdorfer P, Dufrene YF (2006) Nat Methods 3:347
39. Hoh JH (1998) Proteins 32:223
40. Hoh JH, Schoenenberger CA (1994) J Cell Sci 107(Pt 5):1105
41. Horton M, Charras G, Lehenkari P (2002) J Recept Signal Transduct Res 22:169
42. Iwasa Y, Hirono C, Sugita M, Takemoto K, Shiba Y (2001) Cell Physiol Biochem 11:311
43. Janicijevic A, Ristic D, Wyman C (2003) J Microsc 212:264
44. Jena BP (2006) Ultramicroscopy 106:663
45. Jiao Y, Schaffer TE (2004) Langmuir 20:10038
46. Kienberger F, Stroh C, Kada G, Moser R, Baumgartner W, Pastushenko V, Rankl C, Schmidt U, Muller H, Orlova E, LeGrimellec C, Drenckhahn D, Blaas D, Hinterdorfer P (2003) Ultramicroscopy 97:229
47. Korchev YE, Gorelik J, Lab MJ, Sviderskaya EV, Johnston CL, Coombes CR, Vodyanoy I, Edwards CR (2000) Biophys J 78:451
48. Laurent VM, Kasas S, Yersin A, Schaffer TE, Catsicas S, Dietler G, Verkhovsky AB, Meister JJ (2005) Biophys J 89:667
49. Lawrence JC, Saslowsky DE, Edwardson JM, Henderson RM (2003) Biophys J 84:1827
50. Le GC, Lesniewska E, Cachia C, Schreiber JP, de FF, Goudonnet JP (1994) Biophys J 67:36
51. Lehto T, Miaczynska M, Zerial M, Muller DJ, Severin F (2003) FEBS Lett 551:25
52. Lesniewska E, Giocondi MC, Vie V, Finot E, Goudonnet JP, Le GC (1998) Kidney Int Suppl 65:S42
53. Liu W, Murcia NS, Duan Y, Weinbaum S, Yoder BK, Schwiebert E, Satlin LM (2005) Am J Physiol Renal Physiol 289:F978

54. Mackie FE, Campbell DJ, Meyer TW (2001) Kidney Int 59:1458
55. Magaldi AJ, Cesar KR, de Araujo M, Simoes e Silva AC, Santos RA (2003) Pflugers Arch 447:223
56. Malcharek S, Hinz A, Hilterhaus L, Galla HJ (2005) Biophys J 88:2638
57. Maric K, Oksche A, Rosenthal W (1998) Am J Physiol 275:F796
58. Mathur AB, Truskey GA, Reichert WM (2000) Biophys J 78:1725
59. Michel CC, Curry FE (1999) Physiol Rev 79:703
60. Michel CC, Neal CR (1999) Microcirculation 6:45
61. Milhiet PE, Giocondi MC, Baghdadi O, Ronzon F, Roux B, Le Grimellec C (2002) EMBO Rep 3:485
62. Miyazaki H, Hayashi K (1999) Med Biol Eng Comput 37:530
63. Muller DJ, Heymann JB, Oesterhelt F, Moller C, Gaub H, Buldt G, Engel A (2000) Biochim Biophys Acta 1460:27
64. Neal CR, Michel CC (1995) J Physiol 488(Pt 2):427
65. Neher E, Sakmann B, Steinbach JH (1978) Pflugers Arch 375:219
66. Nilius B, Prenen J, Voets T, Eggermont J, Droogmans G (1998) J Physiol 506(Pt 2):353
67. Oberleithner H (2007) Pflugers Arch 454:187
68. Oberleithner H, Schwab A, Wang W, Giebisch G, Hume F, Geibel J (1994) Nephron 66:8
69. Oberleithner H, Schneider S, Larmer J, Henderson RM (1996) Kidney Blood Press Res 19:142
70. Oberleithner H, Schneider SW, Henderson RM (1997) Proc Natl Acad Sci USA 94:14144
71. Oberleithner H, Reinhardt J, Schillers H, Pagel P, Schneider SW (2000) Cell Physiol Biochem 10:429
72. Oberleithner H, Schneider SW, Albermann L, Hillebrand U, Ludwig T, Riethmuller C, Shahin V, Schafer C, Schillers H (2003) J Membr Biol 196:163
73. Oberleithner H, Ludwig T, Riethmuller C, Hillebrand U, Albermann L, Schafer C, Shahin V, Schillers H (2004) Hypertension 43:952
74. Oberleithner H, Riethmuller C, Ludwig T, Shahin V, Stock C, Schwab A, Hausberg M, Kusche K, Schillers H (2006) J Cell Sci 119:1926
75. Ohnesorge FM, Horber JK, Haberle W, Czerny CP, Smith DP, Binnig G (1997) Biophys J 73:2183
76. Olteanu D, Yoder BK, Liu W, Croyle MJ, Welty EA, Rosborough K, Wyss JM, Bell PD, Guay-Woodford LM, Bevensee MO, Satlin LM, Schwiebert EM (2006) Am J Physiol Cell Physiol 290:C952
77. Osada T, Takezawa S, Itoh A, Arakawa H, Ichikawa M, Ikai A (1999) Chem Senses 24:1
78. Pelling AE, Sehati S, Gralla EB, Valentine JS, Gimzewski JK (2004) Science 305:1147
79. Pereira RS (2001) Biochem Pharmacol 62:975
80. Puech PH, Poole K, Knebel D, Muller DJ (2006) Ultramicroscopy 106:637
81. Puntheeranurak T, Stroh C, Zhu R, Angsuthanasombat C, Hinterdorfer P (2005) Ultramicroscopy
82. Rabinovich Y, Esayanur M, Daosukho S, Byer K, El-Shall H, Khan S (2005) J Colloid Interface Sci 285:125
83. Radmacher M (1997) IEEE Eng Med Biol Mag 16:47
84. Radmacher M, Fritz M, Kacher CM, Cleveland JP, Hansma PK (1996) Biophys J 70:556
85. Reddy HK, Campbell SE, Janicki JS, Zhou G, Weber KT (1993) J Lab Clin Med 121:510
86. Riethmuller C, Jungmann P, Wegener J, Oberleithner H (2006) Pflugers Arch 453:157
87. Riethmuller C, Schaeffer TE, Kienberger F, Stracke W, Oberleithner H (2007) Ultramicroscopy 107:895
88. Rotsch C, Radmacher M (2000) Biophys J 78:520
89. Sakmann B, Neher E (1984) Annu Rev Physiol 46:455
90. Sato M, Nagayama K, Kataoka N, Sasaki M, Hane K (2000) J Biomech 33:127

91. Scheuring S, Levy D, Rigaud JL (2005) Biochim Biophys Acta 1712:109
92. Schneider SW, Sritharan KC, Geibel JP, Oberleithner H, Jena BP (1997) Proc Natl Acad Sci USA 94:316
93. Schneider SW, Yano Y, Sumpio BE, Jena BP, Geibel JP, Gekle M, Oberleithner H (1997) Cell Biol Int 21:759
94. Schneider SW, Pagel P, Rotsch C, Danker T, Oberleithner H, Radmacher M, Schwab A (2000) Pflugers Arch 439:297
95. Schneider SW, Matzke R, Radmacher M, Oberleithner H (2004) Methods Mol Biol 242:255
96. Schoenenberger CA, Hoh JH (1994) Biophys J 67:929
97. Schrot S, Weidenfeller C, Schaffer TE, Robenek H, Galla HJ (2005) Biophys J 89:3904
98. Seebach J, Dieterich P, Luo F, Schillers H, Vestweber D, Oberleithner H, Galla HJ, Schnittler HJ (2000) Lab Invest 80:1819
99. Shahin V (2006) Bioessays 28:935
100. Sibug-Aga R, Dunn RC (2004) Photochem Photobiol 80:471
101. Singla V, Reiter JF (2006) Science 313:629
102. Stock C, Allen RD, Naitoh Y (2001) J Exp Biol 204:291
103. Tandon R, Levental I, Huang C, Byfield FJ, Ziembicki J, Schelling JR, Bruggeman LA, Sedor JR, Janmey PA, Miller RT (2007) Am J Physiol Renal Physiol 292:F701
104. Turksen K, Troy TC (2004) J Cell Sci 117:2435
105. Vestweber D (2000) J Pathol 190:281
106. Wagner CA, Huber SM, Warntges S, Zempel G, Kaba NK, Fux R, Orth N, Busch GL, Waldegger S, Lambert I, Nilius B, Heinle H, Lang F (2000) Pflugers Arch 440:295
107. Weidenfeller C, Schrot S, Zozulya A, Galla HJ (2005) Brain Res 1053:162
108. Wolburg H, Lippoldt A (2002) Vascul Pharmacol 38:323
109. Yamane Y, Shiga H, Haga H, Kawabata K, Abe K, Ito E (2000) J Electron Microsc (Tokyo) 49:463
110. Yang L, Froio RM, Sciuto TE, Dvorak AM, Alon R, Luscinskas FW (2005) Blood 106:584
111. Zhang X, Wojcikiewicz EP, Moy VT (2006) Exp Biol Med (Maywood) 231:1306

24 Nanotribological Characterization of Human Hair and Skin Using Atomic Force Microscopy (AFM)

Bharat Bhushan · Carmen LaTorre

From volume IV

Human hair is a nanocomposite biological fiber. Maintaining the health, feel, shine, color, softness, and overall aesthetics of the hair is highly desired. Hair care products such as shampoos and conditioners, along with damaging processes such as chemical dyeing and permanent wave treatments, affect the maintenance and grooming process and are important to study because they alter many hair properties. Nanoscale characterization of the morphological, frictional, and adhesive properties (tribological properties) of hair are essential to evaluate and develop better cosmetic products, and to advance the understanding of biological and cosmetic science. The atomic/friction force microscope (AFM/FFM) has recently become an important tool for studying the micro/nanoscale properties of human hair. This chapter presents a comprehensive review of tribological properties of various hair and skin as a function of ethnicity, damage, conditioning treatment, and various environments. Nanotribological properties such as roughness, friction, adhesion, and wear are presented, as well as investigations of scale effects and directionality dependence on friction and adhesion.

24.1 Introduction

Everybody wants beautiful, healthy hair and skin. For most people, grooming and maintenance of hair and skin is a daily process. The demand for products that improve the look and feel of these surfaces has created a huge industry for hair and skin care. Beauty care technology has advanced the cleaning, protection, and restoration of desirable hair and skin properties by altering the hair surface. For many years, especially in the second half of the twentieth century, scientists have focused on the physical and chemical properties of hair to consistently develop products that alter the health, feel, shine, color, softness, and overall aesthetics of the hair. Table 24.1 displays common products and processes involved in hair care. Hair care products such as shampoos and conditioners aid the maintenance and grooming process. Mechanical processes such as combing, cutting, and blowdrying serve to style the hair. Chemical products and processes such as chemical dyes, colorants, bleaches, and permanent wave treatments enhance the appearance and hue of the hair. Of particular interest is how all these common hair care items deposit onto and change hair properties, since these properties are closely tied to product performance. The fact that companies like Procter & Gamble, L'Oreal, and Unilever have hair care product sales consistently measured in billions of dollars

In: H. Fuchs, B. Bhushan (eds), *Biosystems - Investigated by Scanning Probe Microscopy*,
DOI 10.1007/3-540-26914-2_2, © Springer-Verlag Berlin Heidelberg 2010

Table 24.1. Common hair care products/processes and their functions

Product/process	Functions
Shampoos	Clean the hair and skin of oils
Conditioners	Repair hair damage and make the hair easier to comb; prevent flyaway; add feel, shine, softness
Combing/cutting/blowdrying	Style the hair
Chemical dyes/colorants/bleaches	Enhance or change the color and look of hair
Permanent wave treatments	Change the style and look of the hair

(http://www.pg.com; http://www.loreal.com; http://unilever.com) suggests that understanding the science behind human hair has more than just purely academic benefits.

The tribology of the hair changes as a function of the various hair care products and processes. Figure 24.1 illustrates schematically various functions, along with the macroscale and micro/nanoscale mechanisms behind these interactions, that make surface roughness, friction, and adhesion very important to hair and skin [LaTorre and Bhushan, 2005a]. Desired features and corresponding tribological attributes of conditioners are listed in Table 24.2 [LaTorre et al., 2006]. Friction is the most relevant parameter to hair care. Our perception of a soft, silky feel comes from our ability to glide a comb or skin over the hair fibers. For a smooth wet and dry feel, friction between hair and skin should be minimized in wet and dry environments, respectively. For a good feel with respect to bouncing and shaking of the hair during walking or running, friction between hair fibers and groups of hair fibers should be low. The friction one feels during combing is a result of interactions between hair and the comb material (generally plastic), and this too needs to be low to easily maintain, sculpt, and comb the hair. Adhesion is also important. To minimize entanglement, the adhesive force (the force required to separate the hair fibers) needs to be low. In other cases, a certain level of adhesion may be acceptable and is often a function of the hair style. For individuals seeking "hair alignment", where hair fibers lie flat and parallel to each other, a small amount of adhesive force between fibers may be desired. For more complex and curly styles, even higher adhesion between fibers may be optimal.

Table 24.2. Desired features and corresponding tribological attributes of conditioners

Desired hair feature	Tribological attributes
Smooth feel in wet and dry environments	Low friction between hair and skin in respective environments
Shaking and bouncing during daily activities	Low friction between hair fibers and groups of hair
Easy combing and styling	Low friction between hair and comb (plastic) and low adhesion. Note: more complex styles may require higher adhesion between fibers

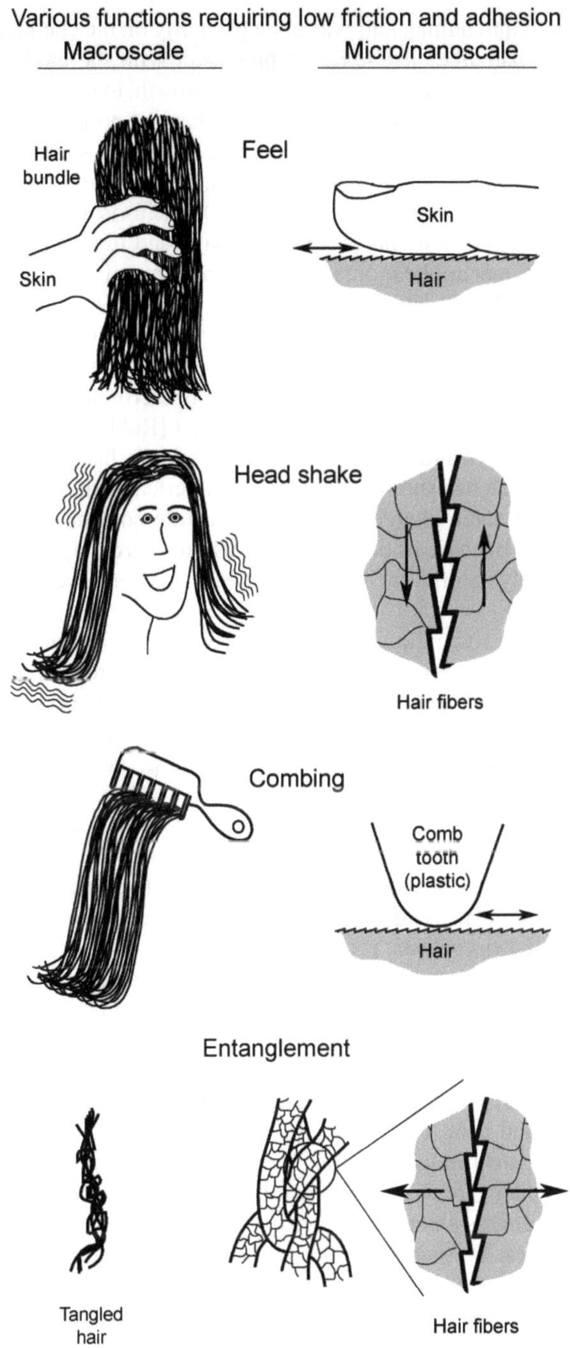

Fig. 24.1. Schematics illustrating various functions with associated macroscale and micro/nanoscale mechanisms of hair and skin friction during feel or touch, shaking and bouncing of the hair, combing, and entanglement [LaTorre and Bhushan, 2005a]

Early research into human hair was done primarily on the chemical and physical properties of the hair fiber itself. Key topics dealt with the analysis of chemical composition in the fiber, microstructure, and hair growth, to name a few. Mechanical properties were also of interest. Most of the mechanical property studies of human hair were on the macroscale and used conventional methods, such as tension, torsion and bending tests [Robbins, 1994; Feughelman, 1997; Swift, 1999, 2000; Barnes and Roberts, 2000; Jachowitz and McMullen, 2002]. Efforts were also made to study the effects of environmental and chemical damage and treatment, such as dyeing, bleaching, and polymer application; these topics have remained a mainstream area of investigation due to the availability and formation of new chemicals and conditioning ingredients. Tribology has generally been studied via the macroscale friction force of hair. As a matter of fact, much of the tribological work performed by the hair care industry today still focuses on the measurement of macroscale friction, particularly between a skin replica and a hair swatch of interest [Robbins, 1994]. The intrinsic differences of the hair as a function of ethnicity eventually became a concern as well. For instance, research has shown that African–American hair has higher resistance to combing, higher static charge, and lower moisture content than Caucasian hair [Syed et al., 1996]. Because of differences like these, a growing number of hair care products specifically tailored for ethnic hair care have been developed and sold with commercial success.

Modern research since the late 1990s has been primarily concerned with using micro/nanoscale experimental methods such as atomic force/friction force microscopies (AFM/FFM) and nanoindentation to answer the complex questions surrounding the structure and behavior of the hair. Nanoscale characterization of the cellular structure, mechanical properties, and tribological properties of hair are essential to evaluate and develop better cosmetic products, and to advance the understanding of composite biological systems, cosmetic science, and dermatology.

AFM/FFM have been used to effectively study the structure of the hair surface and cross-section. AFM provides the potential for being able to see the cellular structure and molecular assembly of hair, for determining various properties of hair, such as elastic modulus and viscoelastic properties, and for investigating the physical behavior of various cellular structures of hair in various environments [Chen and Bhushan, 2005a,b]. As a non-invasive technique, AFM has been used to evaluate the effect of hair treatment and can be operated in ambient conditions in order to study the effect of environment on various physical properties. The nanoindenter has been used to characterize the nanomechanical behavior of the hair surface and cross-section using nanoindentation and nanoscratch techniques [Wei et al., 2005; Wei and Bhushan, 2006]. AFM has also been used to study nanotribological properties of hair and skin and the effects of hair care products on hair [LaTorre and Bhushan, 2005a,b; LaTorre et al., 2006]. Roughness parameters have been measured to compare changes due to damaging processes. The nanoscale friction force has been measured to understand damage or conditioner distribution and its effect on hair tribology. Adhesive force mapping has proven useful to observe the conditioner distribution as well.

It is known that adhesion, friction, and lubrication of many interfaces are scale-dependent [Bhushan, 1999b, 2005; Bhushan et al., 2004]. In the case of hair, the magnitude of the nanoscale coefficient of friction in the paper by LaTorre and Bhushan [2005a] was lower than the macroscale values found in that by Bhushan et al. [2005] for the same hair sets. It was also found that there are both similarities and differences when comparing the trends for various hair types at both scales. Also, directionality effects of friction were observed on both scales, but via slightly different mechanisms due to the size of the contacts involved [LaTorre and Bhushan, 2006]. It is thus recognized that scale effects are an important aspect of studying the tribology of hair. However, no microscale tribological data for hair exists in literature. This is unfortunate because many interactions between hair–skin, hair–comb, and hair-hair contact takes place at microasperities ranging from a few μm to hundreds of μm. For instance, the skin on one's finger is not microscopically smooth, but is instead patterned with bumps and ridges that contact the hair surface. Also, adhesion may be very relevant on the microscale because often many small patches of cuticle scales (5–10 μm long) from two different fibers may be in contact at the same time. Thus, to further the scale effect study and to bridge the gap between the macroscale and nanoscale data, as well as to gain a full understanding of the mechanisms behind the trends, it is worthwhile to look at hair tribology on the microscale as well.

This chapter presents a comprehensive study of various hair and skin nanotribological properties as a function of ethnicity, damage, conditioning treatment, and various environments. Nanotribological properties including surface roughness, friction, adhesion, and wear are presented, as well as investigations of scale effects and directionality dependence on friction and adhesion.

24.2 Human Hair, Skin, and Hair Care Products

24.2.1 Human Hair and Skin

Figure 24.2 shows a schematic of a human hair fiber with its various layers of cellular structure [Zviak, 1986; Robbins, 1994; Feughelman, 1997; Jolles et al., 1997]. Hair fibers (about 50 to 100 μm in diameter) consist of the cuticle and cortex, and in some cases medulla in the central region. All are composed of dead cells, which are mainly filled with keratin protein. Table 24.3 displays a summary of the chemical species of hair [Chen and Bhushan, 2005a]. Depending on its moisture content, human hair consists of approximately 65–95% proteins, which are condensation polymers of amino acids. The remaining constituents are water, lipids (structural and free), pigment, and trace elements. Among numerous amino acids in human hair, cystine is one of the most important amino acids. The distinct cystine content of various cellular structures of human hair results in a significant effect on their physical properties. Cystine has the capacity to cross-link the protein by its intermolecular disulfide linkages. A high cystine content corresponds to rich disulfide cross-links, leading to high mechanical properties. In addition to disulfide bonds, hair is also rich in peptide bonds and the abundant CO- and NH-groups present give rise to hydrogen bonds between groups of neighboring chain molecules. The species responsible for color in hair is the pigment melanin, which is located in the cortex of the hair in granular form.

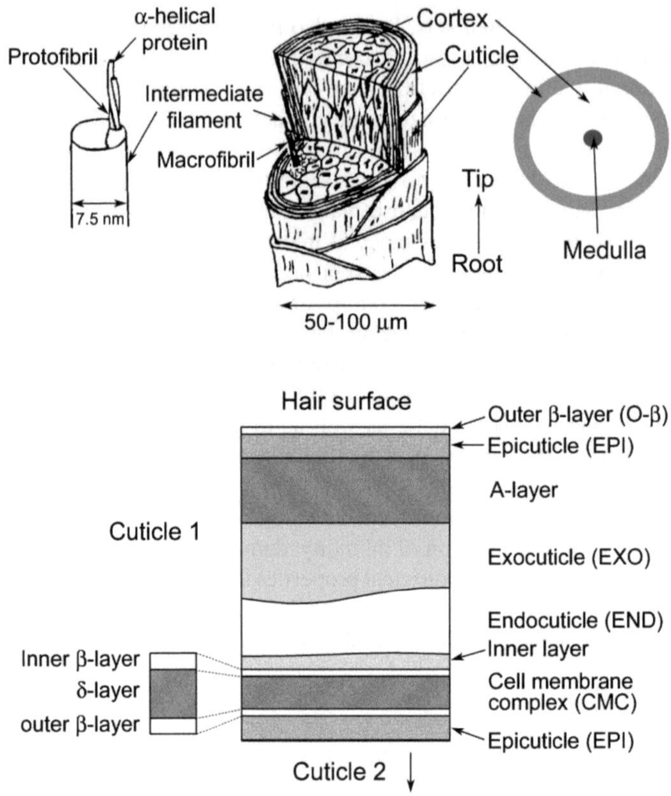

Fig. 24.2. Schematic of hair fiber structure and cuticle sublamellar structure

Table 24.3. Summary of chemical species presented in human hair

Keratin protein	65–95%	
(Amino acids)	$^{\oplus}NH_3-CH(CO_2^{\ominus})-R$	(R: functional group)
Cystine	$^{\oplus}NH_3-CH(CO_2^{\ominus})-CH_2-S-S-CH_2-CH(CO_2^{\ominus})-NH_3^{\oplus}$	
Lipids	Structural and free	
18-MEA	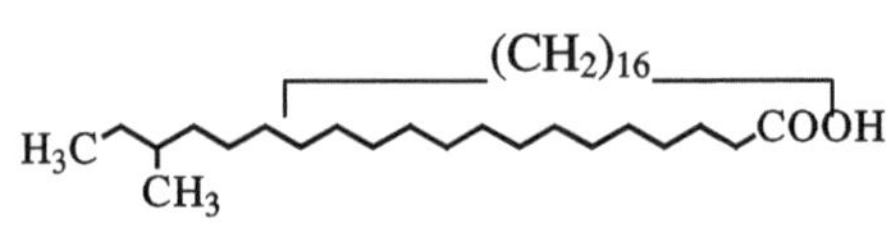	
Water	Up to 30%	
Pigment and trace elements	Melanin	

An average head contains over 100,000 hair follicles, which are the cavities in the skin surface from which hair fibers grow. Each follicle grows about 20 new hair fibers in a lifetime. Each fiber grows for several years until it falls out and is replaced by a new fiber. Hair typically grows at a rate on the order of 10 mm/month.

24.2.1.1
The Cuticle

The cuticle consists of flat overlapping cells (scales). The cuticle cells are attached at the root end and they point forward the tip end of the hair fiber, like tiles on a roof. Each cuticle cell is approximately 0.3 to 0.5 μm thick and the visible length of each cuticle cell is approximately 5 to 10 μm. The cuticle in human hair is generally 5 to 10 scales thick. Each cuticle cell consists of various sublamellar layers (the epicuticle, the A-layer, the exocuticle, the endocuticle and inner layer) and the cell membrane complex (see Fig. 24.2). Table 24.4 displays the various layers of the cuticle, their respective cystine levels [Robbins, 1994], and other details.

The outer layer is the epicuticle, which is covered with a thin layer of covalently attached lipid 18-Methyl Eicosanoic Acid (18-MEA) (see Table 24.3). The A-layer is a component of high cystine content (~30%) and located on the outer-facing aspect of each cell. The A-layer is highly cross-linked,which gives this layer considerable mechanical toughness and chemical resilience, and the swelling in water is presumed to be minimal. The exocuticle, which is immediately adjacent to the A-layer, is also of high cystine content (~15%). On the inner facing aspect of each cuticle cell there is a thin layer of material known as the inner layer. Between the exocuticle and inner layer is the endocuticle, which is low in cystine (~3%). The cell membrane complex itself is a lamellar structure, which consists of the inner β-layer, the δ-layer and the outer β-layer.

Table 24.4. Various layers of the cuticle and their details

Cuticle layer	Cystine component	Details
Epicuticle	~ 12%	18-MEA lipid layer attached to outer epicuticle contributes to lubricity of the hair
A-layer	~ 30%	Highly cross-linked
Exocuticle	~ 15%	Mechanically tough Chemically resilient
Endocuticle	~ 3%	
Inner layer	–	
Cell membrane complex (CMC)	~ 2%	Lamellar structure Consists of inner β-layer, δ-layer, and outer β-layer

Figure 24.3a shows SEM images of virgin Caucasian, Asian and African hair [Wei et al., 2005]. It can be seen that the Asian hair is the thickest (about 100 μm), followed by African hair (about 80 μm) and Caucasian hair (about 50 μm). The visible cuticle cell is about 5 to 10 μm long for the three hair types. A listing of various cross-section dimensional properties is presented in Table 24.5 [Wei et al., 2005]. While Caucasian hair and Asian hair typically have a similar cross-sectional

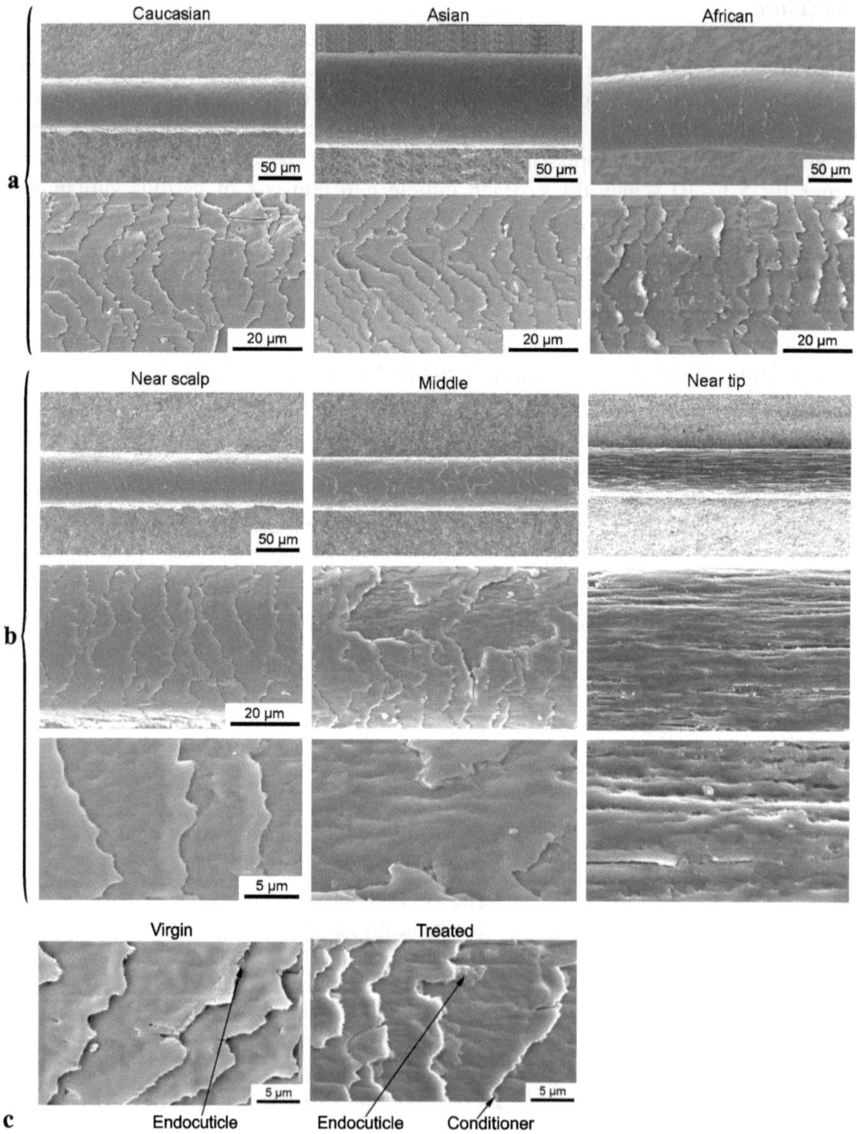

Fig. 24.3. (**a**) SEM images of various hair types. (**b**) SEM images of virgin Caucasian hair at three locations. (**c**) SEM images of Caucasian, virgin and treated hair [Wei et al., 2005]

Table 24.5. Variation in cross-sectional dimensions of human hair

	Shape	Maximum diameter (D_1) (μm)	Minimum diameter (D_2) (μm)	Ratio D_1/D_2	Number of cuticle scales	Cuticle scale thickness (μm)
Caucasian	Nearly oval	74	47	1.6	6–7	0.3–0.5
Asian	Nearly round	92	71	1.3	5–6	0.3–0.5
African	Oval-flat	89	44	2.0	6–7	0.3–0.5

Average length of visible cuticle scale: about 5 to 10 μm

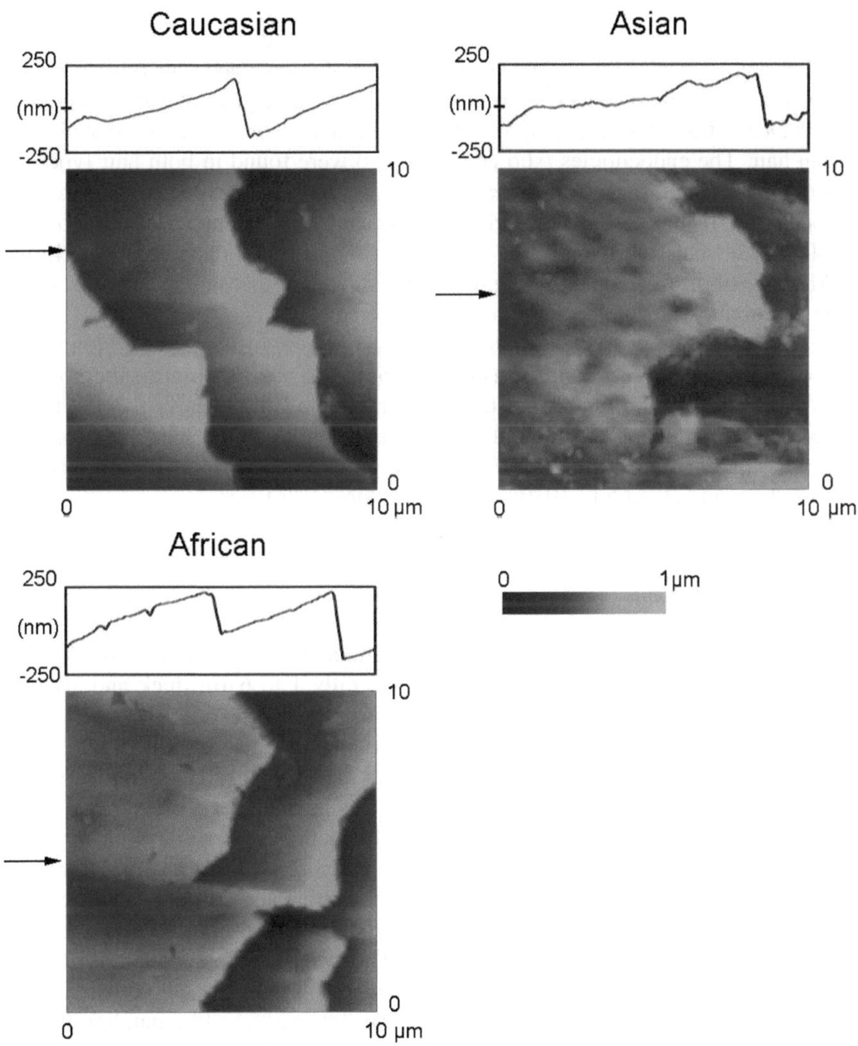

Fig. 24.4. AFM images of various virgin hair types [LaTorre and Bhushan, 2005a]

shape (Asian hair being the most cylindrical), African hair has a highly elliptical shape. African hair is much curlier and wavier along the hair fiber axis than Caucasian or Asian hair. Figure 24.3b shows the SEM images of virgin Caucasian hair at three locations: near scalp, middle and near tip. Three magnifications were used to show the significant differences. The hair near scalp had complete cuticles, while no cuticles were found on the hair near tip. This may be because the hair near the tip experienced more mechanical damage during its life than the hair near the scalp. The hair in the middle experienced intermediate damage, i.e. one or more cuticle scales were worn away, but many cuticles stayed complete. If some substructures of one cuticle scale, like A-layer or A-layer and exocuticle (see Fig. 24.2) are gone, or even worse, one or several cuticle scales are worn away, it is impossible to heal the hair biologically because hair fibers are composed of dead cells.

However, it is possible to physically "repair" the damaged hair by using conditioner, a functions of which is to cover or fill the damaged area of the cuticles. Figure 24.3c shows the high magnification SEM images of virgin and treated Caucasian hair. The endocuticles (shown by arrows) were found in both hair types. In order for the conditioner to physically repair the hair, it is expected to cover the endocuticles. In the case of severely damaged hair, for example, an edge of one whole cuticle scale worn away, the conditioner may fill that damaged edge. In the SEM image of the treated hair in Fig. 24.3c, the substance that stayed near the cuticle edge is probably the conditioner (shown by an arrow).

Figure 24.4 shows the AFM images of various virgin hair types, along with the section plots [LaTorre and Bhushan, 2005a]. The arrows point to the position from where the section plots were taken. Each cuticle cell is nearly parallel to the underlying cuticle cell, and they all have similar angles to the hair axis, forming a tile-like hair surface structure. The visible cuticle cell is approximately 0.3 to 0.5 μm thick and about 5 to 10 μm long for all three hair types.

24.2.1.2
The Cortex and Medulla

The cortex contains cortical cells and intercellular binding material, or the cell membrane complex. The cortical cells are generally 1 to 6 μm thick and 100 μm long, and run longitudinally along the hair fiber axis and take up the majority of the inner hair fiber composition [Randebrook, 1964]. The macrofibrils (about 0.1 to 0.4 μm in diameter) comprise a major portion of the cortical cells. Each macrofibril consists of intermediate filaments (about 7.5 nm in diameter), previously called microfibrils, and the matrix. The intermediate filaments are low in cystine ($\sim 6\%$), and the matrix is rich in cystine ($\sim 21\%$). The cell membrane complex consists of cell membranes and adhesive material that binds the cuticle and cortical cells together. The intercellular cement of the cell membrane complex is primarily nonkeratinous protein, and is low in cystine content ($\sim 2\%$). The medulla of human hair, if present, generally makes up only a small percentage of the mass of the whole hair, and is believed to contribute negligibly to the mechanical properties of human hair fibers.

Figure 24.5a shows the SEM images of a cross-section of virgin hair [Wei et al., 2005] and Fig. 24.5b shows the TEM images of a cross-section of human hair [Swift, 1997].

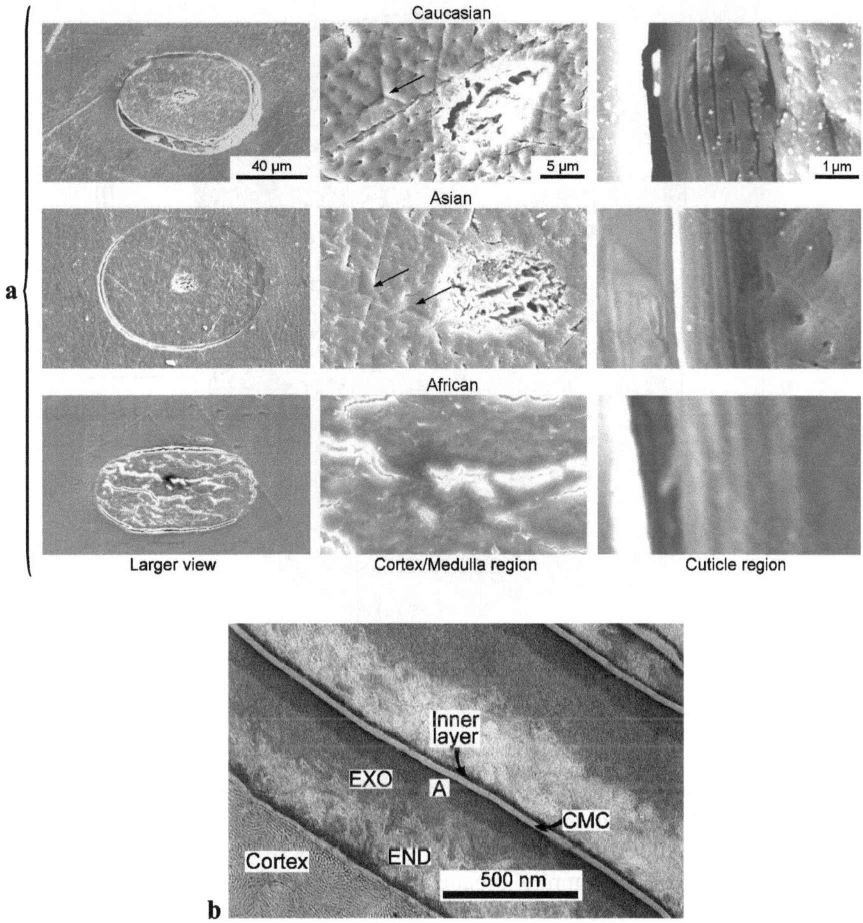

Fig. 24.5. (**a**) SEM images of virgin hair cross-section (in the figure EXO, END, and CMC stand for exocuticle, endocuticle, and cell membrane complex, respectively) [Wei et al., 2005]. (**b**) TEM of hair cross-section

24.2.1.3
Skin

Skin covers and protects our bodies. The skin at the forehead and scalp areas is of most interest when dealing with human hair, since most of the hair care products are developed specifically for head hair. The skin of the hand and fingers is also of importance because the "feel" of hair is often sensed by physically touching the fibers with these regions. In general, skin is composed of three main parts: epidermis, dermis, and subcutaneous tissue (L'Oreal); see Fig. 24.6.

The epidermis contains four distinct cellular layers: basal, spinous, granular, and horny. In the basal layer, melanocytes deliver the pigment melanin to keratinocytes. Keratinocyte cells that have been cornified are referred to as corneocytes [Pugliese,

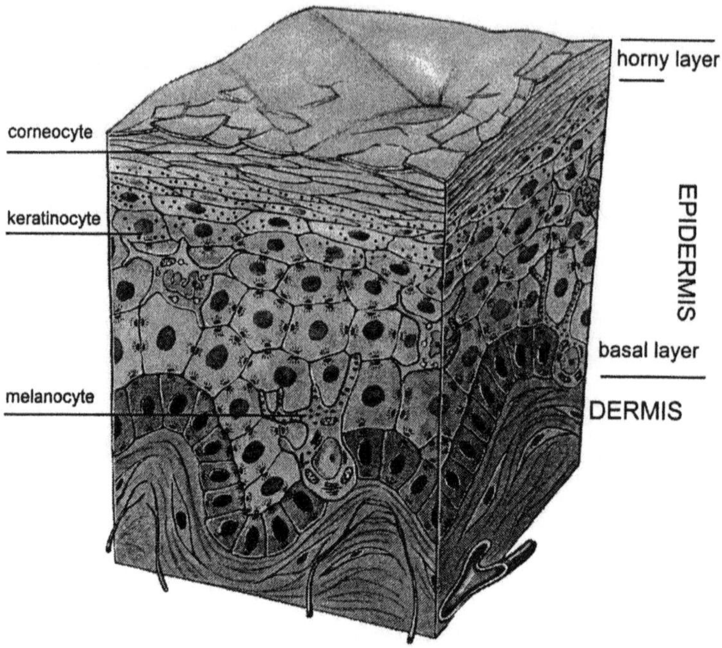

Fig. 24.6. Schematic image of human skin structure with different layers: dermis, epidermis, and horny layer (L'Oreal)

1996]. Hexagonally shaped corneocyte cells compose the horny layer, or stratum corneum. The stratum corneum is the outer layer of the skin; at about 15-μm thick, it acts as a mechanical, thermal, and chemical barrier from environmental factors and contamination. The complex organization of corneocytes and the intercellular matrix contribute to the success of the barrier [Wertz and Downing, 1989]. In fact, Wertz et al. [1989] developed a structural model that observes the matrix as a lamellar phase composition of various lipids that provide a glue-like system to provide a barrier effect.

The dermis structure is known for its ability to handle most of the physical stresses imposed on the skin, and takes up roughly 90% of the mass [Pugliese, 1996]. The dermis is divided into an outermost papillary layer and underlying reticular layer.

24.2.2 Hair Care: Cleaning and Conditioning Treatments, and Damaging Processes

24.2.2.1 Cleaning and Conditioning Treatments: Shampoo and Conditioner

Shampoos are used primarily to clean the hair and scalp of dirt and other greasy residue that can build up after time. Shampoos also have many secondary functions including controlling dandruff, reducing irritation, and even conditioning. Conditioners, on the other hand, are used primarily to give the hair a soft, smooth feel that results in easier hair combing. Secondary functions include preventing "flyaway" hair due to static electricity, giving the hair a shiny appearance, and protecting the hair from further damage by forming a thin coating over the fibers.

Further developments in marketing and aesthetic factors (brand name, fragrance, feel, and color of the shampoos and conditioners) have created new market segments. In many instances, these factors have become primary reasons for use.

24.2.2.1.1
Shampoo: Constitution and Main Functions

The following discussion is based on Gray [2001, 2003]. As stated above, shampoos serve various cleaning functions for the hair and scalp. In the past, typical shampoos were mainly soap-based products. However, soaps did not have very good lathering capability, and often left a residual "scum" layer on the hair that was undesirable and could not be rinsed off. In modern shampoos, advances in chemistry and technology have made it possible to replace the soap bases with complex formulas of cleansing agents, conditioning agents, functional additives, preservatives, aesthetic additives and even medically active ingredients. Table 24.6 shows the most common ingredients of shampoos and their functions.

In most modern shampoos, the primary cleansing agents are surfactants. Dirt and greasy residue are removed from the hair and scalp by these surfactants, making them the most important part of the shampoo. Surfactants have great lathering capabilities and rinse off very easily; see Table 24.6 for a full list of features. Surfactant molecules have two different ends, one that is negatively charged and soluble in water (unable to mix with greasy matter), and another that is soluble in greasy matter (unable to mix with water). In general, surfactants clean the hair by the following process. Surfactant molecules encircle the greasy matter on the hair surface. The molecule end, which is soluble in greasy matter, buries itself in the grease, which leaves the water soluble molecule end to face outward with a negative charge. Since the hair fibers are negatively charged as well, the two negative charges repel each other. Thus, the greasy matter is easily removed from the hair surface and rinsed off.

Table 24.6. Components of common shampoos and their functions

Shampoo component	Functions
Cleansing agents	– Produce lather to trap greasy matter, and prevent redeposition – Remove dirt and grease from hair and scalp – Stabilize the mixture and help keep the ingredient network together – Thicken the shampoo to the desired viscosity
Conditioning agents	Condition the hair
Functional additives	Control the viscosity and pH levels of the shampoo
Preservatives	Prevent decomposition and contamination of the shampoo
Aesthetic additives	Enhance color, scent, and luminescence of the shampoo
Medically active ingredients	Aid treatment of dandruff or hair loss

24.2.2.1.2
Conditioner: Constitution and Main Functions

As stated earlier, many shampoos have certain levels of conditioning agents that mimic the functions of a full conditioner product. Conditioner molecules contain cationic surfactants that give a positive electrical charge to the conditioner. The negative charge of the hair is attracted to the positively charged conditioner molecules, which results in conditioner deposition on the hair; see Fig. 24.7. This is especially true for damaged hair, since damaging processes result in hair fibers being even more negatively charged. The attraction of the conditioner to hair results in a reduction of static electricity on the fiber surfaces, and consequently a reduction in the "fly away" behavior. The conditioner layer also flattens the cuticle scales against each other, which improves shine and color. The smooth feel resulting from conditioner use enables easier combing and detangling in both wet and dry conditions (see Table 24.2).

Conditioner consists of a gel network chassis (cationic surfactants, fatty alcohols, and water) for superior wet feel and a combination of conditioning actives (cationic surfactants, fatty alcohols, and silicones) for superior dry feel. Figure 24.8 shows the transformation of the cationic surfactants and fatty alcohol mixture into the resulting

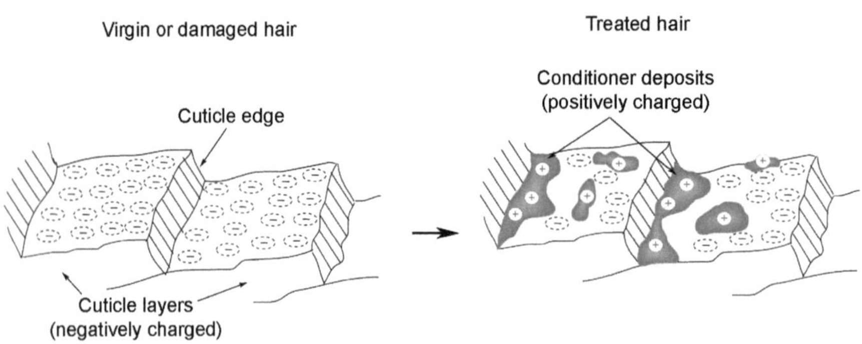

Fig. 24.7. Negatively charged hair and the deposition of positively charged conditioner on the cuticle surface [LaTorre et al., 2006]

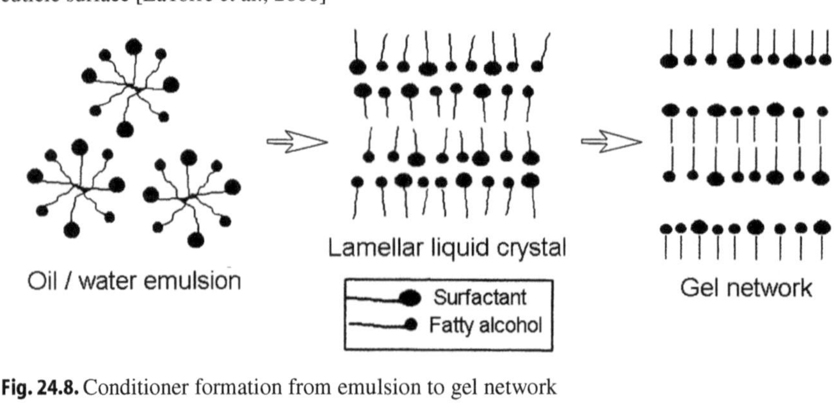

Fig. 24.8. Conditioner formation from emulsion to gel network

gel network, which is a frozen lamellar liquid crystal gel phase. The process starts as an emulsion of the surfactants and alcohols in water. The materials then go through a strictly controlled heating and cooling process: the application of heat causes the solid compounds to melt, and the solidification process enables a setting of the lamellar assembly molecules in a fully extended conformation, creating a lamellar gel network. When this network interacts with the hair surface, the high concentration of fatty alcohols makes them the most deposited ingredient group, followed by the silicones and cationic surfactants. Typical deposition levels for cationic surfactants, fatty alcohol, and silicone are around 500–800 ppm, 1000–2000 ppm, and 200 ppm,

Table 24.7. Combinations of conditioner ingredients and their benefits towards wet and dry feel

Gel network chassis for superior wet feel	
Key ingredients	Benefits
– Cationic surfactant – Fatty alcohols – Water	– Creamy texture – Ease of spreading – Slippery application feel – Soft rinsing feel
Combination of "conditioning actives" for superior dry feel	
Key Ingredients	Benefits
– Silicones – Fatty alcohols – Cationic surfactant	– Moist – Soft – Dry combing ease

Table 24.8. Chemical structure and purpose/function of conditioner ingredients

	Ingredient	Chemical structure	Purpose/Function
Water			
Cationic surfactants	Stearamidopropyl dimethylamine	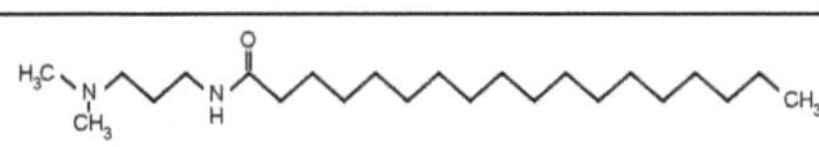	
	Behenyl amidopropyl dimethylamine glutamate (BAPDMA)	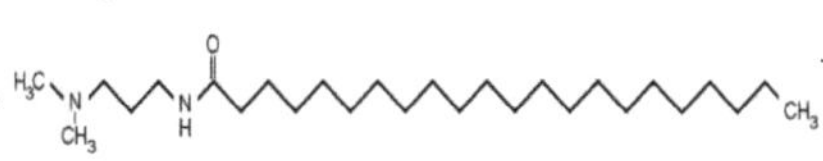	– Aids formation of lamellar gel network – Lubricates and static control agent
	Behentrimonium chloride (BTMAC) $CH_3(CH_2)_{21}N(Cl)(CH_3)_3$	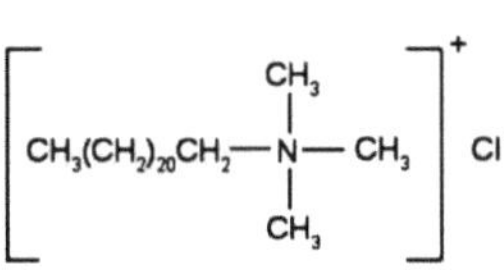	
Fatty alcohols	Stearyl alcohol ($C_{18}OH$)		– Lubricates and moisturizes – Aids formation of lamellar gel network along with cationic surfactant
	Cetyl alcohol ($C_{16}OH$)		
Silicones	PDMS blend (dimethicone)	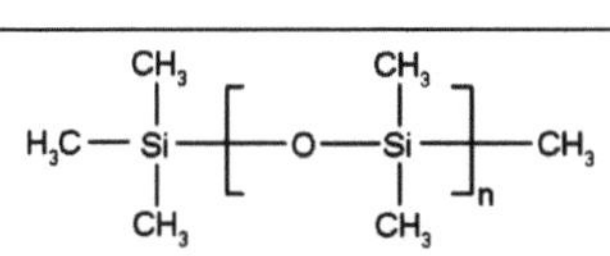	– Primary source of lubrication – Gives hair a soft and smooth feel

respectively. Typical concentrations are approximately 2–5 wt. %, 5–10 wt. %, and 1–10 wt. %, respectively [LaTorre et al., 2006].

The benefits of conditioners are shown in Table 24.7 [LaTorre et al., 2006]. The wet feel benefits are creamy texture, ease of spreading, slippery feel while applying, and soft rinsing feel. The dry feel benefits are moistness, softness, and easier dry combing. Each of the primary conditioner ingredients also have specific functions and roles that affect the performance of the entire product.

Table 24.8 displays the functions of the major conditioner ingredients and their chemical structure [LaTorre et al., 2006]. Cationic surfactants are critical to the forming of the lamellar gel network in conditioners, and also act as a lubricant and static control agent, since their positive charge aids in counteracting the negative charge of the hair fibers. Fatty alcohols are used to lubricate and moisturize the hair surface, along with forming the gel network. Finally, silicones are the main source of lubrication in the conditioner formulation.

24.2.2.2
Damaging Processes

In a previous section, we discussed some of the products that aid in "treating" the hair. There are other hair care products and processes that, while creating a desired look or style to the hair, also bring about significant damage to the fibers (see Table 24.1). Most of these processes occur on some type of periodic schedule, whether it be daily (combing the hair), or monthly (haircut and coloring at a salon). In general, hair fiber damage occurs most readily by mechanical or chemical means, or by a combination of both (chemo-mechanical).

Mechanical Damage

Mechanical damage occurs on a daily basis for many individuals. The damage results from large physical forces or temperatures, which degrade and wear the outer cuticle layers. Common causes are

- combing (scratching and wearing of the cuticle layers)
- scratching (usually with fingernails around the scalp)
- cutting (affects the areas surrounding the fiber tips)
- blowdrying (high temperatures thermally degrade the surface of the hair fibers)

Permanent Wave Treatment

Permanent wave treatments have not changed much with the invention of the Cold Wave. Generally speaking, the Cold Wave uses mercaptans (typically thioglycolic acid) to break down disulfide bridges and style the hair without much user interaction (at least in the period soon after the perm application) [Gray, 2001]. The Cold Wave process does not need increased temperatures (no thermal damage to the hair), but generally consists of a reduction period (whereby molecular reorientation to the cuticle and cortex occurs via a disulfide-mercaptan interchange pathway [Robbins,

1994]), followed by rinsing, setting of the hair to the desired style, and finally neutralization to decrease the mercaptan levels and stabilize the style.

Chemical Relaxation

Commonly used as a means of straightening hair (especially in highly curved, tightly curled African hair), this procedure uses an alkaline agent, an oil phase, and a water phase of a high viscosity emulsion to relax and reform bonds in extremely curly hair. A large part of the ability to sculpt the hair to a desired straightness comes from the breakage of disulfide bonds of the fibers.

Coloring and Dyeing

Hair coloring and dyeing have become extremely successful hair care procedures, due in part to "over-the-counter" style kits that allow home hair care without professional assistance. The most common dyes are para dyes, which contain paraphenylenediamine (PPD) solutions accompanied by conditioners and antioxidants. Hydrogen peroxide (H_2O_2) is combined with the para dyes to effectively create tinted, insoluble molecules that are contained within the cortex and are not small enough to pass through the cuticle layers, leaving a desired color to the hair. Due to the levels of hydrogen peroxide, severe chemical damage can ensue in the cuticle and cortex.

Bleaching

Like dyeing, bleaching consists of using hydrogen peroxide to tint the hair. However, bleaching can only lighten the shade of hair color, as the H_2O_2 releases oxygen to bind hair pigments [Gray, 2001]. Bleaching may also be applied to limited areas of the hair (such as in highlights) to create a desired look. The chemical damage brought on by bleaching leads to high porosity and severe wear of the cuticle layer.

24.3 Experimental Techniques

To date, most information about the detailed structure of human hair was obtained from scanning electron microscope (SEM) and transmission electron microscope (TEM) observations [Swift, 1991; Robbins, 1994; Swift, 1997; Wei et al., 2005]. SEM requires the hair sample to be covered with a very thin layer of a conductive material and needs to be operated under vacuum during both metallization and measurements. Surface metallization and vacuum exposure could potentially induce modifications to the surface details. Figure 24.5a and b shows typical images of human hair obtained by SEM and TEM, respectively. TEM examinations provide fine detailed internal structure of human hair. However, thin sections of 50–100 nm thickness and heavy metal compounds staining treatment are required for TEM examinations. The cutting of these thin sections with the aid of an ultra-microtome is not an easy task. Moreover, since neither SEM nor TEM techniques can measure the physical properties (mechanical, tribological, etc.) of various cellular structures of the human hair of interest, nor do they allow ambient imaging conditions, many outstanding issues remain unanswered.

AFM has been commonly used for characterization of tribological and mechanical properties of surfaces [Bhushan, 1999a,b, 2002, 2004, 2005]. As a non-invasive technique, AFM can evaluate the effect of sample treatment, since it can be used without requiring any specific treatment. Most importantly, it can be operated in ambient conditions in order to study the effect of environment. AFM has the potential of being able to show the cellular structure and molecular assembly of hair, for determining various properties of hair, such as friction, adhesion, wear, elastic stiffness and viscoelastic properties, and for investigating the physical behavior of various cellular structures of hair in various environments. To date, most of AFM studies on hair fibers were focused on the surface topographic imaging [Smith and Swift, 2002; LaTorre and Bhushan, 2005a] and friction, adhesion and wear properties [LaTorre and Bhushan, 2005a,b; LaTorre et al., 2006].

A schematic diagram of an AFM imaging a hair fiber is shown in Fig. 24.9. AFM/FFM uses a sharp tip with a radius of approximately 30–50 nm. This significant reduction in tip-to-sample interaction compared to the macroscale allows the simulation of single asperity contact to give detailed surface information. While there are several AFM operating modes, the so-called "contact mode" is most relevant for studying tribological properties because it allows simultaneous measurement of surface roughness and friction force.

When skin comes in contact with hair, actual contact occurs over a large number of asperities. During relative motion, friction and adhesion are governed by the surface interactions that occur at these asperities. To date, much of the work in

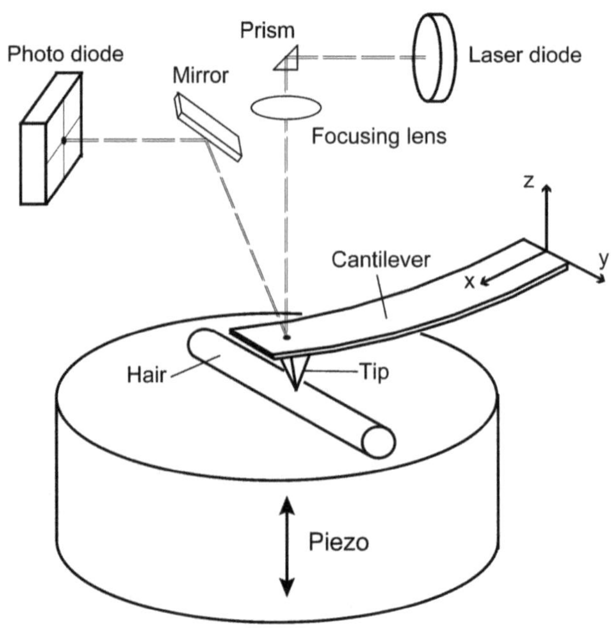

Fig. 24.9. Schematic diagram of AFM operation with human hair sample

the industry has focused on the measurement of macroscale friction, particularly between a skin replica and a hair swatch of interest [Robbins, 1994]. Figure 24.10 shows schematics of typical macroscale and nanoscale test apparatuses.

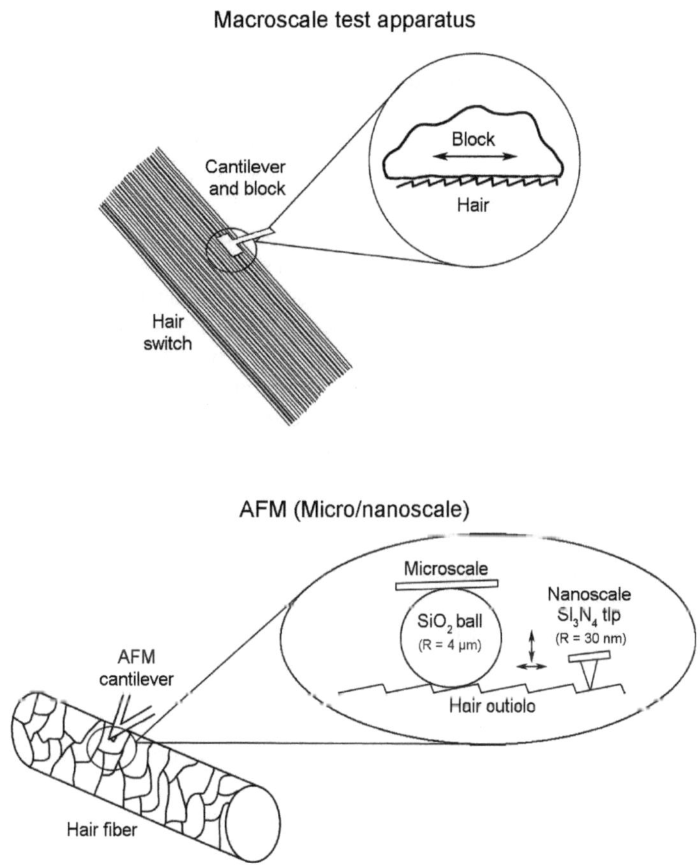

Fig. 24.10. Comparison of macroscale and micro/nanoscale friction test apparatuses

24.3.1 Experimental Procedure

24.3.1.1 Specimen Mounting

Hair specimens were mounted onto AFM sample pucks using Liquid Paper® correction fluid. A thin layer of the fluid was brushed onto the puck, and when the fluid hardened into a tacky state, the hair sample was carefully placed onto it. The Liquid Paper® dries quickly to keep the hair firmly in place. An optical microscope was used to preliminarily image the specimen to ensure none of the Liquid Paper® was deposited on the hair surface.

Synthetic materials were attached to AFM sample pucks using double-sided adhesive tape.

24.3.1.2
Nanotribological Property Measurements

Surface roughness, friction force, and adhesive force measurements were performed using a commercial AFM system (MultiMode Nanoscope IIIa, Digital Instruments, Santa Barbara, CA) in ambient conditions ($22 \pm 1\,°C$, $50 \pm 5\%$ relative humidity) [LaTorre and Bhushan, 2005a,b; LaTorre et al., 2006]. Square pyramidal Si_3N_4 tips of nominal 30–50 nm radius attached to the end of a soft Si_3N_4 cantilever beam (spring constant of 0.06 N/m) were used for surface roughness and nanoscale friction force measurements. This softer cantilever was used to minimize damage to the hair. After engagement of the tip with the cuticle surface, the tip was scanned perpendicular to the longitudinal axis of the fiber. The tip was centered over the cross-section in order to be at the very top of the fiber, so as to negate effects caused by the AFM tip hitting the sides of the hair and adding error to the measurements. In order to minimize scanning artifacts, a scan rate of 1 Hz was used for all measurements. Topographical images to characterize the shape and structure of the various hair types were taken at 5×5, 10×10, and $20 \times 20\,\mu m^2$ scans at a normal load of 5 nN. These scan sizes were suitable for capturing the surface features of multiple scales and scale edges of the cuticle. To characterize roughness, $2 \times 2\,\mu m^2$ scans of the cuticle surface without edges were used. Surface roughness images shown in this chapter were processed using the first-order "planefit" command available in the AFM software, which eliminates tilt in the image. Roughness data, as well as friction force data, were taken after the planefit command was employed. The first-order "flatten" command was also used on friction force mappings to eliminate scanning artifacts and present a cleaner image.

Friction force mapping of the scan area was collected simultaneously with roughness mapping. Figure 24.10 (bottom diagram) and Fig. 24.11 show the AFM tip scanning over the hair surface for untreated hair and hir treated with conditioner, respectively. The effects of the conditioner can be examined by comparing the friction information. A quantitative measure of the friction force was calculated by first calibrating the force based on a method by Bhushan [Bhushan, 1999a, 2002]. The normal load was varied from 5 nN to 45 nN in roughly 5 nN increments, and a friction force measurement was taken at each increment. By plotting the friction force as a function of the normal force, the average coefficient of friction was determined from the slope of the least squares fit line of the data.

Adhesive force measurements were made with square pyramidal Si_3N_4 tips attached to the end of a Si_3N_4 cantilever beam (spring constant of 0.58 N/m), using the force calibration plot technique. In this technique, the AFM tip is brought into contact with the sample by extending the piezo vertically, then retracting the piezo and calculating the force required to separate the tip from the sample. This method is described in detail by Bhushan [Bhushan, 1999a,b, 2002, 2004].

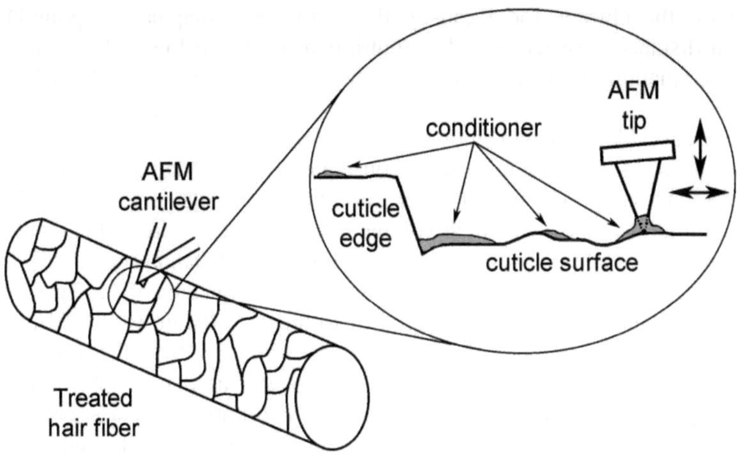

Fig. 24.11. Interactions between the AFM tip and conditioner on the cuticle surface for treated hair [LaTorre and Bhushan, 2005b]

The cantilever deflection is plotted on the vertical axis against the Z-position of the piezo scanner in a force calibration plot, Fig. 24.12. As the piezo extends (from right to left in the graph), it approaches the tip and does not show any deflection while in free air. The tip then approaches within a few nanometers (point A) and becomes attached to the sample via attractive forces at very close range (point B). After initial contact, any extension of the piezo results in a further deflection of the tip. Once the piezo reaches the end of its designated ramp size, it retracts to its starting position (from left to right). The tip goes beyond zero deflection and enters the adhesive regime. At point C, the elastic force of the cantilever becomes

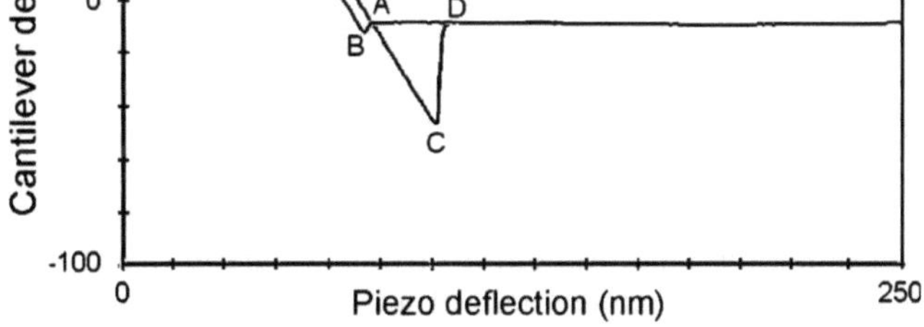

Fig. 24.12. Typical force calibration plot for Caucasian virgin hair. Contact between the tip and hair occurs at *point B*. At *point C*, the elastic force of the cantilever becomes equivalent to the adhesive force, causing the cantilever to snap back to *D*

equivalent to the adhesive force, causing the cantilever to snap back to point D. The horizontal distance between A and D multiplied by the stiffness of the cantilever results in a quantitative measure of the adhesive force [Bhushan, 1999a,b, 2002, 2005].

The force calibration plot allows for the calculation of an adhesive force at a distinct point on the sample surface. Consequently, by taking a force calibration plot at discrete sampling intervals over an entire scan area, a resulting adhesive force mapping (also known as force-volume mapping) can be created to display the variation in adhesive force over the surface [Bhushan and Dandavate, 2000]. In this work, plots were taken at 64×64 distinct points over a scan area of $2 \times 2\ \mu m^2$ for all hair types and ethnicities. Current Digital Instruments software for the Multimode AFM does not allow direct calculation of the adhesive force from the force-volume maps. Thus, a custom program coded in Matlab was used to display the force-volume maps. The adhesive force for each force calibration plot was obtained by multiplying the spring constant with the horizontal distance (in the retract mode) traveled by the piezo from the point of zero applied load to the point where the tip snaps off.

24.3.1.3 Relative Humidity, Temperature, Soaking, and Durability Measurements

A humidity-temperature detector was used to monitor the humidity inside a Plexiglas test chamber enclosing the AFM system. An experimental setup was used to control the humidity inside the chamber. Measurements were taken at nominal relative humidities of 5, 50, and 80%. Hair fibers were placed at each humidity for approximately 2 hours prior to measurements.

A homemade thermal stage was used to conduct temperature effect measurements at 20, 37, 50, and 80 °C. Hair fibers were exposed to each temperature condition for approximately 30 minutes prior to testing.

Soak tests were performed as follows. A dry hair fiber was taken from a swatch, and a sample was cut from the fiber (approximately 7 mm long) for coefficient of friction measurements. An adjacent sample was also taken from the fiber and placed in a small beaker filled with de-ionized water. The sample was subjected to the aqueous environment for 5 minutes, which is representative of a typical exposure time when showering/bathing, then immediately analyzed with AFM. It should be noted that hair becomes saturated when wet in about 1 minute and remains saturated if kept in a humid environment. It was determined from unpublished results that if the wet hair was exposed to the ambient environment for more than 20 minutes while in the AFM, the hair became dry and the coefficient of friction became similar to that of dry hair. Thus, coefficient of friction measurements were made within a 20 minute time frame for each sample.

In order to simulate scratching that can occur on the surface of the hair and its effect on the friction force on the cuticle surface, a durability test was conducted using a stiff silicon tip (spring constant of 40 N/m). A load of 10 μN was used on a 2 μm section of the cuticle. A total of 1000 cycles were performed at 2 Hz. Measurements were conducted using an AFM. The friction force signal was recorded with respect to cycling time.

24.3.1.4
Microscale and Macroscale Friction and Adhesion Measurements for the Study of Scale Effects

Microscale coefficient of friction and adhesive force measurements were performed using a commercial AFM system (MultiMode Nanoscope IIIa, Digital Instruments, Santa Barbara, CA) in ambient conditions (22 ± 1 °C, 50 ± 5% relative humidity). For microscale data, a 4 μm radius silica ball was mounted on a Si_3N_4 cantilever (spring constant of 0.58 N/m). A quantitative measure of friction force was calculated as described previously, except that the normal load was varied over a larger range (0–180 nN with microscale tip, compared to 0–45 nN with nanoscale tip). Adhesive force measurements were made with the microscale AFM tips using the force calibration plot technique, as described previously.

For the macroscale coefficient of friction, the tests were conducted using a flat-on-flat tribometer under reciprocating motion, with measurement techniques similar to those described in Bhushan et al. [2005]. Macroscale coefficient of friction data for virgin and virgin treated hair were taken from Bhushan et al. [2005]. The values for other hair types were taken in part from the indexed coefficient of friction values presented in LaTorre and Bhushan [2005b], which were transformed into actual values by the following procedure. A comparison was made between nanoscale data for chemo-mechanically damaged and chemically damaged hair, and a 15% difference was found. The same % difference was assumed on the macroscale, such that the value for chemo-mechanically damaged hair presented in Bhushan et al. [2005] was used to calculate a value for the chemically damaged hair coefficient of friction. From there, the indexed values for the chemically damaged treated (one cycle and 3 cycle) hair were scaled based on the newly found chemically damaged hair value.

24.3.2
Hair and Skin Samples

For the research reported in this chapter, all hair samples were received from Procter & Gamble (Cincinnati, OH) and Procter & Gamble Far East (Kobe, Japan) and prepared as per Appendix A. Samples arrived as hair swatches approximately 0.3 m long. Although the exact location from the root is unknown, it is estimated that hair samples used for testing were between 0.1 to 0.2 m from the scalp.

Table 24.9 presents a list of all samples used. The main hair categories of interest are: virgin (untreated), virgin (treated with one cycle of commercial conditioner), chemo-mechanically damaged (untreated), chemically damaged (untreated), and chemically damaged (treated with 1 or three cycles of commercial conditioner or a matrix of conditioner ingredients). Virgin samples are considered to be baseline specimens and are defined as having an intact cuticle and absence of chemical damage. Chemo-mechanically damaged hair fibers have been exposed to one or more cycles of coloring and permanent wave treatment, washing, and drying, as well as combing (to contribute mechanical damage), which are representative of common hair management and alteration. In the case of African damaged hair samples, chemical damage occurred only by chemical straightening. Chemically damaged fibers have not been exposed to the combing sequence in their prepa-

Table 24.9. Hair and skin samples

Sample	Type
Caucasian hair	– Virgin – Virgin, treated (1 cycle commercial conditioner) – Chemo-mechanically damaged – Chemically damaged – Chemically damaged, treated (1 cycle commercial conditioner) – Chemically damaged, treated (3 cycles commercial conditioner) – Chemically damaged, treated (various ingredients, see Table 24.10)
Asian hair	– Virgin – Virgin, treated (1 cycle commercial conditioner) – Chemo-mechanically damaged
African hair	– Virgin – Virgin, treated (1 cycle commercial conditioner) – Chemo-mechanically damaged
Synthetic materials	– Artificial collagen film (hair) – Polyurethane film (skin) – Human skin (putty replica)

ration. All treated hair samples were treated with either one or three rinse/wash cycles of a conditioner similar to a Procter & Gamble commercial product, or were treated with various combinations of surfactant, fatty alcohol, and silicone types and deposition levels, presented in the matrix of Table 24.10 [LaTorre et al., 2006]. Two different types of cationic surfactants were used: behentrimonium chloride (BTMAC) and behenyl amidopropyl dimethylamine (BAPDMA). Only one group of fatty alcohols were used for all samples. In the last set of samples, two different silicones were used: a PDMS (blend with high MW) silicone and an amino silicone.

Collagen film is typically used as a synthetic hair material for testing purposes. Polyurethane films represent synthetic human skin. They are cast from human skin and have a similar surface energy, which also makes them suitable test specimens when real skin cannot be used. To characterize the surface roughness of human skin, it was also replicated using a two-part silicone elastomer putty (DMR-503 Replication Putty, Dynamold, Inc., Fortworth, TX). The thickness of the film was approximately 3 mm.

In order to simulate hair conditioner-skin contact in AFM experiments, it is important to have the contact angle and surface energy of an AFM tip close to that of skin. Table 24.11 shows the contact angles and surface energies of materials important to the nanocharacterization of the hair samples: Untreated human hair; PDMS, which is used in conditioners; skin, which comes into contact with hair; and Si_3N_4 film, which in the form of an AFM tip is used for nanotribological measurements [LaTorre et al., 2006]. The static contact angle of Si_3N_4 film with high purity deionized water was measured in air by a sessile-drop method using a contact angle goniometer (Model 100, Rame-Hart Inc., Mountain Lakes, NJ, USA). 5 μL of

Table 24.10. Matrix of hair samples treated with various combinations of ingredients

	Ingredient	Damaged treated (commercial conditioner)	(BTMAC surfactant, no silicone)	(PDMS blend silicone, low deposition)	(PDMS blend silicone, high deposition)	(Amino silicone, low deposition)	(Amino silicone, high deposition)	(BAPDMA surfactant, no silicone)
Cationic surfactants (weight %)	Behentrimonium chloride (BTMAC)	–	×	×	×	×	×	–
	Behenyl amidopropyl dimethylamine (BAPDMA)	–	–	–	–	–	–	×
	Stearamidopropyl dimethylamine (SAPDMA)	×	–	–	–	–	–	–
Fatty alcohols (weight %)	Fatty alcohols	×	×	×	×	×	×	×
Silicones (weight %)	PDMS blend (Dimethicone)	×	–	–	–	–	–	–
	PDMS blend (with blend of high MW)	–	–	Low	High	–	–	–
	Amino silicone	–	–	–	–	Low	High	–
	Deposition levels for conditioner ingredients on hair							
Cationic surfactant deposition level		Similar to commercial conditioner	×	×	×	×	×	×
Fatty alcohol deposition level		Similar to commercial conditioner	×	×	×	×	×	×
Silicone deposition level		Similar to commercial conditioner	None	Low	High	Low	High	None

deionized water was applied on two samples using a micropipette, and three contact angle measurements were taken on each sample and averaged.

24.4 Results and Discussion

24.4.1 Surface Roughness, Friction, and Adhesion for Various Ethnicities of Hair

Topographical images of Caucasian, Asian, and African hair were taken up to scan sizes of $20 \times 20\ \mu m^2$, as shown in Fig. 24.13. Lighter areas of the images correspond to higher topography, and darker areas correspond to lower topography. Only virgin and chemo-mechanically damaged hair are shown in Fig. 24.13, because virgin treated samples closely resemble virgin hair samples. One can see the variation in cuticle structure even in virgin hair. Cracking and miscellaneous damage at the cuticle edges is evident at both virgin and chemo-mechanically damaged conditions.

Table 24.11. Contact angle and surface energy of relevant materials associated with nanotribological characterization of hair

		Contact angle (°)	Surface energy (N/m)
Human hair	– untreated	103[a]	0.024[a]
		100[b]	0.028[b]
	– damaged		
	brown coloring	60[b]	0.038[b]
	blond coloring	55[b]	0.047[b]
PDMS (bulk)		105[c]	0.020[d]
Human skin	– forehead	55[e]	0.043[e]
	– forearm	88[e]	0.038[e]
		84[f]	0.029[f]
	– finger	74[f]	0.024[f]
		58[g] (before soap-washing)	
		104[g] (after soap-washing)	0.027[g]
Si_3N_4 film		35[h]	0.047[i]

[a] Molina et al. [2001]
[b] LaTorre et al. [2006]
[c] Bhushan and Burton [2005]
[d] Jalbert et al. [1993]
[e] Lerebour et al. [2000]
[f] Schott [1971]
[g] Ginn et al. [1968]
[h] LaTorre et al. [2006]
[i] Yanazawa [1984]

In virgin hair, the damage is likely to be caused by mechanical damage resulting from daily activities such as washing, drying, and combing. Most of the virgin cuticle scales that were observed, however, were relatively intact. Long striations similar to scratches, and "scale edge ghosts" (outlines of a former overlying cuticle scale edge left on the underlying scale before it was broken away) were found on the surface. In some instances, the areas surrounding the cuticle edges appeared to show residue or debris on the surface, which were most likely remnants of a previous cuticle or the underside of the cuticle edges that were now exposed (such as the endocuticle). Caucasian and Asian virgin hair displayed a similar surface structure, while the African hair samples showed more signs of endocuticular remains along the scale edges. One can also see more curvature in the cuticle scales of African hair, which is attributed to its elliptical cross-sectional shape and curliness, which can partially uplift the scales in different places. With respect to chemo-mechanically damaged hair, it is observed that several regions seem to exist in these hair samples, ranging from intact cuticle scales to high levels of wear on the surface. In many cases, these regions occur side by side. This wide variation in the chemo-mechanically damaged cuticle structure results in a wider range of tribological properties on the micro/nanoscale for these fibers. Caucasian and Asian chemo-mechanically damaged

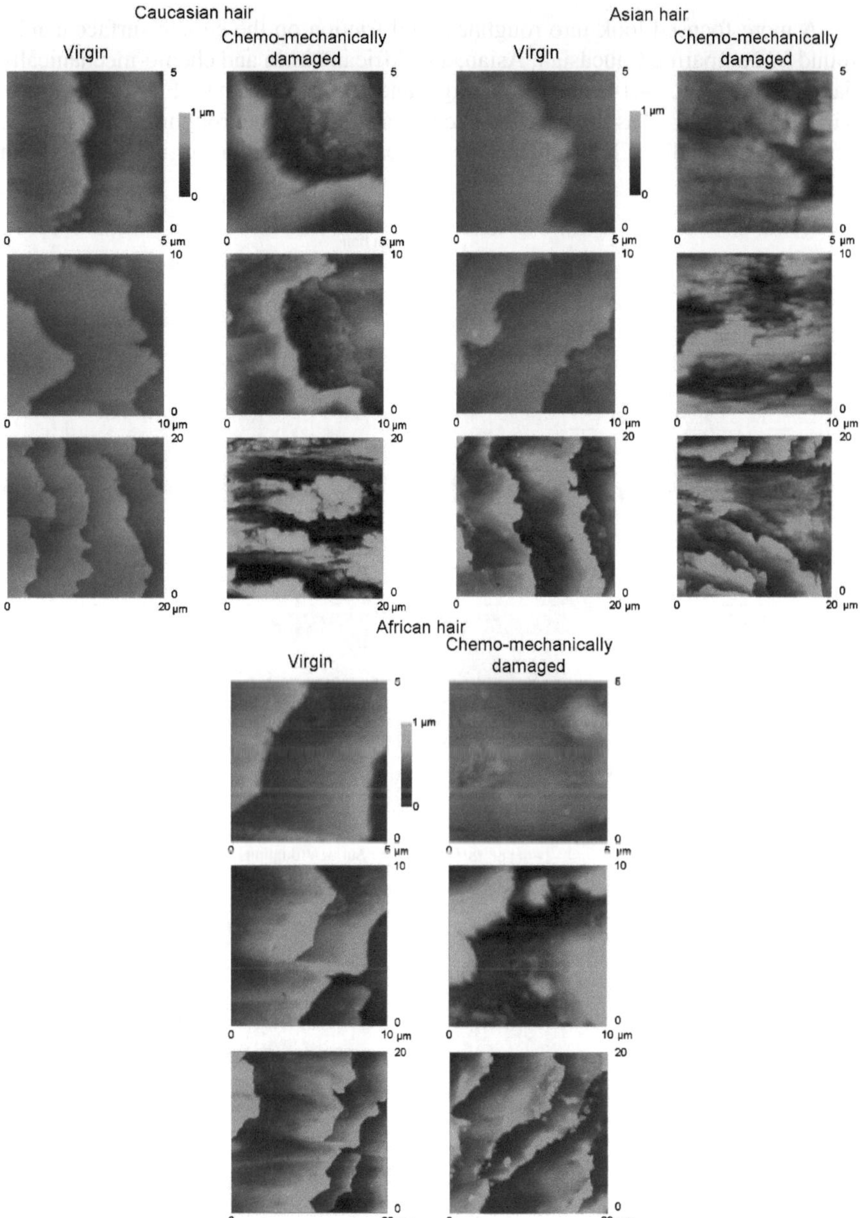

Fig. 24.13. Surface roughness of virgin and chemo-mechanically damaged Caucasian, Asian, and African hair at various scan sizes [LaTorre and Bhushan, 2005a]

hair showed more worn away cuticle scales than in chemo-mechanically damaged African hair, which showed mostly endocuticle remnants. This is most likely due to the different effects that chemical straightening has on the hair versus multiple cycles of perming the hair.

A more focused look into roughness and friction on the cuticle surface can be found by comparing Caucasian, Asian, and African virgin and chemo-mechanically damaged hair, Fig. 24.14a and b, and virgin and virgin treated hair, Fig. 24.15a and b. Virgin hair was used as the baseline to compare variations in roughness and friction force against modified hair (chemo-mechanically damaged or virgin treated). Scan

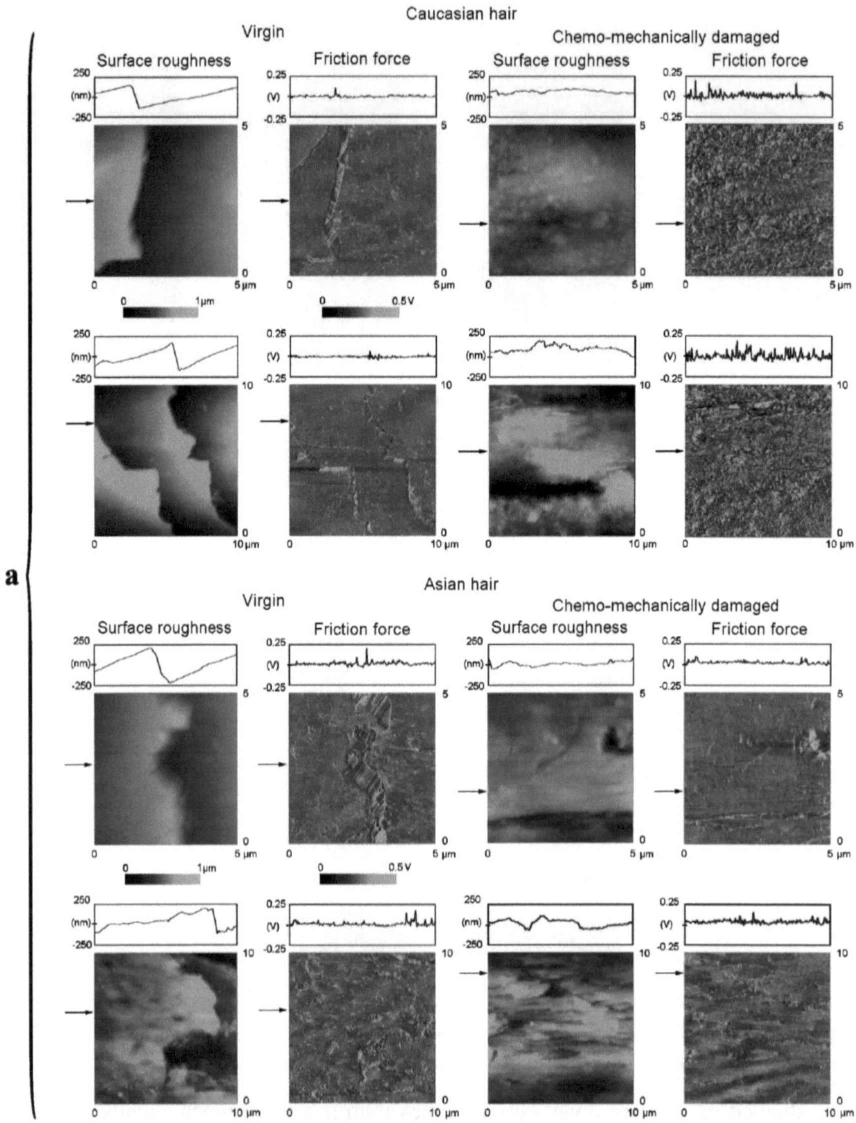

Fig. 24.14a. Surface roughness and friction images for virgin and chemo-mechanically damaged Caucasian, Asian hair at 5 and 10 μm^2 scan sizes. Shown above each image is a cross-section taken at the *corresponding arrows* to demonstrate roughness and friction force information

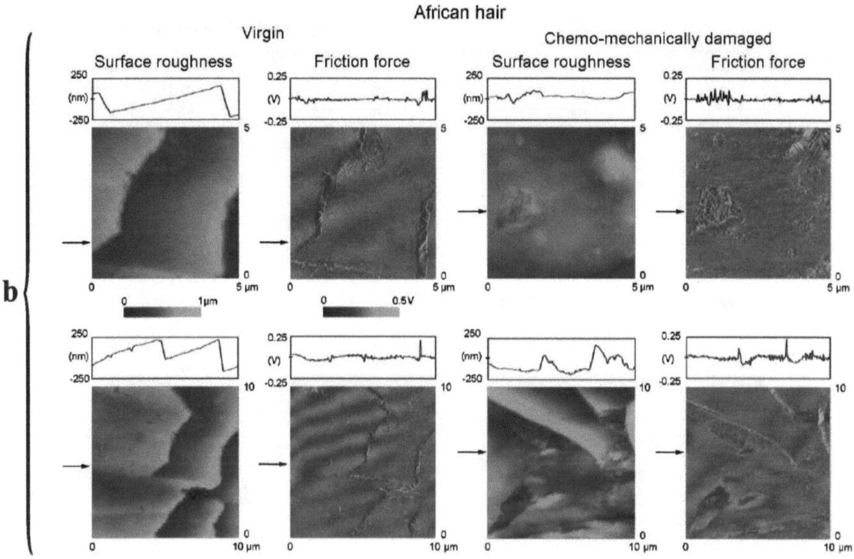

Fig. 24.14b. Surface roughness and friction force images for virgin and chemo-mechanically damaged African hair at 5 and 10 μm^2 scan sizes. Shown above each image is a cross-section taken at the *corresponding arrows* to demonstrate roughness and friction force information [LaTorre and Bhushan, 2005a]

sizes of $5 \times 5\,\mu m^2$ and $10 \times 10\,\mu m^2$ are displayed. Above each AFM and FFM image there are cross-sectional plots of the surface (taken at the accompanying arrows) corresponding to surface roughness or friction force, respectively. From the surface roughness images, the step heights of one or more cuticle edges can be clearly seen. Step heights range from approximately 0.3 to 0.5 μm.

If the surface is assumed to have Gaussian height distribution and an exponential autocorrelation function, then the surface can be statistically characterized by just two parameters: a vertical descriptor, height standard deviation σ, and a spatial descriptor, correlation distance β^* [Bhushan, 1999a,b, 2002]. The standard deviation σ is the square root of the arithmetic mean of the square of the vertical deviation from the mean line. The correlation length can be referred to as the length at which two data points on a surface profile can be regarded as being independent, thus serving as a randomness measure [Bhushan, 2002]. Table 24.12 displays these roughness parameters for each ethnicity as a function of hair type (virgin, chemo-mechanically damaged, and virgin treated) [LaTorre and Bhushan, 2005a]. Virgin hair was shown to generally have the lowest roughness values, with virgin treated hair closely resembling virgin hair. Chemo-mechanically damaged hair showed a significantly higher standard deviation of surface height. This variation is expected because of the nonuniformity of the mechanical and chemical damage that occurs throughout a whole head of hair, as well as in each individual fiber. This is in agreement with the images of chemo-mechanically damaged hair shown previously, where regions of intact cuticle and severe degradation of the surface are intermingled. The trends observed

a

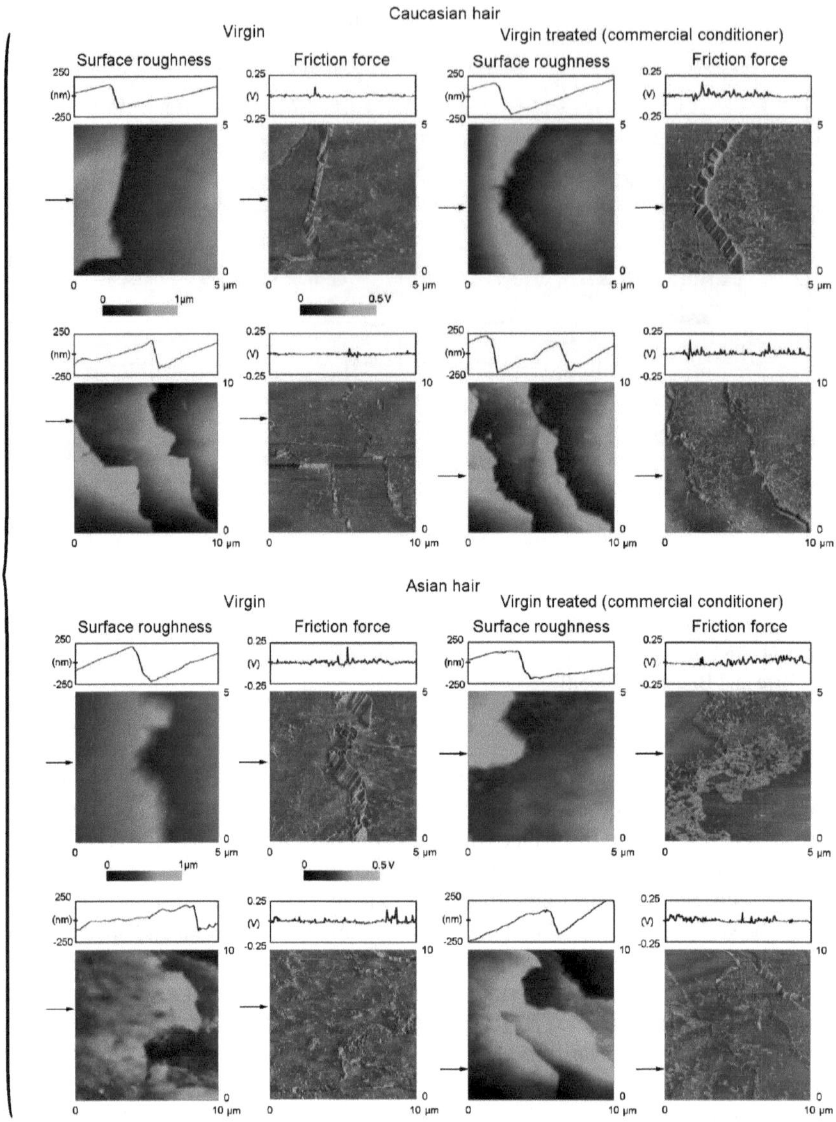

Fig. 24.15a. Surface roughness and friction images for virgin and virgin treated Caucasian, Asian hair at 5 and 10 μm^2 scan sizes. Shown above each image is a cross-section taken at the *corresponding arrows* to show roughness and friction force information

for standard deviation were not as evident for the correlation length β^*. For each ethnicity, chemo-mechanically damaged and virgin treated hair showed similar β^* values.

From Fig. 24.14a and b, friction forces are generally seen to be higher on chemo-mechanically damaged hair than on virgin hair. Although friction forces

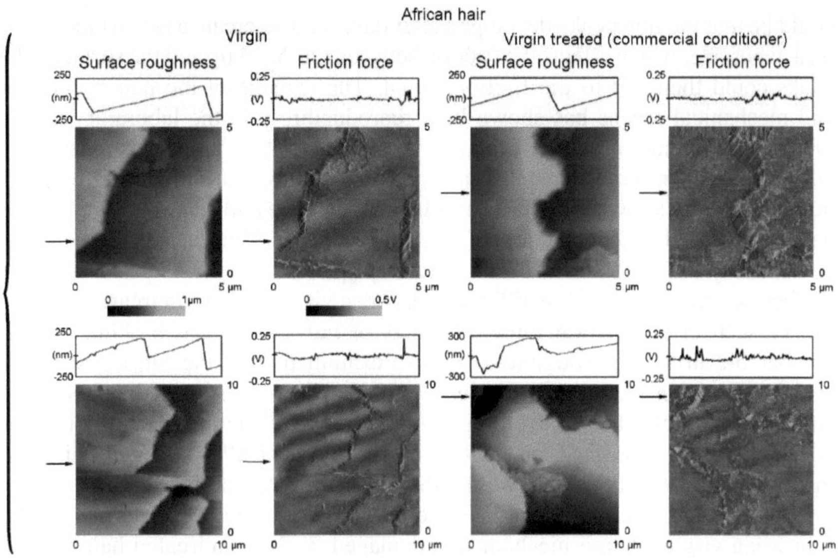

Fig. 24.15b. Surface roughness and friction force images for virgin and virgin treated African hair at 5 and 10 μm^2 scan sizes. Shown above each image is a cross-section taken at the *corresponding arrows* to demonstrate roughness and friction force information [LaTorre and Bhushan, 2005a]

Table 24.12. Surface roughness, coefficient of friction, and adhesive force values of virgin, chemo-mechanically damaged, and virgin treated (1 cycle commercial conditioner) hair at each ethnicity

	Virgin hair		Chemo-mechanically damaged hair		Virgin treated hair (commercial conditioner)	
	Surface roughness parameters (σ (nm), β^* (µm))					
	σ (nm)	β^* (µm)	σ (nm)	β^* (µm)	σ (nm)	β^* (µm)
Caucasian	12 ± 8	0.61 ± 0.3	17 ± 10	1.0 ± 0.3	12 ± 4	0.90 ± 0.3
Asian	9.7 ± 4	0.73 ± 0.3	33 ± 15	0.94 ± 0.3	7.1 ± 0.1	0.97 ± 0.3
African	12 ± 5	0.92 ± 0.3	21 ± 16	0.78 ± 0.3	11 ± 4	0.89 ± 0.2
	Average coefficient of friction μ					
Caucasian	0.02 ± 0.01		0.13 ± 0.05		0.03 ± 0.01	
Asian	0.03 ± 0.01		0.13 ± 0.04		0.06 ± 0.04	
African	0.04 ± 0.02		0.14 ± 0.08		0.05 ± 0.01	
	Adhesive force F_m (nN)					
Caucasian	25		16		32	
Asian	31		18		79	
African	35		38		63	

were similar in magnitude, it was observed that the friction force on the cuticle surface of chemo-mechanically damaged hair showed a much larger variance, which contributed to the higher friction values. Another contribution to the higher friction could be that the tiny peaks developed after damage also create a ratchet mechanism on a nanoscale, which affects the friction between the AFM tip and the surface. These peaks could then add to the friction signal. The damage of the hair by chemical and mechanical means has shown high reproducibility in the laboratory in terms of structure alteration, which explains the similar friction properties regardless of ethnicity for chemo-mechanically damaged hair. With virgin and virgin treated hair, however, it is unknown what prior mechanical damage and sun exposure the fibers have seen, and this depends largely on the individuals. Thus, across ethnicity there is variability in friction force for those hair samples.

Perhaps the most notable difference between virgin and virgin treated hair fibers can be seen in the friction force mappings of Fig. 24.15a and b. Although quite comparable in surface roughness, a close examination of the virgin treated hair surface shows an increase in friction force, usually only surrounding the bottom edge of the cuticle. This was unlike virgin hair, where friction generally remained constant along the surface, and unlike chemo-mechanically damaged hair, where there was large variability that was random over the entire surface.

Figure 24.16 presents friction force curves as a function of normal load for Caucasian virgin, chemo-mechanically damaged, and virgin treated hair to further illustrate the previous discussion. One can see a relatively linear relationship between the data points for each type of hair sample. When plotted in such a way, the coefficient of friction is determined by the slope of the least squares fit line through the data. If this line is extended to intercept the horizontal axis, then a value for adhesive force can also be calculated, since friction force F is governed by the relationship

$$F = \mu(W + F_a)\,, \tag{24.1}$$

where μ is the coefficient of friction, W is the applied normal load, and F_a is the adhesive force [Bhushan, 1999a, 2002].

One explanation for the increase in friction force of virgin treated hair on the micro/nanoscale is that during tip contact, meniscus forces between the tip and the conditioner/cuticle become large as the tip rasters over the surface, causing an increase in the adhesive force. This adhesive force is of the same magnitude as the normal load, which makes the adhesive force contribution to friction rather significant. Thus, at sites where conditioner is accumulated on the surface (namely around the cuticle scale edges), the friction force actually increases. On the macroscale, however, the adhesive force is much lower in magnitude than the applied normal load, so the adhesive force contribution to friction is negligible over the hair swatch. As a result, virgin treated hair shows a decrease in friction force on the macroscale, which is opposite to the micro/nanoscale trend. However, the friction and adhesion data on the micro/nanoscale are useful, because they relate to the presence of conditioner on the cuticle surface and allow one to obtain an estimate of conditioner distribution.

It is also observed from Fig. 24.16 that while chemo-mechanically damaged hair displays a higher friction force on the application of the normal load, and consequently a higher coefficient of friction, chemo-mechanically damaged hair friction is not as strongly dependent on adhesive force contribution as the virgin and virgin

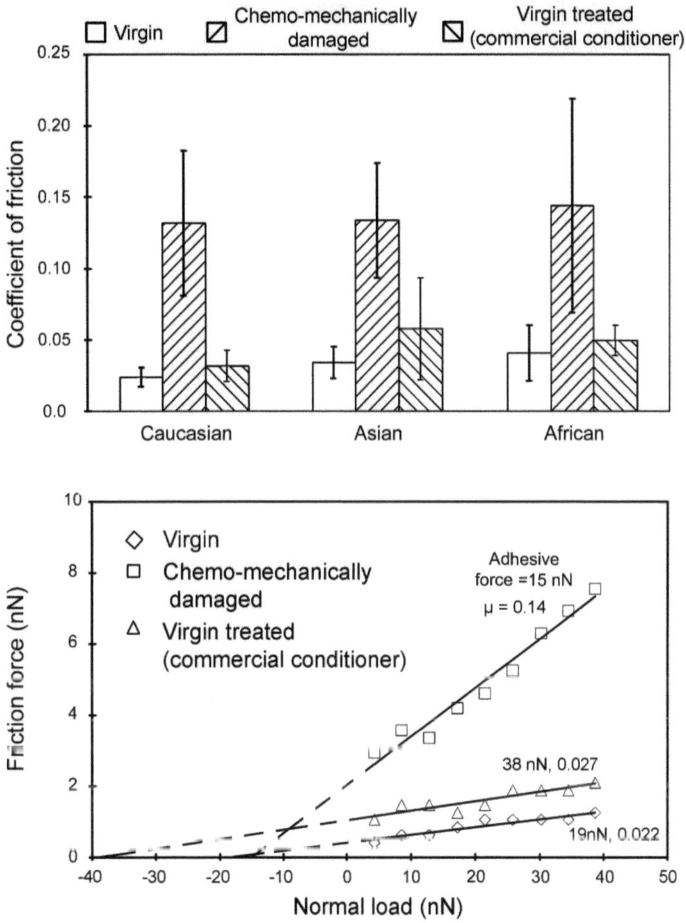

Fig. 24.16. Average coefficient of friction values for virgin, chemo-mechanically damaged, and treated hair at each ethnicity. *Error bars* represent $\pm 1\sigma$ on the average coefficient value. The *bottom plot* shows friction force vs. normal load curves displaying typical values for virgin, damaged, and virgin treated Caucasian hair [LaTorre and Bhushan, 2005a]

treated hair. Average values for μ were calculated and are compiled in Fig. 24.17 and Table 24.12 for all hair ethnicities and types [LaTorre and Bhushan, 2005a]. Error bars represent $\pm 1\sigma$ on the average coefficient value. The coefficients of friction for virgin, chemo-mechanically damaged, and virgin treated Caucasian hair are 0.02, 0.13, and 0.03, respectively. For virgin, chemo-mechanically damaged, and virgin treated Asian hair, the coefficients of friction are 0.03, 0.13, and 0.06, respectively. Finally, virgin, chemo-mechanically damaged, and virgin treated African hair coefficients of friction are 0.04, 0.14, and 0.05, respectively. Chemo-mechanically damaged hair presents the highest coefficient of friction, but also displays the largest standard deviation, due to the large variations in chemical and mechanical damage that each hair or hair bundle experiences. The coefficient of friction of virgin treated

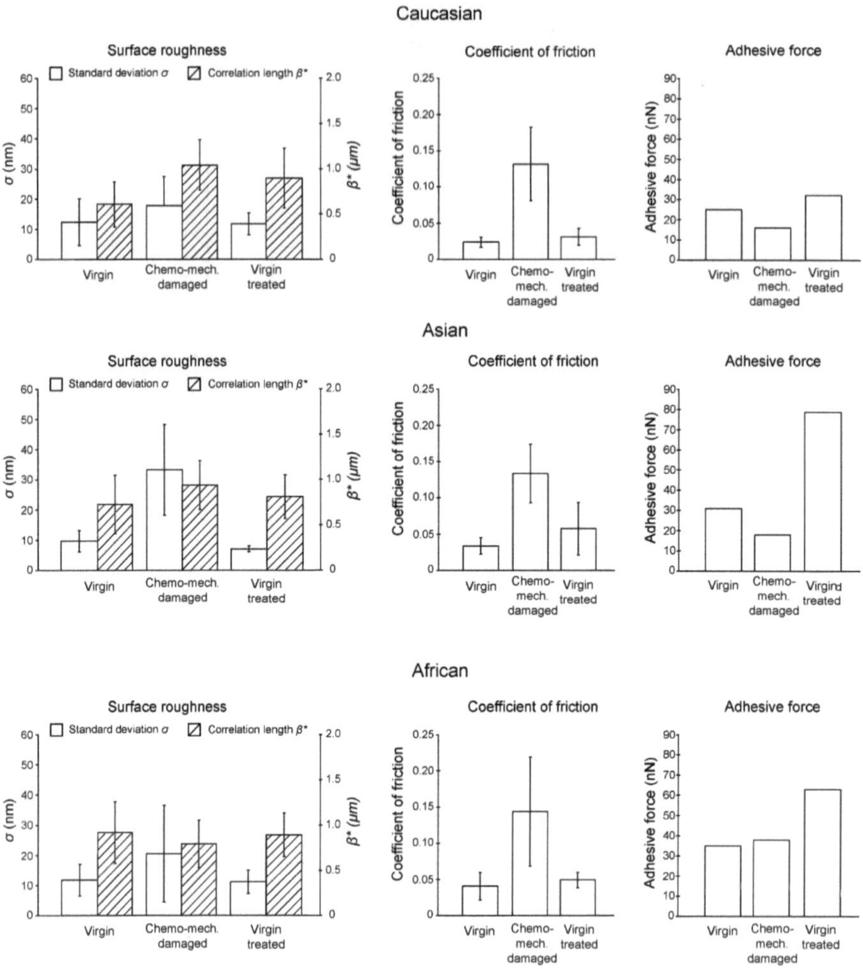

Fig. 24.17. Surface roughness, coefficient of friction, and adhesive force data for virgin, chemo-mechanically damaged, and virgin treated hair at each ethnicity [LaTorre and Bhushan, 2005a]

hair is slightly larger than that of virgin hair for all ethnicities. While the coefficient of friction is similar in virgin and virgin treated hair, the adhesive force contribution to friction for Caucasian virgin treated hair is higher than in Caucasian virgin hair, when calculated according to the method described above. However, this was not always the trend for Asian and African virgin treated hair samples. It should be noted that since in friction force measurement the tip moves laterally over the surface, this might cause a smearing out of the conditioner layer, which accounts for the inconsistent trend. In the adhesive force mappings described in the next section, where determination of the adhesive force does not depend on this lateral movement, all virgin treated hair samples showed higher adhesion than their virgin hair counterparts.

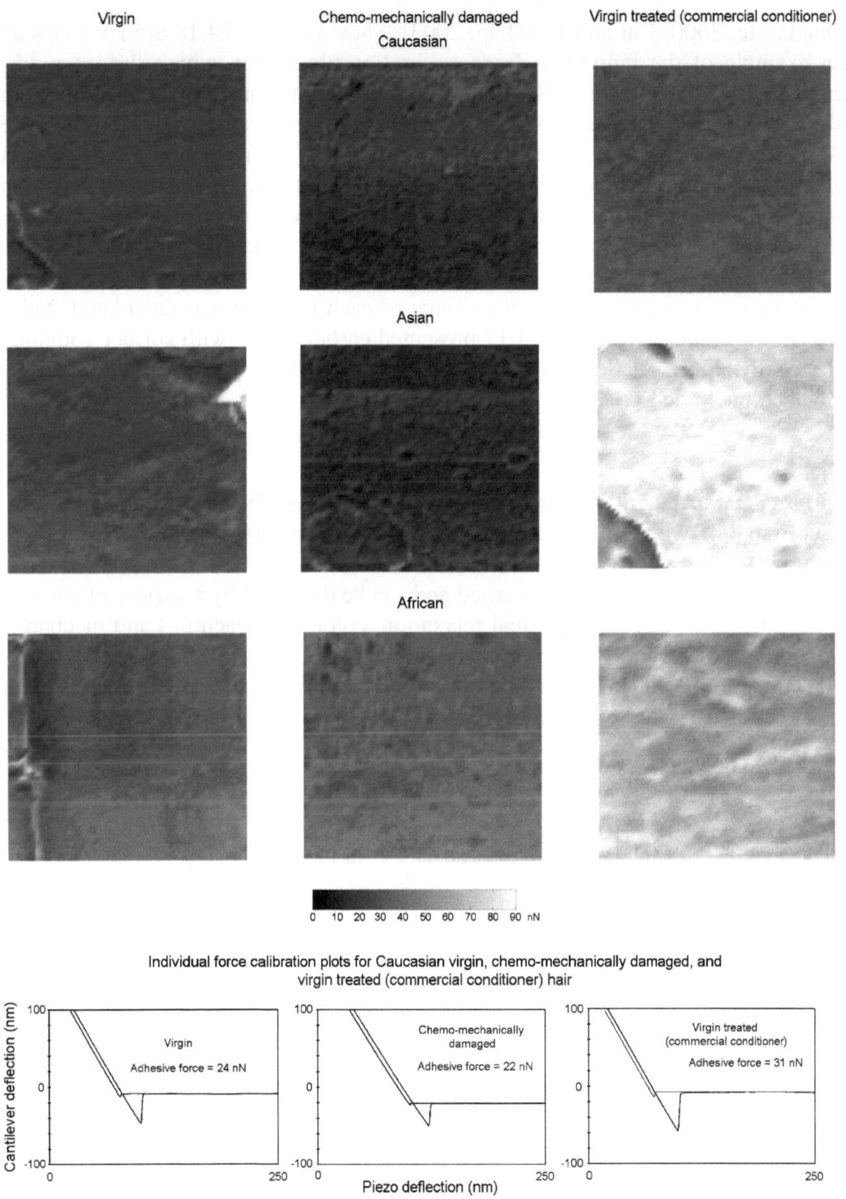

Fig. 24.18. Force-volume maps of virgin, chemo-mechanically damaged, and virgin treated hair at each ethnicity. Examples of the individual force calibration plots, which make up the FV maps, are presented for Caucasian hair of each type [LaTorre and Bhushan, 2005a]

A force calibration plot (FCP) technique and resulting adhesive force maps (commonly called force-volume maps) can be used to understand the adhesive forces between the AFM tip and the sample [Bhushan, 1999b, 2005; Bhushan and

Dandavate, 2000; Liu and Bhushan, 2003]. Shown in Fig. 24.18 are FV maps and an example of the individual force calibration plots from which the maps were created. The adhesive force distribution for chemo-mechanically damaged hair was shown to be comparable to virgin hair adhesive force values, but slightly lower. A significant increase in the adhesive force over the entire mapping was found in all cases of virgin treated hair as compared to virgin hair, especially in Asian and African hair. Conditioner distribution can be seen from these images. This technique shows promise of being very useful in further study of the distribution of materials and hair care products on the surface of the hair.

A typical value for the adhesive force of each FV map was calculated. Values are shown in the plot of Fig. 24.17 presented earlier, along with surface roughness and coefficient of friction data for all hair samples. Adhesive force values are also tabulated in Table 24.12.

24.4.2 Surface Roughness, Friction, and Adhesion for Virgin and Chemically Damaged Caucasian Hair (with and without Commercial Conditioner Treatment)

The hair surface is negatively charged and can be damaged by a variety of chemical (permanent hair waving, chemical relaxation, coloring, bleaching) and mechanical (combing, blowdrying) factors [Robbins, 1994; Bolduc and Shapiro, 2001; Gray, 2001]. Figure 24.19 shows the transformation and wear of the cuticles scales before

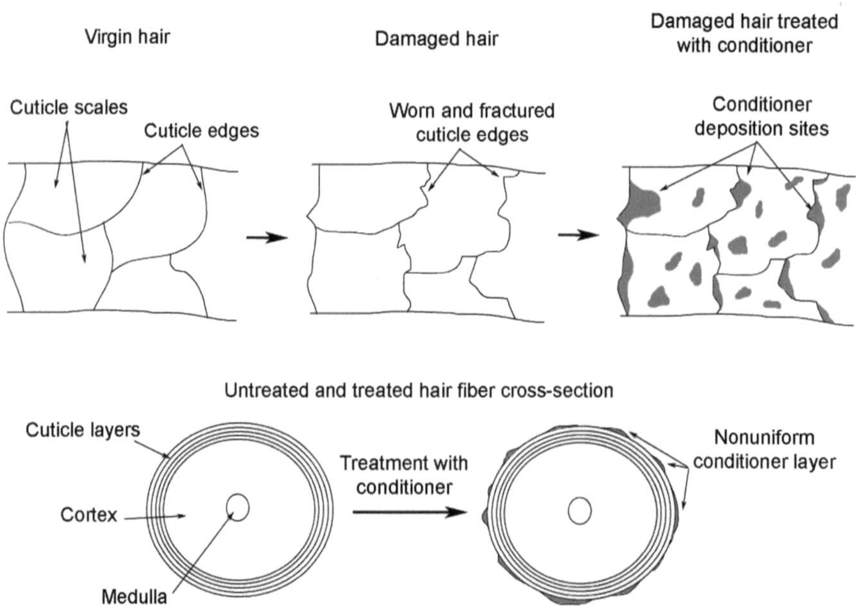

Fig. 24.19. The effect of damage to the cuticle scales and the deposition of conditioner on the cuticle surface. The cross-section of the hair with and without conditioner is shown below [LaTorre et al., 2006]

and after damage. Chemical damage causes parts of the scales to fracture and reveal underlying cuticle remnants. Conditioner application provides a protective coating to the hair surface to prevent future damage.

Shown in Fig. 24.20 are surface roughness and friction force plots for virgin, virgin treated, chemically damaged, chemically damaged treated (one cycle conditioner), and chemically damaged treated (three cycles conditioner). Above each AFM and FFM image are cross-sectional plots of the surface (taken at the accompanying arrows) corresponding to surface roughness and friction force, respectively. Although virgin and virgin treated hair are quite comparable in surface roughness maps, examination of the treated hair surface shows an increase in friction force, especially in the area surrounding the scale edge bottom level. These frictional patterns observed in treated hair were not like anything observed in the virgin or chemically damaged cases. Images of all hair types have shown friction variation due to edge contributions and cuticle mechanical damage that has left only remnants of the cuticle sublayer (such as the endocuticle). Further investigation of the corresponding

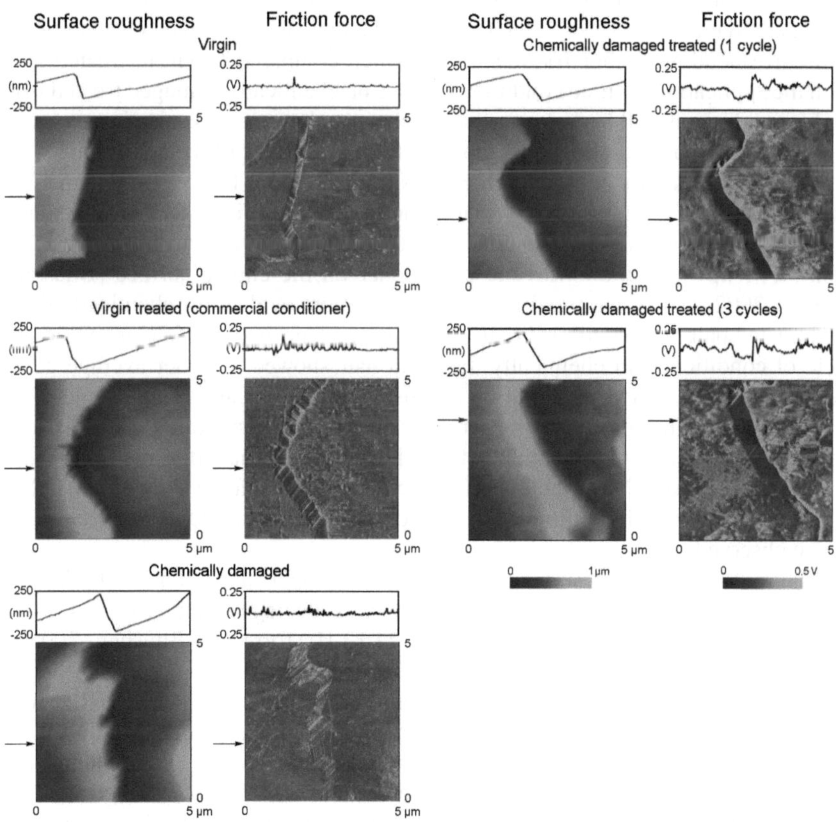

Fig. 24.20. Surface roughness and friction force images for virgin, virgin treated, damaged, damaged treated (one cycle conditioner), and damaged treated (three cycles conditioner) hair at 5 µm scan sizes [LaTorre and Bhushan, 2005b]

treated hair roughness images showed that this increase in friction was not due to a significant change in surface roughness either. One explanation for the increase in friction force of treated hair on the micro/nanoscale is that during tip contact, meniscus forces between the tip and the conditioner/cuticle become large as the tip rasters over the surface, causing an increase in the adhesive force. This adhesive force is of the same magnitude as the normal load, which makes the adhesive force contribution to friction rather significant. Thus, at sites where conditioner is accumulated on the surface (namely around the cuticle scale edges), friction force actually increases.

In general, friction forces are higher on chemically damaged hair than on virgin hair. Although friction forces were similar in magnitude, it was observed that the friction force on the cuticle surface of chemically damaged hair showed a much larger variance, which contributed to the higher friction values. Chemically damaged treated hair shows a much stronger affinity to the conditioner. It is well-known that the cuticle surface of hair is negatively charged. This charge becomes even more negative with the application of chemical damage to the hair. As a result, the positively charged particles of the conditioner have an even stronger attraction to the chemically damaged surface, which explains the increased presence of conditioner (and corresponding higher friction forces) when compared to virgin treated hair. With the application of three conditioner cycles on chemically damaged treated hair, the friction force is still higher near the cuticle edge, however it is also increased all over the cuticle surface, showing a more uniform placement of the conditioner.

Figure 24.21 shows adhesive force maps for the various hair types, which gives a measurement of adhesive force variation on the surface. Virgin treated hair shows a higher adhesive force than virgin hair due to the meniscus effects that come about from AFM tip interaction with the conditioner on the cuticle surface [Bhushan, 1999a,b, 2002]. The same trend is true and even more evident for chemically damaged treated hair compared to chemically damaged hair. A possible reason that one cycle of conditioner on chemically damaged hair showed a higher average adhesive force than three cycles could be because the three cycles generally place the conditioner more uniformly on the surface rather than accumulating it mostly near the bottom surface near the cuticle edge, which is where the adhesive force maps were generally taken. Nevertheless, the increased adhesive force shown in the plots is a clear indication of conditioner present on the hair surface, and its localization can be observed.

Figure 24.22 presents surface roughness, coefficient of friction, and adhesive force plots for the various virgin and chemically damaged hair types discussed above. The data is also presented in Table 24.13 [LaTorre and Bhushan, 2005b]. Surface roughness for human hair is characterized by a vertical descriptor, height standard deviation σ, and a spatial descriptor, correlation distance β^* [Bhushan, 1999a,b, 2002]. Virgin and virgin treated hair showed similar σ values, while β^* was higher in virgin treated hair. Chemically damaged hair and both types of chemically damaged treated hair showed similar roughness values, although σ was higher for the treated cases. The chemically damaged hair presented in this work is different from the chemo-mechanically damaged hair studied in LaTorre and Bhushan [2005a]. It seems that chemically damaging the surface does not lead to as much wear and surface roughness increase as the combination of both chemical and mechanical

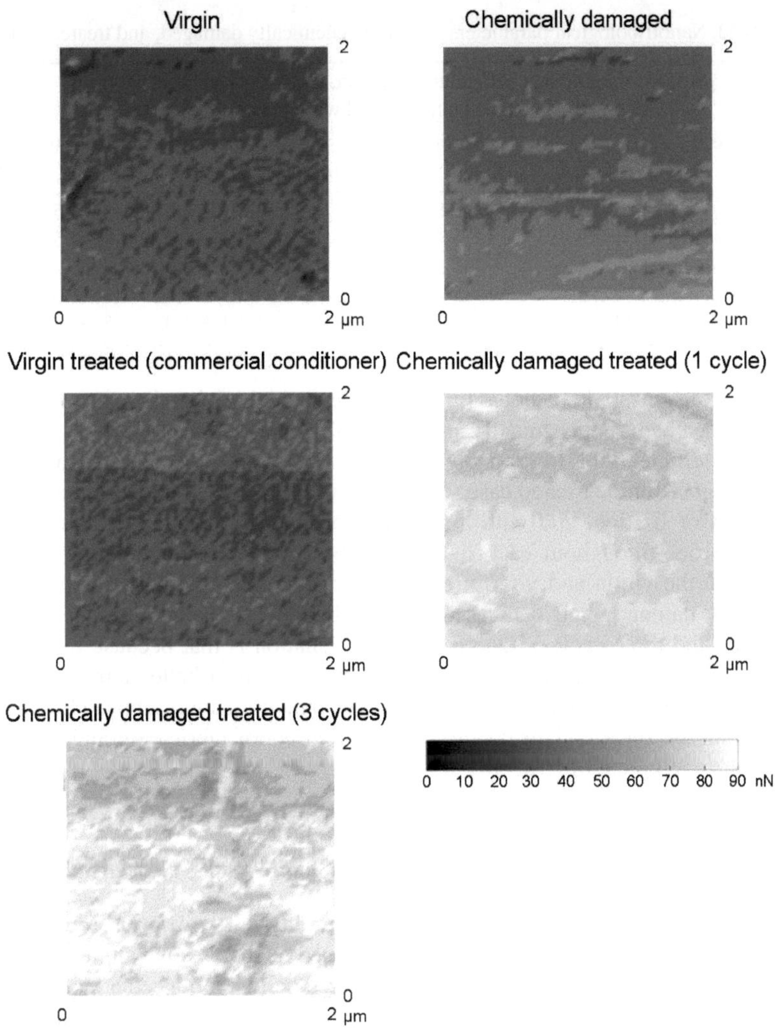

Fig. 24.21. Adhesive force maps displaying variations in the adhesive force on the cuticle surface of various hair types [LaTorre and Bhushan, 2005b]

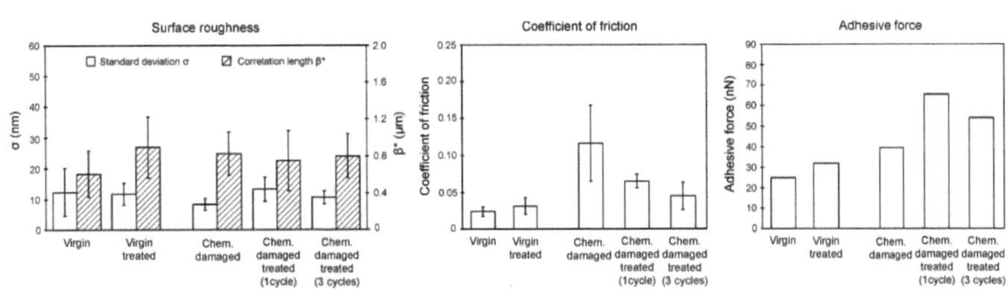

Fig. 24.22. Surface roughness, coefficient of friction, and adhesive force plots for various hair types [LaTorre and Bhushan, 2005b]

Table 24.13. Nanotribological parameters for virgin, chemically damaged, and treated hair

Surface roughness, coefficient of friction, and adhesive force for virgin and chemically damaged hair, with and without conditioner treatment

Hair type	Surface roughness		Coefficient of friction	Adhesive force (nN)
	σ (nm)	β^* (μm)		
Virgin	12 ± 8	0.61 ± 0.2	0.02 ± 0.01	25 ± 5
Virgin, treated	12 ± 4	0.90 ± 0.3	0.03 ± 0.01	32 ± 5
Chemically damaged	8.4 ± 2	0.83 ± 0.2	0.13 ± 0.06	39 ± 0.5
Damaged, treated (1 cycle)	13 ± 4	0.75 ± 0.3	0.05 ± 0.02	66 ± 0.7
Damaged, treated (3 cycles)	11 ± 2	0.80 ± 0.2	0.04 ± 0.02	54 ± 33

damage. Thus, it should be noted (and it is understandable) that there are differences between the chemo-mechanically and chemically damaged hair.

The coefficients of friction of virgin and virgin treated hair are similar, but slightly higher for the treated cases. Chemically damaged hair shows a much higher coefficient of friction, and with more variation in the values, since the chemical damage varies throughout each individual fiber. An interesting finding was that, contrary to the virgin and virgin treated hair results, the coefficient of friction for chemically damaged hair decreased with application of the conditioner treatment (both one and three cycles). One possible explanation is that because the stronger negative charge on chemically damaged hair results in a better attraction of the conditioner, this leads to a higher adhesive force but, more importantly, a lower shear strength on the surface. This creates an overall effect of lubrication and ultimately lowers the coefficient of friction.

24.4.2.1 Effect of Relative Humidity, Temperature, Soaking, and Durability Measurements

Figure 24.23 displays the effect of relative humidity on the friction force and the adhesive force. The coefficient of friction remained relatively constant for virgin and virgin treated hair. However, chemically damaged hair experienced a large increase in the coefficient of friction at high humidity, while chemically damaged treated hair experienced the opposite trend. This clearly shows that heavy moisture in the air plays affects the frictional properties of chemically damaged hair. When combined with conditioner, a lubricating effect once again dominates as the water helps form a liquid layer that is more easily sheared. In terms of the adhesive force, most samples showed a decrease in adhesive force with high humidity. It is expected that as water builds up on a surface, meniscus effects diminish and as a result do not readily contribute to the adhesive force. Thus, the adhesive force is expected to decrease at very high humidity.

Figure 24.24 displays the effect of temperature on the friction force and the adhesive force. The coefficient of friction generally decreased with increasing temperature. As the hair fiber heats up, conditioner present on the surface decreases in viscosity, causing a thinner film and a lower friction force. The lower friction force ultimately leads to lower coefficient of friction values. The adhesive force was shown to decrease with increasing temperature as well. This was especially evident

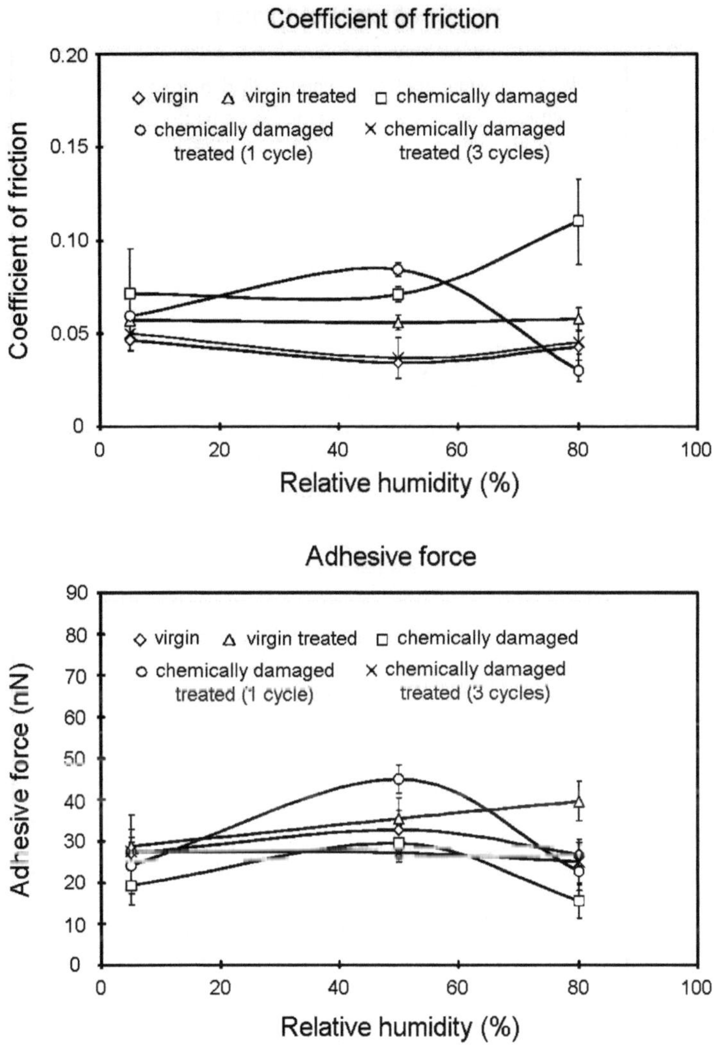

Fig. 24.23. Effect of relative humidity on nanotribological properties of various hair types [LaTorre and Bhushan, 2005b]

for treated hair fibers, whereas a large adhesive force at room temperature decreased rapidly to adhesion values similar to non-treated fibers. It is most likely that at higher temperatures the thinning conditioner layer causes a reduced surface tension, which directly relates to the drop in the adhesive force.

Virgin, chemically damaged, and chemically damaged treated hair samples were soaked in de-ionized water for five minutes. Their corresponding coefficient of friction was measured and compared to coefficient of friction values for dry samples that were adjacent to the wet samples on the respective hair fiber. Figure 24.25 shows the results for two hair samples of each hair type. Virgin hair exhibits a decrease

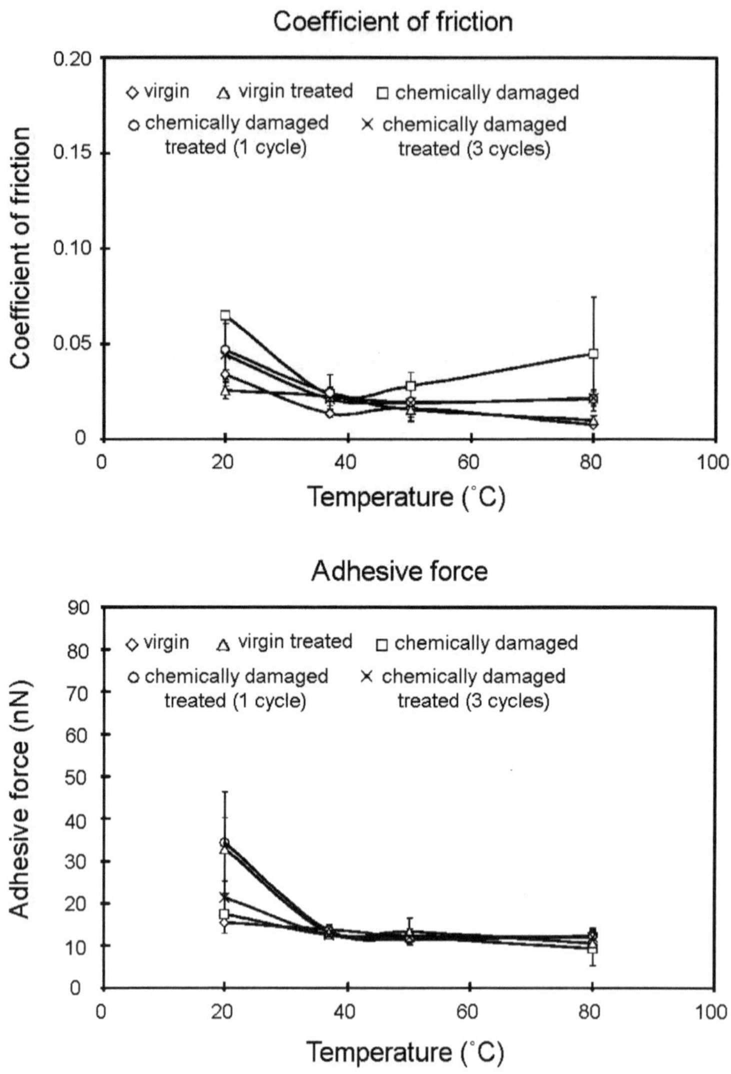

Fig. 24.24. Effect of temperature on nanotribological properties of various hair types [LaTorre and Bhushan, 2005b]

in the coefficient of friction after soaking. Virgin hair is more hydrophobic (see Table 24.11), so more water is present on the surface and this results in a lubrication effect after soaking. Chemically damaged hair tends to be hydrophilic due to the chemical degradation of the cuticle surface, and this results in absorption of water after soaking. This softens the hair, which leads to higher friction, even with conditioner treatment. This is yet another indication that virgin and chemically damaged hair types have significantly different surface properties which in many cases results in opposite trends for their nanoscale tribological properties. The adhesive force for

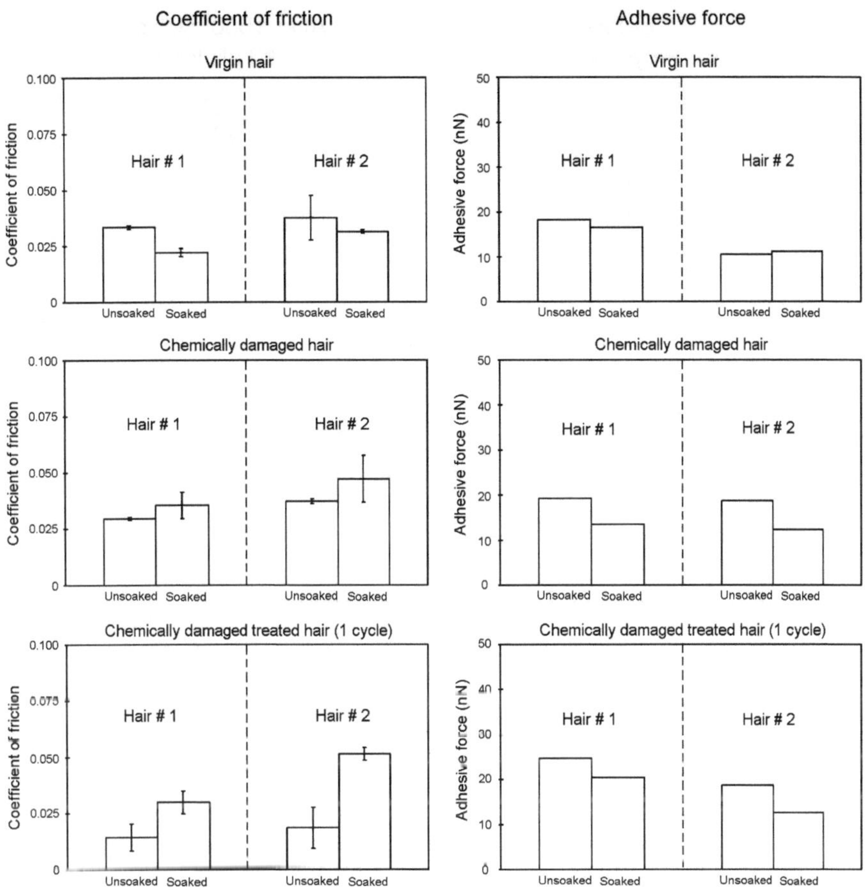

Fig. 24.25. Effect of soaking in de-ionized water on coefficient of friction and adhesive force for virgin, chemically damaged, and chemically damaged treated (one cycle commercial) hair [LaTorre et al., 2006]

virgin hair remained approximately the same before and after soaking, while it decreased for chemically damaged and chemically damaged treated hair after soaking.

Figure 24.26 shows the durability effects on the friction force for various hair types. Above the graph are pictures of unworn and worn virgin hair, with the cuticle edge serving as a reference point. Before testing, the surface is relatively smooth and void of any large debris or wear. After 1000 cycles at approximately 10 μN load with a stiff silicon AFM tip, however, the interaction has caused degradation and wear (scratch) marks on the cuticle scale. This is the type of wear one could potentially see if hair were to come in contact with sand from a day spent at the beach, among other activities. Virgin hair shows an obvious increase in friction force signal as the scratch mark digs further into the surface. By this time, the lubricious lipid layer on the surface of the virgin cuticle has been worn away and the friction force comes close to the magnitude of chemically damaged hair friction force at

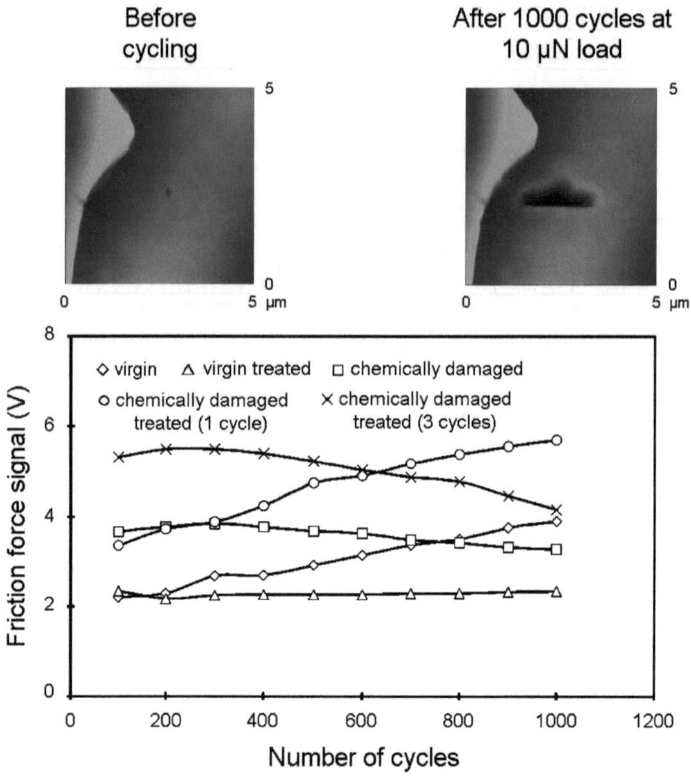

Fig. 24.26. Durability study of friction force change as a function of AFM tip cycling for various hair types. The *images above* the plot signify before and after comparisons of a cuticle surface subjected to cycling at a 10 μN load [LaTorre and Bhushan, 2005b]

the onset of cycling. When conditioner is applied to virgin hair, however, the wear does not show up as an increase in the friction force. Thus, conditioner serves as a protective covering to virgin hair and helps protect the tribological properties when wear ensues.

24.4.3 Surface Roughness, Friction, and Adhesion for Hair Treated with Various Combinations of Conditioner Ingredients

Figure 24.27 displays the representative surface roughness and friction force maps for chemically damaged hair and the seven different applied treatments (see Table 24.10 presented earlier). When conditioner is applied to the surface, a pattern of high friction is shown in the area surrounding the bottom edge of the cuticle. Likewise, the application of a BTMAC surfactant (without silicone deposition) or a BAPDMA surfactant (without silicone deposition) results in similar friction features. This is believed to be an area of conditioner accumulation, which causes increased friction due to meniscus effects. Friction maps for PDMS blend of silicone (at both low and

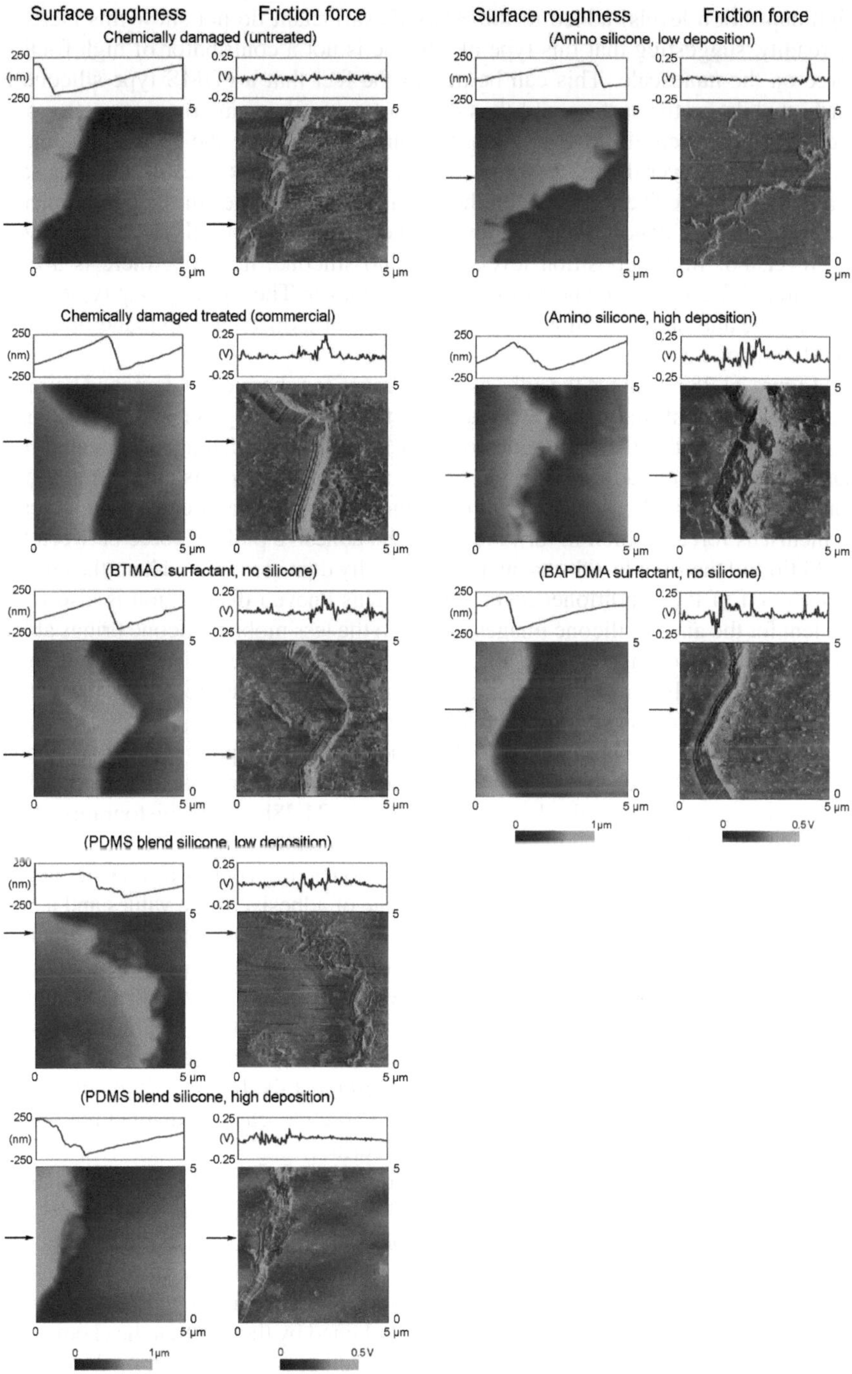

Fig. 24.27. Surface roughness and friction force maps of Caucasian chemically damaged hair with various treatments [LaTorre et al., 2006]

high deposition levels) added to the BTMAC surfactant do not show this increase as readily, suggesting that this type of silicone is not a contributor of high friction force on the nanoscale. This can be due to the fact that a PDMS type silicone is fairly mobile on the surface and thus does not cause the same meniscus effects as the AFM tip rasters through it. Also, the surface energy of the PDMS silicone is believed to be lower than that of the cationic surfactants. As a result, the meniscus force affecting the friction between the AFM tip and silicones in the conditioner is lower than that of conditioner without silicones. Thus, the overall friction force will be lower. For high deposition levels of amino silicone, however, there is a high variation of friction force and a more distributed layer. The amino group typically is less mobile and harder to move, which accounts for a different slip plane flow than PDMS silicone.

Figure 24.28a displays the adhesive force maps for chemically damaged hair and the seven different treatments. As shown in the legend, a lighter area corresponds to a higher tip pull-off force (adhesive force). Chemically damaged untreated hair has a relatively low adhesive force, and is more or less consistent over the hair surface. In nearly all cases, addition of a conditioner treatment caused an increase in meniscus forces, which in turn increased the adhesive pull-off force between the AFM tip and the sample. Observing the chemically damaged treated hair, the uneven distribution of the conditioner layer is seen. This uneven distribution is also most evident for the amino silicone images, in which the less mobile silicone brings about a distinguishable change in adhesion over the surface. For the PDMS blend silicone, it is seen that adhesion over the surface is much more consistent than the amino silicone, where various areas of high adhesion occur.

It is important to note that while the adhesive force maps presented are representative images for each treatment, the adhesive force varies significantly when treatments are applied to the hair surface. Figure 24.28b shows histograms of all adhesive force data for chemically damaged, chemically damaged treated, PDMS blend silicones (high deposition), and amino silicones (high deposition). Chemically damaged treated hair shows a much larger range of adhesive force values and a normal distribution, which suggests that the conditioner layer is normally distributed. The histogram for PDMS blend silicone treatment shows a normal distribution at the larger adhesive force values, but also shows another peak at low adhesion values. Amino silicone treated hair follows a normal distribution, but it is interesting to note the distinct groupings of the adhesion values and the spacing between them. This is further evidence that the amino silicone groups most likely attach immediately to the hair surface and are less mobile than PDMS silicone, causing distinct regions of high and low adhesion values over the cuticle surface.

Figure 24.29 displays a summary of the data collected for all chemically damaged hair samples and their treatments. Table 24.14 reviews some of the observations and corresponding mechanisms that help to explain the trends found [LaTorre et al., 2006]. The application of the commercial conditioner to the chemically damaged hair caused a decreased coefficient of friction and a large increase in adhesive force. The decreased coefficient of friction may be explained by the fact that the chemically damaged hair accumulates much of the positively charged conditioner on the surface due to its highly negative charge, which in turn makes it easier to shear the liquid on the surface, causing a lower coefficient of friction. However, the nanoscale pull-

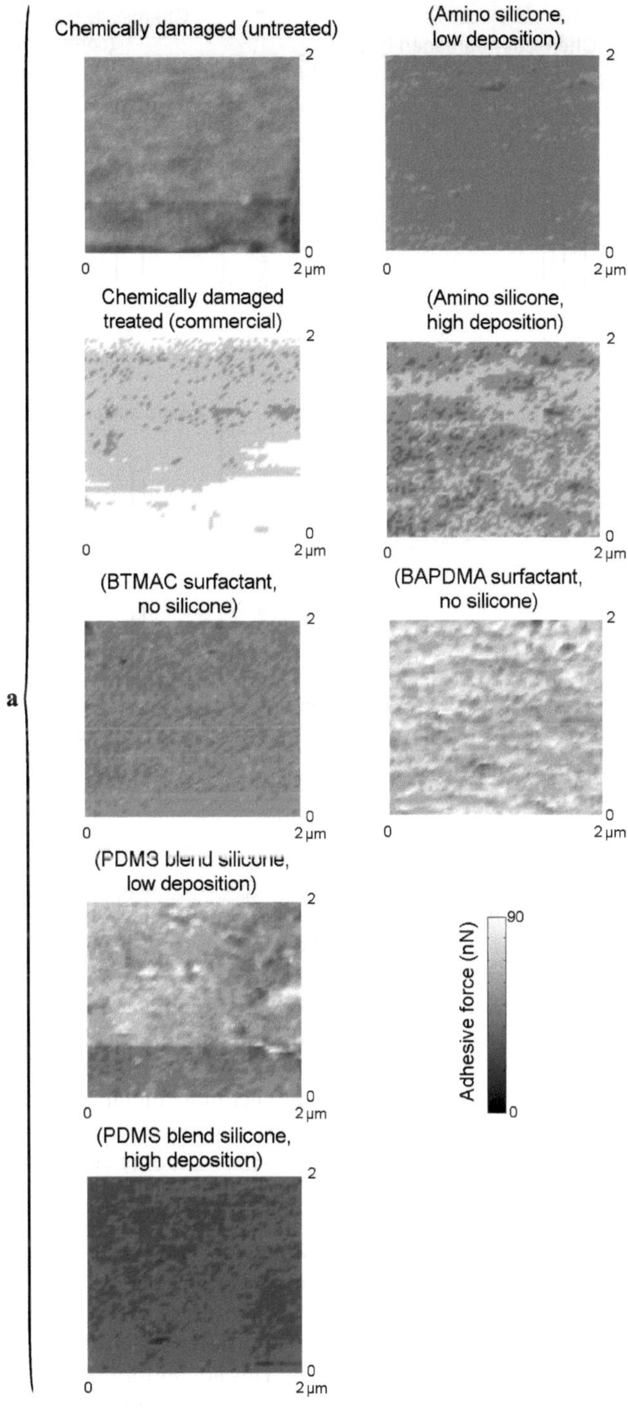

Fig. 24.28a. Adhesive force maps for damaged hair with various treatments

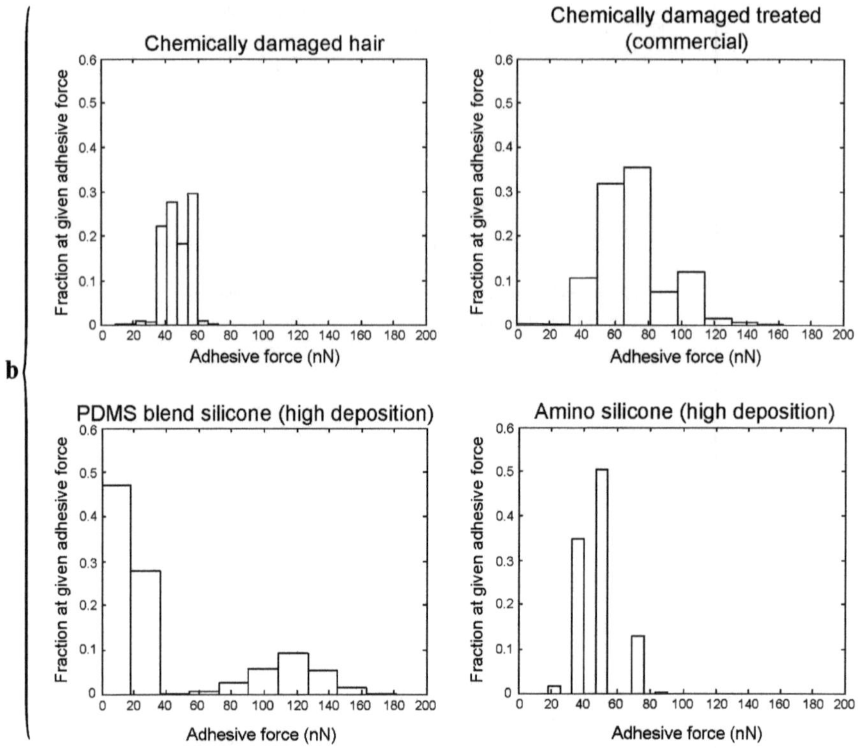

Fig. 24.28b. Adhesive force histograms for various hair types [LaTorre et al., 2006]

off force (adhesive force) is much larger than on the untreated hair because of meniscus effects. In general, the adhesive force varied widely, but typically showed a significant increase with the presence of conditioner. As discussed previously, this is a clear sign that meniscus effects are influencing the pull-off force between the tip and the sample.

In most cases, the macroscale and microscale coefficient of friction followed the same trend, in which a decrease was observed with the addition of the PDMS blend or amino silicones to the BTMAC surfactant. The silicones are typically used as a major source of lubrication and thus give the conditioner more mobility on the hair surface compared to just surfactants and fatty alcohols. The inverse trend was seen only for the amino silicone group. The amino silicones have a strong electrostatic attraction to the negatively charged hair surface, which in turn creates higher binding forces and less mobility. The dampened mobility of the amino silicone at high deposition levels, with respect to hair surface and tip, may account for this wide variation in coefficient of friction and large adhesive force values.

In terms of adhesive force, it was previously observed in Fig. 24.28a that the amino silicone treatments showed much more distinct regions of higher and lower adhesion compared to PDMS blend silicones. This nonuniform amino silicone thickness distribution on hair is also believed to be caused by the inhibited mobility, as the molecules immediately attach to the hair at contact and do not redistribute as

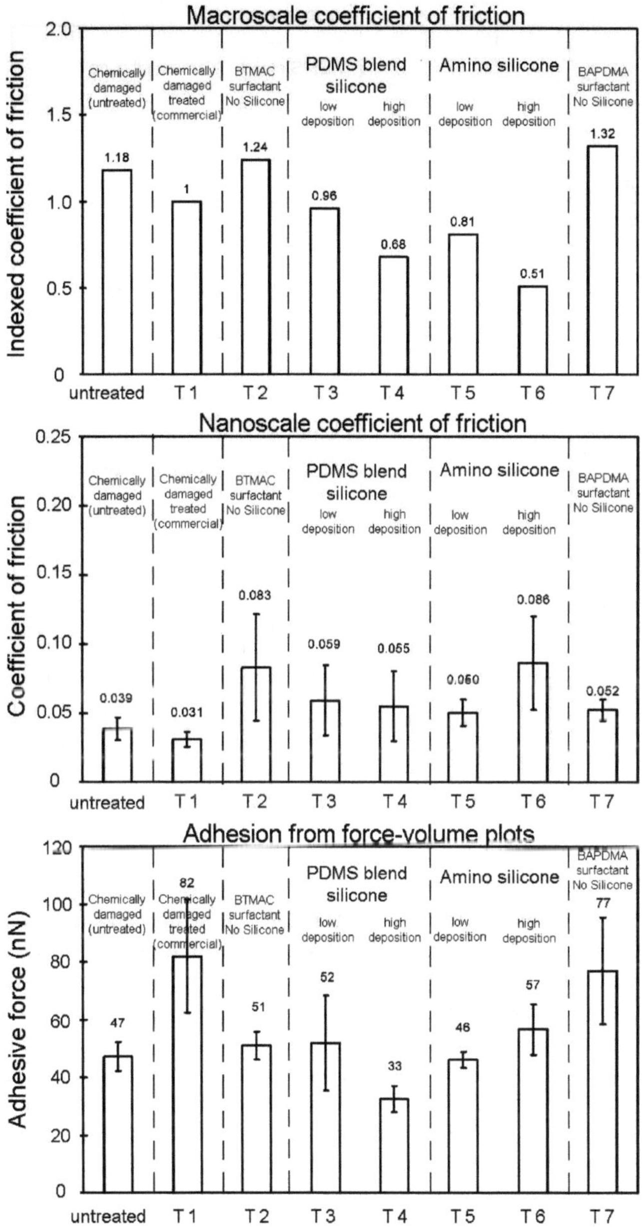

Fig. 24.29. Coefficient of friction, adhesive force, and surface roughness plots for damaged hair with various treatments [LaTorre et al., 2006]

a uniform coating. The increased polarity of the amino silicones compared to the PDMS blend can also be a major contributor of the higher friction and adhesion at high deposition levels.

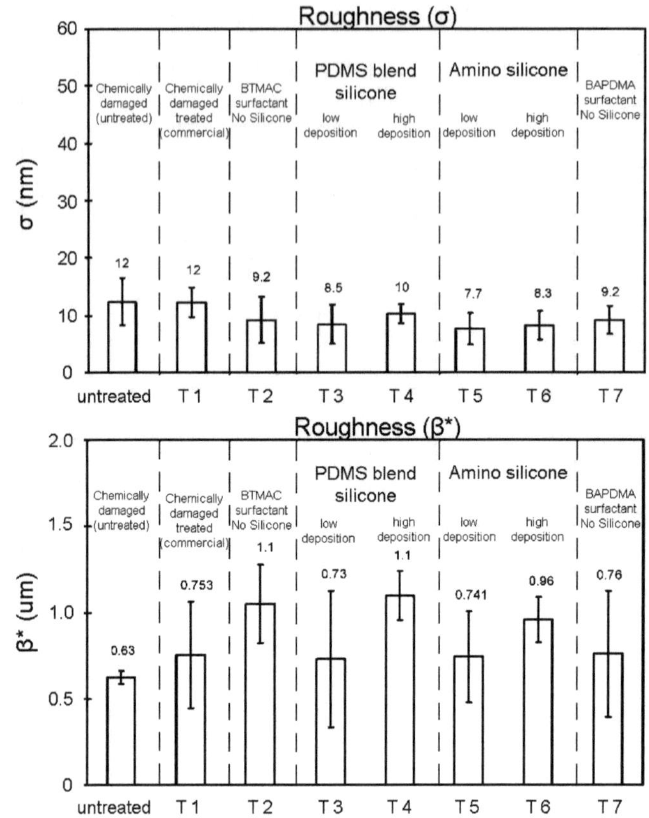

Fig. 24.29. (continued)

If looking at the hair with only the BTMAC surfactant and fatty alcohols, and then adding low deposition levels of PDMS blend (with high MW weight) or amino silicone, it is seen that the coefficient of friction decreases, while the adhesion remains approximately the same. If the deposition level is increased, it is observed that the PDMS blend is still lower, but now the coefficient of friction for the amino silicone is about the same as in the BTMAC only samples. The mobility of the conditioner layer is again a major issue, as this easier mobility accounts for the lower coefficient of friction. However, as is seen when there is a large amount of the amino silicones on the hair surface, the mobility ceases and the coefficient of friction becomes high again.

The BAPDMA surfactant typically showed a higher adhesive force than the BTMAC surfactant. However, there may be an increased accumulation of the BAPDMA surfactant, such that the conditioner layer is more easily sheared by the tip, as is the case for chemically damaged treated hair. The inherent differences in surfactant composition and how they interact with the chemically damaged hair surface most likely account for the coefficient of friction differences.

Table 24.14. Observations and corresponding mechanisms regarding the coefficient of friction and the adhesion for various hair treatments

Observation	Mechanism
Damaged vs. damaged treated hair	
Damaged hair shows a decrease in the coefficient of friction but an increase in the adhesion from the application of commercial conditioner	The conditioner layer deposited on the surface of the damaged hair results in a lower shear strength, which in turn lowers the coefficient of friction, while meniscus effects increase the pull-off (adhesive) force between the tip and hair sample
PDMS blend vs. amino silicone (at high deposition)	
Amino silicones interact strongly with negatively charged hair surface, especially at high deposition levels	A stronger electrostatic attraction exists, which results in stronger binding forces (which leads to higher adhesion) for high deposition amino silicone
Amino silicone thickness distribution on hair is less uniform than with the PDMS blend	Less mobility with amino silicones, so molecules attach to hair at contact and do not redistribute easily
Surfactant vs. surfactant plus addition of silicone	
Adhesion remained approximately the same, while the coefficient of friction decreased with addition of PDMS blend silicone and amino silicone (low deposition). The coefficient of friction remained about the same at high deposition amino silicone	Mobility becomes easier with addition of PDMS blend silicone, which leads to a lower coefficient of friction. At high deposition of amino silicones, mobility ceases and the coefficient of friction becomes high again
BTMAC surfactant vs. BAPDMA surfactant	
The BTMAC surfactant has lower adhesion but a higher coefficient of friction than the BAPDMA surfactant	Interaction of surfactants with damaged hair surface causes inherent differences in the coefficient of friction and adhesion. BAPDMA has both amino and amine groups, which increases polarity

With respect to roughness, the vertical standard deviation decreased slightly with most treatments, although standard deviations were similar. The spatial parameter increased slightly with treatments, but the variation also became extremely high.

24.4.4 Investigation of Directionality Dependence and Scale Effects on Friction and Adhesion of Hair

24.4.4.1 Directionality Dependence

Figure 24.10, shown previously, displays friction test apparatuses used for macroscale and micro/nanoscale coefficient of friction measurements, as discussed earlier. The macroscale friction test apparatu, consists of taking a hair fiber swatch and running a block of synthetic skin over the fibers. AFM coefficient of friction measurements take place on a single fiber. With the microscale AFM tip (4 μm radius), several cuticle scales may come into contact at the same time, depending on their length, causing the scale edges to become a significant factor in the resulting coefficient of friction values. This is especially evident when scanning in the tip-to-root direction of the hair (referred to as "against cuticle"). With the nanoscale AFM tip (30–50 nm radius), single asperity contact can be simulated, and the cuticle edges can be avoided by limiting the scan area to a small section of the surface.

Figure 24.30 displays SEM micrographs and AFM height maps of virgin hair at various scales. The top SEM micrograph shows the hair fiber so that the full diameter is observed. During macroscale friction testing, many hair fibers will come into contact with the upper sample (generally synthetic skin) at the same time. The

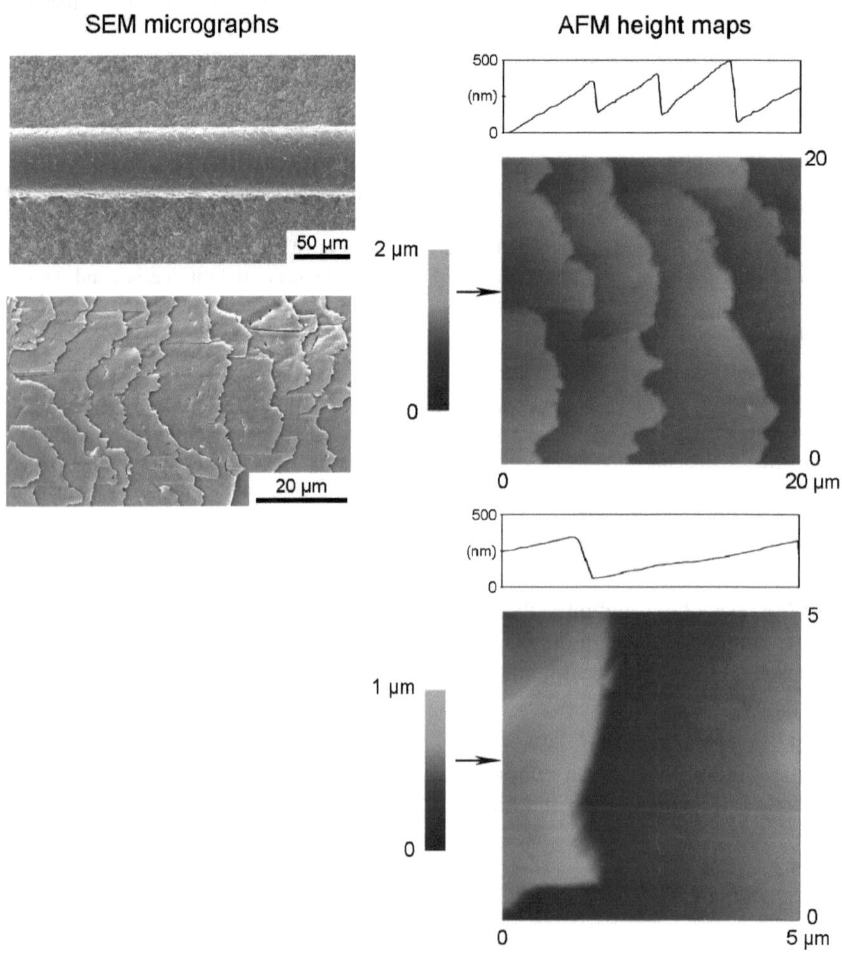

Fig. 24.30. Comparison of hair topography at different scales

next micrograph reveals a better magnification of the cuticle scales. The scan length here is representative of that used for microscale coefficient of friction measurements. The $20 \times 20\,\mu m^2$ AFM height map and the corresponding cross-sectional plot show that cuticle scale heights vary from layer to layer, and range from approximately 300 to 500 nm. With a $5 \times 5\,\mu m^2$ map we can focus on one cuticle scale edge. Most visible cuticle scale lengths are between 5 and 10 μm. For nanoscale coefficient of friction measurements, a 2 μm scan length is generally used, so as to avoid these scale edges affecting the data. However, if directionality dependence is to be studied, a 5 μm scan line is sufficient.

Human hair has been shown in the past to have a directionality friction effect on the macroscale, making it easier to travel over the hair surface from root-to-tip than in the opposite direction due to the tile-like orientation of cuticles [Robbins, 1994]. The outer surface of human hair is composed of numerous cuticle scales running along the fiber axis, generally stacked on top of each other. As previously discussed, the heights of these step changes are approximately 300 nm. These large changes in topography make the cuticle an ideal surface for investigating the directionality effects of friction. The first row of Fig. 24.31a displays a low resolution SEM micrograph and a friction profile for macroscale coefficient of friction measurements. Note that these data are taken for measurements where multiple fibers are in contact with the synthetic skin upper specimen at the same time, and do not correspond directly with the SEM micrograph. With an applied normal load of 50 mN and 3 mm travel, it is observed that the friction force produced when scanning from root-to-tip (referred to as "along cuticle") is lower than that when scanning against cuticle. This is a direct consequence of the literally thousands of scale edges that come into contact with the synthetic skin. When traveling against cuticle, these edges act as tiny resistors to motion as they are forced backwards and uplifted from their interface with the underlying cuticle layers. The resistance to motion of so many cuticle edges at the same time becomes "additive" and results in higher values of friction, corresponding to a higher coefficient of friction than when traveling along cuticle. For along cuticle travel, these edges are forced downward against the underlying cuticle layers, so that the resistance effect of these edges is limited, which results in lower friction values.

The second row of Fig. 24.31a shows an AFM height map corresponding to the microscale friction profile shown to its right. Due to the size of the hair, it was only possible to capture a rectangular height map as shown. It is evident that the 100 μm travel results in the involvement of several cuticle scales. The applied normal load for the microscale friction profile (about 20 nN) is significantly reduced from that of the macroscale value, which consequently yields much lower friction forces. In the along cuticle direction, we can see small fluctuations in the friction data over the scan distance. These are caused by local variations in surface roughness, system noise, and changes due to traveling over the scale edges. However, when scanning against the cuticle, distinctly large spikes in the data are observed at roughly 5–10 μm intervals. This is clearly the effect of the scale edges coming into contact with the AFM microscale tip and causing local collisions and ratcheting of the tip. Because of the sign convention of the AFM that causes a reversal in the sign when traversing the opposite direction, this signal is now observed to be highly negative. These edge effects are hence the primary item responsible for the higher friction and the coefficient of friction observed in the against cuticle direction.

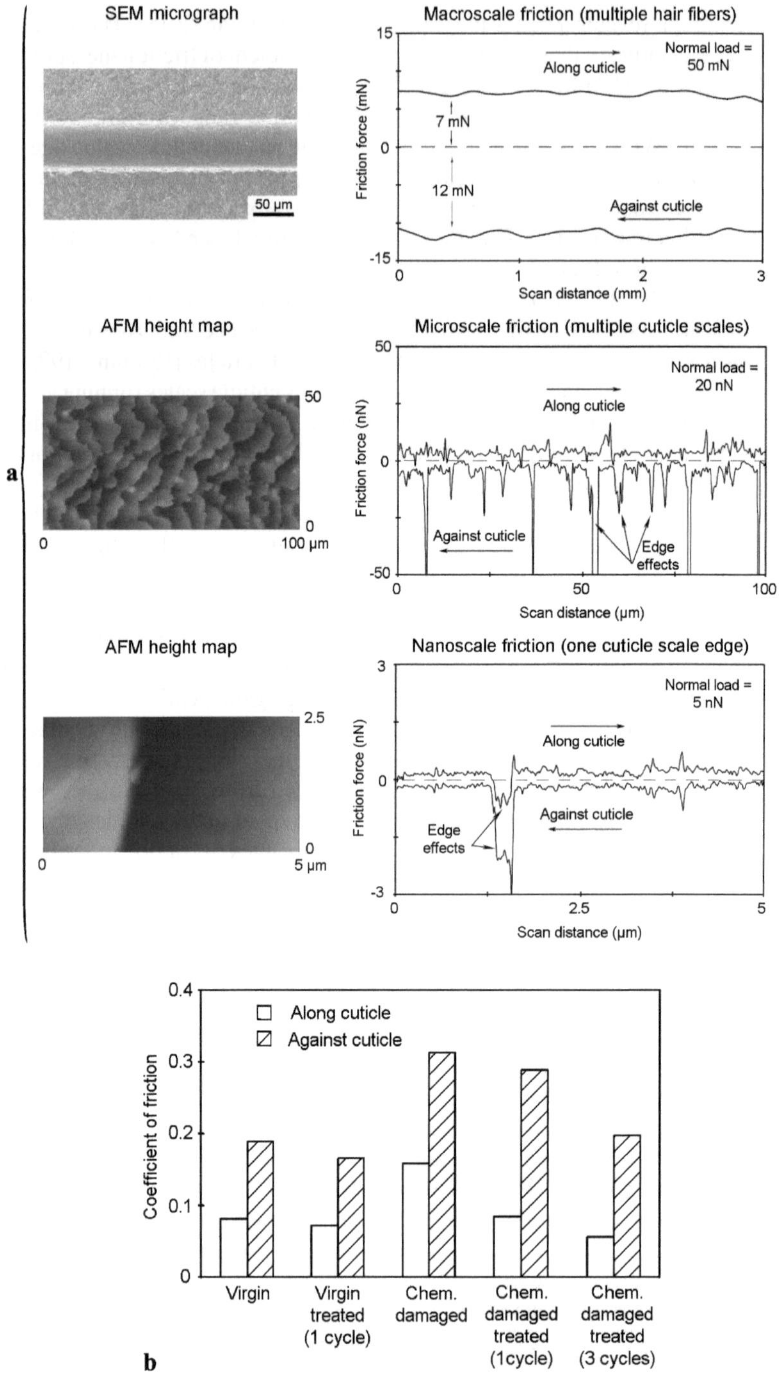

Fig. 24.31. (**a**) Directionality effects of hair friction on various scales. (**b**) Microscale coefficient of friction data for various hair types showing directionality effects

Directionality effects on nanoscale hair friction have previously been reported by LaTorre and Bhushan [2005a]. As shown in the bottom row of Fig. 24.31a, the scan size of 5 × 2.5 μm^2 provides one cuticle scale edge to be studied (the height map is rectangular only to be consistent with the microscale image in the middle row). As the tip rasters in the along cuticle direction, a small decrease in the friction force is observed as the tip follows the scale edge on a downward slope. When the tip comes back in the against cuticle direction, colliding with the scale edge and climbing up the sharp peak results in a high friction signal. As discussed above, because of the sign convention of the AFM/FFM that causes a reversal in the sign when traversing the opposite direction, this signal is now observed to be highly negative. The interesting difference between the two profiles lies in the fact that the magnitude of the decrease in friction when going up the step is much larger than the magnitude of the friction when the tip is going down the step, yet both signals are in the same direction. It is thus shown that the cuticle edge provides a local ratchet and collision mechanism that increases the friction signal at that point and clearly shows the directionality dependence caused by edge effects.

Coefficient of friction data showing directionality dependence on the macroscale for various virgin, damaged, and treated hair types can be found in Bhushan et al. [2005]. Figure 24.31b and Table 24.15 present a summary of the microscale coefficient of friction data for the various hair samples. In most cases, the coefficient of friction has more than doubled when scanning in the against cuticle direction. A more in depth discussion on the coefficient of friction trends between the different hair types will follow. For now, however, it is important to realize that there is a strong directionality dependence on coefficient of friction data for hair, especially on the microscale. On the nanoscale, while directionality dependence of friction force has been studied, actual coefficient of friction data is generally only measured on a small scan area that does not include the cuticle edge, so that only the true cuticle surface is involved [LaTorre and Bhushan, 2005a]. Hence, nanoscale coefficient of friction directionality data similar to Fig. 24.31b are not shown.

Table 24.15. Coefficient of friction values showing directionality dependence on the microscale. This dependence has also been observed on the nanoscale and macroscale

Hair sample	Microscale coefficient of friction	
	Along cuticle	Against cuticle
Virgin (untreated)	0.08	0.19
Virgin treated (one cycle commercial conditioner)	0.07	0.17
Chemically damaged (untreated)	0.16	0.31
Chemically damaged treated (one cycle commercial conditioner)	0.08	0.29
Chemically damaged treated (3 cycle commercial conditioner)	0.06	0.20

24.4.4.2
Scale Effects

Given the fact that the "directionality effect" is now observed to be universal for all types of hair and on all scales, and since the along cuticle direction is more relevant to our daily life (i.e. combing), we shall now focus on coefficient of friction data in the along cuticle direction.

As described earlier, microscale and nanoscale coefficients of friction are taken as the slope of the least squares fit line of a friction force vs. the normal force data curve. Figure 24.32a shows these types of raw data curves for various representative hair samples using both microscale (top plot) and nanoscale (bottom plot) AFM tips. The nanoscale data were taken from the raw data used in LaTorre and Bhushan [2005b].

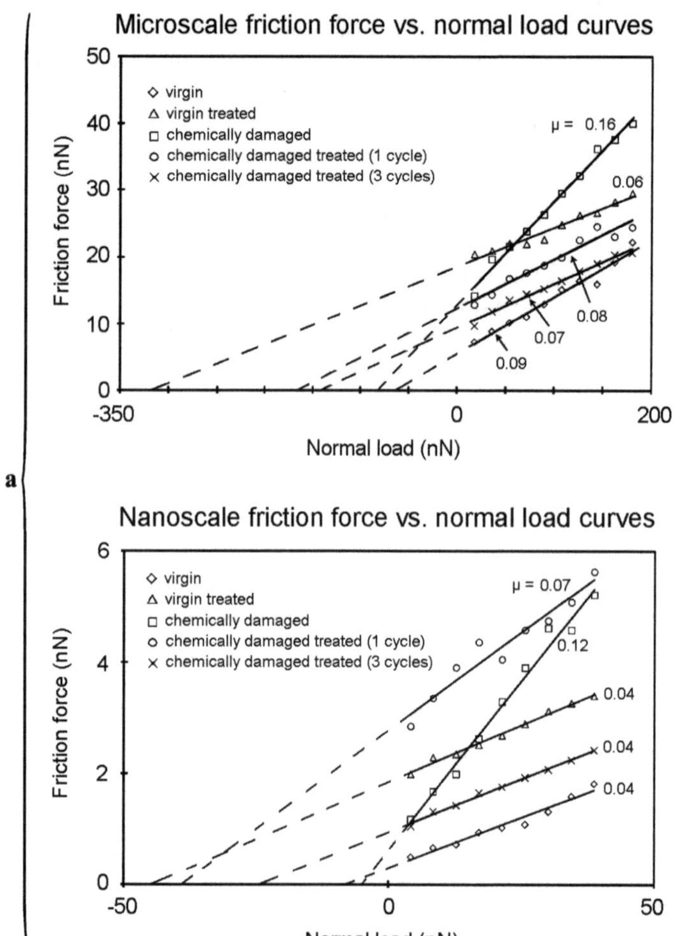

Fig. 24.32a. Friction force vs. normal force curves for microscale and nanoscale coefficients of friction

One can see a relatively linear relationship between the data points for each type of hair sample. If the least squares fit lines of Fig. 24.32a are extended to intercept the horizontal axis (indicated by the dotted lines), then a value for the adhesive force can also be calculated, since friction force F is governed by the relationship of (3.1). These adhesive force values serve as average adhesive force values over the course of the full scan profile, and differ slightly from the force calibration plots of Fig. 24.32b. From Fig. 24.32a, it is observed in both plots that treated hair fibers, whether virgin or chemically damaged, have much higher average adhesive force values as compared to their untreated counterparts. Chemically damaged hair is observed to have the highest coefficient of friction (highest slope) on both the microscale and nanoscale. As explained in LaTorre and Bhushan [2005a,b] and LaTorre et al. [2006], chemical damage to the hair causes the outer lubricious layer of the cuticle to wear off, resulting in an increased coefficient of friction.

Force calibration plots yield adhesive force values at a single point and are considered to be more relevant for measurement of the pull-off force between the tip and hair surface. Since it has been previously shown that treating both virgin and chemically damaged hair with conditioner results in large increases in the adhesive force using both microscale and nanoscale AFM tips, we will focus only on the force calibration plots of the virgin and virgin treated hair to discuss mechanisms for this trend. The first plot in Fig. 24.32b shows a typical virgin hair force calibration plot with the microscale AFM tip. The cantilever deflection is plotted on the vertical axis against the Z-position of the piezo scanner. As the piezo extends (from right to left in the graph), it approaches the tip and does not show any deflection while in free air.

b

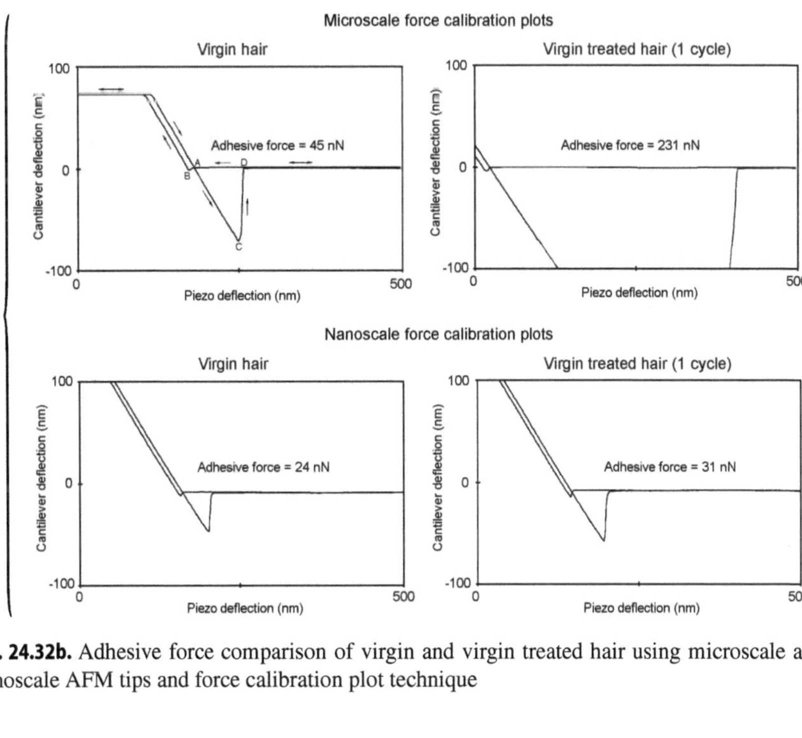

Fig. 24.32b. Adhesive force comparison of virgin and virgin treated hair using microscale and nanoscale AFM tips and force calibration plot technique

The tip then approaches within a few nanometers (point A) and becomes attached to the sample via attractive forces at a very close range (point B). After initial contact, any extension of the piezo results in a further deflection of the cantilever and tip. Once the piezo reaches the end of its designated ramp size, it retracts to its starting position (from left to right). The tip goes beyond zero deflection and enters the adhesive regime. At point C, the elastic force of the cantilever becomes equivalent to the adhesive force, causing the cantilever to snap back to point D. The horizontal distance between A and D multiplied by the stiffness of the cantilever results in a quantitative measure of the adhesive force [Bhushan, 1999a,b, 2002, 2005].

We can see that on the application of one cycle conditioner treatment to the virgin hair, the microscale adhesive force jumps to about 230 nN. This is clearly the effect of the meniscus forces brought about by the presence of the conditioner layer on the cuticle surface that interacts with the tip [LaTorre and Bhushan, 2005a,b; Chen and Bhushan, 2005b]. With the nanoscale tip (bottom row of Fig. 24.32b), an increase in the adhesive force is again seen with conditioner treatment.

It is important to notice that the adhesive force values on the microscale are always larger than those on the nanoscale for a given hair. To explain the scale dependency of the adhesive force, we can model the hair-conditioner-tip interaction as a sphere close to a surface with a continuous liquid film [Chen and Bhushan, 2005b]. The adhesive force F_a, is the force needed to pull the sample away from the tip (which is the same as the adhesive force calculated with force calibration plots). F_a is the sum of van der Waals force F_{vdw} mediated by the adsorbed water or conditioner layer and the meniscus force F_m due to the Laplace pressure ($F_a = F_{vdw} + F_m$). The meniscus force F_m is given by

$$F_m = 2\pi R\gamma(1 + \cos\theta)\,, \tag{24.2}$$

where R is the tip radius, γ is the surface tension of the conditioner, and θ is the contact angle between the tip and conditioner [Bhushan, 2002]. The increase in the adhesive force calculated by force calibration plots with the microscale AFM tip compared to the nanoscale AFM tip is in large part due to the increased radius R of the microscale ball, which consequently induces larger F_m.

Figure 24.33 and Table 24.16 display the coefficient of friction and the adhesive force data on the macroscale, microscale, and nanoscale. Scale dependence is clearly observed. Macroscale data for virgin and virgin treated hair were taken from Bhushan et al. [2005]. The values for other hair were taken in part from the indexed coefficient

Table 24.16. Coefficient of friction and the adhesive force of hair on various scales

Hair sample	Coefficient of friction			Adhesive force (nN)		
	Macroscale	Microscale	Nanoscale	Macroscale	Microscale	Nanoscale
Virgin (untreated)	0.14	0.08 ± 0.01	0.04 ± 0.003	–	53	25
Virgin treated (1 cycle commercial conditioner)	0.12	0.07 ± 0.03	0.04 ± 0.01	–	221	32
Chemically damaged (untreated)	0.24	0.16 ± 0.003	0.12 ± 0.05	–	66	40
Chemically damaged treated (1 cycle commercial conditioner)	0.14	0.08 ± 0.02	0.07 ± 0.01	–	191	66
Chemically damaged treated (3 cycles commercial conditioner)	0.13	0.06 ± 0.02	0.04 ± 0.02	–	97	54

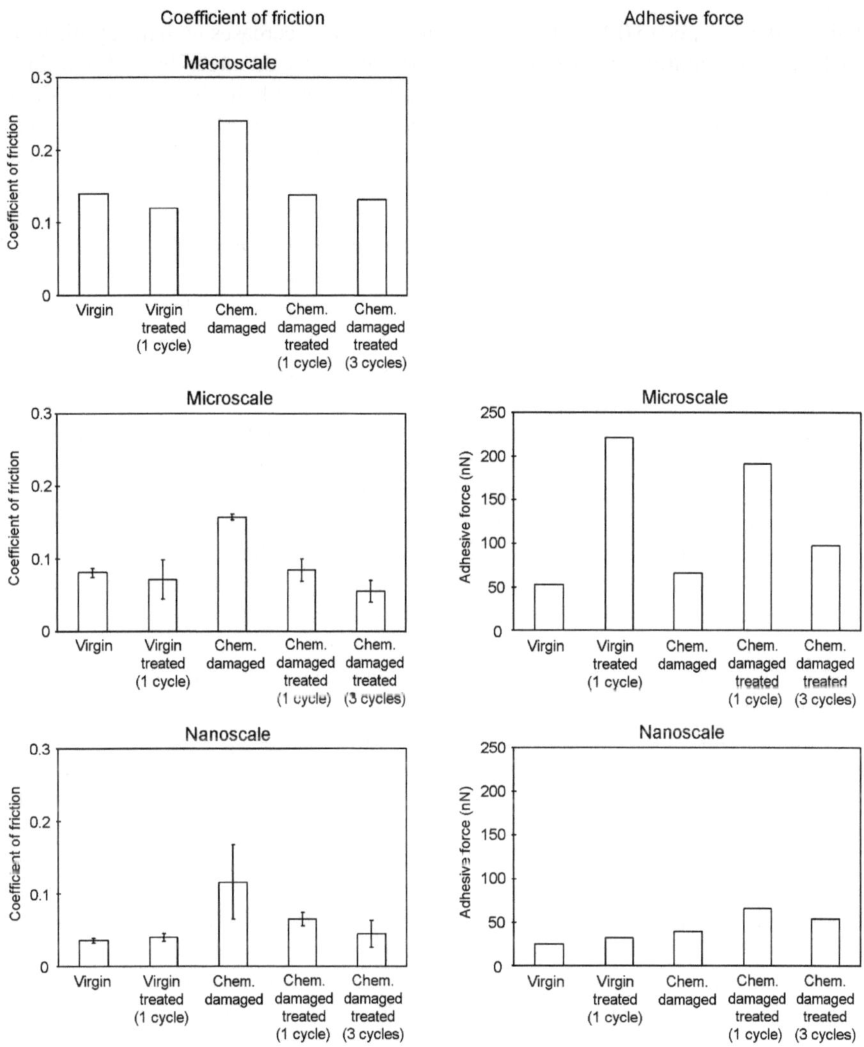

Fig. 24.33. Summary of coefficient of friction and adhesive force data on various scales

of friction values in LaTorre and Bhushan [2005b], which were transformed into actual values (as described previously). No adhesive force data are presented for the macroscale data, because the adhesive force contribution to friction is considered to be negligible compared to the applied normal load. On the microscale and nanoscale, however, the magnitude of the adhesive force is the same as that of the applied normal load, so they have significant contributions on the coefficient of friction data, and are thus presented. Microscale values are taken from the raw data represented in Fig. 24.32a and b. Nanoscale data were taken from LaTorre and Bhushan [2005b].

Macroscale coefficient of friction (COF) data is shown in the top row of Fig. 24.33: Chemically damaged hair has higher coefficient of friction than virgin hair, 0.24 compared to 0.14. The coefficient of friction decreases with the application of 1 cycle of conditioner for both virgin and chemically damaged hair. When three cycles of conditioner are applied to chemically damaged hair, there is only a slight decrease compared to one cycle application, 0.14 to 0.13. Thus, the data readily reveal that conditioner treatment decreases coefficient for friction of hair. The main mechanism for this macroscale trend is that lubrication with a thin conditioner layer occurs over a large contact area, and thus the conditioner layer shears easily to create a lubricious effect [LaTorre and Bhushan, 2005b]. It is important to note that the magnitude of the coefficient of friction values is higher on the macroscale than on the other scales for all hair types. Bhushan et al. [2004] have previously outlined several differences in operating conditions which can responsible for higher macroscale friction values. The one most relevant to our situation is that coefficient of friction increases with an increase in the AFM tip radius. Nanoscale friction data is taken with a sharp AFM tip, while the macroscale and microscale tests have contacts that range from nanoasperities to much larger asperities that may be responsible for larger values of friction force on these scales. The combination of higher normal loads with a larger contact area (due to contact with multiple fibers at the same time) may also be responsible for the increased coefficient of friction on the macroscale.

The coefficient of friction trends are similar on the microscale. On the microscale, the virgin hair coefficient of friction was 0.08, while application of one cycle conditioner decreased the coefficient of friction only slightly to 0.07. From Fig. 24.33 it is important to note the large standard deviation on the virgin treated value. The corresponding adhesive force data in the same row are useful for a better understanding of this behavior. It is shown that the virgin treated hair has a large adhesive force contribution (due to meniscus effects caused by the conditioner layer), which has significant effects on the variation of friction force, and consequently the coefficient of friction. The chemically damaged hair has the largest coefficient of friction of the set, 0.16. Application of 1 conditioner cycle brought the value down to 0.08, while it was even lower for 3 conditioner cycles, 0.06. The adhesive force increased significantly with conditioner application; for virgin hair, the adhesive force increased from about 50 nN to 220 nN due to the meniscus effects that come about from AFM tip interaction with the conditioner on the cuticle surface. Likewise, the adhesive force for chemically damaged hair was about 65 nN and jumped to 190 nN and 100 nN for 1 and three cycles of conditioner treatment, respectively. A possible reason that one cycle of conditioner on damaged hair showed higher average adhesive force than three cycles could be because the three cycles generally place the conditioner more uniformly on the surface rather than accumulating it mostly on the bottom surface near the cuticle edge, which is where the adhesive force maps were generally taken. Nevertheless, the increased adhesive force shown in the plots is a clear indication of conditioner being present on the hair surface [LaTorre and Bhushan, 2005b].

On the nanoscale, the coefficients of friction of virgin and virgin treated hair are similar, but slightly higher for the treated cases. This is opposite to the trend on the macroscale and slightly different to that observed on the microscale. These meniscus bridges require more force to break through than with untreated hair, which causes the coefficient of friction to be similar to the untreated value, instead of experiencing the traditionally expected significant decrease. Damaged hair shows a much higher coefficient of friction with more variation in the values, since the chemical damage varies throughout each individual fiber. Contrary to the virgin and virgin treated hair nanoscale results, the coefficient of friction for damaged hair decreased with application of conditioner treatment (both 1 and 3 cycles), which agrees well with macroscale and microscale trends.

There are several reasons that the nanoscale coefficient of friction trends for virgin treated and chemically damaged treated hair are different. In order to serve as a reference for this discussion, Figs. 24.20 and 24.21 show surface roughness, friction force, and adhesive force maps for the various hair types (taken with a nanoscale AFM tip) [LaTorre and Bhushan, 2005b]. The effects of chemical damage play a large role. It is widely known that the cuticle surface of any hair is negatively charged. This charge becomes even more negative with the application of chemical damage to the hair. As a result, the positively charged particles of conditioner have an even stronger attraction to the chemically damaged surface, which explains the increased presence of conditioner when compared to virgin treated hair. With the application of three conditioner cycles on damaged treated hair, there is an even more uniform placement of the conditioner [LaTorre and Bhushan, 2005b; LaTorre et al., 2006]. This leads to a high adhesive force due to meniscus effects (similar to that of virgin treated hair) but more importantly, a lower shear strength on the surface. This creates an overall effect of lubrication as the tip travels across the cuticle surface and ultimately lowers the coefficient of friction.

Another reason for the difference in nanoscale trends between virgin and chemically damaged hair may have to do with the drastic differences in hydrophobicity of the two hair types. Virgin hair has been shown to be hydrophobic, with a contact angle of around 100° (see Table 24.11). Chemically damaged hair, however, is hydrophilic, with a contact angle of around 60°. The conditioner gel network is primarily composed of water, together with fatty alcohols, cationic surfactants, and silicones. Thus, the hydrophobicity of the hair will be relevant to not only how much conditioner is deposited, but also to how it diffuses into the hair and bonds to the hair surface. For virgin treated hair, the conditioner deposits in certain locations, especially near the cuticle edge, but due to the hydrophobicity of the cuticle does not spread out as readily as with chemically damaged hair. For chemically damaged hair, the conditioner spreads out a bit more uniformly and in more places over the cuticle surface due to both the hydrophilicity and the stronger negative charge that attracts more conditioner deposition. Thus, as the tip scans over the virgin treated surface, the conditioner does not smear as readily, causing the tip to have to break the tiny meniscus bridges formed with the conditioner. This results in an increased adhesive force, which contributes to a higher friction force. This ends ups increasing the coefficient of friction to about the same level as the untreated virgin hair, instead of the reduction that is typically expected for a lubricated surface. In the case of chemically damaged hair, however, the conditioner layer is already more spread out, especially in the case of three cycles of treatment, as shown in the friction force maps of Fig. 24.20. As the tip scans over the surface, the overall effect is one of reduced shear strength, i.e. the conditioner layer, albeit not fully continuous, smears with tip travel and causes a reduced coefficient of friction between the tip and the hair. With three cycles the conditioner thickness increases slightly [Chen and Bhushan, 2005b], and the layer is even more uniformly distributed over the surface [LaTorre and Bhushan, 2005b], which causes a further reduction in the coefficient of friction, much like the results seen on both the microscale and macroscale.

Figure 24.34 schematically shows the mechanisms responsible for the reverse trends seen on the nanoscale for virgin treated hair. Table 24.17 summarizes the observations seen for both directionality dependence and scale effects. In the top diagram of Fig. 24.34, the thin layer of conditioner acts as a lubricant over the hair fiber, limiting the dry contact with the synthetic skin block and creating easier relative motion, which decreases coefficient of friction compared to the untreated hair. This is true on the macroscale for both virgin and chemically damaged hair. On the microscale, the same trend is experienced; that is, the 4 μm radius of the AFM ball comes in contact with multiple cuticle scales at the same time, causing an overall lubrication effect for both virgin treated and chemically damaged treated hair as the thin conditioner layer is sheared to create easier relative motion. It is

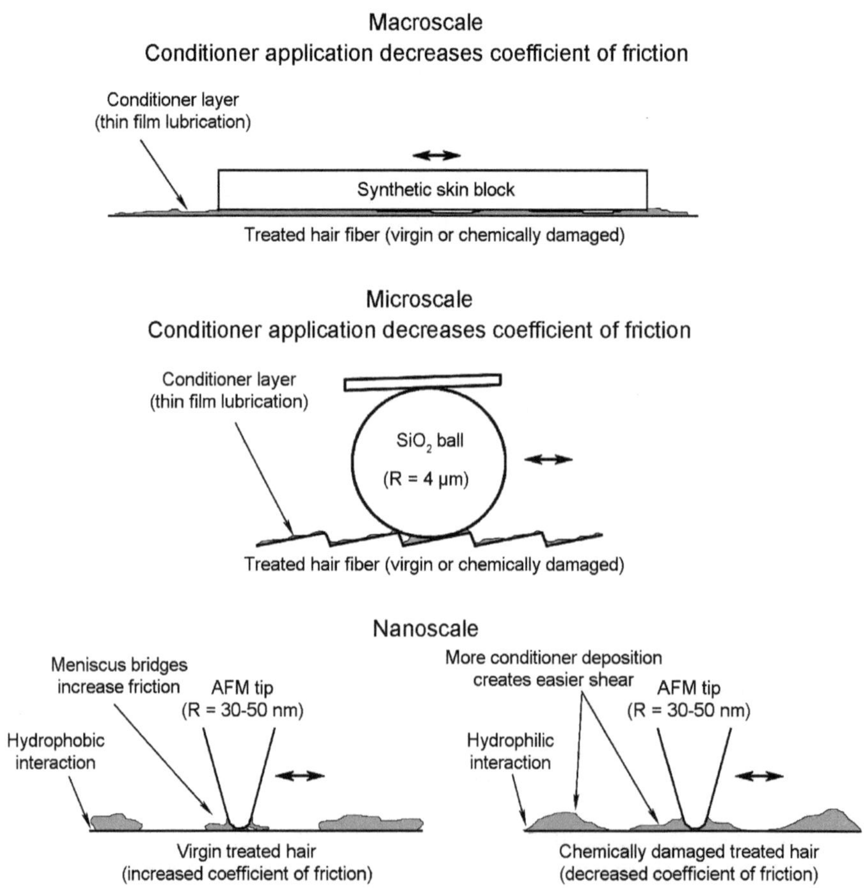

Fig. 24.34. Schematic of various mechanisms responsible for scale dependence of hair friction and adhesion

Table 24.17. Observations and corresponding mechanisms regarding the coefficient of friction and adhesion for various hair treatments

Observation	Mechanism
Directionality dependence of the coefficient of friction	
The coefficient of friction is always higher in the against cuticle direction as compared to the along cuticle direction	Macroscale: Thousands of scale edges resist motion (creating higher friction forces) as they are forced backwards and uplifted. Microscale: Scale edges come in contact with the AFM microscale tip and cause local collisions and ratcheting of the tip, which creates higher friction Nanoscale: AFM nanoscale tip collides with the scale edge and has to climb up the edge, resulting in a high friction signal
	Scale effects
The coefficient of friction values are largest on the macroscale, followed by microscale and then nanoscale values	The coefficient of friction increases with an increase in the AFM tip radius. Nanoscale friction data are taken with a sharp AFM tip, while the macroscale and microscale tests have contacts ranging from nanoasperities to much larger asperities. The macroscale has contact with multiple fibers at the same time
Adhesive force values are larger on the microscale than on the nanoscale	The increase in the adhesive force on the microscale is due to the increased radius of the microscale ball, which consequently induces larger meniscus force contribution to the adhesive force
On all scales, the coefficient of friction decreases for chemically damaged treated hair compared to untreated. The same trend occurs for virgin treated hair on the macroscale and microscale, but the nanoscale value is same as for untreated hair	Macroscale: A thin layer of conditioner acts as a lubricant over the hair fiber (decreases the coefficient of friction) (whether virgin hair or chemically damaged hair) Microscale: The same trend is experienced at macroscale because the AFM microscale tip comes in contact with multiple cuticle scales at the same time, causing an overall lubrication effect Nanoscale: Hydrophobicity of virgin hair causes different deposition of conditioner. The AFM tip has to break the tiny meniscus bridges when scanning the increasing adhesive force coefficient of friction. For chemically damaged hair, a higher negative charge and a hydrophilic surface results in more uniform deposition and better smearing of the conditioner layer and a lower coefficient of friction
On both microscale and nanoscale, the adhesive force increases for treated hair	Meniscus forces brought about by the presence of the conditioner layer increase the adhesive force

important to note, however, that the adhesive forces due to meniscus effects are of the same magnitude as the applied microscale normal load.

On the nanoscale (bottom diagram of Fig. 24.34), we see different trends for virgin treated and chemically damaged treated hair. As discussed earlier, the hydrophobicity of the virgin hair causes a different deposition of conditioner. The AFM tip has to break the tiny meniscus bridges formed with the conditioner as it scans across the hair surface, which increases the adhesive force contribution and results in an increased coefficient of friction. For hydrophilic chemically damaged hair, there is more uniform deposition and better smearing of the conditioner layer, which serves to lower the coefficient of friction between the tip and the cuticle surface.

24.4.5 Surface Roughness and Friction of Skin

Synthetic materials were also studied for surface roughness and friction force information, see Fig. 24.35a. While macroscale dimples could be seen on the surface of collagen film, it was interesting to find similar pits and dimples on the micro/nanoscale, consequently with a large variation in dimple size and depth. Polyurethane films are shown to have quite different topography and friction forces, while their coefficient of friction is very similar. Human skin shows a rougher texture with higher peaks, Fig. 24.35b. The roughness parameters for the collagen and polyurethane films, and also for human skin, are presented in Table 24.18 [LaTorre and Bhushan, 2005a]. The vertical height standard deviation σ was approximately 3 times larger than that of virgin hair for both synthetic materials. However, the correlation length β^* was lower than what was typically observed in hair. The average coefficient of friction for these synthetic materials are shown in Fig. 24.36, plotted next to virgin Caucasian hair as a reference. These values were calculated using the slope of the friction force curves, described previously. Both collagen and polyurethane films displayed similar coefficient of friction values of 0.22 and 0.24, respectively. Virgin hair displays a much lower coefficient of friction than both materials, approximately eight times lower.

Table 24.18. Surface roughness parameters σ, β^* for collagen, polyurethane films, and human skin

	σ (nm)	β^* (µm)
Collagen (synthetic hair)	36 ± 11	0.50 ± 0.1
Polyurethane (synthetic skin)	33 ± 6	0.71 ± 0.1
Human skin	80 ± 28	0.59 ± 0.2

24.5 Closure

The AFM contact mode has been used to perform nanotribological studies on various hair types and skin. The friction force and the resulting coefficient of friction are seen to be higher on chemo-mechanically damaged hair than on virgin hair, due to

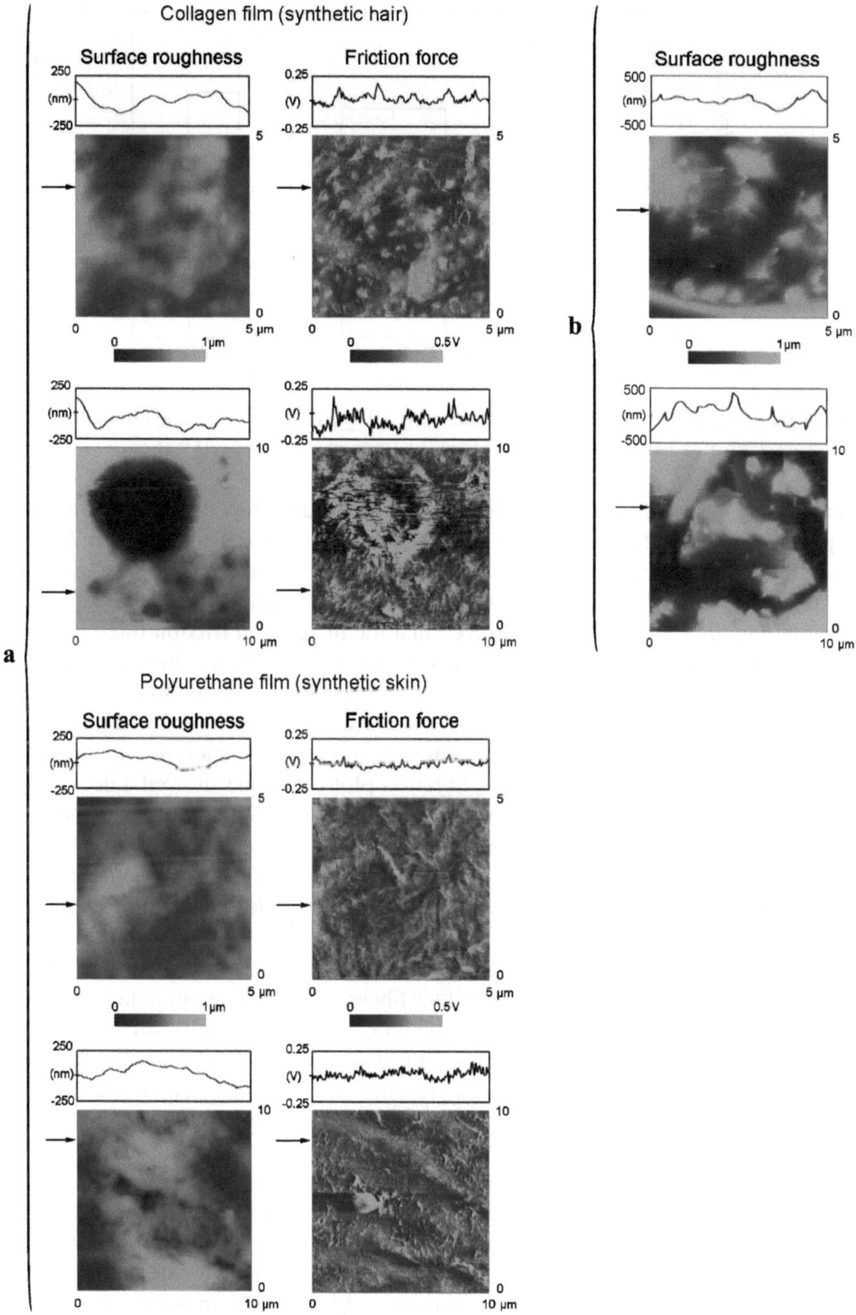

Fig. 24.35. (**a**) Surface roughness and friction force images for collagen and polyurethane films at 5 and 10 μm^2 scan sizes. Shown above each image is a cross-section taken at the *corresponding arrows* to demonstrate roughness and friction force information. (**b**) Surface roughness for human skin at 5 and 10 μm^2 scan sizes [LaTorre and Bhushan, 2005a]

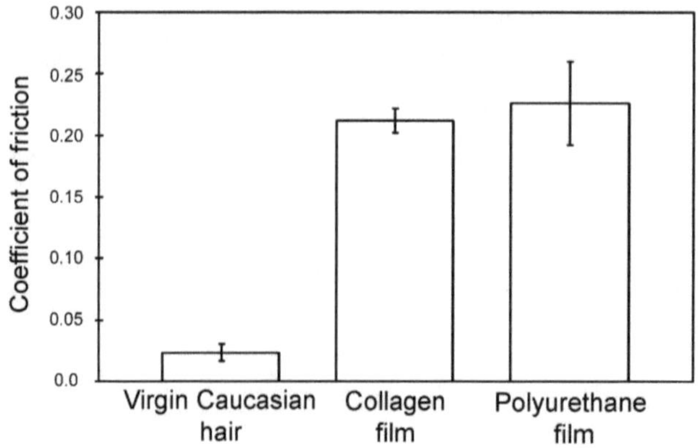

Fig. 24.36. Average coefficient of friction values for collagen and polyurethane films. *Error bars* represent $\pm 1\sigma$ of the average coefficient value [LaTorre and Bhushan, 2005a]

the increase in surface roughness and a change in surface properties that results from exposure to chemical damage. Generally speaking, the average coefficient of friction is similar between virgin and virgin treated hair of each ethnicity. However, in virgin treated hair, there is an increase in friction forces around the cuticle edges and surrounding area. It is currently believed that the increase in friction force is due in part to an increase in meniscus effects, which increase the adhesive force contribution to friction at sites where conditioner is deposited or accumulated on the surface, namely around the cuticle scale edges. This adhesive force is of the same magnitude as the normal load, which makes the adhesive force contribution to friction rather significant. Studies using the force calibration plot technique showed a decrease in the adhesive force with damaged hair, and significantly higher adhesive force for treated hair. This increase on the micro/nanoscale is most likely due to meniscus force contributions from the accumulation and localization of a conditioner layer on the hair surface. Thus, the presence of conditioner can be detected by this increasing adhesive force.

Chemically damaged treated hair shows a much stronger affinity to conditioner than virgin hair. The negative charge of hair fibers becomes even more negative with the application of chemical damage to the hair. As a result, the positively charged particles of conditioner have even stronger attraction to the chemically damaged surface, and this results in an increased presence of conditioner (and corresponding higher friction forces) when compared to virgin treated hair. With the application of three conditioner cycles on chemically damaged treated hair, the friction force increases all over the cuticle surface, showing a more uniform placement of the conditioner. Contrary to the virgin and virgin treated hair results, the coefficient of friction for chemically damaged hair decreased with the application of commercial conditioner treatment (both one and three cycles). One possible explanation is that because the stronger negative charge on damaged hair results in better attraction of conditioner, this leads to a higher adhesive force but, more importantly, a lower shear strength on the surface.

Environmental effects were studied for various hair types. The coefficient of friction generally decreased with increasing temperature. After soaking hair in deionized water, virgin hair exhibits a decrease in the coefficient of friction after soaking. Virgin hair is more hydrophobic (based on the contact angle data), so more water is present on the surface and this results in a lubrication effect after soaking. Chemically damaged hair tends to be hydrophilic due to the chemical degradation of the cuticle surface, and this results in absorption of water after soaking. This softens the hair, which leads to higher friction, even with conditioner treatment. Durability tests show that once conditioner is applied to virgin hair, wear does not show up as an increase in the friction force. Thus, conditioner serves as a protective covering to virgin hair and helps to protect the tribological properties when wear ensues.

In most cases, a decrease in the coefficient of friction was observed on chemically damaged hair with the addition of the PDMS blend or amino silicones to the BTMAC surfactant. The silicones are typically used as a major source of lubrication and thus give the conditioner more mobility on the hair surface compared to just surfactants and fatty alcohols. The inverse trend was seen only for the amino silicone group at high deposition. The dampened mobility of the amino silicone at high deposition levels, with respect to hair surface and tip, may account for this wide variation in the coefficient of friction. At high deposition levels, the amino silicones showed much more distinct regions of high and low friction and adhesion, which shows that there is less mobility of these molecules and much less redistribution as they coat the hair.

Investigations of scale effects and directionality dependence on friction and adhesion have been studied. On the macroscale, when traveling against cuticle, edges resist motion (creating higher friction forces) as they are forced backwards and uplifted from their interface with the underlying cuticle layers. On the microscale, when scanning against the cuticle, distinctly large spikes in the data are observed at roughly 5–10 μm intervals. This is clearly the effect of the scale edges coming in contact with the AFM microscale tip and causing local collisions and ratcheting of the tip, which creates higher friction. On the nanoscale, the AFM nanoscale tip collides with the scale edge and has to climb up the edge, resulting in a high friction signal.

The coefficient of friction values are largest on the macroscale, followed by microscale and then nanoscale values. In general, coefficient of friction increases with an increase in the AFM tip radius. Nanoscale friction data are taken with a sharp AFM tip, while the macroscale and microscale tests have contacts that range from nanoasperities to much larger asperities that may be responsible for larger values of friction force on these scales. Adhesive force values are larger on the microscale than on the nanoscale. The increase in the adhesive force calculated by force calibration plots with the microscale AFM tip compared to the nanoscale AFM tip is in large part due to the increased radius of the microscale ball, which consequently induces a larger meniscus force contribution to the adhesive force.

On all scales, the coefficient of friction decreases for chemically damaged treated hair compared to untreated hair. The same trend occurs for virgin treated hair on the macroscale and microscale, but not on the nanoscale. On the macroscale, the thin layer of conditioner acts as a lubricant over the hair fiber, which decreases the coefficient of friction compared to the untreated hair (whether virgin or chemically damaged). On the microscale, the same trend is experienced because the AFM microscale tip comes into contact with multiple cuticle scales at the same time, causing an overall lubrication effect. On the nanoscale, the hydrophobicity of virgin hair causes different deposition of conditioner. The AFM tip has to break the tiny meniscus bridges formed with the conditioner as it scans across the hair surface, which increases the adhesive force contribution and results in an increased coefficient of friction. For chemically damaged hair, there is a higher negative charge and a hydrophilic surface, which results in more uniform deposition and better smearing of the conditioner layer, which serves to lower the coefficient of friction between the tip and the cuticle surface.

References

1. Barnes HA, Roberts GP (2000) Int J Cosmet Sci 22:259
2. Bhushan B (1999a) Principles and Applications of tribology. Wiley, New York
3. Bhushan B (1999b) Handbook of Micro/Nanotribology. 2nd Ed. CRC Press, Boca Raton
4. Bhushan B (2002) Introduction to Tribology. Wiley, New York
5. Bhushan B (ed) (2004) Springer Handbook of Nanotechnology. Springer, Berlin Heidelberg New York
6. Bhushan B (2005) Nanotribology and Nanomechanics – an Introduction. Springer, Berlin Heidelberg New York
7. Bhushan B, Burton Z (2005) Nanotechnology 16:467
8. Bhushan B, Dandavate C (2000) J Appl Phys 87:1201
9. Bhushan B, Liu H, Hsu SM (2004) ASME J Tribol 126:583
10. Bhushan B, Wei G, Haddad P (2005) Wear 259:1012
11. Bolduc C, Shapiro J (2001) Clinics Dermatology 19:431
12. Chen NH, Bhushan B (2005) J Microscopy 220:96
13. Chen NH, Bhushan B (2006) J Microscopy (in presss)
14. Feughelman A (1997) Mechanical Properties and Structure of Alpha-Keratin Fibres: Wool, Human Hair and Related Fibres. University of New South Wales Press, Sydney
15. Ginn ME, Noyes CM, Jungermann E (1968) J Interface Sci 26:146
16. Gray J (2001) Clinics in Dermatology 19:227
17. Gray J (2003) The World of Hair. http://www.pg.com/science/haircare/hair_twh_toc.htm
18. Jachowicz J, McMullen R (2002) J Cosmet Sci 53:345
19. Jalbert C, Koberstein JT, Yilgor I, Gallagher P, Krukonis V (1993) Macromolecules 26:3069
20. Jollès P, Zahn H, Höcker H (eds) Formation and Structure of Human Hair. Birkhäuser, Berlin
21. LaTorre C, Bhushan B (2005) Ultramicroscopy 105:155
22. LaTorre C, Bhushan B (2005) J Vac Sci Technol A 23:1034
23. LaTorre C, Bhushan B (2006) Ultramicroscopy (in press)
24. LaTorre C, Bhushan B, Torgerson PM, Yang J (2006) J Cosmetic Sci (in press)
25. Lerebour G, Cupferman S, Cohen C, Bellon-Fontaine MN (2000) Skin Res Technol 6:245
26. Liu H, Bhushan B (2003) Ultramicroscopy 97:321
27. Molina R, Comelles F, Julia MR, Erra P (2001) J Colloid Interface Sci 237:40
28. Pugliese PT (1996) Physiology of the Skin. Allured Publishing, Carol Stream
29. Randebrook RJ (1964) Soc Cosmet Chem 15:691
30. Robbins C (1994) Chemical and Physical Behavior of Human Hair. Springer, Berlin Heidelberg New York
31. Schott H (1971) J Pharm Sci 60:1893
32. Smith JR, Swift JA (2002) J Microscopy 206:182
33. Swift JA (1991) Int J Cosmet Sci 13:143
34. Swift JA (1997) In: Jolles P, Zahn H, Hocker H (eds) Formation and Structure of Human Hair. Birkhauser, Berlin, p 149
35. Swift JA (1999) Int J Cosmet Sci 21:227

36. Swift JA (2000) Int J Cosmet Sci 51:37
37. Syed AN, Kuhajda A, Ayoub H, Ahmad K, Frank EM (1996) African–American Hair: Its physical Properties and Differences relative to Caucasian Hair. In: Hair Care (cosmetics & toiletries applied research series). Allured Publishing, Carol Stream
38. Wei G, Bhushan B (2006) Ultramicroscopy (in press)
39. Wei G, Bhushan B, Torgerson PM (2005) Ultramicroscopy 105:248
40. Wertz PW, Downing DT (1989) In: Swarbrick J, Guy RH (eds) Transdermal Drug Delivery. Marcel Dekker, New York
41. Wertz PW, Madison KC, Downing DT (1989) J Invest Dermatol 92:109
42. Yanazawa H (1984) Colloids Surf 9:133
43. Zviak C (ed.) (1986) The Science of Hair Care. Marcel Dekker, New York

Appendix

This appendix outlines the steps involved in washing hair swatches with shampoo and/or conditioner.

– Shampoo treatments

Shampoo treatments consisted of applying a commercial shampoo evenly down a hair swatch with a syringe. Hair was lathered for 30 seconds, rinsed with tap water for 30 seconds, then repeated. The amount of shampoo used for each hair swatch was 0.1 cm^3 shampoo per gram of hair. Swatches were hung to dry in an environmentally controlled laboratory, and then wrapped in aluminum foil.

– Conditioner treatments

0.1 cm^3 of commercial conditioner was applied per gram of hair. The conditioner was applied in a downward direction (scalp to tip) thoroughly throughout the hair swatch for 30 seconds, and then allowed to sit on the hair for another 30 seconds. The swatch was then rinsed thoroughly for 30 seconds. Swatches were hung to dry in an environmentally controlled laboratory, and then wrapped in aluminum foil.

25 Evaluating Tribological Properties of Materials for Total Joint Replacements Using Scanning Probe Microscopy

Sriram Sundararajan · Kanaga Karuppiah Kanaga Subramanian

From volume IX

Abstract. Scanning probe methods have been utilized to reveal useful information regarding the tribological phenomena of total joint replacement materials. In particular, studies on the effect of synovial fluid constituents on the tribological response of polymeric materials and the effect of residual stresses on the fretting response of metallic materials are described. The studies show that an adsorbed layer of proteins causes an increase in friction behavior of hydrophobic ultrahigh molecular weight polyethylene (UHMWPE) owing to denaturation upon adsorption. Changing the crystallinity of UHMWPE can affect the friction behavior and have ramifications for mechanisms of protein adsorption. Studies on fretting behavior of metallic (CoCrMo) implant materials showed that significant plastic deformation of the metal may not be a prerequisite for oxide fracture. In addition, results suggest that nearly neutral boundary lubricants may provide significant protection against the onset of fretting. Finally, the chapter concludes with an outlook on how scanning probe methods can be used to provide greater insight into the design of biomaterials.

Key words: Total joint replacements, Ultrahigh molecular weight polyethylene, Polyethylene, Atomic force microscopy, Friction, Wear, Proteins

25.1 Introduction

This chapter describes examples of scanning probe applications in the field of bioengineering and biomaterials, specifically the area of understanding and improving the tribological behavior and durability of joint replacement materials. The chapter begins with an introduction to total joint replacements (TJRs), including the problems being studied and the far-reaching impact of finding potential solutions. Examples of scanning probe methods used and the results obtained are then described. The chapter concludes with a summary and future outlook.

25.1.1 Total Joint Replacements

TJR is a procedure in which damaged joints are removed and replaced with an artificial device (prosthesis). Each year, more than a million total replacement surgeries are performed worldwide, with a predominant number being hip and knee replacements. A natural hip joint, shown in Fig. 25.1, is a ball-and-socket joint that

In: H. Fuchs, B. Bhushan (eds), *Biosystems - Investigated by Scanning Probe Microscopy*,
DOI 10.1007/978-3-540-74083-4_13,

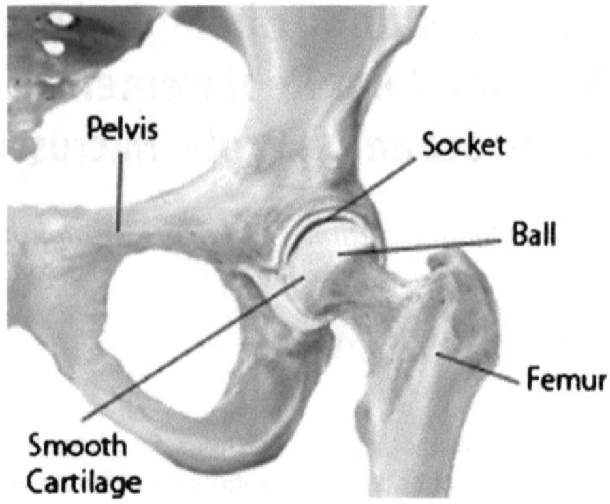

Fig. 25.1. A natural hip joint and related components

allows movement in three planes. The head of the femur mates with the socket of the pelvic bone to form the articulating joint. These bearing surfaces are lined with a 1–3-mm-thick layer of human cartilage and operate in the presence of a natural joint fluid known as the synovial fluid. A typical total replacement hip joint is shown in Fig. 25.2a, in which the bearing surfaces are formed by the acetabular liner (usually a polymer) and the femoral head (usually a ceramic or metallic alloy). Most replacement joints are modular in nature as shown in Fig. 25.2a. The acetabular shell and the hip implant are embedded into the bone using cement or using a press-fit.

25.1.2 Social and Economic Significance

Researchers have estimated that the demand for TJRs is expected to increase dramatically in the next 25 years [1–3]. A study conducted by the American Academy of Orthopedic Surgeons [2] projects the number of yearly procedures for primary (first-time) total knee replacement to jump by 673% to 3.48 million in 2030. The number of primary total hip replacements performed yearly will increase by 174% to 572,000. The research team based its projections on historical procedure rates from 1990 to 2003, combined with population projections from the US Census Bureau. Reasons for the expected increase include an aging population with arthritis requiring joint replacement; the increasing prevalence of obesity, which puts undue stress on the knee and hip joints; and the trend toward baby boomers remaining physically active later in life, which also places demands on the joints.

25.2 Problems Associated with Total Joint Replacements

Though joint replacement procedures are mostly successful, the artificial joints can become loose and unstable during use, which is commonly referred to as aseptic loosening (Fig. 25.2b). The TJRs thus often require revision surgeries to repair or replace the joint. Herbert et al. [4] have reported that a revision surgery costs three to four times more hospital resources than a primary implant. The number of revision surgeries likely will double by 2015 for total knee replacement and by 2026 for total

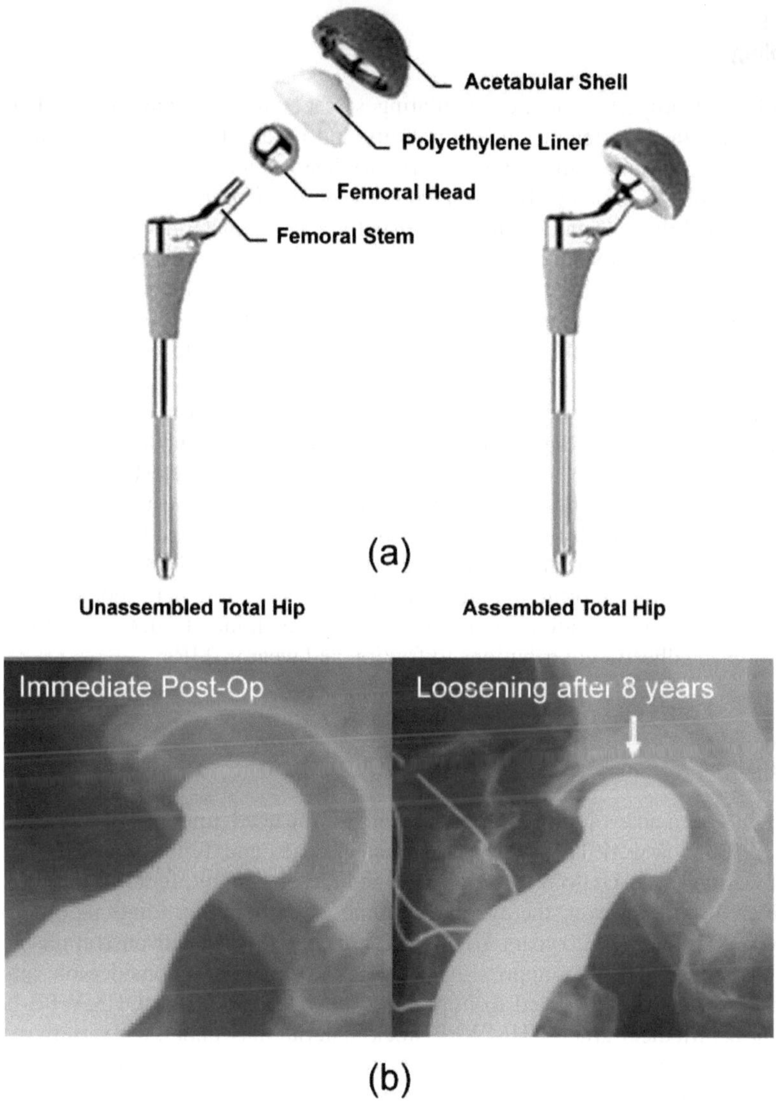

Fig. 25.2. a A typical modular total replacement hip joint showing the unassembled components and the assembled joint. **b** A new implant and the same implant exhibiting aseptic loosening

hip replacement [2,5]. Costs of revision joint replacement are particularly great and must be reduced. The best way is to make the original joint implant last longer. For that to happen, there is a need for more research to improve surgical materials and technologies, which in turn require a better understanding of the processes involved in their in-use failure. Continuing education on best practices for surgeons performing the procedures and new medical approaches to prevent loosening of the implant is also important.

25.2.1 Tribology

Tribological properties of the articulating surfaces in TJRs have been identified as a critical factor affecting their durability and reliability [6–9]. For example, in a total hip replacement there are two primary bearing interfaces that determine the useful life of the implant: the interface between the femoral head (usually a hard material) and the acetabular cup liner (usually a softer material); and the interface between the femoral stem and the femoral head (Fig. 25.2). Sliding wear occurs at the former interface, whereas fretting wear occurs at the latter interface. The modular junctions of hip replacements are especially prone to fretting corrosion [10–13]. The occurrence of wear itself is not catastrophic, but does create wear debris, which causes inflammation of the joint and osteolysis (bone loss). This can cause implant-loosening. In order to understand the debris generation it is necessary to understand the mechanism that governs sliding, damage nucleation and subsequent wear at the material interface. Owing to the high contact area between the cup and the liner in hip joints, the friction and wear mechanisms are generally adhesive in nature at the onset of operation, where maximum principal stresses are much less than the yield strength of the ultrahigh molecular weight polyethylene (UHMWPE). However, at later stages, ductile deformation micromechanisms and particulate debris compound the wear. These modes are also predominant in knee joints. Hence, it is of interest to minimize adhesive contributions to friction and wear in TJRs.

25.2.2 Materials

There has been an evolution of material pairs for the head–liner interface, beginning with Charnley's work [14] with stainless steel heads and Teflon liners. Today, the combination of an UHMWPE liner and metallic or ceramic femoral heads is extensively used. However, the average lifetime of artificial hip joints incorporating UHMWPE is only 15–20 years. UHMWPE is a polyethylene with an average molecular mass of greater than approximately 3.1×10^6 g/mol. A considerable amount of research has been devoted to improving the wear life of the UHMWPE-based prosthesis. Medical-grade UHMWPE stock material may undergo a variety of processing techniques during implant manufacturing [15], including molding, extrusion and milling or turning. These manufacturing processes can affect the surface morphology and mechanical properties of the polymer [16–18], which in turn can affect its friction and wear performance. The current state of the art in the orthopedics community is to fully cross-link UHMWPE using radiation and heat treatment in order to minimize sliding wear [19–22]. Cross-linking of UHMWPE is believed to improve the wear resistance by eliminating the modes of plasticity in the semicrystalline polymer.

The femoral head and stem form part of the modular total hip replacement (Fig. 25.2a). The stem and head have interlocking tapers. Commonly used materials for these components include titanium alloys (Ti-6Al-4V, ASTM F-136) or cobalt alloys (CoCrMo, ASTM F75 cast and ASTM 562, ASTM F799, ASTM F1537 wrought). In addition, the femoral head may also be composed of ceramic alumina or ceramic zirconia [12,13,23–25]. The use of coatings and other surface treatments on the head have been explored to improve the wear behavior of UHMWPE–metal prostheses [26–32].

25.2.3 Lubrication in Joints—the Synovial Fluid

In a natural joint, synovial fluid lubricates the cartilage. Studies suggest that boundary lubrication is dominant in TJRs [33] and it is generally believed that the synovial fluid acts as a boundary lubricant to the joint interface and helps to minimize friction and wear. Fluid film lubrication, including a squeeze film effect, is believed to play a role in knee joint lubrication as well. It is recognized that the synovial fluid contains various kinds of serum proteins, hyaluronic acid and also various kinds of lipids [34, 35] as shown in Table 25.1. It has been found that albumin constitutes almost 60% of the total protein concentration in serum and synovia [35,36]. Hyaluronic acid is a type of polysaccharide which is responsible for the viscous nature of the synovial fluid [36] and is believed to help in fluid film lubrication. Phospholipids are believed to play an active role in boundary lubrication of the joints.

Synovial fluid was obviously not designed to lubricate engineered TJR materials such as polyethylene and metals/ceramics. The interaction of the TJR materials with the synovial fluid constituents is thus an important factor affecting the tribological performance of the materials [18,35,37–44]. The adsorption mechanism of proteins onto the polymeric surface can affect its subsequent friction behavior [45]. Tailoring the hydrophobicity of UHMWPE can affect the protein adsorption and hence the friction behavior of the polymer surface [45,46]. Clearly, it is of importance to understand how the tribological characteristics of UHMWPE change as soon as the TJR comes in contact with the synovial fluid. As a result of numerous studies to identify an alternative physiological fluid for use in experiments in vitro, the American Society for Testing and Materials has established that the use of bovine serum is acceptable for TJR biomaterials testing.

Table 25.1. The constituents of synovial fluid and related information on their function in joint lubrication

Constituent	Typical concentration (g/l)	Function	Reported behavior
Hyaluronic acid	2–4	Viscoelastic properties	Fails to lubricate under load [78,79]
Lubricin	–	Mucinous glycoprotein	Not active ingredient; carrier for surface-active phospholipids [33,80]
Phospholipids	2.4	Highly surface active [33]	Excellent boundary lubricant [81,82]
Proteins	18	Protein adsorbs on foreign surfaces [83]	May enhance [79] or impede lubrication [44,45]

25.3 Conventional Tribological Testing of Material Pairs for Total Joint Replacements

25.3.1 Wear Tests

Joint simulators are the standard for wear testing of TJR materials since they are designed to mimic the orientation and magnitude of joint loads. Joint simulators typically produce wear behavior and particle morphology comparable to those found in clinical cases; however, simulator tests are often difficult to analyze quantitatively and the cost, time and variability of lubrication are some drawbacks to the efficiency of this testing method. Simpler testing geometries such as pin-on-disk and pin-on-flat tests are also commonly employed to isolate and study specific mechanisms and articulating conditions. It has been shown that multidirectional paths yield wear behavior and particle morphologies that are more comparable to those found in clinical cases than unidirectional or bidirectional paths [47–49]. Nonetheless, simple geometry tests have proven to be valuable in screening potential material pairs for TJRs and to understand specific tribological mechanisms.

25.3.2 Friction Tests

Compared with a wear test, a friction measurement is a more rapid tribological assay. Friction tests can alleviate issues with lubricant volume, time and cost concerns with wear testing. Although low-friction behavior does not necessarily imply low wear, friction measurements can provide useful information regarding the tribological behavior. Frictional work is manifested in energy dissipation processes that can damage the surface. Mechanisms associated with high friction (adhesion, abrasion) can have implications for wear as well.

25.4 Scanning Probe Microscopy as a Tool to Study Tribology of Total Joint Replacements

Scanning probe microscopy (SPM) or more specifically atomic force microscopy (AFM) techniques have been extensively used to simulate a single asperity contact and to study the nanotribological properties of material pairs [50]. Techniques to perform adhesion, friction and wear studies using SPM are quite well established. Calibration methods to perform quantitative friction force microscopy are available [51,52]. Techniques to reliably calibrate the normal and lateral spring constants of the cantilevers allow researchers to perform more accurate force measurements [53,54]. The use of probe shape characterizers and reconstruction algorithms allows precise evaluation and monitoring of the tip shape during experiments [44]. Several contact mechanics theories have been applied to adhesion and friction measurements as well [44, 55, 56]. Scanning probe techniques thus allow accurate determination of the probe and contact dimensions during tribological measurements, which provide means to isolate tribological contributions from specific sources of interest such as adsorbed films. Owing to the small interaction volume, SPM techniques also alleviate concerns with lubricant volume requirements. Despite these advantages, few researchers have utilized scanning probe techniques to study the tribological behavior of TJR materials [18,40–44,57–59]. These studies are described in the following sections.

25.4.1 Nanotribology of Ultrahigh Molecular Weight Polyethylene

Ho et al. [59] studied the friction behavior of UHMWPE using an atomic force microscope. They evaluated a friction coefficient of 0.22 between a Si_3N_4 probe and direct compression molded UHMWPE at normal loads of up to 11 nN. Transients in frictional response of the UHMWPE as a function of normal load range (17–340 nN) during sliding were observed. The second transient in the frictional force was observed at a normal load of 288 ± 13 nN, corresponding to a contact pressure of 518 ± 13 MPa. Following the second transition in frictional response, plowing of the UHMWPE surface by the probe was observed. They therefore postulated that the deep scratches noticed in retrieved TJR prostheses could be due to the abrasion from protruding hard carbide asperities that exist within the cast CoCr alloy of TJR prostheses.

Kanaga Karuppiah et al. [40,44,60] have studied the effect of material processing, and the effect of protein adsorption on the nanotribological behavior of medical-grade UHMWPE. Experiments in contact mode were carried out with a Dimension^{TM} 3100 atomic force microscope (Nanoscope IV, Veeco Instruments, Santa Barbara, CA, USA) in controlled low-humidity (10 ± 4% relative humidity) conditions to minimize effects of adsorbed water vapor. Standard V-shaped silicon nitride probes with a quoted normal spring constant of 0.58 N/m and a tip radius of 10–40 nm were used. The normal spring constants of the cantilevers used were calibrated using the technique described by Torii et al. [53] and were found to have actual values of 0.20–0.25 N/m. The pull-off (adhesive) force between the Si_3N_4 tip and the UHMWPE was measured before and after each test from force–displacement curves. The friction force was calibrated using Ruan and Bhushan's method [51]. The radius of the tip was characterized before and after the experiments using a commercially available tip characterizer sample TGT01 (Mikromasch). The images were analyzed using commercial software (Image Metrology) to calculate the tip radius for accurate contact area calculations.

25.4.1.1 Effect of Material Processing

The effect of UHMWPE processing on its friction response was investigated. Specifically, milled and surface-melted samples of comparable roughness were studied. Figure 25.3 shows representative AFM topography maps of the UHMWPE samples. The melt sample was prepared by heating the polymer surface to above its melting temperature and cooling it at room temperature. The melt sample displayed a lamellar type of structure that is indicative of recrystallization and has been reported in

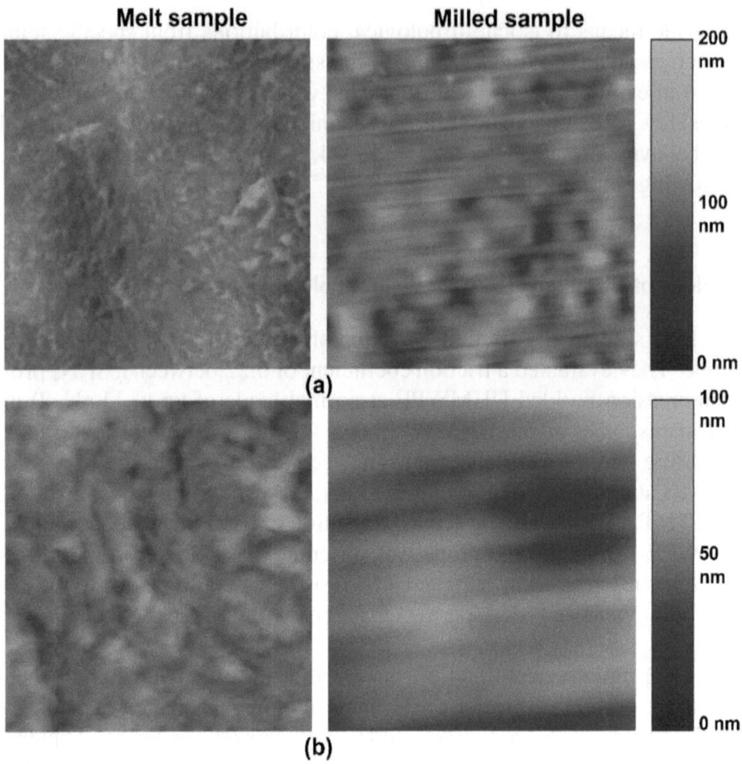

Fig. 25.3. Topography maps of melt and milled ultrahigh molecular weight polyethylene (UHMWPE) samples obtained using atomic force microscopy (AFM) at **a** 5 μm × 5 μm and **b** 1 μm × 1 μm scan sizes [44]. The melt sample exhibits lamellar structures that are indicative of recrystallization

previous studies [15, 61]. Figure 25.4, panel a shows the friction response of the materials. The friction force increased with normal load for both samples and exhibited a slight nonlinear behavior. The melt sample showed higher friction than the milled sample. No discernable wear was observed up to loads of 80 nN. Permanent damage (groove depths on the order of 2–10 nm) was observed at loads beyond 80 nN. In the absence of wear, the predominant mechanism during the friction force microscopy experiments can be assumed to be adhesive. The adhesive friction is then given by [62]

$$F_a = \tau_0 A_r + \beta L \;, \qquad (25.1)$$

where τ_0 is the interfacial shear strength, A_r is the real area of contact, β is a constant that describes the normal load dependency of the shear strength and L is the normal load. The real area of contact can be evaluated using an appropriate contact mechanics theory, while τ_0 and β are typically evaluated using curve fits. The generalized transition analysis by Carpick et al. [63] was used to determine the appropriate contact model, which in this case pointed to the Derjaguin–Muller–Toporov (DMT)

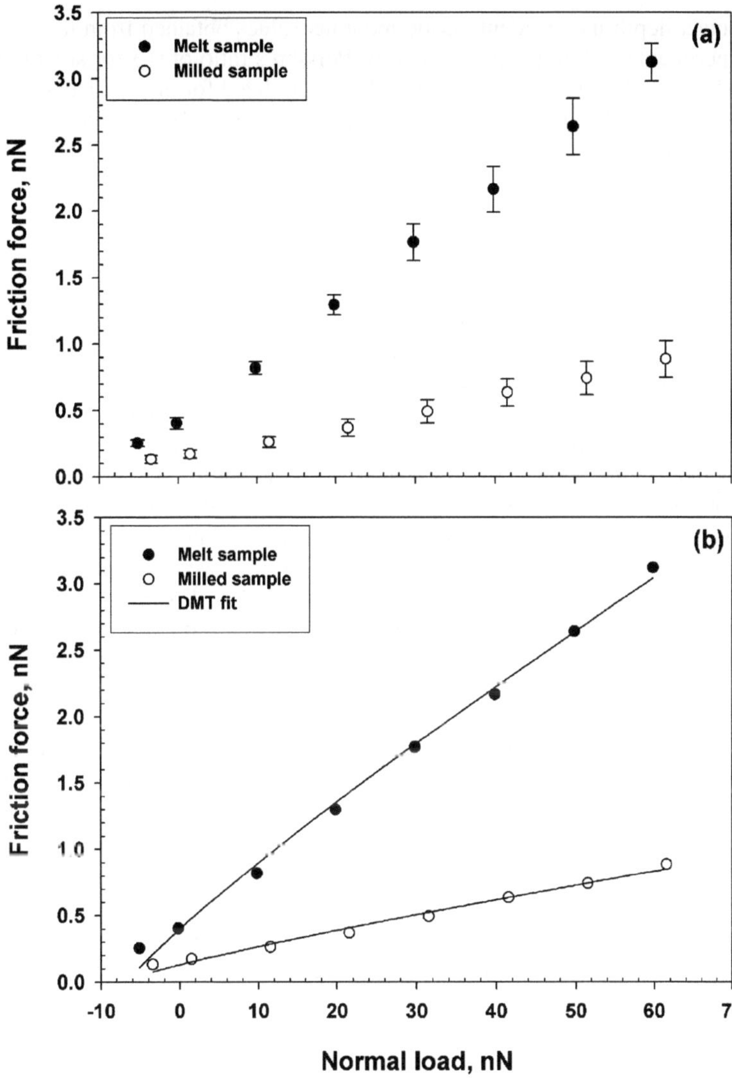

Fig. 25.4. **a** Friction response of the processed UHMWPE samples as a function of normal load obtained using friction force microscopy. **b** Example of a Derjaguin–Muller–Toporov (*DMT*) model fit to the friction data according to (25.1)

model [64]. According to the DMT model, the real area of contact is given by

$$A_{\mathrm{r}} = \pi \left(\frac{3R^*}{4E^*} \left(L + 2\pi R^* W_{12} \right) \right)^{\frac{2}{3}}, \tag{25.2}$$

where R^* is the composite radius and E^* is the composite modulus given by

$$E^* = \left(\frac{1 - \upsilon_1^2}{E_1} + \frac{1 - \upsilon_2^2}{E_2} \right)^{-1}. \tag{25.3}$$

The contact-depth-independent elastic modulus values obtained from nanoindentation experiments and an assumed value of Poisson's ratio of 0.3 for silicon nitride and 0.45 for UHMWPE, respectively [59], were utilized for these calculations. The work of adhesion (W_{12}) was measured from pull-off forces (F_{PO}) for various samples from the following equation derived from the DMT analysis:

$$F_{PO} = -2\pi R^* W_{12} . \quad (25.4)$$

Since the AFM experiments were carried out in dry conditions, it is reasonable to assume that the adhesion component dominates over capillary contributions and that W_{12} can be estimated quite reliably from (25.4). It is noted that the DMT analysis for this interface assumes that no wear occurs during contact and that the probe has a parabolic profile. Figure 25.5 shows that the probe did indeed exhibit a parabolic profile which was maintained during experiments. The parameters τ_0 and β can then be obtained by fitting (25.1) to the friction force data as shown in Fig. 25.4, panel b. The figure shows that the DMT-based equation gave a good fit, which was characteristic of the friction data for all the samples. The values for τ_0 and β obtained are also listed in Table 25.2. The measured interfacial shear strengths of the UHMWPE samples prior to protein adsorption were comparable to those measured using macroscale torsion tests (1–7 MPa) [65]. The values of β obtained were comparable to values reported for high-density and low-density polyethylene (0–0.05) [66]. The β values obtained imply that the melt samples exhibit a slightly higher normal load dependency compared to the milled samples, which displayed values closer to zero (Table 25.2).

The melting and reforming process can affect the degree of crystallinity of the sample [61]. The melt sample exhibited lower water contact angles and hence higher surface energy, which is suggestive of a higher degree of crystallinity [67]. The melt sample also exhibited almost twice the elastic modulus and hardness of the milled sample (Table 25.2), which is also indicative of increased crystallinity [66]. The data

Table 25.2. Summary of surface and interfacial characteristics for the two ultrahigh molecular weight polyethylene (*UHMWPE*) samples described in the study [44]. Measured values indicated are average ± standard deviation

UHMWPE sample	RMS roughness[a] (nm)	Elastic modulus[b] (GPa)	Hardness[b] (GPa)	Contact angle (°)	τ_0^c (MPa)	β^c
Melt sample	3.55 ± 0.90	2.10 ± 0.08	0.151 ± 0.003	66.1 ± 3.57	5.96	0.03
Melt sample with proteins	3.28 ± 0.28	–		58.3 ± 5.24	12.04	0.01
Milled sample	6.98 ± 0.85	0.90 ± 0.07	0.073 ± 0.005	96.3 ± 1.43	0.919	0.005
Milled sample with proteins	7.78 ± 0.75	–		64.3 ± 2.53	3.50	0

RMS root mean square
[a] Measured using atomic force microscopy over a 1 μm × 1 μm scan area
[b] Measured using a nanoindenter at 15-μN peak load
[c] Calculated according to Derjaguin–Muller–Toporov contact mechanics from friction force data

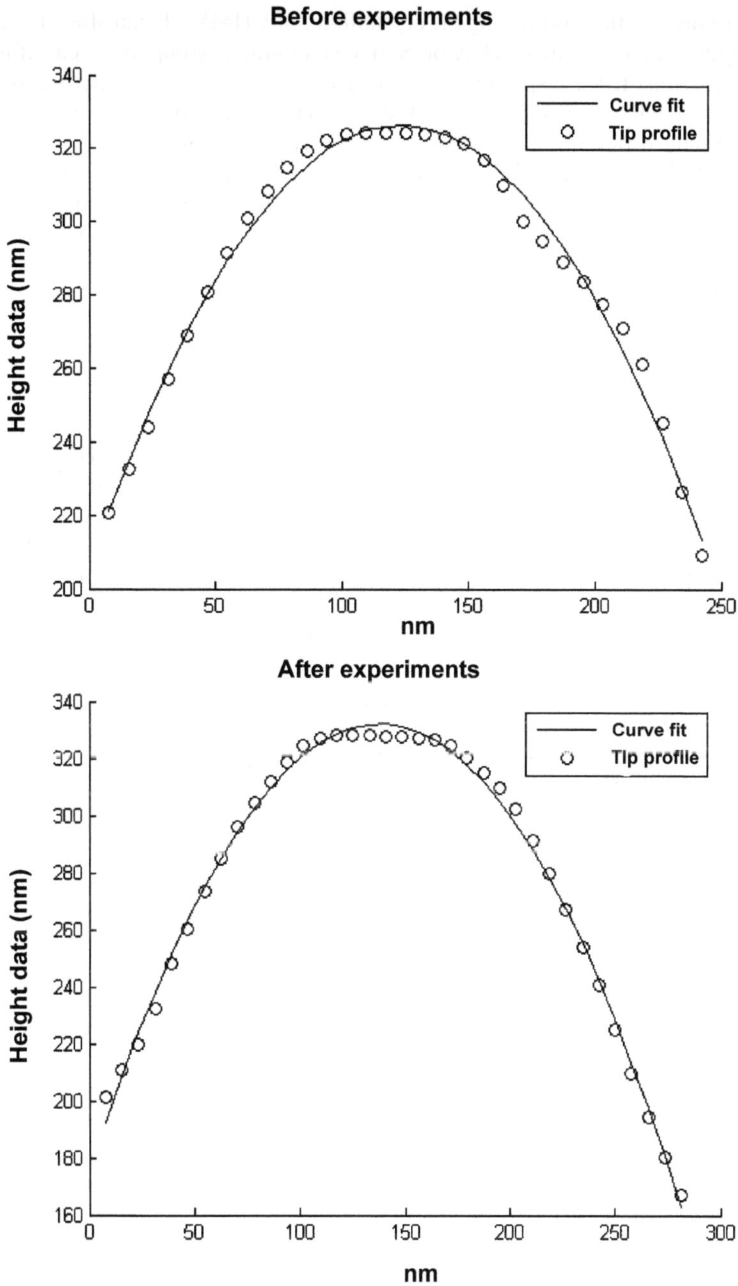

Fig. 25.5. AFM probe profiles before and after experiments obtained using a probe characterization sample and subsequent reconstruction. The profiles are parabolic and there appears to be negligible distortion as a result of experiments [44]

therefore suggest that controlling the crystallinity of UHMWPE can affect its surface tribological and mechanical behavior. Surface topography (roughness) can affect the lateral (friction) force response measured using an atomic force microscope [68]. This is especially true in the case of sharp changes in topography and an order of magnitude difference in roughness levels. The difference in roughness levels between the milled and the melt sample is quite small; hence, it is reasonable to assume the observed differences in friction force were caused by material-based effects (at the surface) rather than topography or roughness. Residual stresses (such as the compressive stresses imparted by milling) may also have an effect on the observed interfacial shear strength behavior.

25.4.1.2 Effect of Protein Adsorption

Although studies suggest that proteins affect the friction and wear of UHMWPE more significantly than the other constituents [35], the role of proteins in lubrication is not fully understood. Protein adsorption has previously been reported to increase the wear rates of UHMWPE [37,69] as well as to decrease them [70,71]. By studying the friction response of a well-characterized nanoscale contact as a function of protein adsorption, one can better understand the role of proteins in boundary lubrication.

Friction response of surface-melted and milled medical-grade UHMWPE as a function of protein adsorption has been investigated using friction force microscopy [44]. Friction response was monitored on indexed areas of the samples before exposure to proteins. The areas were then exposed to protein solution and assessed for protein adsorption before the friction response was measured again. In this study, bovine serum albumin (BSA) tagged with fluorescein isothiocyanate dye was dissolved in phosphate-buffered saline (PBS) solution of pH 7.4 (1×) to make up a concentration of 7.2 mg/ml (BSA) representing 10 vol % dilution. This dilution is comparable to the protein concentration in the human synovial fluid [46]. A 20-μl aliquot of the protein solution was placed on a small indexed area of each UHMWPE sample for 5 min, following which the samples were rinsed with deionized water and dried with nitrogen gas. Protein adsorption onto the samples was verified qualitatively using fluorescence microscopy.

Figure 25.6 shows the friction response of the UHMWPE samples before and after exposure to BSA. Both samples exhibited an increase in friction response upon exposure to BSA. The magnitude of this increase was considerably larger in the case of the milled sample. The corresponding interfacial shear strengths of the samples obtained using DMT contact analysis (Fig. 25.7, panel a) showed that exposure to proteins resulted in a 300% increase in the shear strength in the case of the milled sample, whereas the increase in the shear strength was about 100% for the melt sample. In both the milled and the melt samples, the β values (shown in Table 25.2) decreased upon exposure to proteins. Water contact angle data for the two samples are shown in Fig. 25.7, panel b. The milled sample was initially significantly more hydrophobic than the melt sample. In both cases, adsorption of BSA rendered the surfaces more hydrophilic. The reduction in contact angle was very significant in the case of the milled sample compared with that in the case of the melt sample. Table 25.2 lists all the parameters measured for the samples in this study.

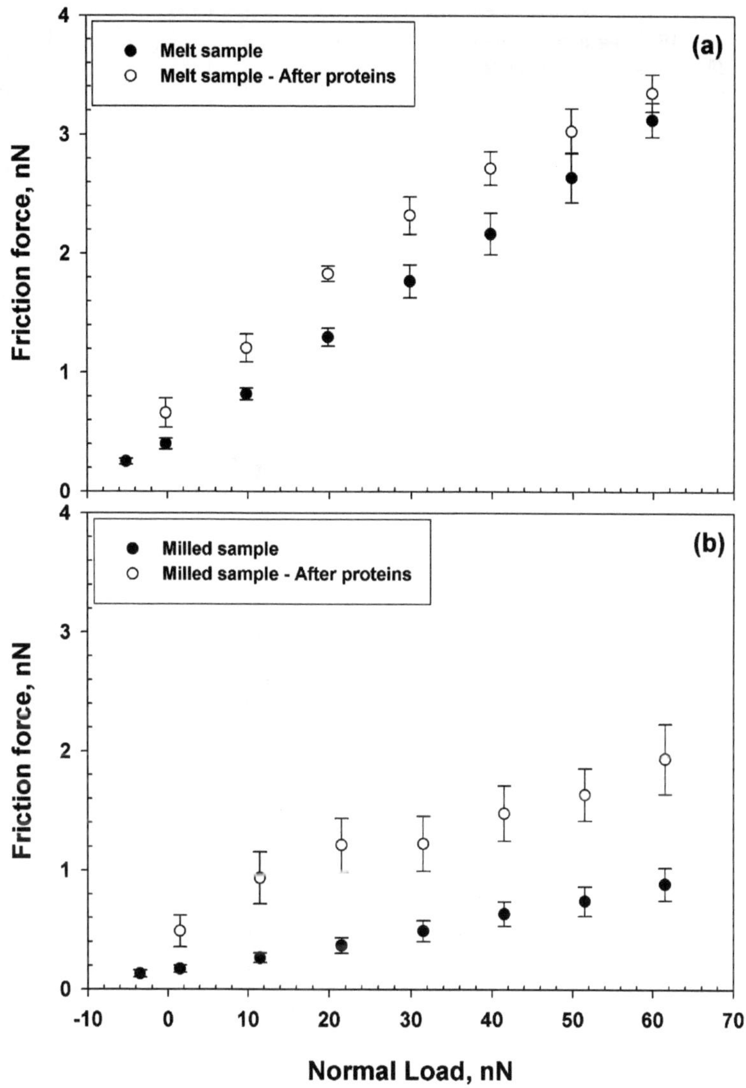

Fig. 25.6. Friction force microscopy for **a** melt and **b** milled UHMWPE samples before and after exposure to protein. The friction response increases upon exposure to proteins in both cases [44]

25.4.1.3
Discussion

The data show that protein adsorption tends to increase the friction behavior and interfacial shear strength of UHMWPE. Widmer et al. [45] showed that hydrophobic polyethylene surfaces exhibit higher friction than hydrophilic ones. They used optical waveguide lightmode spectroscopy to show that proteins adsorbed onto hydrophobic surfaces occupy more surface area than those that adsorb onto hydrophilic surfaces. They therefore proposed that proteins denature during adsorption onto

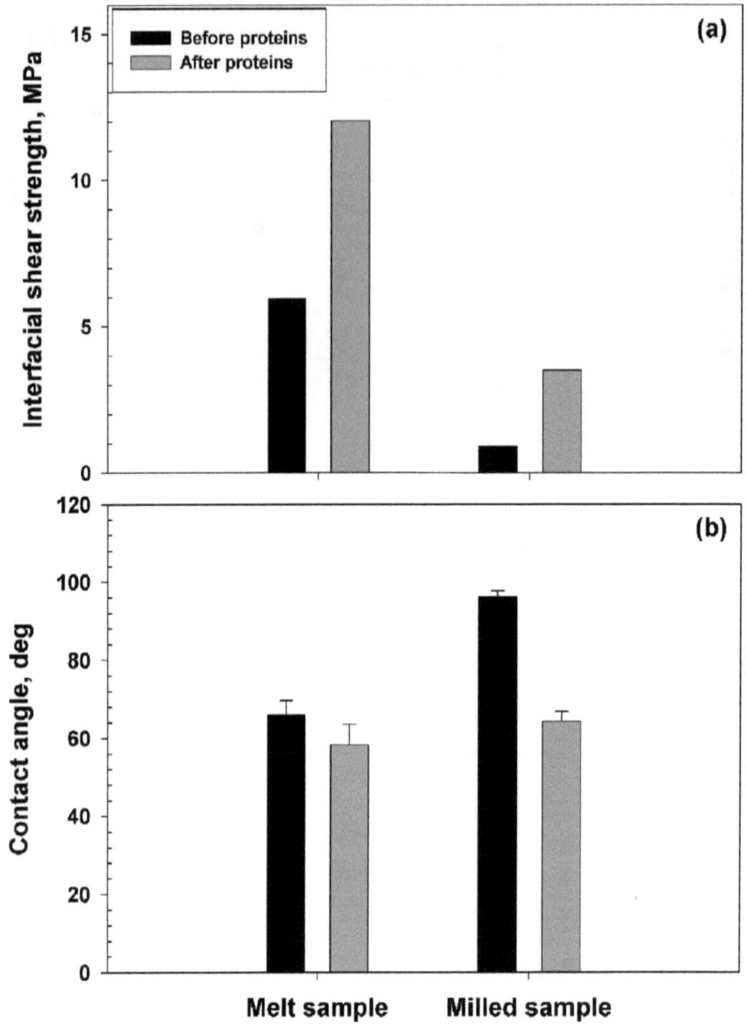

Fig. 25.7. Effect of protein adsorption on **a** interfacial shear strength and **b** water contact angles of the UHMWPE samples (adapted from [44])

a hydrophobic surface. The denaturing of proteins upon adsorption on hydrophobic surfaces has been reported in relation to other applications [72, 73] as well. Denatured proteins occupy large surface area, with hydrophobic parts undergoing adsorption, exposing hydrophilic regions. The contact angle data support the idea that the proteins denature upon adsorption onto the hydrophobic UHMWPE surfaces or that denatured proteins adsorb preferentially onto the hydrophobic surfaces (Fig. 25.8). The data therefore indicate that a denatured protein layer forms a relatively higher shear strength layer than the virgin material, which increases the adhesive friction response. An increase in the interfacial shear strength may have implications for adhesive wear mechanisms. However, higher friction in most

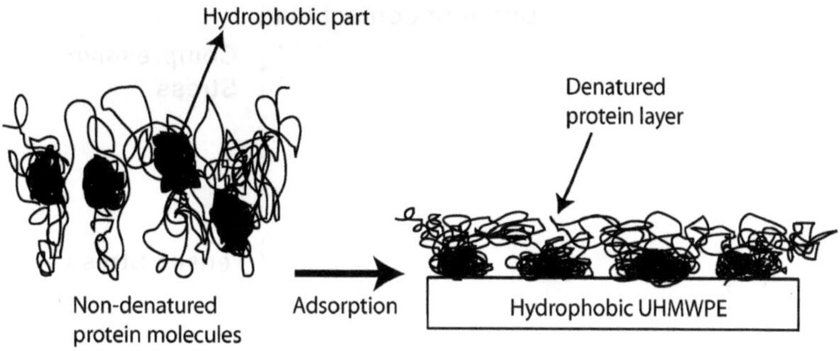

Fig. 25.8. The denaturation of proteins as they adsorb onto hydrophobic UHMWPE. The adsorbed layer exposes hydrophilic regions of the protein molecules. This denatured layer appears to increase the adhesive friction force of the surface. (Adapted from [45])

cases does not necessarily imply higher wear and one should evaluate wear behavior independently from friction experiments. The melt sample showed higher protein adsorption but lower increase in friction upon exposure to protein than the milled sample. This suggests that the conformation of the adsorbed proteins affects the friction performance significantly. The higher surface energy of the melt sample reduces the extent of protein denaturation (or adsorption of denatured proteins) compared with that for the more hydrophobic milled sample and leads to the relatively lower increase in shear strength upon protein adsorption. These studies suggest that changing the crystallinity of UHMWPE can affect its surface energy and protein adsorption and hence the friction behavior in the presence of proteins.

25.4.2 Fretting Wear of Cobalt–Chromium Alloy

Mitchell and Shrotriya [58, 74] have investigated the nanoscale wear response of cobalt–chromium (CoCrMo) alloy (ASTM F-75) as a function of contact load and surface residual stress in order to identify the mechanism governing the onset of surface damage in modular implants.

CoCrMo samples were cut from a canine femoral stem prosthesis by electric discharge machining, which was chosen to minimize processing-induced residual stress on the machined surfaces. The machined samples were mechanically polished to yield a root mean square roughness of less than 2 nm over a 10-μm range.

A unique loading configuration was utilized to apply a range of known in-plane stress states to the specimen surface in order to simulate different residual stress levels. This was achieved through a four-point bending frame as shown in Fig. 25.9. A capacitance gage monitors the deflection of the free end of the c-clamp. Opening of the c-clamp results in application of forces onto the four rollers, leading to four-point bending of the sample. Calibration of the deflection–force relationship allows for the in-plane stress states at any location along the specimen thickness to be determined from the geometry of the sample, the end deflection of the c-clamp, and the distance of the location from the sample mid-plane.

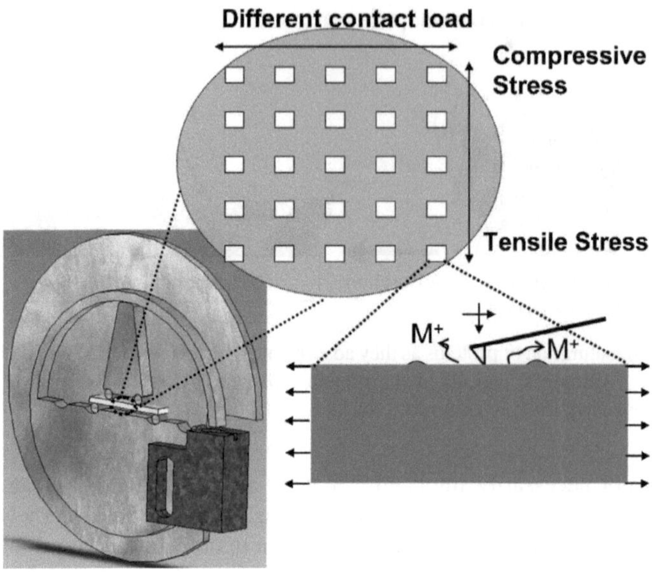

Fig. 25.9. The four-point bending setup used for in situ AFM wear testing of samples subjected to an external tensile or compressive stress state. The main components are a c-clamp, a base, a capacitance probe, rollers, and a rectangular sample. *Top right*: Five contact loading conditions and prestress locations tested. *Bottom right*: The loading action created by the atomic force microscope tip on a tensile residual stress state. (From [74])

Wear experiments were conducted using a Dimension^{TM} 3100 atomic force microscope (Nanoscope IV, Veeco Instruments, Santa Barbara, CA, USA) with a standard V-shaped silicon nitride probe at different normal loads and at different locations of a loaded specimen which correspond to different residual stresses. The wear tests were conducted at ambient conditions (25 °C, approximately 50% relative humidity) as well as in a simulated corrosive physiological environment of PBS solution, augmented with HCl to a pH of 5.01. This lower pH was chosen owing to the fact that the modular junction of a total hip replacement produces differential aeration, creating local drops in pH within the crevice. Additionally, a pH of around 5 is the thermodynamic stability point of chrome oxide (Cr_2O_3), the dominant oxide in the passive layer of CoCrMo. The AFM tests in solution were conducted in the same manner as for the ambient conditions except that a fluid cell was used and 0.5–1 ml of solution was added to cover the exposed surface of the sample. Each wear test was performed over a $2\,\mu m \times 500\,nm$ area. Seventy cycles at 5 Hz were performed and regions were imaged before and after wear. The normal spring constants and probe radii were determined as described earlier, which allowed for a more precise calculation of Hertzian contact pressures. Normal loads used for calculations of the pressure included the observed adhesive force. In order to minimize the effects of initial surface roughness on the calculation of wear depths, the initial and final images obtained for each test were registered and subtracted to determine the amount of material removed. For each experiment, average trench depth was used to estimate the wear volume.

25.4.2.1
Effect of Residual Stresses

Figure 25.10 shows the wear volumes as a function of residual stress state and applied normal load (contact pressure) under ambient conditions. The compressive residual stresses have a more pronounced influence on the wear depth. Whereas the tensile residual stress suppressed the wear, the compressive residual stress exacerbates the wear process. For all contact loads, the wear rate increased linearly as the stress state changed from the maximum tensile to maximum compressive stress state. In addition, the wear increases as the load increases for all residual stress states. Even for lower contact loads (2.3–2.6 GPa) the onset of deformation only occurred on the compressively loaded side of the sample. No wear could be determined from the lowest loading condition (1–2 nN, approximately 0.8 GPa, data not shown). It is noted that the same wear depth was achieved with around 2 GPa of contact stress on the compressive side that required approximately 4 GPa of contact stress on the tensile side. The aqueous environment behaves in a slightly different manner.

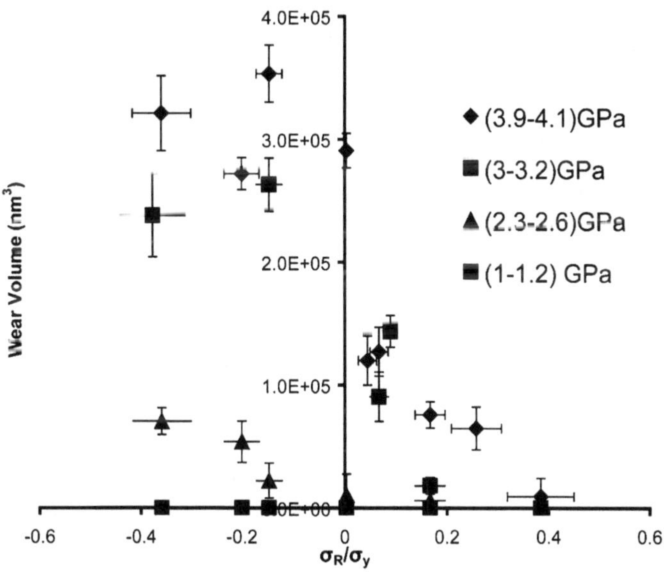

Fig. 25.10. Average wear volume as a function of residual stress (normalized over yield strength) for different contact loads under ambient conditions [58]

25.4.2.2
Effect of Aqueous Solution in Conjunction with Residual Stresses

In contrast to ambient conditions, the rate of wear was increased under both tensile and compressive stresses (Fig. 25.11) in the aqueous solution. The compressive stress still creates larger wear depths than the tensile residual stress; however, in

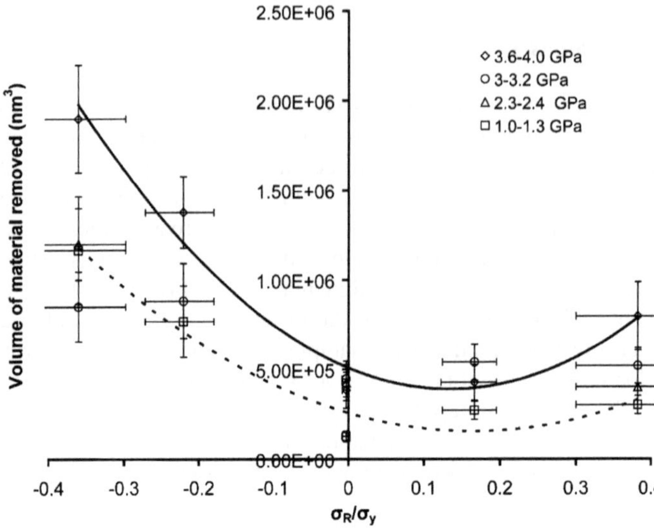

Fig. 25.11. Average wear volume as a function of residual stress (normalized over yield strength) for different contact loads under simulated corrosive physiological solution [58]

all instances the unstressed state produced less wear than the compressive or tensile stressed states. For all contact loads, the wear depth tends to follow a skewed quadratic dependence on the residual stress state (skewed toward the compressive side). The contact stress required to accelerate wear is not as apparent in the aqueous environment as in the ambient environment. However, in contrast to ambient conditions, measurable wear occurred at 0.8–1 GPa of contact stress.

25.4.2.3 Discussion

Since the residual stress is 3 orders of magnitude less than the contact loads, the residual stress should not affect the wear depth so drastically. In addition, it is counterintuitive to suggest that compressive stress is deleterious to the wear properties of the metal since it is well known that surface treatments such as shot peening and nitriding often yield enhanced wear performance in metals. The wear depth values however agree very well with the oxide thickness reported in the literature, which is 1.8 nm in ambient conditions [75]. It may then be thought that the film is fractured and needs to repassivate. The difference in the wear volumes may be explained by a change in the repassivation kinetics caused by the residual stress state. This is highly probable since the timescale for repassivation is several orders of magnitude faster than the time required for an AFM scan (milliseconds/seconds compared with minutes) [76]. Since only the scratched region would be expected to undergo fracture/repassivation, the unscratched surface is already in a passive state and the wear depth may be thought of as a reaction rate snapshot.

The difference between the residual stressed wear rates and that of the unstressed average is an indication of the effect residual stress plays in the mobility of the reaction rate. Yu and Suo [77] argue that residual stress will affect a surface reaction in either a linear (mobility dependent) or a quadratic (driving force dependent) manner. It was demonstrated that troughs on a wavy surface undergoing chemical reaction and subjected to in-plane stress will have different reaction rates from at the crests. Specifically, a compressively stressed solid gaining mass through a reaction, like repassivation, will experience slower growth at the troughs than at the crests, with the opposite being true for tensile stresses. For the present case this would give credence to the marked linear (mobility) dependence on residual stress.

The loading data demonstrate that approximately 2 GPa of average contact stress was required to produce a trench under no residual stress in ambient conditions. Under an aqueous environment, a trench of approximately the same depth is generated with 0.8–1 GPa of contact stress. This suggests that the passive layer may begin to fail by abrasive brittle fracture as would be expected of a ceramic, given that different wear levels were detected for each contact load.

25.5 Summary and Future Outlook

Scanning probe methods can thus reveal useful information regarding the tribological phenomena of TJR materials. In particular, the effect of synovial fluid constituents on the tribological response of polymeric materials and the effect of residual stresses on the fretting response of metallic materials have been described. The studies showed that an adsorbed layer of proteins causes an increase in friction behavior of hydrophobic UHMWPE owing to denaturation upon adsorption. Changing the crystallinity of UHMWPE can affect the friction behavior and have ramifications for mechanisms of protein adsorption.

Studies on fretting behavior of metallic (CoCrMo) implant materials showed that significant plastic deformation of the metal may not be a prerequisite for oxide fracture. The increased wear depth in ambient compared with aqueous environments suggests that fretting of asperities or wear particles (under dry conditions) are a significant damage mechanism that may precede crevice corrosion. In addition, nearly neutral boundary lubricants may provide significant protection against the onset of fretting.

Systematic studies of friction and wear in the presence of the other synovial fluid constituents (i.e., in fluid) can reveal a better picture of how engineered materials will behave in a human joint. These results may have ramifications for design and manufacturing of the implant materials and their surface engineering.

Scanning probe methods can also be applied to natural joint materials as demonstrated by Park et al. [57], who studied both macroscopic and microscopic (AFM) friction measurements on bovine articular cartilages. Scanning probe methods can thus be used to explore the role of boundary lubricants in cartilages and provide greater insight into design of biomaterials.

Acknowledgements. The authors wish to thank Michael Conzemius (University of Minnesota) for his useful discussions and insight into the area of TJR. He provided the medical perspective to our engineering approach to the research problems.

References

1. Dixon T, Shaw M, Ebrahim S, Dieppe P (2004) Ann Rheum Dise 63:825
2. Kurtz S, Mowat F, Ong K, Chan N, Lau E, Halpern M (2005) J Bone Joint Surg Am 87A:1487
3. Mehrotra C, Remington Patrick L, Naimi Timothy S, Washington W, Miller R (2005) Public Health Rep 120:278
4. Hebert CK, Williams RE, Levy RS, Barrack RL (1996) Clin Orthop Relat Res 140
5. Kurtz (2006) http://www6.aaos.org/news/Pemr/press_release.cfm?PRNumber=442
6. Revell PA, Alsaffar N, Kobayashi A (1997) Proc Inst Mech Eng H 211:187
7. McGee MA, Howie DW, Costi K, Haynes DR, Wildenauer CI, Pearcy MJ, McLean JD (2000) Wear 241:158
8. Ingham E, Fisher J (2000) Proc Inst Mech Eng H 214:21
9. El-Warrak AO, Olmstead ML, von Rechenberg B, Auer JA (2001) Vet Comp Orthop Traumatol 14:115
10. Schaaff P (2004) J Appl Biomater Biomech 2:121
11. Hanawa T (2002) Sci Technol Adv Mater 3:289
12. Hallab NJ, Messina C, Skipor A, Jacobs JJ (2004) J Orthop Res 22:250
13. Hallab NJ, Jacobs JJ (2003) Corrosion Reviews 21:183
14. Charnley J (1960) J Bone Joint Surg Br 42:28
15. Kurtz SM, Muratoglu OK, Evans M, Edidin AA (1999) Biomaterials 20:1659
16. Bellare A, Cohen RE (1996) Biomaterials 17:2325
17. Ho SP, Riester L, Drews M, Boland T, LaBerge M (2003) Proc Inst Mech Eng H 217:357
18. Check J, Karuppiah KSK, Sundararajan S (2005) J Biomed Mater Res A 74A:687
19. Shen FW, McKellop HA, Salovey R (1996) J Polym Sci Part B Polym Phys 34:1063
20. Jasty M, Bragdon CR, O'Connor DO, Muratoglu OK, Premnath V, Merrill EW, Harris WH (1997) In: Proceedings of the 43rd annual meeting the of Orthopaedic Research Society, San Francisco, p 785
21. Fisher J, McEwen HMJ, Tipper JL, Galvin AL, Ingram J, Kamali A, Stone MH, Ingham E (2004) Clin Orthop Relat Res 114
22. Shen FW, McKellop H (2005) Clin Orthop Relat Res 80
23. Willmann G (1998) Orthopedics 21:173
24. Katti KS (2004) Colloids Surf B 39:133
25. Cales B (2000) Clin Orthop Rel Res 94
26. Rostami H, Iskandarani B, Kamel I (1992) Polym Composites 13:207
27. Alonso F, Ugarte JJ, Sansom D, Viviente JL, Onate JI (1996) Surf Coat Technol 83:301
28. Lappalainen R, Heinonen H, Anttila A, Santavirta S (1998) Diamond Relat Mater 7:482
29. Raimondi MT, Pietrabissa R (2000) Biomaterials 21:907
30. Onate JI, Comin M, Braceras I, Garcia A, Viviente JL, Brizuela M, Garagorri N, Peris JL, Alava JI (2001) Surf Coat Technol 142:1056
31. Platon F, Fournier P, Rouxel S (2001) Wear 250:227
32. Pompe W, Worch H, Epple M, Friess W, Gelinsky M, Greil P, Hempel U, Scharnweber D, Schulte K (2003) Mater Sci Eng A 362:40
33. Hills BA (1995) Ann Biomed Eng 23:112
34. Brown SS, Clarke IC (2006) Tribol Transa 49:72
35. Sawae Y, Murakami T, Chen J (1998) Wear 216:213
36. Swann DA (1987) Bull Soc Belg Ophtalmol 223(Pt 1):59
37. Chandrasekaran M, Loh NL (2001) Wear 250:237
38. Mazzucco D, Spector M (2004) Clin Orthop Relat Res 17
39. Nakanishi Y, Murakami T, Higaki H, Miyagawa H (1999) JSME Int J Ser C 42:481

40. Kanaga Karuppiah KS, Sundararajan (2006) In: Proceedings of IJTC2006 STLE/ASME international joint tribology conference, San Antonio, October 2006
41. Kanaga Karuppiah KS, Sundararajan Z-H, Xu Li X (2006) In: Proceedings of IMECE2006 2006 ASME international mechanical engineering congress and exposition, Chicago, November 2006
42. Kanaga Karuppiah KS, Sundararajan (2005) In: Proceedings of IMECE2005 2005 ASME international mechanical engineering congress and exposition, Orlando, November 2005
43. Kanaga Karuppiah KS, Sundararajan (2005) In: Proceedings of the 4th international surface engineering congress, St Paul, August 2005
44. Karuppiah KSK, Sundararajan S, Xu ZH, Li XD (2006) Tribol Lett 22:181
45. Widmer MR, Heuberger M, Voros J, Spencer ND (2001) Tribol Lett 10:111
46. Heuberger MP, Widmer MR, Zobeley E, Glockshuber R, Spencer ND (2005) Biomaterials 26:1165
47. Bragdon CR, Jasty M, Kawate K, McGrory BJ, Elder JR, Lowenstein J, Harris WH (1997) J Arthroplasty 12:119
48. Joyce TJ, Monk D, Scholes SC, Unsworth A (2000) Biomed Mater Eng 10:241
49. Bennett D, Orr JF, Beverland DE, Baker R (2002) Proc Inst Mech Eng H 216:393
50. Bhushan B (1999) Handbook of micro/nano tribology. CRC, Boca Raton
51. Ruan J-A, Bhushan B (1994) J Tribol 116:378
52. Ogletree DF, Carpick RW, Salmeron M (1996) Rev Sci Instrum 67:3298
53. Torii A, Sasaki M, Hane K, Okuma S (1996) Meas Sci Technol 7:179
54. Sader JE, Chon JWM, Mulvaney P (1999) Rev Sci Instrum 70:3967
55. Schwarz UD, Zworner O, Koster P, Wiesendanger R (1997) Phys Rev B Condens Matter 56:6987
56. Carpick RW, Agrait N, Ogletree DF, Salmeron M (1996) J Vac Sci Technol B 14:1289
57. Park S, Costa Kevin D, Ateshian Gerard A (2004) J Biomech 37:1679
58. Mitchell AS, Shrotriya P (2005) In: Proceedings of the 4th international surface engineering congress, St Paul, August 2005
59. Ho SP, Carpick RW, Boland T, LaBerge M (2002) Wear 253:1145
60. Kanaga Karuppiah KS, Sundararajan Bruck AL, Lin Z (2006) In: Proceedings of MRS fall 2006 meeting, Boston, November 2006
61. Turell MB, Bellare A (2004) Biomaterials 25:3389
62. Carpick RW, Salmeron M (1997) Chem Rev 97:1163
63. Carpick RW, Ogletree DF, Salmeron M (1999) J Colloid Interface Sci 211:395
64. Derjaguin BV, Muller VM, Toporov YP (1975) J Colloid Interface Sci 53:314
65. Park KD, Kim J, Yang SJ, Yao A, Park JB (2003) J Biomed Mater Res B 65B:272
66. Gracias DH, Somorjai GA (1998) Macromolecules 31:1269
67. Yui N, Suzuki Y, Mori H, Terano M (1995) Polym J 27:614
68. Sundararajan S, Bhushan B (2000) J Appl Phys 88:4825
69. Good VD, Clarke IC, Gustafson GA, Downs B, Anissian L, Sorensen K (2000) Acta Orthop Scand 71:365
70. Cooper JR, Dowson D, Fisher J (1993) Clin Mater 14:295
71. Kernick M, Allen C (1997) Wear 203:537
72. Norde W, Giacomelli CE (2000) J Biotechnol 79:259
73. Ying P, Yu Y, Jin G, Tao Z (2003) Colloids Surf B 32:1
74. Mitchell AS, Shrotriya P (2007) Wear 263:1117
75. Milosev I, Strehblow HH (2003) Electrochim Acta 48:2767
76. Goldberg JR, Gilbert JL (1997) J Biomed Mater Res 37:421
77. Yu HH, Suo Z (2000) J Appl Phys 87:1211
78. Kawano T, Miura H, Mawatari T, Moro-Oka T, Nakanishi Y, Higaki H, Iwamoto Y (2003) Arthritis Rheum 48:1923

79. Swann DA, Radin EL, Nazimiec M, Weisser PA, Curran N, Lewinnek G (1974) Ann Rheum Dis 33:318
80. Mabuchi K, Obara T, Ikegami K, Yamaguchi T, Kanayama T (1999) Clin Biomech 14:352
81. Hills BA, Butler BD (1984) Ann Rheum Dis 43:641
82. Purbach B, Hills BA, Wroblewski BM (2002) Clin Orthop Relat Res 115
83. Adams GA, Feuerstein IA (1981) Trans Am Soc Artif Intern Organs 27:219

12 Atomic Force Microscopy in Nanomedicine

Dessy Nikova · Tobias Lange · Hans Oberleithner · Hermann Schillers · Andreas Ebner · Peter Hinterdorfer

From volume III

12.1 AFM in Biological Sciences

Advances in our understanding of molecular and cellular biology were and still are dictated by the development of new techniques, allowing the structural and functional study of living material. This started with van Leeuwenhoek's "craving after knowledge" leading to the discovery of the first microscope in the 17th century, which made the visualization of individual cells possible. His powerful magnifying glasses have laid the foundations for many revolutions in biology. Since then the development of new microscopes have allowed us to observe smaller and smaller structures, ranging from sub-cellular components to individual molecules. Undoubtedly, elucidating the three dimensional structures of proteins, nucleic acids and assemblies of these macromolecules at a molecular level is critical to gain insight into their biological function. Using techniques such as X-ray diffraction, nuclear magnetic resonance (NMR) and electron microscopy (EM), a wealth of information on the structure and function of individual biomolecules has been gathered. However, those standard structural tools also have limitations, e.g. (i) difficulties in crystallization of multimeric membrane proteins may limit the general use of X-ray diffraction; (ii) solution NMR studies of macromolecular structure and dynamics have size restriction that limits the studied structures; and (iii) the contrast is low in biological EM, thus additional sample preparation is required. It thus seems clear that future progress in revealing the complex machinery of the cell will depend in part on additional and perhaps ground-breaking developments in structural biology techniques. In this regard, recently considerable attention has focused upon the biological and biomedical applications of the atomic force microscopy (AFM), and in particular on its ability to directly "look at" both the biomolecule structure and dynamics at the single molecule level with outstanding signal-to-noise ratio. Initially designed as a technique for the solid-state material sciences [1], nowadays AFM has completely crossed the limits of its traditional areas of application and it is routinely used as a diagnostic probe of biomaterials. The breakthrough came with the development of the liquid cell, which permits the observation of samples in buffer solutions [2]. However, the first images of DNA molecules obtained with contact mode AFM were encouraging, but often very irreproducible as a result of the movement of the molecules under the tip. In contact mode, the tip is scanned over the surface, maintaining a constant deflection (i.e. force). The created large lateral forces can easily sweep away or deform the fragile biomolecules encountered. This problem was greatly reduced upon the invention of the tapping-mode in liquid [3].

In: H. Fuchs, B. Bhushan (eds), *Biosystems - Investigated by Scanning Probe Microscopy*,
DOI 10.1007/3-540-26910-X_1,

Since the tip is oscillating, it only probes (or touches) the sample for a very short time, reducing lateral forces on the molecules enormously. Therefore, this mode allows imaging of very delicate biological samples. The unique potentials of AFM to image biological substrates with molecular or even sub-molecular resolution under physiological conditions in a non-damaging manner, where the native structure and function of the biomolecules under investigation are preserved, make this technique very attractive to biologists. To this end, AFM has been used in a broad scope of biological investigations, ranging from high-resolution of the arrangement of membrane proteins within two-dimensional arrays [4] and, e.g. Fig. 12.1, to determination of the structure of protein–DNA complexes such as nucleosomes [5]; also from molecular manipulation [6] to real-time visualization of DNA degradation [7] and, e.g. Fig. 12.2; or from pulling of nucleic acids [8] to imaging of cells [9] and force mapping to measure properties such as the cell elasticity [10].

The cell, the basic unit of life, is considered as one of the building blocks in all organisms. Early biologists saw cells as simple membranous sacs containing fluid and a few floating particles. It is known today that cells are infinitely more complex than this. There are many different types, sizes, and shapes of cells, each possessing unique functions in the body. Just, e.g. the cell membrane itself, by which every cell is enclosed, is a multifunctional dynamic organization of lipid bilayers and proteins (Fig. 12.3). The role of the membrane is not only to maintain the integrity of the cell but to perform a variety of physiologically important functions as well. The plasma membrane with its embedded membrane proteins is the locus of all communications of the cell with its environment. It acts as receptor sites, controls the transport of molecules into and out of the cell and maintains signal transfer between cells, e.g. of the nervous system by membrane potential.

However, if dysfunctional, those membrane proteins could be the cause of many diseases. The misfolding of the protein three-dimensional macromolecule structure, e.g. is recognized as such a factor in cystic fibrosis (CF) [11], and Parkinson's and Alzheimer's disease [12].

CF is a lethal genetic disease affecting approximately one in 3200 people. CF is mainly the result of an incorrectly folded membrane protein called the cystic fibrosis transmembrane regulator (CFTR). In a healthy organism, CFTR functions as

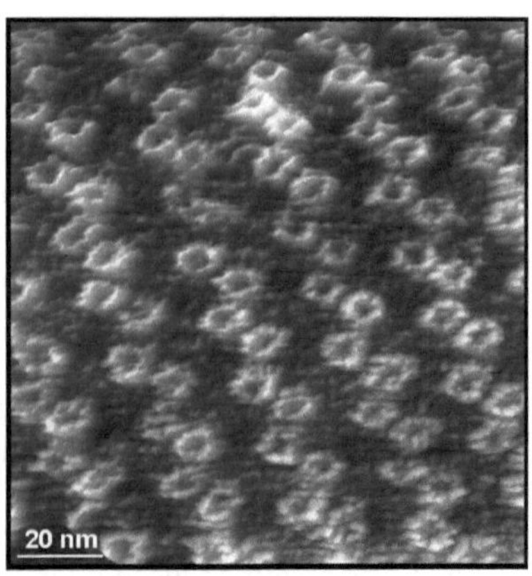

Fig. 12.1. High-resolution AFM image of light harvesting complexes LH2 from photosynthetic purple bacterium assembled in two-dimensional arrays

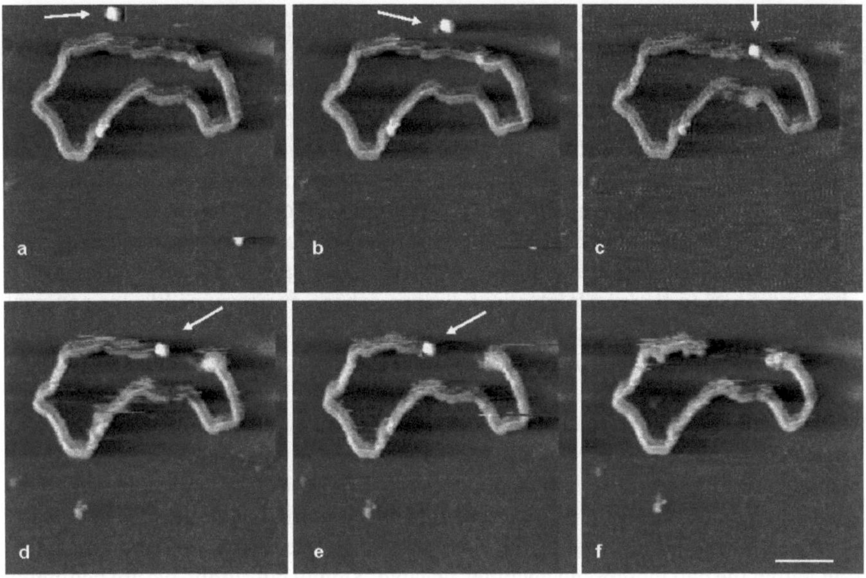

Fig. 12.2. An enzymatic degradation of plasmid DNA. (**a–f**) Consecutive AFM images in buffer solution with a time interval of 30 s, revealing the action of DNase I (indicated by *arrows*). Bar 100 nm, z-range = 6 nm

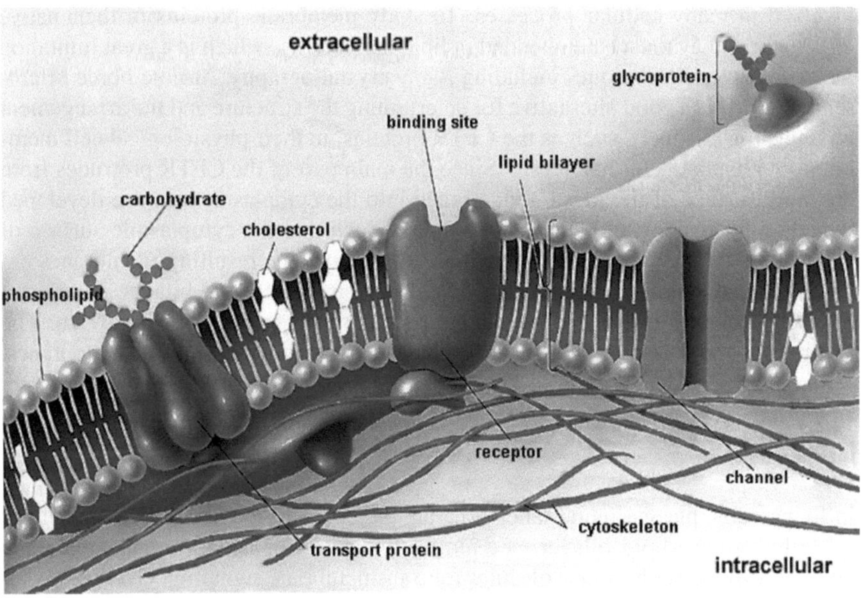

Fig. 12.3. Schematic of a cell membrane composed of a lipid bilayer and proteins

a chloride channel and it controls several other membrane proteins by still unknown mechanisms. CFTR is known to play a crucial role in maintaining the salt and water balance on the epithelium and to regulate processes such as cell volume regulation. A mutation in the gene encoding the protein causes the deletion of the amino acid phenylalanine at position 508 (ΔF508). The resulting protein (ΔF508 CFTR) is misfolded and this leads to an impaired trafficking to the plasma membrane. In most of the cases when misfolded, the protein cannot be inserted into the cell membrane. The loss of CFTR functions on the surface of airway epithelial cells causing enormous obstructive airway problems and hypoxia. CFTR dysfunction is also a problem in the gastrointestinal system, where it is responsible for pancreas insufficiency and malabsorption.

This chapter describes the utilization of AFM for studying cells and particularly cell microelasticity in relation to CF, and the identification of CFTR within the cell membrane at a single molecule level. Such measurement can provide considerable advantages, as it removes the data-averaging drawback inherent in the conventional techniques that record measurements over large ensembles of molecules. With the single molecule approach, the differences between individual molecules are clearly observed. Moreover, elucidating the distribution of single proteins and their organization within the cell membrane becomes feasible. Slowly but surely, we are beginning to learn that the world is a very different place when looked at one molecule at a time.

12.2 Plasma Membrane Preparation for AFM Imaging

Membrane proteins are important components of the cellular machinery and are involved in many cellular processes. To study membrane proteins in their native environment they must be embedded in lipid membranes, which is a great limitation factor for several techniques including X-ray crystallography. Atomic Force Microscopy (AFM) is a good alternative for determining the structure and the arrangement of single ion channels, such as the CFTR proteins, in their physiological cell membranes with nanometer resolution. Since the main part of the CFTR protrudes from the inner surface of the plasma membrane into the cytoplasm, we have developed a method for inside–out membrane isolation, such that the cytoplasmic surface of plasma membrane is accessible for the scanning tip. The resulting membranes exhibit a typical height of ~ 5 nm corresponding to an intact lipid bilayer and a great number of clearly detectable membrane proteins with heights of 5 to 20 nm. The method sets the basis for identifying single proteins within their plasma membranes.

12.2.1 Introduction

In living cells, plasma membranes separate the cell interior from the extracellular space by using a lipid bilayer, ~ 5 nm in thickness, for isolation and membrane proteins as selective barrier molecules for transmembrane signaling. Over the years, AFM has been applied to biological membranes [13] artificial bilayers [14], 2D and 3D protein crystals [15], isolated membranes [16], nuclear envelopes [17], and also to living cells [18]. AFM has permitted an unprecedented level of exploration of different biomolecules because of the ease with which samples are prepared compared to electron microscopy and also because of the freedom to image under solution. However, the flatness of the studied specimen and the supporting surface is a prerequisite for a high-resolution imaging. Spreading a plasma membrane patch on a flat solid support is currently the method of choice for identifying individual protein structures at the cell membrane, since it overcomes the softness and the height variations of living cells and allows single molecule resolution.

12.2.2 Plasma Membrane Preparation

The cells are attached to a poly-L-lysine (PLL) coated mica or glass surface and then depending on the cell type, different approaches for the membrane isolation are applied:

(i) *Xenopus Laevis* oocytes serve as a well-established expression system for membrane proteins in molecular cell biology. Before attaching the cell to a surface, the vitelline membrane, a protein meshwork, has to be removed. The oocyte, a cell with a diameter of more than 1 mm, can be easily brought in contact with the coated glass for a minute and then rolled off with the help of a pipette [19] (Fig. 12.4). The plasma membrane patches remain on the glass exposing their cytoplasmic side.

(ii) Nearly all cells need a surface to grow on. Some of these adherent cells are polarized, which means that they express specific proteins either on the basolateral or apical membrane, such as MDCK or Calu-3 cells. In order to isolate inside out oriented membranes of these cells, we let them grow on a PLL coated mica surface to a certain degree of confluence. Then, another PLL-mica is pressed on top, creating a "sandwich" of cells placed between two mica surfaces. The "sandwich" is rapidly frozen in liquid nitrogen. Then the two mica surfaces are fractured apart while still frozen, resulting in clearly separated apical and basolateral membrane patches attached to the top and base surfaces, respectively (Fig. 12.5).

(iii) Red blood cells (RBC) are cells that exist in suspension and they are too small to be manipulated by a pipette. Therefore, we used a method developed by Swiheard et al. [20] to isolate membrane patches. RBC were attached to PLL coated glass and sheared open by exposing to a jet stream of isotonic phosphate buffered saline (PBS with 0.2 mM EGTA) (Fig. 12.6). The upper membrane is washed away and a membrane patch remains attached to the glass surface.

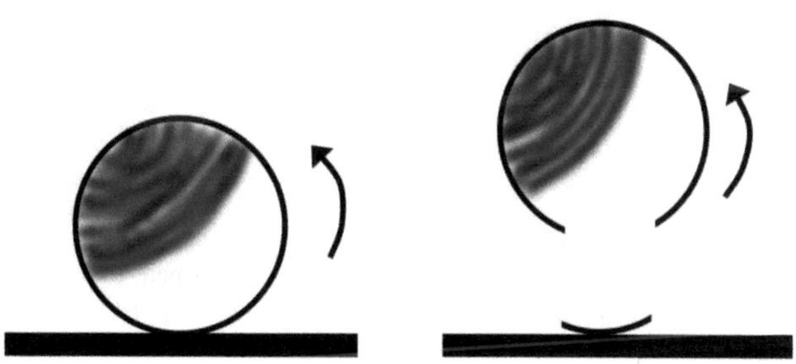

Fig. 12.4. Schematic of the oocyte membrane isolation

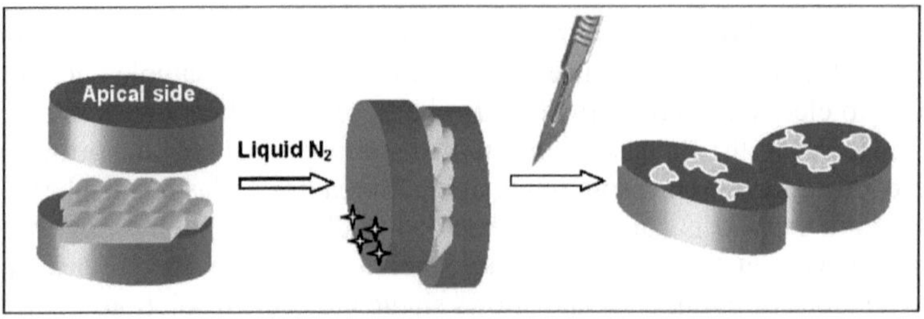

Fig. 12.5. Schematic of membrane isolation from polarized cells. The cells are sandwiched between two mica surfaces. Then the "sandwich" is rapidly frozen in liquid nitrogen. Finally, the two mica surfaces are fractured apart while frozen

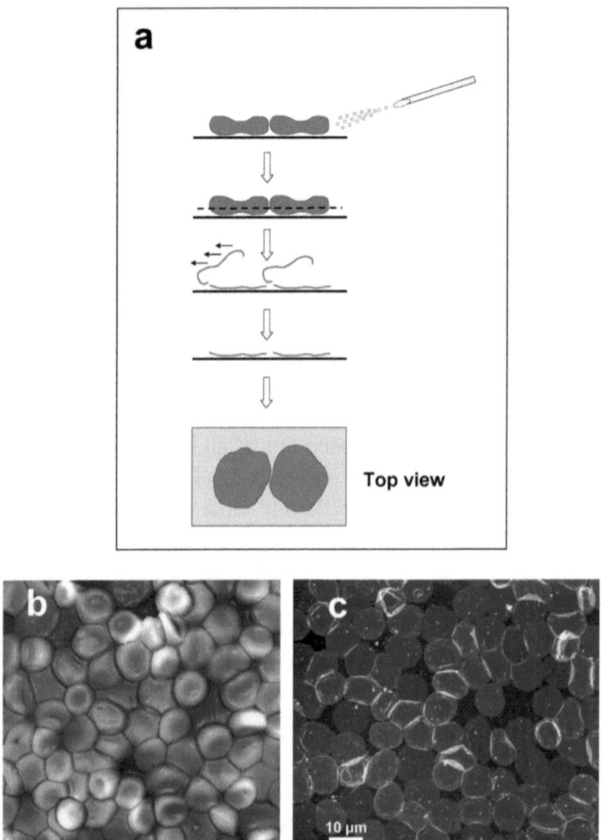

Fig. 12.6. (**a**) Schematics of the RBC membrane preparation. RBCs are exposed to fluid flow-imposed shear stress and as a result the cells are open, exposing their cytoplasmic side of the membrane. AFM images of (**b**) RBCs attached to poly-L-lysine coated glass, (**c**) inside out oriented RBC membrane patches, spread on the glass surface, after shear stress (to be published)

Thus prepared inside–out oriented RBC membrane patches, spread on glass or a mica surface are gently shaken in low salt buffer (0.3 mM NaPi, 0.2 mM EGTA) at 37 °C for 20 min in order to remove cell organelles, remnant cytoskeletal proteins or hemoglobin. Finally, the membranes are fixed with 4% paraformaldehyde in PBS for 35 min at room temperature. Subsequently, the membranes are washed with PBS and used for AFM imaging.

12.2.3 Atomic Force Microscopy

AFM was performed in contact mode using a Nanoscope III Multimode-AFM (Digital Instruments, Santa Babara, California, USA) unless specified. V-shaped oxide sharpened cantilevers with spring constants of 0.06 N/m (Digital Instruments) were used for scanning in air. Images (512 × 512 pixels) were captured with scan sizes between 1 and 25 m^2 at a scan rate of up to12 Hz (12 scan lines/s). Images were processed using the Nanoscope III software (Digital Instruments). Particle counting and 3D presentation were performed with the software SPIP (Scanning probe image processor, Image Metrology, Lyngby, Denmark).

12.2.4 Molecular Volume Measurements of Membrane Proteins

In order to estimate the molecular mass (M_0) of individual membrane proteins, we used a model published by Larmer et al. [21]. The calculation is based on a simplified model imaging a membrane protein as a sphere embedded in the lipid bilayer. The volume of a single protein (V_{Prot}) was calculated using the sphere's volume equation ($V = 4/3\pi r^3$), with the protein radius, r, given by the half height of the protein. The molecular mass M_0 can then be calculated:

$$M_0 = N_A/(V_1 + dV_2) \times V_{Prot}\,. \tag{12.1}$$

In this equation, N_A is the Avogadro constant (6.022×10^{23} mol^{-1}), V_1 is the partial specific volume of the protein (0.74 cm^3/g), V_2 is the specific volume of water (1 cm^3/g) and d is a factor describing the extent of hydration for air-dried proteins (0.4 mol H_2O/mol protein).

12.2.5 AFM Imaging

AFM imaging of thus prepared samples revealed single cell membrane patches with diameters of up to 30 m. Figure 12.7 depicts a typical image of an oocyte plasma membrane fragment spread on a glass surface, revealing a great number of protruding membrane proteins. This area showing the outline of the membrane was chosen so that the total height of the plasma membrane and its structures could be determined. Three different height levels are clearly detectable. The first level has a value of ∼ 1.5 nm and it is associated with the PLL coating. The second level corresponds to

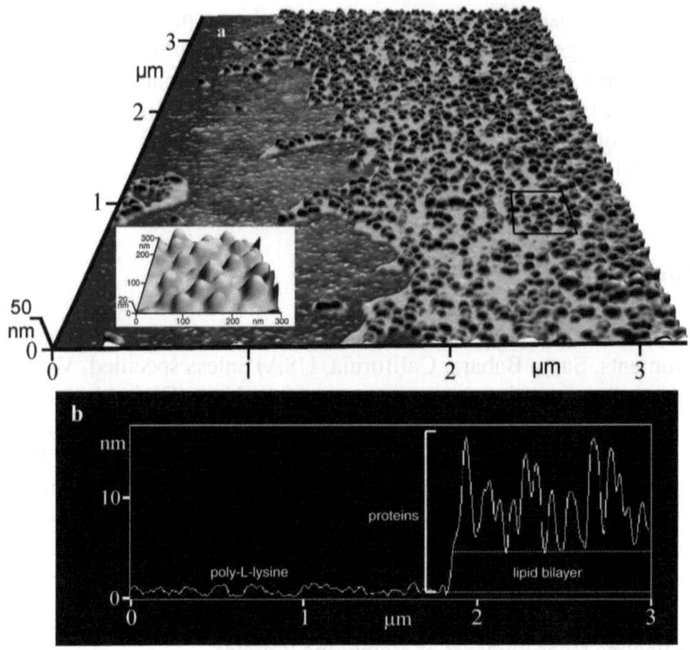

Fig. 12.7. (**a**) AFM image of a membrane fragment attached to a PLL coated glass (3D representation). The image is color-coded: the glass is shown in *blue*, the lipid bilayer membrane is shown in *turquoise* and the membrane proteins are shown in *brown*. The *inset* shows a detail from the *right part* of this image (marked with a *black rectangle*) at higher magnification. (**b**) A height profile taken along the *broken line* in (**a**). In the *lower part* of the figure, the first height level is $\sim$ 1.5 nm corresponding to PLL coating. The second level corresponds to the 5 nm-height of the lipid bilayer. The third level indicates the height of the proteins (up to $\sim$ 15 nm) protruding from the inner surface of the plasma membrane. The figure is adopted from [19]

the 5 nm-height of the lipid bilayer. Finally, the third level indicates the height of the proteins (up to 15 nm) protruding from the inner surface of the plasma membrane. The proteins appear with different heights and shapes. Some proteins are located so close to each other that they overlap or merge into one structure (inset).

Similar membrane fragments and membrane structures are observed when the other two methods for membrane isolation from Calu-3 cells and RBCs are applied. In Fig. 12.8 an image of the apical membrane of a Calu-3 cell is shown. The border of the membrane and the proteins emerging from the lipid bilayer are clearly visible.

The RBC membrane patches prepared by the shear stress exhibit circular shapes with a diameter of $\sim$ 10 m (Fig. 12.6c, Fig. 12.9). The high-resolution image reveals proteins in an extremely high density. Even if they are positioned close to one another, single proteins or groups of proteins are easily distinguishable.

It should be noted here that the apparent shape of the proteins is usually found to be conical, which is most likely due to the fact that lateral dimensions of the base of individual proteins are overestimated as a result of AFM-tip geometry. Please notice that the height measurement of proteins is still precise. The height measurements

Fig. 12.8. AFM image of the apical membrane isolated using the freeze–fracture method from the Calu-3 cell, inside out oriented. The image is gray-level coded: from *dark* (0 nm) to *white* (~ 20 nm). Scan size: 4 μm

(data not shown) of the plasma membranes and the proteins from the Calu-3 and RBCs reveal comparable values to those taken from the oocyte membranes. Clearly, using either of these approaches of isolation results in membranes with similar characteristics and this be further applied for identification of single proteins.

Furthermore, in order to verify whether the structures emerging from the oocyte lipid membrane towards the cytoplasmic space are proteins, the membranes were incubated with the enzyme trypsin. About 5 min after incubation with 0.05% trypsin protruding structures are dramatically decreased in height. Before trypsin treatment, the structures have a wide range of different heights with peak values of ~ 11 nm. After trypsin treatment, a strong shift to the left with a sharp peak appearing at ~ 6 nm is observed (Fig. 12.10). The data indicate that the protrusions are indeed proteins sensitive to trypsin digestion. Modeling plasma membrane proteins as spherical structures protruding from the lipid bilayer allowed an estimation of their possible

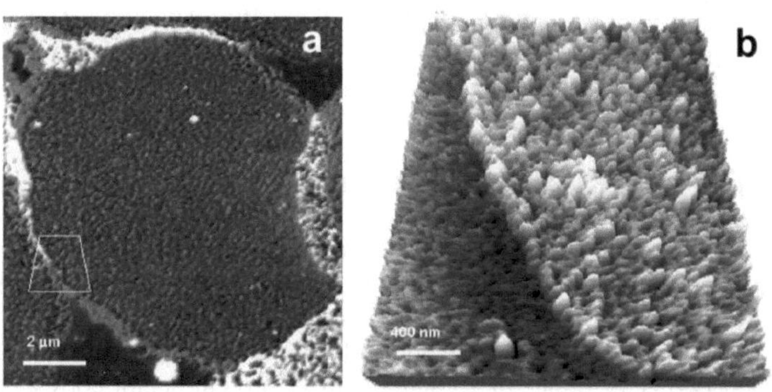

Fig. 12.9. (**a**) AFM image of an inside-out oriented RBC membrane, isolated using shear stress. (**b**) Higher resolution image of the area indicated with a *white rectangle* in (**a**). Gray-level code: *dark* (0 nm), *white* (~ 20 nm)

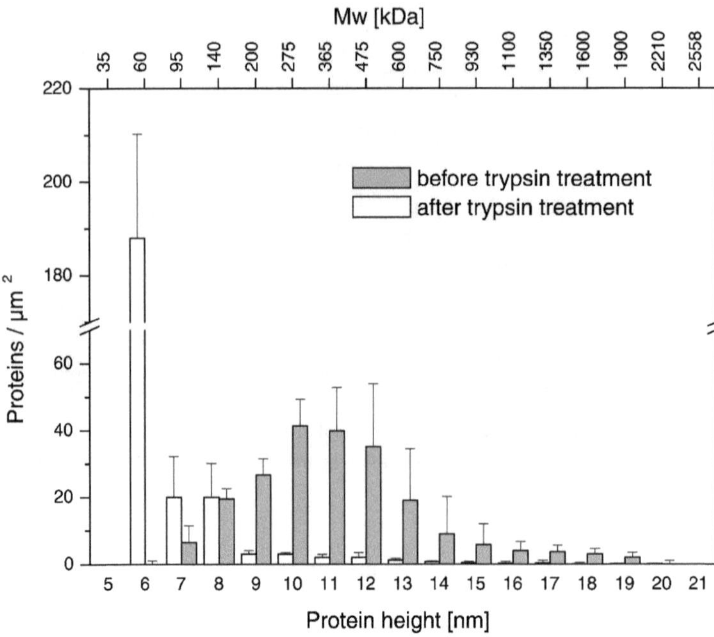

Fig. 12.10. Histogram of the plasma membrane protein distribution (intracellular side) in relation to measured protein heights and calculated molecular weights before and after incubation with trypsin. Molecular weights were calculated from the respective volume measurements. The figure is modified from [19]

molecular weights. From the calculated molecular weights it is assumed that single proteins assemble into multimers or clusters.

In conclusion, the techniques for membrane isolation described here may serve as a suitable method for imaging the inner surface of the plasma membrane from different cell types. Quantification such as determining the number and distribution of membrane proteins within their physiological environment, as well as their height and molecular weight becomes possible. Undoubtedly, this is a valuable step in identifying specific proteins imbedded in native membranes.

12.3 AFM Imaging of CFTR in Oocyte Membranes

Xenopus Laevis Oocytes are a widely used expression system, which means that these cells express a specific protein after injection of the corresponding mRNA. AFM has been applied to study the protein incorporation into plasma membranes of CFTR-expressing oocytes upon cAMP stimulation. Quantification of the protein distribution has revealed two peak values of 10 and 14 nm of protein height corresponding to 275 kDa and 750 kDa, respectively. The results suggest that CFTR causes clustering of plasma membrane proteins after stimulation. In addition, immuno-gold labeling of CFTR in combination with AFM has been used to specifically identify CFTR among the proteins present in the plasma membrane at a single molecule resolution. The frequently observed ring-like structure in the vicinity of a gold particle and the proposed model imply a tail-to-tail dimerization of the CFTR associated with a functional channel.

12.3.1 Introduction

Raising intracellular cAMP concentration stimulates, among other things, CFTR insertion into the plasma membrane and activates CFTR via protein kinase A (PKA) [22]. It is discussed that stimulation of CFTR by cAMP raising agents also initiates several types of protein–protein interactions that might result in a formation of clusters. Clusters are structural and functional units of proteins, which has been shown for numerous plasma membrane proteins, e.g. a potassium channel with members of the ATP-binding cassette (ABC) transporter family [23]. The CFTR, as a member of the ABC-transporter family, is assumed to form clusters with different proteins, such as sodium channels [24], potassium and chloride channels, as well as purinergic receptors [25], in order to perform autocrine cell regulation. Although activation of CFTR has been elegantly examined by electrophysiological and fluorescence microscopy techniques [26,27], little is known about the spatial distribution and organization of this protein within the plasma membrane. Since imaging with AFM has been beneficial in protein counting and protein height measurements essential for the determination of individual molecular masses and protein distribution on the cell surface, this technique would be relevant to the identification of CFTR and its assemblies within the cell membrane.

12.3.2 Does the CFTR Form Functional Assemblies?

In order to visualize CFTR molecules in their natural environment AFM has been employed to inside-out oriented plasma membrane patches of CFTR-expressing oocytes in relation to cAMP stimulation [28]. The AFM topography shown in Fig. 12.11a represents a typical image of a non-stimulated oocyte membrane with regularly distributed proteins. After stimulation (Fig. 12.11b), the density of protein increases clearly.

Quantification of protein distribution (Fig. 12.12) reveals a mean density of 200 proteins per μm^2 in a non-stimulated CFTR-expressing oocyte membrane while in the stimulated ones, the number of proteins increased to 400 per μm^2. The data obtained indicate that the protein covered area increases dramatically in response to cAMP by $\sim$ 110%. This observation strongly suggests protein insertion into the plasma membrane.

Furthermore, the accurate height determination of the non-stimulated oocytes showed distribution with a peak value of 12 nm, corresponding to a molecular mass of 475 kDa (hatched area in Fig. 12.12). In contrast, the height of the proteins from the stimulated membranes exhibited a two-peak distribution with values of 9 nm and 14 nm corresponding to molecular masses of 275 kDa and 750 kDa, respectively.

Fig. 12.11. Membrane patches of CFTR-positive oocytes. The membrane patch shown in (**a**) was isolated before cAMP stimulation, the membrane shown in (**b**) was isolated after cAMP stimulation. In this gray-level coded image the plasma membrane, 5 nm in height, is shown in *"dark"*. Proteins are shown with a gray-level gradient from *dark* to *white*, corresponding to heights from 6 nm to 20 nm. The figure is modified from [28]

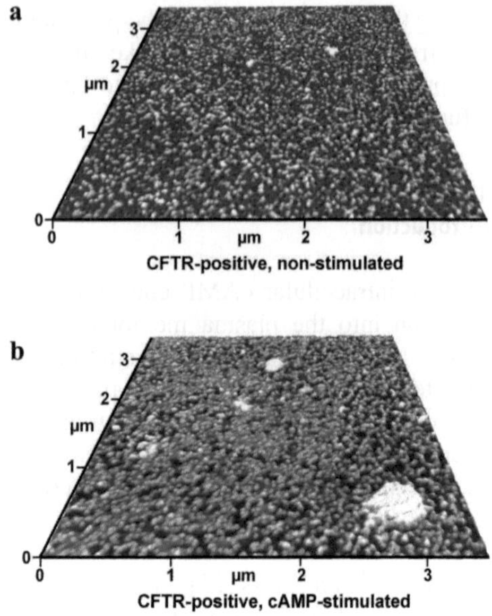

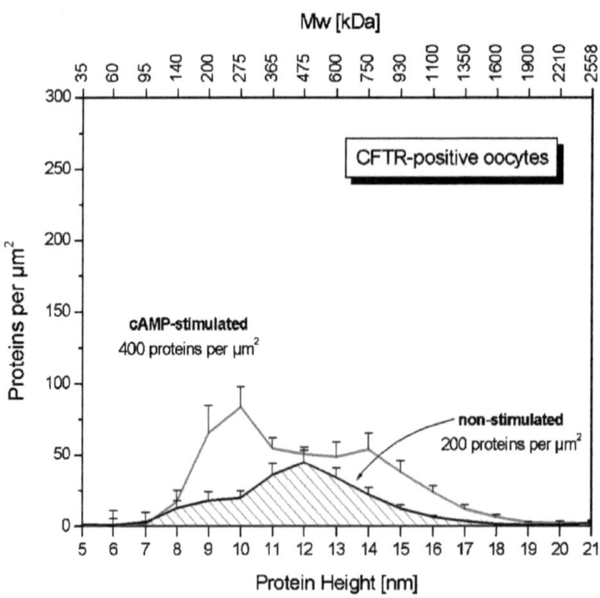

Fig. 12.12. Protein distribution of a CFTR-expressing plasma membrane (mean ± s.e.m., $n = 12$; 4 oocytes, 3 patches per oocyte). The *gray line* represents the protein height distribution of stimulated oocytes, while the *hatched areas* represent the respective height distributions of non-stimulated oocytes. Molecular weights were calculated from the respective volume measurements (see Sect. 12.2). The figure is modified from [28]

Considering the molecular mass of 200 kDa for a CFTR monomer, the peak at 275 kDa and 750 kDa could be multimeric CFTR or CFTR forming clusters with other proteins. It should be noted here that in a control experiment, CFTR negative (native, not CFTR expressing) oocytes do not exhibit any clustering upon cAMP stimulation.

Evidently, using AFM imaging makes it possible to show complex protein–protein interactions within native membranes upon stimulation. These measurements clearly reveal that, at least in the examined cell type, cAMP induces both CFTR insertion into the plasma membrane and leads to the formation of protein clusters with yet unknown stoichiometry. However, these data are consistent with the concept that the physiological function of CFTR involves interactions with other proteins or with itself.

12.3.3 Two CFTRs are Better Than One

The long-standing dilemma of whether the functional CFTR is a monomer or a dimer has not been solved yet. Although recent publications have provided evidence that two CFTR molecules interact together to form a single conductance pore for chloride ions [29,30], structural details of CFTR in native membranes are still unknown. The predicted model of an individual CFTR molecule exhibits two hydrophobic core regions, each with six transmembrane spanning domains (TMs), and two nucleotide binding folds (NBD's). A large cytoplasmic regulatory domain (R-domain) connects the two halves of the CFTR molecule [31] (Fig. 12.13). CFTR-related structures that protrude from the inner side of the membrane towards the cytoplasm are expected to be in the range of only a few nanometers.

Structures with such dimensions cannot be detected with more conventional methods. As we have seen in the section above, AFM is a suitable method to detect proteins within native membranes, since scanning can be performed with a vertical resolution (resolution in height) in the sub-nanometer range. However, it is important now to be able to distinguish and pinpoint a specific single protein, such as the CFTR, among the vast diversity of proteins present in the membrane. In order to identify

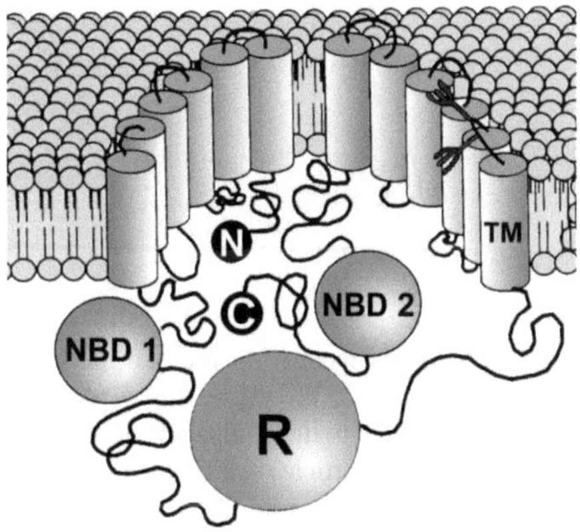

Fig. 12.13. Schematic of the predicted organization of CFTR within the plasma membrane (the crystal structure has not been solved yet). The intracellular domains (R domain, NBD1 and NBD2) are known to be involved in channel gating, while the transmembrane segments (TMs) are presumed to form the anion-conducting pore

CFTR, a novel method of staining CFTR with gold-labeled antibodies was applied and AFM was used for high-resolution single molecule imaging [32].

Figure 12.14 shows a 1 μm^2 patch of plasma membrane excised from a cAMP-stimulated, CFTR-expressing oocyte. The light structures on the surface correspond to the individual gold labels. The height of the gold particles is in the range of 10 to 15 nm, while the respective width is sometimes more than 50 nm. The latter number is an overestimate due to the AFM tip geometry. About 100 gold particles per μm^2 of plasma membrane were detected. Gold particles appear, uniform in height and size, as the most prominent structures on the membrane. Although the density of proteins in the plasma membrane is high and the shape of proteins varies considerably, frequently distinct circular structures in close vicinity to individual gold labels were detected (see inset of Fig. 12.14). Higher resolution images of such ring-like structures (Fig. 12.15) reveal globular subunits correlated with the ring and crater-like structure with a central aperture and a surrounding fringe exhibiting bipartite symmetry. A single CFTR is expected to have three intracellular substructures: one regulatory subunit (R domain) and two nucleotide binding domains (NBD1 and NBD2). However, the AFM images demonstrate up to 6 subunits associated with one pore. On the basis of the substructures a model was developed representing the CFTR as a dimer with a tail-to-tail configuration and a central pore (Fig. 12.16). The C-termini form a tail-to-tail outlet to which the primary antibody is bound. Then, the secondary antibody spans the short distance to the gold label. Starting from the C-terminus, the NBD2 is the first structure detectable as a tiny protrusion. Then, the R-domain and the NBD1 follow. In between coiled hydrophilic parts of the CFTR molecule are expected. The tail-to-tail arrangement indicates a mirror-like

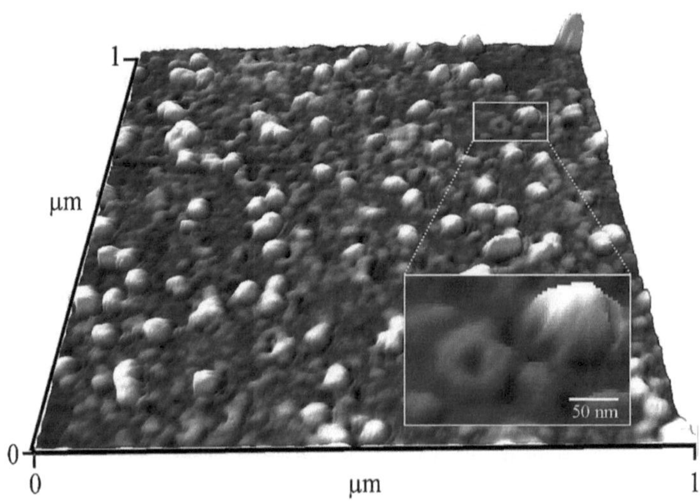

Fig. 12.14. AFM image of inside-out plasma membrane patch excised 5 minutes after cAMP stimulation from CFTR expressing Xenopus laevis oocyte. Light particles (height = 10 to 15 nm) correspond to the gold labels of the secondary antibodies directed against CFTR primary antibodies. Gold particles are frequently found associated with ring-like structures (see *inset*). The figure is modified from [32]

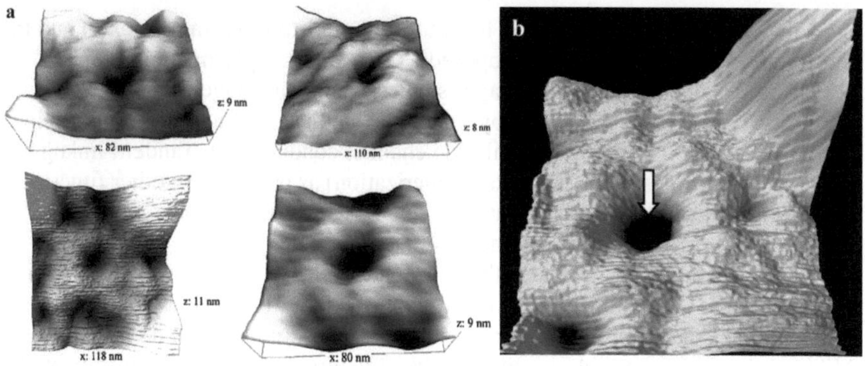

Fig. 12.15. (**a**) AFM images of four different ring-like structures supposed to be the cytoplasmic morphological correlate of CFTR. The structures are found in close vicinity to gold particles. Heights (given as "*z*" in the images) are gray-level coded from *dark* to *light*. (**b**) High-resolution AFM image of a CFTR ring. The *arrow* indicates the cytoplasmic aperture of the central pore. Intracellular substructures of possible nucleotide binding domains (NBD1 and NBD2) and regulatory domains (R domain) are fairly distinguishable. Gray-level code: from *dark* to *light*. The figure is modified from [32]

dimerization of two CFTR molecules. Two such molecules form a CFTR dimer as indicated in Fig. 12.16.

Analysis of several ABC-transporters report homodimeric structures in which monomers interact within the transmembrane domains. They form a central pore or internal chamber open to the inner or outer leaflet between the two subunits constituting the homodimer [33–39]. The findings are consistent with electron microscopy [40] and electrophysiology [41–43] studies showing that CFTR forms a homodimer in plasma membrane. However, no structural data exist that directly

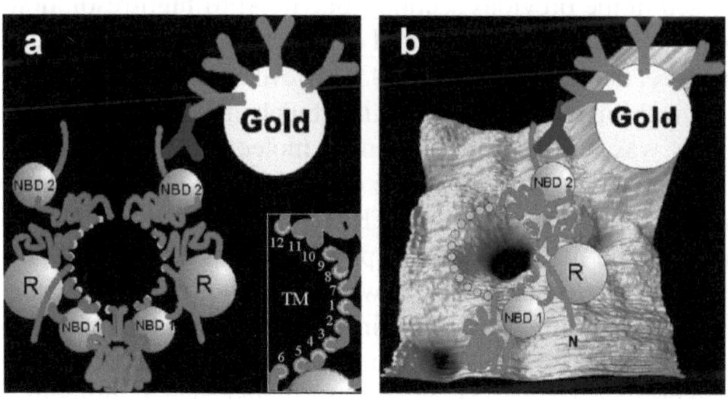

Fig. 12.16. Model of a CFTR dimer in tail-to-tail configuration (**a**) superimposed on the AFM image of a CFTR ring (**b**). The length of the extramembrane loops are drawn to scale. The transmembrane helices (TM) are numbered in the *inset*. The model assumes that the *N*-terminus interacts with the R domain and may be flexible enough to interfer with ion channel gating at the pore entrance. The C-termini form parallel structures at one end of the dimer interacting with the primary monoclonal antibody. The figure is modified from [32]

show the orientation of CFTR (e.g. head-to-tail or tail-to-tail arrangements) and the substructural components of the proposed CFTR-dimer in plasma membrane.

The assessment of the AFM images suggests a mirror-image-like two-fold symmetry of CFTR. This is supported by functional studies in which tail-to-tail dimerization is associated with highest chloride channel conductance. Tandem linkage of two CFTR molecules (i.e. head-to-tail dimerization) as proposed in other functional studies [42, 44] seems less likely, at least in CFTR expressing oocytes, since AFM images show mirror-image-like dimeric structures.

In conclusion, the study shows the advantage of the AFM technique for imaging fully functional CFTR in its native environment without the limitations of artificial systems (e.g. isolated proteins). Since intramolecular domains can be identified by AFM, it should be possible to address the structure–function relationship of the CFTR chloride channel. Imaging CFTR mutations like ΔF508 could be one of the next exciting goals to be addressed in the future.

12.4 Single Antibody–CFTR Recognition Imaging

A novel AFM mode that can visualize specific antigen–antibody interactions while simultaneously recording high-resolution topographical images using "MacMode" dynamic force microscopy was employed. An antibody-modified AFM tip was used to specifically recognize CFTR molecules within red blood cell membranes. The images revealed recognition patterns that were assigned to the topography of single and/or clustered CFTR proteins.

12.4.1 Introduction

As we discussed in the previous sections, due to AFM high-resolution spatial imaging capabilities, a wealth of structural information on DNA-protein complexes, membranes and cells has been gathered. However, AFM has long been criticized because of its inability to provide chemical/biochemical specificity of the imaged structures. This was realized by combining molecular recognition with dynamic force microscopy [45], which led to the development of TREC imaging (topography and recognition; the principles of TREC will be discussed in greater detail in Vol. *II*, Chap. 5). The method vastly improves the ability of the AFM to visualize the chemical composition of a sample while mapping its topographic structure at sub-molecular resolution. For example, investigating both the interactions of single molecules with their specific receptors and their localization becomes feasible.

In order to provide a specific recognition a ligand such as antibody is covalently bound to the scanning AFM tip. The AFM tip approaches the surface that contains cognate antigen and an antibody–antigen bond is formed. During a subsequent retraction of the tip, the bond will cause a tension and with the increased force will eventually break, allowing the estimation of affinity constants [46]. Importantly, the specificity of recognition can be tested by blocking either the antibody on the tip or the antigen on the sample, resulting in a complete abolishment of the recognition signals. TREC imaging has been used to study single lysozyme molecules showing ~ 70% recognition efficiency [47] and to recognize histone H3 within a complex chromatin fiber by an antihistone H3 functionalized tip. Moreover, the authors were able to follow an ATP-dependent nucleosome remodeling process [48].

Here the TREC imaging was used as an alternative method to directly map CFTR sites on RBC membranes as a result of the molecular recognition of the antigen by a CFTR-antibody tethered to an AFM tip.

12.4.2 Tethering of Antibodies to AFM Tips

MacLevers tips (Molecular Imaging Corp., Tempe/AZ, USA) were activated with ethanolamine HCl and subsequently coupled to a heterobifunctional tether of ~ 8 nm length of polyethylene glycol (PEG) derivative as described [49]. A mouse monoclonal antibody against a C-terminal epitope of CFTR (MAB25031 (21-4), R&D Systems Inc., Minneapolis, MN) was then bound to the reactive end of the PEG tether. The flexible linker gives the possibility for the antibody to freely orient, achieving unconstrained binding to its cognate site. The antibody density is adjusted to a value where only one antibody on the tip is expected to have access to the receptor on the surface.

12.4.3 AFM Imaging and Recognition

Magnetized cantilevers are driven by a small solenoid using a MacMode dynamic force microscope (Molecular Imaging). The tip is oscillated by an alternating magnetic field while being scanned across the surface. By passing the raw deflection signal from the AFM-scanning head to a PicoTREC signal-processing system, the oscillation is split into lower and upper parts, resulting in simultaneously acquired topography and recognition images. MacLever cantilevers had a spring constant of 0.1 N/m. Measurements were performed in PBS buffer with 5 nm free tip oscillation amplitude at 5 kHz driving frequency and 20% amplitude reduction. The lateral scanning frequency was of 1 Hz.

12.4.4 A Single Antibody Sees a Single CFTR

In this study of TREC, an anti-CFTR antibody functionalized tip was used to map single CFTR molecules at the cytoplasmic side of RBC membranes (Fig. 12.17).

The topographical image of a part of an RBC membrane together with the map of recognition sites was recorded with nm-resolution. The identification of CFTR molecules at the single molecule level was achieved without compromising its topographic imaging performance, despite the fact that the tip was carrying the tethered antibody (Fig. 12.18). Figure 12.18a revealed structures protruding out of the membrane with 10–15 nm in height, representing the membrane proteins at the cytoplasmic site of the membrane. The observed structures were comparable

Fig. 12.17. Schematic of recognition imaging. When the tip-tethered antibody binds to its antigen in the sample being scanned, there is a transient reduction in the oscillation amplitude. This reduction, corresponding to the recognition event, is then presented in an image as a *dark spot* at its defined position on which it occurred. The figure is modified from Molecular Imaging Corp., Tempe/AZ, USA

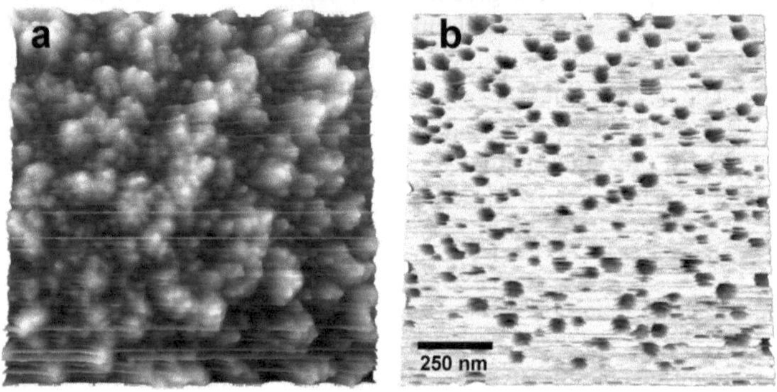

Fig. 12.18. TREC imaging (**a**) topography of an area of a non-CF RBC membrane showing the protruding proteins of the lipid bilayer; (**b**) the same area as in (**a**) but representing the specific interaction between the modified tip (i.e. anti-CFTR antibody tip) and the CFTR at the membrane, identified as darker spots (to be published)

to the topography of membrane proteins obtained with the standard AFM. The simultaneously acquired recognition image showed darker spots, corresponding to the specific interaction between the antibody on the tip and the surface CFTR molecules (Fig. 12.18b). When the tethered antibody binds to an antigen on the surface, the upward extent of the cantilever swing is restricted by the PEG tether. This reduction of the oscillation amplitude on binding is compensated for by the microscope servo. The servo pulls the probe away from the surface to restore the amplitude to its previous value, but now the peak signal is displaced downward by the amount of the amplitude reduction. Thus, the locations of the antibody-binding events are shown as dark patches [50].

Interestingly, the observed recognition patterns here do not appear to be uniform in size and distribution across the membrane surface. This finding supports the hypothesis that CFTR forms functional clusters (as discussed in Sect. 12.3) to fulfill its regulatory functions within the cell membrane. The specificity of the antibody–CFTR recognition process can be tested by blocking the antibody attached to the tip with an antigenic peptide. After blocking, the specific interactions disappear. Clearly, the TREC has an advantage over the labeling methods where the direct visualization of the molecule is not possible since gold dots hinder it. Moreover, with further advances in the TREC methodology (increasing spatial resolution, improving sample preparation), this technique will provide more detailed information on issues such as clustering.

In conclusion, by using TREC it was possible to recognize CFTR molecules in heterogeneous RBC membranes and carry out epitope mapping on a nanometer scale.

12.5 Single Cell Elasticity: Probing for Diseases

As discussed above, the lack of CFTR in human RBC might contribute to the development of pulmonary hypertension and hemolytic anemia in CF patients. In this respect, it was hypothesized that a reason for those symptoms might be altered cell elasticity. In this study, AFM was used to directly quantify the local stiffness of individual CF and non-CF RBCs. According to this method, a very low force (in the order of pico Newtons) was applied via an AFM cantilever to non-fixed RBCs and as a result an indentation with a controlled depth was generated by the AFM tip. The obtained force–distance relations were used to assess the Young's modulus, a measure of the matter's elasticity. Comparison of the Young's moduli of CF and non-CF RBCs revealed a two-fold increase in stiffness for RBCs from patients with CF.

12.5.1 Introduction

Mechanical properties of living cells are very important for a proper physiological functioning, and yet our knowledge in this regard is rather limited. For instance, it is not fully understood how a cell responds structurally and mechanically to external stress or what role the micromechanistics of the cell may play in various physiological or pathological conditions; or how the elasticity of a certain type of cells will be altered in diseased compared to healthy organisms. The red blood cell (RBC) is known for its ability to withstand great deformations when it passes through small capillaries. Because cell cytosol is only composed of a viscous fluid (solution of hemoglobin), the resistance to stress is mainly attributed to the elastic properties of its membrane. The RBC membrane, made of a lipid bilayer and membrane proteins, is reinforced on its inner side by a flexible two-dimensional cytoskeleton. The cytoskeleton elasticity is determined by the association/dissociation of the spectrin and actin filaments, its integral components. It is obvious that an abnormal RBC deformability will affect the ability of the cell to transit in the small blood vessels, resulting in blood circulation disturbances. Interestingly, CF patients have been diagnosed with pulmonary hypertension or hemolytic anemia. Clearly, decreased cell deformability might be a factor contributing to those symptoms in CF patients.

In addition, it has been demonstrated that CFTR associated deformation-induced ATP release of RBCs from patients with CF is abolished [51]. This RBC associated

ATP plays a crucial role in stimulating endothelial cells of the blood vessels, leading, ultimately, to an increase in vascular caliber and thus regulating the vascular resistance. In turn, is not unreasonable to think, that this ATP might act on the RBC itself, initiating different signaling pathways. For example, studies have shown that ATP induces changes in the cell cytoskeleton leading to alteration of the membrane elasticity [52, 53]. In the present work, we investigated the hypothesis that the lack of CFTR in RBC membranes may alter their mechanical properties.

As we have seen in the previous sections, AFM provides detailed surface topography at nm-scale by raster scanning a cantilever tip over the sample. In addition, due to the technique's force detection sensitivity, it has given us the possibility of measuring forces in the pico Newton range. The force felt by the tip as it approaches and retracts from a point on the sample surface is presented as force–distance curves. Thus, local elastic properties of biological matter, in terms of the Young's modulus or elastic modulus, can be quantitatively derived from the force–distance curves obtained at a point of contact on the surface. Furthermore, a two-dimensional spectroscopic matrix of force–distance curves can be collected simultaneously with the topographical data. In this way, elastic properties of the sample can be mapped and correlated with the topographic image. The capability of AFM to provide valuable information on the mechanical properties of living cells has been gaining increasing attention [54–56].

Therefore, we employed AFM as well as a force sensor to quantify the local elasticity of individual RBCs in relation to CF.

12.5.2 Force–Mapping AFM

RBCs were attached to poly-L-lysine coated glass coverslips. AFM measurements were performed in force–mapping mode in PBS buffer using a Nanoscope III Multimode-AFM (Digital Instruments, Santa Babara, CA, USA). We used soft silicon nitride cantilevers with spring constant of 0.01 N/m and pyramidal tip with an opening angle of 35 (Veeco Instruments, USA). Force–mapping was performed with the following parameters: scan sizes of 10 μm^2, scan rate of 14 Hz, samples/line 256, number of samples 64, force/line 64, in relative trigger mode.

To calculate Young's modulus from the force curves, the data were fit with the Hertz model [57] for an indentation of a completely elastic soft sample by a stiff cone. The relation between the indentation, δ (for our calculations we used a value of $\sim$ 100 nm), and the loading force, F, is given by the following:

$$\Delta F = \delta^2 \times (2/\pi) \times (E/[1 - \nu^2]) \times \tan(\alpha)\,, \tag{12.2}$$

where ν is the Poisson ratio (for incompressible materials it is 0.5), E is the elastic modulus (or Young's modulus), and α is the half opening angle of the AFM tip (for Veeco Instruments is 35°). Data were analyzed with self-written macros for the software IGOR PRO (WaveMetrics).

12.5.3
Can One Protein Change Cell Elasticity?

The adhesion of the cells to the support surface is very critical for obtaining reliable data. Poly-L-lysine (PLL) is a polycation commonly adsorbed to surfaces for strong attachment to the negatively charged molecules such as the glycocalyx of RBCs.

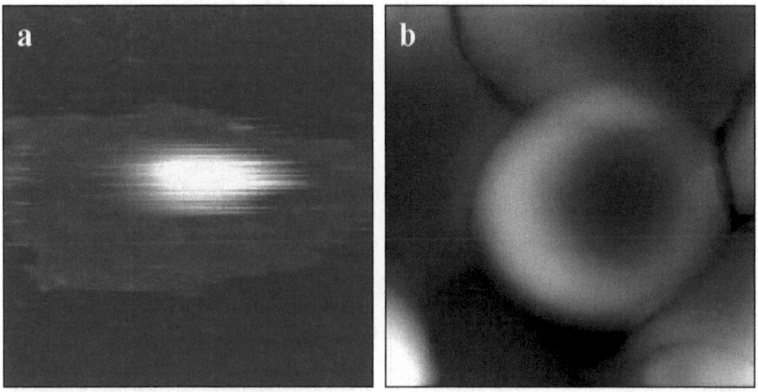

Fig. 12.19. Low force contact mode AFM imaging in PBS buffer: (**a**) topography image of a spread RBC, demonstrating compression of the disc shape due to strong adhesion to the glass surface; (**b**) immobilization at reduced adhesion shows regularly shaped RBCs. Scan size: 12 μm

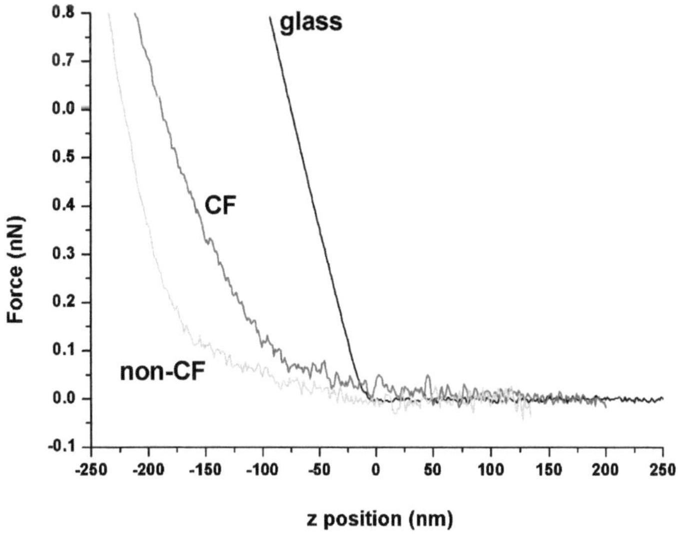

Fig. 12.20. Comparison of typical force curves on soft and stiff samples. The *black trace* represents an approach curve taken on a glass substrate, on CF and non-CF RBCs, respectively. This slope can be used for the determination of the sensitivity of the photodetector. On the stiff sample the force curve shows a sharp bend (the contact point is at the 0 *z* position of the piezo). In contrast, the approach on the cell surface shows a shallow transition

At 1 mg/ml concentrations of PLL preadsorbed to a coverslip, erythrocytes adhere firmly to the substrate but a drastic change of the shape occurs due to the strong adhesion (Fig. 12.19a). With a 10-fold-diluted PLL solution (0.1 mg/ml), erythrocyte immobilization is accompanied without alteration of their shape (Fig. 12.19b). For our force mapping measurements only intact RBCs exhibiting disc shape, which were not under tension due to spreading, were used.

Local micromechanical properties of single non-fixed CF and non-CF RBCs were quantitatively characterized using force–mapping AFM. In this mode force

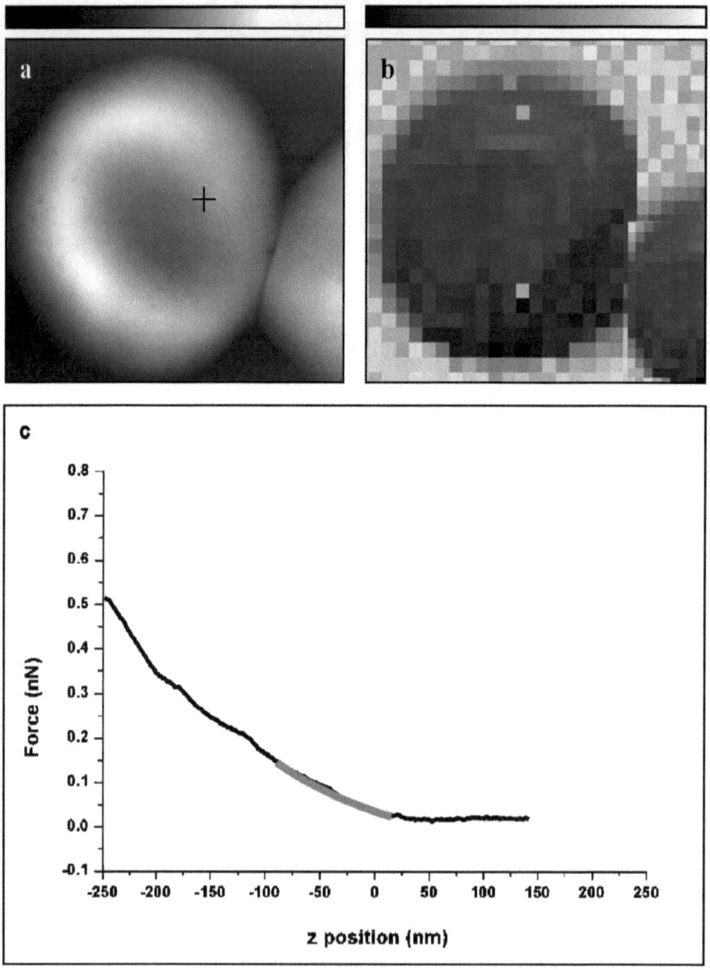

Fig. 12.21. Force map of a non-fixed RBC measured in PBS buffer. (**a**) A typical topography image of an RBC, higher areas are depicted in *lighter color*; (**b**) corresponding elasticity map, softer areas are shown in *gray*; (**c**) a force–distance curve during approach to RBC membrane taken at the point of the RBC indicated with a cross on (**a**). The analysis of the Young's modulus is made for an indentation of ~100 nm (*gray line*). Scan size: 10 μm

curves (Fig. 12.20) are taken while scanning laterally over the sample and both topographic and elasticity maps are recorded. However, in order to obtain reliable local elasticity measurement of the cell, sensitivity of the photodetector needs to be determined accurately. This can be determined from the slope of the force–distance curves taken on the bare surface of glass coverslips where there are no cells present.

In Fig. 12.21a a height image of an RBC, reconstructed from the force measurement is shown, revealing the typical erythrocyte shape of a bi-concave disc. Simultaneously the 2D elastic properties of the cell, in terms of Young's modulus, are obtained (Fig. 12.21b). At every point across the scanned area the corresponding approach curve can be acquired (Fig. 12.21c). For calculating the membrane elasticity we used the loading force required to indent the cell by ~ 100 nm. Since the contact area between tip and cell is not known, and also the tension in the cell center dip is not well determined, only force curves taken on the rim of the "doughnut" were further considered for the calculation of the Young's modulus.

The results of the calculated Young's moduli of CF and non-CF RBCs are presented in Fig. 12.22. As indicated by the histograms, the data show that the RBCs from patients with CF exhibited a two-fold increase in stiffness compared to RBC from healthy donors.

Clearly, the deficiency of CFTR in the CF RBC membranes, directly or indirectly, leads to a reduced deformability.

Considering the fact that CFTR is present in a very low amount even in healthy cells, only ~ 640 copies per cell, it appears rather unlikely that its depletion would lead to a direct alteration of membrane elasticity (based on lipid/protein ratio). It seems more likely that the protein is involved in a secondary pathway, e.g. acting via the cytoskeleton a change in elasticity might occur. For instance, reduction of the elasticity of mouse F9 embryonic carcinoma cells deficient of vinculin, a membrane-associated protein that plays a role in linkage of the cytoskeleton, has been shown recently [58]. Possibly impaired volume regulation as observed in CF cells [59]

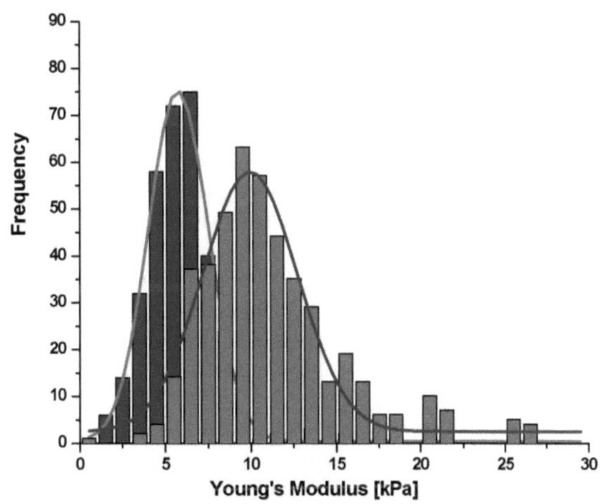

Fig. 12.22. Distribution histograms of the Young's modulus of CF (*light gray*) and non-CF (*dark gray*) RBCs. Each distribution represents data from three different individuals. The Gaussian fits revealed a mean value of 5.8 ± 0.1 kPa ($n = 341$) for the RBCs from the healthy donors and 9.9 ± 0.2 kPa ($n = 458$) for the RBCs from CF patients

changes intracellular pressure leading to an increased membrane tension and thus to stiffening of the membrane.

Further studies are needed to clarify this phenomenon, however, the use of the force–mapping AFM to investigate the elasticity of cells in various pathological conditions certainly presents a new direction for single cell diagnostics and a better understanding of diseases.

12.6 Summary

The combination of AFM with conventional techniques, as well as AFM itself, allows answering biomedical questions of high interest. We could show this clearly for CFTR with single molecule imaging and observation of structural dynamics in native cell membranes. AFM also allows identification and determination of CFTR at single molecule level. The observation that the lack of CFTR influences the mechanical and, therefore, rheological properties of RBC could lead to a novel therapeutic approach for CF treatment. Regarding the fact that the defect of a single protein causes lethal diseases, research at the single molecule level would become vital in the future.

Acknowledgements. We thank Dr. Rainer Matzke, Ludwig Maximilians University Munich, Germany, for the force mapping analysis. We are also grateful to Dr. Johannes Haeberle, Dr. Angelika Duebbers and Dr. Sabine Falk, Department of Pediatrics, University Hospital Muenster, Germany, for the supply with patients' blood. The work was supported by DFG (German Research Foundation) grant SFB 629 (A6) and EU grant Tips4cells.

References

1. Binnig G, Garcia N, Rohrer H (1985) Phys Rev B Condens Matter 32:1336
2. Drake B, Prater CB, Weisenhorn AL, Gould SA, Albrecht TR, Quate CF, Cannell DS, Hansma HG, Hansma PK (1989) Science 243:1586
3. Putman CA, van der Werf KO, de Grooth BG, van Hulst NF, Greve J (1994) Biophys J 67:1749
4. Muller DJ, Fotiadis D, Scheuring S, Muller SA, Engel A (1999) Biophys J 76:1101
5. Nikova DN, Pope LH, Bennink ML, van Leijenhorst-Groener KA, van der WK, Greve J (2004) Biophys J 87:4135
6. Jiang F, Khairy K, Poole K, Howard J, Muller DJ (2004) Microsc Res Tech 64:435
7. Abdelhady HG, Allen S, Davies MC, Roberts CJ, Tendler SJ, Williams PM (2003) Nucleic Acids Res 31:4001
8. Krautbauer R, Pope LH, Schrader TE, Allen S, Gaub HE (2002) FEBS Lett 510:154
9. Oberleithner H, Ludwig T, Riethmuller C, Hillebrand U, Albermann L, Schafer C, Shahin V, Schillers H (2004) Hypertension 43:952
10. Lieber SC, Aubry N, Pain J, Diaz G, Kim SJ, Vatner SF (2004) Am J Physiol Heart Circ Physiol 287:H645
11. Johnston JA, Ward CL, Kopito RR (1998) J Cell Biol 143:1883
12. Paterson D and Nordberg A (2000) Prog Neurobiol 61:75
13. Heinz WF and Hoh JH (1999) Trends Biotechnol 17:143

14. Radmacher M, Tillmann RW, Fritz M, Gaub HE (1992) Science 257:1900
15. Muller DJ, Engel A, Matthey U, Meier T, Dimroth P, Suda K (2003) J Mol Biol 327:925
16. Fotiadis D, Liang Y, Filipek S, Saperstein DA, Engel A, Palczewski K (2004) FEBS Lett 564:281
17. Shahin V, Albermann L, Schillers H, Kastrup L, Schafer C, Ludwig Y, Stock C, Oberleithner H (2005) J Cell Physiol 202:591
18. Pesen D and Hoh JH (2005) Biophys J 88:670
19. Schillers H, Danker T, Schnittler HJ, Lang F, Oberleithner H (2000) Cell Physiol Biochem 10:99
20. Swihart AH, Mikrut JM, Ketterson JB, Macdonald RC (2001) J Microsc 204:212
21. Larmer J, Schneider SW, Danker T, Schwab A, Oberleithner H (1997) Pflugers Arch 434:254
22. Prince LS, Workman RB, Jr, Marchase RB (1994) Proc Natl Acad Sci U S A 91:5192
23. Seino S (1999) Annu Rev Physiol 61:337
24. Kunzelmann K, Schreiber R, Nitschke R, Mall M (2000) Pflugers Arch 440:193
25. al-Awqati Q (1995) Science 269:805
26. Moyer BD, Loffing J, Schwiebert EM, Loffing-Cueni D, Halpin PA, Karlson KH, Ismailov II, Guggino WB, Langford GM, Stanton BA (1998) J Biol Chem 273:21759
27. Weber WM, Cuppens H, Cassiman JJ, Clauss W, Van DW (1999) Pflugers Arch 438:561
28. Schillers H, Danker T, Madeja M, Oberleithner H (2001) J Membr Biol 180:205
29. Zerhusen B, Zhao J, Xie J, Davis PB, Ma J (1999) J Biol Chem 274:7627
30. Wang S, Yue H, Derin RB, Guggino WB, Li M (2000) Cell 103:169
31. Riordan JR, Rommens JM, Kerem B, Alon N, Rozmahel R, Grzelczak Z, Zielenski J, Lok S, Plavsic N, Chou JL (1989) Science 245:1066
32. Schillers H, Shahin V, Albermann L, Schafer C, Oberleithner H (2004) Cell Physiol Biochem 14:1
33. Higgins CF and Linton KJ (2001) Science 293:1782
34. Schmitt L and Tampe R (2002) Curr Opin Struct Biol 12:754
35. Chang G and Roth CB (2001) Science 293:1793
36. Locher KP, Lee AT, Rees DC (2002) Science 296:1091
37. Chami M, Steinfels E, Orelle C, Jault JM, Di Pietro A, Rigaud JL, Marco S (2002) J Mol Biol 315:1075
38. van Veen HW, Margolles A, Muller M, Higgins CF, Konings WN (2000) EMBO J 19:2503
39. Rosenberg MF, Callaghan R, Ford RC, Higgins CF (1997) J Biol Chem 272:10685
40. Eskandari S, Wright EM, Kreman M, Starace DM, Zampighi GA (1998) Proc Natl Acad Sci USA 95:11235
41. Wang S, Yue H, Derin RB, Guggino WB, Li M (2000) Cell 103:169
42. Zerhusen B, Zhao J, Xie J, Davis PB, Ma J (1999) J Biol Chem 274:7627
43. Raghuram V, Mak DD, Foskett JK (2001) Proc Natl Acad Sci USA 98:1300
44. Wang S, Yue H, Derin RB, Guggino WB, Li M (2000) Cell 103:169
45. Hinterdorfer P, Baumgartner W, Gruber HJ, Schilcher K, Schindler H (1996) Proc Natl Acad Sci USA 93:3477
46. Allison DP, Hinterdorfer P, Han W (2002) Curr Opin Biotechnol 13:47
47. Stroh CM, Ebner A, Geretschlager M, Freudenthaler G, Kienberger F, Kamruzzahan AS, Smith-Gill SJ, Gruber HJ, Hinterdorfer P (2004) Biophys J 87:1981
48. Stroh C, Wang H, Bash R, Ashcroft B, Nelson J, Gruber H, Lohr D, Lindsay SM, Hinterdorfer P (2004) Proc Natl Acad Sci USA 101:12503

49. Hinterdorfer P, Baumgartner W, Gruber HJ, Schilcher K, Schindler H (1996) Proc Natl Acad Sci USA 93:3477
50. Stroh C, Wang H, Bash R, Ashcroft B, Nelson J, Gruber H, Lohr D, Lindsay SM, Hinterdorfer P (2004) Proc Natl Acad Sci USA 101:12503
51. Sprague RS, Ellsworth ML, Stephenson AH, Kleinhenz ME, Lonigro AJ (1998) Am J Physiol 275:H1726
52. Marikovsky Y (1996) Mech Ageing Dev 86:137
53. Gov NS and Safran SA (2005) Biophys J 88:1859
54. Ikai A, Afrin R, Sekiguchi H, Okajima T, Alam MT, Nishida S (2003) Curr Protein Pept Sci 4:181
55. Lieber SC, Aubry N, Pain J, Diaz G, Kim SJ, Vatner SF (2004) Am J Physiol Heart Circ Physiol 287:H645
56. Oberleithner H, Reinhardt J, Schillers H, Pagel P, Schneider SW (2000) Cell Physiol Biochem 10:429
57. Radmacher M, Fritz M, Kacher CM, Cleveland JP, Hansma PK (1996) Biophys J 70:556
58. Goldmann WH, Galneder R, Ludwig M, Xu W, Adamson ED, Wang N, Ezzell RM (1998) Exp Cell Res 239:235
59. Braunstein GM, Roman RM, Clancy JP, Kudlow BA, Taylor AL, Shylonsky VG, Jovov B, Peter K, Jilling T, Ismailov II, Benos DJ, Schwiebert LM, Fitz JG, Schwiebert EM (2001) J Biol Chem 276:6621

Part IV
Functional Bio(-inspired) Surfaces

21 Lotus Effect: Roughness-Induced Superhydrophobicity

Michael Nosonovsky · Bharat Bhushan

From volume VII

21.1 Introduction

The rapid advances in nanotechnology since the 1980s, including such applications as microelectromechanical/nanoelectromechanical systems (MEMS/NEMS), have stimulated development of new materials and the design of surfaces, which should meet new sets of requirements [3, 6, 11]. In MEMS/NEMS, the surface-to-volume ratio grows with miniaturization and surface phenomena dominate. One of the crucial surface properties for materials in microscale/nanoscale applications is water-repellency, nonwetting or hydrophobicity. It is also usually desirable to reduce wetting in liquid-flow applications and some conventional applications, such as glass windows, in order for liquid to flow away along their surfaces. Wetting may lead to formation of menisci at the interface between solid bodies during sliding contact, which increases adhesion/friction. As a result of this, the wet friction force is greater than the dry friction force, which is usually undesirable [4, 5, 7]. On the other hand, high adhesion is desirable in some applications, such as adhesive tapes and adhesion of cells to biomaterial surfaces; therefore, enhanced wetting by changing roughness would be desirable in these applications [35, 36]. There is also an analogy between the motion of a droplet on a rough solid surface and the motion of gas bubbles in a liquid near rough walls of a vessel or channel, which allows us to apply the results of droplet motion analysis to study the mobility of nanobubbles [37].

Wetting is characterized by the contact angle, which is defined as the measurable angle which a liquid makes with a solid. The contact angle depends on several factors, such as roughness and the manner of surface preparation, and its cleanliness [1, 25]. If the liquid wets the surface (referred to as wetting liquid or hydrophilic surface), the value of the static contact angle is $0 \leq \theta \leq 90°$, whereas if the liquid does not wet the surface (referred to as nonwetting liquid or hydrophobic surface), the value of the contact angle is $90° < \theta \leq 180°$. The term hydrophobic/hydrophilic, which was originally applied only to water (the Greek word *hydro* means "water"), is often used to describe the contact of a solid surface with any liquid. Surfaces with a contact angle between 150° and 180° are called superhydrophobic. Superhydrophobic surfaces also have very low water contact angle hysteresis. The contact angle hysteresis is the difference between the advancing and the receding contact angles. If additional liquid is added to a sessile drop, the contact line advances and each time motion ceases, the drop exhibits an advancing contact angle. Alternatively, if liquid is removed from the drop, the contact angle decreases to a receding value before the contact retreats. For a droplet moving along the solid surface, the contact angle at the front

In: H. Fuchs, B. Bhushan (eds), *Biosystems - Investigated by Scanning Probe Microscopy*,
DOI 10.1007/978-3-540-37321-6_1,

of the droplet (advancing contact angle) is greater than that at the back of the droplet (receding contact angle), owing to roughness and surface heterogeneity, resulting in the contact angle hysteresis. For liquid-flow applications, in addition to high contact angle another wetting property of interest is a very low water roll-off angle, which denotes the angle to which a surface must be tilted for roll-off of water drops (i.e., very low water contact angle hysteresis) [22, 28].

One of the ways to increase the hydrophobic or hydrophilic properties of the surface is to increase surface roughness; so roughness-induced hydrophobicity has become a subject of extensive investigation. Wenzel (1936) found that the contact angle of a liquid with a rough surface is different from that with a smooth surface. Cassie and Baxter (1944) showed that air (or gas) pockets may be trapped in the cavities of a rough surface, resulting in a composite solid–liquid-air interface, as opposed to a homogeneous solid–liquid interface. Shuttleworth and Bailey (1948) studied spreading of a liquid over a rough solid surface and found that the contact angle at the absolute minimum of the surface energy corresponds to the values predicted by [51] or [15]. Johnson and Dettre (1964) showed that the homogeneous and composite interfaces correspond to the two metastable equilibrium states of a droplet. Bico et al. (2002), Marmur (2003, 2004), Lafuma and Quéré (2003), Patankar (2003, 2004a), He et al. (2003), and other authors investigated recently metastability of artificial superhydrophobic surfaces and showed that whether the interface is homogeneous or composite may depend on the history of the system (in particular, whether the liquid was applied from the top or from the bottom). Extrand (2002) pointed out that whether the interface is homogeneous or composite depends on droplet size, owing to gravity. It was suggested also that the so-called two-tiered (or double) roughness, created by superposition of two roughness patterns at different length scales [24, 41, 47], and fractal roughness [45] may lead to superhydrophobicity. Herminghaus (2000) showed that certain self-affine profiles may result in superhydrophobic surfaces even for wetting liquids; in the case the local equilibrium condition for the triple line (line of contact between solid, liquid, and air) is satisfied. Nosonovsky and Bhushan (2005, 2006a) pointed out that such configurations, although formally possible, are likely to be unstable. Nosonovsky and Bhushan (2006a,b) proposed a stochastic model for wetting of rough surfaces with a certain probability associated with every equilibrium state. According to their model, the overall contact angle with a two-dimensional rough profile is calculated by assuming that the overall configuration of a droplet is a result of superposition of numerous metastable states. The probability-based concept is consistent with the experimental data [29], which suggest that transition between the composite and homogeneous interfaces is gradual, rather than instant.

It has been demonstrated experimentally, that roughness changes the contact angle in accordance with the Wenzel model. Yost et al. (1994) found that roughness enhances wetting of a copper surface with Sn–Pb eutectic solder, which has a contact angle of 15–20° for a smooth surface. Shibushi et al. (1996) measured the contact angle of various liquids (mixtures of water and 1,4-dioxane) on alkylketen dimmer (AKD) substrate (contact angle not larger than 109° for a smooth surface). They found that for wetting liquids the contact angle decreases with increasing roughness, whereas for nonwetting liquids it increases. Semal et al. (1999) investigated the effect of surface roughness on contact angle hysteresis by studying a sessile droplet of squalane spreading dynamically on multilayer substrates (behenic acid on glass) and found that microroughness increases slow the rate of droplet spreading. Erbil et al. (2003) measured the contact angle of polypropylene (contact angle of 104° for a smooth surface) and found that the contact angle increases with increasing roughness. Burton and Bhushan (2005) measured the variation of the contact angle with the roughness of patterned surfaces and found that in the case of hydrophilic surfaces, it decreases with increasing roughness, and for hydrophobic surfaces, it increases with increasing roughness.

In the last decade, material scientists paid attention to natural surfaces, which are extremely hydrophobic. Among them are leaves of water-repellent plants such as *Nelumbo nucifera* (lotus) and *Colocasia esculenta*, which have high contact angles with water [2, 34, 50]. First, the surface of the leaves is usually covered with a range of different waxes made from a mixture of large hydrocarbon molecules, measuring around 1 nm in diameter, that have a strong phobia of being wet. Second, the surface is very rough owing to so-called papillose epidermal cells, which form asperities or papillae (Fig. 21.1). In addition to the microscale roughness of the leaf due to the papillae, the surface of the papillae is also rough with submicron-sized asperities [50]. Thus, they have hierarchical micro- and nano-sized structures. The water droplets on these surfaces readily sit on the apex of nanostructures because air bubbles fill in the valleys of the structure under the droplet; therefore, these leaves exhibit considerable superhydrophobicity. The water droplets on the leaves remove any contaminant particles from there surfaces when they roll off, leading to self-cleaning ability, also referred to as the lotus effect. Other examples of biological surfaces include duck feathers and butterfly wings. Their corrugated surfaces provide air pockets that prevent water from completely touching the surface. Study and simulation of biological objects with desired properties is referred to as "biomimetics", which comes from the Greek word *biomimesis* meaning to mimic life. As far as realization of strongly water repellent artificial surfaces is concerned, they can be constructed by chemically treating surfaces with low-surface-energy substances, such as poly(tetrafluoroethylene), silicon, or wax, or by fabricating extremely rough hydrophobic surfaces directly [23, 28, 33, 45]. Sun et al. (2005) studied an artificial poly(dimethylsiloxane) (PDMS) replica of a lotus leaf surface and found a water contact angle of 160° for the rough surface, whereas for the smooth PDMS surface it is about 105°.

As stated earlier, when two solids come in contact in the presence of a wetting liquid, a meniscus is often formed [4, 5, 7]. A meniscus results in a normal meniscus force, which, in turn, results in an increase in tangential friction force. The magnitude of the meniscus force depends on the number of asperity contacts and asperity radii, which depend on roughness, and on the surface tension of the liquid and the contact angle. The contact angle, as stated before, depends on surface roughness, and thus roughness affects the wet friction force; this effect was neglected earlier [35].

In this chapter, a model to provide a relationship between local roughness and contact angle as well as contact angle hysteresis is presented and experimental data are discussed. Both homogeneous and composite interfaces are studied. The model is used to analyze different possible roughness distributions and to calculate

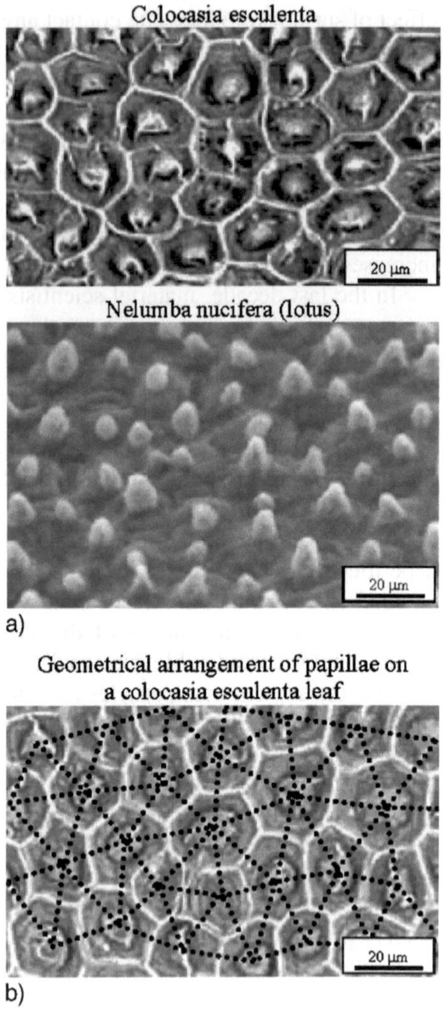

Fig. 21.1. (**a**) Scanning electron microscope (SEM) image of papillae on water-repellent plant leaves of *Colocasia esculenta* and *Nelumbo nucifera* (lotus). (**b**) Distribution of the papillae on the *C. esculenta* leaf surface which forms a pattern, close to hexagonal [35]

the effect of local roughness distribution on the contact angle, which allows us to make recommendations for optimized biomimetic superhydrophobic surfaces. The analysis is also compared with experimental data on natural and artificial water-repellent surfaces. Furthermore, the effect of roughness on the meniscus force and on the friction force during liquid-mediated contact is considered and discussed.

21.2 Contact Angle Analysis

As pointed out in the preceding section, a liquid in contact with a solid surface may form both homogeneous (solid–liquid) and composite (solid–liquid–air) interfaces. Both cases are studied in this section.

21.2.1 Homogeneous Solid–Liquid Interface

In this section, the dependence of the contact angle on the surface tension is considered for a liquid in contact with a smooth and a rough solid surface, forming a homogeneous interface. The effects of the contact area and of sharp edges on the contact angle are discussed.

21.2.1.1 Smooth Surface

It is well known that the surface atoms or molecules of liquids or solids have energy above that of similar atoms and molecules in the interior, which results in surface tension or free surface energy being an important surface property. This property is characterized quantitatively by the surface tension or free surface energy γ, which is equal to the work that is required to create a unit area of the surface at constant volume and temperature. The units of γ are joules per square meter or newtons per meter and it can be interpreted either as energy per unit surface area or as tension force per unit length of a line at the surface. When a solid is in contact with a liquid, the molecular attraction will reduce the energy of the system below that for the two separated surfaces. This may be expressed by the Dupré equation:

$$W_{\mathrm{SL}} = \gamma_{\mathrm{SA}} + \gamma_{\mathrm{LA}} - \gamma_{\mathrm{SL}} \,, \tag{21.1}$$

where W_{SL} is the work of adhesion per unit area between two surfaces, γ_{SA} and γ_{SL} are the surface energies (surface tensions) of the solid against air and liquid, and γ_{LA} is the surface energy (surface tension) of liquid against air [25].

If a droplet of liquid is placed on a solid surface, the liquid and solid surfaces come together under equilibrium at a characteristic angle called the static contact angle θ_0 (Fig. 21.2). The contact angle can be determined from the condition of the total energy of the system being minimized [1,25]. The total energy E_{tot} is given by

$$E_{\mathrm{tot}} = \gamma_{\mathrm{LA}}(A_{\mathrm{LA}} + A_{\mathrm{SL}}) - W_{\mathrm{SL}} A_{\mathrm{SL}} \,, \tag{21.2}$$

where A_{LA} and A_{SL} are the contact areas of the liquid with the solid and air, respectively. It is assumed that the droplet is small enough so that the gravitational potential energy can be neglected. At the equilibrium $\mathrm{d}E_{\mathrm{tot}} = 0$, which yields

$$\gamma_{\mathrm{LA}}(\mathrm{d}A_{\mathrm{LA}} + \mathrm{d}A_{\mathrm{SL}}) - W_{\mathrm{SL}} \mathrm{d}A_{\mathrm{SL}} = 0 \,. \tag{21.3}$$

For a droplet of constant volume, it is easy to show using geometrical considerations that

$$\mathrm{d}A_{\mathrm{LA}} / \mathrm{d}A_{\mathrm{SL}} = \cos\theta_0 \,. \tag{21.4}$$

Combining (21.1), (21.3), and (21.4), we obtain the so-called Young equation for the contact angle:

$$\cos\theta_0 = \frac{\gamma_{\mathrm{SA}} - \gamma_{\mathrm{SL}}}{\gamma_{\mathrm{LA}}} \,. \tag{21.5}$$

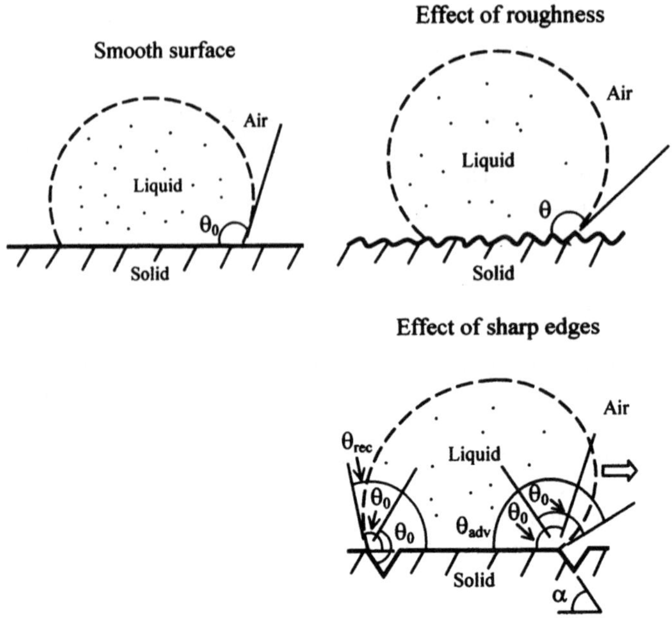

Fig. 21.2. Droplet of liquid in contact with a solid surface – smooth surface, contact angle θ_0; rough surface, contact angle θ – and a surface with sharp edges. For a droplet moving from *left* to *right* on a sharp edge (shown by an *arrow*), the contact angle at a sharp edge may be any value between the contact angle with the horizontal plane and with the inclined plane. This effect results in a difference between advancing ($\theta_{adv} = \theta_0 + \alpha$) and receding ($\theta_{rec} = \theta_0 - \alpha$) contact angles [35]

Equation (21.5) provides with the value of the static contact angle for given surface tensions. Note that although the term "air" is used throughout this chapter, the analysis does not change in the case of another gas, such as a liquid vapor.

For microscale/nanoscale droplets, the value of the contact angle may be different from that for macroscale droplets. The effect of line tension is believed to be responsible for that, since the energy of the molecules in the vicinity of the triple line is different from that in the bulk or at the surface; however, this effect is significant only at the nanometer range [42]. Checco et al. (2003) showed that the atomic force microscopy (AFM) data on the scale dependence of the contact angle for nanosized and microsized alkane droplets may rather be explained by the effect of surface heterogeneity, assuming that tiny droplets first appear at the most wettable spots.

21.2.1.2 Rough Surface

The Young equation (21.5) is valid only in the case of a flat solid surface. If the surface is rough, the roughness affects the contact angle in two ways: owing to the increased contact area A_{SL} and owing to the effect of sharp edges. We consider the effect of the contact area first.

Let us consider a rough solid surface, which consists of asperities and valleys, with a typical size of roughness details smaller than the size of the droplet (Fig. 21.1b). For a droplet of constant volume in contact with a rough surface without air pockets, the cosine of the contact angle is given as the ratio of the differentials of the liquid–air contact area and the area under the droplet, based on the geometrical considerations, which is given as [51]

$$\cos\theta = \mathrm{d}A_{\mathrm{LA}}/\mathrm{d}A_{\mathrm{F}} = \frac{A_{\mathrm{SL}}}{A_{\mathrm{F}}}\mathrm{d}A_{\mathrm{LA}}/\mathrm{d}A_{\mathrm{SL}} = R_{\mathrm{f}}\cos\theta_0 \,, \tag{21.6}$$

where θ is the contact angle for the rough surface and A_{F} is the flat solid–liquid contact area or a projection of the solid–liquid area A_{SL} on the horizontal plane. R_{f} is a roughness factor defined as

$$R_{\mathrm{f}} = \frac{A_{\mathrm{SL}}}{A_{\mathrm{F}}} \,. \tag{21.7}$$

Equation (21.6) shows that if the liquid wets a flat surface ($\cos\theta_0 > 0$), it will wet also the rough surface with a contact angle of $\theta < \theta_0$, since $A_{\mathrm{SL}}/A_{\mathrm{F}} > 1$. Furthermore, for nonwetting liquids ($\cos\theta_0 < 0$), the contact angle with a rough surface will be greater than that with the flat surface, $\theta > \theta_0$. The dependence of the contact angle on the roughness factor is presented in Fig. 21.3 for different values of θ_0 (Nosonovsky and Bhushan 2005).

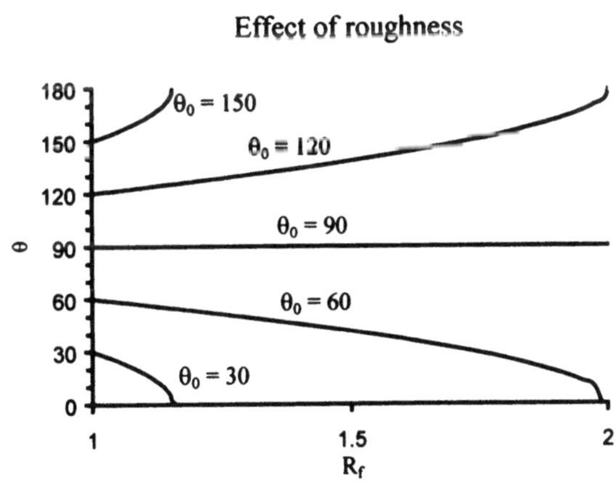

Fig. 21.3. Contact angle for a rough surface (θ) as a function of the roughness factor (R_{f}) for various contact angles for a smooth surface (θ_0) [35]

21.2.1.3 Contact Angle Hysteresis

There are several reasons which may lead to the contact angle hysteresis. These include surface roughness and heterogeneity.

For a rough surface, a sharp edge can pin the line of contact of the solid, liquid, and air (also known as the triple line) at a position far from stable equilibrium, i.e., at contact angles different from θ_0 [19]. This effect is illustrated in the bottom sketch of Fig. 21.2, which shows a droplet propagating along a solid surface with grooves. At the edge point, the contact angle is not defined and can have any value between the values corresponding to the contact with the horizontal and inclined surfaces. For a droplet moving from left to right, the triple line will be pinned at the edge point until it is able to proceed to the inclined plane. As observed from Fig. 21.2, the change of the surface slope (α) at the edge is the reason for the pinning. Because of the pinning, the value of the contact angle at the front of the droplet (dynamic maximum advancing contact angle or $\theta_{adv} = \theta_0 + \alpha$) is greater than θ_0, whereas the value of the contact angle at the back of the droplet (dynamic minimum receding contact angle or $\theta_{rec} = \theta_0 - \alpha$) is smaller than θ_0. This phenomenon is known as the contact angle hysteresis [19,25,26]. A hysteresis domain of the dynamic contact angle is thus defined by the difference $\theta_{adv} - \theta_{rec}$. The liquid can travel easily along the surface if the contact angle hysteresis is small. It is noted that the static contact angle lies within the hysteresis domain; therefore, increasing the static contact angle up to the values of a superhydrophobic surface (approaching 180°) will result also in reduction of the contact angle hysteresis. The contact angle hysteresis can also exist even if the surface slope changes smoothly, without sharp edges. For a chemically heterogeneous surface, the change of local contact angle due to heterogeneity has a similar effect as the change of the surface slope.

21.2.2 Composite Solid–Liquid–Air Interface

In this section, the dependence of the contact angle on the surface tension is considered for a liquid in contact with a smooth and a rough solid surface, forming a composite solid–liquid–air interface with air pockets trapped at the valleys between the asperities.

21.2.2.1 Contact Angle

It is noted that (21.6) is valid only for moderate values of R_f, when $-1 \leq R_f \cos\theta_0 \leq 1$. For high roughness, a wetting liquid will be completely absorbed by the rough surface cavities. As will be shown in this section, a nonwetting liquid cannot penetrate into surface cavities with large slopes, resulting in the formation of air pockets, leading to a composite solid–liquid–air interface, shown for a sawtooth and a smooth profile in Fig. 21.4a. The solid–liquid contact zones are located at the peaks of the asperities, whereas the air pockets and solid–air contact zones are in the valleys. The solid–liquid contact area will not further increase with increasing roughness [19].

Formation of the composite interface is shown in Fig. 21.4a for a sawtooth profile with slope α and for a smooth profile. In order to determine whether the interface is solid–liquid or composite, the change of net energy $\mathrm{d}E_{tot}$ should be considered, which corresponds to the displacement $\mathrm{d}s$ of the liquid–air surface along the inclined groove wall. For the solid–liquid interface, $\mathrm{d}E_{tot} < 0$; therefore

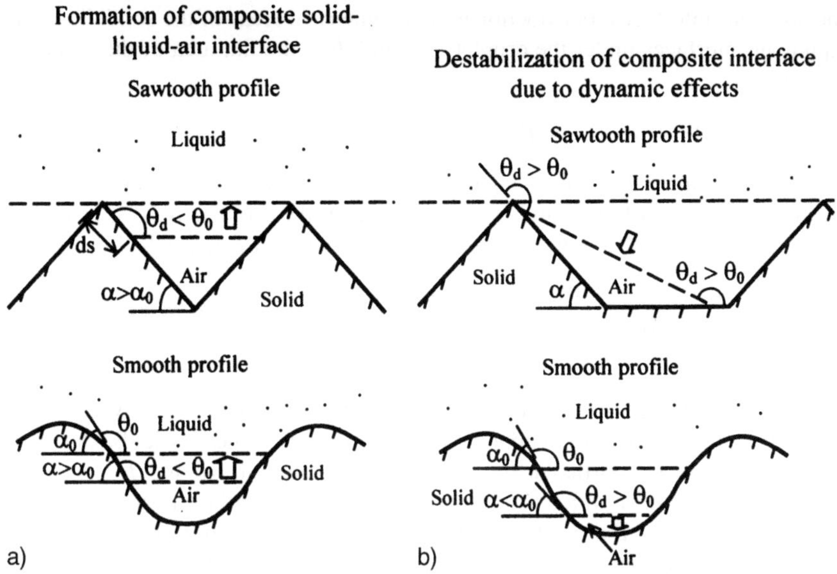

Fig. 21.4. (**a**) Formation of a composite solid–liquid–air interface for sawtooth and smooth profiles and (**b**) destabilization of the composite interface for the sawtooth and smooth profiles owing to dynamic effects. Dynamic contact angle $\theta_d > \theta_0$ corresponds to an advancing liquid–air interface, whereas $\theta_d < \theta_0$ corresponds to a receding interface [35]

it is more energetically profitable for the liquid to advance and fill the groove. For the composite interface, however, $dE_{tot} > 0$; therefore it is more energetically profitable for the liquid to recede and leave the groove. The change of energy depends on slope α, and the critical value of the slope, α_0, can be found by setting $dE_{tot} = 0$:

$$\begin{aligned} dE_{tot} &= dA_{SL}\gamma_{SL} + dA_{LA}\gamma_{LA} = -2\,ds\cos\alpha_0\,ds + 2\,ds(\gamma_{SL} - \gamma_{SA}) \\ &= -2\gamma_{LA}\cos\alpha_0\,ds + 2(\gamma_{SL} - \gamma_{SA})\,ds = 0\,. \end{aligned} \tag{21.8}$$

Substituting (21.5) into (21.8), we obtain $\cos\alpha_0 = -\cos\theta_0$ or

$$\alpha_0 = 180^\circ - \theta_0\,. \tag{21.9}$$

For small slopes ($\alpha < \alpha_0$), $dE_{tot} < 0$ and the interface is solid–liquid, whereas for large slopes ($\alpha > \alpha_0$), $dE_{tot} > 0$ and the interface is composite, as shown in Fig. 21.4a. For a profile of arbitrary smooth shape, a composite interface is possible if the slope exceeds the critical value α_0 at some point. In this case, the liquid would recede and leave space in the groove, as shown in Fig. 21.4a for a smooth profile.

Cassie and Baxter (1944) extended the Wenzel equation, which was originally developed for the homogeneous solid–liquid interface, for the composite interface. For this case, there are two sets of interfaces: a liquid–air interface with the ambient and a flat composite interface under the droplet involving solid–liquid, liquid–air, and solid–air interfaces. For fractional flat geometrical areas of the solid–liquid and liquid–air interfaces under the droplet, f_{SL} and f_{LA}, the flat area of the composite interface is

$$A_C = f_{SL} A_C + f_{LA} A_C = R_f A_{SL} + f_{LA} A_C \, . \tag{21.10}$$

In order to calculate the contact angle in a manner similar to the derivation of (21.6), the differential area of the liquid–air interface under the droplet, $f_{LA}\, dA_C$, should be subtracted from the differential of the total liquid–air area dA_{LA}, which yields

$$\cos\theta = \frac{dA_{LA} - f_{LA}\, dA_C}{dA_C} = \frac{dA_{SL}}{dA_F} \frac{dA_F}{dA_C} \frac{dA_{LA}}{dA_{SL}} - f_{LA} = R_f f_{SL} \cos\theta_0 - f_{LA} \, . \tag{21.11}$$

According to (21.11), in the limit of high R_f, f_{SL} approaches zero, whereas f_{LA} approaches unity, and hence θ approaches 180°. However, the Cassie–Baxter model does not provide any particular form of the dependence of f_{SL} and f_{LA} on R_f and does not explain under which conditions the composite interface forms.

Shuttleworth and Bailey (1948) studied spreading of a liquid over a rough solid surface and found that the contact angle at the absolute minimum of the surface energy corresponds to the values given by (21.2) (for the homogeneous interface) or (21.5) (for the composite interface). According to their analysis, spreading of a liquid continues, until simultaneously (21.1) (Young's equation) is satisfied locally at the triple line and the minimal surface condition is satisfied over the entire liquid–air interface. The minimal surface condition states that the sum of the inverse principal radii of curvature, r_1 and r_2, is constant at any point, and thus governs the shape of the liquid–air interface:

$$\frac{1}{r_1} + \frac{1}{r_2} = \text{const.} \tag{21.12}$$

Johnson and Detre (1964) showed that the homogeneous and composite interfaces correspond to the two states of a droplet. Even though it may be geometrically possible for the system to become composite, it may be energetically profitable for the liquid to penetrate into valleys between asperities and to form the homogeneous interface. Swain and Lipowski (1998) extended the Wenzel and Cassie–Baxter model for the case of a chemically heterogeneous surface. Marmur (2003) formulated geometrical conditions for a surface under which the energy of the system has a local minimum and the composite interface may exist. Patankar (2004a) pointed out that whether the homogeneous or the composite interface exists depends on the system's history, i.e., on whether the droplet was formed at the surface or deposited. However, the aforementioned analyses do not provide us with the answer as to which of the two possible configurations, homogeneous or composite, will actually form.

In order to investigate whether the composite interface can exist, its stability should be analyzed. It was assumed in the preceding paragraphs that the composite interface is stable, and the derivation was based on the assumption that the advancing liquid–air interface remains a horizontal plane. This assumption is hard to justify in the case of large distances between the asperities or in the case when the slope changes slowly. In these cases, the liquid–air interface can easily be destabilized owing to an imperfectness of the profile shape or owing to dynamic effects, such as surface waves. This may result in formation of the homogeneous solid–liquid interface. On the basis of these considerations, Fig. 21.4b shows schematically such destabilization of the composite interface, for the case of a large distance between the asperities and for the case of a smooth profile. The liquid advances, if the liquid–air interface reaches a position at which its local angle with the solid surface is greater than θ_0, and eventually the liquid fills the valley. Therefore, (21.11) should not be used to predict the contact angle, since it ignores the possibility of destabilization of the composite interface. It is suggested that a stochastic approach should be developed with account for possibility of composite interface destabilization and different probabilities of different states of equilibrium [36, 37].

It is noted that the composite interface may normally exist only for hydrophobic materials, because in the case of a hydrophilic material the liquid will advance and fill the valleys between the asperities. It has been reported, however, that the so-called metastable equilibrium states with a composite interface may be experimentally observed in some cases even for hydrophilic materials [29], resulting in high contact angles and effectively making a hydrophilic surface hydrophobic. However, these states are highly unstable and can hardly be practically utilized. Although their nature is not completely understood, they exist apparently owing to pinning of the triple line as a consequence of nanoroughness and surface heterogeneity.

21.2.2.2 Effect of the Composite Interface on the Contact Angle Hysteresis

The transition to the composite solid–liquid–air interface, which was considered earlier, results in a dramatic reduction of the solid–liquid area of contact and, thus, adhesion of the droplet to the solid surface. Therefore, transition to the composite interface can decrease the contact angle hysteresis, so that a droplet would move easily. Reduction of hysteresis due to transition to a composite interface was observed experimentally [26, 29].

21.2.3 Stability of the Composite Interface

As pointed out in the preceding sections, although a composite interface may be possible geometrically, it may be unstable, so the stability of a homogeneous interface should be analyzed. The stability of the homogeneous interface is not considered here, because such an interface is usually stable. While transition from the composite interface to the homogeneous is a well-known phenomenon, the opposite transition has never been observed experimentally. This is due to very high activation energy required to induce the composite state. Stability implies that in addition to satisfying the equilibrium condition for the net energy

$$\mathrm{d}E = 0\,, \tag{21.13}$$

the stable configuration should satisfy the minimum net energy condition

$$\mathrm{d}^2E > 0\,. \tag{21.14}$$

The interface may be destabilized owing to small perturbations caused by various external influences and effects, for example, by capillary or gravitational waves. Furthermore, the configuration may have many stable equilibrium conditions and be transformed from one stable position to another owing to the external effects, with a certain probability of finding the system at a given state. These phenomena are considered in the following section.

21.2.3.1
Destabilization Due to Capillary and Gravitational Waves

A wave may form at the liquid–air interface owing to gravitational or capillary forces

$$z = A\cos(kx - \omega t)\ , \tag{21.15}$$

where z is the vertical displacement, and k and ω are the wavenumber and frequency, which are related to each other by

$$\omega^2 = gk + \frac{\gamma_{\text{LA}}}{\rho}k^3\ , \tag{21.16}$$

where g is the gravity constant, ρ is the liquid density and γ_{LA} is the liquid–air interface energy [30]. For most microscale/nanoscale applications, the effect of gravity is small and the frequency is given by

$$\omega = \sqrt{\frac{\gamma_{\text{LA}}k^3}{\rho}}\ . \tag{21.17}$$

The capillary waves may lead to composite interface destabilization, as will be shown later [36].

It is assumed that the interface energy γ_{LA} is a constant for given materials and that it is size-independent. Generally speaking, this is not true for a very small thickness of liquid comparable with the range of intermolecular forces. However, in the present work we are assuming that the relevant size of the surface roughness, as well as the thickness of the liquid layer, is greater than the range of the intermolecular forces and therefore that γ_{LA} is constant.

Consider a sawtooth profile (Fig. 21.5a) with teeth height $a\tan\alpha/2$ and distance between teeth d. The teeth represent asperities of a rough surface. It is assumed that the model of the sawtooth profile can capture important features of more complicated rough surfaces. The horizontal liquid–air interface is located at a distance z from the valley and has small waves of amplitude A and wavenumber k. The total change of the energy of the system, from the energy of the homogeneous solid–liquid interface, is given by the sum of surface changes throughout the inclined and horizontal portions of the surface and corresponding liquid–air parts, plus the energy of the waves. The surface energy changes at inclined and horizontal portions of the surfaces and at corresponding liquid–air parts are given by corresponding surfaces' lengths times the corresponding interface energies. The length of the inclined portion of the interface is $z/\sin\alpha$ and the length of the corresponding section of the wavy surface is $(z/\tan\alpha)(kL_0/2\pi)$, where L_0 is the length of the liquid–air interface per wave period, given by the integral

$$L_0 = \int_0^{2\pi/k} \sqrt{1+(Ak)^2\sin^2(xk)}\,\mathrm{d}x = 4\sqrt{1+(Ak)^2}\,E\left(\frac{Ak}{\sqrt{1+(Ak)^2}}\right), \tag{21.18}$$

where $E(x)$ is the elliptical integral of the second kind [36]. The length of the horizontal portion of the interface is d, and the length of the corresponding section of the wavy surface is $dkL_0/(2\pi)$. The energy change is given by

$$\begin{aligned} U(z) &= -\frac{2z}{\sin\alpha}(\gamma_{\mathrm{SL}}-\gamma_{\mathrm{SA}}) + \frac{2z}{\tan\alpha}\frac{kL_0}{2\pi}\gamma_{\mathrm{LA}} \\ &\quad - d\left[(\gamma_{\mathrm{SL}}-\gamma_{\mathrm{SA}}) - \gamma_{\mathrm{LA}}\frac{kL_0}{2\pi}\right]H(z) + E_0 \\ &= \frac{2z\gamma_{\mathrm{LA}}}{\sin\alpha}\left(\cos\theta_0 + \cos\alpha\frac{kL_0}{2\pi}\right) + d\gamma_{\mathrm{LA}}\left(\cos\theta_0 + \frac{kL_0}{2\pi}\right)H(z) + E_0 \end{aligned} \tag{21.19}$$

and E_0 is the energy of the waves, γ_{SL} and γ_{SA} are interface energies for the solid–liquid and solid–air interfaces, respectively, and $H(z)$ is the step function, such that $H(z) = 0$ for $z \leq 0$ and $H(z) = 1$ for $z > 0$. Young's equation for the contact angle is used in the derivation of (21.19) [1,25]:

$$\cos\theta_0 = \frac{\gamma_{\mathrm{SA}} - \gamma_{\mathrm{SL}}}{\gamma_{\mathrm{LA}}}. \tag{21.20}$$

It is assumed in (21.19), that $z > A$.

In the limiting case of a flat liquid–air interface ($A = 0$), the surface energy is given by [36]

$$\begin{aligned} U(z) &= -\frac{2z}{\sin\alpha}(\gamma_{\mathrm{SL}}-\gamma_{\mathrm{SA}}) + \frac{2z}{\tan\alpha}\gamma_{\mathrm{LA}} - d\left[(\gamma_{\mathrm{SL}}-\gamma_{\mathrm{SA}}) - \gamma_{\mathrm{LA}}\right]H(z) \\ &= \frac{2z\gamma_{\mathrm{LA}}}{\sin\alpha}(\cos\theta_0 + \cos\alpha) + d\gamma_{\mathrm{LA}}(\cos\theta_0 + 1)H(z) \end{aligned} \tag{21.21}$$

for $A < z$.

For $z > 0$, the energy may increase or decrease with increasing z, depending on the sign of $(\cos\theta_0 + \cos\alpha)$, since both γ_{LA} and $\sin\alpha$ are positive. In particular, if $180° - \alpha > \theta_0$, the energy increases with z, whereas otherwise it decreases. The stable position corresponds to the minimum value of the energy, which is $z = a/(2\tan\alpha)$ (liquid staying at the tops of the asperities) for $180° - \alpha < \theta_0$ and $z = 0$ for $180° - \alpha > \theta_0$ (homogeneous solid–liquid interface).

For the wavy liquid–air interface, based on (21.19), the energy may increases with increasing z, if $\cos\theta_0 + \cos\alpha kL_0/(2\pi) > 0$, for $z > A$. However, for

Fig. 21.5. Sawtooth profile (**a**) with a smooth liquid–air interface, (**b**) with a wavy liquid–air interface and (**c**) with a stochastic distribution of air pockets [36]

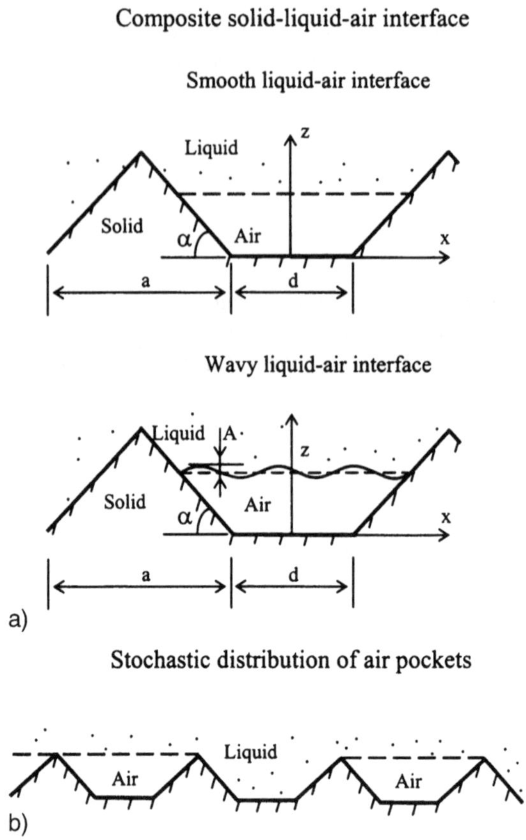

$z < A$, the waves touch the horizontal part of the interface, and only the fraction $[\pi - \arccos(z/A)]/\pi$ of the interface is liquid–air. In this case the energy change is given by

$$U(z) = \frac{z\gamma_{LA}}{\sin\alpha}\left[\cos\theta_0 + \frac{\pi - \arccos(z/A)}{\pi}\left(\frac{kL_{LA}}{2\pi}\right)\cos\alpha\right] + d\gamma_{LA}\left(\cos\theta_0 + \frac{kL_{LA}}{2\pi}\right)\frac{\pi - \arccos(z/A)}{\pi}H(z) + E_0 \qquad (21.22)$$

for $A > z$, where L_{LA} is the length of the liquid–air part of the interface (the wave not touching the solid horizontal part of the interface) given by [36]

$$L_{LA} = \int_{-\arccos(z/A)/k}^{\arccos(z/A)/k} \sqrt{1 + (Ak)^2 \sin^2(xk)}\,\mathrm{d}x\,. \qquad (21.23)$$

The energy change U as a function of liquid–air interface position z is presented in Fig. 21.6 for the smooth interface (21.21) and for the wavy liquid–air interface

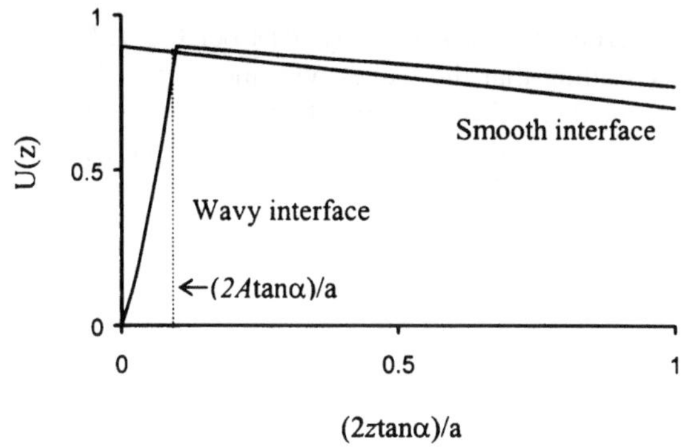

Fig. 21.6. Energy change as a function of interface position for smooth and wavy liquid–air interfaces, $d\gamma_{LA}(\cos\theta_0 + 1) = 0.9$, $\gamma_{LA}(\cos\theta_0 + \cos\alpha)/\sin\alpha = -0.2$, $\gamma_{LA}[\cos\theta_0 + (kL_0/2\pi)\cos\alpha]/\sin\alpha = -0.15$, $E_0 = 0.015$, $A = 0.1$ [36]

((21.19), (21.22)). It is noted that in the case when the waves are introduced, $U(z)$ has a local minimum at $z = 0$, which corresponds to the homogeneous solid–liquid interface, and in the case of $\cos\theta_0 + \cos\alpha kL_0/(2\pi) < 0$ it has another minimum at $z = a/(2\tan\alpha)$, which corresponds to the composite solid–liquid–air interface (liquid staying at the tops of the asperities). The interface position z is normalized in such a manner that the first equilibrium position ($z = 0$) corresponds to zero and the second equilibrium position [$z = a/(2\tan\alpha)$] corresponds to the value of unity in Fig. 21.6. In this case, the system has two equilibriums and may be, with a certain probability, in either one or the other position. The interface consists of many asperities and valley, some of the valleys have a homogeneous interface, whereas others have a composite interface (Fig. 21.5b). It is assumed that the probability p for the interface to be composite depends on geometrical parameters of the interface and the values of the energy which correspond to the equilibrium states [36].

21.2.3.2 Stochastic Model

In this section, the mechanism of destabilization of the composite interface due to liquid–air interface waves will be considered, and a statistical model for the interface destabilization will be discussed [36]. It was shown in the previous section that the interface may have two stable states. The first stable state corresponds to the homogeneous interface with energy level

$$U(0) = 0\,. \tag{21.24}$$

The second state corresponds to the composite interface ($z = a/2\tan\alpha$) with energy level obtained from (21.19)

$$U(a/2\tan\alpha) = \frac{a\gamma_{LA}}{\tan\alpha\sin\alpha}\left(\cos\theta_0 + \cos\alpha\frac{kL_0}{2\pi}\right) + d\gamma_{LA}\left(\cos\theta_0 + \frac{kL_0}{2\pi}\right) + E_0\,. \tag{21.25}$$

A certain probability p may be associated with each of the two stable states of energy. Assuming that the waves with the energy E_0 have a similar effect on the system as the thermal fluctuation of an ideal gas with energy kT, the Maxwell–Boltzmann statistical distribution may be applied [21]. On the basis of the Maxwell–Boltzmann distribution, the probability is exponentially dependent upon the energy level:

$$p = B \exp\left(-\frac{U}{E_0}\right) , \tag{21.26}$$

where B is a normalization constant [36]. Substituting (21.25) into (21.26), we obtain the probability of the composite interface:

$$p(\phi) = C \exp(-\phi/\phi_0) , \tag{21.27}$$

where

$$\phi = d/a , \tag{21.28}$$

$$\phi_0 = \frac{E_0}{a\gamma_{\mathrm{LA}}\left(\cos\theta_0 + \frac{kL_0}{2\pi}\right)} , \tag{21.29}$$

and

$$C = B \exp\left[-\frac{a\gamma_{\mathrm{LA}}}{E_0 \tan\alpha \sin\alpha}\left(\cos\theta_0 + \cos\alpha\frac{kL_0}{2\pi}\right) - 1\right] . \tag{21.30}$$

21.2.3.3 Analysis of Rough Profiles

In this section, a patterned rough surface will be analyzed. Consider a periodic sawtooth profile with a distance between asperities d and width a, as shown in Fig. 21.5. Let us assume that the probability of the interface remaining composite, p, decreases exponentially with distance between asperities according to (21.27) [36]. The roughness factor for the homogeneous interface based on (21.7), is given by

$$R_{\mathrm{f}} = \frac{d + a/\cos\alpha}{d + a} = \frac{\phi + 1/\cos\alpha}{\phi + 1} . \tag{21.31}$$

The total fraction of valleys which are covered with liquid (homogeneous interface) is given by $1 - p$, whereas the fraction of the valleys which have air pockets (composite interface) is given by p, obtained from (21.27). On the basis of this, the fractional areas are given by

$$f_{\mathrm{SL}} = \frac{(1-p)(d + a/\cos\alpha)}{(1-p)(d + a/\cos\alpha) + p(d+a)} = \frac{(1-p)(\phi + 1/\cos\alpha)}{(1-p)(\phi + 1/\cos\alpha) + p(\phi + 1)} \tag{21.32}$$

and

$$f_{\mathrm{LA}} = \frac{p(d+a)}{(1-p)(d+a/\cos\alpha)+p(d+a)} = \frac{p(\phi+1)}{(1-p)(\phi+1/\cos\alpha)+p(\phi+1)} \,. \tag{21.33}$$

Substituting (21.31)–(21.33) into (21.11), we obtain the expression for the contact angle:

$$\cos\theta = \frac{(1-p)(\phi+1/\cos\alpha)^2\cos\theta_0 - p(\phi+1)^2}{(1-p)(\phi+1/\cos\alpha)(\phi+1)+p(\phi+1)^2} \,. \tag{21.34}$$

The results for the contact angle as a function of ϕ are presented in Fig. 21.7a. It is observed that higher roughness (lower ϕ) corresponds to higher contact angles [36].

Comparison of the models, based on (21.6) (homogeneous interface), (21.11) (solid–liquid–air composite interface) and (21.32) (stochastic interface) is presented

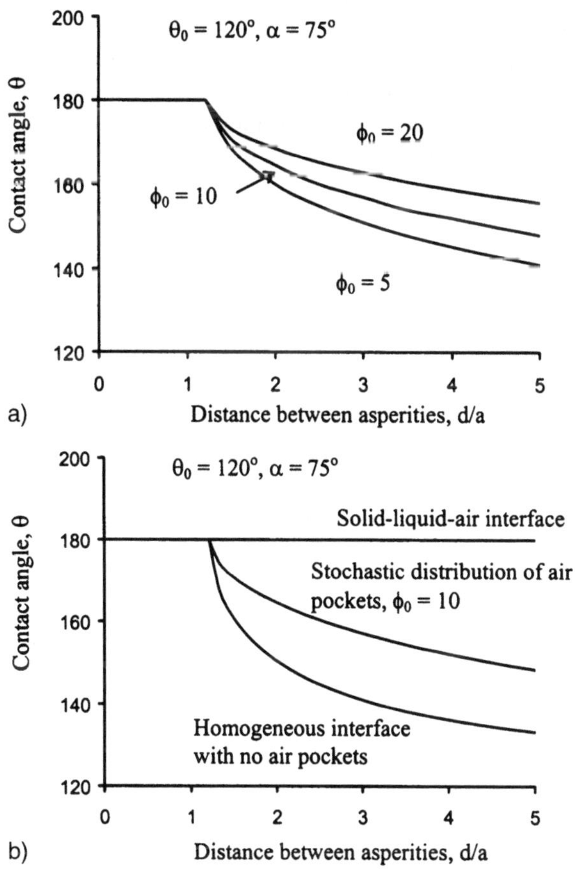

Fig. 21.7. (**a**) Contact angle as a function of distance between asperities for the stochastic model and (**b**) comparison of the interface with no air pockets, a composite liquid–air interface and a stochastic distribution of air pockets [36]

in Fig. 21.7b. It is observed that for high roughness (small ϕ), $|R_f \cos\theta_0| > 1$ and all three models predict $\theta = 180°$. However, for greater distance between the asperities (higher ϕ), the composite interface model, which does not account for the possibility of destabilization, still predicts $\theta = 180°$, if $\theta_0 + \alpha > 180°$, whereas the homogeneous interface model predicts a rapid decrease of θ down to the value of θ_0, owing to a decreasing roughness factor. The stochastic model yields the values of the contact angle close to the composite interface model for short distances between the asperities (small ϕ); however, with increasing ϕ, the probability of destabilization of the composite interface grows, and eventually the values of the contact angle approach those predicted by the homogeneous interface model.

21.2.3.4 Effect of Gravity

In the preceding analysis we ignored the effect of gravity by assuming that the gravity force is small compared with the surface tension forces. However, for big droplets, this assumption may not be correct. If the weight of a droplet exceeds the vertical component of the total surface tension force at the triple line, a droplet suspended at the tops of the asperities will collapse [22]. Thus, a maximum critical size of the droplet R_{max} exits above which the droplet cannot remain suspended on the tops of the asperities. Let us investigate how this maximum size depends on the period of the asperities l. Consider a rough surface with roughness period l and amplitude $\hat{z}$, which corresponds to a maximum droplet size R_{max}. The weight of the droplet is proportional to its volume and to the third power of R_{max}:

$$W \propto R_{max} \,. \tag{21.35}$$

For the maximum value of the droplet radius, the weight is equal to the total vertical component of the surface tension and is proportional to the surface tension times the cosine of the contact angle times the total perimeter of the triple lines:

$$W \propto \gamma_{SL} t N \cos\theta \,, \tag{21.36}$$

where N is the number of asperities under the droplet [37].

Consider another rough surface with period αl and amplitude $\alpha\hat{z}$ which has the same roughness factor and the same contact angle. The length of the triple line for each asperity will be αt. The number of asperities under the droplet is proportional to the second power of R_{max}, divided by the second power of α:

$$N \propto R_{max}^2/\alpha^2 \,. \tag{21.37}$$

Combining (21.35)–(21.37), we obtain

$$R_{max} \propto 1/\alpha \,. \tag{21.38}$$

This result suggests that with increasing size of asperities, the ability of a rough surface to form a composite interface decreases and larger droplets collapse. Therefore, smaller asperities make a composite interface more likely, owing to gravity. Increasing the droplet size has the same effect as increasing the period of roughness [37].

21.2.3.5 Effect of Dual Roughness

It is known that natural superhydrophobic surfaces have hierarchical roughness at two scale ranges: microroughness and nanoroughness (Fig. 21.8), although the reason for this is still obscure [8, 14, 24, 29, 47]. It was suggested by Nosonovsky and Bhushan (2006c) that nanoroughness is required to enhance the stability of the composite interface. A superhydrophobic surface should be able to form a composite interface in order to have a high contact angle and low contact angle hysteresis. However, the composite interface is often fragile and can be irreversibly transformed into the homogeneous interface, thus damaging superhydrophobicity. The mechanisms which lead to destabilization of the composite interface are scale-dependent. To effectively resist these scale-dependent mechanisms, a multiscale (hierarchical) roughness is required. High asperities resist the capillary waves, while nanobumps prevent nanodroplets from filling the valleys between asperities and pin the triple line in the case of a hydrophilic spot. Such multiscale roughness was found in natural and successful artificial superhydrophobic surfaces [8, 14, 20, 24, 34, 37, 45, 47].

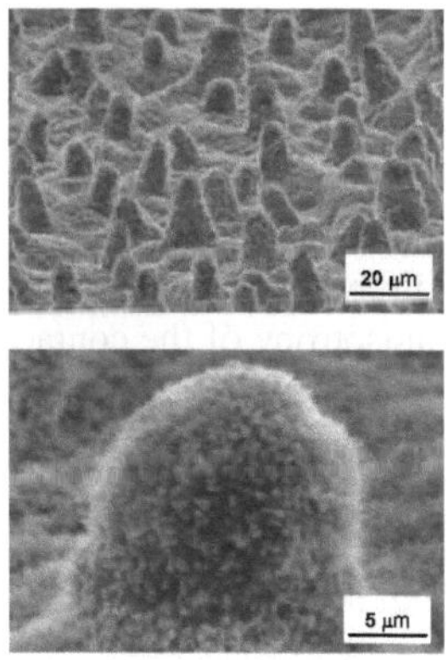

Nelumbo nucifera (lotus)

Fig. 21.8. Scanning electron micrographs of *N. nucifera* (lotus) leaf surface show hierarchical roughness with nanobumps (average peak-to-valley height 780 nm, midwidth 40 nm, peak radius 150 nm) on the top of microbumps (papillae, average peak-to-valley height 13 µm, midwidth 10 µm, peak radius 3 µm). Nanobumps data obtained with atomic force microscopy, scan size 2 × 2 µm; microbumps measured using an optical profiler [8]

21.3 Calculation of the Contact Angle for Selected Rough Surfaces and Surface Optimization

The contact angle of a liquid with a number of rough surfaces is calculated in this section. The model, presented in the preceding sections, combines the effect of surface area, possibility of formation of a composite interface, and the effect of sharp edges. Several selected rough surfaces are considered, shown in Fig. 21.5. First, two-dimensional surface profiles are analyzed, followed by more complex three-dimensional surfaces. On the basis of the analysis, roughness optimization for contact angle is conducted.

21.3.1 Two-Dimensional Periodic Profiles

21.3.1.1 *Sawtooth Periodic Profile*

Let us consider a surface with a sawtooth profile with a tooth angle (or the absolute value of the slope) of α (Fig. 21.9). Using (21.7), we calculate the roughness factor as

$$R_{\mathrm{f}} = \frac{A_{\mathrm{SL}}}{A_{\mathrm{F}}} = (\cos\alpha)^{-1} . \tag{21.39}$$

Using (21.6), we obtain the contact angle as

$$\cos\theta = \frac{\cos\theta_0}{\cos\alpha} . \tag{21.40}$$

An increase of α above the critical level α_0, given by (21.9), will result in a transition from a complete solid–liquid contact to a composite solid–liquid–air interface, and (21.40) cannot be used any further. Substituting the value of slope $\alpha = \alpha_0$ found from (21.9) in (21.40), we can obtain the value of θ, which corresponds to α_0, and this gives that the critical value α_0 corresponds to the contact angle $\theta = 180°$. This means that by increasing the tooth angle toward the critical value, a surface with the contact angle approaching 180° can be produced, for a given θ_0. However, sharp edges, which may lead to pinning of the triple line, make the sawtooth profile undesirable. In addition to this, the sawtooth profile provides roughness only in the direction perpendicular to the grooves, so the anisotropy of the profile will result in anisotropy of the contact angle [17].

21.3.1.2 *General Periodic Profile*

For a general form of the surface $z(x, y)$, the solid–liquid area of contact is equal to

$$A_{\mathrm{SL}} = A_{\mathrm{F}} \iint\limits_{A_{\mathrm{F}}} \sqrt{1 + \left(\frac{\partial z}{\partial x}\right)^2 + \left(\frac{\partial z}{\partial y}\right)^2}\, \mathrm{d}x\,\mathrm{d}y . \tag{21.41}$$

A periodic two-dimensional surface profile with period λ can be presented as a Fourier series:

$$z(x) = \sum_{n=1}^{\infty} A_n \sin\frac{2\pi n x}{\lambda} + B_n \cos\frac{2\pi n x}{\lambda} . \tag{21.42}$$

The derivatives of $z(x)$ are given as

$$\frac{\mathrm{d}z}{\mathrm{d}x} = \frac{2\pi}{\lambda} \sum_{n=1}^{\infty} A_n n \cos\frac{2\pi n x}{\lambda} - B_n n \sin\frac{2\pi n x}{\lambda} ,$$
$$\frac{\mathrm{d}z}{\mathrm{d}y} = 0 . \tag{21.43}$$

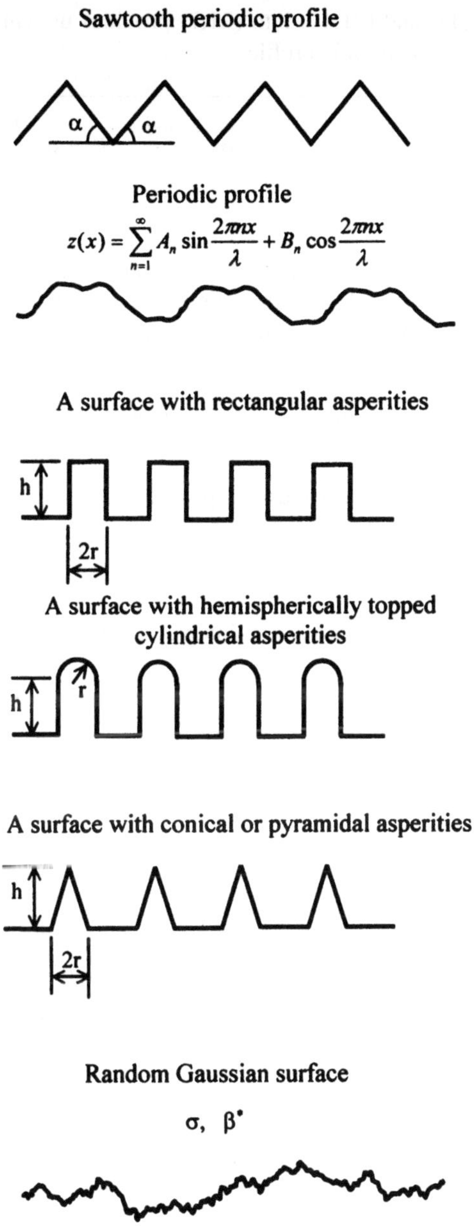

Fig. 21.9. Various rough surfaces [35]

Substitution of (21.43) and (21.41) into (21.7) provides us with an expression for the roughness factor of a periodic profile:

$$R_\mathrm{f} = 1 + \frac{1}{\lambda} \int_0^\lambda \sqrt{1 + \frac{4\pi^2}{\lambda^2} \left(\sum_{n=1}^{\infty} A_n n \cos \frac{2\pi n x}{\lambda} - B_n n \sin \frac{2\pi n x}{\lambda} \right)^2} \, \mathrm{d}x \, . \tag{21.44}$$

It is possible to determine whether the composite interface is possible, by considering the slope of the profile. In order for the composite interface to form, the absolute value of the slope must exceed the critical angle α_0, given by (21.9), at any point:

$$\left| \frac{2\pi}{\lambda} \sum_{n=1}^{\infty} A_n n \cos \frac{2\pi n x}{\lambda} - B_n n \sin \frac{2\pi n x}{\lambda} \right| > \tan(\alpha_0) = \tan(-\theta_0) \, . \tag{21.45}$$

As an example, let us consider a sinusoidal profile

$$z(x) = A_1 \sin \frac{2\pi x}{\lambda} \, . \tag{21.46}$$

By substituting (21.46) into (21.44) and integrating, we can obtain a closed-form solution:

$$\begin{aligned} R_\mathrm{f} &= \frac{1}{\lambda} \int_0^\lambda \sqrt{1 + (2\pi A_1/\lambda)^2 \cos^2(2\pi x/\lambda)} \, \mathrm{d}x \\ &= \frac{1}{2\pi} \int_0^{2\pi} \sqrt{1 + (2\pi A_1/\lambda)^2 \cos^2 x} \, \mathrm{d}x \\ &= \frac{2}{\pi} \sqrt{1 + (2\pi A_1/\lambda)^2} \int_0^{\pi/2} \sqrt{1 - \frac{(2\pi A_1/\lambda)^2}{1 + (2\pi A_1/\lambda)^2} \sin^2 x} \, \mathrm{d}x \\ &= \frac{2}{\pi} \sqrt{1 + (2\pi A_1/\lambda)^2} E \left(\frac{(2\pi A_1/\lambda)}{\sqrt{1 + (2\pi A_1/\lambda)^2}} \right) , \end{aligned} \tag{21.47}$$

where $E(x)$ is the so-called elliptical integral of the second kind, the values of which are tabulated in handbooks:

$$E(k) = \int_0^{\pi/2} \sqrt{1 - k^2 \sin^2 x} \, \mathrm{d}x \, . \tag{21.48}$$

The maximum absolute value of the slope of the sinusoidal profile (21.46) is achieved at $x = 0$ and is equal to $2\pi A_1/\lambda$. With an increase of A_1/λ, the slope increases, and a composite interface may be formed (Fig. 21.4a). For the composite interface to form, the slope at some points should exceed the critical value α_0, given by (21.9). By using (21.45) and setting $x = 0$ and using (21.45), we find the condition for the existence of the composite interfaceas

$$\frac{2\pi A_1}{\lambda} > \tan(-\theta_0) . \tag{21.49}$$

The contact angle can be calculated by substituting R_f from (21.47) into (21.6). The dependence of the contact angle on the amplitude of the sinusoidal profile is presented in Fig. 21.10a. It is observed that lower values of θ correspond to lower values of θ_0 at the transition to the composite interface, unlike in the case of the sawtooth surface, which has critical values corresponding to $\theta_0 = 180°$. For $\theta_0 = 100°$ the critical value of R_f is 5.67 ($\theta = 131°$), for $\theta_0 = 120°$ the critical value of R_f is 1.73 ($\theta = 140°$), and for $\theta_0 = 150°$ the critical value of R_f is 0.58 ($\theta = 159°$). Further increase of A_1/λ may lead to a corresponding increase of R_f and θ, on the basis of (21.38) [54]. However, as discussed earlier, the composite interface can be destabilized and (21.38) cannot be applied; therefore, the sinusoidal interface is not recommended for producing superhydrophobic surfaces. In addition to this, the sinusoidal profile provides roughness only in the direction perpendicular to the grooves, which would result in contact angle anisotropy.

21.3.2 Three-Dimensional Surfaces

The analysis of profiles provides us with critical values of the roughness parameters in the case when the contact line is parallel to the grooves. Three-dimensional surfaces, which constitute a more general case, with various typical shapes of asperities are considered in this section.

21.3.2.1 Array of Asperities of Identical Shape and Size

Let us consider a rough surface with rectangular asperities, which have a square foundation with side $2r$ and height h (Fig. 21.9). For each asperity, the area of the surface is given by

$$A_{\text{asp}} = 8rh + 4r^2 , \tag{21.50}$$

whereas the flat projection area is $4r^2$. Under the assumption that asperities are randomly distributed throughout the surface with a density of η asperities per unit area, the total contact surface area is given by

$$A_{\text{SL}} = A_{\text{F}} + A_{\text{F}}\eta(8rh + 4r^2) - A_{\text{F}}4\eta r^2 = A_{\text{F}}(1 + 8\eta r^2) . \tag{21.51}$$

The roughness factor is found using (21.7) and (21.51):

$$R_f = 1 + 8\eta rh = 1 + 2p^2 h/r , \tag{21.52}$$

where p is a packing parameter for asperities with a square foundation, $p = 2r\sqrt{\eta}$. The packing parameter is equal to the fraction of the surface area which is covered by asperities.

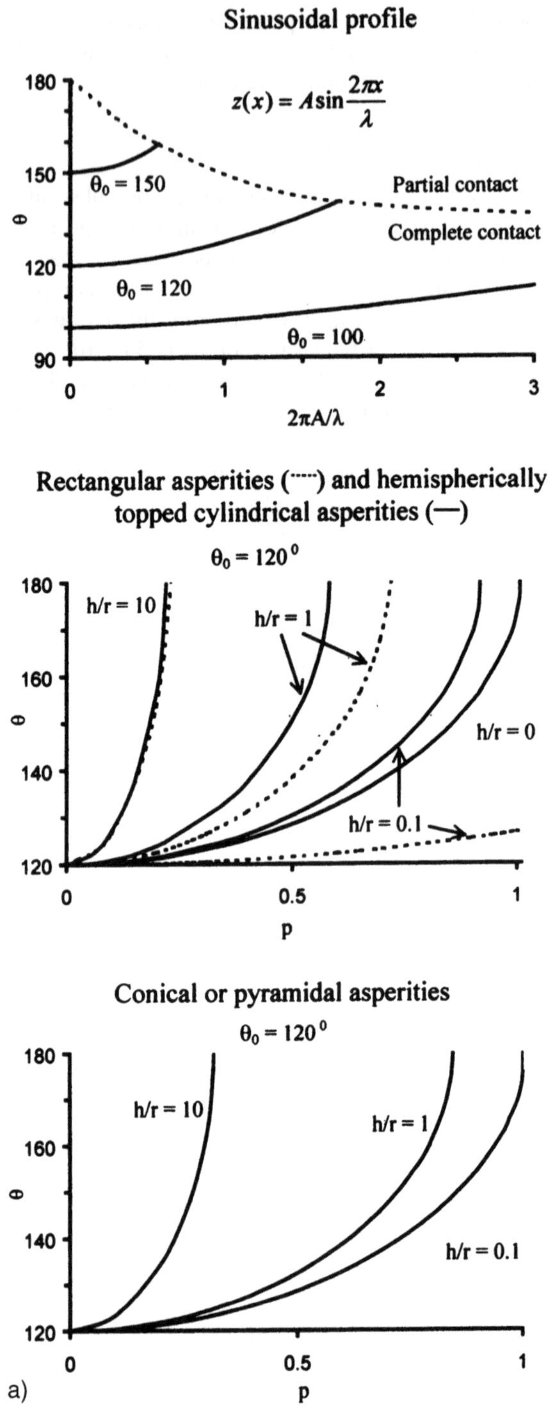

Fig. 21.10. (**a**) Contact angle for a rough surface (θ) as a function of surface parameters for the surface with a sinusoidal profile, rectangular (*dotted line*), hemispherically topped cylindrical (*solid line*), and conical/pyramidal asperities. (**b**) Dependence of the roughness factor (R_f) and the contact angle for a rough surface (θ) on roughness parameters for a Gaussian surface [35]

Fig. 21.10. (continued)

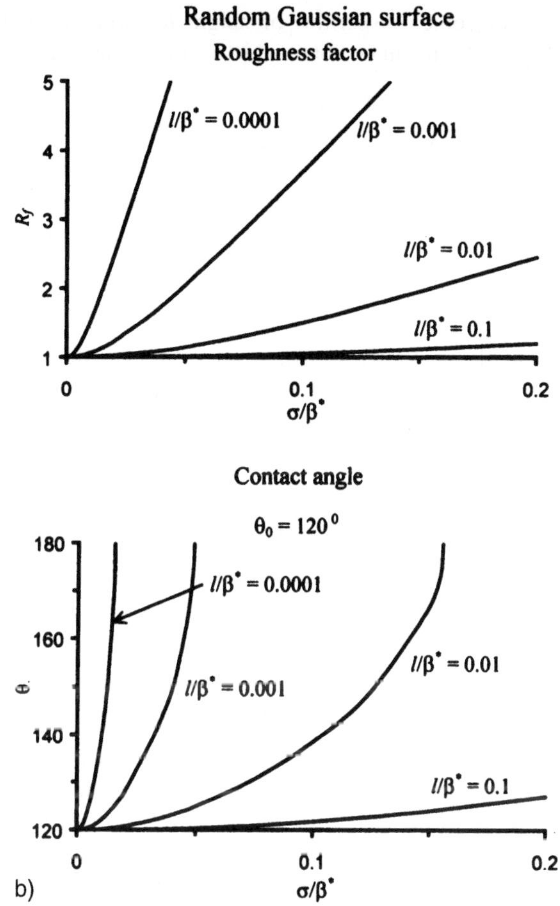

In a similar manner R_f can be calculated for asperities with a cylindrical foundation of height h and hemispherical top of radius r (Fig. 21.9). For each asperity, the area of the surface is given by

$$A_\mathrm{asp} = 2\pi r^2(1 + h/r) , \tag{21.53}$$

whereas the flat projection area is given by πr^2. Under the assumption that asperities are randomly distributed throughout the surface with a density of η asperities per unit area, the total contact surface area is given by

$$A_\mathrm{SL} = A_\mathrm{F} + A_\mathrm{F}\eta 2\pi r^2(1 + h/r) - A_\mathrm{F}\eta\pi r^2 = A_\mathrm{F}\left[1 + \eta\pi r^2(1 + 2h/r)\right] . \tag{21.54}$$

The roughness factor is found using (21.7) and (21.54):

$$R_\mathrm{f} = 1 + \eta\pi r^2(1 + 2h/r) = 1 + p^2(1 + 2h/r) , \tag{21.55}$$

where the packing parameter for asperities with a circular foundation is $p = r\sqrt{\pi\eta}$.

For conical asperities of height h, radius r, and side length $L = \sqrt{h^2 + r^2}$ (Fig. 21.5), it can be obtained in a similar manner:

$$A_{\text{asp}} = \pi r^2(1 + L/r) \,, \tag{21.56}$$

$$\begin{aligned} A_{\text{SL}} &= A_{\text{F}} + A_{\text{F}}\eta\pi r^2(1 + L/r) - A_{\text{F}}\eta\pi r^2 \\ &= A_{\text{F}}(1 + \eta\pi r L) = A_{\text{F}}\left(1 + \eta\pi r^2\sqrt{1 + (h/r)^2}\right) \,, \end{aligned} \tag{21.57}$$

and

$$R_{\text{f}} = 1 + \eta\pi r L = 1 + \eta\pi r^2\sqrt{1 + (h/r)^2} = 1 + p^2\sqrt{1 + (h/r)^2} \,, \tag{21.58}$$

where the packing parameter for asperities with a circular foundation is $p = r\sqrt{\pi\eta}$.

For pyramidal asperities with a square foundation of width $2a$ and height h, the corresponding quantities are given as

$$A_{\text{asp}} = 4r^2\left[1 + \sqrt{1 + (h/r)^2}\right] \,, \tag{21.59}$$

$$\begin{aligned} A_{\text{SL}} &= A_{\text{F}} + 4A_{\text{F}}\eta r^2\left[1 + \sqrt{1 + (h/r)^2}\right] - 4A_{\text{F}}\eta r^2 \\ &= A_{\text{F}}\left(1 + 4\eta r^2\sqrt{1 + (h/r)^2}\right) \,, \end{aligned} \tag{21.60}$$

and

$$R_{\text{f}} = 1 + 4\eta r^2\sqrt{1 + (h/r)^2} = 1 + p^2\sqrt{1 + (h/r)^2} \,, \tag{21.61}$$

where the packing parameter for asperities with a square foundation is $p = 2r\sqrt{\eta}$.

The dependence of the contact angle on the normalized radius of the asperities (taken as p) for $\theta_0 = 120°$ and for different ratios of h/r is presented in Fig. 21.10a, determined on the basis of (21.6), (21.52), (21.55), (21.59), and (21.61), for the rectangular, hemispherically topped, conical and pyramidal asperities. It is observed that with an increase of p, the value of the contact angle increases and reaches 180°. For higher aspect ratios, the increase of θ is faster.

In order to determine the critical values of roughness parameters which correspond to the transition to the composite interface, it should be analyzed whether the local slope can exceed the critical value α_0 and whether the composite interface is likely to remain stable. It is difficult to conduct such an analysis owing to its complexity; however, an estimate can be made using the fact that with increasing average absolute value of the slope of the surface both the local slope increases and the destabilization of the composite interface becomes less likely, since the surface is less smooth. On the basis of this, we assume here that, in a similar manner as for the two-dimensional profiles, the absolute value of the surface slope is responsible for the transition to the composite liquid–solid–air interface, and consider an average absolute value of the slope. For rectangular, hemispherically topped, conical, and pyramidal asperities, the mean absolute value of the slope, m, is equal to the density of the asperities and the flat projection area times the average absolute value of the slope (equal to twice the aspect ratio):

$$m = \eta\pi r^2(h/r) = \eta\pi h r \,. \tag{21.62}$$

The critical value can be found using a similar approach as in derivation of (21.45)

$$m_0 = \eta\pi hr = \tan(180 - \theta_0) = \tan(-\theta_0)\,. \tag{21.63}$$

From (21.52), (21.55), (21.58), (21.61), and (21.63), it may be shown that, for the selected value of $\theta_0 = 120°$, for rectangular, hemispherically topped, conical, and pyramidal asperities, the contact angle may approach 180° before the critical value of the roughness is reached for the values of h/r shown in Fig. 21.10a [35].

The equations developed here are used to calculate the contact angle for a lotus leaf and this is compared with measured data. The lotus leaf has almost hemispherically topped asperities (papillae) which are covered with wax ([34,50], Fig. 21.1). The static contact angle for a water droplet against a paraffin wax surface was reported by [18] as 104° and by [27] as 103°. On the basis of the data reported by [50], the number of asperities (papillae) can be estimated for the lotus leaf as 3400 per square millimeter ($\eta = 0.0034\,\mu\text{m}^{-2}$), the average radius of the hemispherically topped asperities r is 10 µm, and aspect ratio h/r is approximately 1. From (21.6) and (21.27), these values of the parameters correspond to the roughness factor $R_\text{f} \sim 4$ and the contact angle $\theta = 165°$ (using $\theta_0 = 104°$ for wax). During the measurements, conducted in our laboratory, the value of the static contact angle for deionized water on a lotus leaf was found to be $156 \pm 2°$. Neinhuis and Barthlott (1997) reported a contact angle value of 162° for a water droplet on a lotus leaf.

21.3.2.2 Random Rough Surface

A nominally flat random rough surface can be considered as a superposition of a flat plane and a two-dimensional random process which is characterized by a height distribution and an autocorrelation function. Many engineering and natural rough surfaces can be characterized by a Gaussian height distribution and an exponential autocorrelation function [4,5]. In this case, a rough surface is described by only two parameters: the standard deviation of asperity heights, σ, and the correlation length, β^*. The correlation length is a spatial parameter and it can be viewed as a measure of randomness. β^* is responsible for the horizontal scale of the surface, whereas σ is responsible for the vertical scale of the surface. Measured roughness is dependent on the short- and long-wavelength limit of measurement [4,5].

The absolute value of slope of a Gaussian surface also has a Gaussian distribution with the mean

$$m = \frac{\sigma}{l}\sqrt{\frac{1 - \left[\exp(-\beta^*/l)\right]^2}{\pi}}\,, \tag{21.64}$$

where l is the sampling interval or short-wavelength limit, which is a distance between data points during a measurement [52]. For a surface, the sampling interval is given by a low-wavelength limit of the Gaussian roughness, and is comparable with the atomic dimensions [35].

An element of the area of a surface with slopes of $\partial z/\partial x$ and $\partial z/\partial y$ in x- and y-directions is given by

$$dA = \sqrt{1 + (\partial z/\partial x)^2 + (\partial z/\partial y)^2}\,dx\,dy\ . \tag{21.65}$$

The distribution of $\sqrt{1 + (\partial z/\partial x)^2 + (\partial z/\partial y)^2}$ is not Gaussian in general, but in most applications the slope is small and the mean value of slope m can be taken to calculate the mean value of $\sqrt{1 + (\partial z/\partial x)^2 + (\partial z/\partial y)^2}$. It can also be assumed that the slopes in the x- and y-directions are the same. Using (21.63) and substituting m given by (21.64) and integrating, we can calculate [35] the roughness factor as

$$\begin{aligned} R_f &= 1 + \frac{1}{A_{SL}} \int\int_{A_f} \sqrt{1 + (\partial z/\partial x)^2 + (\partial z/\partial y)^2}\,dx\,dy \\ &= 1 + \sqrt{1 + 2m^2} = \sqrt{1 + 2\frac{\sigma^2}{l^2}\frac{1 - \exp(-l/\beta^*)^2}{\pi}}\ . \end{aligned} \tag{21.66}$$

For small l/β^*

$$R_f \approx \sqrt{1 + 2\left(\frac{\sigma}{\beta^*}\right)^2 \frac{2}{\pi(l/\beta^*)}}\ . \tag{21.67}$$

Furthermore, for small values of $\sigma^2/(l\beta^*)$

$$R_f \approx 1 + \frac{2}{\pi}\left(\frac{\sigma}{\beta^*}\right)^2 \left(\frac{\beta^*}{l}\right)\ . \tag{21.68}$$

In order to estimate the critical value of the roughness parameter, we assume, as in the previous subsection, that the average absolute value of the surface slope is responsible for the transition to the composite solid–liquid–air interface. The absolute value of slope is given by (21.64), so, in a similar manner as in the derivation of (21.63), the critical value of the Gaussian surface roughness parameter is

$$m_0 = \left(\frac{\sigma}{l}\sqrt{\frac{1 - \left[\exp(-\beta^*/l)\right]^2}{\pi}}\right)_0 = \tan(-\theta_0)\ . \tag{21.69}$$

The dependence of the roughness factor on σ/β^* is presented in Fig. 21.10b on the basis of (21.66). The dependence of the contact angle on σ/β^* using the roughness factor is presented in Fig. 21.10b. It is observed that both the roughness factor and the contact angle increase with increasing σ/β^*. From (21.69), it may be shown that for the selected value of $\theta_0 = 120°$, the contact angle may approach 180° prior to the critical values of the roughness parameter being reached for the values of σ/β^* and l/β^* shown. It is noted that for most natural and engineering Gaussian surfaces, the ratio σ/β^* is much less than 0.1 and the average value of the slope is small ($m \ll 1$). Therefore, although the roughness is below the critical value, it is difficult to achieve high contact angles with Gaussian random surfaces with a realistic value of σ/β^* [35].

21.3.3 Surface Optimization for Maximum Contact Angle

Among the several types of the surfaces considered in the preceding subsections, the highest contact angles are achieved with the sawtooth profile and rectangular, hemispherically topped, conical, and pyramidal asperities. As stated earlier, the sawtooth profile is undesirable owing to its sharp edges, which may pin the triple line, and because the grooves may reinforce wetting. Therefore, the rectangular, hemispherically topped, conical, and pyramidal asperities should be considered as the most appropriate for producing the highest contact angles. In order to prevent the contact angle hysteresis, it is desirable to avoid asperities with sharp edges, which may cause pinning of the triple line. Therefore, hemispherically topped asperities are the most appropriate. A case will also be made later for pyramidal asperities.

Two-tiered roughness involving two wavelengths has been considered by some authors (e.g., Herminghaus 2000) to decrease wetting, however, it is more likely to involve sharp edges, which are undesirable, and lead to unstable composite solid–liquid–air interface.

On the basis of (21.55), (21.58), and (21.61) and the results shown in Fig. 21.10, the maximum contact angle can be achieved by increasing the aspect ratio h/r and the packing parameter p. The maximum aspect ratio may be achieved by increasing asperity height. The maximum packing parameter may be achieved by packing the asperities as tightly as possible. The square of the packing parameter p^2 is equal to the ratio of the foundation area of the asperities to the total surface area; therefore, a higher value of p corresponds to a higher packing density. For asperities with a circular foundation, the square pattern of the distribution of the asperities results in packing of $1/(2r)$ rows per unit area with $1/(2r)$ asperities per unit length in the row. A higher density of the packing of the asperities can be achieved by a hexagonal distribution of asperities (Fig. 21.11). This distribution pattern results in packing of $1/(\sqrt{3}r)$ rows of asperities per unit length with $1/(2r)$ asperities per unit length in the row, or $\eta = 1/(2\sqrt{3}r^2)$, which yields

$$p = r\sqrt{\pi\eta} = \sqrt{\frac{\pi}{2\sqrt{3}}} \approx 0.952 \,. \tag{21.70}$$

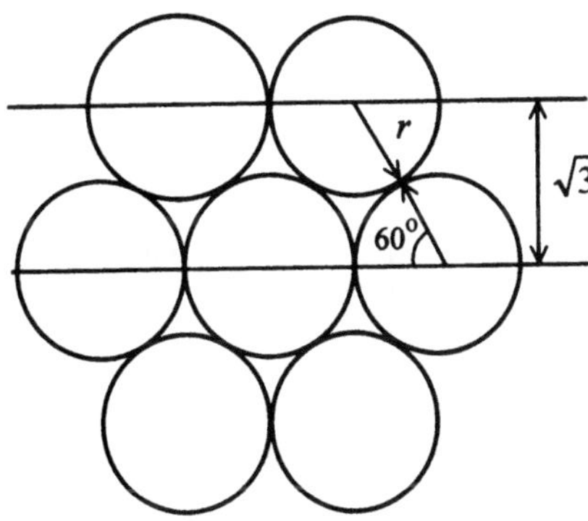

Fig. 21.11. Hexagonal pattern of packing of circular asperities for highest packing density [35]

Therefore, the recommendation for surface optimization is to take hexagonally packed hemispherically topped asperities with a high aspect ratio (needlelike). It is noted that certain leaves tend to have the distribution of the papillae close to hexagonal ([35]; Fig. 21.1).

An alternative shape which provides packing density $p = 1$ is given by pyramidal asperities with a square foundation. In order to avoid pinning due to sharp edges, the tops may be rounded with the hemispheres. Rectangular asperities do not provide space for liquid to penetrate; therefore, in the case of asperities with a square foundation, the pyramidal shape should be used. It should be noted that valleys with rounded edges have the same effect on the contact angle as asperities do [35].

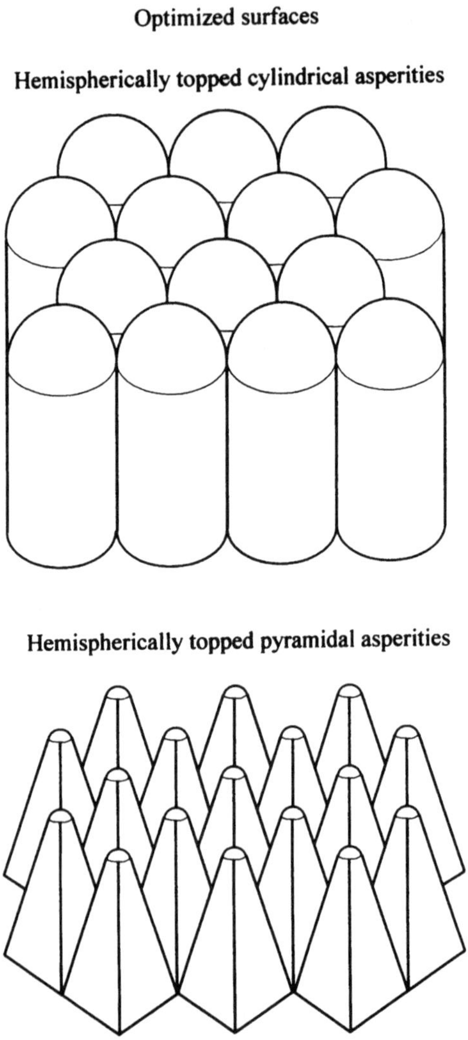

Fig. 21.12. Optimized spaced roughness distribution – hemispherically topped cylindrical asperities and pyramidal asperities with a square foundation and rounded tops. A square base gives higher packing density but introduces undesirable sharp edges [35]

The foundation radius of individual asperities, r (for circular foundation), or foundation side length, $2r$ (for square foundation), should be small compared with typical droplets. The upper limit of the droplet size may be estimated on the basis of the requirement that the gravity effect is small compared with that of the surface tension. The gravitational energy of the droplet is given by its density ρ multiplied by the volume, the gravitational constant $g = 9.81\ \mathrm{m/s^2}$, and the radius r,

$$W_g = \frac{4}{3}\pi r^3 \rho g r\ , \tag{21.71}$$

whereas the energy due to the surface tension can be estimated by the droplet surface area multiplied by the surface tension:

$$W_g = 4\pi r^2 \gamma_{LA}\ . \tag{21.72}$$

From $W_g \ll W_s$, we find the maximum droplet radius which can exist is

$$r_{max} \ll \sqrt{\frac{3\gamma_{LA}}{\rho g}}\ . \tag{21.73}$$

Typical quantities for water of $\rho = 1000\ \mathrm{kg/m^3}$ and $\gamma_{LA} = 72\ \mathrm{mJ/m^2}$ result in $r_{max} \ll 4.7$ mm. Although the small droplets will tend to unite into bigger ones, the minimum droplet radius is limited only by the molecular scale, so it is desirable to have r as small as possible.

To summarize, the highest possible contact angle and the lowest contact angle hysteresis, which is desirable in applications, may be achieved by using hemispherically topped asperities with a hexagonal packing pattern or by using pyramidal asperities with a rounded top. These recommendations can be used for producing superhydrophobic surfaces ([35]; Fig. 21.12).

For wetting liquids, roughness results in a decreased contact angle, in accordance with (21.6). Therefore, in order to create a superhydrophobic surface using the effect of roughness, a hydrophobic film is required. Hydrophobic coating is a well-known method of increasing the water-repellency of a material [43].

21.4 Meniscus Force

When two solids come in contact, a meniscus can form owing to condensation of liquid or because the liquid film may be present initially, if the liquid is wetting. For nonwetting liquids, a meniscus will not be formed. The meniscus causes an increase of the friction force. In this section the effect of surface roughness on the meniscus force will be considered for the case of a sphere in contact with a flat surface (single asperity contact) and for the case of multiple-asperity contact.

21.4.1 Sphere in Contact with a Smooth Surface

Consider a sphere with radius R in contact with a flat surface with a meniscus (Fig. 21.13a). The shape and size of the meniscus, as well as the total energy of the

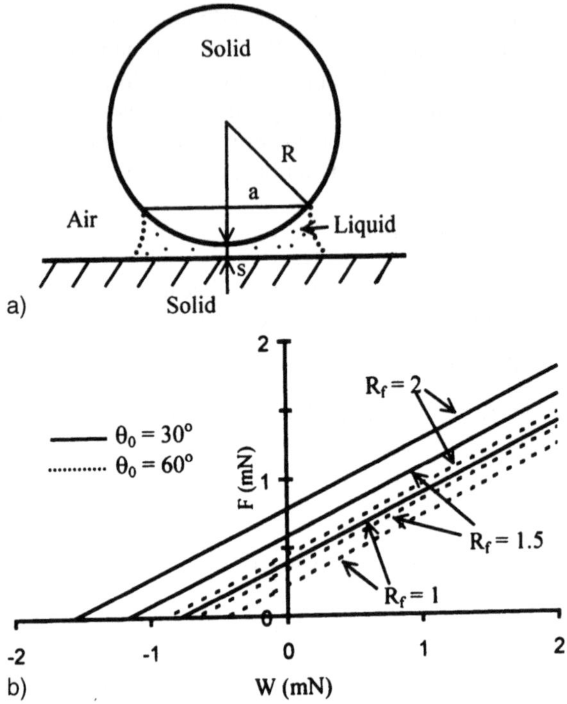

Fig. 21.13. (**a**) Meniscus formation during wet contact of a flat surface with a sphere. (**b**) Dependence of the friction force (F) on the normal load (W) for single-asperity contact, $\gamma_{LA} = 0.073\,\mathrm{J/m^2}$ (water), $R = 1$ mm, $\mu = 0.5$, for different values of $R_f = R_{f1} = R_{f2}$, $\theta_0 = \theta_1 = \theta_2$ [35]

system, depend on the separation distance s between the flat surface and the center of the sphere. The normal meniscus force F_m which acts upon the sphere and the flat surface can be calculated as the derivative of the energy E_{tot} by s:

$$F_m = \frac{dE_{tot}}{ds} . \tag{21.74}$$

There are two solid–liquid interfaces, with the sphere and with the flat surface. The areas of these interfaces are approximately equal to πa^2, where a is the meniscus radius. Under the assumption that the ratio a/R is small and $A_{LA} \ll A_{SL}$, the total energy E_{tot} is equal to the sum of the surface energies at the solid–liquid interface. On the basis of these assumptions, E_{tot} can be calculated, using (21.2), as

$$E_{tot} = \pi a^2(\gamma_{SL1} - \gamma_{SA1} + \gamma_{SL2} - \gamma_{SA2}) = \pi a^2 \gamma_{LA}(\cos\theta_1 + \cos\theta_2) , \tag{21.75}$$

where the indices 1 and 2 correspond to the sphere and the flat surface, and a is defined in Fig. 21.13 [35]. The volume of liquid V is a function of s and a and is given as the sum of the cylindrical volume with height $s + a^2/(2R)$ and cross-sectional area πa^2 minus a volume of the spherical segment of height $a^2/(2R)$:

$$V = \pi a^2 s + \frac{\pi a^4}{4R} . \tag{21.76}$$

The volume of the liquid remains constant during the contact, so (21.76) may be viewed as a quadratic equation for a^2, which is solved as

$$a^2 = -2Rs \pm 2R\sqrt{s^2 + V/(\pi R)} . \tag{21.77}$$

The derivative of a^2 by s, $\mathrm{d}(a^2)/\mathrm{d}s$ for the sphere touching the plane ($s = 0$) is given as

$$\frac{\mathrm{d}a^2}{\mathrm{d}s} = -2R\,. \tag{21.78}$$

By using the derivative of (21.75) and the expression in (21.78) in (21.74), we get

$$F_\mathrm{m} = 2\pi R\gamma_\mathrm{LA}(\cos\theta_1 + \cos\theta_2)\,. \tag{21.79}$$

Equation (21.79) (also known as the equation for the Laplace pressure) provides us with the value of the normal force due to the meniscus. If the sphere and the surface are rough, with roughness factors of R_f1 and R_f2, respectively, the roughness must be taken into account:

$$F_\mathrm{m} = 2\pi R\gamma_\mathrm{LA1}(R_\mathrm{f1}\cos\theta_1 + R_\mathrm{f2}\cos\theta_2)\,. \tag{21.80}$$

In the presence of a meniscus, the friction force is given by [4, 5]

$$F = \mu\,(W + F_\mathrm{m})\,. \tag{21.81}$$

The coefficient of friction in the presence of the meniscus force, μ_wet, is calculated using only the applied normal load, as normally measured in experiments:

$$\mu_\mathrm{wet} = \mu\left(1 + \frac{F_\mathrm{m}}{W}\right)\,. \tag{21.82}$$

Equation (21.55) shows that μ_wet is greater than μ, because F_m is not taken into account for calculation of the normal load in the wet contact.

The effect of a meniscus on friction force for different surface roughness is presented in Fig. 21.13b. It is observed that a roughness factor of $R_\mathrm{f} = 2$ may result in a significant change of the friction force due to the meniscus. In applications, it is usually desirable to decrease the meniscus forces; therefore, a smooth surface is preferable in the case of a single asperity contact.

21.4.2 Multiple-Asperity Contact

In the case of multiple-asperity contact, a statistical approach is used to model the contact. For a random surface with a certain σ and β^*, the average peak radius R_p and the number of contacts N depend on roughness owing to the so-called scale effect [9]. Bhushan and Nosonovsky (2004) showed that the average peak radius $\overline{R_\mathrm{p}}$ is related to σ and β^*, whereas the number of contacts, for moderate loads, is proportional to the load, divided by σ and β^*:

$$\overline{R_\mathrm{p}} \propto \frac{(\beta^*)^2}{\sigma}\,, \tag{21.83}$$

and

$$N \propto \frac{W}{\sigma\beta^*}\,. \tag{21.84}$$

For asperities of equal peak radius R_p, the meniscus force is given by [4,5,49]

$$F_m = 2\pi R_p \gamma_{LA} (\cos\theta_1 + \cos\theta_2)\, N \propto \frac{\beta^* W}{\sigma^2} (\cos\theta_1 + \cos\theta_2)\ . \tag{21.85}$$

The size of menisci is comparable with the size of individual contacts [35].

We consider a rough surface which consists of the short-wavelength roughness superimposed over the long-wavelength roughness with a typical size of the roughness smaller than the meniscus size (Fig. 21.14a). The nanoscale roughness of the two bodies is characterized by the roughness factors R_{f1} and R_{f2}. We further assume that asperities have an average peak radius $\overline{R_p}$. Substituting (21.6) into (21.85), we obtain

$$F_m = 2\pi \overline{R_p} \gamma_{LA} (\cos\theta_1 + \cos\theta_2)\, N \propto \frac{\beta^* W}{\sigma^2} (R_{fn1}\cos\theta_1 + R_{fn2}\cos\theta_2)\ . \tag{21.86}$$

The meniscus force as a function of σ^2/β^*, which is a measure of roughness, is presented in Fig. 21.14b [35]. It is observed that with increasing roughness σ^2/β^*, the meniscus force decreases. A high nanoscale roughness factor may slightly increase the meniscus force. Since it is usually desirable to reduce the meniscus force, a rough surface with high σ^2/β^* is preferable in the case of multiple-asperity contact [35].

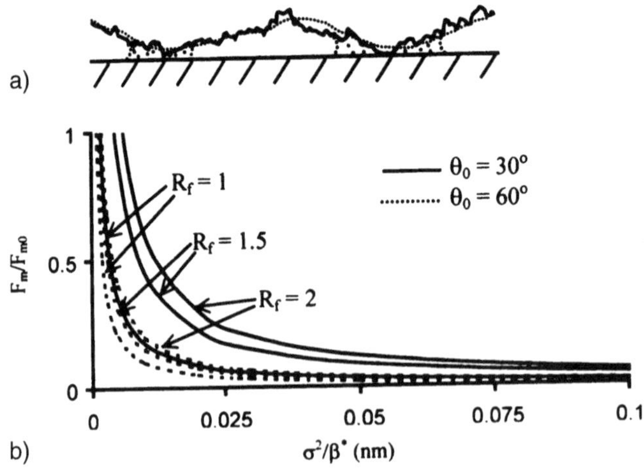

Fig. 21.14. (**a**) Menisci formation during wet contact of a smooth surface with a rough surface with a short-wavelength roughness superimposed over long-wavelength roughness (*dotted line*). (**b**) Dependence of the meniscus force (normalized by F_{m0}, meniscus force value at $\sigma^2/\beta^* = 0.001$ nm) on roughness σ^2/β^*, for different values of $R_f = R_{f1} = R_{f2}$, $\theta_0 = \theta_1 = \theta_2$ [35]

21.5 Experimental Data

Experimental data on natural, artificial and biomimetic rough surfaces which demonstrate the dependence of the contact angle on roughness are discussed in this section.

Burton and Bhushan (2006) measured the roughness of several water-repellent plants, including *N. nucifera* (lotus) and *C. esculenta* using scanning electron microscopy (SEM), optical microscopy, and AFM. Figure 21.15 shows optical profiler images of papillae on the surface of a lotus leaf. It is observed that the papilla surface itself is rough. Asperities on a fresh lotus leaf are too large to be measured by AFM; therefore, measurements have to be made on a dried leaf which does not have tall asperities. Furthermore, the top and the bottom of the leaf have to be measured separately. Figure 21.16 shows AFM scans of a dried lotus leaf. The two-dimensional profiles are taken from the top scan and the bottom scan for each scan size and are spliced together to get the total profile of the leaf [14]. The contact angle with water for lotus was found to be approximately 156°.

Burton and Bhushan (2005) measured the contact angle with artificial patterned poly(methyl methacrylate) (PMMA) surfaces with bumps with a low aspect ratio (LAR; 1:1 height-to-diameter ratio) and a high aspect ratio (HAR; 3:1 height-to-diameter ratio). PMMA was chosen because it is a polymer often used in BioMEMS/BioNEMS devices. The diameter of the asperities near the top is approximately 100 nm and the pitch of the asperities (the distance between each asperity) is approximately 500 nm. Figure 21.17 shows SEM images of the two types of patterned structures, LAR and HAR, on a PMMA surface. According to the model presented earlier, by introducing roughness to a flat surface, the hydrophobicity will either increase or decrease depending on the initial contact angle on a flat surface. The material chosen was initially hydrophilic, so to obtain a sample that is hydrophobic, a self-assembled monolayer (SAM) was deposited on the sample surfaces. The samples chosen for the SAM deposition were the flat film and the HAR for each polymer. The SAM perfluorodecyltriethoxysilane (PFDTES) was deposited on the polymer surface using a vapor phase deposition technique. PFDTES was chosen because of the hydrophobic nature of the surface. It should be noted that the bumps should be as close as possible to provide a large surface area. In order to benefit from an increase in the contact angle and a decrease in the contact area with an increase in surface roughness, the pitch of the bumps should be smaller than the water droplet and the size of the contacting body.

Figure 21.18 shows the static contact angle for various samples. These values correlate well with the model describing roughness with hydrophobicity. The contact

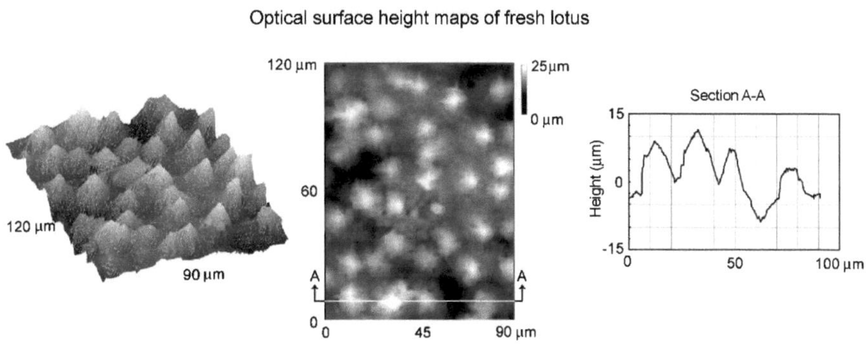

Fig. 21.15. Surface height map and a 2D profile of a lotus leaf using an optical profiler [14]

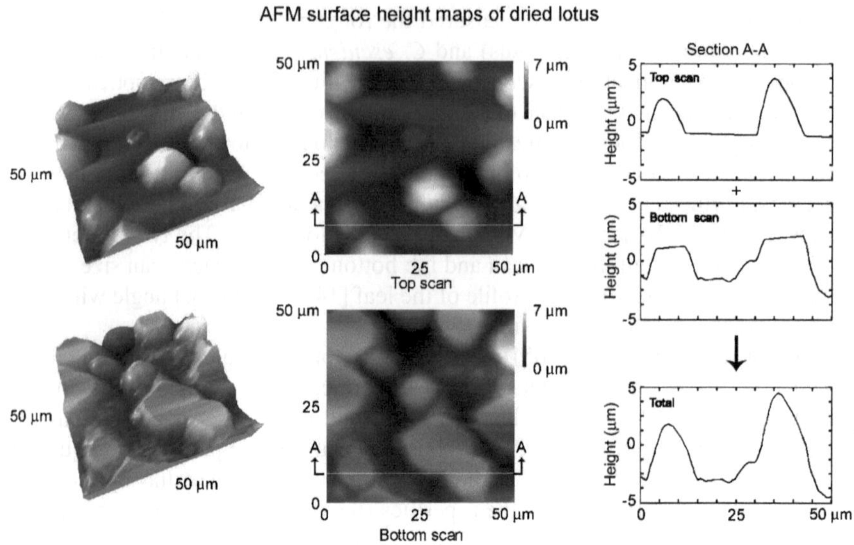

Fig. 21.16. Surface height map and a 2D profile of a lotus leaf using an atomic force microscope (*AFM*) [14]

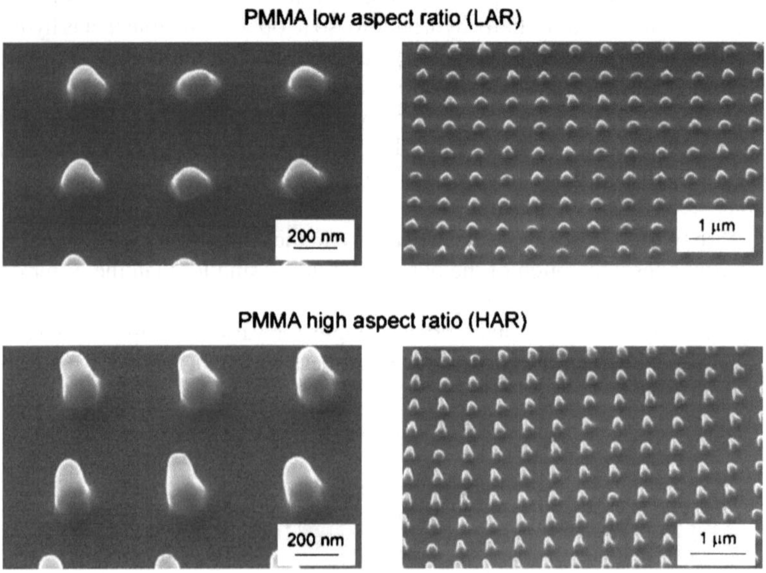

Fig. 21.17. SEM micrograph of the patterned poly(methyl methacrylate) (*PMMA*) surface. Both low aspect ratio (*LAR*) and high aspect ratio (*HAR*) are shown from two magnifications to see the asperity shape and the asperity pattern on the surface [13]

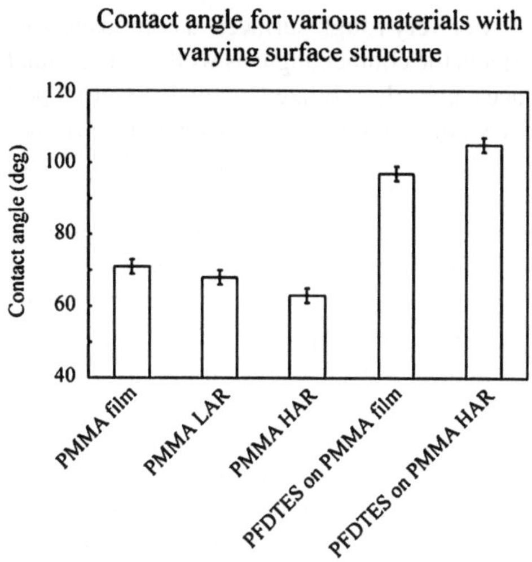

Fig. 21.18. The contact angle for different materials and for different roughness [13]

angle decreased with increased roughness for the hydrophilic PMMA film. For a hydrophobic surface, the model predicts an increase of contact angle with roughness, which is what happens when PMMA HAR is coated with PFDTES.

Sun et al. (2005) produced an artificial lotus leaf from PDMS; by nanocasting, making positive and negative replicas of an actual leaf. Flat PDMS has a contact angle with water of about 105°, comparable with that of wax (104° according to [18] and 103° according to [27]). They found a contact angle of 160° for the positive leaf replica and only 110° for the negative replica. The value for the positive replica is comparable with that for the lotus leaf. Since both positive and negative replicas have the same roughness factor but reversed asperities and valleys, this result suggests that air pockets at the valleys (composite interface) play a main role in increasing the contact angle for lotus leaf.

These experimental data are consistent with the model presented in the preceding sections.

21.6 Closure

Roughness affects the contact angle between a liquid and a solid surface in several ways, such as the increased area of the solid–liquid interface and the effect of sharp edges. For nonwetting liquids, the contact angle with a rough surface is greater than that with a flat surface and may approach 180° in the case of superhydrophobic surfaces. This effect is known for water-repellent leaves of plants, such as lotus, which have high roughness and high contact angles, and can be utilized for producing superhydrophobic surfaces. The effect of sharp edges is in pinning of the triple line and it may result in a difference between advancing and receding contact angles (contact angle hysteresis), which impedes fluid flow. This is not desirable in most applications, which require easy fluid flow.

For very rough surfaces, a composite solid–liquid–air interface can form, which affects the contact angle as well as the contact angle hysteresis. The composite interface can be destabilized owing to external perturbations, such as capillary waves. For a system with multiple states of equilibrium, a stochastic model has been developed, which assumes that a certain probability, depending on the net energy of the system, is associated with every state.

On the basis of an analysis of various modeled roughness distributions, a surface with roughness-induced superhydrophobic properties should satisfy the following requirements. First, asperities must have high aspect ratio to provide a large surface area. Second, sharp edges should be avoided, to prevent pinning of the triple line. Third, asperities should be tightly packed in to minimize the distance between them and to avoid destabilization of the composite interface. Fourth, asperities should be small compared with a typical droplet size. And fifth, in the case of a *hydrophilic* surface, a hydrophobic film must be applied in order to have initial $\theta > 90°$. These recommendations can be utilized for producing superhydrophobic surfaces. Remarkably, all these conditions are satisfied by biological water-repellent surfaces, such as leaves: they have tightly packed hemispherically topped papillae with high (on the order of unity) aspect ratios and a wax coating. Verification of the model was conducted using experimental data for the contact angle of a water droplet on a lotus leaf surface and artificial surfaces.

In addition, for wetting liquids, the effect of roughness on meniscus force was considered and it was found that roughness with typical asperity size greater than of the meniscus may decrease the meniscus force, whereas the nanoscale roughness, with typical asperity size smaller than that of the meniscus, may slightly increase the meniscus force, which is usually undesirable.

Experimental measurements on nanopatterned surfaces and replicas are consistent with the model.

References

1. Adamson AV (1990) Physical Chemistry of Surfaces, Wiley, New York
2. Barthlott W, Neinhuis C (1997) Purity of the Sacred Lotus, or Escape from Contamination in Biological Surfaces, Planta 202:1–8
3. Bhushan B (1998) Tribology Issues and Opportunities in MEMS, Kluwer, Dordrecht
4. Bhushan B (1999) Principles and Applications of Tribology, Wiley, New York
5. Bhushan B (2002) Introduction to Tribology, Wiley, New York
6. Bhushan B (2004) Springer Handbook of Nanotechnology, Springer, Berlin, Heidelberg, New York
7. Bhushan B (2005) Nanotribology and Nanomechanics – An Introduction, Springer, Berlin, Heidelberg, New York
8. Bhushan B, Jung YC (2006) Micro- and Nanoscale Characterization of Hydrophobic and Hydrophilic Leaf Surfaces. Nanotechnology 17:2758–2772
9. Bhushan B, Nosonovsky M (2003) Scale Effects in Friction Using Strain Gradient Plasticity and Dislocation-Assisted Sliding (Microslip), Acta Mater 51:4331–4345
10. Bhushan B, Nosonovsky M (2004) Scale Effects in Dry and Wet Friction, Wear, and Interface Temperature, Nanotechnology 15:749–761
11. Bhushan B, Israelachvili JN, Landman U (1995) Nanotribology: Friction, Wear and Lubrication at the Atomic Scale. Nature 374:607–616

12. Bico J, Thiele U, Quéré D (2002) Wetting of Textured Surfaces. Colloids Surf A 206:41–46
13. Burton Z, Bhushan B (2005) Hydrophobicity, Adhesion, and Friction Properties of Nanopatterned Polymers and Scale Dependence for Micro- and Nanoelectromechanical Systems. Nanoletters 5:1607–1613
14. Burton Z, Bhushan B (2006) Surface Characterization and Adhesion and Friction Properties of Hydrophobic Leaf Surfaces. Ultramicroscopy 106:709–719
15. Cassie A, Baxter S (1944) Wettability of Porous Surfaces. Trans Faraday Soc 40:546–551
16. Checco A, Guenoun P, Daillant J (2003) Nonlinear Dependence of the Contact Angle of Nanodroplets on Contact Line Curvatures. Phys Rev Lett 91:186101
17. Chen Y, He B, Lee J, Patankar NA (2005) Anisotropy in the Wetting of Rough Surfaces. J Colloid Interface Sci 281:458–465
18. Craig RG, Berry GC, Peyton FA (1960) Wetting of Poly-(Methyl Methacrylate) and Polystyrene by Water and Saliva. J Phys Chem 64:541–543
19. Eustathopoulos N, Nicholas MG, Drevet B (1999) Wettability at High Temparatures, Pergamon, Amsterdam
20. Erbil HY, Demirel AL, Avci Y (2003) Transformation of a Simple Plastic into a Superhydrophobic Surface. Science 299:1377–1380
21. Eyring H (1964) Statistical Mechanics and Dynamics, Wiley, New York
22. Extrand CW (2002) Model for Contact Angle and Hysteresis on Rough and Ultraphobic Surfaces. Langmuir 18:7991–7999
23. He B, Patankar NA, Lee J (2003) Multiple Equllibrium Droplet Shapes and Design Criterion for Rough Hydrophobic Surfaces. Langmuir 19:4999–5003
24. Herminghaus S (2000) Roughness-Induced Non-Wetting. Europhys Lett 52:165–170
25. Israelachvili JN (1992) Intermolecular and Surface Forces, 2nd edition, Academic, London
26. Johnson RE, Dettre RH (1964) Contact Angle Hysteresis, In: Fowkes FM (ed) Contact Angle, Wettability, and Adhesion. Advances in Chemistry Series 43. American Chemical Society Washington DC, p 112–135
27. Kamusewitz H, Possart W, Paul D (1999) The Relation Between Young's Equilibrium Contact Angle and the Hysteresis on Rough Paraffin Wax Surfaces. Colloids Surf A 156:271–279
28. Kijlstra J, Reihs K, Klamt A (2002) Roughness and Topology of Ultra-Hydrophobic surfaces. Colloids Surf A 206:521–529
29. Lafuma A, Quéré D (2003) Superhydrophobic States. *Nature Materials* 2:457–460
30. Landau L, Lifshitz E (1959) Fluid Mechanics, Pergamon, London
31. Marmur A (2003) Wetting on Hydrophobic Rough Surfaces: to Be Heterogeneous or Not to Be? Langmuir 19:8343–8348
32. Marmur A (2004) The Lotus Effect: Superhydrophobicity and Metastability, Langmuir 20:3517–3519
33. Miwa M, Nakajima A, Fujishima A, Hashimoto K, Watanabe T (2000) Effects of the Surface Roughness on Sliding Angles of Water Droplets on Superhydrophobic Surfaces. Langmuir 16:5754–5760
34. Neinhuis C, Barthlott W (1997) Characterization and Distribution of Water-Repellent, Self-Cleaning Plant Surfaces. Ann Bot 79:667–677
35. Nosonovsky M, Bhushan B (2005) Roughness Optimization for Biomimetic Superhydrophobic Surfaces. Microsyst Technol 11:535–549
36. Nosonovsky M, Bhushan B (2006a) Stochastic Model for Metastable Wetting of Roughness-Induced Superhydrophobic Surfaces. Microsyst Technol 12:231–237
37. Nosonovsky M, Bhushan B (2006b) Wetting of Rough Three-Dimensional Superhydrophobic Surfaces. Microsyst Technol 12:273–281
38. Nosonovsky M, Bhushan B (2006c) Hierarchical Roughness Makes Superhydrophobic States Stable. Microelectronic Engineering (in press)

39. Patankar NA (2003) On the Modeling of Hydrophobic Contact Angles on Rough Surfaces. Langmuir 19:1249–1253
40. Patankar NA (2004a) Transition Between Superhydrophobic States on Rough Surfaces. Langmuir 20:7097–7102
41. Patankar NA (2004b) Mimicking the Lotus Effect: Influence of Double Roughness Structures and Slender Pillars. Langmuir 20:8209–8213
42. Quéré D (2004) Surface Wetting: Model Droplets. Nate Mater 3:79–80
43. Satas D (ed) (1991) Coating Technology Handbook, Dekker, New York
44. Semal S, Blake TD, Geskin V, de Ruijter ML, Castelein G, De Coninck J (1999) Influence of Surface Roughness on Wetting Dynamics. Langmuir 15:8765–8770
45. Shibuichi S, Onda T, Satoh N, Tsujii K (1996) Super-Water-Repellent Surfaces Resulting from Fractal Structure. J Phys Chem 100:(1951)2–(1951)7
46. Shuttleworth R, Bailey GLJ (1948) The Spreading of a Liquid over a Rough Solid. Discuss Faraday Soc 3:16–22
47. Sun M, Luo C, Xu L, Ji H, Ouyang Q, Yu D, Chen Y (2005) Artificial Lotus Leaf by Nanocasting. Langmuir 21:8978–8981
48. Swain PS, Lipowsky R (1998) Contact Angles on Heterogeneous Surfaces: a New Look at Cassie's and Wenzel's Laws. Langmuir 14:6772–6780
49. Tian X, Bhushan B (1996) The Micro-Meniscus Effect of a Thin Liquid Film on the Static Friction of Rough Surface Contact. J Phys D: Appl Phys 29:163–178
50. Wagner P, Furstner R, Barthlott W, Neinhuis C (2003) Quantitative Assessment to the Structural Basis of Water Repellency in Natural and Technical Surfaces. J Exp Bot 54:1295–1303
51. Wenzel RN (1936) Resistance of Solid Surfaces to Wetting by Water. Ind Eng Chem 28:988–994
52. Whitehouse DJ, Archard JF (1970) The Properties of Random Surfaces of Significance in Their Contact. Proc R Soc Lond Ser A 316:97–121
53. Yost FG, Michael JR, Eisenmann ET (1995) Extensive Wetting due to Roughness. Acta Metall Mater 45:299–305
54. Zhou XB, De Hosson JTM (1995) Influence of Surface Roughness on the Wetting Angle. Acta Metall Mater 45:299–305

27 Gecko Feet: Natural Attachment Systems for Smart Adhesion—Mechanism, Modeling, and Development of Bio-Inspired Materials

Bharat Bhushan · Robert A. Sayer

From volume X

Abstract. Several creatures, including insects, spiders, and lizards, have developed a unique clinging ability that utilizes dry adhesion. Geckos, in particular, have developed the most complex adhesive structures capable of smart adhesion—the ability to cling to different smooth and rough surfaces and detach at will. These animals make use of about three million microscale hairs (setae) (about 14000/mm^2) that branch off into hundreds of nanoscale spatulae (about a billion spatulae). This hierarchical surface construction gives the gecko the adaptability to create a large real area of contact with surfaces. Modeling of the gecko attachment system as a hierarchical spring model has provided insight into adhesion enhancement generated by this system. van der Waals forces are the primary mechanism utilized to adhere to surfaces, and capillary forces are a secondary effect that can further increase adhesion force. Preload applied to the setae increases adhesion force. Although a gecko is capable of producing on the order of 20 N of adhesion force, it retains the ability to remove its feet from an attachment surface at will. The adhesion strength of gecko setae is dependent on the orientation; maximum adhesion occurs at 30°. During walking a gecko is able to peel its foot from surfaces by changing the angle at which its setae contact a surface. A man-made fibrillar structure capable of replicating gecko adhesion has the potential for use in dry, superadhesive tapes that would be of use in a wide range of applications. These adhesives could be created using microfabrication/nanofabrication techniques or self-assembly.

Key words: Gecko feet, Adhesion, Surface energy, Nanostructures, Robots

27.1 Introduction

Almost 2500 years ago, the ability of the gecko to "run up and down a tree in any way, even with the head downwards" was observed by Aristotle (1918). This phenomenon is not limited to geckos, but occurs in several animals and insects as well. This dynamic attachment ability will be referred to as reversible adhesion or smart adhesion (Bhushan et al. 2006). Many insects (e. g., flies and beetles) and spiders have been the subject of investigation. However, the attachment pads of geckos have been the most widely studied owing to the fact that they exhibit the most versatile and effective adhesive known in nature. As a result, the vast majority of this chapter will be concerned with gecko feet.

Although there are over 1000 species of geckos (Kluge 2001; Han et al. 2004) that have attachment pads of varying morphology (Ruibal and Ernst 1965), the Tokay gecko (*Gekko gecko*) has been the main focus of scientific research (Hiller 1968; Irschick et al. 1996; Autumn 2006). The Tokay gecko is the second-largest gecko species, attaining respective lengths of approximately 0.3–0.4 and 0.2–0.3 m for males and females. They have a distinctive blue or gray body with orange or red spots

In: H. Fuchs, B. Bhushan (eds), *Biosystems - Investigated by Scanning Probe Microscopy*,
DOI 10.1007/978-3-540-74085-8_1, © Springer-Verlag Berlin Heidelberg 2010

and can weigh up to 300 g (Tinkle 1992). These geckos have been the most widely investigated species of gecko owing to the availability and size of these creatures.

Even though the adhesive ability of geckos has been known since the time of Aristotle, little was understood about this phenomenon until the late nineteenth century when microscopic hairs covering the toes of the gecko were first noted. The development of electron microscopy in the 1950s enabled scientists to view a complex hierarchical morphology that covers the skin on the gecko's toes. Over the past century and a half, scientific studies have been conducted to determine the factors that allow the gecko to adhere to and detach from surfaces at will, including surface structure (Ruibal and Ernst 1965; Russell 1975, 1986; Williams and Peterson 1982; Schleich and Kästle 1986; Irschick et al. 1996; Autumn and Peattie 2002; Arzt et al. 2003), the mechanisms of adhesion (Wagler 1830; Simmermacher 1884; Schmidt 1904; Hora 1923; Dellit 1934; Ruibal and Ernst 1965; Hiller 1968; Gennaro 1969; Stork 1980; Autumn et al. 2000, 2002; Bergmann and Irschick 2005; Huber et al. 2005b), and adhesion strength (Hiller 1968; Irschick et al. 1996; Autumn et al. 2000; Arzt et al. 2003; Huber et al. 2005a,b). Recent work in modeling the gecko attachment system as a system of springs (Bhushan et al. 2006; Kim and Bhushan 2007a–d) has provided valuable insight into adhesion enhancement. van der Waals forces are widely accepted in the literature as the dominant adhesive mechanism utilized by hierarchical attachment systems. Capillary forces created by humidity naturally present in the air can further increase the adhesive force generated by the spatulae. Both experimental and theoretical work support these adhesive mechanisms.

There is great interest among the scientific community to further study the characteristics of gecko feet in the hope that this information could be applied to the production of microsurfaces/nanosurfaces capable of recreating the adhesion forces generated by these lizards (Bhushan, 2007b). Common man-made adhesives such as tape or glue involve the use of wet adhesives that permanently attach two surfaces. However, replication of the characteristics of gecko feet would enable the development of a superadhesive polymer tape capable of clean, dry adhesion (Geim et al. 2003; Sitti 2003; Sitti and Fearing 2003a; Northen and Turner 2005, 2006; Yurdumakan et al. 2005; Zhao et al. 2006; Bhushan 2007a, Bhushan and Sayer 2007; Gorb et al. 2007). These reusable adhesives have potential for use in everyday objects such as tapes, fasteners, and toys and in high technology such as microelectric and space applications. Replication of the dynamic climbing and peeling ability of geckos could find use in the treads of wall-climbing robots (Sitti and Fearing 2003b; Menon et al. 2004; Autumn et al. 2005).

27.2
Tokay Gecko

27.2.1
Construction of Tokay Gecko

The explanation for the adhesive properties of gecko feet can be found in the surface morphology of the skin on the toes of the gecko. The skin is composed of a complex hierarchical structure of lamellae, setae, branches, and spatulae (Ruibal and Ernst

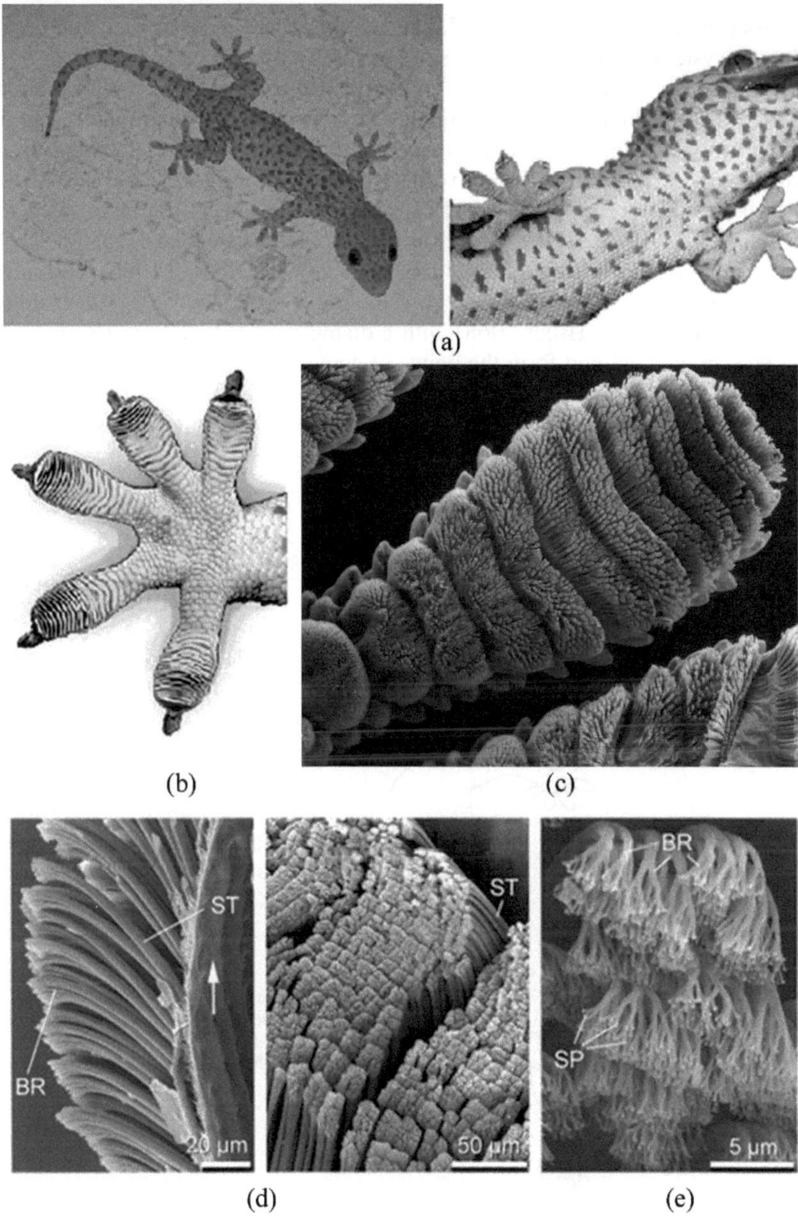

Fig. 27.1. **a** Tokay gecko (Autumn et al. 2000) The hierarchical structures of a gecko foot; **b** a gecko foot (Autumn et al. 2000) and **c** a gecko toe (Autumn 2006). Each toe contains hundreds of thousands of setae and each seta contains hundreds of spatulae. Scanning electron microscope (SEM) micrographs of **d** the setae (Gao et al. 2005) and **e** the spatulae (Gao et al. 2005). *ST* seta, *SP* spatula, *BR* branch

1965). As shown in Figs. 27.1 and 27.2 and summarized in Table 27.1, the gecko attachment system consists of an intricate hierarchy of structures beginning with lamellae, soft ridges that are 1–2 mm in length (Ruibal and Ernst 1965) that are located on the attachment pads (toes) that compress easily so that contact can be made with rough, bumpy surfaces. Tiny curved hairs known as setae extend from the lamellae with a density of approximately 14,000/mm^2 (Schleich and Kästle 1986). These setae are typically 30–130 μm in length and 5–10 μm in diameter (Ruibal and Ernst 1965; Hiller 1968; Russell 1975; Williams and Peterson 1982) and are composed primarily of β-keratin (Maderson 1964; Russell 1986) with some α-keratin components (Rizzo et al. 2006). At the end of each seta, 100–1000 spatulae (Ruibal and Ernst 1965; Hiller 1968) with a diameter of 0.1–0.2 μm (Ruibal and Ernst 1965) branch out and form the points of contact with the surface. The tips of the spatulae are approximately 0.2–0.3 μm in width (Ruibal and Ernst 1965), 0.5 μm in length, and 0.01 μm in thickness (Persson and Gorb 2003) and garner their name from their resemblance to a spatula.

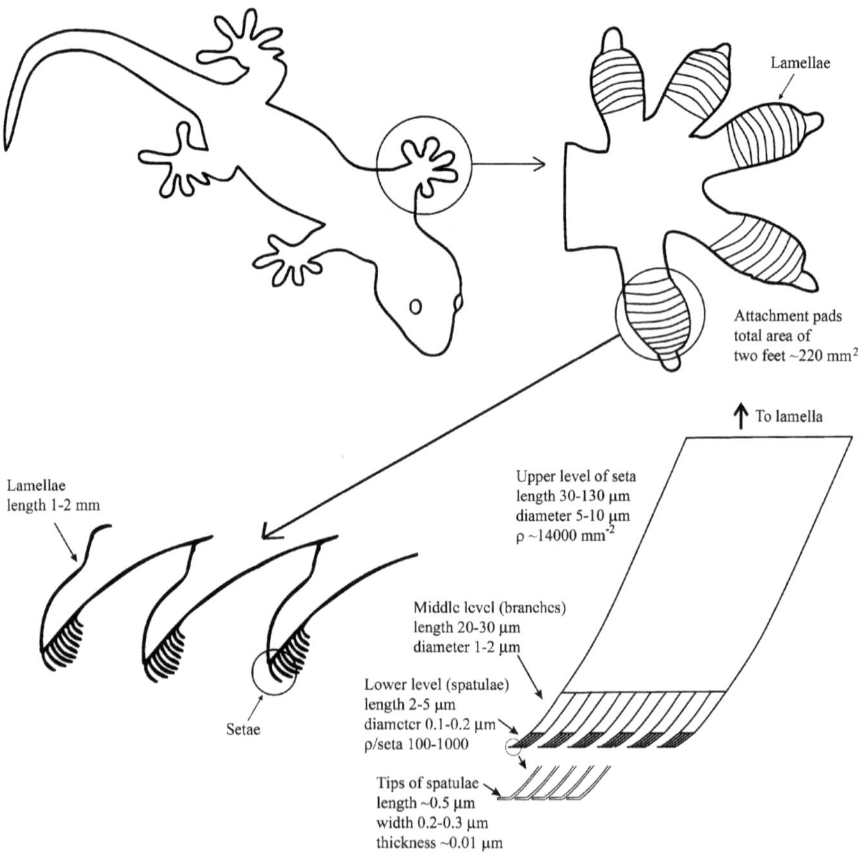

Fig. 27.2. A Tokay gecko including the overall body, one foot, a cross-sectional view of the lamellae, and an individual seta. ρ represents number of spatulae

Table 27.1. Surface characteristics of Tokay gecko feet

Component	Size	Density	Adhesive force
Seta	30–130[a–d]/5–10[a–d] length/diameter (μm)	~ 14000[f,g] setae/mm^2	194 μN[h] (in shear) ~ 20 μN[h] (normal)
Branch	20–30[a]/1–2[a] length/diameter (μm)	–	–
Spatula	2–5[a]/0.1–0.2[a,e] length/diameter (μm)	100–1000[c,d] spatulae per seta	–
Tip of spatula	~ 0.5[a,e]/0.2–0.3[a,d]/ ~ 0.01[e] length/width/thickness (μm)	– –	11 nN[i] (normal)

Young's modulus of surface material, keratin 1–20 GPa (Russell 1986; Bertram and Gosline 1987)
[a]Ruibal and Ernst (1965)
[b]Hiller (1968)
[c]Russell (1975)
[d]Williams and Peterson (1982)
[e]Persson and Gorb (2003)
[f]Schleich and Kästle (1986)
[g]Autumn and Peattie (2002)
[h]Autumn et al. (2000)
[i]Huber et al. (2005a)

The attachment pads on two feet of the Tokay gecko have an area of about 220 mm^2. About three million setae on their toes can produce a clinging ability of about 20 N [vertical force required to pull a lizard down a nearly vertical (85°) surface] (Irschick et al. 1996) and allow them to climb vertical surfaces at speeds over 1 m/s with capability to attach and detach their toes in milliseconds. In isolated setae, a 2.5-μN preload yielded adhesion of 20–40 μN and thus the adhesion coefficient, which represents the strength of adhesion as a function of preload, ranges from 8 to 16 (Autumn et al. 2002).

27.2.2 Other Attachment Systems

Attachment systems in other creatures such as insects and spiders have similar structures to that of gecko skin. The microstructures utilized by beetles, flies, spiders, and geckos can be seen in Fig. 27.3a. As the size (mass) of the creature increases, the radius of the terminal attachment elements decreases. This allows a greater number of setae to be packed into an area, hence increasing the linear dimension of contact and the adhesion strength. Arzt et al. (2003) determined that the density of the terminal attachment elements ρ_A per square meter strongly increases with increasing body mass m in kilograms. In fact, a master curve can be fit for all the different species (Fig. 27.3b):

$$\log \rho_A = 13.8 + 0.669 \log m \ . \tag{27.1}$$

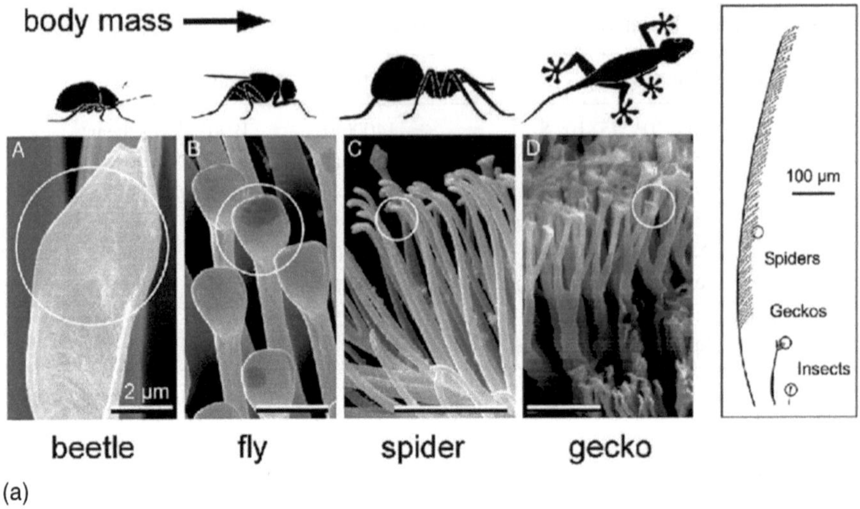

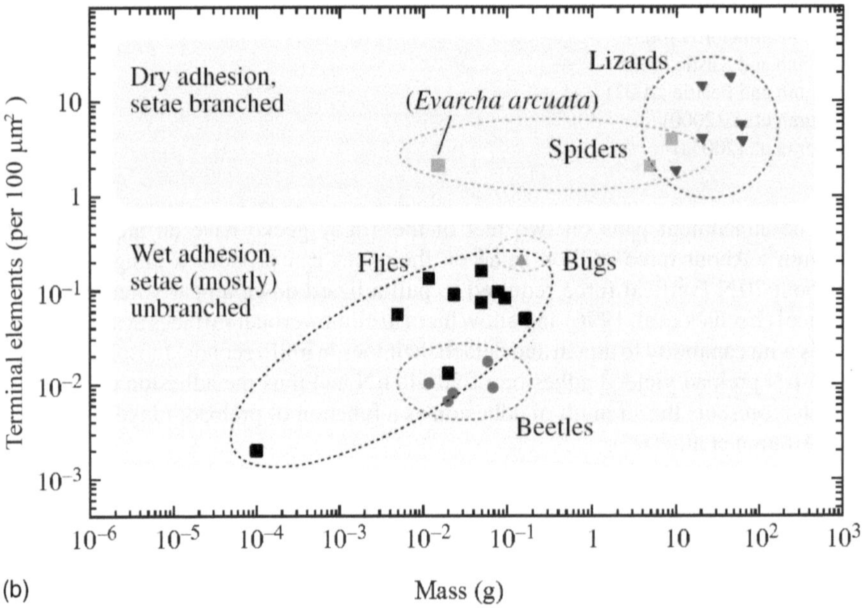

Fig. 27.3. a Terminal elements of the hairy attachment pads of a beetle, fly, spider, and gecko shown at two different scales (Arzt et al. 2003) and **b** the dependence of terminal element density on body mass (Federle 2006). (The data are from Artz et al. 2003 and Kesel et al. 2003)

The correlation coefficient of the master curve is equal to 0.919. Flies and beetles have the largest attachment pads and the lowest density of terminal attachment elements. Spiders have highly refined attachment elements that cover the leg. Geckos have both the highest body mass and the greatest density of terminal elements (spatulae).

Spiders and geckos can generate high dry adhesion, whereas beetles and flies increase adhesion by secreting liquid at the contacting surface.

27.2.3
Adaptation to Surface Roughness

Typical rough, rigid surfaces are only able to make intimate contact with a mating surface over a very small portion of the perceived apparent area of contact. In fact, the real area of contact is typically 2–6 orders of magnitude less than the apparent area of contact (Bhushan 2002, 2005). Autumn et al. (2002) proposed that divided contacts serve as a means for increasing adhesion. A surface energy approach can be used to calculate adhesion force in dry environments in order to calculate the effect of division of contacts. If the tip of a spatula is considered to be a hemisphere with radius R, the adhesion force of a single contact F_{ad} based on the Johnson–Kendall–Roberts (JKR) theory is given as (Johnson et al. 1971):

$$F_{ad} = \frac{3}{2}\pi W_{ad} R \,, \tag{27.2}$$

where W_{ad} is the work of adhesion (units of energy per unit area). Equation (27.2) shows that adhesion force of a single contact is proportional to a linear dimension of the contact. For a constant area divided into a large number of contacts or setae, n, the radius of a divided contact, R_1, is given by $R_1 = R/\sqrt{n}$ (self-similar scaling) (Arzt et al. 2003). Therefore, the adhesion force of (27.2) can be modified for multiple contacts such that

$$F'_{ad} = \frac{3}{2}\pi W_{ad}\left(\frac{R}{\sqrt{n}}\right) n = \sqrt{n} F_{ad} \,, \tag{27.3}$$

where F'_{ad} is the total adhesion force from the divided contacts. Thus, the total adhesion force is simply the adhesion force of a single contact multiplied by the square root of the number of contacts.

For a contact in the humid environment, the meniscus (or capillary) forces further increase the adhesion force (Bhushan 1999, 2002, 2005). The attractive meniscus force (F_m) consists of a contribution from both Laplace pressure and surface tension (Orr et al. 1975; Bhushan 2002). The contribution from Laplace pressure is directly proportional to the meniscus area. The other contribution is from the vertical component of surface tension around the circumference. This force is proportional to the circumference as is the case for the work of adhesion (Bhushan, 2002). Going through the analysis presented earlier, one can show that the contribution from the vertical component of surface tension increases as a surface is divided into a larger number of contacts. It increases linearly with the square root of the number of contacts n (self-similar scaling) (Bhushan 2007b; Kim and Bhushan 2007d):

$$\left(F'_m\right)_{\text{surface tension}} = \sqrt{n}\left(F_m\right)_{\text{surface tension}} \,, \tag{27.4}$$

where F'_m is the force from the divided contacts and F_m is the force of an undivided contact.

The models just presented only consider contact with a flat surface. On natural rough surfaces the compliance and adaptability of setae are the primary sources of high adhesion. Intuitively, the hierarchical structure of gecko setae allows for greater contact with a natural rough surface than a nonbranched attachment system. Modeling of the contact between gecko setae and rough surfaces is discussed in detail in Sect. 27.5.

Material properties also play an important role in adhesion. A soft material is able to achieve greater contact with a mating surface than a rigid material. Although, gecko skin is primarily comprised of β-keratin, a stiff material with a Young's modulus in the range of 1–20 GPa (Russell 1986; Bertman and Gosline 1987), the effective modulus of the setal arrays on gecko feet is about 100 kPa (Autumn et al. 2006), which is approximately 4 orders of magnitude lower than for the bulk material. Nature has selected a relatively stiff material to avoid clinging to adjacent setae. Division of contacts, as discussed earlier, provides high adhesion. By combining optimal surface structure and material properties, mother nature has created an evolutionary superadhesive.

27.2.4 Peeling

Although geckos are capable of producing large adhesion forces, they retain the ability to remove their feet from an attachment surface at will by peeling action. The orientation of the spatulae facilitates peeling. Autumn et al. (2000) were the first to experimentally show that the adhesion force of gecko setae is dependent on the three-dimensional orientation as well as the preload applied during attachment (see Sect. 27.4.1.1). Owing to this fact, geckos have developed a complex foot motion during walking. First the toes are carefully uncurled during attachment. The maximum adhesion occurs at an attachment angle of 30°—the angle between a seta and the mating surface. The gecko is then able to peel its foot from surfaces one row of setae at a time by changing the angle at which its setae contact a surface. At an attachment angle greater than 30° the gecko will detach from the surface.

Shah and Sitti (2004) determined the theoretical preload required for adhesion as well as the adhesion force generated for setal orientations of 30°, 40°, 50°, and 60°. We consider a solid material (elastic modulus E, Poisson's ratio ν) in contact with the rough surface whose profile is given by

$$f(x) = H \sin^2\left(\frac{\pi x}{\chi}\right) , \tag{27.5}$$

where H is the amplitude and χ is the wavelength of the roughness profile. For a solid adhesive block to achieve intimate contact with the rough surface neglecting surface forces, it is necessary to apply a compressive stress, σ_c (Jagota and Bennison 2002):

$$\sigma_c = \frac{\pi E H}{2\chi\left(1-\nu^2\right)} . \tag{27.6}$$

Equation (27.6) can be modified to account for fibers oriented at an angle θ. The preload required for contact is summarized in Fig. 27.4a. As the orientation angle

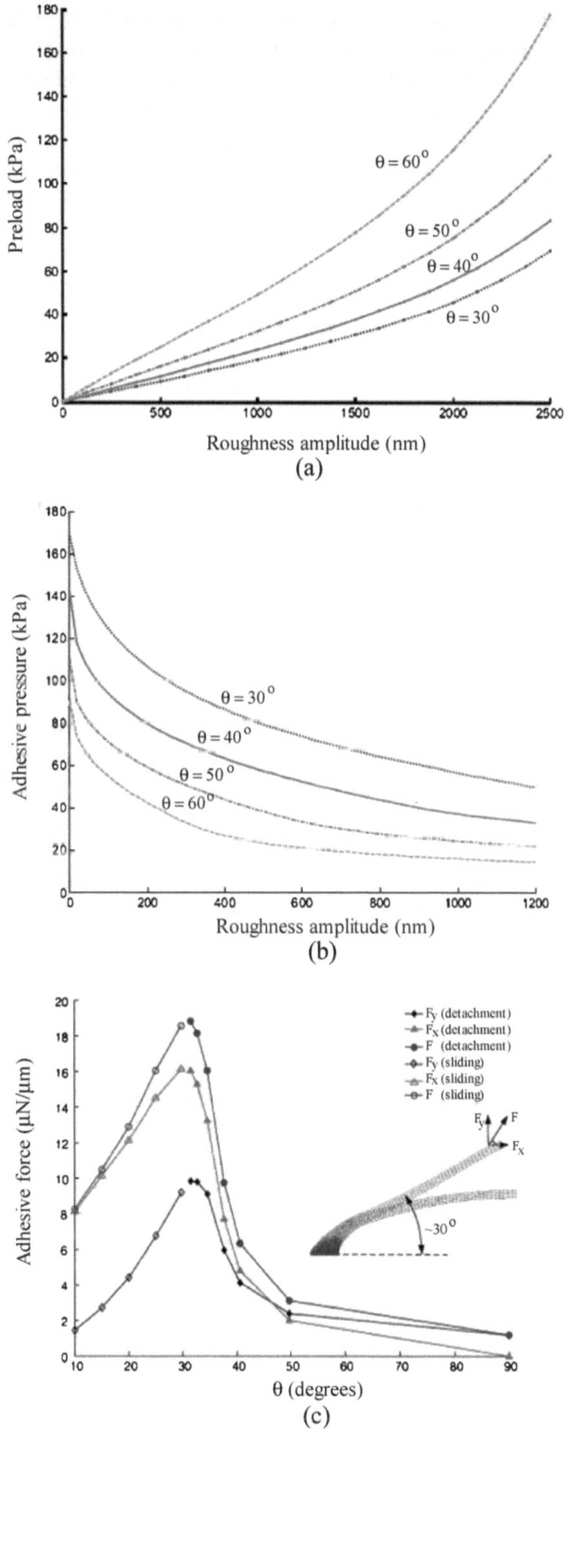

Fig. 27.4. Contact mechanics results for the affect of fiber orientation on **a** preload and **b** adhesive force for roughness amplitudes ranging from 0 to 2500 nm (Shah and Sitti 2004). **c** Finite-element analysis of the adhesive force of a single seta as a function of pull direction (Gao et al. 2005)

decreases, so does the required preload. Similarly, adhesion strength is influenced by fiber orientation. As seen in Fig. 27.4b, the greatest adhesion force occurs at $\theta = 30°$.

Gao et al. (2005) created a finite-element model of a single gecko seta in contact with a surface. A tensile force was applied to the seta at various angles, θ, as shown in Fig. 27.4c. For forces applied at an angle less than 30°, the dominant failure mode was sliding. In contrast, the dominant failure mode for forces applied at angles greater than 30° was detachment. This verifies the results of Autumn et al. (2000) that detachment occurs at attachment angles greater than 30°.

Tian et al. (2006) have suggested that during detachment, the angular dependence of adhesion and that of friction play a role. The pulling force of a spatula along its shaft with an angle between 0 and 90° to the substrate has a normal adhesive force produced at the spatula–substrate bifurcation zone, and a lateral friction force contribution from the part of the spatula still in contact with the substrate. High net friction and adhesion forces on the whole gecko are obtained by rolling down and gripping the toes inward to realize small pulling angles of the large number of spatulae in contact with the substrate. To detach, the high adhesion/friction is rapidly reduced to a very low value by rolling the toes upward and downward, which, mediated by the lever function of the setal shaft, peels the spatula off perpendicularly from the substrate.

27.2.5 Self-Cleaning

Natural contaminants (dirt and dust) as well as man-made pollutants are unavoidable and have the potential to interfere with the clinging ability of geckos. Particles found in the air consist of particulates that are typically less than 10 μm in diameter, while those found on the ground can often be larger (Hinds 1982; Jaenicke 1998). Intuitively, it seems that the great adhesion strength of gecko feet would cause dust and other particles to become trapped in the spatulae and that they would have no way of being removed without some sort of manual cleaning action on behalf of the gecko. However, geckos are not known to groom their feet like beetles (Stork 1983) nor do they secrete sticky fluids to remove adhering particles like ants (Federle et al. 2002) and tree frogs (Hanna and Barnes 1991), yet they retain adhesive properties. One potential source of cleaning is during the time when the lizards undergo molting, or the shedding of the superficial layer of epidermal cells. However, this process only occurs approximately once per month (Van der Kloot 1992). If molting were the sole source of cleaning, the gecko would rapidly lose its adhesive properties as it is exposed to contaminants in nature (Hansen and Autumn 2005).

Hansen and Autumn (2005) tested the hypothesis that gecko setae become cleaner with repeated use—a phenomenon known as self-cleaning. The cleaning ability of gecko feet was first tested experimentally by applying 2.5-μm-radius silica–alumina ceramic microspheres to clean setal arrays. It was found that a significant fraction of the particles was removed from the setal arrays with each step taken by the gecko.

In order to understand this cleaning process, substrate–particle interactions must be examined. The interaction energy between a dust particle and a wall and spatulae can be modeled as shown in Fig. 27.5. The interaction between a spherical dust

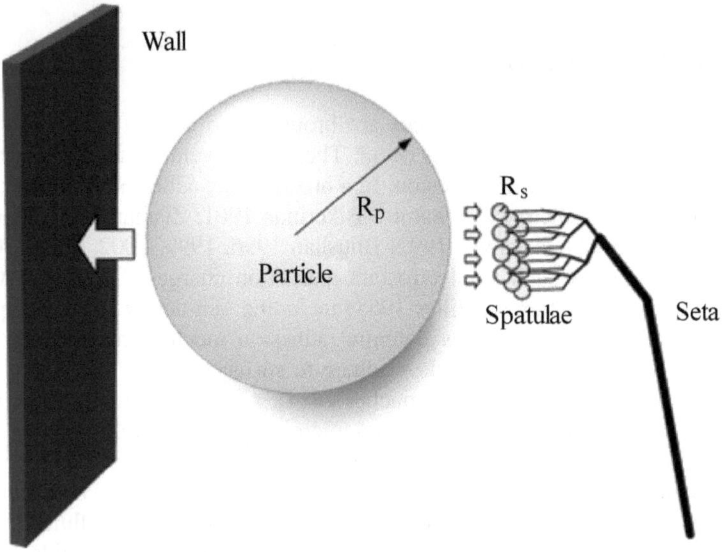

Fig. 27.5. Model of interactions between gecko spatulae of radius R_s, a spherical dirt particle of radius R_p, and a planar wall that enable self-cleaning (Hansen and Autumn 2005)

particle and the wall, W_{pw}, can be expressed as (Israelachvili 1992)

$$W_{pw} = \frac{-H_{pw} R_p}{6D_{pw}} , \tag{27.7}$$

where p and w refer to the particle and wall, respectively. H is the Hamaker constant, R_p is the radius of the particle, and D_{pw} is the separation distance between the particle and the wall. Similarly, the interaction energy between a spherical dust particle and a spatula, s, assuming that the spatula tip is spherical is (Israelachvili 1992)

$$W_{ps} = \frac{-H_{ps} R_p R_s}{6D_{ps}(R_p + R_s)} . \tag{27.8}$$

The ratio of the two interaction energies, N, can be expressed as

$$N = \frac{W_{pw}}{W_{ps}} = \left(1 + \frac{R_p}{R_s}\right) \frac{H_{pw} D_{ps}}{H_{ps} D_{pw}} . \tag{27.9}$$

When the energy required to separate a particle from the wall is greater than that required to separate it from a spatula, self-cleaning will occur. For small contaminants ($R_p < 0.5\,\mu m$), there are not enough spatulae available to adhere to the particle. For larger contaminants, the curvature of the particle makes it impossible for enough spatulae to adhere to it. As a result, Hansen and Autumn (2005) concluded that self-cleaning should occur for all spherical spatulae interacting with all spherical particles.

27.3 Attachment Mechanisms

When asperities of two solid surfaces are brought into contact with each other, chemical and/or physical attractions occur. The force developed that holds the two surfaces together is known as adhesion. In a broad sense, adhesion is considered to be either physical or chemical in nature (Bikerman 1961; Zisman 1963; Houwink and Salomon 1967; Israelachvili 1992; Bhushan 1996, 1999, 2002, 2005, 2007a). Chemical interactions such as electrostatic attraction charges (Schmidt 1904) as well as intermolecular forces (Hiller 1968) including van der Waals and capillary forces have all been proposed as potential adhesion mechanisms in gecko feet. Others have hypothesized that geckos adhere to surfaces through the secretion of sticky fluids (Wagler 1830; Simmermacher 1884), suction (Simmermacher 1884), increased frictional force (Hora 1923), and microinterlocking (Dellit 1934).

Through experimental testing and observations conducted over the last century and a half many potential adhesive mechanisms have been eliminated. Observation has shown that geckos lack glands capable of producing sticky fluids (Wagler 1830; Simmermacher 1884), thus ruling out the secretion of sticky fluids as a potential adhesive mechanism. Furthermore, geckos are able to create large adhesive forces normal to a surface. Since friction only acts parallel to a surface, the attachment mechanism of increased frictional force has been ruled out. Dellit (1934) experimentally ruled out suction and electrostatic attraction as potential adhesive mechanisms. Experiments carried out in a vacuum did not show a difference between the adhesion force at low pressures and that in ambient conditions. Since adhesive forces generated during suction are based on pressure differentials, which are insignificant under vacuum, suction was rejected as an adhesive mechanism (Dellit 1934). Additional testing utilized X-ray bombardment to create ionized air in which electrostatic attraction charges would be eliminated. It was determined that geckos were still able to adhere to surfaces in these conditions and, therefore, electrostatic charges could not be the sole cause of attraction (Dellit 1934). Autumn et al. (2000) demonstrated the ability of a gecko to generate large adhesive forces when in contact with a molecularly smooth SiO_2 microelectromechanical semiconductor. Since surface roughness is necessary for microinterlocking to occur, it has been ruled out as a mechanism of adhesion. Two mechanisms, van der Waals forces and capillary forces, remain as the potential sources of gecko adhesion. These attachment mechanisms are described in detail in the following sections.

27.3.1 Van der Waals Forces

Van der Waals bonds are secondary bonds that are weak in comparison with other physical bonds such as covalent, hydrogen, ionic, and metallic bonds. Unlike other physical bonds, van der Waals forces are always present regardless of separation and are effective from very large separations (approximately 50 nm) down to atomic separation (approximately 0.3 nm). The van der Waals force per unit area between two parallel surfaces, f_{vdW}, is given by (Hamaker 1937; Israelachvili and Tabor 1972; Israelachvili 1992)

$$f_{\mathrm{vdW}} = \frac{H}{6\pi D^3} \quad \text{for } D < 30\,\text{nm}\,, \tag{27.10}$$

where H is the Hamaker constant and D is the separation between surfaces.

Hiller (1968) showed experimentally that the surface energy of a substrate is responsible for gecko adhesion. One potential adhesive mechanism would then be van der Waals forces (Stork 1980; Autumn et al. 2000). Assuming van der Waals forces to be the dominant adhesive mechanism utilized by geckos, one can claculate the adhesion force of a gecko. Typical values of the Hamaker constant range from 4×10^{-20} to 4×10^{-19} J (Israelachvili 1992). In calculation, the Hamaker constant is assumed to be 10^{-19} J, the surface area of a spatula is taken to be $2 \times 10^{-14}\,\text{m}^2$ (Ruibal and Ernst 1965; Williams and Peterson 1982; Autumn and Peattie 2002), and the separation between the spatula and the contact surface is estimated to be 0.6 nm. This equation yields the force of a single spatula to be about 0.5 μN. By applying the surface characteristics of Table 27.1, the maximum adhesive force of a gecko is 150–1500 N for varying spatula density of 100–1000 spatulae per seta. If an average value of 550 spatulae per seta is used, the adhesion force of a single seta is approximately 270 μN, which is in agreement with the experimental value obtained by Autumn et al. (2000), which will be discussed in Sect. 27.4.1.1.

Another approach to calculate adhesive force is to assume that spatulae are cylinders that terminate in hemispherical tips. By using (27.2) and assuming that the radius of each spatula is about 100 nm and that the surface energy is expected to be $50\,\text{mJ/m}^2$ (Arzt et al. 2003), one predicts the the adhesion force of a single spatula to be 0.02 μN. This result is an order of magnitude lower than the first approach calculated for the higher value of A. For a lower value of 10^{-20} J for the Hamaker constant, the adhesion force of a single spatula is comparable to that obtained using the surface-energy approach.

Several experimental results favor van der Waals forces as the dominant adhesive mechanism, including temperature testing (Bergman and Irschick 2005) and adhesion force measurements of a gecko seta with both hydrophilic and hydrophobic surfaces (Autumn et al. 2000). These data will be presented in the Sects. 27.4.2–27.4.4.

27.3.2 Capillary Forces

It has been hypothesized that capillary forces that arise from liquid-mediated contact could be a contributing or even the dominant adhesive mechanism utilized by gecko spatulae (Hiller 1968; Stork 1980). Experimental adhesion measurements (presented in Sects. 27.4.3, 27.4.4) conducted on surfaces with different hydrophobicities and at various humidities (Huber et al. 2005b) as well as numerical simulations (Kim and Bhushan 2007d) support this hypothesis as a contributing mechanism. During contact, any liquid that wets or has a small contact angle on surfaces will condense from vapor in the form of an annular-shaped capillary condensate. Owing to the natural humidity present in the air, water vapor will condense to liquid on the surface of bulk materials. During contact this will cause the formation of adhesive bridges (menisci) owing to the proximity of the two surfaces and the affinity of the surfaces for condensing liquid. (Zimon 1969; Fan and O'Brien 1975; Phipps and Rice 1979).

Capillary force can be divided into two components: the Laplace force F_L and the surface tension force F_s such that the total capillary force F_c is given by the sum of the components:

$$F_\mathrm{c} = F_\mathrm{L} + F_\mathrm{s} \,. \tag{27.11}$$

The Laplace force is caused by the pressure difference across the interface of a curved liquid surface (Fig. 27.6) and depends on pressure difference multiplied by the meniscus area, which can be expressed as (Orr et al. 1975)

$$F_L = -\pi\kappa\gamma R^2 \sin^2\phi \,, \tag{27.12}$$

where γ is the surface tension of the liquid, R is the tip radius, ϕ is the filling angle and κ is the mean curvature of the meniscus. From the Kelvin equation (Israelachvili 1992), which is the thermal equilibrium relation, the mean curvature of the meniscus can be determined as

$$\kappa = \frac{\mathcal{R}T}{V\gamma} \ln\left(\frac{p}{p_\mathrm{o}}\right) , \tag{27.13}$$

where $\mathcal{R}$ is the universal gas constant, T is the absolute temperature, V is the molecular volume, p_o is the saturated vapor pressure of the liquid at T, and p is the ambient pressure acting outside the curved surface (p/p_o is the relative humidity). Orr et al. (1975) formulated the mean curvature of the meniscus between a sphere and a plane in terms of elliptic integrals. The filling angle ϕ can be calculated from the expression just mentioned and (27.13) using an iteration method. Then the Laplace force is calculated at a given environment using (27.12).

The surface tension of the liquid results in the formation of a curved liquid–air interface. The surface tension force acting on the sphere is (Orr et al. 1975)

$$F_\mathrm{s} = 2\pi R\gamma \sin\phi \sin(\theta_1 + \phi) \,. \tag{27.14}$$

Owing to the fact that surface tension force depends on the radius, division would result in an increase of the surface tension force by the square root of N_c upon division (Bhushan et al. 2007b; Kim and Bhushan 2007d).

Hence, the total capillary force on the sphere is

$$F_\mathrm{c} = \pi R\gamma \left[2\sin\phi \sin(\theta_1 + \phi) - \kappa R \sin^2\phi\right] \,. \tag{27.15}$$

27.4 Experimental Adhesion Test Techniques and Data

Experimental measurements of the adhesion force of a single gecko seta (Autumn et al. 2000) and single gecko spatula (Huber et al. 2005a, 2005b) have been made.

The effect of the environment, including temperature (Losos 1990; Bergmann and Irschick 2005) and humidity (Huber et al. 2005b), has been studied. Some of the data have been used to understand the adhesion mechanism utilized by the gecko attachment system—van der Waals or capillary forces. The majority of experimental results point towards van der Waals forces as the dominant mechanism of adhesion (Autumn et al. 2000; Bergmann and Irschick 2005). Recent research suggests that capillary forces can be a contributing adhesive factor (Huber et al. 2005b).

27.4.1 Adhesion Under Ambient Conditions

Two feet of a Tokay gecko are capable of producing about 20 N of adhesion force with a pad area of about 220 mm^2 (Irschick et al. 1996). Assuming that there are about 14,000 setae per square millimeter, the adhesion force from a single hair should be approximately 7 μN. It is likely that the magnitude is actually greater than this value because it is unlikely that all setae are in contact with the mating surface (Autumn et al. 2000). Setal orientation greatly influences adhesive strength. This dependency was first noted by Autumn et al. (2000). It was determined that the greatest adhesion occurs at 30°. In order to determine the adhesion mechanism(s) utilized by gecko feet, it is important to know the adhesion force of a single seta. Hence, the adhesion force of gecko foot-hair has been the focus of several investigations (Autumn et al. 2000; Huber et al. 2005a).

27.4.1.1 Adhesion Force of a Single Seta

Autumn et al. (2000) used both a microelectromechanical force sensor and a wire as a force gauge to determine the adhesion force of a single seta. The microelectromechanical force sensor is a dual-axis atomic force microscope (AFM) cantilever with independent piezoresistive sensors, which allows simultaneous detection of

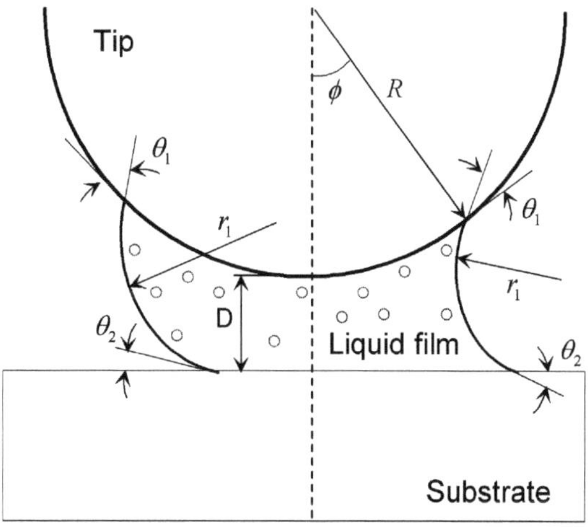

Fig. 27.6. A sphere on a plane at distance D with a liquid film in-between, forming menisci. R is the tip radius, ϕ is the filling angle, θ_1 and θ_2 are contact angles on the sphere and the plane respectively, and r_1 and r_2 are the two principal radii of the curved surface (Kim and Bhushan 2007d)

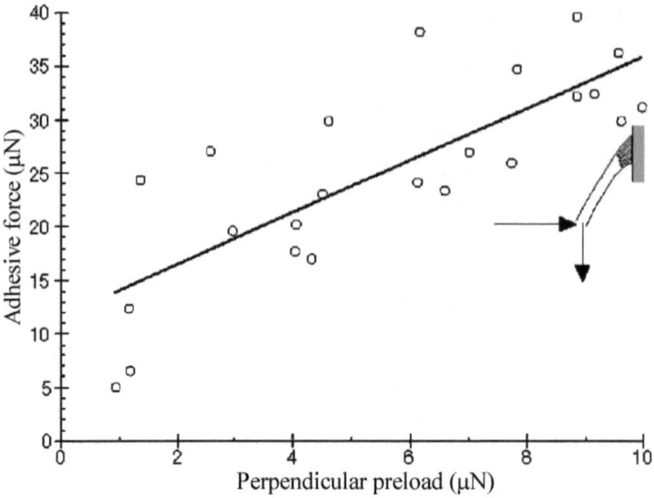

Fig. 27.7. Adhesive force of a single gecko seta as a function of applied preload. The seta was first pushed perpendicularly against the surface and then pulled parallel to the surface (Autumn et al. 2000)

vertical and lateral forces (Chui et al. 1998). The wire force gauge consisted of an aluminum bonding wire that displaced under a perpendicular pull. Autumn et al. (2000) discovered that setal force actually depends on the three-dimensional orientation of the seta as well as the preloading force applied during initial contact. Setae that were preloaded vertically to the surface exhibited only one tenth of the adhesive force ($0.6 \pm 0.7\,\mu$N) compared with setae that were pushed vertically and then pulled horizontally to the surface ($13.6 \pm 2.6\,\mu$N). The dependence of adhesion force of a single gecko spatula on perpendicular preload is illustrated in Fig. 27.7. The adhesion force increases linearly with the preload, as expected (Bhushan 1996, 1999, 2002). The maximum adhesion force of a single gecko foot-hair occurred when the seta was first subjected to a normal preload and then slid 5 μm along the contacting surface. Under these conditions, adhesion force measured $194 \pm 25\,\mu$N (approximately 10 atm adhesive pressure).

27.4.1.2
Adhesion Force of a Single Spatula

Huber et al. (2005a) used atomic force microscopy to determine the adhesion force of individual gecko spatulae. A seta with four spatulae was glued to an AFM tip. The seta was then brought in contact with a surface and a compressive preload of 90 nN was applied. The force required to pull the seta off the surface was then measured. As seen in Fig. 27.8, there are two distinct peaks in the graph—one at 10 nN and the other at 20 nN. The first peak corresponds to one of the four spatulae adhering to the contact surface, while the peak at 20 nN corresponds to two of the four spatulae adhering to the contact surface. The average adhesion force of a single spatula was

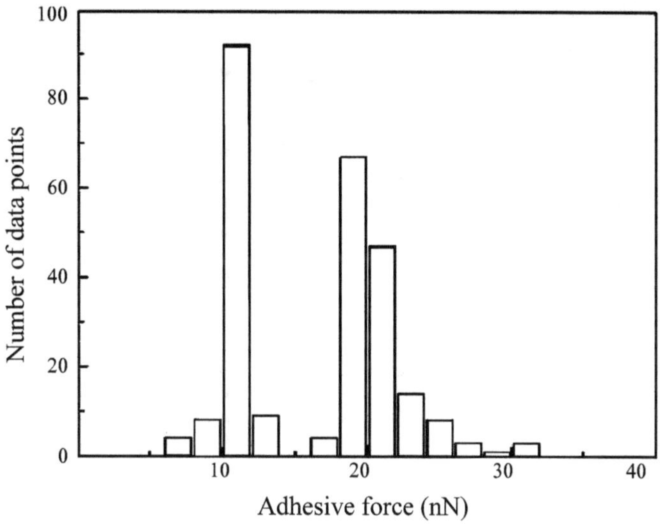

Fig. 27.8. Adhesive force of a single gecko spatula. The peak at 10 nN corresponds to the adhesive force of one spatula and the peak at 20 nN corresponds to the adhesive force of two spatulae (Huber et al. 2005a)

found to be 10.8 ± 1 nN. The measured value is in agreement with the measured adhesive strength of an entire gecko (on the order of 10^9 spatulae on a gecko).

27.4.2 Effects of Temperature

Environmental factors are known to affect several aspects of vertebrate function, including speed of locomotion, digestion rate, and muscle contraction, and as a result several studies have been undertaken to investigate environmental impact on these functions. Relationships between the environment and other properties such as adhesion have been far less studied (Bergmann and Irschick 2005). Only two known studies exist that examine the affect of temperature on the clinging force of the gecko (Losos 1990; Bergmann and Irschick 2005). Losos (1990) examined the adhesion ability of large live geckos at temperatures up to 17 °C. Bergmann and Irschick (2005) expanded upon this research for body temperatures ranging from 15 to 35 °C. The geckos were incubated until their body temperature reached a desired level. The clinging ability of these animals was then determined by measuring the maximum exerted force by the geckos as they were pulled off a custom-built force plate. The clinging force of a gecko for the experimental test range is plotted in Fig. 27.9. It was determined that variation in temperature is not statistically significant for the adhesion force of a gecko. From these results, it was concluded that the temperature independence of adhesion supports the hypothesis of clinging as a passive mechanism (i. e., van der Waals forces). Both studies only measured overall clinging ability on the macroscale. There have not been any investigations into effects of temperature on the clinging ability of a single seta on the microscale and therefore testing in this area would be extremely important.

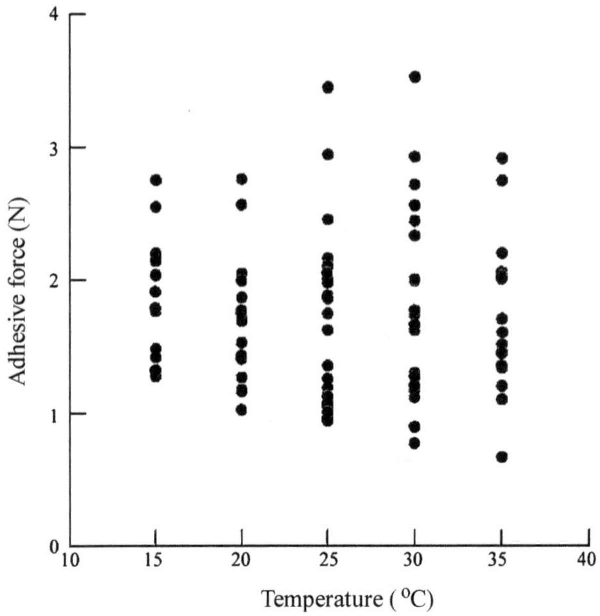

Fig. 27.9. Adhesive force of a gecko as a function of temperature (Bergmann and Irschick 2005)

27.4.3
Effects of Humidity

Huber et al. (2005b) employed similar methods to Huber et al. (2005a) (discussed previously in Sect. 27.4.1.2) in order to determine the adhesion force of a single spatula at varying humidity. Measurements were made using an AFM placed in an airtight chamber. The humidity was adjusted by varying the flow rate of dry nitrogen into the chamber. The air was continuously monitored with a commercially available hygrometer. All tests were conducted at ambient temperature.

As seen in Fig. 27.10, even at low humidity, adhesion force is large. An increase in humidity further increases the overall adhesion force of a gecko spatula. The pull-off force roughly doubled as the humidity was increased from 1.5 to 60%. This humidity effect can be explained in two possible ways: (1) by standard capillarity or (2) by a change of the effective short-range interaction due to absorbed monolayers of water—in other words, the water molecules increase the number of van der Waals bonds that are made. On the basis of these data, van der Waals forces are the primary adhesion mechanism and capillary forces are a secondary adhesion mechanism.

27.4.4
Effects of Hydrophobicity

To further test the hypothesis that capillary forces play a role in gecko adhesion, the spatular pull-off force was determined for contact with both hydrophilic and hydrophobic surfaces. As seen in Fig. 27.11a, the capillary adhesion theory predicts

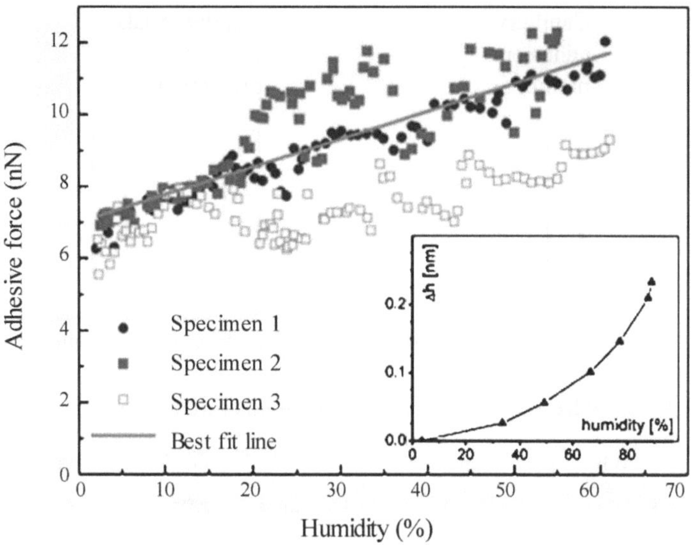

Fig. 27.10. Humidity effects on spatular pull-off force. *Inset*: The increase in water film thickness on a Si wafer with increasing humidity (Huber et al. 2005b)

that a gecko spatula will generate a greater adhesion force when in contact with a hydrophilic surface compared with a hydrophobic surface, while the van der Waals adhesion theory predicts that the adhesion force between a gecko spatula and a surface will be the same regardless of the hydrophobicity of the surface (Autumn et al. 2002). Figure 27.11b shows the adhesion pressure of a whole gecko and the adhesion force of a single seta on hydrophilic and hydrophobic surfaces. The data show that the values of the adhesion force are the same for both surfaces. This supports the van der Waals prediction of Fig. 27.11a. Huber et al. (2005b) found that the hydrophobicity of the attachment surface had an effect on the adhesion force of a single gecko spatula, as shown in Fig. 27.11c. These results show that adhesion force has a finite value for a superhydrophobic surface and increases as the surface becomes hydrophilic. It is concluded that van der Waals forces are the primary mechanism and capillary forces further increase the adhesion force generated.

27.5 Adhesion Modeling

With regard to the natural living conditions of the animals, the mechanics of gecko attachment can be separated into two parts: the mechanics of adhesion of a single contact with a flat surface, and an adaptation of a large number of spatulae to a natural, rough surface. Modeling of the mechanics of adhesion of spatulae to a smooth surface was developed by Autumn et al. (2002), Jagota and Bennison (2002), and Arzt et al. (2003). As discussed previously in Sect. 27.2.3, the adhesion force of multiple contacts F'_{ad} can be increased by dividing the contact into a large

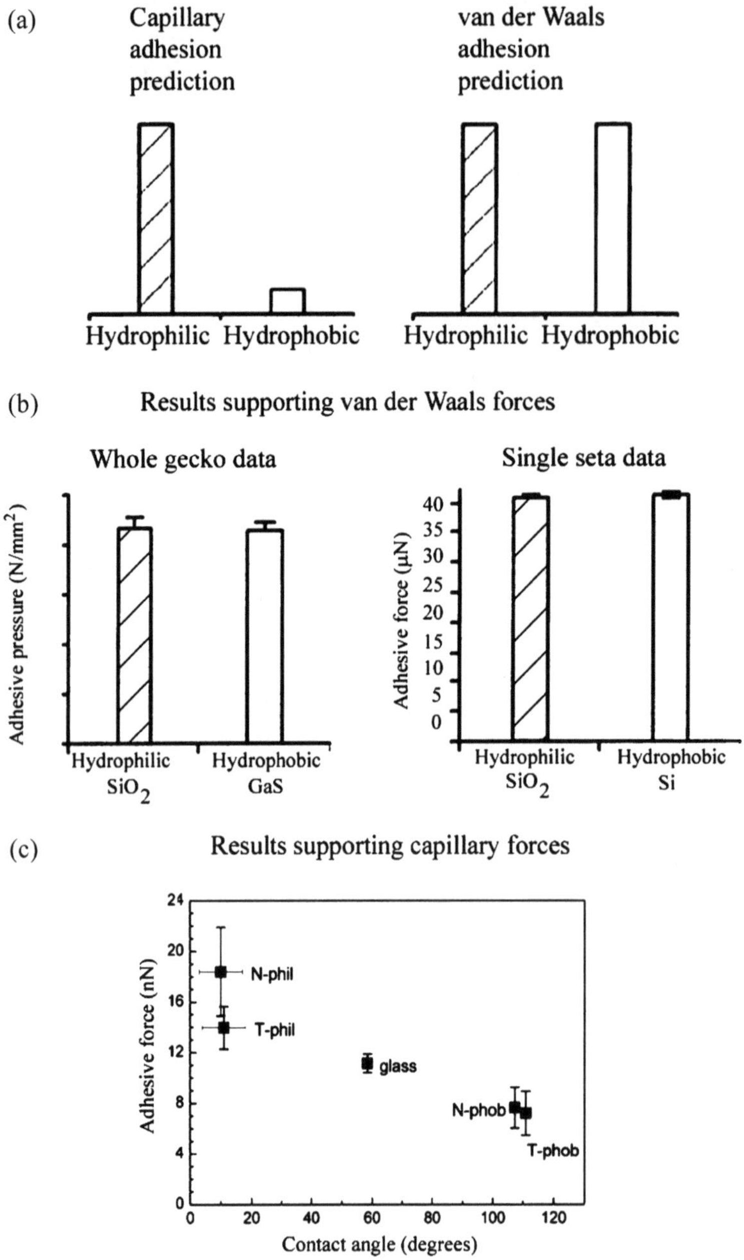

Fig. 27.11. **a** Capillary and van der Waals adhesion predictions for the relative magnitude of the adhesive force of gecko setae with hydrophilic and hydrophobic surfaces (Autumn et al. 2002). **b** Results of adhesion testing for a whole gecko and single seta with hydrophilic and hydrophobic surfaces (Autumn et al. 2002) and **c** results of adhesive force testing with surfaces with different contact angles (Huber et al. 2005b)

number (n) of small contacts, while the nominal area of the contact remains the same, $F'_{ad} \sim \sqrt{n} F_{ad}$. However, this model only considers contact with a flat surface. On natural, rough surfaces, the compliance and adaptability of setae are the primary sources of high adhesion. Intuitively, the hierarchical structure of gecko setae allows for a greater contact with a natural, rough surface than a nonbranched attachment system (Sitti and Fearing 2003a).

27.5.1 Spring Model

Bhushan et al. (2006) and Kim and Bhushan (2007a–d) have recently approximated a gecko seta in contact with random rough surfaces using a hierarchical spring model. Each level of springs in their model corresponds to a level of seta hierarchy. The upper level of springs corresponds to the thicker part of gecko seta, the middle spring level corresponds to the branches, and the lower level of springs corresponds to the spatulae. The upper level is the thickest branch of the seta. It is 75 μm in length and 5 μm in diameter. The middle level, referred to as a branch, has a length of 25 μm and a diameter of 1 μm. The lower level, called a spatula, is the thinnest branch, with a length of 2.5 μm and a diameter of about 0.1 μm (Table 27.2). Autumn et al. (2000) showed that the optimal attachment angle between the substrate and a gecko seta is 30° in the single seta pull-off experiment. This finding is supported by the adhesion models of setae as cantilever beams (Shah and Sitti 2004; Gao et al. 2005) (see Sect. 27.2.4 for more details). Therefore, θ was fixed at 30° in the studies (Bhushan et al. 2006; Kim and Bhushan 2007a–d).

Table 27.2. Geometrical size, calculated stiffness, and typical densities of branches of seta for Tokay gecko (Kim and Bhushan 2007a)

Level of seta	Length (μm)	Diameter (μm)	Bending stiffness[a] (N/m)	Typical density (no./mm^2)
III upper	75	5	2.908	14×10^3
II middle	25	1	0.126	–
I lower	2.5	0.1	0.0126	$1.4–14 \times 10^6$

[a] For an elastic modulus of 10 GPa with load applied at 60° to the spatula long axis

27.5.2 Single Spring Contact Analysis

In their analysis, Bhushan et al. (2006) and Kim and Bhushan (2007a–d) assumed the tip of the spatula in a single contact to be spherical. The springs on every level of hierarchy have the same stiffness as the bending stiffness of the corresponding branches of the seta. If the beam is oriented at an angle θ to the substrate and the contact load F is aligned normal to the substrate, its components along and tangential to the direction of the beam, $F\cos\theta$ and $F\sin\theta$, give rise to bending and compressive deformations, δ_b and δ_c, respectively, as (Young and Budynas 2001)

$$\delta_b = \frac{F\cos\theta l_m^3}{3EI}, \delta_c = \frac{F\sin\theta l_m}{A_C E}, \tag{27.16}$$

where $I = \pi R_m^4/4$ and $A_C = \pi R_m^2$ are the moment of inertia and the cross-sectional area of the beam, respectively, l_m and R_m are the length and the radius of seta branches, respectively, and m is the level number. The net displacement, $\delta_\perp$, normal to the substrate is given by

$$\delta_\perp = \delta_c \sin\theta + \delta_b \cos\theta . \tag{27.17}$$

Using (27.16) and (27.17), one can calculate the stiffness of seta branches k_m as (Glassmaker et al. 2004)

$$k_m = \frac{\pi R_m^2 E}{l_m \sin^2\theta \left(1 + \frac{4l_m^2 \cot^2\theta}{3R_m^2}\right)} . \tag{27.18}$$

For an assumed elastic modulus E of seta material of 10 GPa with a load applied at an angle of 60° to spatulae long axis, Kim and Bhushan (2007a) calculated the stiffness of every level of seta as given in Table 27.2.

In the model, both tips of a spatula and asperity summits of the rough surface are assumed to be spherical with a constant radius (Bhushan et al. 2006). As a result, a single spatula adhering to a rough surface was modeled as the interaction between two spherical tips. Because β-keratin has a high elastic modulus (Russell 1986; Bertram and Gosline 1987), the adhesion force between two round tips was calculated according to the Derjaguin–Muller–Toporov (DMT) theory (Derjaguin et al. 1975) as

$$F_{ad} = 2\pi R_c W_{ad} , \tag{27.19}$$

where R_c is the reduced radius of contact, which is calculated as $R_c = 1/(1/R_1 + 1/R_2)$; R_1 and R_2 are the radii of the contacting surfaces; $R_1 = R_2$, $R_c = R/2$. The work of adhesion W_{ad} is then calculated using the following equation for two flat surfaces separated by a distance D (Israelachvili 1992):

$$W_{ad} = -\frac{H}{12\pi D^2} , \tag{27.20}$$

where H is the Hamakar constant, which depends on the medium between the two surfaces. Typical values of the Hamakar constant for polymers are 10^{-19} J in air and 3.7×10^{-20} J in water (Israelachvili 1992). For a gecko seta, which is composed of β-keratin, the value of H is assumed to be 10^{-19} J. The work of adhesion of two surfaces in contact separated by an atomic distance $D \approx 0.2$ nm is approximately equal to 66 mJ/m^2 (Israelachvili 1992). By assuming that the tip radius R is 50 nm, using (27.19), one can calculate the adhesion force of a single contact to be 10 nN (Kim and Bhushan 2007a). This value is identical to the adhesion force of a single spatula measured by Huber et al. (2005a). This adhesion force is used as a critical force in the model for judging whether the contact between the tip and the surface is broken or not during the pull-off cycle (Bhushan et al. 2006). If the elastic force of a single spring is less than the adhesion force, the spring is regarded as having been detached.

27.5.3 The Multilevel Hierarchical Spring Analysis

In order to study the effect of the number of hierarchical levels in the attachment system on attachment ability, models with one level (Bhushan et al. 2006; Kim and Bhushan 2007a, b), two levels (Bhushan et al. 2006; Kim and Bhushan 2007a, b) and three levels (Kim and Bhushan 2007a, b) of hierarchy were simulated (Fig. 27.12). The one-level model has springs with length $l_{\mathrm{I}} = 2.5\,\mu\mathrm{m}$ and stiffness $k_{\mathrm{I}} = 0.0126\,\mathrm{N/m}$. The length and stiffness of the springs in the two-level model are $l_{\mathrm{I}} = 2.5\,\mu\mathrm{m}$, $k_{\mathrm{I}} = 0.0126\,\mathrm{N/m}$ and $l_{\mathrm{II}} = 25\,\mu\mathrm{m}$, $k_{\mathrm{II}} = 0.126\,\mathrm{N/m}$ for levels I and II, respectively. The three-level model has additional upper-level springs with $l_{\mathrm{III}} = 75\,\mu\mathrm{m}$, $k_{\mathrm{III}} = 2.908\,\mathrm{N/m}$ on the springs of the two-level model, which is identical to gecko setae. The base of the springs and the connecting plate between the levels are assumed to be rigid. The distance S_{I} between neighboring structures of level I is $0.35\,\mu\mathrm{m}$, obtained from the average value of the measured spatula density, $8 \times 10^6\,\mathrm{mm}^{-2}$, calculated by multiplying $14{,}000\,\mathrm{setae/mm}^2$ by an average of 550 spatulae per seta (Schleich and Kästle 1986) (Table 27.2). A 1:10 proportion of the number of springs in the upper level to that in the level below was assumed (Bhushan et al. 2006). This corresponds to one spring at level III being connected to ten springs at level II and each spring at level II also has ten springs at level I. The number of springs at level I considered in the model is calculated by dividing the scan length ($2000\,\mu\mathrm{m}$) with the distance S_{I} ($0.35\,\mu\mathrm{m}$), which corresponds to 5700.

The spring deflection Δl was calculated as

$$\Delta l = h - l_0 - z\,, \tag{27.21}$$

where h is the position of the spring base relative to the mean line of the surface; l_0 is the total length of a spring structure, which is $l_0 = l_{\mathrm{I}}$ for the one-level model, $l_0 = l_{\mathrm{I}} + l_{\mathrm{II}}$ for the two-level model, and $l_0 = l_{\mathrm{I}} + l_{\mathrm{II}} + l_{\mathrm{III}}$ for the three-level model; and z is the profile height of the rough surface. The elastic force F_{el} arising in the springs at a distance h from the surface was calculated for the one-level model as (Bhushan et al. 2006)

$$F_{\mathrm{el}} = -k_{\mathrm{I}} \sum_{i=1}^{p} \Delta l_i u_i \quad u_i = \begin{cases} 1 & \text{if contact} \\ 0 & \text{if no contact}\,, \end{cases} \tag{27.22}$$

where p is the number of springs in level I of the model. For the two-level model, the elastic force was calculated as (Bhushan et al. 2006)

$$F_{\mathrm{el}} = -\sum_{j=1}^{q} \sum_{i=1}^{p} k_{ji} \left(\Delta l_{ji} - \Delta l_j\right) u_{ji} \quad u_{ji} = \begin{cases} 1 & \text{if contact} \\ 0 & \text{if no contact}\,, \end{cases} \tag{27.23}$$

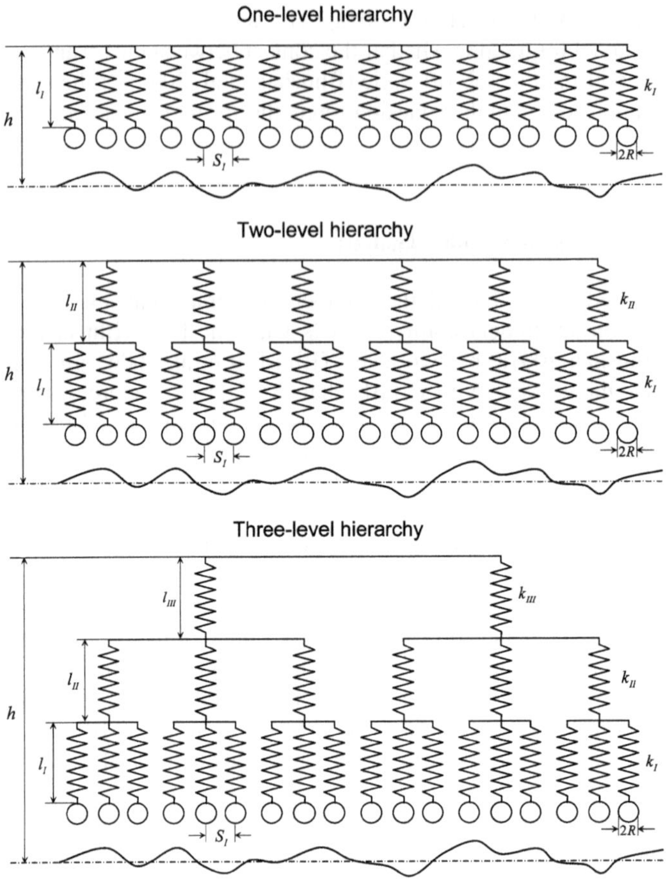

Fig. 27.12. One-, two-, and three-level hierarchical spring models for simulating the effect of hierarchical morphology on the interaction of a seta with a rough surface. l_{I}, l_{II}, and l_{III} are lengths of the structures, S_{I} is the space between spatulae, k_{I}, k_{II}, and k_{III} are stiffnesses of the structures, I, II, and III are level indexes, R is the radius of the tip, and h is the distance between the base of the upper spring of each model and the mean line of the rough profile (Kim and Bhushan 2007a)

where q is the number of springs level II of the model. For the three-level model, the elastic force was calculated as (Kim and Bhushan 2007a)

$$F_{\mathrm{el}} = -\sum_{k=1}^{r}\sum_{j=1}^{q}\sum_{i=1}^{p} k_{kji}\left(\Delta l_{kji} - \Delta l_{kj} - \Delta l_j\right) u_{kji} \quad u_{kji} = \begin{cases} 1 & \text{if contact} \\ 0 & \text{if no contact}\,, \end{cases} \tag{27.24}$$

where r is the number of springs in level III of the model. The spring force when the springs approach the rough surface is calculated using (27.22), (27.23), or (27.24) for the one-, two-, and three-level models, respectively. During pull-off, the same equations are used to calculate the spring force. However, when the applied load is zero, the springs do not detach owing to adhesion attraction given by (27.19). The springs are pulled apart until the net force (pull-off force minus attractive adhesion force) at the interface is zero.

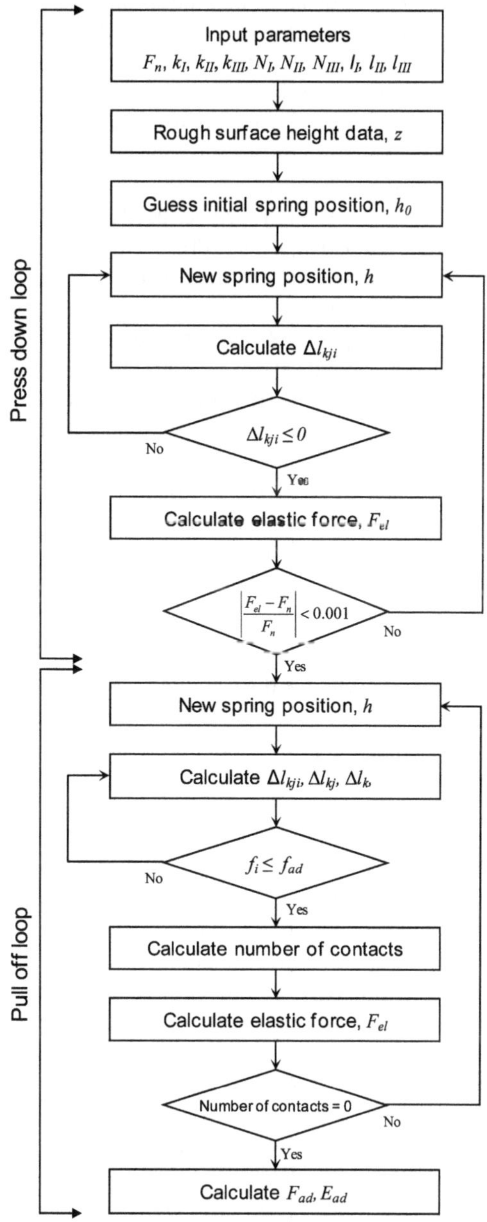

Fig. 27.13. Flow chart for the calculation of the adhesion force (F_{ad}) and the adhesion energy (E_{ad}) for three-level hierarchical spring model. F_n is an applied load, k_{I}, k_{II}, and k_{III} and l_{I}, l_{II}, and l_{III} are stiffnesses and lengths of the structures, Δl_{kji}, Δl_{ki}, and Δl_k are the spring deformations on levels I, II, and III, respectively, i, j, and k are spring indexes on each level, f_i is the elastic force of a single spring and f_{ad} is the adhesion force of a single contact (Kim and Bhushan 2007a)

The adhesion force is the lowest value of the elastic force F_{el} when the seta has detached from the contacting surface. The adhesion energy is calculated as

$$W_{ad} = \int_{\overline{D}}^{\infty} F_{el}(D)\,\mathrm{d}D\,, \tag{27.25}$$

where D is the distance that the spring base moves away from the contacting surface. The lower limit of the distance $\overline{D}$ is the value of D where F_{el} is first zero when the model is pulled away from the contacting surface. Also although the upper limit of the distance is infinity, in practice, the $F_{el}(D)$ curve is integrated to the upper limit where F_{el} increases from a negative value to zero. Figure 27.13 shows the flow chart for the calculation of the adhesion force and the adhesion energy employed by Kim and Bhushan (2007a).

The random rough surfaces used in the simulations were generated by a computer program (Bhushan 1999, 2002). Two-dimensional profiles of surfaces that a gecko may encounter were obtained using a stylus profiler (Bhushan et al. 2006). These profiles along with the surface selection methods and surface roughness parameters [root mean square (RMS) amplitude σ and correlation length β^*] for scan lengths of 80, 400, and 2000 μm are presented in the Appendix. The roughness parameters are scale-dependent, and, therefore, adhesion values also are expected to be scale-dependent. As the scan length was increased, the measured values of RMS amplitude and correlation length both increased. The range of values of σ from 0.01 to 30 μm and a fixed value of $\beta^* = 200$ μm were used for modeling the contact of a seta with random rough surfaces. The range chosen covers values of roughnesses for relatively smooth, artificial surfaces to natural, rough surfaces. A typical scan length of 2000 μm was also chosen, which is comparable to the length of a gecko lamella.

27.5.4 Adhesion Results for the Gecko Attachment System Contacting a Rough Surface

The multilevel spring model produced many useful results. Those obtained by Kim and Bhushan (2007a) will be discussed in detail in this section. Figure 27.14a shows the calculated spring force–distance curves for the one-, two-, and three-level hierarchical models in contact with rough surfaces of different values of RMS amplitude σ ranging from $\sigma = 0.01$ μm to $\sigma = 10$ μm at an applied load of 1.6 μN, which was derived from the gecko's weight. When the spring model is pressed against the rough surface, contact between the spring and the rough surface occurs at point A; as the spring tip presses into the contacting surface, the force increases up to point B, B′, or B″. During pull-off, the spring relaxes, and the spring force passes an equilibrium state (0 N); tips break free of adhesion forces at point C, C′, or C″ as the spring moves away from the surface. The perpendicular distance from C, C′, or C″ to zero is the adhesion force. Adhesion energy stored during contact can be obtained by calculating the area of the triangle during the unloading part of the curves (27.25).

Using the spring force–distance curves, Kim and Bhushan (2007a) calculated the adhesion coefficient, the number of contacts per unit length, and the adhesion energy per unit length of the one-, two-, and three-level models for an applied load of 1.6 μN and a wide range of RMS roughness values as seen in the left graphs in

Fig. 27.14b. The adhesion coefficient, defined as the ratio of the pull-off force to the applied preload, represents the strength of adhesion with respect to the preload. For the applied load of 1.6 μN, which corresponds to the weight of a gecko, the maximum adhesion coefficient is about 36 when σ is smaller than 0.01 μm. This means that a gecko can generate enough adhesion force to support 36 times its bodyweight. However, if σ is increased to 1 μm, the adhesion coefficient for the three-level model is reduced to 4.7. It is noteworthy that the adhesion coefficient falls below 1 when the contacting surface has a RMS roughness σ greater than 10 μm. This implies that the attachment system is no longer capable of supporting the gecko's weight. Autumn et al. (2000, 2002) showed that in isolated gecko setae contacting with the surface of a single crystalline silicon wafer, a 2.5-μN preload yielded adhesion of 20–40 μN and thus a value of the adhesion coefficient of 8–16, which supports the simulation results of Kim and Bhushan (2007a).

Figure 27.14b (top left) shows that the adhesion coefficient for the two-level model is lower than that for the three-level model, but there is only a small difference

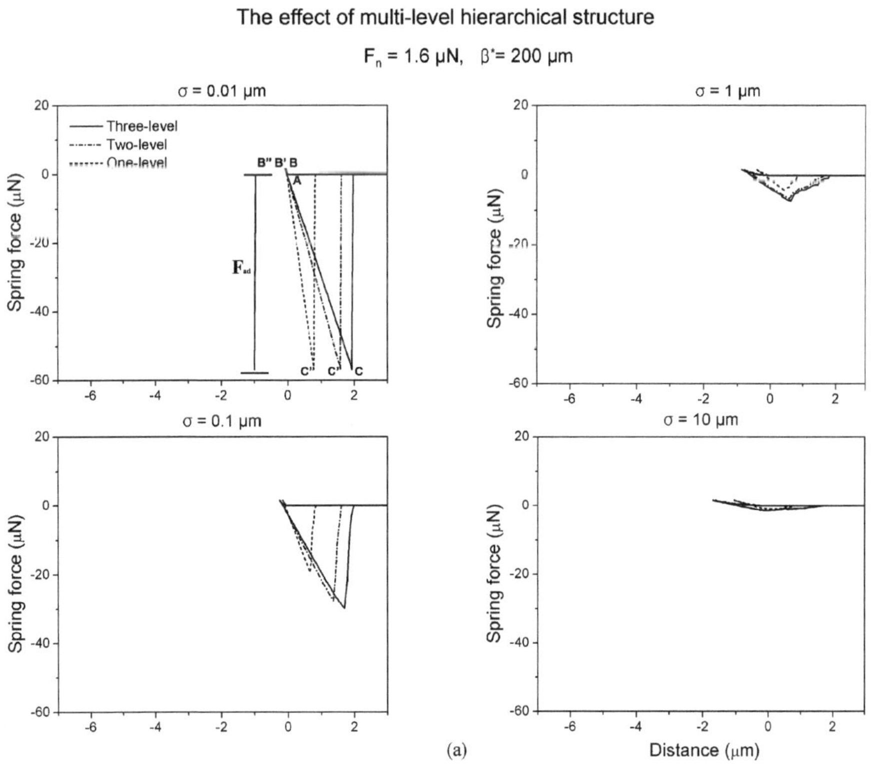

Fig. 27.14. **a** Force–distance curves of one-, two-, and three-level models in contact with rough surfaces with different σ values for an applied load of 1.6 μN. **b** The adhesion coefficient, the number of contacts, and the adhesion energy per unit length of profile for one-level and multilevel models with an increase of σ value (*left*), and relative increases between multilevel and one-level models (*right*) for an applied load of 1.6 μN. The value of k_{III} in the analysis is 2.908 N/m (Kim and Bhushan 2007a)

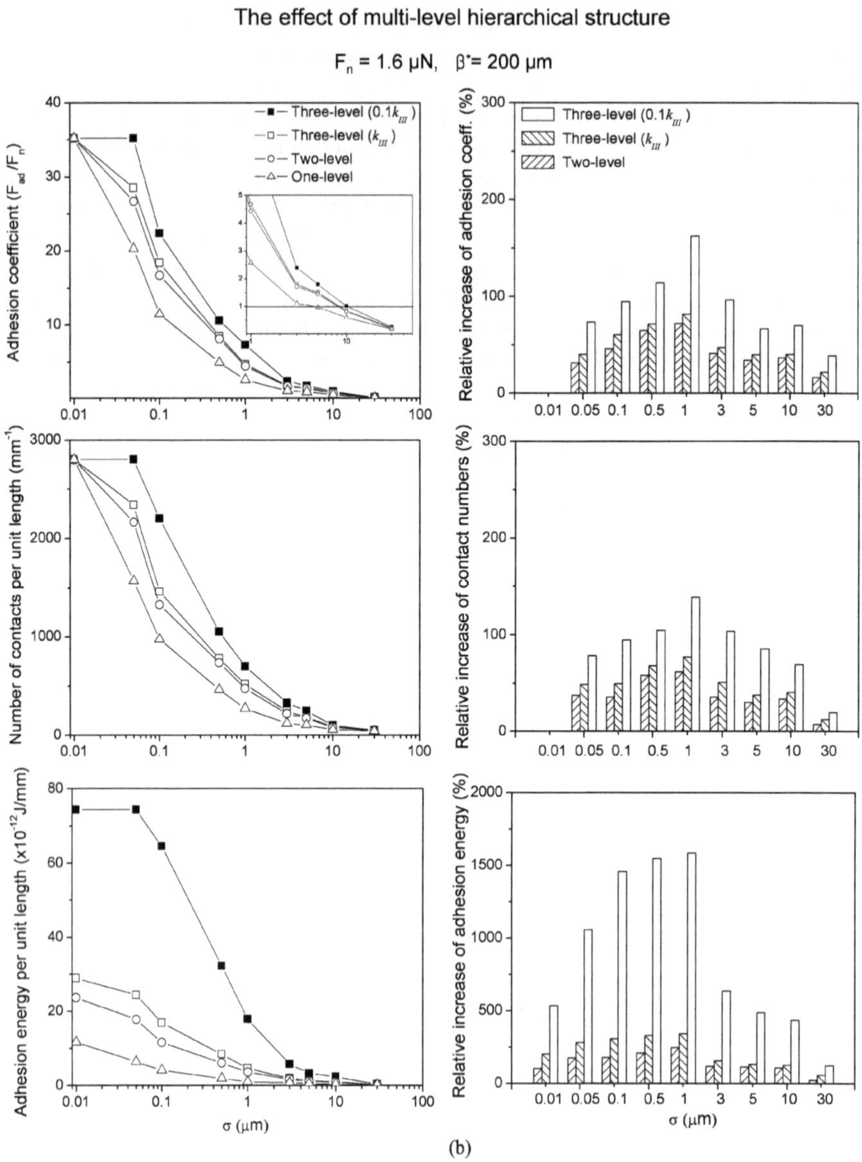

Fig. 27.14. (continued)

between the adhesion forces between the two- and three-level models, because the stiffness of level III for the three-level model is calculated to be higher than those of levels I and II. In order to show the effect of stiffness, the results for the three-level model with springs in level III of which the stiffness is 10 times smaller than that of original level III springs are plotted. It can be seen that the three-level model with a third-level stiffness of $0.1k_{\rm III}$ has a 20–30% higher adhesion coefficient than the three-level model. The results also show that the trends in the number of contacts are similar to that of the adhesion force. The study also investigated the effect of σ on adhesion energy. It was determined that the adhesion energy decreases with an increase of σ. For the smooth surface with $\sigma = 0.01$ μm, the adhesion energies for the two- and three-level hierarchical models are 2 times and 2.4 times larger than that for the one-level model, respectively, but these values decrease rapidly at surfaces with σ greater than 0.05 μm; and in every model the adhesion energy finally decreases to zero at surfaces with σ greater than 10 μm. The adhesion energy for the three-level model with $0.1k_{\mathrm{III}}$ is 2–3 times higher than that for three-level model.

In order to demonstrate the effect of the hierarchical structure on adhesion enhancement, Kim and Bhushan (2007a) calculated the increases in the adhesion coefficient, the number of contacts, and the adhesion energy of the two-, three-, and three-level (with $0.1k_{\mathrm{III}}$) models relative to the one-level model. These results are shown on the right side of Fig. 27.14b. It was found for the two- and three-level models, the relative increase of the adhesion coefficient increases slowly with an increase of σ and has the maximum values of about 70 and 80% at $\sigma = 1$ μm, respectively, and then decreases for surfaces with σ greater than 3 μm. On the whole, at the applied load of 1.6 μN, the effect of the variation of σ on the adhesion enhancement for both two- and three-level models is not so large. However, the relative increase of the adhesion coefficient for the three-level model with $0.1k_{\mathrm{III}}$ has the maximum value of about 170% at $\sigma = 1$ μm, which shows significant adhesion enhancement. Owing to the relative increase of adhesion energy, the three-level model with $0.1k_{\mathrm{III}}$ shows significant adhesion enhancement.

Figure 27.15 shows the variation of adhesion force and adhesion energy as a function of applied load for both one- and three-level models contacting with a surface with $\sigma = 1$ μm. It is shown that as the applied load increases, the adhesion force increases up to a certain applied load and then has a constant value, whereas adhesion energy continues to increase with an increase in the applied load. The one-

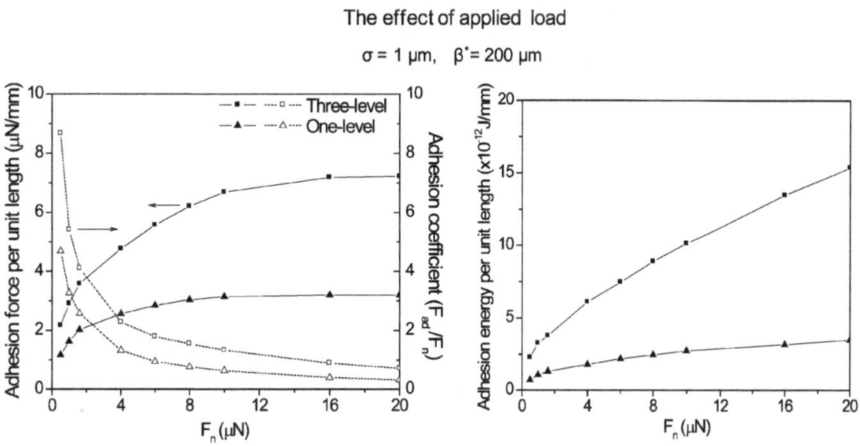

Fig. 27.15. The variation of adhesion force, adhesion coefficient, and adhesion energy as a function of applied loads for both one- and three-level models contacting with surface with $\sigma = 1$ μm. The value of k_{III} in the analysis is 2.908 N/m (Kim and Bhushan 2007a)

level model has a maximum value of adhesive force per unit length of about 3 μN/mm at an applied load of 10 μN, and the three-level model has a maximum value of about 7 μN/mm at an applied load of 16 μN. However, the adhesion coefficient continues to decrease at higher applied loads because adhesion force is constant even if the applied load increases.

The simulation results for the three-level model, which closely models gecko setae, presented in Fig. 27.14 show that roughness reduces the adhesion force. At a surface with σ greater than 10 μm, the adhesion force cannot support the gecko's weight. However, in practice, a gecko can cling or crawl on the surface of a ceiling with higher roughness. Kim and Bhushan (2007a) did not consider the effect of lamellae in their study. The authors state that the lamellae can adapt to the waviness of the surface, while the setae allow for the adaptation to microroughness or nanoroughness and expect that adding the lamellae of gecko skin to the model would lead to higher adhesion over a wider range of roughness. In addition, their hierarchical model considers only normal-to-surface deformation and motion of the seta. It should be noted that measurements of adhesion force of a single gecko seta made by Autumn et al. (2000) demonstrated that a load applied normal to the surface was insufficient for effective attachment. The lateral force required to pull parallel to the surface was observed by sliding the seta approximately 5 μm laterally along the surface under a preload.

27.5.5 Capillarity Effects

Kim and Bhushan (2007d) investigated the effects of capillarity on gecko adhesion by considering capillary force as well as the solid-to-solid interaction. The Laplace and surface tension components of the capillary force are treated according to Sect. 27.3.2. The solid-to-solid adhesion force was calculated by DMT theory according to (27.19) and will be denoted as F_{DMT}.

The work of adhesion was then calculated by (27.20). Kim and Bhushan (2007d) assumed typical values of the Hamaker constant to be 10^{-19} J in air and 6.7×10^{19} J in water (Israelachvili 1992). The work of adhesion of two surfaces in contact separated by an atomic distance $D \approx 0.2$ nm (Israelachvili 1992) is approximately $-66\ \mathrm{mJ/m^2}$ in air and $-44\ \mathrm{mJ/m^2}$ in water. Assuming the tip radius R is 50 nm, the DMT adhesion forces F_{DMT} of a single contact in air and water are $F_{\mathrm{DMT}}^{\mathrm{air}} = 11$ nN and $F_{\mathrm{DMT}}^{\mathrm{water}} = 7.3$ nN, respectively. As the humidity increases from 0 to 100%, the DMT adhesion force will take a value between $F_{\mathrm{DMT}}^{\mathrm{air}}$ and $F_{\mathrm{DMT}}^{\mathrm{water}}$. To calculate the DMT adhesion force for the intermediate humidity, an approximation method by Wan et al. (1992) was used. The work of adhesion W_{ad} for the intermediate humidity can be expressed as

$$W_{\mathrm{ad}} = \int_D^{\infty} \frac{H}{6\pi h^3}\,\mathrm{d}h = \int_D^{h_{\mathrm{f}}} \frac{H_{\mathrm{water}}}{6\pi h^3}\,\mathrm{d}h + \int_{h_{\mathrm{f}}}^{\infty} \frac{H_{\mathrm{air}}}{6\pi h^3}\,\mathrm{d}h\ , \tag{27.26}$$

where h is the separation along the plane. h_{f} is the water film thickness at a filling angle ϕ, which can be calculated as

$$h_{\mathrm{f}} = D + R(1 - \cos\phi)\ . \tag{27.27}$$

Therefore, using (27.19), (27.26), and (27.27), the DMT adhesion force for intermediate humidity is

$$F_{\mathrm{DMT}} = F_{\mathrm{DMT}}^{\mathrm{water}} \left(1 - \frac{1}{[1 + R(1 - \cos\phi)/D]^2}\right) + F_{\mathrm{DMT}}^{\mathrm{air}} \left(\frac{1}{[1 + R(1 - \cos\phi)/D]^2}\right) . \tag{27.28}$$

Finally, Kim and Bhushan (2007d) calculated the total adhesion force F_{ad} as the sum of (27.15) and (27.28):

$$F_{\mathrm{ad}} = F_{\mathrm{c}} + F_{\mathrm{DMT}} . \tag{27.29}$$

Kim and Bhushan (2007d) then used the total adhesion force as a critical force in the three-level hierarchical spring model discussed previously. In the spring model of a gecko seta the spring is regarded as having been detached if the force applied upon spring deformation is greater than the adhesion force.

27.5.6 Adhesion Results that Account for Capillarity Effects

To simulate the capillarity contribution to the adhesion force of a gecko spatula, Kim and Bhushan (2007d) set the contact angle on the gecko spatula tip θ_1 to 128° (Huber et al. 2005b). It was assumed that the spatula tip radius $R = 50\,\mathrm{nm}$, the ambient temperature $T = 25\,°\mathrm{C}$, the surface tension of water $\gamma = 73\,\mathrm{mJ/m^2}$, and the molecular volume of water $V = 0.03\,\mathrm{nm^3}$ (Israelachvili 1992).

Figure 27.16a shows the total adhesion force as a function of relative humidity for a single spatula in contact with surfaces with different contact angles. The total adhesion force decreases with an increase in the contact angle on the substrate, and the difference of the total adhesion force among different contact angles is larger in the intermediate-humidity regime. As the relative humidity increases, the total adhesion force for the surfaces with contact angle less than 60° has a higher value than the DMT adhesive force not considering wet contact, whereas for values above 60°, the total adhesion force has lower values at most relative humidities.

The simulation results of Kim and Bhushan (2007d) are compared with the experimental data of Huber et al. (2005b) in Fig. 27.16b. Huber et al. (2005b) measured the pull-off force of a single spatula in contact with four different types of Si wafer and glass at the ambient temperature of 25 °C and a relative humidity of 52%. According to their description, wafer families "N" and "T" in Fig. 27.16b differ by the thickness of the top amorphous silicon oxide layer. The "Phil" type is the cleaned silicon oxide surface which is hydrophilic with a water contact angle of approximately 10°, whereas the "Phob" type is a silicon wafer covered hydrophobic monolayer causing a water contact angle of more than 100°. The glass has a water contact angle of 58°. Huber et al. (2005b) showed that the adhesive force of a gecko spatula rises significantly for substrates with increasing hydrophilicity (adhesion force increases by a factor of 2 as mating surfaces go from hydrophobic to hydrophylic). As shown in Fig. 27.16b, the simulation results of Kim and Bhushan (2007d) closely match the experimental data of Huber et al. (2005b).

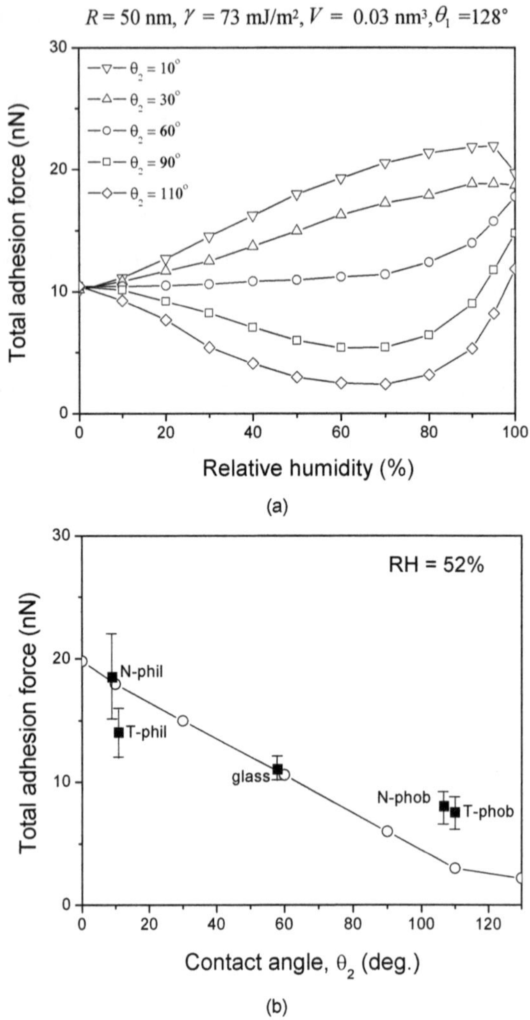

Fig. 27.16. a Total adhesion force as a function of relative humidity for a single spatula in contact with surfaces with different contact angles. **b** Comparison of the simulation results of Kim and Bhushan (2007d) with the measured data obtained by Huber et al. (2005b) for a single spatula in contact with hydrophilic and hydrophobic surfaces (Kim and Bhushan 2007d). *RH* relative humidity

Kim and Bhushan (2007d) performed an adhesion analysis for the three-level hierarchical model for gecko seta. Figure 27.17 shows the adhesion coefficient and number of contacts per unit length for the three-level hierarchical model in contact with rough surfaces with different values of RMS amplitude σ ranging from $\sigma = 0.01\,\mu m$ to $\sigma = 30\,\mu m$ for different relative humidities and contact angles of the surface. It can be seen that for the surface with contact angle $\theta_2 = 10°$ the adhesion coefficient is greatly influenced by relative humidity. At 0% relative humidity the maximum adhesion coefficient is about 36 at a value of σ smaller than $0.01\,\mu m$ compared with 78 for 90% relative humidity with the same surface roughness. As expected, the effect of relative humidity on increasing the adhesion coefficient decreases as the contact angle becomes larger. For hydrophobic surfaces, relative humidity decreases the adhesion coefficient. Similar trends can be noticed in

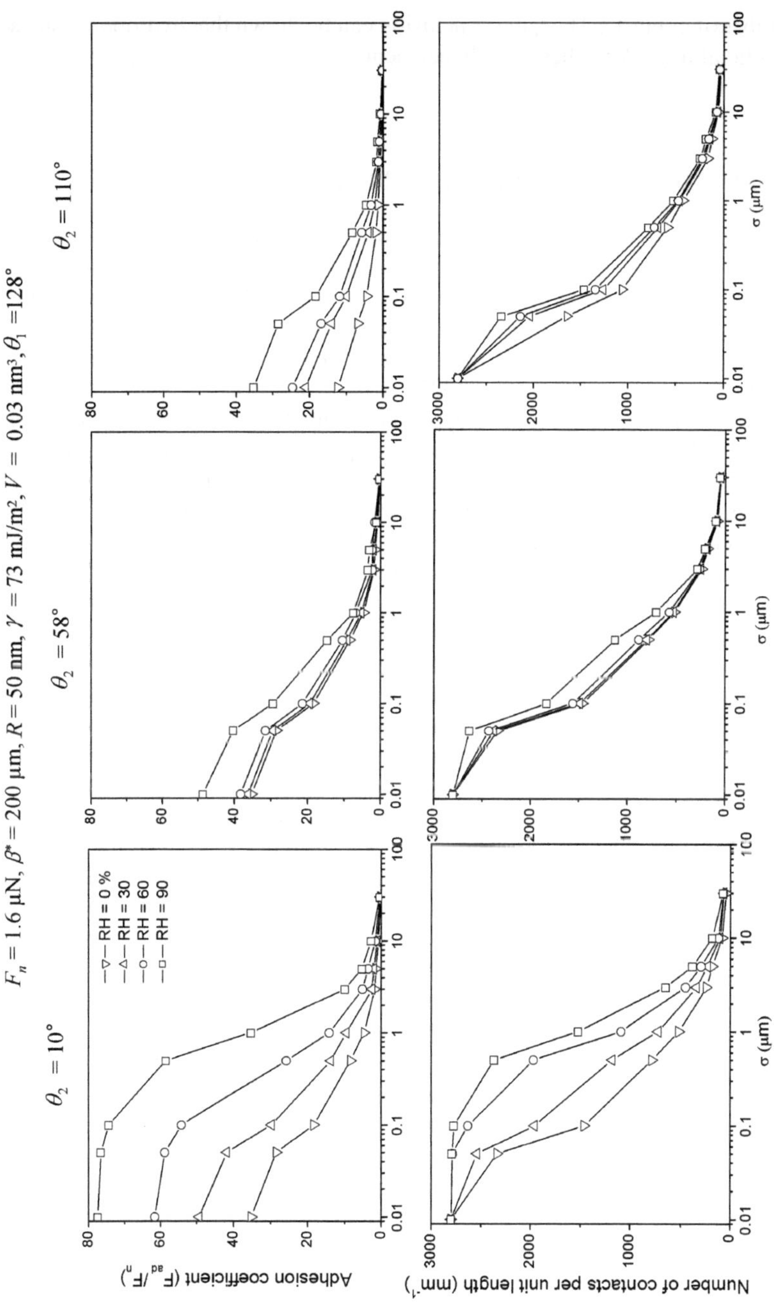

Fig. 27.17a–c. The adhesion coefficient and number of contacts per unit length for the three-level hierarchical model in contact with rough surfaces with different values of root mean square amplitudes σ and contact angles for different relative humidities (Kim and Bhushan 2007d)

the number of contacts. Thus, the conclusion can be drawn that hydrophilic surfaces are beneficial to gecko adhesion enhancement.

27.6 Modeling of Biomimetic Fibrillar Structures

The mechanics of adhesion between a fibrillar structure and a rough surface as it relates to the design of biomimetic structures has been a topic of investigation by many researchers (Jagota and Bennison 2002; Persson 2003; Sitti and Fearing 2003a; Glassmaker et al. 2004, 2005; Gao et al. 2005; Yao and Gao 2006; Kim and Bhushan 2007b,c). In order to better understand the mechanics of adhesion related to design, the approach of Kim and Bhushan (2007c) will be described.

Kim and Bhushan (2007c) developed a convenient, general, and useful guideline for understanding biological systems and for improving biomimetic attachment. This adhesion database was constructed by modeling the fibers as oriented cylindrical cantilever beams with spherical tips. The authors then carried out numerical simulation of the attachment system in contact with random rough surfaces considering three constraint conditions—buckling, fracture, and sticking of fiber structure. For a given applied load, roughness of contacting surface, and fiber material, a procedure to find the optimal fiber radius and aspect ratio for the desired adhesion coefficient was developed.

The model of Kim and Bhushan (2007c) is used to find the design parameters for fibers of a single-level attachment system capable of achieving the desired properties—high adhesion coefficient and durability. The design variables for an attachment system are as follows: fiber geometry (radius and aspect ratio of fibers, tip radius), fiber material, fiber density, and fiber orientation. The optimal values for the design variables to achieve the desired properties should be selected for fabrication of a biomimetic attachment system.

27.6.1 Fiber Model

The fiber model of Kim and Bhushan (2007c) consists of a simple idealized fibrillar structure consisting of a single-level array of microbeams/nanobeams protruding from a backing as shown in Fig. 27.18. The fibers are modeled as oriented cylindrical cantilever beams with spherical tips. In Fig. 27.18, l is the length of the fibers, θ is the fiber orientation, R is the fiber radius, R_t is the tip radius, S is the spacing between fibers, and h is the distance between the upper-spring base of each model and the mean line of the rough profile. The end terminal of the fibers is assumed to be a spherical tip with a constant radius and a constant adhesion force.

27.6.2 Single Fiber Contact Analysis

Kim and Bhushan (2007c) modeled an individual fiber as a beam oriented at an angle θ to the substrate and the contact load F is aligned normal to the substrate.

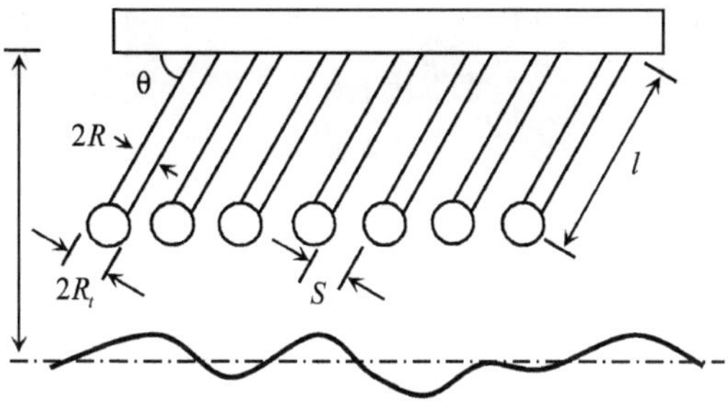

Fig. 27.18. Single-level attachment system with oriented cylindrical cantilever beams with spherical tip. l is the length of the fibers, θ is the fiber orientation, R is the fiber radius, R_t is the tip radius, S is the spacing between fibers, and h is the distance between the base of the model and the mean line of the rough profile (Kim and Bhushan 2007c)

The net displacement normal to the substrate can be calculated according to (27.16) and (27.17). The fiber stiffness ($k = F/\delta_\perp$) is given by (Glassmaker et al. 2004)

$$k = \frac{\pi R^2 E}{l \sin^2\theta \left(1 + \frac{4l^2 \cot^2\theta}{3R^2}\right)} = \frac{\pi R E}{2\lambda \sin^2\theta \left(1 + \frac{16\lambda^2 \cot^2\theta}{3}\right)}, \tag{27.30}$$

where $\lambda = l/2R$ is the aspect ratio of the fiber and θ is fixed at 30°.

Two alternative models dominate the world of contact mechanics—the JKR theory (Johnson et al. 1971) for compliant solids and the DMT theory (Derjaguin et al. 1975) for stiff solids. Although gecko setae are composed of β-keratin with a high elastic modulus (Russell 1986; Bertram and Gosline 1987) which is close to the DMT model, in general the JKR theory prevails for biological or artificial attachment systems. Therefore, the JKR theory was applied in the subsequent analysis of Kim and Bhushan (2007c) to compare the materials with wide ranges of elastic modulus. The adhesion force between a spherical tip and a rigid flat surface is thus calculated using the JKR theory as (Johnson et al. 1971)

$$F_{\mathrm{ad}} = \frac{3}{2}\pi R_{\mathrm{t}} W_{\mathrm{ad}}, \tag{27.31}$$

where R_{t} is the radius of the spherical tip and W_{ad} is the work of adhesion (calculated according to (27.24)). Kim and Bhushan (2007c) used this adhesion force as a critical force. If the elastic force of a single spring is less than the adhesion force, they regarded the spring as having been detached.

27.6.3 Constraints

In the design of fibrillar structures a trade-off exists between the aspect ratio of the fibers and their adaptability to a rough surface. If the aspect ratio of the fibers is too large, they can adhere to each other or even collapse under their own weight as shown in Fig. 27.19a. If the aspect ratio is too small (Fig. 27.19b), the structures will lack the necessary compliance to conform to a rough surface. The spacing between the individual fibers is also important. If the spacing is too small, adjacent

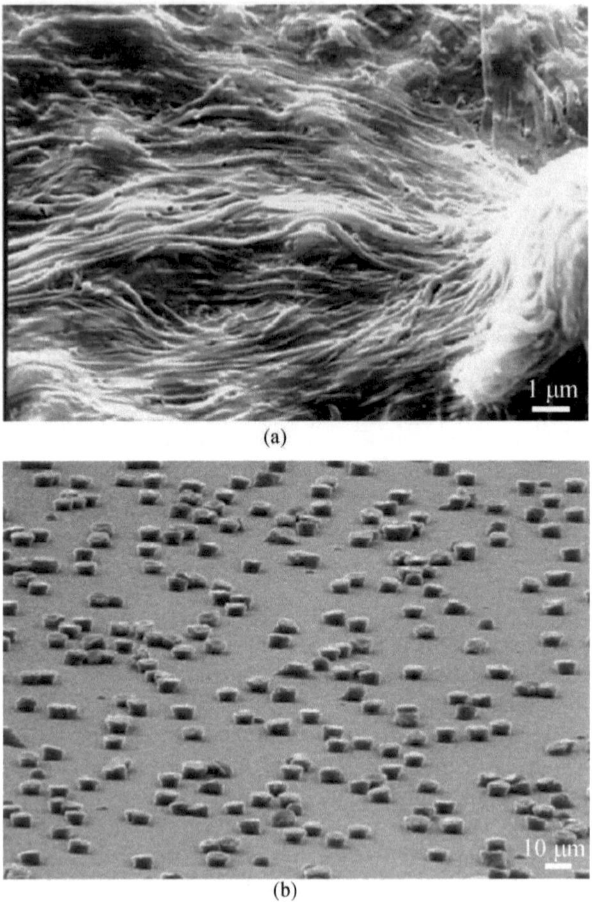

Fig. 27.19. SEM micrographs of **a** high-aspect-ratio polymer fibrils that have collapsed under their own weight and **b** low-aspect-ratio polymer fibrils that are incapable of adapting to rough surfaces (Sitti and Fearing 2003a)

fibers can attract each other through intermolecular forces, which will lead to bunching. Therefore, Kim and Bhushan (2007c) considered three necessary conditions in their analysis, buckling, fracture, and sticking of fiber structure, which constrain the allowed geometry.

27.6.3.1 Nonbuckling Condition

A fibrillar interface can deliver a compliant response while still employing stiff materials because of bending and microbuckling of fibers. Based on classical Euler buckling, Glassmaker et al. (2004) established a stress–strain relationship and a critical compressive strain for buckling ε_{cr} for the fiber oriented at an angle θ to the substrate,

$$\varepsilon_{cr} = -\frac{b_c \pi^2}{3(Al^2/3I)} \left(1 + \frac{A_C l^2}{3I} \cot^2 \theta\right) , \qquad (27.32)$$

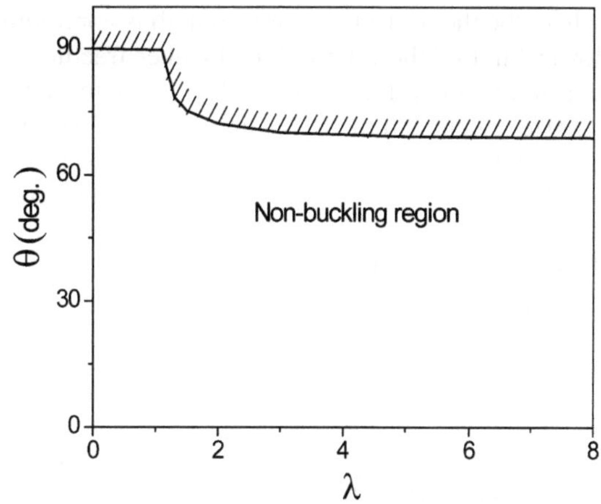

Fig. 27.20. Critical fiber orientation as a function of aspect ratio λ for the nonbuckling condition for pinned-clamped microbeams ($b_c = 2$) (Kim and Bhushan 2007c)

where A_c is the cross-sectional area of the fibril and b_c is a factor that depends on boundary conditions. The factor b_c has a value of 2 for pinned-clamped microbeams. For fibers having a circular cross section, ε_{cr} is calculated as

$$\varepsilon_{cr} = -\frac{b_c \pi^2}{3(4l^2/3R^2)}\left(1 + \frac{4l^2}{3R^2}\cot^2\theta\right) = -b_c\pi^2\left(\frac{1}{16\lambda^2} + \frac{\cot^2\theta}{3}\right) . \quad (27.33)$$

In (27.33), ε_{cr} depends on both the aspect ratio λ and the orientation θ of the fibers. If $\varepsilon_{cr} = 1$, which means the fiber deforms up to the backing, buckling does not occur. Figure 27.20 plots the critical orientation θ as a function of aspect ratio for the case of $\varepsilon_{cr} = 1$. The critical fiber orientation for buckling is 90° at λ less than 1.1. This means that the buckling does not occur regardless of the orientation of the fiber at λ less than 1.1. For λ greater than 1.1, the critical fiber orientation for buckling decreases with an increase in λ, and has a constant value of 69° at λ greater than 3. Kim and Bhushan (2007c) used a fixed value at 30° for θ, because as stated earlier the maximum adhesive force is achieved at this orientation and buckling is not expected to occur.

27.6.3.2 Nonfiber Fracture Condition

For small contacts, the strength of the system will eventually be determined by fracture of the fibers. Spolenak et al. (2005) suggested the limit of fiber fracture as a function of the adhesion force. The axial stress σ_f in a fiber is limited by its theoretical fracture strength σ^f_{th} as

$$\sigma_f = \frac{F_{ad}}{R^2\pi} \le \sigma^f_{th} . \quad (27.34)$$

Using (27.31), one calculates a lower limit for the useful fiber radius R as

$$R \ge \sqrt{\frac{3R_t W_{ad}}{2\sigma^f_{th}}} \approx \sqrt{\frac{15 R_t W_{ad}}{E}} , \quad (27.35)$$

where the theoretical fracture strength is approximated by $E/10$ (Dieter 1988). The lower limit of the fiber radius for fiber fracture by adhesion force depends on the elastic modulus. By assuming $W_{ad} = 66\,\mathrm{mJ/m^2}$ stated earlier, Kim and Bhushan (2007c) calculated the lower limits of the fiber radius for $E = 1$ MPa, 0.1 GPa, and 10 GPa to be 0.32, 0.032, and 0.0032 μm, respectively.

The contact stress cannot exceed the ideal contact strength transmitted through the actual contact area at the instant of tensile instability (Spolenak et al. 2005). Kim and Bhushan (2007c) used this condition (27.34) to extract the limit of the tip radius, R_t:

$$\sigma_c = \frac{F_{ad}}{a_c^2 \pi} \le \sigma_{th} \,, \tag{27.36}$$

where σ_c is the contact stress and σ_{th} is the ideal strength of van der Waals bonds, which is approximately W_{ad}/b, b is the characteristic length of the surface interaction, and a_c is the contact radius. Based on the JKR theory, for the rigid contacting surface, a_c at the instant of pull-off is calculated as

$$a_c = \left(\frac{9\pi W_{ad} R_t^2 (1 - \nu^2)}{8E} \right)^{1/3} , \tag{27.37}$$

where ν is the Poisson ratio. The tip radius can then be calculated by combining (27.36) and (27.37) as

$$R_t \ge \frac{8b^3 E^2}{3\pi^2 (1 - \nu^2)^2 W_{ad}^2} \,. \tag{27.38}$$

The lower limit of the tip radius also depends on the elastic modulus. Assuming $W_{ad} = 66\,\mathrm{mJ/m^2}$ and $b = 2 \times 10^{-10}$ m (Dieter 1988), one calculates the lower limits of the tip radius for $E = 1$ MPa, 0.1 GPa, and 10 GPa as 6×10^{-7}, 6×10^{-3}, and 60 nm, respectively. In this study, Kim and Bhushan (2007c) fixed the tip radius at 100 nm, which satisfies the tip radius condition throughout a wide range of elastic modulus up to 10 GPa.

27.6.3.3 Nonsticking Condition

A high density of fibers is also important for high adhesion. However, if the space S between neighboring fibers is too small, the adhesion forces between them become stronger than the forces required to bend the fibers. Then, fibers might stick to each other and get entangled. Therefore, to prevent fibers from sticking to each other, they must be spaced apart and be stiff enough to prevent sticking or bunching. Several authors (e.g., Sitti and Fearing 2003a) have formulated a nonsticking criterion. Kim and Bhushan (2007c) adopted the approach of Sitti and Fearing (2003a). Both adhesion and elastic forces will act on bent structures. The adhesion force between neighboring two round tips is calculated as

$$F_{ad} = \frac{3}{2} \pi R_t' W_{ad} \,, \tag{27.39}$$

where R_t' is the reduced radius of contact, which is calculated as $R_t' = 1/(1/R_{t1} + 1/R_{t2})$; R_{t1} and R_{t2} are the radii of the contacting tips; for the case of similar tips, $R_{t1} = R_{t2}$, $R_t' = 2/R_t$.

The elastic force of a bent structure can be calculated by multiplying the bending stiffness ($k_b = 3\pi R^4 E/4l^3$) by a given bending displacement δ as

$$F_{el} = \frac{3}{4}\frac{\pi R^4 E\delta}{l^3} . \tag{27.40}$$

The condition for the prevention of sticking is $F_{el} > F_{ad}$. By combining (27.39) and (27.40), a requirement for the minimum distance S between structures which will prevent sticking of the structures is (Kim and Bhushan 2007c)

$$S > 2\delta = 2\left(\frac{4}{3}\frac{W_{ad}l^3}{ER^3}\right) = 2\left(\frac{32}{3}\frac{W_{ad}\lambda^3}{E}\right) . \tag{27.41}$$

The constant 2 takes into account two nearest structures. Using the distance S, the fiber density ρ is

$$\rho = \frac{1}{(S+2R)^2} . \tag{27.42}$$

Equation (27.42) was then used to calculate the allowed minimum density of fibers without sticking or bunching. In (27.41), it is shown that the minimum distance S depends on both the aspect ratio λ and the elastic modulus E. A smaller aspect ratio and a higher elastic modulus allow for greater packing density. However, fibers with a low aspect ratio and a high modulus are not desirable for adhering to rough surfaces owing to lack of compliance.

27.6.4 Numerical Simulation

The simulation of adhesion of an attachment system in contact with random rough surfaces was carried out numerically. In order to conduct two-dimensional simulations it is necessary to calculate the applied load F_n as a function of applied pressure P_n as an input condition. Using ρ calculated for the nonsticking condition, Kim and Bhushan (2007c) calculated F_n as

$$F_n = \frac{P_n p}{\rho} , \tag{27.43}$$

where p is the number of springs in scan length L, which equals $L/(S+2R)$.

Fibers of the attachment system are modeled as one-level-hierarchy elastic springs (Fig. 27.12) (Kim and Bhushan 2007c). The deflection of each spring and the elastic force arising in the springs are calculated according to (27.21) and (27.22), respectively. The adhesion force is the lowest value of elastic force F_{el} when the fiber has detached from the contacting surface. Kim and Bhushan (2007c) used an iterative process to obtain the optimal fiber geometry—fiber radius and aspect ratio. If the applied load, the roughness of the contacting surface and the fiber material are given, the procedure for calculating the adhesion force is repeated iteratively until the desired adhesion force is satisfied. In order to simplify the design problem, the fiber material is regarded as a known variable. The next step is constructing the design database. The left side of Fig. 27.21 shows the flow chart for the construction

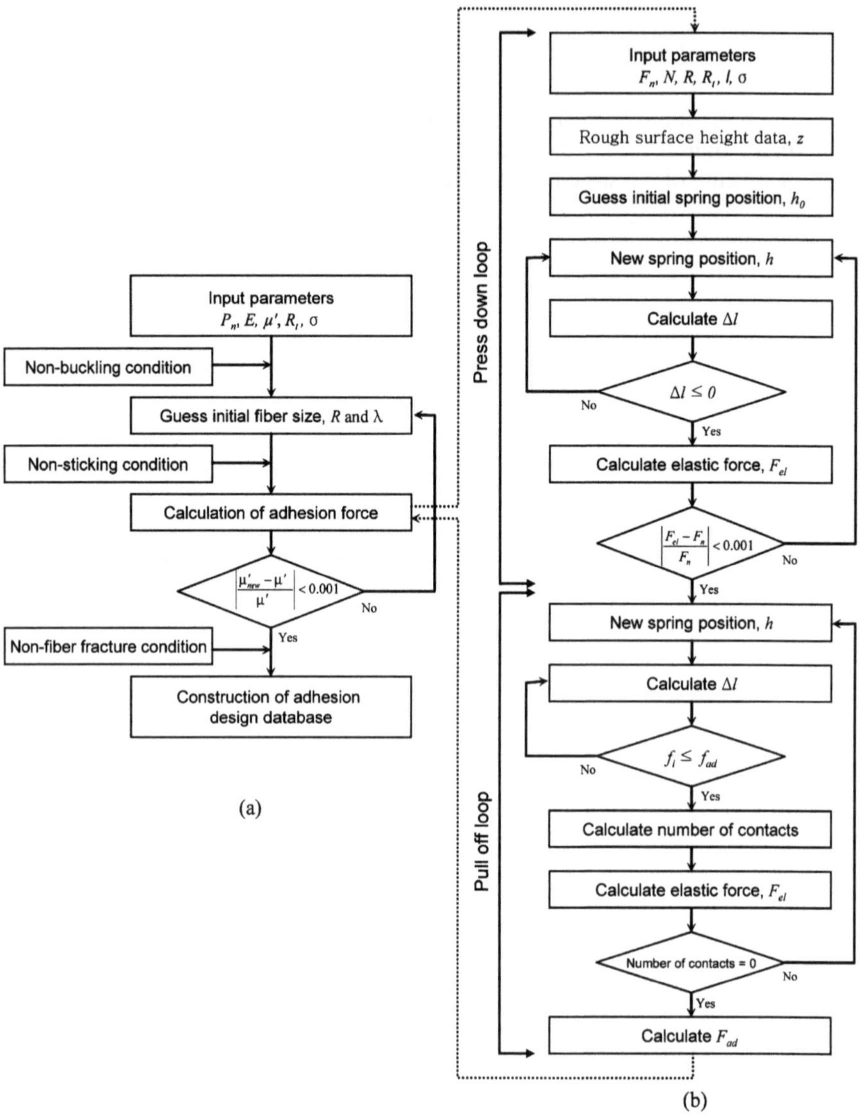

Fig. 27.21. Flow chart for **a** the construction of the adhesion design database and **b** the calculation of the adhesion force. P_n is the applied pressure, E is the elastic modulus, μ' is the adhesion coefficient, R_t is the tip radius, σ is the root mean square amplitude, R is the fiber radius, λ is the aspect ratio of the fiber, F_n is the applied load, N is the number of springs, k and l are the stiffness and the length of the structures, Δl is the spring deformation, f_i is the elastic force of a single spring, and f_{ad} is the adhesion force of a single contact (Kim and Bhushan 2007c)

of the adhesion design database and the right side of Fig. 27.21 shows the calculation of the adhesion force that is a part of the procedure to construct the adhesion design database.

27.6.5
Results and Discussion

Figure 27.22 shows an example of the adhesion design database for biomimetic attachment systems consisting of single-level cylindrical fibers with an orientation angle of 30° and spherical tips of $R_t = 100\,\text{nm}$ constructed by Kim and Bhushan (2007c). The minimum fiber radius calculated for the nonfiber fracture condition, which plays a role in the lower limit of the optimized fiber radius, is also added on the plot. The plots in Fig. 27.22 cover all applicable fiber materials from soft elastomer material such as poly(dimethylsiloxane) to stiffer polymers such as polyimide and β-keratin. The dashed lines in each plot represent the limits of fiber fracture due to the adhesion force. For a soft material with $E = 1\,\text{MPa}$ in Fig. 27.22a, the range of the desirable fiber radius is more than 0.3 μm and that of the aspect ratio is approximately less than 1. As the elastic modulus increases, the feasible ranges of both the fiber radius and the aspect ratio also increase as shown in Fig. 27.22b and c. In Fig. 27.22, the fiber radius has a linear relation with the surface roughness on a logarithmic scale.

If the applied load, the roughness of the contacting surface, and the elastic modulus of a fiber material are specified, the optimal fiber radius and aspect ratio for the desired adhesion coefficient can be selected from this design database. The adhesion databases are useful for understanding biological systems and for guiding the fabrication of biomimetic attachment systems. Two case studies (Kim and Bhushan 2007c) are calculated below.

Case study 1: Select the optimal size of fibrillar adhesive for a wall-climbing robot with the following requirements:

- Material: polymer with $E \approx 100\,\text{MPa}$
- Applied pressure by weight less than 10 kPa
- Adhesion coefficient less than 5
- Surface roughness $\sigma < 1\,\mu\text{m}$

The subplot of the adhesion database that satisfies the requirement is in the second column and second row in Fig. 27.22b. From this subplot, any values on the marked line can be selected to meet the requirements. For example, a fiber radius of 0.4 μm with an aspect ratio of 1 or a fiber radius of 10 μm an with aspect ratio of 0.8 satisfies the specified requirements.

Case study 2: Compare with the adhesion test for a single gecko seta (Autumn et al. 2000, 2002)

- Material: β-keratin with $E \approx 10\,\text{GPa}$
- Applied pressure 57 kPa (2.5 μN on an area of 43.6 μm^2)
- Adhesion coefficient 8–16
- Surface roughness $\sigma < 0.01\,\mu\text{m}$

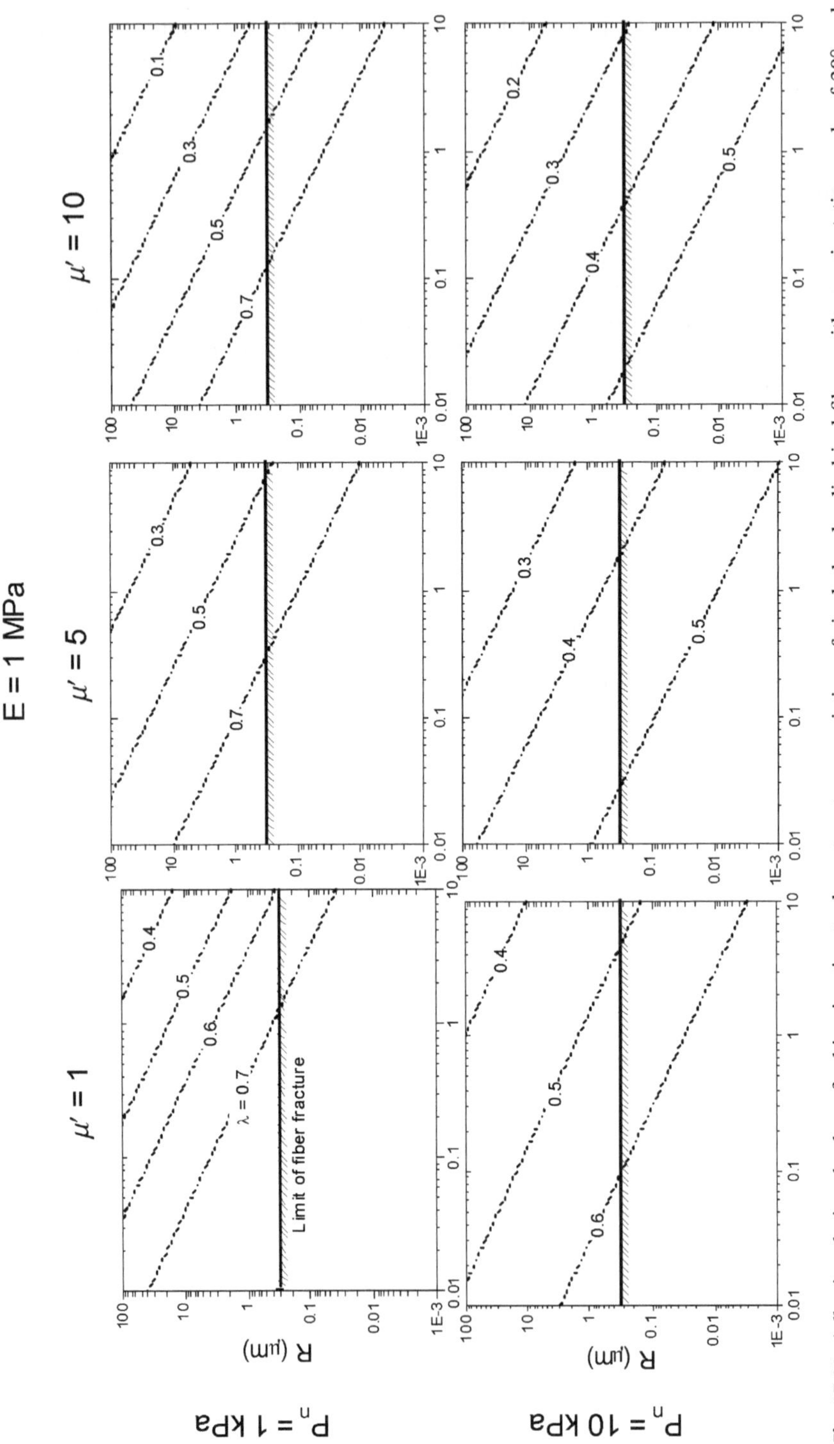

Fig. 27.22. Adhesion design database for biomimetic attachment system consisting of single-level cylindrical fibers with an orientation angle of 30° and spherical tips of 100 nm for an elastic modulus of **a** 1 MPa, **b** 100 MPa, and **c** 10 GPa (Kim and Bhushan 2007c). The *solid lines* shown in **b** and **c** correspond to the case studies I and II, respectively, which satisfy the specified requirements

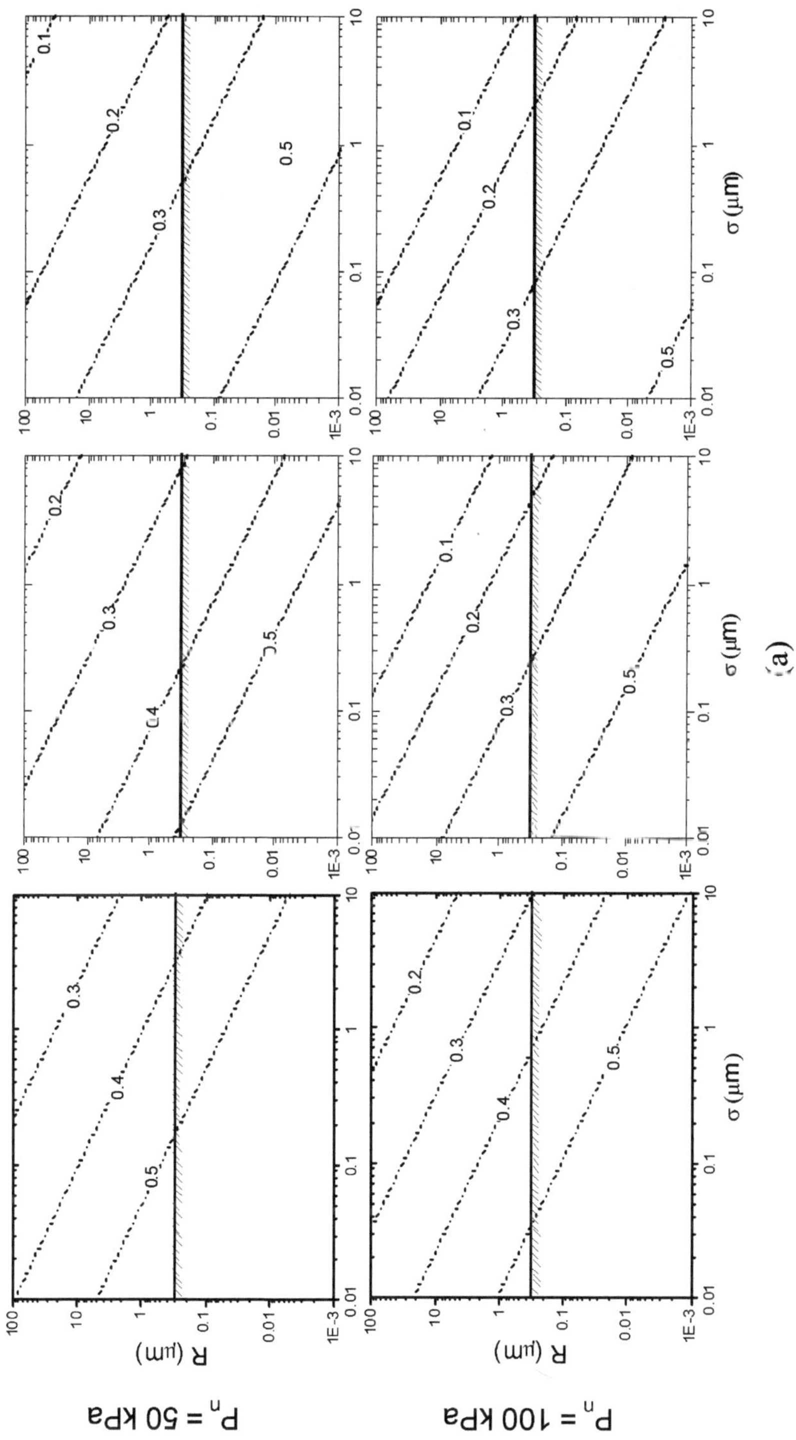

Fig. 27.22. (continued)

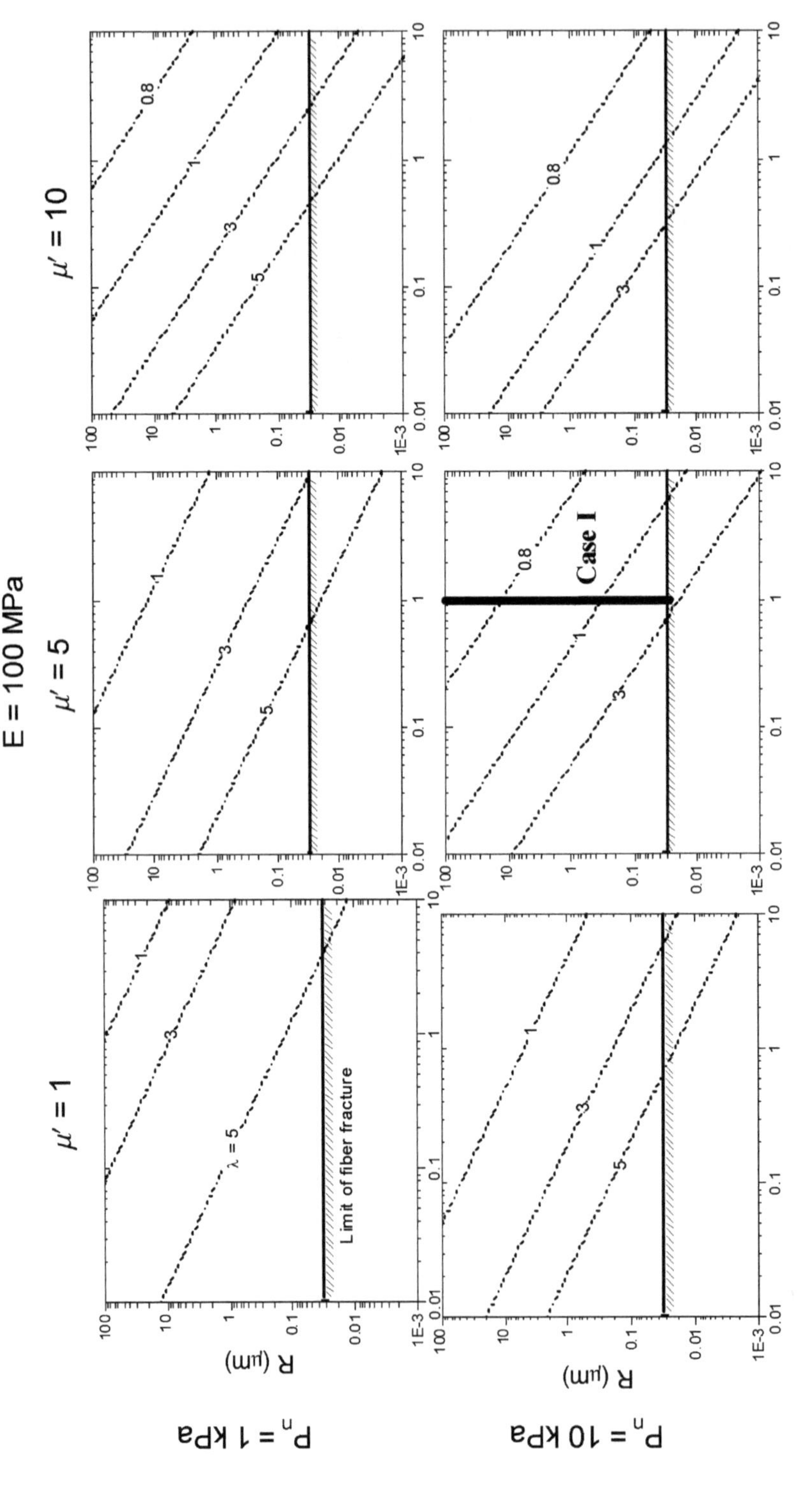

Fig. 27.22. (continued)

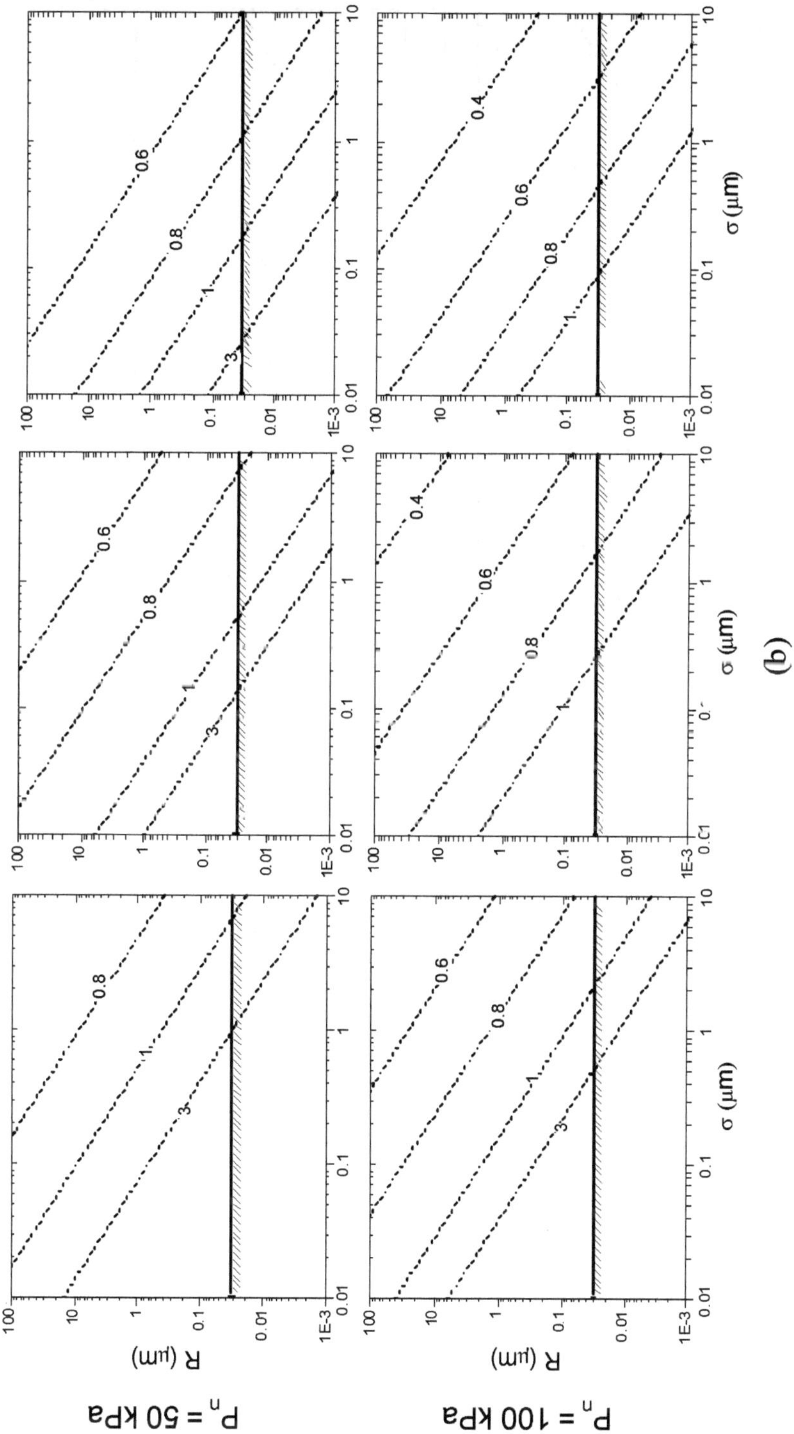

Fig. 27.22. (continued)

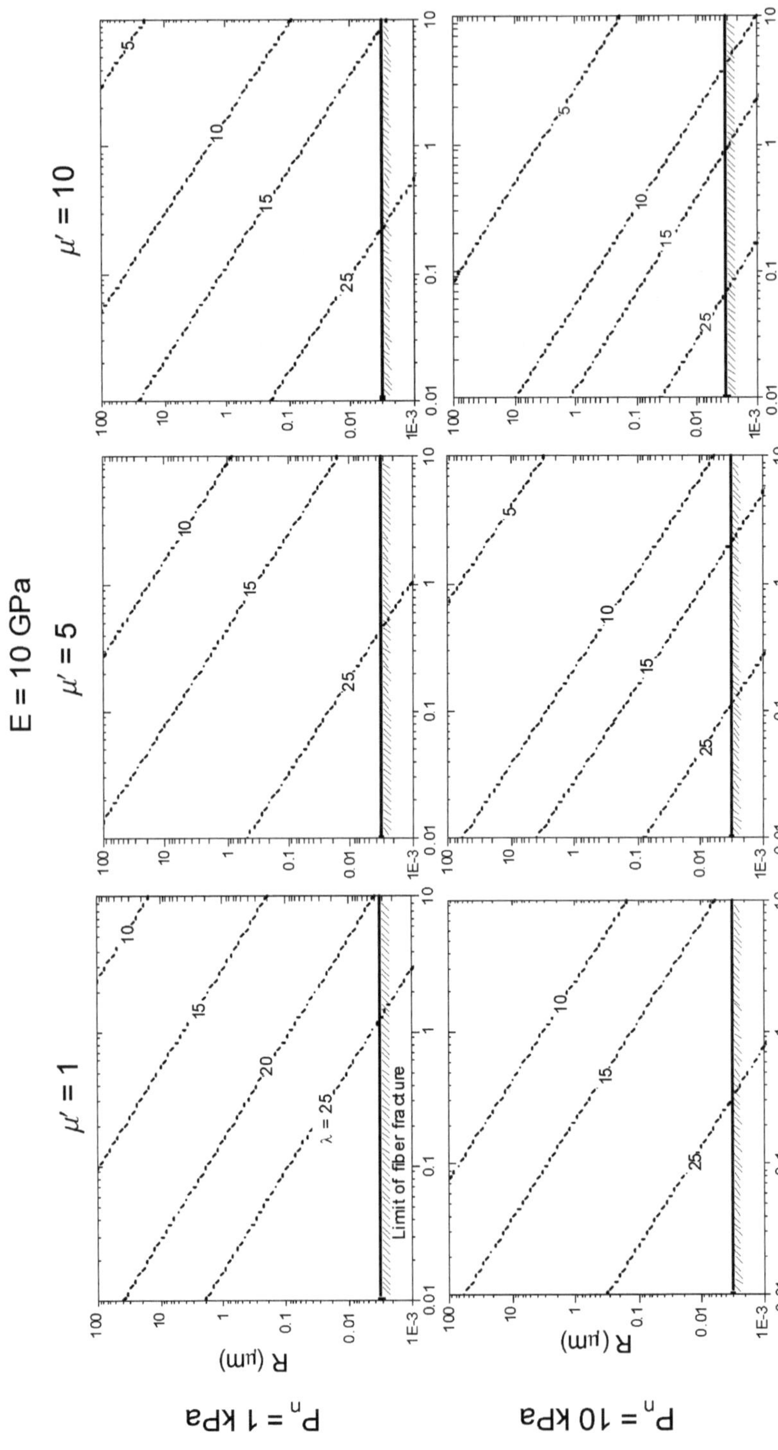

Fig. 27.22. (continued)

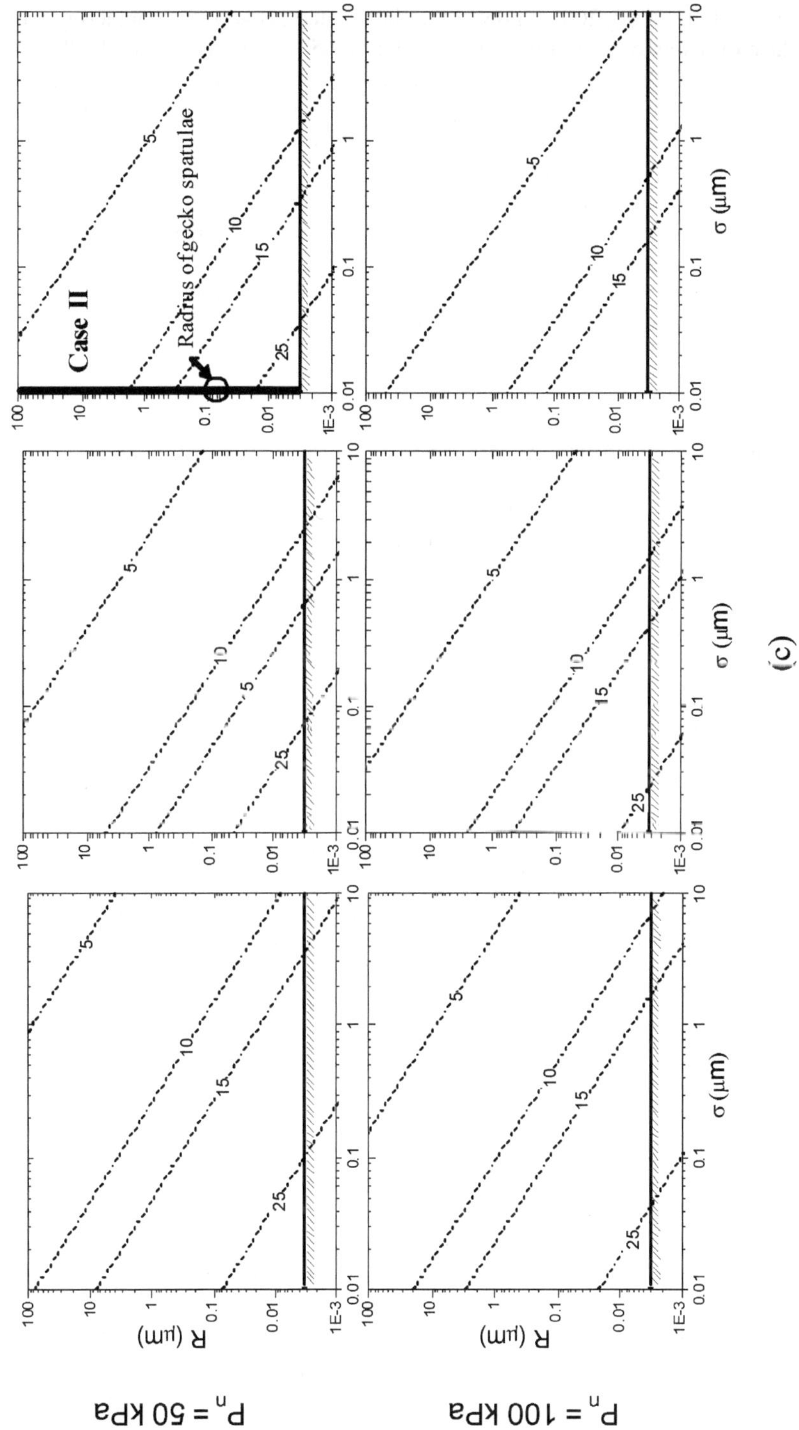

Fig. 27.22. (continued)

Autumn et al. (2000, 2002) showed that in isolated gecko setae contacting with the surface of a single crystalline silicon wafer, a 2.5-μN preload yielded adhesion of 20–40 μN and thus a value of the adhesion coefficient of 8–16. The region that satisfies the above requirements is marked in Fig. 27.22c. The spatulae have an approximate radius of 0.05 μm and an aspect ratio of 25. However, the radius corresponding to $\lambda = 25$ in the marked line is about 0.015 μm. This discrepancy is due to the difference between the simulated fiber model and real gecko setae model. Gecko setae are composed of a three-level hierarchical structure in practice, so higher adhesion can be generated than with a single-level model (Bhushan et al 2006; Kim and Bhushan 2007a, b). Given the simplification in the fiber model, this simulation result is very close to the experimental result.

27.7 Fabrication of Biomimetric Gecko Skin

On the basis of studies found in the literature, the dominant adhesion mechanism utilized by geckos and other spider attachment systems appears to be van der Waals forces. The complex divisions of the gecko skin (lamellae–setae–branches–spatulae) enable a large real area of contact between the gecko skin and the mating surface. Hence, a hierarchical fibrillar microstructure/nanostructure is desirable for dry, superadhesive tapes. The development of nanofabricated surfaces capable of replicating this adhesion force developed in nature is limited by current fabrication methods. Many different techniques have been used in an attempt to create (Geim et al. 2003; Sitti and Fearing 2003a; Northen and Turner 2005, 2006; Yurdumkan et al. 2005; Zhao et al. 2006) and characterize (Peressadko and Gorb 2004; Gorb et al. 2007; Bhushan and Sayer 2007) bio-inspired adhesive tapes.

Gorb et al. (2007) and Bhushan and Sayer (2007) characterized two poly(vinylsiloxane) PVS samples from the Max Planck Institute for Metals Research, Stuttgart, Germany: one consisting of mushroom-shaped pillars (Fig. 27.23a) and the other sample was an unstructured control surface (Fig. 27.23b). The structured sample is composed of pillars that are arranged in a hexagonal order to allow maximum packing density. They are approximately 100 μm in height, 60 μm in base diameter, 35 μm in middle diameter, and 25 μm in diameter at the narrowed region just below the terminal contact plates. These plates were about 40 μm in diameter and 2 μm in thickness at the lip edges. The adhesion force of the two samples in contact with a smooth flat glass substrate was measured by Gorb et al. (2007) using a homemade microtribometer. The results revealed that the structured specimens featured an adhesion force more than twice that of the unstructured specimens. The adhesion force was also found to be independent of the preload. Moreover, it was found that the adhesion force of the structured sample was more tolerant to contamination compared with the control and it could be easily cleaned with a soap solution.

Bhushan and Sayer (2007) characterized the surface roughness, friction force, and contact angle of the structured sample and compared the results with those of an unstructured control. As shown in Fig. 27.24a, the macroscale coefficient of kinetic friction of the structured sample was found to be almost 4 times greater than that of the unstructured sample. This increase was determined to be a result of the structured

Fig. 27.23. SEM micrographs of the **a** structured and **b** unstructured polyvinylsiloxane samples. *SH* shaft, *NR* neck region, *LP* lip (Bhushan and Sayer 2007)

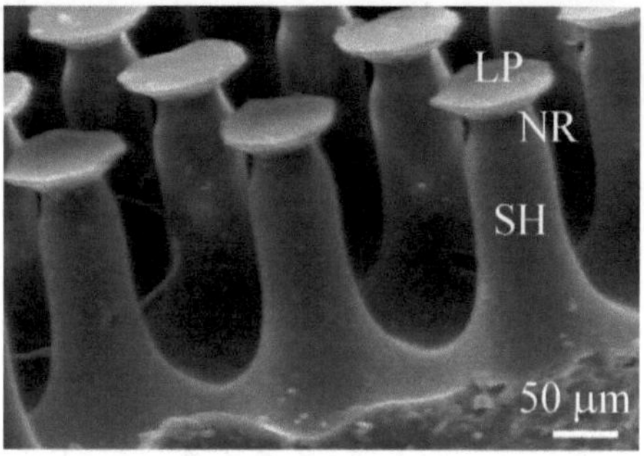

(a)

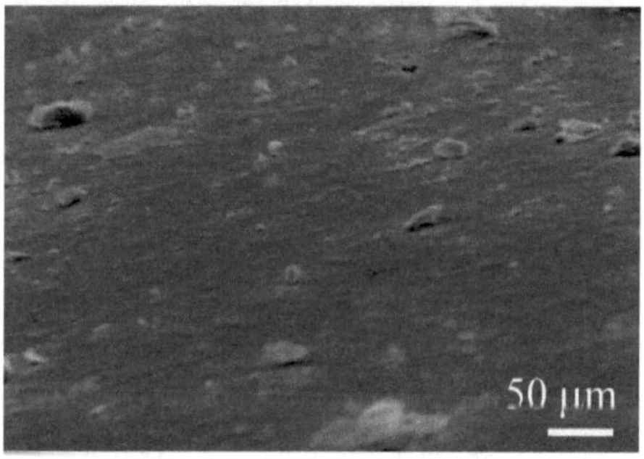

(b)

roughness of the sample and not the random nanoroughness. It is also noteworthy that the static and kinetic coefficients of friction are approximately equal for the structured sample. It is believed that the divided contacts allow the broken contacts of the structured sample to constantly recreate contact. As seen in Fig. 27.24b, the pillars also increased the hydrophobicity of the structured sample in comparison with the unstructured sample as expectedowing to increased surface roughness (Wenzel 1936; Burton and Bhushan 2005). A large contact angle is important for self-cleaning (Barthlott and Neinhius 1997), which agrees with the findings of Gorb et al. (2007) that the structured sample is more tolerant of contamination than the unstructured sample.

27.7.1 Single-Level Hierarchical Structures

One of the simplest approaches to create a single-level hierarchical surface employed an AFM tip to create a set of dimples on a wax surface. These dimples served as a mold for creating the polymer nanopyramids shown in Fig. 27.25a (Sitti and Fearing 2003a). The adhesive force to an individual pyramid was measured using another AFM cantilever. The force was found to be about 200 μN. Although each pyramid of the material is capable of producing large adhesion forces, the surface failed to replicate gecko adhesion on a macroscale. This was due to the lack of flexibility in the

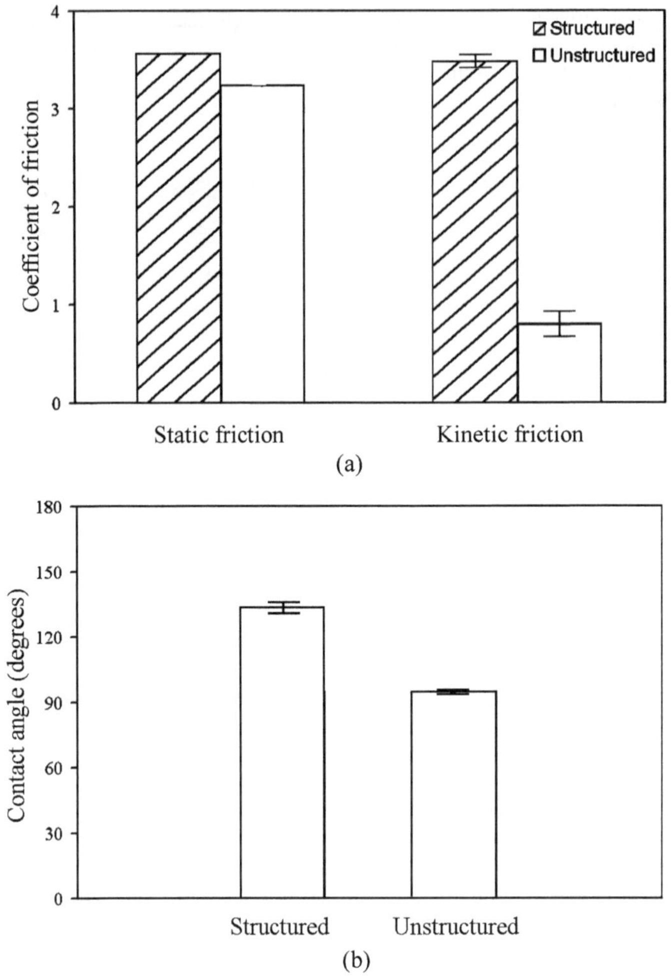

Fig. 27.24. **a** Coefficients of static and kinetic friction for the structured and unstructured samples slid against magnetic tape with a normal load of 130 mN. **b** Water contact angle for the structured and unstructured samples (Bhushan and Sayer 2007)

pyramids. In order to ensure that the largest possible area of contact occurs between the tape and the mating surface, a soft, compliant fibrillar structure would be desired (Jagota and Bennison 2002). As shown in previous calculations, the van der Waals adhesion force for two parallel surfaces is inversely proportional to the cube of the distance between two surfaces. Compliant fibrillar structures enable more fibrils to be in close proximity of a mating surface, thus increasing van der Waals forces.

Geim et al. (2003) created arrays of nanohairs using electron-beam lithography and dry etching in oxygen plasma (Fig. 27.25b, left). The original arrays were created on a rigid silicon wafer. This design was only capable of creating 0.01 N of adhesion force for a 1-cm^2 patch. The nanohairs were then transferred from the silicon wafer

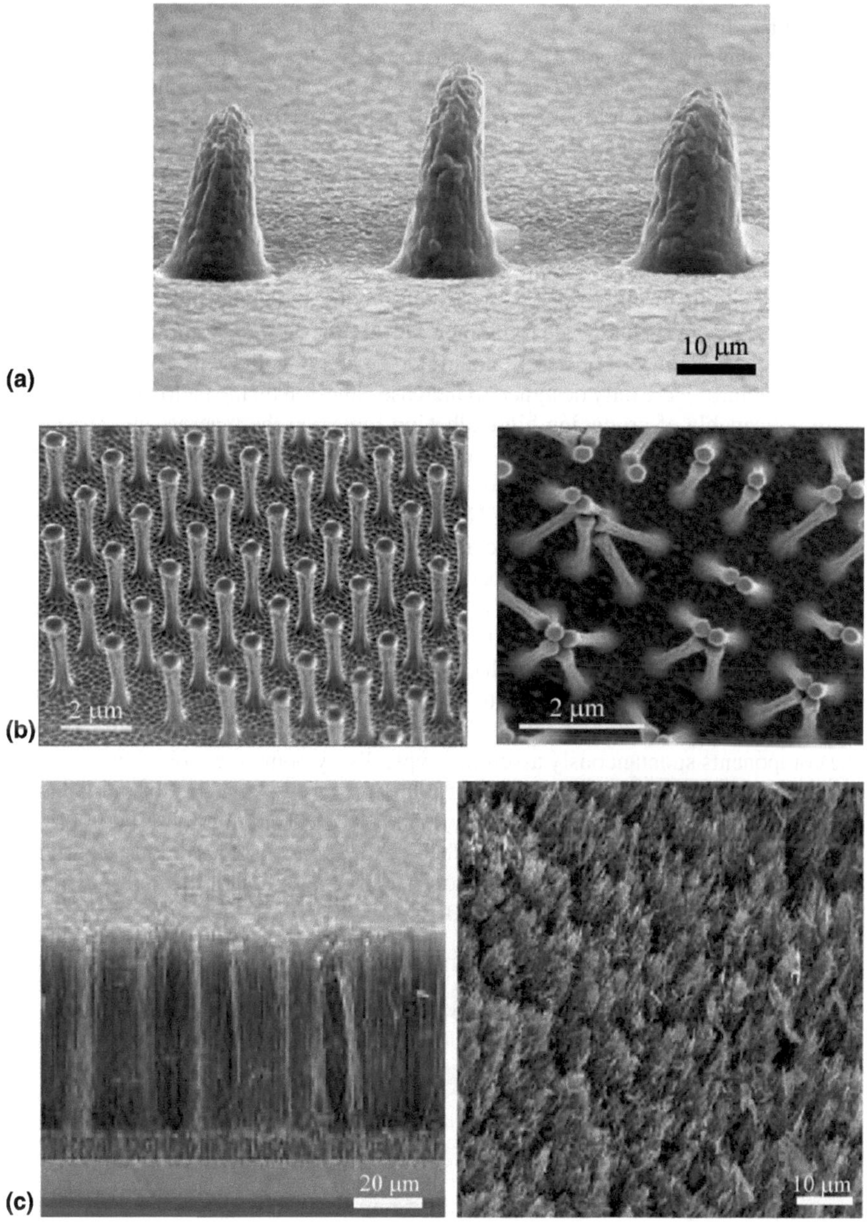

Fig. 27.25. SEM micrographs of **a** three pillars created by nanotip indentation (Sitti 2003), **b** *left* an array of polyimide nanohairs and *right* bunching of the nanohairs, which leads to a reduction in adhesive force (Geim et al. 2003), and **c** multiwalled carbon nanotube structures *left* grown on silicon by vapor deposition and *right* transferred into a poly(methyl methacrylate) matrix and then exposed on the surface after solvent etching (Yurdumakan et al. 2005)

to a soft bonding substrate. A 1-cm^2 sample was able to create 3 N of adhesion force under the new arrangement. This is approximately one third the adhesion strength of a gecko. Bunching (as described earlier) was determined to greatly reduce both the adhesion strength and the durability of the polymer tape. The bunching can be clearly seen in Fig. 27.25b (right).

Multiwalled carbon nanotube (MWCNT) hairs have been used to create super-adhesive tapes. Yurdumakan et al. (2005) used chemical vapor deposition to grow MWCNT that are 50–100 μm in length on quartz or silicon substrates. Patterns were then created using a combination of photolithography and a wet and/or a dry etching. Scanning electron microscope images of the nanotube surfaces can be seen in Fig. 27.25c. On a small scale (nanometer level), the MWCNT surface was able to achieve adhesive forces 2 orders of magnitude greater than those of gecko foot-hairs. These structures were only designed to increase adhesion on the nanometer level and were not capable of producing high adhesion forces on the macroscale. Zhao et al (2006) created MWCNT arrays that are much more capable of macroscale adhesion. MWCNT of 5–10 μm were grown on a Si substrate. This arrangement was able to create an adhesion pressure of $11.7\,\mathrm{N/cm^2}$. The durability of the adhesive tape is an issue as some of the nanotubes detach from the substrate with repeated use. This work shows promise for MWCNT being implemented in bio-inspired tapes.

Directed self-assembly has been proposed as a method to produce regularly spaced fibers (Schäffer et al. 2000; Sitti 2003). In this technique, a thin liquid polymer film is coated on a flat conductive substrate. As demonstrated in Fig. 27.26, a closely spaced metal plate is used to apply a DC electric field on the polymer film. Owing to instabilities, pillars will begin to grow. Self-assembly is desirable because the components spontaneously assemble, typically by bouncing around in a solution or gas phase until a stable structure of minimum energy is reached. This method is crucial in biomolecular nanotechnology, and has the potential to be used in precise devices (Anonymous 2002). These surface coatings have been demonstrated to be both durable and capable of creating superhydrophobic conditions and have been used to form clusters on the nanoscale (Pan et al. 2005).

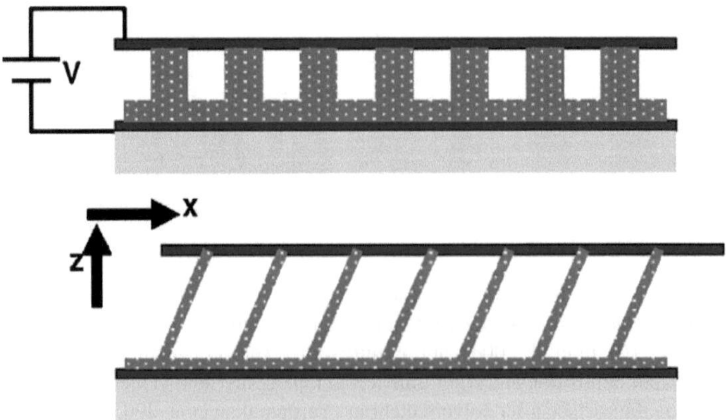

Fig. 27.26. Directed self-assembly based method of producing high-aspect-ratio micro-hairs/nanohairs (Sitti 2003)

27.7.2 Multilevel Hierarchical Structures

The aforementioned fabricated surfaces only have one level of hierarchy. Although these surfaces are capable of producing high adhesion on the microscale/nanoscale, all have failed in producing large-scale adhesion owing to a lack of compliance and bunching. In order to overcome these problems, Northen and Turner (2005, 2006) created a multilevel compliant system by employing a microelectromechanical-based approach. They created a layer of nanorods which they deemed "organorods" (Fig. 27.27a). These organorods are comparable in size to that of gecko spatulae (50–200 nm in diameter and 2 μm tall). They sit atop a silicon dioxide chip (approximately 2 μm thick and 100–150 μm across a side), which was created using photolithography (Fig. 27.27b). Each chip is supported on top of a pillar (1 μm in diameter and 50 μm tall) that attaches to a silicon wafer (Fig. 27.27c). The multilevel structures have been created across a 100-mm wafer (Fig. 27.27d).

Adhesion testing was performed using a nanorod surface on a solid substrate and on the multilevel structures. As seen in Fig. 27.28, adhesive pressure of the multilevel structures was several times higher than that of the surfaces with only one level of hierarchy. The durability of the multilevel structure was also much greater than that of the single-level structure. The adhesion of the multilevel structure did not change between iterations one and five. During the same number of iterations, the adhesive pressure of the single level structure decreased to zero.

Fig. 27.27. Multilevel fabricated adhesive structure composed of **a** organorods, **b** silicon dioxide chips, and **c** support pillars. **d** This structure was repeated multiple times over a silicon wafer (Northen and Turner 2005 2006)

Sitti (2003) proposed a nanomolding technique for creating structures with two levels of structures. In this method two different molds are created—one with pores on the order of magnitude of microns in diameter and a second with pores of nanometer-scale diameter. One potential mold material is porous anodic alumina, which has been demonstrated to produce ordered pores on the nanometer scale of equal size (Maschmann et al. 2006). Pore-widening techniques could be used to cre-

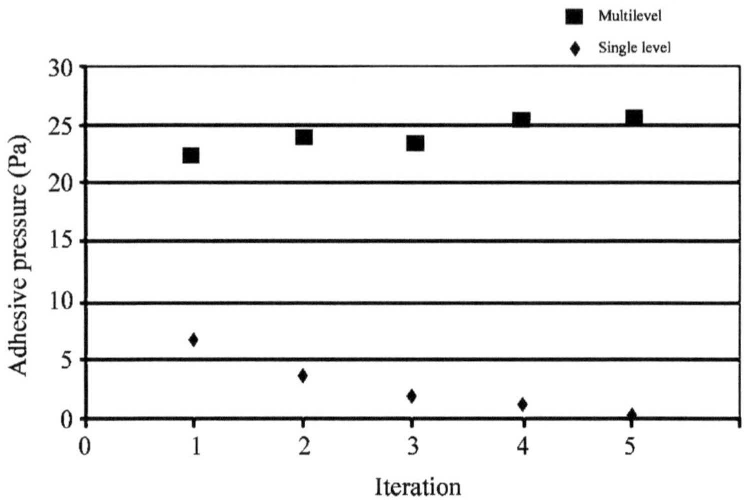

Fig. 27.28. Adhesion test results of a multilevel hierarchical structure and a single-level hierarchical structure repeated for five iterations (Northen and Turner 2006)

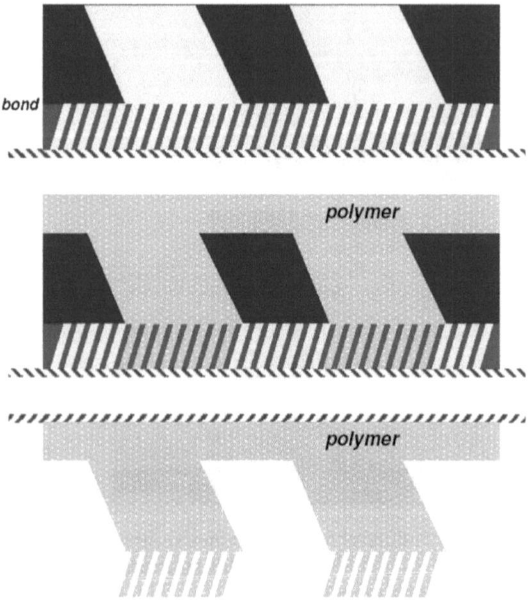

Fig. 27.29. Proposed process of creating multilevel synthetic gecko foot-hairs using nanomolding. Micron- and nanometer-sized pore membranes are bonded together (*top*) and filled with liquid polymer (*middle*). The membranes are then etched away leaving the polymer surface (*bottom*) (Sitti 2003)

ate micron-scale pores. As seen in Fig. 27.29, the two molds would be bonded to each other and then filled with a liquid polymer. According to Sitti (2003), the method would enable the manufacturing of a high volume of synthetic gecko foot-hairs at low cost.

The literature clearly indicates that in order to create a dry superadhesive, a fibrillar surface construction is necessary to maximize the van der Waals forces by decreasing the distance between the two surfaces. It is also desirable to have a superhydrophobic surface in order to utilize self-cleaning. A material must be soft enough to conform to rough surfaces yet hard enough to avoid bunching, which will decrease the adhesion force.

27.8 Closure

The adhesive properties of geckos and other creatures such as flies, beetles, and spiders are due to the hierarchical structures present on each creature's attachment pads. Geckos have developed the most intricate adhesive structures of any of the aforementioned creatures. The attachment system consists of ridges called lamellae that are covered in microscale setae that branch off into nanoscale spatulae. Each structure plays an important role in adapting to surface roughness, bringing the spatulae in close proximity with the mating surface. These structures as well as material properties allow the gecko to obtain a much larger real area of contact between its feet and a mating surface than is possible with a nonfibrillar material. Two feet of a Tokay gecko have about 220 mm^2 of attachment pad area on which the gecko is able to generate approximately 20 N of adhesion force. Although capable of generating high adhesion forces, a gecko is able to detach from a surface at will—an ability known as smart adhesion. Detachment is achieved by a peeling motion of the gecko's feet from a surface.

Experimental results have supported the adhesion theories of intermolecular forces (van der Waals) as a primary adhesion mechanism and capillary forces as a secondary mechanism, and have been used to rule out several other mechanisms of adhesion, including the secretion of sticky fluids, suction, and increased frictional forces. Atomic force microscopy has been employed by several investigators to determine the adhesive strength of gecko foot-hairs. The measured values of the lateral force required to pull parallel to the surface of a single seta (194 μN) and the adhesion force (normal to the surface) of a single spatula (11 nN) are comparable to the van der Waals prediction of 270 and 11 nN for a seta and a spatula, respectively. The adhesion force generated by a seta increases with preload and reaches a maximum when both perpendicular and parallel preloads are applied. Although gecko feet are strong adhesives, they remain free of contaminant particles through self-cleaning. Spatular size along with material properties enable geckos to easily expel any dust particles that come into contact with their feet.

Recent creation of a three-level hierarchical model for a gecko lamella consisting of setae, branches, and spatulae has brought more insight into adhesion of biological attachment systems. One-, two-, and three-level hierarchically structured spring models for simulation of a seta contacting with random rough surfaces were considered. The simulation results show that the multilevel hierarchical structure has a higher adhesion force as well as adhesion energy than the one-level structure for a given applied load, owing to better adaptation and attachment ability. It is concluded that the multilevel hierarchical structure produces adhesion enhancement, and this enhancement increases with an increase in the applied load and a decrease in the stiffness of springs. The condition at which a significant adhesion enhancement occurs appears to be related to the maximum spring deformation. The result shows that significant adhesion enhancement occurs when the maximum spring deformation is greater than 2–3 times larger than the σ value of surface roughness. For the effect of applied load, as the applied load increases, adhesion force increases up to a certain applied load and then has a constant value, whereas adhesion energy continues to increase with an increase in the applied load. Inclusion of capillary forces in the spring model shows that the total adhesion force decreases with an increase in the contact angle of water on the substrate, and the difference of the total adhesion force among contact angles is larger in the intermediate-humidity regime. In addition, the simulation results match measured data for a single spatula in contact with a hydrophilic and a hydrophobic surface, which further supports van der Waals forces as the dominant mechanism of adhesion and capillary forces as a secondary mechanism.

There is great interest among the scientific community to create surfaces that replicate the adhesion strength of gecko feet. These surfaces would be capable of reusable dry adhesion and would have uses in a wide range of applications from everyday objects such as tapes, fasteners, and toys to microelectric and space applications and even wall-climbing robots. In the design of fibrillar structures, it is necessary to ensure that the fibrils are compliant enough to easily deform to the mating surface's roughness profile, yet rigid enough to not collapse under their own weight. The spacing between the individual fibrils is also important. If the spacing is too small, adjacent fibrils can attract each other through intermolecular forces, which will lead to bunching. The adhesion design database developed by Kim and Bhushan (2007c) serves as a reference for choosing design parameters.

Nanoindentation, electron-beam lithography, and growing of carbon nanotube arrays are all methods that have been used to create fibrillar structures. The limitations of current machining methods on the microscale/nanoscale have resulted in the majority of fabricated surfaces consisting of only one level of hierarchy. Although typically capable of producing high adhesion force with an individual fibril, these surfaces have failed to generate high adhesion forces on the macroscale. Bunching, lack of compliance, and lack of durability are all problems that have arisen with the aforementioned structures. Recently, a multilayered compliant system was created using a microelectromechanical-based approach in combination with nanorods. This method as well as other proposed methods of multilevel nanomolding and directed self-assembly show great promise in the creation of adhesive structures with multiple levels of hierarchy, much like those of gecko feet.

Appendix

Several natural (sycamore tree bark and siltstone) and artificial surfaces (dry wall, wood laminate, steel, aluminum, and glass) were chosen to determine the surface parameters of typical rough surfaces that a gecko might encounter. An Alpha-step® 200 (Tencor Instruments, Mountain View, CA, USA) was used to obtain surface profiles for three different scan lengths: 80 μm, which is approximately the size of a single gecko seta, 2000 μm, which is close to the size of a gecko lamella; and an intermediate scan length of 400 μm. The radius of the stylus tip was 1.5–2.5 μm and the applied normal load was 30 μN. The surface profiles were then analyzed using a specialized computer program to determine the RMS amplitude σ, correlation length β^*, peak-to-valley distance P-V, skewness Sk, and kurtosis K.

Samples of surface profiles and their corresponding parameters at a scan length of 2000 μm can be seen in Fig. 27.30a. The roughness amplitude σ, varies from as little as 0.01 μm in glass to as much as 30 μm in tree bark. Similarly, the correlation

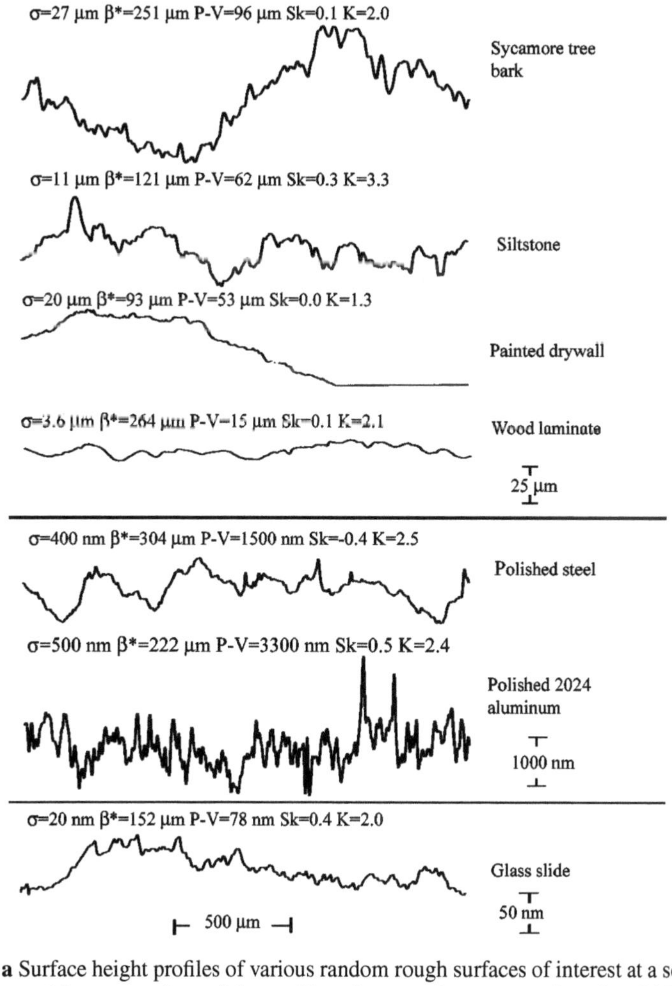

Fig. 27.30. a Surface height profiles of various random rough surfaces of interest at a scan length of 2000 μm and **b** a comparison of the profiles of two surfaces at scan lengths of 80, 400, and 2000 μm. (From Bhushan et al. 2006)

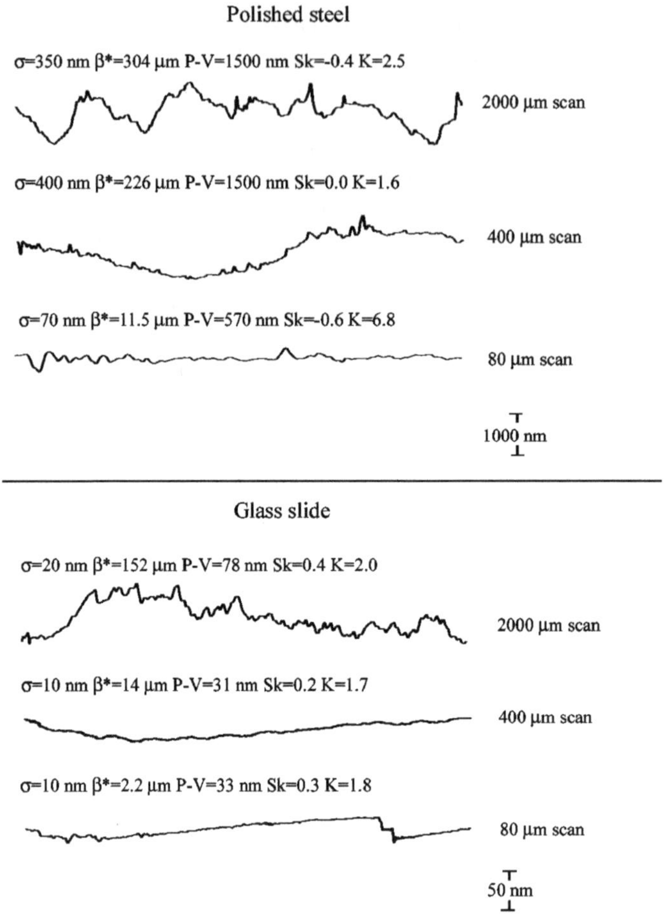

Fig. 27.30. (continued)

Table 27.3. Scale dependence of surface parameters σ and β^* for rough surfaces at scan lengths of 80 and 2000 µm (Bhushan et al. 2006)

Surface	Scan length 80 µm		Scan length 2000 µm	
	σ (µm)	β^* (µm)	σ (µm)	β^* (µm)
Sycamore tree bark	4.4	17	27	251
Siltstone	1.1	4.8	11	268
Painted drywall	1	11	20	93
Wood laminate	0.11	18	3.6	264
Polished steel	0.07	12	0.40	304
Polished 2024 aluminum	0.40	6.5	0.50	222
Glass	0.01	2.2	0.02	152

length varies from 2 to 300 μm. The scale dependency of the surface parameters is illustrated in Fig. 27.30b. As the scan length of the profile increases, so too does the roughness amplitude and the correlation length. Table 27.3 summarizes the scan length dependent factors σ and β^* for all seven sampled surfaces. At a scan length of 80 μm (size of seta), the roughness amplitude does not exceed 5 μm, while at a scan length of 2000 μm (size of lamella), the roughness amplitude is as high as 30 μm. This suggests that setae are responsible for adapting to surfaces with roughness on the order of several microns, while lamellae must adapt to roughness on the order of tens of microns. Larger roughness values would be adapted to by the skin of the gecko. The spring model of Bhushan et al. (2006) verifies that setae are only capable of adapting to roughnesses of a few microns and suggests that lamellae are responsible for adaptation to rougher surfaces.

References

1. Anonymous (2002) Merriam-Webster's medical dictionary, Merriam-Webster, Springfield
2. Aristotle (1918) Historia animalium, book IX, part 9. Translated by Thompson DAW. http://classics.mit.edu/Aristotle/history_anim.html
3. Arzt E, Gorb S, Spolenak R (2003) Proc Natl Acad Sci USA 100:10603–10606
4. Autumn K (2006) Am Sci 94:124–132
5. Autumn K, Peattie AM (2002) Integr Comp Biol 42:1081–1090
6. Autumn K, Liang YA, Hsieh ST, Zesch W, Chan WP, Kenny TW, Fearing R, Full RJ (2000) Nature 405:681–685
7. Autumn K, Sitti M, Liang YA, Peattie AM, Hansen WR, Sponberg S, Kenny TW, Fearing R, Israelachvili JN, Full RJ (2002) Proc Natl Acad Sci USA 99:12252–12256
8. Autumn K, Buehler M, Cutkosky M, Fearing R, Full RJ, Goldman D, Groff R, Provancher W, Rizzi AA, Sranli U, Saunders A, Koditschek DE (2005) Proc SPIE 5804:291–302
9. Autumn K, Majidi C, Groff RE, Dittmore A, Fearing R (2006) J Exp Biol 209:3558–3568
10. Barthlott W, Neinhuis C (1997) Planta 202:1–8
11. Bergmann PJ, Irschick DJ (2005) J Exp Zool 303A:785–791
12. Bertram JEA, Gosline JM (1987) J Exp Biol 130:121–136
13. Bhushan B (1996) Tribology and mechanics of magnetic storage devices, 2nd edn. Springer, New York
14. Bhushan B (1999) Principles and applications of tribology. Wiley, New York
15. Bhushan B (2002) Introduction to tribology. Wiley, New York
16. Bhushan B (ed) (2005) Nanotribology and nanomechanics—an introduction. Springer, Heidelberg
17. Bhushan B (ed) (2007a) Springer handbook of nanotechnology, 2nd edn. Springer, Heidelberg
18. Bhushan B (2007b) J Adhes Sci Technol (in press)
19. Bhushan B, Sayer RA (2007) Microsyst Technol 13:71–78
20. Bhushan B, Peressadko AG, Kim TW (2006) J Adhes Sci Technol 20:1475–1491
21. Bikerman JJ (1961) The science of adhesive joints. Academic, New York
22. Burton Z, Bhushan B (2005) Nano Lett 5:1607–1613
23. Chui BW, Kenny TW, Mamin HJ, Terris BD, Rugar D (1998) Appl Phys Lett 72:1388–1390
24. Dellit WD (1934) Jena Z Naturwissen 68:613–658
25. Derjaguin BV, Muller VM, Toporov YP (1975) J Colloid Interface Sci 53:314–326
26. Dieter GE (1988) Mechanical metallurgy. McGraw-Hill, London
27. Fan PL, O'Brien MJ (1975) In: Lee LH (ed) Adhesion science and technology, vol 9A. Plenum, New York, p 635

28. Federle W (2006) J Exp Biol 209:2611–2621
29. Federle W, Riehle M, Curtis ASG, Full RJ (2002) Integr Comp Biol 42:1100–1106
30. Gao H, Wang X, Yao H, Gorb S, Arzt E (2005) Mech Mater 37:275–285
31. Geim AK, Dubonos SV, Grigorieva IV, Novoselov KS, Zhukov AA, Shapoval SY (2003) Nat Mater 2:461–463
32. Gennaro JGJ (1969) Nat Hist 78:36–43
33. Glassmaker NJ, Jagota A, Hui CY, Kim J (2004) J R Soc Interface 1:23–33
34. Glassmaker NJ, Jagota A, Hui CY (2005) Acta Biomater 1:367–375
35. Gorb S, Varenberg M, Peressadko A, Tuma J (2007) J Royal Soc Interface 4:271–275
36. Hamaker HC (1937) Physica 4:1058
37. Han D, Zhou K, Bauer AM (2004) Biol J Linn Soc 83:353–368
38. Hanna G, Barnes WJP (1991) J Exp Biol 155:103–125
39. Hansen WR, Autumn K (2005) Proc Natl Acad Sci USA 102:385–389
40. Hiller U (1968) Z Morphol Tiere 62:307–362
41. Hinds WC (1982) Aerosol technology: properties, behavior, and measurement of airborne particles. Wiley, New York
42. Hora SL (1923) J Asiat Soc Beng 9:137–145
43. Houwink R, Salomon G (1967) J Appl Phys 38:1896–1904
44. Huber G, Gorb SN, Spolenak R, Arzt E (2005a) Biol Lett 1:2–4
45. Huber G, Mantz H, Spolenak R, Mecke K, Jacobs K, Gorb SN, Arzt E (2005b) Proc Natl Acad Sci USA 102:16293–16296
46. Irschick DJ, Austin CC, Petren K, Fisher RN, Losos JB, Ellers O (1996) Biol J Linn Soc 59:21–35
47. Israelachvili JN, Tabor D (1972) Proc R Soc Lond Ser A 331:19–38
48. Israelachvili JN (1992) Intermolecular and surface Forces, 2nd edn. Academic, San Diego
49. Jaenicke R (1998) In: Harrison RM, van Grieken R (eds) Atmospheric particles. Wiley, New York, pp 1–29
50. Jagota A, Bennison SJ (2002) Integr Comp Biol 42:1140–1145
51. Johnson KL, Kendall K, Roberts AD (1971) Proc R Soc Lond Ser A 324:301–313
52. Kesel AB, Martin A, Seidl T (2003) J Exp Biol 206:2733–2738
53. Kim TW, Bhushan B (2007a) J Adhes Sci Technol 21:1–20
54. Kim TW, Bhushan B (2007b) Ultramicroscopy 107:902–912
55. Kim TW, Bhushan B (2007c) J Vac Sci Technol A 25:1003–1012
56. Kim TW, Bhushan B (2007d) J Royal Soc Interface (in press)
57. Kluge AG (2001) Hamadryad 26:1–209
58. Losos JB (1990) Asiat Herp Res 3:54–59
59. Maderson PFA (1964) Nature 2003:780–781
60. Maschmann MR, Franklin AD, Amama PB, Zakharov DN, Stach EA, Sands TD, Fisher TS (2006) Nanotechnology 17:3925–3929
61. Menon C, Murphy M, Sitti M (2004) In: Proceedings of the IEEE international conference on robotics and biomimetics, 22–26 August, pp 431–436
62. Northen MT, Turner KL (2005) Nanotechnology 16:1159–1166
63. Northen MT, Turner KL (2006) Sens Actuators A 130–131:583–587
64. Orr FM, Scriven LE, Rivas AP (1975) J Fluid Mech 67:723–742
65. Pan B, Gao F, Ao L, Tian H, He R, Cui D (2005) Colloids Surf A 259:89–94
66. Peressadko A, Gorb SN (2004) J Adhes 80:247–261
67. Persson BNJ (2003) J Chem Phys 118:7614–7621
68. Persson BNJ, Gorb S (2003) J Chem Phys 119:11437–11444
69. Phipps PB, Rice DW (1979) In: Brubaker GR, Phipps PB (eds) ACS symposium series no 89. American Chemical Society, Washington

70. Rizzo N, Gardner K, Walls D, Keiper-Hrynko N, Hallahan D (2006) J Royal Soc Interface 3:441–451
71. Ruibal R, Ernst V (1965) J Morphol 117:271–294
72. Russell AP (1975) J Zool Lond 176:437–476
73. Russell AP (1986) Can J Zool 64:948–955
74. Schäffer E, Thurn-Albrecht T, Russell TP, Steiner U (2000) Nature 403:874–877
75. Schleich HH, Kästle W (1986) Amphib Reptil 7:141–166
76. Schmidt HR (1904) Jena Z Naturwissen 39:551
77. Shah GJ, Sitti M (2004) In: Proceedings of the IEEE international conference on robotics and biomimetics, pp 873–878
78. Simmermacher G (1884) Zeitschr Wissen Zool 40:481–556
79. Sitti M (2003) In: Proceedings of the IEEE/ASME advanced mechatronics conference, vol 2, pp 886–890
80. Sitti M, Fearing RS (2003a) J Adhes Sci Technol 17:1055–1073
81. Sitti M, Fearing RS (2003b) In: Proceedings of the IEEE international conference on robotics and automation, vol 1, pp 1164–1170
82. Spolenak R, Gorb S, Arzt E (2005) Acta Biomater 1:5–13
83. Stork NE (1980) J Exp Biol 88:91–107
84. Stork NE (1983) J Nat Hist 17:829–835
85. Tian Y, Pesika N, Zeng H, Rosenberg K, Zhao B, McGuiggan P, Autumn K, Israelachvili J (2006) Proc Natl Acad Sci USA 103:19320–19325
86. Tinkle DW (1992) Encyl Am 12:359
87. Van der Kloot WG (1992) Encyl Am 19:336–337
88. Wagler J (1830) Natürliches System der Amphibien. Cotta'schen, Munich
89. Wan KT, Smith DT, Lawn BR (1992) J Am Ceram Soc 75:667–676
90. Wenzel RN (1936) Ind Eng Chem 28:988–994
91. Williams EE, Peterson JA (1982) designs in the digital adhesive pads of scincid lizards. Science 215:1509–1511
92. Yao H, Gao H (2006) J Mech Phys Solids 54:1120–1146
93. Young WC, Budynas R (2001) Roark's formulas for stress and strain, 7th edn. McGraw-Hill, New York
94. Yurdumakan B, Raravikar NR, Ajayan PM, Dhinojwala A (2005) Chem Commun 3799–3801
95. Zhao Y, Tong T, Delzeit L, Kashani A, Meyyappan M, Majumdar A (2006) J Vac Sci Technol B 24:331–335
96. Zimon AD (1969) Adhesion of dust and powder. Translated from Russian by M Corn. Plenum, New York
97. Zisman WA (1963) Ind Eng Chem 55(10):18–38

MIX
Papier aus verantwortungsvollen Quellen
Paper from responsible sources
FSC® C105338

If you have any concerns about our products,
you can contact us on
ProductSafety@springernature.com

In case Publisher is established outside the EU,
the EU authorized representative is:
Springer Nature Customer Service Center GmbH
Europaplatz 3, 69115 Heidelberg, Germany

Printed by Libri Plureos GmbH
in Hamburg, Germany